Before you begin, review your basic study skills.

To take full advantage of the *Study Skills Workshops* that begin each chapter, you may choose to review them in the early weeks of your course. Each one includes action items, in addition to simple suggestions that can put you on a clear path to success. Below, we have included a table of contents to aid you in locating these:

Study Skills Workshop ▶ Making Homework a

Attending class and taking notes are important, but they are not enough. The only way really going to learn algebra is by doing your homework.

WHEN TO DO YOUR HOMEWORK: Homework should be started on the day it is assigned, when the material is fresh in your mind. It's best to break your homework sessions into 30-minute periods, allowing for short breaks in between.

HOW TO BEGIN YOUR HOMEWORK: Review your notes and the examples in your text before starting your homework assignment.

GETTING HELP WITH YOUR HOMEWORK: It's normal to have some questions when doing homework. Talk to a tutor, a classmate, or your instructor to get those questions answered.

Now Try This ▶

1. Write a one-page paper that describes *when, where,* and *how* you go about completing your algebra homework assignments.
2. For each problem on your next homework assignment, find an example in this book that is similar. Write the example number next to the problem.
3. Make a list of questions that you have while doing your next assignment. Then decide whom you are going to ask to get those questions answered.

5e

Elementary Algebra

5e

Elementary Algebra

▶ **Alan S. Tussy**
Citrus College

▶ **R. David Gustafson**
Rock Valley College

BROOKS/COLE
CENGAGE Learning™

Australia • Brazil • Japan • Korea • Mexico • Singapore • Spain • United Kingdom • United States

Elementary Algebra, **Fifth Edition**
Alan S. Tussy, R. David Gustafson

Publisher: Charlie Van Wagner

Senior Developmental Editor:
  Danielle Derbenti

Assistant Editor: Carrie Jones

Senior Editorial Assistant: Jennifer Cordoba

Media Editors: Heleny Wong,
  Guanglei Zhang

Marketing Manager: Gordon Lee

Marketing Assistant: Shannon Maier

Marketing Communications Manager:
  Darlene Macanan

Content Project Manager: Jennifer Risden

Design Director: Rob Hugel

Art Director: Vernon Boes

Print Buyer: Judy Inouye

Rights Acquisitions Specialist:
  Tom McDonough

Production Service: Chapter Two,
  Ellen Brownstein

Text Designer: Terri Wright

Photo Researcher: Bill Smith Group

Copy Editor: Ellen Brownstein

Illustrator: Lori Heckelman

Cover Designer: Terri Wright

Cover Image: Artie Ng/ww.flickr.com/
  photos/artiephotography

Compositor: Graphic World, Inc.

For product information and technology assistance, contact us at
Cengage Learning Customer & Sales Support, 1-800-354-9706.
For permission to use material from this text or product,
submit all requests online at **www.cengage.com/permissions.**
Further permissions questions can be e-mailed to
permissionrequest@cengage.com.

Library of Congress Control Number: 2010939895

Student Edition:
ISBN-13: 978-1-111-56766-8
ISBN-10: 1-111-56766-2

Loose-leaf Edition:
ISBN-13: 978-1-111-98775-6
ISBN-10: 1-111-98775-0

Brooks/Cole
20 Davis Drive
Belmont, CA 94002-3098
USA

Cengage Learning is a leading provider of customized learning solutions with office locations around the globe, including Singapore, the United Kingdom, Australia, Mexico, Brazil, and Japan. Locate your local office at **www.cengage.com/global.**

Cengage Learning products are represented in Canada by Nelson Education, Ltd.

To learn more about Brooks/Cole, visit **www.cengage.com/brookscole**

Purchase any of our products at your local college store or at our preferred online store **www.cengagebrain.com.**

Printed in China
3 4 5 6 7 18 17 16 15

To my wife, Liz,

Thank you for your faithful help and encouragement.

—AST

To my wife, Carol,

with love and appreciation.

—RDG

ABOUT THE AUTHORS

Alan S. Tussy

Alan Tussy teaches all levels of developmental mathematics at Citrus College in Glendora, California. He has written nine math books—a paperback series and a hard-cover series. A meticulous, creative, and visionary teacher who maintains a keen focus on his students' greatest challenges, Alan Tussy is an extraordinary author, dedicated to his students' success. Alan received his Bachelor of Science degree in Mathematics from the University of Redlands and his Master of Science degree in Applied Mathematics from California State University, Los Angeles. He has taught up and down the curriculum from prealgebra to differential equations. He is currently focusing on the developmental math courses. Professor Tussy is a member of the American Mathematical Association of Two-Year Colleges.

R. David Gustafson

R. David Gustafson is Professor Emeritus of Mathematics at Rock Valley College in Illinois and coauthor of several best-selling math texts, including Gustafson/Frisk's *Beginning Algebra, Intermediate Algebra, Beginning and Intermediate Algebra: A Combined Approach, College Algebra,* and the Tussy/Gustafson developmental mathematics series. His numerous professional honors include Rock Valley Teacher of the Year and Rockford's Outstanding Educator of the Year. He earned a Master of Arts from Rockford College in Illinois, as well as a Master of Science from Northern Illinois University.

CONTENTS

CHAPTER 3 ▶ Graphing Linear Equations and Inequalities in Two Variables; Functions 185

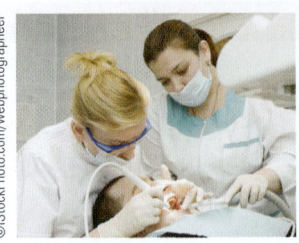

©iStockPhoto.com/webphotographeer

CHAPTER 4 ▶ Systems of Linear Equations and Inequalities 283

©Andrei Contiu/Shutterstock.com

CHAPTER 5 ▶ Exponents and Polynomials 347

©iStockPhoto.com/beetle8

CHAPTER 6 ▶ Factoring and Quadratic Equations 431

CHAPTER 7 ▶ Rational Expressions and Equations 511

CHAPTER 8 ▶ Radical Expressions and Equations 609

©Mark Richards/PhotoEdit

CHAPTER 9 ▶ Quadratic Equations 675

APPENDIXES

PREFACE

We are excited to present the Fifth Edition of *Elementary Algebra*. We believe the revision process has produced an even stronger instructional experience for both students and teachers. First, we have fine-tuned several of the popular features of our series, including the *Strategy* and *Why* example structure, the problem-solving strategy, and the online homework.

Second, we have introduced some new features to further promote student understanding and success. The new *Look Alikes* problems in the *Study Sets* will help students improve their problem recognition. The *Are You Ready?* exercises that begin each section give students the opportunity to review necessary prerequisite skills before they are asked to apply them in the study of new topics. Furthermore, the additional *Campus to Careers* problems inserted in the *Study Sets* expand on the real-life applications of the chapter material.

We want to thank all of you throughout the country who have provided suggestions and input about the previous edition. Your insight has proven invaluable. Throughout this process, our fundamental belief has remained the same: Algebra is a language in its own right. And, as always, the prime objective of this textbook is to teach students how to read, write, speak, and think using the language of algebra.

New to the Fifth Edition

Sections That Begin with Review: *Are You Ready?*

Each section begins with a set of *Are You Ready?* problems. These problems review crucial prerequisite skills that students need to have mastered in order to be successful with the new topics of that section.

> **ARE YOU READY?**
>
> The following problems review some basic skills that are needed to find the slope of a line.
>
> 1. Evaluate: $\dfrac{4-1}{8-3}$
>
> 2. Evaluate: $\dfrac{-10-1}{-4-(-4)}$
>
> 3. Multiply: $-\dfrac{7}{9}\cdot\dfrac{9}{7}$
>
> 4. Simplify: $\dfrac{15}{18}$

Study Sets with More Problem-Recognition Practice: *Look Alikes*

After a poorer than expected performance on a test, students often tell their instructors, "I could do the homework each night, but when it comes to the test, I get confused." The new *Look Alikes* feature builds students' problem-recognition skills. It requires students to distinguish between similar looking problem types and then to select the correct strategy to solve the problem. Encountering such situations in the homework assignments will better prepare students for quizzes and tests.

> **Look Alikes . . .**
>
> *Perform the indicated operations to simplify each expression, if possible.*
>
> 105. a. $(x-2)+(x^2+2x+4)$ b. $(x-2)(x^2+2x+4)$
>
> 106. a. $(a+3)+(a^2-3a+9)$ b. $(a+3)(a^2-3a+9)$
>
> 107. a. $(6x^2z^5)-(-3xz^3)$ b. $(6x^2z^5)(-3xz^3)$
>
> 108. a. $(-5r^4t^2)-(2r^2t)$ b. $(-5r^4t^2)(2r^2t)$
>
> 109. a. $(2x^2-x)-(3x^2-3x)$ b. $(2x^2-x)(3x^2-3x)$
>
> 110. a. $(4.9a-b)-(2a+b)$ b. $(4.9a-b)(2a+b)$

And Even More Problem-Recognition Practice: *Try It Yourself*

Designed to promote problem recognition, instructors and reviewers requested more *Try It Yourself* problems for the Fifth Edition. These problem types are thoroughly mixed, giving students an opportunity to practice decision making and strategy selection as they would when taking a test or quiz. With more than 80% more problems added, students will have even more opportunities to practice this essential skill.

TRY IT YOURSELF

Graph each equation. Solve for y first, when necessary.

53. $y = x$ **54.** $y = 4x$
55. $y = -x - 1$ **56.** $y = -x + 2$
57. $3y = 12x + 15$ **58.** $5y = 20x - 30$
59. $y = \frac{3}{8}x - 6$ **60.** $y = -\frac{3}{2}x + 2$
61. $y = 1.5x - 4$ **62.** $y = 0.5x + 3$
63. $8x + 4y = 16$ **64.** $14x + 7y = 28$
65. $y = -\frac{1}{2}x$ **66.** $y = \frac{3}{4}x$
67. $y = \frac{5}{6}x - 5$ **68.** $y = \frac{2}{3}x - 2$
69. $-6y = 30x + 12$ **70.** $-3y = 9x - 15$

Comprehensive Test Preparation: *Chapter Tests*

Instructors often assign an end-of-chapter test for students to use as a means to study for a classroom exam. However, after taking the exam, students often remark that the classroom exam included problem types that weren't in the *Chapter Test*. To address this issue, we have made sure that each *Chapter Test* is a comprehensive collection of problems that covers *all* of the topics discussed in the chapter. As a result, the *Chapter*

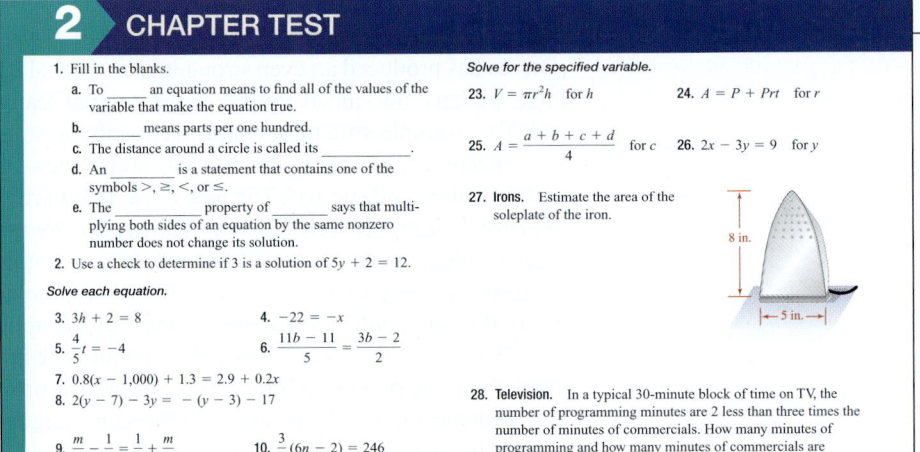

2 ▶ CHAPTER TEST

1. Fill in the blanks.
   a. To _____ an equation means to find all of the values of the variable that make the equation true.
   b. _____ means parts per one hundred.
   c. The distance around a circle is called its _____.
   d. An _____ is a statement that contains one of the symbols $>$, $\geq$, $<$, or $\leq$.
   e. The _____ property of _____ says that multiplying both sides of an equation by the same nonzero number does not change its solution.
2. Use a check to determine if 3 is a solution of $5y + 2 = 12$.

Solve each equation.

3. $3h + 2 = 8$
4. $-22 = -x$
5. $\frac{4}{5}t = -4$
6. $\frac{11b - 11}{5} = \frac{3b - 2}{2}$
7. $0.8(x - 1,000) + 1.3 = 2.9 + 0.2x$
8. $2(y - 7) - 3y = -(y - 3) - 17$
9. $\frac{m}{2} - \frac{1}{3} = \frac{1}{4} + \frac{m}{6}$
10. $\frac{3}{4}(6n - 2) = 246$

Solve for the specified variable.

23. $V = \pi r^2 h$ for h
24. $A = P + Prt$ for r
25. $A = \frac{a + b + c + d}{4}$ for c
26. $2x - 3y = 9$ for y
27. **Irons.** Estimate the area of the soleplate of the iron.

8 in.
← 5 in. →

28. **Television.** In a typical 30-minute block of time on TV, the number of programming minutes are 2 less than three times the number of minutes of commercials. How many minutes of programming and how many minutes of commercials are there?

Tests are lengthy. If students have time to complete a *Chapter Test,* that would be optimal. If, because of time constraints, they are unable to do so, the instructor can assign an appropriate subset of problems that reflects the types of problems that students can see on the exam. This should alleviate the discrepancy between what students practice and what they will see on the test.

Additional Relevant, Motivating Applications in the *Examples* and *Study Sets*

We have included many new applied examples and problems that involve relevant topics such as our environment and sustainability issues, energy savings, technology and social media, and recycling. For a complete list of topics, see the *Index of Applications* following the *Preface.* To see a sampling of new topics added to each chapter, see *Content Changes by Chapter* in this *Preface.*

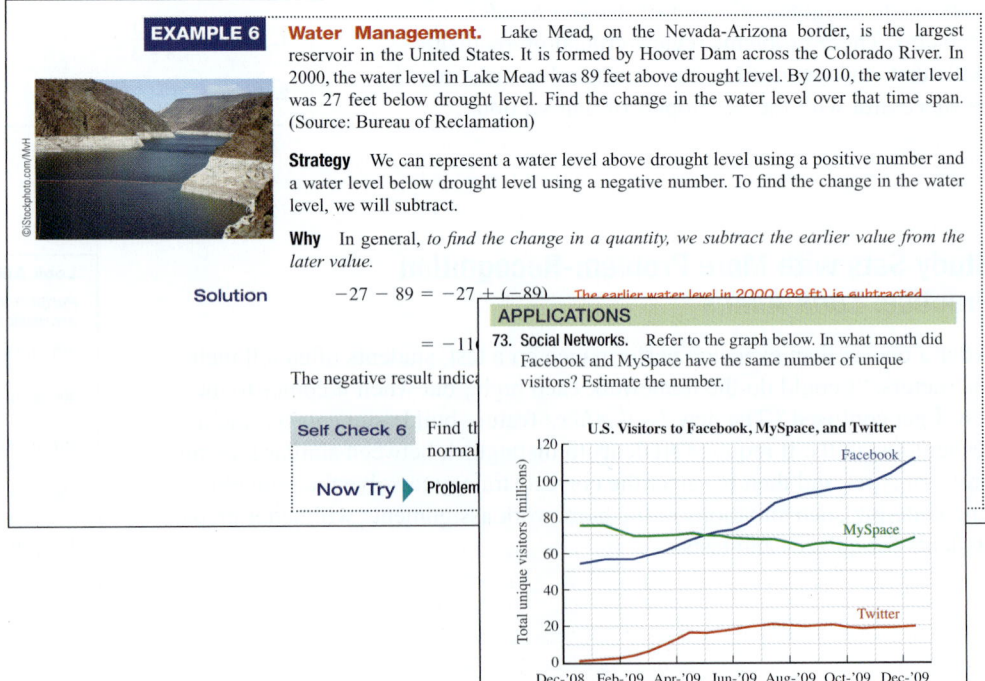

EXAMPLE 6

Water Management. Lake Mead, on the Nevada-Arizona border, is the largest reservoir in the United States. It is formed by Hoover Dam across the Colorado River. In 2000, the water level in Lake Mead was 89 feet above drought level. By 2010, the water level was 27 feet below drought level. Find the change in the water level over that time span. (Source: Bureau of Reclamation)

Strategy We can represent a water level above drought level using a positive number and a water level below drought level using a negative number. To find the change in the water level, we will subtract.

Why In general, *to find the change in a quantity, we subtract the earlier value from the later value.*

Solution $-27 - 89 = -27 + (-89)$ The earlier water level in 2000 (89 ft) is subtracted.

$= -11$

The negative result indica...

Self Check 6 Find th... normal...

Now Try ▶ Problem...

APPLICATIONS

73. **Social Networks.** Refer to the graph below. In what month did Facebook and MySpace have the same number of unique visitors? Estimate the number.

U.S. Visitors to Facebook, MySpace, and Twitter

Facebook
MySpace
Twitter

Total unique visitors (millions): 0, 20, 40, 60, 80, 100, 120

Dec-'08 Feb-'09 Apr-'09 Jun-'09 Aug-'09 Oct-'09 Dec-'09

Source: Comscore Media Metrix

A More Precise Problem-Solving Strategy

In an effort to better describe the problem-solving strategy used in this book, we have inserted a new second step in what was formerly a five-step process. This additional step (*Assign a variable to represent an unknown value in the problem.*) better delineates the thought process students should use as they solve application problems. The six steps of the problem-solving strategy are now: *Analyze the problem*, *Assign a variable*, *Form an equation*, *Solve the equation*, *State the conclusion*, and *Check the result*.

EXAMPLE 11

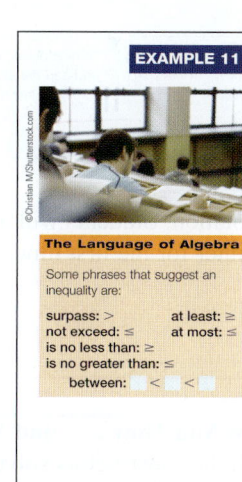
©Christian M/Shutterstock.com

The Language of Algebra

Some phrases that suggest an inequality are:

surpass: $>$ at least: $\geq$
not exceed: $\leq$ at most: $\leq$
is no less than: $\geq$
is no greater than: $\leq$
    between: $\boxed{} < \boxed{} < \boxed{}$

Grades. A student has scores of 72%, 74%, and 78% on three exams. What percent score does he need on the last exam to earn a grade of no less than B (80%)?

Analyze We know three scores. We are to find what the student must score on the last exam to earn a grade of B or higher.

Assign Let x = the score on the fourth (and last) exam.

Form To find the average grade, we add the four scores and divide by 4. To earn a grade of *no less than* B, the student's average must be *greater than or equal to* 80%.

The average of the four grades	must be no less than	80.
$\dfrac{72 + 74 + 78 + x}{4}$	$\geq$	80

Solve

$$\frac{224 + x}{4} \geq 80 \qquad \text{Combine like terms in the numerator: } 72 + 74 + 78 = 224.$$

$$4\left(\frac{224 + x}{4}\right) \geq 4(80) \qquad \text{To clear the inequality of the fraction, multiply both sides by 4.}$$

$$224 + x \geq 320 \qquad \text{Simplify each side.}$$

$$x \geq 96 \qquad \begin{array}{l}\text{To isolate } x\text{, undo the addition of 224 by}\\ \text{subtracting 224 from both sides.}\end{array}$$

State To earn a B, the student must score 96% or better on the last exam.

Check Pick several exam scores that are 96% or better and verify that the student's average will be 80% or greater. For example, a score of 96% gives the student an average that is exactly 80%.

$$\frac{72 + 74 + 78 + 96}{4} = \frac{320}{4} = 80$$

Self Check 11   **Grades.** A student has scores of 78%, 82%, and 76% on three exams. What percent score does he need on the last test to earn a grade of no less than a B (80%)?

Now Try ▶ Problem 103

More Emphasis on "When Will I Use This?"

Each chapter now has three *Campus to Careers* problems that explore the mathematical connections to careers that are presented at the beginning of each chapter.

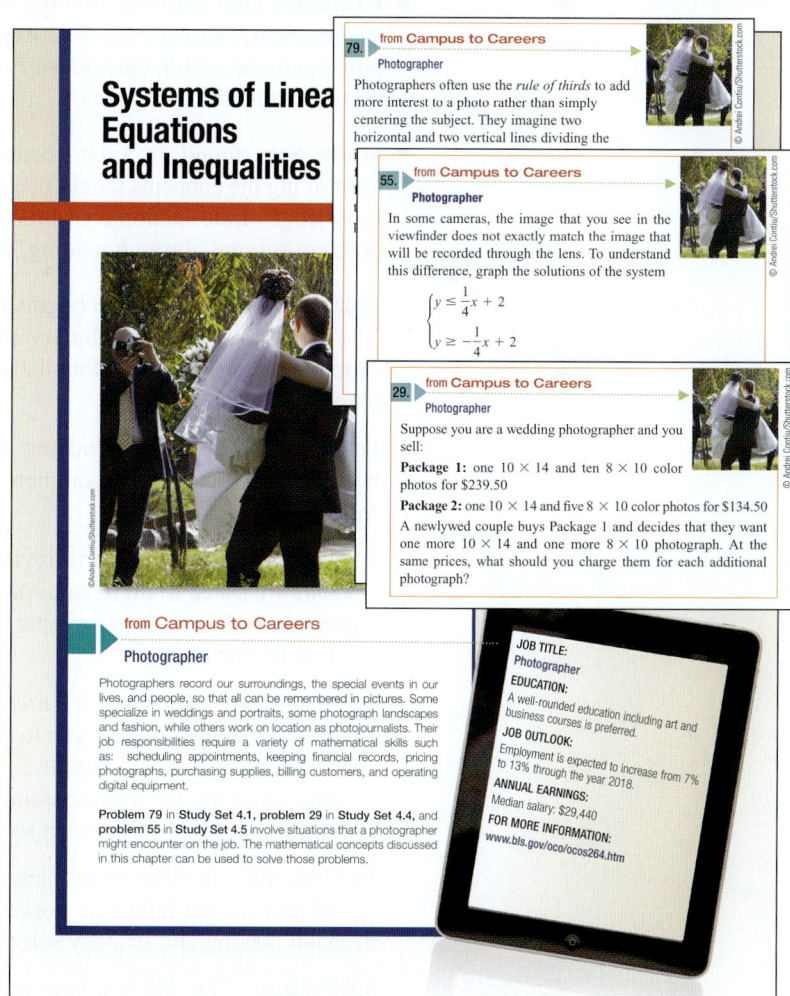

Systems of Linear Equations and Inequalities

79. from **Campus to Careers**
Photographer
Photographers often use the *rule of thirds* to add more interest to a photo rather than simply centering the subject. They imagine two horizontal and two vertical lines dividing the

55. from **Campus to Careers**
Photographer
In some cameras, the image that you see in the viewfinder does not exactly match the image that will be recorded through the lens. To understand this difference, graph the solutions of the system

$$\begin{cases} y \leq \dfrac{1}{4}x + 2 \\ y \geq -\dfrac{1}{4}x + 2 \end{cases}$$

29. from **Campus to Careers**
Photographer
Suppose you are a wedding photographer and you sell:
Package 1: one 10×14 and ten 8×10 color photos for $239.50
Package 2: one 10×14 and five 8×10 color photos for $134.50
A newlywed couple buys Package 1 and decides that they want one more 10×14 and one more 8×10 photograph. At the same prices, what should you charge them for each additional photograph?

from **Campus to Careers**
Photographer

Photographers record our surroundings, the special events in our lives, and people, so that all can be remembered in pictures. Some specialize in weddings and portraits, some photograph landscapes and fashion, while others work on location as photojournalists. Their job responsibilities require a variety of mathematical skills such as: scheduling appointments, keeping financial records, pricing photographs, purchasing supplies, billing customers, and operating digital equipment.

Problem 79 in Study Set 4.1, problem 29 in Study Set 4.4, and problem 55 in Study Set 4.5 involve situations that a photographer might encounter on the job. The mathematical concepts discussed in this chapter can be used to solve those problems.

JOB TITLE:
Photographer
EDUCATION:
A well-rounded education including art and business courses is preferred.
JOB OUTLOOK:
Employment is expected to increase from 7% to 13% through the year 2018.
ANNUAL EARNINGS:
Median salary: $29,440
FOR MORE INFORMATION:
www.bls.gov/oco/ocos264.htm

283

Reading the Language of Algebra

Students often have difficulty reading the
mathematical notation of algebra and, as a
consequence, their understanding suffers. To
provide assistance in this area, we have

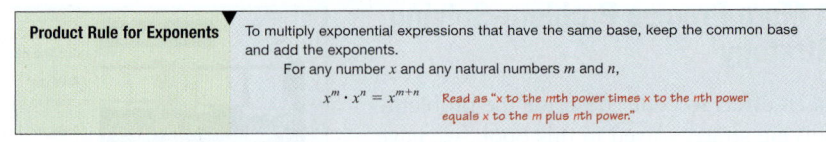

Product Rule for Exponents	To multiply exponential expressions that have the same base, keep the common base and add the exponents.
	For any number x and any natural numbers m and n,
	$x^m \cdot x^n = x^{m+n}$ Read as "x to the mth power times x to the nth power equals x to the m plus nth power."

inserted notes that explain how to read newly introduced notation. Students will appreciate the ever-present "Read as ..."
statements that follow algebraic symbolism that they encounter for the first time.

Trusted Features

- **Examples That Show You How . . . and WHY:** Why? That question is often asked by
  students as they watch their instructors solve problems in class and as they are working on
  problems at home. It's not enough to know how a problem is solved. Students gain a deeper
  understanding of the algebraic concepts if they know why a particular approach was taken.
  This instructional truth was the motivation for adding a Strategy and Why explanation to
  each worked example.

- **Examples That Offer Immediate Feedback:** Each worked example includes a Self Check.
  You can complete these on your own after reading the worked example or your instructor
  may choose to use them as classroom lecture examples. The Self Check answers can be
  found in the *Answers to Selected Exercises* in the back of the text.

- **Examples That Ask You to Work Independently:** Each worked example ends with a Now
  Try problem. These are the final step in the learning process. Each one is linked to a similar
  problem found within the *Guided Practice* sections of the *Study Sets,* offering you a smooth
  transition into the homework.

- **Study Sets** found in each section offer a multifaceted approach to practicing and
  reinforcing the concepts taught in each section. They are designed for students to build their
  knowledge of the section concepts methodically, from basic recall to increasingly complex
  problem solving, through reading, writing, and thinking mathematically.

  Vocabulary—Each *Study Set* begins with the important *Vocabulary* discussed in that section.
  The fill-in-the-blank vocabulary problems emphasize the main concepts taught in the
  chapter and provide the foundation for learning and communicating the language of
  algebra.

  Concepts—In *Concepts,* students are asked about the specific subskills and procedures
  necessary to successfully complete the *Guided Practice* and *Try It Yourself* problems that
  follow.

  Notation—*Notation* problems review the new symbols introduced in a section. Often,
  students are asked to fill in steps of a sample solution. This strengthens their ability to read
  and write mathematics and prepares them for the *Guided Practice* problems by modeling
  solution formats.

  Guided Practice—The *Guided Practice* section of each *Study Set* consistently provides 1-to-
  1 linking for each problem type to a single worked example or objective (*i.e.,* See Example
  1, See Example 2, See Example 3, and so on). Students will appreciate this 1-to-1 linking
  as opposed to the all-encompassing linking statements such as See Examples 2–8 or See
  Example 7–11 that are found in some textbooks.

  Try It Yourself—To promote problem recognition, the *Try It Yourself* problems are thoroughly
  mixed and are *not* linked to worked examples, giving students an opportunity to practice
  decision making and strategy selection as they would when taking a test or quiz.

  Applications—The *Applications* provide students the opportunity to apply their newly
  acquired algebraic skills to relevant and interesting real-life situations.

  Writing—The *Writing* problems help students build mathematical communication skills.

Review—The *Review* problems consist of randomly selected problems from previous chapters. These problems are designed to keep students' successfully mastered skills up-to-date before they move on to the next section.

Challenge Problems—The *Challenge Problems* provide students with an opportunity to stretch themselves and develop their skills beyond the basics. Instructors often find these to be useful as extra-credit problems.

- **Detailed Author Notes** that guide students along in a step-by-step process appear in the solutions to every worked example.

- **The Language of Algebra** boxes draw connections between mathematical terms and everyday references to reinforce the language of algebra thread that runs throughout the text.

- **The Notation, Success Tips, Caution,** and **Calculators** boxes offer helpful tips to reinforce correct mathematical notation, improve students' problem-solving abilities, warn students of potential pitfalls and increase clarity, and offer tips on using scientific calculators.

- **Chapter Tests,** at the end of every chapter, can be used as preparation for the class exam.

- **Cumulative Reviews** follow the end-of-chapter material and keep students' skills current before moving on to the next chapter. Each problem is linked to the associated section from which the problem came for ease of reference. The final *Cumulative Review* often is used by instructors as a final exam review.

- **Using Your Calculator** is an optional feature that is designed for instructors who want to use calculators as part of the instruction in this course. This feature introduces keystrokes and shows how scientific and graphing calculators can be used to solve problems. In the *Study Sets,* icons are used to denote problems that may be solved using a calculator.

Content Changes by Chapter

Based on feedback from colleagues and users of the Fourth Edition, the following changes have been made in an effort to further streamline and update the text.

Chapter 1

- New example and exercise applications include topics such as snowboarding, Lake Mead water levels, melting glaciers, calories burned doing housework, lost luggage, iPhone signal strength, U.S. Federal Budget Deficit/Surplus, and average wait time in airport security lines.

- New instructional features include a screened color 1 that is used when explaining how to build and simplify fractions, additional notes explaining how the notation is read, a worked example demonstrating uses of the commutative and associative properties, a more comprehensive *Chapter Test,* additional cautions, and an upgraded *Group Project.*

Chapter 2

- New example and exercise applications include topics such as Twitter, how tire pressure affects gas mileage, rainforest deforestation, Craigslist ads, *Harry Potter* box office revenue, calculating horsepower, iPhone apps, water usage, consignment shops, signing bonuses, and target heart rates.

- New instructional features include a comparison of linear and nonlinear equations, a new worked example of solving formulas for a specified variable, a new step added to the problem-solving strategy (*Assign a variable.*), additional cautions, additional explanation of consecutive integers, a worked example showing how to clear an inequality of fractions to solve it, and a more comprehensive *Chapter Test.*

Chapter 3

- New example and exercise applications include topics such as the Hollywood sign, René Descartes, endangered species, the U.S. Space Program, Honda Insight gas mileage, renewable energy, dental-assistant programs, firefighting, printing presses, calculating the cost to use an iPad, online games, U.S. credit card debt, the cost of raising a family, managing dental appointment times, and the amount of carbon dioxide in the Earth's atmosphere.

- New instructional features include a comparison of linear and nonlinear equations, tips for constructing a table of solutions for equations of the form $Ax + By = 0$, a comparison of one- and two-dimensional graphs, lines with slopes 1 and -1, additional cautions, a summary table of the forms of linear equations in two variables, and a more comprehensive *Chapter Test.*

Chapter 4

- New example and exercise applications include topics such as social networking, using the *rule of thirds* when taking photographs, the cost of changing CFL light bulbs, the number of women awarded Bachelor's degrees, sources of electricity, newspaper readership, greenhouse gas emissions, and lung cancer statistics.

- New instructional features include additional emphasis on what to write when both variables drop out when using the substitution method, a new step added to the problem-solving strategy (*Assign variables to the unknowns.*), additional cautions, and a comparison of one-variable and two-variable approaches to solving application problems.

Chapter 5

- New example and exercise applications include topics such as threshold hearing, supercomputers, replacing a fan belt, and diabetes diagnoses.

- New instructional features include additional notes explaining how exponential notation is read, additional cautions, a verification that the definitions of zero and negative exponents are consistent with students' previous experience with exponents, a more visible explanation showing how to multiply and divide numbers written in scientific notation, a comparison of polynomials and expressions that are not polynomials, an additional example of polynomial subtraction (vertical form), and more detailed author notes explaining polynomial long division.

Chapter 6

- New example and exercise applications include topics such as making crayons, staining a front door, Grammy nominations, antique shows, and rate of change of Wikipedia entries.

- New instructional features include a worked example in which the terms of a polynomial are rearranged to facilitate factoring by grouping, additional factoring tips, additional insight into factoring using the key number method, more information about how to recognize perfect-square trinomials, alternate factoring approaches, additional *Language of Algebra* boxes, and a more extensive list of types of quadratic equations.

Chapter 7

- New example and exercise applications include topics such as preparing an operating room, number of Tweets per ten seconds, building design, computer hard drives, and exercise equipment depreciation.

- New instructional features include a screened color 1 that is used when explaining how to build and simplify fractions, a worked example in which the common factor $x + 1$ in the numerator and $1 + x$ in the denominator are removed, and increased use of the term *rational expression* in place of the word *fraction.*

Chapter 8

- New example and exercise applications include topics such as marking out a crime scene, curving test scores, and Skype subscriptions.

- New instructional features include a list of natural-number perfect-squares, a list of perfect-square expression forms, a worked example that involves adding unlike radicals, additional examples of squaring a square root, a list of natural-number perfect-cubes, perfect fourth-powers, and perfect fifth-powers, and additional notes explaining how radical notation is read.

Chapter 9

- New example and exercise applications include topics such as the use of the golden triangle in graphic design, sports fishing, and building the St. Louis Arch.

- New instructional features include additional notes explaining how radical notation is read, more emphasis on *approximate* versus *exact* solutions, and checking approximate solutions.

- *Section 9.4: Complex Numbers* and *Section 9.5: Graphing Quadratic Equations,* along with the corresponding *Chapter Summary* and *Chapter Test* material, are now available online at www.cengage.com/math/tussy or www.cengagebrain.com.

Student Resources

Print Ancillaries

Complete Course Notebook (1-133-36391-1)
Ann Ostberg
NEW! The *Complete Course Notebook* is your blueprint for success in your algebra course and in all your future studies! This notebook organizes and guides you through each section of your textbook. Each chapter contains a chapter readiness assessment, structured note-taking guides, questions for reflection, activities, and end-of-chapter test prep. The notebook also contains general tip sheets on note-taking, studying, assessing your test scores, and much more.

Student Workbook (1-111-98831-5)
Maria H. Andersen, *Muskegon Community College*
Get a head-start. The *Student Workbook* contains assessments, activities, and worksheets for classroom discussions, in-class activities, and group work.

Student Solutions Manual (1-111-98902-8)
Alexander Lee, *Hinds Community College*
The *Student Solutions Manual* provides worked-out solutions to the odd-numbered problems in the text.

Enhanced WebAssign: Start Smart Guide for Students (0-495-38479-8)
If your instructor has chosen to package *Enhanced WebAssign* with your text, this manual will help you get up and running quickly with the *Enhanced WebAssign* system so you can study smarter and improve your performance in class.

Electronic Ancillaries

WebAssign Enhanced WebAssign (0-538-73810-3)
Get instant feedback on your homework assignments with *Enhanced WebAssign* (assigned by your instructor). This online homework system is easy to use and includes a multimedia eBook, video examples, and problem-specific tutorials.

Website www.cengagebrain.com or www.cengage.com/math/tussy
Visit us on the web and search by title to access a wealth of learning resources, including online Sections 9.4 and 9.5, *Study Skills Workshop* materials, tutorials, final exams, chapter outlines, chapter reviews, web links, videos, flashcards, study skills handouts, and more!

**Printed Access Card for CourseMate with eBook for *Elementary Algebra*, Fifth Edition
(1-133-50781-6)**

**Instant Access Card for CourseMate with eBook for *Elementary Algebra*, Fifth Edition
(1-133-50693-3)**
The more you study, the better the results. Make the most of your study time by accessing everything you need to succeed in one place: read your textbook, take notes, review flashcards, watch videos, and take practice quizzes—online with *CourseMate*.

Acknowledgments

We want to express our gratitude to our accuracy checkers, Diane Koenig and Steve Odrich, as well as many others for their help with this project: Maria H. Andersen, Sheila Pisa, Alexander Lee, Ed Kavanaugh, Karl Hunsicker, Cathy Gong, Dave Ryba, Terry Damron, Marion Hammond, Lin Humphrey, Doug Keebaugh, Robin Carter, Tanja Rinkel, Bob Billups, Jeff Cleveland, Jo Morrison, Sheila White, Jim McClain, Paul Swatzel, Brandon Tussy, Liz Tussy, Dan Davison, Marshall Dean, Dennis Korn, Matt Greenbeck, Joyce Low, Ralph Tippins, Mohamad Trad, and the Citrus College library staff (including Barbara Rugeley) for their help with this project. Your encouragement, suggestions, and insight have been invaluable to us.

We also would like to express our thanks to the Cengage Learning editorial, marketing, production, and design staff for helping us craft this new edition: Charlie Van Wagner, Danielle Derbenti, Gordon Lee, Carrie Jones, Jennifer Cordoba, Heleny Wong, Guanglei Zhang, Maureen Ross, Sam Subity, Jennifer Risden, Vernon Boes, Terri Wright, Ellen Brownstein, Helen Walden, Lori Heckelman, and Graphic World.

Additionally, we would like to say that authoring a textbook is a tremendous undertaking. A revision of this scale would not have been possible without the thoughtful feedback and support from the colleagues listed below. Their contributions to this edition have shaped this revision in countless ways.

Alan S. Tussy
R. David Gustafson

Advisory Reviewers

Ashish Gupta, *William Paterson University*
Katrina Keating, *Diablo Valley College*
Roger Larson, *Anoka Ramsey Community College*
Lorraine Lopez, *San Antonio College*
Paul J. Vroman, *St. Louis Community College at Florissant Valley*

Reviewers of the Fourth and Fifth Editions

Andrea Adlman, *Ventura College*
Khadija Ahmed, *Monroe Community College*
Rodney Alford, *Calhoun Community College*
Maria Andersen, *Muskegon Community College*
Hamid Attarzadeh, *Jefferson Community and Technical College*
Victoria Baker, *University of Houston–Downtown*
Betty Barks, *Lansing Community College*
Scott Barnett, *Henry Ford Community College*
Susan Beane, *University of Houston–Downtown*
David Behrman, *Somerset Community College*
Chad Bemis, *Riverside Community College*
John F. Beyers, *University of Maryland University College*
Barbara Blass, *Oakland Community College*
Candace Blazek, *Anoka Ramsey Community College*
Jennifer Bluth, *Anoka Ramsey Community College*
A. Elena Bogardus, *Camden Community College*
Carilynn Bouie, *Cuyahoga Community College*
Charles A. Bower, *St. Philip's College*
Jeanne Bowman, *University of Cincinnati*
Kim Brown, *Tarrant Community College*
Kirby Bunas, *Santa Rosa Junior College*
Shawna M. Bynum, *Napa Valley College*
Kim Caldwell, *Volunteer State Community College*
Carole Carney, *Brookdale Community College*
Edythe L. Carter, *Amarillo College*
Joe Castillo, *Broward Community College*
Sandra Chandler, *Tidewater Community College*
Carol Cheshire, *Macon State College*
John Close, *Salt Lake Community College*
Chris Copple, *Northwest State Community College*
Tony Craig, *Paradise Valley Community College*
Patrick Cross, *University of Oklahoma*
Mary Deas, *Johnson County Community College*
Suzanne Doviak, *Old Dominion University*
Archie Earl, *Norfolk State University*

Melody Eldred, *State University of New York at Cobleskill*
Peter Embalabala, *Lincoln Land Community College*
Joan Evans, *Texas Southern University*
Mike Everett, *Santa Ana College*
Betsy Farber, *Bucks County Community College*
Rita Fielder, *University of Central Arkansas*
Maggie Flint, *Northeast State*
Anissa Florence, *Jefferson Community and Technical College*
Pat Foard, *South Plains College*
Nancy Forrest, *Grand Rapids Community College*
Tom Fox, *Cleveland State Community College*
Heng Fu, *Thomas Nelson Community College*
Douglas Furman, *SUNY Ulster Community College*
Abel Gage, *Skagit Valley College*
John Garlow, *Tarrant Community College–Southeast Campus*
Vicki Gearhart, *San Antonio College*
Radu Georgescu, *Prince George's Community College*
Rebecca Giles, *Jefferson State Community College*
Alketa Gjikuria, *Cecil College*
Megan Goodwin, *Anoka Ramsey Community College*
Kim Gregor, *Delaware Technical Community
  College–Wilmington*
Thomas Grogan, *Cincinnati State*
Sally Haas, *Angelina College*
Paula Jean Haigis, *Calhoun Community College*
Haile Kebede Haile, *Minneapolis Community and Technical
  College*
Kelli Jade Hammer, *Broward Community College*
Mehdi Hakim Hashemi, *Normandale Community College*
Julia Hassett, *Oakton Community College*
Jennifer Hastings, *Northeast Mississippi Community College*
Alan Hayashi, *Oxnard College*
Kristy Hill, *Hinds Community College*
Jim Hodge, *Mountain State University*
Amy Hoherz, *Johnson County Community College*
Laura Hoye, *Trident Technical College*
Becki Huffman, *Tyler Junior College*
Jeffrey Hughes, *Hinds Community College*
Vera Hu-Hyneman, *SUNY–Suffolk Community College*
Angela Jahns, *North Idaho College*
Cassandra Johnson, *Robeson Community College*
Cynthia Johnson, *Heartland Community College*
Leslie Johnson, *John C. Calhoun State Community College*
Pete Johnson, *Eastern Connecticut State University*
Shelbra Jones, *Wake Technical Community College*
Ed Kavanaugh, *Schoolcraft College*
Leonid Khazanov, *Borough of Manhattan Community College*
MC Kim, *Suffolk County Community College*
Lynette King, *Gadsden State Community College*
Mike Kirby, *Tidewater Community College*
Alex Kolesnik, *Ventura College*
Patricia Kopf, *Kellogg Community College*
Elena Kravchuk, *University of Alabama–Birmingham*
Marlene Kutesky, *Virginia Commonwealth University*
Fred Lang, *Art Institute of Washington*
Hoat Le, *San Diego Community College*
Alexander Lee, *Hinds Community College, Rankin Campus*
Wayne (Paul) Lee, *Saint Philip's College*

Richard Leedy, *Polk Community College*
Mary Legner, *Riverside Community College*
Lamar Lider-Manuel, *Seminole Community College*
Daniel Lopez, *Brookdale Community College*
Ann Loving, *J. Sargeant Reynolds Community College*
Yixia Lu, *South Suburban College*
Keith Luoma, *Augusta State University*
Julie L. Mays, *Angelina College*
Mikal McDowell, *Cedar Valley College*
Marcus McGuff, *Austin Community College*
Owen Mertens, *Missouri State University*
Susan Meshulam, *Indiana University/Purdue University
  Indianapolis*
James Metz, *Kapi'olani Community College*
Trudy Meyer, *El Camino College*
Pam Miller, *Phoenix College*
Molly Misko, *Gadsden State Community College*
Catherine Moushon, *Elgin Community College*
Tania Munding, *Ohlone College*
Charlie Naffziger, *Central Oregon Community College*
Oscar Neal, *Grand Rapids Community College*
Doug Nelson, *Central Oregon Community College*
Elsie Newman, *Owens Community College*
Charlotte Newsom, *Tidewater Community College*
Katrina Nichols, *Delta College*
Randy Nichols, *Delta College*
Stephen Nicoloff, *Paradise Valley Community College*
Megan Nielsen, *St. Cloud State University*
Charles Odion, *Houston Community College*
Jason Pallett, *Longview Community College*
Mary Beth Pattengale, *Sierra College*
Naeemah Payne, *Los Angeles Community College*
Fred Peskoff, *Borough of Manhattan Community College*
Sheila Pisa, *Riverside Community College–Moreno Valley*
Carol Ann Poore, *Hinds Community College*
Jill Rafael, *Sierra College*
Pamela Reed, *North Harris Montgomery Community College*
Pamelyn Reed, *Cy-Fair College*
Nancy Ressler, *Oakton Community College*
Elaine Richards, *Eastern Michigan University*
Harriette Roadman, *New River Community College*
Lilia Ruvalcaba, *Oxnard College*
Jeffrey Saikali, *San Diego Miramar College*
Fary Sami, *Harford Community College*
Emma Sargent, *Tennessee State University*
Ned Schillow, *Lehigh Carbon Community College*
Joe Sedlacek, *Kirkwood Community College*
Wendiann Sethi, *Seton Hall University*
Debra Shafer, *University of North Carolina*
Hazel Shedd, *Hinds Community College*
Patty Sheeran, *McHenry Community College*
Karen Smith, *Nicholls State University*
Christa Solheid, *Santa Ana College*
Donald Solomon, *University of Wisconsin–Milwaukee*
Frankie Solomon, *University of Houston–Downtown*
Jim Spencer, *Santa Rosa Junior College*
John Squires, *Cleveland State Community College*
Michael Stack, *South Suburban College*

Kristen Starkey, *Rose State College*

Robin Steinberg, *Pima Community College*

Kristin Stoley, *Blinn College*

Eleanor Storey, *Front Range Community College–Westminster Campus*

Teresa Sutcliffe, *Los Angeles Valley College*

Eden Thompson, *Utah Valley State College*

Cindy Thore, *Central Piedmont Community College*

Rose Toering, *Kilian Community College*

Fariheh Towfiq, *Palomar College*

James Vallade, *Monroe County Community College*

Gowribalan "Ana" Vamadeva, *University of Cincinnati*

Maggie Pasqua Viz, *Brookdale Community College*

Beverly Vredevelt, *Spokane Falls Community College*

Andreana Walker, *Calhoun Community College*

Carol Walker, *Hinds Community College*

Cynthia Wallin, *Central Virginia Community College*

John Ward, *Kentucky Community and Technical College–Jefferson Community College*

Richard Watkins, *Tidewater Community College*

Diane Williams, *Northern Kentucky University*

Antoinette Willis, *St. Philip's College*

Jackie Wing, *Angelina College*

Judith Wood, *Central Florida Community College*

Nazar Wright, *Guilford Technical Community College*

Valerie Wright, *Central Piedmont Community College*

Shishen Xie, *University of Houston–Downtown*

Catalina Yang, *Oxnard College*

Heidi Young, *Bryant and Stratton College*

Mary Young, *Brookdale Community College*

Ghidei Zedingle, *Normandale Community College*

Loris Zucca, *Kingwood College*

INDEX OF APPLICATIONS

Examples that are applications are shown with **boldface** page numbers.
Exercises that are applications are shown with lightface page numbers.
Page numbers beginning with 9- refer to online content available at **www.cengage.com/math/tussy** or **www.cengagebrain.com**.

An Introduction to Algebra

©Carolina K. Smith, M.D./Shutterstock.com

from Campus to Careers

Lead Transportation Security Officer

Since 9/11, Homeland Security is one of the fastest-growing career choices in the United States. A lead transportation security officer works in an airport where he or she searches passengers, screens baggage, reviews tickets, and determines staffing requirements. The job description calls for the ability to perform arithmetic computations correctly and solve practical problems by choosing from a variety of mathematical techniques such as formulas and percentages.

Problem 113 in **Study Set 1.2, Problem 99** in **Study Set 1.5,** and **Problem 117** in **Study Set 1.7** involve situations that a lead transportation security officer might encounter on the job. The mathematical concepts discussed in this chapter can be used to solve those problems.

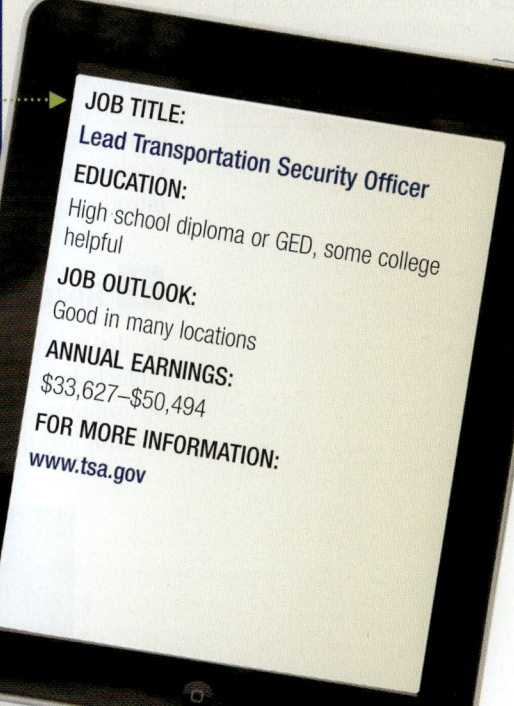

JOB TITLE:
Lead Transportation Security Officer

EDUCATION:
High school diploma or GED, some college helpful

JOB OUTLOOK:
Good in many locations

ANNUAL EARNINGS:
$33,627–$50,494

FOR MORE INFORMATION:
www.tsa.gov

Starting a new course is exciting, but it might also make you a bit nervous. In order to be successful in your algebra class, you need a plan.

MAKE TIME FOR THE COURSE: As a general guideline, 2 hours of independent study time is recommended for every hour in the classroom.

KNOW WHAT IS EXPECTED: Read your instructor's syllabus thoroughly. It lists class policies about attendance, homework, tests, calculators, grading, and so on.

BUILD A SUPPORT SYSTEM: Know where to go for help. Take advantage of your instructor's office hours, your school's tutorial services, the resources that accompany this textbook, and the assistance that you can get from classmates.

Now Try This ▶

Each of the forms referred to below can be found online at: www.cengage.com/math/tussy.
1. To help organize your schedule, fill out the *Weekly Planner Form*.
2. Review the class policies by completing the *Course Information Sheet*.
3. Use the *Support System Worksheet* to build your course support system.

SECTION 1.1

OBJECTIVES

1. Read tables and graphs.
2. Use the basic vocabulary and notation of algebra.
3. Identify expressions and equations.
4. Use equations to construct tables of data.

Introducing the Language of Algebra

ARE YOU READY?

The following problems review some basic arithmetic skills that are needed in this section. Answers to the Are You Ready? *problems are located in Appendix 3 at the back of the book.*

1. Add: $125 + 85$
2. Subtract: $2,400 - 650$
3. Multiply: $78 \cdot 14$
4. Divide: $243 \div 27$

Algebra is the result of contributions from many cultures over thousands of years. The word *algebra* comes from the title of the book *Ihm Al-jabr wa'l muqābalah,* written by an Arabian mathematician around A.D. 800. We can think of algebra as a language with its own vocabulary and notation. In this section, we begin to explore the language of algebra by introducing some of its basic components.

1 Read Tables and Graphs.

In algebra, we often use **tables** to show relationships between quantities. For example, the table below lists the number of calories a 160-pound adult burns during 10, 20, 30, and 40 minutes of snowboarding. For a workout of, say, 30 minutes, we locate 30 in the left column and then scan across the table to see that 300 calories are burned.

Minutes snowboarding	Calories burned
10	100
20	200
30	300
40	400

©Ipatov/Shutterstock.com

The information in the table also can be presented in a **bar graph,** as shown on the next page, on the left. The **horizontal axis** of the graph is labeled "Minutes snowboarding," and it is scaled in units of 10 minutes. The **vertical axis,** labeled "Calories burned," is scaled in units of 50 calories. The height of a bar indicates the number of calories burned. For example, the bar over 40 minutes extends upward to 400. This means 400 calories are burned during a 40-minute snowboarding workout.

Another way to present the snowboarding information is with a **line graph.** Instead of using a bar to represent the number of calories burned, we use a dot drawn at the correct height. After drawing the data points for workouts of 10, 20, 30, and 40 minutes, we connect them with line segments to create the graph shown below, on the right.

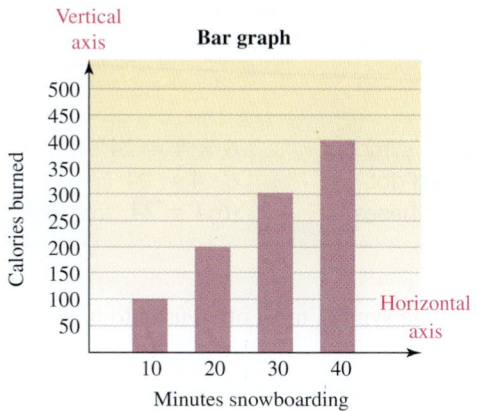
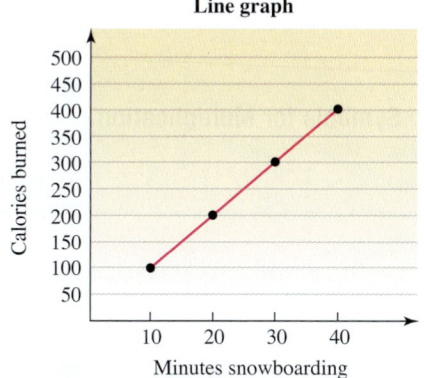

EXAMPLE 1

Fitness.   Use the line graph above to find the number of calories burned during a 25-minute snowboarding workout.

Strategy   We will start at 25 on the horizontal axis of the graph. Then we will scan up to the red line, and over, to read the number of calories burned on the vertical axis.

Why   We start on the horizontal axis because that scale gives the number of minutes of snowboarding. We scan up and over to the vertical axis because that scale gives the number of calories burned.

Solution   We locate 25 minutes (between 20 and 30 minutes) on the horizontal axis and draw a dashed line upward to intersect the red line. From the point of intersection, we then draw a dashed line to the left that points to the vertical axis at 250. This means that a 25-minute snowboarding workout burns 250 calories.

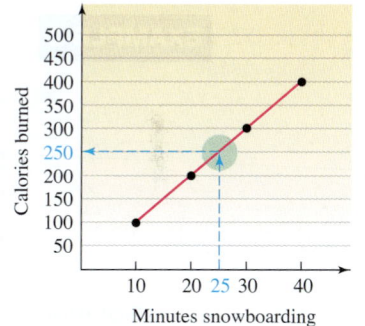

Self Check 1   **Fitness.** Use the graph to find the number of calories burned during a 35-minute snowboarding workout.

Now Try   Problem 29

---

2   Use the Basic Vocabulary and Notation of Algebra.

From the table and graphs, we see that there is a relationship between the number of calories burned and the number of minutes snowboarding. Using words, we can express this relationship as a **verbal model:**

"The number of calories burned is ten times the number of minutes snowboarding."

Since the word **product** indicates the result of a multiplication, we can also write:

"The number of calories burned is the *product* of ten and the number of minutes snowboarding."

To indicate other arithmetic operations, we will use the following words.

- A **sum** is the result of an addition: The sum of 5 and 6 is 11.
- A **difference** is the result of a subtraction: The difference of 3 and 2 is 1.
- A **quotient** is the result of a division: The quotient of 6 and 3 is 2.

Many symbols used in arithmetic are also used in algebra. For example, a + symbol is used to indicate addition, a − symbol is used to indicate subtraction, and an = symbol means *is equal to.*

Since the letter x is often used in algebra and could be confused with the multiplication symbol $\times$, we usually write multiplication using a **raised dot** or **parentheses.**

Symbols for Multiplication			
	$\times$	Times symbol	$6 \times 4 = 24$
	$\cdot$	Raised dot	$6 \cdot 4 = 24$
	$(\ )$	Parentheses	$(6)4 = 24$ or $6(4) = 24$ or $(6)(4) = 24$

In algebra, the symbol most often used to indicate division is the **fraction bar.**

Symbols for Division			
	$\div$	Division symbol	$24 \div 4 = 6$
	$\overline{)}$	Long division	$4\overline{)24}$ with quotient 6
	$—$	Fraction bar	$\dfrac{24}{4} = 6$

EXAMPLE 2   Write each statement in words, using one of the words *sum, product, difference,* or *quotient:*
a. $\dfrac{22}{11} = 2$ **b.** $22 + 11 = 33$

Strategy   We will examine each statement to determine whether addition, subtraction, multiplication, or division is being performed.

Why   The word that we should use (*sum, product, difference,* or *quotient*) depends on the arithmetic operation that we have to describe.

Solution   **a.** Since the fraction bar indicates division, we have: The quotient of 22 and 11 equals 2.
b. The + symbol indicates addition: The sum of 22 and 11 equals 33.

Self Check 2   Write the following statement in words: $22 \cdot 11 = 242$

Now Try ▶ Problems 33 and 35

3 Identify Expressions and Equations.

Another way to describe the relationship between calories burned and snowboarding time uses *variables.* **Variables** are letters that stand for numbers. If we let the letter m represent the number of minutes snowboarding, then the number of calories burned is ten times m, written $10m$. In this notation, the number 10 is an example of a **constant** because it does not change value.

When multiplying a variable by a number, or a variable by another variable, we can omit the symbol for multiplication. For example,

$10m$ means $10 \cdot m$ xy means $x \cdot y$ $8abc$ means $8 \cdot a \cdot b \cdot c$

We call $10m$, xy, and $8abc$ algebraic expressions.

| **Algebraic Expressions** | ▼ | Variables and/or numbers can be combined with the operations of addition, subtraction, multiplication, and division to create **algebraic expressions**. |

Here are some other examples of algebraic expressions.

$4a + 7$ *This expression is a combination of the numbers 4 and 7, the variable a, and the operations of multiplication and addition.*

$\dfrac{10 - y}{3}$ *This expression is a combination of the numbers 10 and 3, the variable y, and the operations of subtraction and division.*

$15mn(2m)$ *This expression is a combination of the numbers 15 and 2, the variables m and n, and the operation of multiplication.*

In the snowboarding example, if we let the letter c stand for the number of calories burned, we can translate the verbal model to mathematical symbols.

The number of calories burned	is	ten	times	the number of minutes snowboarding.
c	$=$	10	$\cdot$	m

The statement $c = 10 \cdot m$, or more simply, $c = 10m$, is called an *equation*. An **equation** is a mathematical sentence that contains an $=$ symbol. The $=$ symbol indicates that the expressions on either side of it have the same value. Other examples of equations are

$$3 + 5 = 8 \qquad x + 5 = 20 \qquad 17 - 2r = 14 + 3r \qquad p = 100 - d$$

©iStockPhoto.com/MvH

EXAMPLE 3

Stormy Weather.   One way to estimate your distance (in miles) from a lightning strike is to count the number of seconds between the flash of lightning and the sound of thunder and divide by five. Translate this verbal model into an equation.

Strategy   We will represent the two unknown quantities using variables and we will use symbols to represent the words *is* and *divided by.*

Why   To translate a verbal (word) model into an equation means to write it using mathematical symbols.

Solution   Let d = your distance (in miles) from the lightning strike and s = the number of seconds between the lightning and the thunder. Then we have:

Your distance (in miles) from the lightning strike	is	the number of seconds between the lightning and thunder	divided by	five.
d	$=$	s	$\div$	5

If we write the division using a fraction bar, then the verbal model translates to the equation $d = \frac{s}{5}$.

Self Check 3   Translate into an equation: The number of unsold tickets is the difference of 500 and the number of tickets that have been purchased.

Now Try ▶ Problems 41 and 45

In the snowboarding example, we have seen that a table, a graph, and an equation can be used to describe the relationship between calories burned and workout time. The equation $c = 10m$ has one major advantage over the other methods. It can be used to accurately determine the number of calories burned during a snowboarding workout of *any* length of time.

EXAMPLE 4

Fitness. Use the equation $c = 10m$ to find the number of calories burned during a 36-minute snowboarding workout.

Strategy In $c = 10m$, we will replace m with 36. Then we will multiply 36 by 10 to obtain the value of c.

Why The equation $c = 10m$ indicates that the number of calories burned is found by multiplying the number of minutes snowboarding by 10.

Solution

$c = 10m$ This is the describing equation.

$c = 10(36)$ Replace *m*, which stands for the number of minutes snowboarding, with 36. Use parentheses to show the multiplication. We also could write 10 · 36.

$c = 360$ Do the multiplication.

A snowboarding workout of 36 minutes will burn 360 calories.

Self Check 4   **Fitness.** Use the equation $c = 10m$ to find the number of calories burned during a 48-minute snowboarding workout.

Now Try ▶ Problem 53

4   Use Equations to Construct Tables of Data.

Equations such as $c = 10m$, which express a relationship between two or more variables, are called **formulas.** Some applications require the repeated use of a formula.

EXAMPLE 5

Fitness. Find the number of calories burned during snowboarding workouts of 18 minutes and 65 minutes. Present the results in a table.

Strategy We need to use the formula $c = 10m$ twice.

Why There are two different workouts: one that is 18 minutes long and another that is 65 minutes long.

Solution   *Step 1:* We construct a two-column table and enter the workout times in the first column, as shown below in red.

The Language of Algebra

To **substitute** means to put or use in place of another, as with a *substitute* teacher. Here, we *substitute* 18 and 65 for *m*.

$c = 10m$

Since *m* represents the number of minutes snowboarding, we use it as the heading of the first column.

m	c
18	180
65	650

Since *c* represents the number of calories burned, we use it as the heading of the second column.

Step 2: We substitute 18 and 65 for m in $c = 10m$ and find each corresponding value of c. The results are entered in the second column of the table, as shown above.

$c = 10m$ $c = 10m$

$c = 10(18)$ $c = 10(65)$

$c = 180$ $c = 650$

Self Check 5   **Fitness.** Find the number of calories burned during snowboarding workouts of 8 minutes and 75 minutes. Present the results in a table.

Success Tip

Answers to the odd-numbered problems in each Study Set can be found at the back of the book in Appendix 3, beginning on page A-7.

Now Try ▶ Problem 55

SECTION 1.1 ▶ STUDY SET

VOCABULARY

Fill in the blanks.

1. A _____ is the result of an addition. A _____ is the result of a subtraction. A _____ is the result of a multiplication. A _____ is the result of a division.

2. _____ are letters (or symbols) that stand for numbers.

3. A number, such as 8, is called a _____ because it does not change.

4. Variables and numbers can be combined with the operations of addition, subtraction, multiplication, and division to create algebraic _____.

5. An _____ is a mathematical sentence that contains an = symbol. An algebraic _____ does not.

6. An equation such as $c = 10m$, which expresses a relationship between two or more variables, is called a _____.

7. The _____ axis of a graph extends left and right and the vertical axis extends up and down.

8. The word _____ comes from the title of a book written by an Arabian mathematician around A.D. 800.

CONCEPTS

Classify each item as an algebraic expression or an equation.

9. a. $m + 18 = 23$ b. $m + 18$

10. a. $30x$ b. $30x = 600$

11. a. $\dfrac{c - 7}{5}$ b. $\dfrac{c - 7}{5} = 7c$

12. a. $r = \dfrac{2}{3}$ b. $\dfrac{2}{3}r$

13. What arithmetic operations does the expression $\frac{12 + 9t}{25}$ contain? What variable does it contain?

14. What arithmetic operations does the equation $4y - 14 = 5(6)$ contain? What variable does it contain?

15. Construct a line graph using the data in the following table.

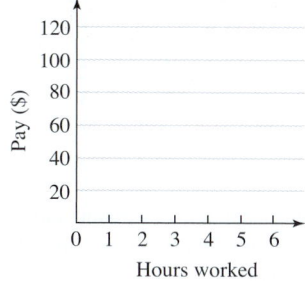

Hours worked	Pay (dollars)
1	20
2	40
3	60
4	80
5	100

16. Use the data in the graph to complete the table.

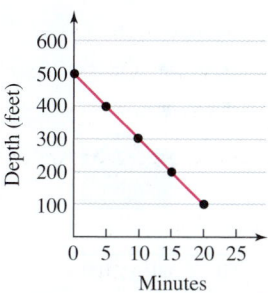

Minutes	Depth (feet)
0	
5	
10	
15	
20	

NOTATION

Fill in the blanks.

17. The symbol $\neq$ means ___ ____ _____ ___.

18. The symbols ( ) are called _____.

19. Write the multiplication 5×6 using a raised dot and then using parentheses.

20. Give four verbs that can be represented by an equal symbol =.

Write each expression without using a multiplication symbol or parentheses.

21. $4 \cdot x$ 22. $P \cdot r \cdot t$

23. $2(w)$ 24. $(x)(y)$

Write each division using a fraction bar.

25. $32 \div x$ 26. $3)\overline{90}$

27. $5)\overline{55}$ 28. $h \div 15$

GUIDED PRACTICE

Use the given line graphs to answer the following questions. See Example 1.

29. **Accounting.** Explain what the dashed lines in the graph below help us find.

30. **Accounting.** What is the value of 35-year-old machinery?

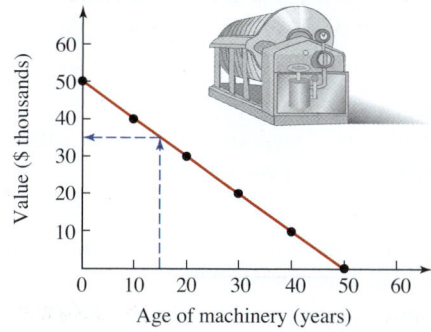

31. Business. Refer to the graph below. Find the income received from 30 customers.

32. Business. Refer to the graph below. Find the income received from 70 customers.

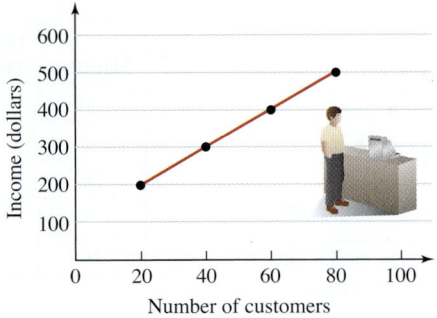

Express each statement using one of the words sum, product, difference, *or* quotient. *See Example 2.*

33. $8(2) = 16$

34. $45 \cdot 12 = 540$

35. $11 - 9 = 2$

36. $65 + 89 = 154$

37. $x + 2 = 10$

38. $16 - t = 4$

39. $\dfrac{66}{11} = 6$

40. $12 \div 3 = 4$

Translate each verbal model into an equation. (Answers may vary, depending on the variables chosen.) See Example 3.

41.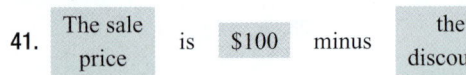

42.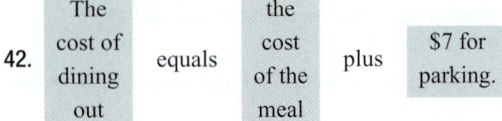

43.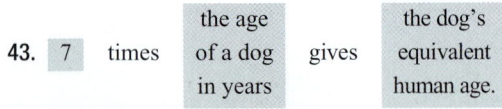

44.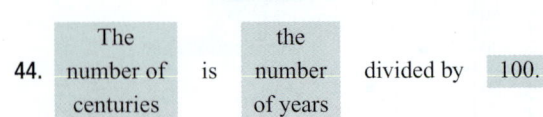

45. The amount of sand that should be used is the product of 3 and the amount of cement used.

46. The number of waiters needed is the quotient of the number of customers and 10.

47. The weight of the truck is the sum of the weight of the engine and 1,200.

48. The number of classes still open is the difference of 150 and the number of classes that are closed.

49. The profit is the difference of the revenue and 600.

50. The distance is the product of the rate and 3.

51. The quotient of the number of laps run and 4 gives the number of miles run.

52. The sum of the tax and 35 gives the total cost.

Use the formula to complete each table. **See Examples 4 and 5.**

53. $d = 360 + L$

Lunch time (minutes) L	School day (minutes) d
30	
40	
45	

54. $b = 1{,}024k$

Kilobytes k	Bytes b
1	
5	
10	

55. $t = 1{,}500 - d$

Deductions d	Take-home pay t
200	
300	
400	

56. $w = \dfrac{s}{12}$

Inches of snow s	Inches of water w
12	
24	
72	

Use the data in the table to complete the formula.

57. $d = \dfrac{e}{\boxed{}}$

Eggs e	Dozens d
24	2
36	3
48	4

58. $p = \boxed{}\,c$

Canoes c	Paddles p
6	12
7	14
8	16

59. $I = \boxed{}\,c$

Couples c	Individuals I
20	40
100	200
200	400

60. $t = \dfrac{p}{\boxed{}}$

Players p	Teams t
5	1
10	2
15	3

APPLICATIONS

61. Exercise. The number of calories that a 125-pound adult burns doing general house cleaning chores is three times the number of minutes spent cleaning.

  a. Write a verbal model using the word *product* that describes the relationship between calories burned and minutes cleaning.

  b. Write a formula using the variables c and m that describes the relationship between calories burned and minutes cleaning.

c. Use your answer to part b to complete the following table.

m	10	20	30	40	50	60
c						

d. Use the data from the table to construct a line graph. Scale the horizontal axis in units of 10 minutes. Scale the vertical axis in units of 30 calories.

62. **Traffic Safety.** As the railroad crossing guard drops, the measure of angle 1 (written $\angle 1$) increases while the measure of $\angle 2$ decreases. At any instant the *sum* of the measures of the two angles is 90°. Complete the table. Then use the data to construct a line graph. Scale each axis in units of 15°.

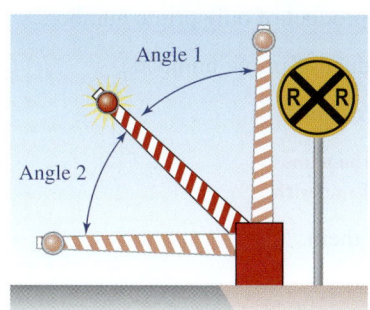

Angle 1 (degrees)	Angle 2 (degrees)
0	
15	
30	
45	
60	
75	
90	

WRITING

63. Many students misuse the word *equation* when discussing mathematics. What is an equation? Give an example.

64. Explain the difference between an algebraic expression and an equation. Give an example of each.

65. In this section, four methods for describing numerical relationships were discussed: tables, verbal models (words), graphs, and equations. Which method do you think is the most useful? Explain why.

66. In your own words, define *horizontal* and *vertical*.

CHALLENGE PROBLEMS

67. Complete the formula. $t = \boxed{}\, s + \boxed{}$

s	t
18	55
33	100
47	142

68. Suppose $h = 4n$ and $n = 2g$. Complete the following formula: $h = \boxed{}\, g$

 SECTION 1.2

Fractions

OBJECTIVES

1. Factor and prime factor natural numbers.
2. Recognize special fraction forms.
3. Multiply and divide fractions.
4. Build equivalent fractions.
5. Simplify fractions.
6. Add and subtract fractions.
7. Simplify answers.
8. Compute with mixed numbers.

The Language of Algebra

When we say "factor 8," we are using the word **factor** as a verb. When we say "2 is a *factor* of 8," we are using the word **factor** as a noun.

ARE YOU READY?

The following problems review some basic skills that are needed when working with fractions.

1. What is the value of $\frac{8}{8}$?
2. Multiply: $2 \cdot 3 \cdot 5 \cdot 5$
3. Is 42 divisible by 3?
4. Write the fraction $\frac{4}{5}$ in words.

In arithmetic, we add, subtract, multiply, and divide **natural numbers:** 1, 2, 3, 4, 5, and so on. Assuming that you have mastered those skills, we will now review the arithmetic of fractions.

1 Factor and Prime Factor Natural Numbers.

To compute with fractions, we need to know how to *factor* natural numbers. To **factor** a number means to express it as a product of two or more numbers. For example, some ways to factor 8 are

$$1 \cdot 8, \qquad 4 \cdot 2, \qquad \text{and} \qquad 2 \cdot 2 \cdot 2$$

The numbers 1, 2, 4, and 8 that were used to write the products are called *factors* of 8. In general, a **factor** is a number being multiplied.

Sometimes a number has only two factors, itself and 1. We call such numbers *prime numbers.*

Prime Numbers and Composite Numbers	A **prime number** is a natural number greater than 1 that has only itself and 1 as factors. The first ten prime numbers are 2, 3, 5, 7, 11, 13, 17, 19, 23, and 29.
	A **composite number** is a natural number, greater than 1, that is not prime. The first ten composite numbers are 4, 6, 8, 9, 10, 12, 14, 15, 16, and 18.

Every composite number can be factored into the product of two or more prime numbers. This product of these prime numbers is called its **prime factorization.**

EXAMPLE 1   Find the prime factorization of 210.

Strategy   We will use a series of steps to express 210 as a product of only prime numbers.

Why   To *prime factor* a number means to write it as a product of prime numbers.

Solution   First, write 210 as the product of two natural numbers other than 1.

$$210 = \mathbf{10 \cdot 21}$$ The resulting prime factorization will be the same no matter which two factors of 210 you begin with.

The Language of Algebra

Prime factors often are written in **ascending** order. To **ascend** means to move upward.

Neither 10 nor 21 are prime numbers, so we factor each of them.

$$210 = \mathbf{2 \cdot 5 \cdot 3 \cdot 7}$$ Factor 10 as 2 · 5 and factor 21 as 3 · 7.

Writing the factors in order, from least to greatest, the **prime-factored form** of 210 is $2 \cdot 3 \cdot 5 \cdot 7$. Two other methods for prime factoring 210 are shown below.

Factor tree   *Division ladder*

Work downward. Factor each number as a product of two numbers (other than 1 and itself) until all factors are prime. Circle prime numbers as they appear at the end of a branch.

```
      210
    7     30
        6    5
      3   2
```

```
      7
   5)35
   3)105
   2)210
```

Work upward. Perform repeated division until the final quotient is a prime number. It is helpful to start with the smallest prime, 2, as a trial divisor. Then, in order, try larger primes as divisors: 3, 5, 7, 11, and so on.

Success Tip

The following divisibility rules are helpful when prime factoring.
A whole number is divisible by
• 2 if it ends in 0, 2, 4, 6, or 8
• 3 if the sum of the digits is divisible by 3
• 5 if it ends in 0 or 5
• 10 if it ends in 0

Either way, the factorization is $2 \cdot 3 \cdot 5 \cdot 7$. To check it, multiply the prime factors. The product should be 210.

Self Check 1   Find the prime factorization of 189.

Now Try   Problems 15 and 23

2 Recognize Special Fraction Forms.

A **fraction** describes the number of equal parts of a whole. For example, consider the figure below with 5 of the 6 equal parts colored red. We say that $\frac{5}{6}$ (five-sixths) of the figure is shaded.

In a fraction, the number above the **fraction bar** is called the **numerator,** and the number below is called the **denominator.**

The Language of Algebra

The word **fraction** comes from the Latin word *fractio* meaning "breaking in pieces."

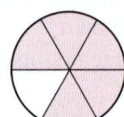

$$\text{Fraction bar} \longrightarrow \frac{5 \leftarrow \text{numerator}}{6 \leftarrow \text{denominator}}$$

Fractions are also used to indicate division. For example, the fraction bar in $\frac{8}{2}$ indicates that the numerator, 8, is to be divided by the denominator, 2:

$$\frac{8}{2} = 8 \div 2 = 4$$ We know that $\frac{8}{2} = 4$ because of its related multiplication statement: $2 \cdot 4 = 8$.

If the numerator and denominator of a fraction are the same nonzero number, the fraction indicates division of a number by itself, and the result is 1. Each of the following fractions is, therefore, a **form of 1.**

$$1 \;=\; \frac{1}{1} \;=\; \frac{2}{2} \;=\; \frac{3}{3} \;=\; \frac{4}{4} \;=\; \frac{5}{5} \;=\; \frac{6}{6} \;=\; \frac{7}{7} \;=\; \frac{8}{8} \;=\; \frac{9}{9} \;=\; \cdots$$

If a denominator is 1, the fraction indicates division by 1, and the result is simply the numerator. For example, $\frac{5}{1} = 5$ and $\frac{24}{1} = 24$.

Special Fraction Forms

For any nonzero number a,

$$\frac{a}{a} = 1 \qquad \text{and} \qquad \frac{a}{1} = a$$

---

3 **Multiply and Divide Fractions.**

The rule for multiplying fractions can be expressed in words and in symbols as follows.

Multiplying Fractions

To multiply two fractions, multiply the numerators and multiply the denominators. For any two fractions $\frac{a}{b}$ and $\frac{c}{d}$,

$$\frac{a}{b} \cdot \frac{c}{d} = \frac{a \cdot c}{b \cdot d}$$

EXAMPLE 2   Multiply: $\dfrac{7}{8} \cdot \dfrac{3}{5}$

Strategy   To find the product, we will multiply the numerators, 7 and 3, and multiply the denominators, 8 and 5.

Why   This is the rule for multiplying two fractions.

Solution
$$\frac{7}{8} \cdot \frac{3}{5} = \frac{7 \cdot 3}{8 \cdot 5}$$ Multiply the numerators.
Multiply the denominators.

$$= \frac{21}{40}$$

Self Check 2   Multiply: $\dfrac{5}{9} \cdot \dfrac{2}{3}$

Now Try ▶   Problem 27

One number is called the **reciprocal** of another if their product is 1. To find the reciprocal of a fraction, we invert its numerator and denominator.

$\frac{3}{4}$ is the reciprocal of $\frac{4}{3}$, because $\frac{3}{4} \cdot \frac{4}{3} = \frac{12}{12} = 1$.

$\frac{1}{10}$ is the reciprocal of 10, because $\frac{1}{10} \cdot 10 = \frac{10}{10} = 1$.

We use reciprocals to divide fractions.

Dividing Fractions

To divide two fractions, multiply the first fraction by the reciprocal of the second.
For any two fractions $\frac{a}{b}$ and $\frac{c}{d}$, where $c \neq 0$,

$$\frac{a}{b} \div \frac{c}{d} = \frac{a}{b} \cdot \frac{d}{c}$$

EXAMPLE 3   Divide: $\dfrac{1}{3} \div \dfrac{4}{5}$

Strategy   We will multiply the first fraction, $\frac{1}{3}$, by the reciprocal of the second fraction, $\frac{4}{5}$.

Why   This is the rule for dividing two fractions.

Solution

$\dfrac{1}{3} \div \dfrac{4}{5} = \dfrac{1}{3} \cdot \dfrac{5}{4}$ Multiply $\frac{1}{3}$ by the reciprocal of $\frac{4}{5}$, which is $\frac{5}{4}$.

$= \dfrac{1 \cdot 5}{3 \cdot 4}$ Use the rule for multiplying fractions.
Multiply the numerators. Multiply the denominators.

$= \dfrac{5}{12}$

Self Check 3   Divide: $\frac{6}{25} \div \frac{1}{2}$

Now Try ▶   Problem 31

4   Build Equivalent Fractions.

The two rectangular regions on the right are identical. The first one is divided into 10 equal parts. Since 6 of those parts are red, $\frac{6}{10}$ of the figure is shaded.

$\frac{6}{10}$

The second figure is divided into 5 equal parts. Since 3 of those parts are red, $\frac{3}{5}$ of the figure is shaded. We can conclude that $\frac{6}{10} = \frac{3}{5}$ because $\frac{6}{10}$ and $\frac{3}{5}$ represent the same shaded portion of the figure. We say that $\frac{6}{10}$ and $\frac{3}{5}$ are *equivalent fractions*.

$\frac{3}{5}$

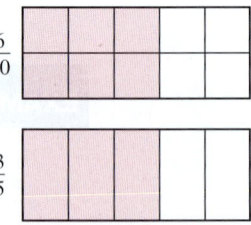

Equivalent Fractions

Two fractions are **equivalent** if they represent the same number. **Equivalent fractions** represent the same portion of a whole.

Writing a fraction as an equivalent fraction with a larger denominator is called **building the fraction**. To build a fraction, we multiply it by a form of 1. Since any number multiplied by 1 remains the same (identical), 1 is called the **multiplicative identity element**.

Multiplication Property of 1	The product of 1 and any number is that number. For any number a, $$1 \cdot a = a \qquad \text{and} \qquad a \cdot 1 = a$$

EXAMPLE 4   Write $\dfrac{3}{5}$ as an equivalent fraction with a denominator of 35.

Strategy   We will compare the given denominator to the required denominator and ask, "By what must we multiply 5 to get 35?"

Why   The answer to that question helps us determine the form of 1 to be used to build an equivalent fraction.

Solution   We need to multiply the denominator of $\dfrac{3}{5}$ by 7 to obtain a denominator of 35. It follows that $\dfrac{7}{7}$ should be the form of 1 that is used to build $\dfrac{3}{5}$. Multiplying $\dfrac{3}{5}$ by $\dfrac{7}{7}$ changes its appearance but does not change its value, because we are multiplying it by 1.

> **Success Tip**
>
> Multiplying $\frac{3}{5}$ by $\frac{7}{7}$ changes its appearance, but does not change its value, because we are multiplying it by a form of 1.

$$\frac{3}{5} = \frac{3}{5} \cdot \frac{7}{7} \qquad \text{\color{red}{Multiply } \frac{3}{5} \text{\color{red}{ by a form of 1: }} \frac{7}{7} = 1}$$

$$= \frac{3 \cdot 7}{5 \cdot 7} \qquad \begin{array}{l} \text{\color{red}{Multiply the numerators.}} \\ \text{\color{red}{Multiply the denominators.}} \end{array}$$

$$= \frac{21}{35}$$

Self Check 4   Write $\dfrac{5}{8}$ as an equivalent fraction with a denominator of 24.

Now Try ▶ Problem 35

Building Fractions	To build a fraction, multiply it by 1 in the form of $\dfrac{c}{c}$, where c is any nonzero number.

To build an equivalent fraction in Example 4, we multiplied $\frac{3}{5}$ by 1 in the form of $\frac{7}{7}$. As a result of that step, the numerator and the denominator of $\frac{3}{5}$ were multiplied by 7:

$$\frac{3 \cdot 7}{5 \cdot 7} \qquad \begin{array}{l} \leftarrow \text{\color{red}{The numerator is multiplied by 7.}} \\ \leftarrow \text{\color{red}{The denominator is multiplied by 7.}} \end{array}$$

This process illustrates the following property of fractions.

The Fundamental Property of Fractions	If the numerator and denominator of a fraction are multiplied by the same nonzero number, the resulting fraction is equivalent to the original fraction.

Since multiplying the numerator and denominator of a fraction by the same nonzero number produces an equivalent fraction, your instructor may allow you to begin your solution to problems like Example 4 as shown above.

5 Simplify Fractions.

Every fraction can be written in infinitely many equivalent forms. For example, some equivalent forms of $\frac{10}{15}$ are:

$$\frac{2}{3} = \frac{4}{6} = \frac{6}{9} = \frac{8}{12} = \mathbf{\frac{10}{15}} = \frac{12}{18} = \frac{14}{21} = \frac{16}{24} = \frac{18}{27} = \frac{20}{30} = \ldots$$

Of all of the equivalent forms in which we can write a fraction, we often need to determine the one that is in *simplest form*.

Simplest Form of a Fraction	A fraction is in **simplest form,** or **lowest terms,** when the numerator and denominator have no common factors other than 1.

To **simplify a fraction,** we write it in simplest form by removing a factor equal to 1. For example, to simplify $\frac{10}{15}$, we note that the greatest factor common to the numerator and denominator is 5 and proceed as follows:

$$\frac{10}{15} = \frac{2 \cdot \mathbf{5}}{3 \cdot \mathbf{5}}$$ To prepare to simplify the fraction, factor 10 and 15. Note the form of 1 highlighted in red.

$$= \frac{2}{3} \cdot \frac{\mathbf{5}}{\mathbf{5}}$$ Use the rule for multiplying fractions in reverse: write $\frac{2 \cdot 5}{3 \cdot 5}$ as the product of two fractions, $\frac{2}{3}$ and $\frac{5}{5}$.

$$= \frac{2}{3} \cdot \mathbf{1}$$ Any nonzero number divided by itself is equal to 1: $\frac{5}{5} = 1$.

$$= \frac{2}{3}$$ Any number multiplied by 1 remains the same.

To simplify $\frac{10}{15}$, we removed a factor equal to 1 in the form of $\frac{5}{5}$. The result, $\frac{2}{3}$, is equivalent to $\frac{10}{15}$.

We can easily identify the greatest common factor of the numerator and the denominator of a fraction if we write them in prime-factored form.

EXAMPLE 5   Simplify each fraction, if possible: **a.** $\frac{63}{42}$ **b.** $\frac{33}{40}$

Strategy   We will begin by prime factoring the numerator and denominator of the fraction. Then, to simplify it, we will remove a factor equal to 1.

Why   We need to make sure that the numerator and denominator have no common factors other than 1. If that is the case, then the fraction is in *simplest form.*

Solution   **a.** After prime factoring 63 and 42, we see that the greatest common factor of the numerator and the denominator is $3 \cdot 7 = 21$.

$$\frac{63}{42} = \frac{3 \cdot \mathbf{3 \cdot 7}}{2 \cdot \mathbf{3 \cdot 7}}$$ To prepare to simplify the fraction, write 63 and 42 in prime-factored form.

$$= \frac{3}{2} \cdot \frac{\mathbf{3 \cdot 7}}{\mathbf{3 \cdot 7}}$$ Write $\frac{3 \cdot 3 \cdot 7}{2 \cdot 3 \cdot 7}$ as the product of two fractions, $\frac{3}{2}$ and $\frac{3 \cdot 7}{3 \cdot 7}$.

$$= \frac{3}{2} \cdot \mathbf{1}$$ Any nonzero number divided by itself is equal to 1: $\frac{3 \cdot 7}{3 \cdot 7} = 1$.

$$= \frac{3}{2}$$ Any number multiplied by 1 remains the same.

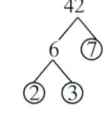

b. To attempt to simplify the fraction, prime factor 33 and 40.

$$\frac{33}{40} = \frac{3 \cdot 11}{2 \cdot 2 \cdot 2 \cdot 5}$$

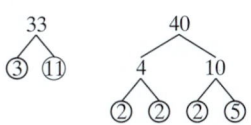

Since the numerator and the denominator have no common factors other than 1, the fraction $\frac{33}{40}$ is in simplest form (lowest terms).

> **Self Check 5** Simplify each fraction, if possible: **a.** $\frac{24}{56}$ **b.** $\frac{16}{125}$
>
> **Now Try ▶** Problems 45 and 49

To streamline the simplifying process, we can replace pairs of factors common to the numerator and denominator with the equivalent fraction $\frac{1}{1}$.

EXAMPLE 6   Simplify: $\dfrac{90}{105}$

Strategy   We will begin by prime factoring the numerator, 90, and denominator, 105. Then we will look for any factors common to the numerator and denominator and remove them.

Why   When the numerator and/or denominator of a fraction are large numbers, such as 90 and 105, writing their prime factorizations is helpful in identifying any common factors.

Solution

$$\frac{90}{105} = \frac{2 \cdot 3 \cdot 3 \cdot 5}{3 \cdot 5 \cdot 7}$$

To prepare to simplify the fraction, write 90 and 105 in prime-factored form.

$$= \frac{2 \cdot \overset{1}{\cancel{3}} \cdot 3 \cdot \overset{1}{\cancel{5}}}{\underset{1}{\cancel{3}} \cdot \underset{1}{\cancel{5}} \cdot 7}$$

Slashes and 1's are used to show that $\frac{3}{3}$ and $\frac{5}{5}$ are replaced by the equivalent fraction $\frac{1}{1}$. A factor equal to 1 in the form of $\frac{3 \cdot 5}{3 \cdot 5} = \frac{15}{15}$ was removed.

$$= \frac{6}{7}$$

Multiply the remaining factors in the numerator: $2 \cdot 1 \cdot 3 \cdot 1 = 6$. Multiply the remaining factors in the denominator: $1 \cdot 1 \cdot 7 = 7$.

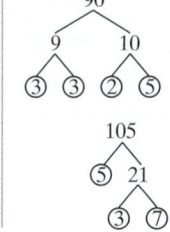

> **Self Check 6** Simplify: $\dfrac{126}{70}$
>
> **Now Try ▶** Problem 53

We can use the following steps to simplify a fraction.

Simplifying Fractions

1. Factor (or prime factor) the numerator and denominator to determine their common factors.
2. Remove factors equal to 1 by replacing each pair of factors common to the numerator and denominator with the equivalent fraction $\frac{1}{1}$.
3. Multiply the remaining factors in the numerator and in the denominator.

6 Add and Subtract Fractions.

In algebra as in everyday life, we can add or subtract only objects that are similar. For example, we can add dollars to dollars, but we cannot add dollars to oranges. This concept is important when adding or subtracting fractions.

Consider the problem $\frac{2}{5} + \frac{1}{5}$. When we write it in words, it is apparent we are adding similar objects.

two-**fifths**   +   one-**fifth**
└─ Similar objects ─┘

Because the denominators of $\frac{2}{5}$ and $\frac{1}{5}$ are the same, we say that they have a **common denominator.**

Adding and Subtracting Fractions That Have the Same Denominator	To add (or subtract) fractions that have the same denominator, add (or subtract) their numerators and write the sum (or difference) over the common denominator. Simplify the result, if possible.

For any fractions $\frac{a}{d}$ and $\frac{b}{d}$,

$$\frac{a}{d} + \frac{b}{d} = \frac{a+b}{d} \qquad \text{and} \qquad \frac{a}{d} - \frac{b}{d} = \frac{a-b}{d}$$

For example,

$$\frac{2}{5} + \frac{1}{5} = \frac{2+1}{5} = \frac{3}{5} \qquad \text{and} \qquad \frac{18}{23} - \frac{9}{23} = \frac{18-9}{23} = \frac{9}{23}$$

Now we consider the problem $\frac{2}{5} + \frac{1}{3}$. Since the denominators are not the same, we cannot add these fractions in their present form.

two-**fifths**   +   one-**third**
└─ Not similar objects ─┘

To add (or subtract) fractions with different denominators, we express them as equivalent fractions that have a common denominator. The smallest common denominator, called the **least** or **lowest common denominator,** is always the easiest common denominator to use.

Least Common Denominator (LCD)	The **least** or **lowest common denominator (LCD)** for a set of fractions is the smallest number each denominator will divide exactly (divide with no remainder).

The denominators of $\frac{2}{5}$ and $\frac{1}{3}$ are 5 and 3. The numbers 5 and 3 divide many numbers exactly (30, 45, and 60, to name a few), but the smallest number that they divide exactly is 15. Thus, 15 is the LCD for $\frac{2}{5}$ and $\frac{1}{3}$.

To find $\frac{2}{5} + \frac{1}{3}$, we build equivalent fractions that have denominators of 15 and we use the rule for adding fractions.

$$\frac{2}{5} + \frac{1}{3} = \frac{2}{5} \cdot \frac{3}{3} + \frac{1}{3} \cdot \frac{5}{5}$$

Multiply $\frac{2}{5}$ by 1 in the form of $\frac{3}{3}$. Multiply $\frac{1}{3}$ by 1 in the form of $\frac{5}{5}$.

We need to multiply this denominator by 5 to obtain 15. It follows that $\frac{5}{5}$ should be the form of 1 used to build $\frac{1}{3}$.

We need to multiply this denominator by 3 to obtain 15. It follows that $\frac{3}{3}$ should be the form of 1 used to build $\frac{2}{5}$.

$$= \frac{6}{15} + \frac{5}{15}$$

Multiply the numerators and multiply the denominators. Note that the denominators are now the same.

$$= \frac{6 + 5}{15}$$

Add the numerators.
Write the sum over the common denominator, 15.

$$= \frac{11}{15}$$

Since 11 and 15 have no common factors other than 1,
this fraction is in simplest form.

When adding (or subtracting) fractions with unlike denominators, the least common denominator is not always obvious. Prime factorization is helpful in determining the LCD.

Finding the LCD Using Prime Factorization

1. Prime factor each denominator.
2. The LCD is a product of prime factors, where each factor is used the greatest number of times it appears in any one factorization found in step 1.

EXAMPLE 7   Subtract: $\dfrac{3}{10} - \dfrac{5}{28}$

Strategy   We will begin by expressing each fraction as an equivalent fraction that has the LCD for its denominator. Then we will use the rule for subtracting fractions with *like* denominators.

Why   To add or subtract fractions, the fractions must have like denominators.

Solution   To find the LCD, we find the prime factorization of both denominators and use each prime factor the *greatest* number of times it appears in any one factorization.

$$\left. \begin{array}{l} 10 = 2 \cdot 5 \\ 28 = 2 \cdot 2 \cdot 7 \end{array} \right\} \mathrm{LCD} = 2 \cdot 2 \cdot 5 \cdot 7 = 140$$

2 appears twice in the factorization of 28.
5 appears once in the factorization of 10.
7 appears once in the factorization of 28.

Since 140 is the smallest number that 10 and 28 divide exactly, we write $\frac{3}{10}$ and $\frac{5}{28}$ as fractions with the LCD 140.

$$\frac{3}{10} - \frac{5}{28} = \frac{3}{10} \cdot \frac{14}{14} - \frac{5}{28} \cdot \frac{5}{5}$$

We must multiply 10 by 14 to obtain 140.
We must multiply 28 by 5 to obtain 140.

$$= \frac{42}{140} - \frac{25}{140}$$

Multiply the numerators and multiply the denominators.
Note that the denominators are now the same.

$$= \frac{42 - 25}{140}$$

Subtract the numerators.
Write the difference over the common denominator, 140.

$$= \frac{17}{140}$$

Since 17 and 140 have no common factors
other than 1, this fraction is in simplest form.

Self Check 7   Subtract: $\frac{11}{48} - \frac{7}{40}$

Now Try ▶ Problem 65

We can use the following steps to add or subtract fractions with different denominators.

Adding and Subtracting Fractions That Have Different Denominators

1. Find the LCD.
2. Rewrite each fraction as an equivalent fraction with the LCD as the denominator. To do so, build each fraction using a form of 1 that involves any factors needed to obtain the LCD.
3. Add or subtract the numerators and write the sum or difference over the LCD.
4. Simplify the result, if possible.

7 Simplify Answers.

When adding, subtracting, multiplying, or dividing fractions, remember to express the answer in simplest form.

EXAMPLE 8 Perform the operations and simplify: **a.** $45\left(\dfrac{4}{9}\right)$ **b.** $\dfrac{5}{12} + \dfrac{3}{2} - \dfrac{1}{4}$

Strategy   We will perform the indicated operations and then make sure that the answer is in simplest form.

Why   Fractional answers should always be given in simplest form.

Solution   **a.**

$$45\left(\frac{4}{9}\right) = \frac{45}{1}\left(\frac{4}{9}\right)$$ Write 45 as a fraction: $45 = \frac{45}{1}$.

$$= \frac{45 \cdot 4}{1 \cdot 9}$$ Multiply the numerators.
Multiply the denominators.

$$= \frac{5 \cdot \overset{1}{\cancel{9}} \cdot 4}{1 \cdot \cancel{9}}$$ To simplify the result, factor 45 as $5 \cdot 9$. Then remove the common factor 9 in the numerator and denominator.

$$= \frac{20}{1}$$ Multiply the remaining factors in the numerator.
Multiply the remaining factors in the denominator.

$$= 20$$ Any number divided by 1 is the number itself.

> **Caution**
>
> Remember that an LCD is **not needed** when multiplying or dividing fractions.

b.   Since the smallest number that 12, 2, and 4 divide exactly is 12, the LCD is 12.

$$\frac{5}{12} + \frac{3}{2} - \frac{1}{4} = \frac{5}{12} + \frac{3}{2} \cdot \frac{\mathbf{6}}{\mathbf{6}} - \frac{1}{4} \cdot \frac{\mathbf{3}}{\mathbf{3}}$$ $\frac{5}{12}$ already has a denominator of 12. Build $\frac{3}{2}$ and $\frac{1}{4}$ so that their denominators are 12.

$$= \frac{5}{12} + \frac{18}{12} - \frac{3}{12}$$ Multiply the numerators and multiply the denominators. The denominators are now the same.

$$= \frac{20}{12}$$ Add the numerators, 5 and 18, to get 23. From that sum, subtract 3. Write that result, 20, over the common denominator.

$$= \frac{\overset{1}{\cancel{4}} \cdot 5}{3 \cdot \underset{1}{\cancel{4}}}$$ To simplify $\frac{20}{12}$, factor 20 and 12, using their greatest common factor, 4. Then remove $\frac{4}{4} = 1$.

$$= \frac{5}{3}$$

> **The Language of Algebra**
>
> Fractions such as $\frac{5}{3}$, with a numerator greater than or equal to the denominator, are called **improper fractions**. In algebra, such fractions are often preferable to their equivalent mixed number form.

Self Check 8   Perform the operations and simplify: **a.** $24\left(\dfrac{7}{6}\right)$
b. $\dfrac{1}{15} + \dfrac{31}{30} - \dfrac{3}{10}$

Now Try ▶   Problems 67 and 71

8 Compute with Mixed Numbers.

A **mixed number** represents the sum of a whole number and a fraction. For example, $5\frac{3}{4}$ means $5 + \frac{3}{4}$ and $179\frac{15}{16}$ means $179 + \frac{15}{16}$.

EXAMPLE 9　Divide:　$5\frac{3}{4} \div 2$

Strategy　We begin by writing the mixed number $5\frac{3}{4}$ and the whole number 2 as fractions. Then we use the rule for dividing two fractions.

Why　To multiply (or divide) with mixed numbers, we first write them as improper fractions, and then multiply (or divide) as usual.

Solution

$$5\frac{3}{4} \div 2 = \frac{23}{4} \div \frac{2}{1}$$

Write $5\frac{3}{4}$ as an improper fraction by multiplying its whole-number part by the denominator: $5 \cdot 4 = 20$. Then add the numerator to that product: $3 + 20 = 23$. Finally, write the result, 23, over the denominator 4. Write 2 as a fraction: $2 = \frac{2}{1}$.

$$= \frac{23}{4} \cdot \frac{1}{2}$$

Multiply by the reciprocal of $\frac{2}{1}$, which is $\frac{1}{2}$.

$$= \frac{23}{8}$$

Multiply the numerators.
Multiply the denominators.

$$= 2\frac{7}{8}$$

Since the original problem involves a mixed number, we will express the answer in mixed-number form. Write $\frac{23}{8}$ as a mixed number by dividing the numerator, 23, by the denominator, 8.

$$\begin{array}{r} 2 \\ 8\overline{)23} \\ -16 \\ \hline 7 \end{array}$$

Self Check 9　Multiply:　$1\frac{1}{8} \cdot 9$

Now Try ▶ Problem 77

EXAMPLE 10　**Freeway Signs.**　How far apart are the Downtown San Diego and Sea World Drive exits?

Strategy　We can find the distance between exits by finding the difference in the mileages on the freeway sign: $6\frac{1}{2} - 1\frac{3}{4}$.

Why　The word *difference* indicates subtraction.

Solution

$$\begin{array}{rcccccc} 6\frac{1}{2} &=& 6\frac{2}{4} &=& 5\frac{2}{4} + \frac{4}{4} &=& 5\frac{6}{4} \\ -1\frac{3}{4} &=& -1\frac{3}{4} &=& -1\frac{3}{4} && -1\frac{3}{4} \\ \hline &&&&&& 4\frac{3}{4} \end{array}$$

Using vertical form, express $\frac{1}{2}$ as an equivalent fraction with denominator 4. Then, borrow 1 in the form of $\frac{4}{4}$ from 6 to subtract the fractional parts of the mixed numbers.

The Downtown San Diego and Sea World Drive exits are $4\frac{3}{4}$ miles apart.

Self Check 10　Subtract:　$9\frac{1}{8} - 2\frac{2}{3}$

Now Try ▶ Problem 114

Success Tip

This problem could also be solved by writing the mixed numbers $6\frac{1}{2}$ and $1\frac{3}{4}$, as improper fractions and subtracting them. However, answers to real-world problems are most often given as mixed numbers instead of improper fractions, because mixed numbers are easier to understand.

SECTION 1.2 ▶ STUDY SET

VOCABULARY

Fill in the blanks.

1. A factor is a number being _____.
2. Numbers that have only 1 and themselves as factors, such as 23, 37, and 41, are called _____ numbers.
3. When we write 60 as $2 \cdot 2 \cdot 3 \cdot 5$, we say that we have written 60 in _____ form.
4. The _____ of the fraction $\frac{3}{4}$ is 3, and the _____ is 4.
5. Two fractions that represent the same number, such as $\frac{1}{2}$ and $\frac{2}{4}$, are called _____ fractions.
6. $\frac{2}{3}$ is the _____ of $\frac{3}{2}$, because their product is 1.
7. The _____ common denominator for a set of fractions is the smallest number each denominator will divide exactly.
8. The _____ number $7\frac{1}{3}$ represents the sum of a whole number and a fraction: $7 + \frac{1}{3}$.

CONCEPTS

Complete each fact about fractions. Assume there are no divisions by 0.

9. a. $\frac{a}{a} = \blacksquare$ b. $\frac{a}{1} = \blacksquare$

    c. $\frac{a}{b} \cdot \frac{c}{d} = \underline{\quad}$ d. $\frac{a}{b} \div \frac{c}{d} = \underline{\quad}$

    e. $\frac{a}{d} + \frac{b}{d} = \underline{\quad}$ f. $\frac{a}{d} - \frac{b}{d} = \underline{\quad}$

10. What two equivalent fractions are shown?

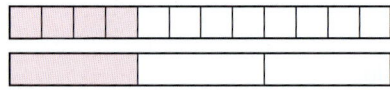

11. Complete each statement.
    a. To simplify a fraction, we remove factors equal to $\blacksquare$ in the form of $\frac{2}{2}$, $\frac{3}{3}$, or $\frac{4}{4}$, and so on.
    b. To build a fraction, we multiply it by $\blacksquare$ in the form of $\frac{2}{2}$, $\frac{3}{3}$, or $\frac{4}{4}$, and so on.

12. What is the LCD for fractions having denominators of:
    a. 3 and 7?
    b. 4 and 6?

NOTATION

Fill in the blanks.

13. a. Multiply $\frac{5}{6}$ by a form of 1 to build an equivalent fraction with denominator 30.

    $$\frac{5}{6} \cdot \frac{\blacksquare}{\blacksquare} = \frac{\blacksquare}{\blacksquare}$$

    b. Remove common factors to simplify $\frac{12}{42}$.

    $$\frac{12}{42} = \frac{2 \cdot \blacksquare \cdot 3}{2 \cdot 3 \cdot \blacksquare} = \frac{\blacksquare}{\blacksquare}$$

14. a. Write $2\frac{15}{16}$ as an improper fraction.
    b. Write $\frac{49}{12}$ as a mixed number.

GUIDED PRACTICE

Find the prime factorization of each number. See Example 1.

15. 75   16. 20
17. 28   18. 54
19. 81   20. 125
21. 117   22. 147
23. 220   24. 270
25. 1,254   26. 1,144

Multiply. See Example 2.

27. $\frac{5}{6} \cdot \frac{1}{8}$ 28. $\frac{2}{3} \cdot \frac{1}{5}$

29. $\frac{7}{11} \cdot \frac{3}{5}$ 30. $\frac{13}{9} \cdot \frac{2}{3}$

Divide. See Example 3.

31. $\frac{3}{4} \div \frac{2}{5}$ 32. $\frac{7}{8} \div \frac{6}{13}$

33. $\frac{6}{5} \div \frac{5}{7}$ 34. $\frac{4}{3} \div \frac{3}{2}$

Build each fraction or whole number to an equivalent fraction with the indicated denominator. See Example 4.

35. $\frac{1}{3}$, denominator 9   36. $\frac{3}{8}$, denominator 24
37. $\frac{4}{9}$, denominator 54   38. $\frac{9}{16}$, denominator 64
39. 7, denominator 5   40. 12, denominator 3
41. 5, denominator 7   42. 6, denominator 8

Simplify each fraction, if possible. See Examples 5 and 6.

43. $\frac{6}{18}$ 44. $\frac{6}{9}$ 45. $\frac{24}{28}$ 46. $\frac{35}{14}$

47. $\frac{15}{40}$ 48. $\frac{22}{77}$ 49. $\frac{33}{56}$ 50. $\frac{26}{21}$

51. $\frac{26}{39}$ 52. $\frac{72}{64}$ 53. $\frac{36}{225}$ 54. $\frac{175}{490}$

Perform the operations and, if possible, simplify. See Objective 6 and Example 7.

55. $\frac{3}{5} + \frac{3}{5}$ 56. $\frac{4}{9} - \frac{1}{9}$

57. $\frac{6}{7} - \frac{2}{7}$ 58. $\frac{5}{13} + \frac{6}{13}$

59. $\frac{1}{6} + \frac{1}{24}$ 60. $\frac{17}{25} - \frac{2}{5}$

61. $\frac{7}{10} - \frac{1}{14}$ 62. $\frac{9}{8} - \frac{5}{6}$

63. $\dfrac{2}{15} + \dfrac{7}{9}$

64. $\dfrac{7}{25} + \dfrac{3}{10}$

65. $\dfrac{13}{28} - \dfrac{1}{21}$

66. $\dfrac{13}{24} - \dfrac{3}{40}$

Perform the operations and, if possible, simplify. See Example 8.

67. $16\left(\dfrac{3}{2}\right)$

68. $30\left(\dfrac{5}{6}\right)$

69. $18 \cdot \dfrac{2}{9}$

70. $14 \cdot \dfrac{3}{7}$

71. $\dfrac{2}{3} + \dfrac{5}{18} - \dfrac{1}{6}$

72. $\dfrac{3}{5} + \dfrac{7}{20} - \dfrac{7}{10}$

73. $\dfrac{5}{12} + \dfrac{1}{3} - \dfrac{2}{5}$

74. $\dfrac{7}{15} + \dfrac{1}{5} - \dfrac{4}{9}$

Perform the operations and, if possible, simplify. See Examples 9 and 10.

75. $4\dfrac{2}{3} \cdot 7$

76. $7 \cdot 1\dfrac{3}{28}$

77. $8 \div 3\dfrac{1}{5}$

78. $15 \div 3\dfrac{1}{3}$

79. $8\dfrac{2}{9} - 7\dfrac{2}{3}$

80. $3\dfrac{4}{5} - 3\dfrac{1}{10}$

81. $3\dfrac{3}{16} + 2\dfrac{5}{24}$

82. $15\dfrac{5}{6} + 11\dfrac{5}{8}$

TRY IT YOURSELF

Perform the operations and, if possible, simplify.

83. $\dfrac{3}{5} + \dfrac{2}{3}$

84. $\dfrac{4}{3} + \dfrac{7}{2}$

85. $21\left(\dfrac{10}{3}\right)$

86. $28\left(\dfrac{4}{7}\right)$

87. $6 \cdot 2\dfrac{7}{24}$

88. $3\dfrac{1}{2} \cdot \dfrac{1}{5}$

89. $\dfrac{2}{3} - \dfrac{1}{4} + \dfrac{1}{12}$

90. $\dfrac{3}{7} - \dfrac{2}{5} + \dfrac{2}{35}$

91. $\dfrac{21}{35} \div \dfrac{3}{14}$

92. $\dfrac{23}{25} \div \dfrac{46}{5}$

93. $\dfrac{4}{3}\left(\dfrac{6}{5}\right)$

94. $\dfrac{21}{8}\left(\dfrac{2}{15}\right)$

95. $\dfrac{4}{63} + \dfrac{1}{45}$

96. $\dfrac{5}{18} + \dfrac{1}{99}$

97. $3 - \dfrac{3}{4}$

98. $4 - \dfrac{7}{3}$

99. $\dfrac{1}{5} \cdot \dfrac{3}{5}$

100. $\dfrac{3}{4} \cdot \dfrac{5}{7}$

101. $3\dfrac{1}{3} \div 1\dfrac{5}{6}$

102. $2\dfrac{1}{2} \div 1\dfrac{5}{8}$

103. $\dfrac{11}{21} - \dfrac{8}{21}$

104. $\dfrac{19}{35} - \dfrac{12}{35}$

105. $\dfrac{7}{30} + \dfrac{1}{50} - \dfrac{19}{75}$

106. $\dfrac{11}{12} - \dfrac{7}{15} - \dfrac{9}{20}$

107. $1\dfrac{31}{32} \cdot 7\dfrac{1}{9}$

108. $3\dfrac{1}{16} \cdot 4\dfrac{4}{7}$

Look Alikes . . .

109. a. $\dfrac{4}{9} + \dfrac{3}{7}$ b. $\dfrac{4}{9} - \dfrac{3}{7}$ c. $\dfrac{4}{9} \cdot \dfrac{3}{7}$ d. $\dfrac{4}{9} \div \dfrac{3}{7}$

110. a. $4\dfrac{1}{8} + 1\dfrac{5}{6}$ b. $4\dfrac{1}{8} - 1\dfrac{5}{6}$

   c. $4\dfrac{1}{8} \cdot 1\dfrac{5}{6}$ d. $4\dfrac{1}{8} \div 1\dfrac{5}{6}$

APPLICATIONS

111. **Forestry.** A ranger cut down a tree and measured the widths of the outer two growth rings.
   a. What was the growth over this 2-year period?
   b. What is the difference in the widths of the rings?

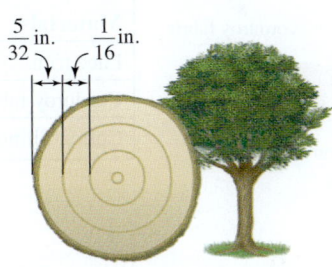

$\dfrac{5}{32}$ in. $\dfrac{1}{16}$ in.

112. **Hardware.** To secure the bracket to the stock, a bolt and a nut are used. How long should the threaded part of the bolt be?

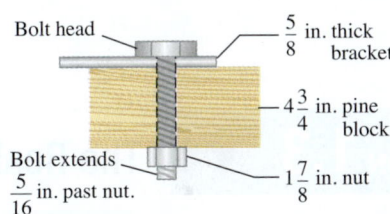

Bolt head

$\dfrac{5}{8}$ in. thick bracket

$4\dfrac{3}{4}$ in. pine block

Bolt extends
$\dfrac{5}{16}$ in. past nut.

$1\dfrac{7}{8}$ in. nut

---

113. ▶ from **Campus to Careers**

Lead Transportation Security Officer

Each year, the Transportation Security Administration (TSA) screens more than 500 million pieces of luggage. On many flights, airlines do not accept luggage whose total dimension (length + width + height) exceeds 62 inches. What is the total dimension figure for the suitcase shown below?

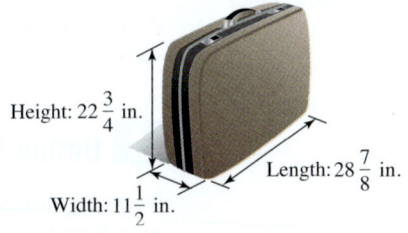

Height: $22\dfrac{3}{4}$ in.

Length: $28\dfrac{7}{8}$ in.

Width: $11\dfrac{1}{2}$ in.

114. **Cooking.** How much butter is left in a $10\dfrac{1}{2}$-pound tub of butter if $4\dfrac{3}{4}$ pounds are used to make a wedding cake?

115. Frames. How many inches of molding are needed to make the square picture frame?

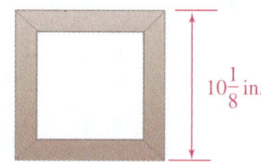

$10\frac{1}{8}$ in.

116. Decorating. The materials used to make a pillow are shown below. Examine the inventory list to decide how many pillows can be manufactured in one production run with the materials in stock.

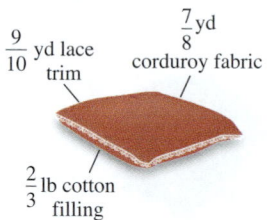

$\frac{9}{10}$ yd lace trim

$\frac{7}{8}$ yd corduroy fabric

$\frac{2}{3}$ lb cotton filling

Materials	Amount in stock
Lace trim	135 yd
Corduroy fabric	154 yd
Cotton filling	98 lb

WRITING

117. Explain the error made below in simplifying $\frac{15}{45}$.

$$\frac{15}{45} = \frac{3 \cdot 5}{3 \cdot 3 \cdot 5} = \frac{0}{3} = 0$$

118. To multiply two fractions, must they have like denominators? Explain.

119. What are equivalent fractions? Give an example.

120. Explain the error in the following addition.

$$\frac{4}{3} + \frac{3}{2} = \frac{4+3}{3+2} = \frac{7}{5}$$

REVIEW

Fill in the blanks.

121. _____ are letters (or symbols) that stand for numbers.

122. A number, such as 10, is called a _____ because it does not change.

CHALLENGE PROBLEMS

123. Which is larger: $\frac{11}{12}$ or $\frac{8}{9}$?

124. If the circle represents a whole, find the missing value.

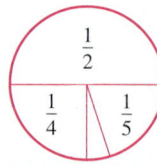

$\frac{1}{2}$ $\frac{1}{4}$ $\frac{1}{5}$

SECTION 1.3

The Real Numbers

OBJECTIVES

1. Define the set of integers.
2. Define the set of rational numbers.
3. Define the set of irrational numbers.
4. Classify real numbers.
5. Graph sets of real numbers on the number line.
6. Find the absolute value of a real number.

ARE YOU READY?

The following problems review several types of numbers that we use in everyday life.

1. Count the number of letters in the word *antidisestablishmentarianism*.

2. What number represents a temperature that is 10 degrees below zero?

3. What type of number is used to express a grade point average (GPA)?

4. Suppose a recipe calls for only part of a full cup of sugar. What type of number is normally used to describe such an amount?

A **set** is a collection of objects, such as a set of golf clubs or a set of dishes. In this section, we will define some important sets of numbers that are used in algebra.

1 Define the Set of Integers.

Natural numbers are the numbers that we use for counting. To write this set, we list its **members** (or **elements**) within **braces** { }.

Natural Numbers   The set of **natural numbers** is {1, 2, 3, 4, 5, . . .}.   Read as "the set containing one, two, three, four, five, and so on."

The natural numbers, together with 0, form the set of **whole numbers.**

| **Whole Numbers** | The set of **whole numbers** is {0, 1, 2, 3, 4, 5, . . . }. |

Notation

The symbol . . . used in the previous definitions is called an **ellipsis** and it indicates that the established pattern continues forever.

Whole numbers are not adequate for describing many real-life situations. For example, if you write a check for more than what's in your account, the account balance will be less than zero.

We can use the **number line** below to visualize numbers less than zero. A number line is straight and has uniform markings. The arrowheads indicate that it extends forever in both directions. For each natural number on the number line, there is a corresponding number, called its *opposite,* to the left of 0. In the diagram, we see that 3 and -3 (negative three) are opposites, as are -5 (negative five) and 5. Note that 0 is its own opposite.

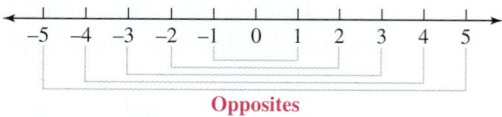

Opposites

| **Opposites** | Two numbers that are the same distance from 0 on the number line, but on opposite sides of it, are called **opposites.** |

The whole numbers, together with their opposites, form the set of **integers.**

| **Integers** | The set of **integers** is {. . . , $-4, -3, -2, -1, 0, 1, 2, 3, 4,$ . . . }. |

The Language of Algebra

The **positive integers** are:
1, 2, 3, 4, 5, . . .
The **negative integers** are:
$-1, -2, -3, -4, -5,$ . . .
The **nonnegative integers** are: 0, 1, 2, 3, 4, 5, . . .

On the number line, numbers greater than 0 are to the right of 0. They are called **positive numbers.** Positive numbers can be written with or without a **positive sign** $+$. For example, $2 = +2$ (positive two). They are used to describe such quantities as an elevation above sea level ($+3,000$ ft) or a pay raise ($25).

Numbers less than 0 are to the left of 0 on the number line. They are called **negative numbers.** Negative numbers are always written with a **negative sign** $-$. They are used to describe such quantities as an overdrawn checking account ($-\$75$) or a below-zero temperature ($-12°$).

Positive and negative numbers are called **signed numbers.**

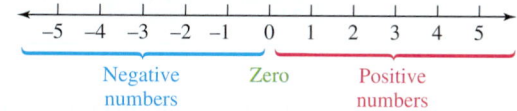

2   Define the Set of Rational Numbers.

We use fractions to describe many situations in daily life. For example, a morning commute might take $\frac{1}{4}$ hour or a recipe might call for $\frac{2}{3}$ cup of sugar. Fractions such as $\frac{1}{4}$ and $\frac{2}{3}$, which are quotients of two integers, are called *rational numbers.*

| **Rational Numbers** | A rational number is any number that can be expressed as a fraction (ratio*) with an integer numerator and a nonzero integer denominator.
 *Ratios are discussed in more detail in Section 7.8. |

The Language of Algebra

Rational numbers are so named because they can be expressed as the *ratio* (quotient) of two integers.

Some other examples of rational numbers are

$$\frac{3}{8}, \quad \frac{41}{100}, \quad \frac{25}{25}, \quad \text{and} \quad \frac{19}{12}$$

To show that negative fractions are rational numbers, we use the following fact.

Negative Fractions

For any numbers a and b where b is not 0,

$$-\frac{a}{b} = \frac{-a}{b} = \frac{a}{-b}$$

To illustrate this rule, we consider $-\frac{11}{16}$ (read as negative eleven-sixteenths). It is a rational number because it can be written as $\frac{-11}{16}$ or as $\frac{11}{-16}$.

Positive and negative mixed numbers are also rational numbers because they can be expressed as fractions. For example,

$$7\frac{5}{8} = \frac{61}{8} \quad \text{and} \quad -6\frac{1}{2} = -\frac{13}{2} = \frac{-13}{2}$$

Any natural number, whole number, or integer can be expressed as a fraction with a denominator of 1. For example, $5 = \frac{5}{1}$, $0 = \frac{0}{1}$, and $-3 = \frac{-3}{1}$. Therefore, every natural number, whole number, and integer is also a rational number.

The Language of Algebra

To **terminate** means to bring to an end. In the movie *The Terminator,* actor Arnold Schwarzenegger plays a heartless machine sent to Earth to bring an end to his enemies.

Many numerical quantities are written in decimal notation. For instance, a candy bar might cost \$0.89, a dragster might travel at 203.156 mph, or a business loss might be −\$4.7 million. These decimals are called **terminating decimals** because their representations terminate (stop). As shown below, terminating decimals can be expressed as fractions. Therefore, terminating decimals are rational numbers.

$$0.89 = \frac{89}{100} \qquad 203.156 = 203\frac{156}{1{,}000} = \frac{203{,}156}{1{,}000} \qquad -4.7 = -4\frac{7}{10} = \frac{-47}{10}$$

Decimals such as 0.3333 . . . and 2.8167167167 . . . , which have a digit (or a block of digits) that repeats, are called **repeating decimals.** Since any repeating decimal can be expressed as a fraction, repeating decimals are rational numbers.

The set of rational numbers cannot be listed in the same way as the natural numbers, the whole numbers, and the integers. Instead, we use **set-builder** notation.

Rational Numbers

The set of rational numbers is

$$\left\{\frac{a}{b} \,\middle|\, a \text{ and } b \text{ are integers, with } b \neq 0.\right\}$$

Read as "the set of all numbers of the form $\frac{a}{b}$ such that a and b are integers, with $b \neq 0$."

A fraction and its **decimal equivalent** are different forms that represent the same value. To find the decimal equivalent for a fraction, we divide its numerator by its denominator. For example, to write $\frac{1}{4}$ and $\frac{5}{22}$ as decimals, we proceed as follows:

$$\begin{array}{r} 0.25 \\ 4\overline{)1.00} \\ -\ 8 \\ \hline 20 \\ -20 \\ \hline 0 \end{array}$$

Write a decimal point and additional zeros to the right of 1.

The remainder is 0.

$$\begin{array}{r} 0.22727\ldots \\ 22\overline{)5.00000} \\ -4\,4 \\ \hline 60 \\ -44 \\ \hline 160 \\ -154 \\ \hline 60 \\ -44 \\ \hline 160 \end{array}$$

Write a decimal point and additional zeros to the right of 5.

60 and 160 continually appear as remainders. Therefore, 2 and 7 will continually appear in the quotient.

The decimal equivalent of $\frac{1}{4}$ is 0.25 and the decimal equivalent of $\frac{5}{22}$ is 0.2272727 . . . . We can use an **overbar** to write repeating decimals in more compact form: $0.2272727\ldots = 0.2\overline{27}$. Here are more fractions and their decimal equivalents.

Terminating decimals	*Repeating decimals*
$\frac{1}{2} = 0.5$	$\frac{1}{6} = 0.166666\ldots$ or $0.1\overline{6}$
$\frac{5}{8} = 0.625$	$\frac{1}{3} = 0.333333\ldots$ or $0.\overline{3}$
$\frac{3}{4} = 0.75$	$\frac{5}{11} = 0.454545\ldots$ or $0.\overline{45}$

3 Define the Set of Irrational Numbers.

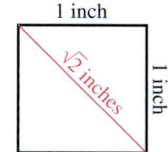

1 inch

$\sqrt{2}$ inches

1 inch

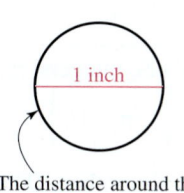

1 inch

The distance around the circle is π inches.

Not all numbers are rational numbers. One example is the square root of 2, written $\sqrt{2}$. It is the number that, when multiplied by itself, gives 2. That is, $\sqrt{2} \cdot \sqrt{2} = 2$. It can be shown that $\sqrt{2}$ *cannot* be written as a fraction with an integer numerator and an integer denominator. Therefore, it is not rational; it is an *irrational number*. It is interesting to note that a square with sides of length 1 inch has a diagonal that is $\sqrt{2}$ inches long.

The number represented by the Greek letter π (pi) is another example of an irrational number. A circle, with a 1-inch diameter, has a circumference of π inches. Expressed in decimal form,

$$\sqrt{2} = 1.414213562\ldots \qquad \text{and} \qquad \pi = 3.141592654\ldots$$

These decimals neither terminate nor repeat.

Irrational Numbers

An **irrational number** is a nonterminating, nonrepeating decimal. An irrational number cannot be expressed as a fraction with an integer numerator and an integer denominator.

We have seen that $\sqrt{2}$ and π are irrational numbers. Other examples of irrational numbers are:

$$\sqrt{3} = 1.732050808\ldots \qquad -\sqrt{5} = -2.236067977\ldots$$
$$-\pi = -3.141592654\ldots \qquad 3\pi = 9.424777961\ldots \quad \textcolor{red}{3\pi \text{ means } 3 \cdot \pi.}$$

We can use a calculator to approximate the decimal value of an irrational number. To approximate $\sqrt{2}$ using a scientific calculator, we use the square root key $\boxed{\sqrt{}}$. To approximate π, we use the *pi* key $\boxed{\pi}$.

$$\sqrt{2} \approx 1.414213562 \quad \text{and} \quad \pi \approx 3.141592654 \quad \textcolor{red}{\text{Read } \approx \text{ as "is approximately equal to."}}$$

Rounded to the nearest thousandth, $\sqrt{2} \approx 1.414$ and $\pi \approx 3.142$.

4 Classify Real Numbers.

The set of **real numbers** is formed by combining the set of rational numbers and the set of irrational numbers. Every real number has a decimal representation. If it is rational, its corresponding decimal terminates or repeats. If it is irrational, its decimal representation is nonterminating and nonrepeating.

The Real Numbers

A **real number** is any number that is a rational number or an irrational number.

The following diagram shows how various sets of numbers are related. Note that a number can belong to more than one set. For example, -6 is an integer, a rational number, and a real number.

The Language of Algebra

The symbol $\mathbb{R}$ is used to represent the set of real numbers.

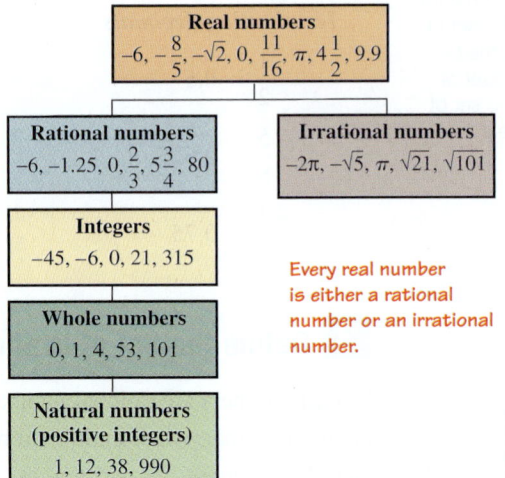

Real numbers
$-6, -\frac{8}{5}, -\sqrt{2}, 0, \frac{11}{16}, \pi, 4\frac{1}{2}, 9.9$

Rational numbers
$-6, -1.25, 0, \frac{2}{3}, 5\frac{3}{4}, 80$

Irrational numbers
$-2\pi, -\sqrt{5}, \pi, \sqrt{21}, \sqrt{101}$

Integers
$-45, -6, 0, 21, 315$

Every real number is either a rational number or an irrational number.

Whole numbers
$0, 1, 4, 53, 101$

Natural numbers (positive integers)
$1, 12, 38, 990$

EXAMPLE 1 Which numbers in the following set are natural numbers, whole numbers, integers, rational numbers, irrational numbers, real numbers? $\left\{ -3.4, \ \frac{2}{5}, \ 0, \ -6, \ 1\frac{3}{4}, \ \pi, \ 16 \right\}$

Strategy We begin by scanning the given set, looking for any natural numbers. Then we scan it five more times, looking for whole numbers, for integers, for rational numbers, for irrational numbers, and finally, for real numbers.

Why We need to scan the given set of numbers six times, because numbers in the set can belong to more than one classification.

Solution
Natural numbers: 16   *16 is a member of {1, 2, 3, 4, 5, . . .}.*

Whole numbers: 0, 16   *0 and 16 are members of {0, 1, 2, 3, 4, 5, . . .}.*

Integers: 0, −6, 16   *0, −6, and 16 are members of { . . . , −3, −2, −1, 0, 1, 2, 3, . . .}.*

Rational numbers:
$-3.4, \frac{2}{5}, 0, -6, 1\frac{3}{4}, 16$ *A rational number can be expressed as a ratio of two integers: $-3.4 = \frac{-34}{10}, 0 = \frac{0}{1}, -6 = \frac{-6}{1}, 1\frac{3}{4} = \frac{7}{4},$ and $16 = \frac{16}{1}$.*

Irrational numbers: π *$\pi = 3.1415 \ldots$ is a nonterminating, nonrepeating decimal.*

Real numbers:
$-3.4, \frac{2}{5}, 0, -6, 1\frac{3}{4}, \pi, 16$ *Every natural number, whole number, integer, rational number, and irrational number is a real number.*

Self Check 1 Use the instructions for Example 1 with: $\left\{ 0.1, \ -\frac{2}{7}, \ 45, \ -2, \ \frac{13}{4}, \ -6\frac{7}{8} \right\}$

Now Try ▶ Problem 27

---

5 Graph Sets of Real Numbers on the Number Line.

Every real number corresponds to a point on the number line, and every point on the number line corresponds to exactly one real number. As we move right on the number line, the values of the numbers increase. As we move left, the values decrease. On the following number line, we see that 5 is greater than -3, because 5 lies to the right of -3. Similarly, -3 is less than 5, because it lies to the left of 5.

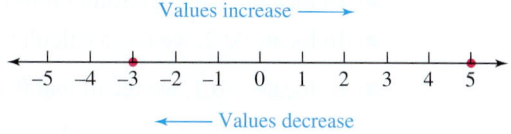

The **inequality symbol** $>$ means "is greater than." It is used to show that one number is greater than another. The inequality symbol $<$ means "is less than." It is used to show that one number is less than another. For example,

$$5 > -3 \qquad \text{Read as "5 is greater than } -3\text{."}$$
$$-3 < 5 \qquad \text{Read as "}-3 \text{ is less than 5."}$$

To distinguish between these inequality symbols, remember that each one points to the smaller of the two numbers involved.

$$5 > -3 \qquad\qquad\qquad -3 < 5$$

Points to the smaller number.

The Language of Algebra

The prefix **in** means *not.* For example:

inaccurate ↔ not accurate
inexpensive ↔ not expensive
inequality ↔ not equal

EXAMPLE 2   Use one of the symbols $>$ or $<$ to make each statement true:

a. $-4 \quad 4$ **b.** $-2 \quad -3$ **c.** $4.47 \quad 12.5$ **d.** $\dfrac{3}{4} \quad \dfrac{5}{8}$

Strategy   To pick the correct inequality symbol to place between a given pair of numbers, we need to determine the position of each number on a number line.

Why   For any two numbers on a number line, the number to the *left* is the smaller number and the number to the *right* is the larger number.

Solution   **a.** Since -4 is to the left of 4 on the number line, we have $-4 < 4$.

b. Since -2 is to the right of -3 on the number line, we have $-2 > -3$.

c. Since 4.47 is to the left of 12.5 on the number line, we have $4.47 < 12.5$.

d. To compare fractions, express them in terms of the same denominator, preferably the LCD. If we write $\dfrac{3}{4}$ as an equivalent fraction with denominator 8, we see that $\dfrac{3}{4} = \dfrac{3}{4} \cdot \dfrac{2}{2} = \dfrac{6}{8}$. Therefore, $\dfrac{3}{4} > \dfrac{5}{8}$.

To compare the fractions, we also could convert each to its decimal equivalent. Since $\dfrac{3}{4} = 0.75$ and $\dfrac{5}{8} = 0.625$, we know that $\dfrac{3}{4} > \dfrac{5}{8}$.

Self Check 2   Use one of the symbols $<$ or $>$ to make each statement true:

a. $1 \quad -1$ **b.** $-5 \quad -4$ **c.** $6.7 \quad 4.999$ **d.** $\dfrac{3}{5} \quad \dfrac{2}{3}$

Now Try ▶   Problems 37 and 42

To **graph a number** means to mark its position on the number line.

EXAMPLE 3   Graph each number in the set: $\left\{ -2.43, \ \sqrt{2}, \ 1, \ -0.\overline{3}, \ 2\dfrac{5}{6}, \ -\dfrac{3}{2} \right\}$

Strategy   We locate the position of each number on the number line, draw a bold dot, and label it.

Why   To *graph a number* means to make a drawing that represents the number.

Solution   It is helpful to approximate the value of a number or to write the number in an equivalent form to determine its location on a number line.

- To locate -2.43, we round it to the nearest tenth: $-2.43 \approx -2.4$.
- To locate $\sqrt{2}$, we use a calculator: $\sqrt{2} \approx 1.4$.
- To locate $-0.\overline{3}$, we recall that $0.\overline{3} = 0.333\ldots = \frac{1}{3}$. Therefore, $-0.\overline{3} = -\frac{1}{3}$.
- In mixed-number form, $-\frac{3}{2} = -1\frac{1}{2}$. This is midway between -1 and -2.

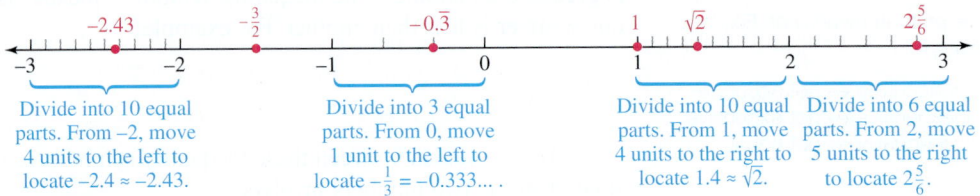

| Divide into 10 equal parts. From −2, move 4 units to the left to locate −2.4 ≈ −2.43. | Divide into 3 equal parts. From 0, move 1 unit to the left to locate $-\frac{1}{3}$ = −0.333... . | Divide into 10 equal parts. From 1, move 4 units to the right to locate 1.4 ≈ √2. | Divide into 6 equal parts. From 2, move 5 units to the right to locate $2\frac{5}{6}$. |

Self Check 3   Graph each number in the set: $\left\{ 1.7, \ \pi, \ -1\frac{3}{4}, \ 0.\overline{6}, \ \frac{5}{2}, \ -3 \right\}$

Now Try ▶ Problem 57

6 Find the Absolute Value of a Real Number.

A number line can be used to measure the distance from one number to another. For example, on the number line below, we see that the distance from 0 to -4 is 4 units and the distance from 0 to 3 is 3 units.

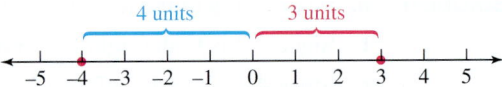

To express the distance that a number is from 0 on a number line, we can use *absolute value.*

Absolute Value | The **absolute value** of a number is its distance from 0 on the number line.

To indicate the absolute value of a number, we write the number between two vertical bars. From the figure above, we see that $|-4| = 4$. This is read as "the absolute value of negative 4 is 4" and it tells us that the distance from 0 to -4 is 4 units. It also follows from the figure that $|3| = 3$.

EXAMPLE 4   Find each absolute value: **a.** $|18|$ **b.** $\left| -\frac{7}{8} \right|$ **c.** $|98.6|$ **d.** $|0|$

Strategy   We need to determine the distance that the number within the vertical absolute value bars is from 0.

Why   The absolute value of a number is the distance between 0 and the number on a number line.

Solution   **a.** Since 18 is a distance of 18 from 0 on the number line, $|18| = 18$.

b. Since $-\frac{7}{8}$ is a distance of $\frac{7}{8}$ from 0 on the number line, $\left| -\frac{7}{8} \right| = \frac{7}{8}$.

c. Since 98.6 is a distance of 98.6 from 0 on the number line, $|98.6| = 98.6$.

d. Since 0 is a distance of 0 from 0 on the number line, $|0| = 0$.

Self Check 4 Find each absolute value: **a.** $|100|$ **b.** $|-4.7|$ **c.** $|\sqrt{2}|$

Now Try ▶ Problems 61 and 67

SECTION 1.3 ▶ STUDY SET

VOCABULARY

Fill in the blanks.

1. The set of _____ numbers is $\{0, 1, 2, 3, 4, 5, \ldots\}$.

2. The set of _____ numbers is $\{1, 2, 3, 4, 5, \ldots\}$.

3. The figure $\begin{array}{c}\overset{|\ |\ |\ |\ |}{-2\ -1\ 0\ 1\ 2}\end{array}$ is called a _____ _____.

4. The set of _____ is $\{\ldots, -2, -1, 0, 1, 2, \ldots\}$.

5. Positive and negative numbers are called _____ numbers.

6.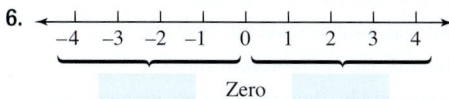
   Zero

7. The symbols $<$ and $>$ are _____ symbols.

8. A _____ number is any number that can be expressed as a fraction with an integer numerator and a nonzero integer denominator.

9. 0.25 is called a _____ decimal and 0.333 . . . is called a _____ decimal.

10. An _____ number cannot be expressed as a quotient of two integers.

11. An irrational number is a nonterminating, nonrepeating _____.

12. The _____ _____ of a number is the distance on the number line between the number and 0.

CONCEPTS

13. Represent each situation using a signed number.
    a. A loss of $15 million
    b. A building foundation $\frac{5}{16}$ inch above grade

14. Show that each of the following numbers is a rational number by expressing it as a fraction with an integer numerator and a nonzero integer denominator: 6, -9, $-\frac{7}{8}$, $3\frac{1}{2}$, -0.3, 2.83.

15. Give the opposite of each number.
    a. 20
    b. $-\dfrac{2}{3}$

16. What two numbers are a distance of 8 away from 5 on the number line?

17. What two numbers are a distance of 5 away from -9 on the number line?

18. Refer to the number line below. Use an inequality symbol, $<$ or $>$, to make each statement true.
    a. a ☐ b **b.** b ☐ a
    c. b ☐ 0 and a ☐ 0   **d.** $|a|$ ☐ $|b|$

    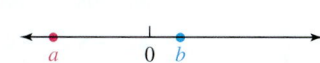

NOTATION

Fill in the blanks.

19. $\sqrt{2}$ is read "the _____ _____ of 2."

20. $|-15|$ is read "the _____ _____ of -15."

21. The symbol $\approx$ means ___ _____ _____ ___.

22. The symbols $\{\ \}$ are called _____.

23. The symbol π is a letter from the _____ alphabet.

24. To find the decimal equivalent for $\frac{2}{3}$, we perform the following long division: ☐$\overline{)\ }$

25. $-\dfrac{4}{5} = \dfrac{\ \boxed{\ }\ }{5} = \dfrac{4}{\boxed{\ }}$

26. Write each repeating decimal using an overbar.
    a. 0.666 . . .   **b.** 0.2444 . . .
    c. 0.717171 . . .   **d.** 0.456456456 . . .

GUIDED PRACTICE

Place check marks in the table to show the set or sets to which each number belongs. For example, the check shows that $\sqrt{2}$ is irrational. See Example 1.

27.

	5	0	-3	$\frac{7}{8}$	0.17	$-9\frac{1}{4}$	$\sqrt{2}$	π
Real								
Irrational							✓	
Rational								
Integer								
Whole								
Natural								

28. Which numbers in the following set are natural numbers, whole numbers, integers, rational numbers, irrational numbers, real numbers? $\left\{67, \frac{4}{13}, -5.9, 11\frac{2}{3}, \sqrt{2}, 0, -3, \pi\right\}$

Determine whether each statement is true or false. See Example 1.

29. Every whole number is an integer.

30. Every integer is a natural number.

31. Every integer is a whole number.

32. Every real number is either a rational number or an irrational number.

33. Irrational numbers are real numbers.

34. Every whole number is a rational number.

35. Every rational number can be written as a fraction (ratio) of two integers.

36. Every rational number is a whole number.

Use one of the symbols $<$ or $>$ to make each statement true. See Example 2.

37. 0 ▢ -4

38. 0 ▢ 32

39. 917 ▢ 971

40. 898 ▢ 889

41. -2 ▢ -3

42. -5 ▢ -4

43. $-\frac{5}{8}$ ▢ $-\frac{3}{8}$

44. $-19\frac{2}{3}$ ▢ $-19\frac{1}{3}$

45. $\frac{2}{3}$ ▢ $\frac{3}{5}$

46. $\frac{3}{4}$ ▢ $\frac{5}{6}$

47. -6.19 ▢ -5.8

48. -2.27 ▢ -5.25

Write each fraction as a decimal. If the result is a repeating decimal, use an overbar. See Objective 2.

49. $\frac{5}{8}$

50. $\frac{3}{32}$

51. $\frac{1}{30}$

52. $\frac{7}{9}$

53. $\frac{1}{60}$

54. $\frac{5}{11}$

55. $\frac{21}{50}$

56. $\frac{2}{125}$

Graph each set of numbers on a number line. See Example 3.

57. $\left\{-\pi, 4.25, -1\frac{1}{2}, -0.333\ldots, \sqrt{2}, -\frac{35}{8}, 3\right\}$

58. $\left\{-2\frac{1}{8}, \pi, 2.75, -\sqrt{2}, \frac{17}{4}, 0.666\ldots, -3\right\}$

59. The integers between -5 and 2

60. The whole numbers less than 4

Find each absolute value. See Example 4.

61. $|83|$

62. $|29|$

63. $\left|\frac{4}{3}\right|$

64. $\left|\frac{9}{16}\right|$

65. $|-11|$

66. $|-14|$

67. $|-6.1|$

68. $|-25.3|$

Insert one of the symbols $>$, $<$, or $=$ in the blank to make each statement true. See Examples 2 and 4.

69. $|3.4|$ ▢ -3

70. 0.08 ▢ 0.079

71. $|-1.1|$ ▢ 1.2

72. -5.5 ▢ $-5\frac{1}{2}$

73. $\left|-\frac{15}{2}\right|$ ▢ 7.5

74. $\left|-2\frac{2}{3}\right|$ ▢ $\frac{7}{3}$

75. $\frac{99}{100}$ ▢ 0.99

76. $|2|$ ▢ $|-2|$

77. 0.3 ▢ $0.333\ldots$

78. $-0.666\ldots$ ▢ -0.6

79. 1 ▢ $\left|-\frac{15}{16}\right|$

80. $\sqrt{2}$ ▢ π

APPLICATIONS

81. **Drafting.** Which dimensions of the aluminum bracket shown below are natural numbers, whole numbers, integers, rational numbers, irrational numbers, and real numbers?

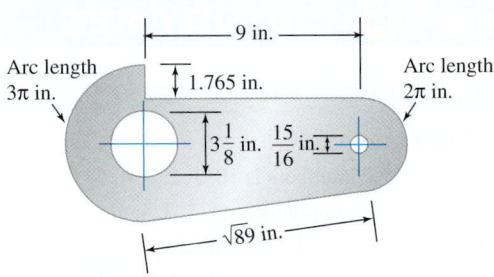

82. **History.** Refer to the time line shown below.

a. What basic unit was used to scale the time line?

b. What symbolism is used to represent zero?

c. Which numbers could be thought of as positive and which as negative?

d. Express the dates for the Maya civilization using positive and negative numbers.

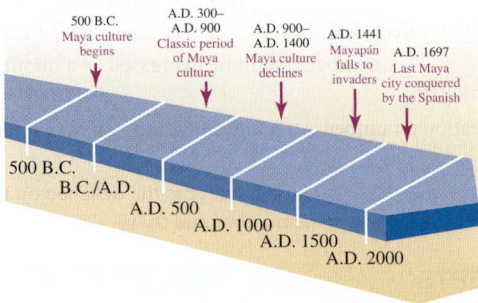

MAYA CIVILIZATION

Based on data from *People in Time and Place, Western Hemisphere* (Silver Burdett & Ginn, 1991), p. 129.

83. **iPhones.** You can get a more accurate reading of an iPhone's signal strength by dialing *3001#12345#*. Field test mode is then activated and the standard signal strength bars (in the upper left corner of the display) are replaced by a negative number. The closer the negative number is to zero, the stronger the signal. Which iPhone shown below is receiving the strongest signal?

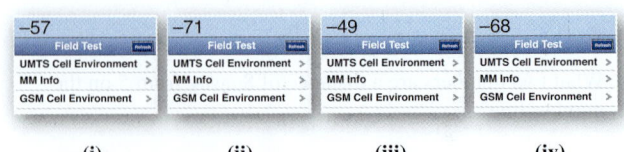

(i)   (ii)   (iii)   (iv)

84. Drafting. On an architect's scale, the edge marked 16 divides each inch into 16 equal parts. Find the decimal form for each fractional part of one inch that is highlighted on the scale.

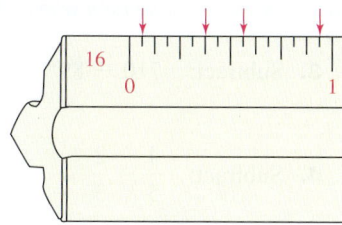

85. Trade. Each year from 1994 through 2009, the United States imported more goods and services from Japan than it exported to Japan. This caused trade *deficits*, which are represented by negative numbers on the following graph.

  a. In which year was the deficit the worst? Express that deficit using a signed number.

  b. In which year was the deficit the smallest? Express that deficit using a signed number.

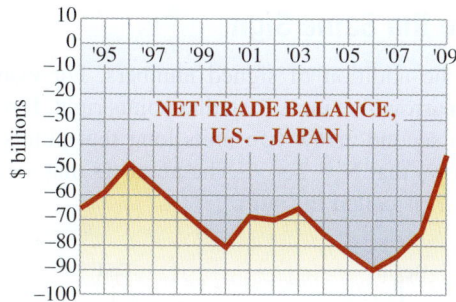

Source: U.S. Bureau of the Census

86. U.S. Budget. A budget *deficit* is a negative number that indicates the government spent more money than it took in that year. A budget *surplus* is a positive number that indicates the government took in more money than it spent that year.

  a. Refer to the graph in the next column that shows the U.S. Federal Budget Deficit/Surplus for the years 1980 through 2009. For how many of those years was there a budget surplus?

  b. Consider the years in which there was a budget deficit. For how many of those years was it smaller than $300 billion?

Federal Budget Deficit/Surplus

Source: U.S. Bureau of the Census

WRITING

87. Explain the difference between a rational and an irrational number.

88. Can two different numbers have the same absolute value? Explain.

89. Explain how to find the decimal equivalent of a fraction.

90. What is a real number?

91. *Pi Day* (or *Pi Approximation Day*) is an unofficial holiday held to celebrate π. Why do you think Pi Day is observed each year on March 14?

92. Explain why $0.1\overline{333}$ is not the simplest way to represent $0.1333\ldots$.

REVIEW

93. Simplify: $\frac{24}{54}$

94. Multiply: $\frac{3}{4}\left(\frac{8}{5}\right)$

95. Divide: $5\frac{2}{3} \div 2\frac{5}{9}$

96. Add: $\frac{3}{10} + \frac{2}{15}$

CHALLENGE PROBLEMS

97. How many integers have an absolute value that is less than 1,000?

98. Is $0.10100100010000\ldots$ a repeating decimal? Explain.

Find a rational number between each pair of numbers.

99. $\frac{1}{8}$ and $\frac{1}{9}$

100. $1.7\overline{1}$ and $1.7\overline{2}$

SECTION 1.4

SECTION 1.4

OBJECTIVES

1. Add two numbers that have the same sign.

2. Add two numbers that have different signs.

3. Use properties of addition.

4. Identify opposites (additive inverses).

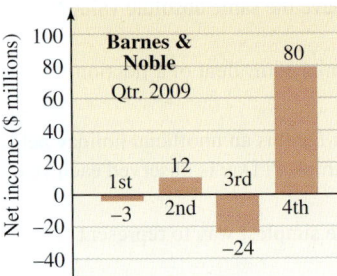

Source: msnbc.msn.com

Adding Real Numbers; Properties of Addition

ARE YOU READY?

The following problems review some basic concepts that are important when adding positive and negative real numbers.

1. Find $|3|$ and $|-5|$. Which number, 3 or -5, has the larger absolute value?

2. Add: $4.37 + 2.8$

3. Subtract: $710 - 89$

4. Subtract: $\dfrac{4}{5} - \dfrac{2}{3}$

In the graph to the left, signed numbers are used to show the financial performance of Barnes and Noble Corporation for the year 2009. Positive numbers indicate *profits* and negative numbers indicate *losses*. To find the company's 2009 net income (in millions of dollars), we need to calculate the following sum:

$$\text{Net income} = -3 + 12 + (-24) + 80$$

In this section, we discuss how to perform this addition and others involving signed numbers.

1   Add Two Numbers That Have the Same Sign.

A number line can be used to explain the addition of signed numbers. For example, to compute $5 + 2$, we begin at 0 and draw an arrow five units long that points right. It represents 5. From the tip of that arrow, we draw a second arrow two units long that points right. It represents 2. Since we end up at 7, it follows that $5 + 2 = 7$. The numbers that we added, 5 and 2, are called **addends,** and the result, 7, is called the **sum.**

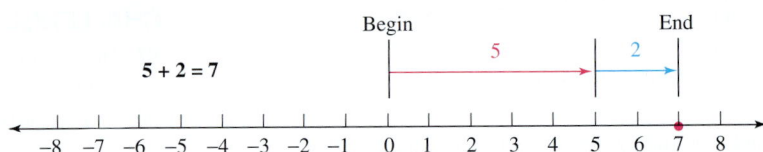

To compute $-5 + (-2)$, we begin at 0 and draw an arrow five units long that points left. It represents -5. From the tip of that arrow, we draw a second arrow two units long that points left. It represents -2. Since we end up at -7, it follows that $-5 + (-2) = -7$.

Notation

To avoid confusion, we write negative numbers within parentheses to separate the negative sign $-$ from the addition symbol $+$.

$$-5 + (-2)$$

To check this result, think of the problem in terms of money. If you lost \$5 (-5) and then lost another \$2 (-2), you would have lost a total of \$7 (-7).

When we use a number line to add numbers with the same sign, the arrows point in the same direction and they build upon each other. Furthermore, the answer has the same sign as the numbers that we added. These observations suggest the following rules.

Adding Two Numbers That Have the Same (Like) Signs

1. To add two positive numbers, add them as usual. The final answer is positive.

2. To add two negative numbers, add their absolute values and make the final answer negative.

EXAMPLE 1 Add: **a.** $-20 + (-15)$ **b.** $-7.89 + (-0.6)$ **c.** $-\frac{1}{3} + \left(-\frac{1}{2}\right)$

Strategy We will use the rule for adding two numbers that have the same sign.

Why In each case, we are asked to add two negative numbers.

Solution **a.** $-20 + (-15) = -35$ Add their absolute values, 20 and 15, to get 35.
Then make the final answer negative.

The Language of Algebra

Two negative numbers, as well as two positive numbers, are said to have **like** signs.

b. To find $-7.89 + (-0.6)$, add their absolute values, 7.89 and 0.6.

$$\begin{array}{r} 7.89 \\ +0.6 \\ \hline 8.49 \end{array}$$ Remember to align the decimal points when adding decimals.

Then make the final answer negative: $-7.89 + (-0.6) = -8.49$.

Success Tip

The sum of two positive numbers is *always* positive. The sum of two negative numbers is *always* negative.

c. To find $-\frac{1}{3} + \left(-\frac{1}{2}\right)$, add their absolute values, $\frac{1}{3}$ and $\frac{1}{2}$.

$$\frac{1}{3} + \frac{1}{2} = \frac{2}{6} + \frac{3}{6}$$ The LCD is 6. Build each fraction: $\frac{1}{3} \cdot \frac{2}{2} = \frac{2}{6}$ and $\frac{1}{2} \cdot \frac{3}{3} = \frac{3}{6}$.

$$= \frac{5}{6}$$ Add the numerators and write the sum over the LCD.

Then make the final answer negative: $-\frac{1}{3} + \left(-\frac{1}{2}\right) = -\frac{5}{6}$.

Self Check 1 Add: **a.** $-51 + (-9)$ **b.** $-12.3 + (-0.88)$
c. $-\frac{1}{4} + \left(-\frac{2}{3}\right)$

Now Try ▶ Problems 15, 21, and 25

2 Add Two Numbers That Have Different Signs.

To compute $5 + (-2)$, we begin at 0 and draw an arrow five units long that points right. From the tip of that arrow, we draw a second arrow two units long that points left. Since we end up at 3, it follows that $5 + (-2) = 3$. In terms of money, if you won \$5 and then lost \$2, you would have \$3 left.

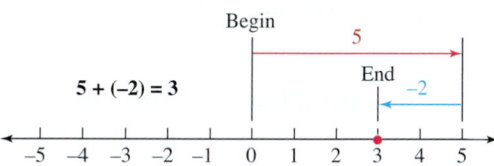

The Language of Algebra

A positive number and a negative number are said to have **unlike** signs.

To compute $-5 + 2$, we begin at 0 and draw an arrow five units long that points left. From the tip of that arrow, we draw a second arrow two units long that points right. Since we end up at -3, it follows that $-5 + 2 = -3$. In terms of money, if you lost \$5 and then won \$2, you have lost \$3.

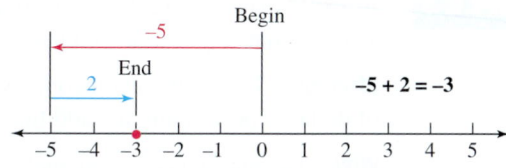

When we use a number line to add numbers with different signs, the arrows point in opposite directions and the longer arrow determines the sign of the answer. If the longer arrow represents a positive number, the sum is positive. If it represents a negative number, the sum is negative. These observations suggest the following rules.

Adding Two Numbers That Have Different (Unlike) Signs	To add a positive number and a negative number, subtract the smaller absolute value from the larger. 1. If the positive number has the larger absolute value, the final answer is positive. 2. If the negative number has the larger absolute value, make the final answer negative.

EXAMPLE 2 Add: **a.** $-20 + 32$ **b.** $5.7 + (-7.4)$ **c.** $-\dfrac{19}{25} + \dfrac{2}{5}$

Strategy We will use the rule for adding two numbers that have different (unlike) signs.

Why In each case, we are asked to add a positive number and a negative number.

Solution **a.** $-20 + 32 = 12$ Subtract the smaller absolute value from the larger: $32 - 20 = 12$. The positive number, 32, has the larger absolute value, so the final answer is positive.

Success Tip

The sum of two numbers with different signs may be positive or negative. The sign of the sum is the sign of the number with the greater absolute value.

b. To find $5.7 + (-7.4)$, subtract the smaller absolute value, 5.7, from the larger, 7.4.

$$\begin{array}{r} 7.4 \\ -5.7 \\ \hline 1.7 \end{array}$$ Remember to align the decimal points when subtracting decimals.

Since the negative decimal, -7.4, has the larger absolute value, make the final answer negative: $5.7 + (-7.4) = -1.7$.

Calculators

Entering negative numbers We don't do anything special to enter positive numbers on a calculator. To enter a negative number, say -7.4, some calculators require the $-$ sign to be entered before entering 7.4 while others require the $-$ sign to be entered after entering 7.4. Consult your owner's manual to determine the proper keystrokes.

c. Since $\dfrac{2}{5} = \dfrac{10}{25}$, the fraction $-\dfrac{19}{25}$ has the larger absolute value. To find $-\dfrac{19}{25} + \dfrac{2}{5}$, we subtract the smaller absolute value from the larger:

$$\frac{19}{25} - \frac{2}{5} = \frac{19}{25} - \frac{10}{25}$$ The LCD is 25. Build $\frac{2}{5}$: $\frac{2}{5} \cdot \frac{5}{5} = \frac{10}{25}$.

$$= \frac{9}{25}$$ Subtract the numerators and write the difference over the LCD.

Since the negative fraction $-\dfrac{19}{25}$ has the larger absolute value, make the final answer negative: $-\dfrac{19}{25} + \dfrac{10}{25} = -\dfrac{9}{25}$.

Self Check 2 Add: **a.** $63 + (-87)$ **b.** $-6.27 + 8$
 c. $-\dfrac{1}{10} + \dfrac{1}{2}$

Now Try ▶ Problems 29, 33, and 35

EXAMPLE 3 **Accounting.** Find the net income of Barnes and Noble Corporation for the year 2009 using the data in the graph on page 32.

Strategy To find the net income, we will add the quarterly profits and losses (in millions of dollars), performing the additions as they occur from left to right.

Why The phrase *net income* means that we should combine (add) the quarterly profits and losses to determine whether there was an overall profit or loss that year.

Solution

$$-3 + 12 + (-24) + 80 = 9 + (-24) + 80 \quad \text{Add: } -3 + 12 = 9.$$
$$= -15 + 80 \quad \text{Add: } 9 + (-24) = -15.$$
$$= 65$$

The Language of Algebra

Net refers to what remains after all the deductions (losses) have been accounted for. *Net income* is a term used in business that often is referred to as the *bottom line*. Net income indicates what a company has earned (or lost) in a given period of time (usually one year).

In 2009, Barnes and Noble's net income was $65 million.

Self Check 3   Add: $650 + (-13) + 87 + (-155)$

Now Try ▶ Problem 43

3   Use Properties of Addition.

The addition of two numbers can be done in any order and the result is the same. For example, $8 + (-1) = 7$ and $-1 + 8 = 7$. This example illustrates that addition is **commutative.**

The Commutative Property of Addition	Changing the order when adding does not affect the answer. For any real numbers a and b, $$a + b = b + a$$

The Language of Algebra

Commutative is a form of the word *commute*, meaning to go back and forth. *Commuter* trains take people to and from work.

In the following example, we add $-3 + 7 + 5$ in two ways. We will use grouping symbols ( ), called **parentheses,** to show this. Standard practice requires that the operation within the parentheses be performed first.

We read $(-3 + 7) + 5$ as "The quantity of –3 plus 7" pause slightly, and then say "plus 5." We read $-3 + (7 + 5)$ as "–3" pause slightly, and then say "plus the quantity of 7 plus 5." The word **quantity** alerts the reader to the parentheses that are used as grouping symbols.

Method 1: Group −3 and 7
$$(-3 + 7) + 5 = 4 + 5$$
$$= 9$$

Method 2: Group 7 and 5
$$-3 + (7 + 5) = -3 + 12$$
$$= 9$$

It doesn't matter how we group the numbers in this addition; the result is 9. This example illustrates that addition is **associative.**

The Associative Property of Addition	Changing the grouping when adding does not affect the answer. For any real numbers a, b, and c, $$(a + b) + c = a + (b + c)$$

Sometimes, an application of the associative property can simplify a computation.

EXAMPLE 4   Find the sum: $98 + (2 + 17)$

Strategy   We will use the associative property to group 2 with 98. Then, we evaluate the expression by performing the addition within the parentheses first.

Why   It is helpful to regroup because 98 and 2 are a pair of numbers that are easily added.

Solution $98 + (2 + 17) = (\mathbf{98 + 2}) + 17$ Use the associative property of addition to regroup. Note that the order of the addends, 98, 2, and 17, is not changed.

$$= \mathbf{100} + 17 \qquad \text{Do the addition within the parentheses first.}$$
$$= 117$$

Self Check 4   Find the sum: $(39 + 25) + 75$

Now Try ▶ Problem 49

EXAMPLE 5

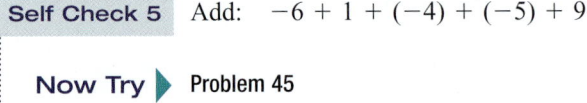

Game Shows.   A contestant on *Jeopardy!* correctly answered the first question to win $100, missed the second to lose $200, correctly answered the third to win $300, and missed the fourth to lose $400. What is her score after answering four questions?

Strategy   We can represent money won by a positive number and money lost by a negative number. Her score is the sum of 100, -200, 300, and -400. Instead of doing the additions from left to right, we will use another approach. Applying the commutative and associative properties, we will add the positives, add the negatives, and then add those results.

Why   It is easier to add numbers that have the same sign than numbers that have different signs. This method minimizes the possibility of an error, because we have to add numbers that have different signs only once.

Solution $100 + (-200) + 300 + (-400)$

$$= (\mathbf{100 + 300}) + [(\mathbf{-200}) + (\mathbf{-400})] \qquad \text{Reorder the numbers. Group the positives together. Group the negatives together using brackets [ ].}$$

$$= \mathbf{400} + (\mathbf{-600}) \qquad \text{Add the positives. Add the negatives.}$$
$$= -200 \qquad \text{Add the results.}$$

After four questions, her score was $-\$200$, which represents a loss of $200.

Self Check 5   Add: $-6 + 1 + (-4) + (-5) + 9$

Now Try ▶ Problem 45

Whenever we add 0 to a number, the result is the number. Therefore, $8 + 0 = 8$, $2.3 + 0 = 2.3$, and $0 + (-16) = -16$. These examples illustrate the **addition property of 0.** Since any number added to 0 remains the same, 0 is called the **identity element** for addition.

Addition Property of 0 (Identity Property of Addition)	When 0 is added to any real number, the result is the same real number. For any real number a, $$a + 0 = a \qquad \text{and} \qquad 0 + a = a$$

4 Identify Opposites (Additive Inverses).

Recall that two numbers that are the same distance from 0 on a number line, but on opposite sides of it, are called **opposites.** To develop a property for adding opposites, we will find $-4 + 4$ using a number line. We begin at 0 and draw an arrow four units long that points left, to represent -4. From the tip of that arrow, we draw a second arrow, four units long that points right, to represent 4. We end up at 0; therefore, $-4 + 4 = 0$.

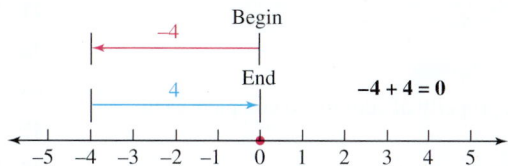

This example illustrates that when we add opposites, the result is 0. Therefore, $1.6 + (-1.6) = 0$ and $-\frac{3}{4} + \frac{3}{4} = 0$. Also, whenever the sum of two numbers is 0, those numbers are opposites. For these reasons, opposites also are called **additive inverses.**

Addition Property of Opposites (Inverse Property of Addition)	The sum of a number and its opposite (additive inverse) is 0. For any real number a and its opposite or additive inverse $-a$, $$a + (-a) = 0 \quad \text{Read } -a \text{ as "the opposite of } a\text{."}$$

EXAMPLE 6   Add: $12 + (-5) + 6 + 5 + (-12)$

Strategy   Instead of working from left to right, we will use the commutative and associative properties of addition to add pairs of opposites.

Why   Since the sum of a number and its opposite is 0, it is helpful to identify such pairs in an addition.

Solution

$$\begin{array}{c} \overbrace{12 + (-5) + 6 + 5 + (-12)}^{\text{opposites}} = 0 + 0 + 6 \\ \underbrace{}_{\text{opposites}} \qquad\qquad = 6 \end{array}$$

Self Check 6   Add: $8 + (-1) + 6 + 5 + (-8) + 1$

Now Try ▶   Problem 73

SECTION 1.4 ▶ STUDY SET

VOCABULARY

Fill in the blanks.

1. In the addition statement $-2 + 5 = 3$, the result, 3, is called the _____.

2. Two numbers that are the same distance from 0 on a number line, but on opposite sides of it, are called _____ or additive _____.

3. The _____ property of addition states that changing the order when adding does not affect the answer. The _____ property of addition states that changing the grouping when adding does not affect the answer.

4. Since any number added to 0 remains the same (is identical), the number 0 is called the _____ element for addition.

CONCEPTS

5. For each pair of numbers, which one has the larger absolute value?
   a. 6 or 5   b. 8.9 or -9.2

6. Determine whether each statement is true or false.
   a. The sum of a number and its opposite is always 0.
   b. The sum of two negative numbers is always negative.
   c. The sum of two numbers with different signs is always negative.

7. For each addition, just determine the sign of the answer.
   a. $39.6 + (-64.9)$ b. $-18.9 + 19.8$

8. Complete each property of addition. Then give its name.

   a. $a + (-a) =$ ▢

   b. $a + 0 =$ ▢

   c. $a + b = b +$ ▢

   d. $(a + b) + c = a +$ ▢

9. Use the commutative property of addition to complete each statement.

   a. $-5 + 1 =$ ▢

   b. $15 + (-80.5) =$ ▢

   c. $-20 + (4 + 20) = -20 + ($ ▢ $)$

   d. $(2.1 + 3) + 6 = ($ ▢ $) + 6$

10. Use the associative property of addition to complete each statement.

   a. $(-6 + 2) + 8 =$ ▢

   b. $-7 + (7 + 3) =$ ▢

11. What properties were used in Step 1 and Step 2 of the solution?

$$(99 + 4) + 1 = (4 + 99) + 1 \quad \text{Step 1}$$
$$= 4 + (99 + 1) \quad \text{Step 2}$$
$$= 4 + 100$$
$$= 104$$

12. Consider: $\quad -3 + 6 + (-9) + 8 + (-4)$

   a. Add all the positives in the expression.

   b. Add all of the negatives.

   c. Add the results from parts **a** and **b**.

NOTATION

13. a. Express the commutative property of addition using the variables x and y.

   b. Express the associative property of addition using the variables x, y, and z.

14. Fill in the blank: We read $-a$ as "the _____ of a."

GUIDED PRACTICE

Add. See Example 1.

15. $-8 + (-1)$

16. $-3 + (-2)$

17. $-5 + (-12)$

18. $-4 + (-14)$

19. $-29 + (-45)$

20. $-23 + (-31)$

21. $-4.2 + (-6.1)$

22. $-5.1 + (-5.1)$

23. $-\dfrac{3}{4} + \left(-\dfrac{2}{3}\right)$

24. $-\dfrac{1}{5} + \left(-\dfrac{3}{4}\right)$

25. $-\dfrac{1}{4} + \left(-\dfrac{1}{10}\right)$

26. $-\dfrac{3}{8} + \left(-\dfrac{1}{3}\right)$

Add. See Example 2.

27. $-7 + 4$

28. $-9 + 7$

29. $50 + (-11)$

30. $27 + (-30)$

31. $15.84 + (-15.84)$

32. $9.19 + (-9.19)$

33. $-6.25 + 8.5$

34. $21.37 + (-12.1)$

35. $-\dfrac{7}{15} + \dfrac{3}{15}$

36. $-\dfrac{8}{11} + \dfrac{3}{11}$

37. $\dfrac{1}{2} + \left(-\dfrac{1}{8}\right)$

38. $\dfrac{5}{6} + \left(-\dfrac{1}{4}\right)$

Add. See Examples 3 and 5.

39. $8 + (-5) + 13$

40. $17 + (-12) + (-23)$

41. $21 + (-27) + (-9)$

42. $-32 + 12 + 17$

43. $-27 + (-3) + (-13) + 22$

44. $53 + (-27) + (-32) + (-7)$

45. $-60 + 70 + (-10) + (-10) + 205$

46. $-100 + 200 + (-300) + (-100) + 200$

Apply the associative property of addition to find the sum. See Example 4.

47. $-99 + (99 + 215)$

48. $67 + (-67 + 127)$

49. $(-112 + 56) + (-56)$

50. $(-67 + 5) + (-5)$

51. $\dfrac{1}{8} + \left(\dfrac{7}{8} + \dfrac{2}{3}\right)$

52. $\left(\dfrac{1}{2} + \dfrac{9}{16}\right) + \dfrac{7}{16}$

53. $(12.4 + 1.9) + 1.1$

54. $87.6 + (2.4 + 1.7)$

Add. See Example 6.

55. $-1 + 9 + 1$

56. $5 + 8 + (-5)$

57. $-8 + 11 + (-11) + 8 + 1$

58. $2 + 15 + (-15) + 8 + (-2)$

TRY IT YOURSELF

Add.

59. $-9 + 81 + (-2)$

60. $11 + (-21) + (-13)$

61. $0 + (-6.6)$

62. $0 + (-2.14)$

63. $-\dfrac{9}{16} + \dfrac{7}{16}$

64. $-\dfrac{3}{4} + \dfrac{1}{4}$

65. $-6 + (-8)$

66. $-4 + (-3)$

67. $-167 + 167$

68. $-25 + 25$

69. $-20 + (-16) + 10$

70. $-13 + (-16) + 4$

71. $19.35 + (-20.21) + 1.53$

72. $33.12 + (-35.7) + 2.98$

73. $-7 + 5 + (-10) + 7$

74. $-3 + 6 + (-9) + (-6)$

75. $19.2 + (-41.3)$

76. $57.93 + (-93.27)$

77. $2{,}345 + (-178)$

78. $-4{,}061 + 5{,}000$

79. $-2.1 + 6.5 + (-8.2) + 2.1$

80. $0.9 + 0.5 + (-0.2) + (-0.9)$

81. $3 + (-6) + (-3) + 74$

82. $4 + (-3) + (-4) + 5$

83. $-\dfrac{1}{4} + \left(-\dfrac{2}{7}\right)$

84. $-\dfrac{3}{32} + \left(-\dfrac{1}{2}\right)$

85. $-0.2 + (-0.3) + (-0.4)$

86. $-0.9 + (-1.9) + (-2.9)$

Look Alikes . . .

87. a. $12 + 15$
   b. $-12 + 15$
   c. $-12 + (-15)$
   d. $12 + (-15)$

88. a. $432 + 67$
   b. $-432 + 67$
   c. $-432 + (-67)$
   d. $432 + (-67)$

89. a. $\dfrac{1}{2} + \dfrac{2}{9}$
   b. $-\dfrac{1}{2} + \dfrac{2}{9}$
   c. $-\dfrac{1}{2} + \left(-\dfrac{2}{9}\right)$
   d. $\dfrac{1}{2} + \left(-\dfrac{2}{9}\right)$

90. a. $0.87 + 0.29$
   b. $-0.87 + 0.29$
   c. $-0.87 + (-0.29)$
   d. $0.87 + (-0.29)$

APPLICATIONS

91. Military Science. During a battle, an army retreated 1,500 meters, regrouped, and advanced 2,400 meters. The next day, it advanced another 1,250 meters. Find the army's net gain.

92. Health. Find the point total for the six risk factors (in blue) on the medical questionnaire. Then use the table at the bottom of the form to determine the patient's risk of contracting heart disease in the next 10 years.

Age		Total Cholesterol	
Age *35*	Points *−4*	Reading *280*	Points *3*
Cholesterol		**Blood Pressure**	
HDL *62*	Points *−3*	Systolic/Diastolic *124/100*	Points *3*
Diabetic		**Smoker**	
	Points		Points
Yes	*4*	*Yes*	*2*
10-Year Heart Disease Risk			
Total Points	**Risk**	Total Points	**Risk**
−2 or less	1%	5	4%
−1 to 1	2%	6	6%
2 to 3	3%	7	6%
4	4%	8	7%

Source: National Heart, Lung, and Blood Institute

93. Golf. The leaderboard below shows the top four finishers from the 2009 PGA Championship Golf Tournament. Scores for each round are compared to *par,* the standard number of strokes necessary to complete the course. A score of -2, for example, indicates that the golfer used two strokes less than par to complete the course. A score of 5 indicates five strokes more than par. Determine the tournament total for each golfer.

Leaderboard

	Round				
	1	**2**	**3**	**4**	**Total**
Y.E. Yang	+1	−2	−5	−2	
Tiger Woods	−5	−2	−1	+3	
Lee Westwood	−2	0	+1	−2	
Rory McIlroy	−1	+1	−1	−2	

94. Submarines. A submarine was cruising at a depth of 1,250 feet. The captain gave the order to climb 550 feet. Compared to sea level, find the new depth of the sub.

95. Credit Cards. Refer to the monthly statement. What is the new balance?

Previous Balance	New Purchases, Fees, Advances & Debts	Payments & Credits	New Balance
3,660.66	1,408.78	3,826.58	
04/21/11 Billing Date	**05/16/11** Date Payment Due		**9,100** Credit Line

96. Politics. The following proposal to limit campaign contributions was on the ballot in a state election, and it passed. What will be the net fiscal impact on the state government?

212 Campaign Spending Limits	YES ☐ NO ☐

Limits contributions to $200 in state campaigns. Fiscal impact: Costs of $4.5 million for implemen– tation and enforcement. Increases state revenue by $6.7 million by eliminating tax deductions for lobbying.

97. Movie Losses. According to the Numbers Box Office Data website, the movie *Stealth,* released in 2005 by Sony Pictures, cost about $176,350,000 to produce, promote, and distribute. It reportedly earned back just $76,700,000 worldwide. Express the dollar loss suffered by Sony as a signed number.

98. Stocks. The last entry on the line for June 12 indicates that one share of Walt Disney Co. stock lost $0.81 in value that day. How much did the value of a share of Disney stock rise or fall over the 5-day period from June 12 through June 16?

June 12	43.88	23.38	Disney	.21	0.5	87	−43	40.75	−.81
June 13	43.88	23.38	Disney	.21	0.5	86	−15	40.19	−.56
June 14	43.88	23.38	Disney	.21	0.5	87	−50	41.00	+.81
June 15	43.88	23.38	Disney	.21	0.5	89	−28	41.81	+.81
June 16	43.88	23.38	Disney				−15	41.19	−.63

Based on data from the *Los Angeles Times*

99. Chemistry. An atom is composed of protons (with a charge of $+1$), neutrons (with no charge), and electrons (with a charge of -1). Two simple models of atoms are shown. What is the overall charge of each atom?

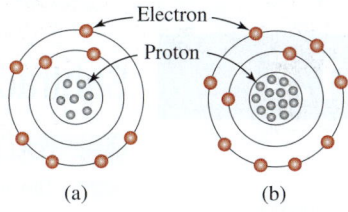

(a)   (b)

100. Physics. In the illustration, arrows show the two forces acting on a lamp hanging from a ceiling. What is the sum of the forces?

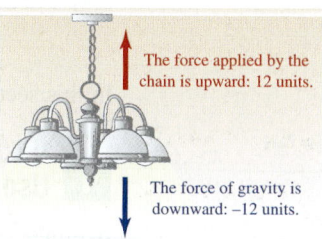

The force applied by the chain is upward: 12 units.

The force of gravity is downward: −12 units.

101. The Big Easy. The city of New Orleans lies, on average, 6 feet below sea level. What is the elevation of the top of an 85-foot tall building in New Orleans?

102. Electronics. A closed circuit contains two batteries and three resistors. The sum of the voltages in the loop must be 0. Is it?

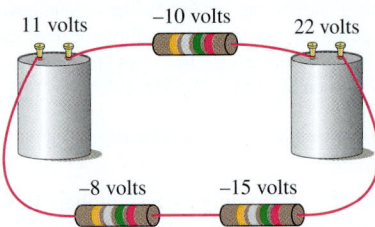

103. Accounting. The 2009 quarterly profits and losses of the Bank of America are shown in the table. Losses are denoted using parentheses. Calculate the company's total net income for 2009.

Quarter	Net income ($ million)
1st	4,247
2nd	3,224
3rd	(1,001)
4th	(194)

Source: www.scottrade.com

104. Politics. Six months before an election, the incumbent trailed the challenger by 18 points. To overtake her opponent, the incumbent decided to use a four-part strategy. Each part of the plan is shown below, with the expected point gain. With these gains, will the incumbent overtake the challenger on election day?

- TV ads   +10 pts
- Voter mailing   +3 pts
- Union endorsement   +2 pts
- Telephone calls   +1 pts

105. Explain why the sum of two positive numbers is always positive and the sum of two negative numbers is always negative.

106. Explain why the sum of a negative number and a positive number is sometimes positive, sometimes negative, and sometimes zero.

REVIEW

107. True or false: Every real number can be expressed as a decimal.

108. Multiply: $\dfrac{1}{3} \cdot \dfrac{1}{3}$

109. What two numbers are a distance of 6 away from -3 on the number line?

110. Graph: $\left\{ -2.5, \ \sqrt{2}, \ \dfrac{11}{3}, \ -0.333 \ldots, \ 0.75 \right\}$

CHALLENGE PROBLEMS

111. A set is said to be *closed under addition* if the sum of any two of its members is also a member of the set. Is the set $\{-1, 0, 1\}$ a closed set under addition? Explain.

112. Think of two numbers. First, add the absolute value of the two numbers, and write your answer. Second, add the two numbers, take the absolute value of that sum, and write that answer. Do the two answers agree? Can you find two numbers that produce different answers? When do you get answers that agree, and when don't you?

SECTION 1.5

Subtracting Real Numbers

OBJECTIVES

1. Use the definition of subtraction.
2. Solve application problems using subtraction.

ARE YOU READY?

The following problems review some basic concepts that are important when subtracting positive and negative real numbers.

1. What is the opposite of 6? What is the opposite of -15?
2. Write *twenty-two minus six* in symbols.
3. If 8 is subtracted from 20, what is the result?
4. Add: $-11 + 2$

In this section, we discuss a rule to use when subtracting signed numbers.

1 Use the Definition of Subtraction.

A minus symbol $-$ is used to indicate subtraction. However, this symbol is also used in two other ways, depending on where it appears in an expression.

$5 - 18$ This is read as "five minus eighteen."

-5 This usually is read as "negative five." It also could be read as "the additive inverse of five" or "the opposite of five."

$-(-5)$ This usually is read as "the opposite of negative five." It also could be read as "the additive inverse of negative five."

In $-(-5)$, parentheses are used to write the opposite of a negative number. When such expressions are encountered in computations, we simplify them by finding the opposite of the number within the parentheses.

$$-(-5) = 5$$ Read as "the opposite of negative five is five."

This observation illustrates the following rule.

Opposite of an Opposite   The opposite of the opposite of a number is that number.
For any real number a,

$$-(-a) = a$$ Read as "the opposite of the opposite of a is a."

EXAMPLE 1   Simplify each expression: **a.** $-(-45)$ **b.** $-(-h)$ **c.** $-|-10|$

Strategy   To simplify each expression, we will use the concept of opposite.

Why   In each case, the outermost $-$ symbol is read as "the opposite."

Solution   **a.** The number within the parentheses is -45. Its opposite is 45. Therefore, $-(-45) = 45$.

b. The opposite of the opposite of h is h. Therefore, $-(-h) = h$.

c. The notation $-|-10|$ means "the opposite of the absolute value of negative ten." Since $|-10| = 10$, we have:

$$-|-10| = -10$$ The absolute value bars do not affect the $-$ symbol outside them. Therefore, the result is negative.

Self Check 1   Simplify each expression: **a.** $-(-1)$ **b.** $-(-y)$
c. $-|-500|$

Now Try ▶   Problems 15, 17, and 19

To develop a rule for subtraction, we consider the following illustration. It represents the subtraction $5 - 2 = 3$.

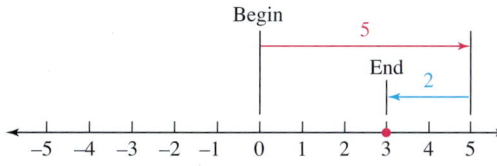

The illustration above also represents the addition $5 + (-2) = 3$. We see that

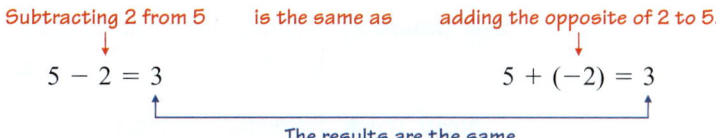

Subtracting 2 from 5   is the same as   adding the opposite of 2 to 5.

$$5 - 2 = 3 \qquad\qquad 5 + (-2) = 3$$

The results are the same.

This observation suggests the following definition.

Subtraction of Real Numbers	To subtract two real numbers, add the first number to the opposite (additive inverse) of the number to be subtracted. For any real numbers a and b, $$a - b = a + (-b)$$ Read as "*a minus b equals a plus the opposite of b.*"

EXAMPLE 2   Subtract and check the result:

a. $-13 - 8$ **b.** $-7.6 - (-4.5)$ **c.** $\dfrac{1}{4} - \left(-\dfrac{1}{8}\right)$

Strategy   To find each difference, we will apply the rule for subtraction: Add the first number to the opposite of the number to be subtracted.

Why   It is easy to make an error when subtracting signed numbers. We will probably be more accurate if we write each subtraction as addition of the opposite.

Solution   **a.** We read $-13 - 8$ as "negative thirteen *minus* eight." Subtracting 8 is the same as adding -8.

The Language of Algebra

When we change a number to its opposite, we say we have **changed** (or *reversed*) its sign.

Change the subtraction to addition.

$$-13 - 8 \quad = \quad -13 + (-8) = -21$$ Use the rule for adding two numbers with like signs.

No change   Change the number being subtracted to its opposite.

To check, we add the *difference*, -21, and the *subtrahend*, 8, to obtain the *minuend*, -13.

Check: $-21 + 8 = -13$

The Language of Algebra

The rule for subtracting real numbers is often summarized as: **Subtracting a number is the same as adding its opposite.**

b. We read $-7.6 - (-4.5)$ as "negative seven point six *minus* negative four point five." Subtracting -4.5 is the same as adding 4.5.

Add . . .

$$-7.6 - (-4.5) \quad = \quad -7.6 + 4.5 = -3.1$$ Use the rule for adding two numbers with unlike signs.

No change   . . . the opposite

Check: $-3.1 + (-4.5) = -7.6$

Calculators

The subtraction key
When using a calculator to subtract signed numbers, be careful to distinguish between the *subtraction* key $-$ and the keys that are used to enter negative values: $+/-$ on a scientific calculator and $(-)$ on a graphing calculator.

c. $\dfrac{1}{4} - \left(-\dfrac{1}{8}\right) = \dfrac{2}{8} - \left(-\dfrac{1}{8}\right)$ The LCD is 8. Build $\frac{1}{4}$ so that the denominator is 8: $\frac{1}{4} \cdot \frac{2}{2} = \frac{2}{8}$.

$\qquad\qquad\qquad = \dfrac{2}{8} + \dfrac{1}{8}$ To subtract, add the opposite. Do not change $\frac{2}{8}$.

$\qquad\qquad\qquad = \dfrac{3}{8}$

Check: $\frac{3}{8} + \left(-\frac{1}{8}\right) = \frac{2}{8} = \frac{1}{4}$

Self Check 2   Subtract and check the result: **a.** $-32 - 25$
b. $1.7 - (-1.2)$ **c.** $-\dfrac{1}{3} - \left(-\dfrac{3}{4}\right)$

Now Try ▶   Problems 25, 39, and 43

EXAMPLE 3   **a.** Subtract 0.5 from 4.6   **b.** Subtract 4.6 from 0.5

Strategy   We will translate each phrase to mathematical symbols and then perform the subtraction. We must be careful when translating the instruction to subtract one number *from* another number.

Why   The order of the numbers in each word phrase must be reversed when we translate it to mathematical symbols.

Solution   **a.** The number to be subtracted is 0.5.

$$\text{Subtract } 0.5 \text{ from } 4.6$$

To translate, reverse the order in which 0.5 and 4.6 appear in the sentence.

$$4.6 - 0.5 = 4.1$$

b. The number to be subtracted is 4.6.

$$\text{Subtract } 4.6 \text{ from } 0.5$$

To translate, reverse the order in which 4.6 and 0.5 appear in the sentence. Add the opposite of 4.6.

$$0.5 - 4.6 = 0.5 + (-4.6)$$
$$= -4.1$$

> **Caution**
>
> Notice from parts **a** and **b** that $4.6 - 0.5 \neq 0.5 - 4.6$. This result illustrates an important fact: Subtraction is *not* commutative. When subtracting two numbers, it is important that we write them in the correct order, because, in general, $a - b \neq b - a$.

Self Check 3   **a.** Subtract 2.2 from 4.9   **b.** Subtract 4.9 from 2.2

Now Try ▶ Problem 47

EXAMPLE 4   Perform the operations: $-9 - 15 + 20 - (-6)$

Strategy   This expression contains addition and subtraction. We will write each subtraction as addition of the opposite and then evaluate the expression.

Why   It is easy to make an error when subtracting signed numbers. We probably will be more accurate if we write each subtraction as addition of the opposite.

Solution $-9 - 15 + 20 - (-6) = -9 + (-15) + 20 + 6$
$$= -24 + 26 \quad \text{Add the negatives. Add the positives.}$$
$$= 2 \quad \text{Add the results.}$$

Self Check 4   Perform the operations: $-40 - (-10) + 7 - (-15)$

Now Try ▶ Problem 51

2   Solve Application Problems Using Subtraction.

Subtraction finds the *difference* between two numbers. When we find the difference between the maximum value and the minimum value of a collection of measurements, we are finding the **range** of the values.

EXAMPLE 5   **U.S. Temperatures.**   The record high temperature in the United States of 134°F was set in Death Valley, California, on July 10, 1913. The record low of −80°F was set at Prospect Creek, Alaska, on January 23, 1971. Find the temperature range for these extremes.

Strategy We will subtract the lowest temperature from the highest temperature.

Why The *range* of a collection of data indicates the spread of the data. It is the difference between the largest and smallest values.

Solution
$$134 - (-80) = 134 + 80 \qquad \textcolor{red}{134° \text{ is the higher temperature and } -80° \text{ is the lower.}}$$
$$= 214$$

The temperature range for these extremes is 214°F.

134° — Record high

Difference in temperature extremes

0°

−80° — Record low

> **Self Check 5** Find the temperature range for a day that had a low of −5°F and a high of 36°F.
>
> **Now Try** ▶ Problem 95

Things are constantly changing in our daily lives. The amount of money we have in the bank, the price of gasoline, and our ages are examples. In mathematics, the operation of subtraction is used to measure change. To find the **change** in a quantity, we subtract the earlier value from the later value.

$$\text{Change} = \text{later value} - \text{earlier value}$$

EXAMPLE 6

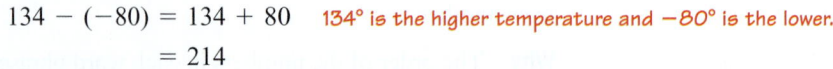

©iStockphoto.com/MvH

Water Management. Lake Mead, on the Nevada-Arizona border, is the largest reservoir in the United States. It is formed by Hoover Dam across the Colorado River. In 2000, the water level in Lake Mead was 89 feet above drought level. By 2010, the water level was 27 feet below drought level. Find the change in the water level over that time span. (Source: Bureau of Reclamation)

Strategy We can represent a water level above drought level using a positive number and a water level below drought level using a negative number. To find the change in the water level, we will subtract.

Why In general, *to find the change in a quantity, we subtract the earlier value from the later value.*

Solution
$$-27 - 89 = -27 + (-89) \qquad \textcolor{red}{\text{The earlier water level in 2000 (89 ft) is subtracted}}$$
$$\textcolor{red}{\text{from the later water level in 2010 (}-27 \text{ ft).}}$$

$$= -116 \qquad \textcolor{red}{\text{Do the addition.}}$$

The negative result indicates that the water level *fell* 116 feet in that time span.

> **Self Check 6** Find the change in water level for a week that started at 4 feet above normal and went to 7 feet below normal level.
>
> **Now Try** ▶ Problem 97

SECTION 1.5 ▶ **STUDY SET**

VOCABULARY

Fill in the blanks.

1. _____ finds the difference between two numbers.
2. In the subtraction $-2 - 5 = -7$, the result of -7 is called the _____.

3. The difference between the maximum and the minimum value of a collection of measurements is called the _____ of the values.
4. To find the _____ in a quantity, subtract the earlier value from the later value.

CONCEPTS

5. Find the opposite (additive inverse) of each number.

   a. 12
   b. $-\dfrac{1}{5}$
   c. 2.71
   d. 0

6. Complete each statement.
   a. $a - b = a +$ ▢

   To subtract two numbers, add the first number to the _____ of the number to be subtracted.

   b. $-(-a) =$ ▢

   The opposite of the opposite of a number is that _____.

7. Apply the rule for subtraction and fill in the blanks.

   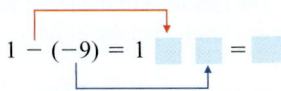

   $1 - (-9) = 1$ ▢ ▢ $=$ ▢

8. Use addition to check this subtraction: $15 - (-8) = 7$. Is the result correct?

9. Write each subtraction in the following expression as addition of the opposite.

   $-10 - 8 + (-23) + 5 - (-34)$

10. For each subtraction, just determine the sign of the answer.
    a. $8.76 - 12.91$
    b. $8.76 - (-12.91)$

11. Circle any minus signs in each expression.
    a. $-6 - (-4)$
    b. $7 + (-3) - 5 - (-2)$

12. In each case, determine what number is being subtracted.
    a. $5 - 8$
    b. $-5 - (-8)$

NOTATION

13. Write each phrase using symbols. Then find its value.
    a. One minus negative seven
    b. The opposite of negative two
    c. The opposite of the absolute value of negative three
    d. Subtract 6 from 2

14. Write each expression in words.
    a. $-(-m)$
    b. $-2 - (-3)$
    c. $x - (-y)$

GUIDED PRACTICE

Simplify each expression. See Example 1.

15. $-(-55)$
16. $-(-27.2)$
17. $-(-x)$
18. $-(-t)$
19. $-|-25|$
20. $-|-100|$
21. $-\left|-\dfrac{3}{16}\right|$
22. $-\left|-\dfrac{4}{3}\right|$

Subtract. See Example 2.

23. $4 - 7$
24. $1 - 6$
25. $-6 - 4$
26. $-3 - 4$
27. $8 - (-3)$
28. $17 - (-21)$
29. $0 - 6$
30. $0 - 9$
31. $-1 - (-3)$
32. $-1 - (-7)$

33. $20 - (-20)$
34. $30 - (-30)$
35. $-2 - (-7)$
36. $-9 - (-1)$
37. $0 - (-12)$
38. $0 - 12$
39. $-1.4 - 5.5$
40. $-1.3 - 4.7$
41. $-1.5 - 0.81$
42. $-1.57 - (-0.8)$
43. $-\dfrac{1}{8} - \dfrac{3}{8}$
44. $-\dfrac{3}{4} - \dfrac{1}{4}$
45. $\dfrac{1}{3} - \dfrac{3}{4}$
46. $\dfrac{1}{6} - \dfrac{5}{8}$

Perform the indicated operation. See Example 3.

47. Subtract -5 from 17.
48. Subtract 45 from -50.
49. Subtract 12 from -13.
50. Subtract -11 from -20.

Perform the operations. See Example 4.

51. $-6 + 8 - (-1) - 10$
52. $-4 + 5 - (-3) - 13$
53. $61 - (-62) + (-64) - 60$
54. $93 - (-92) + (-94) - 95$

TRY IT YOURSELF

Perform the operations.

55. $244 - (-12)$
56. $354 - (-29)$
57. $-20 - (-30) - 50 + 40$
58. $-24 - (-28) - 48 - 44$
59. $-1.2 - 0.9$
60. $-2.52 - 1.72$
61. $\dfrac{1}{8} - \left(-\dfrac{5}{7}\right)$
62. $\dfrac{5}{8} - \left(-\dfrac{2}{9}\right)$
63. $-62 - 71 - (-37) + 99$
64. $-17 - 32 - (-85) - 51$
65. Subtract 47.5 from 0.
66. Subtract 30.3 from 0.
67. Subtract -137 from 12.
68. Subtract 512 from -47.
69. $-1,903 - (-1,732)$
70. $-300 - (-11)$
71. $2.83 - (-1.8)$
72. $4.75 - (-1.9)$
73. $-\dfrac{5}{6} - \dfrac{3}{4}$
74. $-\dfrac{3}{7} - \dfrac{2}{5}$
75. $8 - 9 - 10$
76. $1 - 2 - 3$
77. $-44 - 44$
78. $-33 - 33$
79. $-0.9 - 0.2$
80. $-0.3 - 0.2$
81. $-25 - (-50) - 75$
82. $-33 - (-22) - 44$
83. $6.3 - 9.8$
84. $2.1 - 9.4$
85. $-\dfrac{9}{16} - \left(-\dfrac{1}{4}\right)$
86. $-\dfrac{1}{2} - \left(-\dfrac{1}{4}\right)$
87. $0 - (-1)$
88. $0 - (-8)$
89. $2 - 15$
90. $3 - 14$

Look Alikes . . .

91. a. $-50 + (-3)$ b. $-50 - (-3)$
92. a. $-\dfrac{1}{16} + \dfrac{1}{4}$ b. $-\dfrac{1}{16} - \dfrac{1}{4}$
93. a. $-\dfrac{5}{9} + \left(-\dfrac{1}{6}\right)$ b. $-\dfrac{5}{9} - \left(-\dfrac{1}{6}\right)$
94. a. $2.96 + (-1.78)$ b. $2.96 - (-1.78)$

APPLICATIONS

95. The Empire State.   New York state's record high temperature of 108°F was set in 1926, and the record low of −52°F was set in 1979. What is the range of these temperature extremes?

96. Eyesight.   Nearsightedness, the condition where near objects are clear and far objects are blurry, is measured using negative numbers. Farsightedness, the condition where far objects are clear and near objects are blurry, is measured using positive numbers. Find the range in the measurements shown.

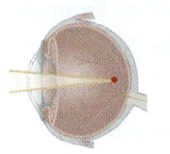 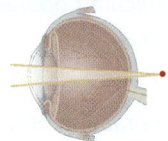

Nearsighted: −2.5   Farsighted: +4.35

97. Law Enforcement.   A burglar scored −18 on a lie detector test, a score that indicates deception. However, on a second test, he scored +3, a score that is inconclusive. Find the change in the scores.

98. Racing.   To improve handling, drivers often adjust the angle of the wheels of their car. When the wheel leans out, the degree measure is considered positive. When the wheel leans in, the degree measure is considered negative. Find the change in the position of the wheel shown below.

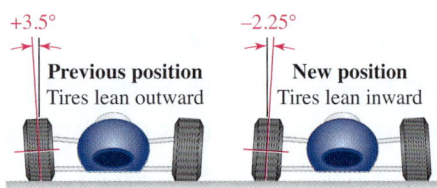

+3.5°   −2.25°

Previous position   **New position**
Tires lean outward   Tires lean inward

99. from **Campus to Careers**

Lead Transportation Security Officer

Determine the change in the number of passengers using each airport in 2009 compared with 2008.

Top 2 Destination Airports in the U.S.
(Number of passengers)

Orlando Int'l Airport Florida	2009	1,296,000
	2008	1,411,000
Ft Lauderdale Airport Florida	2009	1,148,000
	2008	1,198,000

Source: Bureau of Transportation Statistics

© Carolina K. Smith, M.D./Shutterstock.com

100. U.S. Jobs.   The table lists the three occupations that are predicted to have the largest job declines from 2008 to 2018. Complete the column labeled "Change."

Occupation	Number of jobs		
	2008	2018	Change
Farmers/ranchers	985,900	906,700	
Sewing machine operators	212,400	140,900	
Order clerks	245,700	181,500	

Source: Bureau of Labor Statistics

101. Geography.   The elevation of Death Valley, California, is 282 feet below sea level. The elevation of the Dead Sea in Israel is 1,312 feet below sea level. Find the difference in their elevations.

102. Card Games.   Gonzalo won the second round of a card game and earned 50 points. Matt and Hydecki had to deduct the value of each of the cards left in their hands from their score on the first round. Use this information to update the score sheet below. (Face cards are counted as 10 points, aces as 1 point, and all others have the value of the number printed on the card.)

Matt   Hydecki

Running point total	Round 1	Round 2
Matt	+50	
Gonzalo	−15	
Hydecki	−2	

103. World's Coldest Ice Cream.   Dippin' Dots is an ice cream snack that was invented by Curt Jones in 1987. The tiny multi-colored beads are created by flash freezing ice cream mix in liquid nitrogen at a temperature of −355°F. When they come out of the processor, they are stored at a temperature of −40°F. Find the change in temperature of Dippin' Dots from production to storage. (Source: fundinguniverse.com)

104. History.   Plato, a famous Greek philosopher, died in 347 B.C. at the age of 81. When was he born?

105. NASCAR.   Complete the table below to determine how many points the third and fourth place finishers were behind the leader.

2009 Final Driver Standings			
Rank	Driver	Points	Points behind leader
1	Jimmie Johnson	6,652	. . .
2	Mark Martin	6,511	−141
3	Jeff Gordon	6,473	
4	Kurt Busch	6,446	

106. Gauges. With the engine off, the ammeter on a car reads 0. If the headlights, which draw a current of 7 amps, and the radio, which draws a current of 6 amps, are both turned on, what will be the new reading?

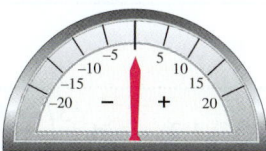

WRITING

107. Explain what it means when we say that subtraction is *not commutative*.

108. Why is addition of signed numbers taught before subtraction of signed numbers?

109. Explain why we know that the answer to $4 - 10$ is negative without having to do any computation.

110. Is the following statement true or false? Explain.

Having a debt of $100 forgiven is equivalent to gaining $100.

REVIEW

111. Find the prime factorization of 30.

112. Write the set of integers.

113. True or false: $-4 > -5$?

114. Use the associative property of addition to simplify the calculation: $-18 + (18 + 89)$

CHALLENGE PROBLEMS

115. Suppose x is positive and y is negative. Determine whether each statement is true or false.
  a. $x - y > 0$ **b.** $y - x < 0$
  c. $|-x| < 0$ **d.** $-|y| < 0$

116. Find:

$$1 - 2 + 3 - 4 + 5 - 6 + \ldots + 99 - 100$$

SECTION 1.6

OBJECTIVES

1 Multiply signed numbers.

2 Use properties of multiplication.

3 Divide signed numbers.

4 Use properties of division.

Multiplying and Dividing Real Numbers; Multiplication and Division Properties

ARE YOU READY?

The following problems review some basic concepts that are important when multiplying positive and negative real numbers.

1. Find $|-14|$ and $|6.75|$.

2. Do the integers -3 and 24 have the same sign or different signs?

3. Multiply: $\dfrac{7}{8} \cdot \dfrac{5}{14}$

4. Divide: $2.22 \div 0.6$

In this section, we will develop rules for multiplying and dividing positive and negative numbers.

1 Multiply Signed Numbers.

Multiplication represents repeated addition. For example, 4(3) is equal to the sum of four 3's.

$$4(3) = 3 + 3 + 3 + 3$$
$$= 12$$

The Language of Algebra

The names of the parts of a multiplication fact are:

Factor Factor Product
$$4(3) = 12$$

This example illustrates that *the product of two positive numbers is positive.*

To develop a rule for multiplying a positive number and a negative number, we will find $4(-3)$, which is equal to the sum of four -3's.

$$4(-3) = -3 + (-3) + (-3) + (-3)$$
$$= -12$$

We see that the result is negative. As a check, think in terms of money. If you lose $3 four times, you have lost a total of $12, which is written $-\$12$. This example illustrates that *the product of a positive number and a negative number is negative.*

| **Multiplying Two Numbers That Have Different (Unlike) Signs** | To multiply a positive real number and a negative real number, multiply their absolute values. Then make the final answer negative. |

EXAMPLE 1 Multiply: **a.** $8(-12)$ **b.** $-151 \cdot 5$ **c.** $(-0.6)(1.2)$ **d.** $\dfrac{3}{4}\left(-\dfrac{4}{15}\right)$

Strategy We will use the rule for multiplying two numbers that have different signs.

Why In each case, we are asked to multiply a positive number and a negative number.

Solution **a.** $8(-12) = -96$ Multiply the absolute values, 8 and 12, to get 96. Since the signs are unlike, make the final answer negative.

b. $-151 \cdot 5 = -755$ Multiply the absolute values, 151 and 5, to get 755. Since the signs are unlike, make the final answer negative.

c. To find the product of these two decimals with unlike signs, first multiply their absolute values, 0.6 and 1.2.

$$\begin{array}{r} 1.2 \\ \times\ 0.6 \\ \hline 0.72 \end{array}$$ Place the decimal point in the result so that the answer has the same number of decimal places as the sum of the number of decimal places in the factors.

Then make the final answer negative: $(-0.6)(1.2) = -0.72$.

Success Tip

The product of two numbers with unlike signs is *always* negative.

d. $\dfrac{3}{4}\left(-\dfrac{4}{15}\right) = -\dfrac{\overset{1}{\cancel{3}} \cdot \overset{1}{\cancel{4}}}{\underset{1}{\cancel{4}} \cdot \underset{1}{\cancel{3}} \cdot 5}$ Multiply the absolute values $\frac{3}{4}$ and $\frac{4}{15}$. Since the signs are unlike, make the final answer negative.

$$= -\frac{1}{5}$$ To simplify the fraction, factor 15 as $3 \cdot 5$. Remove the common factors 3 and 4 in the numerator and denominator.

Self Check 1 Multiply: **a.** $20(-3)$ **b.** $-3 \cdot 5$
c. $4.3(-2.6)$ **d.** $-\dfrac{5}{8} \cdot \dfrac{16}{25}$

Now Try ▶ Problems 21, 27, and 29

To develop a rule for multiplying two negative numbers, consider the following list, where we multiply -4 by factors that decrease by 1. We know how to find the first four products. Graphing those results on a number line is helpful in determining the last three products.

This factor decreases by 1 each time →   Look for a pattern here

$$-4(\mathbf{3}) = -12$$
$$-4(\mathbf{2}) = -8$$
$$-4(\mathbf{1}) = -4$$
$$-4(\mathbf{0}) = \ \ 0$$
$$-4(\mathbf{-1}) = \ \ ?$$
$$-4(\mathbf{-2}) = \ \ ?$$
$$-4(\mathbf{-3}) = \ \ ?$$

A graph of the products

From the pattern, we see that the product increases by 4 each time. Thus,

$$-4(-1) = 4, \qquad -4(-2) = 8, \qquad \text{and} \qquad -4(-3) = 12$$

These results illustrate that *the product of two negative numbers is positive.* As a check, think of losing four debts of $3. This is equivalent to gaining $12. Therefore, $-4(-\$3) = \$12$.

Since the product of two positive numbers is positive, and the product of two negative numbers is also positive, we can summarize the multiplication rule as follows.

Multiplying Two Numbers That Have the Same (Like) Signs	To multiply two real numbers that have the same sign, multiply their absolute values. The final answer is positive.

EXAMPLE 2   Multiply: **a.** $-5(-6)$ **b.** $\left(-\dfrac{1}{2}\right)\left(-\dfrac{5}{8}\right)$

Strategy   We will use the rule for multiplying two numbers that have the same sign.

Why   In each case, we are asked to multiply two negative numbers.

Solution   **a.** $-5(-6) = 30$ Multiply the absolute values, 5 and 6, to get 30. Since both factors are negative, the final answer is positive.

Success Tip

The product of two numbers with like signs is *always* positive.

b. $\left(-\dfrac{1}{2}\right)\left(-\dfrac{5}{8}\right) = \dfrac{5}{16}$ Multiply the absolute values, $\frac{1}{2}$ and $\frac{5}{8}$, to get $\frac{5}{16}$. Since the two factors have the same sign, the final answer is positive.

Self Check 2   Multiply: **a.** $-15(-8)$ **b.** $-\dfrac{1}{4}\left(-\dfrac{1}{3}\right)$

Now Try   Problems 33 and 41

2   Use Properties of Multiplication.

The multiplication of two numbers can be done in any order; the result is the same. For example, $-9(4) = -36$ and $4(-9) = -36$. This illustrates that multiplication is **commutative.**

The Commutative Property of Multiplication	Changing the order when multiplying does not affect the answer. For any real numbers a and b, $$ab = ba$$

In the following example, we multiply $-3 \cdot 7 \cdot 5$ in two ways. Recall that the operation within the parentheses should be performed first. We read $(-3 \cdot 7)5$ as "the *quantity* of -3 times 7," pause slightly, and then say "times 5." We read $-3(7 \cdot 5)$ as "-3 times the *quantity* of 7 times 5." The word *quantity* alerts the reader to the parentheses that are used as grouping symbols.

Method 1: Group -3 *and* 7	*Method 2: Group* 7 *and* 5
$(-3 \cdot 7)5 = (-21)5$	$-3(7 \cdot 5) = -3(35)$
$= -105$	$= -105$

It doesn't matter how we group the numbers in this multiplication; the result is -105. This example illustrates that multiplication is **associative.**

The Associative Property of Multiplication	Changing the grouping when multiplying does not affect the answer. For any real numbers a, b, and c, $$(ab)c = a(bc)$$

EXAMPLE 3 Multiply: **a.** $-5(-37)(-2)$ **b.** $-4(-3)(-2)(-1)$

Strategy First, we will use the commutative and associative properties of multiplication to reorder and regroup the factors. Then we will perform the multiplications.

Why Applying one or both of these properties before multiplying can simplify the computations and lessen the chance of a sign error.

Solution Using the commutative and associative properties of multiplication, we can reorder and regroup the factors to simplify computations.

a. Since it is easy to multiply by 10, we will find $-5(-2)$ first.

$$-5(-37)(-2) = -5(-2)(-37) \quad \text{Use the commutative property of multiplication.}$$
$$= 10(-37)$$
$$= -370$$

b. $-4(-3)(-2)(-1) = 12(2) \quad \text{Multiply the first two factors and multiply the last two factors.}$
$$= 24$$

Self Check 3 Multiply: **a.** $-25(-3)(-4)$ **b.** $-1(-2)(-3)(-3)$

Now Try ▶ Problems 43 and 47

In Example 3a, we multiplied three negative numbers. In Example 3b, we multiplied four negative numbers. The results illustrate the following fact.

Multiplying Negative Numbers	The product of an even number of negative numbers is positive. The product of an odd number of negative numbers is negative.

Recall that the product of 0 and any whole number is 0. The same is true for any real number. Therefore, $-6 \cdot 0 = 0$, $\frac{7}{16} \cdot 0 = 0$, and $0(4.51) = 0$.

Multiplication Property of 0	The product of 0 and any real number is 0. For any real number a, $$0 \cdot a = 0 \quad \text{and} \quad a \cdot 0 = 0$$

Whenever we multiply a number by 1, the number remains the same. Therefore, $1 \cdot 6 = 6$, $4.57 \cdot 1 = 4.57$, and $1(-9) = -9$. Since any number multiplied by 1 remains the same (is identical), the number 1 is called the **identity element** for multiplication.

Multiplication Property of 1 (Identity Property of Multiplication)	The product of 1 and any number is that number. For any real number a, $$1 \cdot a = a \quad \text{and} \quad a \cdot 1 = a$$

Whenever we multiply a number by -1, the result is the opposite of that number. For example, $-1 \cdot 12 = -12$ and $-\frac{5}{8}(-1) = \frac{5}{8}$.

Multiplication Property of -1	The product of -1 and any number is the opposite (or additive inverse) of that number. For any real number a, $$-1 \cdot a = -a \quad \text{and} \quad a(-1) = -a$$

The Language of Algebra

Don't confuse the words **opposite** and **reciprocal**. The opposite of 4 is -4. The reciprocal of 4 is $\frac{1}{4}$.

Two numbers whose product is 1 are **reciprocals** or **multiplicative inverses** of each other. For example, 8 is the multiplicative inverse of $\frac{1}{8}$, and $\frac{1}{8}$ is the multiplicative inverse of 8, because $8 \cdot \frac{1}{8} = 1$. Likewise, $-\frac{3}{4}$ and $-\frac{4}{3}$ are multiplicative inverses because $-\frac{3}{4}\left(-\frac{4}{3}\right) = 1$. All real numbers, except 0, have a multiplicative inverse.

Multiplicative Inverses (Inverse Property of Multiplication)	The product of any number and its multiplicative inverse (reciprocal) is 1. For any nonzero real number a, $$a\left(\frac{1}{a}\right) = 1$$

EXAMPLE 4   Find the reciprocal of each number: **a.** $\dfrac{2}{3}$ **b.** $-\dfrac{2}{3}$ **c.** -11

Strategy   To find the reciprocal of a fraction, we invert the numerator and the denominator.

Why   We want the product of the given number and its reciprocal to be 1.

Solution   **a.** The reciprocal of $\frac{2}{3}$ is $\frac{3}{2}$ because $\frac{2}{3}\left(\frac{3}{2}\right) = 1$.   *To find the reciprocal of a fraction, invert the numerator and denominator.*

Caution

Do not change the sign of a number when finding its reciprocal.

b. The reciprocal of $-\frac{2}{3}$ is $-\frac{3}{2}$ because $-\frac{2}{3}\left(-\frac{3}{2}\right) = 1$.

c. The reciprocal of -11 is $-\frac{1}{11}$ because $-11\left(-\frac{1}{11}\right) = 1$.   *Think of -11 as $\frac{-11}{1}$ to find its reciprocal.*

Self Check 4   Find the reciprocal of each number: **a.** $-\dfrac{15}{16}$ **b.** $\dfrac{15}{16}$ **c.** -27

Now Try ▶ Problems 51 and 53

3 Divide Signed Numbers.

Every division fact can be written as an equivalent multiplication fact.

Division	For any real numbers a, b, and c, where $b \neq 0$, $$\frac{a}{b} = c \qquad \text{provided that} \qquad c \cdot b = a \qquad \text{Quotient} \cdot \text{divisor} = \text{dividend}$$

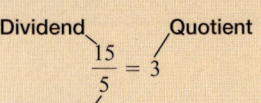
We can use this relationship between multiplication and division to develop rules for dividing signed numbers. For example,

$$\frac{15}{5} = 3 \qquad \text{because} \qquad 3(5) = 15$$

From this example, we see that *the quotient of two positive numbers is positive.*

To determine the quotient of two negative numbers, we consider $\frac{-15}{-5}$.

$$\frac{-15}{-5} = 3 \qquad \text{because} \qquad 3(-5) = -15$$

From this example, we see that the *quotient of two negative numbers is positive.*

To determine the quotient of a positive number and a negative number, we consider $\frac{15}{-5}$.

$$\frac{15}{-5} = -3 \qquad \text{because} \qquad -3(-5) = 15$$

From this example, we see that *the quotient of a positive number and a negative number is negative.*

To determine the quotient of a negative number and a positive number, we consider $\frac{-15}{5}$.

$$\frac{-15}{5} = -3 \qquad \text{because} \qquad -3(5) = -15$$

From this example, we see that *the quotient of a negative number and a positive number is negative.*

We summarize the rules from the previous examples and note that they are similar to the rules for multiplication.

Dividing Two Real Numbers

To divide two real numbers, divide their absolute values.

1. The quotient of two numbers that have the same (*like*) signs is positive.

2. The quotient of two numbers that have different (*unlike*) signs is negative.

EXAMPLE 5   Divide and check the result: **a.** $\dfrac{-81}{-9}$ **b.** $\dfrac{45}{-9}$ **c.** $-2.87 \div 0.7$

d. $-\dfrac{5}{16} \div \left(-\dfrac{1}{2}\right)$

Strategy   We will use the rules for dividing signed numbers. In each case, we need to ask, "Is it a quotient of two numbers with the same sign or different signs?"

Why   The signs of the numbers that we are dividing determine the sign of the result.

Solution   **a.** $\dfrac{-81}{-9} = 9$ Divide the absolute values, 81 by 9, to get 9.
Since the signs are like, the final answer is positive.

Multiply the quotient and the divisor to check the result: $9(-9) = -81$.

b. $\dfrac{45}{-9} = -5$ Divide the absolute values, 45 by 9, to get 5. Since
the signs are unlike, make the final answer negative.

Check: $-5(-9) = 45$ Quotient · divisor = dividend

c. $-2.87 \div 0.7 = -4.1$ Since the signs are unlike, make the final answer negative.

Check: $-4.1(0.7) = -2.87$ Quotient · divisor = dividend

d. $-\dfrac{5}{16} \div \left(-\dfrac{1}{2}\right) = -\dfrac{5}{16}\left(-\dfrac{2}{1}\right)$ Multiply the first fraction by the reciprocal of the second fraction. The reciprocal of $-\frac{1}{2}$ is $-\frac{2}{1}$.

$$= \dfrac{5 \cdot 2}{16 \cdot 1}$$ Multiply the absolute values $\frac{5}{16}$ and $\frac{2}{1}$. Since the signs are like, the final answer is positive.

$$= \dfrac{5 \cdot \overset{1}{\cancel{2}}}{\underset{1}{\cancel{2}} \cdot 8 \cdot 1}$$ To simplify the fraction, factor 16 as $2 \cdot 8$. Then remove the common factor 2.

$$= \dfrac{5}{8}$$

Check: $\dfrac{5}{8}\left(-\dfrac{1}{2}\right) = -\dfrac{5}{16}$ Quotient · divisor = dividend

Self Check 5   Divide and check the result: **a.** $\dfrac{-28}{-4}$ **b.** $\dfrac{75}{-25}$

c. $0.32 \div (-1.6)$ **d.** $\dfrac{3}{4} \div \left(-\dfrac{5}{8}\right)$

Now Try ▶ Problems 55, 59, 63, and 67

EXAMPLE 6

Depreciation.   Over an 8-year period, the value of a $150,000 house fell at a uniform rate to $110,000. Find the amount of depreciation per year.

Strategy   The phrase *uniform rate* means that the value of the house fell the same amount each year, for 8 straight years. We can determine the amount it depreciated in one year (per year) by dividing the total change in value of the house by 8.

Why   The process of separating a quantity into equal parts (in this case, the change in the value of the house) indicates division.

Solution

First, we find the change in the value of the house.

$$110,000 - 150,000 = -40,000$$ Subtract the previous value from the current value.

The negative result represents a drop in value of $40,000. Since the depreciation occurred over 8 years, we divide $-40,000$ by 8.

$$\dfrac{-40,000}{8} = -5,000$$ Divide the absolute values, 40,000 by 8, to get 5,000, and make the quotient negative.

The house depreciated $5,000 per year.

Self Check 6   **Depreciation.**   Over a 6-year period, the value of a $300,000 house fell at a uniform rate to $286,500. Find the amount of depreciation each year.

Now Try ▶ Problem 109

---

4 Use Properties of Division.

Whenever we divide a number by 1, the quotient is that number. Therefore, $\dfrac{12}{1} = 12$, $\dfrac{-80}{1} = -80$, and $7.75 \div 1 = 7.75$. Furthermore, whenever we divide a nonzero number by itself, the quotient is 1. Therefore, $\dfrac{35}{35} = 1$, $\dfrac{-4}{-4} = 1$, and $0.9 \div 0.9 = 1$. These observations suggest the following properties of division.

Division Properties ▼	Any number divided by 1 is the number itself. Any number (except 0) divided by itself is 1. For any real number a, $$\frac{a}{1} = a \quad \text{and} \quad \frac{a}{a} = 1 \quad \text{(where } a \neq 0 \text{)}$$

Caution

Division is *not commutative*. For example, $\frac{6}{3} \neq \frac{3}{6}$ and $\frac{-12}{4} \neq \frac{4}{-12}$. In general, $\frac{a}{b} \neq \frac{b}{a}$.

We will now consider division that involves zero. First, we examine division of zero. Let's look at two examples. We know that

$$\frac{0}{2} = 0 \quad \text{because} \quad 0 \cdot 2 = 0 \quad \text{and} \quad \frac{0}{-5} = 0 \quad \text{because} \quad 0(-5) = 0$$

These examples illustrate that *0 divided by a nonzero number is 0*.

To examine division by zero, let's look at $\frac{2}{0}$ and its related multiplication statement.

$$\frac{2}{0} = ? \quad \text{because} \quad ? \cdot 0 = 2$$

The Language of Algebra

When we say a division by 0, such as $\frac{2}{0}$, is **undefined**, we mean that $\frac{2}{0}$ does not represent a real number.

There is no number that can make $0 \cdot ? = 2$ true because any number multiplied by 0 is equal to 0, not 2. Therefore, $\frac{2}{0}$ does not have an answer. We say that such a division is **undefined.** These results suggest the following division facts.

Division Involving 0 ▼	For any nonzero real number a, $$\frac{0}{a} = 0 \quad \text{and} \quad \frac{a}{0} \text{ is undefined.}$$

EXAMPLE 7   Find each quotient, if possible: **a.** $\dfrac{0}{8}$ **b.** $\dfrac{-24}{0}$

Strategy   In each case, we need to determine if we have division *of* 0 or division *by* 0.

Why   *Division of 0 by a nonzero number is defined, and the result is 0. However, division by 0 is undefined; there is no result.*

Solution   **a.** $\dfrac{0}{8} = 0$ because $0 \cdot 8 = 0$.   This is division of 0 by 8.

The Language of Algebra

Division of 0 by 0, written $\frac{0}{0}$, is called **indeterminate.** This form is studied in advanced mathematics classes.

b. $\dfrac{-24}{0}$ is undefined.   This is division of −24 by 0.

Self Check 7   Find each quotient, if possible: **a.** $\dfrac{4}{0}$ **b.** $\dfrac{0}{17}$

Now Try ▶   Problems 71 and 73

SECTION 1.6 ▸ STUDY SET

VOCABULARY

Fill in the blanks.

1. The answer to a multiplication problem is called a _____.
   The answer to a division problem is called a _____.
2. The _____ property of multiplication states that changing the order when multiplying does not affect the answer.
3. The _____ property of multiplication states that changing the grouping when multiplying does not affect the answer.
4. Division of a nonzero number by 0 is _____.

CONCEPTS

Fill in the blanks.

5. a. The product or quotient of two numbers with like signs is _____.
   b. The product or quotient of two numbers with unlike signs is _____.
6. a. The product of an even number of negative numbers is _____.
   b. The product of an odd number of negative numbers is _____.
7. a. $\frac{-9}{3} = -3$ because ☐ · ☐ = ☐
   b. $\frac{0}{8} = 0$ because ☐ · ☐ = ☐
8. Complete each property of multiplication.
   a. $a \cdot b = b \cdot$ ☐   b. $(ab)c =$ ☐
   c. $0 \cdot a =$ ☐   d. $1 \cdot a =$ ☐
   e. $a\left(\frac{1}{a}\right) =$ ☐   f. $-1 \cdot a =$ ☐
9. Complete each property of division.
   a. $\frac{a}{1} =$ ☐   b. $\frac{a}{a} =$ ☐
   c. $\frac{0}{a} =$ ☐   d. $\frac{a}{0}$ is ☐
10. Which property justifies each statement?
    a. $-5(2 \cdot 17) = (-5 \cdot 2)17$
    b. $-5\left(-\frac{1}{5}\right) = 1$
    c. $-5 \cdot 2 = 2(-5)$
    d. $-5(1) = -5$
    e. $-5 \cdot 0 = 0$
11. For each multiplication or division, just determine the sign of the answer
    a. $-\frac{19}{37}\left(-\frac{51}{75}\right)$ b. $\frac{45.568}{-2.56}$
    c. $-8.2(-4.1)(-6)(-9.3)(-1.5)$
12. Use multiplication to check this division: $\frac{-29.4}{7} = -4.1$. Is the answer correct?

13. Complete each statement using the given property.
    a. $5 \cdot 8 =$ ☐   *Commutative property of multiplication*
    b. $-2(6 \cdot 9) =$ ☐   *Associative property of multiplication*
    c. $5\left(☐\right) = 1$ *Inverse property of multiplication*
    d. ☐$(-20) = -20$ *Multiplication property of 1*
14. Complete the table.

Number	Opposite (additive inverse)	Reciprocal (multiplicative inverse)
2		
$-\frac{4}{5}$		
1.75		

Let POS stand for a positive number and NEG stand for a negative number. Determine the sign of each result, if possible.

15. a. POS · NEG   b. POS + NEG
    c. POS − NEG   d. $\frac{POS}{NEG}$
16. a. NEG · NEG   b. NEG + NEG
    c. NEG − NEG   d. $\frac{NEG}{NEG}$

NOTATION

Write each sentence using symbols.

17. The product of negative four and negative five is twenty.
18. The quotient of sixteen and negative eight is negative two.

GUIDED PRACTICE

Multiply. See Example 1.

19. $4(-1)$ 20. $6(-1)$
21. $-2 \cdot 8$ 22. $-3 \cdot 4$
23. $12(-5)$ 24. $(-9)(11)$
25. $3(-22)$ 26. $-8 \cdot 9$
27. $1.2(-0.4)$ 28. $(-3.6)(0.9)$
29. $\frac{1}{3}\left(-\frac{3}{4}\right)$ 30. $\left(-\frac{3}{4}\right)\left(\frac{4}{5}\right)$

Multiply. See Example 2.

31. $(-1)(-7)$ 32. $(-2)(-5)$
33. $(-6)(-9)$ 34. $(-8)(-7)$

35. $-3(-3)$

36. $-1(-1)$

37. $63(-7)$

38. $43(-6)$

39. $-0.6(-4)$

40. $-0.7(-8)$

41. $\left(-\dfrac{7}{8}\right)\left(-\dfrac{2}{21}\right)$

42. $\left(-\dfrac{5}{6}\right)\left(-\dfrac{2}{15}\right)$

Multiply. See Example 3.

43. $3.3(-4)(-5)$

44. $(-2.2)(-4)(-5)$

45. $-2(-3)(-4)(-5)(-6)$

46. $-9(-7)(-5)(-3)(-1)$

47. $(-41)(3)(-7)(-1)$

48. $56(-3)(-4)(-1)$

49. $(-6)(-6)(-6)$

50. $(-5)(-5)(-5)$

Find the reciprocal of each number. Then find the product of the given number and its reciprocal. See Example 4.

51. $\dfrac{7}{9}$

52. $-\dfrac{8}{9}$

53. -13

54. $\dfrac{1}{8}$

Divide. See Example 5.

55. $-30 \div (-3)$

56. $-12 \div (-2)$

57. $-6 \div (-2)$

58. $-36 \div (-9)$

59. $\dfrac{85}{-5}$

60. $\dfrac{-84}{7}$

61. $\dfrac{-110}{-110}$

62. $\dfrac{-200}{-200}$

63. $\dfrac{-10.8}{1.2}$

64. $\dfrac{-13.5}{-1.5}$

65. $\dfrac{0.5}{-100}$

66. $\dfrac{-1.7}{10}$

67. $-\dfrac{1}{3} \div \dfrac{4}{5}$

68. $-\dfrac{2}{3} \div \dfrac{7}{8}$

69. $-\dfrac{9}{16} \div \left(-\dfrac{3}{20}\right)$

70. $-\dfrac{4}{5} \div \left(-\dfrac{8}{25}\right)$

TRY IT YOURSELF

Perform the operations.

71. $\dfrac{0}{150}$

72. $\dfrac{0}{-12}$

73. $\dfrac{-17}{0}$

74. $\dfrac{225}{0}$

75. $\dfrac{24}{-6}$

76. $\dfrac{-78}{6}$

77. $\dfrac{17}{-17}$

78. $\dfrac{-24}{24}$

79. $(-2)(-2)(-2)(-2)$

80. $(-3)(-3)(-3)(-3)$

81. $-3(-4)(0)$

82. $15(0)(-22)$

83. $\dfrac{-23.5}{5}$

84. $\dfrac{-337.8}{6}$

85. $-5.2 \cdot 100$

86. $-1.17 \cdot 1,000$

87. $\dfrac{1}{2}\left(-\dfrac{1}{3}\right)\left(-\dfrac{1}{4}\right)$

88. $\dfrac{1}{3}\left(-\dfrac{1}{5}\right)\left(-\dfrac{1}{7}\right)$

89. $\dfrac{550}{-50}$

90. $\dfrac{440}{-20}$

91. $-3\dfrac{3}{8} \div \left(-2\dfrac{1}{4}\right)$

92. $-3\dfrac{4}{15} \div \left(-2\dfrac{1}{10}\right)$

93. $7.2(-2.1)(-2)$

94. $4.6(-5.4)(-2)$

95. $\dfrac{1}{2}\left(-\dfrac{3}{4}\right)$

96. $\dfrac{1}{3}\left(-\dfrac{5}{16}\right)$

97. $-\dfrac{16}{25} \div \dfrac{64}{15}$

98. $-\dfrac{15}{16} \div \dfrac{25}{8}$

99. $\dfrac{-24.24}{-0.8}$

100. $\dfrac{-55.02}{-0.7}$

101. $-1\dfrac{1}{4}\left(-\dfrac{3}{4}\right)$

102. $-1\dfrac{1}{8}\left(-\dfrac{3}{8}\right)$

Look Alikes . . .

103. a. $2.7 + (-0.9)$ b. $2.7 - (-0.9)$

c. $2.7(-0.9)$ d. $\dfrac{2.7}{-0.9}$

104. a. $-\dfrac{5}{3} + \left(-\dfrac{9}{25}\right)$ b. $-\dfrac{5}{3} - \left(-\dfrac{9}{25}\right)$

c. $-\dfrac{5}{3}\left(-\dfrac{9}{25}\right)$ d. $-\dfrac{5}{3} \div \left(-\dfrac{9}{25}\right)$

Use the associative property of multiplication to find each product.

105. $-\dfrac{1}{2}(2 \cdot 67)$

106. $\left(-\dfrac{5}{16} \cdot \dfrac{1}{7}\right)7$

107. $-0.2(-10 \cdot 3)$

108. $-1.5(-100 \cdot 4)$

APPLICATIONS

109. **Real Estate.**   Over a 5-year period, the value of a $200,000 lot fell at a uniform rate to $160,000. What signed number indicates the amount of depreciation per year?

110. **Tourism.**   The ocean liner Queen Mary cost $22,500,000 to build in 1936. The ship was purchased by the city of Long Beach, California, in 1967 for $3,450,000. It now serves as a convention center. What signed number indicates the annual average depreciation of the ship over the 31-year period from 1936 to 1967? Round to the nearest dollar.

111. **Fluid Flow.**   In a lab, the temperature of a fluid was decreased 6° per hour for 12 hours. What signed number indicates the change in temperature?

112. **Stress on the Job.**   A health care provider for a company estimates that 75 hours per week are lost by employees suffering from stress-related illness. In one year, how many hours are lost? Use a signed number to answer.

113. **Weight Loss.**   As a result of a diet, Tom has been steadily losing $4\dfrac{1}{2}$ pounds per month.

a. Which expression below can be used to determine how much heavier Tom was 8 months ago?

i. $-4\dfrac{1}{2} \cdot 8$ ii. $-4\dfrac{1}{2}(-8)$

iii. $4\dfrac{1}{2}(-8)$ iv. $-4\dfrac{1}{2} - 8$

b. How much heavier was Tom 8 months ago?

114. **Astronomy.**   The temperature on Pluto gets as low as $-386°F$. This is twice as low as the lowest temperature reached on Jupiter. What is the lowest temperature on Jupiter?

115. Car Radiators. The instructions on a container of antifreeze state, "A 50/50 mixture of antifreeze and water protects against freeze-ups down to $-34°F$, while a 60/40 mix protects against freeze-ups down to one and one-half times that temperature." To what temperature does the 60/40 mixture protect?

116. Accounting. For 2010, the net income of Rite Aid Corporation (the drugstore chain) was about $-\$508$ million. The previous year, the company's net income was even worse, by a factor of about 5.75. What signed number represents Rite Aid's net income in 2009? (Source: moneycentral.msn.com)

117. Airlines. In the 2009 income statement for Delta Air Lines below, numbers within parentheses represent a loss. Complete the statement given these facts. The second quarter loss was about 6.4 times the first quarter loss. The fourth quarter loss was about 5 times the second quarter loss. The third quarter loss was about $\frac{5}{16}$ of the fourth quarter loss.

DELTA INCOME STATEMENT				2009
All amounts in millions of dollars	1st Qtr (25)	2nd Qtr (?)	3rd Qtr (?)	4th Qtr (?)

Source: dailyfinance.com

118. Computers. The formula = A1*B1*C1 in cell D1 of the spreadsheet instructs the computer to multiply the values in cells A1, B1, and C1 and to print the result *in place of the formula* in cell D1. (The symbol * represents multiplication.) What value will be printed in the cell D1? What values will be printed in cells D2 and D3?

	A	B	C	D
1	4	−5	−17	= A1*B1*C1
2	22	−30	14	= A2*B2*C2
3	−60	−20	−34	= A3*B3*C3
4				
5				

File   Edit   View   Insert   Format   Tools   Data   Window

119. Physics. An oscilloscope displays electrical signals as wavy lines on a screen. By switching the magnification dial to ×2, for example, the height of the "peak" and the depth of the "valley" of a graph will be doubled. Use signed numbers to indicate the height and depth of the display for each setting of the dial.

a. normal
b. ×0.5
c. ×1.5
d. ×2

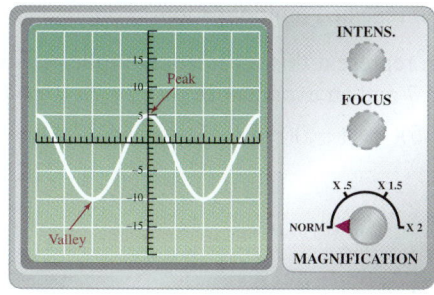

120. Light. Water acts as a selective filter of light. In the illustration, we see that red light waves penetrate water only to a depth of about 5 meters. How many times deeper does

a. yellow light penetrate than red light?
b. green light penetrate than orange light?
c. blue light penetrate than yellow light?

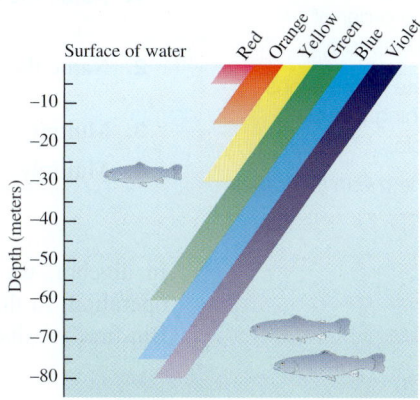

WRITING

121. Explain why $\frac{16}{0}$ is undefined.

122. The commutative property states that changing the order when multiplying does not change the answer. Are the following activities commutative? Explain.

a. Washing a load of clothes; drying a load of clothes
b. Putting on your left sock; putting on your right sock

123. What is wrong with the following statement?
A negative and a positive is a negative.

124. If we multiply two different numbers and the answer is 0, what must be true about one of the numbers? Explain your answer.

REVIEW

125. Add: $-3 + (-4) + (-5) + 4 + 3$

126. Write $-3 - (-5)$ as addition of the opposite.

127. Find $\frac{1}{2} + \frac{1}{4} + \frac{1}{3}$. Answer in decimal form.

128. Which integers have an absolute value equal to 45?

CHALLENGE PROBLEMS

129. If the product of five numbers is negative, how many of them could be negative? Explain.

130. Suppose a is a positive number and b is a negative number. Determine whether the given expression is positive or negative.

a. $-a(-b)$
b. $\dfrac{-a}{b}$
c. $\dfrac{-a}{a}$
d. $\dfrac{1}{b}$

Exponents and Order of Operations

OBJECTIVES

1 Evaluate exponential expressions.

2 Use the order of operations rule.

3 Evaluate expressions containing grouping symbols.

4 Find the mean (average).

ARE YOU READY?

The following problems review some basic concepts that are important when working with numerical expressions.

1. Name the operations that are involved in the expression: $50 - 2(3)$

2. Name the operations that are involved in the expression: $\dfrac{2 + 6 \cdot 8}{4 + 6}$

3. Multiply: $3 \cdot 3 \cdot 3 \cdot 3$

4. Multiply: $(-5)(-5)(-5)$

In algebra, we often have to find the value of expressions that involve more than one operation. In this section, we introduce an order of operations rule to follow in such cases. But first, we discuss a way to write repeated multiplication using *exponents*.

1 Evaluate Exponential Expressions.

In the expression $3 \cdot 3 \cdot 3 \cdot 3 \cdot 3$, the number 3 repeats as a factor five times. We can use **exponential notation** to write this product in a more compact form.

Exponent and Base	An **exponent** is used to indicate repeated multiplication. It is how many times the **base** is used as a factor.

The Language of Algebra

5^2 represents the area of a square with sides 5 units long. 4^3 represents the volume of a cube with sides 4 units long.

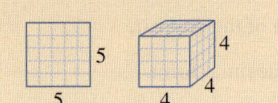

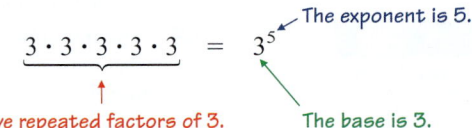

The exponent is 5.

Five repeated factors of 3.   The base is 3.

In the **exponential expression** 3^5, the base is 3, and 5 is the exponent. The expression is called a power of 3. Some other examples of exponential expressions are:

5^2 Read as "5 to the second power" or "5 squared."

4^3 Read as "4 to the third power" or "4 cubed."

$(-2)^5$ Read as "−2 to the fifth power."

EXAMPLE 1

Write each product using exponents: **a.** $7 \cdot 7 \cdot 7$ **b.** $(-5)(-5)(-5)(-5)(-5)$
c. $8 \cdot 8 \cdot 15 \cdot 15 \cdot 15 \cdot 15$ **d.** $a \cdot a \cdot a \cdot a \cdot a \cdot a$ **e.** $4 \cdot \pi \cdot r \cdot r$

Strategy   We need to determine the number of repeated factors in the expression.

Why   An exponent can be used to represent repeated multiplication.

Solution   **a.** The factor 7 is repeated 3 times. We can represent this repeated multiplication with an exponential expression having a base of 7 and an exponent of 3: $7 \cdot 7 \cdot 7 = 7^3$.

b. The factor −5 is repeated five times: $(-5)(-5)(-5)(-5)(-5) = (-5)^5$.

c. $8 \cdot 8 \cdot 15 \cdot 15 \cdot 15 \cdot 15 = 8^2 \cdot 15^4$

d. $a \cdot a \cdot a \cdot a \cdot a \cdot a = a^6$

e. $4 \cdot \pi \cdot r \cdot r = 4\pi r^2$

Self Check 1 Write each product using exponents: **a.** $(12)(12)(12)$
b. $2 \cdot 9 \cdot 9 \cdot 9 \cdot 9$ **c.** $(-30)(-30)$
d. $y \cdot y \cdot y \cdot y \cdot y \cdot y$ **e.** $8 \cdot b \cdot b \cdot b \cdot c$

Now Try ▶ Problems 15 and 21

To **evaluate** (find the value of) an exponential expression, we write the base as a factor the number of times indicated by the exponent. Then we multiply the factors.

EXAMPLE 2 Evaluate each expression: **a.** 5^3 **b.** $\left(-\dfrac{2}{3}\right)^3$ **c.** 10^1 **d.** $(0.6)^2$
e. $(-3)^4$ **f.** $(-3)^5$

Strategy We will rewrite each exponential expression as a product of repeated factors, and then perform the multiplication. This requires that we identify the base and the exponent.

Why The exponent tells the number of times the base is to be written as a factor.

Solution **a.** $5^3 = 5 \cdot 5 \cdot 5$ Write the base, 5, as a factor 3 times.

$= 125$ Multiply, working left to right. We say 125 is the cube of 5.

Because $5^3 = 125$, we say that 125 is a **power** of 5.

Caution

Don't make the common mistake of multiplying the base and the exponent:

$5^3 \neq 5 \cdot 3$

b. $\left(-\dfrac{2}{3}\right)^3 = \left(-\dfrac{2}{3}\right)\left(-\dfrac{2}{3}\right)\left(-\dfrac{2}{3}\right)$ Since $-\frac{2}{3}$ is the base and 3 is the exponent, we write $-\frac{2}{3}$ as a factor 3 times.

$= \dfrac{4}{9}\left(-\dfrac{2}{3}\right)$ Work from left to right: $\left(-\frac{2}{3}\right)\left(-\frac{2}{3}\right) = \frac{4}{9}$.

$= -\dfrac{8}{27}$

Calculators

Finding a power
The squaring key x^2 can be used to find the square of a number. To raise a number to a power, we use the y^x key on a scientific calculator and the $\wedge$ key on a graphing calculator.

c. $10^1 = 10$ The base is 10. Since the exponent is 1, we write the base once.

d. $(0.6)^2 = (0.6)(0.6)$ Write the base, 0.6, as a factor 2 times.
$= 0.36$ We say 0.36 is the square of 0.6.

e. $(-3)^4 = (-3)(-3)(-3)(-3)$ Write the base, -3, as a factor 4 times.
$= 9(-3)(-3)$ Work from left to right.
$= -27(-3)$
$= 81$

The Language of Algebra

A number or a variable has an **understood** exponent of 1. For example,

$8 = 8^1$ and $x = x^1$

f. $(-3)^5 = (-3)(-3)(-3)(-3)(-3)$ Write the base, -3, as a factor 5 times.
$= 9(-3)(-3)(-3)$ Work from left to right.
$= -27(-3)(-3)$
$= 81(-3)$
$= -243$

Self Check 2 Evaluate: **a.** 2^5 **b.** $\left(-\dfrac{3}{4}\right)^3$ **c.** 9^1
d. $(-0.3)^2$ **e.** $(-6)^2$ **f.** $(-5)^3$

Now Try ▶ Problems 23, 29, and 33

In Example 2e, we raised -3 to an even power; the result was positive. In part f, we raised -3 to an odd power; the result was negative. These results illustrate the following rule.

Even and Odd Powers of a Negative Number	When a negative number is raised to an even power, the result is positive.
	When a negative number is raised to an odd power, the result is negative.

The Language of Algebra

Read $(-4)^2$ as "negative four squared" and -4^2 as "the opposite of the square of four."

Although the expressions $(-4)^2$ and -4^2 look alike, they are not. When we find the value of each expression, it becomes clear that they are not equivalent.

$$(-4)^2 = (-4)(-4)$$ The base is -4, the exponent is 2. $$-4^2 = -(4 \cdot 4)$$ The base is 4, the exponent is 2.

$$= 16$$ $$= -16$$

Different results

Any real number can be used as a base. However, the base of an exponential expression *does not include* the negative sign unless parentheses are used.

$$-7^3$$ $$(-7)^3$$

Positive base: 7   Negative base: -7

EXAMPLE 3   Evaluate: -2^4

Strategy   We will rewrite the expression as a product of repeated factors and then perform the multiplication. We must be careful when identifying the base. It is 2, not -2.

Why   Since there are no parentheses around -2, the base is 2.

Solution

$$-2^4 = -(2 \cdot 2 \cdot 2 \cdot 2)$$ Read as "the opposite of the fourth power of two."

$$= -16$$ Do the multiplication within the parentheses to get 16. Then write the opposite of that result.

Self Check 3   Evaluate: -5^4

Now Try ▶ Problem 35

---

2 Use the Order of Operations Rule.

Suppose you have been asked to contact a friend if you see a Rolex watch for sale when you are traveling in Europe. While in Switzerland, you find the watch and send the text message shown on the left. The next day, you get the response shown on the right.

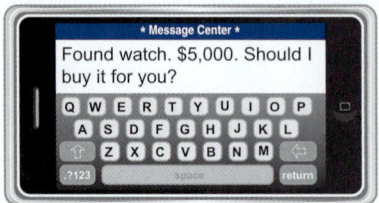

Something is wrong. The first part of the response (No price too high!) says to buy the watch at any price. The second part (No! Price too high.) says not to buy it, because it's too expensive. The placement of the exclamation point makes us read the two parts of the response differently, resulting in different meanings. When reading a mathematical statement, the same kind of confusion is possible. For example, consider the expression

$$2 + 3 \cdot 6$$

We can evaluate this expression in two ways. We can add first, and then multiply. Or we can multiply first, and then add. However, the results are different.

$$2 + 3 \cdot 6 = 5 \cdot 6 \qquad \text{Add 2 and 3 first.} \qquad 2 + 3 \cdot 6 = 2 + 18 \qquad \text{Multiply 3 and 6 first.}$$
$$= 30 \qquad \text{Multiply 5 and 6.} \qquad = 20 \qquad \text{Add 2 and 18.}$$

Different answers

If we don't establish a uniform order of operations, the expression has two different values. To avoid this possibility, we will always use the following set of priority rules.

Order of Operations

1. Perform all calculations within parentheses and other grouping symbols following the order listed in Steps 2–4 below, working from the innermost pair of grouping symbols to the outermost pair.

2. Evaluate all exponential expressions.

3. Perform all multiplications and divisions as they occur from left to right.

4. Perform all additions and subtractions as they occur from left to right.

When grouping symbols have been removed, repeat Steps 2–4 to complete the calculation.

   If a fraction is present, evaluate the expression above and the expression below the bar separately. Then simplify the fraction, if possible.

It isn't necessary to apply all of these steps in every problem. For example, the expression $2 + 3 \cdot 6$ does not contain any parentheses, and there are no exponential expressions. So we look for multiplications and divisions to perform and proceed as follows:

$$2 + 3 \cdot 6 = 2 + 18 \qquad \text{Do the multiplication first.}$$
$$= 20 \qquad \text{Do the addition.}$$

EXAMPLE 4

Evaluate: **a.** $3 \cdot 2^3 - 4$ **b.** $-30 - 4 \cdot 5 + 9$ **c.** $24 \div 6 \cdot 2$
d. $160 - 4 + 6(-2)(-3)$

The Language of Algebra

Sometimes, for problems like these, the instruction **simplify** is used instead of *evaluate*.

Strategy   We will scan the expression to determine what operations need to be performed. Then we will perform those operations, one-at-a-time, following the order of operations rules.

Why   If we don't follow the correct order of operations, the expression can have more than one value.

Solution   **a.** Three operations need to be performed to evaluate this expression: multiplication, raising to a power, and subtraction. By the order of operations rules, we evaluate 2^3 first.

$$3 \cdot 2^3 - 4 = 3 \cdot 8 - 4 \qquad \text{Evaluate the exponential expression: } 2^3 = 8.$$
$$= 24 - 4 \qquad \text{Do the multiplication: } 3 \cdot 8 = 24.$$
$$= 20 \qquad \text{Do the subtraction.}$$

b. This expression involves subtraction, multiplication, and addition. The order of operations rule tells us to multiply first.

$$-30 - \mathbf{4 \cdot 5} + 9 = -30 - \mathbf{20} + 9$$ Do the multiplication: $4 \cdot 5 = 20$.

$$= -50 + 9$$ Working from left to right, do the subtraction: $-30 - 20 = -30 + (-20) = -50$.

$$= -41$$ Do the addition.

Caution

A common mistake is to forget to work from left to right and incorrectly perform the multiplication before the division. This produces the wrong answer, 2.

$$24 \div 6 \cdot 2 = 24 \div 12$$
$$= 2$$

---

c. Since there are no calculations within parentheses nor are there exponents, we perform the multiplications and divisions as they occur from left to right. The division occurs before the multiplication, so it must be performed first.

$$\mathbf{24 \div 6} \cdot 2 = \mathbf{4} \cdot 2$$ Working left to right, do the division: $24 \div 6 = 4$.

$$= 8$$ Do the multiplication.

d. Although this expression contains parentheses, there are no operations to perform within them. Since there are no exponents, we will perform the multiplications as they occur from left to right.

$$160 - 4 + \mathbf{6(-2)}(-3) = 160 - 4 + \mathbf{(-12)}(-3)$$ Do the multiplication, working left to right: $6(-2) = -12$.

$$= 160 - 4 + 36$$ Complete the multiplication: $(-12)(-3) = 36$.

$$= 156 + 36$$ Working left to right, do the subtraction before the addition.

$$= 192$$ Do the addition.

Self Check 4 Evaluate: **a.** $2 \cdot 3^2 + 17$ **b.** $-40 - 9 \cdot 4 + 10$ **c.** $18 \div 2 \cdot 3$ **d.** $240 - 8 + 3(-2)(-4)$

Now Try ▶ Problems 39, 43, 45, and 49

3 Evaluate Expressions Containing Grouping Symbols.

Grouping symbols serve as mathematical punctuation marks. They help determine the order in which an expression is to be evaluated. Examples of grouping symbols are parentheses ( ), brackets [ ], braces { }, absolute value symbols | |, and the fraction bar —.

EXAMPLE 5 Evaluate each expression: **a.** $(6 - 3)^2$ **b.** $5^3 + 2(-8 - 3 \cdot 2)$

Strategy We will perform the operation(s) within the parentheses first. When there is more than one operation to perform within the parentheses, we follow the order of operations rule.

Why This is the first step of the order of operations rule.

Solution **a.** We read $(6 - 3)^2$ as "the square of the quantity of 6 minus 3."

$$(\mathbf{6 - 3})^2 = \mathbf{3}^2$$ Do the subtraction within the parentheses: $6 - 3 = 3$.

$$= 9$$ Evaluate the exponential expression.

Notation

Multiplication is indicated when a number is next to a parenthesis or bracket.

$$5^3 + 2(-8 - 3 \cdot 2)$$

b. We begin by performing the operations within the parentheses in the proper order: multiplication first, and then subtraction.

$$5^3 + 2(-8 - \mathbf{3 \cdot 2}) = 5^3 + 2(-8 - \mathbf{6}) \qquad \text{Do the multiplication: } 3 \cdot 2 = 6.$$
$$= 5^3 + 2(-14) \qquad \text{Do the subtraction: } -8 - 6 = -14.$$
$$= 125 + 2(-14) \qquad \text{Evaluate } 5^3.$$
$$= 125 + (-28) \qquad \text{Do the multiplication: } 2(-14) = -28.$$
$$= 97 \qquad \text{Do the addition.}$$

Self Check 5   Evaluate: **a.** $(12 - 6)^3$ **b.** $1^3 + 6(-6 - 3 \cdot 0)$

Now Try ▶ Problems 53 and 55

Expressions can contain two or more pairs of grouping symbols. To evaluate the following expression, we begin within the innermost pair of grouping symbols, the parentheses. Then we work within the outermost pair, the brackets.

Innermost pair
$$-4[2 + 3(4 - 8^2)] - 2$$
Outermost pair

EXAMPLE 6   Evaluate: $-4[2 + 3(4 - 8^2)] - 2$

Strategy   We will work within the parentheses first and then within the brackets. At each stage, we follow the order of operations rules.

Why   By the order of operations, we must work from the *innermost* pair of grouping symbols to the *outermost*.

Solution

The Language of Algebra

When one pair of grouping symbols is inside another pair, we say that those grouping symbols are **nested**, or **embedded**.

$$-4[2 + 3(4 - \mathbf{8^2})] - 2$$
$$= -4[2 + 3(4 - \mathbf{64})] - 2 \qquad \text{Evaluate the exponential expression within the parentheses: } 8^2 = 64.$$
$$= -4[2 + 3(-60)] - 2 \qquad \text{Do the subtraction within the parentheses: } 4 - 64 = 4 + (-64) = -60.$$
$$= -4[2 + (-180)] - 2 \qquad \text{Do the multiplication within the brackets: } 3(-60) = -180.$$
$$= -4[-178] - 2 \qquad \text{Do the addition within the brackets: } 2 + (-180) = -178.$$
$$= 712 - 2 \qquad \text{Do the multiplication: } -4[-178] = 712.$$
$$= 710 \qquad \text{Do the subtraction.}$$

Self Check 6   Evaluate: $-5[4 + 2(5^2 - 15)] - 10$

Now Try ▶ Problem 61

EXAMPLE 7 Evaluate: $\dfrac{-3(3 + 2) + 5}{17 - 3(-4)}$

Strategy We will evaluate the expression above and the expression below the fraction bar separately. Then we will simplify the fraction, if possible.

Why Fraction bars are grouping symbols. They group the numerator and denominator. The expression could be written $[-3(3 + 2) + 5] \div [17 - 3(-4)]$.

Solution

$$\frac{-3(\mathbf{3 + 2}) + 5}{17 - \mathbf{3(-4)}} = \frac{-3(\mathbf{5}) + 5}{17 - (\mathbf{-12})}$$

In the numerator, do the addition within the parentheses. In the denominator, do the multiplication.

$$= \frac{-15 + 5}{17 + 12}$$

In the numerator, do the multiplication. In the denominator, write the subtraction as the addition of the opposite of -12, which is 12.

$$= \frac{-10}{29}$$

Do the additions.

$$= -\frac{10}{29}$$

Write the $-$ sign in front of the fraction: $\frac{-10}{29} = -\frac{10}{29}$. The fraction does not simplify.

Calculators

Order of operations
Calculators have the order of operations built in. A left parenthesis key (and a right parenthesis key) should be used when grouping symbols, including a fraction bar, are needed.

Self Check 7 Evaluate: $\dfrac{-4(-2 + 8) + 6}{8 - 5(-2)}$

Now Try ▶ Problem 73

EXAMPLE 8 Evaluate: $10|9 - 15| - 2^5$

Strategy The absolute value bars are grouping symbols. We will perform the calculation within them first.

Why By the order of operations, we must perform all calculations within parentheses and other grouping symbols (such as absolute value bars) first.

Solution

$$10|\mathbf{9 - 15}| - 2^5 = 10|\mathbf{-6}| - 2^5 \qquad \text{Subtract: } 9 - 15 = 9 + (-15) = -6.$$
$$= 10(6) - 2^5 \qquad \text{Find the absolute value: } |-6| = 6.$$
$$= 10(6) - 32 \qquad \text{Evaluate the exponential expression: } 2^5 = 32.$$
$$= 60 - 32 \qquad \text{Do the multiplication: } 10(6) = 60.$$
$$= 28 \qquad \text{Do the subtraction.}$$

Notation

Multiplication is indicated when a number is next to an absolute value symbol.

$$10|9 - 15| - 2^5$$

Self Check 8 Evaluate: $10^3 + 3|24 - 25|$

Now Try ▶ Problem 77

---

4 Find the Mean (Average).

The **arithmetic mean** (or simply **mean**) of a set of numbers is a value around which the values of the numbers are grouped. The mean is also commonly called the **average.**

Finding an Arithmetic Mean — To find the **mean** of a set of values, divide the sum of the values by the number of values.

When a value in a set appears more than once, that value has a greater "influence" on the mean than another value that only occurs a single time. To simplify the process of finding a mean, any value that appears more than once can be "weighted" by multiplying it by the number of times it occurs. A mean that is found in this way is called a **weighted mean**.

EXAMPLE 9   **Hotel Reservations.**   In an effort to improve customer service, a hotel electronically recorded the number of times the reservation desk telephone rang before it was answered by a receptionist. The results of the week-long survey are shown in the table. Find the average (mean) number of times the phone rang before a receptionist answered.

Number of rings	Number of calls
1	11
2	46
3	45
4	28
5	20

Strategy   First, we will determine the total number of times the reservation desk telephone rang during the week. Then we will divide that result by the total number of calls received.

Why   To find the *average* value of a set of values, we divide the sum of the values by the number of values.

Solution   To find the total number of rings, we multiply each *number of rings* (1, 2, 3, 4, and 5 rings) by the respective number of occurrences and add those subtotals.

$$\text{Total number of rings} = 11(1) + 46(2) + 45(3) + 28(4) + 20(5)$$

The total number of calls received was $11 + 46 + 45 + 28 + 20$. To find the average, we divide the total number of rings by the total number of calls.

$$\text{Average} = \frac{11(1) + 46(2) + 45(3) + 28(4) + 20(5)}{11 + 46 + 45 + 28 + 20}$$

$$= \frac{11 + 92 + 135 + 112 + 100}{150}$$

In the numerator, do the multiplications.
In the denominator, do the additions.

$$= \frac{450}{150}$$

Do the addition.

$$= 3$$

Simplify the fraction.

The average number of times the phone rang before it was answered was 3.

Self Check 9   **Evaluations.**   On the first question of an instructor's evaluation, 14 students marked 1 for *strongly agree,* 10 students marked 2 for *agree,* 6 students marked 3 for *disagree,* and 4 students marked 4 for *strongly disagree.* What was the average (mean) response for the first question on the evaluation?

Now Try ▶ Problem 119

SECTION 1.7 ▷ STUDY SET

VOCABULARY

Fill in the blanks.

1. In the exponential expression 7^5, 7 is the _____, and 5 is the _____. 7^5 is the fifth _____ of seven.
2. 10^2 can be read as ten _____, and 10^3 can be read as ten _____.
3. An _____ is used to represent repeated multiplication.

4. To _____ the expression $2(-1 + 4^2)$ means to find its value.
5. The rule for the _____ of operations guarantees that an evaluation of a numerical expression will result in a single answer.
6. To find the arithmetic _____ or average of a set of values, divide the sum of the values by the number of values.

CONCEPTS

7. To evaluate each expression, what operation should be performed first?

   a. $24 - 4 + 2$ **b.** $32 \div 8 \cdot 4$

   c. $8 - (3 + 5)^2$ **d.** $65 \cdot 3^3$

8. To evaluate $\dfrac{36 - 4(7)}{2(10 - 8)}$, what operation should be performed first in the numerator? In the denominator?

NOTATION

9. a. Give the name of each grouping symbol: $(\ \ )$, $[\ \ ]$, $\{\ \ \}$, $|\ \ |$, and —.

   b. In the expression $-8 + 2[15 - (-6 + 1)]$, which grouping symbols are innermost, and which are outermost?

10. What operation is indicated?

$$2 + 9 \,\big|\, 5 - (2 + 4) \big|$$

11. a. In the expression $(-5)^2$, what is the base?

   b. In the expression -5^2, what is the base?

12. Write each expression using symbols. Then evaluate it.

   a. Negative two squared

   b. The opposite of the square of two

Complete the evaluation of each expression.

13. $-19 - 2[(1 + 2)^2 \cdot 3] = -19 - 2[\ \ ^2 \cdot 3]$

$= -19 - 2[\ \ \cdot 3]$

$= -19 - 2[\ \ ]$

$= -19 - \ \ $

$= \ \ $

14.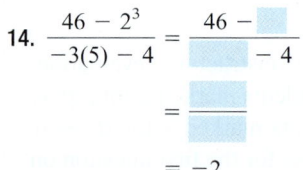

$= \dfrac{46 - \ \ }{\ \ - 4}$

$= \dfrac{\ \ }{\ \ }$

$= -2$

GUIDED PRACTICE

Write each product using exponents. See Example 1.

15. $8 \cdot 8 \cdot 8$

16. $(-4)(-4)(-4)(-4)$

17. $7 \cdot 7 \cdot 7 \cdot 12 \cdot 12$

18. $5 \cdot 5 \cdot 5 \cdot 5 \cdot 5 \cdot 5 \cdot 7 \cdot 7 \cdot 7$

19. $x \cdot x \cdot x$

20. $b \cdot b \cdot b \cdot b$

21. $r \cdot r \cdot r \cdot r \cdot s \cdot s$

22. $m \cdot m \cdot m \cdot n \cdot n \cdot n \cdot n$

Evaluate each expression. See Example 2.

23. 7^2

24. 9^2

25. 6^3

26. 6^4

27. $(-5)^4$

28. $(-5)^3$

29. $(-0.1)^2$

30. $(-0.8)^2$

31. $\left(-\dfrac{1}{4}\right)^3$

32. $\left(-\dfrac{1}{3}\right)^4$

33. $\left(\dfrac{2}{3}\right)^3$

34. $\left(\dfrac{3}{4}\right)^3$

Evaluate each expression. See Example 3.

35. $(-6)^2$ and -6^2

36. $(-4)^2$ and -4^2

37. $(-8)^2$ and -8^2

38. $(-9)^2$ and -9^2

Evaluate each expression. See Example 4.

39. $3 - 5 \cdot 4$

40. $-4 \cdot 6 + 5$

41. $32 - 16 \div 4 + 2$

42. $60 - 20 \div 10 + 5$

43. $3^2 \cdot 5 - 6 \div 3$

44. $2^3 \cdot 5 - 4 \div 2$

45. $12 \div 3 \cdot 2$

46. $18 \div 6 \cdot 3$

47. $-22 - 15 + 3$

48. $-33 - 8 + 10$

49. $-2(9) - 2(5)(10)$

50. $-6(7) - 3(-4)(-2)$

Evaluate each expression. See Example 5.

51. $-4(6 + 5)$

52. $-3(5 - 4)$

53. $(9 - 3)(9 - 9)^2$

54. $-(-8 - 6)(6 - 6)^2$

55. $(-1 - 3^2 \cdot 4)2^2$

56. $-1(28 - 5^2 \cdot 2)3^2$

57. $1 + 5(10 + 2 \cdot 5) - 1$

58. $14 + 3(7 - 5 \cdot 3)$

Evaluate each expression. See Example 6.

59. $(-1)^9[-7^2 - (-2)^2]$

60. $[-9^2 - (-8)^2](-1)^{10}$

61. $64 - 6[15 + 2(-3 + 8)]$

62. $4 - 2[26 + 2(5 - 3)]$

63. $-2[2 + 4^2(8 - 9)]^2$

64. $-3[5 + 3^2(4 - 5)]^2$

65. $3 + 2[-1 - (4 - 5)]$

66. $4 + 2[-7 - (3 - 9)]$

Evaluate each expression. See Example 7.

67. $\dfrac{-2 - 5}{-7 + (-7)}$

68. $\dfrac{-3 - (-1)}{-2 + (-2)}$

69. $\dfrac{2 \cdot 2^5 - 60 + (-4)}{5^4 - (-4)(-5)}$

70. $\dfrac{(6 - 5)^8 - 1}{(-9)(-3) - 4}$

71. $\dfrac{2(-4 - 2 \cdot 2)}{3(-3)(-2)}$

72. $\dfrac{3(-3^2 + 2 \cdot 2^2)}{(5 - 8)(7 - 9)}$

73. $\dfrac{72 - (2 - 2 \cdot 4)}{10^2 - (9 \cdot 10 + 2^2)}$

74. $\dfrac{13^2 - 5^2}{-3(5 - 3^2)}$

Evaluate each expression. See Example 8.

75. $10 - 2|4 - 8|$

76. $45 - 5|1 - 8|$

77. $-|7 - 2^3(4 - 7)|$

78. $-|9 - 5(1 - 2^3)|$

79. $\dfrac{(3 + 5)^2 + |-2|}{-2(5 - 8)}$

80. $\dfrac{|-25| - 8(-5)}{2^4 - 29}$

81. $\dfrac{|6 - 4| + 2|-4|}{226 - 6^3}$

82. $\dfrac{4|9 - 7| + |-7|}{6^3 - 211}$

83. $-(2 \cdot 3 - 2^2)^5$

84. $-(3 \cdot 5 - 2 \cdot 6)^4$

85. $2 \cdot 5^2 + 4 \cdot 3^2$

86. $5 \cdot 3^3 - 4 \cdot 2^3$

87. $-2(-1)^2 + 3(-1) - 3$

88. $-4(-3)^2 + 3(-3) - 1$

89. $8 - 3[5^2 - (7 - 3)^2]$

90. $3 - [3^3 + (3 - 1)^3]$

TRY IT YOURSELF

Evaluate each expression.

91. $[6(5) - 5(5)]^3(-4)$

92. $5 - 2 \cdot 3^4 - (-6 + 5)^3$

93. $8 - 6[(130 - 4^3) - 2]$

94. $91 - 5[(150 - 3^3) - 1]$

95. $-2\left(\dfrac{15}{-5}\right) - \dfrac{6}{2} + 9$

96. $-6\left(\dfrac{25}{-5}\right) - \dfrac{36}{9} + 1$

97. $-5(-2)^3 - |-2 + 1|$

98. $-6(-3)^3 - |-6 + 5|$

99. $\dfrac{18 - [2 + (1 - 6)]}{16 - (-4)^2}$

100. $\dfrac{6 - [6(-1) - 88]}{4 - 2^2}$

101. $-|-5 \cdot 7^2| - 30$

102. $2 + |-3 \cdot 2^2 \cdot 8^2 \cdot 1^2|$

103. $(-3)^3\left(\dfrac{-4}{2}\right)(-1)$

104. $(-2)^3\left(\dfrac{-6}{2}\right)(-1)$

105. $\dfrac{1}{2}\left(\dfrac{1}{8}\right) + \left(-\dfrac{1}{4}\right)^2$

106. $-\dfrac{1}{9}\left(\dfrac{1}{4}\right) + \left(-\dfrac{1}{6}\right)^2$

107. $\dfrac{-5^2 \cdot 10 + 5 \cdot 2^5}{-5 - 3 - 1}$

108. $\dfrac{(-6^2 - 2^4 \cdot 2) + 5}{-4 - 3}$

109. $-\left(\dfrac{40 - 1^3 - 2^4}{3(2 + 5) + 2}\right)$

110. $-\left(\dfrac{8^2 - 10}{2(3)(4) - 5(3)}\right)$

Look Alikes . . .

111. a. $(-7 - 4)(-2)$ b. $(-7 - 4) - 2$

112. a. $2 \cdot 3^3$ b. $(2 \cdot 3)^3$

113. a. $-100 \div 5 \cdot 2$ b. $-100 \div (5 \cdot 2)$

114. a. $8 + 3[-2 - (6 + 1)]$ b. $(8 + 3)[-2 - (6 + 1)]$

APPLICATIONS

115. **Light.** As light energy passes through the first unit of area, 1 yard away from the bulb, it spreads out. How much area does that light energy cover 2 yards, 3 yards, and 4 yards from the bulb? Express each answer using exponents.

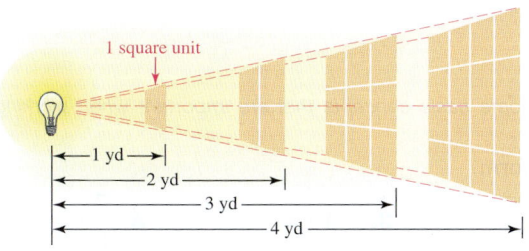

116. **Chain Letters.** A woman sent two friends a letter with the following request: "Please send a copy of this letter to two of your friends." Assume that all those receiving letters responded and that everyone in the chain received just one letter. Complete the table and then determine how many letters will be circulated in the 10th level.

Level	Number of letters circulated
1st	$2 = 2^1$
2nd	$\square = 2^\square$
3rd	$\square = 2^\square$
4th	$\square = 2^\square$

117. ▶ **from Campus to Careers**

Lead Transportation Security Officer

To determine the average afternoon wait time in security lines at an airport, officials monitored four passengers, each at a different gate. The time that each passenger entered a security line and the time the same passenger cleared the checkpoint was recorded, as shown below. Find the average (mean) wait time for these passengers.

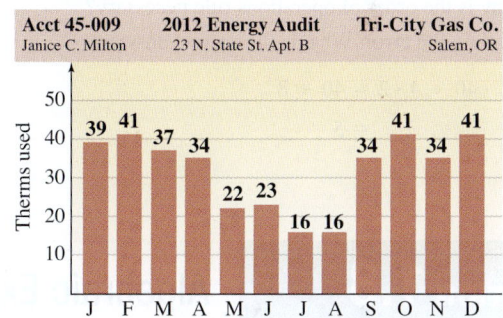

	Time entered	Time cleared
Passenger at Gate A	3:05 pm	3:21 pm
Passenger at Gate B	3:03 pm	3:13 pm
Passenger at Gate C	3:01 pm	3:09 pm
Passenger at Gate D	3:02 pm	3:16 pm

118. **Energy Usage.** Find the average number of therms of natural gas used per month.

Acct 45-009   2012 Energy Audit   Tri-City Gas Co.
Janice C. Milton   23 N. State St. Apt. B   Salem, OR

Therms used: J 39, F 41, M 37, A 34, M 22, J 23, J 16, A 16, S 34, O 41, N 34, D 41

119. **Cash Awards.** A contest is to be part of a promotional kickoff for a new children's cereal. The prizes to be awarded are shown.

a. How much money will be awarded in the promotion?

b. What is the average cash prize?

YouTube Video Contest
Grand prize: Disney World vacation plus $2,500
Four 1st place prizes of $500
Thirty-five 2nd place prizes of $150
Eighty-five 3rd place prizes of $25

120. **Surveys.** Some students were asked to rate their college cafeteria food on a scale from 1 to 5. The responses are shown on the tally sheet. Find the average rating.

Poor		Fair		Excellent
1	2	3	4	5
	III	III	ЖЖ	ЖЖ IIII

121. Wrapping Gifts. How much ribbon is needed to wrap the package if 15 inches of ribbon are needed to make the bow?

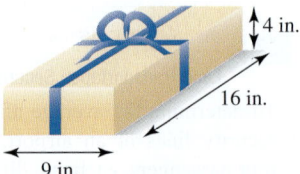

122. Scrabble. Write an expression to determine the number of points received for playing the word QUARTZY and then evaluate it. (The number on each tile gives the point value of the letter.)

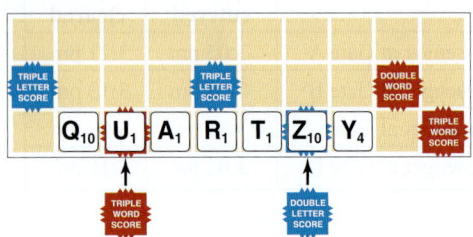

WRITING

123. Explain the difference between 2^3 and 3^2.

124. Why is the order of operations rule necessary?

125. Explain the error. What is the correct answer?

$$40 \div 4 \cdot 2 = 40 \div 8$$
$$= 5$$

126. Explain the error. What is the correct answer?

$$5 + 3(2 - 6) = 5 + 3(-4)$$
$$= 8(-4)$$
$$= -32$$

REVIEW

127. What numbers are a distance of 6 away from -11 on a number line?

128. Fill in the blank with $>$ or $<$: $0.3 \;\rule{0.5cm}{0.15cm}\; \frac{1}{3}$

CHALLENGE PROBLEMS

129. Using each of the numbers 2, 3, and 4 only once, what is the greatest value that the following expression can have?

$$\left(\square^{\square}\right)^{\square}$$

130. Insert a pair of parentheses into $4 \cdot 3^2 - 4 \cdot 2$ so that it has a value of 40.

Translate the set of instructions to an expression and then evaluate it.

131. Subtract the sum of -9 and 8 from the product of the cube of -3 and the opposite of 4.

132. Increase the square of the reciprocal of -2 by the difference of -0.25 and -1.

SECTION 1.8

Algebraic Expressions

OBJECTIVES

1 Identify terms and coefficients of terms.

2 Translate word phrases to algebraic expressions.

3 Analyze problems to determine hidden operations.

4 Evaluate algebraic expressions.

ARE YOU READY?

The following problems review some basic concepts that are important when working with algebraic expressions.

Write each expression in simpler form.

1. $9 \cdot x$ **2.** $1 \cdot m$ **3.** $-1 \cdot t$ **4.** $\frac{7}{8} \cdot y$

Identify each of the following expressions as either a sum, difference, product, or quotient.

5. $\frac{x}{12}$ **6.** $45 - a$ **7.** $2bc$ **8.** $d + 5$

Since problems in algebra are often presented in words, the ability to interpret what you read is important. In this section, we will introduce several strategies that will help you translate English words into algebraic expressions.

1 Identify Terms and Coefficients of Terms.

Recall that variables and/or numbers can be combined with the operations of arithmetic to create **algebraic expressions.** Addition symbols separate expressions into parts called *terms*. For example, the expression $x + 8$ has two terms.

$$\underset{\textcolor{red}{\text{First term}}}{x} \quad + \quad \underset{\textcolor{red}{\text{Second term}}}{8}$$

Since subtraction can be written as addition of the opposite, the expression $a^2 - 3a - 9$ has three terms.

$$a^2 - 3a - 9 = \underset{\text{First term}}{a^2} + \underset{\text{Second term}}{(-3a)} + \underset{\text{Third term}}{(-9)}$$

In general, a **term** is a product or quotient of numbers and/or variables. A single number or variable is also a term. Examples of terms are:

$$4, \quad y, \quad 6r, \quad -w^3, \quad 3.7x^5, \quad \frac{3}{n}, \quad -15ab^2$$

The numerical factor of a term is called the **coefficient** of the term. For instance, the term $6r$ has a coefficient of 6 because $6r = 6 \cdot r$. The coefficient of $-15ab^2$ is -15 because $-15ab^2 = -15 \cdot ab^2$. More examples are shown below.

A term such as 4, that consists of a single number, is called a **constant term.**

Notation

By the commutative property of multiplication, $r6 = 6r$ and $-15b^2a = -15ab^2$. However, we usually write the numerical factor first and the variable factors in alphabetical order.

The Language of Algebra

Terms such as x and y have **implied** coefficients of 1. *Implied* means suggested without being precisely expressed.

Term	Coefficient	
$8y^2$	8	
$-0.9pq$	-0.9	
$\frac{3}{4}b$	$\frac{3}{4}$	This term also could be written as $\frac{3b}{4}$.
$-\frac{x}{6}$	$-\frac{1}{6}$	Because $-\frac{x}{6} = -\frac{1x}{6} = -\frac{1}{6} \cdot x$
x	1	Because $x = 1x$
$-t$	-1	Because $-t = -1t$
27	27	

EXAMPLE 1   Identify the coefficient of each term in the expression: $7x^2 - x + 6$

Strategy   We will begin by writing the subtraction as addition of the opposite. Then we will determine the numerical factor of each term.

Why   Addition symbols separate expressions into terms.

Solution   If we write $7x^2 - x + 6$ as $7x^2 + (-x) + 6$, we see that it has three terms: $7x^2$, $-x$, and 6. The numerical factor of each term is its coefficient.

■ The coefficient of $7x^2$ is **7** because $7x^2$ means $\mathbf{7} \cdot x^2$.

■ The coefficient of $-x$ is **−1** because $-x$ means $\mathbf{-1} \cdot x$.

■ The coefficient of the constant 6 is 6.

Self Check 1   Identify the coefficient of each term in the expression:
$p^3 - 12p^2 + 3p - 4$

Now Try ▶ Problem 19

It is important to be able to distinguish between the *terms* of an expression and the *factors* of a term.

EXAMPLE 2   Is m used as a *factor* or a *term* in each expression?   **a.** $m + 6$ **b.** $8m$

Strategy   We will begin by determining whether m is involved in an addition or a multiplication.

Why   Addition symbols separate expressions into *terms*. A *factor* is a number being multiplied.

Solution   **a.** Since m is added to 6, m is a term of $m + 6$.

b. Since m is multiplied by 8, m is a factor of $8m$.

Self Check 2   Is b used as a *factor* or a *term* in each expression?
    a. $-27b$ **b.** $5a + b$

Now Try ▶   Problems 21 and 23

---

2 Translate Word Phrases to Algebraic Expressions.

The four tables below show how key phrases can be translated into algebraic expressions.

Caution

Be careful when translating subtraction. Order is important. For example, when a translation involves the phrase *less than*, note how the terms are reversed as we translate from English to mathematical symbols.

18 less than w

$w - 18$

Addition	
the sum of a and 8	$a + 8$
4 plus c	$4 + c$
16 added to m	$m + 16$
4 more than t	$t + 4$
20 greater than F	$F + 20$
T increased by r	$T + r$
exceeds y by 35	$y + 35$

Subtraction	
the difference of 23 and P	$23 - P$
550 minus h	$550 - h$
18 less than w	$w - 18$
7 decreased by j	$7 - j$
M reduced by x	$M - x$
12 subtracted from L	$L - 12$
5 less f	$5 - f$

Multiplication	
the product of 4 and x	$4x$
20 times B	$20B$
twice r	$2r$
double the amount a	$2a$
triple the profit P	$3P$
three-fourths of m	$\frac{3}{4}m$

Division	
the quotient of R and 19	$\frac{R}{19}$
s divided by d	$\frac{s}{d}$
the ratio of c to d	$\frac{c}{d}$
k split into 4 equal parts	$\frac{k}{4}$

Be careful when translating division. As with subtraction, order is important. For example, s divided by d is not written $\frac{d}{s}$.

EXAMPLE 3   Write each phrase as an algebraic expression:
    a. one-half of the profit P **b.** 5 less than the capacity c
    c. the product of the weight w and 2,000, increased by 300

Strategy   We will begin by identifying any key phrases.

Why   Key phrases can be translated to mathematical symbols.

Solution   **a.** **Key phrase:** *One-half of*   **Translation:** multiplication by $\frac{1}{2}$

The algebraic expression is: $\frac{1}{2}P$.

Caution

$5 < c$ is the translation of the statement 5 *is less than the capacity c*. It is not the translation of 5 *less than the capacity c*.

b. **Key phrase:** *less than*   **Translation:** subtraction

Sometimes thinking in terms of specific numbers makes translating easier. Suppose the capacity was 100. Then 5 *less than* 100 would be $100 - 5$. If the capacity is c, then we need to make it 5 less. The algebraic expression is: $c - 5$.

c. Key phrase: *product of*   **Translation:** multiplication

Key phrase: *increased by*   **Translation:** addition

In the given wording, the comma after 2,000 means *w* is first multiplied by 2,000; then 300 is added to that product. The algebraic expression is: $2{,}000w + 300$.

If there is no comma, the phrase *the product of the weight w and 2,000 increased by 300* translates to: $w(2{,}000 + 300)$.

Self Check 3   Write each phrase as an algebraic expression:

a. 80 less than the total *t*   **b.** $\frac{2}{3}$ of the time *T*

c. the difference of twice *a* and 15, squared

Now Try   Problems 25, 31, and 35

To solve application problems, we often let a variable represent an unknown quantity.

EXAMPLE 4   **Swimming.**   A pool is to be sectioned into 8 equally wide swimming lanes. Write an algebraic expression that represents the width of each lane.

Strategy   We will begin by letting *x* = the width of the swimming pool in feet. Then we will identify any key phrases.

Why   The width of the pool is unknown.

Solution   The key phrase, *sectioned into 8 equally wide lanes,* indicates division.

Therefore, the width of each lane is $\frac{x}{8}$ feet.

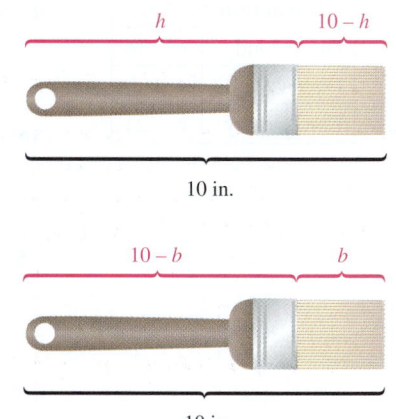

Self Check 4   **Commuting.**   It takes Val *m* minutes to get to work by bus. If she drives her car, her travel time exceeds this by 15 minutes. How many minutes does it take her to get to work by car?

Now Try   Problem 61

EXAMPLE 5   **Painting.**   A 10-inch-long paintbrush has two parts: a handle and bristles. Choose a variable to represent the length of one of the parts. Then write an expression to represent the length of the other part.

Strategy   There are two approaches. We can let *h* = the length of the handle or we can let *b* = the length of the bristles.

Why   Both the length of the handle and the length of the bristles are unknown.

Solution   Refer to the drawing on the top. If we let *h* = the length of the handle (in inches), then the length of the bristles is $10 - h$.

Now refer to the drawing on the bottom. If we let *b* = the length of the bristles (in inches), then the length of the handle is $10 - b$.

Self Check 5   **Scholarships.**   Part of a $900 donation to a college went to the scholarship fund, the rest to the building fund. Choose a variable to represent the amount donated to one of the funds. Then write an expression that represents the amount donated to the other fund.

Now Try ▶ Problem 13

EXAMPLE 6   **Enrollments.**   Second semester enrollment in a nursing program was 32 more than twice that of the first semester. Let x represent the enrollment for one of the semesters. Write an expression that represents the enrollment for the other semester.

Strategy   We will begin by letting x = the enrollment for the first semester.

Why   Because the second-semester enrollment is related to the first-semester enrollment.

Solution   **Key phrase:** *more than*   **Translation:** addition

Key phrase: *twice that*   **Translation:** multiplication by 2

The second semester enrollment was $2x + 32$.

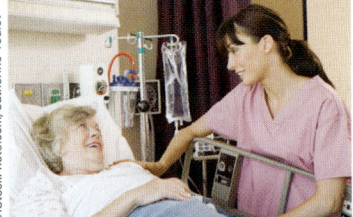

Self Check 6   **Politics.**   In an election, the incumbent received 55 fewer votes than three times the challenger's votes. Let x represent the number of votes received by one candidate. Write an expression that represents the number of votes received by the other.

Now Try ▶ Problem 109

---

3 **Analyze Problems to Determine Hidden Operations.**

Many applied problems require insight and analysis to determine which mathematical operations to use.

EXAMPLE 7   **Vacations.**   Disneyland, in California, was in operation 16 years before the opening of Disney World in Florida. Euro Disney, in France, was constructed 21 years after Disney World. Write algebraic expressions to represent the ages (in years) of each Disney attraction.

Strategy   We will begin by letting x = the age of Disney World.

Why   The ages of Disneyland and Euro Disney are both related to the age of Disney World.

Solution   In carefully reading the problem, we see that Disneyland was built 16 years before Disney World. That makes its age 16 years more than that of Disney World. The key phrase *more than* indicates addition.

Attraction	Age
Disneyland	$x + 16$
Disney World	x
Euro Disney	$x - 21$

$x + 16$ = the age of Disneyland

Euro Disney was built 21 years *after* Disney World. That makes its age 21 years less than that of Disney World. The key phrase *less than* indicates subtraction.

$x - 21$ = the age of Euro Disney

Self Check 7   **Tax Forms.**   Kayla worked 5 more hours preparing her tax return than she did on her daughter's return. Kayla's son's return took her 2 more hours to prepare than her daughter's. Write expressions to represent the hours she spent on each return.

Now Try ▶ Problem 111

EXAMPLE 8   How many months are in x years?

Strategy   There are no key phrases so we must analyze the problem carefully. We will begin by considering some specific cases.

Why   It's often easier to work with specifics first to get a better understanding of the relationship between the two quantities. Then we can generalize using a variable.

Solution   Let's calculate the number of months in 1 year, 2 years, and 3 years. When we write the results in a table, a pattern is apparent.

The number of months in x years is $12 \cdot x$ or $12x$.

Number of years	Number of months
1	12
2	24
3	36
x	$12x$

We multiply the number of years by 12 to find the number of months.

Self Check 8   How many days is h hours?

Now Try   Problems 7 and 67

In some problems, we must distinguish between *the number of* and *the value of* the unknown quantity. For example, to find the value of 3 quarters, we multiply the number of quarters by the value (in cents) of one quarter. Therefore, the value of 3 quarters is $3 \cdot 25$ cents $= 75$ cents.

The same distinction must be made if the number is unknown. For example, the value of n nickels is not n cents. The value of n nickels is $n \cdot 5$ cents $= 5n$ cents. For problems of this type, we will use the relationship

Number $\cdot$ value $=$ total value

EXAMPLE 9   Find the total value of: **a.** five dimes   **b.** q quarters   **c.** $x + 1$ half-dollars

Strategy   To find the total value (in cents) of each collection of coins, we multiply the number of coins by the value (in cents) of one coin, as shown in the table.

Why   Number $\cdot$ value $=$ total value

Solution

Type of coin	Number	Value	Total value	
Dime	5	10	50	Multiply: $5 \cdot 10 = 50$.
Quarter	q	25	$25q$	Multiply: $q \cdot 25$ can be written $25q$.
Half-dollar	$x + 1$	50	$50(x + 1)$	Multiply: $(x + 1) \cdot 50$ can be written $50(x + 1)$.

Self Check 9   Find the value of: **a.** six \$50 savings bonds
b. t \$100 savings bonds
c. $x - 4$ \$1,000 savings bonds

Now Try   Problems 14 and 69

4   Evaluate Algebraic Expressions.

To evaluate an algebraic expression, we substitute given numbers for each variable and perform the necessary calculations in the proper order.

EXAMPLE 10   Evaluate   each   expression   for $x = 3$ and $y = -4$:   **a.** $y^3 + y^2$ **b.** $-y - x$
c. $|5xy - 7|$ **d.** $\dfrac{y - 0}{x - (-1)}$

Strategy   We will replace each x and y in the expression with the given value of the variable, and evaluate the expression using the order of operation rule.

Why   To *evaluate an expression* means to find its numerical value, once we know the value of its variable(s).

Solution   **a.** $y^3 + y^2 = (-4)^3 + (-4)^2$ Substitute -4 for each y. We must write -4 within parentheses so that it is the base of each exponential expression.

$$= -64 + 16 \qquad \text{Evaluate each exponential expression.}$$
$$= -48 \qquad \text{Do the addition.}$$

> **Caution**
>
> When replacing a variable with its numerical value, we must often write the replacement number within parentheses to convey the proper meaning.

b. $-y - x = -(-4) - 3$ Substitute -4 for y and 3 for x. Don't forget to write the $-$ sign in front of (-4).

$$= 4 - 3 \qquad \text{Simplify: } -(-4) = 4.$$
$$= 1 \qquad \text{Do the subtraction.}$$

c. $|5xy - 7| = |5(3)(-4) - 7|$ Substitute 3 for x and -4 for y.

$$= |-60 - 7| \qquad \text{Do the multiplication: } 5(3)(-4) = -60.$$
$$= |-67| \qquad \text{Do the subtraction: } -60 - 7 = -60 + (-7) = -67.$$
$$= 67 \qquad \text{Find the absolute value of } -67.$$

d. $\dfrac{y - 0}{x - (-1)} = \dfrac{-4 - 0}{3 - (-1)}$ Substitute 3 for x and -4 for y.

$$= \dfrac{-4}{4} \qquad \text{In the denominator, do the subtraction: } 3 - (-1) = 3 + 1 = 4.$$
$$= -1 \qquad \text{Simplify the fraction.}$$

Self Check 10   Evaluate each expression for $a = -2$ and $b = 5$:
  a. $|a^3 + b^2|$ **b.** $-a + 2ab$ **c.** $\dfrac{a + 2}{b - 3}$

Now Try ▶ Problems 79 and 91

EXAMPLE 11   **Rocketry.**   If a toy rocket is shot into the air with an initial velocity of 80 feet per second, its height (in feet) after t seconds in flight is approximated by $-16t^2 + 80t$. How many seconds after the launch will it hit the ground?

Strategy   We can substitute positive values for t, the time in flight, until we find the one that gives a height of 0.

Why   When the height of the rocket is 0, it is on the ground.

Solution   We begin by finding the height after the rocket has been in flight for 1 second ($t = 1$).

$$-16t^2 + 80t = -16(1)^2 + 80(1) \qquad \text{Substitute 1 for } t.$$
$$= 64$$

As we evaluate $-16t^2 + 80t$ for several more values of t, we record each result in a table. The columns of the table can also be headed with the terms **input** and **output**. The values of t are the inputs into the expression $-16t^2 + 80t$, and the resulting values are the outputs.

t	$-16t^2 + 80t$
1	64
2	96
3	96
4	64
5	0

Evaluate for $t = 2$: $-16t^2 + 80t = -16(2)^2 + 80(2) = 96$
Evaluate for $t = 3$: $-16t^2 + 80t = -16(3)^2 + 80(3) = 96$
Evaluate for $t = 4$: $-16t^2 + 80t = -16(4)^2 + 80(4) = 64$
Evaluate for $t = 5$: $-16t^2 + 80t = -16(5)^2 + 80(5) = 0$

Input	Output
1	64
2	96
3	96
4	64
5	0

The height of the rocket is 0 when $t = 5$. The rocket will hit the ground 5 seconds after being launched.

Self Check 11 | In Example 11, suppose the height of the rocket is given by $-16t^2 + 112t$. What will be the height of the rocket 6 seconds after launch?

Now Try ▶ Problem 97

VOCABULARY

Fill in the blanks.

1. Variables and/or numbers can be combined with the operations of arithmetic to create algebraic _____.

2. A _____ is a product or quotient of numbers and/or variables. Examples are: $8x$, $\frac{t}{2}$, and $-cd^3$.

3. Addition symbols separate algebraic expressions into parts called _____.

4. A term, such as 27, that consists of a single number is called a _____ term.

5. The _____ of the term $10x$ is 10.

6. To _____ $4x - 3$ for $x = 5$, we substitute 5 for x and perform the necessary calculations.

CONCEPTS

7. Complete the table below on the left to determine the number of days in w weeks.

8. Complete the table below on the right to determine the number of minutes in s seconds.

Number of weeks	Number of days
1	
2	
3	
w	

Number of seconds	Number of minutes
60	
120	
180	
s	

9. The knife shown below is 12 inches long. Write an expression that represents the length (in inches) of the blade.

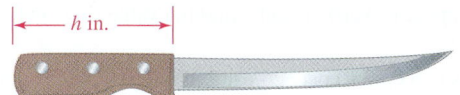

10. A student inherited $5,000 and deposits x dollars in American Savings. Write an expression that represents the number of dollars left to deposit in a City Mutual account.

$5,000

American Savings ← $x   City Mutual $?

11. Solution 2 is poured into solution 1. Write an expression that represents the number of ounces in the mixture.

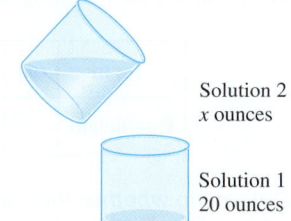

Solution 2
x ounces

Solution 1
20 ounces

12. Peanuts were mixed with c pounds of cashews to make 100 pounds of a mixture. Write an expression that represents the number of pounds of peanuts that were used.

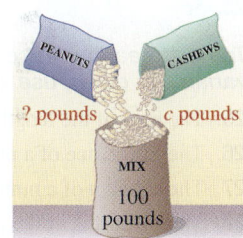

PEANUTS   CASHEWS

? pounds c pounds

MIX
100 pounds

13. **a.** Let b = the length of the beam shown below (in feet). Write an expression that represents the length of the pipe.

b. Let p = the length of the pipe (in feet). Write an expression that represents the length of the beam.

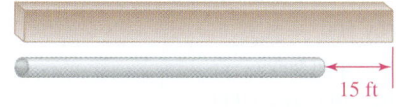

15 ft

14. Complete the table. Give each value in cents.

Coin	Number · Value	= Total value	
Nickel	6		
Dime	d		
Half-dollar	$x + 5$		

NOTATION

Complete each solution. Evaluate each expression for
a = 5, x = −2, and y = 4.

15. $9a - a^2 = 9(\quad) - (5)^2$

$= 9(5) - \boxed{}$

$= \boxed{} - 25$

$= 20$

16. $-x + 6y = -(\boxed{}) + 6(\boxed{})$

$= \boxed{} + 24$

$= 26$

17. Write each term in standard form.

   a. $y8$ **b.** $d2c$

   c. What property of multiplication did you use?

18. Fill in the blanks.

   a. $\dfrac{w}{2} = \boxed{}\,w$ **b.** $\dfrac{2}{3}m = \dfrac{\boxed{}}{3}$

GUIDED PRACTICE

See Example 1.

19. Consider the expression $3x^3 + 11x^2 - x + 9$.

   a. How many terms does the expression have?

   b. What is the coefficient of each term?

20. Complete the following table.

Term	$6m$	$-75t$	w	$\frac{1}{2}bh$	$\frac{x}{5}$	t
Coefficient						

Determine whether the variable c is used as a factor or as a term. See Example 2.

21. $c + 32$ **22.** $-24c + 6$

23. $5c$ **24.** $a + b + c$

Translate each phrase to an algebraic expression. If no variable is given, use x as the variable. See Example 3.

25. The sum of the length l and 15

26. The difference of a number and 10

27. The product of a number and 50

28. Three-fourths of the population p

29. The ratio of the amount won w and lost l

30. The tax t added to c

31. P increased by two-thirds of p

32. 21 less than the total height h

33. The square of k, minus 2,005

34. s subtracted from S

35. 1 less than twice the attendance a

36. J reduced by 500

37. 1,000 split n equal ways

38. Exceeds the cost c by 25,000

39. 90 more than twice the current price p

40. 64 divided by the cube of y

41. 3 times the total of 35, h, and 300

42. Decrease x by -17

43. 680 fewer than the entire population p

44. Triple the number of expected participants

45. The product of d and 4, decreased by 15

46. The quotient of y and 6, cubed

47. Twice the sum of 200 and t

48. The square of the quantity 14 less than x

49. The absolute value of the difference of a and 2

50. The absolute value of a, decreased by 2

51. One-tenth of the distance d

52. Double the difference of x and 18

Translate each algebraic expression into an English phrase. (Answers may vary.) See Example 3.

53. $\dfrac{3}{4}r$ **54.** $\dfrac{2}{3}d$

55. $t - 50$ **56.** $c + 19$

57. xyz **58.** $10ab$

59. $2m + 5$ **60.** $2s - 8$

Answer with an algebraic expression. See Example 4.

61. A model's skirt is x inches long. The designer then lets the hem down 2 inches. What is the length (in inches) of the altered skirt?

62. A soft drink manufacturer produced c cans of cola during the morning shift. Write an expression for how many six-packs of cola can be assembled from the morning shift's production.

63. The tag on a new pair of 36-inch-long jeans warns that after washing, they will shrink x inches in length. What is the length (in inches) of the jeans after they are washed?

64. A caravan of b cars, each carrying 5 people, traveled to the state capital for a political rally. How many people were in the caravan?

Answer with an algebraic expression. See Example 8.

65. How many minutes are there in h hours?

66. How many feet are in y yards?

67. How many feet are in i inches?

68. How many centuries in y years?

Answer with an algebraic expression. See Example 9.

69. A sales clerk earns $\$x$ an hour; how much does he earn in an 8-hour day?

70. A cashier earns $\$d$ an hour; how much does she earn in a 40-hour week?

71. If a car rental agency charges 49¢ a mile, what is the rental fee if a car is driven x miles?

72. If one egg is worth e cents, find the value (in cents) of one dozen eggs.

73. A ticket to a concert costs $\$t$. What would a pair of concert tickets cost?

74. If one apple is worth a cents, find the value (in cents) of 20 apples.

75. Tickets to a circus cost $25 each. What will tickets cost for a family of x people if they also pay for two of their neighbors?

76. A certain type of office desk that used to sell for $\$x$ is now on sale for $50 off. What will a company pay if it purchases 80 of the desks?

Evaluate each expression, for x = 3, y = −2, and z = −4. See Example 10.

77. $-y$ **78.** $-z$

79. $-z + 3x$

80. $-y - 5x$

81. $3y^2 - 6y - 4$

82. $-z^2 - z - 12$

83. $(3 + x)y$

84. $(4 + z)y$

85. $(x + y)^2 - |z + y|$

86. $[(z - 1)(z + 1)]^2$

87. $-\dfrac{2x + y^3}{y + 2z}$

88. $-\dfrac{2z^2 - x}{2x - y^2}$

Evaluate each expression. See Example 10.

89. $b^2 - 4ac$ for $a = -1$, $b = 5$, and $c = -2$

90. $(x - a)^2 + (y - b)^2$ for $x = -2$, $y = 1$, $a = 5$, and $b = -3$

91. $a^2 + 2ab + b^2$ for $a = -5$ and $b = -1$

92. $\dfrac{a - x}{y - b}$ for $x = -2$, $y = 1$, $a = 5$, and $b = 2$

93. $\dfrac{n}{2}[2a + (n - 1)d]$ for $n = 10$, $a = -4.2$, and $d = 6.6$

94. $\dfrac{a(1 - r^n)}{1 - r}$ for $a = -5$, $r = 2$, and $n = 3$

95. $(27c^2 - 4d^2)^3$ for $c = \frac{1}{3}$ and $d = \frac{1}{2}$

96. $\dfrac{-b^2 + 16a^2 + 1}{2}$ for $a = \frac{1}{4}$ and $b = -10$

Complete each table. See Example 11.

97.

x	$x^3 - 1$
0	
-1	
-3	

98.

g	$g^2 - 7g + 1$
0	
7	
-10	

99.

s	$\dfrac{5s + 36}{s}$
1	
6	
-12	

100.

a	$2{,}500a + a^3$
2	
4	
-5	

101.

Input x	Output $2x - \dfrac{x}{2}$
100	
-300	

102.

Input x	Output $\dfrac{x}{3} + \dfrac{x}{4}$
12	
-36	

103.

x	$(x + 1)(x + 5)$
-1	
-5	
-6	

104.

x	$\dfrac{1}{x + 8}$
-7	
-9	
-8	

Look Alikes . . .

Translate each phrase to mathematical symbols. Let x represent the unknown number.

105. **a.** The sum of a number and 7 squared

b. The sum of a number and 7, squared

106. **a.** 19 less than a number

b. 19 is less than a number

107. **a.** 4 times a number increased by 2

b. 4 times a number, increased by 2

108. **a.** Twice a number decreased by 3

b. Twice a number, decreased by 3

109. **Vehicle Weights.** A Hummer H2 weighs 340 pounds less than twice a Honda Element.

a. Let x represent the weight of one of the vehicles. Write an expression for the weight of the other vehicle.

b. If the weight of the Element is 3,370 pounds, what is the weight of the Hummer?

110. **Sod Farms.** The expression $20{,}000 - 3s$ gives the number of square feet of sod that are left in a field after s strips have been removed. Suppose a city orders 7,000 strips of sod. Evaluate the expression and explain the result.

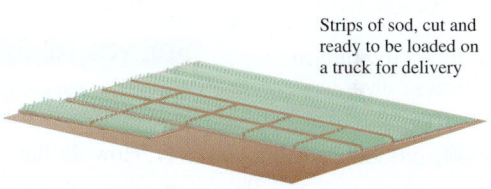

Strips of sod, cut and ready to be loaded on a truck for delivery

111. **Computer Companies.** IBM was founded 80 years before Apple Computer. Dell Computer Corporation was founded 9 years after Apple.

a. Let x represent the age (in years) of one of the companies. Write expressions to represent the ages (in years) of the other two companies.

b. On April 1, 2008, Apple Computer Company was 32 years old. How old were the other two computer companies then?

112. **Thrill Rides.** The distance in feet that an object will fall in t seconds is given by the expression $16t^2$. Find the distance that riders on "Drop Zone" will fall during the times listed in the table.

©sahua d/Shutterstock

Time (seconds)	Distance (feet)
1	
2	
3	
4	

113. What is an algebraic expression? Give some examples.

114. Explain why 2 *less than x* does not translate to $2 < x$.

115. In this section, we substituted a number for a variable. List some other uses of the word *substitute* that you encounter in everyday life.

116. Explain why *d* dimes are not worth *d*¢.

REVIEW

117. Find the LCD for $\frac{5}{12}$ and $\frac{1}{15}$.

118. Simplify: $\frac{3 \cdot 3 \cdot 5}{3 \cdot 5 \cdot 5 \cdot 11}$

119. Evaluate: $\left(\frac{2}{3}\right)^3$

120. Find the result when $\frac{7}{8}$ is multiplied by its reciprocal.

CHALLENGE PROBLEMS

121. Evaluate: $(8 - 1)(8 - 2)(8 - 3) \ldots (8 - 49)(8 - 50)$

122. Translate to an expression: The sum of a number decreased by six, and seven more than the quotient of triple the number and five.

SECTION 1.9

OBJECTIVES

1. Use the commutative and associative properties.
2. Simplify products.
3. Use the distributive property.
4. Identify like terms.
5. Combine like terms.

Simplifying Algebraic Expressions Using Properties of Real Numbers

ARE YOU READY?

The following problems review some basic concepts that are important when simplifying expressions using properties of real numbers.

1. How do the expressions $3 + x$ and $x + 3$ differ?

2. How do the expressions $1 + (7 + x)$ and $(1 + 7) + x$ differ?

3. Evaluate $5(2 + 4)$ and $5 \cdot 2 + 5 \cdot 4$ and compare the results.

4. How do the terms $4x$ and $4y$ differ? What do they have in common?

In algebra, we frequently replace one algebraic expression with another that is equivalent and simpler in form. That process, called *simplifying an algebraic expression,* often involves the use of one or more properties of real numbers.

1 Use the Commutative and Associative Properties.

Recall the commutative and associative properties discussed in Sections 1.4 and 1.6.

Commutative Properties	Changing the order when adding or multiplying does not affect the answer.
	$a + b = b + a$ and $ab = ba$

Associative Properties	Changing the grouping when adding or multiplying does not affect the answer.
	$(a + b) + c = a + (b + c)$ and $(ab)c = a(bc)$

These properties can be applied when working with algebraic expressions that involve addition and multiplication.

EXAMPLE 1   Use the given property to complete each statement.

a. $9 + x =$ _____   Commutative property of addition

b. $t \cdot 5 =$ _____   Commutative property of multiplication

c. $(a + 92) + 8 =$ _____   Associative property of addition

d. $6(10n) =$ _____   Associative property of multiplication

Strategy   For problems like these, it is important to memorize the properties by name. To fill in each blank, we will determine the way in which the property enables us to rewrite the given expression.

Why   We should memorize the properties by name because their names remind us how to use them.

Solution   **a.** To *commute* means to go back and forth. The commutative property of addition enables us to change the order of the terms, 9 and x.

$$9 + x = \underline{\ x + 9\ } \quad \text{Commutative property of addition}$$

b. We change the order of the factors, t and 5.

$$t \cdot 5 = \underline{\ 5 \cdot t\ } \quad \text{Commutative property of multiplication}$$

Success Tip

In part c, note that the order of the terms does not change. In part d, note that the order of the factors does not change.

c. To *associate* means to group together. The associative property of addition enables us to group the terms in a different way.

$$(a + 92) + 8 = \underline{\ a + (92 + 8)\ } \quad \text{Associative property of addition}$$

d. We change the grouping of the factors, 6, 10, and n.

$$6(10n) = \underline{\ (6 \cdot 10)n\ } \quad \text{Associative property of multiplication}$$

Self Check 1   Use the given property to complete each statement.
a. $5x + (3x + 1) =$ _____   Associative property of addition

b. $a(-15) =$ _____   Commutative property of multiplication

Now Try ▶   Problems 15 and 17

2 Simplify Products.

The commutative and associative properties of multiplication can be used to simplify certain products. For example, let's simplify $8(4x)$.

$$8(4x) = 8 \cdot (4 \cdot x) \quad \text{Rewrite } 4x \text{ as } 4 \cdot x.$$
$$= (8 \cdot 4) \cdot x \quad \text{Use the associative property of multiplication to group 4 with 8.}$$
$$= 32x \quad \text{Do the multiplication within the parentheses.}$$

We have found that $8(4x) = 32x$. We say that $8(4x)$ and $32x$ are **equivalent expressions** because for each value of x, they represent the same number. For example, both expressions have the value 320 if $x = 10$, and both have the value -96 if $x = -3$.

If $x = 10$		**If $x = -3$**	
$8(4x) = 8[4(\mathbf{10})]$	$32x = 32(\mathbf{10})$	$8(4x) = 8[4(\mathbf{-3})]$	$32x = 32(\mathbf{-3})$
$= 8(40)$	$= 320$	$= 8(-12)$	$= -96$
$= 320$		$= -96$	

EXAMPLE 2 Multiply: **a.** $-9 \cdot 3b$ **b.** $15a(6)$ **c.** $3(7p)(-5)$ **d.** $\dfrac{8}{3} \cdot \dfrac{3}{8}r$ **e.** $35\left(\dfrac{4}{5}x\right)$

Strategy We will use the commutative and associative properties of multiplication to reorder and regroup the factors in each expression.

Why We want to group all of the numerical factors of an expression together so that we can find their product.

Solution **a.** $-9 \cdot 3b = (-9 \cdot 3)b$ Use the associative property of multiplication to group -9 and 3.

$= -27b$ Do the multiplication within the parentheses.

Success Tip

By the commutative property of multiplication, we can *change* the order of factors. By the associative property of multiplication, we can change the *grouping* of factors.

---

b. $15a(6) = 15(6)a$ Use the commutative property of multiplication to reorder the factors.

$= 90a$ Do the multiplication, working from left to right: $15(6) = 90$.

c. $3(7p)(-5) = [3(7)(-5)]p$ Use the commutative and associative properties of multiplication to reorder and regroup the factors.

$= -105p$ Do the multiplication within the brackets.

d. $\dfrac{8}{3} \cdot \dfrac{3}{8}r = \left(\dfrac{8}{3} \cdot \dfrac{3}{8}\right)r$ Use the associative property of multiplication to regroup the factors.

$= 1r$ Multiply within the parentheses. The product of a number and its reciprocal is 1.

$= r$ The coefficient 1 need not be written.

e. $35\left(\dfrac{4}{5}x\right) = \left(35 \cdot \dfrac{4}{5}\right)x$ Use the associative property of multiplication to regroup the factors.

$= \left(\dfrac{\overset{1}{\cancel{5}} \cdot 7 \cdot 4}{\underset{1}{\cancel{5}}}\right)x$ Factor 35 as $5 \cdot 7$ and then remove the common factor 5.

$= 28x$

Self Check 2 Multiply: **a.** $9 \cdot 6s$ **b.** $-4(6u)(-2)$

c. $\dfrac{2}{3} \cdot \dfrac{3}{2}m$ **d.** $36\left(\dfrac{2}{9}y\right)$

Now Try ▶ Problems 23, 29, 31, and 33

---

3 Use the Distributive Property.

The Language of Algebra

To **distribute** means to give from one to several. You have probably *distributed* candy to children coming to your door on Halloween.

Another property that is often used to simplify algebraic expressions is the **distributive property.** To introduce it, we will evaluate $4(5 + 3)$ in two ways.

Use the order of operations: *Distribute the multiplication:*

$4(\mathbf{5 + 3}) = 4(\mathbf{8})$ $\mathbf{4}(5 + 3) = \mathbf{4} \cdot 5 + \mathbf{4} \cdot 3$

$= 32$ $= 20 + 12$

$= 32$

Each method gives a result of 32. This observation suggests the following property.

The Distributive Property ▼   For any real numbers a, b, and c,

$a(b + c) = ab + ac$ Read as "a times the quantity of b plus c."

The Language of Algebra

Formally, it is called the **distributive property of multiplication over addition.** When we use it to write a product, such as $5(x + 2)$, as a sum, $5x + 10$, we say that we have **removed** or **cleared** the parentheses.

To illustrate one use of the distributive property, let's consider the expression $5(x + 3)$. *Since we are not given the value of x, we cannot add x and 3 within the parentheses.* However, we can distribute the multiplication by the factor of 5 that is outside the parentheses to x and to 3 and add those products.

$$5(x + 3) = \mathbf{5} \cdot x + \mathbf{5} \cdot 3 \qquad \text{Distribute the multiplication by 5.}$$
$$= 5x + 15 \qquad \text{Do the multiplication.}$$

EXAMPLE 3   Multiply: **a.** $8(m + 9)$ **b.** $-12(4t + 1)$ **c.** $6\left(\dfrac{x}{3} + \dfrac{9}{2}\right)$

Strategy   In each case, we will distribute the multiplication by the factor *outside* the parentheses over each term *within* the parentheses.

Why   In each case, we cannot simplify the expression within the parentheses. To multiply, we must use the distributive property.

Solution   **a.** $\mathbf{8}(m + 9) = \mathbf{8} \cdot m + \mathbf{8} \cdot 9$ Read as "8 times the quantity of m plus 9."

Distribute the multiplication by 8.

$$= 8m + 72 \qquad \text{Do the multiplication.}$$

The Language of Algebra

We read $8(m + 9)$ as "eight times the **quantity** of m plus nine." The word *quantity* alerts us to the grouping symbols in the expression.

b. $-\mathbf{12}(4t + 1) = -\mathbf{12} \cdot 4t + -\mathbf{12} \cdot 1$ Distribute the multiplication by -12.

$$= -48t + (-12) \qquad \text{Do the multiplication.}$$
$$= -48t - 12 \qquad \begin{array}{l}\text{Write the result in simpler form. Recall that}\\ \text{adding } -12 \text{ is the same as subtracting 12.}\end{array}$$

c. $\mathbf{6}\left(\dfrac{x}{3} + \dfrac{9}{2}\right) = \mathbf{6} \cdot \dfrac{x}{3} + \mathbf{6} \cdot \dfrac{9}{2}$ Distribute the multiplication by 6.

$$= \frac{2 \cdot \overset{1}{\cancel{3}} \cdot x}{\underset{1}{\cancel{3}}} + \frac{\overset{1}{\cancel{2}} \cdot 3 \cdot 9}{\underset{1}{\cancel{2}}} \qquad \begin{array}{l}\text{Factor 6 as } 2 \cdot 3 \text{ and then remove}\\ \text{the common factors 3 and 2.}\end{array}$$

$$= 2x + 27$$

Self Check 3   Multiply: **a.** $7(m + 2)$ **b.** $-80(8x + 3)$
   c. $24\left(\dfrac{y}{6} + \dfrac{3}{8}\right)$

Now Try ▶   Problems 35, 37, and 39

Since subtraction is the same as adding the opposite, the distributive property also holds for subtraction.

$$a(b - c) = ab - ac$$

EXAMPLE 4   Multiply: **a.** $3(3b - 4)$ **b.** $-6(-3y - 8)$ **c.** $-1(t - 9)$

Strategy   In each case, we will distribute the multiplication by the factor *outside* the parentheses over each term *within* the parentheses.

Why   In each case, we cannot simplify the expression within the parentheses. To multiply, we must use the distributive property.

Solution

a. $3(3b - 4) = 3 \cdot 3b - 3 \cdot 4$ Distribute the multiplication by 3.

           $= 9b - 12$ Do the multiplication.

Caution

A common mistake is to forget to distribute the multiplication over each of the terms within the parentheses.

$3(3b - 4) = 9b - 4$

b. $-6(-3y - 8) = -6(-3y) - (-6)(8)$ Distribute the multiplication by -6.

                $= 18y - (-48)$ Do the multiplication.

                $= 18y + 48$ Write the result in simpler form. Add the opposite of -48.

Another approach is to write the subtraction within the parentheses as addition of the opposite. Then we distribute the multiplication by -6 over the addition.

$-6(-3y - 8) = -6[-3y + (-8)]$ Add the opposite of 8.

              $= -6(-3y) + (-6)(-8)$ Distribute the multiplication by -6.

              $= 18y + 48$ Do the multiplication.

Success Tip

Notice that distributing the multiplication by -1 *changes the sign* of each term within the parentheses.

c. $-1(t - 9) = -1(t) - (-1)(9)$ Distribute the multiplication by -1.

            $= -t - (-9)$ Do the multiplication.

            $= -t + 9$ Write the result in simpler form. Add the opposite of -9.

Self Check 4   Multiply: **a.** $5(2x - 1)$ **b.** $-9(-y - 4)$

                     c. $-1(c - 22)$

Now Try ▶   Problems 43, 47, and 49

Caution

The distributive property applies only to expressions in which multiplication is distributed over addition (or subtraction). For example, to simplify $6(5x)$, do not use the distibutive property.

Correct

$6(5x) = (6 \cdot 5)x = 30x$

Incorrect

$6(5x) = 30 \cdot 6x = 180x$

The distributive property can be extended to several other useful forms. Since multiplication is commutative, we have:

$$(b + c)a = ba + ca \qquad\qquad (b - c)a = ba - ca$$

For situations in which there are more than two terms within parentheses, we have:

$$a(b + c + d) = ab + ac + ad \qquad\qquad a(b - c - d) = ab - ac - ad$$

EXAMPLE 5   Multiply: **a.** $(6x + 4)\frac{1}{2}$ **b.** $2(a - 3b)8$ **c.** $-0.3(3a - 4b + 7)$

Strategy   We will multiply each term within the parentheses by the factor (or factors) outside the parentheses.

Why   In each case, we cannot simplify the expression within the parentheses. To multiply, we must use the distributive property.

Solution

a. $(6x + 4)\frac{1}{2} = (6x)\frac{1}{2} + (4)\frac{1}{2}$ Distribute the multiplication by $\frac{1}{2}$.

            $= 3x + 2$ Do the multiplication.

b. $2(a - 3b)8 = 2 \cdot 8(a - 3b)$ Use the commutative property of multiplication to reorder the factors.

              $= 16(a - 3b)$ Multiply 2 and 8 to get 16.

              $= 16a - 48b$ Distribute the multiplication by 16.

c. $-0.3(3a - 4b + 7) = -0.3(3a) - (-0.3)(4b) + (-0.3)(7)$

$= -0.9a + 1.2b - 2.1$ *Do each multiplication.*

Self Check 5   Multiply: **a.** $(-6x - 24)\frac{1}{3}$ **b.** $6(c - 2d)9$
c. $-0.7(2r + 5s - 8)$

Now Try ▶   **Problems 53, 55, and 57**

We can use the distributive property to find the opposite of a sum. For example, to find $-(x + 10)$, we interpret the $-$ symbol as a factor of -1, and proceed as follows:

$-(x + 10) = -1(x + 10)$ *Replace the $-$ symbol with -1.*

$= -1(x) + (-1)(10)$ *Distribute the multiplication by -1.*

$= -x - 10$ *Do the multiplication.*

In general, we have the following property of real numbers.

The Opposite of a Sum	The opposite of a sum is the sum of the opposites. For any real numbers a and b, $$-(a + b) = -a + (-b)$$

EXAMPLE 6   Simplify: $-(-9s - 3)$

Strategy   We will multiply each term within the parentheses by -1.

Why   The $-$ outside the parentheses represents a factor of -1 that is to be distributed.

Solution

$-(-9s - 3) = -1(-9s - 3)$ *Replace the $-$ symbol in front of the parentheses with -1.*

$= -1(-9s) - (-1)(3)$ *Distribute the multiplication by -1.*

$= 9s + 3$ *Do the multiplication.*

Success Tip

Notice that the $-$ symbol in front of the parentheses changes the sign of each term within the parentheses.

Self Check 6   Simplify: $-(-5x + 18)$

Now Try ▶   **Problem 59**

4 Identify Like Terms.

Before we can discuss methods for simplifying algebraic expressions involving addition and subtraction, we need to introduce some new vocabulary.

Like Terms	**Like terms** are terms containing exactly the same variables raised to exactly the same powers. Any constant terms in an expression are considered to be like terms. Terms that are not like terms are called **unlike terms**.

Success Tip

When looking for like terms, don't look at the coefficients of the terms. Consider only the variable factors of each term. If two terms are like terms, only their coefficients may differ.

Here are several examples.

Like terms		*Unlike terms*	
$4x$ and $7x$	Same variable	$4x$ and $7y$	The variables are not the same.
$-10p^2$ and $25p^2$	Same variable to the same power	$-10p$ and $25p^2$	Same variable, but different powers
$\frac{1}{3}c^3d$ and c^3d	Same variables to the same powers	$\frac{1}{3}c^3d$ and c^3	The variables are not the same.

EXAMPLE 7   List the like terms in each expression: **a.** $7r + 5 + 3r$ **b.** $6x^4 - 6x^2 - 6x$
c. $-17m^3 + 3 - 2 + m^3$

Strategy   First, we will identify the terms of the expression. Then we will look for terms that contain the same variables raised to exactly the same powers.

Why   If two terms contain the same variables raised to the same powers, they are like terms.

Solution   **a.** $7r + 5 + 3r$ contains the like terms $7r$ and $3r$.

b. Since the exponents on x are different, $6x^4 - 6x^2 - 6x$ contains no like terms.

c. $-17m^3 + 3 - 2 + m^3$ contains two pairs of like terms: $-17m^3$ and m^3 are like terms, and the constant terms, 3 and -2, are like terms.

Self Check 7   List the like terms: **a.** $2x - 2y + 7y$
b. $5p^2 - 12 + 17p^2 + 2$

Now Try ▶ Problem 65

---

5 Combine Like Terms.

To add or subtract objects, they must be similar. For example, fractions that are to be added must have a common denominator. When adding decimals, we align columns to be sure to add tenths to tenths, hundredths to hundredths, and so on. The same is true when working with terms of an algebraic expression. They can be added or subtracted only if they are like terms.

This expression can be simplified   This expression cannot be simplified
because it contains like terms.   because its terms are not like terms.

$$3x + 4x \qquad\qquad 3x + 4y$$

Recall that the distributive property can be written in the following forms:

$$(b + c)a = ba + ca \qquad (b - c)a = ba - ca$$

We can use these forms of the distributive property in reverse to simplify a sum or difference of like terms. For example, we can simplify $3x + 4x$ as follows:

The Language of Algebra

Simplifying a sum or difference of like terms is called **combining like terms**.

$$3x + 4x = (3 + 4)x \qquad \text{Use } ba + ca = (b + c)a.$$
$$= 7x$$

We can simplify $15m^2 - 9m^2$ in a similar way:

$$15m^2 - 9m^2 = (15 - 9)m^2 \qquad \text{Use } ba - ca = (b - c)a.$$
$$= 6m^2$$

In each case, we say that we **combined like terms.** These examples suggest the following general rule.

Combining Like Terms   Like terms can be combined by adding or subtracting the coefficients of the terms and keeping the same variables with the same exponents.

EXAMPLE 8   Simplify by combining like terms, if possible: **a.** $2x + 9x$ **b.** $-8p + (-2p) + 4p$ **c.** $0.5s^3 - 0.3s^3$ **d.** $4w + 6$ **e.** $\frac{4}{9}b + \frac{7}{9}b$

Strategy   We will use the distributive property in reverse to add (or subtract) the coefficients of the like terms. We will keep the same variables raised to the same powers.

Why   To *combine like terms* means to add or subtract the like terms in an expression.

Solution   **a.** Since $2x$ and $9x$ are like terms with the common variable x, we can combine them.

$$2x + 9x = 11x \quad \text{Think: } (2 + 9)x = 11x.$$

Success Tip

Just as 2 apples plus 9 apples is 11 apples, $2x + 9x = 11x$.

b. $-8p + (-2p) + 4p = -6p$ Think: $[-8 + (-2) + 4]p = -6p.$

c. $0.5s^3 - 0.3s^3 = 0.2s^3$ Think: $(0.5 - 0.3)s^3 = 0.2s^3.$

d. Since $4w$ and 6 are not like terms, they cannot be combined. $4w + 6$ doesn't simplify.

e. $\frac{4}{9}b + \frac{7}{9}b = \frac{11}{9}b$ Think: $\left(\frac{4}{9} + \frac{7}{9}\right)b = \frac{11}{9}b.$

Self Check 8   Simplify, if possible: **a.** $3x + 5x$ **b.** $-6y + (-6y) + 9y$ **c.** $4.4s^4 - 3.9s^4$ **d.** $4a - 2$ **e.** $\frac{10}{7}c - \frac{4}{7}c$

Now Try   Problems 67, 69, 71, and 73

EXAMPLE 9   Simplify by combining like terms: **a.** $16t - 15t$ **b.** $16t - t$ **c.** $15t - 16t$ **d.** $16t + t$

Strategy   As we combine like terms, we must be careful when working with the terms such as t and $-t$.

Why   Coefficients of 1 and -1 are usually not written.

Solution   **a.** $16t - 15t = t$ Think: $(16 - 15)t = 1t = t.$
b. $16t - t = 15t$ Think: $16t - 1t = (16 - 1)t = 15t.$
c. $15t - 16t = -t$ Think: $(15 - 16)t = -1t = -t.$
d. $16t + t = 17t$ Think: $16t + 1t = (16 + 1)t = 17t.$

Self Check 9   Simplify: **a.** $9h - h$ **b.** $9h + h$ **c.** $9h - 8h$ **d.** $8h - 9h$

Now Try   Problems 75 and 77

EXAMPLE 10   Simplify: $6a^2 + 54a - 4a - 36$

Strategy   First, we will identify any like terms in the expression. Then we will use the distributive property in reverse to combine them.

Why   To *simplify* an expression, we use properties of real numbers to write an equivalent expression in simpler form.

Solution We can combine the like terms that involve the variable a.

$$6a^2 + \mathbf{54a} - \mathbf{4a} - 36 = 6a^2 + \mathbf{50a} - 36 \qquad \text{Think: } (54 - 4)a = 50a.$$

Self Check 10 Simplify: $7y^2 + 21y - 2y - 6$

Now Try ▶ Problem 85

EXAMPLE 11 Simplify: $4(x + 5) - 5 - (2x - 4)$

Strategy First, we will use the distributive property to remove the parentheses. Then we will identify any like terms and combine them.

Why To *simplify* an expression, we use properties of real numbers, such as the distributive property, to write an equivalent expression in simpler form.

Solution

$$4(x + 5) - 5 - (2x - 4) = 4(x + 5) - 5 - \mathbf{1}(2x - 4) \qquad \text{Replace the } - \text{ symbol in front of } (2x - 4) \text{ with } -1.$$

$$= \mathbf{4x + 20 - 5 - 2x + 4} \qquad \text{Distribute the multiplication by 4 and } -1.$$

$$= \mathbf{2x + 19} \qquad \text{Think: } (4 - 2)x = 2x. \text{ Think: } (20 - 5 + 4) = 19.$$

Success Tip

Here, the distributive property is used both *forward* (to remove parentheses) and in *reverse* (to combine like terms).

Self Check 11 Simplify: $6(3y - 1) + 2 - (-3y + 4)$

Now Try ▶ Problem 87

SECTION 1.9 ▶ **STUDY SET**

VOCABULARY

Fill in the blanks.

1. To _____ the expression $5(6x)$ means to write it in simpler form: $5(6x) = 30x$.

2. $5(6x)$ and $30x$ are _____ expressions because for each value of x, they represent the same number.

3. To perform the multiplication $2(x + 8)$, we use the _____ property.

4. We call $-(c + 9)$ the _____ of a sum.

5. Terms such as $7x^2$ and $5x^2$, which have the same variables raised to exactly the same power, are called _____ terms.

6. When we write $9x + x$ as $10x$, we say we have _____ like terms.

CONCEPTS

7. **a.** Fill in the blanks to simplify the expression.

$$4(9t) = (\boxed{} \cdot \boxed{})t = \boxed{}\, t$$

b. What property did you use in part a?

8. **a.** Fill in the blanks to simplify the expression.

$$-6y \cdot 2 = \boxed{} \cdot \boxed{} \cdot y = \boxed{}\, y$$

b. What property did you use in part a?

9. Fill in the blanks.

a. $2(x + 4) = 2x \boxed{} 8$ **b.** $2(x - 4) = 2x \boxed{} 8$

c. $-2(x + 4) = -2x \boxed{} 8$ **d.** $-2(-x - 4) = 2x \boxed{} 8$

10. Fill in the blanks to combine like terms.

a. $4m + 6m = (\boxed{})m = \boxed{}\, m$

b. $30n^2 - 50n^2 = (\boxed{})n^2 = \boxed{}\, n^2$

c. $12 + 32d + 15 = 32d + \boxed{}$

d. Like terms can be combined by adding or subtracting the _____ of the terms and keeping the same _____ with the same exponents.

11. Simplify each expression, if possible.

a. $5(2x)$ **b.** $5 + 2x$

c. $6(-7x)$ **d.** $6 - 7x$

e. $2(3x)(3)$ **f.** $2 + 3x + 3$

12. Fill in the blanks: Distributing multiplication by -1 changes the _____ of each term within the parentheses.

$$-(x + 10) = \boxed{}(x + 10) = -x \boxed{} 10$$

NOTATION

13. Translate to symbols.
   a. Six times the quantity of h minus four.
   b. The opposite of the sum of z and sixteen.
14. Write an equivalent expression for the given expression using fewer symbols.
   a. $1x$ b. $-1d$ c. $0m$
   d. $5x - (-1)$ e. $16t + (-6)$

GUIDED PRACTICE

Use the given property to complete each statement. See Example 1.

15. $8 + (7 + a) =$ _____ Associative property of addition

16. $-2(5b) =$ _____ Associative property of multiplication

17. $y \cdot 11 =$ ____ Commutative property of multiplication

18. $x + x^2 =$ _____ Commutative property of addition

19. $(8d \cdot 2)6 =$ _____ Associative property of multiplication

20. $(-1 + 3a) + 7a =$ _____ Associative property of addition

21. $9t + (4 + t) = 9t + ($ ___ $)$ Commutative property of addition

22. $(x - 2)3 =$ _____ Commutative property of multiplication

Simplify each expression. See Example 2.

23. $3 \cdot 4t$ 24. $9 \cdot 3s$
25. $5(-7q)$ 26. $-7(5t)$
27. $(-5.6x)(-2)$ 28. $(-4.4x)(-3)$
29. $5(4c)(3)$ 30. $9(2h)(2)$
31. $\frac{5}{3} \cdot \frac{3}{5}g$ 32. $\frac{9}{7} \cdot \frac{7}{9}k$
33. $12\left(\frac{5}{12}x\right)$ 34. $15\left(\frac{4}{15}w\right)$

Multiply. See Example 3.

35. $5(x + 3)$ 36. $4(x + 2)$
37. $-3(4x + 9)$ 38. $-5(8x + 9)$
39. $45\left(\frac{x}{5} + \frac{2}{9}\right)$ 40. $35\left(\frac{y}{5} + \frac{8}{7}\right)$
41. $0.4(x + 4)$ 42. $2.2(2q + 1)$

Multiply. See Example 4.

43. $6(6c - 7)$ 44. $9(9d - 3)$
45. $-6(13c - 3)$ 46. $-2(10s - 11)$
47. $-15(-2t - 6)$ 48. $-20(-4z - 5)$
49. $-1(-4a + 1)$ 50. $-1(-2x + 3)$

Multiply. See Example 5.

51. $(3t + 2)8$ 52. $(2q + 1)9$

53. $(3w - 6)\frac{2}{3}$ 54. $(2y - 8)\frac{1}{2}$
55. $4(7y + 4)2$ 56. $8(2a - 3)4$
57. $2.5(2a - 3b + 1)$ 58. $5(9s - 12t - 3)$

Multiply. See Example 6.

59. $-(x - 7)$ 60. $-(y + 1)$
61. $-(-5.6y + 7)$ 62. $-(-4.8a - 3)$

List the like terms in each expression, if any. See Example 7.

63. $3x + 2 - 2x$
64. $3y + 4 - 11y + 6$
65. $-12m^4 - 3m^3 + 2m^2 - m^3$
66. $6x^3 + 3x^2 + 6x$

Simplify by combining like terms. See Example 8.

67. $3x + 7x$ 68. $12y - 15y$
69. $-7b^2 + 27b^2$ 70. $-2c^3 + 12c^3$

Simplify by combining like terms. See Example 9.

71. $36y + y - 9y$ 72. $32a - a + 5a$
73. $\frac{3}{5}t + \frac{1}{5}t$ 74. $\frac{3}{16}x - \frac{5}{16}x$
75. $13r - 12r$ 76. $25s + s$
77. $43s^3 - 44s^3$ 78. $8j^3 - 9j^3$

Simplify by combining like terms. See Example 10.

79. $15y - 10 - y - 20y$ 80. $9z - 7 - z - 19z$
81. $3x + 4 - 5x + 1$ 82. $4b + 9 - 9b + 9$
83. $9m^2 - 6m + 12m - 4$ 84. $6a^2 + 18a - 9a + 5$
85. $4x^2 + 5x - 8x + 9$ 86. $10y^2 - 8y + y - 7$

Simplify. See Example 11.

87. $2z + 5(z - 3) - 10$ 88. $12(m + 11) - 11 + m$
89. $2(s^2 - 7) - (s^2 - 2)$ 90. $4(d^2 - 3) - (d^2 - 1)$

TRY IT YOURSELF

Simplify each expression, if possible.

91. $-\frac{7}{16}x - \frac{3}{16}x$ 92. $-\frac{5}{18}x - \frac{7}{18}x$
93. $-9.8c + 6.2c$ 94. $-5.7m + 4.3m$
95. $-4(-6)(-4m)$ 96. $-5(-9)(-4n)$
97. $-4x + 4x$ 98. $-16y + 16y$
99. $-0.2r - (-0.6r)$ 100. $-1.1m - (-2.4m)$
101. $8\left(\frac{3}{4}y\right)$ 102. $27\left(\frac{2}{3}x\right)$

103. $-9(3r - 9) - 7(2r - 7)$ **104.** $-6(3t - 6) - 3(11t - 3)$

105. $9(7m)$ **106.** $12n(8)$

107. $6 - 4(-3c - 7)$ **108.** $10 - 5(-5g - 1)$

109. $5t \cdot 60$ **110.** $70a \cdot 10$

111. $36\left(\dfrac{2}{9}x - \dfrac{3}{4}\right) + 36\left(\dfrac{1}{2}\right)$ **112.** $40\left(\dfrac{3}{8}y - \dfrac{1}{4}\right) + 40\left(\dfrac{4}{5}\right)$

113. $-4r - 7r + 2r - r$ **114.** $-v - 3v + 6v + 2v$

115. $24\left(-\dfrac{5}{6}r\right)$ **116.** $\dfrac{3}{4} \cdot \dfrac{1}{2}g$

117. $a + a + a$ **118.** $t - t - t - t$

119. $60\left(\dfrac{3}{20}r - \dfrac{4}{15}\right)$ **120.** $72\left(\dfrac{7}{8}f - \dfrac{8}{9}\right)$

121. $4a + 4b + 4c$ **122.** $2x + 2y + 2z$

123. $-(c + 7) + 2(c - 3)$ **124.** $-(z + 2) + 5(3 - z)$

125. $a^3 + 2a^2 + 4a - 2a^2 - 4a - 8$

126. $c^3 - 3c^2 + 9c + 3c^2 - 9c + 27$

Look Alikes . . .

127. a. $2(7x)5$ **b.** $2(7x + 5)$

128. a. $-3(-4a)(-2)$ **b.** $-3(-4a) - 2$

APPLICATIONS

In Exercises 129 and 130, recall that the perimeter of a figure is equal to the sum of the lengths of its sides.

129. First Aid.   Each side of the red cross has length x inches. Write an algebraic expression that represents the perimeter of the cross.

x inches

130. Billiards.   Write an algebraic expression that represents the perimeter of the table.

x ft $2x$ ft

WRITING

131. Explain why the distributive property applies to $2(3 + x)$ but not to $2(3x)$.

132. Explain each error. Then give the correct answer.
   a. $9(4b - 2) = 36b - 2$ **b.** $3(2x) = 6 \cdot 3x = 18x$
   c. $-(23c + 2) = -23c + 2$ **d.** $(5n + 1)2 = 5n + 2$

REVIEW

Evaluate each expression for $x = -3$ and $y = -5$.

133. $\dfrac{x - y^2}{2y - 1 + x}$ **134.** $\dfrac{2y + 1}{x} - x$

CHALLENGE PROBLEMS

135. Fill in the blanks: $\boxed{}(- \boxed{}) = -187x + 119$

Simplify.

136. $2\{-2[x + 4(2x + 1)] - 5[x + 2(3x + 4)]\} + 106x$

---

1 **Summary & Review**

SECTION 1.1 ▶ Introducing the Language of Algebra

DEFINITIONS AND CONCEPTS	EXAMPLES
Tables, bar graphs, and **line graphs** are used to describe numerical relationships.	See pages 2 and 3 for examples of tables and graphs.
A **sum** is the result of an addition. A **difference** is the result of a subtraction. A **product** is the result of a multiplication. A **quotient** is the result of a division.	$3 + 15 = $ **18** $\quad$ $16 - 1 = $ **15** $\quad$ $7 \cdot 8 = $ **56** $\quad$ $\dfrac{63}{9} = $ **7** sum $\qquad$ difference $\qquad$ product $\qquad$ quotient

A **variable** is a letter (or symbol) that stands for a number.	Variables: x, a, and y
Algebraic expressions contain variables and numbers combined with the operations of addition, subtraction, multiplication, and division.	Expressions: $5y + 7$, $\dfrac{12 - x}{5}$, and $8a^2(b - 3)$
An **equation** is a statement that two expressions are equal.	Equations: $3x - 4 = 12$ and $\dfrac{t}{9} = 12$ **Equations contain an = sign. Expressions do not.**
Equations that express a relationship between two or more variables are called **formulas.**	$A = lw$ (The formula for the area of a rectangle)

REVIEW EXERCISES

The line graph shows the number of cars in a parking structure from 6 P.M. to 12 midnight on a Saturday.

1. What units are used to scale the horizontal and vertical axes?

2. How many cars were in the parking structure at 11 P.M.?

3. At what time did the parking structure have 500 cars in it?

4. When was the structure empty of cars?

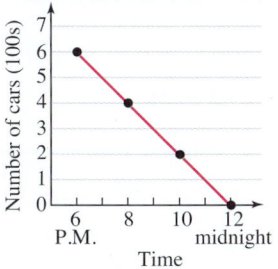

Express each statement in words, using one of these words: sum, difference, product, or quotient.

5. $15 - 3 = 12$

6. $15 + 3 = 18$

7. $15 \div 3 = 5$

8. $15 \cdot 3 = 45$

9. **a.** Write the multiplication 4×9 with a raised dot and then with parentheses.

   b. Write the division $9 \div 3$ using a fraction bar.

10. Write each multiplication without a multiplication symbol.

    a. $8 \cdot b$ **b.** $P \cdot r \cdot t$

11. Classify each item as either an expression or an equation.

    a. $5 = 2x + 3$ **b.** $2x + 3$

12. Use the formula $n = b + 5$ to complete the table.

Brackets (b)	Nails (n)
5	
10	
20	

SECTION 1.2 ▶ Fractions

DEFINITIONS AND CONCEPTS	EXAMPLES
A **factor** is a number being multiplied.	$8 \cdot 9 = 72$ Factor Factor
A **prime number** is a natural number that is greater than 1 that has only itself and 1 as factors. A **composite number** is a natural number, greater than 1, that is not prime.	Primes: $\{2, 3, 5, 7, 11, 13, 17, 19, 23, \ldots\}$ Composites: $\{4, 6, 8, 9, 10, 12, 14, 15, \ldots\}$
Any composite number can be factored into the product of two or more prime factors. The product of these prime numbers is called its **prime factorization.**	Find the prime factorization of 98. $98 = 2 \cdot 49 = 2 \cdot 7 \cdot 7$ 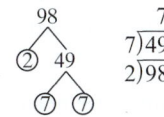
In a fraction, the number above the **fraction bar** is the **numerator** and the number below the fraction bar is called the **denominator.**	$\dfrac{11}{15}$ ← Numerator ← Denominator
Two fractions are **equivalent** if they represent the same number.	Equivalent fractions: $\dfrac{1}{2} = \dfrac{2}{4} = \dfrac{3}{6} = \dfrac{4}{8} = \ldots$
To **multiply two fractions,** multiply their numerators and multiply their denominators.	Multiply: $\dfrac{5}{8} \cdot \dfrac{3}{4} = \dfrac{15}{32}$

One number is the **reciprocal** of another if their product is 1.	The reciprocal of $\frac{4}{5}$ is $\frac{5}{4}$ because $\frac{4}{5} \cdot \frac{5}{4} = 1$.
To **divide two fractions,** multiply the first fraction by the reciprocal of the second fraction.	Divide: $\frac{4}{7} \div \frac{5}{8} = \frac{4}{7} \cdot \frac{8}{5} = \frac{32}{35}$ The reciprocal of $\frac{5}{8}$ is $\frac{8}{5}$.
Multiplication property of 1: The product of 1 and any number is that number.	$1 \cdot 5 = 5$ and $\frac{7}{8} \cdot 1 = \frac{7}{8}$
To **build a fraction,** multiply it by a form of 1 such as $\frac{2}{2}, \frac{3}{3}, \frac{4}{4}, \ldots$.	Write $\frac{3}{4}$ as an equivalent fraction with a denominator of 20. $$\frac{3}{4} = \frac{3}{4} \cdot \frac{5}{5} = \frac{15}{20}$$
To **simplify a fraction,** remove pairs of factors common to the numerator and the denominator. A fraction is in **simplest form,** or **lowest terms,** when the numerator and denominator have no common factors other than 1.	Simplify: $\frac{12}{18} = \frac{2 \cdot \overset{1}{\cancel{6}}}{3 \cdot \underset{1}{\cancel{6}}} = \frac{2}{3}$ Factor 12 and 18. Remove the common factor 6 from the numerator and denominator. Slashes and 1's are used to show that $\frac{6}{6}$ is replaced by the equivalent fraction $\frac{1}{1}$. A factor equal to 1 in the form of $\frac{6}{6}$ was removed.
To find the **LCD** of two fractions, prime factor each denominator and find the product of the prime factors, using each factor the greatest number of times it appears in any one factorization.	Find the LCD of $\frac{5}{12}$ and $\frac{7}{8}$. $$\left.\begin{array}{l} 12 = 2 \cdot 2 \cdot 3 \\ 8 = 2 \cdot 2 \cdot 2 \end{array}\right\} \text{LCD} = 2 \cdot 2 \cdot 2 \cdot 3 = 24$$
To **add (or subtract) fractions that have the same denominator,** add (or subtract) the numerators and write the sum (or difference) over the common denominator. Simplify, if possible. To **add (or subtract) fractions that have different denominators,** rewrite each fraction as an equivalent fraction with the LCD as the denominator. Then add (or subtract) as usual. Simplify, if possible.	Add: $\frac{5}{12} + \frac{7}{8} = \frac{5}{12} \cdot \frac{2}{2} + \frac{7}{8} \cdot \frac{3}{3}$ The LCD is 24. Build each fraction. $$= \frac{10}{24} + \frac{21}{24}$$ The denominators are now the same. $$= \frac{31}{24}$$ This result does not simplify.
A **mixed number** represents the sum of a whole number and a fraction. In some computations, it is necessary to write mixed numbers as improper fractions.	Mixed numbers: $6\frac{1}{3} = 6 + \frac{1}{3}$ and $1\frac{3}{4} = \frac{7}{4}$

REVIEW EXERCISES

13. a. Write 24 as the product of two factors.

   b. Write 24 as the product of three factors.

   c. List the factors of 24.

14. What do we call fractions, such as $\frac{1}{8}$ and $\frac{2}{16}$, that represent the same number?

Give the prime factorization of each number, if possible.

15. 54

16. 147

17. 385

18. 41

Simplify each fraction.

19. $\frac{20}{35}$

20. $\frac{24}{18}$

Build each number to an equivalent fraction with the indicated denominator.

21. $\frac{5}{8}$, denominator 64

22. 12, denominator 3

What is the LCD for fractions having the following denominators?

23. 10 and 18

24. 21 and 70

Perform each operation and simplify, if possible.

25. $\frac{1}{8} \cdot \frac{7}{8}$

26. $\frac{16}{35} \cdot \frac{25}{48}$

27. $\frac{1}{3} \div \frac{15}{16}$

28. $16\frac{1}{4} \div 5$

29. $\frac{17}{25} - \frac{7}{25}$

30. $\frac{8}{11} - \frac{1}{2}$

31. $\dfrac{17}{24} + \dfrac{11}{40}$ **32.** $4\dfrac{1}{9} - 3\dfrac{5}{6}$

33. The Internet.   A popular website averaged $1\dfrac{3}{4}$ million hits per day during a 30-day period. How many hits did it receive during that time?

34. Machine Shops.   How much must be milled off the $\dfrac{17}{24}$-inch-thick steel rod so that the collar will slip over it?

Steel rod

SECTION 1.3 ▶ **The Real Numbers**

DEFINITIONS AND CONCEPTS	EXAMPLES
To write a set, we list its **elements** within **braces** { }.	In the English alphabet, the set of vowels is {a, e, i, o, u}.
The **natural numbers** are the numbers we count with.	Natural numbers: {1, 2, 3, 4, 5, 6, . . .}
The **whole numbers** are the natural numbers together with 0.	Whole numbers: {0, 1, 2, 3, 4, 5, 6, . . .}
Two numbers are called **opposites** if they are the same distance from 0 on the number line but are on opposite sides of it.	The opposite of 3 is -3 and the opposite of -21 is 21.
The **integers** include the whole numbers and their opposites.	Integers: {. . . , $-3, -2, -1, 0, 1, 2, 3, \ldots$}
The **rational numbers** are numbers that can be expressed as fractions with an integer numerator and a nonzero integer denominator.	Rational numbers: $-6, \quad -3.1, \quad -\dfrac{1}{2}, \quad 0, \quad \dfrac{11}{12}, \quad 9\dfrac{4}{5}, \quad \text{and} \quad 87$ $-6 = \dfrac{-6}{1}, \quad -3.1 = \dfrac{-31}{10},$ $0 = \dfrac{0}{1}, \quad 9\dfrac{4}{5} = \dfrac{49}{5}, \quad 87 = \dfrac{87}{1}$
Terminating and **repeating decimals** can be expressed as fractions and are, therefore, rational numbers.	Terminating decimal: $-0.25 = -\dfrac{1}{4}$ Repeating decimal: $0.\overline{6} = \dfrac{2}{3}$
Negative fractions: $-\dfrac{a}{b} = \dfrac{-a}{b} = \dfrac{a}{-b}$	$-\dfrac{3}{4} = \dfrac{-3}{4} = \dfrac{3}{-4}$ and $\dfrac{-1}{8} = \dfrac{1}{-8} = -\dfrac{1}{8}$
An **irrational number** is a nonterminating, nonrepeating decimal. An irrational number cannot be expressed as a fraction with an integer numerator and a nonzero integer denominator. A **real number** is any number that is either a rational or an irrational number. Every real number corresponds to a point on the **number line,** and every point on the number line corresponds to exactly one real number.	Irrational numbers: $\sqrt{2}, \pi,$ and $-\sqrt{7}$ Graph the numbers in the set $\left\{-2, 4, -0.75, 1\dfrac{3}{4}, \pi, 0, \dfrac{7}{8}\right\}$ on a number line and classify them. 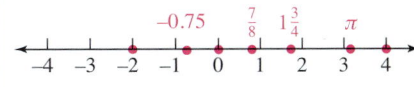 Natural numbers: 4 Whole numbers: 4, 0 Integers: -2, 4, 0 Rational numbers: -2, 4, -0.75, $1\dfrac{3}{4}$, 0, $\dfrac{7}{8}$ Irrational numbers: π Real numbers: all
Inequality symbols: $>$ is greater than $<$ is less than	$25 > 15$ and $-2 > -7$ $3.3 < 9.7$ and $-10 < -9$
The **absolute value** of a number is the distance on the number line between the number and 0.	$|5| = 5, \quad |-7| = 7, \quad \text{and} \quad -\left|\dfrac{5}{9}\right| = -\dfrac{5}{9}$

35. a. Which number is a whole number but not a natural number?

    b. Write the set of integers.

36. Represent 206 feet below sea level with a signed number.

37. Use one of the symbols $>$ or $<$ to make each statement true.

    a. 0 ⬚ 5   **b.** -12 ⬚ -13

38. Show that each of the following numbers is a rational number by expressing it as a ratio (quotient) of two integers.

    a. 0.7   **b.** $4\frac{2}{3}$

Write each fraction as a decimal. Use an overbar if the result is a repeating decimal.

39. $\dfrac{1}{250}$ **40.** $\dfrac{17}{22}$

41. Graph each number on a number line:

    $\left\{ \pi, 0.333\ldots, 3.75, \sqrt{2}, -\frac{17}{4}, \frac{7}{8}, -2 \right\}$

42. Determine which numbers in the given set are natural numbers, whole numbers, integers, rational numbers, irrational numbers, and real numbers. $\left\{ -\frac{4}{5}, 99.99, 0, \sqrt{2}, -12, 4\frac{1}{2}, 0.666\ldots 8 \right\}$

Determine whether each statement is true or false.

43. All integers are whole numbers.

44. π is a rational number.

45. The set of real numbers corresponds to all points on the number line.

46. A real number is either rational or irrational.

Insert one of the symbols $>$, $<$, or $=$ in the blank to make each statement true.

47. $|-6|$ ⬚ $|5|$ **48.** -9 ⬚ $|-10|$

SECTION 1.4 ▶ **Adding Real Numbers; Properties of Addition**

DEFINITIONS AND CONCEPTS	EXAMPLES
To **add two real numbers with like signs:**	
1. To add two positive numbers, add them as usual. The final answer is positive.	Add: $3 + 5 = 8$
2. To add two negative numbers, add their absolute values and make the final answer negative.	Add: $-5 + (-11) = -16$
To **add two real numbers with unlike signs:**	
1. Subtract their absolute values (the smaller from the larger).	Add: $-8 + 6 = -2$ *-8 has the larger absolute value.*
2. To that result, attach the sign of the number with the larger absolute value.	Add: $12 + (-5) = 7$ *12 has the larger absolute value.*
Properties of Addition	
Commutative property: $a + b = b + a$ *Changing the order when adding does not affect the answer.*	$5 + (-9) = -9 + 5$ *Reorder.*
Associative property: $(a + b) + c = a + (b + c)$ *Changing the grouping when adding does not affect the answer.*	$(3 + 7) + 5 = 3 + (7 + 5)$ *Regroup. Note that the order of the addends, 3, 7, and 5, does not change.*
Addition property of 0: $a + 0 = a$ and $0 + a = a$	$-6 + \mathbf{0} = -6$ *0 is the additive identity element.*
Addition property of opposites: $a + (-a) = 0$ and $(-a) + a = 0$	$\mathbf{11} + (\mathbf{-11}) = 0$ *11 and -11 are additive inverses.*

Add.

49. $-45 + (-37)$ **50.** $25 + (-13)$

51. $0 + (-7)$ **52.** $-7 + 7$

53. $12 + (-8) + (-15)$ **54.** $-9.9 + (-2.4)$

55. $\dfrac{5}{16} + \left(-\dfrac{1}{2} \right)$

56. $35 + (-13) + (-17) + 6$

57. Determine what property of addition is shown.

    a. $-2 + 5 = 5 + (-2)$

    b. $(-2 + 5) + 1 = -2 + (5 + 1)$

    c. $80 + (-80) = 0$

    d. $-5.75 + 0 = -5.75$

58. Temperatures. Determine Washington State's record high temperature if it is $166°$ greater than the state's record low temperature of $-48°$F.

SECTION 1.5 ▶ Subtracting Real Numbers

DEFINITIONS AND CONCEPTS	EXAMPLES
The **opposite of the opposite of a number** is that number. For any real number a, $-(-a) = a$.	$-(-13) = 13$ and $-(-x) = x$
To **subtract two real numbers,** add the first to the opposite (additive inverse) of the number to be subtracted. For any real numbers a and b, $$a - b = a + (-b)$$ To **check** a subtraction, the difference plus the subtrahend should equal the minuend.	Subtract: $4 - \mathbf{7} = 4 + (\mathbf{-7}) = -3$ The opposite of 7 is −7. $6 - (\mathbf{-8}) = 6 + \mathbf{8} = 14$ The opposite of −8 is 8. $-1 - (\mathbf{-2}) = -1 + \mathbf{2} = 1$ The opposite of −2 is 2. To check $-6 - 2 = -8$, verify that $-8 + 2 = -6$.

REVIEW EXERCISES

Write the expression in simpler form.

59. a. The opposite of 10

   b. The additive inverse of -3

60. a. $-\left(-\dfrac{9}{16}\right)$ **b.** $-|-4|$

Perform the operations.

61. $45 - 64$ **62.** Subtract $\dfrac{1}{3}$ from $-\dfrac{3}{5}$

63. $-7 - (-12)$ **64.** $3.6 - (-2.1)$

65. $0 - 10$ **66.** $-33 + 7 - 5 - (-2)$

67. Geography. The tallest peak on Earth is Mount Everest, at 29,028 feet, and the greatest ocean depth is the Mariana Trench, at $-36,205$ feet. Find the difference in these elevations. Check the result.

68. History. Archimedes, a famous Greek mathematician, died in 212 B.C. (-212) at the age of 75. When was he born? Check the result.

SECTION 1.6 ▶ Multiplying and Dividing Real Numbers; Multiplication and Division Properties

DEFINITIONS AND CONCEPTS	EXAMPLES
To **multiply two real numbers,** multiply their absolute values. **1.** If the numbers have **like signs,** the final answer is positive. **2.** If the numbers have **unlike signs,** the final answer is negative.	Multiply: $-5(-7) = 35$ and $14(3) = 42$ Multiply: $6(-6) = -36$ and $-11(5) = -55$
Properties of multiplication **Commutative property:** $ab = ba$ *Changing the order when multiplying does not affect the answer.*	$-8(12) = 12(-8)$ Reorder.
Associative property: $(ab)c = a(bc)$ *Changing the grouping when multiplying does not affect the answer.*	$(-4 \cdot 9) \cdot 7 = -4(9 \cdot 7)$ Regroup. Note that the order of the factors, −4, 9, and 7, does not change.
Multiplication property of 0: $0 \cdot a = 0$ and $a \cdot 0 = 0$	$0 \cdot (-7) = 0$ and $6(5)0 = 0$
Multiplication property of 1: $1 \cdot a = a$ and $a \cdot 1 = a$	$1 \cdot 32 = 32$ 1 is the multiplicative identity.
Multiplication property of −1: $-1 \cdot a = -a$ and $a(-1) = -a$	$-1(8) = -8, \quad -1(-32) = 32, \quad -x = -1 \cdot x$
Multiplicative inverse property: $a\left(\dfrac{1}{a}\right) = 1$ and $\dfrac{1}{a}(a) = 1$	$4\left(\dfrac{1}{4}\right) = 1$ 4 and $\frac{1}{4}$ are multiplicative inverses.

To **divide two real numbers,** divide their absolute values.	Divide: $\dfrac{16}{8} = 2$ and $\dfrac{-25}{-5} = 5$
1. If the numbers have **like signs,** the final answer is positive.	
2. If the numbers have **unlike signs,** the final answer is negative.	Divide: $\dfrac{-36}{9} = -4$ and $\dfrac{56}{-7} = -8$
For any real number, $\dfrac{a}{1} = a$ and $\dfrac{a}{a} = 1$, where $a \neq 0$.	$\dfrac{25}{1} = 25$ and $\dfrac{-32}{-32} = 1$
Division of zero by a nonzero number is 0. **Division by zero** is undefined.	$\dfrac{0}{17} = 0$ but $\dfrac{2}{0}$ is undefined because no number multiplied by 0 gives 2.
To **check** the division $\dfrac{a}{b} = c$, verify that $c \cdot b = a$.	To check $\dfrac{6}{-2} = -3$, verify that $-3(-2) = 6$. **Quotient · divisor = dividend.**

REVIEW EXERCISES

Multiply.

69. $-8 \cdot 7$

70. $-9\left(-\dfrac{1}{9}\right)$

71. $2(-3)(-2)$

72. $(-4)(-1)(-3)$

73. $-1.2(-5.3)$

74. $0.002(-1,000)$

75. $-\dfrac{2}{3}\left(\dfrac{1}{5}\right)$

76. $-6(-3)(0)(-1)$

77. Electronics. The picture on the screen can be magnified by switching a setting on the monitor. What would be the new high and low if every value changed by a factor of 1.5?

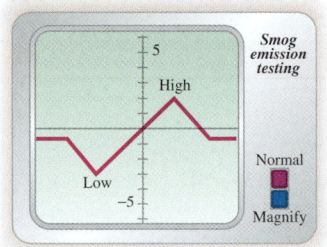

78. Determine what property of multiplication is shown.
   a. $(2 \cdot 3)5 = 2(3 \cdot 5)$
   b. $(-5)(-6) = (-6)(-5)$
   c. $-6 \cdot 1 = -6$
   d. $\dfrac{1}{2}(2) = 1$

Perform each division, if possible.

79. $\dfrac{44}{-44}$

80. $\dfrac{-272}{16}$

81. $\dfrac{-81}{-27}$

82. $-\dfrac{3}{5} \div \dfrac{1}{2}$

83. $\dfrac{-60}{0}$

84. $\dfrac{-4.5}{1}$

85. Fill in the blanks: $\dfrac{0}{18} = 0$ because $\boxed{} \cdot \boxed{} = \boxed{}$.

86. Gemstones. A 3-carat yellow sapphire stone valued at $3,000 five years ago is now worth $1,200. What signed number indicates the average annual depreciation of the sapphire?

SECTION 1.7 ▶ Exponents and Order of Operations

DEFINITIONS AND CONCEPTS	EXAMPLES
An **exponent** represents repeated multiplication.	$8^5 = \mathbf{8 \cdot 8 \cdot 8 \cdot 8 \cdot 8}$ The exponent 5 indicates that 8 is to be used as a factor 5 times.
In a^n, a is the **base** and n is the **exponent.**	In 7^4, the base is 7 and 4 is the exponent.

Order of Operations

1. Perform all calculations within grouping symbols, working from the innermost to the outermost in the following order.

2. Evaluate all exponential expressions.

3. Perform all multiplications and divisions as they occur from left to right.

4. Perform all additions and subtractions as they occur from left to right.

In fractions, evaluate the numerator and denominator separately. Then simplify the fraction.

Grouping Symbols:

Innermost parentheses

$$2[7 + 3(1 - 4)]$$

Outermost brackets

Evaluate:

$$\frac{3(6 - 4^3) - 2^4 + 4}{8 \div 4 \cdot 3} = \frac{3(6 - 64) - 2^4 + 4}{2 \cdot 3}$$ Evaluate: $4^3 = 64$. Divide: $8 \div 4 = 2$.

$$= \frac{3(-58) - 2^4 + 4}{6}$$ Subtract: $6 - 64 = -58$.

$$= \frac{3(-58) - 16 + 4}{6}$$ Evaluate: $2^4 = 16$.

$$= \frac{-174 - 16 + 4}{6}$$ Multiply: $3(-58) = -174$.

$$= \frac{-190 + 4}{6}$$ Subtract: $-174 - 16 = -190$.

$$= \frac{-186}{6}$$ Add: $-190 + 4 = -186$.

$$= -31$$ Divide: -186 by 6.

$$\text{Mean} = \frac{\text{sum of values}}{\text{number of values}}$$ The mean is also called the average.

Find the mean of the test scores of 74, 83, 79, 91, and 73.

$$\text{Mean} = \frac{74 + 83 + 79 + 91 + 73}{5} = 80$$ The mean is a value around which 74, 83, 79, 91, and 73 are grouped.

REVIEW EXERCISES

87. Write each expression using exponents.

    a. $8 \cdot 8 \cdot 8 \cdot 8 \cdot 8$ **b.** $9 \cdot \pi \cdot r \cdot r$

88. Evaluate each expression.

    a. 9^2 **b.** $\left(-\frac{2}{3}\right)^3$

    c. 2^5 **d.** 50^1

Evaluate each expression.

89. $2 + 5 \cdot 3$ **90.** $-24 \div 2 \cdot 3$

91. $-(16 - 3)^2$ **92.** $43 + 2(-6 - 2 \cdot 2)$

93. $10 - 5[-3 - 2(5 - 7^2)] - 5$

94. $\dfrac{-4(4 + 2) - 4}{2|-18 - 4(5)|}$

95. $(-3)^3\left(\dfrac{-8}{2}\right) + 5$

96. $\dfrac{2^4 - (4 - 6)(3 - 6)}{12 + 4[(-1)^8 - 2^2]}$

97. Write each expression in symbols and then evaluate it.

    a. Negative nine squared

    b. The opposite of the square of nine

98. **Walk-A-Thons.** Use the data in the table to find the average (mean) donation to a charity walk-a-thon.

Donation	$5	$10	$20	$50	$100
Number received	20	65	25	5	10

SECTION 1.8 ▶ Algebraic Expressions

DEFINITIONS AND CONCEPTS	EXAMPLES
Addition symbols separate algebraic expressions into **terms**. In a term, the numerical factor is called the **coefficient**. A term that consists of a single number is called a **constant** term.	Since $a^2 + 3a - 5$ can be written as $a^2 + 3a + (-5)$, it has three terms. The coefficient of a^2 is 1, the coefficient of $3a$ is 3, and the coefficient of the constant term -5 is -5.
Key phrases can be translated to algebraic expressions.	*5 more than x* can be expressed as $x + 5$. *25 less than twice y* can be expressed as $2y - 25$. One-half *of c* can be expressed as $\frac{1}{2}c$.
Number · value = total value	The total value (in cents) of n nickels is $n \cdot 5 = 5n$ cents.

To **evaluate algebraic expressions,** we substitute the values of its variables and use the order of operations rule.

Evaluate $\frac{x^2 - y^2}{x + y}$ for $x = 2$ and $y = -3$.

$$\frac{x^2 - y^2}{x + y} = \frac{2^2 - (-3)^2}{2 + (-3)}$$ Substitute 2 for x and −3 for y.

$$= \frac{4 - 9}{-1}$$ Evaluate: $2^2 = 4$ and $(-3)^2 = 9$. Add: $2 + (-3) = -1$.

$$= \frac{-5}{-1}$$ Subtract: $4 - 9 = -5$.

$$= 5$$ Divide −5 by −1.

REVIEW EXERCISES

99. How many terms does each expression have?

    a. $3x^2 + 2x - 5$ **b.** $-12xyz$

100. Identify the coefficient of each term of the given expression.

    a. $16x^2 - 5x + 25$ **b.** $\frac{x}{2} + y$

Write each phrase as an algebraic expression.

101. 25 more than the height h

102. 15 less than triple the cutoff score s

103. 6 less than one-half of the time

104. The absolute value of the difference of 2 and the square of a

105. Hardware. Let n represent the length of the nail in inches. Write an algebraic expression that represents the length of the bolt (in inches).

4 in.

106. Hardware. Let b represent the length of the bolt in inches. Write an algebraic expression that represents the length of the nail (in inches).

107. How many years are in d decades?

108. Five years after a house was constructed, a patio was added. How old, in years, is the patio if the house is x years old?

109. Complete the table below. The units are cents.

Coin	Number	Value	Total value
Nickel	6	5	
Dime	d	10	

110. Complete the table.

x	$20x - x^3$
0	
1	
-4	

Evaluate each algebraic expression for the given values of the variables.

111. $b^2 - 4ac$ for $b = -10$, $a = 3$, and $c = 5$

112. $\frac{x + y}{-x - z}$ for $x = 19$, and $y = 17$, and $z = -18$

SECTION 1.9 ▶ Simplifying Algebraic Expressions Using Properties of Real Numbers

DEFINITIONS AND CONCEPTS	EXAMPLES
We often use the *commutative property of multiplication* to reorder factors and the *associative property of multiplication* to regroup factors when **simplifying expressions.**	Simplify: $-5(3y) = (-5 \cdot 3)y = -15y$ $-5(3y)$ and $-15y$ are called equivalent expressions. $-45b\left(\frac{5}{9}\right) = \left(-45 \cdot \frac{5}{9}\right)b = -25b$
The **distributive property** can be used to *remove parentheses:* $a(b + c) = ab + ac$ $a(b - c) = ab - ac$ $a(b + c + d) = ab + ac + ad$	Multiply: $7(x + 3) = 7 \cdot x + 7 \cdot 3 = 7x + 21$ $-0.2(4m - 5n - 7) = -0.2(4m) - (-0.2)(5n) - (-0.2)(7)$ $= -0.8m + n + 1.4$
The **opposite of a sum:** $-(a + b) = -a + (-b)$	$-(3x + 4) = -3x - 4$ and $-(-2y - 12) = 2y + 12$

Like terms are terms with exactly the same variables raised to exactly the same powers.	$3x$ and $-5x$ are like terms. $-4t^3$ and $3t^2$ are unlike terms because the variable t has different exponents. $0.5xyz$ and $3.7xy$ are unlike terms because they have different variables.
Simplifying the sum or difference of like terms is called **combining like terms**. Like terms can be combined by adding or subtracting the coefficients of the terms and keeping the same variables with the same exponents.	Simplify: $4a + 2a = 6a$ *Think: (4 + 2)a = 6a.* $5p^2 + p - p^2 - 9p = 4p^2 - 8p$ *Think: (5 − 1)p^2 = 4p^2 and* $(1 - 9)p = -8p.$ $2(k - 1) - 3(k + 2) = 2k - 2 - 3k - 6 = -k - 8$

REVIEW EXERCISES

Use the given property to complete each statement.

113. $a \cdot 150 = $ _____   Commutative property of multiplication

114. $9 + (1 + 7y) = $ _____   Associative property of addition

115. $2.7(10b) = $ _____   Associative property of multiplication

116. $x + 2x^2 = $ _____   Commutative property of addition

Simplify each expression.

117. $-4(7w)$ **118.** $3(-2x)(-4)$

119. $0.4(5.2f)$ **120.** $\dfrac{7}{2} \cdot \dfrac{2}{7}r$

Use the distributive property to remove parentheses.

121. $5(x + 3)$ **122.** $-(2x + 3 - y)$

123. $\dfrac{3}{4}(4c - 8)$ **124.** $-2(-3c - 7)(2.1)$

Simplify each expression by combining like terms.

125. $8p + 5p - 4p$ **126.** $-5m + 2 - 2m - 2$

127. $n + n + n + n$ **128.** $5(p - 2) - 2(3p + 4)$

129. $55.7k^2 - 55.6k^2$

130. $8a^3 + 4a^3 + 2a - 4a^3 - 2a - 1$

131. $\dfrac{3}{5}w - \left(-\dfrac{2}{5}w\right)$ **132.** $36\left(\dfrac{1}{9}h - \dfrac{3}{4}\right) + 36\left(\dfrac{1}{3}\right)$

133. $-(7.6t - 1.9) + (1.4t - 1.2)8 + t$

134. Write an equivalent expression for the given expression using fewer symbols.

   a. $1x$ **b.** $-1x$

   c. $4x - (-1)$ **d.** $4x + (-1)$

1 ▶ CHAPTER TEST

1. Fill in the blanks.

  a. Two fractions, such as $\dfrac{1}{2}$ and $\dfrac{5}{10}$, that represent the same number are called _____ fractions.

  b. The result of a multiplication is called a _____.

  c. $\dfrac{8}{7}$ is the _____ of $\dfrac{7}{8}$ because $\dfrac{8}{7} \cdot \dfrac{7}{8} = 1$.

  d. $9x^2$ and $7x^2$ are _____ _____ because they have the same variable raised to exactly the same power.

  e. For any nonzero real number a, $\dfrac{a}{0}$ is _____.

2. Security Guards.

The graph shows the cost to hire a security guard.

  a. What will it cost to hire a security guard for 3 hours?

  b. If a school was billed $40 for hiring a security guard for a dance, for how long did the guard work?

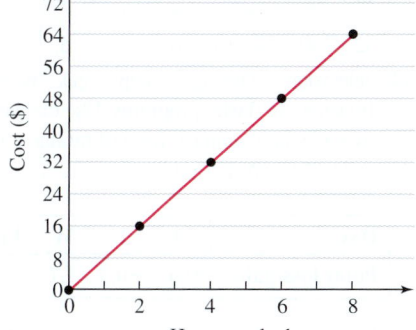

3. Use the formula $f = \frac{a}{5}$ to complete the table.

Square miles (a)	Fire stations (f)
15	
100	
350	

4. Give the prime factorization of 180.

5. Simplify: $\frac{42}{105}$

6. Divide: $\frac{15}{16} \div \frac{5}{8}$

7. Add: $\frac{7}{10} + \frac{1}{14}$

8. Subtract: $8\frac{2}{5} - 1\frac{2}{3}$

9. **Shopping.**

   a. What is the weight of the oranges? Answer using a mixed number.

   b. Find the cost of the oranges. Answer in dollars.

Oranges
84 cents a pound

10. Write $\frac{5}{6}$ as a decimal.

11. a. Graph the set of numbers on a number line.

   $$\left\{-1\frac{1}{4},\ 0,\ \sqrt{2},\ -3.75,\ 2,\ \frac{7}{2},\ 0.5,\ -3\right\}$$

   b. Determine which numbers in the set are natural numbers, whole numbers, integers, rational numbers, irrational numbers, and real numbers.

12. Determine whether each statement is true or false.

   a. Every integer is a rational number.

   b. Every rational number is an integer.

   c. π is an irrational number.

   d. 500 is a whole number.

   e. A real number is any number that is either a rational or an irrational number.

13. Insert the proper symbol, $>$ or $<$, in the blank.

   a. $-2 \quad\boxed{}\quad -3$

   b. $-|-9| \quad\boxed{}\quad 8$

   c. $|-4| \quad\boxed{}\quad -(-5)$

   d. $\left|-\frac{7}{8}\right| \quad\boxed{}\quad 0.5$

14. **Television.** During "sweeps week," networks try to gain viewers by showing flashy programs. Use the data to determine the average daily gain (or loss) of ratings points by a network for the 7-day "sweeps period."

Day	M	T	W	Th	F	Sa	Su
Point loss/gain	0.6	−0.3	1.7	1.5	−0.2	1.1	−0.2

Perform the operations.

15. $-5.6 + (-2)$

16. $(-6) + 8 + (-4)$

17. $-\frac{1}{2} + \frac{7}{8}$

18. a. $-10 - (-4)$

   b. Show a check of the result.

19. a. $\dfrac{-12.6}{-0.9}$

   b. Show a check of the result.

20. $(-2)(-3)(-5)$

21. $-6.1(0.4)$

22. $\dfrac{0}{-3}$

23. $\left(-\dfrac{3}{5}\right)^3$

24. $3 + (-3)$

25. $0 - 3$

26. $-30 + 50 - 10 - (-40)$

27. **Astronomy.** *Magnitude* is a term used in astronomy to describe the brightness of planets and stars. Negative magnitudes are associated with brighter objects. By how many magnitudes do a full moon and the sun differ?

Object	Magnitude
Sun	−26.5
Full moon	−12.5

28. **Glaciers.** In 2005, the South Cascade Glacier in Washington State gained approximately 87 inches in thickness because of snowfall. That same year it lost approximately 165 inches of thickness due to melting. Was there a net gain or loss in the glacier's thickness that year? How much? (Source: U.S. Geological Survey)

29. **Inventory.** It was discovered that fifteen MP3 players were missing from the stockroom of an electronics store. If the players cost $85 each, what signed number represents the store's financial loss?

30. Use the given property to complete each statement.

   a. $(-12 + 97) + 3 =$ _____
      Associative property of addition

   b. $2(x + 7) =$ _____
      Distributive property

   c. $-2(m)5 =$ _____
      Commutative property of multiplication

   d. $\frac{1}{8}(8) =$ ___
      Inverse property of multiplication

   e. $0 + 15x =$ _____
      Identity property of addition

31. Write each product using exponents:

   a. $9(9)(9)(9)(9)$

   b. $3 \cdot x \cdot x \cdot z \cdot z \cdot z$

32. Complete the table.

x	$2x - \frac{30}{x}$
5	
10	
−30	

Evaluate each expression.

33. $8 + 2 \cdot 3^4$

34. $\dfrac{3(40 - 2^3)}{-2(6 - 4)^2}$

35. $-10^2 - 5 + 6$

36. $9 - 3[45 - 5^2(1^5 - 4)]$

37. $|-50 \div 5 \cdot 2|$

38. Evaluate $3(10x - y) - 5(x + y^2)$ for $x = 2$ and $y = -5$.

39. Translate to an algebraic expression: seven less than twice the width w.

40. a. Music. A band recorded x songs for a CD. However, two of the songs were not included in the album because of poor sound quality. Write an algebraic expression that represents the number of songs on the CD.

   b. Money. Find the value of q quarters in cents.

41. How many terms are in the expression $4x^2 + 5x - 7$?

42. What is the coefficient of each term of $a^3 - 6a^2 - a + 10$?

Simplify each expression.

43. $5(-4x)$

44. $-8(-7t)(4)$

45. $\dfrac{4}{5}(15a + 5) - 16a$

46. $-1.1d^3 - 3.8d^3 - d^3$

47. $9x + 2(7x - 3) - 9(x - 1)$

48. $m^2 + 4m^2 + 5m - 2m^2 - 3m - 4$

49. Tell whether each statement is true or false.

   a. $-\dfrac{3}{5} = \dfrac{-3}{5} = \dfrac{3}{-5}$

   b. $-|-8| = -(-8)$

   c. $-5^2 = (-5)^2$

   d. $-(4a + 7) = -4a + 7$

   e. $(-9)^{14}$ is a positive number.

   f. The reciprocal of $-\dfrac{4}{7}$ is $\dfrac{7}{4}$.

Group Project

WRITING FRACTIONS AS DECIMALS

Overview: This is a good activity to try at the beginning of the course. You can become acquainted with other students in your class while you review the process for finding decimal equivalents of fractions.

Instructions: Form groups of 6 students. Select one person from your group to record the group's responses on the questionnaire. Express the results in fraction form and in decimal form.

What fraction (decimal) of the students in your group . . .	Fraction	Decimal
■ have the letter a in their first names?		
■ have a birthday in January or February?		
■ work full-time or part-time?		
■ have ever been on television?		
■ have downloaded music in the last week?		
■ log into Facebook at least once a day?		
■ send 30 or more text messages a day?		
■ have downloaded at least 10 applications to their cell phones?		

Equations, Inequalities, and Problem Solving

2

©barang/Shutterstock.com

from Campus to Careers

Automotive Service Technician

Anyone whose car has ever broken down appreciates the talents of automotive service technicians. To work on today's high-tech cars and trucks, a person needs strong diagnostic and problem-solving skills. Courses in automotive repair, electronics, physics, chemistry, English, computers, and mathematics provide a good educational background for a career as a service technician.

Problem 47 in **Study Set 2.3, problem 85** in **Study Set 2.4,** and **problem 54** in **Study Set 2.5** involve situations that an automotive service technician might encounter on the job. The mathematical concepts discussed in this chapter can be used to solve those problems.

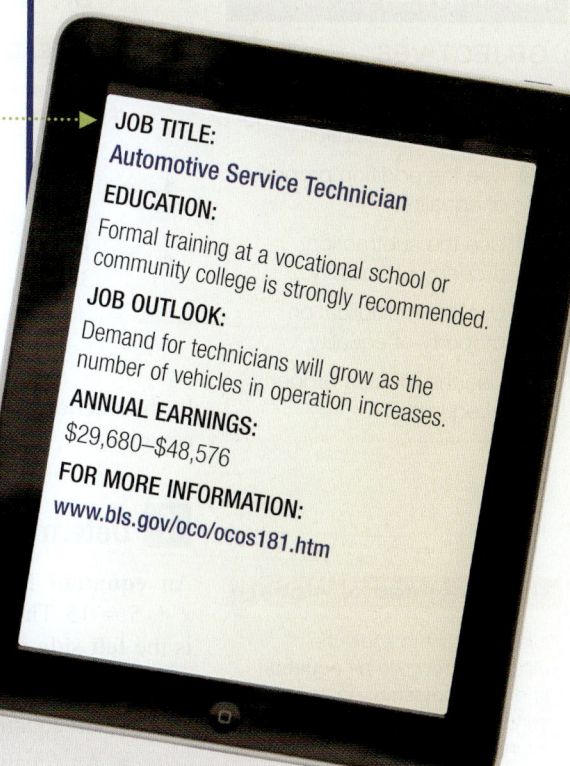

JOB TITLE:
Automotive Service Technician

EDUCATION:
Formal training at a vocational school or community college is strongly recommended.

JOB OUTLOOK:
Demand for technicians will grow as the number of vehicles in operation increases.

ANNUAL EARNINGS:
$29,680–$48,576

FOR MORE INFORMATION:
www.bls.gov/oco/ocos181.htm

Many students think that there are two types of people—those who are good at math and those who are not—and that this cannot be changed. This isn't true! Here are some suggestions that can increase your chances for success in algebra.

DISCOVER YOUR LEARNING STYLE: Are you a visual, verbal, or audio learner? Knowing this will help you determine how best to study.

GET THE MOST OUT OF THE TEXTBOOK: This book and the software that comes with it contain many student-support features. Are you taking advantage of them?

TAKE GOOD NOTES: Are your class notes complete so that they are helpful when doing your homework and studying for tests?

Now Try This ▶

1. To determine what type of learner you are, take the *Learning Style Survey* found online at http://www.metamath.com/multiple/multiple_choice_questions.html. Then, write a one-page paper explaining what you learned from the survey results and how you will use the information to help you succeed in the class.

2. To learn more about the student-support features of this book, take the *Textbook Tour* found online at www.cengage.com/math/tussy.

3. Rewrite a set of your class notes to make them more readable and to clarify the concepts and examples covered. If they are not already, write them in outline form. Fill in any information you didn't have time to copy down in class and complete any phrases or sentence fragments.

SECTION 2.1

Solving Equations Using Properties of Equality

OBJECTIVES

1. Determine whether a number is a solution.
2. Use the addition property of equality.
3. Use the subtraction property of equality.
4. Use the multiplication property of equality.
5. Use the division property of equality.

ARE YOU READY?

The following problems review some basic skills that are needed when solving equations. Fill in the blanks.

1. $8 \boxed{} 8 = 0$

2. $-1.6 \boxed{} 1.6 = 0$

3. $\dfrac{7}{\boxed{}} = 1$

4. $3 \boxed{} \dfrac{1}{3} = 1$

5. $\dfrac{2}{5} - \boxed{} = 0$

6. $-\dfrac{9}{8}\left(\boxed{} \right) = 1$

In this section, we introduce four fundamental properties of equality that are used to solve equations.

1 Determine Whether a Number Is a Solution.

The Language of Algebra

It is important to know the difference between an **equation** and an **expression**. An equation contains an = symbol and an expression does not.

An **equation** is a statement indicating that two expressions are equal. An example is $x + 5 = 15$. The equal symbol = separates the equation into two parts: The expression $x + 5$ is the **left side** and 15 is the **right side.** The letter x is the **variable** (or the **unknown**). The sides of an equation can be reversed, so we can write $x + 5 = 15$ or $15 = x + 5$.

- An equation can be true: $6 + 3 = 9$
- An equation can be false: $2 + 4 = 7$
- An equation can be neither true nor false. For example, $x + 5 = 15$ is neither true nor false because we don't know what number x represents.

An equation that contains a variable is made true or false by substituting a number for the variable. If we substitute 10 for x in $x + 5 = 15$, the resulting equation is true: $10 + 5 = 15$. If we substitute 1 for x, the resulting equation is false: $1 + 5 = 15$. A number that makes an equation true when substituted for the variable is called a **solution** and it is said to **satisfy** the equation. Therefore, 10 is a solution of $x + 5 = 15$, and 1 is not. The **solution set** of an equation is the set of all numbers that make the equation true.

EXAMPLE 1   Check to determine whether 9 is a solution of $3y - 1 = 2y + 7$.

Strategy   We will substitute 9 for each y in the equation and evaluate the expression on the left side and the expression on the right side separately.

Why   If a true statement results, 9 is a solution of the equation. If we obtain a false statement, 9 is not a solution.

Solution

The Language of Algebra

Read $\overset{?}{=}$ as "is possibly equal to."

Evaluate the expression on the left side.

$$3y - 1 = 2y + 7$$
$$3(9) - 1 \overset{?}{=} 2(9) + 7$$
$$27 - 1 \overset{?}{=} 18 + 7$$
$$26 = 25$$

Evaluate the expression on the right side.

Since $26 = 25$ is false, 9 is not a solution of $3y - 1 = 2y + 7$.

Self Check 1   Check to determine whether 25 is a solution of $10 - x = 35 - 2x$.

Now Try ▶ Problem 19

2 Use the Addition Property of Equality.

To **solve an equation** means to find all values of the variable that make the equation true. We can develop an understanding of how to solve equations by referring to the scales shown on the right.

The first scale represents the equation $x - 2 = 3$. The scale is in balance because the weights on the left side and right side are equal. To find x, we must add 2 to the left side. To keep the scale in balance, we must also add 2 to the right side. After doing this, we see that x is balanced by 5. Therefore, x must be 5. We say that we have solved the equation $x - 2 = 3$ and that the solution is 5.

In this example, we solved $x - 2 = 3$ by transforming it to a simpler *equivalent equation*, $x = 5$.

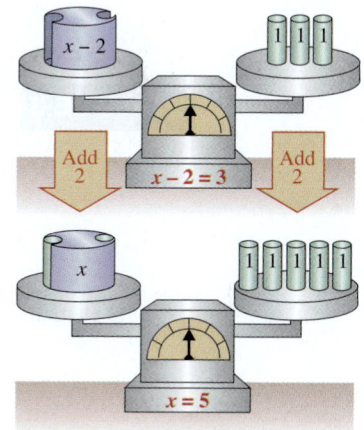

Equivalent Equations	Equations with the same solutions are called **equivalent equations**.

The procedure that we used above suggests the following property of equality.

Addition Property of Equality	Adding the same number to both sides of an equation does not change its solution. For any real numbers a, b, and c, if $a = b$, then $a + c = b + c$

When we use this property, the resulting equation is *equivalent to the original one.* We will now show how it is used to solve $x - 2 = 3$ algebraically.

EXAMPLE 2   Solve: $x - 2 = 3$

Strategy   We will use a property of equality to isolate the variable on one side of the equation.

Why   To solve the original equation, we want to find a simpler equivalent equation of the form **$x = $ a number**, whose solution is obvious.

Solution

The Language of Algebra

We solve equations by writing a series of steps that result in an equivalent equation of the form

$x = $ *a number*

or

a number $= x$

We say the variable is **isolated** on one side of the equation. *Isolated* means alone or by itself.

$x - 2 = 3$	This is the equation to solve.
$x - 2 + 2 = 3 + 2$	Use the addition property of equality to isolate x on the left side of the equation. Undo the subtraction of 2 by adding 2 to both sides.
$x + 0 = 5$	The sum of a number and its opposite is zero: $-2 + 2 = 0$.
$x = 5$	When 0 is added to a number, the result is the same number.

Since 5 is obviously the solution of the equivalent equation $x = 5$, the solution of the original equation, $x - 2 = 3$, is also 5. To check this result, we substitute 5 for x in the original equation and simplify.

$$x - 2 = 3$$
$$5 - 2 \stackrel{?}{=} 3 \qquad \text{Substitute 5 for x.}$$
$$3 = 3 \qquad \text{True}$$

Since the resulting statement is true, 5 is the solution of $x - 2 = 3$. A more formal way to present this result is to write the solution within braces as a solution set: $\{5\}$.

Self Check 2   Solve: $n - 16 = 33$

Now Try ▶ Problem 37

EXAMPLE 3   Solve: **a.** $-19 = y - 7$ **b.** $-27 + y = -3$

Strategy   We will use a property of equality to isolate the variable on one side of the equation.

Why   To solve the original equation, we want to find a simpler equivalent equation of the form **$y = $ a number** or **a number $= y$**, whose solution is obvious.

Solution   **a.**

$-19 = y - 7$	This is the equation to solve.
$-19 + 7 = y - 7 + 7$	To isolate y on the right side, use the addition property of equality. Undo the subtraction of 7 by adding 7 to both sides.
$-12 = y$	The sum of a number and its opposite is zero: $-7 + 7 = 0$.

Notation

We may solve an equation so that the variable is isolated on either side of the equation. Note that $-12 = y$ is equivalent to $y = -12$.

Check: $-19 = y - 7$	This is the original equation.
$-19 \stackrel{?}{=} -12 - 7$	Substitute -12 for y.
$-19 = -19$	True

Since the resulting statement is true, the solution is -12. The solution set is $\{-12\}$.

b.

$-27 + y = -3$	The equation to solve.
$-27 + y + 27 = -3 + 27$	To isolate y, use the addition property of equality. Eliminate -27 on the left side by adding its opposite (additive inverse) to both sides.
$y = 24$	The sum of a number and its opposite is zero: $-27 + 27 = 0$.

Check: $-27 + y = -3$ This is the original equation.

$\qquad -27 + \mathbf{24} \stackrel{?}{=} -3$ Substitute 24 for y.

$\qquad\qquad -3 = -3$ True

The solution is 24. The solution set is $\{24\}$.

Self Check 3 Solve: **a.** $-5 = b - 38$ **b.** $-20 + n = 29$

Now Try ▶ Problems 41 and 43

3 Use the Subtraction Property of Equality.

Since any subtraction can be written as an addition by adding the opposite of the number to be subtracted, the following property is an extension of the addition property of equality.

Subtraction Property of Equality	Subtracting the same number from both sides of an equation does not change its solution. For any real numbers a, b, and c, $\qquad$ if $a = b$, then $\quad a - c = b - c$

When we use this property, the resulting equation is equivalent to the original one.

EXAMPLE 4 Solve: **a.** $x + \dfrac{1}{8} = \dfrac{7}{4}$ **b.** $54.9 + x = 45.2$

Strategy We will use a property of equality to isolate the variable on one side of the equation.

Why To solve the original equation, we want to find a simpler equivalent equation of the form $x = $ **a number**, whose solution is obvious.

Solution

a. $\qquad x + \dfrac{1}{8} = \dfrac{7}{4}$ This is the equation to solve.

$x + \dfrac{1}{8} - \dfrac{\mathbf{1}}{\mathbf{8}} = \dfrac{7}{4} - \dfrac{\mathbf{1}}{\mathbf{8}}$ To isolate x, use the subtraction property of equality. Undo the addition of $\frac{1}{8}$ by subtracting $\frac{1}{8}$ from both sides.

$\qquad\quad x = \dfrac{7}{4} - \dfrac{1}{8}$ On the left side, $\frac{1}{8} - \frac{1}{8} = 0$.

$\qquad\quad x = \dfrac{7}{4} \cdot \dfrac{\mathbf{2}}{\mathbf{2}} - \dfrac{1}{8}$ To prepare to subtract the fractions, build $\frac{7}{4}$ so that it has a denominator of 8.

$\qquad\quad x = \dfrac{14}{8} - \dfrac{1}{8}$ Multiply the numerators and multiply the denominators.

$\qquad\quad x = \dfrac{13}{8}$ Subtract the numerators. Write the result over the common denominator 8. The fraction is in simplest form.

Verify that $\frac{13}{8}$ is the solution by substituting it for x in the original equation and simplifying.

b. $54.9 + x = 45.2$ This is the equation to solve.

$54.9 + x - \mathbf{54.9} = 45.2 - \mathbf{54.9}$ To isolate x, use the subtraction property of equality. Undo the addition of 54.9 by subtracting 54.9 from both sides.

$x = -9.7$ On the left side, $54.9 - 54.9 = 0$.

Check: $54.9 + x = 45.2$ This is the original equation.

$54.9 + (\mathbf{-9.7}) \overset{?}{=} 45.2$ Substitute −9.7 for x.

$45.2 = 45.2$ True

The solution is -9.7. The solution set is $\{-9.7\}$.

> **Self Check 4**   Solve: **a.** $x + \frac{4}{15} = \frac{11}{5}$ **b.** $0.7 + a = 0.2$
>
> **Now Try** ▶ Problems 45 and 47

4 Use the Multiplication Property of Equality.

To develop another property of equality, consider the first scale shown on the right that represents the equation $\frac{x}{3} = 25$. The scale is in balance because the weights on the left side and right side are equal. To find x, we must triple (multiply by 3) the weight on the left side. To keep the scale in balance, we also must triple the weight on the right side. After doing this, we see in the second illustration that x is balanced by 75. Therefore, x must be 75.

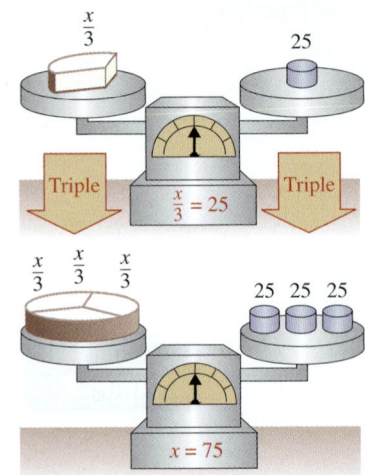

The procedure that we used to keep the scale balanced suggests the following property of equality.

Multiplication Property of Equality	Multiplying both sides of an equation by the same nonzero number does not change its solution.
	For any real numbers a, b, and c, where c is not 0,
	if $a = b$, then $ca = cb$

When we use this property, the resulting equation is equivalent to the original one. We will now show how it is used to solve $\frac{x}{3} = 25$ algebraically.

EXAMPLE 5   Solve: $\dfrac{x}{3} = 25$

Strategy   We will use a property of equality to isolate the variable on one side of the equation.

Why   To solve the original equation, we want to find a simpler equivalent equation of the form *x = a number*, whose solution is obvious.

Solution

$$\frac{x}{3} = 25 \qquad \text{This is the equation to solve.}$$

$$3 \cdot \frac{x}{3} = 3 \cdot 25 \qquad \text{To isolate } x, \text{ use the multiplication property of equality.}$$
Undo the division by 3 by multiplying both sides by 3.

$$\frac{3x}{3} = 75 \qquad \text{Do the multiplication.}$$

$$1x = 75 \qquad \text{Simplify } \frac{3x}{3} \text{ by removing the common factor of 3}$$
in the numerator and denominator: $\frac{3}{3} = 1$.

$$x = 75 \qquad \text{The coefficient 1 need not be written since } 1x = x.$$

If we substitute 75 for x in $\frac{x}{3} = 25$, we obtain the true statement $25 = 25$. This verifies that 75 is the solution. The solution set is $\{75\}$.

Self Check 5 Solve: $\frac{b}{24} = 3$

Now Try ▶ Problem 49

Since the product of a number and its reciprocal (or multiplicative inverse) is 1, we can solve equations such as $\frac{2}{3}x = 6$, where the coefficient of the variable term is a fraction, as follows.

EXAMPLE 6 Solve: **a.** $\frac{2}{3}x = 6$ **b.** $-\frac{5x}{4} = \frac{3}{16}$

Strategy We will use a property of equality to isolate the variable on one side of the equation.

Why To solve the original equation, we want to find a simpler equivalent equation of the form **x = a number**, whose solution is obvious.

Solution **a.** Since the coefficient of x is $\frac{2}{3}$, we can isolate x by multiplying both sides of the equation by the reciprocal of $\frac{2}{3}$, which is $\frac{3}{2}$.

$$\frac{2}{3}x = 6 \qquad \text{This is the equation to solve.}$$

$$\frac{3}{2} \cdot \frac{2}{3}x = \frac{3}{2} \cdot 6 \qquad \text{To undo the multiplication by } \frac{2}{3}, \text{ multiply both sides by the reciprocal of } \frac{2}{3}.$$

$$\left(\frac{3}{2} \cdot \frac{2}{3}\right)x = \frac{3}{2} \cdot 6 \qquad \text{Use the associative property of multiplication to group } \frac{3}{2} \text{ and } \frac{2}{3}.$$

$$1x = 9 \qquad \text{On the left side, } \frac{3}{2} \cdot \frac{2}{3} = 1. \text{ On the right side, } \frac{3}{2} \cdot 6 = \frac{18}{2} = 9.$$

$$x = 9 \qquad \text{The coefficient 1 need not be written since } 1x = x.$$

Check: $\frac{2}{3}x = 6 \qquad$ This is the original equation.

$$\frac{2}{3}(9) \stackrel{?}{=} 6 \qquad \text{Substitute 9 for } x \text{ in the original equation.}$$

$$6 = 6 \qquad \text{On the left side, } \frac{2}{3}(9) = \frac{18}{3} = 6.$$

Since the resulting statement is true, 9 is the solution. The solution set is $\{9\}$.

b.

$$-\frac{5x}{4} = \frac{3}{16}$$ — This is the equation to solve.

$$-\frac{5}{4}x = \frac{3}{16}$$ — Write $-\frac{5x}{4}$ as $-\frac{5}{4}x$.

$$-\frac{4}{5}\left(-\frac{5}{4}x\right) = -\frac{4}{5}\left(\frac{3}{16}\right)$$ — To isolate x, undo the multiplication by $-\frac{5}{4}$ by multiplying both sides by the reciprocal of $-\frac{5}{4}$.

$$1x = -\left(\frac{4 \cdot 3}{5 \cdot 16}\right)$$ — On the left side, $-\frac{4}{5}\left(-\frac{5}{4}\right) = 1$. On the right side, multiply the numerators and multiply the denominators. Since the signs of the fractions are unlike, make the final answer negative.

$$x = -\frac{\overset{1}{\cancel{4}} \cdot 3}{5 \cdot \underset{1}{\cancel{4}} \cdot 4}$$ — On the left side, the coefficient 1 need not be written since $1x = x$. On the right side, simplify the fraction by factoring 16 as $4 \cdot 4$ and removing the common factor 4.

$$x = -\frac{3}{20}$$ — Multiply the remaining factors in the numerator and in the denominator.

Verify that $-\frac{3}{20}$ is the solution by checking.

Self Check 6 Solve: **a.** $\frac{7}{2}x = 21$ **b.** $-\frac{3b}{8} = \frac{11}{56}$

Now Try ▶ Problems 53 and 55

5 Use the Division Property of Equality.

Since any division can be rewritten as a multiplication by multiplying by the reciprocal, the following property is a natural extension of the multiplication property of equality.

Division Property of Equality

Dividing both sides of an equation by the same nonzero number does not change its solution.

For any real numbers a, b, and c, where c is not 0,

$$\text{if } a = b, \text{ then } \frac{a}{c} = \frac{b}{c}$$

When we use this property, the resulting equation is equivalent to the original one.

EXAMPLE 7 Solve: **a.** $2t = 80$ **b.** $-6.02 = -8.6t$

Strategy We will use a property of equality to isolate the variable on one side of the equation.

Why To solve the original equation, we want to find a simpler equivalent equation of the form $t = \textbf{a number}$ or $\textbf{a number} = t$, whose solution is obvious.

Solution

a. $2t = 80$ This is the equation to solve.

$$\dfrac{2t}{2} = \dfrac{80}{2}$$ To isolate t on the left side, use the division property of equality.
Undo the multiplication by 2 by dividing both sides of the equation by 2.

$1t = 40$ Simplify $\frac{2t}{2}$ by removing the common factor of 2
in the numerator and denominator: $\frac{2}{2} = 1$

$t = 40$ The product of 1 and any number is that number: $1t = t$.

If we substitute 40 for t in $2t = 80$, we obtain the true statement $80 = 80$. This verifies that 40 is the solution. The solution set is $\{40\}$.

b. $-6.02 = -8.6t$ This is the equation to solve.

$$\dfrac{-6.02}{-8.6} = \dfrac{-8.6t}{-8.6}$$ To isolate t on the right side, use the division property of equality. Undo the multiplication by -8.6 by dividing both sides by -8.6.

$0.7 = t$ Do the division: $8.6\overline{)6.02}$. The quotient of two negative numbers is positive.

The solution is 0.7. Verify that this is correct by checking.

Success Tip

Since division by 2 is the same as multiplication by $\frac{1}{2}$, we also can solve $2t = 80$ using the multiplication property of equality. We can isolate t by multiplying both sides by the *multiplicative inverse* of 2, which is $\frac{1}{2}$:

$$\tfrac{1}{2} \cdot 2t = \tfrac{1}{2} \cdot 80$$

Success Tip

It is usually easier to multiply on each side if the coefficient of the variable term is a *fraction,* and divide on each side if the coefficient is an *integer* or *decimal.*

Self Check 7   Solve: **a.** $16x = 176$ **b.** $10.04 = -0.4r$

Now Try ▶ Problems 57 and 59

EXAMPLE 8   Solve: $-x = 3$

Strategy   The variable x is not isolated, because there is a $-$ sign in front of it. Since the term $-x$ has an understood coefficient of -1, the equation can be written as $-1x = 3$. We need to select a property of equality and use it to isolate the variable on one side of the equation.

Why   To find the solution of the original equation, we want to find a simpler equivalent equation of the form **x = a number**, whose solution is obvious.

Solution   To isolate x, we can either multiply or divide both sides by -1.

Multiply both sides by -1:		*Divide both sides by* -1:	
$-x = 3$	The equation to solve	$-x = 3$	The equation to solve
$-1x = 3$	Write: $-x = -1x$	$-1x = 3$	Write: $-x = -1x$
$(-1)(-1x) = (-1)3$		$\dfrac{-1x}{-1} = \dfrac{3}{-1}$	
$1x = -3$		$1x = -3$	On the left side, $\frac{-1}{-1} = 1$.
$x = -3$ $1x = x$		$x = -3$ $1x = x$	

Either way, we get the same result, -3.

Check: $-x = 3$ This is the original equation.

$-(-3) \overset{?}{=} 3$ Substitute -3 for x.

$3 = 3$ On the left side, the opposite of -3 is 3.

Since the resulting statement is true, -3 is the solution. The solution set is $\{-3\}$.

Self Check 8   Solve: $-h = -12$

Now Try ▶ Problem 61

SECTION 2.1 ▶ STUDY SET

VOCABULARY

Fill in the blanks.

1. A statement indicating that two expressions are equal, such as $x + 1 = 7$, is called an _____.

2. Any number that makes an equation true when substituted for the variable is said to _____ the equation. Such numbers are called _____.

3. To _____ an equation means to find all values of the variable that make the equation true.

4. To solve an equation, we _____ the variable on one side of the equation.

5. Equations with the same solutions are called _____ equations.

6. To _____ the solution of an equation, we substitute the value for the variable in the original equation and determine whether the result is a true statement.

CONCEPTS

7. Given $x + 6 = 12$,
   a. What is the left side of the equation?
   b. Is this equation true, false, or neither?
   c. Is 5 the solution?
   d. Does 6 satisfy the equation?

8. For each equation, determine what operation is performed on the variable. Then explain how to undo that operation to isolate the variable.
   a. $x - 8 = 24$
   b. $x + 8 = 24$
   c. $\dfrac{x}{8} = 24$
   d. $8x = 24$

9. Complete the following properties of equality. If $a = b$, then
   a. $a + c = b + \boxed{}$ and $a - c = b - \boxed{}$
   b. $ca = \boxed{}\, b$ and $\dfrac{a}{c} = \dfrac{b}{\boxed{}}$ (where $c \neq 0$)

10. a. To solve $\dfrac{h}{10} = 20$, do we multiply both sides of the equation by 10 or 20?
    b. To solve $4k = 16$, do we subtract 4 from both sides of the equation or divide both sides by 4?

11. Simplify each expression.
    a. $x + 7 - 7$
    b. $y - 2 + 2$
    c. $\dfrac{5t}{5}$
    d. $6 \cdot \dfrac{h}{6}$

12. a. To solve $-\dfrac{4}{5}x = 8$, we can multiply both sides by the reciprocal of $-\dfrac{4}{5}$. What is the reciprocal of $-\dfrac{4}{5}$?
    b. What is $-\dfrac{5}{4}\left(-\dfrac{4}{5}\right)$?

NOTATION

Complete each solution to solve the equation.

13.
$$x - 5 = 45$$
$$x - 5 + \boxed{} = 45 + \boxed{}$$
$$x = \boxed{}$$

   Check:
$$x - 5 = 45$$
$$\boxed{} - 5 \boxed{} 45$$
$$\boxed{} = 45 \quad \text{True}$$
   $\boxed{}$ is the solution.

14.
$$8x = 40$$
$$\dfrac{8x}{\boxed{}} = \dfrac{40}{\boxed{}}$$
$$x = \boxed{}$$

   Check:
$$8x = 40$$
$$8(\ \ ) \stackrel{?}{=} 40$$
$$\boxed{} = 40 \quad \text{True}$$
   $\boxed{}$ is the solution.

15. a. What does the symbol $\stackrel{?}{=}$ mean?
    b. If you solve an equation and obtain $50 = x$, can you write $x = 50$?

16. Fill in the blank: $-x = \boxed{}\, x$

GUIDED PRACTICE

Check to determine whether the number in red is a solution of the equation. **See Example 1.**

17. **6**, $x + 12 = 28$
18. **110**, $x - 50 = 60$
19. **−8**, $2b + 3 = -15$
20. **−2**, $5t - 4 = -16$
21. **5**, $0.5x = 2.9$
22. **3.5**, $1.2 + x = 4.7$
23. **−6**, $33 - \dfrac{x}{2} = 30$
24. **−8**, $\dfrac{x}{4} + 98 = 100$
25. **−2**, $|c - 8| = 10$
26. **−45**, $|30 - r| = 15$
27. **12**, $3x - 2 = 4x - 5$
28. **5**, $5y + 8 = 3y - 2$
29. **−3**, $x^2 - x - 6 = 0$
30. **−2**, $y^2 + 5y - 3 = 0$
31. **1**, $\dfrac{2}{a+1} + 5 = \dfrac{12}{a+1}$
32. **4**, $\dfrac{2t}{t-2} - \dfrac{4}{t-2} = 1$
33. **$\dfrac{3}{4}$**, $x - \dfrac{1}{8} = \dfrac{5}{8}$
34. **$\dfrac{7}{3}$**, $-4 = a + \dfrac{5}{3}$
35. **−3**, $(x - 4)(x + 3) = 0$
36. **5**, $(2x + 1)(x - 5) = 0$

Use a property of equality to solve each equation. Then check the result. **See Example 2.**

37. $a - 5 = 66$
38. $x - 34 = 19$
39. $9 = p - 9$
40. $3 = j - 88$

Use a property of equality to solve each equation. Then check the result. **See Example 3.**

41. $-16 = y - 4$
42. $-23 = y - 19$
43. $-3 + a = 0$
44. $-1 + m = 0$

Use a property of equality to solve each equation. Then check the result. **See Example 4.**

45. $x + \dfrac{1}{10} = \dfrac{6}{5}$
46. $x + \dfrac{1}{14} = \dfrac{9}{7}$
47. $3.5 + f = 1.2$
48. $9.4 + h = 8.1$

Use a property of equality to solve each equation. Then check the result. See Example 5.

49. $\dfrac{x}{15} = 3$ **50.** $\dfrac{y}{7} = 12$

51. $\dfrac{d}{8} = -6$ **52.** $\dfrac{n}{9} = -2$

Use a property of equality to solve each equation. Then check the result. See Example 6.

53. $\dfrac{4}{5}t = 16$ **54.** $\dfrac{11}{15}y = 22$

55. $-\dfrac{7r}{2} = \dfrac{5}{12}$ **56.** $-\dfrac{4a}{5} = \dfrac{11}{15}$

Use a property of equality to solve each equation. Then check the result. See Example 7.

57. $4x = 16$ **58.** $5y = 45$

59. $-1.7 = -3.4y$ **60.** $-1.26 = -2.1x$

Use a property of equality to solve each equation. Then check the result. See Example 8.

61. $-x = 18$ **62.** $-y = 50$

63. $-n = \dfrac{4}{21}$ **64.** $-w = \dfrac{11}{16}$

TRY IT YOURSELF

Solve each equation. Then check the result.

65. $63 = 9c$ **66.** $40 = 5t$

67. $d - \dfrac{1}{9} = \dfrac{7}{9}$ **68.** $\dfrac{7}{15} = b - \dfrac{1}{15}$

69. $0 = \dfrac{v}{11}$ **70.** $\dfrac{d}{49} = 0$

71. $x - 1.6 = -2.5$ **72.** $y - 1.2 = -1.3$

73. $\dfrac{2}{3}c = 10$ **74.** $\dfrac{9}{7}d = 81$

75. $-100 = -5g$ **76.** $-80 = -5w$

77. $s + \dfrac{1}{5} = \dfrac{4}{25}$ **78.** $\dfrac{1}{6} = h + \dfrac{4}{3}$

79. $\dfrac{d}{-7} = -3$ **80.** $\dfrac{c}{-2} = -11$

81. $8h = 0$ **82.** $9a = 0$

83. $\dfrac{y}{0.6} = -4.4$ **84.** $\dfrac{y}{0.8} = -2.9$

85. $23b = 23$ **86.** $16 = 16h$

87. $-\dfrac{5}{4}h = -5$ **88.** $-\dfrac{3}{8}t = -3$

89. $8.9 = -4.1 + t$ **90.** $7.7 = -3.2 + s$

91. $-2.5 = -m$ **92.** $-1.8 = -b$

93. $-\dfrac{9}{8}x = 3$ **94.** $-\dfrac{14}{3}c = 7$

95. $\dfrac{2}{3}n = -\dfrac{7}{8}$ **96.** $\dfrac{4}{9}m = -\dfrac{3}{5}$

97. $-10 = n - 5$ **98.** $-8 = t - 2$

99. $\dfrac{h}{-40} = 5$ **100.** $\dfrac{x}{-7} = 12$

101. $-\dfrac{15}{16}a = -\dfrac{5}{4}$ **102.** $-\dfrac{20}{27}b = -\dfrac{4}{9}$

103. $-15x = -60$ **104.** $-14x = -84$

Look Alikes . . .

105. a. $d + \dfrac{1}{10} = \dfrac{3}{4}$ **b.** $d - \dfrac{1}{10} = \dfrac{3}{4}$

   c. $\dfrac{1}{10}d = \dfrac{3}{4}$ **d.** $10d = \dfrac{3}{4}$

106. a. $x + 4.2 = -18.9$ **b.** $x - 4.2 = -18.9$

   c. $4.2x = -18.9$ **d.** $\dfrac{x}{4.2} = -18.9$

APPLICATIONS

107. Synthesizers. To find the unknown angle measure, which is represented by x, solve the equation $x + 115 = 180$.

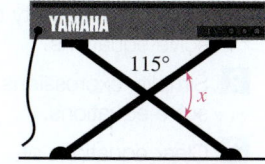

108. Stop Signs. To find the degree measure of one angle of the stop sign, which is represented by x, solve the equation $8x = 1{,}080$.

109. Sharing the Winning Ticket. When a Florida Lotto Jackpot was won by a group of 16 nurses employed at a Southwest Florida Medical Center, each received $375,000. To find the amount of the jackpot, which is represented by x, solve the equation $\dfrac{x}{16} = 375{,}000$.

110. Social Networks. In 2009, the annual revenue generated per employee at Twitter was $142,857. This is $777,143 less than the annual revenue generated per employee at Facebook. To find the annual revenue generated per employee at Facebook, which is represented by x, solve the equation $x - 777{,}143 = 142{,}857$. (Source: businessinsider.com)

WRITING

111. What does it mean to solve an equation?

112. Explain the error in the following work.

   Solve: $-6.4x = 1.1$
   $-6.4x + 6.4 = 1.1 + 6.4$
   $x = 7.5$

113. Explain the error in the following work.

   Solve: $x + 2 = 40$
   $x + 2 - 2 = 40$
   $x = 40$

114. After solving an equation, how do we check the result?

REVIEW

115. Evaluate $-9 - 3x$ for $x = -3$.

116. Evaluate: $-5^2 + (-5)^2$

117. Translate to symbols: Subtract x from 45

118. Evaluate: $\dfrac{2^3 + 3(5 - 3)}{15 - 4 \cdot 2}$

CHALLENGE PROBLEMS

119. If $a + 81 = 49$, what is $a - 81$?

120. Find two solutions of $\dfrac{1}{4}|x + 1| = 25$.

SECTION 2.2

More about Solving Equations

OBJECTIVES

1 Use more than one property of equality to solve equations.

2 Simplify expressions to solve equations.

3 Clear equations of fractions and decimals.

4 Identify identities and contradictions.

ARE YOU READY?

The following problems review some basic skills that are needed when solving equations.

1. Simplify: $4x - 12 - 4x$

2. Simplify: $2a + 2 - 2$

3. Simplify: $5m - 3(4m - 6)$

4. Multiply: $5\left(\dfrac{3}{5}x\right)$

5. Multiply: $18\left(\dfrac{4}{3}n\right)$

6. Multiply: $100 \cdot 0.08$

We have solved simple equations by using properties of equality. We now will expand our equation-solving skills by considering more complicated equations. We want to develop a general strategy that can be used to solve any kind of *linear equation in one variable*.

Linear Equation in One Variable

A **linear equation in one variable** can be written in the form

$$ax + b = c$$

where a, b, and c are real numbers and $a \neq 0$.   Read $\neq$ as "is not equal to 0."

Some examples of linear and nonlinear equations in one variable are shown below.

Linear equations in one variable (x):

$3x + 1 = 4$ Think: $3x^1 + 1 = 4$

$-\dfrac{5}{3}x - 7 = 0$ Think: $-\frac{5}{3}x^1 - 7 = 0$

Not linear equations in one variable:

$x^2 - x - 6 = 0$ The exponent on x is not 1.

$\dfrac{1}{2} + \dfrac{3}{x} = \dfrac{5}{6}$ x is in the denominator.

1   Use More Than One Property of Equality to Solve Equations.

Sometimes we must use several properties of equality to solve an equation. For example, on the left side of $2x + 6 = 10$, the variable x is multiplied by 2, and then 6 is added to that product. To isolate x, we use the order of operations rule in reverse. First, we undo the addition of 6, and then we undo the multiplication by 2.

$2x + 6 = 10$ This is the equation to solve.

$2x + 6 - 6 = 10 - 6$ To undo the addition of 6, subtract 6 from both sides.

$2x = 4$ Do the subtraction.

$\dfrac{2x}{2} = \dfrac{4}{2}$ To undo the multiplication by 2, divide both sides by 2.

$x = 2$ Do the division.

The solution is 2.

Success Tip

Recall that:
- Subtraction undoes addition.
- Addition undoes subtraction.
- Division undoes multiplication.
- Multiplication undoes division.

EXAMPLE 1 Solve: $-12x + 5 = 17$

Strategy First, we will use a property of equality to isolate the *variable term* on one side of the equation. Then we will use a second property of equality to isolate the *variable* itself.

Why To solve the original equation, we want to find a simpler equivalent equation of the form **x = a number**, whose solution is obvious.

Solution On the left side of the equation, x is multiplied by -12, and then 5 is added to that product. To isolate x, we undo the operations in the opposite order.

> **The Language of Algebra**
>
> We subtract 5 from both sides to isolate the **variable term**, $-12x$. Then we divide both sides by -12 to isolate the **variable**, x.

- To isolate the variable term, $-12x$, we subtract 5 from both sides to undo the addition of 5.
- To isolate the variable, x, we divide both sides by -12 to undo the multiplication by -12.

$$-12x + 5 = 17 \qquad \text{This is the equation to solve.}$$

$$-12x + 5 - 5 = 17 - 5 \qquad \begin{array}{l}\text{Use the subtraction property of equality:}\\ \text{Subtract 5 from both sides to undo the}\\ \text{addition and isolate the variable term, } -12x.\end{array}$$

$$-12x = 12 \qquad \text{Do the subtractions: } 5 - 5 = 0 \text{ and } 17 - 5 = 12.$$

$$\frac{-12x}{-12} = \frac{12}{-12} \qquad \begin{array}{l}\text{Use the division property of equality: Divide both sides by } -12\\ \text{to undo the multiplication and isolate the variable } x.\end{array}$$

$$x = -1 \qquad \text{Do the division.}$$

> **Caution**
>
> When checking solutions, always use the original equation.

Check:
$$-12x + 5 = 17 \qquad \text{This is the original equation.}$$
$$-12(-1) + 5 \overset{?}{=} 17 \qquad \text{Substitute } -1 \text{ for } x.$$
$$12 + 5 \overset{?}{=} 17 \qquad \text{Do the multiplication on the left side.}$$
$$17 = 17 \qquad \text{True}$$

Since the resulting statement is true, the solution is -1. The solution set is $\{-1\}$.

Self Check 1 Solve: $8x - 13 = 43$

Now Try ▶ Problem 13

EXAMPLE 2 Solve: $\dfrac{5}{8}m - 2 = -12$

Strategy We will use properties of equality to isolate the variable on one side of the equation.

Why To solve the original equation, we want to find a simpler equivalent equation of the form **m = a number**, whose solution is obvious.

Solution We note that the coefficient of m is $\frac{5}{8}$ and proceed as follows.

- To isolate the variable term, $\frac{5}{8}m$, we add 2 to both sides to undo the subtraction of 2.
- To isolate the variable, m, we multiply both sides by $\frac{8}{5}$ to undo the multiplication by $\frac{5}{8}$.

$$\frac{5}{8}m - 2 = -12$$

This is the equation to solve.

$$\frac{5}{8}m - 2 + 2 = -12 + 2$$

Use the addition property of equality: Add 2 to both sides to undo the subtraction and isolate the variable term, $\frac{5}{8}m$.

$$\frac{5}{8}m = -10$$

Do the additions: $-2 + 2 = 0$ and $-12 + 2 = -10$.

$$\frac{8}{5}\left(\frac{5}{8}m\right) = \frac{8}{5}(-10)$$

Use the multiplication property of equality: Multiply both sides by $\frac{8}{5}$ $\left(\text{which is the reciprocal of } \frac{5}{8}\right)$ to isolate the variable, m.

$$m = -16$$

On the left side: $\frac{8}{5}\left(\frac{5}{8}\right) = 1$ and $1m = m$. On the right side,

multiply: $\frac{8}{5}(-10) = -\dfrac{8 \cdot 2 \cdot \overset{1}{\cancel{5}}}{\cancel{5}} = -16.$

The solution is -16. Verify this by substituting -16 into the original equation. The solution set is $\{-16\}$.

Self Check 2   Solve: $\dfrac{7}{12}a - 6 = -27$

Now Try ▶ Problem 17

EXAMPLE 3   Solve: $-0.2 = -0.8 - y$

Strategy   First, we will use a property of equality to isolate the variable term on one side of the equation. Then we will use a second property of equality to isolate the variable itself.

Why   To solve the original equation, we want to find a simpler equivalent equation of the form **a number $= y$**, whose solution is obvious.

Solution

$$-0.2 = -0.8 - y$$

This is the equation to solve.

$$-0.2 + 0.8 = -0.8 - y + 0.8$$

To isolate the variable term $-y$ on the right side, we eliminate -0.8 by adding 0.8 to both sides.

$$0.6 = -y$$

Do the addition.

Since the term $-y$ has an understood coefficient of -1, the equation can be written as $0.6 = -1y$. To isolate y, we can either multiply both sides or divide both sides by -1.

$$0.6 = -1y$$

If it is helpful, write $-y$ as $-1y$.

$$\frac{0.6}{-1} = \frac{-1y}{-1}$$

To isolate y, undo the multiplication by -1 by dividing both sides by -1.

$$-0.6 = y$$

The solution is -0.6. Verify this by substituting -0.6 into the original equation.

Success Tip

We also can multiply both sides of the equation by -1 to isolate y:
$$-1(0.6) = -1(-1y)$$
$$-0.6 = y$$

Self Check 3   Solve: $-6.6 - m = -2.7$

Now Try ▶ Problem 23

---

2   Simplify Expressions to Solve Equations.

When solving equations, we should simplify the expressions that make up the left and right sides before applying any properties of equality. Often, that involves using the distributive property to remove parentheses and/or combining like terms.

EXAMPLE 4   Solve: **a.** $3(k + 1) - 5k = 0$ **b.** $10a - 2(2a - 7) = 68$

Strategy   We will use the distributive property along with the process of combining like terms to simplify the left side of each equation.

Why   It's best to simplify each side of an equation before using a property of equality.

Solution   **a.** $3(k + 1) - 5k = 0$ This is the equation to solve.

$3k + 3 - 5k = 0$ Distribute the multiplication by 3.

$-2k + 3 = 0$ Combine like terms: $3k - 5k = -2k$.

$-2k + 3 - 3 = 0 - 3$ To isolate the variable term $-2k$, undo the addition of 3 by subtracting 3 from both sides.

$-2k = -3$ Do the subtraction: $3 - 3 = 0$ and $0 - 3 = -3$.

$\dfrac{-2k}{-2} = \dfrac{-3}{-2}$ To isolate the variable k, undo the multiplication by -2 by dividing both sides by -2.

$k = \dfrac{3}{2}$ Simplify the fraction: $\dfrac{-3}{-2} = \dfrac{3}{2}$.

Success Tip

To check a result, evaluate each side of the equation following the order of operations rule.

Check: $3(k + 1) - 5k = 0$ This is the original equation.

$3\left(\dfrac{3}{2} + 1\right) - 5\left(\dfrac{3}{2}\right) \overset{?}{=} 0$ Substitute $\dfrac{3}{2}$ for k.

$3\left(\dfrac{5}{2}\right) - 5\left(\dfrac{3}{2}\right) \overset{?}{=} 0$ Do the addition within the parentheses. Think of 1 as $\dfrac{2}{2}$ and then add: $\dfrac{3}{2} + \dfrac{2}{2} = \dfrac{5}{2}$.

$\dfrac{15}{2} - \dfrac{15}{2} \overset{?}{=} 0$ Do the multiplication.

$0 = 0$ True

The solution is $\dfrac{3}{2}$ and the solution set is $\left\{\dfrac{3}{2}\right\}$.

b. $10a - 2(2a - 7) = 68$ This is the equation to solve.

$10a - 4a + 14 = 68$ Distribute the multiplication by -2.

$6a + 14 = 68$ Combine like terms: $10a - 4a = 6a$.

$6a + 14 - 14 = 68 - 14$ To isolate the variable term $6a$, undo the addition of 14 by subtracting 14 from both sides.

$6a = 54$ Do the subtraction.

$\dfrac{6a}{6} = \dfrac{54}{6}$ To isolate the variable a, undo the multiplication by 6 by dividing both sides by 6.

$a = 9$ Do the division.

Use a check to verify that 9 is the solution.

Self Check 4   Solve: **a.** $4(a + 2) - a = 11$ **b.** $19x - 5(3x - 9) = 1$

Now Try ▶ Problems 25 and 27

When solving an equation, if variables appear on both sides, we can use the addition (or subtraction) property of equality to get all variable terms on one side and all constant terms on the other.

EXAMPLE 5   Solve: $3x - 15 = 4x + 36$

Strategy   There are variable terms ($3x$ and $4x$) on both sides of the equation. We will eliminate $3x$ from the left side of the equation by subtracting $3x$ from both sides.

Why   To solve for x, all the terms containing x must be on the same side of the equation.

Solution

$$3x - 15 = 4x + 36$$ This is the equation to solve.

$$3x - 15 - 3x = 4x + 36 - 3x$$ Subtract 3x from both sides to isolate the variable term on the right side.

$$-15 = x + 36$$ Combine like terms: 3x − 3x = 0 and 4x − 3x = x.

$$-15 - 36 = x + 36 - 36$$ To isolate x, undo the addition of 36 by subtracting 36 from both sides.

$$-51 = x$$ Do the subtraction.

Check:

$$3x - 15 = 4x + 36$$ The original equation.

$$3(-51) - 15 \stackrel{?}{=} 4(-51) + 36$$ Substitute −51 for x.

$$-153 - 15 \stackrel{?}{=} -204 + 36$$ Do the multiplication.

$$-168 = -168$$ True

The solution is -51 and the solution set is $\{-51\}$.

Self Check 5   Solve: $30 + 6n = 4n - 2$

Now Try ▶ Problem 29

Success Tip

We could have eliminated $4x$ from the right side by subtracting $4x$ from both sides. The same solution results.

$$3x - 15 - 4x = 4x + 36 - 4x$$
$$-x - 15 = 36$$

However, it is usually easier to isolate the variable term on the side that will result in a *positive* coefficient.

---

3 Clear Equations of Fractions and Decimals.

Equations are usually easier to solve if they don't involve fractions. We can use the multiplication property of equality to clear an equation of fractions by multiplying both sides of the equation by the least common denominator of all the fractions that appear in the equation.

EXAMPLE 6   Solve: $\dfrac{1}{6}x + \dfrac{5}{2} = \dfrac{1}{3}$

Strategy   To clear the equation of fractions, we will multiply both sides by the LCD of all the fractions in the equation.

Why   It's easier to solve an equation that involves only integers.

Solution

$$\frac{1}{6}x + \frac{5}{2} = \frac{1}{3}$$ This is the equation to solve.

$$6\left(\frac{1}{6}x + \frac{5}{2}\right) = 6\left(\frac{1}{3}\right)$$ Multiply both sides by the LCD of $\frac{1}{6}$, $\frac{5}{2}$, and $\frac{1}{3}$, which is 6. Don't forget the parentheses on the left side.

$$6\left(\frac{1}{6}x\right) + 6\left(\frac{5}{2}\right) = 6\left(\frac{1}{3}\right)$$ On the left side, distribute the multiplication by 6.

$$x + 15 = 2$$ Do each multiplication: $6\left(\frac{1}{6}\right) = 1$, $6\left(\frac{5}{2}\right) = \frac{30}{2} = 15$, and $6\left(\frac{1}{3}\right) = \frac{6}{3} = 2$. The fractions have been cleared.

$$x + 15 - 15 = 2 - 15$$ To undo the addition of 15, subtract 15 from both sides.

$$x = -13$$

Check the solution by substituting -13 for x in $\frac{1}{6}x + \frac{5}{2} = \frac{1}{3}$.

Success Tip

Some students find it helpful to perform each fraction multiplication as shown below:

$$\overset{1}{\cancel{6}}\left(\frac{1}{\cancel{6}}x\right) + \overset{3}{\cancel{6}}\left(\frac{5}{\cancel{2}}\right) = \overset{2}{\cancel{6}}\left(\frac{1}{\cancel{3}}\right)$$

Self Check 6 Solve: $\frac{1}{4}x + \frac{1}{2} = -\frac{1}{8}$

Now Try ▶ Problem 33

If an equation contains decimals, it is often convenient to multiply both sides by a power of 10 to change the decimals in the equation to integers.

EXAMPLE 7 Solve: $0.04(12) + 0.01x = 0.02(12 + x)$

Strategy To clear the equation of decimals, we will multiply both sides by a carefully chosen power of 10.

Why It's easier to solve an equation that involves only integers.

Solution

Success Tip

Recall that multiplying a decimal by 10 moves the decimal point 1 place to the right, multiplying it by 100 moves it 2 places to the right, and so on.

The equation contains the decimals 0.04, 0.01, and 0.02. Since the greatest number of decimal places in any one of these numbers is two, we multiply both sides of the equation by 10^2 or 100. This changes 0.04 to 4, and 0.01 to 1, and 0.02 to 2.

$$0.04(12) + 0.01x = 0.02(12 + x)$$ This is the equation to solve.

$$100[0.04(12) + 0.01x] = 100 \cdot 0.02(12 + x)$$ Multiply both sides by 100. Don't forget the brackets.

$$100 \cdot 0.04(12) + 100 \cdot 0.01x = 100 \cdot 0.02(12 + x)$$ Distribute.

$$4(12) + 1x = 2(12 + x)$$ Multiply each decimal by 100 by moving its decimal point 2 places to the right.

$$48 + x = 24 + 2x$$ Distribute the multiplication by 2.

$$48 + x - 24 - x = 24 + 2x - 24 - x$$ Subtract 24 and x from both sides.

$$24 = x$$ Simplify each side.

$$x = 24$$

The solution is 24. Check by substituting 24 for x in the original equation.

Success Tip

When we write the decimals in the original equation as fractions, it becomes more apparent why it is helpful to multiply both sides by the LCD, 100.

$$\frac{4}{100}(12) + \frac{1}{100}x = \frac{2}{100}(12 + x)$$

Self Check 7 Solve: $0.08x + 0.07(15,000 - x) = 1,110$

Now Try ▶ Problem 37

The previous examples suggest the following strategy for solving equations. It is important to note that not every step is needed to solve every equation.

Strategy for Solving Linear Equations in One Variable

1. **Clear the equation of fractions or decimals:** Multiply both sides by the LCD to clear fractions or multiply both sides by a power of 10 to clear decimals.
2. **Simplify each side of the equation:** Use the distributive property to remove parentheses, and then combine like terms on each side.
3. **Isolate the variable term on one side:** Add (or subtract) to get the variable term on one side of the equation and a number on the other using the addition (or subtraction) property of equality.
4. **Isolate the variable:** Multiply (or divide) to isolate the variable using the multiplication (or division) property of equality.
5. **Check the result:** Substitute the possible solution for the variable in the *original* equation to see if a true statement results.

EXAMPLE 8   Solve: $\dfrac{7m + 5}{5} = -4m + 1$

Strategy   We will follow the steps of the equation-solving strategy to solve the equation.

Why   This is the most efficient way to solve a linear equation in one variable.

$$\frac{7m + 5}{5} = -4m + 1 \qquad \text{This is the equation to solve.}$$

Step 1 $\quad 5\left(\dfrac{7m + 5}{5}\right) = 5(-4m + 1)$ — Clear the equation of the fraction by multiplying both sides by 5.

Step 2 $\qquad 7m + 5 = -20m + 5$ — On the left side, remove the common factor 5 in the numerator and denominator. On the right side, distribute the multiplication by 5.

Step 3 $\quad 7m + 5 + \mathbf{20m} = -20m + 5 + \mathbf{20m}$ — To eliminate the term $-20m$ on the right side, add $20m$ to both sides.

$$27m + 5 = 5 \qquad \text{Combine like terms: } 7m + 20m = 27m \text{ and } -20m + 20m = 0.$$

$$27m + 5 - \mathbf{5} = 5 - \mathbf{5} \qquad \text{To isolate the term } 27m, \text{ undo the addition of 5 by subtracting 5 from both sides.}$$

$$27m = 0 \qquad \text{Do the subtraction.}$$

Step 4 $\qquad \dfrac{27m}{\mathbf{27}} = \dfrac{0}{\mathbf{27}}$ — To isolate m, undo the multiplication by 27 by dividing both sides by 27.

$$m = 0 \qquad \text{0 divided by any nonzero number is 0.}$$

Step 5   Substitute 0 for m in $\dfrac{7m + 5}{5} = -4m + 1$ to check that the solution is 0.

Self Check 8   Solve: $6c + 2 = \dfrac{18 - c}{9}$

Now Try ▶ Problem 41

Success Tip

We can remove the common factor 5 from the numerator and the denominator in the following way:

$$\dfrac{\overset{1}{\cancel{5}}}{1}\left(\dfrac{7m + 5}{\underset{1}{\cancel{5}}}\right)$$

Caution

Remember that when you multiply one side of an equation by a nonzero number, you must multiply the other side of the equation by the same number.

---

4 Identify Identities and Contradictions.

Each of the equations that we solved in Examples 1 through 8 had exactly one solution. However, not every linear equation in one variable has a single solution. Some equations are made true by *any* permissible replacement value for the variable. Such equations are called **identities.** An example of an **identity** is

$\qquad x + x = 2x$ If we substitute -10 for x, we get the true statement $-20 = -20$. If we substitute 7 for x, we get $14 = 14$, and so on.

Since we can replace x with any number and the equation will be true, all real numbers are solutions of $x + x = 2x$. This equation has infinitely many solutions.

Another type of equation, called a **contradiction,** is false for all replacement values for the variable. An example is

$\qquad x = x + 1$ No number is equal to 1 more than itself.

Since this equation is false for any value of x, it has no solution.

EXAMPLE 9   Solve: $3(x + 8) + 5x = 2(12 + 4x)$

Strategy   We will follow the steps of the equation-solving strategy to solve the equation.

Why   This is the most efficient way to solve a linear equation in one variable.

Solution

$3(x + 8) + 5x = 2(12 + 4x)$	This is the equation to solve.
$3x + 24 + 5x = 24 + 8x$	Distribute the multiplication by 3 and by 2.
$8x + 24 = 24 + 8x$	Combine like terms: 3x + 5x = 8x. Note that the sides of the equation are identical.
$8x + 24 - \mathbf{8x} = 24 + 8x - \mathbf{8x}$	To eliminate the term 8x on the right side, subtract 8x from both sides.
$24 = 24$	Combine like terms on both sides: 8x − 8x = 0.

> **Success Tip**
>
> At the step $8x + 24 = 24 + 8x$, we know that the equation is an identity because both sides are exactly the same.

In this case, the terms involving x drop out and the result is true. This means that any number substituted for x in the original equation will give a true statement. Therefore, *all real numbers* are solutions and this equation is an identity. Its solution set is written as {all real numbers} or using the symbol $\mathbb{R}$.

Self Check 9   Solve: $3(x + 5) - 4(x + 4) = -x - 1$

Now Try ▶ Problem 45

EXAMPLE 10   Solve: $3(d + 7) - d = 2(d + 10)$

Strategy   We will follow the steps of the equation-solving strategy to solve the equation.

Why   This is the most efficient way to solve a linear equation in one variable.

Solution

$3(d + 7) - d = 2(d + 10)$	This is the equation to solve.
$3d + 21 - d = 2d + 20$	Distribute the multiplication by 3 and by 2.
$2d + 21 = 2d + 20$	Combine like terms: 3d − d = 2d.
$2d + 21 - \mathbf{2d} = 2d + 20 - \mathbf{2d}$	To eliminate the term 2d on the right side, subtract 2d from both sides.
$21 = 20$	Combine like terms on both sides: 2d − 2d = 0.

> **The Language of Algebra**
>
> **Contradiction** is a form of the word *contradict*, meaning conflicting ideas. During a trial, evidence might be introduced that *contradicts* the testimony of a witness.

In this case, the terms involving d drop out and the result is false. This means that any number that is substituted for d in the original equation will give a false statement. Therefore, this equation has *no solution* and it is a contradiction. Its solution set is the **empty set**, which is written as { } or using the symbol $\varnothing$.

Self Check 10   Solve: $-4(c - 3) + 2c = 2(10 - c)$

Now Try ▶ Problem 47

SECTION 2.2 ▶ STUDY SET

VOCABULARY

Fill in the blanks.

1. $3x + 8 = 10$ is an example of a linear _____ in one variable.

2. To solve $\frac{s}{3} + \frac{1}{4} = -\frac{1}{2}$, we can _____ the equation of the fractions by multiplying both sides by 12.

3. A linear equation that is true for any permissible replacement value for the variable is called an _____.

4. A linear equation that is false for all replacement values for the variable is called a _____.

CONCEPTS

Fill in the blanks.

5. a. To solve $3x - 5 = 1$, we first undo the _____ of 5 by adding 5 to both sides. Then we undo the _____ by 3 by dividing both sides by 3.

    b. To solve $\frac{x}{2} + 3 = 5$, we can undo the _____ of 3 by subtracting 3 from both sides. Then we can undo the _____ by 2 by multiplying both sides by 2.

6. a. Combine like terms on the left side of $6x - 8 - 8x = -24$.

    b. Distribute and then combine like terms on the right side of $-20 = 4(3x - 4) - 9x$.

7. Use a check to determine whether -2 is a solution of $6x + 5 = 7$.

8. Multiply.

    a. $20\left(\frac{3}{5}x\right)$ b. $100 \cdot 0.02x$

9. a. By what must you multiply both sides of $\frac{2}{3} - \frac{1}{2}b = -\frac{4}{3}$ to clear it of fractions?

    b. By what must you multiply both sides of $0.7x + 0.3(x - 1) = 0.5x$ to clear it of decimals?

Look Alikes . . .

10. a. Simplify: $3x + 5 - x$

    b. Solve: $3x + 5 = 9$

    c. Evaluate $3x + 5 - x$ for $x = 9$

    d. Check: Is -1 a solution of $3x + 5 - x = 9$?

NOTATION

Complete the solution.

11. Solve: $2x - 7 = 21$

    $2x - 7 + \boxed{} = 21 + \boxed{}$

    $2x = 28$

    $\dfrac{2x}{\boxed{}} = \dfrac{28}{\boxed{}}$

    $x = 14$

    Check: $2x - 7 = 21$

    $2(\boxed{}) - 7 \overset{?}{=} 21$

    $\boxed{} - 7 \overset{?}{=} 21$

    $\boxed{} = 21$

    $\boxed{}$ is the solution.

12. A student multiplied both sides of $\frac{3}{4}t + \frac{5}{8} = \frac{1}{2}t$ by 8 to clear it of fractions, as shown below. Explain his error in showing this step.

$$8 \cdot \frac{3}{4}t + \frac{5}{8} = 8 \cdot \frac{1}{2}t$$

GUIDED PRACTICE

Solve each equation and check the result. **See Example 1.**

13. $-8x + 1 = 73$ 14. $-7y + 4 = 60$

15. $-5q - 2 = 23$ 16. $-4p + 3 = 43$

Solve each equation and check the result. **See Example 2.**

17. $\frac{5}{6}k - 5 = 10$ 18. $\frac{2}{5}c - 12 = 2$

19. $-\frac{7}{16}h + 28 = 21$ 20. $-\frac{5}{8}h + 25 = 15$

Solve each equation and check the result. **See Example 3.**

21. $-6 - y = -2$ 22. $-1 - h = -9$

23. $-1.7 = 1.2 - x$ 24. $0.6 = 4.1 - x$

Solve each equation and check the result. **See Example 4.**

25. $3(2y - 2) - y = 5$ 26. $2(-3a + 2) + a = 2$

27. $6a - 3(3a - 4) = 30$ 28. $16y - 8(3y - 2) = -24$

Solve each equation and check the result. **See Example 5.**

29. $7a - 12 = 8a + 9$ 30. $10m - 14 = 11m + 13$

31. $60r - 50 = 15r - 5$ 32. $100f - 75 = 50f + 75$

Solve each equation and check the result. **See Example 6.**

33. $\frac{5}{6}x + \frac{2}{9} = \frac{1}{3}$ 34. $\frac{2}{3}x + \frac{2}{3} = \frac{3}{4}$

35. $\frac{1}{8}y - \frac{1}{2} = \frac{1}{4}$ 36. $\frac{1}{15}x - \frac{4}{5} = \frac{2}{3}$

Solve each equation and check the result. **See Example 7.**

37. $0.02(62) - 0.08s = 0.06(s + 9)$

38. $0.04(50) + 0.16x = 0.08(x + 50)$

39. $0.09(t + 50) + 0.15t = 52.5$

40. $0.08(x - 100) = 44.5 - 0.07x$

Solve each equation and check the result. **See Example 8.**

41. $\frac{10 - 5s}{3} = -s + 6$ 42. $\frac{40 - 8s}{5} = -2s + 8$

43. $t = \frac{7t - 9}{16}$ 44. $-3 = \frac{11r + 68}{3}$

Solve each equation, if possible. **See Examples 9 and 10.**

45. $8x + 3(2 - x) = 5x + 6$

46. $5(x + 2) = 5x - 2$

47. $-3(s + 2) = -2(s + 4) - s$

48. $21(b - 1) + 3 = 3(7b - 6)$

TRY IT YOURSELF

Solve each equation, if possible. Check the result.

49. $3x - 8 - 4x - 7x = -2 - 8$

50. $-6t - 7t - 5t - 1 = 12 - 3$

51. $\dfrac{t}{3} + 2 = 6$ 52. $\dfrac{x}{5} - 5 = -12$

53. $4(5b) + 2(6b - 1) = -34$

54. $9(x + 11) + 5(13 - x) = 0$

55. $2x + 5 = 17$ 56. $3x - 5 = 13$

57. $\dfrac{5}{6}(1 - x) = -x + 1$ 58. $\dfrac{3}{8}(14 - u) = -3u + 6$

59. $0.05a + 0.01(90) = 0.02(a + 90)$

60. $0.03x + 0.05(2{,}000 - x) = 99.5$

61. $\dfrac{7}{2} + \dfrac{3}{2}d = -9 + 1.5d$

62. $x + 7 = \dfrac{2x + 6}{2} + 4$

63. $-(19 - 3s) - (8s + 1) = 35$ 64. $2(3x) - 5(3x + 1) = 58$

65. $5x = 4x + 7$ 66. $3x = 2x + 2$

67. $\dfrac{3(b + 2)}{2} = \dfrac{4b - 10}{4}$ 68. $\dfrac{2(5a - 7)}{4} = \dfrac{9(a - 1)}{3}$

69. $8y - 2 = 4y + 16$ 70. $7 + 3w = 4 + 9w$

71. $\dfrac{1}{6}y + \dfrac{1}{4}y = -1$ 72. $\dfrac{1}{3}x + \dfrac{1}{4}x = -2$

73. $0.7 - 4y = 1.7$ 74. $0.3 - 2x = -0.9$

75. $-33 = 5t + 2$ 76. $-55 = 3w + 5$

77. $-3p + 7 = -3$ 78. $-2r + 8 = -1$

79. $2(-3) + 4y = 14$ 80. $4(-1) + 3y = 8$

81. $0.06(a + 200) + 0.1a = 172$

82. $0.03x + 0.05(6{,}000 - x) = 280$

83. $8.6y + 3.4 = 4.2y - 9.8$

84. $9.1y + 3.6 = 6.5y - 30.2$

85. $\dfrac{2}{3}y + 2 = \dfrac{1}{5} + y$ 86. $\dfrac{2}{5}x + 1 = \dfrac{1}{3} + x$

87. $0.4b - 0.1(b - 100) = 70$

88. $0.105x + 0.06(20{,}000 - x) = 1{,}740$

89. $\dfrac{1}{4}(10 - 2y) = 8$ 90. $\dfrac{1}{3}(7 - 7x) = -21$

91. $2 - 3(x - 5) = 4(x - 1)$

92. $2 - (4x + 7) = 3 + 2(x + 2)$

93. $2n - \dfrac{3}{4}n = \dfrac{1}{2}n + \dfrac{13}{3}$ 94. $\dfrac{5}{6}n + 3n = -\dfrac{1}{3}n - \dfrac{11}{9}$

95. $10.08 = 4(0.5x + 2.5)$ 96. $-3.28 = 8(1.5y - 0.5)$

97. $\dfrac{3}{4}(d - 8) = \dfrac{2}{3}(d + 1)$ 98. $\dfrac{3}{2}(c - 2) = \dfrac{2}{5}(2c + 3)$

99. $2d + 5 = 0$ 100. $3c + 8 = 0$

101. $3(A + 2) = 2(A - 7)$

102. $9(T - 1) = 6(T + 2) - T$

103. $4(a - 3) = -2(a - 6) + 6a$

104. $9(t + 2) = -6(t - 3) + 15t$

105. $4(y - 3) - y = 3(y - 4)$

106. $5(x + 3) - 3x = 2(x + 8)$

107. $-(4 - m) = -10$ 108. $-(6 - t) = -12$

109. $-\dfrac{2}{3}z + 4 = 8$ 110. $-\dfrac{7}{5}x + 9 = -5$

Look Alikes . . .

Simplify each expression and solve each equation.

111. a. $-2(9 - 3x) - (5x + 2)$

     b. $-2(9 - 3x) - (5x + 2) = -25$

112. a. $4(x - 5) - 3(12 - x)$

     b. $4(x - 5) - 3(12 - x) = 7$

113. a. $0.6 - 0.2(x + 1)$

     b. $0.6 - 0.2(x + 1) = 0.4$

114. a. $2(6n + 4) + 4(3n + 2) + 2$

     b. $2(6n + 5) = 4(3n + 2) + 2$

WRITING

115. To solve $3x - 4 = 5x + 1$, one student began by subtracting $3x$ from both sides. Another student solved the same equation by first subtracting $5x$ from both sides. Will the students get the same solution? Explain why or why not.

116. What does it mean to *clear* an equation such as $\frac{1}{4} + \frac{1}{2}x = \frac{3}{8}$ of the fractions?

117. Explain the error in the following solution.

Solve: $2x + 4 = 30$

$$\dfrac{2x}{2} + 4 = \dfrac{30}{2}$$

$$x + 4 = 15$$

$$x + 4 - 4 = 15 - 4$$

$$x = 11$$

118. a. Write an equation that is an identity. Explain why every real number is a solution.

     b. Write an equation that is a contradiction. Explain why no real number is a solution.

REVIEW

Name the property that is used.

119. $x \cdot 9 = 9x$

120. $4 \cdot \dfrac{1}{4} = 1$

121. $(x + 1) + 2 = x + (1 + 2)$

122. $2(30y) = (2 \cdot 30)y$

CHALLENGE PROBLEMS

123. Solve: $\dfrac{5}{6}\left(-\dfrac{3}{4}m + 1\right) = -\dfrac{2}{3}\left(\dfrac{1}{2}m - 1\right)$

124. In this section, we discussed equations that have no solution, one solution, and an infinite number of solutions. Do you think an equation could have exactly two solutions? If so, give an example.

SECTION 2.3

Applications of Percent

OBJECTIVES

1 Change percents to decimals and decimals to percents.

2 Solve percent problems by direct translation.

3 Solve applied percent problems.

4 Find percent of increase and decrease.

5 Solve discount and commission problems.

ARE YOU READY?

The following problems review some basic skills that are needed to solve percent problems.

1. Multiply: $100 \cdot 0.61$

2. Multiply: $100 \cdot 0.02$

3. Write $\dfrac{1}{2}$ and $\dfrac{3}{4}$ as decimals.

4. Divide: $\dfrac{27}{100}$

5. Multiply: $124 \cdot 0.03$

6. Divide: $\dfrac{17.82}{0.36}$

In this section, we will use translation skills from Chapter 1 and equation-solving skills from Chapter 2 to solve problems involving percents.

1   Change Percents to Decimals and Decimals to Percents.

93% or $\dfrac{93}{100}$ or 0.93 of the figure is shaded.

The word **percent** means parts per one hundred. We can think of the percent symbol % as representing a denominator of 100. Thus, $93\% = \dfrac{93}{100}$. Since the fraction $\dfrac{93}{100}$ is equal to the decimal 0.93, it is also true that $93\% = 0.93$. In general, $n\% = \dfrac{n}{100}$.

When solving percent problems, we must often convert percents to decimals and decimals to percents. To change a percent to a decimal, we drop the % symbol and *divide the given number by 100 by moving the decimal point 2 places to the left.* For example,

$$31\% = 31.0\% = 0.31$$

To change a decimal to a percent, we *multiply the decimal by 100 by moving the decimal point 2 places to the right, and then inserting a % symbol.* For example,

$$0.678 = 67.8\%$$

2   Solve Percent Problems by Direct Translation.

There are three basic types of percent problems. Examples of these are:

Type 1   What number is 8% of 215?

Type 2   102 is 21.3% of what number?

Type 3   31 is what percent of 500?

Every percent problem has three parts: the *amount,* the *percent,* and the *base.* For example, in the question *What number is 8% of 215?,* the words "what number" represent the **amount,** 8% represents the **percent,** and 215 represents the **base.** In these problems, the word "is" means "is equal to," and the word "of" means "multiplication."

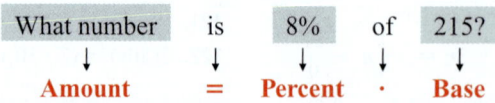

EXAMPLE 1   What number is 8% of 215?

Strategy   We will translate the words of this problem into an equation and then solve the equation.

Why   The variable in the translation equation represents the unknown number that we are asked to find.

Solution

In this problem, the phrase "what number" represents the amount, 8% is the percent, and 215 is the base.

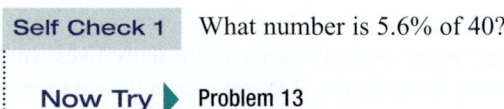

What number	is	8%	of	215?
↓	↓	↓	↓	↓
x	=	0.08	·	215
x	=	17.2		

Change the percent to a decimal: 8% = 0.08.
Do the multiplication.

Thus, 8% of 215 is 17.2.

To check, we note that 17.2 out of 215 is $\frac{17.2}{215} = 0.08 = 8\%$.

Self Check 1   What number is 5.6% of 40?

Now Try ▶ Problem 13

We will illustrate the other two types of percent problems with application problems.

3 Solve Applied Percent Problems.

One method for solving applied percent problems is to use the given facts to write a **percent sentence** of the form

	is		%	of		?

We enter the appropriate numbers in two of the blanks and the words "what number" or "what percent" in the remaining blank. As before, we translate the words into an equation and solve it.

EXAMPLE 2

Aging Populations.   By the year 2050, a study by the Pew Research Center predicts that about 81 million residents of the U.S. will be age 65 or older. The **circle graph** (or **pie chart**) indicates that age group will make up 18.5% of the population. If this prediction is correct, what will the population of the United States be in 2050? (Round to the nearest million.)

Strategy   To find the predicted U.S. population in 2050, we will translate the words of the problem into an equation and then solve the equation.

Why   The variable in the translation equation represents the unknown population in 2050 that we are asked to find.

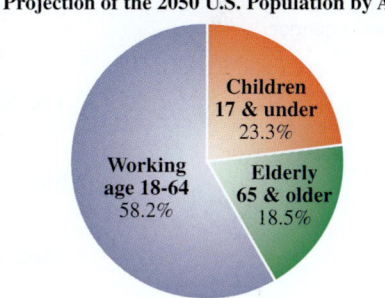

Projection of the 2050 U.S. Population by Age

Children 17 & under 23.3%

Working age 18-64 58.2%

Elderly 65 & older 18.5%

Source: Pew Research Center (2008)

Solution

In this problem, 81 is the amount, 18.5% is the percent, and the words "what number" represent the base. The units are millions of people.

81	is	18.5%	of	what number?
↓	↓	↓	↓	↓
81	=	0.185	·	x

$$\frac{81}{\mathbf{0.185}} = \frac{0.185x}{\mathbf{0.185}}$$

To isolate x, undo the multiplication by 0.185 by dividing both sides by 0.185.

$$437.8 \approx x$$ Do the division.

$$438 \approx x$$ Round 437.8 million to the nearest million.

The U.S. population is predicted to be about 438 million in the year 2050. We can check using estimation: 81 million out of a population of 438 million is about $\frac{80 \text{ million}}{400 \text{ million}}$, or $\frac{1}{5}$, which is 20%. Since this is close to 18.5%, the answer 438 million seems reasonable.

> **Self Check 2**   **Aging Populations.**   By the year 2100, it is predicted that 131 million, or 23%, of the U.S. residents will be age 65 or older. If the prediction is correct, find the population in 2100. (Round to the nearest million.)
>
> **Now Try** ▶ Problem 17

We pay many types of taxes in our daily lives, such as sales tax, gasoline tax, income tax, and Social Security tax. **Tax rates** usually are expressed as percents.

EXAMPLE 3   **Taxes.**   A maid makes $500 a week. One of the deductions from her weekly paycheck is a Social Security tax of $31. Find her Social Security tax rate.

Strategy   To find the tax rate, we will translate the words of the problem into an equation and then solve the equation.

Why   The variable in the translation equation represents the unknown tax rate that we are asked to find.

Solution

31	is	what percent	of	500?
↓	↓	↓	↓	↓
31	=	x	·	500

31 is the amount, x is the percent, and 500 is the base.

> **Success Tip**
>
> One way to check an answer is using *estimation*. If your estimate and your answer are close, you can be reasonably sure that your answer is correct.

$$\frac{31}{500} = \frac{500x}{500}$$ To undo the multiplication by 500, divide both sides by 500.

$0.062 = x$ Do the division.

$6.2\% = x$ Change the decimal 0.062 to a percent.

The Social Security tax rate is 6.2%.

We can use estimation to check: $31 out of $500 is about $\frac{30}{500}$ or $\frac{6}{100}$, which is 6%. Since this is close to 6.2%, the answer seems reasonable.

> **Self Check 3**   **Medicare Tax.**   The maid mentioned in Example 3 also has $7.25 of Medicare tax deducted from her weekly paycheck. Find her Medicare tax rate.
>
> **Now Try** ▶ Problem 23

4 Find Percent of Increase and Decrease.

Percents often are used to describe how a quantity has changed. For example, a health care provider might increase the cost of medical insurance by 3%, or a police department might decrease the number of officers assigned to street patrols by 10%. To describe such changes, we use **percent of increase** or **percent of decrease.**

EXAMPLE 4

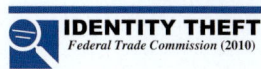

Year	2008	2009
Number of Complaints	314,000	278,000

Identity Theft. The Federal Trade Commission receives complaints involving the theft of someone's identity information, such as a credit card, Social Security number, or cell phone account. Refer to the data in the table. What was the percent of decrease in the number of complaints from 2008 to 2009? (Round to the nearest percent.)

Strategy First, we will subtract to find the *amount of decrease* in the number of complaints. Then we will translate the words of the problem into an equation and solve it.

Why A percent of decrease problem involves finding the *percent of change,* and the change in a quantity is found using subtraction.

Solution To find the *amount of decrease,* we find the difference of 314,000 and 278,000.

$$314,000 - 278,000 = 36,000$$

Now we translate the percent sentence to an equation, and solve it.

Caution

The percent of increase (or decrease) is a percent of the *original* number, that is, the number before the change occurred.

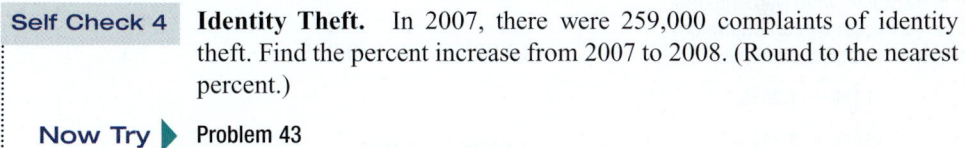

The decrease is what percent of the number of complaints in 2008?

$$36,000 = x \cdot 314,000$$

36,000 is the amount, x is the percent, and 314,000 is the base.

$$\frac{36,000}{314,000} = \frac{314,000x}{314,000}$$

To undo the multiplication by 314,000, divide both sides by 314,000.

$$0.114649681 \approx x$$ Do the division using a calculator.

$$11.4649681\% \approx x$$ Change the decimal to a percent.

Rounding to the nearest percent, we find that the number of identity theft complaints decreased by about 11% from 2008 to 2009.

A 10% decrease would be 0.10(314,000) or 31,400 fewer complaints. It seems reasonable that 36,000 fewer complaints is an 11% decrease.

Self Check 4 **Identity Theft.** In 2007, there were 259,000 complaints of identity theft. Find the percent increase from 2007 to 2008. (Round to the nearest percent.)

Now Try ▶ Problem 43

5 Solve Discount and Commission Problems.

When the price of an item is reduced, we call the amount of the reduction a **discount.** If a discount is expressed as a percent, it is called the **rate of discount.**

EXAMPLE 5 **Health Club Discounts.** A 30% discount on a 1-year membership for a fitness center amounted to a $90 savings. Find the cost of a 1-year membership before the discount.

Strategy We will translate the words of the problem into an equation and then solve the equation.

Why The variable in the translation equation represents the unknown cost of a 1-year membership before the discount.

©barang/Shutterstock.com

Solution   We are told that $90 is 30% of some unknown membership cost.

90	is	30%	of	what number?
↓	↓	↓	↓	↓
90	=	0.30	·	x

90 is the amount, 30% is the percent, and x is the base.

$$\frac{90}{0.30} = \frac{0.30x}{0.30}$$ *To undo the multiplication by 0.30, divide both sides by 0.30.*

$$300 = x$$ *Do the division.*

A one-year membership cost $300 before the discount.

Self Check 5   **Discounts.**   A shopper saved $6 on a pen that was discounted 5%. Find the original cost.

Now Try ▶ Problem 51

Instead of working for a salary or at an hourly rate, many salespeople are paid on **commission.** An employee who is paid a commission is paid a percent of the price of goods or services that he or she sells. We call that percent the **rate of commission.**

EXAMPLE 6   **Commissions.**   A real estate agent earned $14,025 for selling a house. If she received a $5\frac{1}{2}\%$ commission, what was the selling price?

Strategy   We will translate the words of the problem into an equation and then solve the equation.

Why   The variable in the translation equation represents the unknown selling price of the house that we are asked to find.

Solution   We are told that $14,025 is $5\frac{1}{2}\%$ of some unknown selling price of a house.

Success Tip

It is helpful to write percents that involve fractions in an equivalent decimal form. For example,

$1\frac{1}{4}\% = 1.25\%$

$5\frac{1}{2}\% = 5.5\%$

$7\frac{3}{4}\% = 7.75\%$

$14,025	is	5.5%	of	what number?
↓	↓	↓	↓	↓
14,025	=	0.055	·	x

Write $5\frac{1}{2}\%$ as 5.5%.

14,025 is the amount, 5.5% is the percent, and x is the base.

$$\frac{14,025}{0.055} = \frac{0.055x}{0.055}$$ *To undo the multiplication by 0.055, divide both sides by 0.055.*

$$255,000 = x$$ *Do the division.*

The selling price of the house was $255,000.

Self Check 6   **Commissions.**   A jewelry store clerk receives a $4\frac{1}{4}\%$ commission on all sales. What was the price of a gold necklace sold by the clerk if his commission was $25.50?

Now Try ▶ Problem 53

SECTION 2.3 ▶ STUDY SET

VOCABULARY

Fill in the blanks.

1. _____ means parts per one hundred.

2. In the statement "10 is 50% of 20," 10 is the _____, 50% is the percent, and 20 is the _____.

3. In percent questions, the word *of* means _____, and the word ___ means equals.

4. An employee who is paid a _____ is paid a percent of the cost of goods or services that he or she sells.

CONCEPTS

5. Represent the amount of the figure that is shaded using a fraction, a decimal, and a percent.

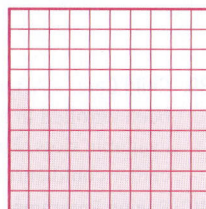

6. Fill in the blanks.

   a. To change a percent to a decimal, drop the % symbol and move the decimal point 2 places to the _____.

   b. To change a decimal to a percent, move the decimal point 2 places to the _____ and insert a % symbol.

7. Fill in the blanks using the words *percent, amount,* and *base.*

   _____ = _____ · _____

8. Translate each sentence into an equation. **Do not solve.**

   a. 12 is 40% of what number?

   b. 99 is what percent of 200?

   c. What is 66% of 3?

9.

Number of Earthquakes in the U.S.	
2008	**2009**
3,618	4,257

   Source: U.S. Geological Survey

   a. Find the *amount* of increase in the number of earthquakes.

   b. Fill in blanks to find the percent of increase in earthquakes: _____ is _____ % of _____?

10. Use estimation to determine if each statement is reasonable.

   a. 18 is 48% of 93.   b. 47 is 6% of 206.

NOTATION

11. Change each percent to a decimal.

   a. 35%   b. 8.5%

   c. 150%   d. $2\frac{3}{4}$%

   e. 9.25%   f. $1\frac{1}{2}$%

12. Change each decimal to a percent.

   a. 0.9   b. 0.99

   c. 0.999   d. 9

GUIDED PRACTICE

See Example 1.

13. What number is 48% of 650?

14. What number is 60% of 200?

15. What number is 92.4% of 50?

16. What number is 2.8% of 220?

See Example 2.

17. 75 is 25% of what number?

18. 78 is 6% of what number?

19. 128.1 is 8.75% of what number?

20. 1.12 is 140% of what number?

See Example 3.

21. 78 is what percent of 300?

22. 143 is what percent of 325?

23. 0.42 is what percent of 16.8?

24. 199.92 is what percent of 2,352?

APPLICATIONS

25. **Antiseptics.** Refer to the label on the bottle in figure (a) below. Find the amount of pure hydrogen peroxide in the bottle.

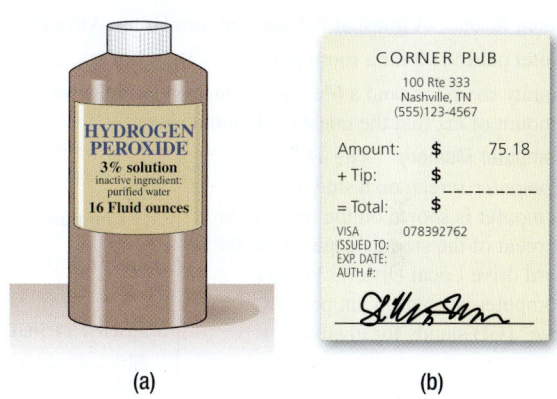

(a)   (b)

26. **Dining Out.** Refer to the sales receipt in figure (b) above. Compute the 15% tip (*rounded up* to the nearest dollar). Then find the total cost of the meal.

27. **U.S. Federal Budget.** The circle graph shows how the government spent $2,980 billion in 2008. How much was spent on

   a. Social Security/Medicare?

   b. Defense/Veterans?

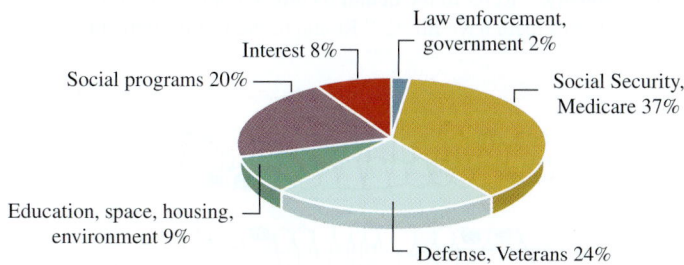

Based on 2009 Federal Income Tax Form 1040

28. Tax Tables. Use the Schedule X table below to compute the amount of federal income tax to be paid on an income of $39,909.

If your income is over—	But not over—	Your income tax is—	of the amount over—
$0	$8,350	.... 10%	$0
8,350	33,950	**$835.00 + 15%**	8,350
33,950	82,250	**4,675.00 + 25%**	33,950

29. PayPal. Many e-commerce businesses use PayPal to perform payment processing for them. For certain transactions, merchants are charged a fee of 2.9% of the selling price of the item plus $0.30. What would PayPal charge an online art store to collect payment on a painting selling for $350?

30. eBay. When a student sold a Play Station 3 on eBay for $195, she was charged a two-part final value fee: 8% of the first $50 of the selling price plus 5% of the remainder of the selling price over $50. Find the fee to sell the Play Station 3 on eBay.

31. Price Guarantees. Home Club offers a "10% Plus" guarantee: If the customer finds the same item selling for less somewhere else, he or she receives the difference in price plus 10% of the difference. A woman bought miniblinds at the Home Club for $120 but later saw the same blinds on sale for $98 at another store. How much can she expect to be reimbursed?

32. Room Taxes. A guest at the San Antonio Hilton Airport Hotel paid $180 for a room plus a 9% city room tax, a $1\frac{3}{4}$% county room tax, and a 6% state room tax. Find the total amount of tax that the guest paid on the room.

33. Computer Memory. The *My Computer* screen on a student's computer is shown on the right. What percent of the storage capacity on the hard drive Local Disk (C:) of his computer is used? What percent is free? (GB stands for gigabytes.)

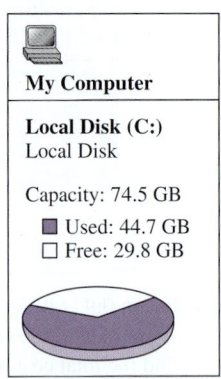

My Computer

Local Disk (C:)
Local Disk

Capacity: 74.5 GB

▪ Used: 44.7 GB
☐ Free: 29.8 GB

34. Genealogy. Through an extensive computer search, a genealogist determined that worldwide, 180 out of every 10 million people had his last name. What percent is this?

35. Dentistry. Refer to the dental record. What percent of the patient's teeth have fillings? Round to the nearest percent.

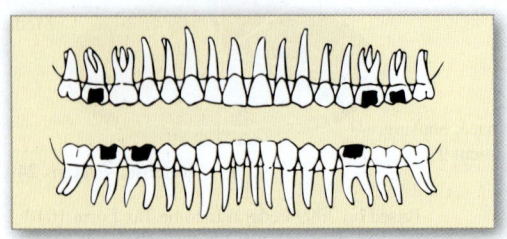

36. Test Scores. The score 175/200 was written by an algebra instructor at the top of a student's test paper. Write the test score as a percent.

37. DMV Written Test. To obtain a learner's permit to drive in Nevada, a score of 80% (or better) on a 50-question multiple-choice test is required. If a teenager answered 33 questions correctly, did he pass the test?

38. iPods. The settings menu screen of an Apple iPod is shown. What percent of the memory capacity is still available? Round to the nearest percent. (GB stands for gigabytes.)

About	
Songs	3639
Videos	32
Photos	0
Capacity	62.5 GB
Available	35.0 GB
Version	1.1.1
S/N	4H534PG7TY1
Model	MA148LL
Format	Windows

39. Child Care. After the first day of registration, 84 children had been enrolled in a day care center. That represented 70% of the available slots. Find the maximum number of children the center could enroll.

40. Racing Programs. One month before a stock car race, the sale of ads for the official race program was slow. Only 12 pages, or just 30% of the available pages, had been sold. Find the total number of pages devoted to advertising in the program.

41. Nutrition. The Nutrition Facts label from a can of clam chowder is shown.

 a. Find the number of grams of saturated fat in one serving. What percent of a person's recommended daily intake is this?

 b. Determine the recommended number of grams of saturated fat that a person should consume daily.

Nutrition Facts

Serving Size 1 cup (240mL)
Servings Per Container about 2

Amount per serving	
Calories 240 Calories from Fat 140	
	% Daily Value*
Total Fat 15 g	**23%**
Saturated Fat 5 g	**25%**
Cholesterol 10 mg	**3%**
Sodium 980 mg	**41%**
Total Carbohydrate 21 g	**7%**
Dietary Fiber 2 g	**8%**
Sugars 1 g	
Protein 7 g	

42. Commercials. Jared Fogle credits his tremendous weight loss to exercise and a diet of low-fat Subway sandwiches. His current weight (about 187 pounds) is 44% of his maximum weight (reached in March of 1998). What did he weigh then?

43. Rainforests. Refer to the graph on the next page, which shows the number of square miles of Brazilian Amazon rainforest that has recently been deforested.

 a. Find the percent of increase in the number of square miles of the Brazilian Amazon rainforest that was cleared from 2007 to 2008. Round to the nearest percent.

 b. Find the percent of decrease in the number of square miles of the Brazilian Amazon rainforest that was cleared from 2008 to 2009. Round to the nearest percent.

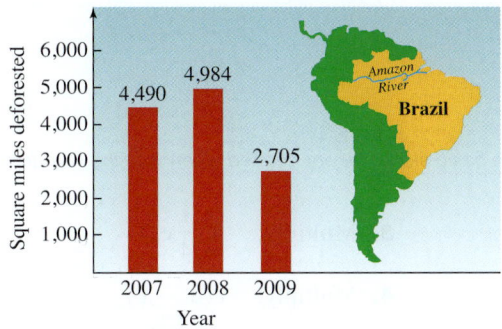

Source: mongabay.com

44. Auctions. A pearl necklace of former First Lady Jacqueline Kennedy Onassis, originally valued at $700, was sold at auction in 1996 for $211,500. Find the percent of increase in the value of the necklace. (Round to the nearest percent.)

45. Insurance Costs. A college student's good grades earned her a student discount on her car insurance premium. Find the percent of decrease to the nearest percent if her annual premium was lowered from $1,050 to $925.

46. U.S. Life Expectancy. Use the following life expectancy data for 1900 and 2010 to find the percent of increase for males and for females. Round to the nearest percent.

Years of life expected at birth		
	Male	**Female**
1900	46.3	48.3
2010	75.8	80.8

Source: *The World Fact Book*, CIA

from **Campus To Careers**

47 ▶ ....

Automotive Service Technician

© barang/Shutterstock.com

A *single* automobile tire that is underinflated by 5 pounds per square inch (psi) decreases the car's gas mileage by about 1.5%. (Source: auto-buying-tips.com and popularmechanics.com)

    a. If all four tires of a car are underinflated by 5 psi, by what percent will its gas mileage be decreased?

    b. A 2010 Chevrolet Camaro with properly inflated tires gets 25 miles per gallon on the highway. If each of its tires is underinflated by 5 psi, what will be its highway mileage?

48. Food Labels. To be labeled "Reduced Fat," foods must contain at least 25% less fat per serving than the regular product. One serving of the original Jif peanut butter has 16 grams of fat per serving. The new Jif Reduced Fat product contains 12 grams of fat per serving. Does it meet the labeling requirement?

49. TV Shopping. Jan bought a toy from the QVC home shopping network that was discounted 20%. If she saved $15, what was the original price of the toy?

50. Discounts. A 12% discount on a watch saved a shopper $48. Find the price of the watch before the discount.

51. Sales. The price of a certain model patio set was reduced 35% because it was being discontinued. A shopper purchased two of them and saved a total of $210. Find the price of a patio set before the discount.

52. Craigslist. A post on the classifieds website Craigslist advertised a mountain bike selling for $800 less than its retail price. If this is a 40% savings, what is the retail price of the bike?

53. Real Estate. The $3\frac{1}{2}$% commission paid to a real estate agent on the sale of a condominium earned her $3,325. Find the selling price of the condo.

54. Consignment. An art gallery agreed to sell an artist's sculpture for a commission of 45%. What must be the selling price of the sculpture if the gallery would like to make $13,500?

55. Stockbrokers. A stockbroker charges a 2.5% commission to sell shares of a stock for a client. Find the value of stock sold by a broker if the commission was $640.

56. Agents. A sports agent made one million dollars by charging a 12.5% commission to negotiate a long-term contract for a professional athlete. Find the amount of the contract.

WRITING

57. Explain the error: What is 5% of 8?

$$x = 5 \cdot 8$$
$$x = 40$$

40 is 5% of 8.

58. Write a real-life situation that could be described by "9 is what percent of 20?"

59. Explain why 150% of a number is more than the number.

60. Why is the problem "What is 9% of 100?" easy to solve?

REVIEW

61. Divide: $-\frac{16}{25} \div \left(-\frac{4}{15}\right)$

62. What two numbers are a distance of 8 away from 4 on the number line?

63. Is -34 a solution of $x + 15 = -49$?

64. Evaluate: $2 + 3[24 - 2(2 - 5)]$

CHALLENGE PROBLEMS

65. Soaps. A soap advertises itself as $99\frac{44}{100}$% pure. First, determine what percent of the soap is impurities. Then express your answer as a decimal.

66. Express $\frac{1}{20}$ of 1% as a percent using decimal notation.

SECTION 2.4

OBJECTIVES

1 Use formulas from business.

2 Use formulas from science.

3 Use formulas from geometry.

4 Solve for a specified variable.

Formulas

ARE YOU READY?

The following problems review some basic skills that are needed when working with formulas.

1. How many variables does each equation contain?

    a. $4x + 3 = 15$

    b. $P = 2l + 2w$

2. Simplify: $a + d - a$

3. Multiply: $4 \cdot \dfrac{1}{4}x$

4. Multiply: $c(8 - x)$

5. If $2 = t$, is it also true that $t = 2$?

6. Simplify: $\dfrac{b}{b}$

A **formula** is an equation that states a mathematical relationship between two or more variables. The variables of a formula represent quantities such as cost, time, and perimeter. Formulas are used in fields such as business, science, and geometry.

1 Use Formulas from Business.

A formula for retail price: To make a profit, a merchant must sell an item for more than he or she paid for it. The price at which the merchant sells the product, called the **retail price,** is the *sum* of what the item cost the merchant plus the **markup.** Using *r* to represent the retail price, *c* the cost, and *m* the markup, we can write this formula as

$$r = c + m \qquad \text{Retail price} = \text{cost} + \text{markup}$$

A formula for profit: The **profit** a business makes is the *difference* between the **revenue** (the money it takes in) and the cost. Using *p* to represent the profit, *r* the revenue, and *c* the cost, we can write this formula as

$$p = r - c \qquad \text{Profit} = \text{revenue} - \text{cost}$$

If we are given the values of all but one of the variables in a formula, we can use our equation-solving skills to find the value of the remaining variable.

EXAMPLE 1

Films. Estimates are that Warner Brothers made a $219 million profit on the film *Harry Potter and the Half-Blood Prince.* If the studio received $469 million in worldwide box office revenue, find the cost to make and distribute the film. (Source: www.thenumbers.com, June 2010)

Strategy To find the cost to make and distribute the film, we will substitute the given values in the formula $p = r - c$ and solve for *c*.

Why The variable *c* in the formula represents the unknown cost.

Solution The film made $219 million (the profit *p*) and the studio took in $469 million (the revenue *r*). To find the cost *c*, we proceed as follows.

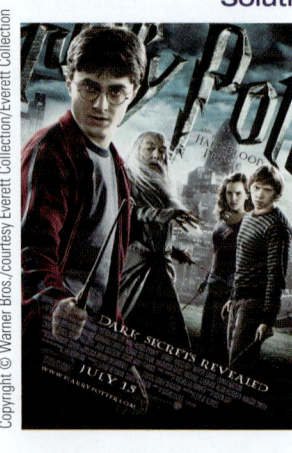

$$p = r - c \qquad \text{This is the formula for profit.}$$

$$219 = 469 - c \qquad \text{Substitute 219 for } p \text{ and 469 for } r.$$

$$219 - 469 = 469 - c - 469 \qquad \begin{array}{l}\text{To eliminate 469 on the right side,}\\ \text{subtract 469 from both sides.}\end{array}$$

$$-250 = -c \qquad \text{Do the subtraction.}$$

$$\frac{-250}{-1} = \frac{-c}{-1} \qquad \text{To solve for } c \text{, divide (or multiply) both sides by } -1.$$

$$250 = c \qquad \text{The units are millions of dollars.}$$

It cost $250 million to make and distribute the film.

Self Check 1   **Fundraisers.**   A PTA spaghetti dinner made a profit of $275.50. If the cost to host the dinner was $1,235, how much revenue did it generate?

Now Try ▶ Problem 13

The Language of Algebra

The word **annual** means occurring once a year. An *annual* interest rate is the interest rate paid per year.

A formula for simple interest: When money is borrowed, the lender expects to be paid back the amount of the loan plus an additional charge for the use of the money, called **interest.** When money is deposited in a bank, the depositor is paid for the use of the money. The money the deposit earns is also called interest.

Interest is calculated in two ways: either as **simple interest** or as **compound interest.** Simple interest is the *product* of the principal (the amount of money that is invested, deposited, or borrowed), the annual interest rate, and the length of time in years. Using I to represent the simple interest, P the principal, r the annual interest rate, and t the time in years, we can write this formula as

$$I = Prt \qquad \text{Interest} = \text{principal} \cdot \text{rate} \cdot \text{time}$$

EXAMPLE 2   **Retirement Income.**   One year after investing $15,000, a retired couple received a check for $1,125 in interest. Find the interest rate their money earned that year.

Strategy   To find the interest rate, we will substitute the given values in the formula $I = Prt$ and solve for r.

Why   The variable r in the formula represents the unknown interest rate.

Solution   The couple invested $15,000 (the principal P) for 1 year (the time t) and made $1,125 (the interest I). To find the annual interest rate r, we proceed as follows.

$$I = Prt \qquad \text{This is the formula for simple interest.}$$

$$1{,}125 = 15{,}000r(1) \qquad \text{Substitute 1,125 for } I, \text{ 15,000 for } P, \text{ and 1 for } t.$$

$$1{,}125 = 15{,}000r \qquad \text{Simplify the right side.}$$

$$\frac{1{,}125}{15{,}000} = \frac{15{,}000r}{15{,}000} \qquad \text{To solve for } r, \text{ undo the multiplication by 15,000 by dividing both sides by 15,000.}$$

$$0.075 = r \qquad \text{Do the division. This is the rate expressed as a decimal.}$$

$$7.5\% = r \qquad \text{To write 0.075 as a percent, multiply 0.075 by 100 by moving the decimal point two places to the right and inserting a \% symbol.}$$

Caution

When using the formula $I = Prt$, always write the interest rate r (which is given as a percent) as a decimal (or fraction) before performing any calculations.

The couple received an annual rate of 7.5% that year on their investment. We can display the facts of the problem in a table as shown on the right.

	P	$\cdot$ r	$\cdot t =$	I
Investment	15,000	0.075	1	1,125

Self Check 2   **Home Loans.**   A father loaned his daughter $12,200 at a 2% annual simple interest rate for a down payment on a house. If the interest on the loan amounted to $610, for how long was the loan?

Now Try ▶ Problem 17

2 Use Formulas from Science.

A formula for distance traveled: If we know the average rate (of speed) at which we will be traveling and the time we will be traveling at that rate, we can find the distance traveled. Using d to represent the distance, r the average rate, and t the time, we can write this formula as

$$d = rt \qquad \text{Distance} = \text{rate} \cdot \text{time}$$

EXAMPLE 3

Whales.   As they migrate from the Bering Sea to Baja California, gray whales swim for about 20 hours each day, covering a distance of approximately 70 miles. Estimate their average swimming rate in miles per hour. (Source: marinebio.net)

Strategy   To find the swimming rate, we will substitute the given values in the formula $d = rt$ and solve for r.

Why   The variable r in the formula represents the unknown average swimming rate.

Solution   The whales swam 70 miles (the distance d) in 20 hours (the time t). To find their average swimming rate r, we proceed as follows.

$d = rt$	This is the formula for distance traveled.
$70 = r(20)$	Substitute 70 for d and 20 for t.
$\dfrac{70}{20} = \dfrac{20r}{20}$	To solve for r, undo the multiplication by 20 by dividing both sides by 20.
$3.5 = r$	Do the division.

> **Caution**
>
> When using the formula $d = rt$, make sure the units are consistent. For example, if the rate is given in miles per hour, the time must be expressed in hours.

The whales' average swimming rate is 3.5 miles per hour (mph). The facts of the problem can be displayed in a table, as shown on the right.

	r	$\cdot$	t	$=$	d
Gray whale	3.5		20		70

Self Check 3   **Elevators.**   An elevator travels at an average rate of 288 feet per minute. How long will it take the elevator to climb 30 stories, a distance of 360 feet?

Now Try ▶ Problem 21

A formula for converting temperatures: In the American system, temperature is measured on the Fahrenheit scale. The Celsius scale is used to measure temperature in the metric system. The formula that relates a Fahrenheit temperature F to a Celsius temperature C is:

$$C = \frac{5}{9}(F - 32)$$

EXAMPLE 4

Temperature Conversion.   Convert the temperature shown on the City Savings sign to degrees Fahrenheit.

CITY SAVINGS
TEMP 30°C

Strategy   To find the temperature in degrees Fahrenheit, we will substitute the given Celsius temperature in the formula $C = \frac{5}{9}(F - 32)$ and solve for F.

Why   The variable F represents the unknown temperature in degrees Fahrenheit.

Solution   The temperature in degrees Celsius is 30°. To find the temperature in degrees Fahrenheit F, we proceed as follows.

$C = \dfrac{5}{9}(F - 32)$	This is the formula for temperature conversion.
$30 = \dfrac{5}{9}(F - 32)$	Substitute 30 for C, the Celsius temperature.
$\dfrac{9}{5} \cdot 30 = \dfrac{9}{5} \cdot \dfrac{5}{9}(F - 32)$	To undo the multiplication by $\frac{5}{9}$, multiply both sides by the reciprocal of $\frac{5}{9}$.
$54 = F - 32$	Do the multiplication.
$54 + 32 = F - 32 + 32$	To isolate F, undo the subtraction of 32 by adding 32 to both sides.
$86 = F$	Do the addition.

> **The Language of Algebra**
>
> In 1724, Daniel Gabriel **Fahrenheit,** a German scientist, introduced the temperature scale that bears his name. The Celsius scale was invented in 1742 by Swedish astronomer Anders **Celsius.**

30°C is equivalent to 86°F.

Planets.   Change $-175°C$, the temperature on Saturn, to degrees Fahrenheit. (Source: universetoday.com)

Problem 27

3 Use Formulas from Geometry.

To find the **perimeter** of a plane (two-dimensional, flat) geometric figure, such as a rectangle or triangle, we find the distance around the figure by computing the sum of the lengths of its sides. Perimeter is measured in American units, such as inches, feet, yards, and in metric units, such as millimeters, meters, and kilometers.

EXAMPLE 5

Perimeter formulas

$P = 2l + 2w$ (rectangle)
$P = 4s$ (square)
$P = a + b + c$ (triangle)

©AP Photo/Las Vegas Review Journal/Carol Ferguson

Flags.   The largest flag ever flown was an American flag that had a perimeter of 1,520 feet and a length of 505 feet. It was hoisted on cables across Hoover Dam to celebrate the 1996 Olympic Torch Relay. Find the width of the flag.

Strategy   To find the width of the flag, we will substitute the given values in the formula $P = 2l + 2w$ and solve for w.

Why   The variable w in the formula represents the unknown width of the flag.

Solution   The perimeter P of the rectangular-shaped flag is 1,520 ft and the length l is 505 ft. To find the width w, we proceed as follows.

$P = 2l + 2w$ This is the formula for the perimeter of a rectangle.

$1{,}520 = 2(505) + 2w$ Substitute 1,520 for P and 505 for l.

$1{,}520 = 1{,}010 + 2w$ Do the multiplication.

$510 = 2w$ To undo the addition of 1,010, subtract 1,010 from both sides. This step is done mentally and is not shown.

$255 = w$ To isolate w, undo the multiplication by 2 by dividing both sides by 2. This step is done mentally and is not shown.

The width of the flag is 255 feet. If its length is 505 feet and its width is 255 feet, its perimeter is $2(505) + 2(255) = 1{,}010 + 510 = 1{,}520$ feet, as given.

Flags.   The largest flag that consistently flies is the flag of Brazil in Brasilia, the country's capital. It has a perimeter 1,116 feet and length 328 feet. Find its width.

Problem 29

Area formulas

$A = lw$ (rectangle)
$A = s^2$ (square)
$A = \frac{1}{2}bh$ (triangle)
$A = \frac{1}{2}h(B + b)$ (trapezoid)

The **area** of a plane (two-dimensional, flat) geometric figure is the amount of surface that it encloses. Area is measured in square units, such as square inches, square feet, square yards, and square meters (written as in.^{2}, ft^2, yd^2, and m^2, respectively).

EXAMPLE 6

a. What is the circumference of a circle with diameter 14 feet? Round to the nearest tenth of a foot.   **b.** What is the area of the circle? Round to the nearest tenth of a square foot.

Strategy   To find the circumference and area of the circle, we will substitute the proper values into the formulas $C = \pi D$ and $A = \pi r^2$ and find C and A.

Why   The variable C in the formula represents the unknown circumference of the circle and A represents the unknown area.

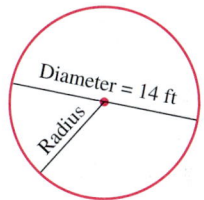

Diameter = 14 ft

Radius

Solution

Circle formulas

$D = 2r$ (diameter)

$r = \dfrac{1}{2}D$ (radius)

$C = 2\pi r = \pi D$ (circumference)

$A = \pi r^2$ (area)

a. Recall that the circumference of a circle is the distance around it. To find the circumference C of a circle with diameter D equal to 14 ft, we proceed as follows.

$C = \pi D$ This is the formula for the circumference of a circle. πD means $\pi \cdot D$.

$C = \pi(14)$ Substitute 14 for D, the diameter of the circle.

$C = 14\pi$ The exact circumference of the circle is 14π.

$C \approx 43.98229715$ To use a calculator to approximate the circumference, enter $\boxed{\pi}\ \boxed{\times}\ 14\ \boxed{=}$. If you do not have a calculator, use 3.14 as an approximation of π. (Answers may vary slightly depending on which approximation of π is used.)

The circumference is exactly 14π ft. Rounded to the nearest tenth, this is 44.0 ft.

Notation

Since π is an irrational number, its decimal representation has an infinite number of decimal places. When an approximation of π is used in a calculation, it produces an approximate answer. Use an *is approximately equal to* symbol $\approx$ in your solution to show that.

b. The radius r of the circle is one-half the diameter, or 7 feet. To find the area A of the circle, we proceed as follows.

$A = \pi r^2$ This is the formula for the area of a circle. πr^2 means $\pi \cdot r^2$.

$A = \pi(7)^2$ Substitute 7 for r, the radius of the circle.

$A = 49\pi$ Evaluate the exponential expression: $7^2 = 49$. The exact area is 49π ft^2.

$A \approx 153.93804$ To use a calculator to approximate the area, enter $49\ \boxed{\times}\ \boxed{\pi}\ \boxed{=}$.

The area is exactly 49π ft^2. To the nearest tenth, the area is 153.9 ft^2.

Self Check 6   Find the circumference of a circle with radius 10 inches. Round to the nearest hundredth of an inch.

Now Try ▶ Problem 30

The **volume** of a three-dimensional geometric solid is the amount of space it encloses. Volume is measured in cubic units, such as cubic inches, cubic feet, and cubic meters (written as in.^{3}, ft^3, and m^3, respectively).

EXAMPLE 7   Find the volume of the cylinder. Round to the nearest tenth of a cubic centimeter.

Success Tip

A complete list of geometric figures and formulas appears on the inside of the back cover of this textbook.

Strategy   To find the volume of the cylinder, we will substitute the proper values into the formula $V = \pi r^2 h$ and find V.

Why   The variable V in the formula represents the unknown volume.

Solution   Since the radius of a circle is one-half its diameter, the radius r of the circular base of the cylinder is $\dfrac{1}{2}(6 \text{ cm}) = 3$ cm. The height h of the cylinder is 12 cm. To find the volume V of the cylinder, we proceed as follows.

Volume formulas

$V = lwh$ (rectangular solid)

$V = s^3$ (cube)

$V = \dfrac{4}{3}\pi r^3$ (sphere)

$V = \pi r^2 h$ (cylinder)

$V = \dfrac{1}{3}\pi r^2 h$ (cone)

$V = \pi r^2 h$ This is the formula for the volume of a cylinder. $\pi r^2 h$ means $\pi \cdot r^2 \cdot h$.

$V = \pi(3)^2(12)$ Substitute 3 for r and 12 for h.

$V = \pi(9)(12)$ Evaluate the exponential expression.

$V = 108\pi$ Multiply. The exact volume is 108π cm^3.

$V \approx 339.2920066$ Use a calculator to approximate the volume.

To the nearest tenth, the volume is 339.3 cubic centimeters. This can be written as 339.3 cm^3.

Self Check 7   Find the volume of a cone whose base has radius 12 meters and whose height is 9 meters. Round to the nearest tenth of a cubic meter. Use the formula $V = \dfrac{1}{3}\pi r^2 h$.

Now Try ▶ Problem 31

4 Solve for a Specified Variable.

The Language of Algebra

The word **specified** is a form of the word *specify,* which means to select something for a purpose. Here, we select a variable for the purpose of solving for it.

Suppose a shopper wants to calculate the markup m on several items, knowing their retail price r and their cost c to the merchant. It would take a lot of time to substitute values for r and c into the formula for retail price $r = c + m$ and then repeatedly solve for m. A better way is to solve the formula for m first, substitute values for r and c, and then compute m directly.

To **solve a formula for a specified variable** means to isolate that variable on one side of the equation, with all other variables and constants on the opposite side.

EXAMPLE 8   Solve the retail-price formula $r = c + m$ for m.

Strategy   To solve for m, we will focus on it as if it is the only variable in the equation. We will use a strategy similar to that used to solve linear equations in one variable to isolate m on one side. (See page 117 if you need to review the strategy.)

Why   We can solve the formula as if it were an equation in one variable because all the other variables are treated as if they were numbers (constants).

Solution

We will isolate m on this side of the equation.

The Language of Algebra

We say that the formula is **solved for** m because m is alone on one side of the equation and the other side does not contain m.

$$r = c + m$$

$$r - c = c + m - c \qquad \text{To isolate } m, \text{ undo the addition of } c \text{ by subtracting } c \text{ from both sides.}$$

$$r - c = m \qquad \text{Simplify the right side: } c - c = 0.$$

$$m = r - c \qquad \text{Reverse the sides of the equation so that } m \text{ is on the left.}$$

The resulting formula, $m = r - c$, indicates that the markup on an item is the difference between its retail price and its cost to the merchant.

Self Check 8   Solve the profit formula $p = r - c$, for c.

Now Try ▶ Problem 33

EXAMPLE 9   Solve the area of a triangle formula $A = \frac{1}{2}bh$ for b.

Strategy   To solve for b, we will treat b as the only variable in the equation and use properties of equality to isolate it on one side. We will treat the other variables as if they were numbers (constants).

Why   To solve for a specified variable means to isolate it on one side of the equation.

Solution   We use the same steps to solve an equation for a specified variable that we use to solve equations with only one variable.

Success Tip

To solve for b, think of it as the only variable in the equation. Treat A and h as if they were numbers. It is often helpful to first circle the variable in the given formula that you are solving for:

$A = \frac{1}{2}\textcircled{b}h$

We will isolate b on this side of the equation.

$$A = \frac{1}{2}bh$$

$$2 \cdot A = 2 \cdot \frac{1}{2}bh \qquad \text{To clear the equation of the fraction, multiply both sides by 2.}$$

$$2A = bh \qquad \text{Simplify the right side: } 2 \cdot \frac{1}{2} = 1.$$

$$\frac{2A}{h} = \frac{bh}{h} \qquad bh \text{ means } b \cdot h. \text{ To isolate } b, \text{ undo the multiplication by } h \text{ by dividing both sides by } h.$$

$$\frac{2A}{h} = b \qquad \text{On the right side, remove the common factor of } h: \frac{b\cancel{h}}{\cancel{h}} = b.$$

$$b = \frac{2A}{h} \qquad \text{Reverse the sides of the equation so that } b \text{ is on the left.}$$

Self Check 9   Solve $A = \frac{1}{2}r^2a$ for a.

Now Try ▶ Problem 37

EXAMPLE 10   Solve the perimeter of a rectangle formula $P = 2l + 2w$ for l.

Strategy   To solve for l, we will treat l as the only variable in the equation and use properties of equality to isolate it on one side. We will treat the other variables as if they were numbers (constants).

Why   To solve for a specified variable means to isolate it on one side of the equation.

Solution

         ┌──────────── We will isolate l on this side of the equation.

$$P = 2l + 2w$$

$$P - 2w = 2l + 2w - 2w \qquad \text{To undo the addition of } 2w, \text{ subtract } 2w \text{ from both sides.}$$

$$P - 2w = 2l \qquad \text{Combine like terms: } 2w - 2w = 0.$$

$$\frac{P - 2w}{2} = \frac{2l}{2} \qquad \text{To isolate } l, \text{ undo the multiplication by 2 by dividing both sides by 2.}$$

$$\frac{P - 2w}{2} = l \qquad \text{Simplify the right side: } \frac{\overset{1}{2}l}{\underset{1}{2}} = l.$$

$$l = \frac{P - 2w}{2} \qquad \text{Reverse the sides of the equation so that } l \text{ is on the left.}$$

> **Caution**
>
> Do not try to simplify the result this way:
>
> $$l = \frac{P - \overset{1}{2}w}{\underset{1}{2}}$$
>
> This step is incorrect because 2 is not a factor of the entire numerator.

Self Check 10   Solve $B = 3c + 4d$ for c.

Now Try ▶ Problem 41

EXAMPLE 11   In Chapter 3, we will work with equations that involve the variables x and y, such as $3x + 5y = 10$. Solve this equation for y.

Strategy   To solve for y, we will treat y as the only variable in the equation and use properties of equality to isolate it on one side.

Why   To solve for a specified variable means to isolate it on one side of the equation.

Solution

       ┌──────────── We will isolate y on this side of the equation.

$$3x + 5y = 10$$

$$3x + 5y - 3x = 10 - 3x \qquad \text{To eliminate } 3x \text{ on the left side, subtract } 3x \text{ from both sides.}$$

$$5y = 10 - 3x \qquad \text{Combine like terms: } 3x - 3x = 0.$$

$$\frac{5y}{5} = \frac{10 - 3x}{5} \qquad \text{To isolate } y, \text{ undo the multiplication by 5 by dividing both sides by 5.}$$

$$y = \frac{10}{5} - \frac{3x}{5} \qquad \text{Write } \frac{10 - 3x}{5} \text{ as the difference of two fractions with like denominators, } \frac{10}{5} \text{ and } \frac{3x}{5}.$$

$$y = 2 - \frac{3}{5}x \qquad \text{Simplify: } \frac{10}{5} = 2. \text{ Write } \frac{3x}{5} \text{ as } \frac{3}{5}x.$$

> **Success Tip**
>
> When solving for a specified variable, there is often more than one way to express the result. Keep this in mind when you are comparing your answers with those in the back of the text.

This result can be written in the following equivalent form:

$$y = -\frac{3}{5}x + 2 \qquad \text{On the right side, write the x-term first.}$$

Self Check 11   Solve $x + 3y = 12$ for y.

Now Try ▶ Problem 45

Sometimes the distributive property is used to solve a formula for a specified variable.

EXAMPLE 12   Solve $S = 2\pi(r^2 + rh)$ for h.

Strategy   To solve for h, we will treat it as the only variable in the equation and isolate it on one side.

Why   To solve for a specified variable means to isolate it on one side of the equation.

Solution   We will isolate h on the right side of the equation. To begin, we use the distributive property to remove the parentheses.

$$S = 2\pi(r^2 + rh) \qquad \text{This is the given formula.}$$

$$S = 2\pi r^2 + 2\pi rh \qquad \text{Distribute the multiplication by } 2\pi.$$

$$S - 2\pi r^2 = 2\pi r^2 + 2\pi rh - 2\pi r^2 \qquad \text{To eliminate } 2\pi r^2 \text{ on the right side, subtract } 2\pi r^2 \text{ from both sides.}$$

$$S - 2\pi r^2 = 2\pi rh \qquad \text{On the right side, combine like terms: } 2\pi r^2 - 2\pi r^2 = 0.$$

$$\frac{S - 2\pi r^2}{2\pi r} = \frac{2\pi rh}{2\pi r} \qquad 2\pi rh \text{ means } 2 \cdot \pi \cdot r \cdot h. \text{ To isolate } h, \text{ undo the multiplication by } 2\pi r \text{ by dividing both sides by } 2\pi r.$$

$$\frac{S - 2\pi r^2}{2\pi r} = h \qquad \text{On the right side, remove the common factors of 2, } \pi, \text{ and r:}$$
$$\overset{1\ 1\ 1}{\frac{2\pi rh}{2\pi r}} = h.$$
$$\underset{1\ 1\ 1}{}$$

$$h = \frac{S - 2\pi r^2}{2\pi r} \qquad \text{Reverse the sides of the equation so that } h \text{ is on the left.}$$

Self Check 12   Solve $A = -xy(3 - s)$ for s.

Now Try ▶ Problem 49

SECTION 2.4 ▶ STUDY SET

VOCABULARY

Fill in the blanks.

1. A _____ is an equation that states a mathematical relationship between two or more variables.

2. The distance around a plane geometric figure is called its _____, and the amount of surface that it encloses is called its _____.

3. The _____ of a three-dimensional geometric solid is the amount of space it encloses.

4. The formula $a = P - b - c$ is _____ for a because a is isolated on one side of the equation and the other side does not contain a.

CONCEPTS

5. Use variables to write the formula relating:
   a. Time, distance, rate
   b. Markup, retail price, cost
   c. Costs, revenue, profit
   d. Interest rate, time, interest, principal

6. Complete the table.

	Principal·	rate ·	time =	interest
Account 1	$2,500	5%	2 yr	
Account 2	$15,000	4.8%	1 yr	

7. Complete the table to find how far light and sound travel in 60 seconds. (*Hint:* mi/sec means miles per second.)

	Rate	· time =	distance
Light	186,282 mi/sec	60 sec	
Sound	1,088 ft/sec	60 sec	

8. Determine which concept (perimeter, area, or volume) should be used to find each of the following. Then determine which unit of measurement, ft, ft^2, or ft^3, would be appropriate.

a. The amount of storage in a freezer

b. The amount of ground covered by a sleeping bag lying on the floor

c. The distance around a dance floor

NOTATION

Complete the solution.

9. Solve $Ax + By = C$ for y.

$$Ax + By = C$$
$$Ax + By - \boxed{} = C - \boxed{}$$
$$By = C - Ax$$
$$\frac{By}{\boxed{}} = \frac{C - Ax}{\boxed{}}$$
$$y = \frac{C - Ax}{\boxed{}}$$

10. Approximate 98π to the nearest hundredth.

11. a. Write $\pi \cdot r^2 \cdot h$ in simpler form.

b. In the formula $V = \pi r^2 h$, what does r represent? What does h represent?

12. a. What does 45°C mean?

b. What does 15°F mean?

GUIDED PRACTICE

Use a formula to solve each problem. See Example 1.

13. Hollywood. As of 2010, the movie *Titanic* had brought in $1,842 million worldwide and made a gross profit of $1,602 million. What did it cost to make the movie? (Source: the numbers.com/movies)

14. Valentine's Day. Find the markup on a dozen roses if a florist buys them wholesale for $12.95 and sells them for $47.50.

15. Service Clubs. After expenses of $55.15 were paid, a Rotary Club donated $875.85 in proceeds from a pancake breakfast to a local health clinic. How much did the pancake breakfast gross?

16. New Cars. The factory invoice for a minivan shows that the dealer paid $16,264.55 for the vehicle. If the sticker price of the van is $18,202, how much over factory invoice is the sticker price?

See Example 2.

17. Entrepreneurs. To start a mobile dog-grooming service, a woman borrowed $2,500. If the loan was for 2 years and the amount of interest was $175, what simple interest rate was she charged?

18. Savings. A man deposited $5,000 in a credit union paying 6% simple interest. How long will the money have to be left on deposit to earn $6,000 in interest?

19. Loans. A student borrowed some money from his father at 2% simple interest to buy a car. If he paid his father $360 in interest after 3 years, how much did he borrow?

20. Banking. Three years after opening an account that paid simple interest of 6.45% annually, a depositor withdrew the $3,483 in interest earned. How much money was left in the account?

See Example 3.

21. Swimming. In 1930, a man swam down the Mississippi River from Minneapolis to New Orleans, a total of 1,826 miles. He was in the water for 742 hours. To the nearest tenth, what was his average swimming rate?

22. Parades. Rose Parade floats travel down the 5.5-mile-long parade route at a rate of 2.5 mph. How long will it take a float to complete the route if there are no delays?

23. Hot-Air Balloons. If a hot-air balloon travels at an average of 37 mph, how long will it take to fly 166.5 miles?

24. Air Travel. An airplane flew from Chicago to San Francisco in 3.75 hours. If the cities are 1,950 miles apart, what was the average speed of the plane?

See Example 4.

25. Frying Foods. One of the most popular cookbooks in U.S. history, *The Joy of Cooking,* recommends frying foods at 365°F for best results. Convert this to degrees Celsius.

26. Freezing Points. Saltwater has a much lower freezing point than freshwater does. For saltwater that is saturated as much as it can possibly get (23.3% salt by weight), the freezing point is −5.8°F. Convert this to degrees Celsius.

27. Biology. Cryobiologists freeze living matter to preserve it for future use. They can work with temperatures as low as −270°C. Change this to degrees Fahrenheit.

28. Metallurgy. Change 2,212°C, the temperature at which silver boils, to degrees Fahrenheit. Round to the nearest degree.

See Examples 5–7. If you do not have a calculator, use 3.14 as an approximation of π. Answers may vary slightly depending on which approximation of π is used.

29. Energy Savings. One hundred inches of foam weather stripping tape was placed around the perimeter of a rectangular-shaped window. If the length of the window is 30 inches, what is its width?

30. Rugs. Find the amount of floor area covered by a circular throw rug that has a radius of 15 inches. Round to the nearest square inch.

31. Straws. Find the volume of a 150 millimeter-long drinking straw that has an inside diameter of 4 millimeters. Round to the nearest cubic millimeter.

32. Rubber Bands. The world's largest rubber band ball is $5\frac{1}{2}$ ft tall and was made in 2006 by Steve Milton of Eugene, Oregon. Find the volume of the ball. Round to the nearest cubic foot. (*Hint:* The formula for the volume of a sphere is $V = \frac{4}{3}\pi r^3$.) (Source: timesunion.com)

Solve for the specified variable. See Example 8.

33. $r = c + m$ for c

34. $p = r - c$ for r

35. $P = a + b + c$ for b

36. $a + b + c = 180$ for a

Solve for the specified variable. See Example 9.

37. $V = \frac{1}{3}Bh$ for h

38. $C = \frac{1}{7}Rt$ for R

39. $E = IR$ for R

40. $d = rt$ for t

Solve for the specified variable. See Example 10.

41. $T = 2r + 2t$ for r

42. $y = mx + b$ for x

43. $Ax + By = C$ for x

44. $A = P + Prt$ for t

Solve for y. See Example 11.

45. $2x + 7y = 21$

46. $3x + 4y = 20$

47. $9x - 2y = -8$

48. $5x - 6y = -12$

Solve for the specified variable. See Example 12.

49. $T = 4b(a + am)$ for m

50. $f = 7n(d + dz)$ for z

51. $G = g(4r - 1)$ for r

52. $F = f(9n - 1)$ for n

TRY IT YOURSELF

Solve for the specified variable or expression.

53. $A = \dfrac{a + b + c}{3}$ for c

54. $x = \dfrac{a + b}{2}$ for b

55. $3x + y = 9$ for y

56. $-5x + y = 4$ for y

57. $K = \frac{1}{2}mv^2$ for m

58. $V = \frac{1}{3}\pi r^2 h$ for h

59. $C = 2\pi r$ for r

60. $V = \pi r^2 h$ for h

61. $\dfrac{M}{2} - 9.9 = 2.1B$ for M

62. $\dfrac{G}{0.5} + 16r = -8t$ for G

63. $w = \dfrac{s}{f}$ for f

64. $P = \dfrac{ab}{c}$ for c

65. $-x + 3y = 9$ for y

66. $5y - x = 25$ for y

67. $A = \frac{1}{2}h(b + d)$ for b

68. $C = \frac{1}{4}s(t - d)$ for t

69. $c^2 = a^2 + b^2$ for a^2

70. $x^2 + y^2 + z^2 = d^2$ for y^2

71. $\frac{7}{8}c + w = 9$ for c

72. $\frac{3}{4}m - t = 5b$ for m

73. $m = 70 + t(a + b)$ for b

74. $B = 50 + r(x + y)$ for y

75. $V = lwh$ for l

76. $I = Prt$ for r

77. $2E = \dfrac{T - t}{9}$ for t

78. $D = \dfrac{C - s}{n}$ for s

79. $s = 4\pi r^2$ for r^2

80. $E = mc^2$ for c^2

Look Alikes . . .

81. Solve $A = R + ab$
  a. for R **b.** for a

82. Solve $m = (a + d)T$
  a. for T **b.** for d

83. Solve $S = 2(2lw + wh)$
  a. for h **b.** for l

84. Solve $t = -40 + 9(r + az)$
  a. for r **b.** for z

APPLICATIONS

85. from **Campus to Careers**

Automotive Service Technician

One of the formulas that is often used by automotive technicians who service engines is:

$$\text{Torque} = \frac{5{,}252 \cdot \text{Horsepower}}{\text{RPM}}$$

RPM stands for revolutions per minute. Solve the formula for horsepower.

86. Properties of Water. Refer to the illustration below. Use the temperature formula from this section to find the boiling point of water in degrees Fahrenheit and the freezing point of water in degrees Celsius.

Water boils
100° C

Water freezes
32° F

87. Avon Products. Complete the financial statement.

Income statement (dollar amounts in millions)	Quarter ending March '09	Quarter ending March '10
Revenue	2,186.9	2,490.4
Cost of goods sold	2,018.5	2,297.6
Operating profit		

Source: Avon Products, Inc.

88. Credit Cards. The finance charge that a student pays on his credit card is 19.8% APR (annual percentage rate). Determine the finance charges (interest) the student would have to pay if the account's average balance for the year was $2,500.

89. Campers. The perimeter of the window of the camper shell is 140 in. Find the length of one of the shorter sides of the window.

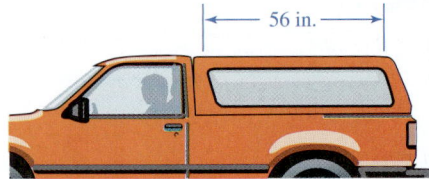

← 56 in. →

90. Flags. The flag of Eritrea, a country in east Africa, is shown. The perimeter of the flag is 160 inches.

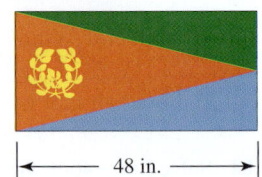

← 48 in. →

a. What is the width of the flag?

b. What is the area of the red triangular region of the flag?

91. Kites. 650 in.² of nylon cloth were used to make the kite shown. If its height is 26 inches, what is the wingspan?

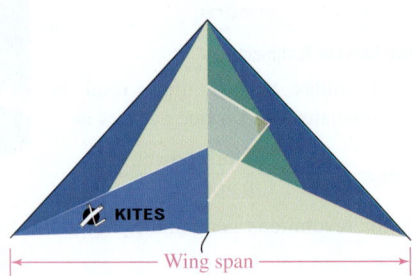

KITES

← Wing span →

92. Memorials. The Vietnam Veterans Memorial is a black granite wall recognizing the more than 58,000 Americans who lost their lives or remain missing. Find the total area of the two triangular-shaped surfaces on which the names are inscribed.

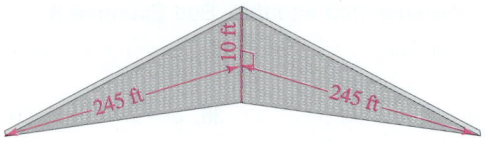

245 ft   10 ft   245 ft

93. Wheelchairs. Find the diameter of the rear wheel and the radius of the front wheel.

12.5 in.   5 in.

94. Archery. The diameter of a standard archery target used in the Olympics is 48.8 inches. Find the area of the target. Round to the nearest square inch.

95. Bulls-Eye. See Exercise 94. The diameter of the center yellow ring of a standard archery target is 4.8 inches. What is the area of the bulls-eye? Round to the nearest tenth of a square inch.

96. Geography. The circumference of the Earth is about 25,000 miles. Find its diameter to the nearest mile.

97. Horses. A horse trots in a circle around its trainer at the end of a 28-foot-long rope. Find the area of the circle that is swept out. Round to the nearest square foot.

98. Yo-Yos. How far does a yo-yo travel during one revolution of the "around the world" trick if the length of the string is 21 inches?

99. History. The Inca Empire (1438–1533) was centered in what is now called Peru. A special feature of Inca architecture was the trapezoid-shaped windows and doorways. A standard Inca window was 70 cm high, 50 cm at the base, and 40 cm at the top. Find the area of a window opening.

©DIOMEDIA/Alamy

100. Hamster Habitats. Find the amount of space in the tube.

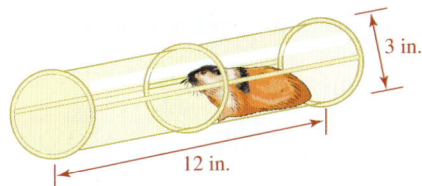
3 in.

12 in.

101. Tires. The road surface footprint of a sport truck tire is approximately rectangular. If the area of the footprint is 45 in.², about how wide is the tire?

$7\frac{1}{2}$ in.

102. Softball. The strike zone in fast-pitch softball is between the batter's armpit and the top of her knees, as shown. If the area of the strike zone for this batter is 442 in.², what is the width of home plate?

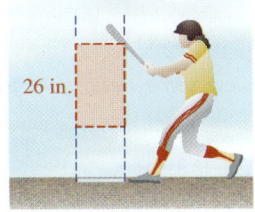
26 in.

103. Firewood. The cord of wood shown occupies a volume of 128 ft³. How long is the stack?

4 ft

4 ft

104. Teepees. The teepees constructed by the Blackfoot Indians were cone-shaped tents about 10 feet high and about 15 feet across at the ground. Estimate the volume of a teepee with these dimensions, to the nearest cubic foot.

105. Igloos. During long journeys, some Canadian Eskimos built winter houses of snow blocks stacked in the dome shape shown. Estimate the volume of an igloo having an interior height of 5.5 feet to the nearest cubic foot.

106. Pyramids. The Great Pyramid at Giza in northern Egypt is one of the most famous works of architecture in the world. Find its volume to the nearest cubic foot. (*Hint:* The formula to use is on the inside back cover.)

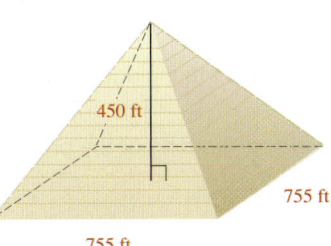
450 ft

755 ft

755 ft

107. Cooking. If the fish shown in the illustration is 18 inches long, what is the area of the grill? Round to the nearest square inch.

108. Skateboarding. A half-pipe ramp is in the shape of a semicircle with a radius of 8 feet. To the nearest tenth of a foot, what is the length of the arc that the rider travels on the ramp?

8 ft

Plywood

109. Pulleys. The approximate length L of a belt joining two pulleys of radii r and R feet with centers D feet apart is given by the formula $L = 2D + 3.25(r + R)$. Solve the formula for R.

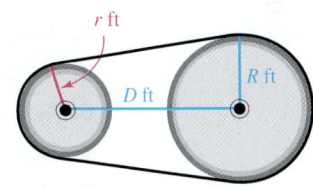
r ft

R ft

D ft

110. Thermodynamics. The Gibbs free-energy formula is given by $G = U - TS + pV$. Solve the formula for the pressure p.

WRITING

111. After solving $A = B + C + D$ for B, a student compared her answer with that at the back of the textbook. Could this problem have two different-looking answers? Explain why or why not.

Student's answer: $B = A - C - D$

Book's answer: $B = A - D - C$

112. A student solved $x + 5c = 3c + a$ for c. His answer was $c = \frac{3c + a - x}{5}$. Explain why the equation is not solved for c.

113. Explain the difference between what perimeter measures and what area measures.

114. Explain the error made below.

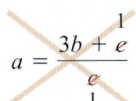
$$a = \frac{\overset{1}{\cancel{3b + e}}}{\underset{1}{\cancel{e}}}$$

115. Find 82% of 168.

116. 29.05 is what percent of 415?

117. What percent of 200 is 30?

118. Shopping. A woman bought a coat for $98.95 and some gloves for $7.95. If the sales tax was 6%, how much did the purchase cost her?

119. In mathematics, letters from the Greek alphabet are often used as variables. Solve the following equation for α (read as "alpha"), the first letter of the Greek alphabet.

$$-7(\alpha - \beta) - (4\alpha - \theta) = \frac{\alpha}{2}$$

120. Solve $B = R - \frac{1}{16}(c - 3D)$ for D.

SECTION 2.5

Problem Solving

OBJECTIVES

1 Apply the steps of a problem-solving strategy.

2 Solve consecutive integer problems.

3 Solve geometry problems.

ARE YOU READY?

The following problems review some basic skills that are needed to solve the application problems in this section.

1. Simplify: $x + x + 1 + 2x + 3$

2. If $x = 8$, find $6x + 1$.

3. If staplers cost $4.35 each, what is the cost of 9 staplers?

4. Simplify: $x - 0.72x$

5. What is the formula for the perimeter of a rectangle?

6. What is the sum of the measures of the angles of a triangle?

7. Translate to symbols: *8 less than twice a number x*

8. Write 6% as a decimal.

In this section, you will see that algebra is a powerful tool that can be used to solve a wide variety of real-world problems.

1 **Apply the Steps of a Problem-Solving Strategy.**

To become a good problem solver, you need a plan to follow, such as the following six-step strategy.

Strategy for Problem Solving

1. **Analyze the problem** by reading it carefully to understand the given facts. What information is given? What are you asked to find? What vocabulary is given? Often, a diagram or table will help you understand the facts of the problem.

2. **Assign a variable** to represent an unknown value in the problem. This means, in most cases, to let $x =$ what you are asked to find. If there are other unknown values, represent each of them using an algebraic expression that involves the variable.

3. **Form an equation** by translating the words of the problem into mathematical symbols.

4. **Solve the equation** formed in step 3.

5. **State the conclusion clearly.** Be sure to include the units (such as feet, seconds, or pounds) in your answer.

6. **Check the result** using the original wording of the problem, not the equation that was formed in step 3.

EXAMPLE 1   **California Coastline.**   The first part of California's magnificent 17-Mile Drive begins at the Pacific Grove entrance and continues to Seal Rock. It is 1 mile longer than the second part of the drive, which extends from Seal Rock to the Lone Cypress as shown in the map below. The third and final part of the drive winds through Pebble Beach, eventually returning to the entrance. This part of the drive is 1 mile longer than four times the length of the second part. How long is each part of 17-Mile Drive?

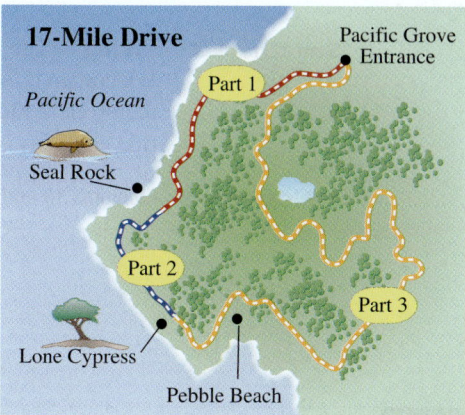

©michalis/Shutterstock.com

Success Tip

When there is more than one unknown value in a problem, let the variable represent the unknown value on which any other unknown values are based.

Analyze   The drive is composed of three parts. We need to find the length of each part. We can straighten out the winding 17-Mile Drive and model it with a line segment.

Assign   Since the lengths of the first part and of the third part of the drive are related to the length of the second part, we will let x represent the length of the second part. We then express the other lengths in terms of x.

$$x = \text{the length of the second part of the drive (in miles)}$$
$$x + 1 = \text{the length of the first part of the drive (in miles)}$$
$$4x + 1 = \text{the length of the third part of the drive (in miles)}$$

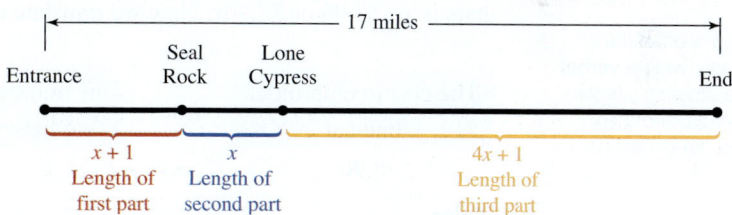

Form   Now we translate the words of the problem to an equation.

The length of part 1	plus	the length of part 2	plus	the length of part 3	equals	the total length of the drive.
$x + 1$	$+$	x	$+$	$4x + 1$	$=$	17

Solve

$$x + 1 + x + 4x + 1 = 17$$
$$6x + 2 = 17 \qquad \text{Combine like terms: } x + x + 4x = 6x \text{ and } 1 + 1 = 2.$$
$$6x = 15 \qquad \text{To undo the addition of 2, subtract 2 from both sides.}$$
$$\frac{6x}{6} = \frac{15}{6} \qquad \text{To isolate } x, \text{ undo the multiplication by 6 by dividing both sides by 6.}$$
$$x = 2.5 \qquad \text{Do the division.}$$

Recall that x represents the length of the second part of the drive. To find the lengths of the first and third parts, we evaluate $x + 1$ and $4x + 1$ for $x = 2.5$.

First part of drive	*Third part of drive*	
$x + 1 = \mathbf{2.5} + 1$	$4x + 1 = 4(\mathbf{2.5}) + 1$	Substitute 2.5 for x.
$= 3.5$	$= 11$	The units are miles.

State   The first part of the drive is 3.5 miles long, the second part is 2.5 miles long, and the third part is 11 miles long.

Check   Since 3.5 mi + 2.5 mi + 11 mi = 17 mi, the results check.

Self Check 1   **Biking.**   The Mountain-Bay State Park Bike Trail in Northeast Wisconsin is 76 miles long. A couple rode the trail in four days. Each day they rode 2 miles more than the previous day. How many miles did they ride each day?

Now Try ▶ Problems 13 and 15

EXAMPLE 2   **Computer Logos.**   A trucking company had its logo embroidered on the front of baseball caps. It was charged $8.90 per hat plus a one time set up fee of $25. If the project cost $559, how many hats were embroidered?

Analyze

- It cost $8.90 to have a logo embroidered on a hat.
- The set up charge was $25.
- The project cost $559.
- We need to find the number of hats that were embroidered.

Assign   Let x = the number of hats that were embroidered.

Form   If x hats are embroidered, at a cost of $8.90 per hat, the cost to embroider all of the hats is $x \cdot \$8.90$ or $\$8.90x$. Now we translate the words of the problem into an equation.

The cost to embroider one hat	times	the number of hats	plus	the set up charge	equals	the total cost.
8.90	·	x	+	25	=	559

Solve

$$8.90x + 25 = 559$$

$$8.90x = 534 \qquad \text{To undo the addition of 25, subtract 25 from both sides.}$$

$$\frac{8.90x}{\mathbf{8.90}} = \frac{534}{\mathbf{8.90}} \qquad \text{To isolate x, undo the multiplication by 8.90 by dividing both sides by 8.90.}$$

$$x = 60 \qquad \text{Do the division.}$$

State   The company had 60 hats embroidered.

Check   The cost to embroider 60 hats is 60($8.90) = $534. When the $25 set up charge is added, we get $534 + $25 = $559. The result checks.

Self Check 2   **T-shirts.**   A school club had their motto screenprinted on the front of T-shirts. They were charged $5 per shirt plus a one-time set up fee of $20. If the project cost $255, how many T-shirts were printed?

Now Try ▶ Problem 23

EXAMPLE 3

©iStockphoto.com/Catherine Yeulet

Auctions. A classic car owner is going to sell his 1959 Chevy Impala at an auction. He wants to make $46,000 after paying an 8% commission to the auctioneer. What should be the selling price (called the "hammer price") for the car owner to make this amount of money?

Analyze When the commission is subtracted from the selling price of the car, the owner wants to have $46,000 left. We need to find the selling price.

Assign Let x = the selling price of the car.

Form The amount of the commission is 8% of x, or $0.08x$. Now we translate the words of the problem to an equation.

The selling price of the car	minus	the auctioneer's commission	should be	$46,000.
x	$-$	$0.08x$	$=$	$46,000$

The Language of Algebra

Here are some words and phrases that often translate to an equal symbol =.

is	are
should be	will be
yields	amounts to
represents	gives
is the same as	was

Solve

$$x - 0.08x = 46,000$$
$$0.92x = 46,000$$

Combine like terms: 1.00x − 0.08x = 0.92x. We could begin with this equation because after the 8% commission is paid, 100%–8% or 92% of the selling price should be $46,000.

$$\frac{0.92x}{0.92} = \frac{46,000}{0.92}$$

To isolate x, undo the multiplication by 0.92 by dividing both sides by 0.92.

$$x = 50,000$$

Do the division. This is the selling price of the car.

State The owner will make $46,000 if the car sells for $50,000.

Check An 8% commission on $50,000 is 0.08($50,000) = $4,000. The owner will keep $50,000 − $4,000 = $46,000. The result checks.

Self Check 3   **Cattle Auction.** A farmer is going to sell one of his Black Angus cattle at an auction and would like to make $2,597 after paying a 6% commission to the auctioneer. For what selling price will the farmer make this amount of money?

Now Try ▶ Problem 29

2 Solve Consecutive Integer Problems.

Integers that follow one another, such as 15 and 16, are called **consecutive integers.** They are 1 unit apart. **Consecutive even integers** are even integers that differ by 2 units, such as 12 and 14. Similarly, **consecutive odd integers** differ by 2 units, such as 9 and 11. When solving consecutive integer problems, if we let x = the first integer, then

- two consecutive integers are x and $x + 1$
- two consecutive even integers are x and $x + 2$
- two consecutive odd integers are x and $x + 2$

- three consecutive integers are x, $x + 1$, and $x + 2$
- three consecutive even integers are x, $x + 2$, and $x + 4$
- three consecutive odd integers are x, $x + 2$, and $x + 4$

EXAMPLE 4

U.S. History. The year George Washington was chosen president and the year the Bill of Rights went into effect are consecutive odd integers whose sum is 3,580. Find the years.

Analyze We need to find two consecutive odd integers whose sum is 3,580. From history, we know that Washington was elected president first and the Bill of Rights went into effect later.

Assign Let x = the first odd integer (the date when Washington was chosen president). The next odd integer is 2 *greater than* x, therefore $x + 2$ = the next larger odd integer (the date when the Bill of Rights went into effect).

The Language of Algebra

Consecutive means following one after the other in order. Elton John holds the record for the most *consecutive* years with a song on the Top 50 music chart: 31 years (1970 to 2000).

Form

The first odd integer	plus	the second odd integer	is	3,580.
x	$+$	$x + 2$	$=$	3,580

Solve

$$x + x + 2 = 3{,}580$$

$2x + 2 = 3{,}580$ Combine like terms: x + x = 2x.

$2x = 3{,}578$ To undo the addition of 2, subtract 2 from both sides.

$x = 1{,}789$ To isolate x, undo the multiplication by 2 by dividing both sides by 2.

State   George Washington was chosen president in the year 1789. The Bill of Rights went into effect in $1789 + 2 = 1791$.

Check   1789 and 1791 are consecutive odd integers whose sum is $1789 + 1791 = 3{,}580$. The answers check.

Self Check 4   **Dictionaries.**   The definitions of the words *little* and *lobby* are on back-to-back pages in a dictionary. If the sum of the page numbers is 1,159, on what page can the definition of *little* be found?

Now Try ▶ **Problem 37**

3 Solve Geometry Problems.

EXAMPLE 5   **Crime Scenes.** Police used 400 feet of yellow tape to fence off a rectangular-shaped lot for an investigation. They used 50 fewer feet of tape for each width than for each length. Find the dimensions of the lot.

Analyze   Since the yellow tape surrounded the lot, the concept of perimeter applies. Recall that the formula for the perimeter of a rectangle is $P = 2l + 2w$. We also know that the width of the lot is 50 feet less than the length.

The Language of Algebra

Dimensions are measurements of length and width. We might speak of the *dimensions* of a dance floor or a TV screen.

Assign   Since the width of the lot is given in terms of the length, we let l = the length of the lot. Then $l - 50$ = the width.

Form   Using the perimeter formula, we have:

2	times	the length	plus	2	times	the width	is	the perimeter.
2	$\cdot$	l	$+$	2	$\cdot$	$(l - 50)$	$=$	400

Success Tip

When solving geometry problems, a sketch is often helpful.

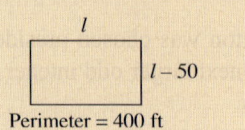

Perimeter = 400 ft

Solve

$2l + 2(l - 50) = 400$ Write the parentheses so that the entire expression *l* − 50 is multiplied by 2.

$2l + 2l - 100 = 400$ Distribute the multiplication by 2.

$4l - 100 = 400$ Combine like terms: 2*l* + 2*l* = 4*l*.

$4l = 500$ To undo the subtraction of 100, add 100 to both sides.

$l = 125$ To isolate *l*, undo the multiplication by 4 by dividing both sides by 4.

State   The length of the lot is 125 feet and width is $125 - 50 = 75$ feet.

Check   The width (75 feet) is 50 less than the length (125 feet). The perimeter of the lot is $2(125) + 2(75) = 250 + 150 = 400$ feet. The results check.

> **Self Check 5**   **Counters.**   A rectangular counter for the customer service department of a store is 6 feet longer than it is wide. If the perimeter is 32 feet, find the outside dimensions of the counter.
>
> **Now Try** ▶ Problem 43

EXAMPLE 6   **Isosceles Triangles.**   If the vertex angle of an isosceles triangle is 56°, find the measure of each base angle.

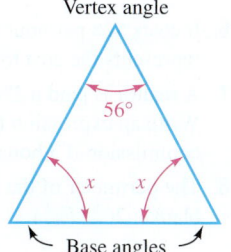

Analyze   An **isosceles triangle** has two sides of equal length, which meet to form the **vertex angle**. In this case, the measurement of the vertex angle is 56°. We can sketch the triangle as shown. The **base angles** opposite the equal sides are also equal. We need to find their measure.

Assign   If we let $x =$ the measure (in degrees) of one base angle, the measure of the other base angle is also x.

Form   Since the sum of the angles of any triangle is 180°, the sum of the base angles and the vertex angle is 180°. We can use this fact to form the equation.

One base angle	plus	the other base angle	plus	the vertex angle	is	180°.
x	$+$	x	$+$	56	$=$	180

Solve

$$x + x + 56 = 180$$
$$2x + 56 = 180 \qquad \text{Combine like terms: } x + x = 2x.$$
$$2x = 124 \qquad \text{To undo the addition of 56, subtract 56 from both sides.}$$
$$x = 62 \qquad \text{To isolate } x, \text{ undo the multiplication by 2 by dividing both sides by 2.}$$

State   The measure of each base angle is 62°.

Check   Since $62° + 62° + 56° = 180°$, the answer checks.

> **Self Check 6**   **Geometry.**   The perimeter of an isosceles triangle is 32 cm. If the base is 8 cm, find the length of each remaining side.
>
> **Now Try** ▶ Problem 47

SECTION 2.5 ▶ STUDY SET

VOCABULARY

Fill in the blanks.

1. Integers that follow one another, such as 7 and 8, are called _____ integers.

2. An _____ triangle is a triangle with two sides of the same length.

3. The equal sides of an isosceles triangle meet to form the _____ angle. The angles opposite the equal sides are called _____ angles, and they have equal measures.

4. When asked to find the dimensions of a rectangle, we are to find its _____ and _____.

CONCEPTS

5. A 17-foot pipe is cut into three sections. The longest section is three times as long as the shortest, and the middle-sized section is 2 feet longer than the shortest. Complete the diagram.

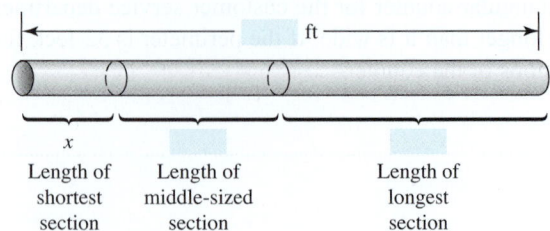

6. It costs $28 per hour to rent a trailer. Write an expression that represents the cost to rent the trailer for x hours.

7. A realtor is paid a 3% commission on the sale of a house. Write an expression that represents the amount of the commission if a house sells for $$x$.

8. The perimeter of the rectangle below is 15 feet. Fill in the blanks: $2(\quad) + 2x = $

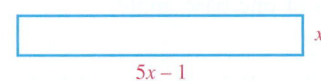

9. What is the sum of the measures of the angles of any triangle?

10. Refer to the isosceles triangle on the right.
  a. Find the missing angle measure.
  b. Find the missing side length.

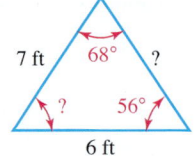

NOTATION

11. a. If x represents an integer, write an expression for the next largest integer.
  b. If x represents an odd integer, write an expression for the next largest odd integer.
  c. If x represents an even integer, write expressions for the next two largest even integers.

12. What does 45° mean?

GUIDED PRACTICE

See Example 1.

13. A 12-foot board has been cut into two sections, one twice as long as the other. How long is each section?

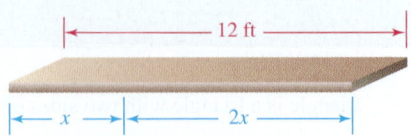

14. The robotic arm will extend a total distance of 18 feet. Find the length of each section.

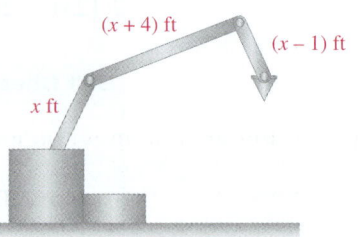

APPLICATIONS

15. National Parks. The Natchez Trace Parkway is a historical 444-mile route from Natchez, Mississippi, to Nashville, Tennessee. A couple drove the Trace in four days. Each day they drove 6 miles more than the previous day. How many miles did they drive each day?

16. Touring. A rock group plans to travel for a total of 38 weeks, making three concert stops. They will be in Japan for 4 more weeks than they will be in Australia. Their stay in Sweden will be 2 weeks shorter than that in Australia. How many weeks will they be in each country?

17. Solar Heating. Two solar panels were installed side-by-side on a roof, as shown below. One panel is 3.4 feet wider than the other. Find the width of each panel.

18. Accounting. Determine the 2010 income of Aeropostale Inc. for each quarter from the data in the graph below.

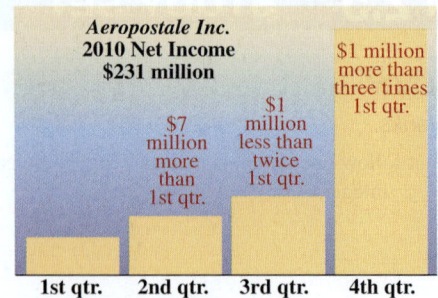

Source: moneycentral.msn.com

19. **iPhone Apps.** A student spent a total of $52.97 on the purchase of three applications in the Apps Store on his iPhone. *Call of Duty: World at War: Zombies II* cost $7 more than *Guitar Hero,* and *Tom Tom USA* cost $4.11 more than twelve times *Guitar Hero.* Find the cost of each application.

20. **Water Usage.** It takes about 3.8 times more gallons of water to produce one pound of grain-fed beef than it does to produce one pound of grain-fed chicken. If a combined total of 3,135 gallons of water are needed to produce one pound of each meat, how many gallons does it take to produce one pound of chicken? How many gallons does it take to produce one pound of beef? (Source: The Sierra Club: *The True Cost of Food*)

21. **Counting Calories.** A slice of pie with a scoop of ice cream has 850 calories. The calories in the pie alone are 100 more than twice the calories in the ice cream alone. How many calories are in each food?

22. **Waste Disposal.** Two tanks hold a total of 45 gallons of a toxic solvent. One tank holds 6 gallons more than twice the amount in the other. How many gallons does each tank hold?

23. **Concerts.** The fee to rent a concert hall is $2,250 plus $150 per hour to pay for the support staff. For how many hours can an orchestra rent the hall and stay within a budget of $3,300?

24. **Truck Mechanics.** An engine repair cost a truck owner $1,185 in parts and labor. If the parts were $690 and the mechanic charged $45 per hour, how many hours did the repair take?

25. **Field Trips.** It costs a school $65 a day plus $0.25 per mile to rent a 15-passenger van. If the van is rented for two days, how many miles can be driven on a $275 budget?

26. **Decorations.** A party supply store charges a set-up fee of $80 plus 35¢ per balloon to make a balloon arch. A business has $150 to spend on decorations for their grand opening. How many balloons can they have in the arch? (*Hint:* 35¢ = $0.35.)

27. **Tutoring.** High school students enrolling in a private tutoring program must first take a placement test (cost $25) before receiving tutoring (cost $18.75 per hour). If a family has set aside $400 to get their child extra help, how many hours of tutoring can they afford?

28. **Data Conversion.** The *Books2Bytes* service converts old print books to Microsoft Word electronic files for $20 per book plus $2.25 per page. If it cost $1,201.25 to convert a novel, how many pages did the novel have?

29. **Cattle Auctions.** A cattle rancher is going to sell one of his prize bulls at an auction and would like to make $45,500 after paying a 9% commission to the auctioneer. For what selling price will the rancher make this amount of money?

30. **Listing Price.** At what price should a home be listed if the owner wants to make $567,000 on its sale after paying a 5.5% real estate commission?

31. **Selling Used Clothing.** A *consignment shop* accepts an item of clothing that no longer fits (or one you have grown tired of) and sells it for you. The shop then charges you an agreed on percent of the selling price as their profit. Suppose the owner of a designer wool coat would like to make $210 on its sale at a consignment shop. If there is a $12\frac{1}{2}$% consignment charge, for what price must the coat be sold?

32. **Finder's Fees.** A *finder's fee* is an amount of money that is paid to someone who brings people together for business purposes. Suppose the owner of a software company needs to sell it and make $9,950,000 to pay back creditors. If he expects to pay a finder's fee of $\frac{1}{2}$% of the selling price to find a qualified buyer, for what price must the company be sold?

33. **Savings Accounts.** The balance in a savings account grew by 5% in one year, to $5,512.50. What was the balance at the beginning of the year?

34. **Aluminum Cans.** Today's aluminum cans are much thinner and lighter than those of the past. From 1972 to 2010, the number of empty cans produced from one pound of aluminum has increased by about 45%. If 32 cans could be produced from one pound of aluminum in 2010, how many cans could be produced from one pound of aluminum in 1972? Round to the nearest can. (Source: cancentral.com)

Consecutive integer problems

35. **Soccer.** Ronaldo of Brazil and Gerd Mueller of Germany rank 1 and 2, respectively, with the most goals scored in World Cup play. The number of goals Ronaldo and Mueller have scored are consecutive integers that total 29. Find the number of goals scored by each man. (Source: planetworldcup.com)

36. **Dictionaries.** The definitions of the words *job* and *join* are on back-to-back pages in a dictionary. If the sum of those page numbers is 1,411, on what page can the definition of *job* be found?

37. **TV History.** *Friends* and *Leave It to Beaver* are two of the most popular television shows of all time. The number of episodes of each show are consecutive even integers whose sum is 470. If there are more episodes of *Friends,* how many episodes of each were there? (Source: angelfire.com)

38. **Time Off.** The table shows the average number of days off an employed adult receives for selected countries. Complete the table. The numbers of days are listed in descending order.

Average Number of Days Off per Year*	
Country	**Days**
Brazil	41
Norway	?
South Africa	?
U.S.	25

Consecutive odd integers whose sum is 68.

* Employee has 10 years of service and works 5 days a week
Source: *The Wall Street Journal,* 2009.

39. **Celebrity Birthdays.** Selena Gomez, Jennifer Lopez, and Sandra Bullock have birthdays (in that order) on consecutive even-numbered days in July. The sum of the calendar dates of their birthdays is 72. Find each birthday.

40. **Locks.** The three numbers of the combination for a lock are consecutive integers, and their sum is 81. Find the combination.

Geometry problems

41. Tennis.   The perimeter of a regulation singles tennis court is 210 feet and the length is 3 feet less than three times the width. What are the dimensions of the court?

42. Swimming Pools.   The seawater Orthlieb Pool in Casablanca, Morocco, is the largest swimming pool in the world. With a perimeter of 1,110 meters, this rectangular-shaped pool is 30 meters longer than 6 times its width. Find its dimensions.

43. Art.   The *Mona Lisa* was completed by Leonardo da Vinci in 1506. The length of the picture is 11.75 inches shorter than twice the width. If the perimeter of the picture is 102.5 inches, find its dimensions.

44. New York City.   Central Park, which lies in the middle of Manhattan, is rectangular-shaped and has a 6-mile perimeter. The length is 5 times the width. What are the dimensions of the park?

45. Engineering.   A truss is in the form of an isosceles triangle. Each of the two equal sides is 4 feet shorter than the third side. If the perimeter is 25 feet, find the lengths of the sides.

46. First Aid.   A sling is in the shape of an isosceles triangle with a perimeter of 144 inches. The longest side of the sling is 18 inches longer than either of the other two sides. Find the lengths of each side.

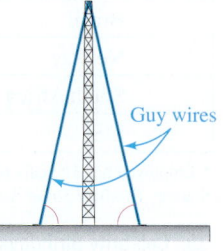

47. TV Towers.   A TV tower is supported by several guy wires. Two of the guy wires are attached to the top of the tower to form an isosceles triangle with the ground, as shown on the right. The measure of each of the base angles of the triangle is 4 times the third angle (the vertex angle). Find the measure of the vertex angle.

Guy wires

48. Clotheslines.   A pair of damp jeans are hung in the middle of a clothesline to dry. Find x, the angle that the clothesline makes with the horizontal.

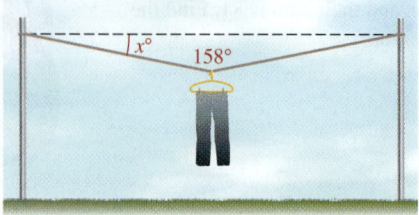

$x°$ 158°

49. Mountain Bicycles.   For the bicycle frame shown, the angle that the horizontal crossbar makes with the seat support is 15° less than twice the angle at the steering column. The angle at the pedal gear is 25° more than the angle at the steering column. Find these three angle measures.

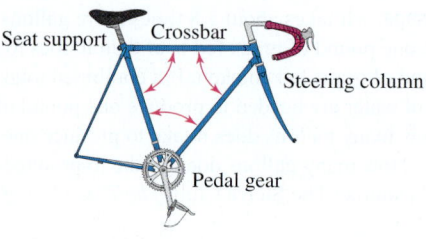

Seat support   Crossbar

Steering column

Pedal gear

50. Triangles.   The measure of $\angle 1$ (read as angle 1) of a triangle is one-half that of $\angle 2$. The measure of $\angle 3$ is equal to the sum of the measures of $\angle 1$ and $\angle 2$. Find each angle measure.

51. Angles.   Two angles are called ***complementary angles*** when the sum of their measures is 90°. Refer to the figure on the right. Find x. Then find the measures of the complementary angles.

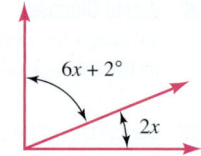

$6x + 2°$

$2x$

52. Angles.   Two angles are called ***supplementary angles*** when the sum of their measures is 180°. Refer to the figure below. Find x. Then find the measures of the supplementary angles.

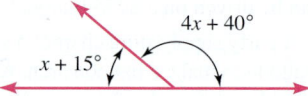

$4x + 40°$

$x + 15°$

53. "Lightning Bolt."   In 2010, Usain Bolt of Jamaica held the world record for the 100 meters and the 200 meters sprints. His *maximum stride angle* shown below is 5° less than 1.5 times its supplement. Find his maximum stride angle. You may need to refer to problem 52 to review the geometry involved. (Source: somaxsports.com)

Maximum stride angle

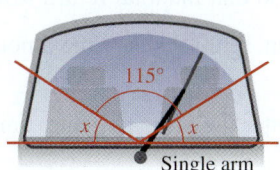

54. from **Campus To Careers**

Automotive Service Technician

The *sweep angle* of a windshield wiper arm is 115° as shown below. Find *x* so that the area that is cleared by the wiper is centered on the car's windshield.

© baranq/Shutterstock.com

115°

x   *x*

Single arm

WRITING

55. Create a geometry problem that could be answered by solving the equation $2w + 2(w + 5) = 26$.

56. What information do you need to know to answer the following question?

A business rented a copy machine for $85 per month plus 4¢ for every copy made. How many copies can be made each month?

57. Make a list of words and phrases that translate to an equal symbol =.

58. Define the word *strategy*.

REVIEW

Solve.

59. $\frac{5}{8}x = -15$

60. $\frac{12x + 24}{13} = 36$

61. $\frac{3}{4}y = \frac{2}{5}y - \frac{3}{2}y - 2$

62. $4.2(y - 4) - 0.6y = -13.2$

CHALLENGE PROBLEMS

63. What concept discussed in this section is illustrated by the following day and time?

Two minutes and three seconds past 1 A.M. on the 5th day of April, 2006

64. **Manufacturing.** A company has two machines that make widgets. The production costs are listed below.

Machine 1: Setup cost $400 and $1.70 per widget

Machine 2: Setup cost $500 and $1.20 per widget

Find the number of widgets for which the cost to manufacture them on either machine is the same.

SECTION 2.6

OBJECTIVES

1 Solve investment problems.

2 Solve uniform motion problems.

3 Solve liquid mixture problems.

4 Solve dry mixture problems.

5 Solve number-value problems.

More about Problem Solving

ARE YOU READY?

 The following problems review basic skills and some formulas that are needed to solve money, motion, and mixture applications.

1. Find the amount of interest earned by $8,000 invested at a 5% annual simple interest rate for 1 year.

2. At 45 miles per hour, how far will a car travel in 3 hours?

3. A 12-gallon mixture of antifreeze and water is 30% antifreeze. How many gallons of the mixture is antifreeze?

4. At $2.45 per pound, what is the value of 8 pounds of ground beef?

5. A couple invested $6,000 of their $20,000 lottery winnings in bonds. How much do they have left to invest in stocks?

6. Multiply: $100(0.03x)$

In this section, we will solve problems that involve money, motion, and mixtures. Tables are a helpful way to organize the information given in these problems.

1 Solve Investment Problems.

To find the amount of *simple interest I* an investment earns, we use the formula **$I = Prt$**, where *P* is the principal (the amount invested), *r* is the annual interest rate, and *t* is the time in years.

EXAMPLE 1   **Paying Tuition.**   A college student wants to invest the $12,000 inheritance he received and use the annual interest earned to pay his tuition cost of $945. The highest rate offered by a bank is 6% annual simple interest. At this rate, he cannot earn the needed $945, so he decides to invest some of the money in a riskier, but more profitable, investment offering a 9% return. How much should he invest at each rate?

Analyze   We know that $12,000 will be invested for 1 year at two rates: 6% and 9%. We are asked to find the amount invested at each rate so that the total return would be $945.

Assign   Let x = the amount invested at 6%. Then $12,000 - x$ = the amount invested at 9%.

Form   To organize the facts of the problem, we enter the principal, rate, time, and interest earned from each account in a table.

Step 1: List each investment in a row of the table.

Bank			
Riskier Investment			

Step 2: Label the columns using $I = Prt$ reversed and also write Total.

	P	$\cdot\ r$	$\cdot\ t =$	I
Bank				
Riskier Investment				
				Total:

Step 3: Enter the rates as decimals, the times, and the total interest.

	P	$\cdot\ r$	$\cdot\ t =$	I
Bank		0.06	1	
Riskier Investment		0.09	1	
			Total: **945**	

Step 4: Enter each unknown principal.

	P	$\cdot\ r$	$\cdot\ t =$	I
Bank	x	0.06	1	
Riskier Investment	$12,000 - x$	0.09	1	
			Total: 945	

Step 5: In the last column, multiply P, r, and t to obtain expressions for the interest earned.

	P	$\cdot\ r$	$\cdot\ t =$	I
Bank	x	0.06	1	**0.06x**
Riskier Investment	$12,000 - x$	0.09	1	**0.09(12,000 − x)**
			Total: 945	

← This is $x \cdot 0.06 \cdot 1$.
← This is $(12,000 - x) \cdot 0.09 \cdot 1$.

Use the information in this column to form an equation.

The interest earned at 6%	plus	the interest earned at 9%	equals	the total interest.
0.06x	+	0.09(12,000 − x)	=	945

Solve

$0.06x + 0.09(12,000 - x) = 945$

$100[0.06x + 0.09(12,000 - x)] = 100(945)$ Multiply both sides by 100 to clear the equation of decimals.

$100(0.06x) + 100(0.09)(12,000 - x) = 100(945)$ Distribute the multiplication by 100.

$6x + 9(12,000 - x) = 94,500$ Do the multiplications by 100.

$6x + 108,000 - 9x = 94,500$ Use the distributive property.

$-3x + 108,000 = 94,500$ Combine like terms.

$-3x = -13,500$ Subtract 108,000 from both sides.

$x = 4,500$ To isolate x, divide both sides by −3.

Success Tip

We can clear an equation of decimals by multiplying both sides by a power of 10. Here, we multiply 0.06 and 0.09 by 100 to move each decimal point two places to the right:

$100(0.06) = 6$ $100(0.09) = 9$

State   The student should invest $4,500 at 6% and $12,000 − $4,500 = $7,500 at 9%.

Check   The first investment will earn 0.06($4,500), or $270. The second will earn 0.09($7,500), or $675. Since the total return will be $270 + $675 = $945, the results check.

> **Self Check 1**   **Investments.**   A student invested a total of $4,200 in certificates of deposit, one at 2% and the other at 3%. Find the amount invested at each rate if the first year combined interest income from the two investments was $102.
>
> **Now Try** ▶ Problem 17

2 Solve Uniform Motion Problems.

If we know the rate r at which we will be traveling and the time t we will be traveling at that rate, we can find the distance d traveled by using the formula **$d = rt$**.

EXAMPLE 2   **Rescues at Sea.**   A cargo ship, heading into port, radios the Coast Guard that it is experiencing engine trouble and that its speed has dropped to 3 knots. (This is 3 sea miles per hour.) Immediately, a Coast Guard cutter leaves port and speeds at a rate of 25 knots directly toward the disabled ship, which is 56 sea miles away. How long will it take the Coast Guard to reach the ship? (Sea miles are also called nautical miles.)

Success Tip

A sketch is helpful when solving uniform motion problems.

Analyze   We know the *rate* of each ship (25 knots and 3 knots), and we know that they must close a *distance* of 56 sea miles between them. We don't know the *time* it will take to do this.

Assign   Let $t =$ the time it takes the Coast Guard to reach the cargo ship. During the rescue, the ships don't travel at the same rate, but they do travel for the same amount of time. Therefore, t also represents the travel time for the cargo ship.

Form   We enter the rates, the variable t for each time, and the total distance traveled by the ships (56 sea miles) in the table. To fill in the last column, we use the formula $r \cdot t = d$ twice to find an expression for each distance traveled: $25 \cdot t = 25t$ and $3 \cdot t = 3t$.

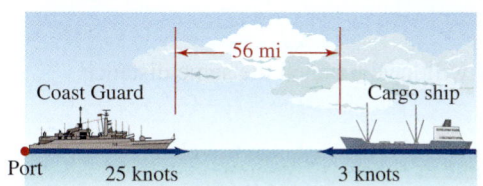

	r	$\cdot$ t	$=$ d
Coast Guard cutter	25	t	$25t$
Cargo ship	3	t	$3t$
		Total:	56

} Multiply $r \cdot t$ to obtain an expression for each distance traveled.

└── Use the information in this column to form an equation.

Caution

A common error is to enter 56 miles as the distance traveled for each ship. However, neither ship traveled 56 miles. Together, they travel 56 miles.

	r	$\cdot$ t	$=$ d
Cutter			56
Cargo			56

The distance the cutter travels	plus	the distance the ship travels	equals	the original distance between the ships.
$25t$	$+$	$3t$	$=$	56

Solve

$$25t + 3t = 56$$

$$28t = 56 \qquad \text{Combine like terms: } 25t + 3t = 28t.$$

$$t = \frac{56}{28} \qquad \text{To isolate } t, \text{ divide both sides by 28.}$$

$$t = 2 \qquad \text{Do the division.}$$

State The ships will meet in 2 hours.

Check In 2 hours, the Coast Guard cutter travels $25 \cdot 2 = 50$ sea miles, and the cargo ship travels $3 \cdot 2 = 6$ sea miles. Together, they travel $50 + 6 = 56$ sea miles. Since this is the original distance between the ships, the result checks.

> **Self Check 2** **Rescues.** Two search-and-rescue teams leave base at the same time looking for a lost boy. The first team, on foot, heads north at 2 mph, and the other, on horseback, heads south at 4 mph. How long will it take them to search a distance of 21 miles between them?
>
> **Now Try** ▶ Problem 29

EXAMPLE 3 **Concert Tours.** While on tour, a country music star travels by bus. Her musical equipment is carried in a truck. How long will it take her bus, traveling 60 mph, to overtake the truck, traveling at 45 mph, if the truck had a $1\frac{1}{2}$-hour head start to her next concert location?

Analyze We know the rate of each vehicle (60 mph and 45 mph) and that the truck began the trip $1\frac{1}{2}$ or 1.5 hours earlier than the bus. We need to determine how long it will take the bus to catch up to the truck.

Assign Let $t =$ the time it takes the bus to overtake the truck. With a 1.5-hour head start, the truck is on the road longer than the bus. Therefore, $t + 1.5 =$ the truck's travel time.

Form We enter each rate and time in the table, and use the formula $r \cdot t = d$ twice to fill in the distance column.

	r $\cdot$	t $=$	d
Bus	60	t	$60t$
Truck	45	$t + 1.5$	$45(t + 1.5)$

Multiply $r \cdot t$ to obtain an expression for each distance traveled.

Enter this information first.

Use the information in this column to form an equation.

When the bus overtakes the truck, they will have traveled the same distance.

The distance traveled by the bus	is the same as	the distance traveled by the truck.
$60t$	$=$	$45(t + 1.5)$

> **Success Tip**
>
> We used 1.5 hrs for the head start because it is easier to solve
> $$60t = 45(t + 1.5)$$
> than
> $$60t = 45\left(t + 1\tfrac{1}{2}\right)$$

Solve

$60t = 45(t + 1.5)$

$60t = 45t + 67.5$ Distribute the multiplication by 45: $45(1.5) = 67.5$.

$15t = 67.5$ Subtract $45t$ from both sides: $60t - 45t = 15t$.

$t = 4.5$ To isolate t, divide both sides by 15: $\frac{67.5}{15} = 4.5$.

State The bus will overtake the truck in 4.5 or $4\frac{1}{2}$ hours.

Check In 4.5 hours, the bus travels $60(4.5) = 270$ miles. The truck travels for $1.5 + 4.5 = 6$ hours at 45 mph, which is $45(6) = 270$ miles. Since the distances traveled are the same, the result checks.

Self Check 3 **Moving Day.**   A moving van, packed with a family's belongings, left their old home for their new home, traveling at 40 miles per hour. Forty-five minutes $\left(\frac{3}{4} \text{ hour}\right)$ later, the family left for their new home, traveling by car at 60 miles per hour. How long did it take the family to overtake the moving van?

Now Try ▶ Problem 33

---

3 Solve Liquid Mixture Problems.

We now discuss how to solve mixture problems. In the first type, a liquid mixture of a desired strength is made from two solutions with different concentrations (strengths).

EXAMPLE 4   **Mixing Solutions.**   A chemistry experiment calls for a 30% sulfuric acid solution. If the lab supply room has only 50% and 20% sulfuric acid solutions, how much of each should be mixed to obtain 12 liters of a 30% acid solution?

Success Tip

The strength *(concentration)* of a mixture is always between the strengths of the two solutions used to make it.

Analyze   The 50% solution is too strong and the 20% solution is too weak. We must find how much of each should be combined to obtain 12 liters of a 30% solution.

Assign   If x = the number of liters of the 50% solution used in the mixture, the remaining $(12 - x)$ liters must be the 20% solution.

Form   The amount of pure sulfuric acid in each solution is given by

Amount of solution · strength of the solution = amount of pure sulfuric acid

A table and sketch are helpful in organizing the facts of the problem.

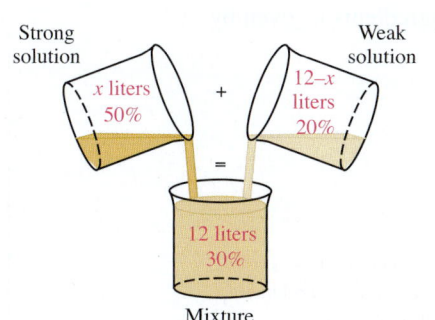

	Amount ·	Strength =	Amount of pure sulfuric acid
Weak	$12 - x$	0.20	$0.20(12 - x)$
Strong	x	0.50	$0.50x$
Mixture	12	0.30	$12(0.30)$

Enter this information first.

Multiply amount · strength three times to fill in this column.

Use the information in this column to form an equation.

The sulfuric acid in the 20% solution	plus	the sulfuric acid in the 50% solution	equals	the sulfuric acid in the mixture.
$0.20(12 - x)$	$+$	$0.50x$	$=$	$12(0.30)$

Solve

$$0.20(12 - x) + 0.50x = 12(0.30)$$

$$2.4 - 0.2x + 0.5x = 3.6 \qquad \text{Distribute the multiplication by 0.20.}$$

$$0.3x + 2.4 = 3.6 \qquad \text{Combine like terms: } -0.2x + 0.5x = 0.3x.$$

$$0.3x = 1.2 \qquad \text{Subtract 2.4 from both sides.}$$

$$x = 4 \qquad \text{To isolate } x, \text{ undo the multiplication by 0.3 by dividing both sides by 0.3: } \frac{1.2}{0.3} = 4.$$

Success Tip

We could begin by multiplying both sides of the equation by 10 to clear it of the decimals.

State   4 liters of 50% solution and $12 - 4 = 8$ liters of 20% solution should be used.

Check   The amount of acid in 4 liters of the 50% solution is $0.50(4) = 2.0$ liters and the amount of acid in 8 liters of the 20% solution is $0.20(8) = 1.6$ liters. Thus, the amount of acid in these two solutions is $2.0 + 1.6 = 3.6$ liters. The amount of acid in 12 liters of the 30% mixture is also $0.30(12) = 3.6$ liters. Since the amounts of acid are equal, the results check.

> **Self Check 4**   **Mixing Solutions.**   How many gallons of a 3% salt solution must be mixed with a 7% salt solution to obtain 25 gallons of a 5.4% salt solution?
>
> **Now Try ▶** Problem 41

4 Solve Dry Mixture Problems.

In another type of mixture problem, a dry mixture of a specified value is created from two differently priced ingredients.

EXAMPLE 5   **Snack Foods.**   Because cashews priced at $9 per pound were not selling, a produce clerk decided to combine them with less expensive peanuts and sell the mixture for $7 per pound. How many pounds of peanuts, selling at $6 per pound, should be mixed with 50 pounds of cashews to obtain such a mixture?

Analyze   We need to determine how many pounds of peanuts to mix with 50 pounds of cashews to obtain a mixture worth $7 per pound.

Assign   Let $x =$ the number of pounds of peanuts to use in the mixture. Since 50 pounds of cashews will be combined with the peanuts, the mixture will weigh $50 + x$ pounds.

Form   The value of the mixture and of each of its ingredients is given by

$$\textbf{Amount} \cdot \textbf{the price} = \textbf{the total value}$$

We can organize the facts of the problem in a table.

Caution

To find the number of pounds in the mixture, add the number of pounds of the ingredients:

$$50 + x$$

It would be incorrect to multiply:

$$50 \cdot x$$

	Amount ·	Price =	Total value
Peanuts	x	6	$6x$
Cashews	50	9	450
Mixture	$50 + x$	7	$7(50 + x)$

Enter this information first.

Multiply amount · price three times to fill in this column.

Use the information in this column to form an equation.

The value of the peanuts	plus	the value of the cashews	equals	the value of the mixture.
$6x$	$+$	450	$=$	$7(50 + x)$

Solve

$$6x + 450 = 7(50 + x)$$
$$6x + 450 = 350 + 7x \quad \text{Distribute the multiplication by 7.}$$
$$450 = 350 + x \quad \text{To eliminate the term } 6x \text{ on the left side, subtract } 6x \text{ from both sides: } 7x - 6x = x.$$
$$100 = x \quad \text{To isolate } x, \text{ subtract 350 from both sides.}$$

State   100 pounds of peanuts should be used in the mixture.

Check The value of 100 pounds of peanuts, at $6 per pound, is 100($6) = $600 and the value of 50 pounds of cashews, at $9 per pound, is 50($9) = $450. Thus, the total value of these two ingredients is $1,050. Since the value of 150 pounds of the mixture, at $7 per pound, is also 150(7) = $1,050, the result checks.

> **Self Check 5**   **Mixing Candy.**   Candy worth $1.90 per pound is to be mixed with 60 lb of a second candy worth $1.20 per pound. How many pounds of the $1.90 per pound candy should be used to make a mixture worth $1.48 per pound?
>
> **Now Try** ▶ Problem 47

5 Solve Number-Value Problems.

When problems deal with collections of different items having different values, we must distinguish between the *number of* and the *value of* the items. For these problems, we will use the fact that

Number · value = total value

EXAMPLE 6   **Dining Area Improvements.**   A restaurant owner needs to purchase some tables, chairs, and dinner plates for the dining area of her establishment. She plans to buy four chairs and four plates for each new table. She also plans to buy 20 additional plates in case of breakage. If a table costs $100, a chair $50, and a plate $5, how many of each can she buy if she takes out a loan for $6,500 to pay for the new items?

Analyze   We know the *value* of each item: Tables cost $100, chairs cost $50, and plates cost $5 each. We need to find the *number* of tables, chairs, and plates she can purchase for $6,500.

Assign   The number of chairs and plates she needs depends on the number of tables she buys. So we let t = the number of tables to be purchased. Since every table requires four chairs and four plates, she needs to order $4t$ chairs. Because 20 additional plates are needed, she should order $(4t + 20)$ plates.

Form   We can organize the facts of the problem in a table.

	Number ·	Value =	Total value
Tables	t	100	$100t$
Chairs	$4t$	50	$50(4t)$
Plates	$4t + 20$	5	$5(4t + 20)$
			Total: 6,500

Multiply number · value three times to fill in this column.

Enter this information first.

Use the information in this column to form an equation.

The value of the tables	plus	the value of the chairs	plus	the value of the plates	equals	the total value of the purchase.
$100t$	$+$	$50(4t)$	$+$	$5(4t + 20)$	$=$	$6,500$

Solve

$$100t + 50(4t) + 5(4t + 20) = 6,500$$

$$100t + 200t + 20t + 100 = 6,500 \qquad \text{Do the multiplications and distribute.}$$

$$320t + 100 = 6,500 \qquad \text{Combine like terms: } 100t + 200t + 20t = 320t.$$

$$320t = 6,400 \qquad \text{Subtract 100 from both sides.}$$

$$t = 20 \qquad \text{To isolate } t, \text{ divide both sides by 320.}$$

To find the number of chairs and plates to buy, we evaluate $4t$ and $4t + 20$ for $t = 20$.

Chairs: $4t = 4(\mathbf{20})$ *Plates:* $4t + 20 = 4(\mathbf{20}) + 20$ Substitute 20 for t.

$= 80$ $= 100$

State   The owner needs to buy 20 tables, 80 chairs, and 100 plates.

Check   The total value of 20 tables is $20(\$100) = \$2,000$, the total value of 80 chairs is $80(\$50) = \$4,000$, and the total value of 100 plates is $100(\$5) = \$500$. Because the total purchase is $\$2,000 + \$4,000 + \$500 = \$6,500$, the results check.

> **Self Check 6**   **Electronics.**   A small electronics store buys iPods for $189, iPod skins for $32, and iTunes cards for $15. If they place an order for three times as many iPods as skins and 20 more iTunes cards than skins, how many of each did they order if the items totaled $2,756?
>
> **Now Try** ▶ Problem 55

SECTION 2.6 ▶ STUDY SET

VOCABULARY

Fill in the blanks.

1. Problems that involve depositing money are called _____ problems, and problems that involve moving vehicles are called uniform _____ problems.

2. Problems that involve combining ingredients are called _____ problems, and problems that involve collections of different items having different values are called _____ problems.

CONCEPTS

3. Complete only the *principal column* given that part of $30,000 is invested in stocks and the rest in art.

	P	$\cdot$	r	$\cdot$	$t = I$
Stocks	x				
Art	?				

4. A man made two investments that earned a combined annual simple interest of $280. Complete the table and then form an equation for this investment problem.

	P	$\cdot$	r	$\cdot t =$	I
Bank	x		0.04	1	
Stocks	$6,000 - x$		0.06	1	
				Total:	

5. Complete only the *rate column* given that the east-bound plane flew 150 mph slower than the west-bound plane.

	r	$\cdot$	$t = d$
West	r		
East	?		

6. a. Complete only the *time column* given that a runner wants to overtake a walker and the walker had a $\frac{1}{2}$-hour head start.

	r	$\cdot$	$t = d$
Runner			t
Walker			?

   b. Complete only the *time column* given that part of a 6-hour drive was in fog and the other part was in clear conditions.

	r	$\cdot$	$t = d$
Foggy			t
Clear			?

7. A husband and wife drive in opposite directions to work. Their drives last the same amount of time and their workplaces are 80 miles apart. Complete the table and then form an equation for this distance problem.

	r	$\cdot$	$t = d$
Husband	35	t	
Wife	45		
		Total:	

8. a. How many gallons of acetic acid are there in barrel 2?

   b. Suppose the contents of the two barrels are poured into an empty third barrel. How many gallons of liquid will the third barrel contain?

   c. Estimate the strength of the solution in the third barrel: 15%, 35%, or 60% acid?

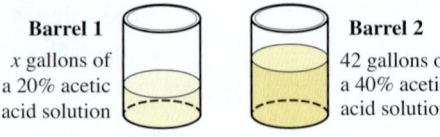

Barrel 1
x gallons of a 20% acetic acid solution

Barrel 2
42 gallons of a 40% acetic acid solution

9. a. Two antifreeze solutions are combined to form a mixture. Complete the table and then form an equation for this mixture problem.

	Amount ·	Strength =	Pure antifreeze
Weak	x	0.25	
Strong	6	0.50	
Mixture		0.30	

b. Two oil-and-vinegar salad dressings are combined to make a new mixture. Complete the table and then form an equation for this mixture problem.

	Amount ·	Strength =	Pure vinegar
Weak		0.03	
Strong	x	0.06	
Mixture	10	0.05	

10. The value of all the nylon brushes that a paint store carries is $670. Complete the table and then form an equation for this number-value problem.

	Number ·	Value =	Total value
1-inch	$2x$	4	
2-inch	x	5	
3-inch	$x + 10$	7	
		Total:	

NOTATION

11. Write 6% and 15.2% in decimal form.
12. By what power of 10 should each decimal be multiplied to make it a whole number?
   a. 0.08   b. 0.162

GUIDED PRACTICE

Solve each equation. See Example 1.

13. $0.18x + 0.45(12 - x) = 0.36(12)$
14. $0.12x + 0.20(4 - x) = 0.6$
15. $0.08x + 0.07(15,000 - x) = 1,110$
16. $0.108x + 0.07(16,000 - x) = 1,500$

APPLICATIONS

Investment problems. See Example 1.

17. **Corporate Investments.** The financial board of a corporation invested $25,000 overseas, part at 4% and part at 7% annual simple interest. Find the amount invested at each rate if the first-year combined income from the two investments was $1,300.

18. **Loans.** A credit union loaned out $50,000, part at an annual simple rate of 5% and the rest at an annual simple rate of 8%. They collected combined interest of $3,400 from the loans that year. How much was loaned out at each rate?

19. **Old Coins.** A salesperson used her $3,500 year-end bonus to purchase some old gold and silver coins. She earned 15% annual simple interest on the gold coins and 12% annual simple interest on the silver coins. If she saw a return on her investment of $480 the first year, how much did she invest in each type of coin?

20. **High-Risk Companies.** An investment club used funds totaling $200,000 to invest in a bio-tech company and in an ethanol plant, with hopes of earning 11% and 14% annual simple interest, respectively. Their hunch paid off. The club made a total of $24,250 interest the first year. How much was invested at each rate?

21. **Retirement.** A professor wants to supplement her pension with investment interest. If she invests $28,000 at 6% annual simple interest, how much would she have to invest at 7% annual simple interest to achieve a goal of $3,500 per year in supplemental income?

22. **Extra Income.** An investor wants to receive $1,000 annually from two investments. He has put $4,500 in a money market account paying 4% annual simple interest. How much should he invest in a stock fund that pays 10% annual simple interest to achieve his goal?

23. **1099 Forms.** The form below shows the interest income Terrell Washington earned in 2011 from two savings accounts. He deposited a total of $15,000 at the first of that year, and made no further deposits or withdrawals. How much money did he deposit in account 822 and in account 721?

RECIPIENT'S name	USA HOME SAVINGS	2011
TERRELL WASHINGTON	This is important tax information and is being furnished to the Internal Revenue Service.	

Account Number	Annual Percent Yield	Interest earned
822	5%	?
721	4.5%	?
FORM 1099		Total Interest Income $720.00

24. **Investment Plans.** A financial planner recommends a plan for a client who has $65,000 to invest. (See the chart at the right.) At the end of the presentation, the client asks, "How much will be invested at each rate?" Answer this question using the given information.

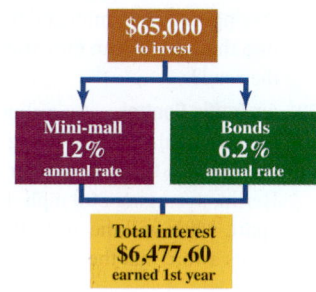

25. **Investments.** Equal amounts are invested in each of three accounts paying 7%, 8%, and 10.5% annual simple interest. If one year's combined interest income is $1,249.50, how much is invested in each account?

26. **Personal Loans.** Maggy lent her brother some money at 2% annual simple interest. She lent her sister twice as much money at half of the interest rate. In one year, Maggy collected combined interest of $200 from her brother and sister. How much did she lend each of them?

27. Bad Investments. A newly hired MBA graduate received an $18,000 signing bonus from her employer. She invested part of it in a credit union account earning 3% annual simple interest and the rest in utility stocks that suffered a 7% loss of value. The net income from both investments for the first year was only $90. How much of her bonus was originally placed in each investment?

28. Losses. A financial planner invested a portion of his client's $190,000 in a high-yield mutual fund that earned 11% annual simple interest. The remainder of the money was invested in a mini-mall development. Unfortunately, that investment lost 25% of its value the first year. Find the amount originally made in each investment if the first-year net income was $6,500.

Uniform motion problems. **See Example 2.**

29. Tornadoes. During a storm, two teams of scientists leave a university at the same time in vans to search for tornadoes. The first team travels east at 20 mph and the second travels west at 25 mph. If their radios have a range of up to 90 miles, how long will it be before they lose radio contact?

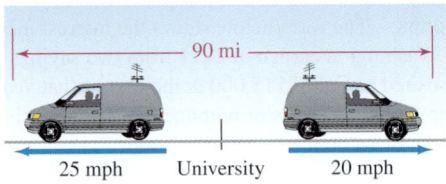

30. Unmanned Aircraft. Two remotely controlled unmanned aircraft are launched in opposite directions. One flies east at 78 mph and the other west at 82 mph. How long will it take the aircraft to fly a combined distance of 560 miles?

31. Hello/Goodbye. A husband and wife work different shifts at the same plant. When the husband leaves from work to make the 20-mile trip home, the wife leaves their home and drives to work. They travel on the same road. The husband's driving rate is 45 mph and the wife's is 35 mph. How long into their drives can they wave at each other when passing on the road?

32. Air Traffic Control. An airliner leaves Berlin, Germany, headed for Montreal, Canada, flying at an average speed of 450 mph. At the same time, an airliner leaves Montreal headed for Berlin, averaging 500 mph. If the airports are 3,800 miles apart, when will the air traffic controllers have to make the pilots aware that the planes are passing each other?

33. Cycling. A cyclist leaves his training base for a morning workout, riding at the rate of 18 mph. One and one-half hours later, his support staff leaves the base in a car going 45 mph in the same direction. How long will it take the support staff to catch up with the cyclist?

34. Parenting. How long will it take a mother, running at 4 feet per second, to catch up with her toddler, running down the sidewalk at 2 feet per second, if the child had a 5-second head start?

35. Road Trips. A car averaged 40 mph for part of a trip and 50 mph for the remainder. If the 5-hour trip covered 210 miles, for how long did the car average 40 mph?

36. Cross-Training. An athlete runs up a set of stadium stairs at a rate of 2 stairs per second, immediately turns around, and then descends the same stairs at a rate of 3 stairs per second. If the workout takes 90 seconds, how long does it take him to run up the stairs?

37. Winter Driving. A trucker drove for 4 hours before he encountered icy road conditions. He reduced his speed by 20 mph and continued driving for 3 more hours. Find his average speed during the first part of the trip if the entire trip was 325 miles.

38. Speed of Trains. Two trains are 330 miles apart, and their speeds differ by 20 mph. Find the speed of each train if they are traveling toward each other and will meet in 3 hours.

Liquid mixture problems. **See Example 3.**

39. Salt Solutions. How many gallons of a 3% salt solution must be mixed with 50 gallons of a 7% solution to obtain a 5% solution?

40. Photography. A photographer wishes to mix 2 liters of a 5% acetic acid solution with a 10% solution to get a 7% solution. How many liters of 10% solution must be added?

41. Making Cheese. To make low-fat cottage cheese, milk containing 4% butterfat is mixed with milk containing 1% butterfat to obtain 15 gallons of a mixture containing 2% butterfat. How many gallons of each milk must be used?

42. Antifreeze. How many quarts of a 10% antifreeze solution must be mixed with 16 quarts of a 40% antifreeze solution to make a 30% solution?

43. Printing. A printer has ink that is 8% cobalt blue color and ink that is 22% cobalt blue color. How many ounces of each ink are needed to make one-half gallon (64 ounces) of ink that is 15% cobalt blue color?

44. Flood Damage. One website recommends a 6% chlorine bleach-water solution to remove mildew. A chemical lab has 3% and 15% chlorine bleach-water solutions in stock. How many gallons of each should be mixed to obtain 100 gallons of the mildew spray?

45. Interior Decorating. The colors on the paint chip card below are created by adding different amounts of orange tint to a white latex base. How many gallons of Desert Sunrise should be mixed with 1 gallon of Bright Pumpkin to obtain Cool Cantaloupe?

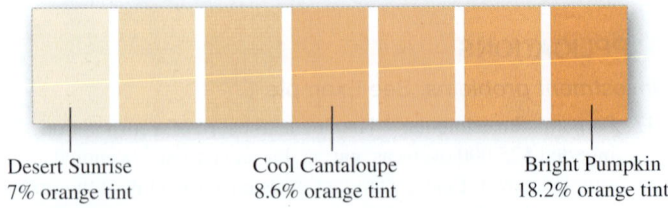

Desert Sunrise
7% orange tint

Cool Cantaloupe
8.6% orange tint

Bright Pumpkin
18.2% orange tint

46. Antiseptics. A nurse wants to add water to 30 ounces of a 10% solution of benzalkonium chloride to dilute it to an 8% solution. How much water must she add? (*Hint:* Water is 0% benzalkonium chloride.)

Dry mixture problems. See Example 4.

47. Lawn Seed. A store sells bluegrass seed for $6 per pound and ryegrass seed for $3 per pound. How much ryegrass must be mixed with 100 pounds of bluegrass to obtain a blend that will sell for $5 per pound?

48. Coffee Blends. A store sells regular coffee for $8 a pound and gourmet coffee for $14 a pound. To get rid of 40 pounds of the gourmet coffee, a shopkeeper makes a blend to put on sale for $10 a pound. How many pounds of regular coffee should he use?

49. Raisins. How many scoops of natural seedless raisins costing $3.45 per scoop must be mixed with 20 scoops of golden seedless raisins costing $2.55 per scoop to obtain a mixture costing $3 per scoop?

50. Fertilizer. Fertilizer with weed control costing $38 per 50-pound bag is to be mixed with a less expensive fertilizer costing $6 per 50-pound bag to make 16 bags of fertilizer that can be sold for $28 per bag. How many bags of cheaper fertilizer should be used?

51. Packaged Salad. How many 10-ounce bags of Romaine lettuce must be mixed with fifty 10-ounce bags of Iceberg lettuce to obtain a blend that sells for $2.50 per ten-ounce bag?

52. Mixing Candy. Lemon drops worth $3.80 per pound are to be mixed with jelly beans that cost $2.40 per pound to make 300 pounds of a mixture worth $2.96 per pound. How many pounds of each candy should be used?

53. Bronze. A pound of tin is worth $1 more than a pound of copper. Four pounds of tin are mixed with 6 pounds of copper to make bronze that sells for $3.65 per pound. How much is a pound of tin worth?

54. Snack Foods. A bag of peanuts is worth $0.30 less than a bag of cashews. Equal amounts of peanuts and cashews are used to make 40 bags of a mixture that sells for $1.05 per bag. How much is a bag of cashews worth?

Number-value problems. See Example 5.

55. Rentals. The owners of an apartment building rent equal numbers of 1-, 2-, and 3-bedroom units. The monthly rent for a 1-bedroom is $550, a 2-bedroom is $700, and a 3-bedroom is $900. If the total monthly income is $36,550, how many of each type of unit are there?

56. Warehousing. A store warehouses 40 more portables than big-screen TV sets, and 15 more consoles than big-screen sets. The monthly storage cost for a portable is $1.50, a console is $4.00, and a big-screen is $7.50. If storage for all the televisions costs $276 per month, how many big-screen sets are in stock?

57. Software. Three software applications are priced as shown. Spreadsheet and database programs sold in equal numbers, but 15 more word processing applications were sold than the other two combined. If the three applications generated sales of $72,000, how many spreadsheets were sold?

Software	Price
Spreadsheet	$150
Database	$195
Word processing	$210

58. Inventories. With summer approaching, the number of air conditioners sold is expected to be double that of stoves and refrigerators combined. Stoves sell for $350, refrigerators for $450, and air conditioners for $500, and sales of $56,000 are expected. If stoves and refrigerators sell in equal numbers, how many of each appliance should be stocked?

59. Piggy Banks. When a child emptied his coin bank, he had a collection of pennies, nickels, and dimes. There were 20 more pennies than dimes and the number of nickels was triple the number of dimes. If the coins had a value of $5.40, how many of each type coin were in the bank?

60. Wishing Wells. A scuba diver, hired by an amusement park, collected $121 in nickels, dimes, and quarters at the bottom of a wishing well. There were 500 nickels, and 90 more quarters than dimes. How many quarters and dimes were thrown into the wishing well?

61. Basketball. Epiphanny Prince, of New York, scored 113 points in a high school game on February 1, 2006, breaking a national prep record that was held by Cheryl Miller. Prince made 46 more 2-point baskets than 3-point baskets, and only 1 free throw. How many 2-point and 3-point baskets did she make?

62. Museum Tours. The ticket prices for the Coca-Cola Museum in Atlanta are shown. A family purchased 3 more children's tickets than adult tickets, and 1 less senior ticket than adult tickets. The total cost of the tickets was $131. How many of each type did they purchase?

WRITING

63. Create a mixture problem of your own, and solve it.

64. Write an investment problem to fit the following equation, and then solve it.

$$0.08x + 0.07(10,000 - x) = 770$$

65. Explain the error in each statement.

    a. If 3 pounds of Mocha coffee are mixed with x pounds of Java coffee, there will be $3x$ pounds of the Mocha-Java blend.

    b. A financial manager has a total of $5,000 to invest in two accounts. If $x is invested in the first account, then $(x − 5,000) is left to be invested in the second account.

66. Is it possible to mix a 10% sugar solution with a 20% sugar solution to get a 30% sugar solution? Explain.

REVIEW

Multiply.

67. $-12(3a + 4b - 32)$

68. $\frac{1}{2}(4b - 8)$

69. $3(5t + 1)2$

70. $2.9(4c - 12)$

CHALLENGE PROBLEMS

71. **Evaporation.** How much water must be boiled away to increase the concentration of 300 milliliters of a 2% salt solution to a 3% salt solution?

72. **Diluting Solutions.** How much water should be added to 20 ounces of a 15% solution of alcohol to dilute it to a 10% alcohol solution?

73. **Financial Planning.** A plumber has a choice of two investment plans:

    ■ An insured fund that pays 11% interest

    ■ A risky investment that pays a 13% return

    If the same amount invested at the higher rate would generate an extra $150 per year, how much does the plumber have to invest?

74. **Investments.** The amount of annual interest earned by $8,000 invested at a certain rate is $200 less than $12,000 would earn at a rate 1% lower. At what rate is the $8,000 invested?

SECTION 2.7

Solving Inequalities

OBJECTIVES

1 Determine whether a number is a solution of an inequality.

2 Graph solution sets and use interval notation.

3 Solve linear inequalities.

4 Solve compound inequalities.

5 Solve inequality applications.

ARE YOU READY?

The following problems review some basic skills that are needed to solve inequalities.

1. Fill in the blanks: The symbol $<$ means "___ ____ ____."

2. Is $-5 > -6$ a true or false statement?

3. Graph each number in the set $\left\{ -4, -1.7, 2, \frac{13}{4} \right\}$ on a number line.

4. Express the fact that $10 > 0$ using an $<$ symbol.

In our daily lives, we often speak of one value being *greater than* or *less than* another. For example, a sick child might have a temperature *greater than* 98.6°F or a granola bar might contain *less than* 2 grams of fat. In mathematics, we use *inequalities* to show that one expression is greater than or is less than another expression.

1 ## Determine Whether a Number Is a Solution of an Inequality.

An **inequality** is a statement that contains one or more of the following symbols.

Inequality Symbols		
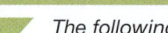 $<$ is less than $\leq$ is less than or equal to	$>$ is greater than $\geq$ is greater than or equal to	$\neq$ is not equal to

An inequality can be true, false, or neither true nor false. For example,

■ $9 \geq 9$ is true because $9 = 9$.

■ $37 < 24$ is false.

■ $x + 1 > 5$ is neither true nor false because we don't know what number x represents.

The Language of Algebra

Because < requires one number to be strictly less than another number and > requires one number to be strictly greater than another number, < and > are called **strict inequalities**.

An inequality that contains a variable can be made true or false depending on the number that is substituted for the variable. If we substitute 10 for x in $x + 1 > 5$, the resulting inequality is true: $10 + 1 > 5$. If we substitute 1 for x, the resulting inequality is false: $1 + 1 > 5$. A number that makes an inequality true is called a **solution** of the inequality, and we say that the number *satisfies* the inequality. Thus, 10 is a solution of $x + 1 > 5$ and 1 is not.

In this section, we will find the solutions of *linear inequalities in one variable*.

Linear Inequality in One Variable	A linear inequality in one variable can be written in one of the following forms where a, b, and c are real numbers and $a \neq 0$.
	$ax + b > c \qquad ax + b \geq c \qquad ax + b < c \qquad ax + b \leq c$

EXAMPLE 1   Is 9 a solution of $2x + 4 \leq 21$?

Strategy   We will substitute 9 for x and evaluate the expression on the left side.

Why   If a true statement results, 9 is a solution of the inequality. If we obtain a false statement, 9 is not a solution.

Solution

The Language of Algebra

A **linear inequality** in one variable is similar to a linear equation in one variable except that the equal symbol is replaced with an inequality symbol.

Equation	Inequality
$2x + 1 = 9$	$2x + 1 > 9$

$$2x + 4 \leq 21$$
$$2(9) + 4 \overset{?}{\leq} 21 \qquad \textcolor{red}{\text{Substitute 9 for } x. \text{ Read } \overset{?}{\leq} \text{ as "is possibly less than or equal to."}}$$
$$18 + 4 \overset{?}{\leq} 21 \qquad \textcolor{red}{\text{Do the multiplication.}}$$
$$22 \leq 21 \qquad \textcolor{red}{\text{This inequality is false.}}$$

The statement $22 \leq 21$ is false because neither $22 < 21$ nor $22 = 21$ is true. Therefore, 9 is not a solution of $2x + 4 \leq 21$.

Self Check 1   Is 2 a solution of $3x - 1 \geq 0$?

Now Try ▶   Problem 15

2 Graph Solution Sets and Use Interval Notation.

The **solution set** of an inequality is the set of all numbers that make the inequality true. Some solution sets are easy to find. For example, if we replace the variable in $x > -3$ with a number greater than -3, the resulting inequality will be true. Because there are infinitely many real numbers greater than -3, it follows that $x > -3$ has infinitely many solutions. Since there are too many solutions to list, we use **set-builder notation** to describe the solutions set.

$$\{x \mid x > -3\}$$

Read as "the set of all x   such that   x is greater than −3."

We can illustrate the solution set by **graphing the inequality** on a number line. To graph $x > -3$, a **parenthesis** or **open circle** is drawn on the endpoint -3 to indicate that -3 is not part of the graph. Then we shade all of the points on the number line to the right of -3. The right arrowhead is also shaded to show that the solutions continue forever to the right.

Notation

The parenthesis (opens in the direction of the shading and indicates that an endpoint is not included in the shaded interval.

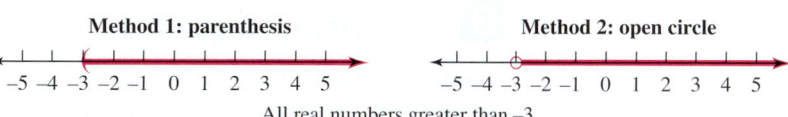

Method 1: parenthesis   Method 2: open circle

All real numbers greater than −3

The graph of $x > -3$ is an example of an **interval** on the number line. We can write intervals in a compact form called **interval notation.**

The interval notation that represents the graph of $x > -3$ is $(-3, \infty)$. As on the number line, a left parenthesis is written next to -3 to indicate that -3 is not included in the interval. The **positive infinity symbol** ∞ that follows indicates that the interval continues without end to the right. With this notation, *a parenthesis is always used next to an infinity symbol.*

The illustration below shows the relationship between the symbols used to graph an interval and the corresponding interval notation. If we begin at -3 and move to the right, the shaded arrowhead on the graph indicates that the interval approaches positive infinity ∞.

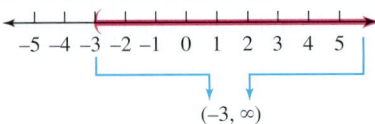

We now have three ways to describe the solution set of an inequality.

Set-builder notation	*Number line graph*	*Interval notation*
$\{x \mid x > -3\}$		$(-3, \infty)$

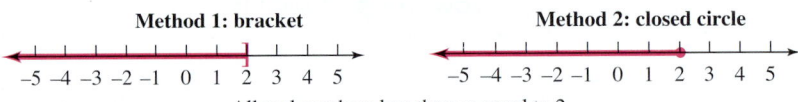

EXAMPLE 2   Graph: $x \le 2$

Strategy   We need to determine which real numbers, when substituted for x, would make $x \le 2$ a true statement.

Why   To graph $x \le 2$ means to draw a "picture" of all of the values of x that make the inequality true.

Solution   If we replace x with a number less than or equal to 2, the resulting inequality will be true. To graph the solution set, a **bracket** or a **closed circle** is drawn at the endpoint 2 to indicate that 2 is part of the graph. Then we shade all of the points on the number line to the left of 2 as well as the left arrowhead.

Method 1: bracket   Method 2: closed circle

All real numbers less than or equal to 2

The interval is written as $(-\infty, 2]$. The right bracket indicates that 2 is included in the interval. The **negative infinity symbol** $-\infty$ shows that the interval continues forever to the left. The illustration below shows the relationship between the symbols used to graph the interval and the corresponding interval notation.

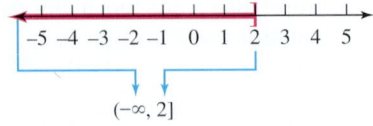

Self Check 2   Graph: $x \ge 0$

Now Try ▶   Problem 19

---

3   Solve Linear Inequalities.

To **solve an inequality** means to find all values of the variable that make the inequality true. As with equations, there are properties that we can use to solve inequalities.

Addition and Subtraction Properties of Inequality	Adding the same number to, or subtracting the same number from, both sides of an inequality does not change its solutions.

For any real numbers a, b, and c,

$$\text{If } a < b, \text{ then } \quad a + c < b + c. \qquad \text{If } a < b, \text{ then } \quad a - c < b - c.$$

Similar statements can be made for the symbols $\leq$, $>$, and $\geq$.

After applying one of these properties, the resulting inequality is equivalent to the original one. **Equivalent inequalities** have the same solution set.

Like equations, inequalities are solved by isolating the variable on one side.

EXAMPLE 3 Solve $x + 3 > 2$. Write the solution set in interval notation and graph it.

Strategy We will use a property of inequality to isolate the variable on one side.

Why To solve the original inequality, we want to find a simpler equivalent inequality of the form $x >$ **a number** or $x <$ **a number**, whose solution is obvious.

Solution We will use the subtraction property of inequality to isolate x on the left side of the inequality. We can undo the addition of 3 by subtracting 3 from both sides.

$x + 3 > 2$	This is the inequality to solve.
$x + 3 - 3 > 2 - 3$	Subtract 3 from both sides.
$x > -1$	Do the subtraction: $3 - 3 = 0$ and $2 - 3 = -1$.

All real numbers greater than -1 are solutions of $x + 3 > 2$. The solution set can be written in set-builder notation as $\{x \mid x > -1\}$ and in interval notation as $(-1, \infty)$. The graph of the solution set is shown below.

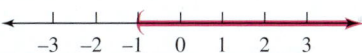

Since there are infinitely many solutions, we cannot check all of them. As an informal check, we can pick some numbers in the graph, say 0 and 30, substitute each number for x in the original inequality, and see whether true statements result.

Check:	$x + 3 > 2$		$x + 3 > 2$	
	$0 + 3 \overset{?}{>} 2$	Substitute 0 for x.	$30 + 3 \overset{?}{>} 2$	Substitute 30 for x.
	$3 > 2$	True	$33 > 2$	True

The solution set appears to be correct.

Self Check 3 Solve $x - 3 < -2$. Write the solution set in interval notation and graph it.

Now Try ▶ Problem 23

As with equations, there are properties for multiplying and dividing both sides of an inequality by the same number. To develop what is called *the multiplication property of inequality*, we consider the true statement $2 < 5$. If both sides are multiplied by a positive number, such as 3, another true inequality results.

$2 < 5$	This inequality is true.
$3 \cdot 2 < 3 \cdot 5$	Multiply both sides by 3.
$6 < 15$	This inequality is true.

However, if we multiply both sides of $2 < 5$ by a negative number, such as -3, the direction of the inequality symbol must be reversed to produce another true inequality.

$$2 < 5 \qquad \text{This inequality is true.}$$
$$-3 \cdot 2 > -3 \cdot 5 \qquad \text{Multiply both sides by } -3 \text{ and reverse the direction of the inequality.}$$
$$-6 > -15 \qquad \text{This inequality is true.}$$

The inequality $-6 > -15$ is true because -6 is to the right of -15 on the number line.

Dividing both sides of an inequality by the same negative number also requires that the direction of the inequality symbol be reversed.

$$-4 < 6 \qquad \text{This inequality is true.}$$
$$\frac{-4}{-2} > \frac{6}{-2} \qquad \text{Divide both sides by } -2 \text{ and change} < \text{to} >.$$
$$2 > -3 \qquad \text{This inequality is true.}$$

These examples illustrate the **multiplication and division properties of inequality.**

Multiplication and Division Properties of Inequality

Multiplying or dividing both sides of an inequality by the same **positive number** does not change its solutions.

For any real numbers a, b, and c, where c is **positive**,

$$\text{If } a < b, \quad \text{then } ac < bc. \qquad \text{If } a < b, \quad \text{then } \frac{a}{c} < \frac{b}{c}.$$

If we multiply or divide both sides of an inequality by the same **negative number**, the direction of the inequality symbol must be reversed for the inequalities to have the same solutions.

For any real numbers a, b, and c, where c is **negative**,

$$\text{If } a < b, \quad \text{then } ac > bc. \qquad \text{If } a < b, \quad \text{then } \frac{a}{c} > \frac{b}{c}.$$

Similar statements can be made for the symbols $\leq$, $>$, and $\geq$.

EXAMPLE 4   Solve each inequality. Write the solution set in interval notation and graph it.

a. $-\frac{3}{2}t \geq -12$ **b.** $-5t < 55$

Strategy   We will use a property of inequality to isolate the variable on one side.

Why   To solve the original inequality, we want to find a simpler equivalent inequality, whose solution is obvious.

Solution   **a.** To undo the multiplication by $-\frac{3}{2}$, we multiply both sides by the reciprocal, which is $-\frac{2}{3}$.

$$-\frac{3}{2}t \geq -12 \qquad \text{This is the inequality to solve.}$$

$$-\frac{2}{3}\left(-\frac{3}{2}t\right) \leq -\frac{2}{3}(-12) \qquad \text{Multiply both sides by } -\frac{2}{3}. \text{ Since we are multiplying both sides by a negative number, reverse the direction of the } \geq \text{ symbol.}$$

$$t \leq 8 \qquad \text{Do the multiplication: } -\frac{2}{3}\left(-\frac{3}{2}\right) = 1 \text{ and } -\frac{2}{\cancel{3}}(-\overset{4}{\cancel{12}}) = 8.$$

The solution set is $(-\infty, 8]$ and it is graphed as shown.

You may find it easier to graph the solution set first and then use the symbols that occur in that form of the result to write the interval notation.

b. To undo the multiplication by -5, we divide both sides by -5.

$$-5t < 55 \qquad \text{This is the inequality to solve.}$$

$$\frac{-5t}{-5} > \frac{55}{-5} \qquad \text{To isolate } t, \text{ undo the multiplication by } -5 \text{ by dividing both sides by } -5.$$
Since we are dividing both sides by a negative number, reverse the direction of the $<$ symbol.

$$t > -11 \qquad \text{Do the division.}$$

The solution set is $(-11, \infty)$ and it is graphed as shown.

Self Check 4   Solve each inequality. Write the solution set in interval notation and graph it.   **a.** $-\frac{h}{20} \le 10$ **b.** $-12a > -144$

Now Try ▶ Problems 27 and 31

EXAMPLE 5   Solve $-5 > 3x + 7$. Write the solution set in interval notation and graph it.

Strategy   First we will use a property of inequality to isolate the *variable term* on one side. Then we will use a second property of inequality to isolate the *variable* itself.

Why   To solve the original inequality, we want to find a simpler equivalent inequality of the form $x >$ **a number** or $x <$ **a number**, whose solution is obvious.

Solution

Success Tip

Don't be confused by the negative number on the left side. We didn't reverse the $>$ symbol because we divided both sides by *positive* 3.

$$\frac{-12}{3} > \frac{3x}{3}$$

$$-5 > 3x + 7 \qquad \text{This is the inequality to solve.}$$

$$-5 - 7 > 3x + 7 - 7 \qquad \text{To isolate the variable term } 3x \text{ on the right side, undo the addition of 7 by subtracting 7 from both sides.}$$

$$-12 > 3x \qquad \text{Do the subtraction: } -5 - 7 = -12 \text{ and } 7 - 7 = 0.$$

$$\frac{-12}{3} > \frac{3x}{3} \qquad \text{To isolate } x, \text{ undo the multiplication by 3 by dividing both sides by 3.}$$

$$-4 > x \qquad \text{Do the division.}$$

To determine the solution set, it is useful to rewrite the inequality $-4 > x$ in an equivalent form with the variable on the left side.

$$x < -4 \qquad \text{If } -4 \text{ is greater than } x, \text{ it follows that } x \text{ must be less than } -4.$$

The solution set is $(-\infty, -4)$ and it is graphed as shown.

Self Check 5   Solve $-13 < 2r - 7$. Write the solution set in interval notation and graph it.

Now Try ▶ Problem 35

EXAMPLE 6   Solve $5.1 - 3k < 19.5$. Write the solution set in interval notation and graph it.

Strategy   We will use properties of inequality to isolate the variable on one side.

Why   To solve the original inequality, we want to find a simpler equivalent inequality of the form $k >$ **a number** or $k <$ **a number**, whose solution is obvious.

Solution

$$5.1 - 3k < 19.5$$ This is the inequality to solve.

$$5.1 - 3k - \mathbf{5.1} < 19.5 - \mathbf{5.1}$$ To isolate $-3k$ on the left side, subtract 5.1 from both sides.

$$-3k < 14.4$$ Do the subtraction.

$$\frac{-3k}{-3} > \frac{14.4}{-3}$$ To isolate k, undo the multiplication by -3 by dividing both sides by -3 and reverse the direction of the $<$ symbol.

$$k > -4.8$$ Do the division.

The solution set is $(-4.8, \infty)$, and it is graphed as shown.

Self Check 6   Solve $-9n + 1.8 > -17.1$. Write the solution set in interval notation and graph it.

Now Try ▶ Problem 39

The equation-solving strategy in Section 2.2 can be applied to inequalities. However, when solving inequalities, we must remember to *reverse the direction of the inequality symbol when multiplying or dividing both sides by a negative number.*

EXAMPLE 7   Solve $8(y + 1) \geq 2(y - 4) + y$. Write the solution set in interval notation and graph it.

Strategy   We will follow the steps of the equation-solving strategy (adapted to inequalities) to solve the inequality.

Why   This is the most efficient way to solve a linear inequality in one variable.

Solution

$$8(y + 1) \geq 2(y - 4) + y$$ This is the inequality to solve.

$$8y + 8 \geq 2y - 8 + y$$ Distribute the multiplication by 8 and by 2.

$$8y + 8 \geq 3y - 8$$ Combine like terms: $2y + y = 3y$.

$$8y + 8 - \mathbf{3y} \geq 3y - 8 - \mathbf{3y}$$ To eliminate $3y$ from the right side, subtract $3y$ from both sides.

$$5y + 8 \geq -8$$ Combine like terms: $8y - 3y = 5y$ and $3y - 3y = 0$.

$$5y + 8 - \mathbf{8} \geq -8 - \mathbf{8}$$ To isolate the variable term $5y$ on the left side, undo the addition of 8 by subtracting 8 from both sides.

$$5y \geq -16$$ Do the subtraction: $8 - 8 = 0$ and $-8 - 8 = -16$.

$$\frac{5y}{5} \geq \frac{-16}{5}$$ To isolate y, undo the multiplication by 5 by dividing both sides by 5. Do not reverse the direction of the $\geq$ symbol.

$$y \geq -\frac{16}{5}$$

> **Success Tip**
>
> As an informal check, substitute a number on the graph that is shaded, such as -3, into $8(y + 1) \geq 2(y - 4) + y$. A true statement should result. Then substitute a number on the graph that is not shaded, such as -4, into the inequality. A false statement should result.

The solution set is $\left[-\frac{16}{5}, \infty\right)$ and it is graphed as shown.

A solution involving an improper fraction is perfectly acceptable. Just make sure the fraction is in simplified form. To locate the endpoint of the interval on the graph, it is helpful to note that $-\frac{16}{5} = -3\frac{1}{5}$.

Self Check 7   Solve $5(b - 2) \geq -(b - 3) + 2b$. Write the solution set in interval notation and graph it.

Now Try ▶ Problem 45

EXAMPLE 8   Solve $\dfrac{3}{4} + \dfrac{x}{2} > \dfrac{6}{7}$. Write the solution set in interval notation and graph it.

Strategy   The first step of the equation-solving strategy (adapted to inequalities) is to clear the inequality of fractions by multiplying both sides by the LCD.

Why   It's easier to solve an inequality that involves only integers.

Solution

$$\dfrac{3}{4} + \dfrac{x}{2} > \dfrac{6}{7}$$ This is the inequality to solve.

$$28\left(\dfrac{3}{4} + \dfrac{x}{2}\right) > 28\left(\dfrac{6}{7}\right)$$ Clear the inequality of fractions by multiplying both sides by the LCD of $\frac{3}{4}, \frac{x}{2}$, and $\frac{6}{7}$, which is 28.

$$28\left(\dfrac{3}{4}\right) + 28\left(\dfrac{x}{2}\right) > 28\left(\dfrac{6}{7}\right)$$ On the left side, distribute the multiplication by 28.

$$21 + 14x > 24$$ Multiply: $\overset{7}{28}\left(\frac{3}{4}\right) = 21$, $\overset{14}{28}\left(\frac{x}{2}\right) = 14x$, and $\overset{4}{28}\left(\frac{6}{7}\right) = 24$.

$$21 + 14x - \mathbf{21} > 24 - \mathbf{21}$$ To isolate the variable term $14x$ on the left side, undo the addition of 21 by subtracting 21 from both sides.

$$14x > 3$$ Do the subtraction: $21 - 21 = 0$ and $24 - 21 = 3$.

$$\dfrac{14x}{\mathbf{14}} > \dfrac{3}{\mathbf{14}}$$ To isolate x, undo the multiplication by 14 by dividing both sides by 14.

$$x > \dfrac{3}{14}$$

The solution set is $\left(\dfrac{3}{14}, \infty\right)$ and it is graphed as shown.

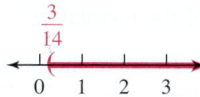

Self Check 8   Solve $\dfrac{1}{6} + \dfrac{2a}{9} < \dfrac{5}{2}$. Write the solution set in interval notation and graph it.

Now Try ▶ Problem 47

4 Solve Compound Inequalities.

Two inequalities can be combined into a **compound inequality** to show that an expression lies between two fixed values. For example, $-2 < x < 3$ is a combination of

$$-2 < x \qquad \text{and} \qquad x < 3$$

It indicates that x is greater than -2 and that x is also less than 3. The solution set of $-2 < x < 3$ consists of all numbers that lie between -2 and 3, and we write it as the interval $(-2, 3)$. The graph of the compound inequality is shown below.

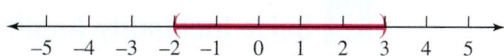

EXAMPLE 9   Graph $-4 \leq x < 0$ and write the solution set in interval notation.

Strategy   We need to determine which real numbers, when substituted for x, would make $-4 \leq x < 0$ a true statement.

Why   To graph $-4 \leq x < 0$ means to draw a "picture" of all of the values of x that make the compound inequality true.

Solution   If we replace the variable in $-4 \leq x < 0$ with a number between -4 and 0, including -4, the resulting compound inequality will be true. Therefore, the solution set is the interval $[-4, 0)$.

Notation

Note that the two inequality symbols in $-4 \leq x < 0$ point in the same direction and both point to the smaller number.

To graph the interval, we draw a bracket at -4, a parenthesis at 0, and shade in between, as shown.

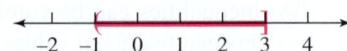

To check, we pick a number in the graph, such as -2, and see whether it satisfies the inequality. Since $-4 \leq -2 < 0$ is true, the answer appears to be correct.

Self Check 9   Graph $-2 \leq x < 1$ and write the solution set in interval notation.

Now Try ▶ Problem 51

To solve these types of compound inequalities, we isolate the variable in the middle part of the inequality. To do this, we apply the properties of inequality to all *three* parts of the inequality: the left, the middle, and the right.

EXAMPLE 10   Solve $-4 < 2(x - 1) \leq 4$. Write the solution set in interval notation and graph it.

Strategy   We will use properties of inequality to isolate the variable by itself as the middle part of the inequality.

Why   To solve the original inequality, we want to find a simpler equivalent inequality of the form **a number $< x \leq$ a number**, whose solution is obvious.

Solution

$-4 < 2(x - 1) \leq 4$	This is the compound inequality to solve.
$-4 < 2x - 2 \leq 4$	In the middle, distribute the multiplication by 2.
$-4 + 2 < 2x - 2 + 2 \leq 4 + 2$	To isolate the variable term $2x$, undo the subtraction of 2 by adding 2 to all three parts.
$-2 < 2x \leq 6$	Do the addition: $-4 + 2 = -2$, $-2 + 2 = 0$, and $4 + 2 = 6$.
$\dfrac{-2}{2} < \dfrac{2x}{2} \leq \dfrac{6}{2}$	To isolate x, we undo the multiplication by 2 by dividing all three parts by 2.
$-1 < x \leq 3$	Do the division.

The solution set is $(-1, 3]$ and it is graphed as shown.

Self Check 10   Solve $-6 \leq 3(t + 2) \leq 6$. Write the solution set in interval notation and graph it.

Now Try ▶ Problem 59

5 Solve Inequality Applications.

When solving application problems, phrases such as "not more than," or "should exceed" suggest that the problem involves an inequality rather than an equation.

©Christian M/Shutterstock.com

EXAMPLE 11

Grades. A student has scores of 72%, 74%, and 78% on three exams. What percent score does he need on the last exam to earn a grade of no less than B (80%)?

Analyze We know three scores. We are to find what the student must score on the last exam to earn a grade of B or higher.

Assign Let x = the score on the fourth (and last) exam.

Form To find the average grade, we add the four scores and divide by 4. To earn a grade of *no less than* B, the student's average must be *greater than or equal to* 80%.

The average of the four grades	must be no less than	80.
$\dfrac{72 + 74 + 78 + x}{4}$	$\geq$	80

Solve

$$\frac{224 + x}{4} \geq 80 \qquad \text{Combine like terms in the numerator: } 72 + 74 + 78 = 224.$$

$$4\left(\frac{224 + x}{4}\right) \geq 4(80) \qquad \text{To clear the inequality of the fraction, multiply both sides by 4.}$$

$$224 + x \geq 320 \qquad \text{Simplify each side.}$$

$$x \geq 96 \qquad \text{To isolate } x, \text{ undo the addition of 224 by subtracting 224 from both sides.}$$

State To earn a B, the student must score 96% or better on the last exam.

Check Pick several exam scores that are 96% or better and verify that the student's average will be 80% or greater. For example, a score of 96% gives the student an average that is exactly 80%.

$$\frac{72 + 74 + 78 + 96}{4} = \frac{320}{4} = 80$$

Self Check 11

Grades. A student has scores of 78%, 82%, and 76% on three exams. What percent score does he need on the last test to earn a grade of no less than a B (80%)?

Now Try ▶ Problem 103

The Language of Algebra

Some phrases that suggest an inequality are:

surpass: $>$ at least: $\geq$
not exceed: $\leq$ at most: $\leq$
is no less than: $\geq$
is no greater than: $\leq$
    between: ☐ $<$ ☐ $<$ ☐

SECTION 2.7 ▶ STUDY SET

VOCABULARY

Fill in the blanks.

1. An _____ is a statement that contains one of the symbols: $>$, $\geq$, $<$, or $\leq$. An equation is a statement that contains an _____ symbol.

2. To _____ an inequality means to find all the values of the variable that make the inequality true.

3. The solution set of $x > 2$ can be expressed in _____ notation as $(2, \infty)$.

4. The inequality $-4 < x \leq 10$ is an example of a _____ inequality.

CONCEPTS

Fill in the blanks.

5. **a.** Adding the same number to _____ sides of an inequality does not change the solutions.

   b. Multiplying or dividing both sides of an inequality by the same _____ number does not change the solutions.

   c. If we multiply or divide both sides of an inequality by a _____ number, the direction of the inequality symbol must be reversed for the inequalities to have the same solutions.

6. To solve $-4 \leq 2x + 1 < 3$, properties of inequality are applied to all _____ parts of the inequality: left, middle, and right.

7. Rewrite the inequality $32 < x$ in an equivalent form with the variable on the left side.

8. The solution set of an inequality is graphed below. Which of the four numbers, 3, −3, 2, and 4.5, when substituted for the variable in that inequality, would make it true?

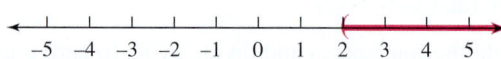

Write the inequality that is represented by each graph. Then describe the graph using interval notation.

9. a. 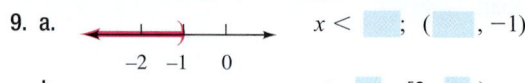 $x < \boxed{}$; $(\boxed{}, -1)$

   b. $x \geq \boxed{}$; $[2, \boxed{})$

10. a. 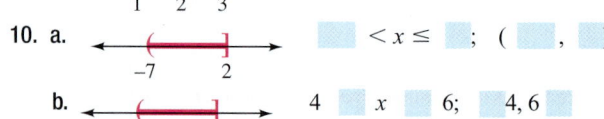 $\boxed{} < x \leq \boxed{}$; $(\boxed{}, \boxed{}]$

    b. $4 \boxed{} x \boxed{} 6$; $\boxed{} 4, 6 \boxed{}$

NOTATION

11. Write each symbol.
    a. is less than or equal to   b. infinity
    c. bracket   d. is greater than

12. Consider the graph of the interval [4, 8).
    a. Is 4 included in the graph?   b. Is 8 included in the graph?

Complete the solution to solve each inequality.

13. $4x - 5 \geq 7$

   $4x - 5 + \boxed{} \geq 7 + \boxed{}$

   $4x \geq \boxed{}$

   $\dfrac{4x}{\boxed{}} \geq \dfrac{12}{\boxed{}}$

   $x \geq 3$ Solution set: $[\boxed{}, \infty)$

14. $-6x > 12$

   $\dfrac{-6x}{\boxed{}} \boxed{} \dfrac{12}{-6}$

   $x < \boxed{}$ Solution set: $(\boxed{}, -2)$

GUIDED PRACTICE

See Example 1.

15. Determine whether each number is a solution of $3x - 2 > 5$.
    a. 5   b. −4

16. Determine whether each number is a solution of $3x + 7 < 4x - 2$.
    a. 12   b. 9

17. Determine whether each number is a solution of $-5(x - 1) \geq 2x + 12$.
    a. 1   b. −1

18. Determine whether each number is a solution of $\frac{4}{5}a \geq -2$.
    a. $-\dfrac{5}{4}$ b. −15

Graph each inequality and describe the graph using interval notation. See Example 2.

19. $x < 5$ 20. $x \geq -2$
21. $-3 < x \leq 1$ 22. $-4 \leq x \leq 2$

Solve each inequality. Write the solution set in interval notation and graph it. See Example 3.

23. $x + 2 > 5$ 24. $x + 5 \geq 2$
25. $g - 30 \geq -20$ 26. $h - 18 \leq -3$

Solve each inequality. Write the solution set in interval notation and graph it. See Example 4.

27. $-\dfrac{3}{16}x \geq -9$ 28. $-\dfrac{7}{8}x \leq 21$

29. $\dfrac{2}{3}x \geq 2$ 30. $\dfrac{3}{4}x < 3$

31. $-3y \leq -6$ 32. $-6y \geq -6$

33. $8h < 48$ 34. $2t > 22$

Solve each inequality. Write the solution set in interval notation and graph it. See Example 5.

35. $64 < 9x + 1$ 36. $32 > 4x + 8$

37. $-20 \geq 3m - 5$ 38. $-29 \leq 7t - 1$

Solve each inequality. Write the solution set in interval notation and graph it. See Example 6.

39. $1.3 - 2x \geq 0.5$ 40. $1.04 - 7x > 0.2$

41. $24.9 - 12a < -3.9$ 42. $37.5 - 16t \leq 99.9$

Solve each inequality. Write the solution set in interval notation and graph it. See Example 7.

43. $9a + 4 > 5a - 16$ 44. $8t + 1 < 4t - 19$

45. $8(2n + 1) \leq 4(6n + 7) + 4n$

46. $5(2 - d) \leq 3(d - 5) + 3d$

Solve each inequality. Write the solution set in interval notation and graph it. See Example 8.

47. $\dfrac{1}{2} + \dfrac{n}{5} > \dfrac{3}{4}$ 48. $\dfrac{1}{3} + \dfrac{c}{5} > -\dfrac{3}{2}$

49. $\dfrac{1}{2} - \dfrac{x}{24} \geq -\dfrac{1}{8}$ 50. $\dfrac{4}{3} - \dfrac{x}{5} \geq \dfrac{4}{15}$

Graph each compound inequality and describe the graph using interval notation.

51. $-2 \leq x < 3$ 52. $-1 < x \leq 4$

53. $-\dfrac{7}{4} < x < 2$ 54. $0 \leq x \leq \dfrac{11}{3}$

Solve each compound inequality. Write the solution set in interval notation and graph it. See Example 10.

55. $2 < x - 5 < 5$

56. $-8 < t - 8 < 8$

57. $0 \leq x + 10 \leq 10$

58. $-9 \leq x + 8 < 1$

59. $3 \leq 2x - 1 < 5$

60. $4 < 3x - 5 \leq 7$

61. $-9 < 6x + 9 \leq 45$

62. $-30 \leq 10d + 20 < 90$

TRY IT YOURSELF

Solve each inequality or compound inequality. Write the solution set in interval notation and graph it.

63. $\dfrac{6x + 1}{4} \leq x + 1$

64. $\dfrac{3x - 10}{5} \leq x + 4$

65. $17(3 - x) \geq 3 - 13x$

66. $7x + 6 \geq -(x - 6)$

67. $0 < 5(x + 2) \leq 15$

68. $-18 \leq 9(x - 5) < 27$

69. $0.4x \leq 0.1x + 0.45$

70. $0.9s \leq 0.3s + 0.54$

71. $-\dfrac{2}{3} \geq \dfrac{2y}{3} - \dfrac{3}{4}$

72. $-\dfrac{2}{9} \geq \dfrac{5x}{6} - \dfrac{1}{3}$

73. $\dfrac{m}{-42} - 1 > -1$

74. $\dfrac{a}{-25} + 3 < 3$

75. $6 - x \leq 3(x - 1)$

76. $3(3 - x) \geq 6 + x$

77. $6 < -2(x - 1) < 12$

78. $4 \leq -4(x - 2) < 20$

79. $-1 \leq -\dfrac{1}{2}n$

80. $-3 \geq -\dfrac{1}{3}t$

81. $-m - 12 > 15$

82. $-5x + 7 \leq 12$

83. $y - \dfrac{1}{7} \leq \dfrac{2}{3}$

84. $m - \dfrac{1}{9} \geq \dfrac{4}{5}$

85. $9x + 13 \geq 2x + 6x$

86. $7x - 16 < 2x + 4x$

87. $7 < \dfrac{5}{3}a + (-3)$

88. $5 < \dfrac{7}{2}a + (-9)$

89. $-8 \leq \dfrac{y}{8} - 4 \leq 2$

90. $-12 < \dfrac{b}{3} < 0$

91. $0.04x + 1.04 \leq 0.01x + 1.085$

92. $0.005 + 2.08x \leq 2.05x - 0.07$

93. $\dfrac{5}{3}(x + 1) \geq -x + \dfrac{2}{3}$

94. $\dfrac{5}{2}(7x - 15) \geq \dfrac{11}{2}x - \dfrac{3}{2}$

95. $\dfrac{4}{5}x < \dfrac{2}{5}$

96. $\dfrac{11}{9}x > \dfrac{5}{9}$

97. $2x + 3(2x + 3) \leq 7(x + 1) + 1$

98. $3(3x + 3) - 2 \leq 2(2x - 1) + 6x$

Look Alikes . . .

Solve each equation and inequality. Write the solution set of each inequality in interval notation and graph it.

99. a. $\dfrac{3}{8} + \dfrac{b}{3} > \dfrac{5}{12}$ **b.** $\dfrac{3}{8} + \dfrac{b}{3} = \dfrac{5}{12}$

100. a. $7(a - 3) < 2(5a - 8)$ **b.** $7(a - 3) = 2(5a - 8)$

101. a. $4 \leq 2x - 6$ **b.** $4 \leq 2x - 6 < 18$

102. a. $-16 < 4(x + 8) \leq 8$ **b.** $-16 = 4(x + 8) + 8$

APPLICATIONS

103. Grades.   A student has test scores of 68%, 75%, and 79% in a government class. What must she score on the last exam to earn a B (80% or better) in the course?

104. Occupational Testing.   An employment agency requires applicants average at least 70% on a battery of four job skills tests. If an applicant scored 70%, 74%, and 84% on the first three exams, what must he score on the fourth test to maintain a 70% or better average?

105. Gas Mileage.   A car manufacturer produces three models in equal quantities. One model has an economy rating of 17 miles per gallon, and the second model is rated for 19 mpg. If government regulations require the manufacturer to have a fleet average that exceeds 21 mpg, what economy rating is required for the third model?

106. Service Charges.   When the average daily balance of a customer's checking account falls below $500 in any week, the bank assesses a $5 service charge. The table shows the daily balances of one customer. What must Friday's balance be to avoid the service charge?

Day	Balance
Monday	$540.00
Tuesday	$435.50
Wednesday	$345.30
Thursday	$310.00

107. Geometry.   The perimeter of an equilateral triangle is at most 57 feet. What could the length of a side be? (*Hint:* All three sides of an equilateral triangle are equal.)

108. Geometry.   The perimeter of a square is no less than 68 centimeters. How long can a side be?

109. Counter Space. A rectangular counter is being built for the customer service department of a store. Designers have determined that the outside perimeter of the counter (shown in red) needs to exceed 30 feet. Determine the acceptable values for x.

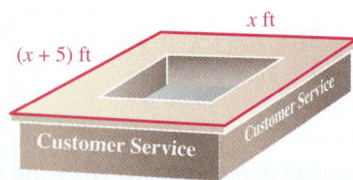

110. Number Puzzles. What numbers satisfy the condition: Four more than three times the number is at most 10?

111. Graduations. It costs a student $18 to rent a cap and gown and 80 cents for each graduation announcement that she orders. If she doesn't want her spending on these graduation costs to exceed $50, how many announcements can she order?

112. Telephones. A cellular telephone company has currently enrolled 36,000 customers in a new calling plan. If an average of 1,200 people are signing up for the plan each day, in how many days will the company surpass their goal of having 150,000 customers enrolled?

113. Windows. An architect needs to design a triangular-shaped bathroom window that has an area no greater than 100 in.^{2}. If the base of the window must be 16 inches long, what window heights will meet this condition?

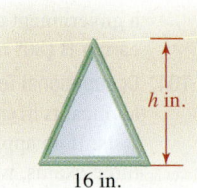

114. Room Temperatures. To hold the temperature of a room between 19° and 22° Celsius, what Fahrenheit temperatures must be maintained? *Hint:* Use the formula $C = \frac{5}{9}(F - 32)$.

115. Number Puzzles. What *whole* numbers satisfy the condition: Twice the number decreased by 1 is between 50 and 60?

116. Exercise. The graph in the next column shows the target heart beat range for different ages and exercise intensity levels. If we let b represent the number of beats per minute, then the compound inequality that estimates the heart beat rate range for a 30-year-old involved in a high-intensity workout is about $168 \leq b \leq 198$. Use a compound inequality to estimate the heart beat rate range for the following ages and zones.

a. 45-year-old, fat-burning zone

b. 70-year-old, high-intensity zone

c. 25-year-old, intermediate zone

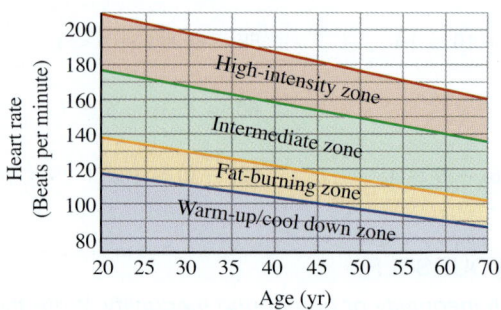

Source: elitefitness.co.nz

WRITING

117. Explain why multiplying both sides of an inequality by a negative number reverses the direction of the inequality.

118. a. What number is a solution of $3x - 26 \geq 4$ but is not a solution of $3x - 26 > 4$? Explain your reasoning.

b. What numbers are solutions of $3x - 26 \geq 4$ but not solutions of $3x - 26 = 4$? Explain your reasoning.

REVIEW

Complete each table.

119.

x	$x^2 - 3$
-2	
0	
3	

120.

x	$\frac{x}{3} + 2$
-6	
0	
12	

CHALLENGE PROBLEMS

Solve each inequality. Write the solution set in interval notation and graph it.

121. $3 - x < 5 < 7 - x$

122. $\frac{1}{x} > 1$. (*Hint:* Use a guess-and-check approach.)

123. $2(5x - 6) > 4x - 15 + 6x$

124. $\dfrac{3a - 4}{-5} < \dfrac{3a + 15}{-5}$

2 ▶ Summary & Review

SECTION 2.1 ▶ Solving Equations Using Properties of Equality

DEFINITIONS AND CONCEPTS	EXAMPLES
An **equation** is a statement indicating that two expressions are equal. The equal symbol $=$ separates an equation into two parts: the *left side* and the *right side*.	Equations: $2x + 4 = 10$ $-5(a + 4) = -11a$ $\dfrac{3}{2}t + 6 = t - \dfrac{1}{3}$

A number that makes an equation a true statement when substituted for the variable is called a **solution** of the equation.	Use a check to determine whether 2 is a solution of $x + 4 = 3x$.
	
	Check: $x + 4 = 3x$
	    $2 + 4 \overset{?}{=} 3(2)$ Substitute 2 for each x.
	    $6 = 6$ True
	
	Since the resulting statement $6 = 6$ is true, 2 is a solution of $x + 4 = 3x$.

Equivalent equations have the same solutions.	$x - 2 = 6$ and $x = 8$ are equivalent equations because they have the same solution, 8.

To **solve an equation** isolate the variable on one side of the equation by undoing the operations performed on it using properties of equality.
 	
Addition (Subtraction) property of equality: If the same number is added to (or subtracted from) both sides of an equation, the result is an equivalent equation.
	Solve: $x - 5 = 7$ Solve: $c + 9 = 16$
	  $x - 5 + 5 = 7 + 5$ $c + 9 - 9 = 16 - 9$
	     $x = 12$ $c = 7$
	The solution is 12. The solution is 7.
	The solution set is $\{12\}$. The solution set is $\{7\}$.

Multiplication (Division) property of equality: If both sides of an equation are multiplied (or divided) by the same nonzero number, the result is an equivalent equation.	Solve: $\dfrac{1}{3}m = 2$ Solve: $10y = 50$
	
	                        $\dfrac{10y}{10} = \dfrac{50}{10}$
	  $3\left(\dfrac{1}{3}m\right) = 3(2)$ $y = 5$ The solution is 5.
	
	    $m = 6$ The solution is 6.

REVIEW EXERCISES

Use a check to determine whether the given number is a solution of the equation.

1. $84, x - 34 = 50$ **2.** $3, 5y + 2 = 12$

3. $-30, \dfrac{x}{5} = 6$ **4.** $2, |a^2 - a - 1| = 0$

5. $-3, 5b - 2 = 3b - 8$ **6.** $1, \dfrac{2}{y + 1} = \dfrac{12}{y + 1} - 5$

Fill in the blanks.

7. An _____ is a statement indicating that two expressions are equal.

8. To solve $x - 8 = 10$ means to find all the values of the variable that make the equation a _____ statement.

Solve each equation and check the result.

9. $x - 9 = 12$ **10.** $-y = -32$

11. $a + 3.7 = -16.9$ **12.** $100 = -7 + r$

13. $120 = 5c$ **14.** $t - \dfrac{1}{3} = \dfrac{3}{7}$

15. $\dfrac{4}{3}t = -12$ **16.** $3 = \dfrac{q}{-2.6}$

17. $6b = 0$ **18.** $\dfrac{15}{16}s = -3$

SECTION 2.2 ▶ More about Solving Equations

DEFINITIONS AND CONCEPTS	EXAMPLES
A five-step **strategy for solving linear equations:** 1. *Clear* the equation of fractions or decimals. 2. *Simplify* each side. Use the distributive property and combine like terms when necessary. 3. *Isolate the variable term.* Use the addition and subtraction properties of equality. 4. *Isolate the variable.* Use the multiplication and division properties of equality. 5. *Check* the result in the original equation.	Solve: $2(y + 2) + 4y = 11 - y$ $\quad 2y + 4 + 4y = 11 - y$ Distribute the multiplication by 2. $\qquad 6y + 4 = 11 - y$ Combine like terms: 2y + 4y = 6y. $\quad 6y + 4 + y = 11 - y + y$ To eliminate −y on the right, add y to both sides. $\qquad 7y + 4 = 11$ Combine like terms: 6y + y = 7y and −y + y = 0. $\quad 7y + 4 - 4 = 11 - 4$ To isolate the variable term 7y, undo the addition of 4 by subtracting 4 from both sides. $\qquad 7y = 7$ Simplify each side of the equation. $\qquad \dfrac{7y}{7} = \dfrac{7}{7}$ To isolate y, undo the multiplication by 7 by dividing both sides by 7. $\qquad y = 1$ The solution is 1.
It is easier to solve an equation that involves only integers. **To clear an equation of fractions,** multiply both sides of an equation by the LCD. **To clear an equation of decimals,** multiply both sides by a power of 10 to change the decimals in the equation to integers.	To solve $\dfrac{1}{2} + \dfrac{x}{3} = \dfrac{3}{4}$, first clear the fractions by multiplying both sides by 12: $12\left(\dfrac{1}{2} + \dfrac{x}{3}\right) = 12\left(\dfrac{3}{4}\right)$ The LCD of $\frac{1}{2}$, $\frac{x}{3}$, and $\frac{3}{4}$ is 12. To solve $0.5(x - 4) = 0.1x + 0.2$, first clear the decimals by multiplying both sides by 10: $10[0.5(x - 4)] = 10(0.1x + 0.2)$
Not every equation in one variable has a single solution. Some equations are made true by *any* permissible replacement value for the variable. Such equations are called **identities.** An equation that is not true for any value of its variable is called a **contradiction.**	When we solve $x + 5 + x = 2x + 5$, the variable drops out and we obtain a true statement $5 = 5$. All real numbers are solutions. When we solve $y + 2 = y$, the variable drops out and we obtain a false statement $2 = 0$. The equation has no solutions.

REVIEW EXERCISES

Solve each equation. Check the result.

19. $5x + 4 = 14$

20. $98.6 - t = 129.2$

21. $\dfrac{n}{5} + (-2) = 4$

22. $\dfrac{b - 5}{4} = -6$

23. $5(2x - 4) - 5x = 0$

24. $-2(x - 5) = 5(-3x + 4) + 3$

25. $\dfrac{3}{4} = \dfrac{1}{2} + \dfrac{d}{5}$

26. $\dfrac{5(7 - x)}{4} = 2x - 3$

27. $\dfrac{3(2 - c)}{2} = \dfrac{-2(2c + 3)}{5}$

28. $\dfrac{b}{3} + \dfrac{11}{9} + 3b = -\dfrac{5}{6}b$

29. $0.15(x + 2) + 0.3 = 0.35x - 0.4$

30. $0.5 - 0.02(y - 2) = 0.16 + 0.36y$

31. $3(a + 8) = 6(a + 4) - 3a$

32. $2(y + 10) + y = 3(y + 8)$

SECTION 2.3 ▶ Applications of Percent

DEFINITIONS AND CONCEPTS	EXAMPLES
To solve **percent problems,** use the facts of the problem to write a sentence of the form: ☐ is ☐ % of ☐ ? Translate the sentence to mathematical symbols: *is* translates to an = symbol and *of* means multiply. Then solve the equation.	648 is 30% of what number? ↓ ↓ ↓ ↓ ↓ 648 = 30% · *x* Translate. $648 = 0.30x$ Change 30% to a decimal: 30% = 0.30. $\dfrac{648}{0.30} = \dfrac{0.30x}{0.30}$ To isolate *x*, undo the multiplication by 0.30 by dividing both sides by 0.30. $2{,}160 = x$ Do the division. Thus, 648 is 30% of 2,160.
To find the **percent of increase** or **the percent of decrease,** find what percent the increase or decrease is of the original amount.	**Sales Prices.** To find the percent of decrease when ground beef prices are reduced from $4.89 to $4.25 per pound, we first find the amount of decrease: $4.89 - 4.25 = 0.64$. Then we determine what percent 0.64 is of 4.89 (the original price). 0.64 is what % of 4.89? ↓ ↓ ↓ ↓ ↓ 0.64 = *x* · 4.89 Translate. $0.64 = 4.89x$ $\dfrac{0.64}{4.89} = \dfrac{4.89x}{4.89}$ To isolate x, undo the multiplication by 4.89 by dividing both sides by 4.89. $0.130879346 \approx x$ Do the division. $013.0879346\% \approx x$ Write the decimal as a percent. To the nearest tenth of a percent, the percent of decrease is 13.1%.

REVIEW EXERCISES

33. Fill in the blanks.

  a. _____ means parts per one hundred.

  b. When the price of an item is reduced, we call the amount of the reduction a _____.

  c. An employee who is paid a _____ is paid a percent of the goods or services that he or she sells.

34. 4.81 is 2.5% of what number?

35. What number is 15% of 950?

36. What percent of 410 is 49.2?

37. **Internet Users.** The circle graph below shows the percent of the U.S. population that used the Internet in 2010.

  a. What percent of the population did not use the Internet?

  b. If the U.S. population in 2010 was about 310 million, how many people used the Internet that year? Round to the nearest million.

U.S. Internet Usage,* 2010

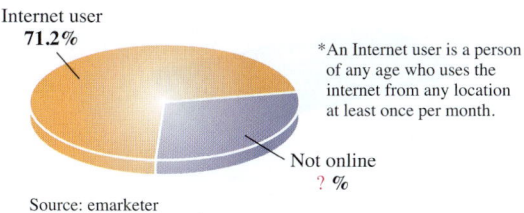

Internet user
71.2%

*An Internet user is a person of any age who uses the internet from any location at least once per month.

Not online
? %

Source: emarketer

38. **Cost of Living.** A retired trucker receives a monthly Social Security check of $764. If she is to receive a 3.5% cost-of-living increase soon, how much larger will her check be?

39. **Family Budgets.** It is recommended that a family pay no more than 30% of its monthly income (after taxes) on housing. If a family has an after-tax income of $1,890 per month and pays $625 in housing costs each month, are they within the recommended range?

40. **Discounts.** A shopper saved $148.50 on a food processor that was discounted 33%. What did it originally cost?

41. **Tupperware.** The hostess of a Tupperware party is paid a 25% commission on her in-home party's sales. What would the hostess earn if sales totaled $600?

42. **Collectibles.** A collector of football trading cards paid $6 for a 1984 Dan Marino rookie card several years ago. If the card is now worth $100, what is the percent of increase in the card's value? (Round to the nearest percent.)

SECTION 2.4 ▶ Formulas

DEFINITIONS AND CONCEPTS	EXAMPLES
A **formula** is an equation that states a relationship between two or more variables.	Retail price: $r = c + m$ Profit: $p = r - c$ Simple Interest: $I = Prt$ Distance: $d = rt$ Temperature: $C = \dfrac{5}{9}(F - 32)$
The **perimeter** of a plane geometric figure is the distance around it. The **area** of a plane geometric figure is the amount of surface that it encloses. The **volume** of a three-dimensional geometric solid is the amount of space it encloses.	Rectangle: $P = 2l + 2w$ Circle: $C = \pi D = 2\pi r$ $A = lw$ $A = \pi r^2$ Rectangular solid: $V = lwh$ Cylinder: $V = \pi r^2 h$ *See inside the back cover of the text for more geometric formulas.
If we are given the values of all but one of the variables in a formula, we can use our equation-solving skills to find the value of the remaining variable.	**Bedding.** The area of a standard queen-size bed sheet is 9,180 in.^{2}. If the length is 102 inches, what is the width? $A = lw$ This is the formula for the area of a rectangle. $9{,}180 = 102w$ Substitute 9,180 for the area A and 102 for the length l. $\dfrac{9{,}180}{102} = \dfrac{102w}{102}$ To isolate w, undo the multiplication by 102 by dividing both sides by 102. $90 = w$ Do the division. The width of a standard queen-size bed sheet is 90 inches.
To solve a formula for a specific variable means to isolate that variable on one side of the equation, with all other variables and constants on the opposite side. Treat the specified variable as if it is the only variable in the equation. Treat the other variables as if they were numbers (constants).	Solve the formula for the volume of a cone for h. $V = \dfrac{1}{3}\pi r^2 h$ This is the formula for the volume of a cone. $3(V) = 3\left(\dfrac{1}{3}\pi r^2 h\right)$ To clear the equation of the fraction, multiply both sides by 3. $3V = \pi r^2 h$ Simplify. $\dfrac{3V}{\pi r^2} = \dfrac{\pi r^2 h}{\pi r^2}$ To isolate h, undo the multiplication by πr^2 by dividing both sides by πr^2. $\dfrac{3V}{\pi r^2} = h$ or $h = \dfrac{3V}{\pi r^2}$

REVIEW EXERCISES

43. **Shopping.** Find the markup on a CD player whose wholesale cost is $219 and whose retail price is $395.

44. **Restaurants.** One month, a restaurant had sales of $13,500 and made a profit of $1,700. Find the expenses for the month.

45. **Snails.** A typical garden snail travels at an average rate of 2.5 feet per minute. How long would it take a snail to cross a 20-foot-long flower bed?

46. **Certificates of Deposit.** A $26,000 investment in a CD earned $1,170 in interest the first year. What was the annual interest rate?

47. **Jewelry.** Gold melts at about 1,065°C. Change this temperature to degrees Fahrenheit.

48. **Camping.**
   a. Find the perimeter of the air mattress.
   b. Find the amount of sleeping area on the top surface of the air mattress.
   c. Find the volume of the air mattress if it is 3 inches thick.

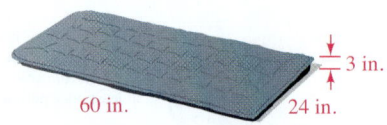

60 in.   24 in.   3 in.

49. Find the area of a triangle with a base 17 meters long and a height of 9 meters.

50. Find the area of a trapezoid with bases 11 inches and 13 inches long and a height of 12 inches.

In Problems 51–53, the answers may vary slightly depending on which approximation of π is used.

51. a. Find the circumference of a circle with a radius of 8 centimeters. Round to the nearest hundredth.

   b. Find the area of the circle. Round to the nearest square centimeter.

52. Find the volume of a 12-foot tall cylinder whose circular base has a radius of 0.5 feet. Give the result to the nearest tenth.

53. Halloween. After being cleaned out, a spherical-shaped pumpkin has an inside diameter of 9 inches. To the nearest hundredth, what is its volume?

54. Find the volume of a pyramid that has a square base, measuring 6 feet on a side, and a height of 10 feet.

Solve each formula for the specified variable.

55. $A = 2\pi rh$ for h

56. $A - BC = \dfrac{G - K}{3}$ for G

57. $C = \dfrac{1}{4}s(t - d)$ for t

58. $4y - 3x = 16$ for y

SECTION 2.5 ▶ Problem Solving

DEFINITIONS AND CONCEPTS	EXAMPLES

To solve application problems, use the six-step problem-solving strategy.

1. **Analyze** the problem.

2. **Assign** a variable.

3. **Form** an equation.

4. **Solve** the equation.

5. **State** the conclusion.

6. **Check** the result.

(See page 142 for the more detailed list of steps.)

In this section, we solved application problems involving:

- More than one unknown (Example 1)
- Set-up fees (Example 2)
- Commissions (Example 3)
- Consecutive integers (Example 4)
- Perimeter (Example 5)
- Isoceles triangles (Example 6)

Income Taxes.   After taxes, an author kept $85,340 of her total annual earnings. If her earnings were taxed at a 15% rate, how much did she earn that year?

Analyze   The author earned some unknown amount of money. On that amount, she paid 15% in taxes. The difference between her total earnings and the taxes paid was $85,340.

Assign   If we let x = the author's total earnings, the amount of taxes that she paid was 15% of x or $0.15x$.

Form   We can use the words of the problem to form an equation.

Her total earnings	minus	the taxes that she paid	equals	the money that she kept.
x	$-$	$0.15x$	$=$	$85,340$

Solve

$x - 0.15x = 85,340$

$0.85x = 85,340$ *Combine like terms: 1x − 0.15x = 0.85x.*

$x = 100,400$ *To isolate x, divide both sides by 0.85.*

State   The author earned $100,400 that year.

Check   The taxes were 15% of $100,400 or $15,060. If we subtract the taxes from her total earnings, we get $100,400 − $15,060 = $85,340. The answer checks.

REVIEW EXERCISES

59. Sound Systems.   A 45-foot-long speaker wire is to be cut into three pieces. One piece is to be 15 feet long. Of the remaining pieces, one must be 2 feet less than 3 times the length of the other. Find the length of the shorter piece.

60. Signing Petitions.   A professional signature collector is paid $50 a day plus $2.25 for each verified signature he gets from a registered voter. How many signatures are needed to earn $500 a day?

61. Lottery Winnings.   After taxes, a lottery winner was left with a lump sum of $1,800,000. If 28% of the original prize was withheld to pay federal income taxes, what was the original cash prize?

62. NASCAR.   The car numbers of drivers Bobby Labonte and Kyle Petty are consecutive odd integers whose sum is 88. If Labonte's number is the smaller, find the numbers of each car.

63. Art History. *American Gothic* was painted in 1930 by Grant Wood. The length of the rectangular painting is 5 inches more than the width. Find the dimensions of the painting if it has a perimeter of 109.5 inches.

©SuperStock/SuperStock

64. Geometry. Find the missing angle measures of the triangle.

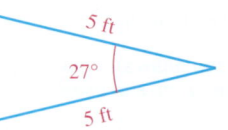
5 ft
27°
5 ft

SECTION 2.6 ▶ More on Problem Solving

DEFINITIONS AND CONCEPTS	EXAMPLES

To solve application problems, use the six-step problem-solving strategy.

1. **Analyze** the problem.

2. **Assign** a variable.

3. **Form** an equation.

4. **Solve** the equation.

5. **State** the conclusion.

6. **Check** the result.

(See page 142 for the more detailed list of steps.)

Tables are a helpful way to organize the facts of a problem.

In this section, we solved application problems involving:

■ Money/investments (Example 1)

■ Motion (Example 2 and Example 3)

■ Liquid mixture (Example 4)

■ Dry mixture (Example 5)

■ Number-value (Example 6)

Trucking. Two trucks leave from the same place at the same time traveling in opposite directions. One travels at a rate of 60 mph and the other at 50 mph. How long will it take them to be 165 miles apart?

Analyze We know that one truck travels at 60 mph and the other at 50 mph. Together, the trucks will travel a distance of 165 miles.

Assign Let t = the number of hours until the distance between the trucks is 165 miles.

Form We enter each rate in the table under the heading r. Since the trucks travel for the same length of time, say t hours, we enter t for each truck under the heading t. We enter the distances traveled under the heading d in the table.

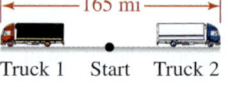
← 165 mi →
Truck 1   Start   Truck 2

	r · t = d		
Truck 1	60	t	$60t$
Truck 2	50	t	$50t$
	Total: 165		

Multiply $r \cdot t$ to obtain an expression for the distance traveled by each truck.

↑___ Use the information in this column to form an equation.

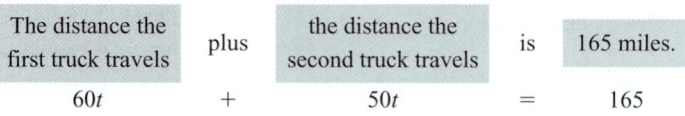

The distance the first truck travels	plus	the distance the second truck travels	is	165 miles.
$60t$	+	$50t$	=	165

Solve $60t + 50t = 165$

$110t = 165$ Combine like terms: 60t + 50t = 110t.

$\dfrac{110t}{110} = \dfrac{165}{110}$ To isolate t, undo the multiplication by 110 by dividing both sides by 110.

$t = 1.5$ Do the division.

State The trucks will be 165 miles apart in 1.5 hours.

Check If the first truck travels 60 mph for 1.5 hours, it will go $60(1.5) = 90$ miles. If the second truck travels 50 mph for 1.5 hours, it will go $50(1.5) = 75$ miles. Since 90 miles + 75 miles = 165 miles, the result checks.

REVIEW EXERCISES

65. Investment Income. A woman has $27,000. Part is invested for 1 year in a certificate of deposit paying 7% interest, and the remaining amount in a cash management fund paying 9%. After 1 year, the total interest on the two investments is $2,110. How much is invested at each rate?

66. Walking and Bicycling. A bicycle path is 5 miles long. A man walks from one end at the rate of 3 mph. At the same time, a friend bicycles from the other end, traveling at 12 mph. In how many minutes will they meet?

67. Airplanes. How long will it take a jet plane, flying at 450 mph, to overtake a propeller plane, flying at 180 mph, if the propeller plane had a $2\frac{1}{2}$-hour head start?

68. Autographs. Kesha collected the autographs of 8 more television celebrities than she has of movie stars. Each TV celebrity autograph is worth $75 and each movie star autograph is worth $250. If her collection is valued at $1,900, how many of each type of autograph does she have?

69. Mixtures. A store manager mixes candy worth 90¢ per pound with gumdrops worth $1.50 per pound to make 20 pounds of a mixture worth $1.20 per pound. How many pounds of each kind of candy does he use?

70. Eliminating Mildew. How many gallons of a 2% fungicide solution must be mixed with 4 gallons of a 5% fungicide solution to get a 4% fungicide solution?

SECTION 2.7 ▶ Solving Inequalities

DEFINITIONS AND CONCEPTS	EXAMPLES
An **inequality** is a mathematical statement that contains an $>$, $<$, $\geq$, or $\leq$ symbol.	Inequalities: $3x < 8$ $\frac{1}{2}y - 4 \geq 12$ $2z + 4 \leq z - 5$
A **solution of an inequality** is any number that makes the inequality true.	Determine whether 3 is a solution of $2x - 7 < 5$. **Check:** $2x - 7 < 5$ $2(3) - 7 \overset{?}{<} 5$ Substitute 3 for x. $-1 < 5$ True Since the resulting statement $-1 < 5$ is true, 3 is a solution of $2x - 7 < 5$.
We **solve inequalities** as we solve equations. However, if we **multiply or divide both sides by a negative number,** we must *reverse* the inequality symbol.	Solve: $-3(z - 1) \geq -6$ $-3z + 3 \geq -6$ Distribute the multiplication by -3. $-3z \geq -9$ To isolate the variable term $-3z$, undo the addition of 3 by subtracting 3 from both sides. $\dfrac{-3z}{-3} \leq \dfrac{-9}{-3}$ To isolate z, undo the multiplication by dividing both sides by -3. Reverse the $\geq$ symbol. $z \leq 3$ Do the division.
Interval notation can be used to describe the solution set of an inequality. A **parenthesis** indicates that a number is not included in the solution set of an inequality. A **bracket** indicates that a number is included in the solution set.	The solution set is $(-\infty, 3]$ and it is graphed as shown.

REVIEW EXERCISES

Solve each inequality. Write the solution set in interval notation and graph it.

71. $3x + 2 < 5$

72. $-\frac{3}{4}x \geq -9$

73. $\frac{3}{4} < \frac{d}{5} + \frac{1}{2}$

74. $5(3 - x) \leq 3(x - 3)$

75. $\frac{t}{-5} - (-1.8) \geq -6.2$

76. $a + 5 - 2(10 - a) > 6$

77. $24 < 3(x + 2) < 39$

78. $0 \leq 3 - 2x < 10$

79. Sports Equipment. The acceptable weight w of Ping-Pong balls used in competition can range from 2.40 to 2.53 grams. Express this range using a compound inequality.

80. Signs. A large office complex has a strict policy about signs. Any sign to be posted in the building must be rectangular in shape, its width must be 18 inches, and its perimeter is not to exceed 132 inches. What possible sign lengths meet these specifications?

2 ▶ CHAPTER TEST

1. Fill in the blanks.

   a. To _____ an equation means to find all of the values of the variable that make the equation true.

   b. _____ means parts per one hundred.

   c. The distance around a circle is called its _____.

   d. An _____ is a statement that contains one of the symbols $>$, $\geq$, $<$, or $\leq$.

   e. The _____ property of _____ says that multiplying both sides of an equation by the same nonzero number does not change its solution.

2. Use a check to determine if 3 is a solution of $5y + 2 = 12$.

Solve each equation.

3. $3h + 2 = 8$

4. $-22 = -x$

5. $\dfrac{4}{5}t = -4$

6. $\dfrac{11b - 11}{5} = \dfrac{3b - 2}{2}$

7. $0.8(x - 1{,}000) + 1.3 = 2.9 + 0.2x$

8. $2(y - 7) - 3y = -(y - 3) - 17$

9. $\dfrac{m}{2} - \dfrac{1}{3} = \dfrac{1}{4} + \dfrac{m}{6}$

10. $\dfrac{3}{4}(6n - 2) = 246$

11. $5x = 0$

12. $6a + (-7) = 3a - 7 + 2a$

13. $9 - 5(2x + 10) = -1$

14. $24t = -6(8 - 4t)$

15. What is 15.2% of 80?

16. **Down Payments.** To buy a house, a woman was required to make a down payment of $11,400. What did the house sell for if this was 15% of the purchase price?

17. **Body Temperatures.** Suppose a person's body temperature rises from 98.6°F to a dangerous 105°F. What is the percent increase? Round to the nearest percent.

18. **Commissions.** An appliance store salesperson receives a commission of 5% of the price of every item that she sells. What will she make if she sells a $599.99 refrigerator?

19. **Grand Openings.** On its first night of business, a pizza parlor brought in $445. The owner estimated his profits that night to be $150. What were the costs?

20. Find the Celsius temperature reading if the Fahrenheit reading is 14°.

21. **Sound.** The speed of sound at sea level is about 1,108 feet per second. How far will sound travel in 1 minute?

22. **Pets.** The spherical fishbowl is three-quarters full of water. To the nearest cubic inch, find the volume of water in the bowl.

    $\left(\text{\textit{Hint:} The volume of a sphere is given by } V = \tfrac{4}{3}\pi r^3.\right)$

10 in.

Solve for the specified variable.

23. $V = \pi r^2 h$ for h

24. $A = P + Prt$ for r

25. $A = \dfrac{a + b + c + d}{4}$ for c

26. $2x - 3y = 9$ for y

27. **Irons.** Estimate the area of the soleplate of the iron.

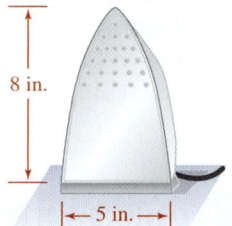

8 in.

5 in.

28. **Television.** In a typical 30-minute block of time on TV, the number of programming minutes are 2 less than three times the number of minutes of commercials. How many minutes of programming and how many minutes of commercials are there?

29. **Plumbing Bills.** A part of the invoice for plumbing work shown below is torn away. What is the cost per hour for labor?

Carter Plumbing 100.W. Dalton Ave.		Invoice #210
Standard service charge	$	25.75
Parts	$	38.75
Labor: 4 hours at $ per hour		
Total charges	$	226.70

30. **Concert Seating.** Two types of seating were sold for a concert. Floor seats cost $12.50 a ticket and balcony seats cost $20.50. Ten times as many floor seats were sold as balcony seats. If the total receipts from the sale of both types of tickets were $11,640, how many of each type of ticket were sold?

31. **Home Sales.** A condominium owner cleared $114,600 on the sale of his condo, after paying a 4.5% real estate commission. What was the selling price?

32. **Colorado.** The state of Colorado is approximately rectangular-shaped with perimeter 1,320 miles. Find the length (east to west) and width (north to south), if the length is 100 miles longer than the width.

33. **Tea.** How many pounds of green tea, worth $40 a pound, should be mixed with herbal tea, worth $50 a pound, to produce 20 pounds of a blend worth $42 a pound?

34. Reading. A bookmark is inserted between two page numbers whose sum is 825. What are the page numbers?

35. Travel Times. A car leaves Rockford, Illinois, at the rate of 65 mph, bound for Madison, Wisconsin. At the same time, a truck leaves Madison at the rate of 55 mph, bound for Rockford. If the cities are 72 miles apart, how long will it take for the car and the truck to meet?

36. Pickles. To make pickles, fresh cucumbers are soaked in a salt water solution called *brine*. How many liters of a 2% brine solution must be added to 30 liters of a 10% brine solution to dilute it to an 8% solution?

37. Exercise. How long will it take a bicyclist, traveling at 20 mph, to catch up with a jogger, traveling at 8 mph, if the jogger had a half-hour head start?

38. Geometry. If the vertex angle of an isosceles triangle is 44°, find the measure of each base angle.

39. Investments. Part of $13,750 is invested at 9% annual interest, and the rest is invested at 8%. After one year, the accounts paid $1,185 in interest. How much was invested at the lower rate?

40. Use a check to determine whether -3 is a solution of $4 - 9w < -4w + 19$.

Solve each inequality. Write the solution set in interval notation and graph it.

41. $-8x - 20 \leq 4$

42. $-8.1 > \dfrac{t}{2} + (-11.3)$

43. $-12 \leq 2(x + 1) < 10$

44. $\dfrac{1}{3}(a - 5) > \dfrac{1}{2}(a + 1)$

45. $-9(h - 3) + 2h \leq 8(4 - h)$

46. Awards. A city honors its citizen of the year with a framed certificate. An artist charges $15 for the frame and 75 cents per word for writing out the proclamation. If a city regulation does not allow gifts in excess of $150, what is the maximum number of words that can be written on the certificate?

Group Project

TRANSLATING KEY WORDS AND PHRASES

▶ *Overview:* Students often say that the most challenging step of the six-step problem-solving strategy is forming an equation. This activity is designed to make that step easier by improving your translating skills.

Instructions: Form groups of 3 or 4 students. Select one person from your group to record the group's responses. Determine whether addition, subtraction, multiplication, or division is suggested by each of the following words or phrases. Then use the word or phrase in a sentence to illustrate its meaning. (If a word stumps everyone in your group, a dictionary can be helpful.)

deflate	recede	partition	evaporate	amplify
bisect	augment	hike	erode	boost
annexed	diminish	plummet	upsurge	wane
quadruple	corrode	taper off	trisect	broaden

COMPUTER SPREADSHEETS

▶ *Overview:* In this activity, you will get some experience working with a spreadsheet.

Instructions: Form groups of 3 or 4 students. Examine the following spreadsheet, which consists of cells named by column and row. For example, 7 is entered in cell B3. In any cell you may enter data or a formula. For each formula in cells D1–D4 and E1–E4, the computer performs a calculation using values entered in other cells and prints the result in place of the formula. Find the value that will be printed in each formula cell. The symbol * means multiply, / means divide, and ^ means raise to a power.

	A	B	C	D	E
1	-8	20	-6	$= 2*B1 - 3*C1 + 4$	$= B1 - 3*A1^2$
2	39	2	-1	$= A2/(B2 - C2)$	$= B3*B2*C2*2$
3	50	7	3	$= A3/5 + C3^3$	$= 65 - 2*(B3 - 5)^5$
4	6.8	-2.8	-0.5	$= 100*A4 + B4*C4$	$= A4/10 + A3/2*5$

Graphing Linear Equations and Inequalities in Two Variables; Functions

3

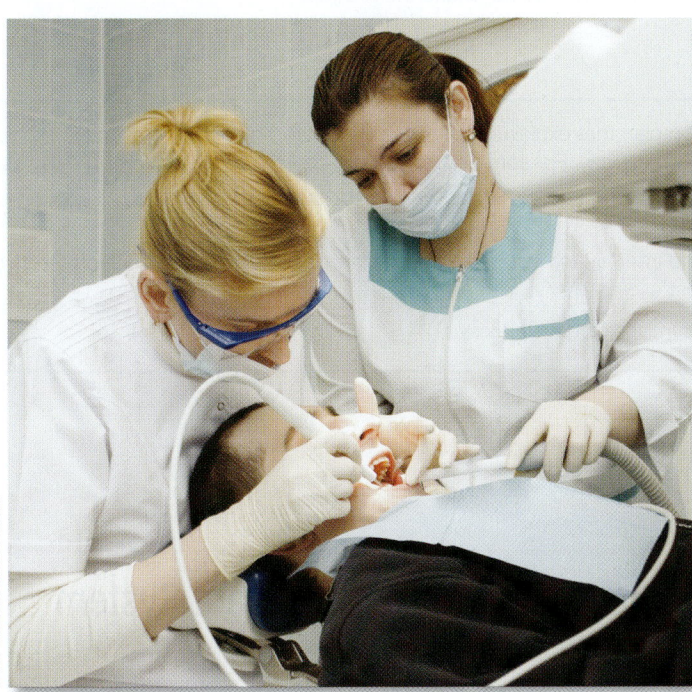

©Yuri Shirokov/Shutterstock.com

from Campus to Careers

Dental Assistant

A dental assistant is a valuable member of the dental health team who prepares patients for treatment, takes x-rays, sterilizes instruments, and keeps records. Part of the training of a dental assistant includes learning about a *coordinate system* that is used to identify the location of teeth in the mouth. This coordinate system is much like one used in algebra to graph points, lines, and curves.

Problem 33 in **Study Set 3.1, problem 105** in **Study Set 3.4,** and **problem 77** in **Study Set 3.7** involve situations that a dental assistant might encounter on the job. The mathematical concepts discussed in this chapter can be used to solve those problems.

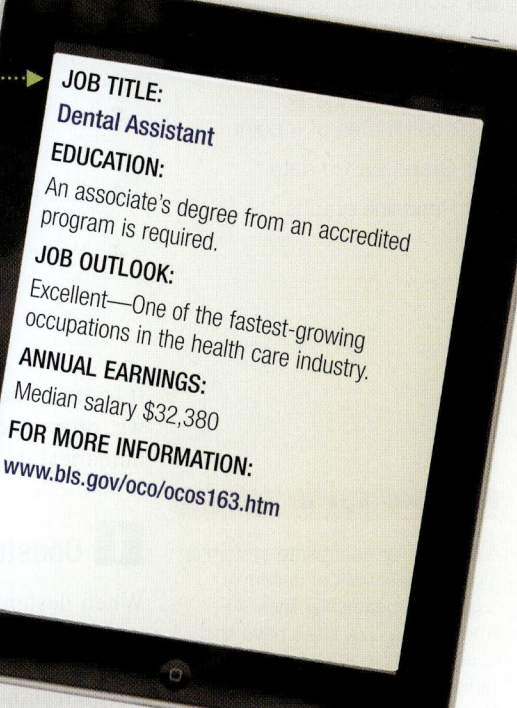

JOB TITLE:
Dental Assistant

EDUCATION:
An associate's degree from an accredited program is required.

JOB OUTLOOK:
Excellent—One of the fastest-growing occupations in the health care industry.

ANNUAL EARNINGS:
Median salary $32,380

FOR MORE INFORMATION:
www.bls.gov/oco/ocos163.htm

Taking a math test doesn't have to be an unpleasant experience. Here are some suggestions that can make it more enjoyable and also improve your score.

PREPARING FOR THE TEST: Begin studying several days before the test rather than cramming your studying into one marathon session the night before.

TAKING THE TEST: Follow a test-taking strategy so you can maximize your score by using the testing time wisely.

EVALUATING YOUR PERFORMANCE: After your graded test is returned, classify the types of errors that you made on the test so that you do not make them again.

Now Try This ▶

1. Write a study session plan that explains how you will prepare on each of the 4 days before the test, as well as on test day. For some suggestions, see *Preparing for a Test.**

2. Develop your own test-taking strategy by answering the survey questions found in *How to Take a Math Test.**

3. Use the outline found in *Analyzing Your Test Results** to classify the errors that you made on your most recent test.

*Found online at: www.cengage.com/math/tussy

SECTION 3.1

Graphing Using the Rectangular Coordinate System

OBJECTIVES

1. Construct a rectangular coordinate system.
2. Plot ordered pairs and determine the coordinates of a point.
3. Graph paired data.
4. Read line graphs.

ARE YOU READY?

▼ *The following problems review some basic skills that are needed when graphing ordered pairs.*

1. Graph each number in the set $\left\{\dfrac{7}{3}, -3, 0, 4, -1.5\right\}$ on a number line.

2. **a.** What number is 8 units to the right of 0 on a number line?

   b. What number is 3.5 units to the left of 0 on a number line?

3. List the first four Roman numerals.

4. Write $\dfrac{9}{2}$ and $-\dfrac{11}{3}$ in mixed-number form.

It is often said, "A picture is worth a thousand words." This is certainly true in algebra, where we often use mathematical pictures called *rectangular coordinate graphs* to illustrate numerical relationships.

The Language of Algebra

A rectangular coordinate system is a **grid**—a network of uniformly spaced perpendicular lines. At times, some U.S. cities have such horrible traffic congestion that vehicles can barely move, if at all. The condition is called *gridlock*.

1 Construct a Rectangular Coordinate System.

When designing the Gateway Arch in St. Louis, architects created a mathematical model called a **rectangular coordinate graph.** This graph, shown on the next page, is drawn on a grid called a **rectangular coordinate system.** This coordinate system also is called a **Cartesian coordinate system,** after the 17th-century French mathematician René Descartes.

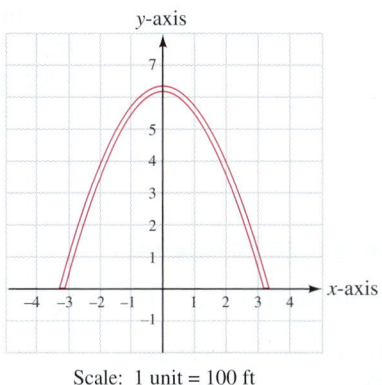

Scale: 1 unit = 100 ft

The Language of Algebra

The word **axis** is used in mathematics and science. For example, Earth rotates on its *axis* once every 24 hours. The plural of *axis* is **axes,** which is pronounced ak-seas.

A rectangular coordinate system is formed by two perpendicular number lines. The horizontal number line is usually called the **x-axis,** and the vertical number line is usually called the **y-axis.** On the *x*-axis, the positive direction is to the right. On the *y*-axis, the positive direction is upward. Each axis should be scaled to fit the data. For example, the axes of the graph of the arch are scaled in units of 100 feet.

The point where the axes intersect is called the **origin.** This is the zero point on each axis. The axes form a **coordinate plane,** and they divide it into four regions called **quadrants,** which are numbered counterclockwise using Roman numerals.

Points in quadrant II have a negative x- and positive y-coordinate.

Points in quadrant I have a positive x- and positive y-coordinate.

The Language of Algebra

A **coordinate plane** can be thought of as a perfectly flat surface extending infinitely far in every direction.

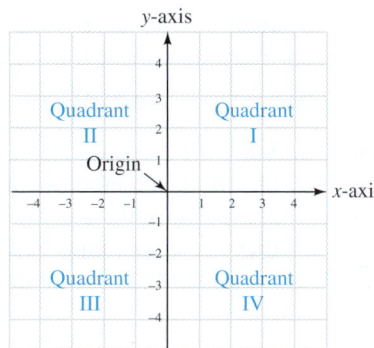

Points in quadrant III have a negative x- and negative y-coordinate.

Points in quadrant IV have a positive x- and negative y-coordinate.

Notation

Don't be confused by this new use of parentheses. $(3, -4)$ represents a point on the coordinate plane, whereas $3(-4)$ indicates multiplication. Also, don't confuse the ordered pair with interval notation.

Each point in a coordinate plane can be identified by an **ordered pair** of real numbers x and y written in the form (x, y). The first number, x, in the pair is called the **x-coordinate,** and the second number, y, is called the **y-coordinate.** Some examples of such pairs are $(3, -4)$, $\left(-1, -\frac{3}{2}\right)$, and $(0, 2.5)$.

$$(3, -4)$$

Read as "the point three, negative four" or as "the ordered pair three, negative four."

The x-coordinate is listed first.

The y-coordinate is listed second.

2 Plot Ordered Pairs and Determine the Coordinates of a Point.

The process of locating a point in the coordinate plane is called **graphing** or **plotting** the point. On the next page, we use blue arrows to show how to graph the point with coordinates $(3, -4)$. Since the *x*-coordinate, 3, is positive, we start at the origin and move 3 units to the *right* along the *x*-axis. Since the *y*-coordinate, -4, is negative, we then move *down* 4 units and draw a dot. This locates the point $(3, -4)$.

In the figure, red arrows are used to show how to plot the point $(-4, 3)$. We start at the origin, move 4 units to the *left* along the x-axis, then move *up* 3 units and draw a dot. This locates the point $(-4, 3)$.

The Language of Algebra

Note that the points $(3, -4)$ and $(-4, 3)$ have different locations. Since the order of the coordinates of a point is important, we call them **ordered pairs.**

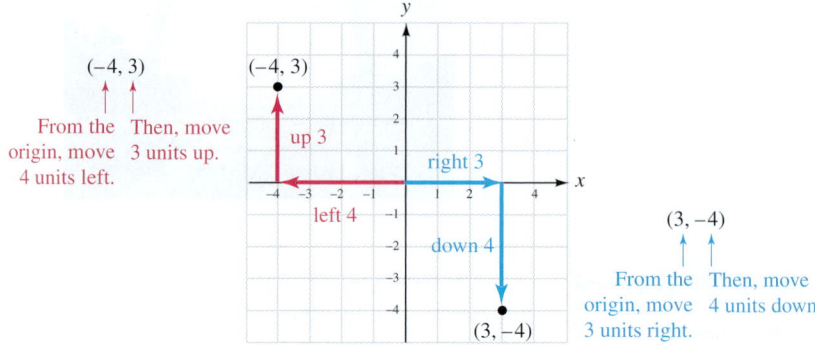

EXAMPLE 1   Plot each point. Then state the quadrant in which it lies or the axis on which it lies.

a. $(4, 4)$ **b.** $\left(-1, -\dfrac{7}{2}\right)$ **c.** $(0, 2.5)$ **d.** $(-3, 0)$ **e.** $(0, 0)$

Strategy   After identifying the x- and y-coordinates of the ordered pair, we will move the corresponding number of units left, right, up, or down to locate the point.

Why   The coordinates of a point determine its location on the coordinate plane.

Solution

Caution

When drawing a rectangular coordinate system, always label each axis with the appropriate letter (or title) and **scale** each axis. Here, the scaling shows that each square in the grid is 1 unit long and 1 unit wide.

Success Tip

Points with an x-coordinate that is 0 lie on the y-axis. Points with a y-coordinate that is 0 lie on the x-axis. Points that lie on an axis are not considered to be in any quadrant.

a. Since the x-coordinate, 4, is positive, we start at the origin and move 4 units to the *right* along the x-axis. Since the y-coordinate, 4, is positive, we then move *up* 4 units and draw a dot. This locates the point $(4, 4)$. The point lies in quadrant I.

b. To plot $\left(-1, -\dfrac{7}{2}\right)$, we begin at the origin and move 1 unit to the *left,* because the x-coordinate is -1. Then, since the y-coordinate is negative, we move $\dfrac{7}{2}$ units, or $3\dfrac{1}{2}$ units, *down.* The point lies in quadrant III.

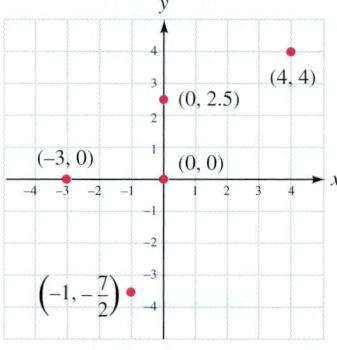

c. To plot $(0, 2.5)$, we begin at the origin and do not move right or left, because the x-coordinate is 0. Since the y-coordinate is positive, we move 2.5 units *up.* The point lies on the y-axis.

d. To plot $(-3, 0)$, we begin at the origin and move 3 units to the *left,* because the x-coordinate is -3. Since the y-coordinate is 0, we do not move up or down. The point lies on the x-axis.

e. To plot $(0, 0)$, we begin at the origin, and we remain there because both coordinates are 0. The point with coordinates $(0, 0)$ is the origin.

Self Check 1   Plot each point: $(2, -2)$, $(-4, 0)$, $\left(1.5, \dfrac{5}{2}\right)$, and $(0, 5)$

Now Try ▶ Problem 17

EXAMPLE 2   Find the coordinates of points *A*, *B*, *C*, *D*, *E*, and *F* plotted in figure (a) below.

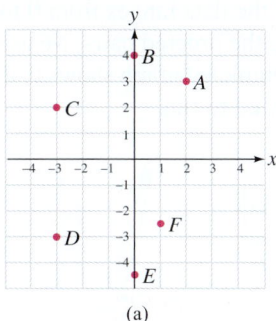

(a)

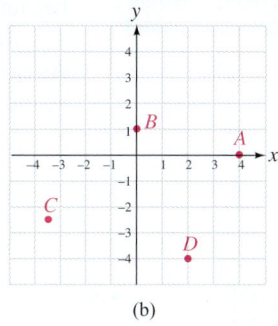

(b)

Strategy   We will start at the origin and count to the left or right on the *x*-axis, and then up or down to reach each point.

Why   The movement left or right gives the *x*-coordinate of the ordered pair and the movement up or down gives the *y*-coordinate.

Solution   To locate point *A*, we start at the origin, move 2 units to the right on the *x*-axis, and then 3 units up. Its coordinates are (2, 3). The coordinates of the other points are found in the same manner.

$$B(0, 4) \qquad C(-3, 2) \qquad D(-3, -3) \qquad E(0, -4.5) \qquad F(1, -2.5)$$

The coordinates of points *E* and *F* could also be given as $E\left(0, -4\frac{1}{2}\right)$ and $F\left(1, -2\frac{1}{2}\right)$.

Self Check 2   Find the coordinates of each point in figure (b) above.

Now Try ▶   Problem 20

3 Graph Paired Data.

Every day, we deal with quantities that are related:

- The time it takes to cook a roast depends on the weight of the roast.
- The money we earn depends on the number of hours we work.
- The sales tax that we pay depends on the price of the item purchased.

We can use graphs to visualize such relationships. For example, suppose a tub is filling with water, as shown below. Obviously, the amount of water in the tub depends on how long the water has been running. To graph this relationship, we can use the measurements that were taken as the tub began to fill.

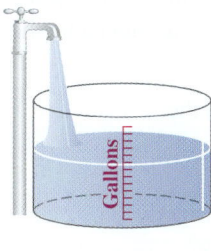

Time (min)	Water in tub (gal)	
0	0	← (0, 0)
1	8	← (1, 8)
3	24	← (3, 24)
4	32	← (4, 32)

The data in the table can be expressed as ordered pairs (x, y).

x-coordinate   y-coordinate

The data in each row of the table can be written as an ordered pair and plotted on a rectangular coordinate system. Since the first coordinate of each ordered pair is a time, we

The Language of Algebra

Data that can be represented as an ordered pair are called **paired data**.

label the *x*-axis *Time (min)*. The second coordinate is an amount of water, so we label the *y*-axis *Amount of water (gal)*. The *y*-axis is scaled in larger units (multiples of 4 gallons) because the size of the data ranges from 0 to 32 gallons.

After plotting the ordered pairs, we use a straightedge to draw a line through the points. As expected, the completed graph shows that the amount of water in the tub increases steadily as the water is allowed to run.

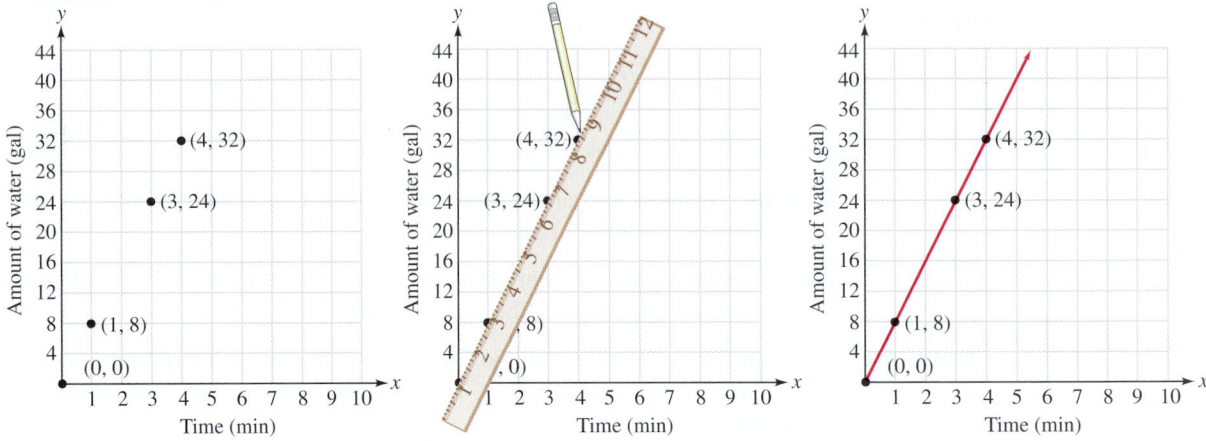

We can use the graph to determine the amount of water in the tub at various times. For example, the green dashed line on the graph at the right shows that in 2 minutes, the tub will contain 16 gallons of water. This process, called **interpolation,** uses known information to predict values that are not known but are *within* the range of the data. The blue dashed line on the graph shows that in 5 minutes, the tub will contain 40 gallons of water. This process, called **extrapolation,** uses known information to predict values that are not known and are *outside* the range of the data.

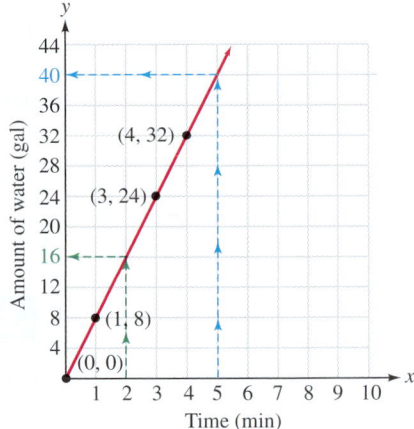

4 Read Line Graphs.

Since graphs are a popular way to present information, the ability to read and interpret them is very important.

EXAMPLE 3   **TV Shows.**   The following graph shows the number of people in an audience before, during, and after the taping of a television show. Use the graph to answer the following questions.

a. How many people were in the audience when the taping began?

b. At what times were there exactly 100 people in the audience?

c. How long did it take the audience to leave after the taping ended?

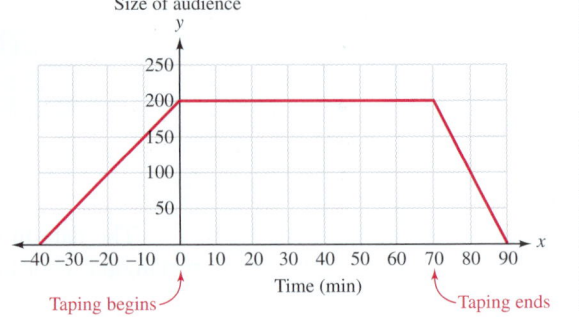

Strategy We will use an ordered pair of the form *(time, size of audience)* to describe each situation mentioned in parts (a), (b), and (c).

Why The coordinates of specific points on the graph can be used to answer each of these questions.

Solution

a. The time when the taping began is represented by 0 on the *x*-axis. The point on the graph directly above 0 is (0, **200**). The *y*-coordinate indicates that 200 people were in the audience when the taping began. This result is shown in the first row of the table in the margin.

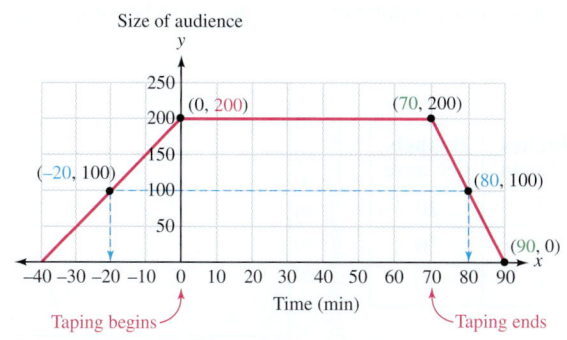

Time	Audience
0	200
−20	100
80	100
70	200
90	0

b. We can draw a horizontal line passing through 100 on the *y*-axis. Since the line intersects the graph twice, at (**−20**, 100) and at (**80**, 100), there are two times when 100 people were in the audience. These results are shown in the second and third rows of the table. The *x*-coordinates of the points tell us those times: 20 minutes before the taping began, and 80 minutes after.

c. The *x*-coordinate of the point (**70**, 200) tells us when the audience began to leave. The *x*-coordinate of (**90**, 0) tells when the exiting was completed. Subtracting the *x*-coordinates, we see that it took 90 − 70 = 20 minutes for the audience to leave.

Self Check 3 Use the graph in Example 3 to answer the following questions.
a. At what times were there exactly 50 people in the audience?
b. How many people were in the audience when the taping took place?
c. When were the first audience members allowed into the taping session?

Now Try ▶ Problems 21 and 23

SECTION 3.1 ▶ **STUDY SET**

VOCABULARY

Fill in the blanks.

1. (7, 1) is called an _____ pair.
2. In the ordered pair (2, −5), the -coordinate is 2 and the *y*-_____ is −5.
3. A rectangular coordinate system is formed by two perpendicular number lines called the *x*-_____ and the *y*-_____. The point where the axes cross is called the _____.
4. The *x*- and *y*-axes divide the coordinate plane into four regions called _____.
5. The point with coordinates (4, 2) can be graphed on a _____ coordinate system.
6. The process of locating the position of a point on a coordinate plane is called _____ the point.

CONCEPTS

Fill in the blanks.

7. **a.** To plot (−5, 4), we start at the _____ and move 5 units to the _____ and then move 4 units ____.
   b. To plot $\left(6, -\frac{3}{2}\right)$, we start at the _____ and move 6 units to the _____ and then move $\frac{3}{2}$ units _____.
8. In which quadrant is each point located?
   a. (−2, 7)   **b.** $\left(\frac{1}{2}, \frac{15}{16}\right)$
   c. (−1, −2.75)   **d.** (50, −16)
9. **a.** In which quadrants are the second coordinates of points positive?
   b. In which quadrants are the first coordinates of points negative?
   c. In which quadrant do points with a positive *x*-coordinate and a negative *y*-coordinate lie?
   d. On what axis are the first coordinates of points zero?

10. Farming. The number of bushels of wheat produced per acre depends on the amount of water it receives. Plot the data in the table as ordered pairs and draw a straight line through the points. Use the graph to determine how many bushels per acre will be produced if

    a. 6 inches of rain fall.   **b.** 10 inches of rain fall.

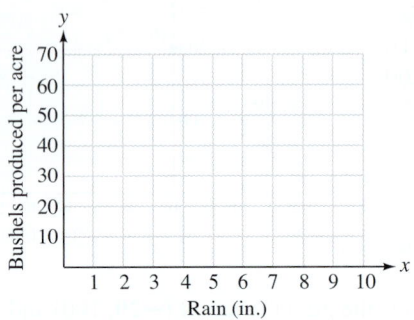

Inches of rain	Bushels per acre
2	20
4	30
8	50

NOTATION

11. Explain the difference between (3, 5) and 3(5).

12. In a paired-data table, does the *first* or *second* column contain values associated with the vertical axis of a graph?

13. Do these ordered pairs name the same point?

$$\left(2.5, -\tfrac{7}{2}\right), \left(2\tfrac{1}{2}, -3.5\right), \left(2.5, -3\tfrac{1}{2}\right)$$

14. Do (3, 2) and (2, 3) represent the same point?

15. In the ordered pair (4, 5), is the number 4 associated with the horizontal or the vertical axis?

16. Fill in the blank: In the notation $P(4, 5)$, the capital letter P is used to name a _____.

GUIDED PRACTICE

See Examples 1 and 2.

17. Plot each point:
$(-3, 4), (4, 3.5), \left(-2, -\tfrac{5}{2}\right), (0, -4), \left(\tfrac{3}{2}, 0\right), (2.7, -4.1)$

18. Plot each point:
$(4, 4), (0.5, -3), (-3.9, -3.2), (0, -1), (0, 0), (0, 3), (-2, 0)$

19. Complete the coordinates for each point in figure (a) below.

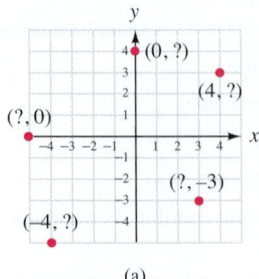

(a)

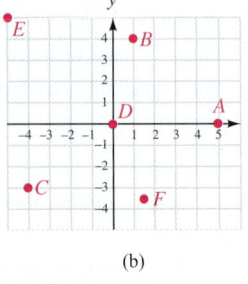

(b)

20. Find the coordinates of points $A, B, C, D, E,$ and F in figure (b) above.

The following graph gives the heart rate of a woman before, during, and after an aerobic workout. Use it to answer Problems 21–24. **See Example 3.**

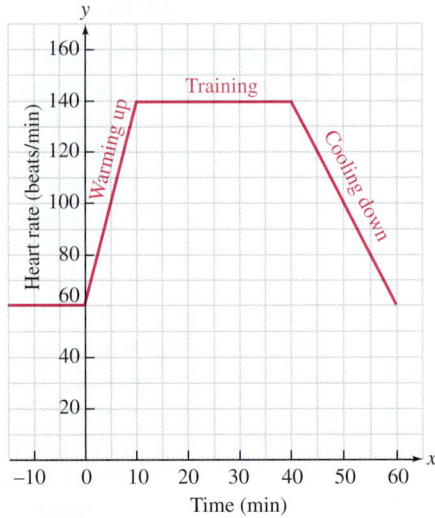

21. a. What was her heart rate before beginning the workout?

    b. After beginning her workout, how long did it take the woman to reach her training-zone heart rate?

22. a. What was the woman's heart rate half an hour after beginning the workout?

    b. For how long did the woman work out at her training zone?

23. a. At what time was her heart rate 100 beats per minute?

    b. How long was her cool-down period?

24. a. What was the difference in the woman's heart rate before the workout and after the cool-down period?

    b. What was her approximate heart rate 8 minutes after beginning?

The following graph shows the depths of a submarine at certain times after it leaves port. Use the graph to answer Problems 25–28. **See Example 3.**

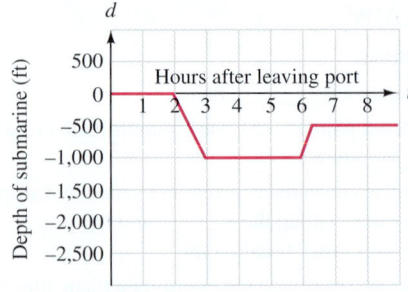

25. a. For how long does the sub travel at sea level?

    b. What is the depth of the sub 5 hours after leaving port?

26. a. Once the sub begins to dive, how long does it take to reach $-1,000$ feet in depth?

    b. For how long does the sub travel at a depth of 1,000 feet?

27. **a.** Explain what happens 6 hours after the sub leaves port.

   b. What is the depth of the sub 8 hours after leaving port?

28. **a.** How long does it take the sub to first reach −500 feet in depth?

   b. Approximate the time when the sub reaches −500 feet in depth for the second time.

APPLICATIONS

29. **Bridge Construction.** Find the coordinates of each rivet, weld, and anchor.

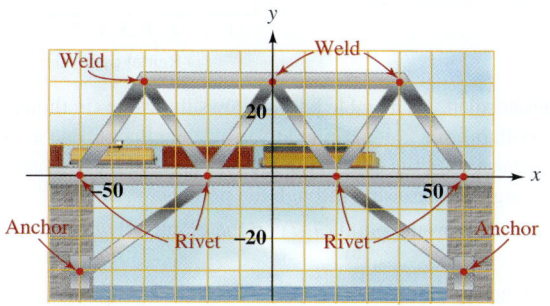

30. **Golf.** A golfer is videotaped and then has her swing displayed on a computer monitor so that it can be analyzed. Give the coordinates of the three highlighted points in red.

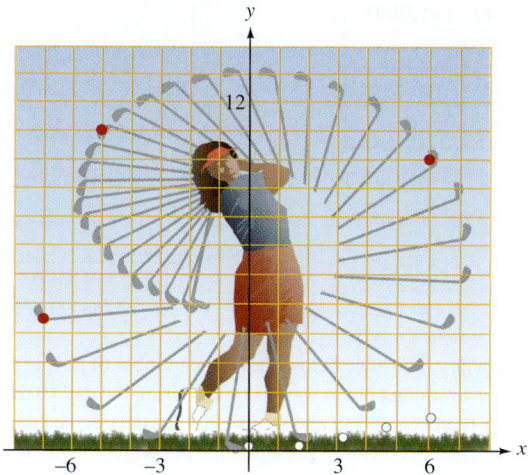

31. **Games.** In the game *Battleship,* players use coordinates to drop depth charges from their ships to hit submarines. What coordinates should be used to make three hits on the submarine seen here? Express each answer in the form (letter, number).

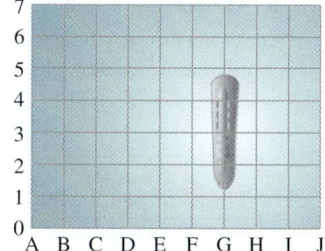

32. **Maps.** Use coordinates of the form (number, letter) to locate each of the following on the map: Tempe, Glendale, Paradise Valley, Sky Harbor Airport, and the intersection of Camelback Road and 7th Avenue.

33. **from Campus to Careers**
Dental Assistant

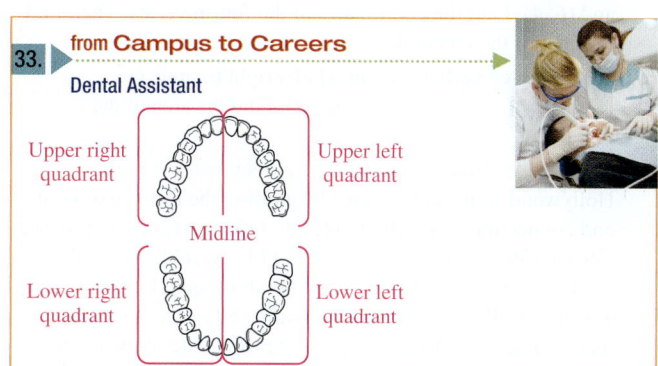

Dentists describe teeth as being located in one of four *quadrants* as shown above.

a. How many teeth are in each quadrant?

b. Why would the upper left quadrant appear on the right in the illustration?

34. **Geography.** A coordinate system that describes the location of any place on the surface of the Earth uses a series of *latitude* and *longitude* lines, as shown below. Estimate the location of the Deep Water Horizon (the oil drilling rig in the Gulf of Mexico that exploded in 2010) using an ordered pair of the form (latitude, longitude). (Source: sailwx.info)

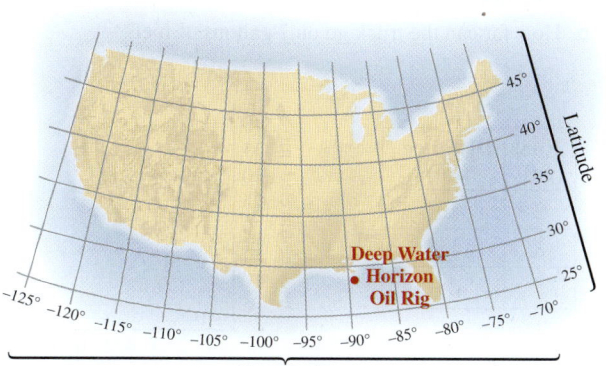

35. Water Pressure. The graphs show how the path of a stream of water changes when the hose is held at two different angles.

a. At which angle does the stream of water shoot up higher? How much higher?

b. At which angle does the stream of water shoot out farther? How much farther?

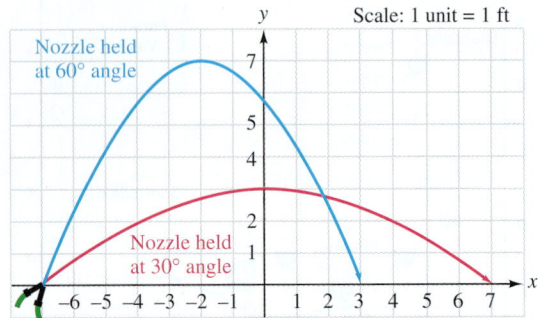

36. Area. Three vertices (corners) of a rectangle are (2, 1), (6, 1), and (6, 4). Find the coordinates of the fourth vertex. Then find the area of the rectangle.

37. Area. Three vertices (corners) of a right triangle are (−1, −7), (−5, −7), and (−5, −2). Find the area of the triangle.

38. Landmarks. A scale model of the block letter H in the Hollywood sign can be drawn by plotting the following points and connecting them: (0, 0), (13, 0), (13, 16), (26, 16), (26, 0), (39, 0), (39, 45), (26, 45), (26, 29), (13, 29), (13, 45), and (0, 45). The scale is 1 unit on the graph is equal to 1 foot on the actual sign. If a gallon of paint covers 350 square feet, how many gallons are needed to paint the front side of the letter H? Round to the nearest gallon.

39. Trucks. The table below shows the number of miles that an 18-wheel truck can be driven on a given number of gallons of diesel fuel. Plot the data in the table as ordered pairs. Then draw a straight line through the points.

a. How far can the truck go on 4 gallons of fuel?

b. How many gallons of fuel are needed to travel a distance of 30 miles?

c. How far can the truck go on 7 gallons of fuel?

Fuel (gal)	Distance (mi)
2	10
3	15
5	25

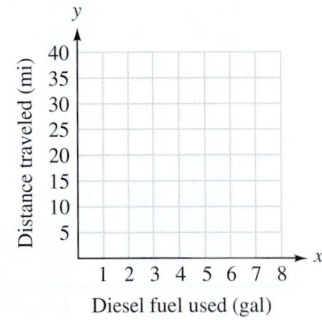

40. Boating. The table below shows the cost to rent a sailboat for a given number of hours. Plot the data in the table as ordered pairs. Then draw a straight line through the points.

a. What does it cost to rent the boat for 3 hours?

b. For how long can the boat be rented for $60?

c. What does it cost to rent the boat for 9 hours?

Rental time (hr)	Cost ($)
2	20
4	30
6	40

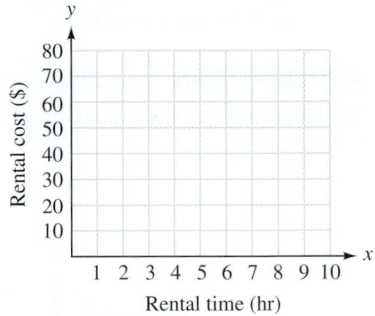

41. Depreciation. The table below shows the value (in thousands of dollars) of a color copier at various lengths of time after its purchase. Plot the data in the table as ordered pairs. Then draw a straight line passing through the points.

a. What does the point (3, 7) on the graph tell you?

b. Find the value of the copier when it is 7 years old.

c. After how many years will the copier be worth $2,500?

Age (yr)	Value ($1,000)
3	7
4	5.5
5	4

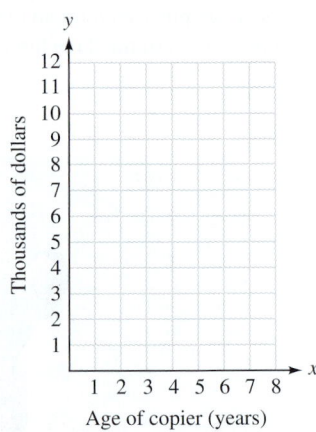

42. Swimming. The table below shows the number of people at a public swimming pool at various times during the day. (0 represents noon, 1 represents 1 P.M., and so on.) Plot the data in the table as ordered pairs. Then draw a straight line passing through the points.

a. How many people will be at the pool at 6 P.M.?

b. At what time will there be 250 people at the pool?

c. At what time will the number of people at the pool be half of what it was at noon?

Time	Number of people
0	350
3	200
5	100

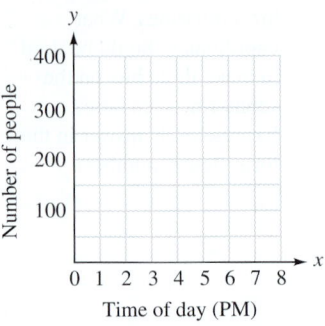

WRITING

43. Explain why the point $(-3, 3)$ is not the same as the point $(3, -3)$.

44. Explain how to plot the point $(-2, 5)$.

45. Use the Internet to perform a search of the name René Descartes. After reading about him, explain how a fly on his bedroom ceiling provided the inspiration for the concept of a rectangular coordinate system.

46. Explain this diagram.

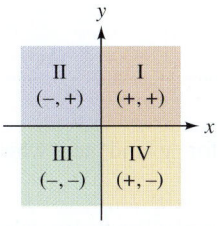

REVIEW

47. Solve $AC = \frac{2}{3}h - T$ for h.

48. Solve $5(x + 1) \le 2(x - 3)$. Write the solution set in interval notation and graph it.

49. Evaluate: $\dfrac{-4(4 + 2) - 2^3}{|-12 - 4(5)|}$

50. Simplify: $\dfrac{24}{54}$

CHALLENGE PROBLEMS

51. In what quadrant does a point lie if the *sum* of its coordinates is negative and the *product* of its coordinates is positive?

52. Draw line segment $\overline{AB}$ with endpoints $A(6, 5)$ and $B(-4, 5)$. Suppose that the x-coordinate of a point C is the average of the x-coordinates of points A and B, and the y-coordinate of point C is the average of the y-coordinates of points A and B. Find the coordinates of point C. Why is C called the midpoint of $\overline{AB}$?

SECTION 3.2

OBJECTIVES

1 Determine whether an ordered pair is a solution of an equation.

2 Complete ordered-pair solutions of equations.

3 Construct a table of solutions.

4 Graph linear equations by plotting points.

5 Use graphs of linear equations to solve applied problems.

Graphing Linear Equations

ARE YOU READY?

The following problems review some basic skills that are needed when graphing linear equations.

1. Is $2(4) + 3(-1) = 4$ a true or false statement?

2. Evaluate $-6x - 7$ for $x = -2$.

3. Solve: $4(2) + 3y = 14$

4. Solve for y: $4x + 5y = -15$

5. Multiply: $\dfrac{3}{2}(2)$

6. Multiply: $-\dfrac{5}{9}(-9)$

In this section, we will discuss equations that contain two variables. Such equations are often used to describe algebraic relationships between two quantities. To see a mathematical picture of these relationships, we will construct graphs of their equations.

1 **Determine Whether an Ordered Pair Is a Solution of an Equation.**

We have previously solved **equations in one variable.** For example, $x + 3 = 9$ is an equation in x. If we subtract 3 from both sides, we see that 6 is the solution. To verify this, we replace x with 6 and note that the result is a true statement: $9 = 9$.

In this chapter, we extend our equation-solving skills to find solutions of **equations in two variables.** To begin, let's consider $y = x - 1$, an equation in x and y.

A solution of $y = x - 1$ is a pair of values, one for x and one for y, that make the equation true. To illustrate, suppose x is 5 and y is 4. Then we have:

$y = x - 1$ This is the given equation.

$4 \overset{?}{=} 5 - 1$ Substitute 5 for x and 4 for y.

$4 = 4$ True

Since the result is a true statement, $x = 5$ and $y = 4$ is a solution of $y = x - 1$. We write the solution as the ordered pair (5, 4), with the value of x listed first. We say that (5, 4) **satisfies** the equation.

In general, a **solution of an equation in two variables** is an ordered pair of numbers that makes the equation a true statement.

EXAMPLE 1   Is $(-1, -3)$ a solution of $y = x - 1$?

Strategy   We will substitute -1 for x and -3 for y and see whether the resulting equation is true.

Why   An ordered pair is a solution of $y = x - 1$ if replacing the variables with the values of the ordered pair results in a true statement.

Solution $y = x - 1$ This is the given equation.

$-3 \overset{?}{=} -1 - 1$ Substitute −1 for x and −3 for y.

$-3 = -2$ False

Since $-3 = -2$ is false, $(-1, -3)$ is not a solution of $y = x - 1$.

Self Check 1   Is (9, 8) a solution of $y = x - 1$?

Now Try ▶   Problem 17

2 Complete Ordered-Pair Solutions of Equations.

If only one of the values of an ordered-pair solution is known, we can substitute it into the equation to determine the other value.

EXAMPLE 2   Complete the solution $(-5, \quad)$ of the equation $y = -2x + 3$.

Strategy   We will substitute the known x-coordinate of the solution into the given equation.

Why   We can use the resulting equation in one variable to find the unknown y-coordinate of the solution.

Solution   In the ordered pair $(-5, \quad)$, the x-value is -5; the y-value is not known. To find y, we substitute -5 for x in the equation and evaluate the right side.

$y = -2x + 3$ This is the given equation.

$y = -2(-5) + 3$ Substitute −5 for x.

$y = 10 + 3$ Do the multiplication.

$y = 13$ This is the missing y-coordinate of the solution.

The completed ordered pair is $(-5, 13)$.

Self Check 2   Complete the solution $(-2, \quad)$ of the equation $y = 4x - 2$.

Now Try ▶   Problem 29

Solutions of equations in two variables are often listed in a **table of solutions** (or **table of values**).

EXAMPLE 3   Complete the table of solutions for $3x + 2y = 5$.

x	y	(x, y)
7		(7,)
	4	(, 4)

Strategy   In each case we will substitute the known coordinate of the solution into the given equation.

Why   We can solve the resulting equation in one variable to find the unknown coordinate of the solution.

Solution   In the first row, we are given an x-value of 7. To find the corresponding y-value, we substitute 7 for x and solve for y.

$$3x + 2y = 5 \qquad \text{This is the given equation.}$$
$$3(7) + 2y = 5 \qquad \text{Substitute 7 for x.}$$
$$21 + 2y = 5 \qquad \text{Do the multiplication.}$$
$$2y = -16 \qquad \text{To isolate the variable term 2y, subtract 21 from both sides.}$$
$$y = -8 \qquad \text{To isolate y, divide both sides by 2.}$$
$$\text{This is the missing y-coordinate of the solution.}$$

x	y	(x, y)
7	-8	$(7, -8)$

A solution of $3x + 2y = 5$ is $(7, -8)$. It is entered in the table on the left.

In the second row, we are given a y-value of 4. To find the corresponding x-value, we substitute 4 for y and solve for x.

$$3x + 2y = 5 \qquad \text{This is the given equation.}$$
$$3x + 2(4) = 5 \qquad \text{Substitute 4 for y.}$$
$$3x + 8 = 5 \qquad \text{Do the multiplication.}$$
$$3x = -3 \qquad \text{To isolate the variable term 3x, subtract 8 from both sides.}$$
$$x = -1 \qquad \text{To isolate x, divide both sides by 3.}$$
$$\text{This is the missing x-coordinate of the solution.}$$

x	y	(x, y)
7	-8	$(7, -8)$
-1	4	$(-1, 4)$

Another solution is $(-1, 4)$. It is entered in the table on the left.

Self Check 3   Complete the table of solutions for $3x + 2y = 5$.

x	y	(x, y)
	-2	(, -2)
5		(5,)

Now Try ▶ Problem 37

---

3   Construct a Table of Solutions.

To find a solution of an equation in two variables, we can select a number, substitute it for one of the variables, and find the corresponding value of the other variable. For example, to find a solution of $y = x - 1$, we can select a value for x, say, -4, substitute -4 for x in the equation, and find y.

$$y = x - 1$$
$$y = -4 - 1 \qquad \text{Substitute} -4 \text{ for x.}$$
$$y = -5 \qquad \text{Do the subtraction.}$$

x	y	(x, y)
-4	-5	$(-4, -5)$

The ordered pair $(-4, -5)$ is a solution. We list it in the table on the left.

To find another solution of $y = x - 1$, we select another value for x, say, -2, and find the corresponding y-value.

x	y	(x, y)
-4	-5	$(-4, -5)$
-2	-3	$(-2, -3)$

$$y = x - 1$$
$$y = -2 - 1 \quad \text{Substitute } -2 \text{ for } x.$$
$$y = -3 \quad \text{Do the subtraction.}$$

A second solution is $(-2, -3)$, and we list it in the table of solutions.

If we let $x = 0$, we can find a third ordered pair that satisfies $y = x - 1$.

x	y	(x, y)
-4	-5	$(-4, -5)$
-2	-3	$(-2, -3)$
0	-1	$(0, -1)$

$$y = x - 1$$
$$y = 0 - 1 \quad \text{Substitute } 0 \text{ for } x.$$
$$y = -1 \quad \text{Do the subtraction.}$$

A third solution is $(0, -1)$, which we also add to the table of solutions.

We can find a fourth solution by letting $x = 2$, and a fifth solution by letting $x = 4$.

x	y	(x, y)
-4	-5	$(-4, -5)$
-2	-3	$(-2, -3)$
0	-1	$(0, -1)$
2	1	$(2, 1)$
4	3	$(4, 3)$

$$y = x - 1$$
$$y = 2 - 1 \quad \text{Substitute } 2 \text{ for } x.$$
$$y = 1 \quad \text{Do the subtraction.}$$

$$y = x - 1$$
$$y = 4 - 1 \quad \text{Substitute } 4 \text{ for } x.$$
$$y = 3 \quad \text{Do the subtraction.}$$

A fourth solution is $(2, 1)$ and a fifth solution is $(4, 3)$. We add them to the table.

Since we can choose any real number for x, and since any choice of x will give a corresponding value of y, it is apparent that the equation $y = x - 1$ has *infinitely many solutions*. We have found five of them: $(-4, -5)$, $(-2, -3)$, $(0, -1)$, $(2, 1)$, and $(4, 3)$.

4 Graph Linear Equations by Plotting Points.

It is impossible to list the infinitely many solutions of the equation $y = x - 1$. However, to show all of its solutions, we can draw a mathematical "picture" of them. We call this picture the *graph of the equation.*

Notation

The graph only shows a part of the line. The arrowheads indicate that it extends indefinitely in both directions.

To graph $y = x - 1$, we plot the ordered pairs shown in the table on a rectangular coordinate system. Then we draw a straight line through the points, because the graph of any solution of $y = x - 1$ will lie on this line. We also draw arrowheads on either end of the line to indicate that the solutions continue indefinitely in both directions, beyond what we can see on the coordinate grid. We call the line the **graph of the equation.** It represents all of the solutions of $y = x - 1$.

$$y = x - 1$$

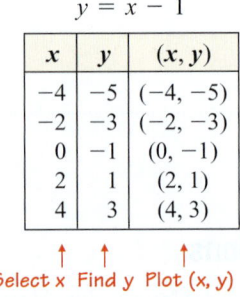

x	y	(x, y)
-4	-5	$(-4, -5)$
-2	-3	$(-2, -3)$
0	-1	$(0, -1)$
2	1	$(2, 1)$
4	3	$(4, 3)$

↑   ↑   ↑
Select x   Find y   Plot (x, y)

Construct a table of solutions.

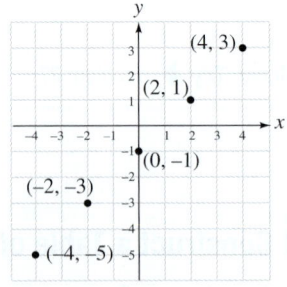

Plot the ordered pairs.

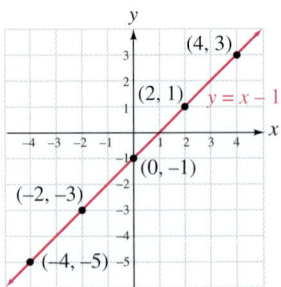

Draw a straight line through the points.
This is the *graph of the equation.*

The equation $y = x - 1$ is said to be *linear* and its graph is a line. By definition, a linear equation in two variables is any equation that can be written in the following form, where the variable terms appear on one side of an equal symbol and a constant appears on the other.

Linear Equations

A **linear equation in two variables** is an equation that can be written in the form

$$Ax + By = C$$

where A, B, and C are real numbers and A and B are not both 0. This form is called **standard form.***

*In some textbooks, the definition of the standard form of a linear equation in two variables contains additional requirements, such as: A, B, and C are integers, $A > 0$, and the greatest common factor of A, B, and C is 1.

Every linear equation in two variables has an infinite number of ordered-pair solutions. The graph of a linear equation in two variables is a straight line. Every point on the line corresponds to a solution.

Some more examples of linear equations are

$$y = 2x + 4, \qquad 2x + 3y = 12, \qquad \text{and} \qquad 3x = 5y$$

The exponent on each variable of a linear equation in two variables is an understood 1. For example, $y = 2x + 4$ can be thought of as $y^1 = 2x^1 + 4$ and $2x + 3y = 12$ can be thought of as $2x^1 + 3y^1 = 12$.

Some examples of equations in two variables that are *not* linear are shown below. You will see later in this course (and in more advanced courses) that the graphs of these equations are not straight lines.

$$y = x^2 + 3, \qquad y = \sqrt{x}, \qquad y = 4^x, \qquad x^2 + y^2 = 25, \qquad \text{and} \qquad y = \frac{1}{x}$$

Linear equations can be graphed in several ways. Generally, the form in which an equation is written determines the method that we use to graph it. To graph linear equations solved for y, such as $y = 2x + 4$, we can use the following **point-plotting method.**

Graphing Linear Equations Solved for y by Plotting Points

1. Find three ordered pairs that are solutions of the equation by selecting three values for x and calculating the corresponding values of y.
2. Plot the solutions on a rectangular coordinate system.
3. Draw a straight line passing through the points. If the points do not lie on a line, check your calculations.

EXAMPLE 4   Graph: $y = 2x + 4$

Strategy   We will find three solutions of the equation, plot them on a rectangular coordinate system, and then draw a straight line passing through the points.

Why   To *graph* a linear equation in two variables means to make a drawing that represents all of its solutions.

Solution   To find three solutions of this linear equation, we select three values for x that will make the calculations easy. Then we find each corresponding value of y.

Success Tip

When selecting x-values for a table of solutions, a rule of thumb is to choose a negative number, a positive number, and 0. When $x = 0$, the calculations to find y are usually quite simple.

If $x = -2$:	If $x = 0$:	If $x = 2$:
$y = 2x + 4$	$y = 2x + 4$	$y = 2x + 4$
$y = 2(-2) + 4$	$y = 2(0) + 4$	$y = 2(2) + 4$
$y = -4 + 4$	$y = 0 + 4$	$y = 4 + 4$
$y = 0$	$y = 4$	$y = 8$
$(-2, 0)$ is a solution.	$(0, 4)$ is a solution.	$(2, 8)$ is a solution.

We enter the results in a table of solutions and plot the points. Then we draw a straight line through the points and label it $y = 2x + 4$.

$$y = 2x + 4$$

x	y	(x, y)
-2	0	$(-2, 0)$
0	4	$(0, 4)$
2	8	$(2, 8)$

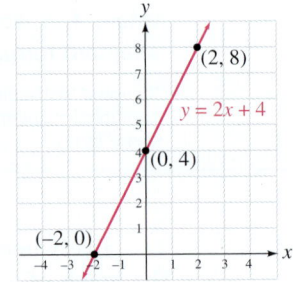

As a check, we can pick two points that the line appears to pass through, such as $(1, 6)$ and $(-1, 2)$. When we substitute their coordinates into the given equation, the two true statements that result indicate that $(1, 6)$ and $(-1, 2)$ are solutions and that the graph of the line is correctly drawn.

Check $(1, 6)$: $y = 2x + 4$ **Check $(-1, 2)$:** $y = 2x + 4$
$\qquad\qquad\quad 6 \stackrel{?}{=} 2(1) + 4$ $\qquad\qquad\qquad\quad 2 \stackrel{?}{=} 2(-1) + 4$
$\qquad\qquad\quad 6 \stackrel{?}{=} 2 + 4$ $\qquad\qquad\qquad\quad 2 \stackrel{?}{=} -2 + 4$
$\qquad\qquad\quad 6 = 6$ True $\qquad\qquad\qquad\quad 2 = 2$ True

Self Check 4   Graph: $y = 2x - 2$

Now Try ▶ Problem 41

EXAMPLE 5   Graph: $y = -3x$

Strategy   We will find three solutions of the equation, plot them on a rectangular coordinate system, and then draw a straight line passing through the points.

Why   To *graph* a linear equation in two variables means to make a drawing that represents all of its solutions.

Solution   To find three solutions, we begin by selecting three x-values: $-1, 0,$ and 1. Then we find the corresponding values of y. If $x = -1$, we have

$\qquad y = -3x$ This is the equation to graph.
$\qquad y = -3(-1)$ Substitute -1 for x.
$\qquad y = 3$ Do the multiplication.

$(-1, 3)$ is a solution.

In a similar manner, we find the y-values for x-values of 0 and 1, and record the results in a table of solutions. After plotting the ordered pairs, we draw a straight line through the points and label it $y = -3x$.

$$y = -3x$$

x	y	(x, y)
-1	3	$(-1, 3)$
0	0	$(0, 0)$
1	-3	$(1, -3)$

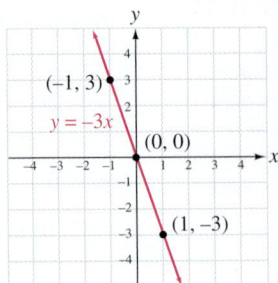

Self Check 5 Graph: $y = -4x$

Now Try ▸ Problem 45

To graph linear equations in x and y using the method discussed in this section, the variable y must be isolated on one side of the equation.

EXAMPLE 6 Graph $2x + 3y = -12$ by first solving for y.

Strategy We will use properties of equality to solve the given equation for y. Then we will use the point-plotting method of this section to graph the resulting equivalent equation.

Why The calculations to find several solutions of a linear equation in two variables are usually easier when the equation is solved for y.

Solution To solve for y, we proceed as follows.

$$2x + 3y = -12 \qquad \text{This is the given equation.}$$

$$2x + 3y - 2x = -2x - 12 \qquad \text{To isolate the variable term } 3y \text{ on the left side, subtract } 2x$$
from both sides. When solving for y, it is common practice to write the subtraction (or addition) of a variable term before the constant term.

$$3y = -2x - 12 \qquad \text{On the left side, combine like terms: } 2x - 2x = 0.$$

$$\frac{3y}{3} = \frac{-2x}{3} - \frac{12}{3} \qquad \text{To isolate the variable } y, \text{ undo the multiplication by } 3$$
by dividing both sides, term-by-term, by 3.

$$y = -\frac{2}{3}x - 4 \qquad \text{Write } \frac{-2x}{3} \text{ as } -\frac{2}{3}x. \text{ Simplify: } \frac{12}{3} = 4.$$

Success Tip

The division by 3 on the right side of the equation is done term-by-term instead of with a single fraction bar. We write:

$$\frac{-2x}{3} - \frac{12}{3} \quad \text{not} \quad \frac{-2x - 12}{3}$$

Since $y = -\frac{2}{3}x - 4$ is equivalent to $2x + 3y = -12$, we can use it to draw the graph of $2x + 3y = -12$.

To find solutions of $y = -\frac{2}{3}x - 4$, each value of x must be multiplied by $-\frac{2}{3}$. This calculation is made easier if we select x-values that are *multiples of the denominator 3*, such as -3, 0, and 6. For example, if $x = -3$, we have

$$y = -\frac{2}{3}x - 4 \qquad \text{This is the equation to graph.}$$

$$y = -\frac{2}{3}(-3) - 4 \qquad \text{Substitute } -3 \text{ for } x.$$

$$y = 2 - 4 \qquad \text{Multiply: } -\frac{2}{\overset{1}{3}}(\overset{1}{-3}) = 2.$$

$$y = -2 \qquad \text{Do the subtraction.}$$

Success Tip

When we chose x-values that are multiples of the denominator 3, the corresponding y-values are integers, and not difficult-to-plot fractions.

Thus, $(-3, -2)$ is a solution.

Two more solutions, one for $x = 0$ and one for $x = 6$, can be found in a similar way, and entered in a table. We plot the ordered pairs, draw a straight line through the points, and label the line as $y = -\frac{2}{3}x - 4$ or as $2x + 3y = -12$.

$2x + 3y = -12$
or
$y = -\frac{2}{3}x - 4$

x	y	(x, y)
-3	-2	$(-3, -2)$
0	-4	$(0, -4)$
6	-8	$(6, -8)$

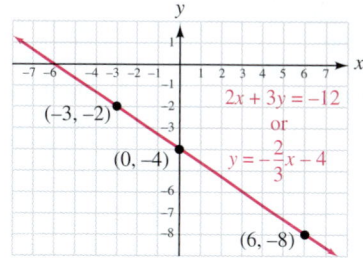

> **Self Check 6**   Graph $5x - 2y = -2$ by first solving for y.
>
> **Now Try** ▶ Problem 49

5 Use Graphs of Linear Equations to Solve Applied Problems.

When linear equations are used to model real-life situations, they are often written in variables other than x and y. In such cases, we must make the appropriate changes when labeling the table of solutions and the graph of the equation.

EXAMPLE 7   **Cleaning Windows.**   The linear equation $A = -0.03n + 32$ estimates the amount A of glass-cleaning solution (in ounces) that is left in the bottle after the sprayer trigger has been pulled a total of n times. Graph the equation and use the graph to estimate the amount of solution that is left after 500 sprays.

Strategy   We will find three solutions of the equation, plot them on a rectangular coordinate system, and then draw a straight line passing through the points.

Why   We can use the graph to estimate the amount of solution left after any number of sprays.

Solution   Since A depends on n in the equation $A = -0.03n + 32$, solutions will have the form (n, A). To find three solutions, we begin by selecting three values of n. Because the number of sprays cannot be negative, and the calculations to find A involve decimal multiplication, we select 0, 100, and 1,000. For example, if $n = 100$, we have

$A = -0.03n + 32$	This is the equation to graph.
$A = -0.03(100) + 32$	Substitute 100 for n.
$A = -3 + 32$	Multiply by moving the decimal point in -0.03 two places to the right: $-0.03(100) = -3$.
$A = 29$	Do the addition.

Thus, $(100, 29)$ is a solution. It indicates that after 100 sprays, 29 ounces of cleaner will be left in the bottle.

In the same way, solutions are found for $n = 0$ and $n = 1,000$ and listed in the table. Then the ordered pairs are plotted and a straight line is drawn through the points.

To graphically estimate the amount of solution that is left after 500 sprays, we draw the dashed blue lines, as shown. Reading on the vertical A-axis, we see that after 500 sprays, about 17 ounces of glass cleaning solution would be left.

> **Success Tip**
>
> It is often helpful, especially with appications, to scale the axes differently. Since we selected large n-values such as 100 and 1,000, the horizontal n-axis was scaled in units of 100. Since the corresponding A-values range from 2 to 32, the vertical A-axis was scaled in units of 4.

$A = -0.03n + 32$

n	A	(n, A)
0	32	$(0, 32)$
100	29	$(100, 29)$
1,000	2	$(1,000, 2)$

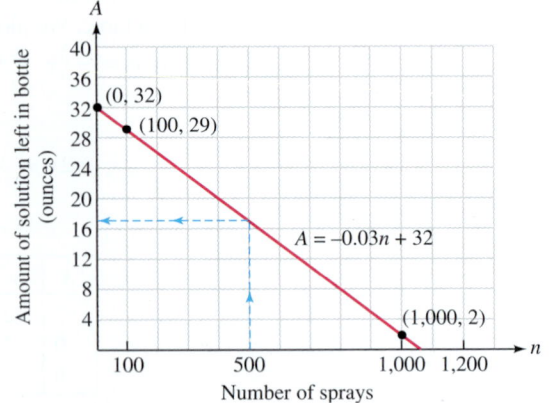

©Matthew Benoit/Shutterstock.com

Self Check 7 **Parties.** A laser tag business offers a party package that includes invitations, a party room, and 2 rounds of laser tag. The cost is $15 plus $10 per child. Write a linear equation that will give the cost for a party of any size, and then graph the equation.

Now Try ▶ Problems 87 and 89

SECTION 3.2 ▷ STUDY SET

VOCABULARY

Fill in the blanks.

1. $y = 9x + 5$ is an equation in _____ variables, x and y.
2. A _____ of an equation in two variables is an ordered pair of numbers that makes the equation a true statement.
3. Solutions of equations in two variables are often listed in a _____ of solutions.
4. The line that represents all of the solutions of a linear equation is called the _____ of the equation.
5. $y = 3x + 8$ is a _____ equation and its graph is a line.
6. The _____ form of a linear equation in two variables is $Ax + By = C$.

CONCEPTS

7. Consider: $y = -3x + 6$
   a. How many variables does the equation contain?
   b. Does $(4, -6)$ satisfy the equation?
   c. Is $(-2, 0)$ a solution of the equation?
   d. How many solutions does this equation have?
8. To graph a linear equation, three solutions were found, they were plotted (in black), and a straight line was drawn through them, as shown below.
   a. Looking at the graph, complete the table of solutions.
   b. From the graph, determine three other solutions of the equation.

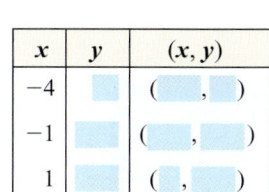

x	y	(x, y)
-4		(,)
-1		(,)
1		(,)

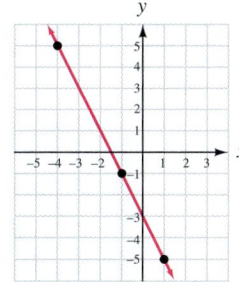

9. The graph of $y = -2x - 3$ is shown in Problem 8. Fill in the blanks: Every point on the graph represents an ordered-pair _____ of $y = -2x - 3$ and every ordered-pair solution is a _____ on the graph.
10. The graph of a linear equation is shown.
    a. If the coordinates of point M are substituted into the equation, will the result be true or false?

    b. If the coordinates of point N are substituted into the equation, will the result be true or false?

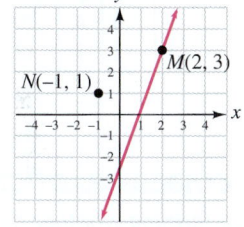

11. Suppose you are making a table of solutions for each given equation. What three x-values would you select to make the calculations for finding the corresponding y-values the easiest?
    a. $y = \frac{4}{5}x + 2$ b. $y = 0.6x + 500$

12. A table of solutions for a linear equation is shown below. When constructing the graph of the equation, how would you scale the x-axis and the y-axis?

x	y	(x, y)
-20	600	$(-20, 600)$
5	100	$(5, 100)$
35	-500	$(35, -500)$

NOTATION

Complete each solution.

13. Verify that $(-2, 6)$ is a solution of $y = -x + 4$.

$$y = -x + 4$$
$$\boxed{} \stackrel{?}{=} -(\boxed{}) + 4$$
$$6 \stackrel{?}{=} \boxed{} + 4$$
$$6 = \boxed{}$$

14. Solve $5x + 3y = 15$ for y.

$$5x + 3y - 5x = \boxed{} + 15$$

$$\boxed{} = -5x + 15$$

$$\frac{3y}{\boxed{}} = \frac{-5x}{\boxed{}} + \frac{15}{\boxed{}}$$

$$y = \boxed{}\, x + \boxed{}$$

15. a. In the linear equation $y = \frac{1}{2}x + 7$ what are the understood exponents on the variables?

 b. Explain why $y = x^2 + 2$ and $y = x^3 - 4$ are not linear equations.

16. Complete the labeling of the table of solutions and the axes of the graph of $c = -a + 4$.

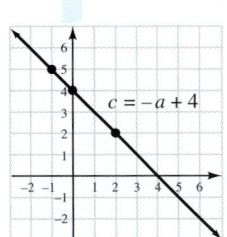

$\boxed{}$	$\boxed{}$	$(\ ,\ )$
-1	5	$(-1, 5)$
0	4	$(0, 4)$
2	2	$(2, 2)$

GUIDED PRACTICE

Determine whether each equation has the given ordered pair as a solution. See Example 1.

17. $y = 5x - 4;\ (1, 1)$
18. $y = -2x + 3;\ (2, -1)$
19. $7x - 2y = 3;\ (2, 6)$
20. $10x - y = 10;\ (0, 0)$
21. $x + 12y = -12;\ (0, -1)$
22. $-2x + 3y = 0;\ (-3, -2)$
23. $3x - 6y = 12;\ (-3.6, -3.8)$
24. $8x + 4y = 10;\ (-0.5, 3.5)$
25. $y - 6x = 12;\ \left(\frac{5}{6}, 7\right)$
26. $y + 8x = 4;\ \left(\frac{3}{4}, 2\right)$
27. $y = -\frac{3}{4}x + 8;\ (-8, 12)$
28. $y = \frac{1}{6}x - 2;\ (-12, 4)$

For each equation, complete the solution. See Example 2.

29. $y = -5x - 4;\ (-3, \boxed{\ })$
30. $y = 8x + 30;\ (-6, \boxed{\ })$
31. $4x - 5y = -4;\ (\boxed{\ }, 4)$
32. $7x + y = -12;\ (\boxed{\ }, 2)$
33. $y = \frac{x}{4} + 9;\ (16, \boxed{\ })$
34. $y = \frac{x}{6} - 8;\ (48, \boxed{\ })$
35. $7x = 4y;\ \left(\boxed{\ }, -2\right)$
36. $11x = 16y;\ \left(\boxed{\ }, -3\right)$

Complete each table of solutions. See Example 3.

37. $y = 2x - 4$

x	y	(x, y)
8		
	8	

38. $y = 3x + 1$

x	y	(x, y)
-3		
	-2	

39. $3x - y = -2$

x	y	(x, y)
-5		
	-1	

40. $5x - 2y = -15$

x	y	(x, y)
5		
	0	

Construct a table of solutions and then graph each equation. See Example 4.

41. $y = 2x - 3$
42. $y = 3x + 1$
43. $y = 5x - 4$
44. $y = 6x - 3$

Construct a table of solutions and then graph each equation. See Example 5.

45. $y = -6x$
46. $y = -2x$
47. $y = -7x$
48. $y = -8x$

Solve each equation for y and then graph it. See Example 6.

49. $2x + 3y = -3$
50. $2x + 3y = 9$
51. $5y - x = 20$
52. $4y - x = 8$

TRY IT YOURSELF

Graph each equation. Solve for y first, when necessary.

53. $y = x$
54. $y = 4x$
55. $y = -x - 1$
56. $y = -x + 2$
57. $3y = 12x + 15$
58. $5y = 20x - 30$
59. $y = \frac{3}{8}x - 6$
60. $y = -\frac{3}{2}x + 2$
61. $y = 1.5x - 4$
62. $y = 0.5x + 3$
63. $8x + 4y = 16$
64. $14x + 7y = 28$
65. $y = -\frac{1}{2}x$
66. $y = \frac{3}{4}x$
67. $y = \frac{5}{6}x - 5$
68. $y = \frac{2}{3}x - 2$
69. $-6y = 30x + 12$
70. $-3y = 9x - 15$
71. $y = \frac{x}{3}$
72. $y = -\frac{x}{3} - 1$
73. $y = -2x + 1$
74. $y = -3x + 2$
75. $7x - y = 1$
76. $2x - y = -3$
77. $7y = -2x$
78. $6y = -4x$
79. $y = -2.5x + 5$
80. $y = -3.5x + 4$

APPLICATIONS

81. Billiards. The path traveled by the black 8-ball is described by the equations $y = 2x - 4$ and $y = -2x + 12$. Construct a table of solutions for $y = 2x - 4$ using the x-values 1, 2, and 4. Do the same for $y = -2x + 12$, using the x-values 4, 6, and 8. Then graph the path of the 8-ball.

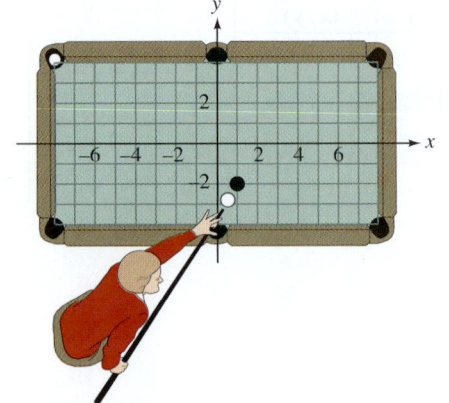

82. **Ping-Pong.** The path traveled by a Ping-Pong ball is described by the equations $y = \frac{1}{2}x + \frac{3}{2}$ and $y = -\frac{1}{2}x - \frac{3}{2}$. Construct a table of solutions for $y = \frac{1}{2}x + \frac{3}{2}$ using the x-values 7, 3, and -3. Do the same for $y = -\frac{1}{2}x - \frac{3}{2}$, using the x-values -3, -5, and -7. Then graph the path of the ball.

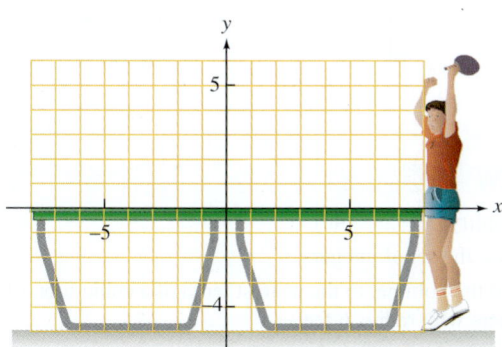

83. **Defrosting Poultry.** The number of hours h needed to defrost a turkey weighing p pounds in the refrigerator can be estimated by $h = 5p$. Graph the equation and use the graph to estimate the time needed to defrost a 25-pound turkey. (Source: helpwithcooking.com.)

84. **Owning a Car.** In 2010, the average cost c (in dollars) to own and operate a car was estimated by $c = 0.57m$, where m represents the number of miles driven. Graph the equation and use the graph to estimate the cost in 2010 of operating a car that is driven 25,000 miles. (Source: AAA Auto Club)

85. **Housekeeping.** The linear equation $A = -0.02n + 16$ estimates the amount A of furniture polish (in ounces) that is left in the bottle after the sprayer trigger has been pulled a total of n times. Graph the equation and use the graph to estimate the amount of polish that is left after 650 sprays.

86. **Sharpening Pencils.** The linear equation $L = -0.04t + 8$ estimates the length L (in inches) of a pencil after it has been inserted into a sharpener and the handle turned a total of t times. Graph the equation and use the graph to estimate the length of the pencil after 75 turns of the handle.

t turns of the handle

87. **NFL Tickets.** The average ticket price p to a National Football League game during the years 1990–2009 is approximated by $p = 2.7t + 20$, where t is the number of years after 1990. Graph this equation and use the graph to predict the average ticket price in 2020. (Source: Team Marketing Report, NFL.)

88. **U.S. Automobile Accidents.** The number n of lives saved by seat belts during the years 2000–2009 is estimated by $n = 170t + 13,800$, where t is the number of years after 2000. Graph this equation and use the graph to predict the number of lives that will be saved by seat belts in 2015. (Source: NHTSA National Center for Statistics.)

89. **Raffles.** A private school is going to sell raffle tickets as a fund raiser. Suppose the number n of raffle tickets that will be sold is predicted by the equation $n = -20p + 300$, where p is the price of a raffle ticket in dollars. Graph the equation and use the graph to predict the number of raffle tickets that will be sold at a price of $6.

90. **Endangered Species.** The number n of endangered plant and animal species in the U.S. during the years 2000–2010 is estimated by $n = 9t + 960$, where t is the number of years after 2000. Graph this equation and use the graph to predict the number of endangered species in the U.S. in 2022. (Source: *The World Almanac and Book of Facts, 2010*).

91. **U.S. Space Program.** Since 1980, the Gallup Poll organization has surveyed Americans to see whether they think the space program has brought enough benefits to the country to justify its cost. The percent p responding "yes" is estimated by $p = \frac{3}{5}t + 40$, where t is the number of years after 1980. Graph the equation. If the polling trend continues, when will the percent that respond "yes" reach 70%? (Source: galluppoll.com)

92. **Gas Mileage.** The mileage for a Honda Insight traveling between 55 mph and 75 mph is estimated by the equation $m = -\frac{3}{4}s + 95$, where s is the speed of the car (in mph) and m is the mileage (in miles per gallon). Graph the equation for s between 55 and 75. Estimate the speed at which the mileage of the car drops below 40 miles per gallon. (Source: *Consumer Reports* 9/10/2009)

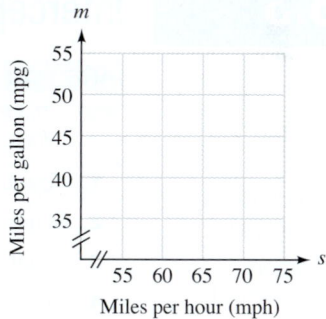

The symbol $\not\equiv$ is used to show the break in the scale on an axis. Such a break enables us to omit large portions of empty space on a graph.

WRITING

93. When we say that $(-2, -6)$ is a solution of $y = x - 4$, what do we mean?

94. What is a table of solutions?

95. What does it mean when we say that a linear equation in two variables has infinitely many solutions?

96. A linear equation and a graph are two ways of describing a relationship between two quantities. Which do you think is more informative and why?

97. From geometry, we know that two points determine a line. Why is it a good practice when graphing linear equations to find and plot three solutions instead of just two?

98. A student found three solutions of a linear equation in two variables and plotted them as shown. What conclusion can be made about the location of the points?

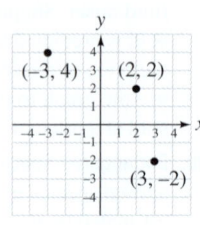

99. To graph $y = 3x - 1$, one student made the table of solutions on the left. Another student made the table on the right. The tables are different. Could they both be correct? Explain.

x	y	(x, y)
0	-1	$(0, -1)$
2	5	$(2, 5)$
3	8	$(3, 8)$

x	y	(x, y)
-2	-7	$(-2, -7)$
-1	-4	$(-1, -4)$
1	2	$(1, 2)$

100. Both graphs below are of the same linear equation $y = 10x$. Why do the graphs have a different appearance?

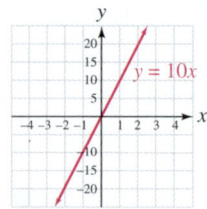 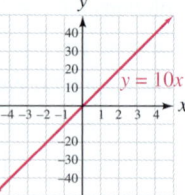

REVIEW

101. Simplify: $-(-5 - 4c)$

102. Write the set of integers.

103. Find the volume, to the nearest tenth of a cubic foot, of a sphere with radius 6 feet.

104. Solve: $-2(a + 3) = 3(a - 5)$

CHALLENGE PROBLEMS

*Graph each of the following **nonlinear** equations in two variables by constructing a table of solutions consisting of seven ordered pairs. These equations are called nonlinear, because their graphs are not straight lines.*

105. $y = x^2 + 1$ **106.** $y = x^3 - 2$

107. $y = |x| - 2$ **108.** $y = (x + 2)^2$

SECTION **3.3**

OBJECTIVES

1 Identify intercepts of a graph.

2 Graph linear equations by finding intercepts.

3 Identify and graph horizontal and vertical lines.

4 Obtain information from intercepts.

Intercepts

ARE YOU READY?

  The following problems review some basic skills that are needed when finding the intercepts of lines.

1. Graph the points $(2, 0)$, $(-4, 0)$, $(0, 1)$ and $(0, -3)$.

2. What is the x-coordinate of any point that lies on the y-axis?

3. What point lies on both the x-axis and the y-axis?

4. Solve: $3(0) + 2y = 10$

In this section, we will graph linear equations by determining the points where their graphs intersect the x-axis and the y-axis. These points are called the *intercepts* of the graph.

The Language of Algebra

Note the difference in spelling. The point where a line **intersects** the x- or y-axis is called an **intercept**.

1 Identify Intercepts of a Graph.

The graph of $y = 2x - 4$ is shown on the next page. We see that the graph intersects (crosses) the y-axis at the point $(0, -4)$; this point is called the **y-intercept** of the graph. The graph intersects (crosses) the x-axis at the point $(2, 0)$; this point is called the **x-intercept** of the graph.

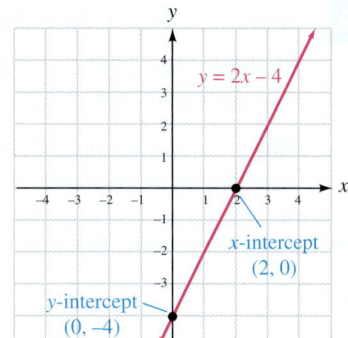

Recall that the y-coordinate of any point on the x-axis is 0.

Recall that the x-coordinate of any point on the y-axis is 0.

y-intercept (0, −4)

x-intercept (2, 0)

$y = 2x - 4$

EXAMPLE 1 For the graphs in figures (a) and (b), identify the x- and y-intercepts.

a. 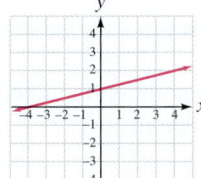 **b.** 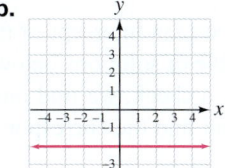 **c.**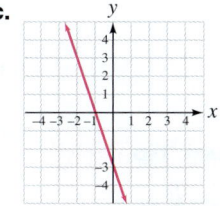

Strategy We will determine where each graph (shown in red) crosses the x-axis and the y-axis.

Why The point at which a graph crosses the x-axis is the x-intercept and the point at which a graph crosses the y-axis is the y-intercept.

Solution

a. In figure (a), the graph crosses the x-axis at (−4, 0). This is the x-intercept. The graph crosses the y-axis at (0, 1). This is the y-intercept.

b. In figure (b), the horizontal line does not cross the x-axis; there is no x-intercept. The graph crosses the y-axis at (0, −2). This is the y-intercept.

> **Self Check 1** Identify the x- and y-intercepts of the graph in figure (c).
>
> **Now Try** ▶ Problem 11

From the previous examples, we see that a y-intercept has an x-coordinate of 0, and an x-intercept has a y-coordinate of 0. These observations suggest the following procedures for finding the intercepts of a graph from its equation.

Finding Intercepts

To find the y-intercept, substitute 0 for x in the given equation and solve for y.

To find the x-intercept, substitute 0 for y in the given equation and solve for x.

---

2 Graph Linear Equations by Finding Intercepts.

Plotting the x- and y-intercepts of a graph and drawing a line through them is called the **intercept method of graphing a line.** This method is useful when graphing linear equations written in the standard form $Ax + By = C$.

EXAMPLE 2   Graph $x - 3y = 6$ by finding the y- and x-intercepts.

Strategy   We will let $x = 0$ to find the y-intercept of the graph. We will then let $y = 0$ to find the x-intercept.

Why   Since two points determine a line, the y-intercept and x-intercept are enough information to graph this linear equation.

Solution

y-intercept: *Let x = 0*

$x - 3y = 6$

$0 - 3y = 6$ Substitute 0 for x.

$-3y = 6$ Subtract.

$y = -2$ Divide both sides by −3.

The y-intercept is $(0, -2)$.

x-intercept: *Let y = 0*

$x - 3y = 6$

$x - 3(0) = 6$ Substitute 0 for y.

$x - 0 = 6$ Multiply.

$x = 6$ Subtract.

The x-intercept is $(6, 0)$.

Since each intercept of the graph is a solution of the equation, we enter the intercepts in the table of solutions below.

As a check, we find one more point on the line. We select a convenient value for x, say, 3, and find the corresponding value of y.

$x - 3y = 6$ This is the equation to graph.

$3 - 3y = 6$ Substitute 3 for x.

$-3y = 3$ To isolate the variable term −3y, subtract 3 from both sides.

$y = -1$ To isolate y, divide both sides by −3.

Therefore, $(3, -1)$ is a solution. It is also entered in the table.

We plot the intercepts and the check point, draw a straight line through them, and label the line as $x - 3y = 6$.

Success Tip

The check point should lie on the same line as the x- and y-intercepts. If it does not, check your work to find the incorrect coordinate or coordinates.

The Language of Algebra

Points that lie on the same line are said to be **collinear**.

$x - 3y = 6$

x	y	(x, y)	
0	−2	$(0, -2)$	← y-intercept
6	0	$(6, 0)$	← x-intercept
3	−1	$(3, -1)$	← Check point

Self Check 2   Graph $x - 2y = 2$ by finding the intercepts.

Now Try ▶ Problem 27

The calculations for finding intercepts can be simplified if we realize what occurs when we substitute 0 for y or 0 for x in an equation written in the form $Ax + By = C$.

EXAMPLE 3   Graph $40x + 3y = -120$ by finding the y- and x-intercepts.

Strategy   We will let $x = 0$ to find the y-intercept of the graph. We will then let $y = 0$ to find the x-intercept.

Why   Since two points determine a line, the y-intercept and x-intercept are enough information to graph this linear equation.

Solution

When we substitute 0 for x, it follows that the term $40x$ will be equal to 0. Therefore, to find the y-intercept, we can cover over the $40x$ and solve the remaining equation for y.

$$\cancel{40x} + 3y = -120 \qquad \text{If } x = 0, \text{ then } 40x = 40(0) = 0. \text{ Cover over the } 40x \text{ term.}$$
$$y = -40 \qquad \text{To solve } 3y = -120, \text{ divide both sides by 3.}$$

The y-intercept is $(0, -40)$. This is entered in the table below.

When we substitute 0 for y, it follows that the term $3y$ will be equal to 0. Therefore, to find the x-intercept, we can cover over the $3y$ and solve the remaining equation for x.

$$40x + \cancel{3y} = -120 \qquad \text{If } y = 0, \text{ then } 3y = 3(0) = 0. \text{ Cover over the } 3y \text{ term.}$$
$$x = -3 \qquad \text{To solve } 40x = -120, \text{ divide both sides by 40.}$$

The x-intercept is $(-3, 0)$. This is entered in the table below.

We can find a third solution by selecting a convenient value for x and finding the corresponding value for y. If we choose $x = -6$, we find that $y = 40$. The solution $(-6, 40)$ is entered in the table, and the equation is graphed as shown.

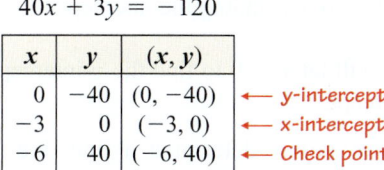

$40x + 3y = -120$

x	y	(x, y)	
0	-40	$(0, -40)$	← y-intercept
-3	0	$(-3, 0)$	← x-intercept
-6	40	$(-6, 40)$	← Check point

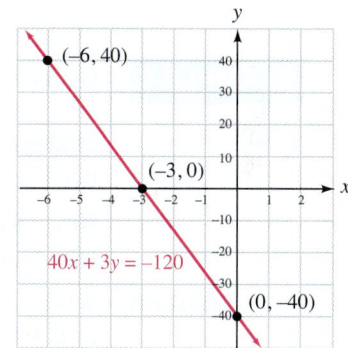

If a different scale is used on the y-axis, the same equation ($40x + 30y = -120$) will produce a graph with a somewhat different look.

Self Check 3 Graph $32x + 5y = -160$ by finding the intercepts.

Now Try ▶ Problem 35

EXAMPLE 4 Graph $3x = -5y + 8$ by finding the intercepts.

Strategy We will let $x = 0$ to find the y-intercept of the graph. We will then let $y = 0$ to find the x-intercept.

Why Since two points determine a line, the y-intercept and x-intercept are enough information to graph this linear equation.

Solution We find the intercepts and select $x = 1$ to find a check point.

y-intercept: Let $x = 0$

$$3x = -5y + 8$$
$$3(0) = -5y + 8$$
$$0 = -5y + 8$$
$$-8 = -5y$$
$$\frac{8}{5} = y$$
$$1\frac{3}{5} = y$$

The y-intercept is $\left(0, 1\frac{3}{5}\right)$.

x-intercept: Let $y = 0$

$$3x = -5y + 8$$
$$3x = -5(0) + 8$$
$$3x = 8$$
$$x = \frac{8}{3}$$
$$x = 2\frac{2}{3}$$

The x-intercept is $\left(2\frac{2}{3}, 0\right)$.

Check point: Let $x = 1$

$$3x = -5y + 8$$
$$3(1) = -5y + 8$$
$$3 = -5y + 8$$
$$-5 = -5y$$
$$1 = y$$

A check point is $(1, 1)$.

The ordered pairs are plotted as shown, and a straight line is then drawn through them.

$$3x = -5y + 8$$

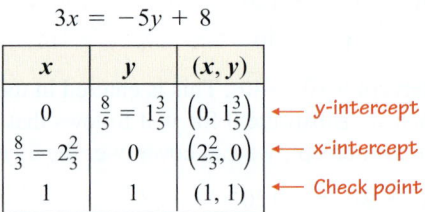

x	y	(x, y)	
0	$\tfrac{8}{5} = 1\tfrac{3}{5}$	$\left(0, 1\tfrac{3}{5}\right)$	← y-intercept
$\tfrac{8}{3} = 2\tfrac{2}{3}$	0	$\left(2\tfrac{2}{3}, 0\right)$	← x-intercept
1	1	$(1, 1)$	← Check point

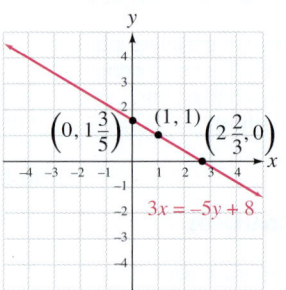

Self Check 4 Graph $8x = -4y + 15$ by finding the intercepts.

Now Try ▶ Problem 39

EXAMPLE 5 Graph $2x + 3y = 0$ by finding the intercepts.

Strategy We will let $x = 0$ to find the y-intercept of the graph. We will then let $y = 0$ to find the x-intercept.

Why Since two points determine a line, the y-intercept and x-intercept are enough information to graph this linear equation.

Solution When we find the y- and x-intercepts (shown below), we see that they are both $(0, 0)$. In this case, the line passes through the origin. Since we are using two points and a check point to graph lines, we need to find two more ordered-pair solutions.

If $x = 3$, we see that $(3, -2)$ is a solution. And if $x = -3$, we see that $(-3, 2)$ is also a solution. These two solutions and the origin are plotted and a straight line is drawn through them to give the graph of $2x + 3y = 0$.

y-intercept: Let $x = 0$	*x*-intercept: Let $y = 0$	Let $x = 3$	Let $x = -3$
$2x + 3y = 0$	$2x + 3y = 0$	$2x + 3y = 0$	$2x + 3y = 0$
$2(0) + 3y = 0$	$2x + 3(0) = 0$	$2(3) + 3y = 0$	$2(-3) + 3y = 0$
$3y = 0$	$2x = 0$	$6 + 3y = 0$	$-6 + 3y = 0$
$y = 0$	$x = 0$	$3y = -6$	$3y = 6$
		$y = -2$	$y = 2$
The y-intercept is $(0, 0)$.	The x-intercept is $(0, 0)$.	$(3, -2)$ is a solution.	$(-3, 2)$ is a solution.

The intercepts are the same.

$$2x + 3y = 0$$

x	y	(x, y)	
0	0	$(0, 0)$	← The x-intercept and y-intercept.
3	−2	$(3, -2)$	← A solution.
−3	2	$(-3, 2)$	← This solution serves as a check point.

Equations that can be written in the form $Ax + By = 0$, such as $2x + 3y = 0$, have graphs that pass through the origin. To find another point on such a line that has integer coordinates, select an x-value equal to the coefficient of y or the opposite of the coefficient of y. Then substitute that x-value into the equation and solve for y.

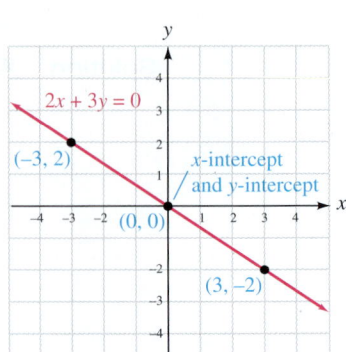

Self Check 5   Graph $5x - 2y = 0$ by finding the intercepts.

Now Try ▶ Problem 47

3   Identify and Graph Horizontal and Vertical Lines.

Equations such as $y = 4$ and $x = -3$ are linear equations in two variables, because they can be written in the standard form $Ax + By = C$. For example, $y = 4$ is equivalent to $0x + 1y = 4$ and $x = -3$ is equivalent to $1x + 0y = -3$. We can graph these types of equations using point-plotting.

| **EXAMPLE 6** | Graph: $y = 4$ |

Strategy   To find three ordered-pair solutions of this equation to plot, we will select three values for x and use 4 for y each time.

Why   The equation requires that $y = 4$.

Solution   We can write the equation in standard form as $0x + y = 4$. Since the coefficient of x is 0, the numbers selected for x have no effect on y. The value of y is always 4. For example, if $x = 2$, we have

$$0x + y = 4 \qquad \textcolor{red}{\text{This is the given equation, } y = 4, \text{ written in standard form: } Ax + By = C.}$$
$$0(2) + y = 4 \qquad \textcolor{red}{\text{Substitute 2 for } x.}$$
$$y = 4 \qquad \textcolor{red}{\text{Simplify the left side.}}$$

One solution is $(2, 4)$. To find two more solutions, we select $x = 0$ and $x = -3$. For any x-value, the y-value is always 4, so we enter $(0, 4)$ and $(-3, 4)$ in the table. If we plot the ordered pairs and draw a straight line through the points, the result is a horizontal line. The y-intercept is $(0, 4)$ and there is no x-intercept.

Success Tip

An equation with only the variable y is a horizontal line that intersects the y-axis.

$y = 4$

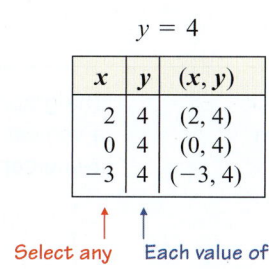

x	y	(x, y)
2	4	$(2, 4)$
0	4	$(0, 4)$
-3	4	$(-3, 4)$

↑ Select any number for x.   ↑ Each value of y must be 4.

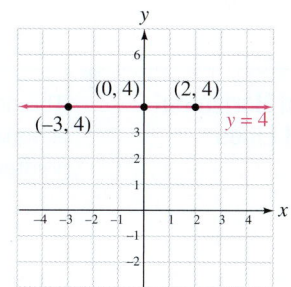

Self Check 6   Graph: $y = -2$

Now Try ▶ Problem 51

| **EXAMPLE 7** | Graph: $x = -3$ |

Strategy   To find three ordered-pair solutions of this equation to plot, we must select -3 for x each time.

Why   The equation requires that $x = -3$.

Solution   We can write the equation in standard form as $x + 0y = -3$. Since the coefficient of y is 0, the value of y has no effect on x. The value of x is always -3. For example, if $y = -2$, we have

$$x + 0y = -3 \quad \text{\color{red}This is the given equation, } x = -3, \text{ written in standard form: } Ax + By = C.$$
$$x + 0(\mathbf{-2}) = -3 \quad \text{\color{red}Substitute } -2 \text{ for } y.$$
$$x = -3 \quad \text{\color{red}Simplify the left side.}$$

One solution is $(-3, -2)$. To find more solutions, we must again select -3 for x. Any number can be used for y. If $y = 0$, then a second solution is $(-3, 0)$. If $y = 3$, a third solution is $(-3, 3)$. The three solutions are entered in the table below. When we plot the ordered pairs and draw a straight line through the points, the result is a vertical line. The x-intercept is $(-3, 0)$ and there is no y-intercept.

$$x = -3$$

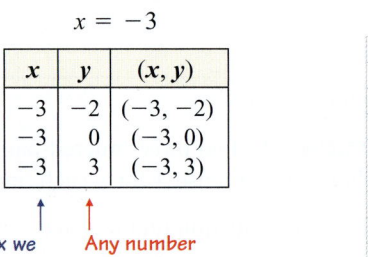

x	y	(x, y)
-3	-2	$(-3, -2)$
-3	0	$(-3, 0)$
-3	3	$(-3, 3)$

↑ Each value of x we select must be -3.   ↑ Any number can be used for y.

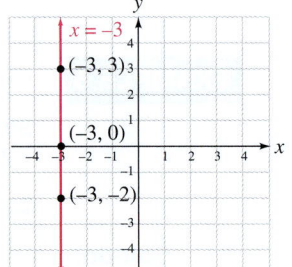

Self Check 7   Graph: $x = 4$

Now Try ▶   Problem 55

From the results of Examples 6 and 7, we have the following facts.

Equations of Horizontal and Vertical Lines

The graph of $y = b$ is a horizontal line with y-intercept $(0, b)$.

The graph of $x = a$ is a vertical line with x-intercept $(a, 0)$.

The graph of $y = 0$ is the x-axis. The graph of $x = 0$ is the y-axis.

---

4   Obtain Information from Intercepts.

The ability to read and interpret graphs is a valuable skill. When analyzing a graph, we should locate and examine the intercepts. As the following example illustrates, the coordinates of the intercepts can give useful information.

©Michael Doolittle/Alamy

EXAMPLE 8

Hybrid Mileage. Figure (a) shows mileage data for a 2010 Toyota Prius Hybrid. What information do the intercepts give about the car?

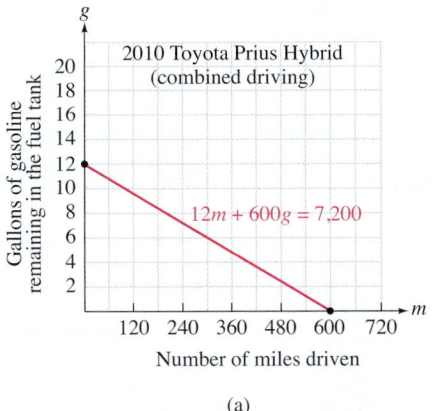

(a)

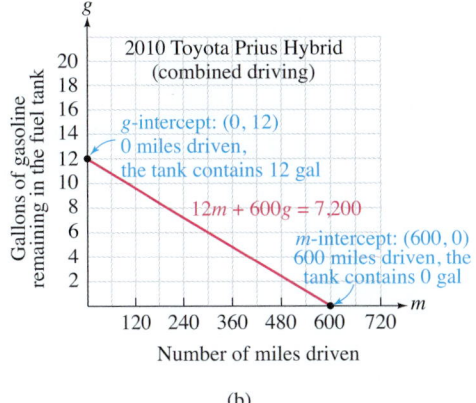

(b)

Strategy We will determine where the graph (the line in red) intersects the g-axis and where it intersects the m-axis.

Why Once we know the intercepts, we can interpret their meaning.

Solution See figure (b). The g-intercept $(0, 12)$ indicates that when the car has been driven 0 miles, the fuel tank contains 12 gallons of gasoline. That is, the Prius has a 12-gallon fuel tank.

The m-intercept $(600, 0)$ indicates that after 600 miles of combined driving, the fuel tank contains 0 gallons of gasoline. Thus, 600 miles of combined driving can be done on 1 tank of gas in a Prius.

Now Try ▶ Problems 79 and 81

Using Your Calculator ▶ **Use a Calculator to Graph Linear Equations. (Optional)**

Courtesy of Texas Instruments Incorporated

So far, we have graphed linear equations by making tables of solutions and plotting points. A graphing calculator can make the task of graphing much easier. However, a graphing calculator does not take the place of a working knowledge of the topics discussed in this chapter. It should serve as an aid to enhance your study of algebra.

The Viewing Window The screen on which a graph is displayed is called the **viewing window.** The **standard window** has settings of

$$\text{Xmin} = -10, \qquad \text{Xmax} = 10, \qquad \text{Ymin} = -10, \qquad \text{and} \qquad \text{Ymax} = 10$$

which indicate that the minimum x- and y-coordinates used in the graph will be -10, and that the maximum x- and y-coordinates will be 10.

Graphing an Equation To graph $y = x - 1$ using a graphing calculator, we press the $\boxed{\mathbf{Y=}}$ key and enter $x - 1$ after the symbol Y_1. Then we press the $\boxed{\mathbf{GRAPH}}$ key to see the graph.

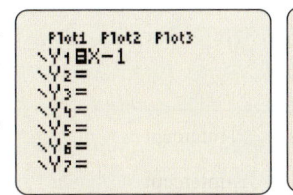
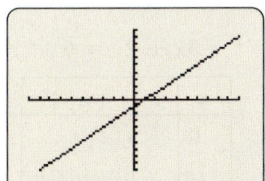

Change the Viewing Window We can change the viewing window by pressing the $\boxed{\mathbf{WINDOW}}$ key and entering -4 for the minimum x- and y-coordinates and 4 for the maximum x- and y-coordinates. Then we press the $\boxed{\mathbf{GRAPH}}$ key to see the graph of $y = x - 1$ in more detail.

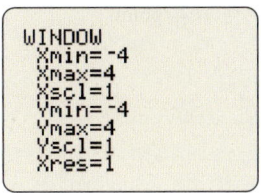
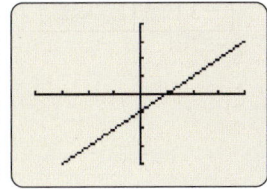

Solving an Equation for _y_   To graph $3x + 2y = 12$, we must first solve the equation for _y_.

$$3x + 2y = 12$$
$$2y = -3x + 12 \qquad \text{Subtract 3x from both sides.}$$
$$y = -\frac{3}{2}x + 6 \qquad \text{Divide both sides by 2.}$$

Next, we press the **WINDOW** key to reenter the standard window settings, press **Y =** and enter $-\frac{3}{2}x + 6$, as shown below. Then press **GRAPH** to see the graph.

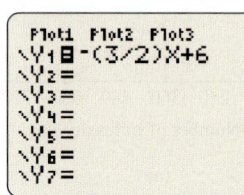
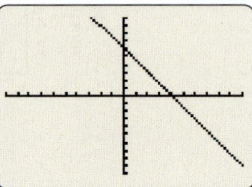

SECTION 3.3 ▶ STUDY SET

VOCABULARY

Fill in the blanks.

1. The _____ of a line is the point where the line intersects/crosses the _x_-axis.

2. The _y_-intercept of a line is the point where the line _____ the _y_-axis.

3. The graph of $y = 4$ is a _____ line and the graph of $x = 6$ is a _____ line.

4. The intercept method is useful when graphing linear equations written in the _____ form $Ax + By = C$.

CONCEPTS

5. Fill in the blanks.

   a. To find the _y_-intercept of the graph of a line, substitute ▢ for _x_ in the equation and solve for ▢.

   b. To find the _x_-intercept of the graph of a line, substitute ▢ for _y_ in the equation and solve for ▢.

6. Complete the table of solutions and fill in the blanks.

   $$3x + 2y = 6$$

_x_	_y_	(_x_, _y_)
0			← ___-intercept
	0		← ___-intercept
−2			← _____ point

7. **a.** Refer to the graph. Which intercept tells the purchase price of the machinery? What was that price?

   b. Which intercept indicates when the machinery will have lost all of its value? When is that?

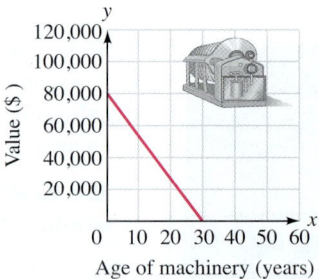

Age of machinery (years)

8. Match each graph with its equation.

   a. $x = 2$ **b.** $y = 2$ **c.** $y = 2x$

   d. $2x - y = 2$ **e.** $y = 2x + 2$ **f.** $y = -2x$

   i.
   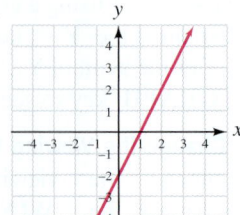

   ii.
   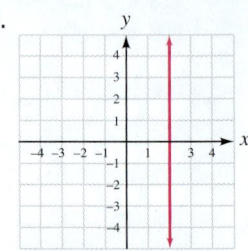

iii. 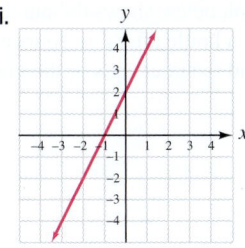 iv.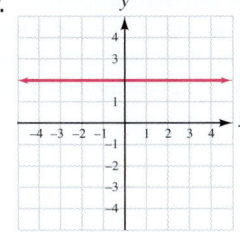

Estimate the coordinates of the intercepts of each graph. (Some are not integers.) See Example 1. (Answers may vary.)

17. 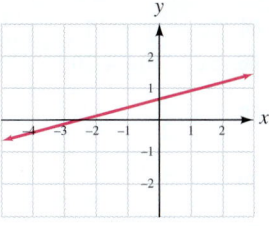 18.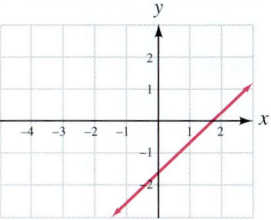

v. 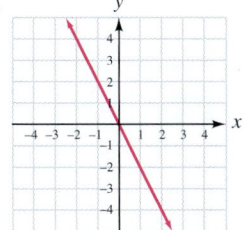 vi.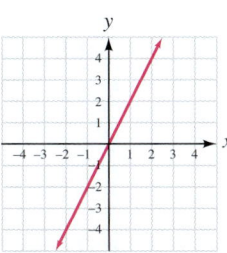

Find the x- and y-intercepts of the graph of each equation. Do not graph the line. See Example 2.

19. $8x + 3y = 24$ 20. $5x + 6y = 30$

21. $7x - 2y = 28$ 22. $2x - 9y = 36$

23. $-5x - 3y = 10$ 24. $-9x - 5y = 25$

25. $6x + y = 9$ 26. $x + 8y = 14$

NOTATION

9. What is the equation of the *x*-axis? What is the equation of the *y*-axis?

10. Write the coordinates that are improper fractions as mixed numbers.

   a. $\left(\frac{7}{2}, 0\right)$ b. $\left(0, -\frac{17}{3}\right)$

Use the intercept method to graph each equation. See Example 2.

27. $4x + 5y = 20$ 28. $3x + 4y = 12$
29. $5x + 15y = -15$ 30. $8x + 4y = -24$
31. $x - y = -3$ 32. $x - y = 3$
33. $x + 2y = -2$ 34. $x + 2y = -4$

Use the intercept method to graph each equation. See Example 3.

35. $30x + y = -30$ 36. $20x - y = -20$
37. $4x - 20y = 60$ 38. $6x - 30y = 30$

GUIDED PRACTICE

Give the coordinates of the intercepts of each graph. See Example 1.

11. 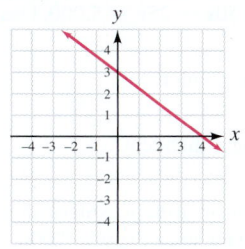 12.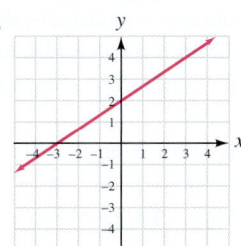

Use the intercept method to graph each equation. See Example 4.

39. $3x + 4y = 8$ 40. $2x + 3y = 9$
41. $-9x + 4y = 9$ 42. $-5x + 4y = 15$
43. $3x - 4y = 11$ 44. $5x - 4y = 13$
45. $9x + 3y = 10$ 46. $4x + 4y = 5$

13. 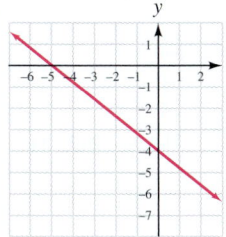 14.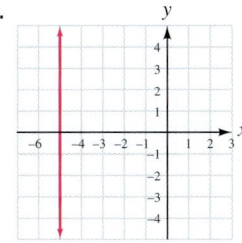

Use the intercept method to graph each equation. See Example 5.

47. $3x + 5y = 0$ 48. $4x + 3y = 0$
49. $2x - 7y = 0$ 50. $6x - 5y = 0$

Graph each equation. See Examples 6 and 7.

15. 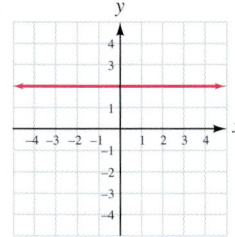 16.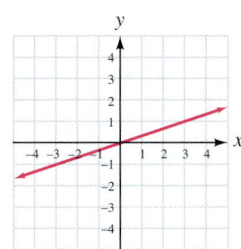

51. $y = 5$ 52. $y = -3$
53. $y = 0$ 54. $x = 0$
55. $x = -2$ 56. $x = 5$
57. $x = \dfrac{4}{3}$ 58. $y = -\dfrac{1}{2}$
59. $y - 2 = 0$ (*Hint:* Solve for *y* first.)
60. $x + 1 = 0$ (*Hint:* Solve for *x* first.)
61. $5x = 7.5$ (*Hint:* Solve for *x* first.)
62. $3y = 4.5$ (*Hint:* Solve for *y* first.)

TRY IT YOURSELF

Graph each equation.

63. $7x = 4y - 12$

64. $7x = 5y - 15$

65. $4x - 3y = 12$

66. $5x - 10y = 20$

67. $x = -\dfrac{5}{3}$

68. $y = \dfrac{5}{2}$

69. $y - 3x = -\dfrac{4}{3}$

70. $y - 2x = -\dfrac{9}{8}$

71. $7x + 3y = 0$

72. $4x - 5y = 0$

73. $-4x = 8 - 2y$

74. $-5x = 10 + 5y$

75. $3x = -150 - 5y$

76. $x = 50 - 5y$

77. $-3y = 3$

78. $-2x = 8$

APPLICATIONS

79. Chemistry. The relationship between the temperature T and volume V of a gas kept in a sealed container at a constant pressure is graphed below. The T-intercept of this graph is a very important scientific fact. It represents the lowest possible temperature, called **absolute zero.**

   a. Estimate absolute zero.

   b. What is the volume of the gas when the temperature is absolute zero?

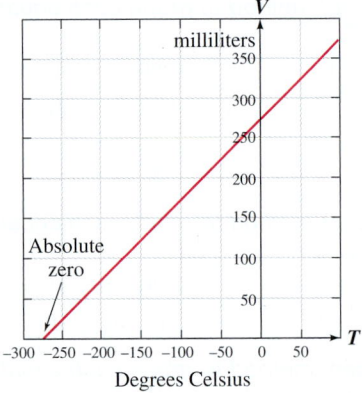

80. Physics. The graph shows the length L of a stretched spring (in inches) as different weights w (in pounds) are attached to it. What information about the spring does the L-intercept give?

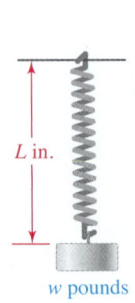

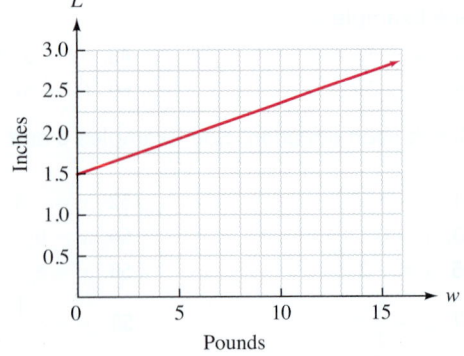

81. Bottled Water Dispenser. The graph below shows the number of gallons g of water remaining in a bottle after c six-ounce cups have been served from it. Find the intercepts of the graph. What information do they give?

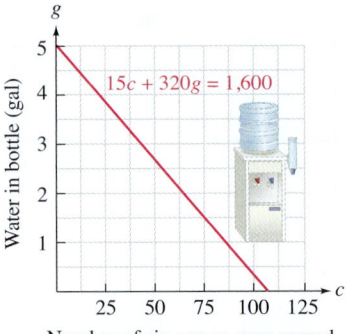

$15c + 320g = 1,600$

Number of six-ounce cups served

82. Renewable Energy. The equation $p = 50s - 300$ estimates the power output in watts from a propeller blade turbine driven by a wind of speed s miles per hour. What information does the s-intercept of the graph of the equation give? (Source: otherpower.com)

83. Landscaping. A developer is going to purchase x trees and y shrubs to landscape a new office complex. The trees cost $50 each and the shrubs cost $25 each. His budget is $5,000. This situation is modeled by the equation $50x + 25y = 5,000$. Use the intercept method to graph it.

   a. What information is given by the y-intercept?

   b. What information is given by the x-intercept?

84. Eggs. The number of eggs eaten by an average American in one year has remained almost constant since the year 2000. See the graph below. Draw a horizontal line that passes through, or near, the data points. What is the equation of the line?

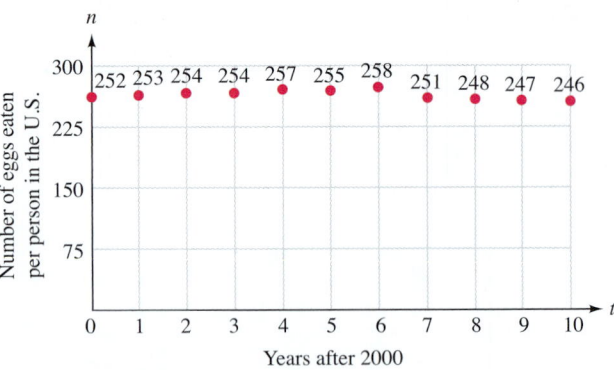

Source: United Egg

WRITING

85. To graph $3x + 2y = 12$, a student found the intercepts and a check point, and graphed them, as shown in figure (a). Instead of drawing a crooked line through the points, what should he have done?

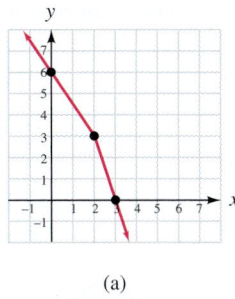

(a)

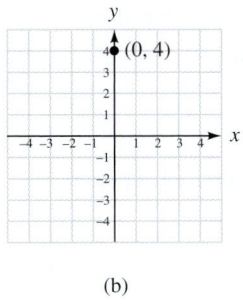

(b)

86. A student graphed the linear equation $y = 4$, as shown above in figure (b). Explain her error.

87. How do we find the intercepts of the graph of an equation without having to graph the equation?

88. In Section 3.2, we discussed a method to graph $y = 2x - 3$. In Section 3.3, we discussed a method to graph $2x + 3y = 6$. Briefly explain the steps involved in each method.

REVIEW

89. Simplify: $\dfrac{3 \cdot 5 \cdot 5}{3 \cdot 5 \cdot 5 \cdot 5}$

90. Simplify: $4\left(\dfrac{d}{2} - 3\right) - 5\left(\dfrac{2}{5}d - 1\right)$

91. Translate: Six less than twice x

92. Is -5 a solution of $2(3x + 10) = 5x + 6$?

CHALLENGE PROBLEMS

93. Where will the line $y = b$ intersect the line $x = a$?

94. Write an equation of the line that has an x-intercept of $(4, 0)$ and a y-intercept of $(0, 3)$.

95. What is the least number of intercepts a line can have? What is the greatest number a line can have?

96. On a rectangular coordinate system, draw a circle that has exactly two intercepts.

SECTION 3.4

Slope and Rate of Change

OBJECTIVES

1 Find the slope of a line from its graph.

2 Find the slope of a line given two points.

3 Find slopes of horizontal and vertical lines.

4 Solve applications of slope.

5 Calculate rates of change.

6 Determine whether lines are parallel or perpendicular using slope.

ARE YOU READY?

The following problems review some basic skills that are needed to find the slope of a line.

1. Evaluate: $\dfrac{4 - 1}{8 - 3}$

2. Evaluate: $\dfrac{-10 - 1}{-4 - (-4)}$

3. Multiply: $-\dfrac{7}{9} \cdot \dfrac{9}{7}$

4. Simplify: $\dfrac{15}{18}$

In this section, we introduce a method to measure the steepness (or slant) of a line. We call this measure the *slope of the line,* and it can be found in several ways.

1 Find the Slope of a Line from Its Graph.

The **slope of a line** is a ratio that compares the vertical change with the corresponding horizontal change as we move along the line from one point to another.

As an example, let's find the slope of the line graphed on the right. To begin, we select two points on the line and call them P and Q. One way to move from P to Q is to start at point P and count upward 5 grid squares. Then, moving to the right, we count 6 grid squares to reach point Q. The vertical change in this movement is called the **rise.** The horizontal change is called the **run.** Notice that a right triangle, called a **slope triangle,** is created by this process.

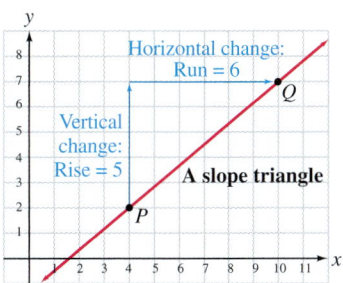

The Language of Algebra

A **ratio** is a comparison of two numbers using a quotient. Ratios are used in many settings. Mechanics speak of gear ratios. Colleges like to advertise their low student-to-teacher ratios.

The slope of a line is defined to be *the ratio of the vertical change to the horizontal change.* By tradition, the letter *m* is used to represent slope. For the line graphed on the previous page, we have

$$m = \text{slope} = \frac{\text{vertical change}}{\text{horizontal change}} = \frac{\text{rise}}{\text{run}} = \frac{5}{6}$$ This ratio is a comparison of the rise and the run using a quotient.

The slope of the line is $\frac{5}{6}$. This indicates that there is a rise (vertical change) of 5 units for each run (horizontal change) of 6 units.

EXAMPLE 1   Find the slope of the line graphed in figure (a) below.

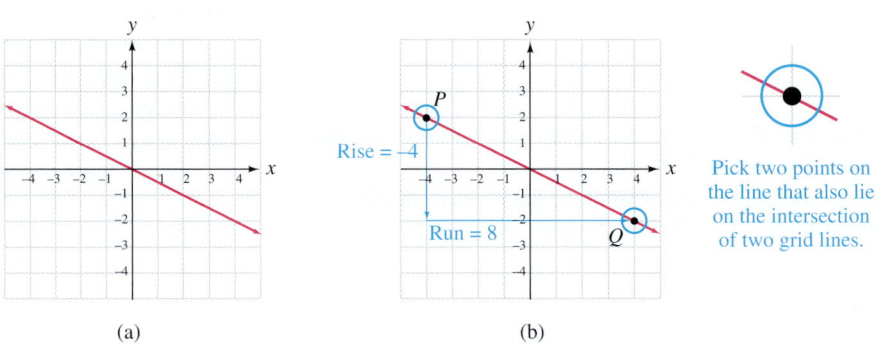

(a)   (b)

Strategy   We will pick two points on the line, construct a slope triangle, and find the rise and run. Then we will write the ratio of the rise to the run.

Why   The slope of a line is the ratio of the rise to the run.

Solution   We begin by choosing two points on the line, *P* and *Q*, as shown in figure (b). One way to move from *P* to *Q* is to start at point *P* and count *downward* 4 grid squares. Because this movement is downward, the rise is -4. Then, moving right, we count 8 grid squares to reach *Q*. This indicates that the run is 8.
    To find the slope of the line, we write a ratio of the rise to the run in simplified form.

$$m = \frac{\text{rise}}{\text{run}} = \frac{-4}{8} = -\frac{1}{2}$$ Always simplify slope fractions.

The slope of the line is $-\frac{1}{2}$.

The movement from *P* to *Q* can be reversed. Starting at *P*, we can move to the right, a run of 8; and then downward, a rise of -4, to reach *Q*. With this approach, the slope triangle is above the line. When we form the ratio to find the slope, we get the *same result* as before:

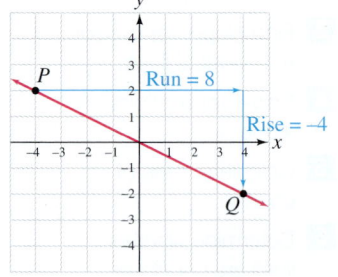

$$m = \frac{\text{rise}}{\text{run}} = \frac{-4}{8} = -\frac{1}{2}$$

Arrowheads are used to show horizontal movement (left or right) and vertical movement (up or down).

Self Check 1   Find the slope of the line shown above using two points different from those used in the solution of Example 1.

Now Try ▶   Problem 21

The identical answers from Example 1 and its Self Check illustrate an important fact: *For any line, the same value will be obtained no matter which two points on the line are used to find the slope.*

2 Find the Slope of a Line Given Two Points.

We can generalize the graphic method for finding slope to develop a slope formula. To begin, we select points P and Q on the line shown in the figure on the right. To distinguish between the coordinates of these two points, we use **subscript notation.** Point P has coordinates (x_1, y_1), which are read as "x sub 1 and y sub 1." Point Q has coordinates (x_2, y_2), which are read as "x sub 2 and y sub 2."

As we move from point P to point Q, the rise is the difference of the y-coordinates: $y_2 - y_1$. We call this difference the **change in y.** The run is the difference of the x-coordinates: $x_2 - x_1$. This difference is called the **change in x.** Since the slope is the ratio $\frac{\text{rise}}{\text{run}}$, we have the following formula for calculating slope.

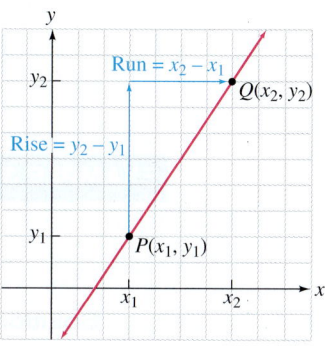

The Language of Algebra

The prefix **sub** means below or beneath, as in submarine or subway. In x_1, x_2, y_1, and y_2, the subscripts 1 and 2 are written lower than the variable. They are not exponents, and they do not represent a calculation.

Success Tip

Recall that subtraction is used to measure change.

Slope of a Line

The **slope** m of a line that passes through points (x_1, y_1) and (x_2, y_2) is

$$m = \frac{\text{vertical change}}{\text{horizontal change}} = \frac{\text{rise}}{\text{run}} = \frac{\text{change in } y}{\text{change in } x} = \frac{y_2 - y_1}{x_2 - x_1} \quad \text{if } x_2 \neq x_1$$

EXAMPLE 2   Find the slope of the line that passes through $(1, 2)$ and $(3, 8)$.

Strategy   We will use the slope formula to find the slope of the line.

Why   We know the coordinates of two points on the line.

Solution   When using the slope formula, it makes no difference which point you call (x_1, y_1) and which point you call (x_2, y_2). If we let (x_1, y_1) be $(1, 2)$ and (x_2, y_2) be $(3, 8)$, then

Success Tip

When using the slope formula, you may find it helpful to begin your solution by labeling the coordinates of the points in this way:

$$\begin{array}{cc} (x_1, y_1) & (x_2, y_2) \\ \downarrow\downarrow & \downarrow\downarrow \\ (1, 2) & (3, 8) \end{array}$$

$m = \dfrac{y_2 - y_1}{x_2 - x_1}$ This is the slope formula.

$m = \dfrac{8 - 2}{3 - 1}$ Substitute 8 for y_2, 2 for y_1, 3 for x_2, and 1 for x_1.

$m = \dfrac{6}{2}$ Do the subtraction.

$m = 3$ Simplify. Think of this as a $\frac{3}{1}$ rise-to-run ratio.

The slope of the line is 3. Note that we obtain the same value for the slope if we let $(x_1, y_1) = (3, 8)$ and $(x_2, y_2) = (1, 2)$.

$$m = \frac{y_2 - y_1}{x_2 - x_1} = \frac{2 - 8}{1 - 3} = \frac{-6}{-2} = 3$$

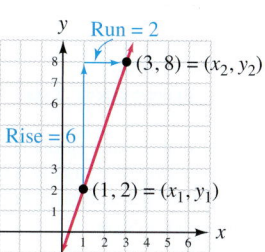

Success Tip

The slope formula is a valuable tool because it allows us to calculate the slope of a line without having to view its graph.

Although it is not necessary, the line passing through $(1, 2)$ and $(3, 8)$ has been graphed on the right. The graph of the line, including the slope triangle, verifies that the rise is 6 and the run is 2, and that $m = \frac{6}{2} = 3$.

CAUTION   When using the slope formula, always subtract the y-coordinates and their corresponding x-coordinates in the same order. Otherwise, your answer will have the wrong sign. If we let (x_1, y_1) be $(1, 2)$ and (x_2, y_2) be $(3, 8)$:

$$m = \frac{\cancel{y_2 - y_1}}{\cancel{x_1 - x_2}} = \frac{8 - 2}{1 - 3} = \frac{6}{-2} = -3 \quad \text{and} \quad m = \frac{\cancel{y_1 - y_2}}{\cancel{x_2 - x_1}} = \frac{2 - 8}{3 - 1} = \frac{-6}{2} = -3$$

Self Check 2   Find the slope of the line that passes through (2, 1) and (4, 11).

Now Try ▶ Problem 33

EXAMPLE 3   Find the slope of the line that passes through $(-2, 4)$ and $(5, -6)$.

Strategy   We will use the slope formula to find the slope of the line.

Why   We know the coordinates of two points on the line.

Solution   If we let (x_1, y_1) be $(-2, 4)$ and (x_2, y_2) be $(5, -6)$, then

> **Caution**
>
> Slopes normally are written as fractions, sometimes as decimals, but never as mixed numbers.
>   As with any fractional answer, always express slope in simplified form.

$$m = \frac{y_2 - y_1}{x_2 - x_1} \qquad \text{This is the slope formula.}$$

$$m = \frac{-6 - 4}{5 - (-2)} \qquad \begin{array}{l}\text{Substitute } -6 \text{ for } y_2, 4 \text{ for } y_1,\\ 5 \text{ for } x_2, \text{ and } -2 \text{ for } x_1.\end{array}$$

$$m = -\frac{10}{7} \qquad \begin{array}{l}\text{Do the subtraction. We can write the}\\ \text{result as } \frac{-10}{7} \text{ or } -\frac{10}{7}.\end{array}$$

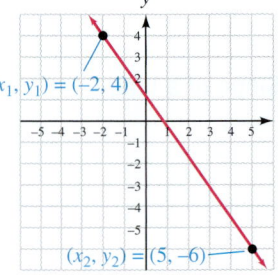

The slope of the line is $-\frac{10}{7}$.

   If we graph the line by plotting the two points, we see that the line falls from left to right—a fact indicated by its negative slope.

Self Check 3   Find the slope of the line that passes through $(-1, -2)$ and $(1, -7)$.

Now Try ▶ Problem 39

   In Example 2, the slope of the line was positive. In Examples 1 and 3, the slopes of the lines were negative. In general, lines that rise from left to right have a positive slope. Lines that fall from left to right have a negative slope.

> **Success Tip**
>
> To classify the slope of a line as positive or negative, follow it from left to right, as you would read a sentence in a book.

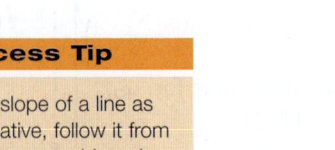

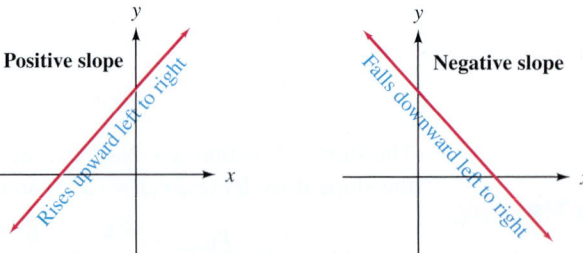

   In figure (a) on the next page, we see that a line with slope 3 is steeper than a line with slope $\frac{5}{6}$, and a line with slope $\frac{5}{6}$ is steeper than a line with slope $\frac{1}{4}$. In general, *the larger the absolute value of the slope, the steeper the line.*
   Lines with slopes of 1 and -1 are graphed in figure (b) on the next page. In each case, there is a special relationship between the rise and the run. When $m = 1$, the rise and run are, of course, the same number. When $m = -1$ the rise and run are opposites. Note that both lines create a $45°$ angle with the horizontal x-axis.

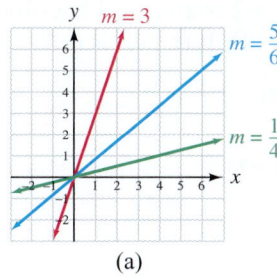

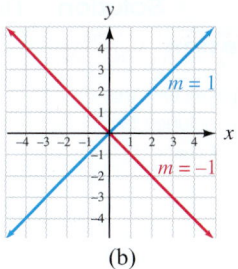

(a)   (b)

3 Find Slopes of Horizontal and Vertical Lines.

In the next two examples, we calculate the slope of a horizontal line and we show that a vertical line has no defined slope.

EXAMPLE 4   Find the slope of the line $y = 3$.

Strategy   We will find the coordinates of two points on the line.

Why   We can then use the slope formula to find the slope of the line.

Solution   The graph of $y = 3$ is a horizontal line. To find its slope, we select two points on the line: $(-2, 3)$ and $(3, 3)$. If (x_1, y_1) is $(-2, 3)$ and (x_2, y_2) is $(3, 3)$, we have

$$m = \frac{y_2 - y_1}{x_2 - x_1} \qquad \text{This is the slope formula.}$$

$$m = \frac{3 - 3}{3 - (-2)} \qquad \text{Substitute 3 for } y_2, \text{ 3 for } y_1, \text{ 3 for } x_2, \text{ and } -2 \text{ for } x_1.$$

$$m = \frac{0}{5} \qquad \text{Simplify the numerator and the denominator.}$$

$$m = 0 \qquad \text{0 divided by any nonzero number is equal to 0.}$$

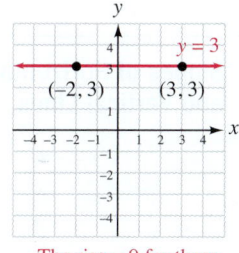

The rise = 0 for these two points.

The slope of the line $y = 3$ is 0.

Self Check 4   Find the slope of the line $y = -10$.

Now Try ▶   Problem 61

The y-coordinates of any two points on a horizontal line will be the same, and the x-coordinates will be different. Thus, the numerator of $\frac{y_2 - y_1}{x_2 - x_1}$ will always be zero, and the denominator will always be nonzero. Therefore, the slope of a horizontal line is 0.

EXAMPLE 5   Find the slope of the line $x = -2$.

Strategy   We will find the coordinates of two points on the line.

Why   We can then use the slope formula to find the slope of the line.

Solution

Notation

This example explains why the definition of slope includes the restriction that $x_1 \neq x_2$.

The graph of $x = -2$ is a vertical line. To find its slope, we select two points on the line: $(-2, 3)$ and $(-2, -1)$. If (x_1, y_1) is $(-2, -1)$ and (x_2, y_2) is $(-2, 3)$, we have

$$m = \frac{y_2 - y_1}{x_2 - x_1}$$ This is the slope formula.

$$m = \frac{3 - (-1)}{-2 - (-2)}$$ Substitute 3 for y_2, -1 for y_1, -2 for x_2, and -2 for x_1. Note that $x_1 = x_2$.

$$m = \frac{4}{0}$$ Simplify the numerator and the denominator.

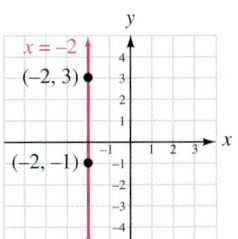

The run = 0 for these two points.

Since division by zero is undefined, $\frac{4}{0}$ has no meaning. The slope of the line $x = -2$ is undefined.

Self Check 5　Find the slope of the line $x = 12$.

Now Try ▶ Problem 67

The Language of Algebra

Undefined and **0** do not mean the same thing. A horizontal line has a defined slope; it is 0. A vertical line does not have a defined slope; we say its slope is *undefined*.

The y-coordinates of any two points on a vertical line will be different, and the x-coordinates will be the same. Thus, the numerator of $\frac{y_2 - y_1}{x_2 - x_1}$ will always be nonzero, and the denominator will always be 0. Therefore, the slope of a vertical line is undefined.

We now summarize the results from Examples 4 and 5.

Slopes of Horizontal and Vertical Lines	Horizontal lines (lines with equations of the form $y = b$) have **slope 0**.	Vertical lines (lines with equations of the form $x = a$) have **undefined slope**.
	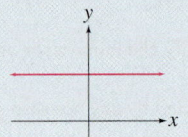	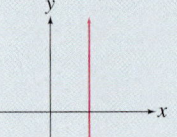

---

4 Solve Applications of Slope.

The concept of slope has many applications. For example, architects use slope when designing ramps and roofs. Truckers must be aware of the slope, or **grade,** of the roads they travel. Mountain bikers ride up rocky trails and snow skiers speed down steep slopes.

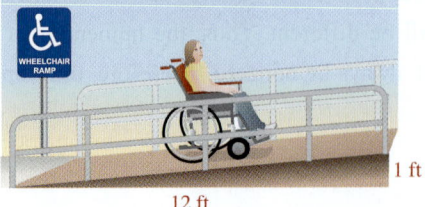

The Americans with Disabilities Act provides a guideline for the steepness of a ramp. The maximum slope for a wheelchair ramp is 1 foot of rise for every 12 feet of run: $m = \frac{1}{12}$.

The grade of an incline is its slope expressed as a percent: A 15% grade means a rise of 15 feet for every run of 100 feet: $m = \frac{15}{100}$, which simplifies to $\frac{3}{20}$. A grade is always expressed as a positive percent.

EXAMPLE 6 **Architecture.** Pitch is the incline of a roof written as a ratio of the vertical rise to the horizontal run. It is always expressed as a positive number. Find the pitch of the roof shown in the illustration.

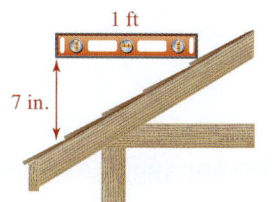

Strategy We will determine the rise and the run of the roof from the illustration. Then we will write the ratio of the rise to the run.

Why The pitch of a roof is its slope, and the slope of a line is the ratio of the rise to the run.

Solution A level is used to create a slope triangle. The rise of the slope triangle is given as 7 inches. Since a ratio is a quotient of two quantities with the *same* units, we will express the length of the one-foot-long level as 12 inches. Therefore, the run of the slope triangle is 12 inches.

$$m = \frac{\text{rise}}{\text{run}} = \frac{7}{12}$$ The roof has a $\frac{7}{12}$ pitch. This means that the roof rises 7 units for every 12 units in the horizontal direction.

Self Check 6 **Roofing.** Find the pitch of the roof.

Now Try ▶ Problem 99

5 Calculate Rates of Change.

We have seen that the slope of a line compares the change in y to the change in x. This is called the **rate of change** of y with respect to x. In our daily lives, we often make many such comparisons of the change in one quantity with respect to another. For example, we might speak of snow melting at the rate of 6 inches per day or a tourist exchanging money at the rate of 12 pesos per dollar.

When finding rates of change in application problems, we attach units to the numerator and denominator in the slope calculation.

EXAMPLE 7 **Banking.** A bank offers a business account with a fixed monthly fee, plus a service charge for each check written. The relationship between the monthly cost y and the number x of checks written is graphed below. At what rate does the monthly cost change?

Notation

In the graph, the symbol ≠ indicates a break in the labeling of the vertical axis. The break enables us to omit a large portion of the grid that would not be used.

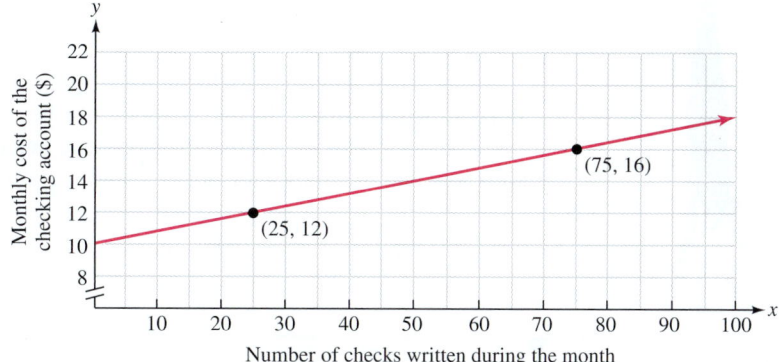

Strategy We will use the slope formula to calculate the slope of the line and attach the proper units to the numerator and denominator.

Why We know the coordinates of two points on the line.

Solution   Two points on the line are $(25, 12)$ and $(75, 16)$. We will let $(x_1, y_1) = (25, 12)$ and $(x_2, y_2) = (75, 16)$, so we have

$$\frac{\text{Rate of}}{\text{change}} = \frac{(y_2 - y_1) \text{ dollars}}{(x_2 - x_1) \text{ checks}} = \frac{(16 - 12) \text{ dollars}}{(75 - 25) \text{ checks}} = \frac{4 \text{ dollars}}{50 \text{ checks}} = \frac{2 \text{ dollars}}{25 \text{ checks}}$$

Since the y-axis is scaled in dollars, we attach the units of "dollars" to the calculation of $y_2 - y_1$ in the numerator. Since x-axis indicates the number of checks written, we attach the units of "checks" to the calculation of $x_2 - x_1$ in the denominator. The monthly cost of the checking account increases $2 for every 25 checks written.

We can express $\frac{2}{25}$ in decimal form by dividing the numerator by the denominator. Then we can write the rate of change in two other ways, using the word *per*, which indicates division.

> Rate of change = $0.08 per check   or   Rate of change = 8¢ per check

---

Self Check 7   **Ethanol.**   In 2005, the U.S. produced about 1,600 million bushels of corn for ethanol. By 2009, production had risen to 4,200 million bushels. Find the rate of change in the number of bushels of corn produced for ethanol over that time span. (Source: USDA, ERS Feed Outlook)

Now Try ▶ Problem 103

---

6 Determine Whether Lines Are Parallel or Perpendicular Using Slope.

Two lines that lie in the same plane but do not intersect are called **parallel lines.** Parallel lines have the same slope and different y-intercepts. For example, the lines graphed in figure (a) are parallel because they both have slope $-\frac{2}{3}$.

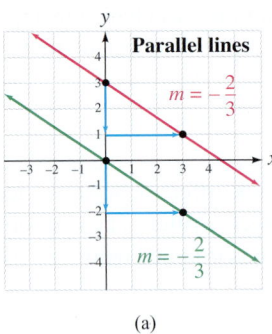

(a)

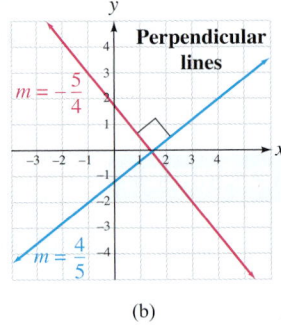

(b)

Lines that intersect to form four right angles (angles with measure 90°) are called **perpendicular lines.** If the product of the slopes of two lines is -1, the lines are perpendicular. This means that the slopes are **negative** (or **opposite**) **reciprocals.** In figure (b), we know that the lines with slopes $\frac{4}{5}$ and $-\frac{5}{4}$ are perpendicular because

$$\frac{4}{5}\left(-\frac{5}{4}\right) = -\frac{20}{20} = -1 \qquad \text{$\frac{4}{5}$ and $-\frac{5}{4}$ are negative reciprocals.}$$

Slopes of Parallel and Perpendicular Lines

1. Two lines with the same slope are parallel.

2. Two lines are perpendicular if the product of their slopes is -1; that is, if their slopes are negative reciprocals.

3. A horizontal line is perpendicular to any vertical line, and vice versa.

EXAMPLE 8 Determine whether the line that passes through $(7, -9)$ and $(10, 2)$ and the line that passes through $(0, 1)$ and $(3, 12)$ are parallel, perpendicular, or neither.

Strategy We will use the slope formula to find the slope of each line.

Why If the slopes are equal, the lines are parallel. If the slopes are negative reciprocals, the lines are perpendicular. Otherwise, the lines are neither parallel nor perpendicular.

Solution To calculate the slope of each line, we use the slope formula.

Success Tip

An example of two lines that are neither parallel nor perpendicular would be lines with slopes of $\frac{5}{2}$ and -2. When graphed, the lines intersect, but they do not form right angles.

The line through $(7, -9)$ and $(10, 2)$:

$$m = \frac{y_2 - y_1}{x_2 - x_1} = \frac{2 - (-9)}{10 - 7} = \frac{11}{3}$$

The line through $(0, 1)$ and $(3, 12)$:

$$m = \frac{y_2 - y_1}{x_2 - x_1} = \frac{12 - 1}{3 - 0} = \frac{11}{3}$$

Since the slopes are the same, the lines are parallel.

Self Check 8 Determine whether the line that passes through $(2, 1)$ and $(6, 8)$ and the line that passes through $(-1, 0)$ and $(-5, 7)$ are parallel, perpendicular, or neither.

Now Try Problems 73 and 75

EXAMPLE 9 Find the slope of a line perpendicular to the line that passes through $(1, -4)$ and $(8, 4)$.

Strategy We will use the slope formula to find the slope of the line passing through $(1, -4)$ and $(8, 4)$.

Why We can then form the negative reciprocal of the result to produce the slope of a line perpendicular to the given line.

Solution The slope of the line that passes through $(1, -4)$ and $(8, 4)$ is

$$m = \frac{y_2 - y_1}{x_2 - x_1} = \frac{4 - (-4)}{8 - 1} = \frac{8}{7} \qquad \text{We let } (x_1, y_1) = (1, -4) \text{ and } (x_2, y_2) = (8, 4).$$

The slope of a line perpendicular to the given line has slope that is the negative (or opposite) reciprocal of $\frac{8}{7}$, which is $-\frac{7}{8}$.

Self Check 9 Find the slope of a line perpendicular to the line that passes through $(-4, 1)$ and $(9, 5)$.

Now Try Problem 85

SECTION 3.4 STUDY SET

VOCABULARY

Fill in the blanks.

1. The _____ of a line is a measure of the line's steepness. It is the _____ of the vertical change to the horizontal change.

2. $m = \dfrac{}{\text{horizontal change}} = \dfrac{\text{rise}}{} = \dfrac{\text{change in } y}{}$

3. The rate of _____ of a linear relationship can be found by finding the slope of the graph of the line and attaching the proper units.

4. _____ lines do not intersect. _____ lines intersect to form four right angles.

CONCEPTS

5. Which line graphed has
   a. a positive slope?
   b. a negative slope?
   c. zero slope?
   d. undefined slope?

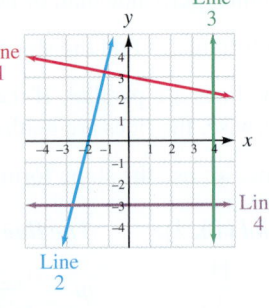

6. Consider each graph of a line and the slope triangle. What is the rise? What is the run? What is the slope of the line?

a.
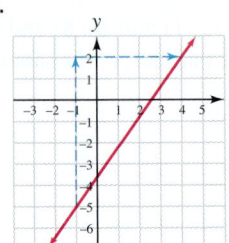

b.
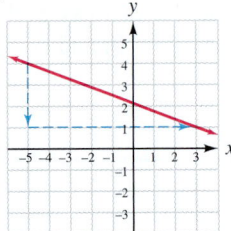

7. For each graph, determine which line has the greater slope.

a.
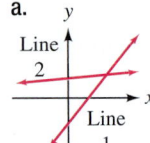

b.
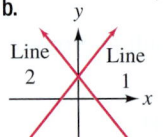

c.
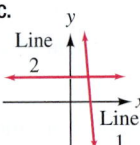

8. Which two labeled points should be used to find the slope of the line?

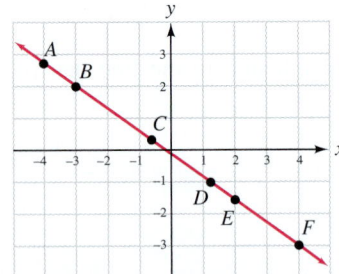

9. a. Find the slope of the line below using the points in black.
   b. Find the slope of the line using the points in green.
   c. Fill in the blank: When finding the slope of a line, the _____ value will be obtained no matter which two points on the line are used.

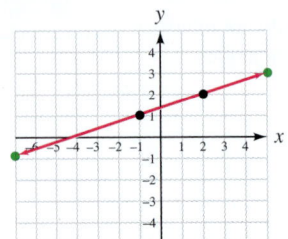

10. Evaluate each expression.
    a. $\dfrac{10 - 4}{6 - 5}$

    b. $\dfrac{-1 - 1}{-2 - (-7)}$

11. Write each slope in a better way.

    a. $m = \dfrac{0}{6}$

    b. $m = \dfrac{8}{0}$

    c. $m = \dfrac{3}{12}$

    d. $m = \dfrac{-10}{-5}$

12. Fill in the blanks: _____ lines have a slope of 0. Vertical lines have _____ slope.

13. The *grade* of an incline is its slope expressed as a percent. Express the slope $\frac{2}{5}$ as a grade.

14. **Growth Rates.** The graph shows how a child's height increased from ages 2 through 5. Fill in the correct units to find the rate of change in the child's height.

$$\frac{\text{Rate of}}{\text{change}} = \frac{(40 - 31)\ \boxed{}}{(5 - 2)\ \boxed{}}$$

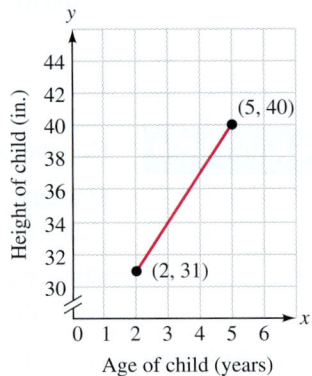

15. Find the negative reciprocal of each number.

    a. 6

    b. $-\dfrac{7}{8}$

    c. -1

16. Fill in the blanks.
    a. Two different lines with the same slope are _____.
    b. If the slopes of two lines are negative reciprocals, the lines are _____.
    c. The product of the slopes of perpendicular lines is $\boxed{}$.

NOTATION

17. a. What is the formula used to find the slope of a line passing through (x_1, y_1) and (x_2, y_2)?
    b. Fill in the blanks to state the slope formula in words: m equals y _____ two minus y _____ one _____ x sub _____ minus x sub _____.

18. Explain the difference between y^2 and y_2.

19. Consider the points $(7, 2)$ and $(-4, 1)$. If we let $x_1 = 7$, then what is y_2?

20. The symbol $\rightleftharpoons$ is used when graphing to indicate a _____ in the labeling of an axis.

GUIDED PRACTICE

Find the slope of each line. See Example 1.

21.
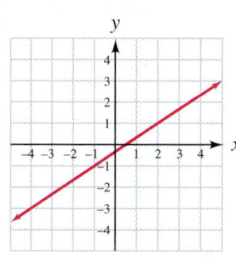

22.
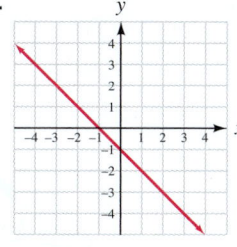

23.
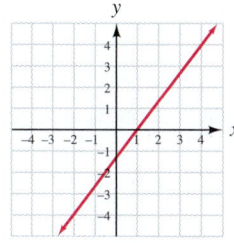

24.
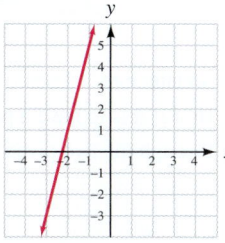

25.
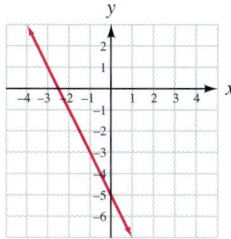

26.
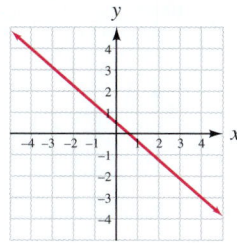

27.
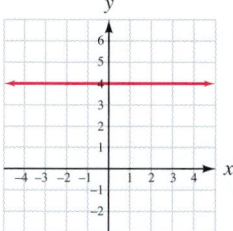

28.
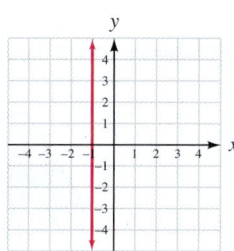

29.
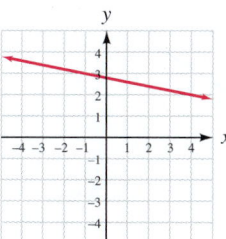

30.
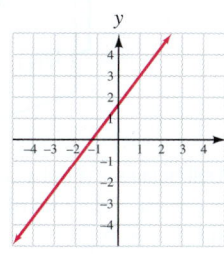

31.
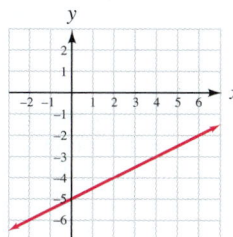

32.
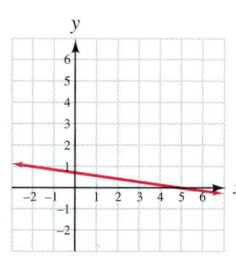

Find the slope of the line passing through the given points.
See Examples 2 and 3.

33. $(1, 3)$ and $(2, 4)$

34. $(1, 3)$ and $(2, 5)$

35. $(3, 4)$ and $(2, 7)$

36. $(3, 6)$ and $(5, 2)$

37. $(0, 0)$ and $(4, 5)$

38. $(4, 3)$ and $(7, 8)$

39. $(-3, 5)$ and $(-5, 6)$

40. $(6, -2)$ and $(-3, 2)$

41. $(-2, -2)$ and $(-12, -8)$

42. $(-1, -2)$ and $(-10, -5)$

43. $(5, 7)$ and $(-4, 7)$

44. $(-1, -12)$ and $(6, -12)$

45. $(8, -4)$ and $(8, -3)$

46. $(-2, 8)$ and $(-2, 15)$

47. $(-6, 0)$ and $(0, -4)$

48. $(0, -9)$ and $(-6, 0)$

49. $(-2.5, 1.75)$ and $(-0.5, -7.75)$

50. $(6.4, -7.2)$ and $(-8.8, 4.2)$

51. $(-2.2, 18.6)$ and $(-1.7, 18.6)$

52. $(4.6, 3.2)$ and $(4.6, -4.8)$

53. $\left(-\frac{4}{7}, -\frac{1}{5}\right)$ and $\left(\frac{3}{7}, \frac{6}{5}\right)$

54. $\left(-\frac{4}{9}, -\frac{1}{8}\right)$ and $\left(\frac{5}{9}, \frac{3}{8}\right)$

55. $\left(-\frac{3}{4}, \frac{2}{3}\right)$ and $\left(\frac{4}{3}, -\frac{1}{6}\right)$

56. $\left(\frac{1}{2}, \frac{3}{4}\right)$ and $\left(-\frac{11}{16}, -\frac{1}{2}\right)$

Determine the slope of the graph of the line that has the given table of solutions. See Examples 2 and 3.

57.

x	y	(x, y)
-3	-1	$(-3, -1)$
1	2	$(1, 2)$

58.

x	y	(x, y)
-3	6	$(-3, 6)$
0	2	$(0, 2)$

59.

x	y	(x, y)
-3	6	$(-3, 6)$
0	6	$(0, 6)$

60.

x	y	(x, y)
4	-5	$(4, -5)$
4	0	$(4, 0)$

Find the slope of each line. See Examples 4 and 5.

61. $y = -11$

62. $y = -2$

63. $y = 0$

64. $x = 0$

65. $x = 6$

66. $x = 4$

67. $x = -10$

68. $y = 8$

69. $y - 9 = 0$

70. $x + 14 = 0$

71. $3x = -12$

72. $2y + 2 = -6$

Determine whether the lines through each pair of points are parallel, perpendicular, or neither. See Example 8.

73. $(5, 3)$ and $(1, 4)$
    $(-3, -4)$ and $(1, -5)$

74. $(2, 4)$ and $(-1, -1)$
    $(8, 0)$ and $(11, 5)$

75. $(-4, -2)$ and $(2, -3)$
    $(7, 1)$ and $(8, 7)$

76. $(-2, 4)$ and $(6, -7)$
    $(-6, 4)$ and $(5, 12)$

77. $(2, 2)$ and $(4, -3)$
    $(-3, 4)$ and $(-1, 9)$

78. $(-1, -3)$ and $(2, 4)$
    $(5, 2)$ and $(8, -5)$

79. $(-1, 8)$ and $(-6, 8)$
    $(3, 3)$ and $(3, 7)$

80. $(11, 0)$ and $(11, -5)$
    $(14, 6)$ and $(25, 6)$

81. $(6, 4)$ and $(2, 5)$
    $(-2, -3)$ and $(2, -4)$

82. $(-3, -1)$ and $(3, -2)$
    $(8, 2)$ and $(9, 8)$

83. $(4, 2)$ and $(5, -3)$
    $(-5, 3)$ and $(-2, 9)$

84. $(8, -3)$ and $(8, -8)$
    $(11, 3)$ and $(22, 3)$

Find the slope of a line perpendicular to the line passing through the given two points. See Example 9.

85. $(0, 0)$ and $(5, -9)$ **86.** $(0, 0)$ and $(5, 12)$

87. $(-1, 7)$ and $(1, 10)$ **88.** $(-7, 6)$ and $(0, 4)$

89. $\left(-2, \frac{1}{2}\right)$ and $\left(-1, \frac{3}{2}\right)$ **90.** $\left(\frac{1}{3}, -1\right)$ and $\left(\frac{4}{3}, -2\right)$

91. $(-1, 2)$ and $(-3, 6)$ **92.** $(5, -4)$ and $(-1, -7)$

APPLICATIONS

93. Pools.   Find the slope of the bottom of the swimming pool as it drops off from the shallow end to the deep end.

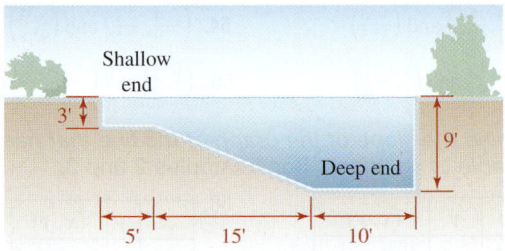

94. Drainage.   Find the slope of the concrete patio slab using the 1-foot ruler, level, and 10-foot-long board shown in the illustration. (*Hint:* 10 feet = 120 in.)

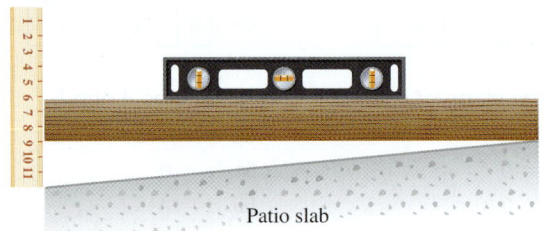

95. Grade of a Road.   Refer to the illustration below. Find the slope of the decline and use that information to find the grade of the road.

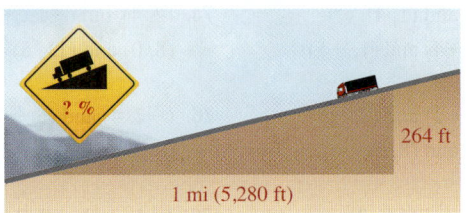

96. Streets.   One of the steepest streets in the United States is Eldred Street in Highland Park, California (near Los Angeles). It rises approximately 220 feet over a horizontal distance of 665 feet. What is the grade of the street?

97. Treadmills.   Find the slope of the jogging surface of the treadmill for a height setting of 6 inches. Then express the incline as a percent.

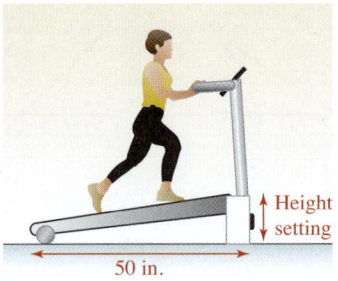

98. Architecture.   Locate the coordinates of the peak of the roof if it is to have a pitch of $\frac{2}{5}$ and the roof line is to pass through the two given points in black.

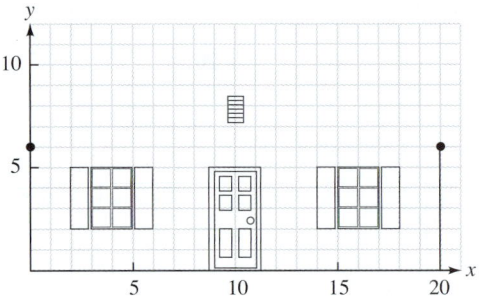

99. Carpentry.   Find the pitch of each roof.

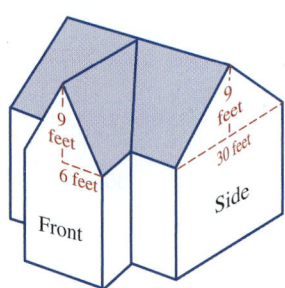

100. Doll Houses.   Find x so that the pitch of the roof of the doll house is $\frac{4}{3}$.

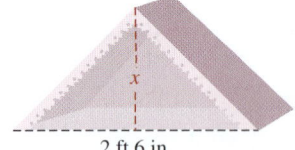

101. Irrigation.   The graph on the next page shows the number of gallons of water remaining in a reservoir as water is used from it to irrigate a field. Find the rate of change in the number of gallons of water in the reservoir.

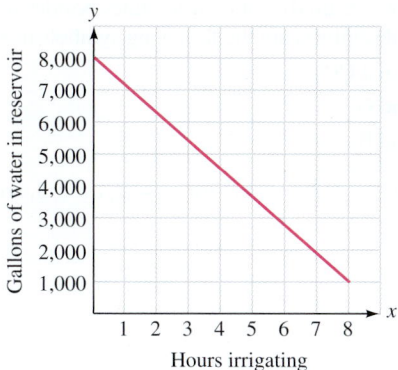

102. Commercial Jets. Examine the graph and consider trips of more than 7,000 miles by a Boeing 777. Use a rate of change to estimate how the maximum payload decreases as the distance traveled increases.

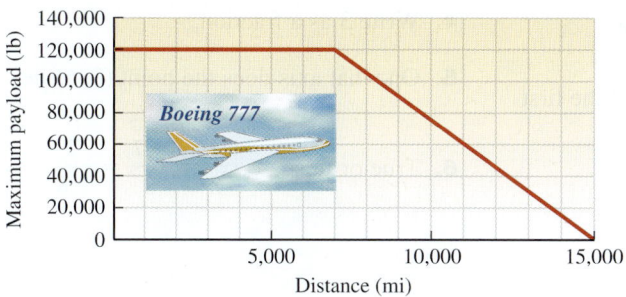

Based on data from Lawrence Livermore National Laboratory and *Los Angeles Times* (October 22, 1998).

103. Milk Production. The following graph approximates the amount of milk produced per cow in the United States for the years 1996–2009. Find the rate of change.

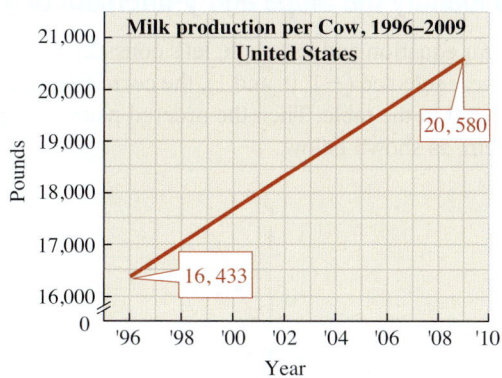

Source: USDA; Agricultural and Applied Economics, UW Madison

104. Wal-Mart. The graph in the next column approximates the sales revenue of Wal-Mart for the years 1991–2009. Find the rate of change in revenue for the years

  a. 1991–1998

  b. 1998–2009

Source: wikinvest.com

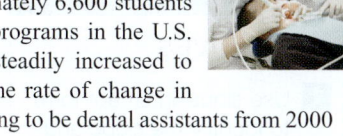

105. **from Campus to Careers**
Dental Assistant
In 2000, there were approximately 6,600 students enrolled in dental assisting programs in the U.S. By 2008, that number had steadily increased to about 9,200 students. Find the rate of change in the number of students studying to be dental assistants from 2000 to 2008. (Source: American Dental Education Association)

106. Firefighting. When flames are tilted due to effects of wind, firefighters measure what is called the **slope percent** of the flames. Calculate the slope percent of the flame shown below by expressing its "slope" as a percent. (Source math.fire.org)

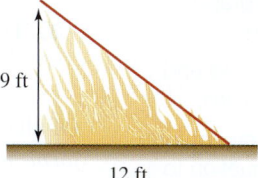

9 ft

12 ft

WRITING

107. Explain why the slope of a vertical line is undefined.

108. How do we distinguish between a line with positive slope and a line with negative slope?

109. Explain the error in the following solution: *Find the slope of the line that passes through* (6, 4) *and* (3, 1).

$$m = \frac{1-4}{6-3} = \frac{-3}{3} = -1$$

110. Explain the difference between a rate of change that is positive and one that is negative. Give an example of each.

REVIEW

111. Halloween Candy. A candy maker wants to make a 60-pound mixture of two candies to sell for $2 per pound. If black licorice bits sell for $1.90 per pound and orange gumdrops sell for $2.20 per pound, how many pounds of each should be used?

112. Medications. A doctor prescribes an ointment that is 2% hydrocortisone. A pharmacist has 1% and 5% concentrations in stock. How many ounces of each should the pharmacist use to make a 1-ounce tube?

CHALLENGE PROBLEMS

113. Use the concept of slope to determine whether $A(-50, -10)$, $B(20, 0)$, and $C(34, 2)$ all lie on the same straight line.

114. A line having slope $\frac{2}{3}$ passes through the point $(10, -12)$. What is the y-coordinate of another point on the line whose x-coordinate is 16?

115. Subscripts are used in other disciplines besides mathematics. In what disciplines are the following symbols used?
   a. H_2O and CO_2
   b. C_7 and G_7
   c. B_6 and B_{12}

116. Evaluate $2a_2^2 + 3a_3^3 + 4a_4^4$ for $a_2 = 2$, $a_3 = 3$, and $a_4 = 4$.

SECTION 3.5

OBJECTIVES

1. Use slope–intercept form to identify the slope and y-intercept of a line.
2. Write a linear equation in slope–intercept form.
3. Write an equation of a line given its slope and y-intercept.
4. Use the slope and y-intercept to graph a linear equation.
5. Recognize parallel and perpendicular lines.
6. Use slope–intercept form to write an equation to model data.

Slope–Intercept Form

ARE YOU READY?

The following problems review some basic skills that are needed when working with equations of lines in slope–intercept form.

1. a. Identify each term in the expression $3x - 6$.
   b. What is the coefficient of the first term?

2. Solve for y: $2x + 5y = 15$

3. True or false: $\dfrac{x}{4} = \dfrac{1}{4}x$

4. Write 3 as a fraction.

5. On what axis does the point $(0, 6)$ lie?

6. True or false: $-\dfrac{7}{8} = \dfrac{-7}{8} = \dfrac{7}{-8}$

Of all of the ways in which a linear equation can be written, one form, called *slope–intercept form,* is probably the most useful. When an equation is written in this form, two important features of its graph are evident.

1 Use Slope–Intercept Form to Identify the Slope and *y*-Intercept of a Line.

To explore the relationship between a linear equation and its graph, let's consider $y = 2x + 1$. To graph this equation, three values of x were selected (-1, 0, and 1), the corresponding values of y were found, and the results were entered in the table. Then the ordered pairs were plotted and a straight line was drawn through them, as shown below.

$$y = 2x + 1$$

x	y	(x, y)
-1	-1	$(-1, -1)$
0	1	$(0, 1)$
1	3	$(1, 3)$

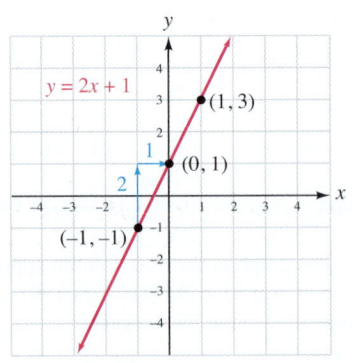

To find the slope of the line, we pick two points on the line, $(-1, -1)$ and $(0, 1)$, and draw a slope triangle and count grid squares:

$$\text{Slope} = \frac{\text{rise}}{\text{run}} = \frac{2}{1} = 2$$

From the equation and the graph, we can make two observations:

- The graph crosses the y-axis at 1. This is the same as the constant term in $y = 2x + \mathbf{1}$.
- The slope of the line is 2. This is the same as the coefficient of x in $y = \mathbf{2}x + 1$.

The Language of Algebra

Recall that a term that contains only a number is called a **constant term**.

This illustrates that the slope and y-intercept of the graph of $y = 2x + 1$ can be determined from the equation.

$$y = \mathbf{2}x + \mathbf{1}$$

The slope of the line is 2.   The y-intercept is (0, 1).

These observations suggest the following form of an equation of a line.

Slope–Intercept Form of the Equation of a Line

If a linear equation is written in the form

$$y = mx + b$$

the graph of the equation is a line with slope m and y-intercept $(0, b)$.

When an equation of a line is written in slope–intercept form, the coefficient of the x-term is the line's slope and the constant term gives the y-coordinate of the y-intercept.

$$y = \mathbf{m}x + \mathbf{b}$$

Slope y-intercept: (0, b)

Caution

For equations in $y = mx + b$ form, the slope of the line is the *coefficient* of x, not the x-term. For the graph of $y = 6x - 2$:

$$m = 6$$
not
$$m = 6x$$

Linear equation	Equation written in slope–intercept form	Slope	y-intercept
$y = 6x - 2$	$y = \mathbf{6}x + (\mathbf{-2})$	6	$(0, -2)$
$y = -\dfrac{5}{4}x$	$y = -\dfrac{5}{4}x + \mathbf{0}$	$-\dfrac{5}{4}$	$(0, 0)$
$y = \dfrac{x}{2} + 3$	$y = \dfrac{1}{2}x + 3$	$\dfrac{1}{2}$	$(0, 3)$
$y = -\dfrac{7}{8} - x$	$y = -x + \left(-\dfrac{7}{8}\right)$	-1	$\left(0, -\dfrac{7}{8}\right)$

2   Write a Linear Equation in Slope–Intercept Form.

The equation of any nonvertical line can be written in slope–intercept form.

Slope-Intercept form

To write a linear equation in two variables in slope–intercept form, solve the equation for y.

EXAMPLE 1   Find the slope and y-intercept of the line with the given equation.
a. $8x + y = 9$ **b.** $x + 4y = 16$ **c.** $-9x - 3y = 11$

Strategy   We will write each equation in slope–intercept form by solving for y.

Why   When the equations are written in slope–intercept form, the slope and y-intercept of their graphs become apparent.

Solution **a.** The slope and y-intercept of the graph of $8x + y = 9$ are not obvious because the equation is not in slope–intercept form. To write it in $y = mx + b$ form, we isolate y.

<table>
<tr><td>$8x + y = 9$</td><td>This is the given equation.</td></tr>
<tr><td>$8x + y - \mathbf{8x} = -\mathbf{8x} + 9$</td><td>To isolate y on the left side, subtract 8x from both sides. Since we want the right side of the equation to have the form mx + b, we show the subtraction from that side as −8x + 9 rather than 9 − 8x.</td></tr>
<tr><td>$y = -8x + \boxed{9}$</td><td>On the left side, combine like terms: 8x − 8x = 0.</td></tr>
</table>

The Language of Algebra

When working with the slope–intercept form, use precise language: b is not the y-intercept of the graph of the equation $8x + y = 9$. b is the y-coordinate of the y-intercept.

Since $m = -8$ and $b = 9$, the slope is -8 and the y-intercept is $(0, 9)$.

b. To write the equation in slope–intercept form, we solve for y.

<table>
<tr><td>$x + 4y = 16$</td><td>This is the given equation.</td></tr>
<tr><td>$x + 4y - \mathbf{x} = -\mathbf{x} + 16$</td><td>To isolate the term 4y on the left side, subtract x from both sides. Write the subtraction before the constant term 16.</td></tr>
<tr><td>$4y = -x + 16$</td><td>On the left side, combine like terms: x − x = 0.</td></tr>
<tr><td>$y = \dfrac{-x}{4} + \dfrac{16}{4}$</td><td>To isolate y, undo the multiplication by 4 by dividing both sides by 4, term-by-term.</td></tr>
<tr><td>$y = -\dfrac{1}{4}x + \boxed{4}$</td><td>Write $\frac{-x}{4}$ as $-\frac{1}{4}x$. Simplify: $\frac{16}{4} = 4$.</td></tr>
</table>

Success Tip

Since we want the right side of the equation to have the form $mx + b$, we show the division by 4 as

$-\dfrac{x}{4} + \dfrac{16}{4}$ not $\dfrac{-x + 16}{4}$

Since $m = -\frac{1}{4}$ and $b = 4$, the slope is $-\frac{1}{4}$ and the y-intercept is $(0, 4)$.

c. To write the equation in $y = mx + b$ form, we isolate y on the left side.

<table>
<tr><td>$-9x - 3y = 11$</td><td>This is the given equation.</td></tr>
<tr><td>$-3y = 9x + 11$</td><td>To isolate the term −3y on the left side, add 9x to both sides. Write the addition before the constant term 11.</td></tr>
<tr><td>$\dfrac{-3y}{-3} = \dfrac{9x}{-3} + \dfrac{11}{-3}$</td><td>To isolate y, undo the multiplication by −3 by dividing both sides by −3, term-by-term.</td></tr>
<tr><td>$y = -3x - \boxed{\dfrac{11}{3}}$</td><td>Simplify.</td></tr>
</table>

Since $m = -3$ and $b = -\frac{11}{3}$, the slope is -3 and the y-intercept is $\left(0, -\frac{11}{3}\right)$.

Self Check 1 Find the slope and y-intercept of the line with the given equation.
a. $9x + y = 4$ **b.** $-x + 11y = -22$ **c.** $-10x - 2y = 7$

Now Try ▶ Problems 11, 31, and 39

3 Write an Equation of a Line Given Its Slope and y-Intercept.

If we are given the slope and y-intercept of a line, we can write an equation of the line by substituting for m and b in the slope–intercept form.

EXAMPLE 2 Write an equation of the line with slope -1 and y-intercept $(0, 9)$.

Strategy We will use the slope–intercept form, $y = mx + b$, to write an equation of the line.

Why We know the slope of the line and its y-intercept.

Solution   If the slope is -1 and the y-intercept is $(0, 9)$, then $m = -1$ and $b = 9$.

$$y = mx + b \qquad \text{This is the slope–intercept form.}$$
$$y = -1x + 9 \qquad \text{Substitute } -1 \text{ for } m \text{ and } 9 \text{ for } b.$$
$$y = -x + 9 \qquad \text{Simplify: } -1x = -x.$$

The equation of the line with slope -1 and y-intercept $(0, 9)$ is $y = -x + 9$.

> **Self Check 2**   Write an equation of the line with slope 1 and y-intercept $(0, -12)$.
>
> **Now Try** ▶ Problem 43

EXAMPLE 3   Write an equation of the line graphed in figure (a).

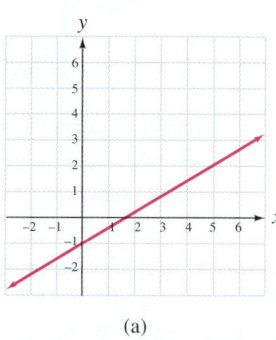 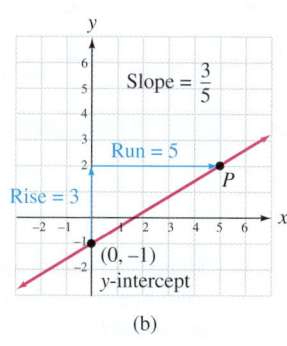 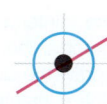

Pick a second point on the line that lies on the intersection of two grid lines.

(a)   (b)

Strategy   We will use the slope–intercept form, $y = mx + b$, to write an equation of the line.

Why   We can determine the slope and y-intercept of the line from the given graph.

Solution   In figure (b), we highlight and label the y-intercept of the line, $(0, -1)$. Then we pick a convenient second point on the line and label it point P. Moving from the y-intercept to point P, we draw a slope triangle to find that the slope of the line is $\frac{3}{5}$. When we substitute $\frac{3}{5}$ for m and -1 for b into the slope–intercept form $y = mx + b$, we obtain an equation of the line:

$$y = \frac{3}{5}x - 1.$$

> **Self Check 3**   Write an equation of the line graphed here.
>
>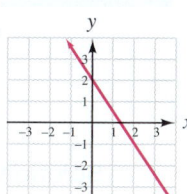
>
> **Now Try** ▶ Problem 55

---

4 Use the Slope and *y*-Intercept to Graph a Linear Equation.

If we know the slope and y-intercept of a line, we can graph the line.

EXAMPLE 4   Use the slope and y-intercept to graph $y = 5x - 4$.

Strategy   We will examine the equation to identify the slope and the y-intercept of the line to be graphed. Then we will plot the y-intercept and use the slope to determine a second point on the line.

Why   Once we locate two points on the line, we can draw the graph of the line.

Solution

Since $y = 5x - 4$ is written in $y = mx + b$ form, we know that its graph is a line with a slope of 5 and a y-intercept of $(0, -4)$. To draw the graph, we begin by plotting the y-intercept. The slope can be used to find another point on the line.

If we write the slope as the fraction $\frac{5}{1}$, the rise is 5 and the run is 1. From $(0, -4)$, we move 5 units *upward* (because the numerator, 5, is positive) and 1 unit to the right (because the denominator, 1, is positive). This locates a second point on the line, $(1, 1)$. The line through $(0, -4)$ and $(1, 1)$ is the graph of $y = 5x - 4$.

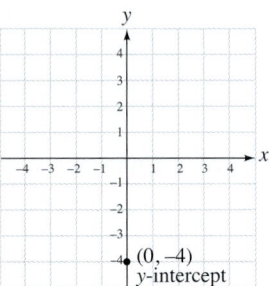

Plot the y-intercept.

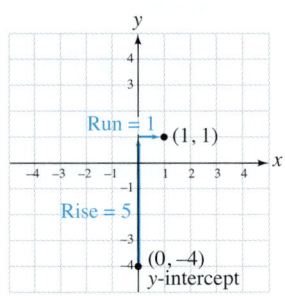

From $(0, -4)$, draw the rise and run parts of the slope triangle for $m = \frac{5}{1}$ to find another point on the line.

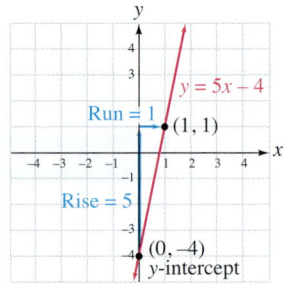

Use a straightedge to draw a line through the two points.

Since $\frac{5}{1} = \frac{-5}{-1}$, an alternate way to find another point on the line is to write the slope as $\frac{-5}{-1}$. Like before, we begin at the y-intercept $(0, -4)$. Since the rise is negative, we move 5 units *downward,* and since the run is negative, we then move 1 unit to the *left.* We arrive at $(-1, -9)$, another point on the graph of $y = 5x - 4$.

Self Check 4   Use the slope and y-intercept to graph $y = 2x - 3$.

Now Try ▶ Problem 63

EXAMPLE 5   Use the slope and y-intercept to graph $4x + 3y = 6$.

Strategy   We will write the equation of the line in slope–intercept form, $y = mx + b$. Then we will identify the slope and y-intercept of its graph.

Why   We can use that information to plot two points that the line passes through.

Solution   To write $4x + 3y = 6$ in slope–intercept form, we isolate y on the left side.

$$4x + 3y = 6 \qquad \text{This is the given equation.}$$

$$3y = -4x + 6 \qquad \text{To isolate the term 3y on the left side, subtract 4x from both sides. Write the subtraction before the constant term 6.}$$

$$\frac{3y}{3} = \frac{-4x}{3} + \frac{6}{3} \qquad \text{To isolate y, undo the multiplication by 3 by dividing both sides by 3, term-by-term.}$$

$$y = -\frac{4}{3}x + 2 \qquad m = -\frac{4}{3} \text{ and } b = 2.$$

The slope of the line is $-\frac{4}{3}$ and the y-intercept is $(0, 2)$. To draw the graph, we begin by plotting the y-intercept. If we write the slope as $\frac{-4}{3}$, the rise is -4 and the run is 3. From $(0, 2)$, we then move 4 units *downward* (because the numerator is negative) and 3 units to the *right* (because the denominator is positive). This locates a second point on the line, $(3, -2)$.

Since $\frac{-4}{3} = \frac{4}{-3}$, we can find another point on the graph by writing the slope as $\frac{4}{-3}$. In this case, the rise is 4 and the run is -3. Again, we begin at the y-intercept $(0, 2)$, but this time, we move 4 units *upward* because the rise is positive. Then we move 3 units to the *left*, because the run is negative, and arrive at the point $(-3, 6)$. The line that passes through $(0, 2)$, $(3, -2)$, and $(-3, 6)$ is the graph of $4x + 3y = 6$.

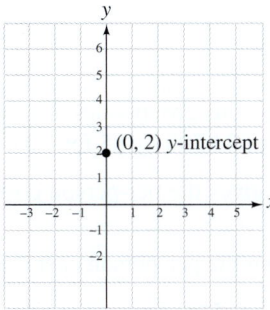
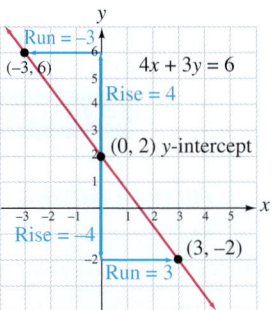

Plot the y-intercept.

From $(0, 2)$, draw the rise and run parts of the slope triangle for $m = \frac{-4}{3}$ $\left(\text{or } m = \frac{4}{-3}\right)$ to find another point on the line.

Use a straightedge to draw a line through the points.

To check the graph, verify that $(3, -2)$ and $(-3, 6)$ satisfy $4x + 3y = 6$.

Self Check 5 Use the slope and y-intercept to graph $5x + 6y = 12$.

Now Try Problem 71

5 Recognize Parallel and Perpendicular Lines.

The slope–intercept form enables us to quickly identify parallel and perpendicular lines.

EXAMPLE 6 Are the graphs of $y = -5x + 6$ and $x - 5y = -10$ parallel, perpendicular, or neither?

Strategy We will find the slope of each line and then compare the slopes.

Why If the slopes are equal, the lines are parallel. If the slopes are negative reciprocals, the lines are perpendicular. Otherwise, the lines are neither parallel nor perpendicular.

Solution The graph of $y = -5x + 6$ is a line with slope -5. To find the slope of the graph of $x - 5y = -10$, we will write the equation in slope–intercept form.

Success Tip

Graphs are not necessary to determine if two lines are parallel, perpendicular, or neither. We simply examine the slopes of the lines.

$$x - 5y = -10 \qquad \text{This is the second given equation.}$$
$$-5y = -x - 10 \qquad \text{To isolate the term } -5y \text{ on the left side, subtract } x \text{ from both sides.}$$
$$\frac{-5y}{-5} = \frac{-x}{-5} - \frac{10}{-5} \qquad \text{To isolate } y, \text{ undo the multiplication by } -5 \text{ by dividing both sides by } -5 \text{ term-by-term.}$$
$$y = \frac{x}{5} + 2 \qquad m = \tfrac{1}{5} \text{ because } \tfrac{x}{5} = \tfrac{1}{5}x.$$

The graph of $y = \frac{x}{5} + 2$ is a line with slope $\frac{1}{5}$. Since the slopes -5 and $\frac{1}{5}$ are negative reciprocals, the lines are perpendicular.

Self Check 6 Determine whether the graphs of $y = 4x + 6$ and $x - 4y = -8$ are parallel, perpendicular, or neither.

Now Try Problem 81

6 Use Slope–Intercept Form to Write an Equation to Model Data.

The concepts that we have studied in the first four sections of this chapter can be used to write equations that mathematically describe, or **model,** many real-world situations. To make the equation more descriptive of the given situation, we can replace the variables x and y in $y = mx + b$ with other letters.

EXAMPLE 7

Cruise to Alaska

$4,500 per person

Group discounts available*

*For groups of up to 100

Group Discounts.   To promote group sales for an Alaskan cruise, a travel agency reduces the regular ticket price of $4,500 by $5 for each person traveling in the group.

a. Write a linear equation that gives the per-person cost c of the cruise, if p people travel together.

b. Use the equation to determine the per-person cost if 55 teachers travel together.

Strategy   We will determine the slope and the y-intercept of the graph of the equation from the given facts about the cruise.

Why   If we know the slope and y-intercept, we can use the slope–intercept form, $y = mx + b$, to write an equation to model the situation.

Solution   **a.** We will let p represent the number of people traveling in the group and c represent the per-person cost of the cruise. Since the cost depends on the number of people in the group, the linear equation that models this situation is

$$c = mp + b$$ This is the slope-intercept form $y = mx + b$ with the variable c in place of y and the variable p in place of x.

Since the per-person cost of the cruise steadily decreases as the number of people in the group increases, the rate of change of $-\$5$ per person is the slope of the graph of the equation. Thus, m is -5.

The Language of Algebra

To determine the slope m for modeling problems like this, look for a phrase that describes a **rate of change,** such as:

■ $5 for each person
■ $20 per unit
■ 5° every minute
■ 12 feet a year

If 0 people take the cruise, there will be no discount and the per-person cost of the cruise will be $4,500. Written as an ordered pair of the form (p, c), we have $(0, 4,500)$. When graphed, this point would be the c-intercept. Thus, b is 4,500.

Substituting for m and b in the slope–intercept form $c = mp + b$, we obtain the linear equation that models the pricing arrangement.

A graph of the equation for groups of up to 100 $(c \le 100)$ is shown on the right.

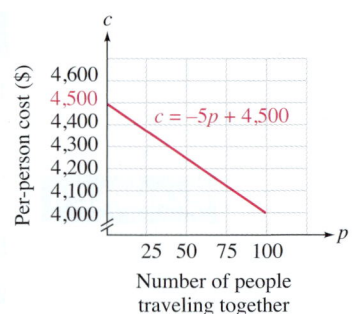

$$c = -5p + 4,500 \quad m = -5 \text{ and } b = 4,500.$$

b. To find the per-person cost of the cruise for a group of 55 people, we substitute 55 for p and evaluate the right side of the equation.

$$c = -5\boldsymbol{p} + 4,500$$
$$c = -5(\boldsymbol{55}) + 4,500 \quad \text{Substitute 55 for } p.$$
$$c = -275 + 4,500 \quad \text{Do the multiplication.}$$
$$c = 4,225 \quad \text{Do the addition.}$$

If a group of 55 people travel together, the cruise will cost each person $4,225.

Self Check 7   **Group Discounts.** Write a linear equation in slope–intercept form that gives the cost of the cruise of Example 7 if a $10.50-per-person discount is offered for groups.

Now Try ▶   Problem 91

SECTION 3.5 ▶ STUDY SET

VOCABULARY

Fill in the blanks.

1. The equation $y = mx + b$ is called the _____ form of the equation of a line.

2. The graph of the linear equation $y = mx + b$ has _____ $(0, b)$ and _____ m.

CONCEPTS

3. Determine whether each equation is in slope–intercept form.
   a. $7x + 4y = 2$ b. $5y = 2x - 3$
   c. $y = 6x + 1$ d. $x = 4y - 8$

4. a. Fill in the blank: To write a linear equation in two variables in slope–intercept form, solve the equation for ▢.
   b. Solve $4x + y = 9$ for y.

5. Simplify the right side of each equation.
   a. $y = \dfrac{4x}{2} + \dfrac{16}{2}$ b. $y = \dfrac{15x}{-3} + \dfrac{9}{-3}$

   c. $y = \dfrac{2x}{6} - \dfrac{6}{6}$ d. $y = \dfrac{-9x}{-5} - \dfrac{20}{-5}$

6. Find the slope and y-intercept of each line graphed below. Then use that information to write an equation for that line.

   a. 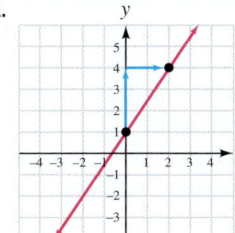   b.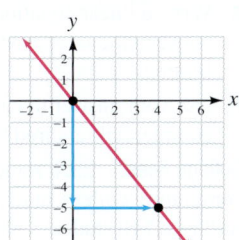

NOTATION

Complete the solution by solving the equation for y. Then find the slope and the y-intercept of its graph.

7. $2x + 5y = 15$

$2x + 5y - 2x = \boxed{} + 15$

$\boxed{} = -2x + 15$

$\dfrac{5y}{\boxed{}} = \dfrac{-2x}{\boxed{}} + \dfrac{15}{\boxed{}}$

$y = \boxed{}\,x + \boxed{}$

The slope is $\boxed{}$ and the y-intercept is $(\boxed{}, \boxed{})$.

8. What is the slope–intercept form of the equation of a line?

9. Fill in the blanks: $-\dfrac{3}{2} = \dfrac{3}{\boxed{}} = \dfrac{\boxed{}}{2}$

10. Determine whether each statement is true or false.
    a. $\dfrac{x}{6} = \dfrac{1}{6}x$ b. $\dfrac{5}{3}x = \dfrac{5x}{3}$

GUIDED PRACTICE

Find the slope and the y-intercept of the line with the given equation. See Example 1.

11. $y = 4x + 2$ 12. $y = 7x + 3$
13. $y = -5x - 8$ 14. $y = -4x - 2$
15. $y = 25x - 9$ 16. $y = 6x - 1$
17. $y = 11 - x$ 18. $y = 12 - 4x$
19. $y = \dfrac{1}{2}x + 6$ 20. $y = \dfrac{4}{5}x - 9$
21. $y = \dfrac{x}{4} - \dfrac{1}{2}$ 22. $y = \dfrac{x}{15} - \dfrac{3}{4}$
23. $y = -5x$ 24. $y = 14x$
25. $y = x$ 26. $y = -x$
27. $y = -2$ 28. $y = 30$
29. $-5y - 2 = 0$ 30. $3y - 13 = 0$
31. $x + y = 8$ 32. $x - y = -30$
33. $6y = x - 6$ 34. $2y = x + 20$
35. $-4y = 6x - 4$ 36. $-6y = 8x + 6$

37. $2x + 3y = 6$ 38. $4x + 5y = 25$

39. $3x - 5y = 15$ 40. $x - 6y = 6$

41. $-6x + 6y = -11$ 42. $-4x + 4y = -9$

Write an equation of the line with the given slope and y-intercept and graph it. See Example 2.

43. Slope 5, y-intercept $(0, -3)$ 44. Slope -2, y-intercept $(0, 1)$

45. Slope -3, y-intercept $(0, 6)$ 46. Slope 4, y-intercept $(0, -1)$

47. Slope $\dfrac{1}{4}$, y-intercept $(0, -2)$ 48. Slope $\dfrac{1}{3}$, y-intercept $(0, -5)$

49. Slope $-\dfrac{8}{3}$, y-intercept $(0, 5)$ 50. Slope $-\dfrac{7}{6}$, y-intercept $(0, 2)$

51. Slope $\dfrac{6}{5}$, y-intercept $(0, 0)$ 52. Slope $\dfrac{5}{4}$, y-intercept $(0, 0)$

53. Slope -2, y-intercept $\left(0, \dfrac{1}{2}\right)$ 54. Slope -3, y-intercept $\left(0, -\dfrac{1}{2}\right)$

Write an equation for each line shown below.
See Example 3.

55.

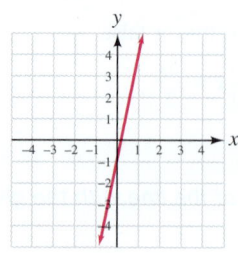

56.

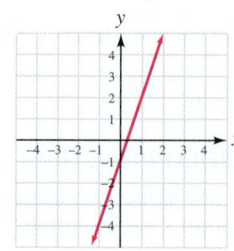

57.

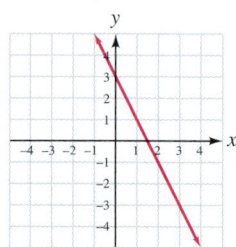

58.

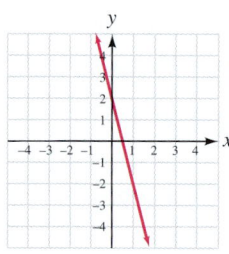

59.

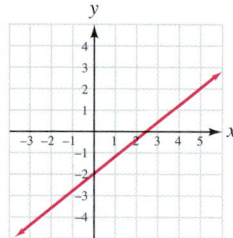

60.

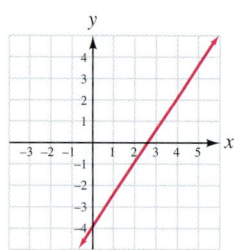

61.

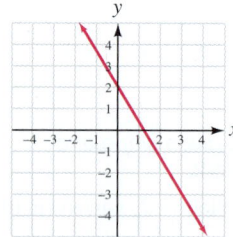

62.

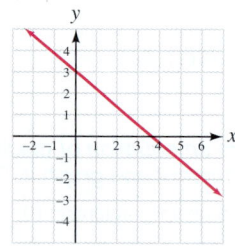

Find the slope and the y-intercept of the graph of each
equation and graph it. See Examples 4 and 5.

63. $y = 3x + 3$

64. $y = -3x + 5$

65. $y = \dfrac{1}{2}x + 2$

66. $y = \dfrac{x}{3}$

67. $y = -3x$

68. $y = -4x$

69. $4x + y = -4$

70. $2x + y = -6$

71. $3x + 4y = 16$

72. $2x + 3y = 9$

73. $10x - 5y = 5$

74. $4x - 2y = 6$

For each pair of equations, determine whether their
graphs are parallel, perpendicular, or neither.
See Example 6.

75. $y = 6x + 8$
$y = 6x$

76. $y = -9x - 3$
$y = -9x$

77. $y = x$
$y = -x$

78. $y = 3x$
$y = 4x$

79. $y = \dfrac{1}{2}x - \dfrac{4}{5}$
$y = 0.5x + 3$

80. $y = 3x - 15$
$y = -\dfrac{1}{3}x + 4$

81. $y = -2x - 9$
$2x - y = 9$

82. $y = \dfrac{3}{4}x + 1$
$4x - 3y = 15$

83. $3x = 5y - 10$
$5x = 1 - 3y$

84. $-2y = 2 - x$
$2x - 3 = 4y$

85. $x - y = 12$
$-2x + 2y = -23$

86. $y = -3x + 1$
$3y = x - 5$

87. $x = 9$
$y = 8$

88. $-x + 4y = 10$
$2y + 16 = -8x$

89. $-4x + 3y = -12$
$8x + 6y = 54$

90. $-5x + 2y = -8$
$15x + 6y = -48$

APPLICATIONS

91. Production Costs. A television production company charges a basic fee of $5,000 and then $2,000 an hour when filming a commercial.

   a. Write a linear equation that describes the relationship between the total production costs c and the hours h of filming.

   b. Use your answer to part a to find the production costs if a commercial required 8 hours of filming.

92. College Fees. Each semester, students enrolling at a community college must pay tuition costs of $20 per unit as well as a $40 student services fee.

   a. Write a linear equation that gives the total fees t to be paid by a student enrolling at the college and taking x units.

   b. Use your answer to part a to find the enrollment cost for a student taking 12 units.

93. Chemistry. A portion of a student's chemistry lab manual is shown on the right. Use the information to write a linear equation relating the temperature F in degrees Fahrenheit of the compound to the time t (in minutes) elapsed during the lab procedure.

> Chem. Lab #1   Aug. 13
> **Step 1:** Removed compound from freezer @ –10° F.
>
> **Step 2:** Used heating unit to raise temperature of compound 5° F every minute.

94. Rentals. Use the information in the newspaper advertisement to write a linear equation that gives the amount of income A (in dollars) the apartment owner will receive when the unit is rented for m months.

> **APARTMENT FOR RENT**
> 1 bedroom/1 bath, with garage
> $500 per month +
> $250 nonrefundable one-time security fee.

95. Employment Services. A policy statement of LIZCO, Inc., is shown below. Suppose a secretary had to pay an employment service $500 to get placed in a new job at LIZCO. Write a linear equation that tells the secretary the actual cost c of the employment service to her m months after being hired.

> **Policy no. 23452**—A new hire will be reimbursed by LIZCO for any employment service fees paid by the employee at the rate of $20 per month.

96. Printing Presses. Every three minutes, 100 feet of paper is used off of an 8,000 foot-roll to print the pages of a magazine. Write a linear equation that relates the number of feet of paper that remain on the roll and the number of minutes the printing press has been operating.

97. Sewing Costs. A tailor charges a basic fee of $20 plus $5 per letter to sew an athlete's name on the back of a jacket. Write a linear equation that will find the cost to have a name containing x letters sewn on the back of a jacket.

98. Salad Bars. For lunch, a delicatessen offers a "Salad and Soda" special where customers serve themselves at a well-stocked salad bar. The cost is $2.00 for the drink and 42¢ an ounce for the salad. Write a linear equation that will find the cost of a "Salad and Soda" lunch when a salad weighing x ounces is purchased.

99. iPads. When a student purchased an Apple iPad with Wi-Fi + 3G for $629.99, he also enrolled in a 250 MB data plan that cost $14.95 per month.

   a. Write a linear equation that gives the cost for him to purchase and use the iPad for m months.

   b. Use your answer to part a to find the cost to purchase and use the iPad for 2 years.

100. Online Games. A new Playstation 3 costs $310.50 and membership in an online videogame multiplayer network cost $18.49 per month.

   a. Write a linear equation that gives the cost for someone to buy the machine and belong to the online network for m months.

   b. Use your answer to part a to find the cost to buy the machine and belong to the network for 3 years.

101. Navigation. The graph below shows the recommended speed at which a ship should proceed into head waves of various heights.

   a. What information does the y-intercept of the line give?

   b. What is the rate of change in the recommended speed of the ship as the wave height increases?

   c. Write an equation of the line.

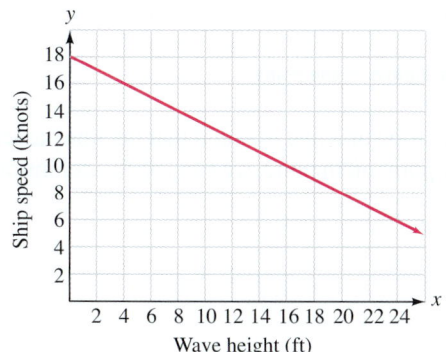

102. Debt. The graph below estimates the amount of credit card debt per U.S. household for the years 1990–2010.

   a. What information does the d-intercept of the line give?

   b. What was the rate of change in the amount of debt per household?

   c. Write an equation of the line, where d is the approximate credit card debt and t is the number of years since 1990.

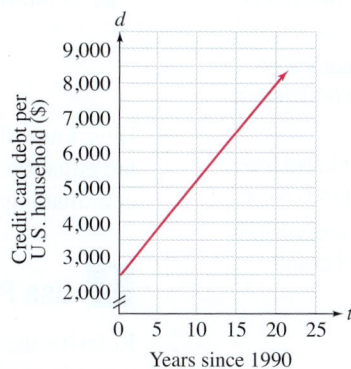

Source: mybudget 360.com

WRITING

103. Why is $y = mx + b$ called the slope–intercept form of the equation of a line?

104. On a quiz, a student was asked to find the slope of the graph of $y = 2x + 3$. She answered: $m = 2x$. Her instructor marked it wrong. Explain why the answer is incorrect.

REVIEW

105. Cable TV. A 186-foot television cable is to be cut into four pieces. Find the length of each piece if each successive piece is 3 feet longer than the previous one.

106. Investments. Joni received $25,000 as part of a settlement in a class action lawsuit. She invested some money at 10% and the rest at a 9% simple interest rate. If her total annual income from these two investments was $2,430, how much did she invest at each rate?

CHALLENGE PROBLEMS

107. If the graph of $y = mx + b$ passes through quadrants I, II, and IV, what can be known about the constants m and b?

108. Which of the following equations has the steeper graph, $103x - 200y = -400$ or $17x - 33y = -66$?

SECTION 3.6

OBJECTIVES

1 Use point–slope form to write an equation of a line.

2 Write an equation of a line given two points on the line.

3 Write equations of horizontal and vertical lines.

4 Use a point and the slope to graph a line.

5 Write linear equations that model data.

Point–Slope Form

ARE YOU READY?

The following problems review some basic skills that are needed when working with equations of lines in point–slope form.

1. Find the slope of the line that passes through $(-2, 0)$ and $(-12, -8)$.

2. Simplify: $x - (-5)$

3. Solve $y + 2 = 6(x - 7)$ for y.

4. Add: $-\dfrac{3}{4} + 8$

If we know the slope of a line and its y-intercept, we can use the slope–intercept form to write the equation of the line. The question that now arises is, can *any* point on the line be used in combination with its slope to write its equation? In this section, we answer this question.

1 Use Point–Slope Form to Write an Equation of a Line.

Refer to the line graphed on the left, with slope 3 and passing through the point $(2,1)$. To develop a new form for the equation of a line, we will find the slope of this line in another way.

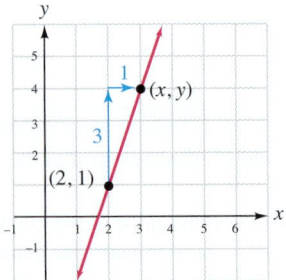

If we pick another point on the line with coordinates (x, y), we can find the slope of the line by substituting the coordinates of the points (x, y) and $(2, 1)$ into the slope formula.

$$\frac{y_2 - y_1}{x_2 - x_1} = m$$

$$\frac{y - 1}{x - 2} = m \qquad \text{Let } (x_1, y_1) \text{ be } (2, 1) \text{ and } (x_2, y_2) \text{ be } (x, y).$$
$$\text{Substitute } y \text{ for } y_2, 1 \text{ for } y_1, x \text{ for } x_2, \text{ and } 2 \text{ for } x_1.$$

Since the slope of the line is 3, we can substitute 3 for m in the previous equation.

$$\frac{y - 1}{x - 2} = 3$$

We then multiply both sides by $x - 2$ to clear the equation of the fraction.

$$\frac{y - 1}{x - 2}(x - 2) = 3(x - 2)$$

$$y - 1 = 3(x - 2) \qquad \text{Simplify the left side. Remove the common factor } x - 2$$
$$\text{in the numerator and denominator: } \frac{y-1}{x-2} \cdot \frac{x-2}{1}.$$

The resulting equation displays the slope of the line and the coordinates of one point on the line:

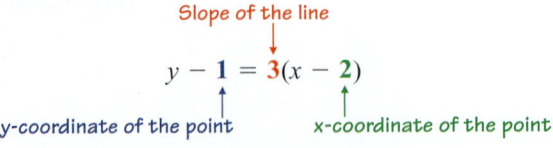

In general, suppose we know that the slope of a line is m and that the line passes through the point (x_1, y_1). Then if (x, y) is any other point on the line, we can use the definition of slope to write

$$\frac{y - y_1}{x - x_1} = m$$

If we multiply both sides by $x - x_1$ to clear the equation of the fraction, we have

$$y - y_1 = m(x - x_1)$$

This form of a linear equation is called **point–slope form.** It can be used to write the equation of a line when the slope and one point on the line are known.

Point–Slope Form of the Equation of a Line	If a line with slope m passes through the point (x_1, y_1), the equation of the line is $$y - y_1 = m(x - x_1)$$ Read as "y minus y sub 1 equals m times the quantity x minus x sub 1."

EXAMPLE 1

Find an equation of a line that has slope -8 and passes through $(-1, 5)$. Write the answer in slope–intercept form.

Strategy Although the problem asks for an answer in slope–intercept form, we will begin with the point–slope form to write an equation of the line.

Why We know the slope of the line and the coordinates of a point that it passes through.

Solution

The given point is $(-1, 5)$, so $x_1 = -1$ and $y_1 = 5$. The given slope is -8, so $m = -8$. We will substitute these values into the point–slope form and simplify the right side of the equation.

Caution

When using the point–slope form, never substitute values for x or y.

$$y - y_1 = m(x - x_1)$$

Only substitute values for x_1, y_1, and m.

$$y - y_1 = m(x - x_1) \quad \text{This is the point–slope form.}$$
$$y - 5 = -8[x - (-1)] \quad \text{Substitute } -8 \text{ for } m, -1 \text{ for } x_1, \text{ and } 5 \text{ for } y_1.$$
$$\text{Brackets are used to enclose } x - (-1).$$
$$y - 5 = -8(x + 1) \quad \text{Simplify the expression within the brackets.}$$

To write an equivalent equation in slope–intercept form, we solve for y.

$$y - 5 = -8(x + 1) \quad \text{This is the simplified point–slope form.}$$
$$y - 5 = -8x - 8 \quad \text{Distribute the multiplication by } -8.$$
$$y - 5 + 5 = -8x - 8 + 5 \quad \text{To isolate } y, \text{ undo the subtraction of 5 by adding 5 to both sides.}$$
$$y = -8x - 3 \quad \text{This is the requested slope–intercept form.}$$

In slope–intercept form, the equation is $y = -8x - 3$.

To verify this result, we note that $m = -8$. Therefore, the slope of the line is -8, as required. To see whether the line passes through $(-1, 5)$, we substitute -1 for x and 5 for y in the equation. If this point is on the line, a true statement should result.

$$y = -8x - 3$$
$$5 \stackrel{?}{=} -8(-1) - 3$$
$$5 \stackrel{?}{=} 8 - 3$$
$$5 = 5 \quad \text{True}$$

Self Check 1 Find an equation of the line that has slope -2 and passes through $(4, -3)$. Write the answer in slope–intercept form.

Now Try ▶ Problems 13 and 19

2 Write an Equation of a Line Given Two Points on the Line.

In the next example, we show that it is possible to write the equation of a line when we know the coordinates of two points on the line.

EXAMPLE 2   Find an equation of the line that passes through $(-2, 6)$ and $(4, 7)$. Write the equation in slope–intercept form.

Strategy   We will use the point–slope form, $y - y_1 = m(x - x_1)$, to write an equation of the line.

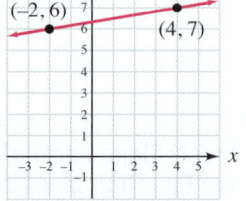

Why   We know the coordinates of a point that the line passes through and we can calculate the slope of the line using the slope formula.

Solution   To find the slope of the line, we use the slope formula.

> **Success Tip**
>
> In Example 2, either of the given points can be used as (x_1, y_1) when writing the point–slope equation. The results will be the same.
>
> Looking ahead, we usually choose the point whose coordinates will make the calculations the easiest.

$$m = \frac{y_2 - y_1}{x_2 - x_1} = \frac{7 - 6}{4 - (-2)} = \frac{1}{6}$$ Substitute 7 for y_2, 6 for y_1, 4 for x_2, and -2 for x_1.

Either point on the line can serve as (x_1, y_1). If we use $(4, 7)$, we have

$$y - y_1 = m(x - x_1)$$ This is the point–slope form.

$$y - 7 = \frac{1}{6}(x - 4)$$ Substitute $\frac{1}{6}$ for m, 7 for y_1, and 4 for x_1.

To write an equivalent equation in slope–intercept form, we solve for y.

$$y - 7 = \frac{1}{6}x - \frac{2}{3}$$ Distribute the multiplication by $\frac{1}{6}$: $\frac{1}{6}(-4) = -\frac{4}{6} = -\frac{2}{3}$.

$$y - 7 + 7 = \frac{1}{6}x - \frac{2}{3} + 7$$ To isolate y, add 7 to both sides.

> **Success Tip**
>
> To check this result, verify that $(-2, 6)$ and $(4, 7)$ satisfy $y = \frac{1}{6}x + \frac{19}{3}$ using substitution.

$$y = \frac{1}{6}x - \frac{2}{3} + \frac{21}{3}$$ Simplify the left side. On the right side, express 7 as $\frac{21}{3}$ to prepare to add the fractions with the common denominator 3.

$$y = \frac{1}{6}x + \frac{19}{3}$$ Add the fractions: $-\frac{2}{3} + \frac{21}{3} = \frac{19}{3}$. This is slope–intercept form.

An equation of the line that passes through $(-2, 6)$ and $(4, 7)$ is $y = \frac{1}{6}x + \frac{19}{3}$.

Self Check 2   Find an equation of the line that passes through $(-5, 4)$ and $(8, -6)$. Write the equation in slope–intercept form.

Now Try ▶   Problem 29

3 Write Equations of Horizontal and Vertical Lines.

We have previously graphed horizontal and vertical lines. We will now discuss how to write their equations.

EXAMPLE 3   Write an equation of each line and graph it.   **a.** A horizontal line that passes through $(-2, -4)$
b. A vertical line that passes through $(1, 3)$

Strategy   We will use the appropriate form, either $y = b$ or $x = a$, to write an equation of each line.

Why   These are the standard forms for the equations of a horizontal and a vertical line.

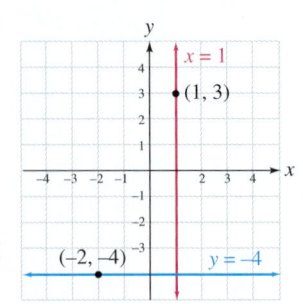

Solution   **a.** The equation of a horizontal line can be written in the form $y = b$. Since the y-coordinate of $(-2, -4)$ is -4, the equation of the line is $y = -4$. The graph is shown in the figure.

b. The equation of a vertical line can be written in the form $x = a$. Since the x-coordinate of $(1, 3)$ is 1, the equation of the line is $x = 1$. The graph is shown in the figure.

Self Check 3 Write an equation of each line. **a.** A horizontal line that passes through (3, 2)   **b.** A vertical line passing through (−1, −3)

Now Try ▶ Problems 41 and 43

4 Use a Point and the Slope to Graph a Line.

If we know the coordinates of a point on a line, and if we know the slope of the line, we can use the slope to determine a second point on the line.

EXAMPLE 4 Graph the line with slope $\frac{2}{5}$ that passes through (−1, −3).

Strategy First, we will plot the given point (−1, −3). Then we will use the slope to find a second point that the line passes through.

Why Once we determine two points that the line passes through, we can draw the graph of the line.

Solution To draw the graph, we begin by plotting the point (−1, −3). From there, we move 2 units up (rise) and then 5 units to the right (run), since the slope is $\frac{2}{5}$. This locates a second point on the line, (4, −1). We then draw a straight line through the two points.

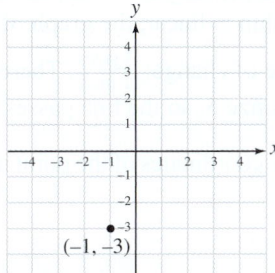
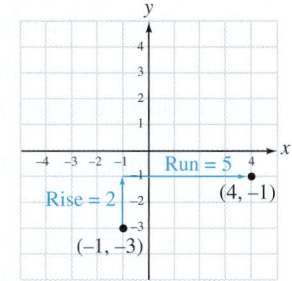
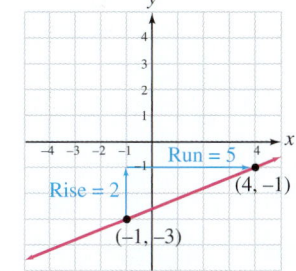

Plot the given point (−1, −3).

From (−1, −3) draw the rise and run parts of the slope triangle for $m = \frac{2}{5}$ to find another point on the line.

Use a straightedge to draw a line through the points.

Self Check 4 Graph the line with slope −4 that passes through (−4, 2).

Now Try ▶ Problem 45

The following table summarizes what you should know about each form of an equation of a line.

Form	Example	Comments
Standard form $Ax + By = C$	$2x + 5y = 9$	■ To graph, find the x- and y-intercepts by letting $y = 0$ and finding x, and letting $x = 0$ and finding y. Also find a checkpoint.
Slope–intercept form $y = mx + b$	$y = \frac{5}{3}x + 4$	■ To graph, plot the y-intercept $(0, b)$. From there, draw a slope triangle using the rise and run to locate another point. ■ Use this form to write a line's equation if you know its slope and y-intercept.
Point–slope form $y - y_1 = m(x - x_1)$	$y - 1 = 6(x - 8)$	■ To graph, plot the point (x_1, y_1). From there, draw a slope triangle using the rise and run of the slope m to locate another point. ■ Use this form to write a line's equation if you know a point on the line and the slope. If two points on the line are known, find the slope and use it with one of the points to write the equation.
Horizontal line $y = b$	$y = 7$	■ To graph, draw a horizontal line ($m = 0$) with y-intercept $(0, b)$.
Vertical line $x = a$	$x = -1$	■ To graph, draw a vertical line (undefined slope) with x-intercept $(a, 0)$.

5 Write Linear Equations That Model Data.

In the next two examples, we will see how the point–slope form can be used to write linear equations that model certain real–world situations.

EXAMPLE 5

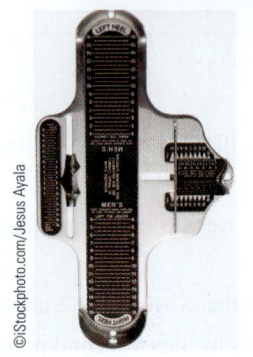

©iStockphoto.com/Jesus Ayala

Men's Shoe Sizes.   The length (in inches) of a man's foot is not his shoe size. For example, the smallest adult men's shoe size is 5, and it fits a 9-inch-long foot. There is, however, a linear relationship between the two. It can be stated this way: Shoe size increases by 3 sizes for each 1-inch increase in foot length.

a. Write a linear equation that relates shoe size s to foot length L.

b. Shaquille O'Neal, a famous basketball player, has a foot that is about 14.6 inches long. Find his shoe size.

Strategy   We will first find the slope of the line that describes the linear relationship between shoe size and the length of a foot. Then we will determine the coordinates of a point on that line.

Why   Once we know the slope and the coordinates of one point on the line, we can use the point–slope form to write the equation of the line.

Solution

a. Since shoe size s depends on the length L of the foot, ordered pairs have the form (L, s). Because the relationship is linear, the graph of the desired equation is a line.

- The line's slope is the rate of change: $\frac{3 \text{ sizes}}{1 \text{ inch}}$. Therefore, $m = 3$.
- A 9-inch-long foot wears size 5, so the line passes through $(9, 5)$.

We substitute 3 for m and the coordinates of the point into the point–slope form and solve for s.

$s - s_1 = m(L - L_1)$	This is the point–slope form using the variables L and s.
$s - 5 = 3(L - 9)$	Substitute 3 for m, 9 for L_1, and 5 for s_1.
$s - 5 = 3L - 27$	Distribute the multiplication by 3.
$s = 3L - 22$	To isolate s, add 5 to both sides: $-27 + 5 = -22$.

The equation relating men's shoe size and foot length is $s = 3L - 22$.

b. To find Shaquille's shoe size, we substitute 14.6 inches for L in the equation.

$s = 3L - 22$	
$s = 3(14.6) - 22$	
$s = 43.8 - 22$	Do the multiplication.
$s = 21.8$	Do the subtraction.

Since men's shoes come in only full- and half-sizes, we round 21.8 up to 22. Shaquille O'Neal wears size 22 shoes.

Self Check 5   **Comparing Temperature Scales.**   Celsius and Fahrenheit measures of temperature are not the same. There is, however, a linear relationship between the two. Degrees Fahrenheit increase by 9° for each 5° increase in Celsius. If a 212° Fahrenheit temperature measure is the same as 100° Celsius, write a linear equation that relates Fahrenheit measure to Celsius measure.

Now Try ▶ Problem 77

| **EXAMPLE 6** | **Studying Learning.**   In a series of 40 trials, a rat was released in a maze to search for food. Researchers recorded the trial number and the time that it took the rat to complete the maze as ordered pairs on a **scatter diagram** shown below. All of the points fell on or near the line drawn in red. Write an equation of the line in slope–intercept form. |

The Language of Algebra

The term *scatter diagram* is somewhat misleading. Often, the data points are not scattered loosely about. In this case, they fall, more or less, along an imaginary straight line, indicating a linear relationship.

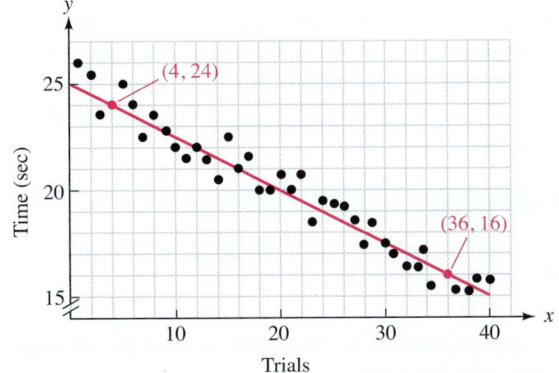

Strategy   From the graph, we will determine the coordinates of two points on the line.

Why   We can write an equation of a line when we know the coordinates of two points on the line. (See Example 2.)

Solution   To write a point–slope equation, we need to know the slope of the line. The line passes through several points; we will use (4, 24) and (36, 16) to find its slope.

$$m = \frac{y_2 - y_1}{x_2 - x_1} = \frac{16 - 24}{36 - 4} = \frac{-8}{32} = -\frac{1}{4}$$

Any point on the line can serve as (x_1, y_1). We will use (4, 24).

$$y - y_1 = m(x - x_1) \qquad \text{This is the point–slope form.}$$

$$y - 24 = -\frac{1}{4}(x - 4) \qquad \text{Substitute } -\tfrac{1}{4} \text{ for } m, 4 \text{ for } x_1, \text{ and } 24 \text{ for } y_1.$$

To write this equation in slope–intercept form, solve for y.

$$y - 24 = -\frac{1}{4}x + 1 \qquad \text{Distribute the multiplication by } -\tfrac{1}{4}: -\tfrac{1}{4}(-4) = 1.$$

$$y = -\frac{1}{4}x + 25 \qquad \text{To isolate } y, \text{ add 24 to both sides: } 1 + 24 = 25.$$

A linear equation that models the rat's performance on the maze is $y = -\frac{1}{4}x + 25$, where x is the number of the trial and y is the time it took, in seconds.

| **Self Check 6** | **Awards.**   Orders for awards to be given to math team members were placed on two separate occasions. The first order of 32 awards cost $172 and the second order of 5 awards cost $37. Write a linear equation that gives the cost for an order of any number of awards. |

Now Try ▶ Problem 85

SECTION 3.6 ▶ STUDY SET

VOCABULARY

Fill in the blanks.

1. $y - y_1 = m(x - x_1)$ is called the _____ form of the equation of a line. In words, we read this as y minus y _____ one equals m _____ the quantity of x _____ x sub _____.

2. $y = mx + b$ is called the _____ form of the equation of a line.

CONCEPTS

3. Determine in what form each equation is written.
   a. $y - 4 = 2(x - 5)$
   b. $y = 2x + 15$

4. What point does the graph of each equation pass through, and what is the line's slope?
   a. $y - 2 = 6(x - 7)$
   b. $y + 3 = -8(x + 1)$

5. Refer to the following graph of a line.
   a. What highlighted point does the line pass through?
   b. What is the slope of the line?
   c. Write an equation of the line in point–slope form.

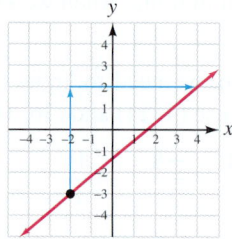

6. On a quiz, a student was asked to write the equation of a line with slope 4 that passes through $(-1, 3)$. Explain how the student can check her answer, $y = 4x + 7$.

7. Suppose you are asked to write an equation of the line in the scatter diagram below. What two points would you use to write the point–slope equation?

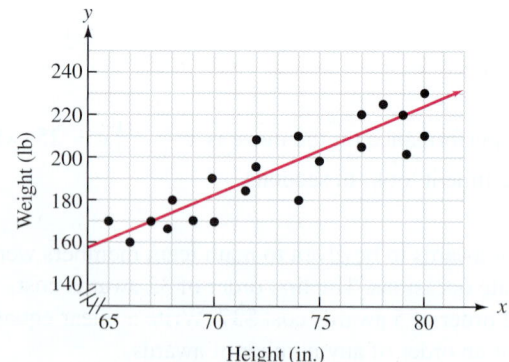

Height (in.)

8. In each case, a linear relationship between two quantities is described. If the relationship were graphed, what would be the slope of the line?
   a. The sales of new cars increased by 15 every 2 months.
   b. There were 35 fewer robberies for each dozen police officers added to the force.
   c. One acre of forest is being destroyed every 30 seconds.

NOTATION

Complete the solution.

9. Find an equation of the line with slope -2 that passes through the point $(-1, 5)$. Write the answer in slope–intercept form.

$$y - y_1 = m(x - x_1)$$
$$y - \boxed{} = -2[x - (\boxed{})]$$
$$y - 5 = -2[x \boxed{} 1]$$
$$y - 5 = -2x - \boxed{}$$
$$y = -2x + \boxed{}$$

10. What is the point–slope form of the equation of a line?

11. Consider the steps below and then fill in the blanks:

$$y - 3 = 2(x + 1)$$
$$y - 3 = 2x + 2$$
$$y = 2x + 5$$

The original equation was in _____ form. After solving for y, we obtain an equation in _____ form.

12. Fill in the blanks: The equation of a horizontal line has the form $\boxed{} = b$ and the equation of a vertical line has the form $\boxed{} = a$.

GUIDED PRACTICE

Use the point–slope form to write an equation of the line with the given slope and point. Leave the equation in that form. See Example 1.

13. Slope 3, passes through $(2, 1)$

14. Slope 2, passes through $(4, 3)$

15. Slope $\dfrac{4}{5}$, passes through $(-5, -1)$

16. Slope $\dfrac{7}{8}$, passes through $(-2, -9)$

Use the point–slope form to find an equation of the line with the given slope and point. Then write the equation in slope–intercept form. See Example 1.

17. Slope 2, passes through $(3, 5)$

18. Slope 8, passes through $(2, 6)$

19. Slope -5, passes through $(-9, 8)$

20. Slope -4, passes through $(-2, 10)$

21. Slope -3, passes through the origin

22. Slope -1, passes through the origin

23. Slope $\dfrac{1}{5}$, passes through $(10, 1)$

24. Slope $\dfrac{1}{4}$, passes through $(8, 1)$

25. Slope $-\dfrac{4}{3}$,

x	y
6	-4

26. Slope $-\dfrac{3}{2}$,

x	y
-2	1

27. Slope $-\dfrac{11}{6}$, passes through $(2, -6)$

28. Slope $-\dfrac{5}{4}$, passes through $(2, 0)$

Find an equation of the line that passes through the two given points. Write the equation in slope–intercept form, if possible. See Example 2.

29. Passes through $(1, 7)$ and $(-2, 1)$

30. Passes through $(-2, 2)$ and $(2, -8)$

31.

x	y
−4	3
2	0

32.

x	y
−1	−4
1	−2

33. Passes through $(5, 5)$ and $(7, 5)$

34. Passes through $(-2, 1)$ and $(-2, 15)$

35. Passes through $(5, 1)$ and $(-5, 0)$

36. Passes through $(-3, 0)$ and $(3, 1)$

37. Passes through $(-8, 2)$ and $(-8, 17)$

38. Passes through $\left(\dfrac{2}{3}, 2\right)$ and $(0, 2)$

39. Passes through $\left(\dfrac{2}{3}, \dfrac{1}{3}\right)$ and $(0, 0)$

40. Passes through $\left(\dfrac{1}{2}, \dfrac{3}{4}\right)$ and $(0, 0)$

Write an equation of each line. See Example 3.

41. Vertical, passes through $(4, 5)$

42. Vertical, passes through $(-2, -5)$

43. Horizontal, passes through $(4, 5)$

44. Horizontal, passes through $(-2, -5)$

Graph the line that passes through the given point and has the given slope. See Example 4.

45. $(1, -2)$, $m = -1$

46. $(-4, 1)$, $m = -3$

47. $(5, -3)$, $m = \dfrac{3}{4}$

48. $(2, -4)$, $m = \dfrac{2}{3}$

49. $(-2, -3)$, slope 2

50. $(-3, -3)$, slope 4

51. $(4, -3)$, slope $-\dfrac{7}{8}$

52. $(4, 2)$, slope $-\dfrac{1}{5}$

TRY IT YOURSELF

Use either the slope–intercept form (from Section 3.5) or the point–slope form (from Section 3.6) to find an equation of each line. Write each result in slope–intercept form, if possible.

53. Passes through $(5, 0)$ and $(-11, -4)$

54. Passes through $(7, -3)$ and $(-5, 1)$

55. Horizontal, passes through $(-8, 12)$

56. Horizontal, passes through $(9, -32)$

57. Slope $-\dfrac{1}{4}$, y-intercept $\left(0, \dfrac{7}{8}\right)$

58. Slope $-\dfrac{9}{5}$, y-intercept $\left(0, \dfrac{11}{3}\right)$

59. Slope $-\dfrac{2}{3}$, passes through $(3, 0)$

60. Slope $-\dfrac{2}{5}$, passes through $(15, 0)$

61. Slope 8, passes through $(2, 20)$

62. Slope 6, passes through $(1, -2)$

63. Vertical, passes through $(-3, 7)$

64. Vertical, passes through $(12, -23)$

65. Slope 7 and y-intercept $(0, -11)$

66. Slope 3 and y-intercept $(0, 4)$

67. Passes through $(-2, -1)$ and $(-1, -5)$

68. Passes through $(-3, 6)$ and $(-1, -4)$

69. x-intercept $(7, 0)$ and y-intercept $(0, -2)$

70. x-intercept $(-3, 0)$ and y-intercept $(0, 7)$

71. Slope $\dfrac{1}{10}$, passes through the origin

72. Slope $\dfrac{9}{8}$, passes through the origin

73. Undefined slope, passes through $\left(-\dfrac{1}{8}, 12\right)$

74. Undefined slope, passes through $\left(\dfrac{2}{5}, -\dfrac{5}{6}\right)$

75. Slope 1.7, y-intercept $(0, -2.8)$

76. Slope 9.5, y-intercept $(0, -14.3)$

APPLICATIONS

77. **Anatomy.** There is a linear relationship between a woman's height and the length of her radius bone. It can be stated this way: Height increases by 3.9 inches for each 1-inch increase in the length of the radius. Suppose a 64-inch-tall woman has a 9-inch-long radius bone. Use this information to find a linear equation that relates height h to the length r of the radius. Write the equation in slope–intercept form.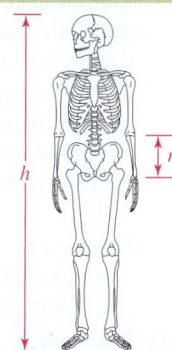

78. **Automation.** An automated production line uses distilled water at a rate of 300 gallons every 2 hours to make shampoo. After the line had run for 7 hours, planners noted that 2,500 gallons of distilled water remained in the storage tank. Find a linear equation relating the time t in hours since the production line began and the number g of gallons of distilled water in the storage tank. Write the equation in slope–intercept form.

79. **Pole Vaulting.** Find the equations of the lines that describe the positions of the pole for parts 1, 3, and 4 of the jump. Write the equations in slope–intercept form, if possible.

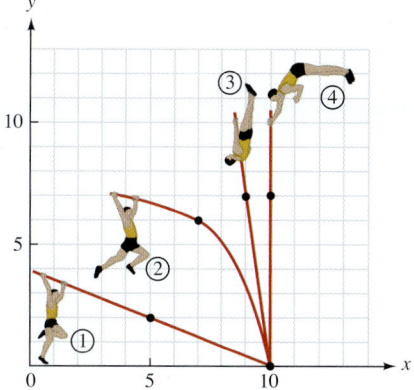

80. **Freeway Design.**   The graph below shows the route of a proposed freeway.

    a. Give the coordinates of the points where the proposed Freeway 133 will join Interstate 25 and Highway 40.

    b. Find an equation of the line that describes the route of the proposed freeway. Write the equation in slope–intercept form.

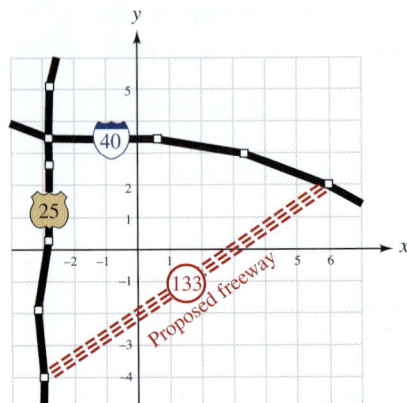

81. **Toxic Cleanup.**   Three months after cleanup began at a dump site, 800 cubic yards of toxic waste had yet to be removed. Two months later, that number had been lowered to 720 cubic yards.

    a. Find an equation that describes the linear relationship between the length of time m (in months) the cleanup crew has been working and the number of cubic yards y of toxic waste remaining. Write the equation in slope–intercept form.

    b. Use your answer to part (a) to predict the number of cubic yards of waste that will still be on the site one year after the cleanup project began.

82. **Depreciation.**   To lower its corporate income tax, accountants of a company depreciated a word processing system over several years using a linear model, as shown in the worksheet.

    a. Find a linear equation relating the years since the system was purchased, x, and its value, y, in dollars. Write the equation in slope–intercept form.

    b. Find the purchase price of the system.

Tax Worksheet Method of depreciation: *Linear*

Property	Years after purchase	Value
Word processing system	2	$60,000
	4	$30,000

83. **Trampolines.**   There is a linear relationship between the length of the protective pad that wraps around a trampoline and the radius of the trampoline. Use the data in the table to find an equation that gives the length l of pad needed for any trampoline with radius r. Write the equation in slope–intercept form. Use units of feet for both l and r.

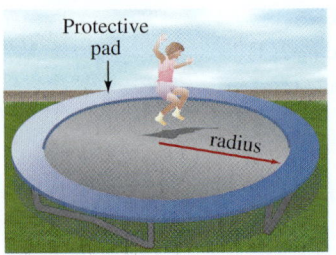

Radius	Pad length
3 ft	19 ft
7 ft	44 ft

84. **Raising a Family.**   In the report *"Expenditures on Children by Families,"* the U.S. Department of Agriculture projected the yearly child-rearing expenditures on children from birth through age 17. For a child born in 2010 to a two-parent middle-income family, the report estimated annual expenditures of $10,808 when the child is 6 years old, and $14,570 when the child is 15 years old.

    a. Write two ordered pairs of the form (child's age, annual expenditure).

    b. Assume the relationship between the child's age a and the annual expenditures E is linear. Use your answers to part (a) to write an equation in slope–intercept form that models this relationship.

    c. What are the projected child-rearing expenses when the child is 17 years old?

85. **Got Milk?**   The scatter diagram shows the amount of milk that an average American drank in one year for the years 1980–2008. A straight line can be used to model the data.

    a. Use the two highlighted points on the line to find its equation. Write the equation in slope–intercept form.

    b. Use your answer to part (a) to predict the amount of milk that an average American will drink in 2020.

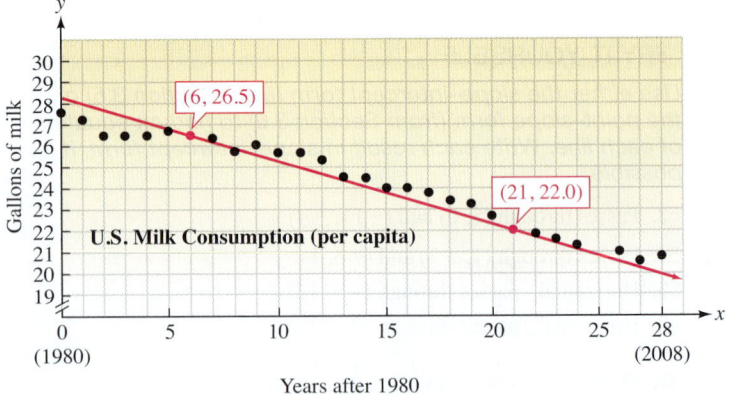

Source: United States Department of Agriculture.

86. **Engine Output.** The horsepower produced by an automobile engine was recorded for various engine speeds in the range of 2,400 to 4,800 revolutions per minute (rpm). The data were recorded on the following scatter diagram. Find an equation of the line that models the relationship between engine speed s and horsepower h. Write the equation in slope–intercept form.

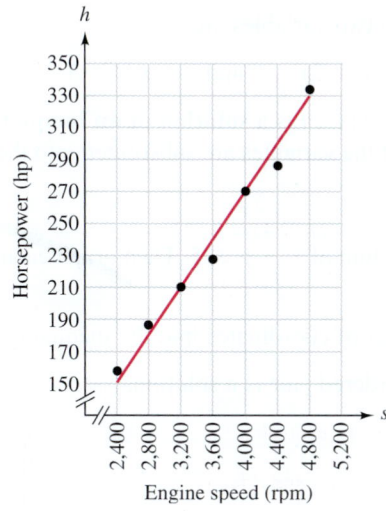

Engine speed (rpm)

WRITING

87. Why is $y - y_1 = m(x - x_1)$ called the point–slope form of the equation of a line?

88. If we know two points that a line passes through, we can write its equation. Explain how this is done.

89. Explain the steps involved in writing $y - 6 = 4(x - 1)$ in slope–intercept form.

90. Think of the points on the graph of the horizontal line $y = 4$. What do the points have in common? How do they differ?

REVIEW

91. **Frames.** The length of a rectangular picture is 5 inches greater than twice the width. If the perimeter is 112 inches, find the dimensions of the frame.

92. **Speed of an Airplane.** Two planes are 6,000 miles apart, and their speeds differ by 200 mph. They travel toward each other and meet in 5 hours. Find the speed of the slower plane.

CHALLENGE PROBLEMS

93. Find an equation of the line that passes through (2, 5) and is parallel to the line $y = 4x - 7$. Write the equation in slope–intercept form.

94. Find an equation of the line that passes through $(-6, 3)$ and is perpendicular to the line $y = -3x - 12$. Write the equation in slope–intercept form.

SECTION 3.7

Graphing Linear Inequalities

OBJECTIVES

1. Determine whether an ordered pair is a solution of an inequality.

2. Graph a linear inequality in two variables.

3. Graph inequalities with a boundary through the origin.

4. Solve applied problems involving linear inequalities in two variables.

Solution set for: $x + 6 < 8$

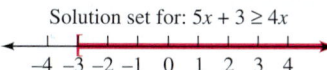

Solution set for: $5x + 3 \geq 4x$

ARE YOU READY?

The following problems review some basic skills that are needed when graphing linear inequalities.

1. True or false: $2(-3) + 1 > -4$

2. True or false: $8 \leq 8$

3. Graph: $2x - 3y = 6$

4. Determine whether each of the following points lies *above, below,* or *on* the line graphed in problem 3.

   a. $(2, -4)$ **b.** $(-3, -4)$ **c.** $(0, 0)$

Recall that an **inequality** is a statement that contains one of the symbols $<$, $\leq$, $>$, or $\geq$. Inequalities in one variable, such as $x + 6 < 8$ and $5x + 3 \geq 4x$, were solved in Section 2.7. Because they have an infinite number of solutions, we represented their solution sets graphically, by shading intervals on a number line. Two examples of this type of shading are shown in the left margin.

We now extend that concept to linear inequalities *in two variables,* as we introduce a procedure that is used to graph their solution sets.

1 Determine Whether an Ordered Pair Is a Solution of an Inequality.

If the $=$ symbol in a linear equation in two variables is replaced with an inequality symbol, we have a **linear inequality in two variables.**

Linear Inequalities	A **linear inequality in two variables** is an inequality that can be written in one of the forms
	$$Ax + By > C, \qquad Ax + By < C, \qquad Ax + By \geq C, \qquad \text{or} \qquad Ax + By \leq C$$
	where A, B, and C are real numbers and A and B are not both 0.

Some examples of linear inequalities in two variables are

$$x - y \leq 5, \qquad 4x + 3y < -6, \qquad y > 2x \qquad \text{and} \qquad x < -3$$

As with linear equations, an ordered pair (x, y) is a **solution of an inequality** in x and y if a true statement results when the values of the variables are substituted into the inequality.

EXAMPLE 1   Determine whether each ordered pair is a solution of $x - y \leq 5$. Then graph each solution:
a. $(4, 2)$ **b.** $(0, -6)$ **c.** $(1, -4)$

Strategy   We will substitute each ordered pair of coordinates into the inequality.

Why   If the resulting statement is true, the ordered pair is a solution.

Solution   **a.** For $(4, 2)$:

$$x - y \leq 5 \qquad \text{This is the given inequality.}$$

$$4 - 2 \overset{?}{\leq} 5 \qquad \text{Substitute 4 for x and 2 for y.}$$

$$2 \leq 5 \qquad \text{True}$$

Notation

The symbol $\overset{?}{\leq}$ is read as "is possibly less than or equal to."

Because $2 \leq 5$ is true, $(4, 2)$ is a solution of $x - y \leq 5$. We say that $(4, 2)$ *satisfies* the inequality. This solution is graphed as shown on the right.

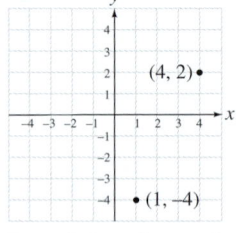

Two solutions of $x - y \leq 5$.

b. For $(0, -6)$:

$$x - y \leq 5 \qquad \text{This is the given inequality.}$$

$$0 - (-6) \overset{?}{\leq} 5 \qquad \text{Substitute 0 for x and -6 for y.}$$

$$6 \leq 5 \qquad \text{False}$$

Because $6 \leq 5$ is false, $(0, -6)$ is not a solution.

c. For $(1, -4)$:

$$x - y \leq 5 \qquad \text{This is the given inequality.}$$

$$1 - (-4) \overset{?}{\leq} 5 \qquad \text{Substitute 1 for x and -4 for y.}$$

$$5 \leq 5 \qquad \text{True}$$

Because $5 \leq 5$ is true, $(1, -4)$ is a solution, and we graph it as shown.

Self Check 1   Using the inequality in Example 1, determine whether each ordered pair is a solution: **a.** $(8, 2)$ **b.** $(4, -1)$
c. $(-2, 4)$ **d.** $(-3, -5)$

Now Try ▶ Problem 19

In Example 1, we graphed two of the solutions of $x - y \leq 5$. Since there are infinitely more ordered pairs (x, y) that make the inequality true, it would not be reasonable to plot all of them. Fortunately, there is an easier way to show all of the solutions.

2   Graph a Linear Inequality in Two Variables.

The graph of a linear inequality is a picture that represents the set of all points whose coordinates satisfy the inequality. In general, such graphs are regions bounded by a line. We call those regions **half-planes,** and we use a two-step procedure to find them.

EXAMPLE 2 Graph: $x - y \leq 5$

Strategy We will graph the **related equation** $x - y = 5$ to establish a boundary line between two regions of the coordinate plane. Then we will determine which region contains points whose coordinates satisfy the given inequality.

Why The graph of a linear inequality in two variables is a region of the coordinate plane on one side of a boundary line.

Solution Since the inequality symbol $\leq$ includes an equal symbol, the graph of $x - y \leq 5$ includes the graph of $x - y = 5$.

Notation

The inequality $x - y \leq 5$ means

$$x - y = 5 \quad \text{or} \quad x - y < 5$$

A table of solutions to graph the boundary

$$x - y = 5$$

x	y	(x, y)
0	-5	$(0, -5)$
5	0	$(5, 0)$
6	1	$(6, 1)$

Let x = 0 and find y.
Let y = 0 and find x.
As a check, let x = 6 and find y.

Step 1: To graph $x - y = 5$, we use the intercept method, as shown in part (a) of the illustration below. The resulting line, called a **boundary line,** divides the coordinate plane into two half-planes. To show that the points on the boundary line are solutions of $x - y \leq 5$, we draw it as a solid line.

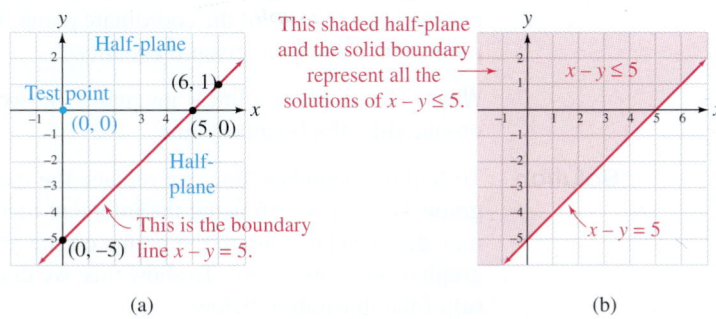

(a) (b)

Step 2: Since the inequality $x - y \leq 5$ also allows $x - y$ to be less than 5, other ordered pairs, besides those on the boundary, satisfy the inequality. For example, consider the origin, with coordinates $(0, 0)$. If we substitute 0 for x and 0 for y in the given inequality, we have

$$x - y \leq 5 \quad \text{This is the given inequality.}$$
$$0 - 0 \stackrel{?}{\leq} 5 \quad \text{Substitute.}$$
$$0 \leq 5 \quad \text{True}$$

Because $0 \leq 5$, the coordinates of the origin satisfy $x - y \leq 5$. In fact, the coordinates of every point on the same side of the boundary as the origin satisfy the inequality. To indicate this, we shade in red the half-plane that contains the test point $(0, 0)$, as shown in part (b). Every point in the shaded half-plane and every point on the boundary line satisfies $x - y \leq 5$. On the other hand, the points in the unshaded half-plane *do not* satisfy $x - y \leq 5$.

As an informal check, we can pick an ordered pair that lies in the shaded region and one that does not lie in the shaded region. When we substitute their coordinates into the inequality, we should obtain a true statement and then a false statement.

Success Tip

All the points in the unshaded region below the boundary line have coordinates that satisfy $x - y > 5$.

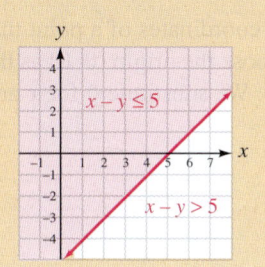

For (3, 1), in the shaded region:

$$x - y \leq 5$$
$$3 - 1 \stackrel{?}{\leq} 5 \quad \text{Substitute.}$$
$$2 \leq 5 \quad \text{True}$$

For (5, −4), not in the shaded region:

$$x - y \leq 5$$
$$5 - (-4) \stackrel{?}{\leq} 5 \quad \text{Substitute.}$$
$$9 \leq 5 \quad \text{False}$$

Self Check 2 Graph: $x - y \leq 2$

Now Try ▶ Problem 35

The previous example suggests the following **test-point method** to graph linear inequalities in two variables.

Graphing Linear Inequalities in Two Variables	1. Replace the inequality symbol with an equal symbol $=$ and graph the boundary line of the region. If the original inequality allows the possibility of equality (the symbol is either $\leq$ or $\geq$), draw the boundary line as a solid line. If equality is not allowed ($<$ or $>$), draw the boundary line as a dashed line. 2. Pick a test point that is on one side of the boundary line. (Use the origin if possible.) Replace x and y in the inequality with the coordinates of that point. If a true statement results, shade the side that contains that point. If a false statement results, shade the other side of the boundary.

EXAMPLE 3   Graph: $4x + 3y < -6$

Strategy   We will graph the related equation $4x + 3y = -6$ to establish the boundary line between two regions of the coordinate plane. Then we will determine which region contains points that satisfy the given inequality.

Why   The graph of a linear inequality in two variables is a region of the coordinate plane on one side of a boundary line.

Solution   To find the boundary line, we replace the inequality symbol with an equal symbol $=$ and graph $4x + 3y = -6$ using the intercept method. Since the inequality symbol $<$ does not include an equal symbol, the points on the graph of $4x + 3y = -6$ will not be part of the graph of $4x + 3y < -6$. To show this, we draw the boundary line as a dashed line. See part (a) of the illustration below.

A table of solutions to graph the boundary

$$4x + 3y = -6$$

x	y	(x, y)
0	-2	$(0, -2)$
$-\frac{3}{2}$	0	$\left(-\frac{3}{2}, 0\right)$
-3	2	$(-3, 2)$

Let x = 0 and find y.
Let y = 0 and find x.
As a check, let x = −3 and find y.

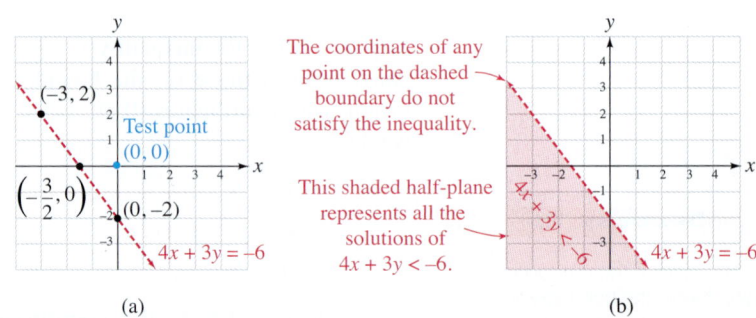

(a)   (b)

To determine which half-plane to shade, we substitute the coordinates of a point that lies on one side of the boundary line into $4x + 3y < -6$. We choose the origin $(0, 0)$ as the test point because the calculations are easy when they involve 0. We substitute 0 for x and 0 for y in the inequality.

Caution

When using a test point to determine which half-plane to shade, remember to substitute the coordinates into the given inequality, not the equation for the boundary.

$$4(0) + 3(0) = -6$$

$$4x + 3y < -6 \qquad \text{This is the given inequality.}$$
$$4(\mathbf{0}) + 3(\mathbf{0}) \overset{?}{<} -6 \qquad \text{The symbol } \overset{?}{<} \text{ is read as "is possibly less than."}$$
$$0 + 0 \overset{?}{<} -6$$
$$0 < -6 \qquad \text{False}$$

Since $0 < -6$ is a false statement, the point $(0, 0)$ does not satisfy the inequality. In fact, no point in the half-plane containing $(0, 0)$ is a solution. Therefore, we shade the other side of the boundary line—the half-plane that does not contain $(0, 0)$. The graph of the solution set of $4x + 3y < -6$ is the half-plane below the dashed line, as shown in part (b).

Self Check 3   Graph: $5x + 6y < -15$

Now Try ▶   Problem 37

3 Graph Inequalities with a Boundary through the Origin.

In the next example, the boundary line passes through the origin. In such cases, the ordered pair (0, 0) should not be used as a test point to determine which half-plane to shade.

EXAMPLE 4   Graph: $y > 2x$

Strategy   We will graph the related equation $y = 2x$ to establish the boundary line between two regions of the coordinate plane. Then we will determine which region contains points that satisfy the given inequality.

Why   The graph of a linear inequality in two variables is a region of the coordinate plane on one side of a boundary line.

Solution   To find the boundary line, we graph $y = 2x$. Since the symbol $>$ does *not* include an equal symbol, the points on the graph of $y = 2x$ are not part of the graph of $y > 2x$. Therefore, the boundary line should be dashed, as shown in part (a) of the illustration below.

> **Success Tip**
>
> Draw a solid boundary line if the inequality has $\leq$ or $\geq$. Draw a dashed line if the inequality has $<$ or $>$.

A table of solutions to graph the boundary

$y = 2x$

x	y	(x, y)
0	0	$(0, 0)$
-1	-2	$(-1, -2)$
1	2	$(1, 2)$

Select three values for x and find the corresponding values of y.

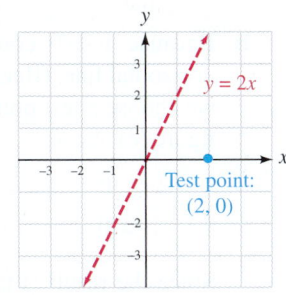

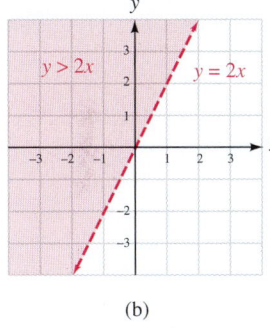

(a)   (b)

> **Success Tip**
>
> The origin (0, 0) is a smart choice for a test point because calculations involving 0 are usually easy. If the origin is on the boundary, choose a test point not on the boundary that has one coordinate that is 0, such as (0, 1) or (2, 0).

To determine which half-plane to shade, we substitute the coordinates of a point that lies on one side of the boundary line into $y > 2x$. Since the origin is on the boundary, it cannot serve as a test point. One of the many possible choices for a test point is (2, 0), because it does not lie on the boundary line. To see whether it satisfies $y > 2x$, we substitute 2 for x and 0 for y in the inequality.

$y > 2x$ This is the given inequality.

$0 \overset{?}{>} 2(2)$ The symbol $\overset{?}{>}$ is read as "is possibly greater than."

$0 > 4$ False

Since $0 > 4$ is a false statement, the point (2, 0) does not satisfy the inequality. We shade the half-plane that does not contain (2, 0), as shown in part (b).

Self Check 4   Graph: $y < 3x$

Now Try ▶ Problem 55

The graphs of some linear inequalities in two variables have boundary lines that are horizontal or vertical.

EXAMPLE 5   Graph each linear inequality: **a.** $x < -3$ **b.** $y \geq 0$

Strategy   We will use the procedure for graphing linear inequalities in two variables.

Why   Since the inequalities can be written as $x + 0y < -3$ and $0x + y \geq 0$, they are linear inequalities in two variables.

Solution   **a.** Because $x < -3$ contains an $<$ symbol, we draw the boundary, $x = -3$, as a dashed vertical line. See figure (a) below. We can use $(0, 0)$ as the test point.

> $x < -3$ This is the given inequality.
>
> $0 < -3$ Substitute 0 for x. The y-coordinate of the test point (0, 0) is not used.

Since the result is false, we shade the half-plane that does not contain $(0, 0)$, as shown in figure (b) below. Note that the solution consists of all points that have an x-coordinate that is less than -3.

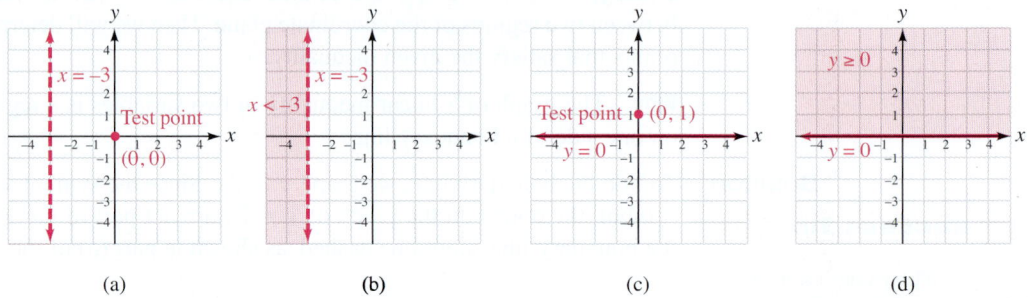

(a)   (b)   (c)   (d)

b. Because $y \geq 0$ contains an $\geq$ symbol, we draw the boundary, $y = 0$, as a solid horizontal line. (Recall that the graph of $y = 0$ is the x-axis. See figure (c) above.) Next, we choose a test point not on the boundary. The point $(0, 1)$ is a convenient choice. See figure (c) above.

> $y \geq 0$ This is the given inequality.
>
> $1 \geq 0$ Substitute 1 for y. The x-coordinate of the test point (0, 1) is not used.

Since the result is true, we shade the half-plane that contains $(0, 1)$, as shown in part (d) above. Note that the solution consists of all points that have a y-coordinate that is greater than or equal to 0.

Self Check 5   Graph each linear inequality: **a.** $x \geq 2$ **b.** $y < 4$

Now Try ▶ Problems 63 and 65

4 Solve Applied Problems Involving Linear Inequalities in Two Variables.

When solving applied problems, phrases such as *at least, at most,* and *should not exceed* indicate that an inequality should be used.

EXAMPLE 6   **Working Two Jobs.** Carlos has two part-time jobs, one paying $10 per hour and another paying $12 per hour. If x represents the number of hours he works on the first job, and y represents the number of hours he works on the second, the graph of $10x + 12y \geq 240$ shows the possible ways he can schedule his time to earn at least $240 per week to pay his college expenses. Find four possible combinations of hours he can work to achieve his financial goal.

Strategy We will graph the inequality and find four points whose coordinates satisfy the inequality.

Why The coordinates of these points will give four possible combinations.

Solution   The graph of the inequality is shown below in part (a) of the illustration. Any point in the shaded region represents a possible way Carlos can schedule his time and earn $240 or more per week. If each shift is a whole number of hours long, the red highlighted points in part (b) represent four of the many acceptable combinations.

> **Success Tip**
>
> Since Carlos cannot work a negative number of hours on job 1 or on job 2, we need to graph only the inequality in quadrant I.

(6, 24):   6 hours on the first job, 24 hours on the second job

(12, 12):   12 hours on the first job, 12 hours on the second job

(22, 4):   22 hours on the first job, 4 hours on the second job

(26, 20):   26 hours on the first job, 20 hours on the second job

To verify one combination, suppose Carlos works 22 hours on the first job and 4 hours on the second job. He will earn

$$\$10(\mathbf{22}) + \$12(\mathbf{4}) = \$220 + \$48$$
$$= \$268 \qquad \text{This is at least \$240 per week.}$$

A table of solutions to graph the boundary

$$10x + 12y = 240$$

x	y	(x, y)
0	20	(0, 20)
24	0	(24, 0)

Let $x = 0$ and find y.
Let $y = 0$ and find x.

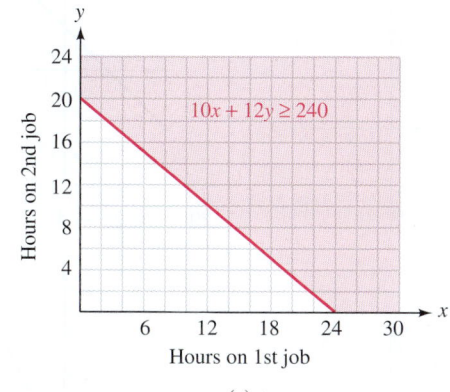

(a)

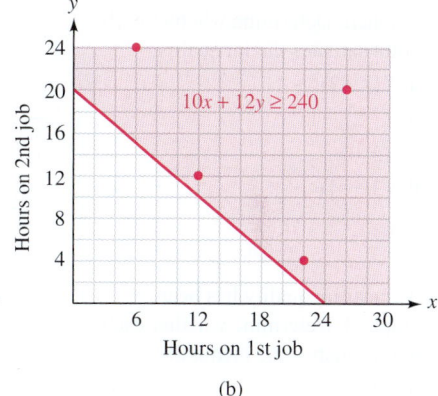

(b)

Self Check 6   **iTunes.**   Brianna and Ashley pool their money to purchase some songs and movies on iTunes. If songs cost $1 and movies cost $15, write an inequality to represent the number of songs and movies they can buy if they want to spend $150 or less.

Now Try ▶   Problem 79

SECTION 3.7   STUDY SET

VOCABULARY

Fill in the blanks.

1. $2x - y \le 4$ is a linear _____ in two variables.

2. An ordered pair (x, y) is a _____ of a linear inequality in two variables if a true statement results when the values of the variables are substituted into the inequality.

3. $(7, 2)$ is a solution of $x - y > 1$. We say that $(7, 2)$ _____ the inequality.

4. In the graph, the line $2x - y = 4$ is the _____ line.

5. In the graph, the line $2x - y = 4$ divides the coordinate plane into two _____.

6. When graphing a linear inequality, we determine which half-plane to shade by substituting the coordinates of a test _____ into the inequality.

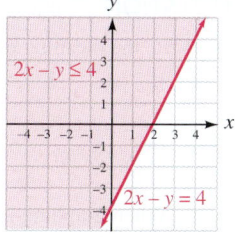

CONCEPTS

7. Determine whether $(-3, -5)$ is a solution of $5x - 3y \ge 0$.

8. Determine whether $(3, -1)$ is a solution of $x + 4y < -1$.

9. Fill in the blanks: A _____ line indicates that points on the boundary are not solutions and a _____ line indicates that points on the boundary are solutions.

10. The boundary for the graph of a linear inequality is shown. Why can't the origin be used as a test point to decide which side to shade?

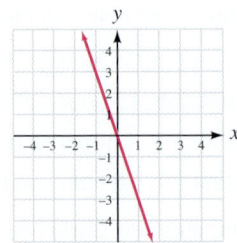

11. If a false statement results when the coordinates of a test point are substituted into a linear inequality, which half-plane should be shaded to represent the solution of the inequality?

12. A linear inequality has been graphed. Determine whether each point satisfies the inequality.

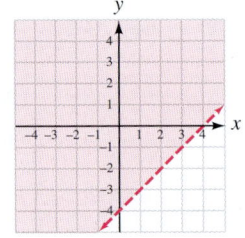

   a. $(1, -3)$
   b. $(-2, -1)$
   c. $(2, 3)$
   d. $(3, -4)$

13. A linear inequality has been graphed. Determine whether each point satisfies the inequality.

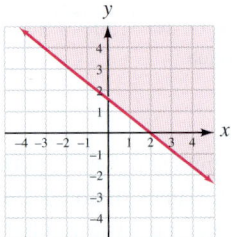

   a. $(2, 1)$
   b. $(-2, -4)$
   c. $(4, -2)$
   d. $(-3, 4)$

14. To graph linear inequalities, we must be able to graph boundary lines. Complete the table of solutions for each given boundary line.

   a. $5x - 3y = 15$

x	y	(x, y)
0		
	0	
1		

   b. $y = 3x - 2$

x	y	(x, y)
-1		
0		
2		

NOTATION

15. Write the meaning of each symbol in words.
   a. $<$
   b. $\geq$

   c. $\leq$
   d. $\overset{?}{>}$

16. a. When graphing linear inequalities, which inequality symbols are associated with a dashed boundary line?
   b. When graphing linear inequalities, which inequality symbols are associated with a solid boundary line?

17. Fill in the blanks: The inequality $4x + 2y \leq 9$ means $4x + 2y$ ▢ 9   or $4x + 2y$ ▢ 9.

18. Fill in the blanks: The inequality $-x + 8y \geq 1$ means $-x + 8y$ ▢ 1   or $-x + 8y$ ▢ 1.

Determine whether each ordered pair is a solution of the given inequality. See Example 1.

19. $2x + y > 6; (3, 2)$

20. $4x - 2y \geq -6; (-2, 1)$

21. $-5x - 8y < 8; (-8, 4)$

22. $x + 3y > 14; (-3, 8)$

23. $4x - y \leq 0; \left(\dfrac{1}{2}, 1\right)$

24. $9x - y \leq 2; \left(\dfrac{1}{3}, 1\right)$

25. $-5x + 2y > -4; (0.8, 0.6)$

26. $6x - 2y < -7; (-0.2, 1.5)$

Complete the graph by shading the correct side of the boundary. See Example 2.

27. $x - y \geq -2$

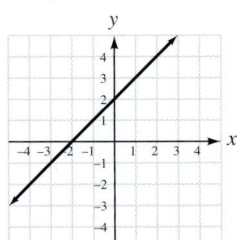

28. $x - y < 3$

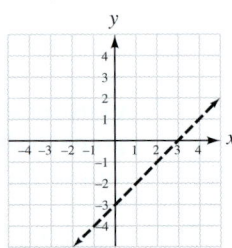

29. $y > 2x - 4$

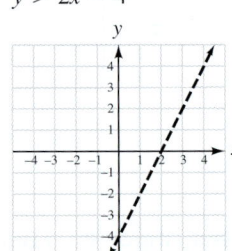

30. $y \leq -x + 1$

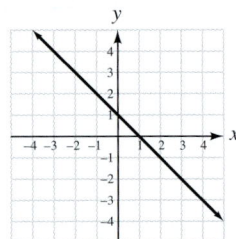

31. $x - 2y \geq 4$

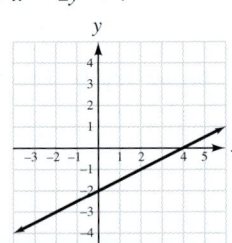

32. $3x + 2y > 12$

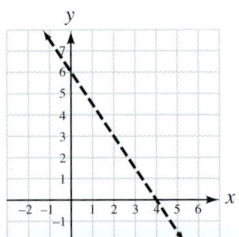

33. $y \leq 4x$

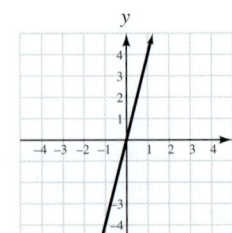

34. $y + 2x < 0$

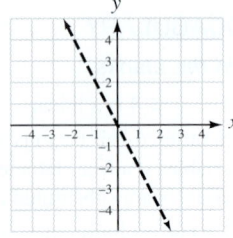

Graph each inequality. See Examples 2 and 3.

35. $x + y \geq 3$

36. $x + y < 2$

37. $3x - 4y > 12$

38. $5x + 4y \geq 20$

39. $2x + 3y \leq -12$

40. $3x - 2y > 6$

41. $y < 2x - 1$

42. $y > x + 1$

43. $y < -3x + 2$

44. $y \geq -2x + 5$

45. $y \geq -\dfrac{3}{2}x + 1$ **46.** $y < \dfrac{x}{3} - 1$

47. $x - 2y \geq 4$ **48.** $4x + y \geq -4$

49. $2y - x < 8$ **50.** $y + 9x \geq 3$

51. $7x - 2y < 21$ **52.** $3x - 3y \geq -10$

53. $2x - 3y \geq 4$ **54.** $4x + 3y < 6$

Graph each inequality. See Example 4.

55. $y \geq 2x$ **56.** $y < 3x$

57. $y < -\dfrac{x}{2}$ **58.** $y \geq x$

59. $y + x < 0$ **60.** $y - x < 0$

61. $5x + 3y < 0$ **62.** $2x + 5y > 0$

Graph each inequality. See Example 5.

63. $x < 2$ **64.** $y > -3$

65. $y \leq 1$ **66.** $x \geq -4$

67. $y + 2.5 > 0$ **68.** $x - 1.5 \leq 0$

69. $x \leq 0$ **70.** $y < 0$

TRY IT YOURSELF

Look Alikes . . .

Graph the given inequality in part a. Then use your answer to part a to help you quickly graph the associated inequality in part b. (Hint: If you spot the relationship between the inequalities, the graph in part b can be completed without having to use the test-point method.)

71. a. $5x - 3y \geq -15$ **b.** $5x - 3y < -15$

72. a. $y > -\dfrac{2}{3}x + 2$ **b.** $y \leq -\dfrac{2}{3}x + 2$

73. a. $y + 2x < 0$ **b.** $y + 2x \geq 0$

74. a. $y \leq \dfrac{1}{4}x$ **b.** $y > \dfrac{1}{4}x$

APPLICATIONS

75. Deliveries. To decide the number x of pallets and the number y of barrels that a truck can hold, a driver refers to the graph below. Can a truck make a delivery of 4 pallets and 10 barrels in one trip?

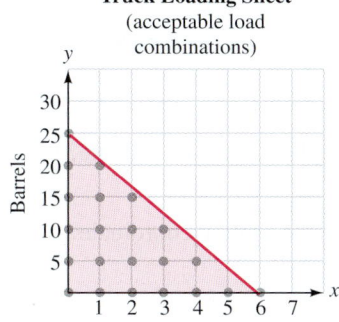

Truck Loading Sheet
(acceptable load combinations)

76. Zoos. To determine the allowable number of juvenile chimpanzees x and adult chimpanzees y that can live in an

enclosure, a zookeeper refers to the graph. Can 6 juvenile and 4 adult chimps be kept in the enclosure?

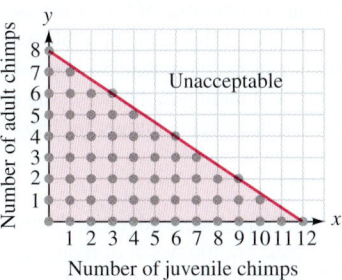

77. ▶ from **Campus to Careers**

Dental Assistant

A dentist's office schedules 1-hour long appointments for adults and $\frac{3}{4}$-hour long appointments for children. The appointment times do not overlap. Let c represent the number of appointments scheduled for children and a represent the number of appointments scheduled for adults. The graph of $\frac{3}{4}c + a \leq 9$ shows the possible ways the time for seeing patients can be scheduled so that it does not exceed 9 hours per day. Graph the inequality. Label the horizontal axis c and the vertical axis a. Then find three possible combinations of children/adult appointments.

78. Rolling Dice. The points on the graph represent all of the possible outcomes when two fair dice are rolled a single time. For example, (5, 2), shown in red, represents a 5 on the first die and a 2 on the second. Which of the following sentences best describes the outcomes that lie in the shaded area?

  (i) Their sum is at most 6.   **(ii)** Their sum exceeds 6.

  (iii) Their sum does not exceed 6.   **(iv)** Their sum is at least 6.

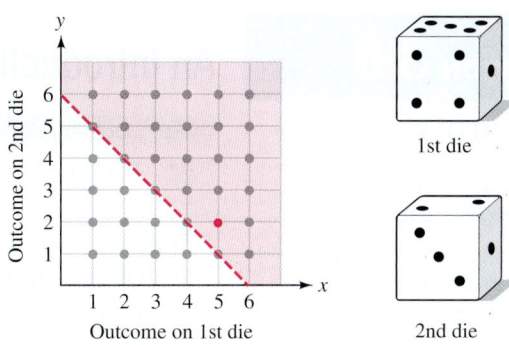

79. Production Planning. It costs a bakery $3 to make a cake and $4 to make a pie. If x represents the number of cakes made, and y represents the number of pies made, the graph of $3x + 4y \leq 120$ shows the possible combinations of cakes and pies that can be produced so that costs do not exceed $120 per day. Graph the inequality. Then find three possible combinations of cakes and pies that can be made so that the daily costs are not exceeded.

80. **Hiring Babysitters.** Mrs. Cansino has a choice of two babysitters. Sitter 1 charges $6 per hour, and Sitter 2 charges $7 per hour. If x represents the number of hours she uses Sitter 1 and y represents the number of hours she uses Sitter 2, the graph of $6x + 7y \le 42$ shows the possible ways she can hire the sitters and not spend more than $42 per week. Graph the inequality. Then find three possible ways she can hire the babysitters so that her weekly budget for babysitting is not exceeded.

81. **Inventories.** A clothing store advertises that it maintains an inventory of at least $4,400 worth of men's jackets at all times. At the store, leather jackets cost $100 and nylon jackets cost $88. If x represents the number of leather jackets in stock and y represents the number of nylon jackets in stock, the graph of $100x + 88y \ge 4,400$ shows the possible ways the jackets can be stocked. Graph the inequality. Then find three possible combinations of leather and nylon jackets so that the store lives up to its advertising claim.

82. **Making Sporting Goods.** A sporting goods manufacturer allocates at least 2,400 units of production time per day to make baseballs and footballs. It takes 20 units of time to make a baseball and 30 units of time to make a football. If x represents the number of baseballs made and y represents the number of footballs made, the graph of $20x + 30y \ge 2,400$ shows the possible ways to schedule the production time. Graph the inequality. Then find three possible combinations of production time for the company to make baseballs and footballs.

WRITING

83. Explain how to decide which side of the boundary line to shade when graphing a linear inequality in two variables.

84. Why is the origin usually a good test point to choose when graphing a linear inequality?

85. Why is (0, 0) not an acceptable choice for a test point when graphing a linear inequality whose boundary passes through the origin?

86. Explain the difference between the graph of the solution set of $x + 1 > 8$, an inequality in one variable, and the graph of $x + y > 8$, an inequality in two variables.

REVIEW

87. Solve $A = P + Prt$ for t.

88. What is the sum of the measures of the three angles of any triangle?

89. Simplify: $40\left(\dfrac{3}{8}x - \dfrac{1}{4}\right) + 40\left(\dfrac{4}{5}\right)$

90. Evaluate: $-4 + 5 - (-3) - 13$

CHALLENGE PROBLEMS

91. Find a linear inequality that has the graph shown.

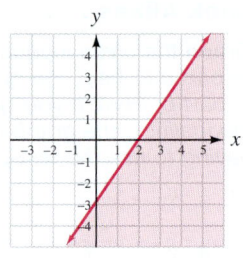

92. Graph the inequality: $4x - 3(x + 2y - 1) \ge -6\left(y - \dfrac{1}{2}\right)$

| SECTION **3.8** | An Introduction to Functions |

OBJECTIVES

1 Find the domain and range of a relation.

2 Identify functions and their domains and ranges.

3 Use function notation.

4 Graph functions.

5 Use the vertical line test.

6 Solve applications involving functions.

ARE YOU READY?

The following problems review some basic skills that are needed when working with functions.

1. If $y = 3x - 1$, find the value for y when $x = 6$.

2. Which of the following ordered pairs have the same x-coordinate?
   $(3, 5), (2, 9), (3, 0), (-1, 5)$

3. Which of the following ordered pairs have the same y-coordinate?
   $(8, 4), (-7, 6), (0, 0), (-3, 4)$

4. Find each absolute value:
   a. $|-5|$ b. $|2|$

In this section, we will discuss *relations* and *functions*. These two concepts are included in our study of graphing because they involve ordered pairs.

1 Find the Domain and Range of a Relation.

The following table shows the number of medals won by American athletes at several recent Winter Olympics.

Year	1988	1992	1994*	1998	2002	2006	2010
Medals	6	11	13	13	34	25	37
	Calgary CAN	Albertville FRA	Lillehammer NOR	Nagano JPN	Salt Lake City USA	Turin ITA	Vancouver CAN

* The Winter Olympics were moved ahead two years so that the winter and summer games would alternate every two years.

We can display the data in the table as a set of ordered pairs, where the **first component** represents the year and the **second component** represents the number of medals won by American athletes:

{(1988, 6), (1992, 11), (1994, 13), (1998, 13), (2002, 34), (2006, 25), (2010, 37)}

A set of ordered pairs, such as this, is called a **relation.** The set of all first components is called the **domain** of the relation and the set of all second components is called the **range** of the relation.

EXAMPLE 1   Find the domain and range of the relation {(1, 7), (4, −6), (−3, 1), (2, 7)}.

Strategy   We will examine the first and second components of the ordered pairs.

Why   The set of first components is the domain and the set of second components is the range.

Solution   The relation {(**1, 7**), (**4, −6**), (**−3, 1**), (**2, 7**)} has the domain {**−3, 1, 2, 4**} and the range is {**−6, 1, 7**}. The elements of the domain and range are usually listed in increasing order, and if a value is repeated, it is listed only once.

Self Check 1   Find the domain and range of the relation {(8, 2), (−1, 10), (6, 2), (−5, −5)}.

Now Try ▶ Problem 15

---

2 Identify Functions and Their Domains and Ranges.

An **arrow** or **mapping diagram** can be used to define a relation. The data from the Winter Olympics example are shown on the right in that form. Relations are also often defined using **two-column tables.**

Notice that for each year, there corresponds exactly one medal count. That is, this relation assigns to each member of the domain exactly one member of the range. Relations that have this characteristic are called *functions.*

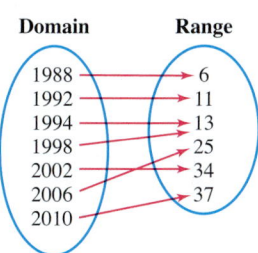

Function	A **function** is a set of ordered pairs (a relation) in which to each first component there corresponds exactly one second component.

We may also think of a function as a rule that assigns to each value of one variable exactly one value of another variable. Since we often worked with sets of ordered pairs of the form (x, y), it is helpful to define a function in an alternate way using the variables x and y.

y Is a Function of *x*	Given a relation in x and y, if to each value of x in the domain there is assigned exactly one value of y in the range, then y is said to be a function of x.

EXAMPLE 2   Determine whether each relation defines y to be a function of x. If a function is defined, give its domain and range.

a.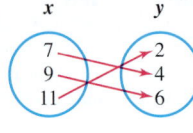

b.

x	y
2	3
5	7
2	1
6	5

c. $\{(0, 8), (3, 8), (4, 8), (9, 8)\}$

d.

x → y diagram: $-1 \to 5$, $-2 \to 0$, $-3 \to 6$, $-4 \to 6$, $-4 \to 11$

Strategy   We will check to see whether each value of x is assigned exactly one value of y.

Why   If this is true, then y is a function of x.

Solution   a. The arrow diagram defines a function because to each value of x there is assigned exactly one value of y: $7 \to 4$, $9 \to 6$, and $11 \to 2$.

The domain of the function is $\{7, 9, 11\}$ and the range is $\{2, 4, 6\}$.

b. The table does not define a function, because to the x value 2 there is assigned more than one value of y: $2 \to 3$ and $2 \to 1$.

c. Since to each number x exactly one value y is assigned, the set of ordered pairs defines y to be a function of x. It also illustrates an important fact about functions: *The same value of y can be assigned to different values of x.* In this case, each number x is assigned the y-value 8.

The domain of the function is $\{0, 3, 4, 9\}$ and the range is $\{8\}$.

d. The arrow diagram does not define a function, because to the x value -4 there is assigned more than one value of y: $-4 \to 6$ and $-4 \to 11$.

Self Check 2   Determine whether each relation defines y to be a function of x. If a function is defined, give its domain and range.

a.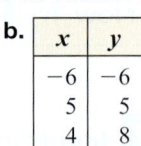

b.

x	y
−6	−6
5	5
4	8

c. $\{(1, 4), (4, 9), (9, 4), (3, 9)\}$

Now Try ▶   Problems 19, 25, and 31

3 Use Function Notation.

A function can be defined by an equation. For example, $y = 2x - 3$ is a rule that assigns to each value of x exactly one value of y. To find the y-value that is assigned to the x-value 4, we substitute 4 for x and evaluate the right side of the equation.

$$y = 2\textbf{x} - 3$$
$$y = 2(\textbf{4}) - 3 \qquad \text{Substitute 4 for x.}$$
$$= 8 - 3 \qquad \text{Evaluate the right side.}$$
$$= 5$$

The function $y = 2x - 3$ assigns the y-value 5 to an x-value of 4. When making such calculations, the value of x is called an **input** and its corresponding value of y is called an **output.**

A special notation is used to name functions that are defined by equations.

Function Notation	The notation $y = f(x)$ indicates that the variable y is a function of x.

Since $y = f(x)$, the equations $y = 2x - 3$ and $f(x) = 2x - 3$ are equivalent. We read $f(x) = 2x - 3$ as "f of x is equal to $2x$ minus 3."

Notation

If y is a function of x, then y and $f(x)$ are interchangeable.

This is the variable used to represent the input value.

$$f(x) = 2x - 3$$

This is the name of the function.

This expression shows how to obtain an output from a given input.

Function notation provides a compact way of representing the value that is assigned to some number x. For example, if $f(x) = 2x - 3$, the value that is assigned to an x-value 5 is represented by $f(5)$.

$$f(x) = 2x - 3$$
$$f(5) = 2(5) - 3 \quad \text{Substitute the input 5 for each } x.$$
$$= 10 - 3 \quad \text{Evaluate the right side.}$$
$$= 7 \quad \text{The output is 7.}$$

Caution

The symbol $f(x)$ **does not** mean $f \cdot x$. We read $f(x)$ as "f of x" or as "the value of f at x."

Thus, $f(5) = 7$. We read this as "f of 5 is 7." The output 7 is called a **function value.**

To see why function notation is helpful, consider these two sentences, which ask you to do the same thing:

1. If $y = 2x - 3$, find the value of y when x is 5.
2. If $f(x) = 2x - 3$, find $f(5)$.

Sentence 2, which uses $f(x)$ notation, is much more compact.

EXAMPLE 3 For $f(x) = 5x + 7$, find each of the following function values:
a. $f(2)$ **b.** $f(-4)$ **c.** $f(0)$

Strategy We will substitute 2, -4, and 0 for x in the expression $5x + 7$ and then evaluate it.

Why The notation $f(x) = 5x + 7$ indicates that we are to multiply each input (each number written within the parentheses) by 5 and then add 7 to that product.

Solution **a.** To find $f(2)$, we substitute the number within the parentheses, 2, for each x in $f(x) = 5x + 7$, and evaluate the right side of the equation.

The Language of Algebra

Another way to read $f(2) = 17$ is to say "the value of f at 2 is 17."

$$f(x) = 5x + 7$$
$$f(2) = 5(2) + 7 \quad \text{Substitute the input 2 for each } x.$$
$$= 10 + 7 \quad \text{Evaluate the right side.}$$
$$= 17 \quad \text{The output is 17.}$$

Thus, $f(2) = 17$.

b. $f(x) = 5x + 7$

$f(-4) = 5(-4) + 7$ Substitute the input −4 for each x.

$\qquad = -20 + 7$ Evaluate the right side.

$\qquad = -13$ The output is −13.

Thus, $f(-4) = -13$.

c. $f(x) = 5x + 7$

$f(0) = 5(0) + 7$ Substitute the input 0 for each x.

$\qquad = 0 + 7$ Evaluate the right side.

$\qquad = 7$ The output is 7.

Thus, $f(0) = 7$.

Self Check 3   For $f(x) = -2x + 3$, find each of the following function values:

   a. $f(4)$ **b.** $f(-1)$ **c.** $f(0)$

Now Try ▶ Problem 35

We can think of a function as a machine that takes some input x and turns it into some output $f(x)$, as shown in part (a) of the figure below. In part (b), the function machine for $f(x) = x^2 + 2x$ turns the input **4** into the output $4^2 + 2(4) = 24$, and we have $f(4) = 24$.

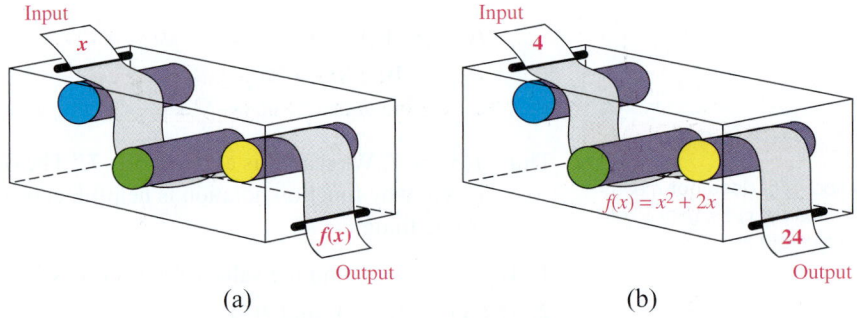

(a)   (b)

The letter f used in the notation $y = f(x)$ represents the word *function*. However, other letters, such as g and h, can also be used to name functions.

EXAMPLE 4   For $g(x) = 3 - 2x$ and $h(x) = x^3 + x^2 - 1$, find: **a.** $g(3)$ **b.** $h(-2)$

Strategy   We will substitute 3 for x in $3 - 2x$ and substitute -2 for x in $x^3 + x^2 - 1$, and then evaluate each expression.

Why   The numbers 3 and -2, which are within the parentheses, are inputs that should be substituted for the variable x.

Solution   **a.** To find $g(3)$, we use the function rule $g(x) = 3 - 2x$ and replace x with 3.

$g(x) = 3 - 2x$ Read g(x) as "g of x."

$g(3) = 3 - 2(3)$ Substitute the input 3 for each x.

$\qquad = 3 - 6$ Evaluate the right side.

$\qquad = -3$ The output is −3.

Thus, $g(3) = -3$.

b. To find $h(-2)$, we use the function rule $h(x) = x^3 + x^2 - 1$ and replace x with -2.

$$h(x) = x^3 + x^2 - 1 \qquad \text{Read h(x) as "h of x."}$$
$$h(-2) = (-2)^3 + (-2)^2 - 1 \qquad \text{Substitute the input } -2 \text{ for each x.}$$
$$= -8 + 4 - 1 \qquad \text{Evaluate the right side.}$$
$$= -5 \qquad \text{This is the output.}$$

Thus, $h(-2) = -5$.

Self Check 4   Find $g(0)$ and $h(4)$ for the functions in Example 4.

Now Try   Problem 41

4   Graph Functions.

We have seen that a function such as $f(x) = 4x + 1$ assigns to each value of x a single value $f(x)$. The input-output pairs generated by a function can be written in the form $(x, f(x))$. These ordered pairs can be plotted on a rectangular coordinate system to give the **graph of the function**.

EXAMPLE 5   Graph: $f(x) = 4x + 1$

Strategy   We can graph the function by creating a table of function values and plotting the corresponding ordered pairs.

Why   After drawing a line through the plotted points, we will have the graph of the function.

Solution   To make a table, we choose several values for x and find the corresponding values of $f(x)$. If x is -1, we have

$$f(x) = 4x + 1 \qquad \text{This is the function to graph.}$$
$$f(-1) = 4(-1) + 1 \qquad \text{Substitute the input } -1 \text{ for each x.}$$
$$= -4 + 1 \qquad \text{Evaluate the right side.}$$
$$= -3 \qquad \text{This is the output.}$$

Thus, $f(-1) = -3$. This means when x is -1, $f(x)$ or y is -3, and that the ordered pair $(-1, -3)$ lies on the graph of $f(x)$.

$$\textit{Function notation} \qquad \textit{Ordered-pair notation}$$
$$f(-1) = -3 \qquad (-1, -3)$$

Similarly, we find the corresponding values of $f(x)$ for x-values of 0 and 1. Then we plot the resulting ordered pairs and draw a straight line through them to get the graph of $f(x) = 4x + 1$. Since $y = f(x)$, the graph of $f(x) = 4x + 1$ is the same as the graph of the equation $y = 4x + 1$.

Notation

A table of function values is similar to a table of solutions, except that the second column is usually labeled $f(x)$ instead of y.

$f(x) = 4x + 1$

x	$f(x)$	
-1	-3	$\rightarrow (-1, -3)$
0	1	$\rightarrow (0, 1)$
1	5	$\rightarrow (1, 5)$

The function generates these ordered pairs.

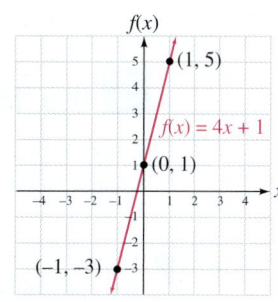

The vertical axis can be labeled y or $f(x)$.

Self Check 5   Graph: $f(x) = -3x - 2$

Now Try ▶   Problem 47

We call $f(x) = 4x + 1$ from Example 5 a **linear function** because its graph is a nonvertical line. Any linear equation, except those of the form $x = a$, can be written using function notation by writing it in slope–intercept form ($y = mx + b$) and then replacing y with $f(x)$.

The graphs of some functions are not straight lines.

EXAMPLE 6   Graph: $f(x) = |x|$

Strategy   We can graph the function by creating a table of function values and plotting the corresponding ordered pairs.

Why   After drawing a "V" shape through the plotted points, we will have the graph of the function.

Solution   To create a table of function values, we choose values for x and find the corresponding values of $f(x)$. For $x = -4$ and $x = 3$, we have

$$f(x) = |x| \qquad\qquad f(x) = |x|$$
$$f(-4) = |-4| \qquad\qquad f(3) = |3|$$
$$= 4 \qquad\qquad\qquad = 3$$

The results $f(-4) = 4$ and $f(3) = 3$ produce the ordered pairs $(-4, 4)$ and $(3, 3)$.

Similarly, we find the corresponding values of $f(x)$ for several other x-values. When we plot the resulting ordered pairs, we see that they lie in a "V" shape. We join the points to complete the graph as shown. We call $f(x) = |x|$ an **absolute value function.**

$f(x) = |x|$

x	$f(x)$	
-4	4	→ $(-4, 4)$
-3	3	→ $(-3, 3)$
-2	2	→ $(-2, 2)$
-1	1	→ $(-1, 1)$
0	0	→ $(0, 0)$
1	1	→ $(1, 1)$
2	2	→ $(2, 2)$
3	3	→ $(3, 3)$
4	4	→ $(4, 4)$

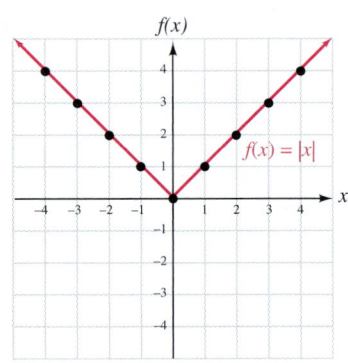

Self Check 6   Graph: $f(x) = |x| + 2$

Now Try ▶   Problem 49

5 Use the Vertical Line Test.

If any vertical line intersects a graph more than once, the graph cannot represent a function, because to one value of x there would correspond more than one value of y.

The Vertical Line Test	If a vertical line intersects a graph in more than one point, the graph is not the graph of a function.

The graph shown on the right in red does not represent a function, because a vertical line intersects the graph at more than one point. The points of intersection indicate that the x-value -1 corresponds to two different y-values, 3 and -1.

When the coordinates of the two points of intersection are listed in a table, it is easy to see that the x-value of -1 is assigned two different y-values. Thus, this is not the graph of a function.

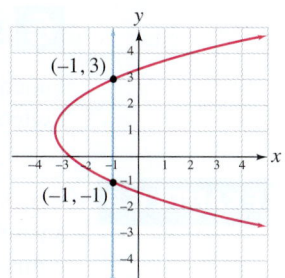

Not a function

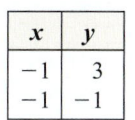

x	y
-1	3
-1	-1

EXAMPLE 7   Determine whether each of the graphs shown in red is the graph of a function.

a. 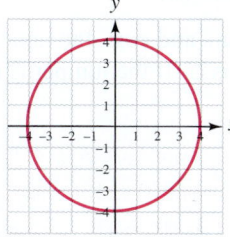   **b.** 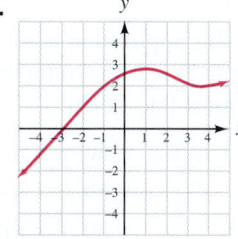   **c.**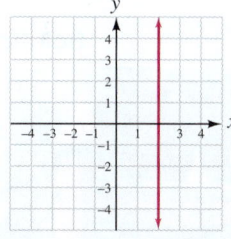

Strategy   We will check to see whether any vertical line intersects the graph more than once.

Why   If any vertical line does intersect the graph more than once, the graph is not a function.

Solution   **a.** Refer to figure (a) below. The graph shown in red is not the graph of a function, because a vertical line can be drawn that intersects the graph at more than one point. The points of intersection of the graph and the line reveal that the x-value 3 is assigned two different y-values, 2.5 and -2.5.

b. Refer to figure (b) below. The graph shown in red is a graph of a function, because no vertical line can be drawn that intersects the graph at more than one point. Several vertical lines are drawn in blue to illustrate this.

c. Refer to figure (c) below. The graph shown in red is not the graph of a function, because a vertical line can be drawn that intersects the graph at more than one point. In fact, it intersects it at infinitely many points. From this example, we can conclude that any vertical line will fail the vertical line test. Thus, vertical lines are not functions.

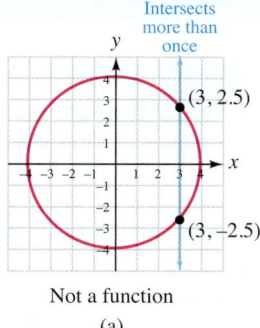

Not a function
(a)

A function
(b)

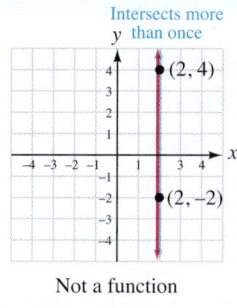

Not a function
(c)

Self Check 7   Determine whether each of the following is the graph of a function.

a. 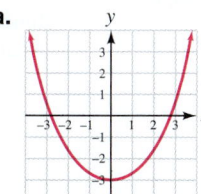   **b.**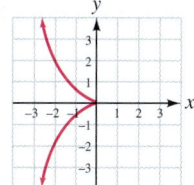

Now Try   Problems 55 and 57

6 Solve Applications Involving Functions.

Functions are used to describe certain relationships where one quantity depends on another. Letters other than f and x are often chosen to more clearly describe these situations.

EXAMPLE 8   **Bounce Houses.**   The function $C(h) = 80 + 15(h - 4)$ gives the cost in dollars to rent an inflatable jumper for h hours. (The terms of the rental agreement require a 4-hour minimum.) Find the cost of renting the jumper for 10 hours.

Strategy   To find the cost to rent the jumper for 10 hours, we will substitute 10 for each h in $C(h) = 80 + 15(h - 4)$ and evaluate the right side.

Why   In $C(h) = 80 + 15(h - 4)$, the variable h represents the number of hours that the jumper is rented. We need to find $C(10)$.

Solution   For this application involving hours and cost, the notation $C(h)$ is used. The input variable is h and the name of the function is C. If the jumper is rented for 10 hours, then h is 10 and we must find $C(10)$.

$$C(h) = 80 + 15(h - 4) \qquad \text{Read } C(h) \text{ as "C of h."}$$
$$C(10) = 80 + 15(10 - 4) \qquad \text{Substitute the input 10 for each } h.$$
$$= 80 + 15(6) \qquad \text{Evaluate the right side.}$$
$$= 80 + 90 \qquad \text{Do the multiplication.}$$
$$= 170 \qquad \text{This is the output.}$$

It costs $170 to rent the jumper for 10 hours.

Self Check 8   **Bounce Houses.**   Find the cost of renting the jumper for 8 hours.

Now Try ▶ Problem 65

SECTION 3.8 ▶ STUDY SET

VOCABULARY

Fill in the blanks.

1. A set of ordered pairs is called a _____.
2. A _____ is a rule that assigns to each x-value exactly one y-value.
3. The set of all input values for a function is called the _____, and the set of all output values is called the _____.
4. We can think of a function as a machine that takes some _____ x and turns it into some output _____.
5. If $f(2) = -3$, we call -3 a function _____.
6. The graph of a _____ function is a straight line and the graph of an _____ value function is V-shaped.

CONCEPTS

7. **Federal Minimum Hourly Wage.**   The following table is an example of a function. Use an arrow diagram to illustrate this.

Year	1992	1994	1996	1998	2000	2002	2004	2006	2008	2010
Minimum wage ($)	4.25	4.25	4.75	5.15	5.15	5.15	5.15	5.15	6.55	7.25

Source: infoplease.com

8. The arrow diagram describes a function. What is the domain and what is the range of the function?

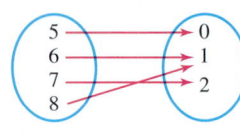

9. For the given input, what value will the function machine output?

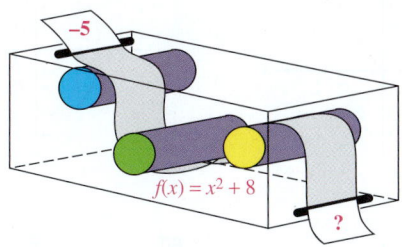

10. a. Fill in the blank: If a _____ line intersects a graph in more than one point, the graph is not the graph of a function.

 b. Give the coordinates of the points where the given vertical line intersects the graph.

 c. Is this the graph of a function?

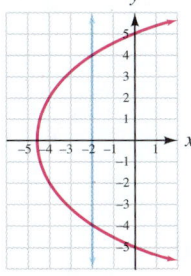

NOTATION

Fill in the blanks.

11. We read $f(x) = 5x - 6$ as "f ____ x is $5x$ minus 6."

12. Since $y =$ _____, the following two equations are equivalent:

$$y = 3x + 2 \quad \text{and} \quad f(x) = 3x + 2$$

13. The notation $f(4) = 5$ indicates that when the x-value ▢ is input into a function rule, the output is ▢. This fact can be shown graphically by plotting the ordered pair (▢, ▢).

14. When graphing the function $f(x) = -x + 5$, the vertical axis of the coordinate system can be labeled ▢ or ▢.

GUIDED PRACTICE

Find the domain and range of each relation. See Example 1.

15. $\{(6, -1), (-1, -10), (-6, 2), (8, -5)\}$

16. $\{(11, -3), (0, 0), (4, 5), (-3, -7)\}$

17. $\{(0, 9), (-8, 50), (6, 9)\}$

18. $\{(1, -12), (-6, 8), (5, 8)\}$

Determine whether the relation defines y to be a function of x. If a function is defined, give its domain and range. If it does not define a function, find two ordered pairs that show a value of x that is assigned more than one value of y. See Example 2.

19.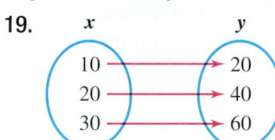

20.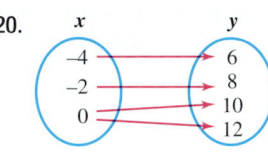

21.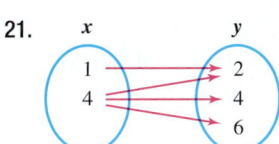

22.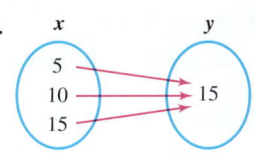

23.

x	y
1	7
2	15
3	23
4	16
5	8

24.

x	y
30	2
30	4
30	6
30	8
30	10

25.

x	y
-4	6
-1	0
0	-3
2	4
-1	2

26.

x	y
1	1
2	2
3	3
4	4

27.

x	y
3	4
3	-4
4	3
4	-3

28.

x	y
-1	1
-3	1
-5	1
-7	1
-9	1

29.

x	y
6	0
-3	-8
1	9
5	4

30.

x	y
1.6	0
-3	-1
2.5	20
-7	0.1
1.6	19

31. $\{(3, 4), (3, -4), (4, 3), (4, -3)\}$

32. $\{(-1, 1), (-3, 1), (-5, 1), (-7, 1), (-9, 1)\}$

33. $\{(-2, 7), (-1, 10), (0, 13), (1, 16)\}$

34. $\{(-2, 4), (-3, 8), (-3, 12), (-4, 16)\}$

Find each function value. See Examples 3 and 4.

35. $f(x) = 4x - 1$

 a. $f(1)$ **b.** $f(-2)$ **c.** $f\left(\dfrac{1}{4}\right)$ **d.** $f(50)$

36. $f(x) = 1 - 5x$

 a. $f(0)$ **b.** $f(-75)$ **c.** $f(0.2)$ **d.** $f\left(-\dfrac{4}{5}\right)$

37. $f(x) = 2x^2$

 a. $f(0.4)$ **b.** $f(-3)$ **c.** $f(1,000)$ **d.** $f\left(\dfrac{1}{8}\right)$

38. $g(x) = 6 - x^2$

    a. $g(30)$ **b.** $g(6)$ **c.** $g(-1)$ **d.** $g(0.5)$

39. $h(x) = |x - 7|$

    a. $h(0)$ **b.** $h(-7)$ **c.** $h(7)$ **d.** $h(8)$

40. $f(x) = |2 + x|$

    a. $f(0)$ **b.** $f(2)$ **c.** $f(-2)$ **d.** $f(-99)$

41. $g(x) = x^3 - x$

    a. $g(1)$ **b.** $g(10)$ **c.** $g(-3)$ **d.** $g(6)$

42. $g(x) = x^4 + x$

    a. $g(1)$ **b.** $g(-2)$ **c.** $g(0)$ **d.** $g(10)$

43. $s(x) = (x + 3)^2$

    a. $s(3)$ **b.** $s(-3)$ **c.** $s(0)$ **d.** $s(-5)$

44. $s(x) = (x - 8)^2$

    a. $s(8)$ **b.** $s(-8)$ **c.** $s(1)$ **d.** $s(12)$

45. If $f(x) = 3.4x^2 - 1.2x + 0.5$, find $f(-0.3)$.

46. If $g(x) = x^4 - x^3 + x^2 - x$, find $g(-12)$.

Complete each table of function values and then graph each function. See Examples 5 and 6.

47. $f(x) = -3x - 2$

x	f(x)
-2	
-1	
0	
1	

48. $f(x) = -2x + 8$

x	f(x)
-1	
0	
1	
2	

49. $h(x) = |1 - x|$

x	h(x)
-2	
-1	
0	
1	
2	
3	
4	

50. $h(x) = |x + 2|$

x	h(x)
-5	
-4	
-3	
-2	
-1	
0	
1	

Graph each function. See Examples 5 and 6.

51. $f(x) = \dfrac{1}{2}x - 2$ **52.** $f(x) = -\dfrac{2}{3}x + 3$

53. $h(x) = -|x|$ **54.** $g(x) = |x| - 2$

Determine whether each graph is the graph of a function. If it is not, find ordered pairs that show a value of x that is assigned more than one value of y. See Example 7.

55.

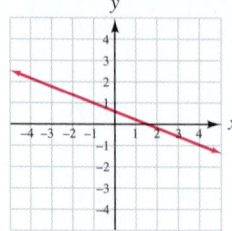

56.

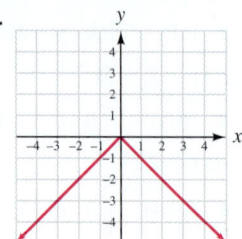

57.

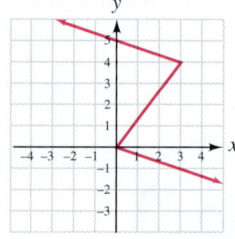

58.

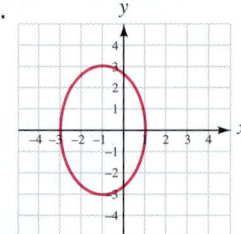

59.

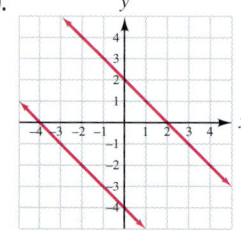

60.

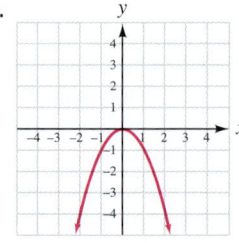

61.

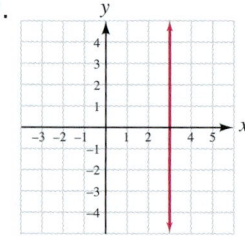

62.

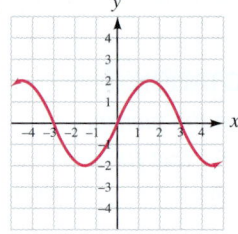

APPLICATIONS

63. Reflections. When a beam of light hits a mirror, it is reflected off the mirror at the same angle that the incoming beam struck the mirror. What type of function could serve as a mathematical model for the path of the light beam shown here?

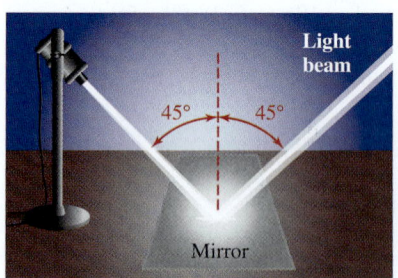

64. Lightning. The function $D(t) = \dfrac{t}{5}$ gives the approximate distance in miles that you are from a lightning strike, where t is the number of seconds between seeing the lightning and hearing the thunder. Find $D(5)$ and explain what it means.

65. Vacationing. The function $C(d) = 500 + 100(d - 3)$ gives the cost in dollars to rent an RV motor home for d days. (The terms of the rental agreement require a 3-day minimum.) Find the cost of renting the RV for a vacation that will last 7 days.

Rent This RV!

(3 day minimum)

66. Structural Engineering. The maximum safe load in pounds of the rectangular beam shown in the figure is given by the function $S(t) = \dfrac{1,875t^2}{8}$, where t is the thickness of the beam, in inches. Find the maximum safe load if the beam is 4 inches thick.

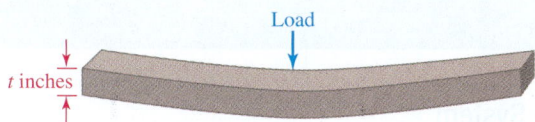

Load

t inches

67. Lawn Sprinklers. The function $A(r) = \pi r^2$ can be used to determine the area in square feet that will be watered by a rotating sprinkler that sprays out a stream of water. Find $A(5)$ and $A(20)$. Round to the nearest tenth.

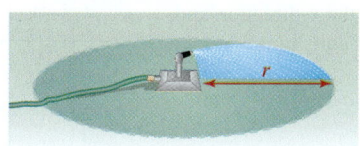

r

68. Parts Lists. The function $f(r) = 2.30 + 3.25(r + 0.40)$ approximates the length (in feet) of the belt that joins the two pulleys, where r is the radius (in feet) of the smaller pulley. Find the belt length needed for each pulley in the parts list.

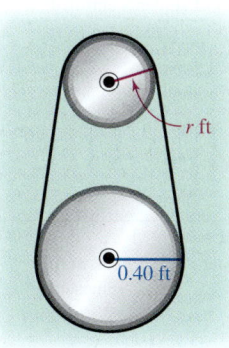

r ft

0.40 ft

Parts list		
Pulley	r	**Belt length**
P-45M	0.32	
P-08D	0.24	

69. Postage. The **step graph** below shows how the cost of a first class U.S. postage stamp increased from 1990 through 2010. An open circle at the end of a line segment means the endpoint of the segment is not included. Is this the graph of a function?

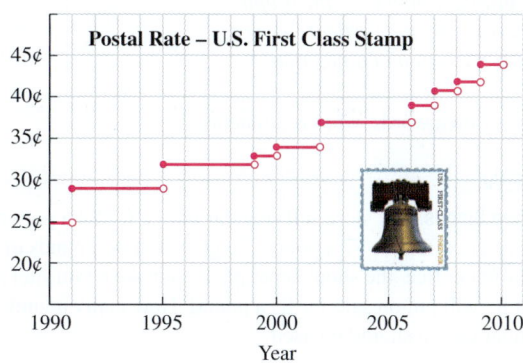

Postal Rate – U.S. First Class Stamp

Source: U.S. Postal Service

70. Sound. We cannot see sound waves, but certain scientific instruments are used to draw mathematical models of them. Is the graph of a sound wave shown below a function?

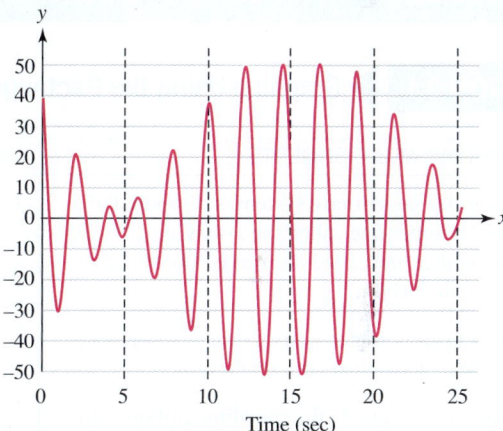

Time (sec)

WRITING

71. In the function $y = -5x + 2$, why do you think the value of x is called the *input* and the corresponding value of y the *output*?

72. Explain what a politician meant when she said, "The speed at which the downtown area will be redeveloped is a function of the number of low-interest loans made available to the property owners."

73. Explain the following diagram:

$$f(4) = 11 \qquad\qquad (4, 11)$$

74. Explain the error in the following solution.
If $f(x) = x^2 + 7x + 1$, find $f(10)$.

$$f(10) = 10^2 + 7x + 1$$
$$= 100 + 7x + 1$$
$$= 101 + 7x$$

75. A student was asked to determine whether the graph on the right is the graph of a function. What is wrong with the following reasoning?

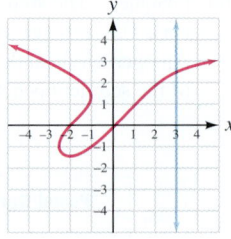

When I draw a vertical line through the graph, it intersects the graph only once. By the vertical line test, this is the graph of a function.

76. In your own words, what is a function?

REVIEW

77. Coffee Blends. A store sells regular coffee for $4 a pound and gourmet coffee for $7 a pound. To get rid of 40 pounds of the gourmet coffee, the shopkeeper plans to make a gourmet blend that he will put on sale for $5 a pound. How many pounds of regular coffee should be used?

78. Photographic Chemicals. A photographer wishes to mix 2 liters of a 5% acetic acid solution with a 10% acetic solution to get a 7% acetic solution. How many liters of 10% acetic solution must be added?

CHALLENGE PROBLEMS

79. Is the graph of $y \geq 3 - x$ a function? Explain.

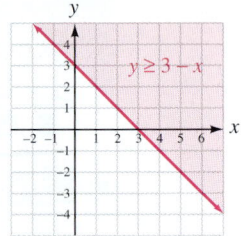

80. If $f(x) = x^2 + x$, find: $f\left(\frac{4}{5}r\right)$

81. Let $f(x) = -2x + 5$. For what value of x is $f(x) = -7$?

82. Let $f(x) = x - 2$ and $g(x) = 3x$. Find $f(g(6))$.

---

3 ▶ Summary & Review

SECTION 3.1 ▶ Graphing Using the Rectangular Coordinate System

DEFINITIONS AND CONCEPTS	EXAMPLES
A **rectangular coordinate system** is composed of a horizontal number line called the **x-axis** and a vertical number line called the **y-axis.** The two axes intersect at the **origin.** To **plot** or **graph** ordered pairs means to locate their position on a rectangular coordinate system. The x- and y-axes divide the **coordinate plane** into four regions called **quadrants.**	Plot the points: $(2, 3), (-4, 2), (-3, -1), (0, -2.5), (4, -2)$ To graph each point, start at the origin and count the appropriate number of units in the x-direction and then the appropriate number of units in the y-direction. 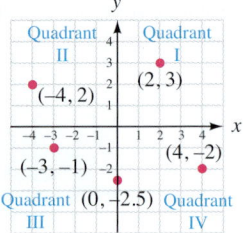

REVIEW EXERCISES

1. Graph the points with coordinates $(-1, 3), (0, 1.5), (-4, -4),$ $\left(2, \frac{7}{2}\right),$ and $(4, 0)$.

2. Hawaii. Estimate the coordinates of Oahu using an ordered pair of the form (longitude, latitude).

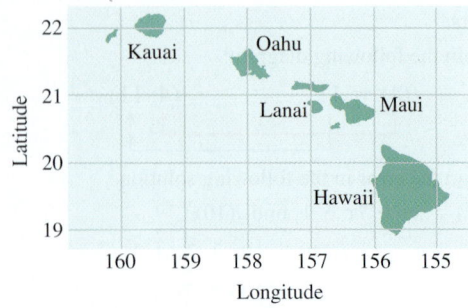

3. In what quadrant does the point $(-3, -4)$ lie?

4. What are the coordinates of the origin?

5. Geometry. Three vertices (corners) of a square are $(-5, 4),$ $(-5, -2),$ and $(1, -2)$. Find the coordinates of the fourth vertex and then find the area of the square.

6. College Enrollments. The graph gives the number of students enrolled at a college for the period from 4 weeks before to 5 weeks after the semester began.

a. What was the maximum enrollment and when did it occur?

b. How many students were enrolled 2 weeks before the semester began?

c. When was the enrollment 2,250?

SECTION 3.2	▶ Graphing Linear Equations

DEFINITIONS AND CONCEPTS	EXAMPLES
A **solution of an equation in two variables** is an ordered pair of numbers that makes the equation a true statement when the numbers are substituted for the variables. The **standard form** of a linear equation in two variables is $Ax + By = C$, where A, B, and C are real numbers and A and B are not both zero.	Determine whether $(2, -3)$ is a solution of $2x - y = 7$. $2x - y = 7$ This is the given equation. $2(2) - (-3) \overset{?}{=} 7$ Substitute 2 for x and −3 for y. $4 + 3 \overset{?}{=} 7$ Evaluate the left side. $7 = 7$ True Since the result $7 = 7$ is true, $(2, -3)$ is a solution of $2x - y = 7$.
If only one coordinate of an ordered-pair solution is known: 1. Substitute it into the equation for the appropriate variable. 2. Solve the resulting equation to find the unknown coordinate.	To complete the solution $(\ \ , 8)$ for $3x + y = -1$, we substitute 8 for y and solve the resulting equation for x. $3x + y = -1$ This is the given equation. $3x + 8 = -1$ Substitute 8 for y. $3x = -9$ Subtract 8 from both sides. $x = -3$ To isolate x, divide both sides by 3. The solution is $(-3, 8)$.
To **graph a linear equation** solved for y: 1. Find three solutions by **selecting three values** of x and finding the corresponding values of y. 2. **Plot** each ordered-pair solution. 3. **Draw** a line through the points.	Graph: $y = -2x + 1$ $y = -2x + 1$ | x	y	(x, y)
---	---	---
-1	3	$(-1, 3)$
0	1	$(0, 1)$
2	-3	$(2, -3)$	 We construct a table of solutions, plot the points, and draw the line. 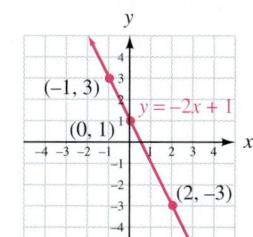

REVIEW EXERCISES

7. Is $(-3, -2)$ a solution of $y = 2x + 4$?

8. Complete the table of solutions.

$$3x + 2y = -18$$

x	y	(x, y)
-2		
	3	

9. Which of the following equations are not linear equations?

$8x - 2y = 6$ $y = x^2 + 1$ $y = x$ $y - x^3 = 0$

10. The graph of a linear equation is shown.

a. When the coordinates of point A are substituted into the equation, will a true or false statement result?

b. When the coordinates of point B are substituted into the equation, will a true or false statement result?

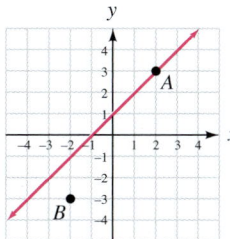

Graph each equation by constructing a table of solutions.

11. $y = 4x - 2$

12. $y = \dfrac{3}{4}x$

13. $5y = -5x + 15$ (*Hint:* Solve for y first.)

14. $6y = -4x$ (*Hint:* Solve for y first.)

15. **Birthday Parties.**   A restaurant offers a party package for children that includes everything: food, drinks, cake, and favors. The cost c, in dollars, is given by the equation $c = 8n + 50$, where n is the number of children attending the party. Graph the equation and use the graph to estimate the cost of a party if 18 children attend.

16. Determine whether each statement is true or false.

a. It takes three or more points to determine a line.

b. A linear equation in two variables has infinitely many solutions.

SECTION 3.3 ▶ Intercepts

DEFINITIONS AND CONCEPTS	EXAMPLES
The point where a line intersects the x-axis is called the **x-intercept**. The point where a line intersects the y-axis is called the **y-intercept**. To **find the y-intercept**, substitute 0 for x in the given equation and solve for y. To **find the x-intercept**, substitute 0 for y and solve for x. Plotting the x- and y-intercepts of a graph and drawing a line through them is called the **intercept method for graphing a line**.	Use the y- and x-intercepts to graph $3x + 4y = -6$. **y-intercept: $x = 0$** **x-intercept: $y = 0$** $3x + 4y = -6$ $3x + 4y = -6$ $3(0) + 4y = -6$ $3x + 4(0) = -6$ $\quad\quad 4y = -6$ $\quad\quad 3x = -6$ $\quad\quad y = -\dfrac{3}{2}$ $\quad\quad x = -2$ 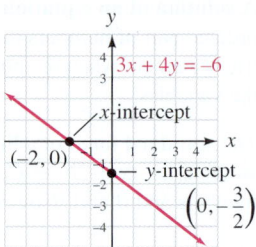 The y-intercept is $\left(0, -\dfrac{3}{2}\right)$ and the x-intercept is $(-2, 0)$.
The equation $y = b$ represents the **horizontal line** that intersects the y-axis at $(0, b)$. The equation $x = a$ represents the **vertical line** that intersects the x-axis at $(a, 0)$.	Graph: $y = 2$ and $x = -1$ on the same rectangular coordinate system. 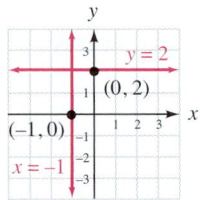

REVIEW EXERCISES

17. Identify the x- and y-intercepts of the graph shown in figure (a) below.

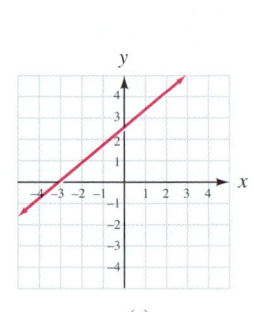

(a)

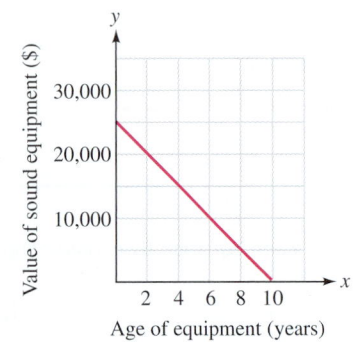

(b)

18. Depreciation. The graph in figure (b) above shows how the value of some sound equipment decreased over the years. Find the intercepts of the graph. What information do the intercepts give about the equipment?

Use the intercept method to graph each equation.

19. $-4x + 2y = 8$

20. $5x - 4y = 13$

21. Graph: $y = 4$

22. Graph: $x = -1$

SECTION 3.4 ▶ Slope and Rate of Change

DEFINITIONS AND CONCEPTS	EXAMPLES
The **slope** m of a line is a ratio that compares the vertical and horizontal change as we move along the line from one point to another. We can find the slope of a line graphically using the ratio $m = \frac{\text{rise}}{\text{run}}$.	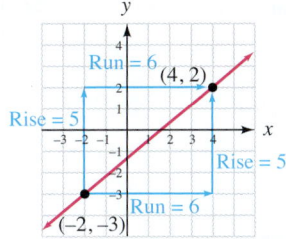 To find the slope of the line graphed on the left, we can use the slope triangle method: $$m = \frac{\text{rise}}{\text{run}} = \frac{5}{6}$$
We also can find the slope of a line using the **slope formula**: $$m = \frac{y_2 - y_1}{x_2 - x_1} \quad \text{if } x_1 \neq x_2$$	
Lines that rise from left to right have a **positive slope,** and lines that fall from left to right have a **negative slope.**	To find the slope of the line that passes through the points $(-2, -3)$ and $(4, 2)$, we substitute into the **slope formula**: $$m = \frac{y_2 - y_1}{x_2 - x_1} = \frac{2 - (-3)}{4 - (-2)} = \frac{5}{6}$$
Horizontal lines have **zero slope** and vertical lines have **undefined slope.**	
When units are attached to a slope, the slope is called a **rate of change.**	An example of a rate of change is:
The **grade** of an incline is its slope expressed as a percent.	$\dfrac{300 \text{ pounds}}{1 \text{ year}}$ Read as "300 pounds per year."
Parallel lines have the same slope. The slopes of **perpendicular lines** are negative reciprocals. The product of their slopes is -1.	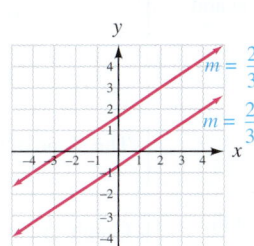 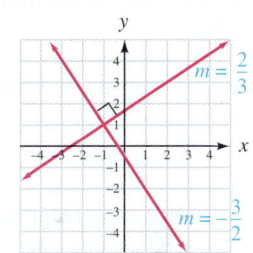 Parallel lines Perpendicular lines

REVIEW EXERCISES

In each case, find the slope of the line.

23.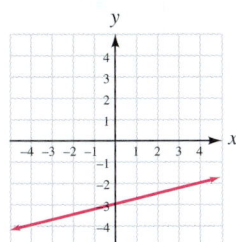

24.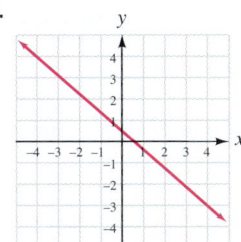

25. The line with this table of solutions:

x	y	(x, y)
2	-3	$(2, -3)$
4	-17	$(4, -17)$

26. The line passing through the points $(1, -4)$ and $(3, -7)$

27. Draw a line having a slope that is
   a. Positive   **b.** Negative   **c.** 0   **d.** Undefined

28. Carpentry.   If a truss like the one shown below is used to build the roof of a shed, find the slope (pitch) of the roof.

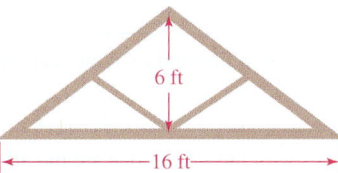

29. Ramps.   Find the grade of the ramp shown below. Round to the nearest tenth of a percent.

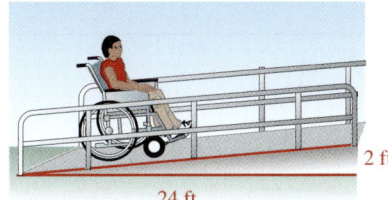

30. Bottled Water. Refer to the graph below. Find the rate of change in the amount of bottled water consumed per person in the U.S. from 2000 through 2008.

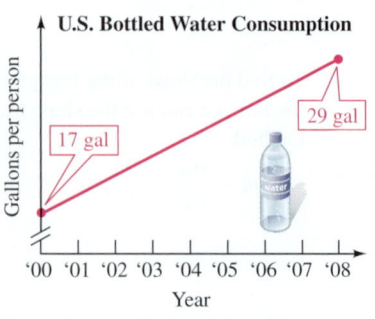

U.S. Bottled Water Consumption

17 gal

29 gal

Gallons per person

'00 '01 '02 '03 '04 '05 '06 '07 '08
Year

Source: Beverage Marketing Corporation

31. Without graphing, determine whether the line that passes through (6, 6) and (4, 2) and the line that passes through (2, −10) and (−2, −2) are parallel, perpendicular, or neither.

32. Find the slope of a line perpendicular to the line passing through (−1, 9) and (−8, 4).

SECTION 3.5 ▶ Slope–Intercept Form

DEFINITIONS AND CONCEPTS	EXAMPLES
If a linear equation is written in **slope–intercept form** $y = mx + b$ the graph of the equation is a line with slope m and y-intercept $(0, b)$. To write a linear equation in two variables in slope–intercept form, solve the equation for y.	Find the slope and y-intercept of the line whose equation is $5x + 3y = 3$. To find the slope and y-intercept, we solve the equation for y. $5x + 3y = 3$ $3y = -5x + 3$ Subtract 5x from both sides. $y = -\dfrac{5}{3}x + 1$ To isolate y, divide both sides by 3. $m = -\dfrac{5}{3}$ and $b = 1$. The slope of the line is $-\dfrac{5}{3}$ and the y-intercept is $(0, \mathbf{1})$.
To **graph a line in slope–intercept form,** plot the y-intercept and use the slope to determine a second point on the line.	Graph: $y = -\frac{5}{3}x + 1$ $y = -\dfrac{5}{3}x + 1$ $m = \dfrac{\text{rise}}{\text{run}} = \dfrac{-5}{3}$ $b = 1$ y-intercept: $(0, 1)$ 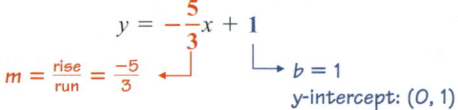 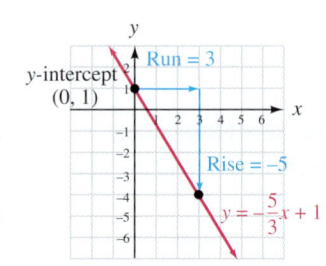
If we know the slope of a line and its y-intercept, we can write its equation.	The equation of a line with slope $\frac{1}{8}$ and y-intercept $(0, -5)$ is $y = \frac{1}{8}x - 5$.
Two different lines with the same slope are **parallel**.	Lines with equations $y = 3x + 4$ and $y = 3x - 12$ are parallel because each line has slope 3.
If the slopes of two lines are negative reciprocals, the product of their slopes is -1 and the lines are **perpendicular.**	Lines with equations $y = 3x + 4$ and $y = -\frac{1}{3}x - 12$ are perpendicular because their slopes, 3 and $-\frac{1}{3}$, are negative reciprocals.

REVIEW EXERCISES

Find the slope and the y-intercept of each line.

33. $y = \frac{3}{4}x - 2$

34. $y = -4x$

35. $y = \frac{x}{8} + 10$

36. $7x + 5y = -21$

37. Graph the line with slope -4 and y-intercept $(0, -1)$. Write an equation of the line.

38. Write an equation for the line shown here.

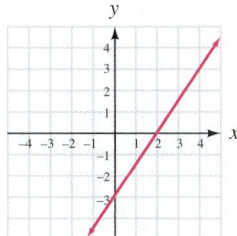

39. Find the slope and the y-intercept of the line whose equation is $9x - 3y = 15$. Then graph it.

40. Copiers. A business buys a used copy machine that has already produced 75,000 copies.

    a. If the business plans to run 300 copies a week, write a linear equation that would find the number of copies c the machine has made in its lifetime after the business has used it for w weeks.

    b. Use your result in part a to predict the total number of copies that will have been made on the machine 1 year, or 52 weeks, after being purchased by the business.

Without graphing, determine whether graphs of the given pairs of lines are parallel, perpendicular, or neither.

41. $y = -\frac{2}{3}x + 6$

    $y = -\frac{2}{3}x - 6$

42. $x + 5y = -10$

    $y - 5x = 0$

SECTION 3.6 ▶ Point–Slope Form

DEFINITIONS AND CONCEPTS	EXAMPLES
If a line with slope m passes through the point with coordinates (x_1, y_1), the equation of the line in **point–slope form** is $$y - y_1 = m(x - x_1)$$	Find an equation of the line with slope -3 that passes through $(-2, 4)$. Write the equation in slope–intercept form. We substitute the slope and the coordinates of the point into the point–slope form. $y - y_1 = m(x - x_1)$ This is point–slope form. $y - 4 = -3[x - (-2)]$ Substitute. $y - 4 = -3(x + 2)$ Simplify within the brackets. $y - 4 = -3x - 6$ Distribute. $y = -3x - 2$ To isolate y, add 4 to both sides. This is slope–intercept form.
If we know **two points that a line passes through,** we can write its equation.	Find an equation of the line that passes through $(2, 5)$ and $(3, 7)$. Write the equation in slope–intercept form. The slope of the line is: $$m = \frac{y_2 - y_1}{x_2 - x_1} = \frac{7 - 5}{3 - 2} = 2$$ Either point on the line can serve as (x_1, y_1). If we use $(2, 5)$, we have: $y - y_1 = m(x - x_1)$ This is point–slope form. $y - 5 = 2(x - 2)$ Substitute: $x_1 = 2$, $y_1 = 5$, and $m = 2$. $y - 5 = 2x - 4$ Distribute. $y = 2x + 1$ To isolate y, add 5 to both sides. This is the slope–intercept form.

REVIEW EXERCISES

Find an equation of the line with the given slope that passes through the given point. Write the equation in slope–intercept form and graph the equation.

43. $m = 3, (1, 5)$

44. $m = -\dfrac{1}{2}, (-4, -1)$

Find an equation of the line with the following characteristics. Write the equation in slope–intercept form.

45. Passing through $(3, 7)$ and $(-6, 1)$

46. Horizontal, passing through $(6, -8)$

47. Car Registration.   When it was 2 years old, the annual registration fee for a Dodge Caravan was \$380. When it was 4 years old, the registration fee dropped to \$310. If the relationship is linear, write an equation that gives the registration fee f in dollars for the van when it is x years old.

48. The Atmosphere.   The scatter diagram below shows the amount of carbon dioxide in the Earth's atmosphere as measured at Hawaii's Mauna Loa Observatory from 1960 through 2010. A straight line can be used to model the data.

  a. Use the two highlighted points in red to write the equation of the line. Write the answer in slope–intercept form.

  b. Use your answer to part a to predict the amount of carbon dioxide in the atmosphere in 2020.

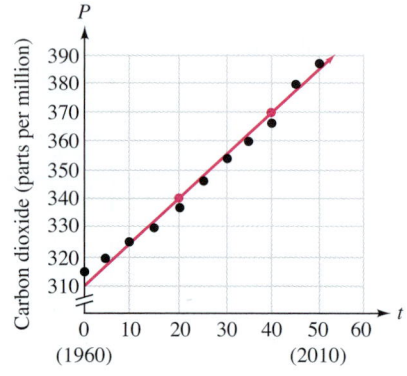

Years after 1960

Source: National Oceanic and Atmospheric Administration

SECTION 3.7 ▶ Graphing Linear Inequalities

DEFINITIONS AND CONCEPTS	EXAMPLES
An ordered pair (x, y) is a **solution of an inequality** in x and y if a true statement results when the variables are replaced by the coordinates of the ordered pair.	Determine whether $(-2, 5)$ is a solution of $x + 3y > -6$. We substitute the coordinates into the inequality and see if a true statement results. $\quad x + 3y > -6$ This is the given inequality. $\quad -2 + 3(5) \overset{?}{>} -6$ Substitute. $\quad\quad\quad 13 > -6$ True Since the result $13 > -6$ is true, $(-2, 5)$ is a solution.
To graph a linear inequality: **1.** Replace the inequality symbol with an $=$ symbol and graph the **boundary line.** Draw a solid line if the inequality contains $\le$ or $\ge$ and a dashed line if it contains $<$ or $>$. **2.** Pick a **test point** not on the boundary. Substitute its coordinates into the inequality. If the inequality is satisfied, shade the half-plane that contains the test point. If the inequality is not satisfied, shade the other half-plane.	Graph: $2x - y \le 4$ **1.** Graph the boundary line $2x - y = 4$ and draw it as a solid line because the inequality symbol is $\le$. **2.** Test the point $(0, 0)$: $\quad 2x - y \le 4$ This is the given inequality. $\quad 2(0) - 0 \overset{?}{\le} 4$ Substitute. $\quad\quad\quad 0 \le 4$ True 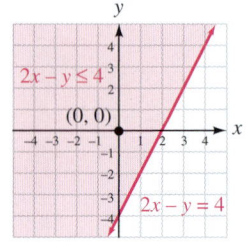 Since the coordinates of the test point satisfy the inequality, we shade the side of the boundary line that contains $(0, 0)$.

REVIEW EXERCISES

49. Determine whether each ordered pair is a solution of
$2x - y \leq -4$.

   a. $(0, 5)$ **b.** $(2, 8)$

   c. $(-3, -2)$ **d.** $\left(\frac{1}{2}, -5\right)$

50. Fill in the blanks: $2x - 3y \geq 6$ means $2x - 3y$ ___ 6 or
$2x - 3y$ ___ 6.

Graph each inequality.

51. $x - y < 5$ **52.** $2x - 3y \geq 6$

53. $y \leq -2x$ **54.** $y < -4$

55. The graph of a linear inequality is shown in the next column. Would a true or a false statement result if the coordinates of

   a. point *A* were substituted into the inequality?

   b. point *B* were substituted into the inequality?

   c. point *C* were substituted into the inequality?

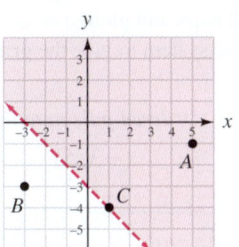

56. Work Schedules. A student told her employer that during the school year, she would be available for up to 30 hours a week, working either 3- or 5-hour shifts. If *x* represents the number of 3-hour shifts she works and *y* represents the number of 5-hour shifts she works, the inequality $3x + 5y \leq 30$ shows the possible combinations of shifts she can work. Graph the inequality and find three possible combinations.

SECTION 3.8 ▶ An Introduction to Functions

DEFINITIONS AND CONCEPTS	EXAMPLES
A **relation** is a set of ordered pairs. The set of all **first components** is called the **domain** of the relation and the set of all **second components** is called the **range** of a relation.	The relation $\{(4, 7), (0, -3), (-3, 8), (1, 7)\}$ has the domain $\{-3, 0, 1, 4\}$ and the range is $\{-3, 7, 8\}$.
A **function** is a set of ordered pairs (a relation) in which to each first component there corresponds exactly one second component. If to each value of *x* in the domain there is assigned exactly one value of *y* in the range, then **y is a function of x**.	 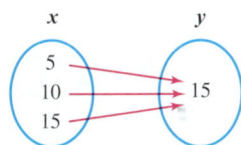 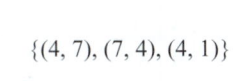 y is not a function of x: y is a function of x Not a function: $5 \rightarrow 3$ and $5 \rightarrow 9$ $4 \rightarrow 7$ and $4 \rightarrow 1$
A function can be defined by an equation. The notation $y = f(x)$ indicates that the variable *y* is a function of *x*. It is read as "*f* of *x*." We can think of a function as a machine that takes some **input** *x* and turns it into some **output** $f(x)$, called a **function value**.	For the function $f(x) = 8x + 5$, $f(-2)$ is the value of $f(x)$ when $x = -2$. $f(x) = 8x + 5$ $f(-2) = 8(-2) + 5$ Substitute the input −2 for each *x*. $\qquad\quad = -16 + 5$ Evaluate the right side. $\qquad\quad = -11$ Thus, $f(-2) = -11$.
The **vertical line test:** If a vertical line intersects a graph in more than one point, the graph is not the graph of a function.	Vertical lines intersect only once 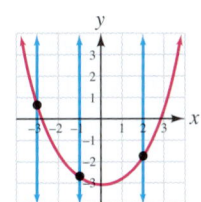 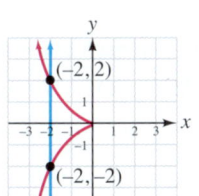 A vertical line intersects more than once A function Not a function 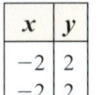

The input-output pairs that a function generates can be written as ordered pairs and plotted on a rectangular coordinate system to give the **graph of a function.**

Function notation   *Ordered-pair notation*

$$f(-3) = 5 \qquad (-3, 5)$$

Graph the function: $f(x) = -\frac{2}{3}x + 3$

We make a table of function values, plot the points, and draw the graph.

$$f(x) = -\frac{2}{3}x + 3$$

x	f(x)	
-3	5	→ (-3, 5)
0	3	→ (0, 3)
3	1	→ (3, 1)

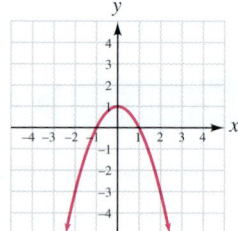

$f(x) = -\frac{2}{3}x + 3$

REVIEW EXERCISES

Find the domain and range of each relation.

57. $\{(7, -3), (-5, 9), (4, 4), (0, -11)\}$

58. $\{(2, -2), (15, -8), (-6, 9), (1, -8)\}$

Determine whether each relation defines y to be function of x. If a function is defined, give its domain and range. If it does not define a function, find ordered pairs that show a value of x that corresponds to more than one value of y.

59. 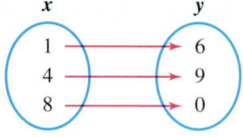   **60.**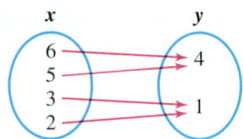

61.

x	y
9	81
7	49
5	25
3	9

62.

x	y
-1	2
0	3
-1	4
1	5

63. $\{(-1, 6), (0, 6), (1, 6), (2, 6)\}$
64. $\{(4, 4), (6, 4), (4, 6)\}$

Fill in the blanks.

65. The set of all input values for a function is called the _____, and the set of all output values is called the _____.

66. Fill in the blank: Since $y = $ ____ , the equations $y = 2x - 8$ and $f(x) = 2x - 8$ are equivalent.

For f(x) = x² − 4x, find each of the following function values.

67. $f(1)$ **68.** $f(0)$ **69.** $f(-3)$ **70.** $f\left(\frac{1}{2}\right)$

For g(x) = 1 − 6x, find each of the following function values.

71. $g(1)$ **72.** $g(-6)$ **73.** $g(0.5)$ **74.** $g\left(\frac{3}{2}\right)$

Determine whether each graph is the graph of a function. If it is not, find two ordered pairs that show a value of x that corresponds to more than one value of y.

75. 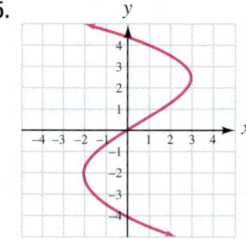   **76.**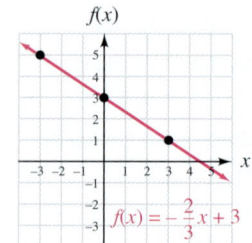

77. Complete the table of function values for $f(x) = 1 - |x|$. Then graph the function.

x	f(x)
0	
1	
3	
-1	
-3	

78. Aluminum Cans. The function $V(r) = 15.7r^2$ estimates the volume in cubic inches of a can 5 inches tall with a radius of r inches. Find the volume of the can shown in the illustration.

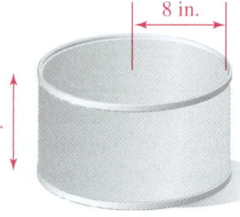

8 in.

5 in.

3 ▸ CHAPTER TEST

1. Fill in the blanks.

   a. A rectangular coordinate system is formed by two perpendicular number lines called the *x*-_____ and the *y*-_____.

   b. A _____ of an equation in two variables is an ordered pair of numbers that makes the equation a true statement.

   c. $3x + y = 10$ is a _____ equation in two variables and its graph is a line.

   d. The _____ of a line is a measure of steepness.

   e. A _____ is a set of ordered pairs in which to each first component there corresponds exactly one second component.

The graph shows the number of dogs being boarded in a kennel over a 3-day holiday weekend.

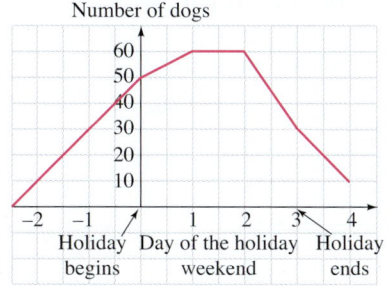

2. How many dogs were in the kennel 2 days before the holiday?

3. What is the maximum number of dogs that were boarded on the holiday weekend at any one time?

4. When were there 30 dogs in the kennel?

5. What information does the *y*-intercept of the graph give?

6. Plot each point on a rectangular coordinate system:
   $(1, 3), (-2, 4), (-3, -2), (3, -2), (-1, 0), (0, -1)$, and $\left(-\frac{1}{2}, \frac{7}{2}\right)$.

7. Find the coordinates of each point shown in the graph.

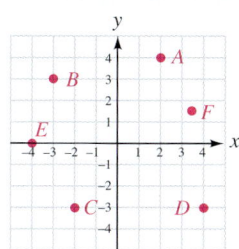

8. In which quadrant is each point located?

   a. $(-1, -5)$ b. $\left(6, -2\frac{3}{4}\right)$

9. Is $(-3, -4)$ a solution of $3x - 4y = 7$?

10. Complete the table of solutions for the linear equation $x + 4y = 6$.

x	y	(x, y)
2		
	3	

11. The graph of a linear equation is shown.

    a. If the coordinates of point *C* are substituted into the equation, will the result be true or false?

    b. If the coordinates of point *D* are substituted into the equation, will the result be true or false?

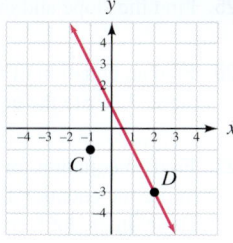

12. Graph: $y = \dfrac{x}{3}$

13. What are the *x*- and *y*-intercepts of the graph of $2x - 3y = 6$?

14. Graph: $8x + 4y = -24$

15. Find the slope of the line.

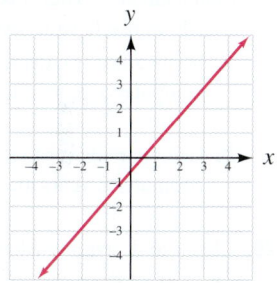

16. Find the slope of the line passing through $(-1, 3)$ and $(3, -1)$.

17. What is the slope of a horizontal line?

18. **Ramps.**   Find the grade of a ramp that rises 2 feet over a horizontal distance of 20 feet.

19. One line passes through $(9, 2)$ and $(6, 4)$. Another line passes through $(0, 7)$ and $(2, 10)$. Without graphing, determine whether the lines are parallel, perpendicular, or neither.

20. When graphed, are the lines $y = 2x + 6$ and $2x - y = 0$ parallel, perpendicular, or neither?

In Problems 21 and 22, refer to the illustration that shows the elevation changes for part of a 26-mile marathon course.

21. Find the rate of change of the decline on which the woman is running.

22. Find the rate of change of the incline on which the man is running.

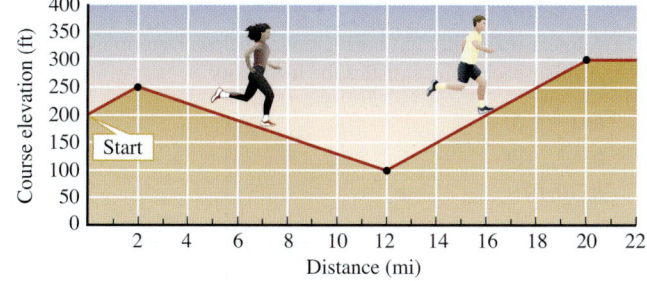

23. Graph: $x = -4$

24. Graph the line passing through $(-2, -4)$ having slope $\frac{2}{3}$.

25. Find the slope and the y-intercept of the graph of $x + 2y = 8$.

26. Find an equation of the line passing through $(-2, 5)$ with slope 7. Write the equation in slope–intercept form.

27. Find an equation for the line shown. Write the equation in slope–intercept form.

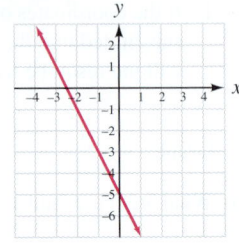

28. **Depreciation.**   After it is purchased, a $15,000 computer loses $1,500 in resale value every year.

    a. Write a linear equation that gives the resale value v of the computer x years after being purchased.

    b. Use your answer to part (a) to predict the value of the computer 8 years after it is purchased.

29. Check to determine whether $(6, 1)$ is a solution of $2x - 4y \geq 8$.

30. **Water Heaters.**   The scatter diagram shows how excessively high temperatures affect the life of a water heater. Write an equation of the line that models the data for water temperatures between $140°$ and $180°$. Let T represent the temperature of the water in degrees Fahrenheit and y represent the expected life of the heater in years. Give the answer in slope–intercept form.

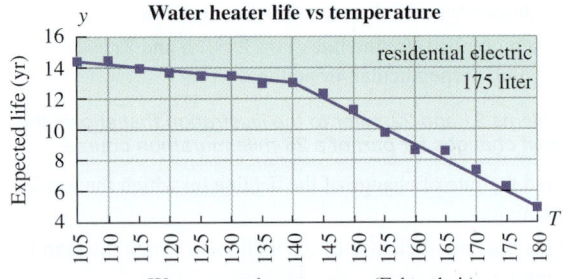

Water heater life vs temperature

residential electric
175 liter

Water: stored temperature (Fahrenheit)

Source: www.uniongas.com/WaterHeating

31. A linear inequality has been graphed below. Determine whether each point satisfies the inequality.

    a. $(-2, 3)$
    b. $(3, -4)$
    c. $(0, 0)$

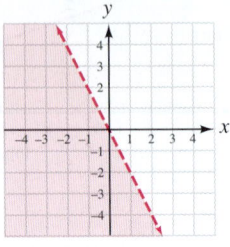

32. Is $(-20, -2)$ a solution of $\frac{1}{2}x - 3y \geq -4$?

33. Graph the inequality: $2x - 5y \leq -10$

34. Find the domain and range of the relation:
    $\{(5, 3), (1, 12), (-4, 3), (0, -8)\}$

Determine whether the relation defines y to be function of x. If a function is defined, give its domain and range. If it does not define a function, find ordered pairs that show a value of x that corresponds to more than one value of y.

35.

x	y
1	4
2	3
3	2
4	1

36.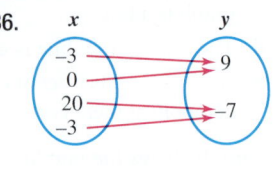

37. $\{(6, 5), (7, 5), (8, 5), (9, 5), (10, 5)\}$

38. $\{(-9, 41), (2, 6), (4, -9), (2, 2)\}$

39. 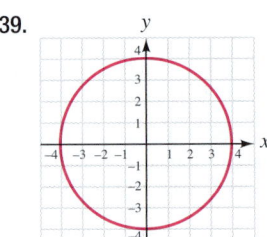   40.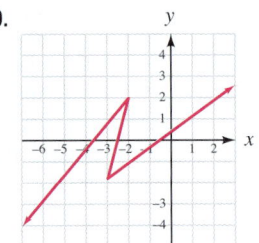

41. If $f(x) = 2x - 7$, find: $f(-3)$

42. If $g(x) = 3.5x^3$ find: $g(6)$

43. **Telephone Calls.**   The function $C(n) = 0.30n + 15$ gives the cost C per month in dollars for making n phone calls. Find $C(45)$ and explain what it means.

44. Graph: $f(x) = |x| - 1$

Group Project

▶ **Overview:** In this activity, you will explore the relationship between a person's height and arm span. Arm span is defined to be the distance between the tips of a person's fingers when his or her arms are held out to the side.

▶ **Instructions:** Form groups of 5 or 6 students. Measure the height and arm span of each person in your group, and record the results in a table like the one shown below.

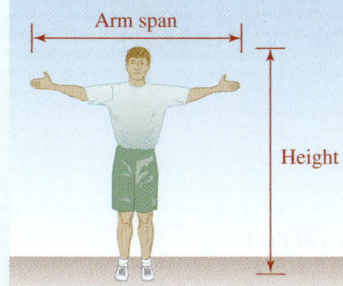

Name	Height (in.)	Arm span (in.)
1.		
2.		
3.		
4.		
5.		
6.		

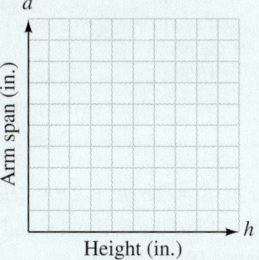

Plot the data in the table as ordered pairs of the form (height, arm span) on a graph like the one shown above. Then draw a straight-line model that best fits the data points.

Pick two convenient points on the line and find its slope. Use the point–slope form $a - a_1 = m(h - h_1)$ to find an equation of the line. Then, write the equation in slope–intercept form.

Ask a person from another group for his or her height measurement. Substitute that value into your linear model to predict that person's arm span. How close is your prediction to the person's actual arm span?

(From *Activities for Beginning and Intermediate Algebra* by Debbie Garrison, Judy Jones, and Jolene Rhodes)

CUMULATIVE REVIEW ▶▶ Chapters 1–3

1. Find the prime factorization of 108. [Section 1.2]
2. Write $\frac{1}{250}$ as a decimal. [Section 1.3]
3. Determine whether each statement is true or false. [Section 1.3]
   a. Every whole number is an integer.
   b. Every integer is a real number.
   c. 0 is a whole number, an integer, and a rational number.

Perform the operations.

4. $-27 + 21 + (-9)$ [Section 1.4]
5. $-1.57 - (-0.8)$ [Section 1.5]
6. $-9(-7)(5)(-3)$ [Section 1.6]
7. $\dfrac{-180}{-6}$ [Section 1.6]
8. Evaluate: $\left| \dfrac{(6-5)^4 - (-21)}{-27 + 4^2} \right|$ [Section 1.7]
9. Evaluate $b^2 - 4ac$ for $a = 2$, $b = -8$, and $c = 4$. [Section 1.8]
10. Suppose x sheets from a 500-sheet ream of paper have been used. How many sheets are left? [Section 1.8]
11. How many terms does the algebraic expression $3x^2 - 2x + 1$ have? What is the coefficient of the second term? [Section 1.8]

12. Multiply. [Section 1.9]
    a. $2(x + 4)$ b. $-2(x - 4)$

Simplify each expression. [Section 1.9]

13. $5a + 10 - a$
14. $-7(9t)$
15. $-2b^2 + 6b^2$
16. $5(-17)(0)(2)$
17. $(a + 2) - (a - 2)$
18. $-4(-5)(-8a)$
19. $-y - y - y$
20. $\frac{3}{2}(4x - 8) + x$

Solve each equation. [Sections 2.1 and 2.2]

21. $3x - 5 = 13$
22. $1.2 - x = -1.7$
23. $\frac{2x}{3} - 2 = 4$
24. $\frac{y - 2}{7} = -3$
25. $-3(2y - 2) - y = 5$
26. $9y - 3 = 6y$
27. $\frac{1}{3} + \frac{c}{5} = -\frac{3}{2}$
28. $5(x + 2) = 5x - 2$
29. $-x = 99$
30. $3c - 2 = \dfrac{11(c - 1)}{5}$

31. Pennies.   A 2010 telephone survey of adults asked whether the penny should be discontinued from the national currency. The results are shown in the circle graph. If 869 people favored keeping the penny, how many took part in the survey? [Section 2.3]

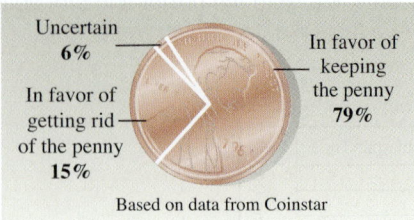

Based on data from Coinstar

32. Solve for h: $S = 2\pi rh + 2\pi r^2$ [Section 2.4]

33. Band Aids.   Find the perimeter and the area of the gauze pad of the bandage. [Section 2.4]

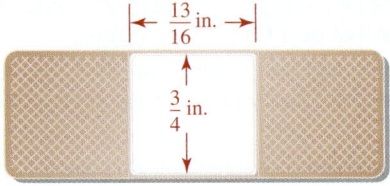

34. High Heels.   Find the unknown angle measure represented by x. [Section 2.5]

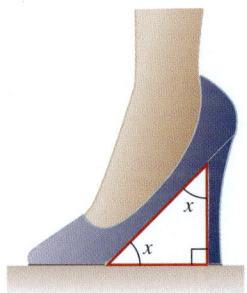

35. Complete the table. [Section 2.6]

	% acid	Liters	Amount of acid
50% solution	0.50	x	
25% solution	0.25	$13 - x$	
30% mixture	0.30	13	

36. Road Trips.   A bus, carrying the members of a marching band, and a truck, carrying their instruments, leave a high school at the same time. The bus travels at 60 mph and the truck at 50 mph. In how many hours will they be 75 miles apart? [Section 2.6]

37. Mixing Candy.   Candy corn worth $2.85 per pound is to be mixed with black gumdrops that cost $1.80 per pound to make 200 pounds of a mixture worth $2.22 per pound. How many pounds of each candy should be used? [Section 2.6]

Solve each inequality. Write the solution set in interval notation and graph it. [Section 2.7]

38. $-\dfrac{3}{16}x \geq -9$

39. $8x + 4 > 3x + 4$

40. In which quadrants are the second coordinates of ordered pairs positive? [Section 3.1]

41. Is $(-2, 4)$ a solution of $y = 2x - 8$? [Section 3.2]

Graph each equation.

42. $y = x$ [Section 3.2]

43. $2x + 4y = -8$ [Section 3.3]

44. What is the slope of the graph of the line $y = 5$? [Section 3.4]

45. What is the slope of the line passing through $(-2, 4)$ and $(5, -6)$? [Section 3.4]

46. Roofing.   Find the pitch of the roof. [Section 3.4]

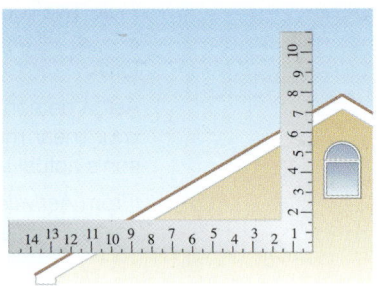

47. Find the slope and the y-intercept of the graph of the line described by $4x - 6y = -12$. [Section 3.5]

48. Write an equation of the line that has slope -2 and y-intercept $(0, 1)$. [Section 3.5]

49. Find an equation of the line that has slope $-\dfrac{7}{8}$ and passes through $(2, -9)$. Write the equation in point–slope form and in slope–intercept form. [Section 3.6]

50. Is $(-2, -4)$ a solution of $x + y \leq -6$? [Section 3.7]

51. Graph: $y \geq x + 1$ [Section 3.7]

52. Graph $x < 4$ on a rectangular coordinate system. [Section 3.7]

53. If $f(x) = x^4 + x$, find: $f(-3)$ [Section 3.8]

54. Is this the graph of a function? [Section 3.8]

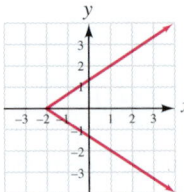

Systems of Linear Equations and Inequalities

4

©Andrei Contiu/Shutterstock.com

from Campus to Careers

Photographer

Photographers record our surroundings, the special events in our lives, and people, so that all can be remembered in pictures. Some specialize in weddings and portraits, some photograph landscapes and fashion, while others work on location as photojournalists. Their job responsibilities require a variety of mathematical skills such as: scheduling appointments, keeping financial records, pricing photographs, purchasing supplies, billing customers, and operating digital equipment.

Problem 79 in **Study Set 4.1, problem 29** in **Study Set 4.4,** and **problem 55** in **Study Set 4.5** involve situations that a photographer might encounter on the job. The mathematical concepts discussed in this chapter can be used to solve those problems.

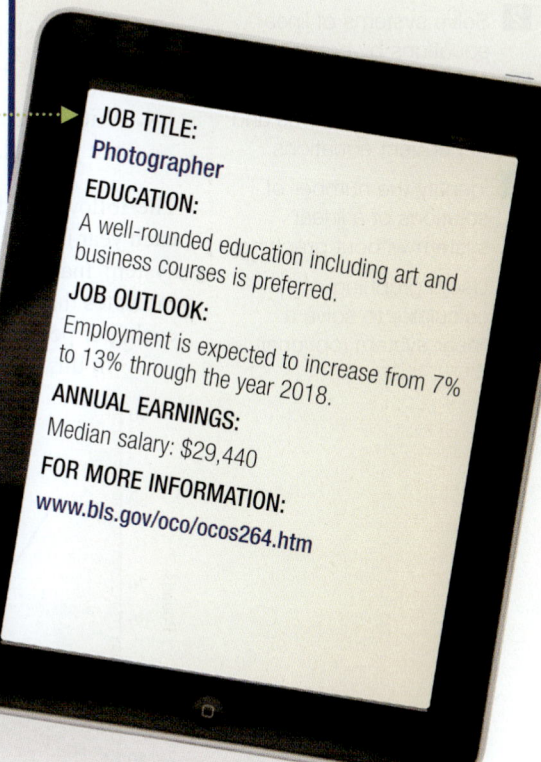

JOB TITLE:
Photographer

EDUCATION:
A well-rounded education including art and business courses is preferred.

JOB OUTLOOK:
Employment is expected to increase from 7% to 13% through the year 2018.

ANNUAL EARNINGS:
Median salary: $29,440

FOR MORE INFORMATION:
www.bls.gov/oco/ocos264.htm

Attending class and taking notes are important, but they are not enough. The only way that you are really going to learn algebra is by doing your homework.

WHEN TO DO YOUR HOMEWORK: Homework should be started on the day it is assigned, when the material is fresh in your mind. It's best to break your homework sessions into 30-minute periods, allowing for short breaks in between.

HOW TO BEGIN YOUR HOMEWORK: Review your notes and the examples in your text before starting your homework assignment.

GETTING HELP WITH YOUR HOMEWORK: It's normal to have some questions when doing homework. Talk to a tutor, a classmate, or your instructor to get those questions answered.

Now Try This ▶

1. Write a one-page paper that describes *when, where,* and *how* you go about completing your algebra homework assignments.
2. For each problem on your next homework assignment, find an example in this book that is similar. Write the example number next to the problem.
3. Make a list of questions that you have while doing your next assignment. Then decide whom you are going to ask to get those questions answered.

SECTION 4.1

OBJECTIVES

1 Determine whether a given ordered pair is a solution of a system.

2 Solve systems of linear equations by graphing.

3 Use graphing to identify inconsistent systems and dependent equations.

4 Identify the number of solutions of a linear system without graphing.

5 Use a graphing calculator to solve a linear system (optional).

Solving Systems of Equations by Graphing

ARE YOU READY?

The following problems review some basic skills that are needed when solving systems of equations by graphing.

1. Is $(-2, -6)$ a solution of $y = 3x - 1$?
2. Use the slope and the y-intercept to graph $y = -4x + 2$.
3. Graph $3x + 4y = 12$ by finding the x- and y-intercepts.
4. Without graphing, determine whether the graphs of $y = \frac{1}{2}x - 3$ and $x - 2y = 2$ are parallel, perpendicular, or neither.

The following illustration shows the average amounts of chicken and beef eaten per person each year in the United States from 1985 to 2010. Plotting both graphs on the same coordinate system makes it easy to compare recent trends. The point of intersection of the graphs indicates that Americans ate equal amounts of chicken and beef in 1992—about 66 pounds of each, per person.

In this section, we will use a similar graphical approach to solve systems of equations.

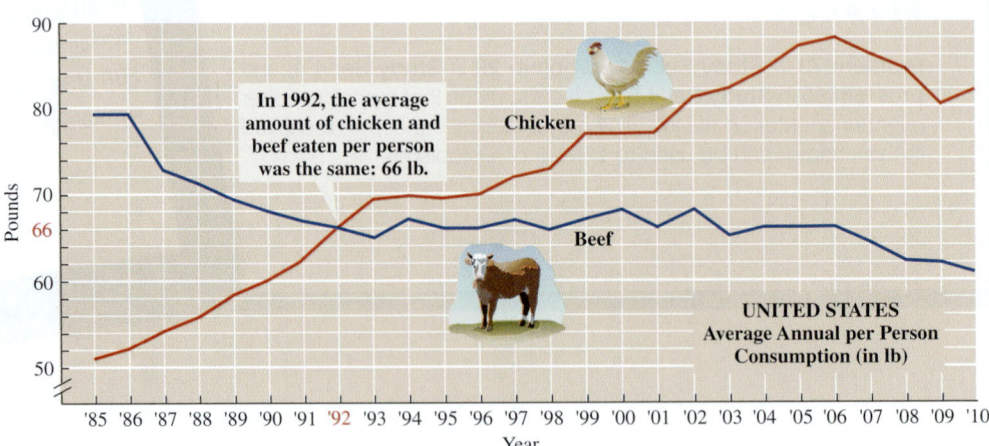

In 1992, the average amount of chicken and beef eaten per person was the same: 66 lb.

Chicken

Beef

UNITED STATES
Average Annual per Person
Consumption (in lb)

Source: U.S. Department of Agriculture

1   Determine Whether a Given Ordered Pair Is a Solution of a System.

We previously have discussed equations in two variables, such as $x + y = 3$. Because there are infinitely many pairs of numbers whose sum is 3, there are infinitely many pairs (x, y) that satisfy this equation. Some of these pairs are listed in table (a).

Now consider the equation $x - y = 1$. Because there are infinitely many pairs of numbers whose difference is 1, there are infinitely many pairs (x, y) that satisfy $x - y = 1$. Some of these pairs are listed in table (b).

The Language of Algebra

We say that $(2, 1)$ **satisfies** $x + y = 3$, because the x-coordinate, 2, and the y-coordinate, 1, make the equation true when substituted for x and y: $2 + 1 = 3$. To *satisfy* means to make content, as in *satisfy* your thirst or a *satisfied* customer.

$x + y = 3$

x	y	(x, y)
0	3	$(0, 3)$
1	2	$(1, 2)$
2	**1**	**$(2, 1)$**
3	0	$(3, 0)$

(a)

$x - y = 1$

x	y	(x, y)
0	-1	$(0, -1)$
1	0	$(1, 0)$
2	**1**	**$(2, 1)$**
3	2	$(3, 2)$

(b)

From the two tables, we see that $(2, 1)$ satisfies both equations.

When two equations with the same variables are considered simultaneously (at the same time), we say that they form a **system of equations.** Using a left brace { , we can write the equations from the previous example as a system:

$$\begin{cases} x + y = 3 \\ x - y = 1 \end{cases}$$ Read as "the system of equations $x + y = 3$ and $x - y = 1$."

Because the ordered pair $(2, 1)$ satisfies both of these equations, it is called a **solution of the system.** In general, a system of linear equations can have exactly one solution, no solution, or infinitely many solutions.

EXAMPLE 1   Determine whether $(-2, 5)$ is a solution of each system of equations.

a. $\begin{cases} 3x + 2y = 4 \\ x - y = -7 \end{cases}$ **b.** $\begin{cases} 4y = 18 - x \\ y = 2x \end{cases}$

Strategy   We will substitute the x- and y-coordinates of $(-2, 5)$ for the corresponding variables in both equations of the system.

Why   If both equations are satisfied (made true) by the x- and y-coordinates, then the ordered pair is a solution of the system.

Solution   **a.** Recall that in an ordered pair, the first number is the x-coordinate and the second number is the y-coordinate. To determine whether $(-2, 5)$ is a solution, we substitute -2 for x and 5 for y in each equation.

The Language of Algebra

A system of equations is two (or more) equations that we consider **simultaneously**—at the same time. On June 3, 2007, in Kansas City, more than 1,680 guitarists set a world record for the most people playing the same song *simultaneously.* The song was Deep Purple's *Smoke on the Water.*

Check: $3x + 2y = 4$ The first equation. $x - y = -7$ The second equation.

$3(-2) + 2(5) \overset{?}{=} 4$ $-2 - 5 \overset{?}{=} -7$

$-6 + 10 \overset{?}{=} 4$ $-7 = -7$ True

$4 = 4$ True

Since $(-2, 5)$ satisfies both equations, it is a solution of the system.

b. Again, we substitute -2 for x and 5 for y in each equation.

Check: $4y = 18 - x$ The first equation. $y = 2x$ The second equation.

$4(5) \overset{?}{=} 18 - (-2)$ $5 \overset{?}{=} 2(-2)$

$20 \overset{?}{=} 18 + 2$ $5 = -4$ False

$20 = 20$ True

Although $(-2, 5)$ satisfies the first equation, it does not satisfy the second. Because it does not satisfy both equations, $(-2, 5)$ is not a solution of the system.

2 Solve Systems of Linear Equations by Graphing.

To **solve a system of equations** means to find all of the solutions of the system. One way to solve a system of linear equations is to graph the equations on the same rectangular coordinate system.

EXAMPLE 2   Solve the system of equations by graphing: $\begin{cases} 2x + 3y = 2 \\ 3x - 2y = 16 \end{cases}$

Strategy   We will graph both equations on the same coordinate system.

Why   Recall that the graph of a linear equation is a "picture" of its solutions. If both equations are graphed on the same coordinate system, we can see whether they have any common solutions.

Solution   The intercept-method is a convenient way to graph equations such as $2x + 3y = 2$ and $3x - 2y = 16$, because they are in standard $Ax + By = C$ form.

Success Tip

Accuracy is crucial when using the graphing method to solve a system. Here are some suggestions for improving your accuracy:

■ Use graph paper.
■ Use a sharp pencil.
■ Use a ruler or straightedge.

$2x + 3y = 2$

x	y	(x, y)
0	$\frac{2}{3}$	$\left(0, \frac{2}{3}\right)$
1	0	$(1, 0)$
-2	2	$(-2, 2)$

$3x - 2y = 16$

x	y	(x, y)
0	-8	$(0, -8)$
$\frac{16}{3}$	0	$\left(\frac{16}{3}, 0\right)$
2	-5	$(2, -5)$

To find the y-intercept, let x = 0 and solve for y.
To find the x-intercept, let y = 0 and solve for x.
As a check, pick another x-value, such as -2 or 2, and find y.

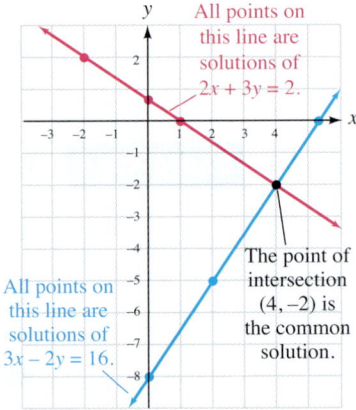

All points on this line are solutions of $2x + 3y = 2$.

All points on this line are solutions of $3x - 2y = 16$.

The point of intersection $(4, -2)$ is the common solution.

Success Tip

When determining the coordinates of a point of intersection from a graph, realize that they are simply estimates. Only after algebraically checking a proposed solution can we be sure that it is an actual solution.

The coordinates of each point on the line graphed in red satisfy $2x + 3y = 2$ and the coordinates of each point on the line graphed in blue satisfy $3x - 2y = 16$. Because the point of intersection is on both graphs, its coordinates satisfy both equations.

It appears that the graphs intersect at the point $(4, -2)$. To verify that it is the solution of the system, we substitute 4 for x and -2 for y in each equation.

Check: $2x + 3y = 2$ *The first equation.* $3x - 2y = 16$ *The second equation.*

$2(4) + 3(-2) \stackrel{?}{=} 2$ $3(4) - 2(-2) \stackrel{?}{=} 16$

$8 + (-6) \stackrel{?}{=} 2$ $12 - (-4) \stackrel{?}{=} 16$

$2 = 2$ *True* $16 = 16$ *True*

Since $(4, -2)$ makes both equations true, it is the solution of the system. The solution set is written as $\{(4, -2)\}$.

To solve a system of linear equations in two variables by graphing, follow these steps.

The Graphing Method	1. Carefully graph each equation on the same rectangular coordinate system. 2. If the lines intersect, determine the coordinates of the point of intersection of the graphs. That ordered pair is the solution of the system. 3. Check the proposed solution in each equation of the original system.

3 Use Graphing to Identify Inconsistent Systems and Dependent Equations.

A system of equations that has at least one solution, like that in Example 2, is called a **consistent system.** A system with no solution is called an **inconsistent system.**

EXAMPLE 3   Solve the system of equations by graphing: $\begin{cases} y = -2x - 6 \\ 4x + 2y = 8 \end{cases}$

Strategy   We will graph both equations on the same coordinate system.

Why   If both equations are graphed on the same coordinate system, we can see whether they have any common solutions.

Solution   Since $y = -2x - 6$ is written in slope–intercept form, we can graph it by plotting the y-intercept $(0, -6)$ and then drawing a slope triangle whose rise is -2 and whose run is 1. We can graph $4x + 2y = 8$ using the intercept method.

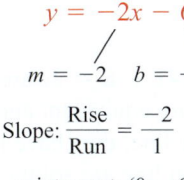

$$y = -2x - 6$$

$$m = -2 \quad b = -6$$

Slope: $\dfrac{\text{Rise}}{\text{Run}} = \dfrac{-2}{1}$

y-intercept: $(0, -6)$

$4x + 2y = 8$

x	y	(x, y)
0	4	$(0, 4)$
2	0	$(2, 0)$
1	2	$(1, 2)$

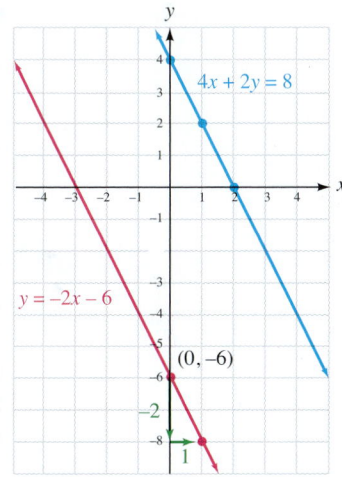

The lines in the graph appear to be parallel. We can verify this by writing the second equation in slope–intercept form and observing that the lines have the same slope, -2, and different y-intercepts, $(0, -6)$ and $(0, 4)$.

$$y = -2x - 6 \qquad 4x + 2y = 8$$

$$2y = -4x + 8 \qquad \text{Subtract 4x from both sides.}$$

$$y = -2x + 4 \qquad \text{To isolate y, divide both sides by 2.}$$

Different y-intercepts

Same slope

Caution

A common error is to graph the parallel lines, but forget to answer with the words *no solution*.

Because the lines are parallel, there is no point of intersection. Such a system has *no solution.* The solution set is the empty set, which is written as { } or as $\varnothing$.

Self Check 3   Solve the system of equations by graphing: $\begin{cases} y = \dfrac{3}{2}x \\ 3x - 2y = 6 \end{cases}$

Now Try ▶ Problem 29

Some systems of equations have infinitely many solutions.

EXAMPLE 4   Solve the system of equations by graphing: $\begin{cases} y = 2x + 4 \\ 4x + 8 = 2y \end{cases}$

Strategy   We will graph both equations on the same coordinate system.

Why   If both equations are graphed on the same coordinate system, we will be able to see if they have any solutions in common.

Solution   To graph $y = 2x + 4$, we use the slope and y-intercept, and to graph $4x + 8 = 2y$, we use the intercept method.

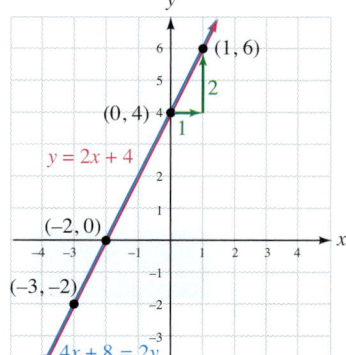

The Language of Algebra

The graphs of these lines **coincide.** That is, they occupy the same location. To illustrate this concept, think of a clock. At noon and at midnight, the hands of the clock coincide.

$$y = 2x + 4$$
$$m = 2 \quad b = 4$$
$$\text{Slope: } \frac{\text{Rise}}{\text{Run}} = \frac{2}{1}$$
$$y\text{-intercept: } (0, 4)$$

$$4x + 8 = 2y$$

x	y	(x, y)
0	4	$(0, 4)$
-2	0	$(-2, 0)$
-3	-2	$(-3, -2)$

The graphs appear to be identical. We can verify this by writing the second equation in slope–intercept form and observing that it is the same as the first equation.

$y = 2x + 4$ The first equation. $\quad 4x + 8 = 2y$ The second equation.

$\qquad\qquad\qquad\qquad\qquad\qquad\qquad\qquad 2y = 4x + 8$ Reverse the sides.

$\qquad\qquad\qquad\qquad\qquad\qquad\qquad\qquad \dfrac{2y}{2} = \dfrac{4x}{2} + \dfrac{8}{2}$ To isolate y, divide both sides by 2.

$\qquad\qquad\qquad\qquad\qquad\qquad\qquad\qquad y = 2x + 4$

This confirms that $y = 2x + 4$ and $4x + 8 = 2y$ are different forms of the same equation. Thus, the equations of this system are equivalent and their graphs are indeed identical.

Since the graphs are the same line, they have infinitely many points in common. All of the points that lie on the common line are solutions because the coordinates of each of those points satisfy both equations of the system. In cases like this, we say that there are *infinitely many solutions.*

From the graph, it appears that four of the infinitely many solutions are $(-3, -2)$, $(-2, 0)$, $(0, 4)$, and $(1, 6)$. Checks for two of these ordered pairs follow.

Caution

A common error is to graph the identical lines, but forget to answer with the words *infinitely many solutions.*

Check $(-3, -2)$:

$4x + 8 = 2y$	$y = 2x + 4$
$4(-3) + 8 \overset{?}{=} 2(-2)$	$-2 \overset{?}{=} 2(-3) + 4$
$-12 + 8 \overset{?}{=} -4$	$-2 \overset{?}{=} -6 + 4$
$-4 = -4$	$-2 = -2$

Check $(0, 4)$:

$4x + 8 = 2y$	$y = 2x + 4$
$4(0) + 8 \overset{?}{=} 2(4)$	$4 \overset{?}{=} 2(0) + 4$
$0 + 8 \overset{?}{=} 8$	$4 \overset{?}{=} 0 + 4$
$8 = 8$	$4 = 4$

Self Check 4   Solve the system of equations by graphing: $\begin{cases} 6x - 4 = 2y \\ y = 3x - 2 \end{cases}$

Now Try ▶ Problem 33

In Examples 2 and 3, the graphs of the equations of the system were different lines. We call equations with different graphs **independent equations.** The equations in Example 4 have the same graph and are equivalent. Because they are different forms of the same equation, they are called **dependent equations.**

There are three possible outcomes when we solve a system of two linear equations using the graphing method.

Solving Systems by Graphing	The two lines intersect at one point.	The two lines are parallel.	The two lines are identical.
	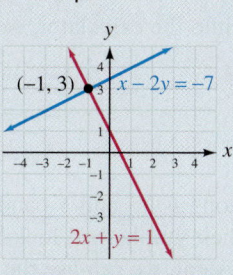	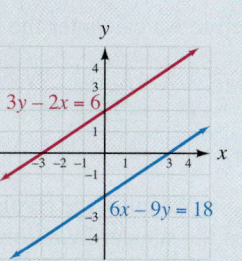	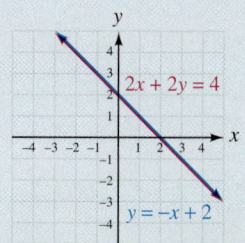
	Exactly one solution (the point of intersection)	No solution	Infinitely many solutions (any point on the line is a solution)
	Consistent system *Independent equations*	*Inconsistent system* *Independent equations*	*Consistent system* *Dependent equations*

4 Identify the Number of Solutions of a Linear System Without Graphing.

We can determine the number of solutions that a system of two linear equations has by writing each equation in slope–intercept form.

- If the lines have different slopes, they intersect, and the system has one solution. (See Example 2.)
- If the lines have the same slope and different y-intercepts, they are parallel, and the system has no solution. (See Example 3.)
- If the lines have the same slope and same y-intercept, they are the same line, and the system has infinitely many solutions. (See Example 4.)

EXAMPLE 5   Without graphing, determine the number of solutions of: $\begin{cases} 5x + y = 5 \\ 3x + 2y = 8 \end{cases}$

Strategy   We will write both equations in slope–intercept form.

Why   We can determine the number of solutions of a linear system by comparing the slopes and y-intercepts of the graphs of the equations.

Solution   To write each equation in slope–intercept form, we solve for y.

$$5x + y = 5 \quad \text{The first equation.} \qquad 3x + 2y = 8 \quad \text{The second equation.}$$
$$y = -5x + 5 \qquad\qquad\qquad 2y = -3x + 8$$
$$y = -\frac{3}{2}x + 4$$

Different slopes

Since the slopes are different, the lines are neither parallel nor identical. Therefore, they will intersect at one point and the system has one solution.

Self Check 5   Without graphing, determine the number of solutions of:
$$\begin{cases} 3x + 6y = 1 \\ 2x + 4y = 0 \end{cases}$$

Now Try ▶ Problem 41

5 Use a Graphing Calculator to Solve a Linear System (Optional).

A graphing calculator can be used to solve systems of equations, such as

$$\begin{cases} 2x + y = 12 \\ 2x - y = -2 \end{cases}$$

Before we can enter the equations into the calculator, we must solve them for y.

$2x + y = 12$ *The first equation.* $2x - y = -2$ *The second equation.*

     $y = -2x + 12$ $-y = -2x - 2$

                                      $y = 2x + 2$

We enter the resulting equations as Y_1 and Y_2 and graph them on the same axes. If we use the standard window setting, their graphs will look like figure (a).

To find the solution of the system, we can use the INTERSECT feature found on most graphing calculators. With this feature, after pushing enter three times to identify each graph and a guess for the point, the cursor automatically moves to the point of intersection of the graphs and displays the coordinates of that point. In figure (b), we see that the solution is $(2.5, 7)$.

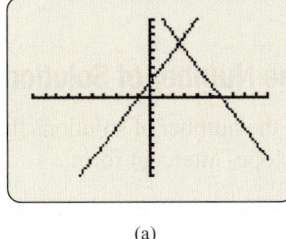 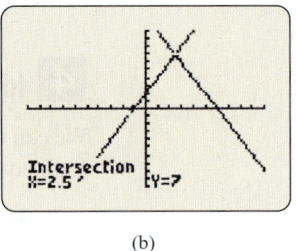

                (a)   (b)

SECTION 4.1   STUDY SET

VOCABULARY

Fill in the blanks.

1. The pair of equations $\begin{cases} x - y = -1 \\ 2x - y = 1 \end{cases}$ is called a _____ of linear equations.

2. Because the ordered pair $(2, 3)$ satisfies both equations in Problem 1, it is called a _____ of the system of equations.

3. The point of _____ of the lines graphed in part (a) below is $(1, 2)$.

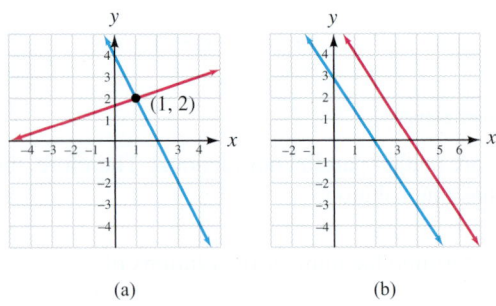

       (a)   (b)

4. The lines graphed in part (b) above do not intersect. They are _____ lines.

5. A system of equations that has at least one solution is called a _____ system. A system with no solution is called an _____ system.

6. We call equations with different graphs _____ equations. Because _____ equations are different forms of the same equation, they have the same graph.

CONCEPTS

7. Refer to the illustration below.
   a. If the coordinates of point A are substituted into the equation for Line 1, will the result be true or false?
   b. If the coordinates of point C are substituted into the equation for Line 1, will the result be true or false?

8. Refer to the illustration on the right.
   a. If the coordinates of point C are substituted into the equation for Line 2, will the result be true or false?
   b. If the coordinates of point B are substituted into the equation for Line 1, will the result be true or false?

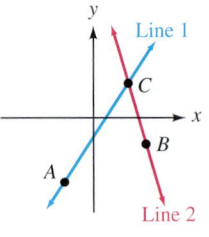

9. a. To graph $5x - 2y = 10$, we can use the intercept method. Complete the table.

x	y
0	
	0

b. To graph $y = 3x - 2$, we can use the slope and y-intercept. Fill in the blanks.

Slope: $\blacksquare = \dfrac{\blacksquare}{1}$ y-intercept: $\blacksquare$

10. What is the apparent solution of the system graphed in figure (a) below? Is the system consistent or inconsistent?

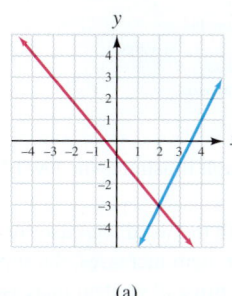 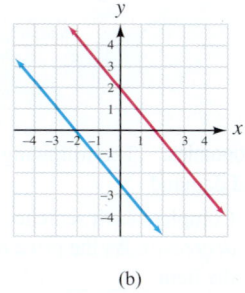

(a)   (b)

11. How many solutions does the system graphed in figure (b) above have? Are the equations dependent or independent?

12. How many solutions does the system graphed on the right have? Give three of the solutions. Is the system consistent or inconsistent?

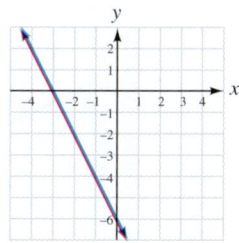

GUIDED PRACTICE

Determine whether the ordered pair is a solution of the given system of equations. See Example 1.

13. $(1, 1)$, $\begin{cases} x + y = 2 \\ 2x - y = 1 \end{cases}$

14. $(1, 3)$, $\begin{cases} 2x + y = 5 \\ 3x - y = 0 \end{cases}$

15. $(3, -2)$, $\begin{cases} 2x + y = 4 \\ y = 1 - x \end{cases}$

16. $(-2, 4)$, $\begin{cases} 2x + 2y = 4 \\ 3y = 10 - x \end{cases}$

17. $(12, 0)$, $\begin{cases} x - 9y = 12 \\ y = 10 - x \end{cases}$

18. $(15, 0)$, $\begin{cases} x - 2y = 15 \\ y = 16 - x \end{cases}$

19. $(-2, -4)$, $\begin{cases} 4x + 5y = -23 \\ -3x + 2y = 0 \end{cases}$

20. $(-5, 2)$, $\begin{cases} -2x + 7y = 17 \\ 3x - 4y = -19 \end{cases}$

21. $\left(\dfrac{1}{2}, 3\right)$, $\begin{cases} 2x + y = 4 \\ 4x - 11 = 3y \end{cases}$

22. $\left(2, \dfrac{1}{3}\right)$, $\begin{cases} x - 3y = 1 \\ -2x + 6 = -6y \end{cases}$

23. $(2.5, 3.5)$, $\begin{cases} 4x - 3 = 2y \\ 4y + 1 = 6x \end{cases}$

24. $(0.2, 0.3)$, $\begin{cases} 20x + 10y = 7 \\ 20y = 15x + 3 \end{cases}$

Solve each system of equations by graphing. See Example 2.

25. $\begin{cases} 2x + 3y = 12 \\ 2x - y = 4 \end{cases}$

26. $\begin{cases} 5x + y = 5 \\ 5x + 3y = 15 \end{cases}$

27. $\begin{cases} x + y = 4 \\ x - y = -6 \end{cases}$

28. $\begin{cases} x + y = 4 \\ x - y = -2 \end{cases}$

Solve each system by graphing. If a system has no solution or infinitely many solutions, so state. See Example 3.

29. $\begin{cases} y = -\dfrac{1}{3}x - 4 \\ x + 3y = 6 \end{cases}$

30. $\begin{cases} y = -\dfrac{1}{2}x - 3 \\ x + 2y = 2 \end{cases}$

31. $\begin{cases} y = 3x \\ y - 3x = -3 \end{cases}$

32. $\begin{cases} y = -2x \\ 2x + y = -2 \end{cases}$

Solve each system by graphing. If a system has no solution or infinitely many solutions, so state. See Example 4.

33. $\begin{cases} y = x - 1 \\ 3x - 3y = 3 \end{cases}$

34. $\begin{cases} y = -x + 1 \\ 4x + 4y = 4 \end{cases}$

35. $\begin{cases} 4x + 6y = 12 \\ 2x + 3y = 6 \end{cases}$

36. $\begin{cases} 2x - y = 0 \\ 2y - 4x = 0 \end{cases}$

Find the slope and the y-intercept of the graph of each line in the system of equations. Then, use that information to determine the number of solutions of the system. See Example 5.

37. $\begin{cases} y = 6x - 7 \\ y = -2x + 1 \end{cases}$

38. $\begin{cases} y = \dfrac{1}{2}x + 8 \\ y = 4x - 10 \end{cases}$

39. $\begin{cases} 3x - y = -3 \\ y - 3x = 3 \end{cases}$

40. $\begin{cases} x + 4y = 4 \\ 12y = 12 - 3x \end{cases}$

41. $\begin{cases} x + y = 6 \\ x + y = 8 \end{cases}$

42. $\begin{cases} 5x + y = 0 \\ 5x + y = 6 \end{cases}$

43. $\begin{cases} 6x + y = 0 \\ 2x + 2y = 0 \end{cases}$

44. $\begin{cases} x + y = 1 \\ 2x - 2y = 5 \end{cases}$

Use a graphing calculator to solve each system. See Objective 5.

45. $\begin{cases} y = 4 - x \\ y = 2 + x \end{cases}$

46. $\begin{cases} 3x - 6y = 4 \\ 2x + y = 1 \end{cases}$

47. $\begin{cases} 6x - 2y = 5 \\ 3x = y + 10 \end{cases}$

48. $\begin{cases} x - 3y = -2 \\ 5x + y = 10 \end{cases}$

TRY IT YOURSELF

Solve each system of equations by graphing.

49. $\begin{cases} y = 3x + 6 \\ y = -2x - 4 \end{cases}$

50. $\begin{cases} y = x + 3 \\ y = -2x - 3 \end{cases}$

51. $\begin{cases} 2y = 3x + 2 \\ 3x - 2y = 6 \end{cases}$

52. $\begin{cases} 3x - 6y = 18 \\ x = 2y + 3 \end{cases}$

53. $\begin{cases} x + y = 2 \\ y = x - 4 \end{cases}$

54. $\begin{cases} x + y = 1 \\ y = x + 5 \end{cases}$

55. $\begin{cases} x = 3 \\ 3y = 6 - 2x \end{cases}$

56. $\begin{cases} x = 4 \\ 2y = 12 - 4x \end{cases}$

57. $\begin{cases} y = \dfrac{3}{4}x + 3 \\ y = -\dfrac{x}{4} - 1 \end{cases}$

58. $\begin{cases} y = \dfrac{2}{3}x + 4 \\ y = -\dfrac{x}{3} + 7 \end{cases}$

59. $\begin{cases} y = -x - 2 \\ y = -3x + 6 \end{cases}$

60. $\begin{cases} y = 2x - 4 \\ y = -5x + 3 \end{cases}$

61. $\begin{cases} -x + 3y = -11 \\ 3x - y = 17 \end{cases}$

62. $\begin{cases} 2x - 3y = -18 \\ 3x + 2y = -1 \end{cases}$

63. $\begin{cases} x + y = 2 \\ y = x \end{cases}$

64. $\begin{cases} x + y = 4 \\ y = x \end{cases}$

65. $\begin{cases} 4x - 2y = 8 \\ y = 2x - 4 \end{cases}$

66. $\begin{cases} 2y = -6x - 12 \\ 3x + y = -6 \end{cases}$

67. $\begin{cases} x + 4y = -2 \\ y = -x - 5 \end{cases}$

68. $\begin{cases} 3x + 2y = -8 \\ 2x - 3y = -1 \end{cases}$

69. $\begin{cases} y = -3 \\ -x + 2y = -4 \end{cases}$

70. $\begin{cases} y = -4 \\ -2x - y = 8 \end{cases}$

71. $\begin{cases} x + 2y = -4 \\ x - \dfrac{1}{2}y = 6 \end{cases}$

72. $\begin{cases} \dfrac{2}{3}x - y = -3 \\ 3x + y = 3 \end{cases}$

APPLICATIONS

73. Social Networks. Refer to the graph below. In what month did Facebook and MySpace have the same number of unique visitors? Estimate the number.

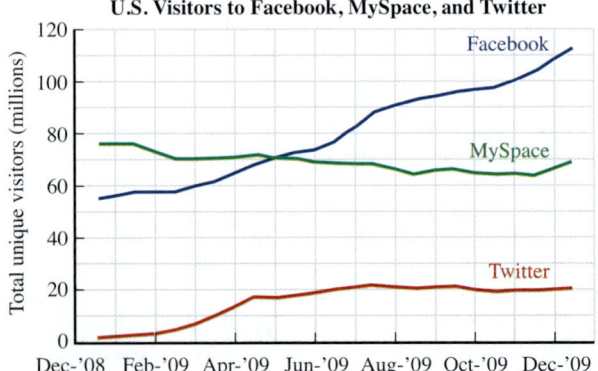

U.S. Visitors to Facebook, MySpace, and Twitter

Source: Comscore Media Metrix

74. Beverages. Refer to the graph below. In what year was average number of gallons of milk and carbonated soft drinks consumed per person the same? Estimate the number of gallons.

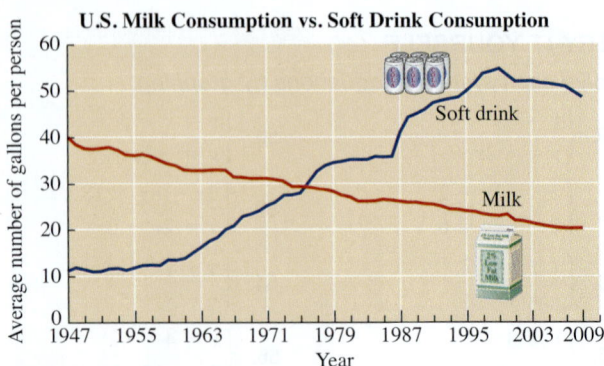

U.S. Milk Consumption vs. Soft Drink Consumption

Source: USDA. Economic Research Service

75. Latitude and Longitude. Refer to the following map.
   a. Name three American cities that lie on a latitude line of 30°.
   b. Name three American cities that lie on a longitude line of −90°.
   c. What city lies on both lines?

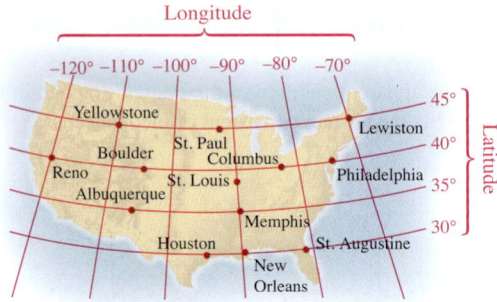

76. Economics. The following graph illustrates the law of supply and demand.
   a. Complete each sentence with the word *increases* or *decreases*. As the price of an item increases, the supply of the item _____. As the price of an item increases, the demand for the item _____.
   b. For what price will the supply equal the demand? How many items will be supplied for this price?

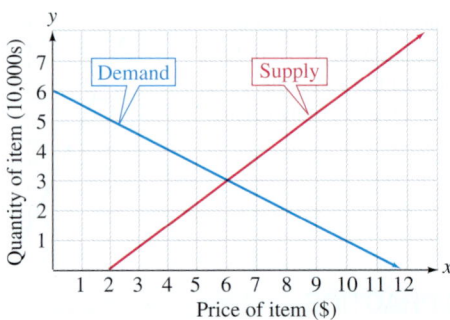

77. Daily Tracking Polls. Refer to the graph below.
   a. Which political candidate was ahead on October 28 and by how much?
   b. On what day did the challenger pull even with the incumbent?
   c. If the election was held November 4, who did the poll predict would win, and by how many percentage points?

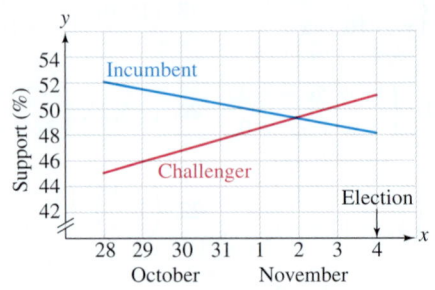

78. Air Traffic Control. The equations describing the paths of two airplanes are $y = -\frac{1}{2}x + 3$ and $3y = 2x + 2$. Graph each equation on the radar screen shown. Is there a possibility of a midair collision? If so, where?

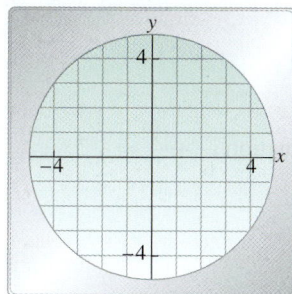

79. from **Campus to Careers**

Photographer

Photographers often use the *rule of thirds* to add more interest to a photo rather than simply centering the subject. They imagine two horizontal and two vertical lines dividing the image into a grid of 9 equal parts (like a tic-tac-toe game). The four points where the lines intersect are possible "sweet spots" for placing the subject of the picture. Give the coordinates of the four sweet spots for the image below. Is the sea gull properly placed?

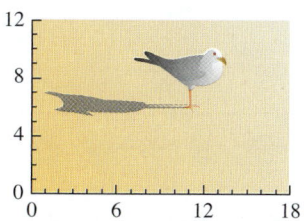

© Andrei Contiu/Shutterstock.com

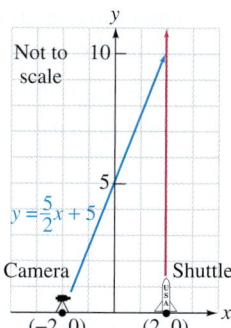

80. TV Coverage. A television camera is located at $(-2, 0)$ and will follow the launch of a space shuttle, as shown here. (Each unit in the illustration is 1 mile.) As the shuttle rises vertically on a path described by $x = 2$, the farthest the camera can tilt back is a line of sight given by $y = \frac{5}{2}x + 5$. For how many miles of the shuttle's flight will it be in view of the camera?

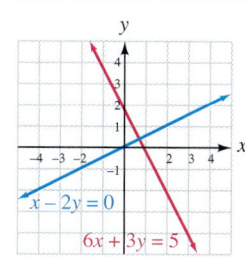

WRITING

81. Explain why it is difficult to determine the solution of the system in the graph.

82. Without graphing, how can you tell that the graphs of $y = 2x + 1$ and $y = 3x + 2$ intersect?

83. Could a system of two linear equations have exactly two solutions? Explain why or why not.

84. What is an inconsistent system? What are dependent equations?

85. Suppose the graphs of the two linear equations of a system are the same line. What is wrong with the following statement? *The system has infinitely many solutions. Any ordered pair is a solution of the system.*

86. The Swine Flu. The graph below is for the month of January, 2010. It shows the total number of doses of the H1N1 Swine Flu vaccine that had been produced to date. It also shows the total number of doses that U.S. health officials had ordered to date. Did the total number of doses ordered ever equal or surpass the total number of doses produced? How can you tell?

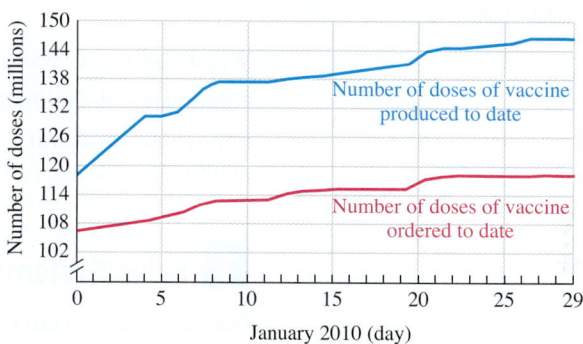

REVIEW

Solve each inequality. Write the solution set in interval notation and graph it.

87. $-4(2y + 2) \leq 4y + 28$

88. $-5 < 3t + 4 \leq 13$

89. $-1 \leq -\frac{1}{2}n$

90. $\frac{1}{3} + \frac{c}{5} > -\frac{3}{2}$

CHALLENGE PROBLEMS

91. Can a system of two linear equations in two variables be inconsistent but have dependent equations? Explain.

92. Construct a system of two linear equations that has a solution of $(-2, 6)$.

93. Write a system of two linear equations such that $(2, 3)$ is a solution of the first equation but is not a solution of the second equation.

94. Solve by graphing: $\begin{cases} \dfrac{1}{3}x - \dfrac{1}{2}y = \dfrac{1}{6} \\ \dfrac{2x}{5} + \dfrac{y}{2} = \dfrac{13}{10} \end{cases}$

SECTION 4.2

Solving Systems of Equations by Substitution

OBJECTIVES

1 Solve systems of linear equations by substitution.

2 Find a substitution equation.

3 Solve systems of linear equations that contain fractions.

4 Use substitution to identify inconsistent systems and dependent equations.

ARE YOU READY?

The following problems review some basic skills that are needed when solving systems of equations by substitution.

1. In $8x + y = 2$, what is the coefficient of y?

2. Solve $2x - y = -3$ for y.

3. What is the LCD for the fractions in the equation $\dfrac{x}{6} + \dfrac{y}{9} = \dfrac{3}{2}$?

4. Substitute 4 for x in $y = 5x - 3$ and find y.

5. Multiply: $8\left(\dfrac{3x}{8}\right)$

When solving a system of equations by graphing, it is often difficult to determine the coordinates of the intersection point. For example, a solution of $\left(\frac{7}{8}, \frac{3}{5}\right)$ would be almost impossible to identify accurately. In this section, we will discuss a second, more precise method for solving systems that does not involve graphing.

1 Solve Systems of Linear Equations by Substitution.

One algebraic method for solving a system of equations is the **substitution method.** It is introduced in the following example.

EXAMPLE 1   Solve the system: $\begin{cases} y = 3x - 2 \\ 2x + y = 8 \end{cases}$

Strategy   Note that the first equation is solved for y. Because y and $3x - 2$ are equal (represent the same value), we will substitute $3x - 2$ for y in the second equation.

Why   The objective is to obtain one equation containing only one unknown. When $3x - 2$ is substituted for y in the second equation, the result will be just that—an equation in one variable, x.

Solution   Since the right side of $y = 3x - 2$ is used to make a substitution, $y = 3x - 2$ is called the **substitution equation.**

Success Tip

Throughout the course, we have been substituting numbers for variables. With this method, we substitute a *variable expression for a variable.*

$$\begin{cases} y = \boxed{3x - 2} \\ 2x + \boxed{y} = 8 \end{cases}$$

To find the solution of the system, we proceed as follows:

$2x + y = 8$ This is the second equation of the system. It has two variables.

$2x + \mathbf{3x - 2} = 8$ Substitute $3x - 2$ for y. This equation has only one variable.

The resulting equation can be solved for x.

Caution

When using the substitution method, a common error is to find the value of one of the variables, say x, and forget to find the value of the other. Remember that a solution of a linear system of two equations is an ordered pair (x, y).

$2x + 3x - 2 = 8$

$5x - 2 = 8$ Combine like terms: $2x + 3x = 5x$.

$5x = 10$ To isolate the variable term, $5x$, add 2 to both sides.

$x = 2$ Divide both sides by 5. This is the x-value of the solution.

We can find the y-value of the solution by substituting 2 for x in either equation of the original system. We will use the substitution equation because it is already solved for y.

$y = 3x - 2$ *This is the substitution equation.*

$y = 3(2) - 2$ *Substitute 2 for x.*

$y = 6 - 2$ *Do the multiplication.*

$y = 4$ *This is the y-value of the solution. We would have obtained the same result if we had substituted 2 for x in 2x + y = 8 and solved for y.*

The ordered pair $(2, 4)$ appears to be the solution of the system. To check, we substitute 2 for x and 4 for y in each equation.

Check: $y = 3x - 2$ *The first equation.* $2x + y = 8$ *The second equation.*

$4 \stackrel{?}{=} 3(2) - 2$ $2(2) + 4 \stackrel{?}{=} 8$

$4 \stackrel{?}{=} 6 - 2$ $4 + 4 \stackrel{?}{=} 8$

$4 = 4$ *True* $8 = 8$ *True*

Since $(2, 4)$ satisfies both equations, it is the solution. The solution set is written as $\{(2, 4)\}$. A graph of the equations of the system shows an intersection point of $(2, 4)$. This illustrates an important fact: *The solution found using the substitution method will be the same as the solution found using the graphing method.*

Self Check 1	Solve the system: $\begin{cases} x + 4y = 7 \\ x = 6y - 3 \end{cases}$
Now Try ▶ Problem 15	

The substitution method works well for solving systems where one equation is solved, or can be easily solved, for one of the variables. To solve a system of equations in x and y by the substitution method, follow these steps.

The Substitution Method ▼

1. Solve one of the equations for either x or y. If this is already done, go to step 2. (We call this equation the **substitution equation.**)

2. Substitute the expression for x or for y obtained in step 1 into the other equation and solve that equation.

3. Substitute the value of the variable found in step 2 into the substitution equation to find the value of the remaining variable.

4. Check the proposed solution in each equation of the original system. Write the solution as an ordered pair.

EXAMPLE 2   Solve the system: $\begin{cases} 4x + 27 = 7y \\ x = -5y \end{cases}$

Strategy   We will use the substitution method to solve this system.

Why   The substitution method works well when one of the equations of the system (in this case, $x = -5y$) is solved for a variable.

Solution   **Step 1:**   Because x and $-5y$ represent the same value, we can substitute $-5y$ for x in the first equation.

Success Tip

The basic objective of this method is to use an appropriate substitution to obtain one equation in one variable.

$\begin{cases} 4x + 27 = 7y \\ x = \boxed{-5y} \end{cases}$ *This is the substitution equation.*

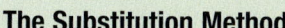

Step 2:   When we substitute $-5y$ for x in the first equation, the resulting equation contains only one variable, and it can be solved for y.

$$4x + 27 = 7y \qquad \text{This is the first equation of the system. It has two variables.}$$

$$4(\mathbf{-5y}) + 27 = 7y \qquad \text{Substitute } -5y \text{ for } x. \text{ Don't forget the parentheses.}$$
$$\text{This equation has only one variable.}$$

$$-20y + 27 = 7y \qquad \text{Do the multiplication.}$$

$$27 = 27y \qquad \text{To eliminate } -20y \text{ on the left side, add } 20y \text{ to both sides.}$$

$$1 = y \qquad \text{Divide both sides by 27. This is the } y\text{-value of the solution.}$$

Caution

We don't have to find the values of the variables in alphabetical order. Here, we found y first.

Step 3:   To find x, substitute 1 for y in the equation $x = -5y$.

$$x = -5\mathbf{y} \qquad \text{This is the substitution equation.}$$

$$x = -5(\mathbf{1}) \qquad \text{Substitute 1 for } y.$$

$$x = -5 \qquad \text{This is the } x\text{-value of the solution.}$$

Step 4:   The following check verifies that the solution is $(-5, 1)$.

Check:
$$4x + 27 = 7y \qquad \text{The first equation.} \qquad\qquad x = -5y \qquad \text{The second equation.}$$
$$4(\mathbf{-5}) + 27 \stackrel{?}{=} 7(\mathbf{1}) \qquad\qquad\qquad\qquad\qquad -5 \stackrel{?}{=} -5(\mathbf{1})$$
$$-20 + 27 \stackrel{?}{=} 7 \qquad\qquad\qquad\qquad\qquad\qquad -5 = -5 \qquad \text{True}$$
$$7 = 7 \qquad \text{True}$$

Self Check 2   Solve the system: $\begin{cases} 3x + 40 = 8y \\ x = -4y \end{cases}$

Now Try ▶ Problem 17

---

2 Find a Substitution Equation.

Sometimes neither equation of a system is solved for a variable. In such cases, we can find a substitution equation by solving one of the equations for one of its variables.

EXAMPLE 3   Solve the system: $\begin{cases} 4x + y = 3 \\ 3x + 5y = 15 \end{cases}$

Strategy   Since the system does not contain an equation solved for x or y, we must choose an equation and solve it for x or y. We will solve for y in the first equation, because y has a coefficient of 1. Then we will use the substitution method to solve the system.

Why   Solving $4x + y = 3$ for x, or solving $3x + 5y = 15$ for x or y, would involve working with cumbersome fractions.

Solution   **Step 1:**   To find a substitution equation, we proceed as follows:

Success Tip

To find a substitution equation, solve one of the equations of the system for one of its variables. If possible, solve for a variable whose coefficient is 1 or -1 to avoid working with fractions.

$$4x + y = 3 \qquad \text{This is the first equation of the system. Think: } 4x + 1y = 3.$$

$$y = 3 - 4x \qquad \text{To isolate } y, \text{ subtract } 4x \text{ from both sides.}$$
$$\text{This is the substitution equation.}$$

Because y and $3 - 4x$ are equal, we can substitute $3 - 4x$ for y in the second equation of the system.

$$\begin{cases} 4x + y = 3 \quad \rightarrow \quad y = \boxed{3 - 4x} \\ 3x + 5y = 15 \end{cases}$$

Step 2: When we substitute for y in the second equation, the resulting equation contains only one variable and can be solved for x.

$$3x + 5y = 15 \qquad \text{This is the second equation of the system. It has two variables.}$$
$$3x + 5(\mathbf{3 - 4x}) = 15 \qquad \text{Substitute } 3 - 4x \text{ for } y. \text{ Don't forget the parentheses.}$$
$$3x + 15 - 20x = 15 \qquad \text{Distribute the multiplication by 5.}$$
$$15 - 17x = 15 \qquad \text{Combine like terms.}$$
$$-17x = 0 \qquad \text{To isolate the variable term, } -17x, \text{ subtract 15 from both sides.}$$
$$x = 0 \qquad \text{Divide both sides by } -17. \text{ This is the x-value of the solution.}$$

Step 3: To find y, substitute 0 for x in the equation $y = 3 - 4x$.

$$y = 3 - 4\mathbf{x} \qquad \text{This is the substitution equation.}$$
$$y = 3 - 4(\mathbf{0}) \qquad \text{Substitute 0 for x.}$$
$$y = 3 - 0$$
$$y = 3 \qquad \text{This is the y-value of the solution.}$$

Step 4: The solution appears to be $(0, 3)$. Check it in the original equations.

Check:

$4x + y = 3$ The first equation.	$3x + 5y = 15$ The second equation.
$4(\mathbf{0}) + \mathbf{3} \stackrel{?}{=} 3$	$3(\mathbf{0}) + 5(\mathbf{3}) \stackrel{?}{=} 15$
$0 + 3 \stackrel{?}{=} 3$	$0 + 15 \stackrel{?}{=} 15$
$3 = 3$ True	$15 = 15$ True

Self Check 3 Solve the system: $\begin{cases} 2x - 3y = 10 \\ 3x + y = 15 \end{cases}$

Now Try ▶ Problem 21

EXAMPLE 4 Solve the system: $\begin{cases} 3a - 3b = 5 \\ 3 - a = -2b \end{cases}$

Strategy Since the coefficient of a in the second equation is -1, we will solve that equation for a. Then we will use the substitution method to solve the system.

Why If we solve for the variable with a numerical coefficient of -1, we can avoid having to work with fractions.

Solution **Step 1:** To find a substitution equation, we proceed as follows:

$$3 - a = -2b \qquad \text{This is the second equation of the system. Think: } 3 - 1a = -2b.$$
$$-a = -2b - 3 \qquad \text{To isolate the variable term, } -a, \text{ subtract 3 from both sides.}$$

To obtain a on the left side, multiply both sides of the equation by -1.

$$\mathbf{-1}(-a) = \mathbf{-1}(-2b - 3) \qquad \text{Don't forget the parentheses.}$$
$$a = 2b + 3 \qquad \text{Do the multiplication. This is the substitution equation.}$$

Because a and $2b + 3$ represent the same value, we can substitute $2b + 3$ for a in the first equation.

$$\begin{cases} 3\mathbf{a} - 3b = 5 \\ 3 - a = -2b \rightarrow a = \boxed{2b + 3} \end{cases}$$

Step 2: Substitute $2b + 3$ for a in the first equation and solve for b.

$$3a - 3b = 5 \qquad \text{This is the first equation of the system. It has two variables.}$$

$$3(\boldsymbol{2b + 3}) - 3b = 5 \qquad \text{Substitute } 2b + 3 \text{ for } a. \text{ Don't forget the parentheses.}$$

$$6b + 9 - 3b = 5 \qquad \text{Distribute the multiplication by 3.}$$

$$3b + 9 = 5 \qquad \text{Combine like terms: } 6b - 3b = 3b.$$

$$3b = -4 \qquad \text{To isolate the variable term, } 3b, \text{ subtract 9 from both sides.}$$

$$b = -\frac{4}{3} \qquad \text{Divide both sides by 3. This is the } b\text{-value of the solution.}$$

Step 3: To find a, substitute $-\frac{4}{3}$ for b in the equation $a = 2b + 3$.

$$a = 2\boldsymbol{b} + 3 \qquad \text{This is the substitution equation.}$$

$$a = 2\left(-\frac{4}{3}\right) + 3 \qquad \text{Substitute } -\frac{4}{3} \text{ for } b.$$

$$a = -\frac{8}{3} + \frac{9}{3} \qquad \begin{array}{l} \text{Do the multiplication. To add, we must write 3} \\ \text{as a fraction with a denominator of 3: } 3 = \frac{9}{3}. \end{array}$$

$$a = \frac{1}{3} \qquad \text{Add the fractions. This is the } a\text{-value of the solution.}$$

Step 4: The solution is $\left(\frac{1}{3}, -\frac{4}{3}\right)$. Check it in the original equations.

Self Check 4 Solve the system: $\begin{cases} 2s - t = 4 \\ 3s - 5t = 2 \end{cases}$

Now Try ▶ Problem 25

3 Solve Systems of Linear Equations that Contain Fractions.

It is usually helpful to clear any equations of fractions and combine any like terms before performing a substitution.

EXAMPLE 5 Solve the system: $\begin{cases} \dfrac{y}{4} = -\dfrac{x}{2} - \dfrac{3}{4} \\ 2x - y = -1 + y - x \end{cases}$

Strategy We will use properties of equality to write each equation of the system in simpler, equivalent form. Then we will use the substitution method to solve the resulting system.

Why The first equation will be easier to work with if we clear it of fractions. The second equation will be easier to work with if we eliminate the variable terms on the right side.

Solution We can clear the first equation of fractions by multiplying both sides by the LCD.

$$\frac{y}{4} = -\frac{x}{2} - \frac{3}{4} \qquad \text{This is the first equation of the system.}$$

$$4\left(\frac{y}{4}\right) = 4\left(-\frac{x}{2} - \frac{3}{4}\right) \qquad \begin{array}{l} \text{Multiply both sides by the LCD, 4.} \\ \text{Don't forget the parentheses.} \end{array}$$

$$4\left(\frac{y}{4}\right) = 4\left(-\frac{x}{2}\right) - 4\left(\frac{3}{4}\right) \qquad \text{Distribute the multiplication by 4.}$$

$$\textbf{(1)} \qquad y = -2x - 3 \qquad \text{Simplify. Call this equation 1.}$$

We can write the second equation of the system in standard $Ax + By = C$ form by adding x and subtracting y from both sides.

$$2x - y = -1 + y - x \qquad \text{\color{red}This is the second equation of the system.}$$
$$2x - y + x - y = -1 + y - x + x - y$$
$$\textbf{(2)} \qquad 3x - 2y = -1 \qquad \text{\color{red}Combine like terms. Call this equation 2.}$$

Step 1: Equations 1 and 2 form an equivalent system, which has the same solution as the original one. To find x, we proceed as follows:

$$\textbf{(1)} \quad \begin{cases} y = -2x - 3 & \text{\color{red}This is the substitution equation.} \\ \textbf{(2)} \quad 3x - 2y = -1 \end{cases}$$

Step 2: To find x, substitute $-2x - 3$ for y in equation 2 and proceed as follows:

$$3x - 2y = -1 \qquad \text{\color{red}This is equation 2. It has two variables.}$$
$$3x - 2(-2x - 3) = -1 \qquad \text{\color{red}Substitute } -2x - 3 \text{ for y. Don't forget the parentheses.}$$
$$3x + 4x + 6 = -1 \qquad \text{\color{red}Distribute the multiplication by } -2.$$
$$7x + 6 = -1 \qquad \text{\color{red}Combine like terms: } 3x + 4x = 7x.$$
$$7x = -7 \qquad \text{\color{red}To isolate the variable term, 7x, subtract 6 from both sides.}$$
$$x = -1 \qquad \text{\color{red}Divide both sides by 7. This is the x-value of the solution.}$$

Step 3: To find y, we substitute -1 for x in equation 1.

$$y = -2x - 3 \qquad \text{\color{red}This is equation 1.}$$
$$y = -2(-1) - 3 \qquad \text{\color{red}Substitute } -1 \text{ for x.}$$
$$y = 2 - 3 \qquad \text{\color{red}Do the multiplication.}$$
$$y = -1 \qquad \text{\color{red}Subtract. This is the y-value of the solution.}$$

Step 4: The solution is $(-1, -1)$. Check it in the original system.

Self Check 5 Solve the system: $\begin{cases} \dfrac{y}{6} = \dfrac{x}{3} + \dfrac{1}{2} \\ 2x - y = -3 + y - x \end{cases}$

Now Try ▶ Problem 29

4 Use Substitution to Identify Inconsistent Systems and Dependent Equations.

In the previous section, we solved inconsistent systems and systems of dependent equations graphically. We also can solve these systems using the substitution method.

EXAMPLE 6   Solve the system: $\begin{cases} 4y - 12 = x \\ y = \dfrac{1}{4}x \end{cases}$

Strategy   We will use the substitution method to solve this system.

Why   The substitution method works well when one of the equations of the system $\left(\text{in this case, } y = \frac{1}{4}x\right)$ is solved for a variable.

Solution   To try to solve this system, substitute $\frac{1}{4}x$ for y in the first equation and solve for x.

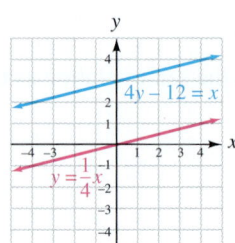

$$4y - 12 = x \qquad \text{\color{red}{This is the first equation of the system.}}$$

$$4\left(\frac{1}{4}x\right) - 12 = x \qquad \text{\color{red}{Substitute $\frac{1}{4}x$ for y.}}$$

$$x - 12 = x \qquad \text{\color{red}{Do the multiplication: $4\left(\frac{1}{4}\right) = 1$.}}$$

$$x - 12 - x = x - x \qquad \text{\color{red}{To eliminate x on the right side, subtract x from both sides.}}$$

$$-12 = 0 \qquad \text{\color{red}{False}}$$

Here, the terms involving x drop out, and we get $-12 = 0$. This false statement indicates that the system has **no solution** and is inconsistent. The solution set is the empty set, $\varnothing$. The graphs of the equations of the system help to verify this; they are parallel lines.

Self Check 6   Solve the system: $\begin{cases} x - 4 = y \\ -2y = 4 - 2x \end{cases}$

Now Try ▶ Problem 33

EXAMPLE 7   Solve the system: $\begin{cases} x = -3y + 6 \\ 2x + 6y = 12 \end{cases}$

Strategy   We will use the substitution method to solve this system.

Why   The substitution method works well when one of the equations of the system (in this case, $x = -3y + 6$) is solved for a variable.

Solution   To solve this system, substitute $-3y + 6$ for x in the second equation and solve for y.

$$2x + 6y = 12 \qquad \text{\color{red}{This is the second equation of the system.}}$$

$$2(-3y + 6) + 6y = 12 \qquad \text{\color{red}{Substitute $-3y + 6$ for x.}}$$
$$\text{\color{red}{Don't forget the parentheses.}}$$

$$-6y + 12 + 6y = 12 \qquad \text{\color{red}{Distribute the multiplication by 2.}}$$

$$12 = 12 \qquad \text{\color{red}{True}}$$

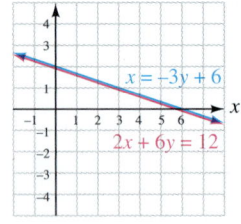

Here, the terms involving y drop out, and we get $12 = 12$. This true statement indicates that the two equations of the system are equivalent. Therefore, they are dependent equations and the system has **infinitely many solutions**. The graphs of the equations help to verify this; they are the same line.

Any ordered pair that satisfies one equation of this system also satisfies the other. To find several of the infinitely many solutions, we can substitute some values of x, say 0, 3, and 6, in either equation and solve for y. The results are: $(0, 2)$, $(3, 1)$, and $(6, 0)$.

Self Check 7   Solve the system: $\begin{cases} y = 2 - x \\ 3x + 3y = 6 \end{cases}$

Now Try ▶ Problem 37

Now we summarize the results from Examples 6 and 7. If both variables are eliminated in the substitution process, there are two possible outcomes.

- If the resulting equation is false, write *"no solution"* as the answer. The system is inconsistent. (*Caution:* Do not answer with "false.")

- If the resulting equation is true, write *"infinitely many solutions"* as the answer. The system's equations are dependent. (*Caution:* Do not answer with "true.")

SECTION 4.2 ▶ **STUDY SET**

VOCABULARY

Fill in the blanks.

1. To solve the system $\begin{cases} x = y + 1 \\ 3x + 2y = 8 \end{cases}$ using the method discussed in this section, we begin by _____ $y + 1$ for x in the second equation.

2. We say that the equation $y = 2x + 4$ is solved for ___.

CONCEPTS

3. If the substitution method is used to solve $\begin{cases} 5x + y = 2 \\ y = -3x \end{cases}$, which equation should be used as the substitution equation?

4. Suppose the substitution method will be used to solve $\begin{cases} x - 2y = 2 \\ 2x + 3y = 11 \end{cases}$. Find a substitution equation by solving one of the equations for one of the variables.

5. Suppose $x - 4$ is substituted for y in the equation $x + 3y = 8$. Insert parentheses in $x + 3x - 4 = 8$ to show the substitution.

6. Fill in the blank. With the substitution method, the objective is to use an appropriate substitution to obtain one equation in _____ variable.

7. A student uses the substitution method to solve the system $\begin{cases} 4a + 5b = 2 \\ b = 3a - 11 \end{cases}$ and finds that $a = 3$. What is the easiest way for her to determine the value of b?

8. a. Clear $\frac{x}{5} + \frac{2y}{3} = 1$ of fractions.
   b. Write $2x + y = x - 5y + 3$ in the form $Ax + By = C$.

9. Suppose $-2 = 1$ is obtained when a system is solved by the substitution method.
   a. Does the system have a solution?
   b. Which of the following is a possible graph of the system?

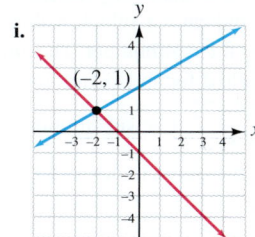

10. Suppose $2 = 2$ is obtained when a system is solved by the substitution method.
   a. Does the system have a solution?
   b. Which graph below is a possible graph of the system?

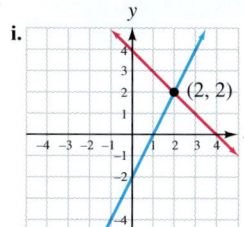

NOTATION

Complete the solution to solve the system.

11. Solve: $\begin{cases} y = 3x \\ x - y = 4 \end{cases}$

   $x - y = 4$ This is the second equation.

   $x - \left(\boxed{}\right) = 4$

   $-2x = \boxed{}$

   $x = \boxed{}$ This is the x-value of the solution.

   $y = 3x$ This is the first equation.

   $y = 3\left(\boxed{}\right)$

   $y = \boxed{}$ This is the y-value of the solution.

   The solution is $\left(\boxed{}, \boxed{}\right)$.

12. The system $\begin{cases} a = 3b + 2 \\ a + 3b = 8 \end{cases}$ was solved, and it was found that $b = 1$ and $a = 5$. Write the solution as an ordered pair.

GUIDED PRACTICE

Solve each system by substitution. See Example 1.

13. $\begin{cases} y = 2x \\ x + y = 6 \end{cases}$

14. $\begin{cases} y = 3x \\ x + y = 4 \end{cases}$

15. $\begin{cases} y = 2x - 6 \\ 2x + y = 6 \end{cases}$

16. $\begin{cases} 2x + y = -7 \\ y = 3x + 8 \end{cases}$

Solve each system by substitution. See Example 2.

17. $\begin{cases} x + 3y = -4 \\ x = -5y \end{cases}$

18. $\begin{cases} x + 5y = -3 \\ x = -4y \end{cases}$

19. $\begin{cases} y = -5x \\ x + 3y = -28 \end{cases}$

20. $\begin{cases} y = -2x \\ 3x + 2y = -1 \end{cases}$

Solve each system by substitution. See Example 3.

21. $\begin{cases} r + 3s = 9 \\ 3r + 2s = 13 \end{cases}$

22. $\begin{cases} x - 2y = 2 \\ 2x + 3y = 11 \end{cases}$

23. $\begin{cases} 4x + y = -15 \\ 2x + 3y = 5 \end{cases}$

24. $\begin{cases} 4x + y = -5 \\ 2x - 3y = -13 \end{cases}$

Solve each system by substitution. See Example 4.

25. $\begin{cases} 8x - 6y = 4 \\ 2x - y = -2 \end{cases}$

26. $\begin{cases} 5x + 4y = 0 \\ 2x - y = 0 \end{cases}$

27. $\begin{cases} 4x + 5y = 2 \\ 3x - y = 11 \end{cases}$

28. $\begin{cases} 5u + 3v = 5 \\ 4u - v = 4 \end{cases}$

Solve each system by substitution. See Example 5.

29. $\begin{cases} \dfrac{x}{4} + \dfrac{y}{4} = -\dfrac{1}{2} \\ x - 2y = y - 24 - x \end{cases}$

30. $\begin{cases} 5a + 2b = 15 - 5a - b \\ \dfrac{a}{3} - \dfrac{2}{3}b = \dfrac{13}{3} \end{cases}$

31. $\begin{cases} x - 5y = 20 - 4x - y \\ \dfrac{y}{3} = \dfrac{x}{2} - \dfrac{5}{2} \end{cases}$

32. $\begin{cases} x - 6y = 9 - 5x - 2y \\ \dfrac{x}{2} - \dfrac{3}{4} = 2y \end{cases}$

Solve each system by substitution. See Example 6.

33. $\begin{cases} 2a + 4b = -24 \\ a = 20 - 2b \end{cases}$

34. $\begin{cases} 8y = 15 - 4x \\ x + 2y = 4 \end{cases}$

35. $\begin{cases} 6 - y = 4x \\ 2y = -8x - 20 \end{cases}$

36. $\begin{cases} 2x - y = x + y \\ -2x + 4y = 6 \end{cases}$

Solve each system by substitution. See Example 7.

37. $\begin{cases} y - 3x = -5 \\ 21x = 7y + 35 \end{cases}$

38. $\begin{cases} 3a + 6b = -15 \\ a = -2b - 5 \end{cases}$

39. $\begin{cases} x = -3y + 6 \\ 2x + 4y = 6 + x + y \end{cases}$

40. $\begin{cases} 9x = 3y + 12 \\ 4 = 3x - y \end{cases}$

TRY IT YOURSELF

Solve each system by substitution. If a system has no solution or infinitely many solutions, so state.

41. $\begin{cases} -y = 11 - 3x \\ 2x + 5y = -4 \end{cases}$

42. $\begin{cases} -x = 10 - 3y \\ 2x + 8y = -6 \end{cases}$

43. $\begin{cases} \dfrac{x}{2} + \dfrac{y}{6} = \dfrac{2}{3} \\ \dfrac{x}{3} - \dfrac{y}{4} = \dfrac{1}{12} \end{cases}$

44. $\begin{cases} \dfrac{c}{2} + \dfrac{d}{14} = 1 \\ \dfrac{c}{5} - \dfrac{d}{2} = -\dfrac{33}{10} \end{cases}$

45. $\begin{cases} y + x = 2x + 2 \\ 6x - 4y = 21 - y \end{cases}$

46. $\begin{cases} y - x = 3x \\ 2x + 2y = 14 - y \end{cases}$

47. $\begin{cases} y - 4 = 2x \\ y = 2x + 2 \end{cases}$

48. $\begin{cases} x + 3y = 6 \\ x = -3y + 6 \end{cases}$

49. $\begin{cases} a + b = 1 \\ a - 2b = -1 \end{cases}$

50. $\begin{cases} 2b - a = -1 \\ 3a + 10b = -1 \end{cases}$

51. $\begin{cases} 5x = \dfrac{1}{2}y - 1 \\ \dfrac{1}{4}y = 10x - 1 \end{cases}$

52. $\begin{cases} \dfrac{x}{4} + y = \dfrac{1}{4} \\ \dfrac{y}{2} + \dfrac{11}{20} = \dfrac{x}{10} \end{cases}$

53. $\begin{cases} x + 2y = -6 \\ x = y \end{cases}$

54. $\begin{cases} y = 2x - 9 \\ x + 3y = 8 \end{cases}$

55. $\begin{cases} b = \dfrac{2}{3}a \\ 8a - 3b = 3 \end{cases}$

56. $\begin{cases} a = \dfrac{2}{3}b \\ 9a + 4b = 5 \end{cases}$

57. $\begin{cases} 6x - 3y = 5 \\ x + 2y = 0 \end{cases}$

58. $\begin{cases} 5s + 10t = 3 \\ 2s + t = 0 \end{cases}$

59. $\begin{cases} 2x + 3 = -4y \\ x - 6 = -8y \end{cases}$

60. $\begin{cases} 5y + 2 = -4x \\ x + 2y = -2 \end{cases}$

61. $\begin{cases} 2x + 5y = -2 \\ y = -\dfrac{x}{2} \end{cases}$

62. $\begin{cases} y = -\dfrac{x}{2} \\ 2x - 3y = -7 \end{cases}$

63. $\begin{cases} 3x + 4y = -19 \\ 2y - x = 3 \end{cases}$

64. $\begin{cases} 5x - 2y = -7 \\ 5 - y = -3x \end{cases}$

65. $\begin{cases} 3(x - 1) + 3 = 8 + 2y \\ 2(x + 1) = 8 + y \end{cases}$

66. $\begin{cases} 4(x - 2) = 19 - 5y \\ 3(x - 2) - 2y = -y \end{cases}$

67. $\begin{cases} x = \dfrac{1}{3}y - 1 \\ x = y + 5 \end{cases}$

68. $\begin{cases} x = \dfrac{1}{2}y + 2 \\ x = y - 6 \end{cases}$

69. $\begin{cases} 2a - 3b = -13 \\ -b = -2a - 7 \end{cases}$

70. $\begin{cases} a - 3b = -1 \\ -b = -2a - 2 \end{cases}$

71. $\begin{cases} x = 7y - 10 \\ 2x - 14y + 20 = 0 \end{cases}$

72. $\begin{cases} y - 1 = 5x \\ 10x - 2y = 2 \end{cases}$

73. $\begin{cases} 4x + 1 = 2x + 5 + y \\ 2x + 2y = 5x + y + 6 \end{cases}$

74. $\begin{cases} 6x = 2(y + 20) + 5x \\ 5(x - 1) = 3y + 4(x + 10) \end{cases}$

75. $\begin{cases} 2a + 3b = 7 \\ 6a - b = 1 \end{cases}$

76. $\begin{cases} 3a + 5b = -6 \\ 5b - a = -3 \end{cases}$

77. $\begin{cases} 2x - 3y = -4 \\ x = -\dfrac{3}{2}y \end{cases}$

78. $\begin{cases} x = -\dfrac{3}{8}y \\ 8x - 3y = 4 \end{cases}$

79. $\begin{cases} \dfrac{9x}{7} - \dfrac{3y}{7} = \dfrac{12}{7} \\ y - 3x = -4 \end{cases}$

80. $\begin{cases} a = 9 - 2b \\ 2(a + b) = 13 + a \end{cases}$

81. $\begin{cases} 4x + 5y + 1 = -8 + 3x \\ x - 3y + 2 = -3 - x \end{cases}$

82. $\begin{cases} 6x + y = -8 + 3x - y \\ 3x + y = 2y + x - 3 \end{cases}$

APPLICATIONS

83. **Offroading.** The *angle of approach* indicates how steep of an incline a vehicle can drive up without damaging the front bumper. The *angle of departure* indicates a vehicle's ability to exit an incline without damaging the rear bumper. The angle of approach *a* and the departure angle *d* for an H3 Hummer are described by the system $\begin{cases} a + d = 77 \\ a = d + 3 \end{cases}$. Use substitution to solve the system. (Each angle is measured in degrees.)

Angle of approach $a°$ Angle of departure $d°$

84. High School Sports. The equations shown in the following graph model the number of boys and girls taking part in high school soccer programs. In both models, x is the number of years after 2000, and y is the number of participants. If the trends continue, the graphs will intersect. Use the substitution method to predict the year when the number of boys and girls participating in high school soccer will be the same.

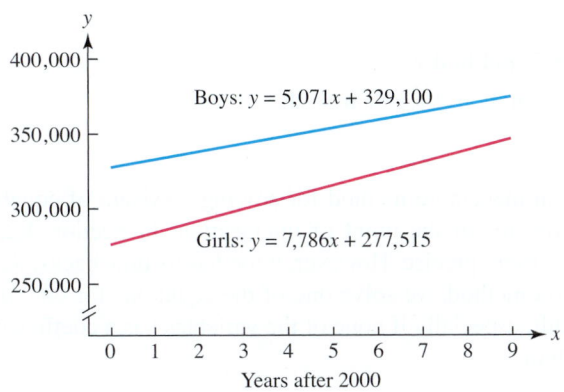

Source: National Federation of State High School Associations

85. Geometry. In the illustration, x and y represent the degree measures of angles. If $x + y = 90$ and $y = 3x$, find x and y.

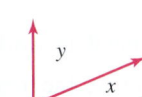

86. Geometry. In the illustration, x and y represent the degree measures of angles. If $x + y = 180$ and $x = 4y$, find x and y.

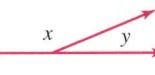

WRITING

87. What concept does this diagram illustrate?

$$\begin{cases} 6x + 5y = 11 \\ y = 3x - 2 \end{cases}$$

88. Explain the error.

Solve: $\begin{cases} 3a + 4b = 1 \\ a + 2b = 9 \end{cases}$

$a = 9 - 2b$ Solve for a in the second equation.

$9 - 2b + 2b = 9$ Substitute $9 - 2b$ for a.

$9 = 9$ The b-terms drop out. True

The system has infinitely many solutions.

89. When using the substitution method, how can you tell whether
   a. a system of linear equations has no solution?
   b. a system of linear equations has infinitely many solutions?

90. When solving a system, what advantages are there with the substitution method compared with the graphing method?

91. Consider the equation $5x + y = 12$. Explain why it is easier to solve for y than it is for x.

92. Could the substitution method be used to solve the following system? Explain why or why not. If not, what method could be used?

$$\begin{cases} y = -2 \\ x = 5 \end{cases}$$

REVIEW

93. Find the prime factorization of 189.

94. Complete each statement. For any nonzero number a,
   a. $\dfrac{0}{a} =$ ▮   **b.** $\dfrac{a}{0}$ is ▮

95. Add: $\dfrac{5}{12} + \dfrac{1}{4}$ **96.** Divide: $\dfrac{1}{3} \div \dfrac{4}{5}$

CHALLENGE PROBLEMS

Use the substitution method to solve each system.

97. $\begin{cases} \dfrac{6x - 1}{3} - \dfrac{5}{3} = \dfrac{3y + 1}{2} \\ \dfrac{1 + 5y}{4} + \dfrac{x + 3}{4} = \dfrac{17}{2} \end{cases}$

98. $\begin{cases} 0.5x + 0.5y = 6 \\ 0.001x - 0.001y = -0.004 \end{cases}$

99. $\begin{cases} 2(2x + 3y) = 5 \\ 8x = 3(1 + 3y) \end{cases}$

100. The system $\begin{cases} \dfrac{1}{2}x = y + 3 \\ x - 2y = 6 \end{cases}$ has infinitely many solutions. Find three of them.

SECTION 4.3

Solving Systems of Equations by Elimination (Addition)

OBJECTIVES

1. Solve systems of linear equations by the elimination method.

2. Use multiplication to eliminate a variable.

3. Use the elimination method twice to solve a system.

4. Use elimination to identify inconsistent systems and dependent equations.

5. Determine the most efficient method to use to solve a linear system.

ARE YOU READY?

 The following problems review some basic skills that are needed when solving systems of equations by elimination (addition).

1. In $4x - 6y = 7$, what is the *opposite* of the coefficient of y?

2. Add: $8x + (-8x)$

3. Substitute 2 for y in $5x + 6y = 7$ and find x.

4. Multiply both sides of the equation $3x + 9y = 1$ by -4.

We have seen that graphing can be an inaccurate method for solving a system of equations because we must estimate the coordinates of the point of intersection. In Section 4.2, we learned that the substitution method is more precise. However, it too has its drawbacks. Recall that in the first step of the substitution method, we solve one of the equations for one of the variables. At times, this can be difficult, especially if none of the variables has a coefficient of 1 or -1. This is the case for the system

$$\begin{cases} 2x + 5y = 11 \\ 7x - 5y = 16 \end{cases}$$

Solving either equation for x or y involves working with cumbersome fractions. For example, if we solve the first equation for x, the resulting substitution equation is $x = \frac{11 - 5y}{2}$. Fortunately, we can solve systems like this one using an easier algebraic method called the **elimination** or the **addition method.**

1   Solve Systems of Linear Equations by the Elimination Method.

The elimination method for solving a system is based on the **addition property of equality:** *When equal quantities are added to both sides of an equation, the results are equal.* In symbols, if $A = B$ and $C = D$, then adding the left sides and the right sides of these equations, we have $A + C = B + D$. This procedure is called *adding the equations.*

Add the terms on the left sides.
$$\begin{array}{r} A = B \\ C = D \\ \hline A + C = B + D \end{array}$$
Add the terms on the right sides.

EXAMPLE 1   Solve the system: $\begin{cases} 2x + 5y = 11 \\ 6x - 5y = 13 \end{cases}$

Strategy   Since the coefficients of the y-terms are opposites (5 and -5), we will add the left sides and the right sides of the given equations.

Why   When we add the equations in this way, the result will be an equation that contains only one variable, x.

Solution   Since $6x - 5y$ and 13 are equal quantities, we can add $6x - 5y$ to the left side and 13 to the right side of the first equation, $2x + 5y = 11$.

The Language of Algebra

The **elimination** method, or **addition** method as it is also known, is so named because one of the variables is eliminated using addition.

$$\begin{array}{r} 2x + 5y = 11 \\ 6x - 5y = 13 \\ \hline 8x \qquad = 24 \end{array}$$

To add the equations, add the like terms, column by column.

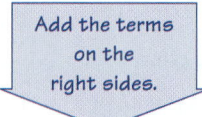

$\llcorner$ 11 + 13 = 24

5y + (−5y) = 0

2x + 6x = 8x

Because the sum of the terms $5y$ and $-5y$ is 0, we say that the variable y has been eliminated. Since the resulting equation has only one variable, we can solve it for x.

$$8x = 24$$

$$x = 3 \qquad \text{Divide both sides by 8. This is the x-value of the solution.}$$

To find the y-value of the solution, substitute 3 for x in *either* equation of the original system.

$$2x + 5y = 11 \qquad \text{This is the first equation of the system.}$$

$$2(3) + 5y = 11 \qquad \text{Substitute 3 for x.}$$

$$6 + 5y = 11 \qquad \text{Multiply.}$$

$$5y = 5 \qquad \text{Subtract 6 from both sides.}$$

$$y = 1 \qquad \text{Divide both sides by 5. This is the y-value of the solution.}$$

Now we check the proposed solution (3, 1) in the equations of the original system.

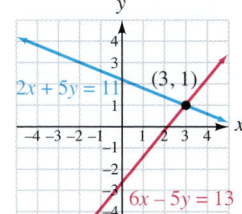

Check:
$2x + 5y = 11$ The first equation.	$6x - 5y = 13$ The second equation.
$2(3) + 5(1) \stackrel{?}{=} 11$	$6(3) - 5(1) \stackrel{?}{=} 13$
$6 + 5 \stackrel{?}{=} 11$	$18 - 5 \stackrel{?}{=} 13$
$11 = 11$ True	$13 = 13$ True

Since (3, 1) satisfies both equations, it is the solution. The graph on the left helps to verify this and it illustrates an important fact: *The solution found using the elimination (addition) method will be the same as the solution found using the graphing method.* The solution set is written $\{(3, 1)\}$.

Self Check 1   Solve the system: $\begin{cases} -4x + 3y = 4 \\ 4x + 5y = 28 \end{cases}$

Now Try ▶ Problem 17

To solve a system of equations in x and y by the elimination (addition) method, follow these steps.

The Elimination (Addition) Method

1. Write both equations of the system in standard $Ax + By = C$ form.

2. If necessary, multiply one or both of the equations by a nonzero number chosen to make the coefficients of x (or the coefficients of y) opposites.

3. Add the equations to eliminate the terms involving x (or y).

4. Solve the equation resulting from step 3.

5. Find the value of the remaining variable by substituting the solution found in step 4 into any equation containing both variables. Or, repeat steps 2–4 to eliminate the other variable.

6. Check the proposed solution in each equation of the original system. Write the solution as an ordered pair.

2 Use Multiplication to Eliminate a Variable.

In Example 1, the coefficients of the terms $5y$ in the first equation and $-5y$ in the second equation were opposites. When we added the equations, the variable y was eliminated. For many systems, however, we are not able to eliminate a variable immediately by adding. In such cases, we use the multiplication property of equality to create coefficients of x or y that are opposites.

EXAMPLE 2   Solve the system: $\begin{cases} 2x + 7y = -18 \\ 2x + 3y = -10 \end{cases}$

Strategy   We will use the elimination method to solve this system.

Why   Since none of the variables has a coefficient of 1 or -1, it would be difficult to solve this system using substitution.

Solution   **Step 1:**   Both equations are in standard $Ax + By = C$ form. We see that neither the coefficients of x nor the coefficients of y are opposites. Adding these equations as written does not eliminate a variable.

Step 2:   To eliminate x, we can multiply both sides of the second equation by -1. This creates the term $-2x$, whose coefficient is opposite that of the $2x$ term in the first equation.

$$\begin{cases} 2x + 7y = -18 \\ 2x + 3y = -10 \end{cases} \xrightarrow[\text{Multiply by } -1]{\text{Unchanged}} \quad \begin{array}{l} 2x + 7y = -18 \\ -1(2x + 3y) = -1(-10) \end{array} \xrightarrow[\text{Simplify}]{\text{Unchanged}} \begin{cases} 2x + 7y = -18 \\ -2x - 3y = 10 \end{cases}$$

Step 3:   When the equations are added, x is eliminated.

$$\begin{array}{r} 2x + 7y = -18 \\ -2x - 3y = 10 \\ \hline 4y = -8 \end{array}$$

In the left column: $2x + (-2x) = 0$.

This equation has only one variable.

Step 4:   Solve the resulting equation to find y.

$$4y = -8$$
$$y = -2 \quad \text{Divide both sides by 4. This is the y-value of the solution.}$$

Step 5:   To find x, we can substitute -2 for y in either of the equations of the original system, or in $-2x - 3y = 10$. It appears the calculations will be simplest if we use $2x + 3y = -10$.

$$2x + 3y = -10 \quad \text{This is the second equation of the original system.}$$
$$2x + 3(-2) = -10 \quad \text{Substitute -2 for y.}$$
$$2x - 6 = -10 \quad \text{Multiply.}$$
$$2x = -4 \quad \text{Add 6 to both sides.}$$
$$x = -2 \quad \text{Divide both sides by 2. This is the x-value of the solution.}$$

Step 6:   The solution is $(-2, -2)$. Check this result in the original equations.

Self Check 2   Solve the system: $\begin{cases} 2x + 7y = -27 \\ 3x + 7y = -30 \end{cases}$

Now Try ▶ **Problem 21**

EXAMPLE 3   Solve the system: $\begin{cases} 7x + 2y - 14 = 0 \\ 9x = 4y - 28 \end{cases}$

Strategy   We will use the elimination method to solve this system.

Why   Since none of the variables has coefficient 1 or -1, it would be difficult to solve this system using substitution.

Solution   **Step 1:**   To compare coefficients, write each equation in the standard $Ax + By = C$ form. Since each of the original equations will be written in an equivalent form, the resulting system will have the same solution as the original system.

$$\begin{cases} 7x + 2y = 14 \quad \text{Add 14 to both sides of $7x + 2y - 14 = 0$.} \\ 9x - 4y = -28 \quad \text{Subtract $4y$ from both sides of $9x = 4y - 28$.} \end{cases}$$

Success Tip

We choose to eliminate y because the coefficient -4 is a *multiple* of the coefficient 2. The same cannot be said for 7 and 9, the coefficients of x.

Caution

When using the elimination method, don't forget to multiply both sides of an equation by the appropriate number.

Multiply both sides by 2.
$$2(7x + 2y) = 2(14)$$

Step 2: Neither the coefficients of x nor the coefficients of y are opposites. To eliminate y, we can multiply both sides of the first equation by 2. This creates the term $4y$, whose coefficient is opposite that of the $-4y$ term in the second equation.

$$\begin{cases} 7x + 2y = 14 \\ 9x - 4y = -28 \end{cases} \xrightarrow[\text{Unchanged}]{\text{Multiply by 2}} \begin{aligned} 2(7x + 2y) = 2(14) \\ 9x - 4y = -28 \end{aligned} \xrightarrow[\text{Unchanged}]{\text{Simplify}} \begin{cases} 14x + 4y = 28 \\ 9x - 4y = -28 \end{cases}$$

Step 3: When the equations are added, y is eliminated.

$$\begin{aligned} 14x + 4y &= 28 \\ 9x - 4y &= -28 \\ \hline 23x \phantom{{}+4y} &= 0 \end{aligned}$$

In the middle column: $4y + (-4y) = 0$.

This equation has only one variable.

Step 4: Solve the resulting equation to find x.

$$23x = 0$$
$$x = 0 \qquad \text{Divide both sides by 23. This is the x-value of the solution.}$$

Step 5: To find y, we can substitute 0 for x in any equation that contains both variables. It appears the computations will be simplest if we use $7x + 2y = 14$.

$$7x + 2y = 14 \qquad \text{This is the first equation of the original system.}$$
$$7(0) + 2y = 14 \qquad \text{Substitute 0 for x.}$$
$$0 + 2y = 14 \qquad \text{Multiply.}$$
$$2y = 14 \qquad \text{Simplify the left side.}$$
$$y = 7 \qquad \text{Divide both sides by 2. This is the y-value of the solution.}$$

Step 6: The solution is $(0, 7)$. Check this result in the original equations.

Self Check 3   Solve the system: $\begin{cases} 3x = 10 - 2y \\ 5x - 6y + 30 = 0 \end{cases}$

Now Try ▶ Problem 25

Sometimes we must apply the multiplication property of equality to both equations to create coefficients of one variable that are opposites.

EXAMPLE 4   Solve the system: $\begin{cases} 4a + 7b = -8 \\ 5a + 6b = 1 \end{cases}$

Strategy   We will use the elimination method to solve this system.

Why   Since none of the variables has coefficient 1 or -1, it would be difficult to solve this system using substitution.

Solution   **Step 1:** Both equations are written in standard $Ax + By = C$ form.

Step 2: In this example, we must write *both* equations in equivalent forms to obtain like terms that are opposites. To eliminate a, we can multiply the first equation by 5 to create the term $20a$, and we can multiply the second equation by -4 to create the term $-20a$.

$$\begin{cases} 4a + 7b = -8 \\ 5a + 6b = 1 \end{cases} \xrightarrow[\text{Multiply by } -4]{\text{Multiply by 5}} \begin{aligned} 5(4a + 7b) = 5(-8) \\ -4(5a + 6b) = -4(1) \end{aligned} \xrightarrow[\text{Simplify}]{\text{Simplify}} \begin{cases} 20a + 35b = -40 \\ -20a - 24b = -4 \end{cases}$$

Success Tip

We create the term $20a$ from $4a$ and the term $-20a$ from $5a$. Note that the *least common multiple* of 4 and 5 is 20:

4, 8, 12, 16, **20**, 24, 28, . . .
5, 10, 15, **20**, 25, 30, . . .

Step 3: When we add the resulting equations, a is eliminated.

$$\begin{array}{r} 20a + 35b = -40 \\ -20a - 24b = -4 \\ \hline 11b = -44 \end{array}$$

In the left column: $20a + (-20a) = 0$.

This equation has only one variable.

Step 4: Solve the resulting equation to find b.

$$11b = -44$$
$$b = -4 \quad \text{Divide both sides by 11. This is the } b\text{-value of the solution.}$$

Success Tip

With this method, it doesn't matter which variable is eliminated. We could have created terms of $42b$ and $-42b$ to eliminate b. We will get the same solution, $(5, -4)$.

Step 5: To find a, we can substitute -4 for b in any equation that contains both variables. It appears the calculations will be simplest if we use $5a + 6b = 1$.

$$5a + 6b = 1 \quad \text{This is the second equation of the original system.}$$
$$5a + 6(-4) = 1 \quad \text{Substitute } -4 \text{ for } b.$$
$$5a - 24 = 1 \quad \text{Multiply.}$$
$$5a = 25 \quad \text{Add 24 to both sides.}$$
$$a = 5 \quad \text{Divide both sides by 5. This is the } a\text{-value of the solution.}$$

Step 6: Written in (a, b) form, the solution is $(5, -4)$. Check it in the original equations.

Self Check 4 Solve the system: $\begin{cases} 5a + 3b = -7 \\ 3a + 4b = 9 \end{cases}$

Now Try ▶ Problem 29

3 Use the Elimination Method Twice to Solve a System.

Sometimes it is easier to find the value of the second variable of a solution by using elimination a second time.

EXAMPLE 5 Solve the system: $\begin{cases} \dfrac{1}{6}x + \dfrac{1}{2}y = \dfrac{1}{3} \\ -\dfrac{x}{9} + y = \dfrac{5}{9} \end{cases}$

Strategy We will begin by clearing each equation of fractions. Then we will use the elimination method to solve the resulting equivalent system.

Why It is easier to create a pair of terms that are opposites if their coefficients are integers rather than fractions.

Solution **Step 1:** To clear the equations of the fractions, multiply both sides of the first equation by 6 and both sides of the second equation by 9.

$$\begin{cases} \dfrac{1}{6}x + \dfrac{1}{2}y = \dfrac{1}{3} \xrightarrow{\text{Multiply by 6}} 6\left(\dfrac{1}{6}x + \dfrac{1}{2}y\right) = 6\left(\dfrac{1}{3}\right) \xrightarrow{\text{Simplify}} \\ -\dfrac{x}{9} + y = \dfrac{5}{9} \xrightarrow{\text{Multiply by 9}} 9\left(-\dfrac{x}{9} + y\right) = 9\left(\dfrac{5}{9}\right) \xrightarrow{\text{Simplify}} \end{cases} \begin{cases} x + 3y = 2 \\ -x + 9y = 5 \end{cases}$$

Success Tip

Some students find it helpful to cross out the terms that are eliminated:

$$\begin{array}{r} x + 3y = 2 \\ -x + 9y = 5 \\ \hline 12y = 7 \end{array}$$

Step 2: The coefficients of x are opposites.

Step 3: The variable x is eliminated when we add the resulting equations.

$$\begin{array}{r} x + 3y = 2 \\ -x + 9y = 5 \\ \hline 12y = 7 \end{array}$$

In the left column: $x + (-x) = 0$.

This equation has only one variable.

Step 4: Solve the resulting equation to find y.

$$12y = 7$$

$$y = \frac{7}{12}$$ Divide both sides by 12. This is the y-value of the solution.

Step 5: We can find x by substituting $\frac{7}{12}$ for y in any equation containing both variables. However, that calculation could be complicated, because $\frac{7}{12}$ is a fraction. Instead, we can begin again with the system that is cleared of fractions, but this time, eliminate y. If we multiply both sides of the first equation by -3, this creates the term $-9y$, whose coefficient is opposite that of the $9y$ term in the second equation.

$$\begin{cases} x + 3y = 2 \\ -x + 9y = 5 \end{cases} \xrightarrow[\text{Unchanged}]{\text{Multiply by } -3} \begin{array}{l} -3(x + 3y) = -3(2) \\ -x + 9y = 5 \end{array} \xrightarrow[\text{Unchanged}]{\text{Simplify}} \begin{cases} -3x - 9y = -6 \\ -x + 9y = 5 \end{cases}$$

When we add the resulting equations, y is eliminated.

$$\begin{array}{r} -3x - 9y = -6 \\ -x + 9y = 5 \\ \hline -4x = -1 \end{array}$$ In the middle column: $9y + (-9y) = 0$.

This equation has only one variable.

Now we solve the resulting equation to find x.

$$-4x = -1$$

$$x = \frac{1}{4}$$ Divide both sides by -4. This is the x-value of the solution.

Step 6: The solution is $\left(\frac{1}{4}, \frac{7}{12}\right)$. To verify this, check it in the original equations.

> **Self Check 5**   Solve the system: $\begin{cases} -\frac{1}{5}x + y = \frac{8}{5} \\ \frac{x}{8} + \frac{y}{2} = \frac{1}{4} \end{cases}$
>
> **Now Try** ▶ Problem 33

4 Use Elimination to Identify Inconsistent Systems and Dependent Equations.

We have solved inconsistent systems and systems of dependent equations by substitution and by graphing. We also can solve these systems using the elimination method.

EXAMPLE 6   Solve the system: $\begin{cases} 3x - 2y = 2 \\ -3x + 2y = -12 \end{cases}$

Strategy   We will use the elimination method to solve this system.

Why   The terms $3x$ and $-3x$ are immediately eliminated.

Solution

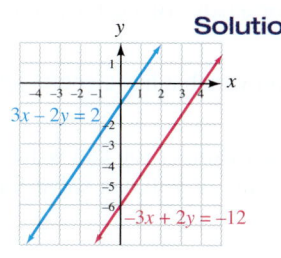

$$\begin{array}{r} 3x - 2y = 2 \\ -3x + 2y = -12 \\ \hline 0 = -10 \end{array}$$

In the left column: $3x + (-3x) = 0$.

In the middle column: $-2y + 2y = 0$.

In the right column: $2 + (-12) = -10$.

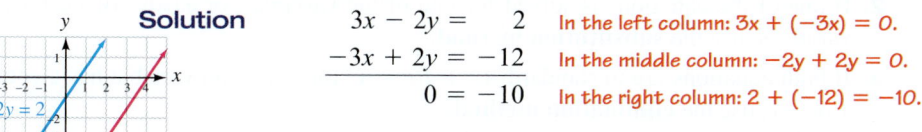

In eliminating x, the variable y is eliminated as well. The resulting false statement, $0 = -10$, indicates that the system has no solution and is inconsistent. The graphs of the equations help to verify this; they are parallel lines.

Self Check 6 Solve the system: $\begin{cases} 2x - 7y = 5 \\ -2x + 7y = 3 \end{cases}$

Now Try ▶ Problem 37

EXAMPLE 7 Solve the system: $\begin{cases} \dfrac{2x - 5y}{15} = \dfrac{8}{15} \\ -0.2x + 0.5y = -0.8 \end{cases}$

Strategy We will begin by clearing the equations of fractions and decimals. Then we will use the elimination method to solve the resulting equivalent system.

Why In this form, the equations do not contain terms with coefficients that are opposites.

Solution We can multiply both sides of the first equation by **15** to clear it of fractions and both sides of the second equation by **10** to clear it of decimals.

$$\begin{cases} \dfrac{2x - 5y}{15} = \dfrac{8}{15} \\ -0.2x + 0.5y = -0.8 \end{cases} \longrightarrow \begin{array}{l} \mathbf{15}\left(\dfrac{2x - 5y}{15}\right) = \mathbf{15}\left(\dfrac{8}{15}\right) \\ \mathbf{10}(-0.2x + 0.5y) = \mathbf{10}(-0.8) \end{array} \longrightarrow \begin{cases} 2x - 5y = 8 \\ -2x + 5y = -8 \end{cases}$$

We add the resulting equations to get

$$
\begin{array}{ll}
2x - 5y = 8 & \text{\small In the left column: } 2x + (-2x) = 0. \\
\underline{-2x + 5y = -8} & \text{\small In the middle column: } -5y + 5y = 0. \\
0 = 0 & \text{\small In the right column: } 8 + (-8) = 0.
\end{array}
$$

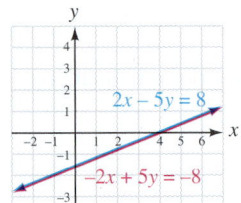

As in Example 6, both variables are eliminated. However, this time a true statement, $0 = 0$, is obtained. It indicates that the equations are dependent and that the system has infinitely many solutions. The graphs of the equations help to verify this; they are identical.

To find several of the infinitely many solutions, we can substitute some values of x, say -1, 4, and 9, in either equation and solve for y. The results are: $(-1, -2)$, $(4, 0)$, and $(9, 2)$.

Self Check 7 Solve the system: $\begin{cases} \dfrac{3x + y}{6} = \dfrac{1}{3} \\ -0.3x - 0.1y = -0.2 \end{cases}$

Now Try ▶ Problem 41

5 Determine the Most Efficient Method to Use to Solve a Linear System.

If no method is specified for solving a particular system of two linear equations, the following guidelines can be helpful in determining whether to use graphing, substitution, or elimination.

1. If you want to show trends and see the point that the two graphs have in common, then use the **graphing method**. However, this method is not exact and can be lengthy.

2. If one of the equations is solved for one of the variables, or easily solved for one of the variables, use the **substitution method**.

3. If both equations are in standard $Ax + By = C$ form, and no variable has a coefficient of 1 or -1, use the **elimination method**.

4. If the coefficient of one of the variables is 1 or -1, you have a choice. You can write each equation in standard $(Ax + By = C)$ form and use elimination, or you can solve for the variable with coefficient 1 or -1 and use substitution.

Here are some examples of suggested approaches:

$$\begin{cases} 2x + 3y = 1 \\ y = 4x - 3 \end{cases} \quad \begin{cases} 5x + 3y = 9 \\ 8x + 4y = 3 \end{cases} \quad \begin{cases} 4x - y = -6 \\ 3x + 2y = 1 \end{cases} \quad \begin{cases} x - 23 = 6y \\ 7x - 9y = -3 \end{cases}$$

Substitution   Elimination   Elimination   Substitution

Each method that we use to solve systems of equations has advantages and disadvantages.

Method	Advantages	Disadvantages
Graphing	■ You see the solution(s). ■ The graphs allow you to observe trends.	■ Inaccurate when the solutions are not integers or are large numbers off the graph
Substitution	■ Always gives the exact solutions ■ Works well if one of the equations is solved for one of the variables, or if it is easy to solve for one of the variables	■ You do not see the solution. ■ If no variable has a coefficient of 1 or -1, solving for one of the variables often involves fractions.
Elimination	■ Always gives the exact solutions ■ Works well if no variable has a coefficient of 1 or -1	■ You do not see the solution. ■ The equations must be written in the form $Ax + By = C$.

SECTION 4.3 ▶ STUDY SET

VOCABULARY

Fill in the blanks.

1. The coefficients of $3x$ and $-3x$ are _____.

2. When the given equations are added, the variable y will be _____.

$$\begin{aligned} 5x - 6y &= 10 \\ -3x + 6y &= 24 \end{aligned}$$

CONCEPTS

3. In the given system, which terms have coefficients that are opposites?

$$\begin{cases} 3x + 7y = -25 \\ 4x - 7y = 12 \end{cases}$$

4. Fill in the blank. The objective of the elimination method is to obtain two equations whose sum will be one equation in one _____.

5. Add each pair of equations.

   a. $2a + 2b = -6$
      $\underline{3a - 2b = 2}$

   b. $x - 3y = 15$
      $\underline{-x - y = -14}$

6. a. Multiply both sides of $4x + y = 2$ by 3.

   b. Multiply both sides of $x - 3y = 4$ by -2.

7. If the elimination method is used to solve

$$\begin{cases} 3x + 12y = 4 \\ 6x - 4y = 7 \end{cases}$$

   a. By what would we multiply the first equation to eliminate the variable x?

   b. By what would we multiply the second equation to eliminate the variable y?

8. Suppose the following system is solved using the elimination method and it is found that x is 2. Find the value of y.

$$\begin{cases} 4x + 3y = 11 \\ 3x - 2y = 4 \end{cases}$$

9. What algebraic step should be performed to

   a. Clear $\frac{2}{3}x + 4y = -\frac{4}{5}$ of fractions?

   b. Clear $0.2x - 0.9y = 6.4$ of decimals?

10. a. Suppose $0 = 0$ is obtained when a system is solved by the elimination method. Does the system have a solution? Which of the following is a possible graph of the system?

    b. Suppose $0 = 2$ is obtained when a system is solved by the elimination method. Does the system have a solution? Which of the following is a possible graph of the system?

    i 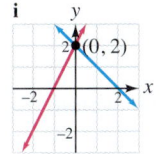   ii 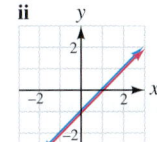   iii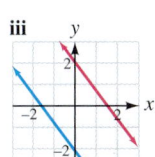

NOTATION

Complete the solution to solve the system.

11. Solve: $\begin{cases} x + y = 5 \\ x - y = -3 \end{cases}$

$$x + y = \quad 5 \qquad \text{Add the equations.}$$
$$\underline{x - y = -3}$$
$$ = 2$$
$$x = \quad$$

$$x + y = 5 \qquad \text{This is the first equation.}$$
$$\boxed{} + y = 5$$
$$y = \boxed{}$$

The solution is (,).

12. Write each equation in $Ax + By = C$ form:

$$\begin{cases} 7x + y + 3 = 0 \rightarrow \\ 8x + 4 = -y \rightarrow \end{cases} \begin{cases} \\ \end{cases}$$

GUIDED PRACTICE

Use the elimination method to solve each system. See Example 1.

13. $\begin{cases} x + y = 5 \\ x - y = 1 \end{cases}$

14. $\begin{cases} x - y = 4 \\ x + y = 8 \end{cases}$

15. $\begin{cases} x + y = -5 \\ -x + y = -1 \end{cases}$

16. $\begin{cases} -x + y = -3 \\ x + y = 1 \end{cases}$

17. $\begin{cases} 4x + 3y = 24 \\ 4x - 3y = -24 \end{cases}$

18. $\begin{cases} -9x + 5y = -9 \\ -9x - 5y = -9 \end{cases}$

19. $\begin{cases} 2s + t = -2 \\ -2s - 3t = -6 \end{cases}$

20. $\begin{cases} -2x + 4y = 12 \\ 2x + 4y = 28 \end{cases}$

Use the elimination method to solve each system. See Example 2.

21. $\begin{cases} x + 3y = -9 \\ x + 8y = -4 \end{cases}$

22. $\begin{cases} x + 7y = -22 \\ x + 9y = -24 \end{cases}$

23. $\begin{cases} 7x - y = 10 \\ 8x - y = 13 \end{cases}$

24. $\begin{cases} 6x - y = 4 \\ 9x - y = 10 \end{cases}$

Use the elimination method to solve each system. See Example 3.

25. $\begin{cases} 7x + 4y - 14 = 0 \\ 3x = 2y - 20 \end{cases}$

26. $\begin{cases} 5x - 14y - 32 = 0 \\ -x = 6y + 20 \end{cases}$

27. $\begin{cases} 7x - 50y + 43 = 0 \\ x = 4 - 3y \end{cases}$

28. $\begin{cases} x - 2y + 1 = 0 \\ 12x = 23 - 11y \end{cases}$

Use the elimination method to solve each system. See Example 4.

29. $\begin{cases} 4x + 3y = 7 \\ 3x - 2y = -16 \end{cases}$

30. $\begin{cases} 3x - 2y = 20 \\ 2x + 7y = 5 \end{cases}$

31. $\begin{cases} 5a + 8b = 2 \\ 11a - 3b = 25 \end{cases}$

32. $\begin{cases} 7a - 5b = 24 \\ 12a + 8b = 8 \end{cases}$

Use the elimination method to solve each system. See Example 5.

33. $\begin{cases} \dfrac{1}{8}x - \dfrac{1}{8}y = \dfrac{3}{8} \\ \dfrac{x}{4} + \dfrac{y}{4} = \dfrac{1}{2} \end{cases}$

34. $\begin{cases} \dfrac{1}{8}x + \dfrac{1}{4}y = \dfrac{1}{4} \\ \dfrac{x}{2} + \dfrac{y}{4} = \dfrac{1}{2} \end{cases}$

35. $\begin{cases} \dfrac{3}{4}x - \dfrac{5}{8}y = \dfrac{1}{24} \\ \dfrac{5x}{6} - y = \dfrac{1}{4} \end{cases}$

36. $\begin{cases} \dfrac{1}{3}x + \dfrac{1}{2}y = \dfrac{5}{3} \\ \dfrac{x}{7} - \dfrac{y}{7} = -\dfrac{1}{7} \end{cases}$

Use the elimination method to solve each system. If there is no solution, or infinitely many solutions, so state. See Example 6.

37. $\begin{cases} 3x - 5y = -29 \\ 3x - 5y = 15 \end{cases}$

38. $\begin{cases} 2a - 3b = -6 \\ 2a - 3b = 8 \end{cases}$

39. $\begin{cases} 3x - 16 = 5y \\ -3x + 5y - 33 = 0 \end{cases}$

40. $\begin{cases} \dfrac{-18x + y}{2} = \dfrac{7}{2} \\ 18x = y \end{cases}$

Use elimination to solve each system. If there is no solution, or infinitely many solutions, so state. See Example 7.

41. $\begin{cases} 0.4x - 0.7y = -1.9 \\ -x + \dfrac{7y}{4} = \dfrac{19}{4} \end{cases}$

42. $\begin{cases} 0.1x + 2y + 0.2 = 0 \\ -\dfrac{x}{4} - 5y = \dfrac{1}{2} \end{cases}$

43. $\begin{cases} \dfrac{x - 6y}{2} = 7 \\ -x + 6y + 14 = 0 \end{cases}$

44. $\begin{cases} 2x + 5y - 13 = 0 \\ -2x + 13 = 5y \end{cases}$

TRY IT YOURSELF

Solve the system by either the substitution or the elimination method.

45. $\begin{cases} y = -3x + 9 \\ y = x + 1 \end{cases}$

46. $\begin{cases} x = 5y - 4 \\ x = 9y - 8 \end{cases}$

47. $\begin{cases} 4x + 6y = 5 \\ 8x - 9y = 3 \end{cases}$

48. $\begin{cases} 3a + 4b = 36 \\ 6a - 2b = -21 \end{cases}$

49. $\begin{cases} 6x - 3y = -7 \\ y + 9x = 6 \end{cases}$

50. $\begin{cases} 9x + 4y = 31 \\ y - 5 = 6x \end{cases}$

51. $\begin{cases} x + y = 1 \\ x - y = 5 \end{cases}$

52. $\begin{cases} x - y = -5 \\ x + y = 1 \end{cases}$

53. $\begin{cases} 4(x - 2y) = 36 \\ 3x - 6y = 27 \end{cases}$

54. $\begin{cases} 2(x + 2y) = 15 \\ 3x = 8 - 6y \end{cases}$

55. $\begin{cases} x = y \\ 0.1x + 0.2y = 1.0 \end{cases}$

56. $\begin{cases} x = y \\ 0.4x - 0.8y = -0.5 \end{cases}$

57. $\begin{cases} 2x + 11y = -10 \\ 5x + 4y = 22 \end{cases}$

58. $\begin{cases} 3x + 4y = 12 \\ 4x + 5y = 17 \end{cases}$

59. $\begin{cases} 7x = 21 - 6y \\ 4x + 5y = 12 \end{cases}$

60. $\begin{cases} -4x = -3y - 13 \\ -6x + 8y = -16 \end{cases}$

61. $\begin{cases} 9x - 10y = 0 \\ \dfrac{9x - 3y}{63} = 1 \end{cases}$

62. $\begin{cases} 8x - 9y = 0 \\ \dfrac{2x - 3y}{6} = -1 \end{cases}$

63. $\begin{cases} \dfrac{m}{4} + \dfrac{n}{3} = -\dfrac{1}{12} \\ \dfrac{m}{2} - \dfrac{5}{4}n = \dfrac{7}{4} \end{cases}$

64. $\begin{cases} \dfrac{x}{2} - \dfrac{y}{3} = -2 \\ \dfrac{x}{3} + \dfrac{2}{3}y = \dfrac{4}{3} \end{cases}$

65. $\begin{cases} x - \dfrac{4}{3}y = \dfrac{1}{3} \\ 2x + \dfrac{3}{2}y = \dfrac{1}{2} \end{cases}$ 66. $\begin{cases} x + y = -\dfrac{1}{4} \\ x - \dfrac{y}{2} = -\dfrac{3}{2} \end{cases}$

67. $\begin{cases} 4x - 7y + 32 = 0 \\ 5x = 4y - 2 \end{cases}$ 68. $\begin{cases} 6x = -3y \\ 5x + 15 = 5y \end{cases}$

69. $\begin{cases} 3(x + 4y) = -12 \\ x = 3y + 10 \end{cases}$ 70. $\begin{cases} 3x + 2y = 3 \\ y = 2(x - 8) \end{cases}$

71. $\begin{cases} 4a + 7b = 2 \\ 9a - 3b = 1 \end{cases}$ 72. $\begin{cases} 5a - 7b = 6 \\ 7a - 6b = 8 \end{cases}$

73. $\begin{cases} 3a - b = 12.3 \\ 4a - b = 14.9 \end{cases}$ 74. $\begin{cases} -7x - y = 8.5 \\ 4x - y = -12.4 \end{cases}$

75. $\begin{cases} 5x - 4y = 8 \\ -5x - 4y = 8 \end{cases}$ 76. $\begin{cases} 2r + s = -8 \\ -2r + 4s = 28 \end{cases}$

77. $\begin{cases} 9a + 16b = -36 \\ 7a + 4b = 48 \end{cases}$ 78. $\begin{cases} 4a + 7b = -24 \\ 9a + b = 64 \end{cases}$

79. $\begin{cases} 8x + 12y = -22 \\ 3x - 2y = 8 \end{cases}$ 80. $\begin{cases} 3x + 2y = 45 \\ 5x - 4y = 20 \end{cases}$

81. $\begin{cases} 6x + 5y + 29 = 0 \\ 0.02x = 0.03y - 0.05 \end{cases}$

82. $\begin{cases} 3x = 20y + 1 \\ 0.04x + 0.05y - 0.33 = 0 \end{cases}$

83. $\begin{cases} c = d - 9 \\ 5c = 3d - 35 \end{cases}$ 84. $\begin{cases} a = b + 7 \\ 3a - 15 = 5b \end{cases}$

85. $\begin{cases} 0.9x + 2.1 = 0.3y \\ 0.4x = 0.7y + 1.9 \end{cases}$ 86. $\begin{cases} 0.7x + 1.1 = 0.4y \\ 0.4x = 0.7y + 2.2 \end{cases}$

87. $\begin{cases} 5c + 2d = -5 \\ 6c + 2d = -10 \end{cases}$ 88. $\begin{cases} 11c + 3d = -68 \\ 10c + 3d = -64 \end{cases}$

89. $\begin{cases} \dfrac{2}{15}x - \dfrac{1}{5}y = \dfrac{1}{3} \\ \dfrac{2}{15}x - \dfrac{1}{5}y = \dfrac{1}{10} \end{cases}$ 90. $\begin{cases} \dfrac{1}{5}x + \dfrac{3}{5}y = \dfrac{4}{5} \\ \dfrac{1}{6}x + \dfrac{1}{2}y = \dfrac{2}{3} \end{cases}$

APPLICATIONS

91. **Education.** The graph shows educational trends during the years 1980–2009 for persons 25 years or older in the United States. The equation $9x + 11y = 352$ approximates the percent y that had less than high school completion. The equation $5x - 11y = -198$ approximates the percent y that had a Bachelor's or higher degree. In each case, x is the number of years since 1980. Use the elimination method to determine in what year the percents were equal.

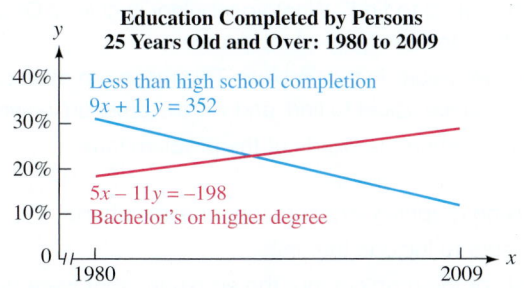

Education Completed by Persons 25 Years Old and Over: 1980 to 2009

Less than high school completion
$9x + 11y = 352$

$5x - 11y = -198$
Bachelor's or higher degree

Source: U.S. Department of Commerce, Census Bureau

92. **Newspapers.** The graph shows the trends in the newspaper publishing industry during the years 1990–2008 in the United States. The equation $37x - 2y = -1{,}128$ models the number y of morning newspapers published and $31x + y = 1{,}059$ models the number y of evening newspapers published. In each case, x is the number of years since 1990. Use the elimination method to determine in what year there were an equal number of morning and evening newspapers being published.

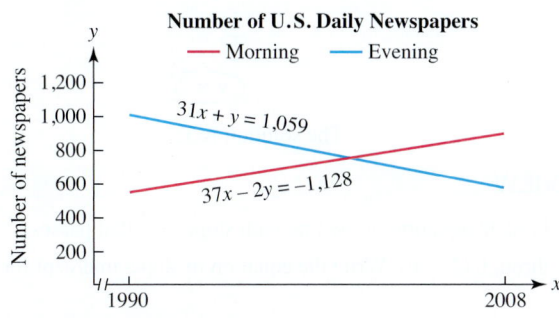

Number of U.S. Daily Newspapers
— Morning — Evening

$31x + y = 1{,}059$
$37x - 2y = -1{,}128$

Source: Editor and Publisher Yearbook data

93. **CFL Bulbs.** The graph below shows how a more expensive, but more energy-efficient, compact fluorescent light bulb eventually costs less to use than an incandescent light bulb. The equation $60c - d = 96$ approximates the cost c (in dollars) to purchase and use a CFL bulb 8 hours a day for d days. The equation $15c - d = 6$ does the same for an incandescent bulb. Use the elimination method to determine after how many days the upgrade to a CFL bulb begins to save money.

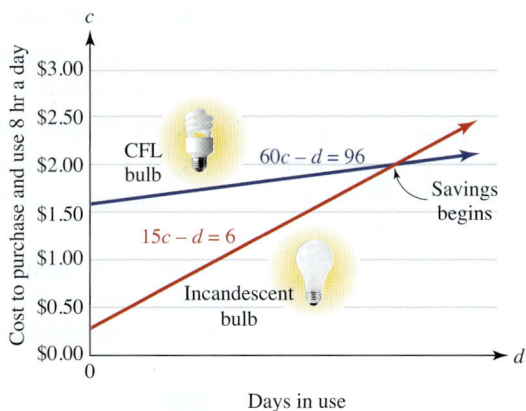

CFL bulb $60c - d = 96$
Savings begins
$15c - d = 6$
Incandescent bulb

Days in use

Source: Whitesites.com

94. **The Human Skeleton.** The equation $h + f = 53$ models the fact that the number of bones in the hand and foot totals 53. The equation $h - f = 1$ models the fact that the difference between the number of bones in the hand and foot is just 1. Use the elimination method to find h and f.

WRITING

95. Why is the method for solving systems that is discussed in this section called the *elimination method*? Why is it also referred to as the *addition method*?

96. If the elimination method is to be used to solve this system, what is wrong with the form in which it is written?

$$\begin{cases} 2x - 5y = -3 \\ -2y - 10 = -5x \end{cases}$$

97. Can the system $\begin{cases} 2x + 3y = 13 \\ 7x - 3y = -5 \end{cases}$ be solved more easily using the elimination method or the substitution method? Explain.

98. Explain the error in the following work.

Solve: $\begin{cases} x + y = 1 \\ x - y = 5 \end{cases}$

$$\begin{array}{r} x + y = 1 \\ +x - y = 5 \\ \hline 2x \quad\quad = 6 \end{array}$$

$$\frac{2x}{2} = \frac{6}{2}$$

$$\boxed{x = 3}$$

The solution is 3.

REVIEW

99. Find an equation of the line with slope $-\frac{11}{6}$ that passes through $(2, -6)$. Write the equation in slope–intercept form.

100. Solve $S = 2\pi rh + 2\pi r^2$ for h.

101. Evaluate: $-10(18 - 4^2)^3$

102. Evaluate: -5^2

CHALLENGE PROBLEMS

Use the elimination method to solve each system.

103. $\begin{cases} \dfrac{x - 3}{2} = \dfrac{11}{6} - \dfrac{y + 5}{3} \\ \dfrac{x + 3}{3} - \dfrac{y + 3}{4} = \dfrac{5}{12} \end{cases}$

104. $\begin{cases} \dfrac{4(x + 1)}{34} = \dfrac{1}{2} - \dfrac{3(y - 1)}{34} \\ 0.2(x + 0.2) + 0.3(y - 0.3) = 0.75 \end{cases}$

Problem Solving Using Systems of Equations

OBJECTIVES

1 Assign variables to two unknowns.

2 Use systems to solve geometry problems.

3 Use systems to solve number-value problems.

4 Use systems to solve interest, uniform motion, and mixture problems.

ARE YOU READY?

The following problems review some basic skills that are needed when solving application problems using systems of equations.

1. What is the formula for the perimeter of a rectangle?

2. At $12.95 per pound, what is the value of 3 pounds of crab legs?

3. Find the amount of interest earned by $55,000 invested at a 4% annual simple interest rate for 1 year.

4. At 35 miles per hour, how far will a Canada goose travel in 12 hours?

5. A 12-ounce mixture of window cleaner is 8% alcohol. How many ounces of the mixture is alcohol?

In previous chapters, many applied problems were modeled and solved with an equation in one variable. In this section, the application problems involve two unknowns. It is often easier to solve such problems using a two-variable approach.

1 Assign Variables to Two Unknowns.

The following steps are helpful when solving problems involving two unknown quantities.

Problem-Solving Strategy

1. **Analyze the problem** by reading it carefully to understand the given facts. What information is given? What are you asked to find? What vocabulary is given? Often a diagram or table will help you understand the facts of the problem.

2. **Assign variables** to represent unknown values in the problem. This means, in most cases, to let $x =$ one of the unknowns that you are asked to find, and $y =$ the other unknown.

3. **Form a system of equations** by translating the words of the problem into mathematical symbols.

4. **Solve the system** of equations using graphing, substitution, or elimination.

5. **State the conclusion clearly.** Be sure to include the units.

6. **Check the results** using the words of the problem, not the equations that were formed in step 3.

EXAMPLE 1   **Motion Pictures.**   Each year, Academy Award winners are presented with Oscars. The 13.5-inch statuette has a base on which a gold-plated figure stands. The figure itself is 7.5 inches taller than its base. Find the height of the figure and the height of the base.

Analyze

13.5 in.

- The statuette is a total of 13.5 inches tall.
- The figure is 7.5 inches taller than the base.
- Find the height of the figure and the height of the base.

Assign   Let x = the height of the figure, in inches, and y = the height of the base, in inches.

Form   We can translate the words of the problem into two equations, each involving x and y.

The height of the figure	plus	the height of the base	is	13.5 inches.
x	$+$	y	$=$	13.5

The height of the figure	is	the height of the base	plus	7.5 inches.
x	$=$	y	$+$	7.5

The resulting system is: $\begin{cases} x + y = 13.5 \\ x = y + 7.5 \end{cases}$

Solve   Since the second equation is solved for x, we will use substitution to solve the system.

$\begin{cases} x + y = 13.5 \\ x = y + 7.5 \end{cases}$

$x + y = 13.5$ This is the first equation of the system.

$y + 7.5 + y = 13.5$ Substitute y + 7.5 for x.

$2y + 7.5 = 13.5$ Combine like terms: y + y = 2y.

$2y = 6$ Subtract 7.5 from both sides.

$y = 3$ Divide both sides by 2. This is the height of the base.

To find x, substitute 3 for y in the second equation of the system.

$x = y + 7.5$ This is the substitution equation.

$x = 3 + 7.5$ Substitute 3 for y.

$x = 10.5$ This is the height of the figure.

State   The height of the figure is 10.5 inches and the height of the base is 3 inches.

Check   The sum of 10.5 inches and 3 inches is 13.5 inches, and the 10.5-inch figure is 7.5 inches taller than the 3-inch base. The results check.

> **Caution**
>
> If two variables are used to represent two unknown quantities, we must form a system of two equations to find the unknowns.

> **Caution**
>
> In this problem we are to find two unknowns, the height of the figure and height of the base. Remember to give both in the *State* step of the solution.

Self Check 1   **Woodworking.**   A carpenter wants to cut a 12-foot board into two pieces. The longer piece is to be twice as long as the shorter piece. Find the length of each piece.

Now Try ▶   Problem 17

2 Use Systems to Solve Geometry Problems.

Two angles are said to be **complementary** if the sum of their measures is 90°. Two angles are said to be **supplementary** if the sum of their measures is 180°.

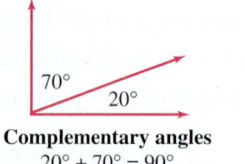

Complementary angles
20° + 70° = 90°

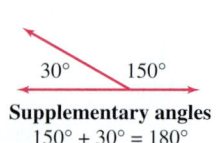

Supplementary angles
150° + 30° = 180°

EXAMPLE 2   **Angles.**   The difference of the measures of two complementary angles is 6°. Find the measure of each angle.

Analyze

- Since the angles are complementary, the sum of their measures is 90°.
- The word *difference* indicates subtraction. If the measure of the smaller angle is subtracted from the measure of the larger angle, the result will be 6°.
- Find the measure of the larger angle and the measure of the smaller angle.

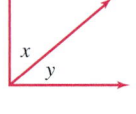

Assign   Let x = the measure of the larger angle and y = the measure of the smaller angle.

Form   We can translate the words of the problem into two equations, each involving x and y.

The measure of the larger angle	plus	the measure of the smaller angle	is	90°.
x	$+$	y	$=$	90

The measure of the larger angle	minus	the measure of the smaller angle	is	6°.
x	$-$	y	$=$	6

The resulting system is: $\begin{cases} x + y = 90 \\ x - y = 6 \end{cases}$

Solve   Since the coefficients of y are opposites, we will use elimination to solve the system.

$$\begin{array}{r} x + y = 90 \\ \underline{x - y = 6} \\ 2x = 96 \end{array}$$ Add the equations to eliminate y.

$$x = 48$$ Divide both sides by 2. This is the measure of the larger angle.

To find y, substitute 48 for x in the first equation of the system.

$$x + y = 90$$
$$48 + y = 90$$ Substitute 48 for x.
$$y = 42$$ Subtract 48 from both sides. This is the measure of the smaller angle.

State   The measure of the larger angle is 48° and the measure of the smaller angle is 42°.

Check   The sum of 48° and 42° is 90°, and the difference is 6°. The results check.

Self Check 2   **Angles.**   The difference of the measures of two supplementary angles is 22°. Find the measure of each angle.

Now Try ▶ Problem 13

Success Tip

The substitution method also could be used to solve this system. If we solve $x + y = 90$ for x, we obtain the substitution equation $x = 90 - y$.

Success Tip

When solving problems using the two-variable approach, be sure to check the results in two ways.

EXAMPLE 3

History.　In 1917, James Montgomery Flagg created the classic *I Want You* poster to help recruiting for World War I. The perimeter of the poster is 114 inches, and its length is 9 inches less than twice its width. Find the length and the width of the poster.

Analyze

- The perimeter of the rectangular poster is 114 inches.
- The length is 9 inches less than twice the width.
- Find the length and the width of the poster.

Assign　Let l = the length of the poster, in inches, and w = the width of the poster, in inches.

Form　The perimeter of a rectangle is the sum of two lengths and two widths, as given by the formula $P = 2l + 2w$, so we have

2	times	the length of the poster	plus	2	times	the width of the poster	is	114 inches.
2	$\cdot$	l	$+$	2	$\cdot$	w	$=$	114

If the length of the poster is 9 inches less than twice the width, we have

The length of the poster	is	2	times	the width of the poster	minus	9 inches.
l	$=$	2	$\cdot$	w	$-$	9

The resulting system is: $\begin{cases} 2l + 2w = 114 \\ l = 2w - 9 \end{cases}$

Solve　Since the second equation is solved for l, we will use substitution to solve the system.

$$2l + 2w = 114 \qquad \text{This is the first equation of the system.}$$
$$2(2w - 9) + 2w = 114 \qquad \text{Substitute } 2w - 9 \text{ for } l. \text{ Don't forget the parentheses.}$$
$$4w - 18 + 2w = 114 \qquad \text{Distribute the multiplication by 2.}$$
$$6w - 18 = 114 \qquad \text{Combine like terms: } 4w + 2w = 6w.$$
$$6w = 132 \qquad \text{Add 18 to both sides.}$$
$$w = 22 \qquad \text{Divide both sides by 6. This is the width of the poster.}$$

To find l, substitute 22 for w in the second equation of the system.

$$l = 2w - 9$$
$$l = 2(22) - 9$$
$$l = 44 - 9$$
$$l = 35 \qquad \text{This is the length of the poster.}$$

State　The length of the poster is 35 inches and the width is 22 inches.

Check　The perimeter is $2(35) + 2(22) = 70 + 44 = 114$ inches, and 35 inches is 9 inches less than twice 22 inches. The results check.

Success Tip

In this section, to solve the application problems that involve two unknowns, we:

- Assign **two** variables.
- Form **two** equations.
- Solve a system of **two** equations.
- State **two** answers.
- Perform **two** checks.

Self Check 3

Gardening.　Tom has 150 feet of fencing to enclose a rectangular garden. If the garden's length is to be 5 feet less than 3 times its width, find the length and width of the garden.

Now Try ▶ Problem 23

3 Use Systems to Solve Number-Value Problems.

EXAMPLE 4   **Photography.** At a school, two picture packages are available, as shown in the illustration. Find the cost of a class picture and the cost of an individual wallet-size picture.

Analyze

- Package 1 contains 1 class picture and 10 wallet-size pictures.
- Package 2 contains 2 class pictures and 15 wallet-size pictures.
- Find the cost of a class picture and the cost of a wallet-size picture.

Assign   Let c = the cost of one class picture and w = the cost of one wallet-size picture.

Form   We can use the fact that **number · value = total value** to write an equation that models the first package. We note that (in dollars) the cost of 1 class picture is $1 \cdot c = c$ and the cost of 10 wallet-size pictures is $10 \cdot w = 10w$.

The cost of 1 class picture	plus	the cost of 10 wallet-size pictures	is	$19.
c	+	$10w$	=	19

To write an equation that models the second package, we note that (in dollars) the cost of 2 class pictures is $2 \cdot c = 2c$, and the cost of 15 wallet-size pictures is $15 \cdot w = 15w$.

The cost of 2 class pictures	plus	the cost of 15 wallet-size pictures	is	$31.
$2c$	+	$15w$	=	31

The resulting system is: $\begin{cases} c + 10w = 19 \\ 2c + 15w = 31 \end{cases}$

Solve   We can use substitution or elimination to solve this system. If we use elimination, we can eliminate c as follows.

$$\begin{array}{rl} -2c - 20w = -38 & \text{Multiply both sides of } c + 10w = 19 \text{ by } -2. \\ \underline{2c + 15w = 31} & \\ -5w = -7 & \text{Add the equations to eliminate } c. \\ w = 1.4 & \text{Divide both sides by } -5. \text{ This is the cost of a wallet-size picture.} \end{array}$$

To find c, substitute 1.4 for w in the first equation of the original system.

$$\begin{array}{rl} c + 10w = 19 & \\ c + 10(1.4) = 19 & \text{Substitute 1.4 for } w. \\ c + 14 = 19 & \text{Multiply.} \\ c = 5 & \text{Subtract 14 from both sides. This is the cost of a class picture.} \end{array}$$

State   A class picture costs $5 and a wallet-size picture costs $1.40.

Check   Package 1 has 1 class picture and 10 wallets: $5 + 10($1.40) = $5 + $14 = $19. Package 2 has 2 class pictures and 15 wallets: 2($5) + 15($1.40) = $10 + $21 = $31. The results check.

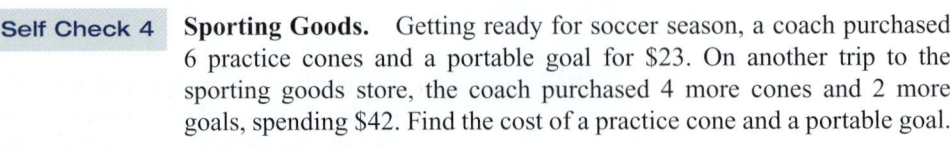

Self Check 4   **Sporting Goods.** Getting ready for soccer season, a coach purchased 6 practice cones and a portable goal for $23. On another trip to the sporting goods store, the coach purchased 4 more cones and 2 more goals, spending $42. Find the cost of a practice cone and a portable goal.

Now Try ▶ Problem 27

4 Use Systems to Solve Interest, Uniform Motion, and Mixture Problems.

We can solve investment problems like those in Section 2.6 using a two-variable approach.

EXAMPLE 5

White-Collar Crime. Investigators discovered that a small business secretly moved $150,000 out of the country to avoid paying income tax. Some of the money was invested in a Swiss bank account that paid 8% annual simple interest. The remainder was deposited in a Cayman Islands account, paying 7% annual simple interest. The investigation also revealed that the combined interest earned the first year was $11,500. How much money was invested in each account?

Analyze We are told that an unknown part of the $150,000 was invested at an annual rate of 8% and the rest at 7%. Together, the accounts earned $11,500 in interest.

> **Caution**
>
> It is incorrect to let
>
> x = the amount invested in each account
>
> This implies that *equal amounts* were invested in the Swiss and Cayman Island accounts. We do not know that.

Assign Let x = the amount invested in the Swiss account and y = the amount invested in the Cayman Islands account.

Form Because the total investment was $150,000, we have

The amount invested in the Swiss account	plus	the amount invested in the Cayman Islands account	is	$150,000
x	$+$	y	$=$	150,000

We can use the formula $I = Prt$ to determine that x dollars invested for 1 year at 8% earns $x \cdot 0.08 \cdot 1 = 0.08x$ dollars. Similarly, y dollars invested for 1 year at 7% earns $y \cdot 0.07 \cdot 1 = 0.07y$ dollars. If the total combined interest earned was $11,500, we have

The income on the 8% investment	plus	the income on the 7% investment	is	$11,500.
$0.08x$	$+$	$0.07y$	$=$	11,500

The resulting system is: $\begin{cases} x + y = 150,000 \\ 0.08x + 0.07y = 11,500 \end{cases}$

Solve First, clear the second equation of decimals. Then we can use substitution or elimination to solve the system. If we use elimination, we can eliminate x as follows.

$$
\begin{array}{rl}
-8x - 8y = -1,200,000 & \text{\color{orange}Multiply both sides of x + y = 150,000 by −8.} \\
\underline{8x + 7y = 1,150,000} & \text{\color{orange}Multiply both sides of 0.08x + 0.07y = 11,500 by 100.} \\
-y = -50,000 & \\
y = 50,000 & \text{\color{orange}Multiply both sides by −1.}
\end{array}
$$

To find x, substitute 50,000 for y in the first equation of the original system.

$$
\begin{aligned}
x + y &= 150,000 \\
x + \mathbf{50,000} &= 150,000 && \text{\color{orange}Substitute 50,000 for y.} \\
x &= 100,000 && \text{\color{orange}Subtract 50,000 from both sides.}
\end{aligned}
$$

State $100,000 was invested in the Swiss bank account, and $50,000 was invested in the Cayman Islands account.

Check

$$
\begin{aligned}
\$100,000 + \$50,000 &= \$150,000 && \text{\color{orange}The two investments total \$150,000.} \\
0.08(\$100,000) &= \$8,000 && \text{\color{orange}The Swiss bank account earned \$8,000.} \\
0.07(\$50,000) &= \$3,500 && \text{\color{orange}The Cayman Islands account earned \$3,500.}
\end{aligned}
$$

The combined interest is $8,000 + $3,500 = $11,500. The results check.

Self Check 5 **Investments.** A woman invested $10,000, some at 9% and the remainder at 10% annual simple interest. The annual income from these two investments was $975. How much was invested at each rate?

Now Try ▶ Problem 35

EXAMPLE 6 **Boating.** A boat traveled 30 miles downstream in 3 hours and made the return trip in 5 hours. Find the speed of the boat in still water and the speed of the current.

Analyze Traveling downstream, the speed of the boat will be faster than it would be in still water. Traveling upstream, the speed of the boat will be slower than it would be in still water.

Traveling downstream with the current

Traveling upstream against the current

Assign Let s = the speed of the boat in still water and c = the speed of the current.

Form The speed of the boat going downstream is $s + c$ and the speed of the boat going upstream is $s - c$. Using the formula **$d = rt$**, we find that $3(s + c)$ represents the distance traveled downstream and $5(s - c)$ represents the distance traveled upstream. We can organize the facts of the problem in a table.

	Rate ·	Time =	Distance
Downstream	$s + c$	3	$3(s + c)$
Upstream	$s - c$	5	$5(s - c)$

Multiply $r \cdot t$ to obtain an expression for each distance traveled.

Enter this information first.

Set each of these expressions for distance traveled equal to 30 to form the system of equations.

Since each trip is 30 miles long, the Distance column of the table helps us to write two equations in two variables. To write each equation in standard form, use the distributive property.

$$\begin{cases} 3(s + c) = 30 \\ 5(s - c) = 30 \end{cases} \xrightarrow{\text{Distribute}} \begin{cases} 3s + 3c = 30 \\ 5s - 5c = 30 \end{cases}$$

Success Tip

At this stage, you also could divide both sides of $3(s + c) = 30$ by 3 and divide both sides of $5(s - c) = 30$ by 5.

Solve To eliminate c, we proceed as follows.

$$\begin{array}{r} 15s + 15c = 150 \\ \underline{15s - 15c = 90} \\ 30s \quad\quad = 240 \end{array}$$

Multiply both sides of $3s + 3c = 30$ by 5.

Multiply both sides of $5s - 5c = 30$ by 3.

$$s = 8$$ Divide both sides by 30. This is the speed of the boat in still water.

To find c, it appears that the calculations will be easiest if we use $3s + 3c = 30$.

$$3s + 3c = 30$$
$$3(8) + 3c = 30$$ Substitute 8 for s.
$$24 + 3c = 30$$ Multiply.
$$3c = 6$$ Subtract 24 from both sides.
$$c = 2$$ Divide both sides by 3. This is the speed of the current.

State The speed of the boat in still water is 8 mph and the speed of the current is 2 mph.

Check   With a 2-mph current, the boat's downstream speed will be $8 + 2 = 10$ mph. In 3 hours, it will travel $10 \cdot 3 = 30$ miles. With a 2-mph current, the boat's upstream speed will be $8 - 2 = 6$ mph. In 5 hours, it will cover $6 \cdot 5 = 30$ miles. The results check.

> **Self Check 6**   **Boating.**   A boat traveled 24 miles downstream in 2 hours and made the return trip in 3 hours. Find the speed of the boat in still water and the speed of the current.
>
> **Now Try** ▶ Problem 41

The liquid and dry mixture problems that we studied in Section 2.6 can be solved with a two-variable approach.

EXAMPLE 7   **Medical Technology.**   A laboratory technician has one batch of antiseptic that is 40% alcohol and a second batch that is 60% alcohol. She would like to make 8 fluid ounces of solution that is 55% alcohol. How many fluid ounces of each batch should she use?

Analyze   Some 60% solution must be added to some 40% solution to make a 55% solution.

Assign   Let x = the number of ounces to be used from batch 1 and y = the number of ounces to be used from batch 2.

Form   The amount of alcohol in each solution is given by

$$\text{Amount of solution} \cdot \text{strength of solution} = \text{amount of alcohol}$$

We can organize the facts of the problem in a table.

Batch 1: Weak solution   Batch 2: Strong solution

x oz 40%   + y oz 60%

=

Mixture   8 oz 55%

	Amount	·	Strength	=	Amount of alcohol
Batch 1 (too weak)	x		0.40		$0.40x$
Batch 2 (too strong)	y		0.60		$0.60y$
Mixture	8		0.55		$0.55(8)$

↑ One equation comes from information in this column.   ↑ 40%, 60%, and 55% have been expressed as decimals.   ↑ Another equation comes from information in this column.

The information in the table provides two equations.

$$\begin{cases} x + y = 8 \\ 0.40x + 0.60y = 0.55(8) \end{cases}$$

The number of ounces of batch 1 plus the number of ounces of batch 2 equals the total number of ounces in the mixture.

The amount of alcohol in batch 1 plus the amount of alcohol in batch 2 equals the total amount of alcohol in the mixture.

Solve   We can solve this system by elimination. To eliminate x, we proceed as follows.

$$\begin{aligned} -40x - 40y &= -320 \\ 40x + 60y &= 440 \\ \hline 20y &= 120 \end{aligned}$$

Multiply both sides of the first equation by −40.
Multiply both sides of the second equation by 100.

$$y = 6$$

Divide both sides by 20. This is the number of ounces of batch 2 needed.

To find x, we substitute 6 for y in the first equation of the original system.

$$x + \textcolor{red}{y} = 8$$
$$x + \textcolor{red}{6} = 8 \qquad \textcolor{red}{\text{Substitute.}}$$
$$x = 2 \qquad \textcolor{red}{\text{Subtract 6 from both sides. This is the number of ounces of batch 1 needed.}}$$

State   The technician should use 2 fluid ounces of the 40% solution and 6 fluid ounces of the 60% solution.

Check   Note that 2 ounces + 6 ounces = 8 ounces, the required number. Also, the amount of alcohol in the two solutions is equal to the amount of alcohol in the mixture.

Alcohol in batch 1: $0.40x = 0.40(2) = 0.8$ ounces ⎫
Alcohol in batch 2: $0.60y = 0.60(6) = 3.6$ ounces ⎬ Total: 4.4 ounces
Alcohol in the mixture: $0.55(8) = 4.4$ ounces ⎭

The results check.

> **Self Check 7**   **Dairy Products.**   How much 1% milk and how much 4% milk must be combined to obtain 60 liters of 2% milk?
>
> **Now Try ▶**   Problem 45

EXAMPLE 8

Breakfast Cereal.   One ounce of raisins (by weight) sells for 22¢ and one ounce of bran flakes (by weight) sells for 12¢. How many ounces of each should be used to create a 20-ounce box of raisin bran cereal that can be sold for 15¢ an ounce?

Analyze   We will use a two-variable approach to solve this dry mixture problem.

Assign   Let $x =$ the number of ounces of raisins and $y =$ the number of ounces of bran flakes that should be mixed.

Form   The value of the mixture and the value of each of its components is given by

$$\textcolor{red}{\textbf{Amount} \cdot \textbf{price} = \textbf{total value}}$$

Thus, the value of x ounces of raisins is $x \cdot 22$¢ or $22x$¢ and the value of y ounces of bran flakes is $y \cdot 12$¢ or $12y$¢. The sum of these values is also equal to the total value of the final mixture, that is $20 \cdot 15$¢ or 300¢. This information is shown in the table.

	Amount	**· Price**	**=Total value**
Raisins	x	22	$22x$
Bran flakes	y	12	$12y$
Mixture	20	15	$20(15)$

One equation comes from this column. ↑   One equation comes from this column. ↑

Multiply amount · price three times to fill in this column.

The facts of the problem give the following two equations:

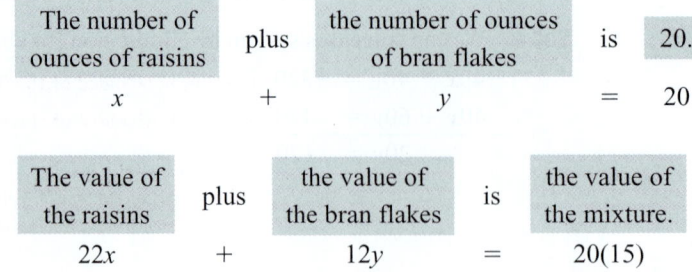

Solve   To find out how many ounces of raisins and bran flakes are needed we solve the following system:

$$\begin{cases} x + y = 20 \\ 22x + 12y = 300 \end{cases}$$ Multiply: 20(15) = 300.

To solve this system by substitution, we can solve the first equation for x:

$$x + y = 20$$
$$x = 20 - y$$ This is the substitution equation.

Then we substitute $20 - y$ for x in the second equation of the system and solve for y.

$$22x + 12y = 300$$
$$22(20 - y) + 12y = 300$$ Substitute 20 − y for x.
$$440 - 22y + 12y = 300$$ Distribute the multiplication by 22.
$$440 - 10y = 300$$ Combine like terms: −22y + 12y = −10y.
$$-10y = -140$$ Subtract 440 from both sides.
$$y = 14$$ Divide both sides by −10. This is the number of ounces of bran flakes needed.

To find x, we substitute 14 for y in the substitution equation and simplify the right side.

$$x = 20 - y$$
$$= 20 - 14$$ Substitute 14 for y.
$$= 6$$ This is the number of ounces of raisins needed.

State   To obtain 20 ounces of raisin bran cereal, 6 ounces of raisins and 14 ounces of bran flakes should be combined.

Check   When 6 ounces of raisins and 14 ounces of bran flakes are combined, the result is 20 ounces of raisin bran cereal. The 6 ounces of raisins are valued at $6 \cdot 22¢ = 132¢$ and the 14 ounces of bran flakes are valued at $14 \cdot 12¢ = 168¢$. The sum of those values, $132¢ + 168¢ = 300¢$, is the same as the value of the mixture, $20 \cdot 15¢ = 300¢$. The results check.

Self Check 8   **Gardening.**   How much planting mix, worth $36 per cubic yard, and how much topsoil, worth $24 per cubic yard, should be combined to make 200 cubic yards of a mixture worth $31.20 per cubic yard?

Now Try   ▶   Problem 49

SECTION 4.4   ▶   **STUDY SET**

VOCABULARY

Fill in the blanks.

1. Two angles are said to be _____ if the sum of their measures is 90°. Two angles are said to be _____ if the sum of their measures is 180°.

2. Problems that involve moving vehicles are called uniform _____ problems. Problems that involve combining ingredients are called _____ problems. Problems that involve collections of different items having different values are called number-_____ problems.

CONCEPTS

3. A length of pipe is to be cut into two pieces. The longer piece is to be 1 foot less than twice the shorter piece. Write two equations that model the situation.

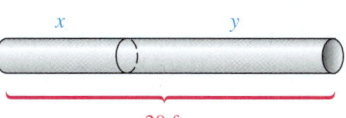

20 ft

4. Two angles are complementary. The measure of the larger angle is four times the measure of the smaller angle. Write two equations that model the situation.

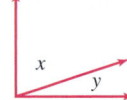

5. Two angles are supplementary. The measure of the smaller angle is 25° less than the measure of the larger angle. Write two equations that model the situation.

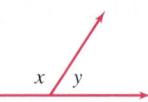

6. The perimeter of the following Ping-Pong table is 28 feet. The length is 4 feet more than the width. Write two equations that model the situation.

7. Let x = the cost of a chicken taco, in dollars, and y = the cost of a beef taco, in dollars. Write an equation that models the offer shown in the advertisement.

TUESDAY TACO SPECIAL

5 CHICKEN TACOS   *2 BEEF TACOS*

only $15

8. **a.** Complete the following table.

	Principal ·	Rate ·	Time =	Interest
City Bank	x	5%	1 yr	
USA Savings	y	11%	1 yr	

b. A total of $50,000 was deposited in the two accounts. Use that information to write an equation about the principal.

c. A total of $4,300 was earned by the two accounts. Use that information to write an equation about the interest.

9. For each case on the right, write an algebraic expression that represents the speed of the canoe in miles per hour if its speed in still water is x mph.

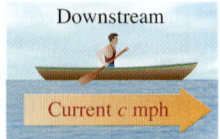

Downstream

Current c mph

Upstream

Current c mph

10. Complete the table, which contains information about an airplane flying in windy conditions.

	Rate ·	Time =	Distance
With wind	$x + y$	3	
Against wind	$x - y$	5	

11. **a.** If the contents of the two test tubes are poured into a third tube, how much solution will the third tube contain? (mL stands for milliliter. A milliliter is about 15 drops from an eyedropper.)

b. Which of the following strengths could the mixture possibly be: 27%, 33%, or 44% acid solution?

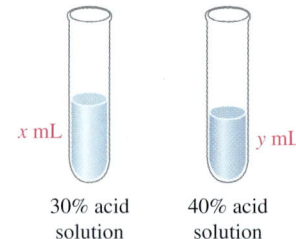

x mL y mL

30% acid solution   40% acid solution

12. **a.** Complete the table, which contains information about mixing two salt solutions to get 12 gallons of a 3% salt solution.

	Amount ·	Strength =	Amount of salt
Weak	x	0.01	
Strong	y	0.06	
Mix			

b. Use the information from the Amount column to write an equation.

c. Use the information from the Amount of salt column to write an equation.

GUIDED PRACTICE

See Example 2.

13. **Complementary Angles.** Two angles are complementary. The measure of one angle is 10° more than three times the measure of the other. Find the measure of each angle.

14. **Supplementary Angles.** Two angles are supplementary. The measure of one angle is 20° less than 19 times the measure of the other. Find the measure of each angle.

15. **Supplementary Angles.** The difference of the measures of two supplementary angles is 80°. Find the measure of each angle.

16. **Complementary Angles.** Two angles are complementary. The measure of one angle is 15° more than one-half of the measure of the other. Find the measure of each angle.

APPLICATIONS

Write a system of two equations in two variables to solve each problem.

17. **Tree Trimming.** When fully extended, the arm on a tree service truck is 51 feet long. If the upper part of the arm is 7 feet shorter than the lower part, how long is each part of the arm?

Upper part

Lower part

TREE SERVICE

18. **Alaska.** Most of the 1,422-mile-long Alaskan Highway is actually in Canada. Find the length of the highway that is in Alaska and the length of the highway that is in Canada if it is known that the difference in the lengths is 1,020 miles.

19. **Government.** The salaries of the president and vice president of the United States total $627,300 a year. If the president makes $172,700 more than the vice president, find each of their salaries.

20. **Causes of Death.** According to the *National Vital Statistics Reports,* in 2007, the number of Americans who died from motor vehicle accidents was about twice the number who died from falls. If the total number of deaths from these two causes was approximately 66,000, how many Americans died from each cause in 2007?

21. **Monuments.** The Marine Corps War Memorial in Arlington, Virginia, portrays the raising of the U.S. flag on Iwo Jima during World War II. Find the measures of the two angles shown if the measure of ∠1 is 15° less than twice the measure of ∠2.

22. **Physical Therapy.** To rehabilitate her knee, an athlete does leg extensions. Her goal is to regain a full 90° range of motion in this exercise. Use the information in the illustration to determine her current range of motion in degrees and the number of degrees of improvement she still needs to make.

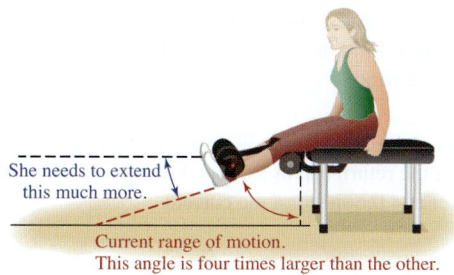

23. **Theater Screens.** At an IMAX theater, the giant rectangular movie screen has a width 26 feet less than its length. If its perimeter is 332 feet, find the length and the width of the screen.

24. **Art.** In 1770, Thomas Gainsborough painted *The Blue Boy.* The sum of the length and width of the painting is 118 inches. The difference of the length and width is 22 inches. Find the length and width.

25. **Geometry.** A 50-meter path surrounds a rectangular garden. The width of the garden is two-thirds its length. Find the length and width.

26. **Ballroom Dancing.** A rectangular-shaped dance floor has a perimeter of 200 feet. If the floor were 20 feet wider, its width would equal its length. Find the length and width of the dance floor.

27. **Empty Cartridges.** A bank recycles its empty printer and copier cartridges. In January, the bank received $40 for recycling 5 printer and 2 copier cartridges. In February, the bank received $57 for recycling 6 printer and 3 copier cartridges. How much is the bank paid for an empty printer cartridge and for an empty copier cartridge?

28. **Thanksgiving Dinner.** There are a total of 510 calories in 6 ounces of turkey and one slice of pumpkin pie. There are a total of 580 calories in 4 ounces of turkey and two slices of pumpkin pie. How many calories are there in 1 ounce of turkey and in one slice of pumpkin pie?

29. **from Campus to Careers**

    Photographer

    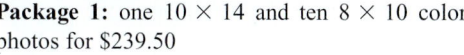

    Suppose you are a wedding photographer and you sell:

    Package 1: one 10 × 14 and ten 8 × 10 color photos for $239.50

    Package 2: one 10 × 14 and five 8 × 10 color photos for $134.50

    A newlywed couple buys Package 1 and decides that they want one more 10 × 14 and one more 8 × 10 photograph. At the same prices, what should you charge them for each additional photograph?

30. **Buying Painting Supplies.** Two partial receipts for paint supplies are shown. (Assume no sales tax was charged.) Find the cost of one gallon of latex paint and the cost of one paint brush.

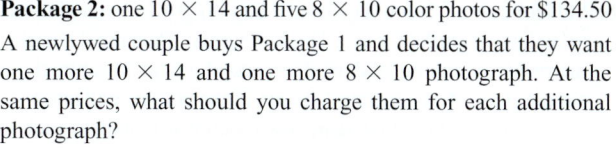

31. Collecting Stamps. Determine the price of an Elvis Presley stamp and a Statue of Liberty stamp given the following information.

 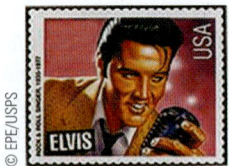

- One Elvis stamp and one Liberty stamp cost a total of 63¢.
- A sheet of 40 Elvis stamps and a sheet of 20 Liberty stamps cost a total of $18.40. (*Hint:* $18.40 = 1,840¢.)

32. Recycling. A boy scout troop earned $24 by recycling a total of 330 beverage containers. The recycling rates are shown below. How many of the small capacity containers and how many of the large capacity containers did they recycle? (*Hint:* 5¢ = $.05 and 10¢ = $0.10)

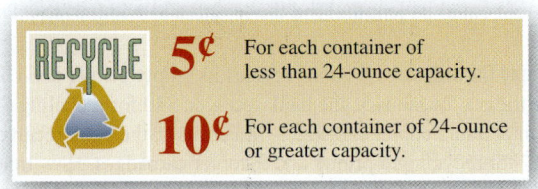

RECYCLE **5¢** For each container of less than 24-ounce capacity.

10¢ For each container of 24-ounce or greater capacity.

33. Selling Ice Cream. At a store, ice cream cones cost $1.80 and sundaes cost $3.30. One day the receipts for a total of 148 cones and sundaes were $360.90. How many cones were sold? How many sundaes?

34. Buying Tickets. The ticket prices for a movie are shown. Receipts for one showing were $1,740 for an audience of 190 people. How many general admission tickets and how many senior citizen tickets were sold?

TICKETS
General Admission: $10
Seniors: $6
Showtimes:
5:00 | 8:00 | 11:00

35. Student Loans. A college used a $5,000 gift from an alumnus to make two student loans. The first was at 5% annual simple interest to a nursing student. The second was at 7% annual simple interest to a business major. If the college collected $310 in interest the first year, how much was loaned to each student?

36. Financial Planning. In investing $6,000 of a couple's money, a financial planner put some of it into a savings account paying 6% annual simple interest. The rest was invested in a riskier mini-mall development plan paying 12% annual simple interest. The combined interest earned for the first year was $540. How much money was invested at each rate?

37. Investing a Bonus. A businessman invested part of his $40,000 end-of-the-year bonus in an international fund that paid an annual yield of 8%. The rest of the bonus was invested in an offshore bank that paid an annual yield of 9%. Find the amount of each investment if he made a total of $3,415 in interest from them the first year.

38. Pension Funds. A state employees' pension fund invested a total of one million dollars in two accounts that earned 3.5% and 4.5% annual simple interest. At the end of the year, the total interest earned from the two investments was $39,000. How much was invested at each rate?

39. Losses. A CEO deposited part of $22,000 in an account paying 4% annual simple interest. The rest of the money was invested in a biotech company that, after only one year, caused him to lose 3% of his initial investment in it. Find the amount of each investment if the net interest he earned the first year was only $110.

40. Lottery Winnings. After winning $60,000 in the lottery, a retired teacher gave $10,000 of it to her grandchildren. She invested part of the remainder in a growth fund that earned 4.4% annually and the rest in certificates of deposit paying a 5.8% annual percentage yield. The interest that she received on these two investments totaled $2,732 at the end of the first year. Find the amount of each investment.

41. The Gulf Stream. The gulf stream is a warm ocean current of the North Atlantic Ocean that flows northward, as shown below. Heading north with the gulf stream, a cruise ship traveled 300 miles in 10 hours. Against the current, it took 15 hours to make the return trip. Find the speed of the ship in still water and the speed of the current.

42. The Jet Stream. The jet stream is a strong wind current that flows across the United States, as shown above. Flying with the jet stream, a plane flew 3,000 miles in 5 hours. Against the same wind, the trip took 6 hours. Find the speed of the plane in still air and the speed of the wind current.

43. Aviation. An airplane can fly with the wind a distance of 800 miles in 4 hours. However, the return trip against the wind takes 5 hours. Find the speed of the plane in still air and the speed of the wind.

44. Boating. A boat can travel 24 miles downstream in 2 hours and can make the return trip in 3 hours. Find the speed of the boat in still water and the speed of the current.

45. Marine Biology. A marine biologist wants to set up an aquarium containing 3% salt water. He has two tanks on hand that contain 6% and 2% salt water. How much water from each tank must he use to fill a 32-gallon aquarium with a 3% saltwater mixture?

46. Commemorative Coins. A foundry has been commissioned to make souvenir coins. The coins are to be made from an alloy that is 40% silver. The foundry has on hand two alloys, one with 50% silver content and one with a 25% silver content. How many kilograms of each alloy should be used to make 20 kilograms of the 40% silver alloy?

47. Cleaning Floors. A custodian is going to mix a 4% ammonia solution and a 12% ammonia solution to get 1 gallon (128 fluid ounces) of a 9% ammonia solution. How many fluid ounces of the 4% solution and the 12% solution should be used?

48. Mouthwash. A pharmacist has a mouthwash solution that is 6% ethanol alcohol and another that is 18% ethanol alcohol. How many milliliters of each must be mixed to make 750 milliliters of a mouthwash that is 10% ethanol alcohol?

49. Coffee Sales. A coffee supply store waits until the orders for its special blend reach 100 pounds before making up a batch. Columbian coffee selling for $8.75 a pound is blended with Brazilian coffee selling for $3.75 a pound to make a product that sells for $6.35 a pound. How much of each type of coffee should be used to make the blend that will fill the orders?

50. Mixing Nuts. A merchant wants to mix peanuts with cashews, as shown in the illustration, to get 48 pounds of mixed nuts that will be sold at $6 per pound. How many pounds of each should the merchant use?

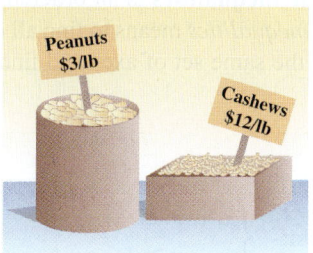

Peanuts $3/lb

Cashews $12/lb

51. Gourmet Foods. A New York delicatessen sells marinated mushrooms for $12 a pint and stuffed Kalamata olives for $9 a pint. How many pints of each should be used to get 20 pints of a mixture that will sell for $10 a pint?

52. Herbs. Ginger root powder sells for $6.50 a pound and ginkgo leaf powder sells for $9.50 a pound. How many pounds of each should be used to make 15 pounds of a mixture that sells for $7 a pound?

WRITING

53. Explain why a table is helpful in solving uniform motion and mixture problems.

54. A man paid $89 for two shirts and four pairs of socks. If we let x = the cost of a shirt, in dollars, and y = the cost of a pair of socks, in dollars, an equation modeling the purchase is $2x + 4y = 89$. Explain why there is not enough information to determine the cost of a shirt or the cost of a pair of socks.

REVIEW

Graph each inequality. Then describe the graph using interval notation.

55. $x < 4$ **56.** $x \geq -3$

57. $-1 < x \leq 2$ **58.** $-2 \leq x \leq 0$

CHALLENGE PROBLEMS

59. Three types of hardware items, nails, bolts, and nuts, are placed on a scale as shown below. On the last scale, how many nails will it take to balance 1 nut?

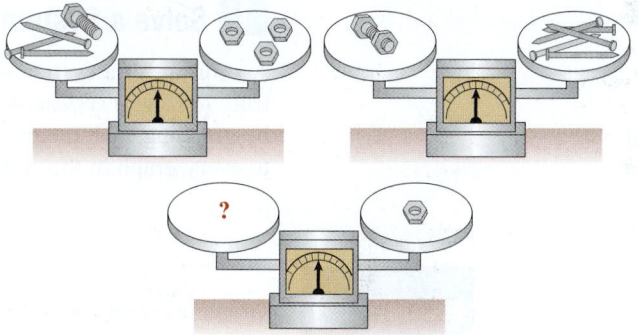

60. Farming. In a pen of goats and chickens, there are 40 heads and 130 feet. How many goats and chickens are in the pen?

SECTION 4.5

Solving Systems of Linear Inequalities

OBJECTIVES

1 Solve a system of linear inequalities by graphing.

2 Solve application problems involving systems of linear inequalities.

ARE YOU READY?

The following problems review some basic skills that are needed when solving systems of linear inequalities.

1. True or false: $3(-4) + 7 > -9$

2. Graph: $4x - 3y = -12$

3. Determine whether each of the following points lies *above*, *below*, or *on* the line graphed in problem 2.

    a. $(-2, 3)$ **b.** $(-4, -5)$ **c.** $(-3, 0)$

4. Is the boundary of the graph of $y < 3x - 1$ solid or dashed?

In Section 4.1, we solved systems of linear *equations* graphically by finding the point of intersection of two lines. Now we consider **systems of linear inequalities**, such as

$$\begin{cases} x + y \geq -1 \\ x - y \geq 1 \end{cases}$$

To solve systems of linear inequalities, we again find the points of intersection of graphs. In this case, however, we are not looking for an intersection of two lines, but an intersection of two regions.

1 Solve a System of Linear Inequalities by Graphing.

A **solution of a system of linear inequalities** is an ordered pair that makes each inequality true. *To solve a system of linear inequalities* means to find all of its solutions. This can be done by graphing each inequality on the same set of axes and finding the points that are common to every graph in the system.

EXAMPLE 1

Graph the solutions of the system: $\begin{cases} x + y \geq -1 \\ x - y \geq 1 \end{cases}$

Strategy We will graph the solutions of $x + y \geq -1$ in one color and the solutions of $x - y \geq 1$ in another color on the same coordinate system.

Why We need to see where the graphs of the two inequalities intersect (overlap).

Solution To graph $x + y \geq -1$, we begin by graphing the boundary line $x + y = -1$. Since the inequality contains an $\geq$ symbol, the boundary is a solid line. Because the coordinates of the test point $(0, 0)$ satisfy $x + y \geq -1$, we shade (in red) the side of the boundary that contains $(0, 0)$. See part (a) of the figure on the next page.

Graph the boundary: The intercept method

$x + y = -1$

x	y	(x, y)
0	-1	$(0, -1)$
-1	0	$(-1, 0)$

Shading: Check the test point $(0, 0)$

$x + y \geq -1$

$0 + 0 \overset{?}{\geq} -1$ Substitute.

$0 \geq -1$ True

$(0, 0)$ is a solution of $x + y \geq -1$.

The Language of Algebra

To solve a system of linear inequalities, we **superimpose** the graphs of the inequalities. That is, we place one graph over the other. Most video camcorders can *superimpose* the date and time over the picture being recorded.

In part (b) of the figure, we superimpose the graph of $x - y \geq 1$ on the graph of $x + y \geq -1$ so that we can determine the points that the graphs have in common. To graph $x - y \geq 1$, we graph the boundary $x - y = 1$ as a solid line. Since the test point $(0, 0)$ does not satisfy $x - y \geq 1$, we shade (in blue) the half-plane that does not contain $(0, 0)$.

Graph the boundary: The intercept method

$$x - y = 1$$

x	y	(x, y)
0	-1	$(0, -1)$
1	0	$(1, 0)$

Shading: Check the test point (0, 0)

$$x - y \geq 1$$

$$0 - 0 \overset{?}{\geq} 1 \quad \text{Substitute.}$$

$$0 \geq 1 \quad \text{False}$$

$(0, 0)$ is not a solution of $x - y \geq 1$.

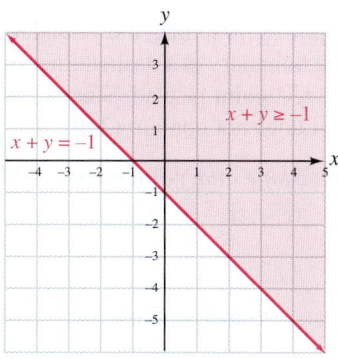

The graph of $x + y \geq -1$ is shaded in red.

(a)

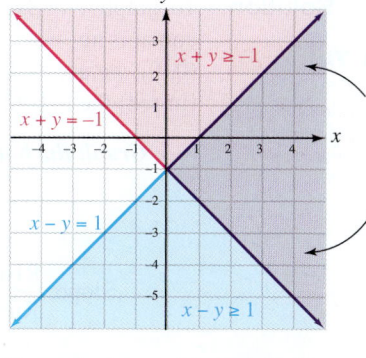

The solutions of the system are shaded in purple. The purple region is the intersection or overlap of the red and blue shaded regions. It includes portions of each boundary.

The graph of $x - y \geq 1$ is shaded in blue. It is drawn over the graph of $x + y \geq -1$.

(b)

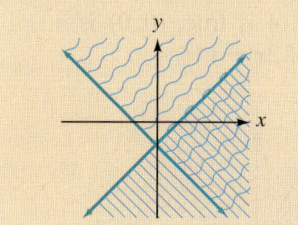

In part (b) of the figure, the area that is shaded twice represents the solutions of the given system. Any point in the **doubly shaded region** in purple (including the purple portions of each boundary) has coordinates that satisfy *both* inequalities.

Since there are infinitely many solutions, we cannot check each of them. However, as an informal check, we can select one ordered pair, say (4, 1), that lies in the doubly shaded region and show that its coordinates satisfy both inequalities of the system.

Check: $x + y \geq -1$ The first inequality. $x - y \geq 1$ The second inequality.

$4 + 1 \overset{?}{\geq} -1$ $4 - 1 \overset{?}{\geq} 1$

$5 \geq -1$ True $3 \geq 1$ True

The resulting true statements verify that (4, 1) is a solution of the system. If we pick a point that is not in the doubly shaded region, such as (1, 3), $(-2, -2)$, or $(0, -4)$, the coordinates of that point will fail to satisfy one or both of the inequalities.

Self Check 1   Graph the solutions of the system: $\begin{cases} x - y \leq 2 \\ x + y \geq -1 \end{cases}$

Now Try ▶ Problem 15

In general, to solve systems of linear inequalities, we will follow these steps.

Solving Systems of Linear Inequalities

1. Graph each inequality on the same rectangular coordinate system.

2. Use shading to highlight the intersection of the graphs (the region where the graphs overlap). The points in this region are the solutions of the system.

3. As an informal check, pick a point from the region where the graphs intersect and verify that its coordinates satisfy each inequality of the original system.

EXAMPLE 2   Graph the solutions of the system: $\begin{cases} y > 3x \\ 2x + y < 4 \end{cases}$

Strategy   We will graph the solutions of $y > 3x$ in one color and the solutions of $2x + y < 4$ in another color on the same coordinate system to see where the graphs intersect.

Why   The solution set of the system is the set of all points in the intersection of the two graphs.

Solution   To graph $y > 3x$, we begin by graphing the boundary line $y = 3x$. Since the inequality contains an $>$ symbol, the boundary is a dashed line. Because the boundary passes through $(0, 0)$, we use $(2, 0)$ as the test point instead. Since $(2, 0)$ does not satisfy $y > 3x$, we shade (in red) the half-plane that does not contain $(2, 0)$. See part (a) of the following figure.

> **Success Tip**
>
> You do not have to use the same test point when graphing the two inequalities.

Graph the boundary: Slope and y-intercept

$$y = 3x + 0$$

$$m = 3 \qquad b = 0$$

$$\text{Slope: } \frac{\text{Rise}}{\text{Run}} = \frac{3}{1} \qquad \text{y-intercept: } (0, 0)$$

Shading: Check the test point (2, 0)

$$y > 3x$$
$$0 \overset{?}{>} 3(2) \qquad \text{Substitute.}$$
$$0 > 6 \qquad \text{False}$$

Since $0 > 6$ is false, $(2, 0)$ is not a solution of $y > 3x$.

In part (b) of the figure, we superimpose the graph of $2x + y < 4$ on the graph of $y > 3x$ to determine the points that the graphs have in common. To graph $2x + y < 4$, we graph the boundary $2x + y = 4$ as a dashed line. Then we shade (in blue) the half-plane that contains $(0, 0)$, because the coordinates of the test point satisfy $2x + y < 4$.

> **Success Tip**
>
> The ordered pairs that lie on a dashed boundary line are never part of the solution of a system of linear inequalities.

Graph the boundary: The intercept method

$$2x + y = 4$$

x	y	(x, y)
0	4	$(0, 4)$
2	0	$(2, 0)$

Shading: Use the test point (0, 0)

$$2x + y < 4$$
$$2(0) + 0 \overset{?}{<} 4 \qquad \text{Substitute.}$$
$$0 < 4 \qquad \text{True}$$

Since $0 < 4$ is true, $(0, 0)$ is a solution of $2x + y < 4$.

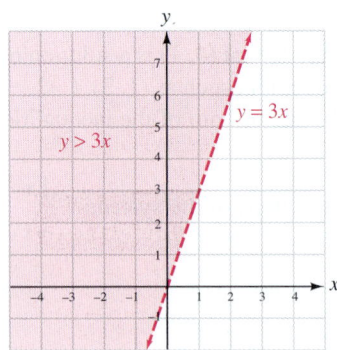

The graph of $y > 3x$ is shaded in red.

(a)

The solutions of the system are shaded in purple. Points on the boundaries are not solutions.

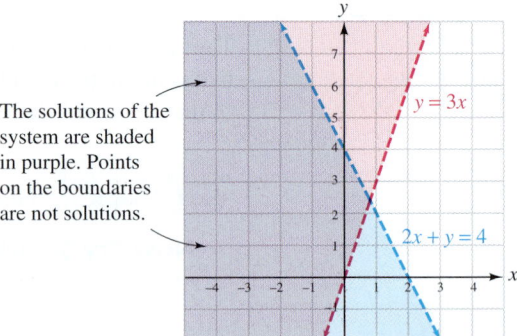

The graph of $2x + y < 4$ is shaded in blue. It is drawn over the graph of $y > 3x$.

(b)

In part (b) of the figure, the area that is shaded twice represents the solutions of the given system. Any point in the doubly shaded region in purple has coordinates that satisfy both inequalities. Pick a point in the region and show that this is true. Note that the region does not include either boundary; points on the boundaries are not solutions of the system.

Self Check 2   Graph the solutions of the system: $\begin{cases} x + 3y < 3 \\ y > \dfrac{1}{3}x \end{cases}$

Now Try ▶ Problem 19

EXAMPLE 3   Graph the solutions of the system: $\begin{cases} x \le 2 \\ y > 3 \end{cases}$

Strategy   We will graph the solutions of $x \le 2$ in one color and the solutions of $y > 3$ in another color on the same coordinate system to see where the graphs of the two inequalities intersect.

Why   The solution set of the system is the set of all points in the intersection of the two graphs.

Solution   The boundary of the graph of $x \le 2$ is the line $x = 2$. Since the inequality contains the symbol $\le$, we draw the boundary as a solid line. The test point $(0, 0)$ makes $x \le 2$ true, so we shade the side of the boundary that contains $(0, 0)$. See part (a) of the figure below.

Graph the boundary: A table of solutions   *Shading: Check the test point (0, 0)*

$$x = 2$$ $x \le 2$

x	y	(x, y)
2	0	$(2, 0)$
2	2	$(2, 2)$
2	4	$(2, 4)$

$0 \le 2$ True

Since $0 \le 2$ is true, $(0, 0)$ is a solution of $x \le 2$.

In part (b) of the figure, the graph of $y > 3$ is superimposed over the graph of $x \le 2$. The boundary of the graph of $y > 3$ is the line $y = 3$. Since the inequality contains the symbol $>$, we draw the boundary as a dashed line. The test point $(0, 0)$ makes $y > 3$ false, so we shade the side of the boundary that does not contain $(0, 0)$.

Graph the boundary: A table of solutions   *Shading: Check the test point (0, 0)*

$$y = 3$$ $y > 3$

x	y	(x, y)
0	3	$(0, 3)$
1	3	$(1, 3)$
4	3	$(4, 3)$

$0 > 3$ False

Since $0 > 3$ is false, $(0, 0)$ is not a solution of $y > 3$.

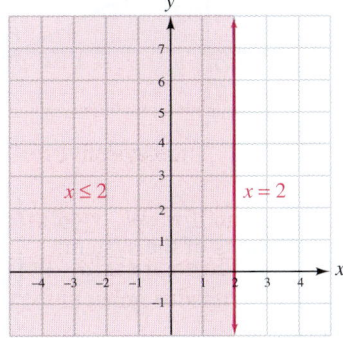

The solutions of the system are shaded in purple. Points on the purple portion of $x = 2$ are solutions. Points on the dashed boundary line are not.

The graph of $x \le 2$ is shaded in red.   The graph of $y > 3$ is shaded in blue. It is drawn over the graph of $x \le 2$.

(a)   (b)

The area that is shaded twice represents the solutions of the system of inequalities. Any point in the doubly shaded region in purple has coordinates that satisfy both inequalities, including the purple portion of the $x = 2$ boundary. Pick a point in the region and show that this is true.

Self Check 3 Graph the solutions of the system: $\begin{cases} y \leq 1 \\ x > 2 \end{cases}$

Now Try ▶ Problem 23

EXAMPLE 4 Graph the solutions of the system: $\begin{cases} x \geq 0 \\ y \geq 0 \\ x + 2y \leq 6 \end{cases}$

Strategy We will graph the solutions of $x \geq 0$, $y \geq 0$, and $x + 2y \leq 6$ on the same coordinate system to see where all three graphs intersect (overlap).

Why The solution set of the system is the set of all points in the intersection of the three graphs.

Solution This is a system of three linear inequalities. If shading is used to graph them on the same set of axes, it can become difficult to interpret the results. Instead, we can draw directional arrows attached to each boundary line in place of the shading.

- The graph of $x \geq 0$ has the boundary $x = 0$ and includes all points on the y-axis and to the right.
- The graph of $y \geq 0$ has the boundary $y = 0$ and includes all points on the x-axis and above.
- The graph of $x + 2y \leq 6$ has the boundary $x + 2y = 6$. Because the coordinates of the origin satisfy $x + 2y \leq 6$, the graph includes all points on and below the boundary.

The solutions of the system are the points that lie on triangle OPQ and the shaded triangular region that it encloses.

$x + 2y = 6$

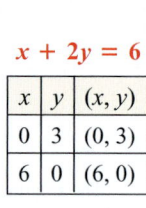

x	y	(x, y)
0	3	(0, 3)
6	0	(6, 0)

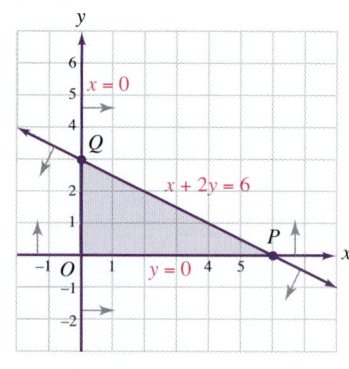

The solutions of the system are shaded purple.

Self Check 4 Graph the solutions of the system: $\begin{cases} x \leq 1 \\ y \leq 2 \\ 2x - y \leq 4 \end{cases}$

Now Try ▶ Problem 27

2 Solve Application Problems Involving Systems of Linear Inequalities.

©iStockphoto/Chris Bernard

EXAMPLE 5

Landscaping. A homeowner budgets from $300 to $600 for trees and bushes to landscape his yard. After shopping around, he finds that good trees cost $150 each and mature bushes cost $75 each. What combinations of trees and bushes can he buy?

Analyze

- At least $300 but not more than $600 is to be spent for trees and bushes.
- Trees cost $150 each and bushes cost $75 each.
- What combination of trees and bushes can he buy?

Assign Let x = the number of trees purchased and y = the number of bushes purchased.

Form We can form the following system of inequalities:

The cost of a tree	times	the number of trees purchased	plus	the cost of a bush	times	the number of bushes purchased	should at least be	$300.
$150	·	x	+	$75	·	y	≥	$300

The cost of a tree	times	the number of trees purchased	plus	the cost of a bush	times	the number of bushes purchased	should not be more than	$600.
$150	·	x	+	$75	·	y	≤	$600

Solve To solve the following system of linear inequalities

$$\begin{cases} 150x + 75y \geq 300 \\ 150x + 75y \leq 600 \end{cases}$$

we use the graphing methods discussed in this section. Neither a negative number of trees nor a negative number of bushes can be purchased, so we restrict the graph to Quadrant I.

State The coordinates of each point highlighted in the graph give a possible combination of the number of trees, x, and the number of bushes, y, that can be purchased. Written as ordered pairs, these possibilities are

(0, 4), (0, 5), (0, 6), (0, 7), (0, 8),
(1, 2), (1, 3), (1, 4), (1, 5), (1, 6),
(2, 0), (2, 1), (2, 2), (2, 3), (2, 4),
(3, 0), (3, 1), (3, 2), (4, 0)

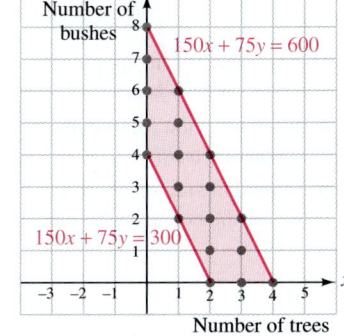

Check Suppose the homeowner picks the combination of 3 trees and 2 bushes, as represented by (3, 2). Show that this point satisfies both inequalities of the system.

Self Check 5 **Buying CDs and DVDs.** An electronics store sells CDs for $10 and DVDs for $20. Donna wants to spend at least $100 but no more than $200 on ($x$) CDs and ($y$) DVDs. What combinations of CDs and DVDs can she afford to buy?

Now Try ▶ Problem 51

SECTION 4.5 ▶ STUDY SET

VOCABULARY

Fill in the blanks.

1. $\begin{cases} x + y > 2 \\ x + y < 4 \end{cases}$ is a system of linear _____.

2. To graph the linear inequality $x + y > 2$, first graph the _____ $x + y = 2$. Then pick the test _____ $(0, 0)$ to determine which half-plane to shade.

3. To find the solutions of a system of two linear inequalities graphically, look for the _____, or overlap, of the two shaded regions.

4. The phrase *should not surpass* can be represented by the inequality symbol _____ and the phrase *must be at least* can be represented by the inequality symbol _____.

CONCEPTS

5. a. What is the equation of the boundary line of the graph of $3x - y < 5$?

   b. Is the boundary a solid or dashed line?

6. a. What is the equation of the boundary line of the graph of $y \geq 4x$?

   b. Is the boundary a solid or dashed line?

   c. Why can't $(0, 0)$ be used as a test point to determine what to shade?

7. Find the slope and the y-intercept of the line whose equation is $y = 4x - 3$.

8. Complete the table to find the x- and y-intercepts of the line whose equation is $8x - 3y = -24$.

x	y
0	
	0

9. The boundary of the graph of $2x + y > 4$ is shown.

   a. Does the point $(0, 0)$ make the inequality true?

   b. Should the region above or below the boundary be shaded?

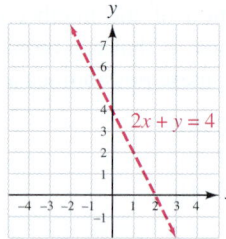

10. Linear inequality 1 is graphed in red below and linear inequality 2 is graphed in blue. Determine whether a true or false statement results when

    a. The coordinates of point A are substituted into inequality 1

    b. The coordinates of point A are substituted into inequality 2

    c. The coordinates of point B are substituted into inequality 1

    d. The coordinates of point B are substituted into inequality 2

    e. The coordinates of point C are substituted into inequality 1

    f. The coordinates of point C are substituted into inequality 2

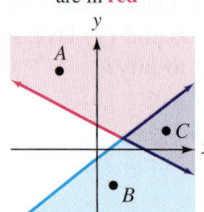

Inequality 1 solutions are in **red**

Inequality 2 solutions are in **blue**

11. The graph of a system of two linear inequalities is shown. Determine whether each point is a solution of the system.

    a. $(4, -2)$

    b. $(1, 3)$

    c. the origin

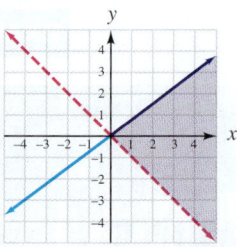

12. Use a check to determine whether each ordered pair is a solution of the system.

$$\begin{cases} x + 2y \geq -1 \\ x - y < 2 \end{cases}$$

    a. $(1, 4)$ b. $(-2, 0)$

13. Match each equation, inequality, or system with the graph of its solution.

    a. $x + y = 2$ b. $x + y \geq 2$

    c. $\begin{cases} x + y = 2 \\ x - y = 2 \end{cases}$ d. $\begin{cases} x + y \geq 2 \\ x - y \leq 2 \end{cases}$

i. 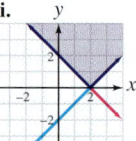   ii. 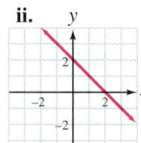   iii. 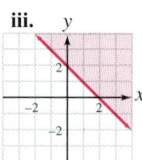   iv.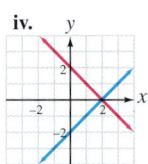

14. Match the system of inequalities with the correct graph.

    a. $\begin{cases} x \geq 2 \\ y < 1 \end{cases}$ b. $\begin{cases} x > 2 \\ y \leq 1 \end{cases}$

    c. $\begin{cases} x \geq 2 \\ y \geq 1 \end{cases}$ d. $\begin{cases} x > 2 \\ y > -1 \end{cases}$

i. 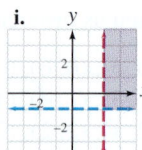   ii. 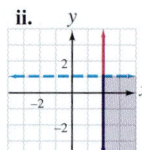   iii. 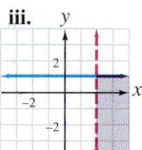   iv.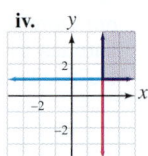

GUIDED PRACTICE

Graph the solutions of each system. See Example 1.

15. $\begin{cases} x + 2y \leq 3 \\ 2x - y \geq 1 \end{cases}$ 16. $\begin{cases} 2x + y \geq 3 \\ x - 2y \leq -1 \end{cases}$

17. $\begin{cases} x + y < -1 \\ x - y > -1 \end{cases}$ 18. $\begin{cases} x + y > 2 \\ x - y < -2 \end{cases}$

Graph the solutions of each system. See Example 2.

19. $\begin{cases} y > 2x \\ x + 2y < 6 \end{cases}$ 20. $\begin{cases} y \leq 2x \\ x + y < 4 \end{cases}$

21. $\begin{cases} y \geq x \\ y \leq \dfrac{1}{3}x + 1 \end{cases}$ 22. $\begin{cases} y > 3x \\ y \leq -x - 1 \end{cases}$

Graph the solutions of each system. See Example 3.

23. $\begin{cases} x \geq 2 \\ y \leq 3 \end{cases}$

24. $\begin{cases} x \geq -1 \\ y > -2 \end{cases}$

25. $\begin{cases} x > 0 \\ y > 0 \end{cases}$

26. $\begin{cases} x \leq 0 \\ y < 0 \end{cases}$

Graph the solutions of each system. See Example 4.

27. $\begin{cases} x \geq 0 \\ y \geq 0 \\ x + y \leq 3 \end{cases}$

28. $\begin{cases} x - y \leq 6 \\ x + 2y \leq 6 \\ x \geq 0 \end{cases}$

29. $\begin{cases} x - y < 4 \\ y \leq 0 \\ x \geq 0 \end{cases}$

30. $\begin{cases} 2x + y \leq 2 \\ y > x \\ x \geq 0 \end{cases}$

TRY IT YOURSELF

Graph the solutions of each system.

31. $\begin{cases} 2x - 3y \leq 0 \\ y \geq x - 1 \end{cases}$

32. $\begin{cases} y > 2x - 4 \\ y \geq -x - 1 \end{cases}$

33. $\begin{cases} x + y < 2 \\ x + y \leq 1 \end{cases}$

34. $\begin{cases} y > -x + 2 \\ y < -x + 4 \end{cases}$

35. $\begin{cases} 3x + 4y \geq -7 \\ 2x - 3y \geq 1 \end{cases}$

36. $\begin{cases} 3x + y \leq 1 \\ 4x - y \geq -8 \end{cases}$

37. $\begin{cases} 2x + y < 7 \\ y > 2 - 2x \end{cases}$

38. $\begin{cases} 2x + y \geq 6 \\ y \leq 2(2x - 3) \end{cases}$

39. $\begin{cases} 2(x - 2y) > -6 \\ 3x + y \geq 5 \end{cases}$

40. $\begin{cases} 2x - 3y < 0 \\ 2x + 3y \geq 12 \end{cases}$

41. $\begin{cases} 3x - y + 4 \leq 0 \\ 3y > -2x - 10 \end{cases}$

42. $\begin{cases} 3x + 2y - 12 \geq 0 \\ x < -2 + y \end{cases}$

43. $\begin{cases} x \geq -1 \\ y \leq -x \\ x - y \leq 3 \end{cases}$

44. $\begin{cases} y > -2.5 \\ 2x - y \geq 2 \\ x \leq 2 \end{cases}$

45. $\begin{cases} x + y > 0 \\ y - x < -2 \end{cases}$

46. $\begin{cases} y + 2x \leq 0 \\ y \leq \frac{1}{2}x + 2 \end{cases}$

Look Alikes . . .

In part a, graph the solution of each system. Use your answer to part a to determine the solution of the system of equations in part b. (No new work is needed.)

47. a. $\begin{cases} x + y > -1 \\ y \geq x - 3 \end{cases}$ b. $\begin{cases} x + y = -1 \\ y = x - 3 \end{cases}$

48. a. $\begin{cases} x - 2y \geq 6 \\ y < -\frac{1}{2}x + 1 \end{cases}$ b. $\begin{cases} x - 2y = 6 \\ y = -\frac{1}{2}x + 1 \end{cases}$

APPLICATIONS

49. **Birds of Prey.** Parts (a) and (b) of the illustration show the fields of vision for each eye of an owl. In part (c), shade the area where the fields of vision overlap—that is, the area that is seen by both eyes.

(a)   (b)

Right eye   Left eye

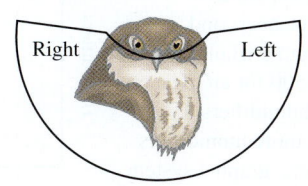

(c)

Right   Left

50. **Earth Science.** Shade the area of the earth's surface that is north of the Tropic of Capricorn and south of the Tropic of Cancer.

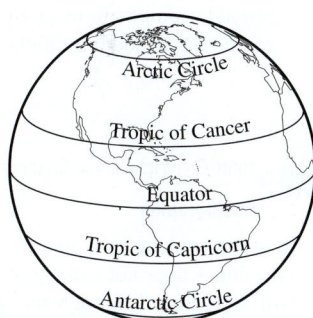

Arctic Circle
Tropic of Cancer
Equator
Tropic of Capricorn
Antarctic Circle

In Problems 51–54, graph each system of inequalities and give two possible solutions. See Example 5.

51. **Buying Compact Discs.** Melodic Music has compact discs on sale for either $10 or $15. If a customer wants to spend at least $30 but no more than $60 on CDs, graph a system of inequalities showing the possible combinations of $10 CDs ($x$) and $15 CDs ($y$) that the customer can buy.

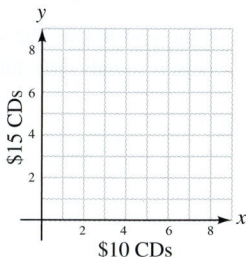

$15 CDs
$10 CDs

52. **Boats.** Boatworks wholesales aluminum boats for $800 and fiberglass boats for $600. Northland Marina wants to make a purchase totaling at least $2,400 but no more than $4,800. Graph a system of inequalities showing the possible combinations of aluminum boats (x) and fiberglass boats (y) that can be ordered.

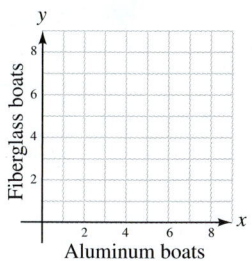

Fiberglass boats
Aluminum boats

53. Furniture. A distributor wholesales desk chairs for $150 and side chairs for $100. Best Furniture wants its order to total no more than $900; Best also wants to order more side chairs than desk chairs. Graph a system of inequalities showing the possible combinations of desk chairs (x) and side chairs (y) that can be ordered.

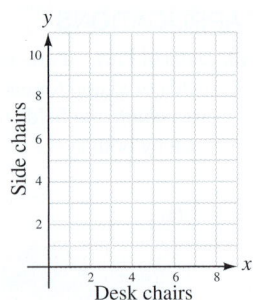

54. Air Quality. J. Bolden Heating Company wants to order no more than $2,000 worth of electronic air cleaners and humidifiers from a wholesaler that charges $500 for air cleaners and $200 for humidifiers. If Bolden wants more humidifiers than air cleaners, graph a system of inequalities showing the possible combinations of air cleaners (x) and humidifiers (y) that can be ordered.

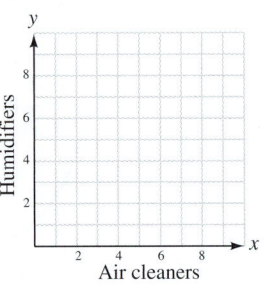

55. from **Campus to Careers**

Photographer

In some cameras, the image that you see in the viewfinder does not exactly match the image that will be recorded through the lens. To understand this difference, graph the solutions of the system

$$\begin{cases} y \le \dfrac{1}{4}x + 2 \\ y \ge -\dfrac{1}{4}x + 2 \end{cases}$$

on the grid below. The shaded solution shows how the viewfinder image is slightly different from the lens image for extreme close-ups.

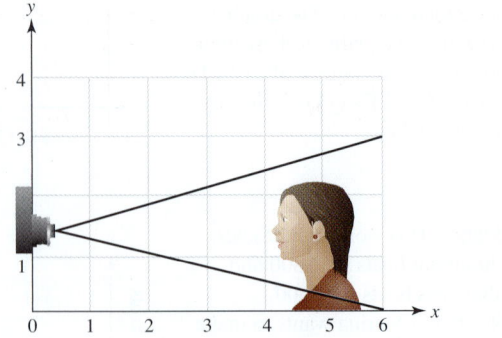

56. Pesticides. To get rid of fruit flies, helicopters sprayed an area of a city that can be described by $y \ge -2x + 1$ (within the city limits). Two weeks later, more spraying was ordered over the area described by $y \ge \frac{1}{4}x - 4$ (within the city limits). Show the part of the city that was sprayed twice.

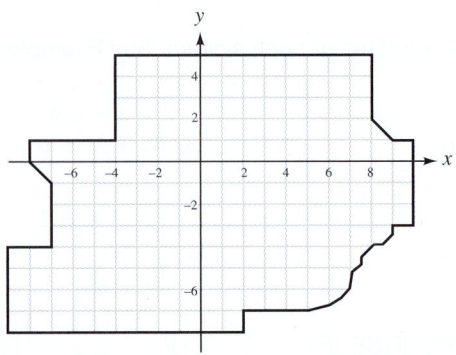

WRITING

57. Explain the error.

Graph: $\begin{cases} y > x \\ x + y < 1 \end{cases}$

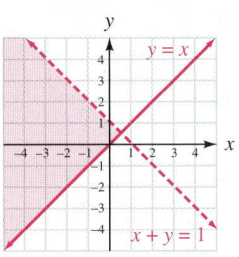

58. When a solution of a system of linear inequalities is graphed, what does the shading represent?

59. Describe how the graphs of the solutions of these systems are similar and how they differ.

$$\begin{cases} x + y = 4 \\ x - y = 4 \end{cases} \quad \text{and} \quad \begin{cases} x + y \ge 4 \\ x - y \ge 4 \end{cases}$$

60. Explain when a system of inequalities will have no solutions.

REVIEW

Simplify each expression.

61. $8\left(\dfrac{3}{4}t\right)$

62. $-\dfrac{2}{3}(3w - 6)$

63. $-\dfrac{7}{16}x - \dfrac{3}{16}x$

64. $60\left(\dfrac{3}{20}r - \dfrac{4}{15}\right)$

CHALLENGE PROBLEMS

Graph the solutions of each system.

65. $\begin{cases} \dfrac{x}{3} - \dfrac{y}{2} < -3 \\ \dfrac{x}{3} + \dfrac{y}{2} > -1 \end{cases}$

66. $\begin{cases} 3x + y < -2 \\ y > 3(1 - x) \end{cases}$

67. $\begin{cases} 2x + 3y \le 6 \\ 3x + y \le 1 \\ x \le 0 \end{cases}$

68. $\begin{cases} x \ge 0 \\ y \ge 0 \\ 9x + 3y \le 18 \\ 3x + 6y \le 18 \end{cases}$

4 ► Summary & Review

DEFINITIONS AND CONCEPTS	EXAMPLES

When two equations are considered at the same time, we say that they form a **system of equations.**

A **solution of a system** of equations in two variables is an ordered pair that satisfies both equations of the system.

Is (4, 3) a solution of the system $\begin{cases} x + y = 7 \\ x - y = 5 \end{cases}$?

To answer this question, we substitute 4 for x and 3 for y in each equation.

$x + y = 7$	The first equation.	$x - y = 5$	The second equation.
$4 + 3 \overset{?}{=} 7$	Substitute.	$4 - 3 \overset{?}{=} 5$	Substitute.
$7 = 7$	True	$1 = 5$	False

Although (4, 3) satisfies the first equation, it does not satisfy the second. Because it does not satisfy both equations, (4, 3) is not a solution of the system.

To **solve a system graphically:**

1. Graph each equation on the same coordinate system.

2. Determine the coordinates of the **point of intersection** of the graphs. That ordered pair is the solution.

3. Check the solution in each equation of the original system.

Use graphing to solve the system: $\begin{cases} y = -2x + 3 \\ x - 2y = 4 \end{cases}$

Step 1: Graph each equation.

$y = -2x + 3$

$m = \dfrac{\text{Rise}}{\text{Run}} = \dfrac{-2}{1}$

$b = 3$

y-intercept: (0, 3)

$x - 2y = 4$

x	y
0	-2
4	0

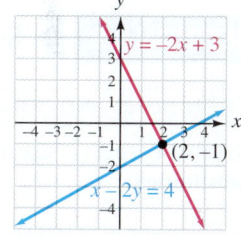

Step 2: It appears that the graphs intersect at the point $(2, -1)$. To verify that this is the solution of the system, substitute 2 for x and -1 for y in each equation.

Step 3: Check

$y = -2x + 3$	The first equation.	$x - 2y = 4$	The second equation.
$-1 \overset{?}{=} -2(2) + 3$	Substitute.	$2 - 2(-1) \overset{?}{=} 4$	Substitute.
$-1 = -1$	True	$4 = 4$	True

Since $(2, -1)$ makes both equations true, it is the solution of the system.

A system of equations that has at least one solution is called a **consistent system.** If the graphs of the equations of the system are parallel lines, the system has no solution and is called an **inconsistent system.**

Equations with different graphs are called **independent equations.** If the graphs of the equations in a system are the same line, the system has infinitely many solutions. The equations are called **dependent equations.**

We can determine the **number of solutions** that a system of two linear equations has by writing each equation in slope-intercept form, $y = mx + b$, and comparing the slopes and y-intercepts.

There are three possible outcomes when solving a system by graphing.

Consistent system
Independent equations

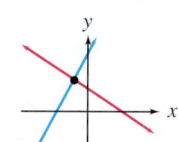

- Exactly one solution
- The lines have different slopes.

Inconsistent system
Independent equations

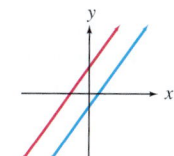

- No solution
- The lines have the same slope but different y-intercepts.

Consistent system
Dependent equations

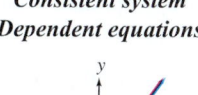
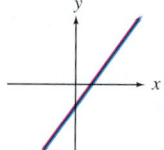

- Infinitely many solutions
- The lines have the same slope and same y-intercept.

REVIEW EXERCISES

Check to determine whether the ordered pair is a solution of the system.

1. $(2, -3)$, $\begin{cases} 3x - 2y = 12 \\ 2x + 3y = -5 \end{cases}$ **2.** $\left(\dfrac{7}{2}, -\dfrac{2}{3}\right)$, $\begin{cases} 3y = 2x - 9 \\ 2x + 3y = 6 \end{cases}$

Use the graphing method to solve each system.

3. $\begin{cases} x + y = 7 \\ 2x - y = 5 \end{cases}$ **4.** $\begin{cases} 2x + y = 5 \\ y = -\dfrac{x}{3} \end{cases}$

5. $\begin{cases} 3x + 6y = 6 \\ x + 2y - 2 = 0 \end{cases}$ **6.** $\begin{cases} 6x + 3y = 12 \\ y = -2x + 2 \end{cases}$

7. Find the slope and the y-intercept of the graph of each line in the system $\begin{cases} y = -2x + 1 \\ 8x + 4y = 3 \end{cases}$. Then, use that information to determine the number of solutions of the system.

8. Bachelor's Degrees. Estimate the point of intersection of the graphs below. Explain its significance.

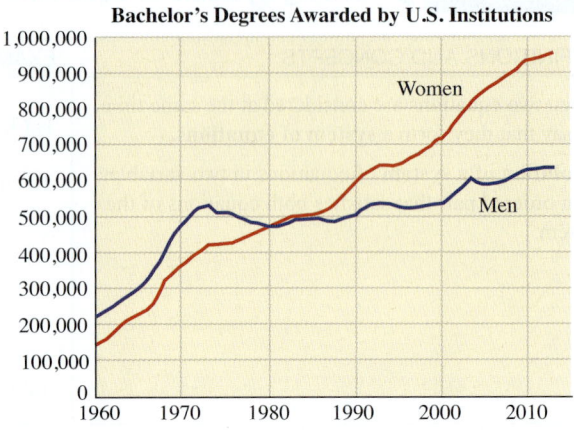

Bachelor's Degrees Awarded by U.S. Institutions

SECTION 4.2 ▶ Solving Systems of Equations by Substitution

DEFINITIONS AND CONCEPTS	EXAMPLES

DEFINITIONS AND CONCEPTS

To solve a system of equations in x and y by **substitution:**

1. Solve one of the equations for either x or y. If this is already done, go to step 2. (We call this equation the **substitution equation.**)

2. Substitute the expression for x (or for y) obtained in step 1 into the other equation and solve the equation.

3. Substitute the value of the variable found in step 2 into the substitution equation to find the value of the remaining variable.

4. Check the proposed solution in the equations of the original system.

With the substitution method, the objective is to use an appropriate substitution to obtain *one equation in one variable.*

If in step 2 the variable drops out and a false statement results, the system has **no solution.** If a true statement results, the system has **infinitely many solutions.**

EXAMPLES

Use substitution to solve the system: 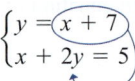 $\begin{cases} y = x + 7 \\ x + 2y = 5 \end{cases}$

Step 1: The first equation is already solved for y.

Step 2: Substitute $x + 7$ for y in the second equation.

$x + 2y = 5$	
$x + 2(x + 7) = 5$	Substitute $x + 7$ for y.
$x + 2x + 14 = 5$	Distribute the multiplication by 2.
$3x + 14 = 5$	Combine like terms.
$3x = -9$	Subtract 14 from both sides.
$x = -3$	Divide both sides by 3. This is the x-value of the solution.

Step 3: $y = x + 7$ This is the substitution equation.

$y = -3 + 7$ Substitute −3 for x.

$y = 4$ This is the y-value of the solution.

Step 4: The following check verifies that the solution is $(-3, 4)$.

$y = x + 7$	The first equation.	$x + 2y = 5$	The second equation.
$4 \overset{?}{=} -3 + 7$	Substitute.	$-3 + 2(4) \overset{?}{=} 5$	Substitute.
$4 = 4$	True	$5 = 5$	True

REVIEW EXERCISES

Use the substitution method to solve each system.

9. $\begin{cases} y = 15 - 3x \\ 7y + 3x = 15 \end{cases}$ **10.** $\begin{cases} x = y \\ 5x - 4y = 3 \end{cases}$

11. $\begin{cases} 6x + 2y = 8 - y + x \\ 3x = 2 - y \end{cases}$ **12.** $\begin{cases} r = 3s + 7 \\ r = 2s + 5 \end{cases}$

13. $\begin{cases} 9x + 3y - 5 = 0 \\ 3x + y = \dfrac{5}{3} \end{cases}$

14. $\begin{cases} \dfrac{x}{2} + \dfrac{y}{2} = 11 \\ \dfrac{5x}{16} - \dfrac{3y}{16} = \dfrac{15}{8} \end{cases}$

15. When solving a system using the substitution method, suppose you obtain the result $8 = 9$.

a. How many solutions does the system have?

b. Describe the graph of the system.

c. What term is used to describe the system?

16. Fill in the blank. With the substitution method, the objective is to use an appropriate substitution to obtain one equation in _____ variable.

SECTION 4.3 ▶ Solving Systems of Equations by Elimination (Addition)

DEFINITIONS AND CONCEPTS	EXAMPLES

To solve a system of equations in x and y using elimination (addition):

1. Write each equation in the standard $Ax + By = C$ form.

2. Multiply one (or both) equations by nonzero quantities to make the coefficients of x (or y) opposites.

3. Add the equations to eliminate the terms involving x (or y).

4. Solve the equation obtained in step 3.

5. Find the value of the other variable by substituting the value of the variable found in step 4 into any equation containing both variables.

6. Check the solution in the equations of the original system.

With the elimination method, the basic objective is to obtain two equations whose sum will be one equation in one variable.

If in step 3 both variables drop out and a false statement results, the system has **no solution.** If a true statement results, the system has **infinitely many solutions.**

Use elimination to solve: $\begin{cases} 2x - 3y = 4 \\ 3x + y = -5 \end{cases}$

Step 1: Both equations are written in $Ax + By = C$ form.

Step 2: Multiply the second equation by 3 so that the coefficients of y are opposites, -3 and 3.

Step 3:

$$\begin{array}{l} 2x - 3y = 4 \\ \underline{9x + 3y = -15} \\ 11x = -11 \end{array}$$

In the middle column, $-3y + 3y = 0$.
Add the like terms, column by column.

Step 4: Solve for x.

$$11x = -11$$
$$x = -1$$

Divide both sides by 11.
This is the x-value of the solution.

Step 5: Find y.

$$3x + y = -5 \quad \text{This is the second equation.}$$
$$3(-1) + y = -5 \quad \text{Substitute } -1 \text{ for } x.$$
$$y = -2 \quad \text{This is the y-value of the solution.}$$

Step 6: The following check verifies that the solution is $(-1, -2)$.

$2x - 3y = 4$ First equation. $3x + y = -5$ Second equation.

$2(-1) - 3(-2) \stackrel{?}{=} 4$ Substitute. $3(-1) + (-2) \stackrel{?}{=} -5$ Substitute.

$-2 + 6 \stackrel{?}{=} 4$ $-3 - 2 \stackrel{?}{=} -5$

$4 = 4$ True $-5 = -5$ True

REVIEW EXERCISES

17. Write each equation of the system $\begin{cases} 4x + 2y - 7 = 0 \\ 3y = 5x + 6 \end{cases}$ in standard $Ax + By = C$ form.

18. Fill in the blank. With the elimination method, the basic objective is to obtain two equations whose sum will be one equation in _____ variable.

Solve each system using the elimination (addition) method.

19. $\begin{cases} 2x + y = 1 \\ 5x - y = 20 \end{cases}$

20. $\begin{cases} x + 8y = 7 \\ x - 4y = 1 \end{cases}$

21. $\begin{cases} 5a + b = 2 \\ 3a + 2b = 11 \end{cases}$

22. $\begin{cases} 11x + 3y = 27 \\ 8x + 4y = 36 \end{cases}$

23. $\begin{cases} 9x + 3y = 15 \\ 3x = 5 - y \end{cases}$

24. $\begin{cases} 0.02x + 0.05y = 0 \\ 0.3x - 0.2y = -1.9 \end{cases}$

25. $\begin{cases} -\dfrac{a}{4} - \dfrac{b}{3} = \dfrac{1}{12} \\ \dfrac{a}{2} - \dfrac{5b}{4} = \dfrac{7}{4} \end{cases}$

26. $\begin{cases} -\dfrac{1}{4}x = 1 - \dfrac{2}{3}y \\ 6x - 18y = 5 - 2y \end{cases}$

For each system, determine which method, substitution or elimination (addition), would be easier to use to solve the system and explain why.

27. $\begin{cases} 6x + 2y = 5 \\ 3x - 3y = -4 \end{cases}$

28. $\begin{cases} x = 5 - 7y \\ 3x - 3y = -4 \end{cases}$

SECTION 4.4 ▶ **Problem Solving Using Systems of Equations**

DEFINITIONS AND CONCEPTS	EXAMPLES
We can solve many types of problems using a system of two linear equations in two variables:	The difference of the measures of two supplementary angles is 40°. Find the measure of each angle.

We can solve many types of problems using a system of two linear equations in two variables:

- Geometry problems
- Number-value problems
- Interest problems
- Uniform motion problems
- Liquid and dry mixture problems

To solve problems involving two unknown quantities:

1. **Analyze** the facts of the problem. Make a table or a diagram if it is helpful.

2. **Assign** different variables to represent the two unknown quantities.

3. Translate the words of the problem to **form two equations** involving those variables.

4. **Solve** the system of equations using graphing, substitution, or elimination.

5. **State** the conclusion.

6. **Check** the results.

Two angles are said to be **complementary** if the sum of their measures is 90°. Two angles are said to be **supplementary** if the sum of their measures is 180°.

EXAMPLES

The difference of the measures of two supplementary angles is 40°. Find the measure of each angle.

Analyze   Since the angles are supplementary, the sum of their measures is 180°. If we subtract the smaller angle from the larger, the result should be 40°.

Assign   Let x = the measure (in degrees) of the larger angle and y = the measure (in degrees) of the smaller angle.

Form $\begin{cases} x + y = 180 & \text{Their sum is 180°.} \\ x - y = 40 & \text{Their difference is 40°.} \end{cases}$

Solve   If we add the equations, we get

$$\begin{array}{r} x + y = 180 \\ \underline{x - y = 40} \\ 2x = 220 \end{array}$$

$\qquad\qquad x = 110$ Divide both sides by 2. This is the measure of the larger angle.

We can use the first equation of the system to find y.

$\qquad\quad x + y = 180$

$\qquad\; 110 + y = 180$ Substitute 110 for x.

$\qquad\qquad\quad y = 70$ Subtract 110 from both sides. This is the measure of the smaller angle.

State   The angles measure 110° and 70°.

Check   Angles with measures of 110° and 70° are supplementary (their sum is 180°) and their difference is 40°. The results check.

REVIEW EXERCISES

Write a system of two equations in two variables to solve each problem.

29. Elevations. The elevation of Las Vegas, Nevada, is 20 times greater than that of Baltimore, Maryland. The sum of their elevations is 2,100 feet. Find the elevation of each city.

30. Painting Equipment. When fully extended, a ladder is 35 feet in length. If the extension is 7 feet shorter than the base, how long is each part of the ladder?

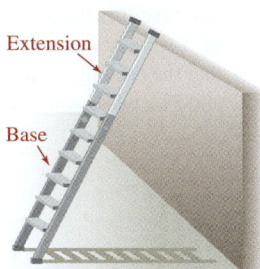
Extension
Base

31. Geometry. Two angles are complementary. The measure of one is 15° more than twice the measure of the other. Find the measure of each angle.

32. Crash Investigation. In an effort to protect evidence, investigators used 420 yards of yellow "Police Line—Do Not Cross" tape to seal off a large rectangular-shaped area around an airplane crash site. How much area will the investigators have to search if the width of the rectangle is three-fourths of the length?

33. Complete each table.

a.

	Amount	· Strength	= Amount of pesticide
Weak	x	0.02	
Strong	y	0.09	
Mixture	100	0.08	

b.

	Rate	· Time	= Distance
With the wind	$s + w$	5	
Against the wind	$s - w$	7	

c.

	$P \cdot$	r	$\cdot t =$	I
Mack Financial	x	0.11	1	
Union Savings	y	0.06	1	

d.

	Amount	· Price	= Total value
Caramel corn	x	4	
Peanuts	y	8	
Mixture	10	5	

34. Candy Store. A merchant wants to mix gummy worms worth $6 per pound and gummy bears worth $3 per pound to make 30 pounds of a mixture worth $4.20 per pound. How many pounds of each type of candy should he use?

35. Boating. It takes a motorboat 4 hours to travel 56 miles down a river, and 3 hours longer to make the return trip. Find the speed of the current.

36. Shopping. Packages containing two bottles of contact lens cleaner and three bottles of soaking solution cost $63.40, and packages containing three bottles of cleaner and two bottles of soaking solution cost $69.60. Find the cost of a bottle of cleaner and a bottle of soaking solution.

37. Investing. Carlos invested part of $3,000 in a 10% certificate account and the rest in a 6% passbook account. The total annual interest from both accounts is $270. How much did he invest at 6%?

38. Antifreeze. How much of a 40% antifreeze solution must a mechanic mix with a 70% antifreeze solution if she needs 20 gallons of a 50% antifreeze solution?

SECTION 4.5 ▶ Solving Systems of Linear Inequalities

DEFINITIONS AND CONCEPTS	EXAMPLES

A solution of a **system of linear inequalities** is an ordered pair that satisfies each inequality.

To **solve a system of linear inequalities:**

1. Graph each inequality on the same coordinate system.

2. Use shading to highlight the intersection of the graphs. The points in this region are the solutions of the system.

3. As an informal check, pick a point from the region and verify that its coordinates satisfy each inequality of the original system.

Graph the solutions of the system: $\begin{cases} y \le x + 1 \\ y > -1 \end{cases}$

Step 1: Graph each inequality on the same coordinate system as shown.

Step 2: Use shading to highlight where the graphs intersect.

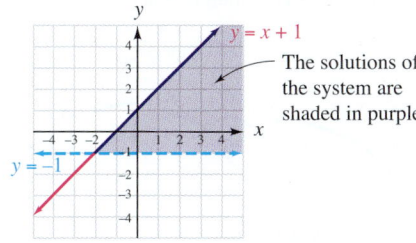

The solutions of the system are shaded in purple.

Step 3: Pick a point from the solution region such as (1, 0) and verify that it satisfies both inequalities.

REVIEW EXERCISES

Solve each system of inequalities.

39. $\begin{cases} 5x + 3y < 15 \\ 3x - y > 3 \end{cases}$ **40.** $\begin{cases} 3y \leq x \\ y > 3x \end{cases}$

41. $\begin{cases} x \leq 0 \\ y < 0 \end{cases}$ **42.** $\begin{cases} y \geq x \\ y \leq \dfrac{1}{3}x + 1 \\ x > -3 \end{cases}$

43. Use a check to determine whether each ordered pair is a solution of the system: $\begin{cases} x + 2y \leq 3 \\ 2x - y > 1 \end{cases}$

    a. $(5, -4)$ **b.** $(-1, -3)$

44. Gift Shopping. A grandmother wants to spend at least $40 but no more than $60 on school clothes for her grandson. If T-shirts sell for $10 each and pants sell for $20 each, write a system of inequalities that describes the possible numbers of T-shirts x and pairs of pants y that she can buy. Graph the system and give two possible solutions.

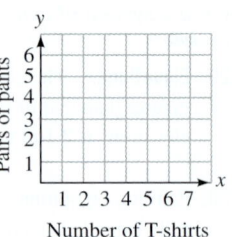

Pairs of pants

Number of T-shirts

---

4 ▶ CHAPTER TEST

Use a check to determine whether the ordered pair is a solution of the system.

1. $(5, 3)$, $\begin{cases} 3x + 2y = 21 \\ x + y = 8 \end{cases}$ **2.** $(-2, -1)$, $\begin{cases} 4x + y = -9 \\ 2x - 3y = -7 \end{cases}$

3. Fill in the blanks.

    a. A _____ of a system of linear equations is an ordered pair that satisfies each equation.

    b. A system of equations that has at least one solution is called a _____ system.

    c. A system of equations that has no solution is called an _____ system.

    d. Equations with different graphs are called _____ equations.

    e. A system of _____ equations has an infinite number of solutions.

    f. Two angles are said to be _____ if the sum of their measures is 180°.

4. Energy. The graphs below show U.S. electricity generation from natural gas and nuclear sources from 1995 through 2009. What is the intersection point of the graphs? Explain its significance.

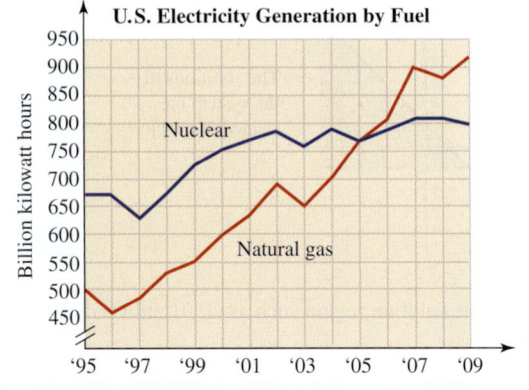

U.S. Electricity Generation by Fuel

Billion kilowatt hours

Nuclear

Natural gas

Solve each system by graphing.

5. $\begin{cases} y = 2x - 1 \\ x - 2y = -4 \end{cases}$ **6.** $\begin{cases} x + y = 5 \\ y = -x \end{cases}$

7. Find the slope and the y-intercept of the graph of each line in the system $\begin{cases} y = 4x - 10 \\ x - 2y = -16 \end{cases}$. Then, use that information to determine the *number of solutions* of the system. **Do not solve the system.**

8. How many solutions does the system of two linear equations graphed on the right have? Give three of the solutions.

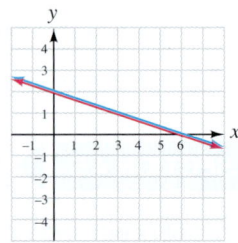

Solve each system by substitution.

9. $\begin{cases} y = x - 1 \\ 2x + y = -7 \end{cases}$ **10.** $\begin{cases} 3x + 6y = -15 \\ x + 2y = -5 \end{cases}$

Solve each system using elimination.

11. $\begin{cases} 3x - y = 2 \\ 2x + y = 8 \end{cases}$ **12.** $\begin{cases} 4x + 3y = -3 \\ -3x = -4y + 21 \end{cases}$

Solve each system using substitution or elimination.

13. $\begin{cases} 3x - 5y - 16 = 0 \\ \dfrac{x}{2} - \dfrac{5}{6}y = \dfrac{1}{3} \end{cases}$ **14.** $\begin{cases} 3a + 4b = -7 \\ 2b - a = -1 \end{cases}$

15. $\begin{cases} y = 3x - 1 \\ y = 2x + 4 \end{cases}$ **16.** $\begin{cases} 0.6c + 0.5d = 0 \\ 0.02c + 0.09d = 0 \end{cases}$

17. $\begin{cases} a - 1 = 2b \\ 3a + 1 = -10b \end{cases}$

18. $\begin{cases} \dfrac{x + 2}{6} = \dfrac{y + 3}{4} \\ \dfrac{x}{5} = \dfrac{3y - 3}{6} \end{cases}$

19. $\begin{cases} 4(a - 2) + 5y = 19 \\ 3(a - 2) - y = 0 \end{cases}$

20. $\begin{cases} 3x + 1 = 2x - 4y + 2 \\ 7x - y - 1 = 9y + 5x + 10 \end{cases}$

Write a system of two equations in two variables to solve each problem.

21. **Child Care.** On a mother's 22-mile commute to work, she drops her daughter off at a child care center. The first part of the trip is 6 miles less than the second part. How long is each part of her morning commute?

22. **Vacationing.** It cost a family of 7 a total of $219 for general admission tickets to the San Diego Zoo. How many adult tickets and how many child tickets were purchased?

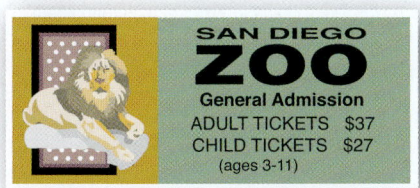

SAN DIEGO
ZOO
General Admission
ADULT TICKETS $37
CHILD TICKETS $27
(ages 3-11)

23. **Financial Planning.** A woman invested some money at 8% and some at 9% annual simple interest. The interest for 1 year on the combined investment of $10,000 was $840. How much was invested at each rate?

24. **Tailwinds/Headwinds.** Flying with a tailwind, a pilot flew an airplane 450 miles in 2.5 hours. Flying into a headwind, the return trip took 3 hours. Find the speed of the plane in calm air and the speed of the wind.

25. **Tether Ball.** The angles shown in the illustration are complementary. The measure of the larger angle is 10° more than three times the measure of the smaller angle. Find the measure of each angle.

26. **Antifreeze.** How many pints of a 5% antifreeze solution and how many pints of a 20% antifreeze solution must be mixed to obtain 12 pints of a 15% solution?

27. **Sunscreen.** A sunscreen selling for $1.50 per ounce is to be combined with another sunscreen selling for $0.80 per ounce. How many ounces of each are needed to make 10 ounces of a sunscreen mix that sells for $1.01 per ounce?

28. Use a check to determine whether (3, 1) is a solution of the system: $\begin{cases} y \le 2x - 1 \\ x + 3y > 6 \end{cases}$

Solve each system by graphing.

29. $\begin{cases} 3x + 2y < 6 \\ y \ge x + 1 \end{cases}$

30. $\begin{cases} x - y < 3 \\ y \le 0 \\ x \ge 0 \end{cases}$

31. **Clothes Shopping.** This system of inequalities describes the number of $20 shirts, x, and $40 pairs of pants, y, a person can buy if he or she plans to spend not less than $80 but not more than $120. Graph the system. Then give three solutions.

$$\begin{cases} 20x + 40y \ge 80 \\ 20x + 40y \le 120 \end{cases}$$

32. Match each equation, inequality, or system with the graph of its solution.

   a. $2x + y = 2$

   b. $2x + y \ge 2$

   c. $\begin{cases} 2x + y = 2 \\ 2x - y = 2 \end{cases}$

   d. $\begin{cases} 2x + y \ge 2 \\ 2x - y \le 2 \end{cases}$

   i.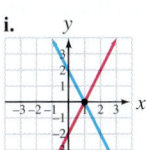

   ii.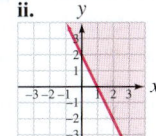

   iii.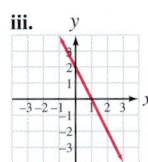

   iv.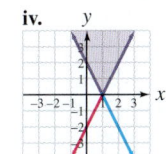

Group Project

WRITING APPLICATION PROBLEMS

▶ **Overview:** In Section 4.4, you solved application problems by translating the words of the problem into a system of two equations. In this activity, you will reverse these steps.

Instructions: Form groups of 2 or 3 students. For each type of application, write a problem that could be solved using the given equations. If you need help getting started, refer to the specific problem types in the text. When finished writing the five applications, pick one problem and solve it completely.

A rectangle problem:
$$\begin{cases} 2l + 2w = 320 \\ l = w + 40 \end{cases}$$

A number-value problem:
$$\begin{cases} 5x + 2y = 23 \\ 3x + 7y = 37 \end{cases}$$

An interest problem:
$$\begin{cases} x + y = 75{,}000 \\ 0.03x + 0.05y = 2{,}750 \end{cases}$$

A with–against the wind problem:
$$\begin{cases} 2(x + y) = 600 \\ 3(x - y) = 600 \end{cases}$$

A liquid mixture problem:
$$\begin{cases} x + y = 36 \\ 0.50x + 0.20y = 0.30(36) \end{cases}$$

CUMULATIVE REVIEW ▶▶ Chapters 1–4

1. Fill in the blanks. The answer to an addition problem is called a _____. The answer to a subtraction problem is called a _____. The answer to a multiplication problem is called a _____. The answer to a division problem is called a _____. [Section 1.1]

2. Give the prime factorization of 100. [Section 1.2]

3. Divide: $\dfrac{3}{4} \div \dfrac{6}{5}$ [Section 1.2]

4. Subtract: $\dfrac{7}{10} - \dfrac{1}{14}$ [Section 1.2]

5. Is π a rational or irrational number? [Section 1.3]

6. Graph each member of the set on the number line. [Section 1.3]

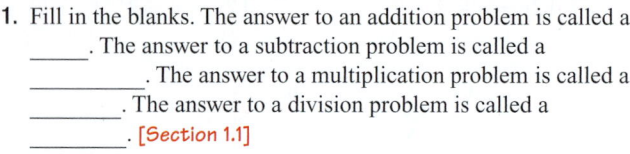

$$\left\{ -2\frac{1}{4}, \ \sqrt{2}, \ -1.75, \ \frac{7}{2}, \ 0.5 \right\}$$

7. Write $\dfrac{2}{3}$ as a decimal. [Section 1.3]

8. What property of real numbers is illustrated? [Section 1.6]

$$3(2x) = (3 \cdot 2)x$$

Evaluate each expression.

9. $-3^2 + |4^2 - 5^2|$ [Section 1.7]

10. $(4 - 5)^{20}$ [Section 1.7]

11. $\dfrac{-3 - (-7)}{2^2 - 3}$ [Section 1.7]

12. $12 - 2[1 - (-8 + 2)]$ [Section 1.7]

13. **Racing.** Suppose a driver has completed x laps of a 250-lap race. Write an expression for how many more laps he must make to finish the race. [Section 1.8]

14. What is the value of d dimes in cents? [Section 1.8]

Simplify each expression.

15. $13r - 12r$ [Section 1.9]

16. $27\left(\dfrac{2}{3}x\right)$ [Section 1.9]

17. $4(d - 3) - (d - 1)$ [Section 1.9]

18. $(13c - 3)(-6)$ [Section 1.9]

Solve each equation. Check each result.

19. $3(x - 5) + 2 = 2x$ [Section 2.2]

20. $\dfrac{x - 5}{3} - 5 = 7$ [Section 2.2]

21. $\dfrac{2}{5}x + 1 = \dfrac{1}{3} + x$ [Section 2.2]

22. $-\dfrac{5}{8}h = 15$ [Section 2.2]

23. **Gymnastics.** After the first day of registration, 119 children had been enrolled in a Gymboree class. That represented 85% of the available slots. Find the maximum number of children the center could enroll. [Section 2.3]

24. Greenhouse Gases. In 2008, the total U.S. greenhouse gas emissions were approximately 6,957 teragrams of carbon dioxide equivalent. Use the graph below to determine the greenhouse emissions that come from transportation sources. Round to the nearest teragram. [Section 2.3]

U.S. Greenhouse Gas Emissions, by source

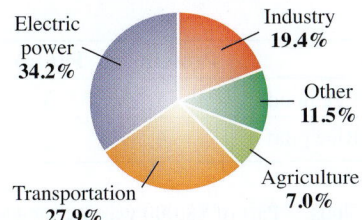

Electric power **34.2%**

Industry **19.4%**

Other **11.5%**

Agriculture **7.0%**

Transportation **27.9%**

Source: U.S. Environmental Protection Agency

25. Solve $A = \dfrac{1}{2}h(b + B)$ for h. [Section 2.4]

26. Cancer. According to the National Lung Cancer Partnership, approximately 219,000 people are diagnosed with lung cancer in the U.S. each year, with 13,000 more of them being men than women. Approximately how many men and how many women are diagnosed with lung cancer each year? [Section 2.5]

27. Mixing Candy. The owner of a candy store wants to make a 30-pound mixture of two candies to sell for $4 per pound. If red licorice bits sell for $3.80 per pound and lemon gumdrops sell for $4.40 per pound, how many pounds of each should be used? [Section 2.6]

28. Solve $8(4 + x) > 10(6 + x)$. Write the solution set in interval notation and graph it. [Section 2.7]

29. New York. In Manhattan, avenues run north and south, and streets run east and west to form a grid, as shown below. Give the location of the Chipotle Mexican Grill as an ordered pair of the form (avenue, street). [Section 3.1]

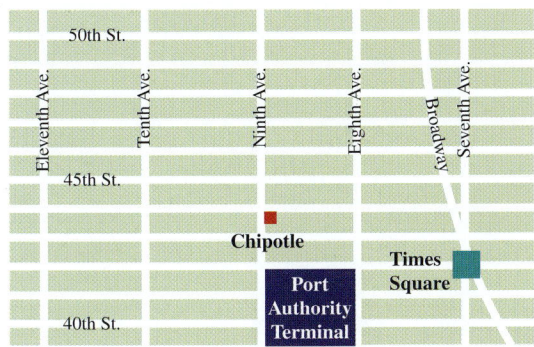

30. Perimeter. The eight vertices (corners) of a figure are $(-4, 6)$, $(4, 6)$, $(4, -3)$, $(2, -3)$, $(2, -5)$, $(-2, -5)$, $(-2, -3)$, and $(-4, -3)$. Find the perimeter of the figure. [Section 3.1]

31. In what quadrant does $(-3.5, 6)$ lie? [Section 3.1]

32. Is $(-2, 8)$ a solution of $y = -2x + 3$? [Section 3.2]

Graph each equation.

33. $x = 4$ [Section 3.3]

34. $4x - 3y = 12$ [Section 3.3]

Find the slope of the line with the given properties.

35. Passing through $(-2, 4)$ and $(6, 8)$ [Section 3.4]

36. A line that is horizontal [Section 3.4]

37. An equation of $2x - 3y = 12$ [Section 3.5]

38. Are the graphs of the lines whose equations are given below parallel or perpendicular? [Section 3.5]

$$y = -\frac{3}{4}x + \frac{15}{4} \qquad 4x - 3y = 25$$

39. Find the slope and the x- and y-intercepts of the line graphed on the right. [Sections 3.3 and 3.4]

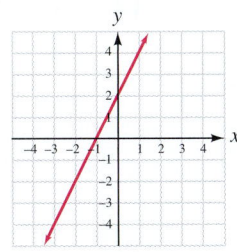

40. Newspapers. The line graph below approximates the percent of adults ages 25–34 who read a newspaper on a regular basis. Find the rate of change in newspaper readership for that age group. [Section 3.4]

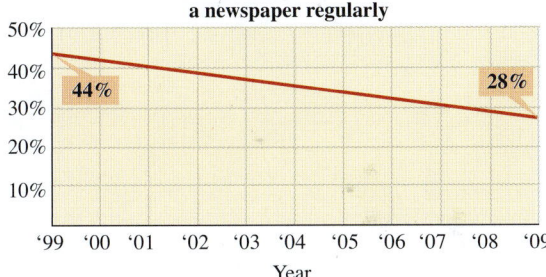

Percent of those ages 25–34 who read a newspaper regularly

44%   28%

Source: The State of the News Media, 2010

Find an equation of the line with the following properties. Write the equation in slope–intercept form.

41. Slope $= \dfrac{2}{3}$, y-intercept $= (0, 5)$ [Section 3.5]

42. Passing through $(-2, 4)$ and $(6, 10)$ [Section 3.6]

43. A horizontal line passing through $(2, 4)$ [Section 3.6]

44. Graph: $y < \dfrac{x}{3} - 1$ [Section 3.7]

45. If $f(x) = -2x^2 - 3x^3$, find $f(-1)$. [Section 3.8]

46. Refer to the graph on the right. Is this the graph of a function? If it is, give the domain and range. If it is not, find ordered pairs that show a value of x that is assigned more than one value of y. [Section 3.8]

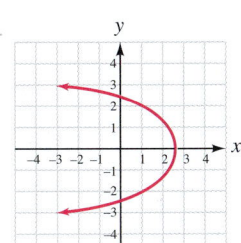

Solve each system by graphing.

47. $\begin{cases} x + 4y = -2 \\ y = -x - 5 \end{cases}$ [Section 4.1]

48. $\begin{cases} 2x - 3y < 0 \\ y > x - 1 \end{cases}$ [Section 4.5]

49. Solve $\begin{cases} x - 2y = 2 \\ 2x + 3y = 11 \end{cases}$ by substitution. [Section 4.2]

50. Solve $\begin{cases} \dfrac{3}{2}x - \dfrac{2}{3}y = 0 \\ \dfrac{3}{4}x + \dfrac{4}{3}y = \dfrac{5}{2} \end{cases}$ by elimination. [Section 4.3]

Write a system of two equations in two variables to solve each problem.

51. **Nutrition.** The table shows per serving nutritional information for egg noodles and rice pilaf. How many servings of each food should be eaten to consume exactly 22 grams of protein and 21 grams of fat? [Section 4.4]

	Protein (g)	Fat (g)
Egg noodles	5	3
Rice pilaf	4	5

52. **Investment Clubs.** Part of $8,000 was invested by an investment club at 10% interest and the rest at 12%. If the annual income from these investments is $900, how much was invested at each rate? [Section 4.4]

Exponents and Polynomials

5

©iStockphoto.com/beetle8

from Campus to Careers

Sound Engineering Technician

Today's digitally recorded music is crystal clear thanks to the talents of sound engineering technicians. They operate console mixing boards and microphones to record the music, voices, and sound effects that are so much a part of our media-filled lives. The job requires strong mathematical skills and an aptitude for working with electronic equipment. Sound technicians constantly work with numbers as they read meters and graphs, adjust dials and switches, and keep written logs.

Problem 125 in Study Set 5.2, problem 89 in Study Set 5.3, and problem 85 in Study Set 5.4 involve situations that a sound engineering technician might encounter on the job. The mathematical concepts discussed in this chapter can be used to solve those problems.

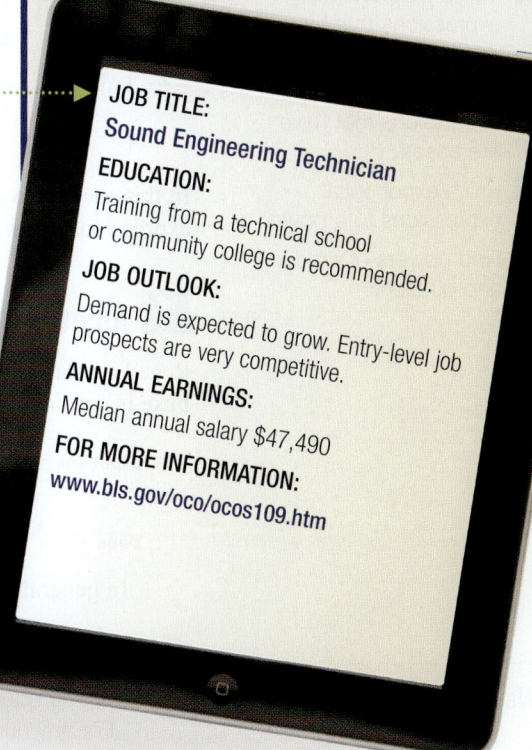

JOB TITLE:
Sound Engineering Technician

EDUCATION:
Training from a technical school or community college is recommended.

JOB OUTLOOK:
Demand is expected to grow. Entry-level job prospects are very competitive.

ANNUAL EARNINGS:
Median annual salary $47,490

FOR MORE INFORMATION:
www.bls.gov/oco/ocos109.htm

It's not uncommon for students' enthusiasm to lessen toward the middle of the term. Sometimes their effort and attendance begin to slip. Realize that missing even one class can have a great effect on your grade. Being tardy takes its toll as well. If you are just a few minutes late, or miss an entire class, you risk getting behind. So, keep the following tips in mind.

ARRIVE ON TIME, OR A LITTLE EARLY: When you arrive, get out your note-taking materials and homework. Identify any questions that you plan to ask your instructor once the class starts.

IF YOU MUST MISS A CLASS: Get a set of notes, the homework assignments, and any handouts that the instructor may have provided for the day(s) that you missed.

STUDY THE MATERIAL YOU MISSED: Take advantage of the online resources that are available with this textbook, such as video examples and problem-specific tutorials. Watch the explanations of the material from the section(s) that you missed.

Now Try This ▶

1. Plan ahead! List five possible situations that could cause you to be late to class or miss a class. (Some examples are parking/traffic delays, lack of a babysitter, oversleeping, or job responsibilities.) What can you do ahead of time so that these situations won't cause you to be tardy or absent?

2. Watch one section from the video series that accompanies this book. Take notes as you watch the explanations.

SECTION 5.1

Rules for Exponents

OBJECTIVES

1. Identify bases and exponents.
2. Multiply exponential expressions that have like bases.
3. Divide exponential expressions that have like bases.
4. Raise exponential expressions to a power.
5. Find powers of products and quotients.

ARE YOU READY?

The following problems review some basic skills that are needed when working with exponents.

1. Evaluate: **a.** $5 + 5 + 5$
   b. $5 \cdot 5 \cdot 5$

2. Evaluate: **a.** 2^6
   b. $2 \cdot 6$

3. Simplify: $\dfrac{x \cdot x \cdot x \cdot x}{x \cdot x \cdot x}$

4. Evaluate: $\dfrac{4^3}{4^2}$

In this section, we will use the definition of exponent to develop some rules for simplifying expressions that contain exponents.

1 Identify Bases and Exponents.

Recall that an **exponent** indicates repeated multiplication. It indicates how many times the **base** is used as a factor. For example, 3^5 represents the product of five 3's.

$$\text{Exponent} \rightarrow 3^5 = \overbrace{3 \cdot 3 \cdot 3 \cdot 3 \cdot 3}^{\text{5 factors of 3}}$$
$$\text{Base} \rightarrow$$

In general, we have the following definition.

Natural-Number Exponents

A natural-number exponent tells how many times its base is to be used as a factor. For any number x and any natural number n,

$$x^n = \overbrace{x \cdot x \cdot x \cdot \,\cdots\, \cdot x}^{n \text{ factors of } x}$$

Expressions of the form x^n are called **exponential expressions.** The base of an exponential expression can be a number, a variable, or a combination of numbers and variables. Some examples are:

$$10^5 = 10 \cdot 10 \cdot 10 \cdot 10 \cdot 10$$

The base is 10. The exponent is 5. Read as "10 to the fifth power" or simply as "10 to the fifth."

$$y^2 = y \cdot y$$

The base is y. The exponent is 2. Read as "y squared."

$$(-2s)^3 = (-2s)(-2s)(-2s)$$

The base is $-2s$. The exponent is 3. Read as "negative 2s raised to the third power" or "negative 2s cubed."

$$-8^4 = -(8 \cdot 8 \cdot 8 \cdot 8)$$

Since the $-$ sign is not written within parentheses, the base is 8. The exponent is 4. Read as "the opposite (or the negative) of 8 to the fourth power."

> ### Notation
>
> Bases that contain a $-$ sign *must* be written within parentheses.
>
> $(-2s)^3 \leftarrow$ Exponent
>
> $\overset{\text{Base}}{}$

When an exponent is 1, it is usually not written. For example, $4 = 4^1$ and $x = x^1$.

EXAMPLE 1   Identify the base and the exponent in each expression: **a.** 9^5 **b.** $7a^3$ **c.** $(7a)^3$ **d.** $-t^{10}$

Strategy   To identify the base and exponent, we will look for the form ▪.

Why   The exponent is the small raised number to the right of the base.

Solution **a.** In 9^5, the base is 9 and the exponent is 5.

b. $7a^3$ means $7 \cdot a^3$. Thus, the base is a, not $7a$. The exponent is 3.

c. Because of the parentheses in $(7a)^3$, the base is $7a$ and the exponent is 3.

d. Since the $-$ symbol is not written within parentheses, the base in $-t^{10}$ is t and the exponent is 10.

> **Self Check 1**   Identify the base and the exponent: **a.** 16^2
> **b.** $3y^4$ **c.** $(3y)^4$ **d.** $-m^{15}$
>
> **Now Try** ▶ Problems 13 and 17

EXAMPLE 2   Write as an exponential expression: **a.** $5 \cdot t \cdot t \cdot t$ **b.** $5t \cdot 5t \cdot 5t$ **c.** $\dfrac{p}{3} \cdot \dfrac{p}{3} \cdot \dfrac{p}{3} \cdot \dfrac{p}{3}$ **d.** $(a + 1)(a + 1)$

Strategy   We will look for repeated factors and count the number of times each appears.

Why   We can use an exponent to represent repeated multiplication.

Solution **a.** There are three repeated factors of t in $5 \cdot t \cdot t \cdot t$. The expression can be written $5t^3$.

b. There are three repeated factors of $5t$ in $5t \cdot 5t \cdot 5t$. The expression can be written $(5t)^3$.

c. There are four repeated factors of $\frac{p}{3}$ in $\frac{p}{3} \cdot \frac{p}{3} \cdot \frac{p}{3} \cdot \frac{p}{3}$. The expression can be written $\left(\frac{p}{3}\right)^4$.

d. There are two repeated factors of $(a + 1)$. The expression can be written $(a + 1)^2$.

> **Self Check 2**   Write as an exponential expression: **a.** $9 \cdot a \cdot a \cdot b \cdot b \cdot b \cdot b$
> **b.** $9a \cdot 9a$ **c.** $(x + y)(x + y)(x + y)(x + y)(x + y)$
>
> **Now Try** ▶ Problems 19 and 21

2 Multiply Exponential Expressions That Have Like Bases.

To develop a rule for multiplying exponential expressions that have the same base, we consider the product $6^2 \cdot 6^3$. Since 6^2 means that 6 is to be used as a factor two times, and 6^3 means that 6 is to be used as a factor three times, we have

$$6^2 \cdot 6^3 = \overbrace{6 \cdot 6}^{\text{2 factors of 6}} \cdot \overbrace{6 \cdot 6 \cdot 6}^{\text{3 factors of 6}}$$

$$= \overbrace{6 \cdot 6 \cdot 6 \cdot 6 \cdot 6}^{\text{5 factors of 6}}$$

$$= 6^5$$

We can find this result quickly if we keep the common base of 6 and add the exponents on 6^2 and 6^3.

$$6^2 \cdot 6^3 = 6^{2+3} = 6^5$$

This example illustrates the following rule for exponents.

Product Rule for Exponents	To multiply exponential expressions that have the same base, keep the common base and add the exponents.

For any number x and any natural numbers m and n,

$$x^m \cdot x^n = x^{m+n}$$ Read as "x to the mth power times x to the nth power equals x to the m plus nth power."

EXAMPLE 3   Simplify: **a.** $9^5(9^6)$ **b.** $x^3 \cdot x^4$ **c.** $y^2 y^4 y$ **d.** $(x + 2)^8 (x + 2)^7$
e. $(c^2 d^3)(c^4 d^5)$

Strategy   In each case, we want to write an equivalent expression using each base only once. We will use the product rule for exponents to do this.

Why   The product rule for exponents is used to multiply exponential expressions that have the same base.

Solution

a. $9^5(9^6) = 9^{5+6} = 9^{11}$ Read as "9 to the fifth power times 9 to the sixth power."
Keep the common base, 9, and add the exponents.
Since 9^{11} is a very large number, we will leave the answer in this form. We won't evaluate it.

> **Caution**
>
> Don't make the mistake of multiplying the bases when using the product rule. Keep the same base.
>
> $$9^5(9^6) \neq 81^{11}$$

b. $x^3 \cdot x^4 = x^{3+4} = x^7$ Keep the common base, x, and add the exponents.

c. $y^2 y^4 y = y^2 y^4 y^1$ Read as "y squared times y to the fourth power times y." Write y as y^1.
$= y^{2+4+1}$ Keep the common base, y, and add the exponents.
$= y^7$

> **Caution**
>
> Don't make the mistake of "distributing" an exponent over a sum (or difference). There is no such rule.
>
> $$(x + 2)^{15} \neq x^{15} + 2^{15}$$

d. $(x + 2)^8 (x + 2)^7 = (x + 2)^{8+7}$ Read as "the quantity of x + 2, raised to the eighth power, times the quantity of x + 2, raised to the seventh power."
Keep the common base, x + 2, and add the exponents.
$= (x + 2)^{15}$

e. $(c^2 d^3)(c^4 d^5) = (c^2 c^4)(d^3 d^5)$ Read as "the quantity of $c^2 d^3$ times the quantity of $c^4 d^5$." Use the commutative and associative properties of multiplication to group like bases together.
$= (c^{2+4})(d^{3+5})$ Keep the common base, c, and add the exponents.
Keep the common base, d, and add the exponents.
$= c^6 d^8$

Simplify: **a.** $7^8(7^7)$ **b.** x^2x^3x
c. $(y-1)^5(y-1)^5$ **d.** $(s^4t^3)(s^4t^4)$

Now Try ▶ Problems 29 and 33

CAUTION We cannot use the product rule to simplify expressions like $3^2 \cdot 2^3$, where the bases are different. However, we can simplify this expression by doing the arithmetic:

$$3^2 \cdot 2^3 = 9 \cdot 8 = 72$$

EXAMPLE 4 **Geometry.** Find an expression that represents the area of the rectangle.

Strategy We will multiply the length of the rectangle by its width.

Why The area of a rectangle is equal to the product of its length and width.

x^3 feet

x^5 feet

Solution
Area = **length · width**   This is the formula for the area of a rectangle.
$= x^5 \cdot x^3$ Substitute x^5 for the length and x^3 for the width.
$= x^{5+3}$ Use the product rule: Keep the common base, x, and add the exponents.
$= x^8$

The area of the rectangle is x^8 square feet, which can be written as x^8 ft^2.

Self Check 4 **Geometry.** Find an expression that represents the area of a rectangle with length a^8 in. and width a^4 in.

Now Try ▶ Problem 35

3 Divide Exponential Expressions That Have Like Bases.

To develop a rule for dividing exponential expressions that have the same base, we consider the quotient $\frac{4^5}{4^2}$, where the exponent in the numerator is greater than the exponent in the denominator. We can simplify this fraction by removing the common factors of 4 in the numerator and denominator:

$$\frac{4^5}{4^2} = \frac{4 \cdot 4 \cdot 4 \cdot 4 \cdot 4}{4 \cdot 4} = \frac{\overset{1}{\cancel{4}} \cdot \overset{1}{\cancel{4}} \cdot 4 \cdot 4 \cdot 4}{\underset{1}{\cancel{4}} \cdot \underset{1}{\cancel{4}}} = 4^3$$

We can find this result quickly if we keep the common base, 4, and subtract the exponents on 4^5 and 4^2.

$$\frac{4^5}{4^2} = 4^{5-2} = 4^3$$

This example illustrates another rule for exponents.

Quotient Rule for Exponents

To divide exponential expressions that have the same base, keep the common base and subtract the exponents.

For any nonzero number x and any natural numbers m and n, where $m > n$,

$$\frac{x^m}{x^n} = x^{m-n}$$ Read as "x to the mth power divided by x to the nth power equals x to the m minus nth power."

EXAMPLE 5   Simplify: **a.** $\dfrac{20^{16}}{20^9}$ **b.** $\dfrac{x^9}{x^3}$ **c.** $\dfrac{(7.5n)^{12}}{(7.5n)^{11}}$ **d.** $\dfrac{a^3b^8}{ab^5}$

Strategy   In each case, we want to write an equivalent expression using each base only once. We will use the quotient rule for exponents to do this.

Why   The quotient rule for exponents is used to divide exponential expressions that have the same base.

Solution   **a.** $\dfrac{20^{16}}{20^9} = 20^{16-9}$ Read as "20 to the sixteenth power divided by 20 to the ninth power." Keep the common base, 20, and subtract the exponents.

$\qquad\qquad = 20^7$ Since 20^7 is a very large number, we will leave the answer in this form. We won't evaluate it.

Caution

Don't make the mistake of "removing the common bases" when using the quotient rule. Keep the same base.

$$\dfrac{20^{16}}{20^9} \ne 1^7$$

b. $\dfrac{x^9}{x^3} = x^{9-3}$ Keep the common base, x, and subtract the exponents.

$\qquad\quad = x^6$

c. $\dfrac{(7.5n)^{12}}{(7.5n)^{11}} = (7.5n)^{12-11}$ Keep the common base, 7.5n, and subtract the exponents.

$\qquad\qquad\quad = (7.5n)^1$

$\qquad\qquad\quad = 7.5n$ Any number raised to the first power is simply that number.

The Language of Algebra

In this chapter, variables often appear in a denominator. In such cases, we will assume that the variables do not equal 0. That is, we will assume that **there are no divisions by 0**.

d. $\dfrac{a^3b^8}{ab^5} = \dfrac{a^3}{a^1} \cdot \dfrac{b^8}{b^5}$ Group the common bases together. It is helpful to write a as a^1.

$\qquad\quad = a^{3-1}b^{8-5}$ Keep the common base a and subtract the exponents. Keep the common base b and subtract the exponents.

$\qquad\quad = a^2b^3$

Self Check 5   Simplify: **a.** $\dfrac{55^{30}}{55^5}$ **b.** $\dfrac{a^5}{a^3}$ **c.** $\dfrac{(8.9t)^8}{(8.9t)^7}$ **d.** $\dfrac{b^{15}c^4}{b^4c}$

Now Try ▶ Problems 41 and 45

EXAMPLE 6   Simplify: $\dfrac{a^3a^5a^7}{a^4a}$

Strategy   We want to write an equivalent expression using one base and one exponent. First, we will use the product rule to simplify the numerator and the denominator. Then, we will use the quotient rule to simplify that result.

Why   The expression involves multiplication and division of exponential expressions that have the same base.

Solution   We simplify the numerator and denominator separately and proceed as follows.

Success Tip

Sometimes, more than one rule for exponents is needed to simplify an expression.

$\dfrac{a^3a^5a^7}{a^4a} = \dfrac{a^{15}}{a^5}$ In the numerator, keep the common base, a, and add the exponents. In the denominator, keep the common base, a, and add the exponents: $4 + 1 = 5$.

$\qquad\quad = a^{15-5}$ Keep the common base, a, and subtract the exponents.

$\qquad\quad = a^{10}$

Self Check 6   Simplify: $\dfrac{b^2b^6b}{b^4b^4}$

Now Try ▶ Problem 49

Recall that like terms are terms with exactly the same variables raised to exactly the same powers. To add or subtract exponential expressions, they must be like terms. To multiply or divide exponential expressions, only the bases need to be the same.

$$x^5 + x^2 \qquad \text{These are not like terms; the exponents are different. We cannot add.}$$

$$x^2 + x^2 = 2x^2 \qquad \text{These are like terms; we can add. Recall that } x^2 = 1x^2.$$

$$x^5 \cdot x^2 = x^7 \qquad \text{The bases are the same; we can multiply.}$$

$$\frac{x^5}{x^2} = x^3 \qquad \text{The bases are the same; we can divide.}$$

4 Raise Exponential Expressions to a Power.

To develop another rule for exponents, we consider $(5^3)^4$. Here, an exponential expression, 5^3, is raised to a power. Since 5^3 is the base and 4 is the exponent, $(5^3)^4$ can be written as $5^3 \cdot 5^3 \cdot 5^3 \cdot 5^3$. Because each of the four factors of 5^3 contains three factors of 5, there are $4 \cdot 3$ or 12 factors of 5.

$$(5^3)^4 = 5^3 \cdot 5^3 \cdot 5^3 \cdot 5^3 = \overbrace{5 \cdot 5 \cdot 5 \cdot 5 \cdot 5 \cdot 5 \cdot 5 \cdot 5 \cdot 5 \cdot 5 \cdot 5 \cdot 5}^{12 \text{ factors of } x} = 5^{12}$$

We can find this result quickly if we keep the common base of 5 and multiply the exponents.

$$(5^3)^4 = 5^{3 \cdot 4} = 5^{12}$$

This example illustrates the following rule for exponents.

Power Rule for Exponents	To raise an exponential expression to a power, keep the base and multiply the exponents. For any number x and any natural numbers m and n, $$(x^m)^n = x^{m \cdot n} = x^{mn}$$ Read as "x to the mth power raised to the nth power equals x to the mnth power."

EXAMPLE 7   Simplify: **a.** $(2^3)^7$ **b.** $[(-6)^2]^5$ **c.** $(z^8)^8$

Strategy   In each case, we want to write an equivalent expression using one base and one exponent. We will use the power rule for exponents to do this.

Why   Each expression is a power of a power.

Solution   **a.** $(2^3)^7 = 2^{3 \cdot 7} = 2^{21}$ Read as "2 cubed raised to the seventh power." Keep the base, 2, and multiply the exponents.

b. $[(-6)^2]^5 = (-6)^{2 \cdot 5} = (-6)^{10}$ Read as "negative six squared raised to the fifth power." Keep the base, -6, and multiply the exponents. Since $(-6)^{10}$ is a very large number, we will leave the answer in this form.

c. $(z^8)^8 = z^{8 \cdot 8} = z^{64}$ Keep the base, z, and multiply the exponents.

Self Check 7   Simplify: **a.** $(4^6)^5$ **b.** $(y^5)^2$

Now Try ▶   Problems 53 and 55

EXAMPLE 8 Simplify: **a.** $(x^2x^5)^8$ **b.** $(z^2)^4(z^3)^3$

Strategy In each case, we want to write an equivalent expression using one base and one exponent. We will use the product and power rules for exponents to do this.

Why The expressions involve multiplication of exponential expressions that have the same base and they involve powers of powers.

Solution **a.** $(x^2x^5)^8 = (x^7)^8$ Read as "the quantity of x squared times x to the fifth power, raised to the eighth power." Within the parentheses, keep the common base, x, and add the exponents.

$\qquad\qquad = x^{56}$ Keep the base, x, and multiply the exponents.

b. $(z^2)^4(z^3)^3 = z^8z^9$ For each power of z raised to a power, keep the base and multiply the exponents.

$\qquad\qquad = z^{17}$ Keep the common base, z, and add the exponents.

> **Self Check 8** Simplify: **a.** $(a^4a^3)^3$ **b.** $(a^3)^3(a^4)^2$
>
> **Now Try** Problems 59 and 63

5 Find Powers of Products and Quotients.

To develop more rules for exponents, we consider the expression $(2x)^3$, which is a *power of the product* of 2 and x, and the expression $\left(\frac{2}{x}\right)^3$, which is a *power of the quotient* of 2 and x.

$$(2x)^3 = 2x \cdot 2x \cdot 2x$$
$$= (2 \cdot 2 \cdot 2)(x \cdot x \cdot x)$$
$$= 2^3x^3$$
$$= 8x^3$$

$$\left(\frac{2}{x}\right)^3 = \frac{2}{x} \cdot \frac{2}{x} \cdot \frac{2}{x}$$ Assume $x \neq 0$.
$$= \frac{2 \cdot 2 \cdot 2}{x \cdot x \cdot x}$$ Multiply the numerators. Multiply the denominators.
$$= \frac{2^3}{x^3}$$
$$= \frac{8}{x^3}$$ Evaluate: $2^3 = 8$.

These examples illustrate the following rules for exponents.

Powers of a Product and a Quotient	To raise a product to a power, raise each factor of the product to that power. To raise a quotient to a power, raise the numerator and the denominator to that power. For any numbers x and y, and any natural number n, $$(xy)^n = x^ny^n \qquad \text{and} \qquad \left(\frac{x}{y}\right)^n = \frac{x^n}{y^n}, \quad \text{where } y \neq 0$$

EXAMPLE 9 Simplify: **a.** $(3c)^4$ **b.** $(x^2y^3)^5$ **c.** $\left(-\frac{1}{4}a^3b\right)^2$

Strategy In each case, we want to write the expression in an equivalent form in which each base is raised to a single power. We will use the power of a product rule for exponents to do this.

Why Within each set of parentheses is a product, and each of those products is raised to a power.

Solution   **a.** $(3c)^4 = 3^4c^4$ Raise each factor of the product 3c to the 4th power.

$= 81c^4$ Evaluate: $3^4 = 81$.

b. $(x^2y^3)^5 = (x^2)^5(y^3)^5$ Raise each factor of the product x^2y^3 to the 5th power.

$= x^{10}y^{15}$ For each power of a power, keep each base, x and y, and multiply the exponents.

c. $\left(-\frac{1}{4}a^3b\right)^2 = \left(-\frac{1}{4}\right)^2(a^3)^2b^2$ Raise each factor of the product $-\frac{1}{4}a^3b$ to the 2nd power.

$= \frac{1}{16}a^6b^2$ Evaluate: $\left(-\frac{1}{4}\right)^2 = \frac{1}{16}$. Keep the base a and multiply the exponents.

Self Check 9   Simplify: **a.** $(2t)^4$ **b.** $(c^3d^4)^6$ **c.** $\left(-\frac{1}{3}ab^5\right)^3$

Now Try   Problems 67 and 73

EXAMPLE 10   Simplify: $\dfrac{(a^3b^4)^2}{ab^5}$

Strategy   We want to write the expression in an equivalent form in which each base is raised to a single power. We will use the power of a product rule and the quotient rule for exponents to do this.

Why   The expression involves a power of a product and it is the quotient of exponential expressions that have the same base.

Solution $\dfrac{(a^3b^4)^2}{ab^5} = \dfrac{(a^3)^2(b^4)^2}{a^1b^5}$ In the numerator, raise each factor within the parentheses to the second power. In the denominator, it is helpful to write a as a^1.

$= \dfrac{a^6b^8}{a^1b^5}$ In the numerator, for each power of a power, keep each base, a and b, and multiply the exponents.

$= a^{6-1}b^{8-5}$ Keep each of the bases, a and b, and subtract the exponents.

$= a^5b^3$

Self Check 10   Simplify: $\dfrac{(c^4d^5)^3}{c^2d^3}$

Now Try   Problem 75

EXAMPLE 11   Simplify: $\dfrac{(5b)^9}{(5b)^6}$

Strategy   We want to write the expression in an equivalent form in which the base is raised to a single power. We will begin by using the quotient rule for exponents.

Why   The expression involves division of exponential expressions that have the same base, $5b$.

Solution $\dfrac{(5b)^9}{(5b)^6} = (5b)^{9-6}$ Keep the common base, 5b, and subtract the exponents.

$= (5b)^3$

$= 5^3b^3$ Use the power of a product rule: Raise each factor within the parentheses to the 3rd power. Do not multiply the exponent 3 and the base 5.

$= 125b^3$ Evaluate 5^3.

Self Check 11   Simplify: $\dfrac{(-2h)^{20}}{(-2h)^{14}}$

Now Try ▶ Problem 79

EXAMPLE 12   Simplify: **a.** $\left(\dfrac{4}{k}\right)^3$ **b.** $\left(\dfrac{3x^2}{2y^3}\right)^5$

Strategy   We want to write each expression in an equivalent form using each base raised to a single power. We will use the power of a quotient rule for exponents to do this.

Why   Within each set of parentheses is a quotient, and each of those quotients is raised to a power.

Solution   **a.** Since $\frac{4}{k}$ is the quotient of 4 and k, the expression $\left(\frac{4}{k}\right)^3$ is a power of a quotient.

$$\left(\dfrac{4}{k}\right)^3 = \dfrac{4^{\mathbf{3}}}{k^{\mathbf{3}}} = \dfrac{64}{k^3}$$ Raise the numerator and denominator to the 3rd power. Then evaluate: $4^3 = 64$. Do not multiply the exponent 3 and the base 4.

b. $\left(\dfrac{3x^2}{2y^3}\right)^5 = \dfrac{(3x^2)^5}{(2y^3)^5}$ Raise the numerator and the denominator to the 5th power.

$$= \dfrac{3^5(x^2)^5}{2^5(y^3)^5}$$ Raise each factor within parentheses to the 5th power. Do not multiply the base 3 and the exponent 5 or the base 2 and the exponent 5.

$$= \dfrac{243x^{10}}{32y^{15}}$$ Evaluate: $3^5 = 243$ and $2^5 = 32$. For each power of a power, keep the base and multiply the exponents.

Self Check 12   Simplify: **a.** $\left(\dfrac{x}{7}\right)^3$ **b.** $\left(\dfrac{2x^3}{3y^2}\right)^4$

Now Try ▶ Problems 83 and 85

The following rules for exponents are used so often in this course, you need to memorize them.

Rules for Exponents	If m and n represent natural numbers and there are no divisions by zero, then

Exponent of 1

$x^1 = x$

Product rule

$x^m x^n = x^{m+n}$

Power rule

$(x^m)^n = x^{mn}$

Quotient rule

$\dfrac{x^m}{x^n} = x^{m-n}$

Power of a product

$(xy)^n = x^n y^n$

Power of a quotient

$\left(\dfrac{x}{y}\right)^n = \dfrac{x^n}{y^n}$

SECTION 5.1 ▶ **STUDY SET**

VOCABULARY

Fill in the blank.

1. Expressions such as x^4, 10^3, and $(5t)^2$ are called _____ expressions.

2. Match each expression below with the proper description on the next page.

$$\dfrac{a^8}{a^2} \qquad (a^4b^2)^5 \qquad \left(\dfrac{a^6}{a}\right)^3 \qquad (a^8)^4 \qquad a^5 \cdot a^3$$

a. Product of exponential expressions with the same base

b. Quotient of exponential expressions with the same base

c. Power of an exponential expression

d. Power of a product

e. Power of a quotient

CONCEPTS

Fill in the blanks.

3. a. $(3x)^4 = $ ☐ · ☐ · ☐ · ☐

   b. $(-5y)(-5y)(-5y) = $ ☐

4. a. $x = x$☐

   b. $x^m x^n = $ ☐

   c. $(xy)^n = $ ☐

   d. $(a^b)^c = $ ☐

   e. $\dfrac{x^m}{x^n} = $ ☐

   f. $\left(\dfrac{a}{b}\right)^n = $ ☐

5. To simplify each expression, determine whether you add, subtract, multiply, or divide the exponents.

   a. $\dfrac{x^8}{x^2}$

   b. $b^6 \cdot b^9$

   c. $(n^8)^4$

   d. $(a^4 b^2)^5$

6. a. To simplify $(2y^3 z^2)^4$, what factors within the parentheses must be raised to the fourth power?

   b. To simplify $\left(\dfrac{y^3}{z^2}\right)^4$, what two expressions must be raised to the fourth power?

Simplify each expression, if possible.

7. a. $x^2 + x^2$

   b. $x^2 \cdot x^2$

   c. $x^2 + x$

   d. $x^2 \cdot x$

8. a. $x^3 - x^2$

   b. $\dfrac{x^3}{x^2}$

   c. $4^2 \cdot 2^4$

   d. $\dfrac{x^3}{y^2}$

NOTATION

Complete each solution to simplify each expression.

9. $(x^4 x^2)^3 = ($ ☐ $)^3 = x$☐

10. $\dfrac{a^3 a^4}{a^2} = \dfrac{☐}{a^2} = a$☐$^{-2} = a$☐

Fill in the blanks.

11. a. We read 9^4 as "nine to the fourth _____."

    b. We read $(a^2 b^6)(a^4 b^5)$ as "the _____ of $a^2 b^6$ times the _____ of $a^4 b^5$."

    c. We read $(3^6)^9$ as "3 to the _____ power, raised to the _____ power."

12. a. We read $n^2 n^3 n$ as "n _____ times n _____ times n."

    b. We read $\dfrac{x^7}{x^5}$ as "x to the seventh power _____ by x to the _____ power."

    c. We read $(b + 5)^6 (b + 5)^8$ as "the _____ of $b + 5$, _____ to the sixth power, times the _____ of $b + 5$, raised to the _____ power."

GUIDED PRACTICE

Identify the base and the exponent in each expression. See Example 1.

Look Alikes . . .

13. a. 4^3 b. -4^3 c. $(-4)^3$

14. a. x^5 b. $-x^5$ c. $(-x)^5$

15. a. $(-3x)^2$ b. $-3x^2$ c. $-(-3x)^2$

16. a. $-\dfrac{1}{3}x^6$ b. $\left(-\dfrac{1}{3}x\right)^6$ c. $-\left(-\dfrac{1}{3}x\right)^6$

17. a. $9m^{12}$ b. $(9m)^{12}$ c. $-9m^{12}$

18. a. $(y + 9)^4$ b. $y + 9^4$ c. $8(y + 9)^4$

Write each expression in an equivalent form using an exponent. See Example 2.

19. $4t \cdot 4t \cdot 4t \cdot 4t$

20. $-5u(-5u)(-5u)(-5u)(-5u)$

21. $-4 \cdot t \cdot t \cdot t \cdot t \cdot t$

22. $-5 \cdot u \cdot u \cdot u$

23. $\dfrac{t}{2} \cdot \dfrac{t}{2} \cdot \dfrac{t}{2}$

24. $\dfrac{x}{c} \cdot \dfrac{x}{c} \cdot \dfrac{x}{c} \cdot \dfrac{x}{c}$

25. $(x - y)(x - y)$

26. $(m + 4)(m + 4)$

Use the product rule for exponents to simplify each expression. Write the results using exponents. See Example 3.

27. $5^3 \cdot 5^4$

28. $3^4 \cdot 3^6$

29. $bb^2 b^3$

30. $aa^3 a^5$

31. $(y - 2)^5 (y - 2)^2$

32. $(t + 1)^5 (t + 1)^3$

33. $(a^2 b^3)(a^3 b^3)$

34. $(u^3 v^5)(u^4 v^5)$

Find an expression that represents the area or volume of each figure. Recall that the formula for the volume of a rectangular solid is V = length · width · height. See Example 4.

35.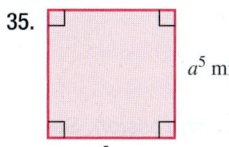

a^5 mi

a^5 mi

36.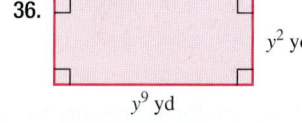

y^2 yd

y^9 yd

37.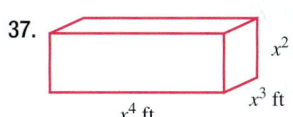

x^2 ft

x^3 ft

x^4 ft

38.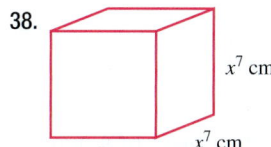

x^7 cm

x^7 cm

x^7 cm

Use the quotient rule for exponents to simplify each expression. Write the results using exponents. See Example 5.

39. $\dfrac{8^{12}}{8^4}$

40. $\dfrac{10^4}{10^2}$

41. $\dfrac{x^{15}}{x^3}$

42. $\dfrac{y^6}{y^3}$

43. $\dfrac{(3.7p)^7}{(3.7p)^2}$

44. $\dfrac{(0.25y)^9}{(0.25y)^3}$

45. $\dfrac{c^3 d^7}{cd}$

46. $\dfrac{r^8 s^9}{rs}$

Use the product and quotient rules for exponents to simplify each expression. See Example 6.

47. $\dfrac{y^3 y^4}{yy^2}$

48. $\dfrac{b^4 b^5}{b^2 b^3}$

49. $\dfrac{a^2 a^3 a^4}{a^8}$

50. $\dfrac{h^3 h^6 h}{h^9}$

Use the power rule for exponents to simplify each expression. Write the results using exponents. See Example 7.

51. $(3^2)^4$

52. $(4^3)^3$

53. $[(-4.3)^3]^8$

54. $[(-1.7)^9]^8$

55. $(m^{50})^{10}$

56. $(n^{25})^4$

57. $(y^5)^3$

58. $(b^3)^6$

Use the product and power rules for exponents to simplify each expression. See Example 8.

59. $(x^2 x^3)^5$

60. $(y^3 y^4)^4$

61. $(p^2 p^3)^5$

62. $(r^3 r^4)^2$

63. $(t^3)^4 (t^2)^3$

64. $(b^2)^5 (b^3)^2$

65. $(u^4)^2 (u^3)^2$

66. $(v^5)^2 (v^3)^4$

Use the power of a product rule for exponents to simplify each expression. See Example 9.

67. $(6a)^2$

68. $(3b)^3$

69. $(5y)^4$

70. $(4t)^4$

71. $(-2r^2 s^3)^3$

72. $(-2x^2 y^4)^5$

73. $\left(-\dfrac{1}{3} y^2 z^4\right)^5$

74. $\left(-\dfrac{1}{4} t^3 u^8\right)^2$

Use rules for exponents to simplify each expression. See Example 10.

75. $\dfrac{(ab^2)^3}{a^2 b^2}$

76. $\dfrac{(m^3 n^4)^3}{m^3 n^6}$

77. $\dfrac{(r^4 s^3)^4}{r^3 s^9}$

78. $\dfrac{(x^2 y^5)^5}{x^6 y^2}$

Use rules for exponents to simplify each expression. See Example 11.

79. $\dfrac{(6k)^7}{(6k)^4}$

80. $\dfrac{(-3a)^{12}}{(-3a)^{10}}$

81. $\dfrac{(3q)^5}{(3q)^3}$

82. $\dfrac{(ab)^8}{(ab)^4}$

Use the power of a quotient rule for exponents to simplify each expression. See Example 12.

83. $\left(\dfrac{a}{b}\right)^3$

84. $\left(\dfrac{r}{s}\right)^4$

85. $\left(\dfrac{8a^2}{11b^5}\right)^2$

86. $\left(\dfrac{7g^4}{6h^3}\right)^2$

TRY IT YOURSELF

Simplify each expression, if possible.

87. $\left(\dfrac{x^2}{y^3}\right)^5$

88. $\left(\dfrac{u^4}{v^2}\right)^6$

89. $y^3 y^2 y^4$

90. $y^4 yy^6$

91. $\dfrac{15^9}{15^6}$

92. $\dfrac{25^{13}}{25^7}$

93. $\dfrac{t^5 t^6 t}{t^2 t^3}$

94. $\dfrac{m^5 m^{12} m}{m^7 m^4}$

95. $\dfrac{(k-2)^{15}}{(k-2)}$

96. $\dfrac{(m+8)^{20}}{(m+8)}$

97. $cd^4 \cdot cd$

98. $ab^3 \cdot ab^4$

99. $\left(\dfrac{y^3 y^5}{yy^2}\right)^3$

100. $\left(\dfrac{s^5 s^6}{s^2 s^2}\right)^4$

101. $\dfrac{s^2 s^2 s^2}{s^3 s}$

102. $\dfrac{w^4 w^4 w^4}{w^2 w}$

103. $(-6a^3 b^2)^3$

104. $(-10r^3 s^2)^2$

105. $\left(\dfrac{3m^4}{2n^5}\right)^5$

106. $\left(\dfrac{2s^2}{3t^5}\right)^5$

107. $\dfrac{(a^2 b^2)^{15}}{(ab)^9}$

108. $\dfrac{(s^3 t^3)^4}{(st)^2}$

109. $(n^4 n)^3 (n^3)^6$

110. $(y^3 y)^2 (y^2)^2$

111. $\dfrac{(6h)^8}{(6h)^6}$

112. $\dfrac{(-7r)^{10}}{(-7r)^8}$

113. $\dfrac{x^4 y^7}{xy^3}$

114. $\dfrac{p^7 q^{10}}{p^2 q^7}$

115. $\left(\dfrac{m}{3}\right)^4$

116. $\left(\dfrac{n}{5}\right)^3$

Look Alikes . . .

117. a. $a^3 \cdot a^3$ b. $(a^3)^3$ c. $a^3 + a^3$

118. a. $(m^5)^7$ b. $m^5 \cdot m^7$ c. $m^5 - m^7$

119. a. $b^3 b^2 b^4$ b. $(b^3 b^2)^4$ c. $\dfrac{b^3 b^2}{b^4}$

120. a. $(2n^4 n)^5$ b. $2n^4 n$ c. $\left(\dfrac{2n^4}{n}\right)^5$

APPLICATIONS

121. **Art History.** Leonardo da Vinci's drawing relating a human figure to a square and a circle is shown. Find an expression for the following:

a. The area of the square if the man's height is $5x$ feet

b. The area of the circle if the waist-to-feet distance is $3a$ feet. Leave π in your answer.

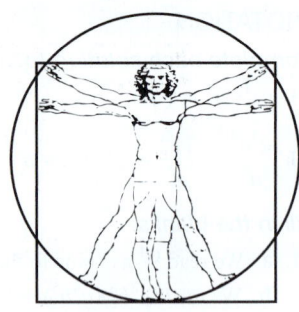

122. **Packaging.** A bowling ball fits tightly against all sides of a cardboard box that it is packaged in. Find expressions for the volume of the ball and box. Leave π in your answer.

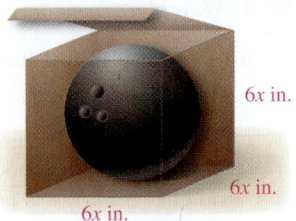

$6x$ in.

$6x$ in.

$6x$ in.

123. **Childbirth.** Mr. and Mrs. Emory Harrison, of Johnson City, Tennessee, had 13 sons in a row during the 1940s and 1950s. The **probability** of a family of 13 children all being male is $\left(\frac{1}{2}\right)^{13}$. Evaluate this expression.

124. **Toys.** A Super Ball is dropped from a height of 1 foot and always rebounds to four-fifths of its previous height. The rebound height of the ball after the third bounce is $\left(\frac{4}{5}\right)^3$ feet. Evaluate this expression. Is the third bounce more or less than $\frac{1}{2}$ foot high?

WRITING

125. Explain the mistake in the following work.

   a. $2^3 \cdot 2^2 = 4^5 = 1{,}024$ b. $(5d^2)^3 = 15d^6$

126. Explain why we can simplify $x^4 \cdot x^5$, but cannot simplify $x^4 + x^5$.

REVIEW

Match each equation with its graph below.

127. $y = 2x - 1$ 128. $y = 3x - 1$
129. $y = 3$ 130. $x = 3$

a. 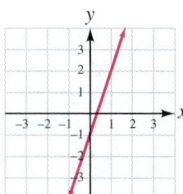   b.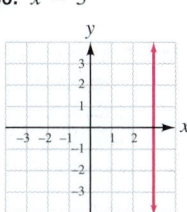

c. 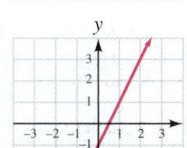   d.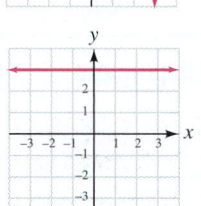

CHALLENGE PROBLEMS

131. Simplify each expression. The variables represent natural numbers.

   a. $x^{2m}x^{3m}$ b. $(y^{5c})^4$

   c. $\dfrac{m^{8x}}{m^{4x}}$ d. $(2a^{6y})^4$

132. Evaluate the following expression without using a calculator:

$$\frac{(108{,}642)^4}{(54{,}321)^4}$$

SECTION 5.2

Zero and Negative Exponents

OBJECTIVES

1. Use the zero exponent rule.

2. Use the negative integer exponent rule.

3. Use exponent rules to change negative exponents in fractions to positive exponents.

4. Use all exponent rules to simplify expressions.

ARE YOU READY?

 The following problems review some basic skills that are needed when working with zero and negative exponents.

1. Simplify: $\dfrac{2 \cdot 2}{2 \cdot 2 \cdot 2}$

2. Simplify: $\dfrac{x \cdot x}{x \cdot x \cdot x \cdot x}$

3. Find the reciprocal: a. $\dfrac{1}{5}$ b. w

4. Identify the base and exponent:
   a. m^0 b. 7^{-2}

We now extend the discussion of natural-number exponents to include exponents that are zero and exponents that are negative integers.

1 Use the Zero Exponent Rule.

To develop the definition of a zero exponent, we will simplify the expression $\frac{5^3}{5^3}$ in two ways and compare the results. First, we apply the quotient rule for exponents, where we subtract the equal exponents in the numerator and denominator. The result is 5^0, which is read as "5 to the zero power." In the second approach, we write 5^3 as $5 \cdot 5 \cdot 5$ and remove the common factors of 5 in the numerator and denominator. The result is 1.

$$\frac{5^3}{5^3} = 5^{3-3} = \mathbf{5^0} \qquad\qquad \frac{5^3}{5^3} = \frac{\overset{1}{\cancel{5}} \cdot \overset{1}{\cancel{5}} \cdot \overset{1}{\cancel{5}}}{\underset{1}{\cancel{5}} \cdot \underset{1}{\cancel{5}} \cdot \underset{1}{\cancel{5}}} = \mathbf{1}$$

These results must be equal.

Since $\frac{5^3}{5^3} = 5^0$ and $\frac{5^3}{5^3} = 1$, we conclude that $5^0 = 1$. This observation illustrates the following definition.

Zero Exponents	Any nonzero base raised to the 0 power is 1. For any nonzero real number x, $\qquad x^0 = 1$ Read as "x to the zero power equals 1."

EXAMPLE 1   Simplify. Assume $a \neq 0$:   **a.** $(-8)^0$ **b.** $\left(\dfrac{14}{15}\right)^0$ **c.** $(3a)^0$ **d.** $3a^0$

Strategy   We note that each exponent is 0. To simplify the expressions, we will identify the base and use the zero-exponent rule.

Why   If an expression contains a nonzero base raised to the 0 power, we can replace it with 1.

Solution   **a.** $(-8)^0 = 1$ Because the base is -8 and the exponent is 0.

b. $\left(\dfrac{14}{15}\right)^0 = 1$ Because the base is $\frac{14}{15}$ and the exponent is 0.

The Language of Algebra

The zero exponent definition does not define 0^0. This expression is called an **indeterminate form,** which is beyond the scope of this book.

c. $(3a)^0 = 1$ Because of the parentheses, the base is 3a. The exponent is 0.

d. $3a^0 = 3 \cdot \mathbf{a^0}$ Read as "3 times a to the zero power." Since there are no parentheses, the base is a, not $3a$. The exponent is 0.

$\qquad = 3 \cdot \mathbf{1}$ Simplify: $a^0 = 1$.

$\qquad = 3$

Self Check 1   Simplify each expression: **a.** $(0.75)^0$ **b.** $-5c^0 d$
 c. $(5c)^0$

Now Try ▶   Problems 13 and 17

---

2   Use the Negative Integer Exponent Rule.

The Language of Algebra

The **negative integers** are:
 $-1, -2, -3, -4, -5, \ldots$

To develop the definition of a negative exponent, we will simplify $\dfrac{6^2}{6^5}$ in two ways and compare the results.

If we apply the quotient rule for exponents, where we subtract the greater exponent in the denominator from the lesser exponent in the numerator, we get 6^{-3}. In the second approach, we remove the two common factors of 6 to get $\frac{1}{6^3}$.

$$\frac{6^2}{6^5} = 6^{2-5} = \mathbf{6^{-3}} \qquad\qquad \frac{6^2}{6^5} = \frac{\overset{1}{\cancel{6}} \cdot \overset{1}{\cancel{6}}}{\underset{1}{\cancel{6}} \cdot \underset{1}{\cancel{6}} \cdot 6 \cdot 6 \cdot 6} = \frac{\mathbf{1}}{\mathbf{6^3}}$$

These must be equal.

Since $\frac{6^2}{6^5} = 6^{-3}$ and $\frac{6^2}{6^5} = \frac{1}{6^3}$, we conclude that $6^{-3} = \frac{1}{6^3}$. Note that 6^{-3} is equal to the reciprocal of 6^3. This observation illustrates the following definition.

Negative Exponents	For any nonzero real number x and any integer n,

$$x^{-n} = \frac{1}{x^n}$$ Read as "x to the negative nth power equals 1 over x to the nth power."

In words, x^{-n} is the reciprocal of x^n.

From the definition, we see that another way to write x^{-n} is to write its reciprocal and change the sign of the exponent. For example,

$$5^{-4} = \frac{1}{5^4}$$ Think of the reciprocal of 5^{-4}. Then change the sign of the exponent.

The definitions of a zero exponent and a negative integer exponent have been written in such a way as to remain consistent with the definition for a natural-number exponent that we learned earlier. This can be seen in the following list.

$$2^4 = 16$$
$$2^3 = 8$$ Divide by 2.   For natural number exponents, each time that we decrease the exponent by 1, the value of the exponential expression is divided by 2.
$$2^2 = 4$$ Divide by 2.
$$2^1 = 2$$ Divide by 2.

$$2^0 = 1$$ For the pattern to continue, we define $2^0 = 1$.

$$2^{-1} = \frac{1}{2}$$ For the pattern to continue, we define $2^{-1} = \frac{1}{2^1} = \frac{1}{2}$.

$$2^{-2} = \frac{1}{4}$$ For the pattern to continue, we define $2^{-2} = \frac{1}{2^2} = \frac{1}{4}$.

$$2^{-3} = \frac{1}{8}$$ For the pattern to continue, we define $2^{-3} = \frac{1}{2^3} = \frac{1}{8}$.

EXAMPLE 2   Express using positive exponents and simplify, if possible: **a.** 3^{-2} **b.** y^{-1} **c.** $(-2)^{-3}$ **d.** $5^{-2} - 10^{-2}$

Strategy   Since each exponent is a negative number, we will use the negative exponent rule.

Why   This rule enables us to write an exponential expression that has a negative exponent in an equivalent form using a positive exponent.

Solution   **a.** $3^{-2} = \frac{1}{3^2} = \frac{1}{9}$ Read as "3 to the negative second power."
Write the reciprocal of 3^{-2} and change the sign of the exponent.

Caution

A negative exponent **does not,** itself, make the simplified expression negative. It indicates a reciprocal. Avoid these common mistakes:

$$3^{-2} = -9 \qquad 3^{-2} = -\frac{1}{9}$$

b. $y^{-1} = \frac{1}{y^1} = \frac{1}{y}$ Read as "y to the negative first power."
Write the reciprocal of y^{-1} and change the sign of the exponent.

c. $(-2)^{-3} = \frac{1}{(-2)^3}$ Read as "−2 to the negative third power."
Because of the parentheses, the base is −2.
Write the reciprocal of $(-2)^{-3}$ and change the exponent.

$$= -\frac{1}{8}$$ Evaluate: $(-2)^3 = -8$. Write the − symbol in front of the fraction.

d. $5^{-2} - 10^{-2} = \dfrac{1}{5^2} - \dfrac{1}{10^2}$ Write the reciprocal of 5^{-2} and 10^{-2} and change the sign of each exponent.

$= \dfrac{1}{25} - \dfrac{1}{100}$ Evaluate: $5^2 = 25$ and $10^2 = 100$.

$= \dfrac{4}{100} - \dfrac{1}{100}$ Build $\frac{1}{25}$ to have a denominator of 100 so that the fractions can be subtracted: $\frac{1}{25} \cdot \frac{4}{4} = \frac{4}{100}$.

$= \dfrac{3}{100}$

Self Check 2   Express using positive exponents and simplify, if possible:
a. 8^{-2} **b.** x^{-5} **c.** $(-3)^{-3}$ **d.** $2^{-3} - 4^{-2}$

Now Try ▶ Problems 21, 25, and 29

EXAMPLE 3   Simplify. Do not use negative exponents in the answer.   **a.** $9m^{-3}$ **b.** -5^{-2}

Strategy   We note that each exponent is a negative number. We will identify the base for each negative exponent and then use the negative exponent rule.

Why   This rule enables us to write an exponential expression that has a negative exponent in an equivalent form using a positive exponent.

Solution   **a.** $9m^{-3} = 9 \cdot m^{-3}$ Read as "9 times m to the negative third power." The base is m. The exponent is -3.

$= 9 \cdot \dfrac{1}{m^3}$ Write the reciprocal of m^{-3} and change the sign of the exponent. Since 9 is not part of the base, it is not part of the reciprocal.

$= \dfrac{9}{m^3}$ Multiply.

b. $-5^{-2} = -1 \cdot 5^{-2}$ Read as "the opposite of 5 to the negative second power." The base is 5. The exponent is -2.

$= -1 \cdot \dfrac{1}{5^2}$ Write the reciprocal of 5^{-2} and change the sign of the exponent. Since -1 is not part of the base, it is not part of the reciprocal.

$= -\dfrac{1}{25}$ Evaluate 5^2 and multiply. The result is negative because of the $-$ sign in front of -5^{-2}.

Self Check 3   Simplify. Do not use negative exponents in the answer.   **a.** $12h^{-9}$
b. -2^{-4}

Now Try ▶ Problems 33 and 37

---

3   Use Exponent Rules to Change Negative Exponents in Fractions to Positive Exponents.

Negative exponents can appear in the numerator and/or the denominator of a fraction. To develop rules for such situations, we consider the following example.

$$\frac{a^{-4}}{b^{-3}} = \frac{\dfrac{1}{a^4}}{\dfrac{1}{b^3}} = \frac{1}{a^4} \div \frac{1}{b^3} = \frac{1}{a^4} \cdot \frac{b^3}{1} = \frac{b^3}{a^4}$$

Read $\frac{a^{-4}}{b^{-3}}$ as "a to the negative fourth power over (or divided by) b to the negative third power."

We can obtain this result in a simpler way. In $\frac{a^{-4}}{b^{-3}}$, we can move a^{-4} from the numerator to the denominator and change the sign of the exponent, and we can move b^{-3} from the denominator to the numerator and change the sign of the exponent.

The Language of Algebra

Factors of a numerator or denominator may be moved **across the fraction bar** if we change the sign of their exponent.

$$\frac{a^{-4}}{b^{-3}} \; = \; \frac{b^{3}}{a^{4}}$$

This example illustrates the following rules.

Changing from Negative to Positive Exponents

A factor can be moved from the denominator to the numerator or from the numerator to the denominator of a fraction if the sign of its exponent is changed.
For any nonzero real numbers x and y, and any integers m and n,

$$\frac{1}{x^{-n}} = x^{n} \qquad \text{and} \qquad \frac{x^{-m}}{y^{-n}} = \frac{y^{n}}{x^{m}}$$

These rules streamline the process when simplifying fractions involving negative exponents.

EXAMPLE 4   Simplify. Do not use negative exponents in the answer.   **a.** $\frac{1}{d^{-10}}$ **b.** $\frac{2^{-3}}{3^{-4}}$ **c.** $\frac{-6s^{-2}}{t^{-9}}$

Strategy   We will move any factors in the numerator that have a negative exponent to the denominator. Then we will move any factors in the denominator that have a negative exponent to the numerator.

Why   In this process, the sign of a negative exponent changes to positive.

Solution   **a.** $\frac{1}{d^{-10}} = d^{10}$ Read as "1 over d to the negative tenth power."
Move d^{-10} to the numerator and change the sign of the exponent.

Caution

A common error is to mistake the $-$ sign in -6 for a negative exponent. *It is not an exponent.* The -6 should not be moved to the denominator and have its sign changed.

$$\frac{-6s^{-2}}{t^{-9}} \neq \frac{t^{9}}{6s^{2}}$$

b. $\frac{2^{-3}}{3^{-4}} = \frac{3^{4}}{2^{3}}$ Move 2^{-3} to the denominator and change the sign of the exponent.
Move 3^{-4} to the numerator and change the sign of the exponent.

$= \frac{81}{8}$ Evaluate: $3^{4} = 81$ and $2^{3} = 8$.

c. $\frac{-6s^{-2}}{t^{-9}} = \frac{-6t^{9}}{s^{2}}$ Since $-6s^{-2}$ has no parentheses, s is the base. Move only s^{-2} to the denominator and change the sign of the exponent. Do not move -6.
Move t^{-9} to the numerator and change the sign of the exponent.

Self Check 4   Simplify. Do not use negative exponents in the answer.
a. $\frac{1}{w^{-5}}$ **b.** $\frac{5^{-2}}{4^{-3}}$ **c.** $\frac{-8h^{-6}}{a^{-7}}$

Now Try ▶   Problems 41, 45, and 47

When a fraction is raised to a negative power, we can use rules for exponents to change the sign of the exponent. For example, we see that

The exponent is the opposite of -3.

$$\left(\frac{a}{2}\right)^{-3} = \frac{a^{-3}}{2^{-3}} = \frac{2^{3}}{a^{3}} = \left(\frac{2}{a}\right)^{3}$$

The base is the reciprocal of $\frac{a}{2}$.

This process can be streamlined using the following rule.

Negative Exponents and Reciprocals	A fraction raised to a power is equal to the reciprocal of the fraction raised to the opposite power. For any nonzero real numbers x and y, and any integer n, $$\left(\frac{x}{y}\right)^{-n} = \left(\frac{y}{x}\right)^{n}$$

EXAMPLE 5 Simplify: $\left(\dfrac{4}{m}\right)^{-2}$

Strategy We want to write this fraction that is raised to a negative power in an equivalent form that involves a positive power. We will use the negative exponent and reciprocal rules to do this.

Why It is usually easier to simplify exponential expressions if the exponents are positive.

Solution

$$\left(\frac{4}{m}\right)^{-2} = \left(\frac{m}{4}\right)^{2}$$ Read as "the quantity of 4 over m, raised to the negative second power." The base is the fraction $\frac{4}{m}$ and the exponent is -2. Write the reciprocal of the base and change the sign of the exponent.

$$= \frac{m^2}{4^2}$$ Use the power of a quotient rule: Raise the numerator m and denominator 4 to the second power.

$$= \frac{m^2}{16}$$ Evaluate: $4^2 = 16$.

Self Check 5 Simplify: $\left(\dfrac{c}{9}\right)^{-2}$

Now Try ▶ Problem 53

4 Use All Exponent Rules to Simplify Expressions.

The rules for exponents involving products, powers, and quotients are also true for zero and negative exponents.

Summary of Exponent Rules	If m and n represent integers and there are no divisions by zero, then

Product rule
$$x^m \cdot x^n = x^{m+n}$$

Power rule
$$(x^m)^n = x^{mn}$$

Power of a product
$$(xy)^n = x^n y^n$$

Quotient rule
$$\frac{x^m}{x^n} = x^{m-n}$$

Power of a quotient
$$\left(\frac{x}{y}\right)^n = \frac{x^n}{y^n}$$

Exponents of 0 and 1
$$x^0 = 1 \text{ and } x^1 = x$$

Negative exponent
$$x^{-n} = \frac{1}{x^n}$$

Negative exponents appearing in fractions
$$\frac{1}{x^{-n}} = x^n \qquad \frac{x^{-m}}{y^{-n}} = \frac{y^n}{x^m} \qquad \left(\frac{x}{y}\right)^{-n} = \left(\frac{y}{x}\right)^n$$

The rules for exponents are used to simplify expressions involving products, quotients, and powers. In general, an expression involving exponents is simplified when

- Each base occurs only once
- No powers are raised to powers
- There are no parentheses
- There are no negative or zero exponents

EXAMPLE 6   Simplify. Do not use negative exponents in the answer.   **a.** $x^5 \cdot x^{-3}$ **b.** $\dfrac{x^3}{x^7}$

c. $(x^3)^{-2}$ **d.** $(2a^3 b^{-5})^3$ **e.** $\left(\dfrac{3}{b^5}\right)^{-4}$

Strategy In each case, we want to write an equivalent expression such that each base is used only once and is raised to a positive power. We will use rules for exponents to do this.

Why   These expressions are not in simplest form. In parts a and b, the base x occurs more than once. In parts a, c, d, and e, there is a negative exponent.

Solution   **a.** $x^5 \cdot x^{-3} = x^{5+(-3)} = x^2$ Use the product rule: Keep the base, x, and add exponents.

b. $\dfrac{x^3}{x^7} = x^{3-7}$ Use the quotient rule: Keep the base, x, and subtract the exponents.

$\quad = x^{-4}$ Do the subtraction: $3 - 7 = -4$.

$\quad = \dfrac{1}{x^4}$ Write the reciprocal of x^{-4} and change the sign of the exponent.

c. $(x^3)^{-2} = x^{-6}$ Use the power rule: Keep the base, x, and multiply exponents.

$\quad = \dfrac{1}{x^6}$ Write the reciprocal of x^{-6} and change the sign of the exponent.

d. $(2a^3 b^{-5})^3 = 2^3 (a^3)^3 (b^{-5})^3$ Raise each factor of the product $2a^3 b^{-5}$ to the 3rd power.

$\quad = 8a^9 b^{-15}$ Use the power rule: Multiply exponents.

$\quad = \dfrac{8a^9}{b^{15}}$ Move b^{-15} to the denominator and change the sign of the exponent.

e. $\left(\dfrac{3}{b^5}\right)^{-4} = \left(\dfrac{b^5}{3}\right)^4$ The base is the fraction $\frac{3}{b^5}$ and its exponent is -4. Write the reciprocal of the base and change the sign of the exponent.

$\quad = \dfrac{(b^5)^4}{(3)^4}$ Use the power of a quotient rule: Raise the numerator and denominator to the 4th power.

$\quad = \dfrac{b^{20}}{81}$ Use the power rule: Keep the base, b, and multiply the exponents 5 and 4. Evaluate: $3^4 = 81$.

Success Tip

We can use the negative exponent rule to simplify $(x^3)^{-2}$ in an alternate way:

$$(x^3)^{-2} = \dfrac{1}{(x^3)^2} = \dfrac{1}{x^6}$$

Self Check 6   Simplify. Do not use negative exponents in the answer.

a. $t^8 \cdot t^{-4}$ **b.** $\dfrac{a^3}{a^8}$ **c.** $(n^4)^{-5}$

d. $(4c^2 d^{-1})^3$ **e.** $\left(\dfrac{c^4}{2}\right)^{-3}$

Now Try ▶ Problems 57, 61, 65, 69, and 71

EXAMPLE 7 Simplify. Do not use negative exponents in the answer.

a. $\dfrac{y^{-4}y^{-3}}{y^{-20}}$ b. $\dfrac{7^{-1}a^3b^4}{6^{-2}a^5b^2}$ c. $\left(\dfrac{x^{-3}y^2}{xy^{-3}}\right)^2$

Strategy In each case, we want to write an equivalent expression such that each base is used only once and is raised to a positive power.

Why These expressions are not in simplest form. The bases occur more than once and the expressions contain negative exponents.

Solution a. $\dfrac{y^{-4}y^{-3}}{y^{-20}} = \dfrac{y^{-7}}{y^{-20}}$ In the numerator, use the product rule: Keep the common base, y, and add exponents: $-4 + (-3) = -7$.

$= y^{-7-(-20)}$ Use the quotient rule: Keep the common base, y, and subtract exponents.

$= y^{13}$ Do the subtraction: $-7 - (-20) = -7 + 20 = 13$.

> **Caution**
>
> We cannot use the alternate approach if the negative exponents occur in a sum or difference of terms. For example:
>
> $$\dfrac{y^{-4} + y^{-3}}{y^{-20}} \neq \dfrac{y^{20}}{y^4 + y^3}$$
>
> You will study this situation in more detail in your next algebra course.

Alternate solution: To avoid working with negative numbers, we could move each factor across the fraction bar and change the sign of its exponent.

$$\dfrac{y^{-4}y^{-3}}{y^{-20}} = \dfrac{y^{20}}{y^4y^3} = \dfrac{y^{20}}{y^7} = y^{13}$$

b. $\dfrac{7^{-1}a^3b^4}{6^{-2}a^5b^2} = \dfrac{6^2a^3b^4}{7^1a^5b^2}$ Move 7^{-1} to the denominator. Change the sign of the exponent. Move 6^{-2} to the numerator. Change the sign of the exponent.

$= \dfrac{36a^{3-5}b^{4-2}}{7}$ Use the quotient rule twice: Keep each base, a and b, and subtract exponents.

$= \dfrac{36a^{-2}b^2}{7}$ Do the subtractions: $3 - 5 = -2$ and $4 - 2 = 2$.

$= \dfrac{36b^2}{7a^2}$ Move a^{-2} to the denominator and change the sign of the exponent.

c. $\left(\dfrac{x^{-3}y^2}{xy^{-3}}\right)^2 = [x^{-3-1}y^{2-(-3)}]^2$ Within the parentheses, use the quotient rule twice: Keep each base, x and y, and subtract exponents.

$= (x^{-4}y^5)^2$ Do the subtractions: $-3 - 1 = -4$ and $2 - (-3) = 5$.

$= x^{-8}y^{10}$ Raise each factor within the parentheses to the second power.

$= \dfrac{y^{10}}{x^8}$ Move x^{-8} to the denominator and change the sign of its exponent. y^{10} does not move.

> **Success Tip**
>
> Since the rules for exponents can be applied in different orders, there are several equally valid ways to simplify these expressions.

Alternate solution: To simplify the expression, we can begin on the "outside" by using the power of a quotient rule first.

$$\left(\dfrac{x^{-3}y^2}{xy^{-3}}\right)^2 = \dfrac{(x^{-3})^2(y^2)^2}{x^2(y^{-3})^2} = \dfrac{x^{-6}y^4}{x^2y^{-6}} = \dfrac{y^4y^6}{x^6x^2} = \dfrac{y^{10}}{x^8}$$

Self Check 7 Simplify. Do not use negative exponents in the answer.

a. $\dfrac{a^{-4}a^{-5}}{a^{-3}}$ b. $\dfrac{1^{-4}x^5y^3}{9^{-2}x^3y^6}$ c. $\left(\dfrac{c^{-2}d^2}{c^4d^{-3}}\right)^3$

Now Try ▶ Problems 73, 77, and 79

SECTION 5.2 ▶ STUDY SET

VOCABULARY

Fill in the blanks.

1. In the expression 5^{-1}, the exponent is a _____ integer.
2. x^{-n} is the _____ of x^n.
3. We read a^0 as "a to the _____ power."
4. We read 3^{-4} as "3 to the _____ _____ power."

CONCEPTS

5. Complete the table.

Expression	Base	Exponent
4^{-2}		
$6x^{-5}$		
$\left(\frac{3}{y}\right)^{-8}$		
-7^{-1}		
$(-2)^{-3}$		
$10a^0$		

6. Complete each rule for exponents.

   a. $x^m \cdot x^n =$
   b. $x^0 =$
   c. $(x^m)^n =$
   d. $(xy)^n =$
   e. $\left(\dfrac{x}{y}\right)^n =$
   f. $x^{-n} =$
   g. $\dfrac{1}{x^{-n}} =$
   h. $\dfrac{x^{-m}}{y^{-n}} =$
   i. $\dfrac{x^m}{x^n} =$
   j. $\left(\dfrac{x}{y}\right)^{-n} =$

Complete each table.

7.

x	3^x
2	
1	
0	
-1	
-2	

8.

x	$(-9)^x$
2	
1	
0	
-1	
-2	

9. Fill in the blanks.

   a. $2^{-3} = \dfrac{1}{2^{\square}}$
   b. $\dfrac{1}{t^{-6}} = t^{\square}$
   c. A factor can be moved from the denominator to the numerator or from the numerator to the denominator of a fraction if the _____ of its exponent is changed.

   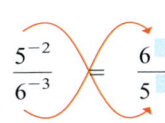
   $$\frac{5^{-2}}{6^{-3}} = \frac{6^{\square}}{5^{\square}}$$

   d. A fraction raised to a power is equal to the _____ of the fraction raised to the opposite power.
   $$\left(\frac{3}{d}\right)^{-2} = \left(\frac{d}{3}\right)^{\square}$$

10. Determine whether each statement is true or false.

   a. $6^{-2} = -36$
   b. $6^{-2} = \dfrac{1}{36}$
   c. $\dfrac{x^3}{y^{-2}} = \dfrac{y^2}{x^3}$
   d. $\dfrac{-6x^{-5}}{y^{-6}} = \dfrac{y^6}{6x^5}$

NOTATION

Complete each solution to simplify each expression.

11. $(y^5 y^3)^{-5} = \left(\boxed{}\right)^{-5} = y^{\boxed{}} = \dfrac{1}{y^{\boxed{}}}$

12. $\left(\dfrac{a^2 b}{a^{-3}b^3}\right)^3 = \left(a^{2-\boxed{}} b^{1-\boxed{}}\right)^3$

   $= (a^{\boxed{}} b^{\boxed{}})^3$

   $= a^{\boxed{}} b^{\boxed{}}$

   $= \dfrac{a^{15}}{b^{\boxed{}}}$

GUIDED PRACTICE

Simplify each expression. See Example 1.

13. 7^0
14. 9^0
15. $\left(\dfrac{1}{4}\right)^0$
16. $\left(\dfrac{3}{8}\right)^0$
17. $2x^0$
18. $8t^0$
19. $\dfrac{5}{2x^0}$
20. $\dfrac{4}{3a^0}$

Express using positive exponents and simplify, if possible. See Example 2.

21. 2^{-2}
22. 7^{-2}
23. 6^{-1}
24. 5^{-1}
25. b^{-5}
26. c^{-4}
27. $(-5)^{-1}$
28. $(-8)^{-1}$
29. $2^{-2} + 4^{-1}$
30. $-9^{-1} + 9^{-2}$
31. $9^0 - 9^{-1}$
32. $7^{-1} - 7^0$

Simplify. Do not use negative exponents in the answer. See Example 3.

33. $15g^{-6}$
34. $16t^{-3}$
35. $5x^{-3}$
36. $27m^{-3}$
37. -3^{-3}
38. -6^{-3}
39. -8^{-2}
40. -4^{-2}

Simplify. Do not use negative exponents in the answer. See Example 4.

41. $\dfrac{1}{5^{-3}}$
42. $\dfrac{1}{3^{-3}}$
43. $\dfrac{8}{s^{-1}}$
44. $\dfrac{6}{k^{-2}}$

45. $\dfrac{2^{-4}}{3^{-1}}$

46. $\dfrac{7^{-2}}{2^{-3}}$

47. $\dfrac{-4d^{-1}}{p^{-10}}$

48. $\dfrac{-9m^{-1}}{n^{-30}}$

Simplify. See Example 5.

49. $\left(\dfrac{1}{6}\right)^{-2}$

50. $\left(\dfrac{1}{7}\right)^{-2}$

51. $\left(\dfrac{1}{2}\right)^{-3}$

52. $\left(\dfrac{1}{5}\right)^{-3}$

53. $\left(\dfrac{c}{d}\right)^{-8}$

54. $\left(\dfrac{a}{x}\right)^{-10}$

55. $\left(\dfrac{3}{m}\right)^{-4}$

56. $\left(\dfrac{2}{t}\right)^{-4}$

Simplify. Do not use negative exponents in the answer. See Example 6.

57. $y^8 \cdot y^{-2}$

58. $m^{10} \cdot m^{-6}$

59. $b^{-7} \cdot b^{14}$

60. $c^{-9} \cdot c^{14}$

61. $\dfrac{y^4}{y^5}$

62. $\dfrac{t^7}{t^{10}}$

63. $\dfrac{h^{-5}}{h^2}$

64. $\dfrac{y^{-3}}{y^4}$

65. $(x^4)^{-3}$

66. $(y^{-3})^2$

67. $(b^2)^{-4}$

68. $(n^3)^{-5}$

69. $(6s^4t^{-7})^2$

70. $(11r^{10}s^{-3})^2$

71. $\left(\dfrac{4}{x^3}\right)^{-3}$

72. $\left(\dfrac{2}{b^5}\right)^{-2}$

Simplify. Do not use negative exponents in the answer. See Example 7.

73. $\dfrac{y^{-3}}{y^{-4}y^{-2}}$

74. $\dfrac{x^{-12}}{x^{-3}x^{-4}}$

75. $\dfrac{a^{-5}a^{-9}}{a^{-8}}$

76. $\dfrac{b^{-2}b^{-3}}{b^{-9}}$

77. $\dfrac{2^{-1}a^4b^2}{3^{-2}a^2b^4}$

78. $\dfrac{6^{-2}b^9c^3}{5^{-3}b^4c^8}$

79. $\left(\dfrac{y^3z^{-2}}{y^{-4}z^3}\right)^2$

80. $\left(\dfrac{xy^3}{x^{-1}y^{-1}}\right)^3$

TRY IT YOURSELF

Simplify. Do not use negative exponents in the answer. Assume that no variables are 0.

81. $\left(\dfrac{a^4}{2b}\right)^{-3}$

82. $\left(\dfrac{n^8}{9m}\right)^{-2}$

83. $\dfrac{r^{-50}}{r^{-70}}$

84. $\dfrac{m^{-30}}{m^{-40}}$

85. $\dfrac{a^{-5}}{b^{-2}}$

86. $\dfrac{r^{-6}}{s^{-1}}$

87. $(-10)^{-3}$

88. $(-9)^{-2}$

89. $\left(\dfrac{a^2b^3}{ab^4}\right)^0$

90. $\left(\dfrac{xyz}{x^2y}\right)^0$

91. $\dfrac{9^{-2}s^6t}{4^{-3}s^4t^5}$

92. $\dfrac{2^{-5}m^{10}n^6}{5^{-2}m^6n^{10}}$

93. $(2u^{-2}v^5)^5$

94. $(3w^{-8}x^3)^4$

95. $\left(\dfrac{y^4}{3}\right)^{-2}$

96. $\left(\dfrac{p^3}{2}\right)^{-2}$

97. $-15x^0y$

98. $24g^0h^2$

99. $\left(\dfrac{4}{h^{10}}\right)^{-2}$

100. $\left(\dfrac{x^4}{3}\right)^{-4}$

101. $x^{-3} \cdot x^{-3}$

102. $y^{-2} \cdot y^{-2}$

103. $\left(\dfrac{c^3d^{-4}}{c^{-1}d^5}\right)^3$

104. $\left(\dfrac{s^2t^{-8}}{s^{-9}t^2}\right)^4$

105. $15(-6x)^0$

106. $4(-12y)^0$

107. $\dfrac{2^{-2}g^{-2}h^{-3}}{9^{-1}h^{-3}}$

108. $\dfrac{5^{-1}x^{-2}y^{-3}}{8^{-2}x^{-11}}$

109. $(5d^{-2})^3$

110. $(9s^{-6})^2$

111. $\left(\dfrac{x^2y^{-2}}{x^{-5}y^3}\right)^4$

112. $\left(\dfrac{r^4s^{-3}}{r^{-3}s^7}\right)^3$

113. $(2x^3y^{-2})^5$

114. $(3u^{-2}v^3)^3$

115. $\dfrac{t(t^{-2})^{-2}}{t^{-5}}$

116. $\dfrac{d(d^{-3})^{-3}}{d^{-7}}$

117. $\dfrac{-4s^{-5}}{t^{-2}}$

118. $\dfrac{-9k^{-8}}{m^{-2}}$

119. $(x^{-4}x^3)^3$

120. $(y^{-2}y)^3$

Look Alikes . . .

121. a. 8^{-1} b. -8^{-1} c. $(-8)^{-1}$ d. $-(-8)^{-1}$

122. a. 9^{-2} b. -9^{-2} c. $(-9)^{-2}$ d. $-(-9)^{-2}$

123. a. $4xy^{-2}$ b. $(4xy)^{-2}$ c. $4x^{-2}y$ d. $4^{-2}xy$

124. a. $\left(\dfrac{r}{s}\right)^{-1}$ b. $\dfrac{r^{-1}}{s}$ c. $\dfrac{r}{s^{-1}}$ d. $\dfrac{r^{-1}}{s^{-1}}$

APPLICATIONS

125.

from **Campus to Careers**

Sound Engineering Technician

The faintest sound that the typical human ear can detect (called the *threshold of hearing*) has an intensity of 10^{-12} units. A sound intensity of 10^4 units will cause an instant rupture of the eardrum. Consider the types of sounds in the table below. In the intensity column, write the most appropriate exponential expression from the following list. Each expression is used only once. (Source: physicsclassroom.com)

$$10^{-10} \qquad 10^{-6} \qquad 10^{-4} \qquad 10^{-1} \qquad 10^2$$

Lowest intensity   **Greatest intensity**

Type of sound	Intensity
Front row rock concert	
Normal conversation	
Vacuum cleaner	
Military jet takeoff	
Whisper	

126. Electronics. The total resistance R of a certain circuit is given by

$$R = \left(\frac{1}{R_1} + \frac{1}{R_2} \right)^{-1} + R_3$$

Find R if $R_1 = 4$, $R_2 = 2$, and $R_3 = 1$.

WRITING

127. Explain how you would help a friend understand that 2^{-3} is not equal to -8.

128. Explain each error.

a. $\dfrac{-5x^{-2}}{y^{-2}} = \dfrac{y^2}{5x^2}$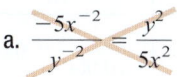

b. $4^{-2} = -\dfrac{1}{16}$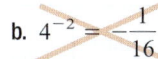

REVIEW

Find the slope of the line that passes through the given points.

129. $(1, -4)$ and $(3, -7)$

130. $(1, 3)$ and $(3, -1)$

131. Write an equation of the line having slope $\frac{3}{4}$ and y-intercept -5.

132. Find an equation of the line that passes through $(4, 4)$ and $(-6, -6)$. Write the answer in slope–intercept form.

CHALLENGE PROBLEMS

133. Simplify each expression. Do not use negative exponents in the answer. The variable m represents a positive integer.

a. $r^{5m} r^{-6m}$

b. $\dfrac{x^{3m}}{x^{6m}}$

134. Write an expression equivalent to $\left(\dfrac{2x^3 y^7}{3z^5} \right)^9$ that involves only *negative* exponents.

SECTION 5.3

Scientific Notation

OBJECTIVES

1 Convert from scientific to standard notation.

2 Write numbers in scientific notation.

3 Perform calculations with scientific notation.

ARE YOU READY?

The following problems review some basic skills that are needed when working with scientific notation.

1. Evaluate: 10^2

2. Multiply: $1,000 \cdot 4.528$

3. Evaluate: 10^{-1}

4. Multiply: $0.01 \cdot 6.22$

Scientists often deal with extremely large and small numbers. For example, the distance from the Earth to the sun is approximately 150,000,000 kilometers. The influenza virus, which causes flu symptoms of cough, sore throat, and headache, has a diameter of 0.00000256 inch.

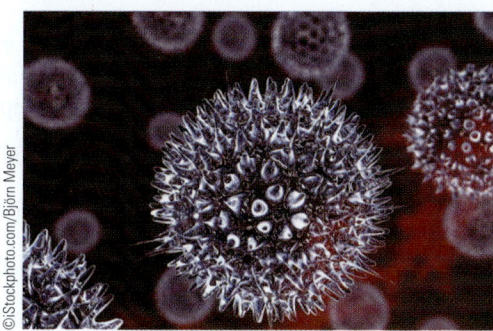

The numbers 150,000,000 and 0.00000256 are written in **standard notation,** which is also called **decimal notation.** Because they contain many zeros, they are difficult to read and cumbersome to work with in calculations. In this section, we will discuss a more convenient form in which we can write such numbers.

1 Convert from Scientific to Standard Notation.

Scientific notation provides a compact way of writing very large or very small numbers.

Scientific Notation	A positive number is written in **scientific notation** when it is written in the form $N \times 10^n$, where $1 \le N < 10$ and n is an integer.

To write numbers in scientific notation, you need to be familiar with **powers of 10,** like those listed in the table below.

Power of 10	10^4	10^3	10^2	10^1	10^0	10^{-1}	10^{-2}	10^{-3}	10^{-4}
Value	10,000	1,000	100	10	1	$\frac{1}{10} = 0.1$	$\frac{1}{100} = 0.01$	$\frac{1}{1{,}000} = 0.001$	$\frac{1}{10{,}000} = 0.0001$

Three examples of numbers written in scientific notation are shown below. Note that each of them is the product of a decimal number (between 1 and 10) and a power of 10.

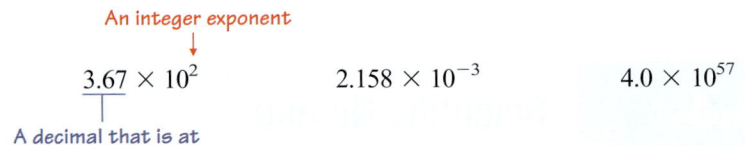

$$3.67 \times 10^2 \qquad 2.158 \times 10^{-3} \qquad 4.0 \times 10^{57}$$

An integer exponent

A decimal that is at least 1, but less than 10

A number written in scientific notation can be converted to standard notation by performing the indicated multiplication. For example, to convert 3.67×10^2, we recall that multiplying a decimal by 100 moves the decimal point 2 places to the right.

$$3.67 \times \mathbf{10^2} = 3.67 \times \mathbf{100} = 3\,6\,7.$$

To convert 2.158×10^{-3} to standard notation, we recall that dividing a decimal by 1,000 moves the decimal point 3 places to the left.

$$2.158 \times \mathbf{10^{-3}} = 2.158 \times \frac{1}{\mathbf{10^3}} = 2.158 \times \frac{1}{\mathbf{1{,}000}} = \frac{2.158}{\mathbf{1{,}000}} = 0.0\,0\,2\,1\,5\,8$$

In 3.67×10^2 and 2.158×10^{-3}, the exponent gives the number of decimal places that the decimal point moves, and the sign of the exponent indicates the direction in which it moves. Applying this observation to several other examples, we have

$5.32 \times 10^6 = 5\,3\,2\,0\,0\,0\,0.$ *Move the decimal point 6 places to the right.*

$1.95 \times 10^{-5} = 0.0\,0\,0\,0\,1\,9\,5$ *Move the decimal point $|-5| = 5$ places to the left.*

$9.7 \times 10^0 = 9.7$ *There is no movement of the decimal point.*

The following procedure summarizes our observations.

Converting from Scientific to Standard Notation	1. If the exponent is positive, move the decimal point the same number of places to the right as the exponent.
	2. If the exponent is negative, move the decimal point the same number of places to the left as the absolute value of the exponent.

EXAMPLE 1   Convert to standard notation: **a.** 3.467×10^5 **b.** 8.9×10^{-4}

Strategy   In each case, we need to identify the exponent on the power of 10 and consider its sign.

Why   The exponent gives the number of decimal places that we should move the decimal point. The sign of the exponent indicates whether it should be moved to the right or the left.

Solution   **a.** Since the exponent in 10^5 is 5, the decimal point moves 5 places to the right.

$$3\,4\,6\,7\,0\,0.$$ To move 5 places to the right, two placeholder zeros must be written.

Thus, $3.467 \times 10^5 = 346{,}700$.

b. Since the exponent in 10^{-4} is -4, the decimal point moves 4 places to the left.

$$0.0\,0\,0\,8\,9$$ To move 4 places to the left, three placeholder zeros must be written.

Thus, $8.9 \times 10^{-4} = 0.00089$.

> **The Language of Algebra**
>
> **Standard notation** is also called **decimal notation**.

> **Self Check 1**   Convert to standard notation: **a.** 4.88×10^6
> **b.** 9.8×10^{-3}
>
> **Now Try** ▶   Problems 13 and 17

2 Write Numbers in Scientific Notation.

To write a number in scientific notation ($N \times 10^n$) we first determine N and then n.

EXAMPLE 2   Write each number in scientific notation: **a.** 150,000,000   **b.** 0.00000256
c. 432×10^5

Strategy   We will write each number as the product of a number between 1 and 10 and a power of 10.

Why   Numbers written in scientific notation have the form $N \times 10^n$.

Solution   **a.** We must write 150,000,000 (the distance in kilometers from the Earth to the sun) as the product of a number between 1 and 10 and a power of 10. We note that 1.5 lies between 1 and 10. To obtain 150,000,000, we must move the decimal point in 1.5 exactly 8 places to the right.

$$1.5\,0\,0\,0\,0\,0\,0\,0$$

This will happen if we multiply 1.5 by 10^8. Therefore,

$$150{,}000{,}000 = 1.5 \times 10^8$$ This is the distance (in kilometers) from the Earth to the sun.

b. We must write 0.00000256 (the diameter in inches of a flu virus) as the product of a number between 1 and 10 and a power of 10. We note that 2.56 lies between 1 and 10. To obtain 0.00000256, the decimal point in 2.56 must be moved 6 places to the left.

$$0\,0\,0\,0\,0\,0\,2.56$$

This will happen if we multiply 2.56 by 10^{-6}. Therefore,

$$0.00000256 = 2.56 \times 10^{-6}$$ This is the diameter (in inches) of a flu virus.

> **Caution**
>
> At first glance, the following numbers might look like they are written in scientific notation, but they are not. In each case, the first factor is not a number between 1 and 10.
>
> 16.38×10^{24} 0.39×10^{-11}

> **Caution**
>
> Don't apply the negative exponent rule when writing numbers in scientific notation.
>
> 2.56×10^{-6} $2.56 \times \dfrac{1}{10^6}$

c. The number 432×10^5 is not written in scientific notation because 432 is not a number between 1 and 10. To write this number in scientific notation, we proceed as follows:

$$432 \times 10^5 = \mathbf{4.32 \times 10^2} \times 10^5 \qquad \text{Write 432 in scientific notation.}$$
$$= 4.32 \times 10^7 \qquad \text{Use the product rule to find } 10^2 \times 10^5.$$
$$\text{Keep the base, 10, and add the exponents.}$$

Written in scientific notation, 432×10^5 is 4.32×10^7.

Self Check 2 Write each number in scientific notation: **a.** 93,000,000 **b.** 0.00009055 **c.** 85×10^{-3}

Now Try ▶ Problems 31, 35, and 55

The results from Example 2 illustrate the following forms to use when converting numbers from standard to scientific notation.

For real numbers between 0 and 1: ▨ $\times 10^{\text{negative integer}}$

For real numbers at least 1, but less than 10: ▨ $\times 10^0$

For real numbers greater than or equal to 10: ▨ $\times 10^{\text{positive integer}}$

3 Perform Calculations with Scientific Notation.

Another advantage of scientific notation becomes apparent when we evaluate products or quotients that involve very large or small numbers. If we express those numbers in scientific notation, we can use rules for exponents to make the calculations easier.

To multiply two numbers written in scientific notation, use the following rule:

$$(a \times 10^m)(b \times 10^n) = (a \cdot b) \times 10^{m+n}$$

EXAMPLE 3 **Astronomy.** Except for the sun, the nearest star visible to the naked eye from most parts of the United States is Sirius. Light from Sirius reaches Earth in about 70,000 hours. If light travels at approximately 670,000,000 mph, how far from Earth is Sirius?

Strategy We can use the formula $d = rt$ to find the distance from Sirius to Earth.

Why We know the *rate* at which light travels and the *time* it takes to travel from Sirius to the Earth.

Solution The rate at which light travels is 670,000,000 mph and the time it takes the light to travel from Sirius to Earth is 70,000 hr. To find the distance from Sirius to Earth, we proceed as follows:

$$d = \mathbf{\textit{rt}} \qquad \text{This is the formula for distance traveled.}$$
$$d = \mathbf{670,000,000(70,000)} \qquad \text{Substitute 670,000,000 for } r \text{ and 70,000 for } t.$$
$$= (6.7 \times 10^8)(7.0 \times 10^4) \qquad \text{Write each number in scientific notation.}$$
$$= (6.7 \cdot 7.0) \times \mathbf{(10^8 \times 10^4)} \qquad \text{Group the decimals together and the powers of 10 together.}$$
$$= (6.7 \cdot 7.0) \times \mathbf{10^{8+4}} \qquad \text{Use the product rule to find } 10^8 \times 10^4.$$
$$\text{Keep the base, 10, and add exponents.}$$
$$= \mathbf{46.9} \times 10^{12} \qquad \text{Do the multiplication and the addition.}$$

We note that 46.9 is not between 0 and 1, so 46.9×10^{12} is not written in scientific notation. To answer in scientific notation, we proceed as follows.

$$= \mathbf{4.69 \times 10^1} \times 10^{12} \qquad \text{Write 46.9 in scientific notation as } 4.69 \times 10^1.$$

$$= 4.69 \times 10^{13} \qquad \text{Use the product rule to find } 10^1 \times 10^{12}.$$
$$\text{Keep the base, 10, and add the exponents.}$$

Sirius is approximately 4.69×10^{13} or 46,900,000,000,000 miles from Earth.

> **Self Check 3**   Use scientific notation to evaluate: (2,540,000,000,000)(0.00041)
>
> **Now Try** ▶ Problem 61

To divide two numbers written in scientific notation, use the following rule:

$$\frac{a \times 10^m}{b \times 10^n} = \frac{a}{b} \times 10^{m-n}$$

EXAMPLE 4   **Atoms.**   As an example of how scientific notation is used in chemistry, we can approximate the weight (in grams) of one atom of the element uranium by evaluating the following expression.

$$\frac{2.4 \times 10^2}{6 \times 10^{23}}$$

Strategy   To simplify, we will divide the numbers and powers of 10 separately.

Why   We can then use the quotient rule for exponents to simplify the calculations.

Solution

$$\frac{2.4 \times 10^2}{6 \times 10^{23}} = \frac{2.4}{6} \times \frac{10^2}{10^{23}} \qquad \text{Divide the decimals and the powers of 10 separately.}$$

$$= \frac{2.4}{6} \times 10^{2-23} \qquad \text{For the powers of 10, use the quotient rule.}$$
$$\text{Keep the base, 10, and subtract the exponents.}$$

$$= \mathbf{0.4} \times 10^{-21} \qquad \text{Divide the decimals. Subtract the exponents.}$$
$$\text{The result is not in scientific notation form.}$$

$$= \mathbf{4 \times 10^{-1}} \times 10^{-21} \qquad \text{Write 0.4 in scientific notation as } 4 \times 10^{-1}.$$

$$= 4 \times 10^{-22} \qquad \text{Use the product rule to find } 10^{-1} \times 10^{-21}.$$
$$\text{Keep the base, 10, and add the exponents.}$$

One atom of uranium weighs 4×10^{-22} gram or 0.0000000000000000000004 g.

> **Self Check 4**   Find the approximate weight (in grams) of one atom of gold by evaluating: $\dfrac{1.98 \times 10^2}{6 \times 10^{23}}$
>
> **Now Try** ▶ Problem 65

SECTION 5.3 ▶ STUDY SET

VOCABULARY

Fill in the blanks.

1. 4.84×10^5 is written in _____ notation. 484,000 is written in _____ notation.
2. 10^3, 10^{50}, and 10^{-4} are _____ of 10.

CONCEPTS

Fill in the blanks.

3. When we multiply a decimal by 10^5, the decimal point moves 5 places to the _____. When we multiply a decimal by 10^{-7}, the decimal point moves 7 places to the _____.
4. Describe the procedure for converting a number from scientific notation to standard form.
   a. If the exponent on the base of 10 is positive, move the decimal point the same number of places to the _____ as the exponent.
   b. If the exponent on the base of 10 is negative, move the decimal point the same number of places to the _____ as the absolute value of the exponent.
5. a. When a real number greater than or equal to 10 is written in scientific notation, the exponent on 10 is a _____ integer.
   b. When a real number between 0 and 1 is written in scientific notation, the exponent on 10 is a _____ integer.
6. The arrows show the movement of a decimal point. By what power of 10 was each decimal multiplied?
   a. 0.0 0 0 0 0 0 5 5 6
   b. 8,0 4 1,0 0 0,0 0 0.

Fill in the blanks to write each number in scientific notation.

7. a. $7{,}700 = \boxed{} \times 10^3$ b. $500{,}000 = \boxed{} \times 10^5$
   c. $114{,}000{,}000 = 1.14 \times 10^{\boxed{}}$

8. a. $0.0082 = \boxed{} \times 10^{-3}$
   b. $0.0000001 = \boxed{} \times 10^{-7}$
   c. $0.00003457 = 3.457 \times 10^{\boxed{}}$

9. Write each expression so that the decimal numbers are grouped together and the powers of ten are grouped together.
   a. $(5.1 \times 10^9)(1.5 \times 10^{22})$
   b. $\dfrac{8.8 \times 10^{30}}{2.2 \times 10^{19}}$

10. Simplify each expression.
    a. $10^{24} \times 10^{33}$ b. $\dfrac{10^{50}}{10^{36}}$ c. $\dfrac{10^{15} \times 10^{27}}{10^{40}}$

NOTATION

11. Fill in the blanks. A positive number is written in scientific notation when it is written in the form $N \times 10^n$, where $\boxed{} \le N < \boxed{}$ and n is an _____.
12. Express each power of 10 in fraction form and decimal form.
    a. 10^{-3} b. 10^{-6}

GUIDED PRACTICE

Convert each number to standard notation. See Example 1.

13. 2.3×10^2 14. 3.75×10^4
15. 8.12×10^5 16. 1.2×10^3
17. 1.15×10^{-3} 18. 4.9×10^{-2}
19. 9.76×10^{-4} 20. 7.63×10^{-5}
21. 6.001×10^6 22. 9.998×10^5
23. 2.718×10^0 24. 3.14×10^0
25. 6.789×10^{-2} 26. 4.321×10^{-1}
27. 2.0×10^{-5} 28. 7.0×10^{-6}

Write each number in scientific notation. See Example 2.

29. 23,000   30. 4,750
31. 1,700,000   32. 290,000
33. 0.062   34. 0.00073
35. 0.0000051   36. 0.04
37. 5,000,000,000   38. 7,000,000
39. 0.0000003   40. 0.0001
41. 909,000,000   42. 7,007,000,000
43. 0.0345   44. 0.000000567
45. 9   46. 2
47. 11   48. 55
49. 1,718,000,000,000,000,000
50. 44,180,000,000,000,000
51. 0.0000000000000123
52. 0.0000000000000000555
53. 73×10^4 54. 99×10^5
55. 201.8×10^{15} 56. 154.3×10^{17}
57. 0.073×10^{-3} 58. 0.0017×10^{-4}
59. 36.02×10^{-20} 60. 56.29×10^{-30}

Use scientific notation to perform the calculations. Give all answers in scientific notation and standard notation. See Examples 3 and 4.

61. $(3.4 \times 10^2)(2.1 \times 10^3)$
62. $(4.1 \times 10^{-3})(3.4 \times 10^4)$
63. $(8.4 \times 10^{-13})(4.8 \times 10^9)$
64. $(5.5 \times 10^{-15})(2.2 \times 10^{13})$
65. $\dfrac{2.24 \times 10^4}{5.6 \times 10^7}$ 66. $\dfrac{2.47 \times 10^5}{3.8 \times 10^{-5}}$
67. $\dfrac{9.3 \times 10^2}{3.1 \times 10^{-2}}$ 68. $\dfrac{7.2 \times 10^6}{1.2 \times 10^8}$
69. $\dfrac{0.00000129}{0.0003}$ 70. $\dfrac{169{,}000{,}000{,}000}{26{,}000{,}000}$

71. $(0.0000000056)(5{,}500{,}000)$
72. $(0.000000061)(3{,}500{,}000{,}000)$

73. $\dfrac{96,000}{(12,000)(0.00004)}$ **74.** $\dfrac{(0.48)(14,400,000)}{96,000,000}$

75. $\dfrac{2,475}{(132,000,000,000,000)(0.25)}$

76. $\dfrac{147,000,000,000,000}{(0.000049)(25)}$

Find each power.

77. $(456.4)^6$

78. $(0.009)^{-6}$

79. 225^{-5}

80. $\left(\dfrac{1}{3}\right)^{-55}$

APPLICATIONS

81. Astronomy. The distance from Earth to Alpha Centauri (the nearest star outside our solar system) is about 25,700,000,000,000 miles. Write this number in scientific notation.

82. Water. According to the U.S. Geological Survey, the total water supply of the world is 366,000,000,000,000,000,000 gallons. Write this number in scientific notation.

83. Earth, Sun, Moon. The surface area of Earth is 1.97×10^8 square miles, the surface area of the sun is 1.09×10^{17} square miles, and the surface area of the moon is 1.46×10^7 square miles. Convert each number to standard notation.

84. Atoms. The number of atoms in 1 gram of iron is approximately 1.08×10^{22}. Convert this number to standard notation.

85. Sand. The mass of one grain of beach sand is approximately 0.00000000045 ounce. Write this number in scientific notation.

86. Molecules. The mass of a water molecule is approximately 0.00000000000000000000001056 ounce. Write this number in scientific notation.

87. Wavelengths. Examples of the most common types of electromagnetic waves are given in the table in the next column. List the wavelengths in order from shortest to longest.

This distance between the two crests of the wave is called the wavelength.

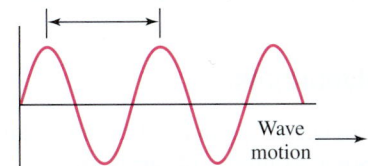

Wave motion

Type	Use	Wavelength (in meters)
visible light	lighting	9.3×10^{-6}
infrared	photography	3.7×10^{-5}
x-ray	medical	2.3×10^{-11}
radio wave	communication	3.0×10^{2}
gamma ray	treating cancer	8.9×10^{-14}
microwave	cooking	1.1×10^{-2}
ultraviolet	sun lamp	6.1×10^{-8}

88. Exploration. On July 4, 1997, the *Pathfinder*, carrying the rover vehicle called Sojourner, landed on Mars. The distance from Mars to Earth is approximately 3.5×10^7 miles. Use scientific notation to express this distance in feet. (*Hint:* 5,280 feet = 1 mile.)

89. ▶ from **Campus to Careers**
Sound Engineering Technician

The speed of sound in air is approximately 3.3×10^4 centimeters per second. Use scientific notation to express this speed in kilometers per second. (*Hint:* 100 centimeters = 1 meter and 1,000 meters = 1 kilometer.)

90. Protons. The mass of one proton is approximately 1.7×10^{-24} gram. Use scientific notation to express the mass of 1 million protons.

91. Light Years. One light year is about 5.87×10^{12} miles. Use scientific notation to express this distance in feet. (*Hint:* 5,280 feet = 1 mile.)

92. Oil. As of 2009, Saudi Arabia was believed to have crude oil reserves of about 2.65×10^{11} barrels. A barrel contains 42 gallons of oil. Use scientific notation to express Saudi Arabia's oil reserves in gallons. (Source: British Petroleum)

93. Insured Deposits. As of June 2009, the total insured deposits in U.S. banks and savings and loans was approximately 7.6×10^{12} dollars. If this money was invested at a rate of 4% simple annual interest, how much would it earn in 1 year? Use scientific notation to express the answer. (Source: Federal Deposit Insurance Corporation.)

94. Currency. As of December 2009, the number of $20 bills in circulation was approximately 6.4×10^9. What was the total value of the currency? Express the answer in scientific notation and standard notation. (Source: The Federal Reserve.)

95. Powers of 10. In the United States, we use Latin prefixes in front of "illion" to name extremely large numbers. Write each number in scientific notation.

One million: 1,000,000

One billion: 1,000,000,000

One trillion: 1,000,000,000,000

One quadrillion: 1,000,000,000,000,000

One quintillion: 1,000,000,000,000,000,000

96. Supercomputers. As of June 2009, the world's fastest computer was the Cray Jaguar, owned by the Oak Ridge National Laboratory in Tennessee. If it could make 1.75×10^{15} calculations in one second, how many could it make in one minute? Answer in scientific notation. (Source: datacenterknowledge.com)

WRITING

97. In what situations would scientific notation be more convenient than standard notation?

98. To multiply a number by a power of 10, we move the decimal point. Which way, and how far? Explain.

99. 2.3×10^{-3} contains a negative sign but represents a positive number. Explain.

100. Explain why 237.8×10^8 is not written in scientific notation.

REVIEW

101. If $y = -1$, find the value of $-5y^{55}$.

102. What is the y-intercept of the graph of $y = -3x - 5$?

103. Counseling. At the end of her first year of practice, a family counselor had 75 clients. At the end of her second year, she had 105 clients. If a linear trend continues, write an equation that gives the number of clients c the counselor will have at the end of t years.

104. Is $(0, -5)$ a solution of $2x + 3y \geq -14$?

CHALLENGE PROBLEMS

105. Consider 2.5×10^{-4}. Answer the following questions in scientific notation form.

   a. What is its opposite?

   b. What is its reciprocal?

106. a. Write the numbers one million and one millionth in scientific notation.

   b. By what number must we multiply one millionth to get one million?

SECTION 5.4

Polynomials

OBJECTIVES

1 Know the vocabulary for polynomials.

2 Evaluate polynomials.

3 Graph equations defined by polynomials.

ARE YOU READY?

 The following problems review some basic skills that are needed when working with polynomials.

1. How many terms does the expression $2x^2 - 5x + 8$ have?

2. What is the coefficient of the term $6a^3$?

3. In $4b^3 + 5b^4 - 16b + b^2$, which term has the largest exponent?

4. For $y = x^3 + 1$, what is the value of y when $x = -3$?

In this section, we will discuss a special type of algebraic expression called a *polynomial*.

1 Know the Vocabulary for Polynomials.

Recall from Chapter 1 that a *term* is a product or quotient of numbers and/or variables. A single number or variable is also a term. Some examples of terms are:

$$14, \qquad x, \qquad -6y^3, \qquad 9cd^2, \qquad \text{and} \qquad \frac{5}{y}$$

Polynomials	A **polynomial** is a single term or a sum of terms in which all variables have whole-number exponents and no variable appears in a denominator.

The Language of Algebra

The prefix **poly** means many. Some other words that begin with this prefix are *poly*gon, *poly*ester, and *poly*unsaturated.

Here are some examples of polynomials. Note that polynomials are expressions, not equations.

$$3x + 2, \qquad 4y^2 - 2y - 3, \qquad a^3 + 3a^2b + 3ab^2 + b^3, \qquad \text{and} \qquad -8xy^2z$$

Some examples of expressions that are *not polynomials* are:

$$6x^3 + 4x^{-2}$$ The variable x has an exponent that is not a whole number.

$$y^2 + \frac{5}{y} + 1$$ The variable y appears in the denominator.

The polynomial $3x + 2$ is the sum of two terms, $3x$ and 2, and we say it is a **polynomial in one variable, *x*.** A single number is called a **constant,** and so its last term, 2, is called the **constant term.**

A polynomial is defined as a single term or the sum of several terms. Since $4y^2 - 2y - 3$ can be written as the sum $4y^2 + (-2y) + (-3)$, it has three terms, $4y^2$, $-2y$, and -3. It is written in **descending powers** of y, because the exponents on y decrease from left to right. When a polynomial is written in descending powers, the first term, in this case $4y^2$, is called the **leading term.** The coefficient of the leading term, in this case 4, is called the **leading coefficient.**

A polynomial can have more than one variable. For example, $a^3 + 3a^2b + 3ab^2 + b^3$ is a **polynomial in two variables,** a and b. It has four terms and is written in descending powers of a and **ascending powers** of b. The polynomial $-8xy^2z$ is a polynomial in three variables, $x, y,$ and z, and has only one term.

Polynomials are classified according to the number of terms they have. A polynomial with exactly one term is called a **monomial;** exactly two terms, a **binomial;** and exactly three terms, a **trinomial.** Polynomials with four or more terms have no special names.

The Language of Algebra

The prefix **mono** means one; Jay Leno begins the *Tonight Show* with a monologue. The prefix **bi** means two, as in bicycle or binoculars. The prefix **tri** means three, as in triangle or the *Lord of the Rings Trilogy.*

Polynomials			
Monomials	**Binomials**	**Trinomials**	**No special name**
$-6x$	$9u - 4$	$5t^2 + 4t + 3$	$a^4 + 2b^2 - 3b - 9$
$5.5x^3y^2$	$-29z^4 - z^2$	$27x^3 - 6x^2 - 2x$	$2.1 - 6.2t + 5.9t^2 + t^3$
11	$18a^2b + 4ab$	$\frac{1}{2}a^2 + 2ab + b^2$	$mn^2 - mn + m - n$

Polynomials and their terms can be described by the exponents on their variables.

Degree of a Term of a Polynomial

The **degree of a term** of a polynomial in one variable is the value of the exponent on the variable. If a polynomial is in more than one variable, the **degree of a term** is the sum of the exponents on the variables in that term. The **degree of a nonzero constant** is 0.

Think of the *degree of a term* as the number of variable factors in that term.

- $9x^6$ has degree **6**. x^6 represents 6 variable factors: $x \cdot x \cdot x \cdot x \cdot x \cdot x$.
- $-2a^4$ has degree **4**. a^4 represents 4 variable factors: $a \cdot a \cdot a \cdot a$.
- $47x^2y^{11}$ has degree **13**. Because $2 + 11 = 13$.
- 8 has degree **0** since it can be written as $8x^0$. There is no variable factor.

We determine the *degree of a polynomial* by considering the degrees of each of its terms.

Degree of a Polynomial

The **degree of a polynomial** is the same as the highest degree of any term of the polynomial.

EXAMPLE 1 Use the vocabulary of this section to describe each polynomial: **a.** $d^4 + 9d^2 - 16$
b. $\frac{1}{2}x^2 - x$ **c.** $-6y^{14} - 1.5y^9z^9 + 2.5y^8z^{10} + yz^{11}$

Strategy First, we will identify the variable(s) in the polynomial and determine whether it is written in ascending or descending powers. Then we will count the number of terms in the polynomial and determine the degree of each term.

Why The number of terms determines the type of polynomial. The highest degree of any term of the polynomial determines its degree.

Solution **a.** $d^4 + 9d^2 - 16$ is a polynomial in one variable that is written in descending powers of d. If we write the subtraction as addition of the opposite, we see that it has 3 terms, d^4, $9d^2$, and -16, and is therefore a trinomial. The leading term is d^4 and the leading coefficient is 1. The highest degree of any of its terms is 4, so it is of degree 4.

$$d^4 + 9d^2 - 16 = \quad d^4 \quad + \quad 9d^2 \quad + \quad (-16)$$
$$\uparrow \qquad\qquad \uparrow \qquad\qquad \uparrow$$
$$\text{1st Term} \quad\quad \text{2nd Term} \quad\quad \text{3rd Term}$$

Term	Coefficient	Degree
d^4	1	**4**
$9d^2$	9	2
-16	-16	0

Degree of the polynomial: **4**

b. $\frac{1}{2}x^2 - x$ is a polynomial in one variable. It is written in descending powers of x. Since it has two terms, it is a binomial. The leading term is $\frac{1}{2}x^2$ and the leading coefficient is $\frac{1}{2}$. The highest degree of any of its terms is 2, so it is of degree 2.

Term	Coefficient	Degree
$\frac{1}{2}x^2$	$\frac{1}{2}$	**2**
$-x$	-1	1

Degree of the polynomial: **2**

c. $-6y^{14} - 1.5y^9z^9 + 2.5y^8z^{10} + yz^{11}$ is a polynomial in two variables, y and z. It is written in descending powers of y and ascending powers of z. It has 4 terms, and therefore has no special name. The leading term is $-6y^{14}$ and the leading coefficient is -6. The highest degree of any term is 18, so it is of degree 18.

Term	Coefficient	Degree
$-6y^{14}$	-6	14
$-1.5y^9z^9$	-1.5	**18**
$2.5y^8z^{10}$	2.5	**18**
yz^{11}	1	12

Degree of the polynomial: **18**

Self Check 1 Describe each polynomial: **a.** $x^2 + 4x - 16$ **b.** $-14s^5t + s^4t^3$

Now Try ▶ Problems 17 and 35

2 Evaluate Polynomials.

A polynomial can have different values depending on the number that is substituted for its variable (or variables).

EXAMPLE 2 Evaluate $3x^2 + 4x - 5$ for $x = 0$ and $x = -2$.

Strategy We will substitute the given value for each x in the polynomial and follow the rules for the order of operations.

Why To *evaluate a polynomial* means to find its numerical value, once we know the value of its variable.

Solution

For x = 0:

$$3x^2 + 4x - 5 = 3(0)^2 + 4(0) - 5$$
$$= 3(0) + 4(0) - 5$$
$$= 0 + 0 - 5$$
$$= -5$$

For x = -2:

$$3x^2 + 4x - 5 = 3(-2)^2 + 4(-2) - 5$$
$$= 3(4) + 4(-2) - 5$$
$$= 12 + (-8) - 5$$
$$= -1$$

Self Check 2 Evaluate $-x^3 + x - 2x + 3$ for $x = -3$.

Now Try ▶ Problem 55

EXAMPLE 3 **Supermarket Displays.** The polynomial $\frac{1}{3}c^3 + \frac{1}{2}c^2 + \frac{1}{6}c$ gives the number of cans used in a display shaped like a square pyramid, having a square base formed by c cans per side. Find the number of cans used in the display.

Strategy We will evaluate the polynomial for $c = 4$.

Why From the illustration, we see that each side of the square base is formed by 4 cans.

Solution

$$\frac{1}{3}c^3 + \frac{1}{2}c^2 + \frac{1}{6}c = \frac{1}{3}(4)^3 + \frac{1}{2}(4)^2 + \frac{1}{6}(4)$$ Substitute 4 for c.

$$= \frac{1}{3}(64) + \frac{1}{2}(16) + \frac{1}{6}(4)$$ Evaluate the exponential expressions first.

$$= \frac{64}{3} + 8 + \frac{2}{3}$$ Do the multiplication, and then simplify: $\frac{4}{6} = \frac{2}{3}$.

$$= 30$$ Add the fractions: $\frac{64}{3} + \frac{2}{3} = \frac{66}{3} = 22$.

There are 30 cans of soup in the display.

Self Check 3 Find the number of cans used in a display having a square base formed by 5 cans per side.

Now Try ▶ Problem 81

In the following example, we evaluate a polynomial in two variables.

EXAMPLE 4 Evaluate $3p^2q - 4pq^2$ for $p = 2$ and $q = -3$.

Strategy We will substitute the given values for each p and q in the polynomial and follow the rules for the order of operations.

Why To evaluate a polynomial means to find its numerical value, once we know the value of its variables.

Solution

$$3p^2q - 4pq^2 = 3(2)^2(-3) - 4(2)(-3)^2$$ Substitute 2 for p and -3 for q.

$$= 3(4)(-3) - 4(2)(9)$$ Find the powers.

$$= -36 - 72$$ Do the multiplication.

$$= -108$$ Do the subtraction.

Self Check 4 Evaluate $3a^3b^2 + 2a^2b$ for $a = 2$ and $b = -1$.

Now Try ▶ Problem 61

3 Graph Equations Defined by Polynomials.

In Chapter 3, we graphed equations in two variables such as $y = x$ and $y = 2x - 3$ using the point-plotting method. Recall that these equations are called *linear equations* and that their graphs are straight lines. Note that the right side of the first two equations shown below is a polynomial of degree 1.

$$y = x \qquad y = 2x - 3 \qquad\qquad y = x^2 \qquad\qquad y = x^3 + 1$$

The degree of each polynomial is 1.   The degree of this polynomial is 2.   The degree of this polynomial is 3.

We also can graph equations defined by polynomials with degrees greater than 1 such as $y = x^2$ and $y = x^3 + 1$ by plotting points.

EXAMPLE 5 Graph: $y = x^2$

Strategy We will find several solutions of the equation, plot them on a rectangular coordinate system, and then draw a smooth curve passing through the points.

Why To *graph* an equation in two variables means to make a drawing that represents all of its solutions.

Solution To find some solutions of this equation, we select several values of x that will make the calculations easy. Then we find each corresponding value of y. If $x = -3$, we substitute -3 for x in $y = x^2$ and find y.

$$y = x^2 = (-3)^2 = 9$$

Thus, $(-3, 9)$ is a solution. In a similar manner, we find the corresponding y-values for x-values of $-2, -1, 0, 1, 2,$ and 3. If we plot the ordered pairs listed in the table and join the points with a smooth curve, we get the graph shown below, which is called a **parabola.**

Success Tip

When constructing a table of solutions, it is wise to select some positive and negative integer-values for x, as well as 0.

The Language of Algebra

The cup-like shape of a **parabola** has many real-life applications. The word *parabolic* (pronounced par·a·bol·ic) means having the form of a parabola. Did you know that a satellite TV dish is more formally known in the electronics industry as a *parabolic* dish?

$y = x^2$

x	y	(x, y)
-3	9	$(-3, 9)$
-2	4	$(-2, 4)$
-1	1	$(-1, 1)$
0	0	$(0, 0)$
1	1	$(1, 1)$
2	4	$(2, 4)$
3	9	$(3, 9)$

Select x. Find y. Plot (x, y).

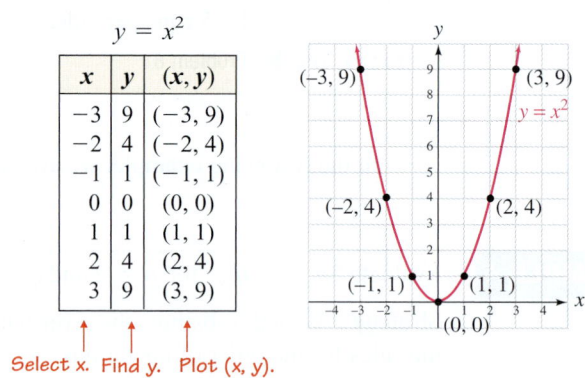

Self Check 5 Graph: $y = x^2 - 2$

Now Try Problem 69

Because the variable x is squared, $y = x^2$ is not a linear equation in two variables. We call it a **nonlinear equation.** In Examples 6 and 7, we will graph other nonlinear equations.

EXAMPLE 6   Graph: $y = -x^2 + 2$

Strategy   We will find several solutions of the equation, plot them on a rectangular coordinate system, and then draw a smooth curve passing through the points.

Why   To *graph* an equation in two variables means to make a drawing that represents all of its solutions.

Solution   To make a table of solutions, we select x-values of $-3, -2, -1, 0, 1, 2,$ and 3 and find each corresponding y-value. For example, if $x = -3$, we have

$y = -x^2 + 2$	This is the equation to graph.
$y = -(-3)^2 + 2$	Substitute -3 for x.
$y = -(9) + 2$	Evaluate the exponential expression first: $(-3)^2 = 9$.
$y = -7$	Do the addition: $-9 + 2 = -7$.

The ordered pair $(-3, -7)$ is a solution. Six other solutions appear in the table. After plotting each pair, we join the points with a smooth curve to obtain the graph, a parabola opening downward.

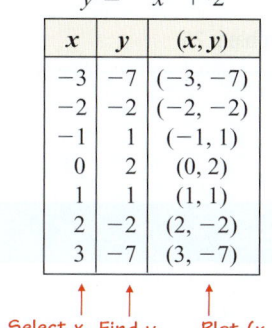

$$y = -x^2 + 2$$

x	y	(x, y)
-3	-7	$(-3, -7)$
-2	-2	$(-2, -2)$
-1	1	$(-1, 1)$
0	2	$(0, 2)$
1	1	$(1, 1)$
2	-2	$(2, -2)$
3	-7	$(3, -7)$

Select x. Find y.   Plot (x, y).

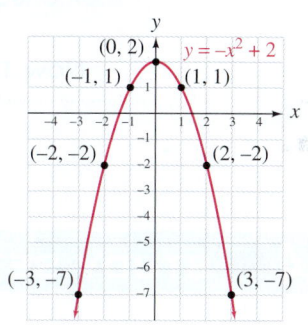

Self Check 6   Graph: $y = -x^2$

Now Try ▶ Problem 71

Success Tip

To graph equations of lines, we must find at least two points on the line to draw its graph. Because the graphs of these equations are more complicated, more work is required. We must find a sufficient number of points on the graph so that its entire shape is revealed.

EXAMPLE 7   Graph: $y = x^3 + 1$

Strategy   We will find several solutions of the equation, plot them on a rectangular coordinate system, and then draw a smooth curve passing through the points.

Why   To *graph* an equation in two variables means to make a drawing that represents all of its solutions.

Solution   If we let $x = -2$, we have

$y = x^3 + 1$	This is the equation to graph.
$y = (-2)^3 + 1$	Substitute -2 for x.
$y = -8 + 1$	Evaluate the exponential expression first: $(-2)^3 = -8$.
$y = -7$	Do the addition.

The ordered pair $(-2, -7)$ is a solution. This pair and others that satisfy the equation are listed in the table. Plotting the ordered pairs and joining the points with a smooth curve gives us the graph.

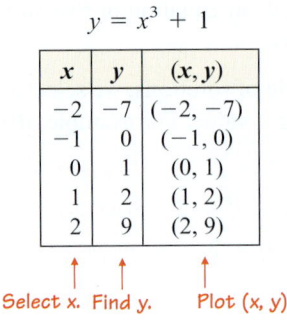

$$y = x^3 + 1$$

x	y	(x, y)
-2	-7	$(-2, -7)$
-1	0	$(-1, 0)$
0	1	$(0, 1)$
1	2	$(1, 2)$
2	9	$(2, 9)$

Select x.   Find y.   Plot (x, y).

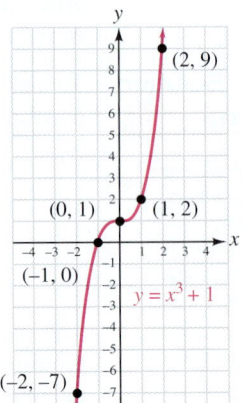

Self Check 7   Graph: $y = x^3 - 1$

Now Try ▶ Problem 75

SECTION 5.4 ▶ STUDY SET

VOCABULARY

Fill in the blanks.

1. A _____ is a term or a sum of terms in which all variables have whole-number exponents and no variable appears in a denominator.

2. The _____ of a polynomial are separated by + symbols.

3. $x^3 - 6x^2 + 9x - 2$ is a polynomial in _____ variable, and is written in _____ powers of x, and $c^3 + 2c^2d - d^2$ is a polynomial in _____ variables and is written in _____ powers of d.

4. For the polynomial $6x^2 + 3x - 1$, the _____ term is $6x^2$, and the leading _____ is 6. The _____ term is -1.

5. A _____ is a polynomial with exactly one term. A _____ is a polynomial with exactly two terms. A _____ is a polynomial with exactly three terms.

6. The _____ of the term $3x^7$ is 7 because x appears as a factor 7 times: $3 \cdot x \cdot x \cdot x \cdot x \cdot x \cdot x \cdot x$.

7. To _____ the polynomial $x^2 - 2x + 1$ for $x = 6$, we substitute 6 for x and follow the rules for the order of operations.

8. The graph of $y = x^2$ is a cup-shaped curve called a _____.

CONCEPTS

Determine whether each expression is a polynomial.

9. a. $x^3 - 5x^2 - 2$ b. $x^{-4} - 5x$

   c. $x^2 - \dfrac{1}{2x} + 3$ d. $x^3 - 1$

   e. $x^2 - y^2$ f. $a^4 + a^3 + a^2 + a$

10. Fill in the blank so that the term has degree 5.

   a. $9x$ ▢   b. $-\dfrac{2}{3}xy$ ▢

Make a term-coefficient-degree table like that shown in Example 1 for each polynomial.

11. $8x^2 + x - 7$

Term	Coefficient	Degree

Degree of the polynomial: ▢

12. $y^4 - y^3 + 16y^2 + 3y$

Term	Coefficient	Degree

Degree of the polynomial: ▢

13. $8a^6b^3 - 27ab$

Term	Coefficient	Degree

Degree of the polynomial: ▢

14. $-1.2c^4 + 2.4c^2d^2 - 3.6d^4$

Term	Coefficient	Degree

Degree of the polynomial: ▢

15. a. Write $x - 9 + 3x^2 + 5x^3$ in descending powers of x.

b. Write $-2xy + y^2 + x^2$ in ascending powers of y.

16. Complete the solution. Evaluate $-2x^2 + 3x - 1$ for $x = -2$.

$$-2x^2 + 3x - 1 = -2(\quad)^2 + 3(\quad) - 1$$
$$= -2(\quad) + 3(\quad) - 1$$
$$= \quad + (-6) - 1$$
$$= \quad$$

GUIDED PRACTICE

Classify each polynomial as a monomial, a binomial, a trinomial, or none of these. See Example 1.

17. $3x + 7$

18. $3y - 5$

19. $y^2 + 4y + 3$

20. $9xy$

21. $\frac{3}{2}z^2$

22. $\frac{3}{5}x^4 - \frac{2}{5}x^3 + \frac{3}{5}x - 1$

23. $t - 32$

24. $12z^4$

25. $s^2 - 23s + 31$

26. $2x^3 - 5x^2 + 6x - 3$

27. $6x^5 - x^4 - 3x^3 + 7$

28. x^3

29. $3m^3n - 4m^2n^2 + mn - 1$

30. $4p^3q^2 + 7p^2q^3 + pq^4 - q^5$

31. $2a^2 - 3ab + b^2$

32. $a^3b - ab^3$

Find the degree of each polynomial. See Example 1.

33. $3x^4$

34. $3x^5$

35. $-2x^2 + 3x + 1$

36. $-5x^4 + 3x^2 - 3x$

37. $\frac{1}{3}x - 5$

38. $\frac{1}{2}y^3 + 4y^2$

39. $-5r^2s^2 - r^3s + 3$

40. $4r^2s^3 - 5r^2s^8$

41. $x^{12} + 3x^2y^3$

42. $17ab^5 - 12a^3b$

43. 38

44. -24

45. $\frac{3}{2}m^7 - \frac{3}{4}m^{18}$

46. $\frac{7}{8}t^{10} - \frac{1}{8}t^{16}$

47. $5.5tw - 6.5t^2w - 7.5t^3$

48. $0.4h + 0.6h^4c + 0.6h^5$

Evaluate each expression. See Example 2 and 3.

49. $x^2 - x + 1$ for
  a. $x = 2$
  b. $x = -3$

50. $x^2 - x + 7$ for
  a. $x = 6$
  b. $x = -2$

51. $4t^2 + 2t - 8$ for
  a. $t = -1$
  b. $t = 0$

52. $3s^2 - 2s + 8$ for
  a. $s = 1$
  b. $s = 0$

53. $\frac{1}{2}a^2 - \frac{1}{4}a$ for
  a. $a = 4$
  b. $a = -8$

54. $\frac{1}{3}b^2 - \frac{1}{9}b$ for
  a. $b = 9$
  b. $b = -9$

55. $-9.2x^2 + x - 1.4$ for
  a. $x = -1$
  b. $x = -2$

56. $-10.3x^2 - x + 6.5$ for
  a. $x = -1$
  b. $x = -2$

57. $x^3 + 3x^2 + 2x + 4$ for
  a. $x = 2$
  b. $x = -2$

58. $x^3 - 3x^2 - x + 9$ for
  a. $x = 3$
  b. $x = -3$

59. $y^4 - y^3 + y^2 + 2y - 1$ for
  a. $y = 1$
  b. $y = -1$

60. $-y^4 + y^3 + y^2 + y + 1$ for
  a. $y = 1$
  b. $y = -1$

Evaluate each polynomial for $a = -2$ and $b = 3$. See Example 4.

61. $6a^2b$

62. $4ab^2$

63. $a^3 + b^3$

64. $a^3 - b^3$

65. $a^2 + 5ab - b^2$

66. $a^3 - 2ab + b^3$

67. $5ab^3 - ab - b + 10$

68. $-a^3b + ab - a - 21$

Construct a table of solutions and then graph the equation. See Examples 5–7.

69. $y = x^2 + 1$

70. $y = x^2 - 4$

71. $y = -x^2 - 2$

72. $y = -x^2 + 1$

73. $y = 2x^2 - 3$

74. $y = -2x^2 + 2$

75. $y = x^3 + 2$

76. $y = x^3 + 4$

77. $y = x^3 - 3$

78. $y = x^3 - 2$

79. $y = -x^3 - 1$

80. $y = -x^3$

APPLICATIONS

81. Supermarkets. A grocer plans to set up a pyramid-shaped display of cantaloupes like that shown in Example 3. If each side of the square base of the display is made of six cantaloupes, how many will be used in the display?

82. Packaging. The polynomial $4x^3 - 44x^2 + 120x$ gives the volume (in cubic inches) of the resulting box when a square with sides x inches long is cut from each corner of a 10 in. × 12 in. piece of cardboard. Find the volume of a box if 3-inch squares are cut out.

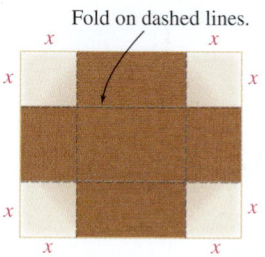
Fold on dashed lines.

83. Stopping Distance. The number of feet that a car travels before stopping depends on the driver's reaction time and the braking distance, as shown in the illustration. For one driver, the stopping distance is given by the polynomial $0.04v^2 + 0.9v$ where v is the velocity of the car. Find the stopping distance when the driver is traveling at 30 mph.

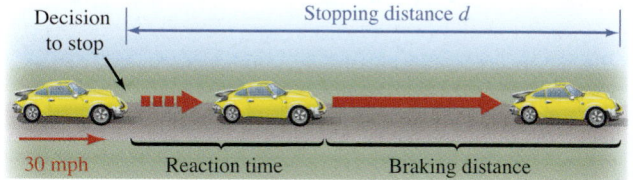

84. Suspension Bridges. The following polynomial $-0.0000001s^4 + 0.0066667s^2 + 400$ approximates the length of the cable between the two vertical towers of a bridge, where s is the sag in the cable (in feet). Estimate the length of the cable if the sag is 24.6 feet.

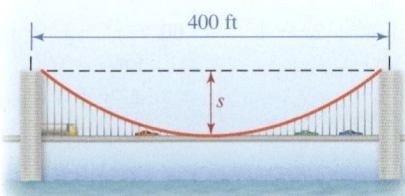

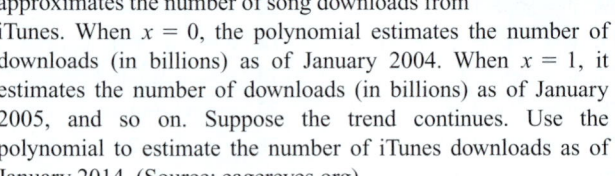

85. from **Campus to Careers**
Sound Engineering Technician

Many people in the recording industry have been impressed by the success of Apple's iTunes Music Store. The polynomial $0.32x^2 - 0.36x + 0.21$ approximates the number of song downloads from iTunes. When $x = 0$, the polynomial estimates the number of downloads (in billions) as of January 2004. When $x = 1$, it estimates the number of downloads (in billions) as of January 2005, and so on. Suppose the trend continues. Use the polynomial to estimate the number of iTunes downloads as of January 2014. (Source: eagereyes.org)

86. Twitter. When $x = 1$, the polynomial $4.4x^2 + 36.2x + 42.5$ approximates the number of Tweets (in millions) on the social network Twitter for January, 2009. When $x = 2$, it approximates the number of Tweets (in millions) for the month of February, 2009, and so on. Use the polynomial to find the number of Tweets (in millions) for the month of October, 2009. (Source: pingdom.com)

87. Science History. The Italian scientist Galileo Galilei (1564–1642) built an incline plane like that shown to study falling objects. As the ball rolled down, he measured the time it took the ball to travel different distances. Graph the data and then connect the points with a smooth curve.

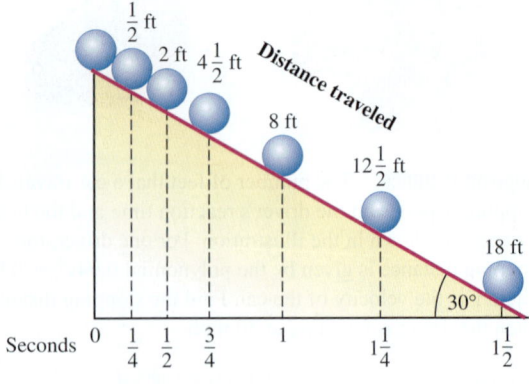

88. Dolphins. At a marine park, three trained dolphins jump in unison over an arching stream of water whose path can be described by the equation $y = -0.05x^2 + 2x$. Given the takeoff points for each dolphin, how high must each jump to clear the stream of water?

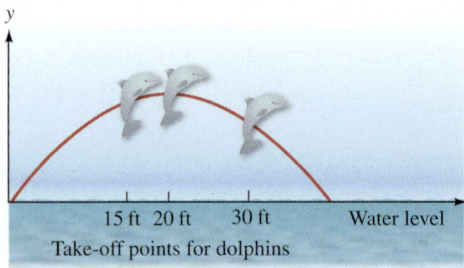

WRITING

89. Describe how to determine the degree of a polynomial.

90. List some words that contain the prefixes *mono, bi,* or *tri.*

91. To graph $y = x^2 - 4$, a table of solutions is constructed and a graph is drawn. Explain the error.

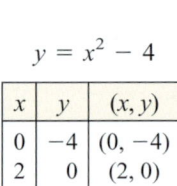 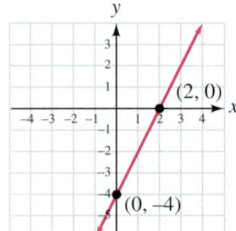

$y = x^2 - 4$

x	y	(x, y)
0	-4	$(0, -4)$
2	0	$(2, 0)$

92. The expression $x + y$ is a binomial. Is xy also a binomial? Explain.

REVIEW

Solve each inequality. Write the solution set in interval notation and graph it.

93. $-4(3y + 2) \leq 28$ **94.** $-5 < 3t + 4 \leq 13$

Simplify each expression. Do not use negative exponents in the answer.

95. $(x^2x^4)^3$ **96.** $(a^2)^3(a^3)^2$

97. $\left(\dfrac{y^2y^5}{y^4}\right)^3$ **98.** $\left(\dfrac{2t^3}{t}\right)^{-4}$

CHALLENGE PROBLEMS

99. Find a three-term polynomial of degree 2 whose value will be 1 when it is evaluated for $x = 2$.

100. Graph: $y = 2x^3 - 3x^2 - 11x + 6$

SECTION 5.5

OBJECTIVES

1 Simplify polynomials by combining like terms.

2 Add polynomials.

3 Subtract polynomials.

Adding and Subtracting Polynomials

ARE YOU READY?

The following problems review some basic skills that are needed when adding and subtracting polynomials.

Combine like terms, if possible.

1. $8x + 9x$

2. $2.5a^3 - 1.3a^3$

3. $2t^2u - (-3t^2u)$

4. $8b^2 + 7b$

5. Simplify: $-(4b^2 - 9b + 1)$

6. Subtract 10 from 2.

If we are to add (or subtract) objects, they must be similar. For example, we can add dollars to dollars and inches to inches, but we can't add dollars to inches. If you keep this concept in mind, then adding and subtracting polynomials will be easy. It simply involves combining like terms.

1 Simplify Polynomials by Combining Like Terms.

Recall that **like terms** have the same variables with the same exponents. Only the coefficients may differ.

Like terms	*Unlike terms*	
$-7x$ and $15x$	$-7x$ and $15a$	Different variables
$9.4y^3$ and $1.6y^3$	$9.4y^3$ and $1.6y^2$	Different exponents on the same variable
$\frac{1}{2}x^5y^2$ and $-\frac{1}{3}x^5y^2$	$\frac{1}{2}x^5y^2$ and $-\frac{1}{3}x^2y^5$	Different exponents on different variables

The Language of Algebra

Simplifying the sum or difference of like terms is called **combining like terms.**

Also recall that to **combine like terms,** we combine their coefficients and keep the same variables with the same exponents. For example,

$$4y + 5y = (4 + 5)y \qquad 8x^2 - x^2 = (8 - 1)x^2$$
$$= 9y \qquad\qquad\qquad = 7x^2$$

Polynomials with like terms can be simplified by combining like terms.

EXAMPLE 1

Simplify each polynomial by combining like terms:

a. $4x^4 + 81x^4$ **b.** $-0.3r - 0.4r + 0.6r$

c. $17x^2y^2 + 2x^2y - 6x^2y^2$ **d.** $\frac{3}{4}p^2 + \frac{1}{2}q^2 - 7 + \frac{1}{3}p^2 - \frac{5}{4}q^2 + 4$

Strategy We will use the distributive property in reverse to add (or subtract) the coefficients of the like terms. We will keep the same variables raised to the same powers.

Why To *combine like terms* means to add or subtract the like terms in an expression.

Solution

a. $4x^4 + 81x^4 = 85x^4$ Think: $(4 + 81)x^4 = 85x^4$.

b. $-0.3r - 0.4r + 0.6r = -0.1r$ Think: $(-0.3 - 0.4 + 0.6)r = -0.1r$.

c. The first and third terms are like terms.

$$17x^2y^2 + 2x^2y - 6x^2y^2 = 11x^2y^2 + 2x^2y \qquad \text{Think: } (17 - 6)x^2y^2 = 11x^2y^2.$$

Caution

When combining like terms, simply combine their coefficients. Don't incorrectly add the exponents. The exponents on the variables *stay the same.*

d. $\dfrac{3}{4}p^2 + \dfrac{1}{2}q^2 - 7 + \dfrac{1}{3}p^2 - \dfrac{5}{4}q^2 + 4$

$= \left(\dfrac{3}{4} + \dfrac{1}{3}\right)p^2 + \left(\dfrac{1}{2} - \dfrac{5}{4}\right)q^2 - 7 + 4$ Combine like terms.

$= \left(\dfrac{9}{12} + \dfrac{4}{12}\right)p^2 + \left(\dfrac{2}{4} - \dfrac{5}{4}\right)q^2 - 7 + 4$ Build equivalent fractions: $\dfrac{3}{4} \cdot \dfrac{3}{3} = \dfrac{9}{12}, \dfrac{1}{3} \cdot \dfrac{4}{4} = \dfrac{4}{12},$ and $\dfrac{1}{2} \cdot \dfrac{2}{2} = \dfrac{2}{4}.$

$= \dfrac{13}{12}p^2 - \dfrac{3}{4}q^2 - 3$ Do the addition and the subtraction.

CAUTION Do not try to clear this expression of fractions by multiplying it by the LCD 12. That strategy works only when we multiply *both sides of an equation* by the LCD.

$$\cancel{12}\left(\dfrac{3}{4}p^2 + \dfrac{1}{2}q^2 - 7 + \dfrac{1}{3}p^2 - \dfrac{5}{4}q^2 + 4\right)$$

Self Check 1   Simplify each polynomial: **a.** $6m^4 + 3m^4$
b. $-19x + 21x - x$ **c.** $1.7s^3t + 0.3s^2t - 0.6s^3t$
d. $\dfrac{1}{8}c^5 + \dfrac{1}{3}d^5 - 9 + \dfrac{5}{4}c^5 - \dfrac{3}{5}d^5 + 1$

Now Try ▶ Problems 13, 23, and 27

Notice that the terms of each solution in Example 1 are written in **descending powers** of one variable. When working with polynomials, answers are almost always presented in this form because it makes them easy to compare.

2   Add Polynomials.

When adding polynomials horizontally, each polynomial is usually enclosed within parentheses. For example,

$$(3x^2 + 6x + 7) \mathbf{+} (2x - 5)$$

is the sum of a trinomial and a binomial. To find the sum, we reorder and regroup the terms using the commutative and associative properties of addition so that like terms are together.

$(3x^2 + 6x + 7) + (2x - 5) = 3x^2 + (\mathbf{6x + 2x}) + (\mathbf{7 - 5})$ The *x*-terms are together. The constant terms are together.

$= 3x^2 + \mathbf{8x} + \mathbf{2}$ Combine like terms.

This example illustrates the following rule.

Adding Polynomials	To add polynomials, combine their like terms.

EXAMPLE 2   Add the polynomials: **a.** $(-6a^3 + 5a^2 - 7a + 9) + (4a^3 - 5a^2 - a - 8)$
b. $\left(\dfrac{1}{2}m^2 + \dfrac{2}{3}m + 1\right) + \left(\dfrac{3}{4}m^2 - \dfrac{7}{9}m - 4\right)$ **c.** $(16g^2 - h^2) + (4g^2 + 2gh + 10h^2)$

Strategy   We will reorder and regroup to get the like terms together. Then we will combine like terms.

Why   To add polynomials means to combine their like terms.

Solution

Success Tip

Combine the like terms in order: a^3-terms first, a^2-terms second, a-terms third, and constants last. Then the answer will be in descending powers of a.

a. $(-6a^3 + 5a^2 - 7a + 9) + (4a^3 - 5a^2 - a - 8)$

$= (\mathbf{-6a^3 + 4a^3}) + (\mathbf{5a^2 - 5a^2}) + (\mathbf{-7a - a}) + (\mathbf{9 - 8})$ Group like terms together.

$= \mathbf{-2a^3 + 0a^2 + (-8a) + 1}$ Combine like terms.

$= -2a^3 - 8a + 1$ It is not necessary to write $0a^2$.

b. $\left(\dfrac{1}{2}m^2 + \dfrac{2}{3}m + 1\right) + \left(\dfrac{3}{4}m^2 - \dfrac{7}{9}m - 4\right)$

$= \left(\dfrac{1}{2}m^2 + \dfrac{3}{4}m^2\right) + \left(\dfrac{2}{3}m - \dfrac{7}{9}m\right) + (1 - 4)$ Group like terms together.

$= \left(\dfrac{2}{4}m^2 + \dfrac{3}{4}m^2\right) + \left(\dfrac{6}{9}m - \dfrac{7}{9}m\right) + (1 - 4)$ Build equivalent fractions: $\frac{1}{2} \cdot \frac{2}{2} = \frac{2}{4}$ and $\frac{2}{3} \cdot \frac{3}{3} = \frac{6}{9}$.

$= \dfrac{5}{4}m^2 - \dfrac{1}{9}m - 3$ Combine like terms.

c. $(16g^2 - h^2) + (4g^2 + 2gh + 10h^2)$

$= (\mathbf{16g^2 + 4g^2}) + \mathbf{2gh} + (\mathbf{-h^2 + 10h^2})$ Group like terms together.

$= \mathbf{20g^2 + 2gh + 9h^2}$ Combine like terms.

Self Check 2   Add the polynomials:

a. $(2a^2 - a + 4) + (5a^2 + 6a - 5)$

b. $\left(\frac{3}{2}b^3 + \frac{4}{5}b + 7\right) + \left(\frac{3}{4}b^3 - \frac{11}{10}b - 10\right)$

c. $(7x^2 - 2xy - y^2) + (4x^2 - y^2)$

Now Try ▶ Problems 29, 31, and 35

EXAMPLE 3   Find a polynomial that represents the perimeter of the trapezoid.

Strategy   We will add the polynomials that represent the lengths of the sides of the trapezoid.

Why   To find the perimeter of a figure, we find the distance around the figure by finding the sum of the lengths of its sides.

$(2a^2 - 5)$ ft

$(a^2 + a)$ ft $3a^2$ ft

$(4a^2 - 2a + 1)$ ft

Solution   To add the four polynomials that represent the lengths of the sides of the trapezoid, we combine their like terms.

The Language of Algebra

A **trapezoid** is a four-sided figure with exactly two sides parallel.

$(2a^2 - 5) + (a^2 + a) + (4a^2 - 2a + 1) + 3a^2$

$= (\mathbf{2a^2 + a^2 + 4a^2 + 3a^2}) + (\mathbf{a - 2a}) + (\mathbf{-5 + 1})$ Reorder and regroup terms.

$= \mathbf{10a^2 - a - 4}$ Combine like terms. Think: $2 + 1 + 4 + 3 = 10$, $1 - 2 = -1$, and $-5 + 1 = -4$.

The perimeter of the trapezoid is $(10a^2 - a - 4)$ ft.

Self Check 3   Find a polynomial that represents the perimeter of a triangle with sides of length $(19h^2 + 11h - 2)$ in., $(4h^2 - 4h + 4)$ in., and $(6h^2 - 22h - 2)$ in.

Now Try ▶ Problem 39

Polynomials also can be added vertically by aligning like terms in columns.

EXAMPLE 4   Add $4x^2 - 3$ and $3x^2 - 8x + 8$ using vertical form.

Strategy   First, we will write one polynomial underneath the other and draw a horizontal line beneath them. Then we will add the like terms, column by column, and write each result under the line.

Why   *Vertical form* means to use an approach similar to that used in arithmetic to add two numbers.

Solution   When performing vertical addition, any missing term may be written with a coefficient of 0. Since the first polynomial does not have an x-term, we insert a **placeholder** term $0x$ in the second column so that the constant terms line up in the third column.

x^2-terms ⎯⎯⎯⎯ x-terms   ⎾⎯ Constants

$$
\begin{array}{rrr}
4x^2 & +\ 0x & -\ 3 \\
3x^2 & -\ 8x & +\ 8 \\
\hline
7x^2 & -8x & +5
\end{array}
$$

In the x^2-column, find $4x^2 + 3x^2$.
In the x-column, find $0x + (-8x)$.
In the constant column, find $-3 + 8$.

The sum is $7x^2 - 8x + 5$.

Self Check 4   Add $4q^2 - 7$ and $2q^2 - 8q + 9$ using vertical form.

Now Try ▶ Problem 41

3 Subtract Polynomials.

Recall from Chapter 1 that we can use the distributive property to find the opposite of several terms enclosed within parentheses. For example, consider $-(2a^2 - a + 9)$.

$$-(2a^2 - a + 9) = -\mathbf{1}(2a^2 - a + 9)$$ Replace the $-$ symbol in front of the parentheses with -1.

$$= -2a^2 + a - 9$$ Use the distributive property to remove parentheses.

This example illustrates the following method of subtracting polynomials.

Subtracting Polynomials ▼   To subtract two polynomials, change the signs of the terms of the polynomial being subtracted, drop the parentheses, and combine like terms.

EXAMPLE 5   Subtract the polynomials: **a.** $(3a^2 - 4a - 6) - (2a^2 - a + 9)$
b. $(-t^3u + 2t^2u - u + 1) - (-3t^2u - u + 8)$

Strategy   In each case, we will change the signs of the terms of the polynomial being subtracted, drop the parentheses, and combine like terms.

Why   This is the method for subtracting two polynomials.

Solution   **a.** $(3a^2 - 4a - 6) \mathbf{- (2a^2 - a + 9)}$

$$= 3a^2 - 4a - 6 \mathbf{- 2a^2 + a - 9}$$ Change the sign of each term of $2a^2 - a + 9$ and drop the parentheses.

$$= a^2 - 3a - 15$$ Combine like terms.

b. $(-t^3u + 2t^2u - u + 1) - (-3t^2u - u + 8)$

$= -t^3u + 2t^2u - u + 1 + 3t^2u + u - 8$ *Change the sign of each term of $-3t^2u - u + 8$ and drop the parentheses.*

$= -t^3u + 5t^2u - 7$ *Combine like terms.*

Self Check 5   Subtract the polynomials:
a. $(8a^3 - 5a^2 + 5) - (a^3 - a^2 - 7)$
b. $(x^2y - 2x + y - 2) - (6x + 9y - 2)$

Now Try Problems 49 and 55

Polynomials can also be subtracted vertically by aligning like terms in columns.

EXAMPLE 6   **a.** Subtract $3x^2 - 2x + 3$ from $2x^2 + 4x - 1$ using vertical form.
b. Subtract $4x^3 - 6x^2 + x$ from $7x^3 - 2x$ using vertical form.

Strategy   Since the first polynomial is to be subtracted from the second, we will write the first underneath the second, change the sign of each of its terms and add, column-by-column.

Why   *Vertical form* means to arrange the like terms in columns.

Solution   **a.**
$$\begin{array}{r} 2x^2 + 4x - 1 \\ -(3x^2 - 2x + 3) \end{array} \xrightarrow{\substack{Change\ signs\\ and\ add}} \begin{array}{r} 2x^2 + 4x - 1 \\ -3x^2 + 2x - 3 \\ \hline -x^2 + 6x - 4 \end{array}$$

In the x^2-column, find $2x^2 + (-3x^2)$.
In the x-column, find $4x + 2x$.
In the constant column, find $-1 + (-3)$.

The difference is $-x^2 + 6x - 4$.

b. Since $7x^3 - 2x$ is missing an x^2-term, we will insert the placeholder term $0x^2$ to ensure that like terms are in the same column.

$$\begin{array}{r} 7x^3 + 0x^2 - 2x \\ -(4x^3 - 6x^2 + x) \end{array} \xrightarrow{\substack{Change\ signs\\ and\ add}} \begin{array}{r} 7x^3 + 0x^2 - 2x \\ -4x^3 + 6x^2 - x \\ \hline 3x^3 + 6x^2 - 3x \end{array}$$

In the x^3-column, find $7x^3 + (-4x^3)$.
In the x^2-column, find $0x^2 + 6x^2$.
In the x-column, find $-2x + (-x)$.

The difference is $3x^3 + 6x^2 - 3x$.

Self Check 6   **a.** Subtract $2p^2 + 2p - 8$ from $5p^2 - 6p + 7$ using vertical form.
b. Subtract $4m^3 - 6m^2 + 7m$ from $-8m^3 + 16m$.

Now Try Problem 57

EXAMPLE 7   Subtract $1.2a^4 - 0.7a$ from the sum of $0.6a^4 + 1.5a$ and $0.4a^4 - 1.1a$.

Strategy   First, we will translate the words of the problem into mathematical symbols. Then we will perform the indicated operations.

Why   The words of the problem contain the key phrases *subtract from* and *sum*.

Solution   Since $1.2a^4 - 0.7a$ is to be subtracted from the sum, the order must be reversed when we translate to mathematical symbols.

Subtract $1.2a^4 - 0.7a$ from   the sum of $0.6a^4 + 1.5a$ and $0.4a^4 - 1.1a$.

$[(0.6a^4 + 1.5a) + (0.4a^4 - 1.1a)] - (1.2a^4 - 0.7a)$ *Use brackets [ ] to enclose the sum.*

Next, we change the sign of each term within $(1.2a^4 - 0.7a)$ and drop the parentheses.

$$= 0.6a^4 + 1.5a + 0.4a^4 - 1.1a \mathbf{- 1.2a^4 + 0.7a}$$
$$= -0.2a^4 + 1.1a \quad \textit{Combine like terms.}$$

Self Check 7 Subtract $-0.2q^2 - 0.2q$ from the sum of $0.1q^2 - 0.6q$ and $0.3q^2 + 0.1q$.

Now Try ▶ Problem 65

EXAMPLE 8

Fireworks. Two firework shells are fired upward at the same time from different platforms. The height, after t seconds, of the first shell is $(-16t^2 + 160t + 3)$ feet. The height, after t seconds, of a higher-flying second shell is $(-16t^2 + 200t + 1)$ feet.

a. Find a polynomial that represents the difference in the heights of the shells.

b. In 5 seconds, the first shell reaches its peak and explodes. How much higher is the second shell at that time?

Strategy To find the difference in their heights, we will subtract the height of the first shell from the height of the higher-flying second shell.

Why The key word *difference* indicates that we should subtract the polynomials.

Solution **a.** Since the height of the higher flying second shell is represented by $-16t^2 + 200t + 1$ and the height of the lower flying shell is represented by $-16t^2 + 160t + 3$, we can find their difference by performing the following subtraction.

$$(-16t^2 + 200t + 1) - (-16t^2 + 160t + 3)$$
$$= -16t^2 + 200t + 1 + 16t^2 - 160t - 3 \quad \begin{array}{l}\textit{Change the sign of each term of} \\ -16t^2 + 160t + 3 \textit{ and drop the} \\ \textit{parentheses.}\end{array}$$

$$= 40t - 2 \quad \textit{Combine like terms.}$$

The difference in the heights of the shells t seconds after being fired is $(40t - 2)$ feet.

b. To find the difference in their heights after 5 seconds, we will evaluate the polynomial found in part (a) at a value of 5 seconds. If we substitute 5 for t, we have

$$40t - 2 = 40(\mathbf{5}) - 2 = 200 - 2 = 198$$

When the first shell explodes, the second shell will be 198 feet higher than the first shell.

Self Check 8 **Property values.**   A real estate investor purchased two houses on the same day. The value of the first house, x years after its purchase, is given by $\$(2{,}500x + 95{,}000)$. The value of the second house, x years after its purchase, is given by $\$(4{,}500x + 125{,}000)$. Find a polynomial that represents the total value of the houses after x years.

Now Try ▶ Problem 109

SECTION 5.5 ▶ STUDY SET

VOCABULARY

Fill in the blanks.

1. $(b^3 - b^2 - 9b + 1) + (b^3 - b^2 - 9b + 1)$ is the sum of two _____.

2. $(b^2 - 9b + 11) - (4b^2 - 14b)$ is the _____ of a trinomial and a binomial.

3. _____ terms have the same variables with the same exponents.

4. The polynomial $2t^4 + 3t^3 - 4t^2 + 5t - 6$ is written in _____ powers of t.

CONCEPTS

Fill in the blanks.

5. To add polynomials, _____ their like terms.

6. To subtract polynomials, _____ the signs of the terms of the polynomial being subtracted, drop parentheses, and combine like terms.

7. Simplify each polynomial, if possible.
   a. $2x^2 + 3x^2$ b. $15m^3 - m^3$
   c. $8a^3b - a^3b$ d. $6cd + 4c^2d$

8. What is the result of the addition in the x-column?

   $$4x^2 + \quad x - 12$$
   $$\underline{5x^2 - 8x + 23}$$

9. Write without parentheses.
   a. $-(5x^2 - 8x + 23)$ b. $-(-5y^4 + 3y^2 - 7)$

10. What is the result of the subtraction in the x-column?

   $$8x^2 - 7x - 1 \qquad\qquad 8x^2 - 7x - 1$$
   $$\underline{-(4x^2 + 6x - 9)} \longrightarrow \underline{-4x^2 - 6x + 9}$$

NOTATION

Fill in the blanks to add (subtract) the polynomials.

11. $(6x^2 + 2x + 3) + (4x^2 - 7x + 1)$
   $$= (6x^2 + \boxed{}) + (\boxed{} - 7x) + (3 + \boxed{})$$
   $$= \boxed{} - 5x + \boxed{}$$

12. $(6x^2 + 2x + 3) - (4x^2 - 7x + 1)$
   $$= 6x^2 + 2x + 3 \boxed{} 4x^2 \boxed{} 7x - 1$$
   $$= \boxed{} + 9x + \boxed{}$$

GUIDED PRACTICE

Simplify each polynomial and write it in descending powers of one variable. See Example 1.

13. $8t^2 + 4t^2$

14. $15x^2 + 10x^2$

15. $18x^2 - 19x + 2x^2$

16. $17y^2 - 22y - y^2$

17. $10x^2 - 8x + 9x - 9x^2$

18. $-3y^2 - y - 6y^2 + 7y$

19. $\frac{1}{5}x^2 - \frac{3}{8}x + \frac{2}{3}x^2 + \frac{1}{4}x$

20. $\frac{6}{7}y^2 + \frac{1}{2}y - \frac{2}{3}y^2 + \frac{1}{5}y$

21. $0.6x^3 + 0.8x^4 + 0.7x^3 + (-0.8x^4)$

22. $1.9m^4 - 2.4m^6 - 3.7m^4 + 2.8m^6$

23. $\frac{1}{2}st + \frac{3}{2}st$ 24. $\frac{2}{5}at + \frac{1}{5}at$

25. $-4ab + 4ab - ab$ 26. $xy - 4xy - 2xy$

27. $4x^2y + 5 - 6x^3y - 3x^2y + 2x^3y$

28. $5b - 9ab^2 + 10a^3b - 8ab^2 - 9a^3b$

Add the polynomials. See Example 2.

29. $(3q^2 - 5q + 7) + (2q^2 + q - 12)$

30. $(2t^2 + 11t - 15) + (-5t^2 - 13t + 10)$

31. $\left(\frac{2}{3}y^3 + \frac{3}{4}y^2 + \frac{1}{2}\right) + \left(\frac{1}{3}y^3 + \frac{1}{5}y^2 - \frac{1}{6}\right)$

32. $\left(\frac{1}{16}r^6 + \frac{1}{2}r^3 - \frac{11}{12}\right) + \left(\frac{9}{16}r^6 + \frac{9}{4}r^3 + \frac{1}{12}\right)$

33. $(0.3p + 2.1q) + (0.4p - 3q)$

34. $(-0.3r - 5.2s) + (0.8r - 5.2s)$

35. $(2x^2 + xy + 3y^2) + (5x^2 - y^2)$

36. $(-4a^2 - ab + 15b^2) + (5a^2 - b^2)$

Find a polynomial that represents the perimeter of the figure. See Example 3.

37.

$(x^2 + 3x + 1)$ yd; $70°$ $70°$; $(x^2 - 4)$ yd

38.

$(2y^5 - 3y^3)$ inches; $(y^5 + y^3)$ inches; $(5y^5 - 5y^3)$ inches

39.

$(2x^2 - 7)$ mi; $(x + 6)$ mi; $45°$; $45°$; $(x + 6)$ mi; $(5x^2 + 3x + 1)$ mi

40.

$(9a^2 + a - 3)$ ft; $(4a^2 + 1)$ ft; $(5a^2 - 6a)$ ft; $(11a^2 - 5a + 1)$ ft

Use vertical form to add the polynomials. See Example 4.

41. $3x^2 + 4x + 5$
   $\underline{2x^2 - 3x + 6}$

42. $6x^3 - 4x^2 + 7$
   $\underline{7x^3 + 9x^2 + 12}$

43. $6a^2 + 7a + 9$
          $\underline{-9a^2 \qquad - 2}$

44. $-2c^2 - 3c - 5$
          $\underline{14c^2 \qquad - 1}$

45. $z^3 + 6z^2 - 7z + 16$
          $\underline{9z^3 - 6z^2 + 8z - 18}$

46. $3x^3 + 4x^2 - 3x + 5$
          $\underline{3x^3 - 4x^2 - \ x - 7}$

47. $-3x^3y^2 + 4x^2y - 4x + 9$
          $\underline{\ 2x^3y^2 \qquad + 9x - 3}$

48. $3x^2y^2 + 4xy + 25$
          $\underline{5x^2y^2 \qquad - 12}$

Subtract the polynomials. **See Example 5.**

49. $(3a^2 - 2a + 4) - (a^2 - 3a + 7)$

50. $(2b^2 + 3b - 5) - (2b^2 - 4b - 9)$

51. $(-4h^3 + 5h^2 + 15) - (h^3 - 15)$

52. $(-c^5 + 5c^4 - 12) - (2c^5 - c^4)$

53. $\left(\dfrac{3}{8}s^8 - \dfrac{3}{4}s^7\right) - \left(\dfrac{1}{3}s^8 + \dfrac{1}{5}s^7\right)$

54. $\left(\dfrac{5}{6}q^9 - \dfrac{4}{5}q^8\right) - \left(\dfrac{1}{4}q^9 + \dfrac{3}{8}q^8\right)$

55. $(5ab + 2b^2) - (2 + ab + b^2)$

56. $(mn + 8n^2) - (6 - 5mn + n^2)$

Use vertical form to subtract the polynomials. **See Example 6.**

57. Subtract $2x^2 - 2x + 3$ from $3x^2 + 4x + 5$.

58. Subtract $3y^2 - 6y + 7$ from $6y^2 + 4y + 13$.

59. Subtract $(s^2 + 4s + 2)$ from $(5s^2 + 9)$.

60. Subtract $(4p^2 - 4p - 40)$ from $(10p^2 - 30)$.

61. Subtract $8a^3 + 8a^2 - 3a + 1$ from $17a^3 + 25a - 10$.

62. Subtract $m^3 + 20m^2 - 15m + 39$ from $-4m^3 - m + 22$.

63. $0.8x^3 \qquad \ - 2.3x + 0.6$
          $\underline{-(0.2x^3 - 1.2x^2 - 3.6x + 0.9)}$

64. $9.7y^3 \qquad + \quad y + 1.1$
          $\underline{-(6.3y^3 - 4.4y^2 + 2.7y + 8.8)}$

Perform the operations. **See Example 7.**

65. Subtract $(3x^2 + 4x - 7)$ from the sum of $(-2x^2 - 7x + 1)$ and $(-4x^2 + 8x - 7)$.

66. Subtract $(32x^2 - 17x + 45)$ from the sum of $(23x^2 - 12x - 7)$ and $(-11x^2 + 12x + 7)$.

67. Subtract $t^3 - 2t^2 + 2$ from the sum of $3t^3 + t^2$ and $-t^3 + 6t - 3$.

68. Subtract $-3z^3 - 4z + 7$ from the sum of $2z^2 + 3z - 7$ and $-4z^3 - 2z - 3$.

Perform the operations.

69. $(9a^2 + 3a) - (2a - 4a^2)$

70. $(4b^2 + 3b) - (7b - b^2)$

71. Subtract $(-y^5 + 5y^4 - 1.2)$ from $(2y^5 - y^4)$.

72. Subtract $(-4w^3 + 5w^2 + 7.6)$ from $(w^3 - 15w^2)$.

73. $3r^4 - 4r + 7r^4$

74. $-2b^4 + 7b - 3b^4$

75. $(0.03f^2 + 0.25f + 0.91) - (0.17f^2 - 1.18)$

76. $(0.05r^2 - 0.33r) - (0.48\,r^2 + 0.15r + 2.14)$

77. $\left(\dfrac{7}{8}r^4 + \dfrac{5}{9}r^2 - \dfrac{9}{4}\right) - \left(-\dfrac{3}{8}r^4 - \dfrac{2}{3}r^2 - \dfrac{1}{4}\right)$

78. $\left(\dfrac{4}{5}t^4 - \dfrac{1}{3}t^2 + \dfrac{1}{2}\right) - \left(-\dfrac{1}{2}t^4 + \dfrac{3}{8}t^2 - \dfrac{1}{16}\right)$

79. $8c^2 - 4c - 5$
          $\underline{-(-c^2 + 2c + 9)}$

80. $3t^3 - 4t^2 - 3t + 5$
          $\underline{+11t^3 \qquad - 8t - 2}$

81. $(12.1h^3 + 9.9h^2) + (7.3h^3 + 1.1h^2)$

82. $(5.7n^3 - 2.1n) + (-6.2n^3 - 3.9n)$

83. $(20 - 4rt - 5r^2t) + (10 - 5rt)$

84. $(5m^2 - 8m + 8) - (-20m^2 + m)$

85. $(3x^2 - 3x - 2) + (3x^2 + 4x - 3)$

86. $(4c^2 + 3c - 2) + (3c^2 + 4c + 2)$

87. $\dfrac{2}{3}d^2 - \dfrac{1}{4}c^2 + \dfrac{5}{6}c^2 - \dfrac{1}{2}cd + \dfrac{1}{3}d^2$

88. $\dfrac{3}{5}s^2 - \dfrac{2}{5}t^2 - \dfrac{1}{2}s^2 - \dfrac{7}{10}st - \dfrac{3}{10}st$

89. $(3x + 7) + (4x - 3)$

90. $(2y - 3) + (4y + 7)$

91. Subtract $1.7t^2 - 1.1t$ from the sum of $-2.7t^2 + 2.1t - 1.7$ and $3.1t^2 - 2.5t + 2.3$.

92. Subtract $1.07x^2 - 2.07x$ from the sum of $1.04x^2 - 5.01$ and $1.33x - 1.9x^2 + 5.02$.

93. $-32u^3 - 16u^3$

94. $-25x^3 - 7x^3$

95. $(9d^2 + 6d) + (8d - 4d^2)$

96. $(2c^2 - 4c) + (8c - c^2)$

97. $3x^3y^2 + 4x^2y + 7x + 12$
          $\underline{-(-4x^3y^2 + 6x^2y + 9x - 3)}$

98. $-2x^2y^2 \qquad + 12y^2$
          $\underline{-(10x^2y^2 + 9xy - 24y^2)}$

99. $(2x^2 - 3x + 1) - (4x^2 - 3x + 2) + (2x^2 + 3x + 2)$

100. $(-3z^2 - 4z + 7) + (2z^2 + 2z - 1) - (2z^2 - 3z + 7)$

101. $\begin{array}{r} 4x^3 + 4x^2 - 3x + 10 \\ +(5x^3 - 2x^2 - 4x - 4) \\ \hline \end{array}$ **102.** $\begin{array}{r} 7m^5 + m^3 + 9m^2 - m \\ -(8m^5 - 2m^3 + m^2 + m) \\ \hline \end{array}$

Look Alikes . . .

103. **a.** $(-8x^2 - 3x) + (-11x^2 + 6x + 10)$
   b. $(-8x^2 - 3x) - (-11x^2 + 6x + 10)$

104. **a.** $(10 - 2st - 3s^2t) + (4 - 6st)$
   b. $(10 - 2st - 3s^2t) - (4 - 6st)$

APPLICATIONS

105. **Greek Architecture.**

   a. Find a polynomial that represents the difference in the heights of the columns.

   b. If the columns were stacked one atop the other, to what height would they reach?

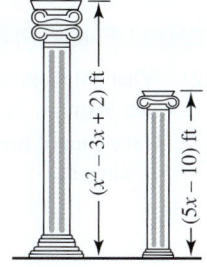

106. **Jets.** Find a polynomial that represents the length of the larger jet.

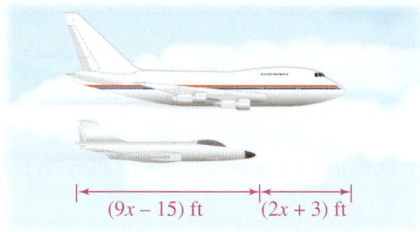

$(9x - 15)$ ft $(2x + 3)$ ft

107. **Piñatas.** Find a polynomial that represents the length of the rope used to hold up the piñata.

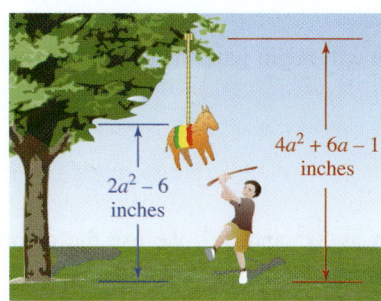

108. **Reading Blueprints.** Find a polynomial that represents

   a. the difference in the length and width of the one-bedroom apartment.

   b. the perimeter of the apartment.

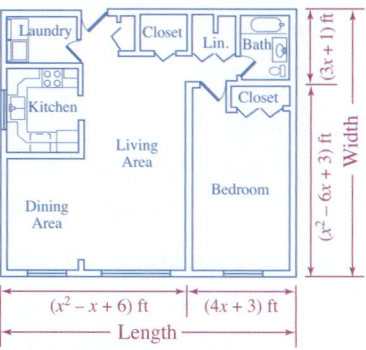

109. **Naval Operations.** Two warning flares are fired upward at the same time from different parts of a ship. The height of the first flare is $(-16t^2 + 128t + 20)$ feet and the height of the higher-traveling second flare is $(-16t^2 + 150t + 40)$ feet, after t seconds.

   a. Find a polynomial that represents the difference in the heights of the flares.

   b. In 4 seconds, the first flare reaches its peak, explodes, and lights up the sky. How much higher is the second flare at that time?

110. **Auto Mechanics.** The length of a fan belt that wraps around three pulleys is $(3x^2 + 11x + 4.5\pi)$ in. Find a polynomial that represents the unknown length of a part of the belt shown in the illustration below.

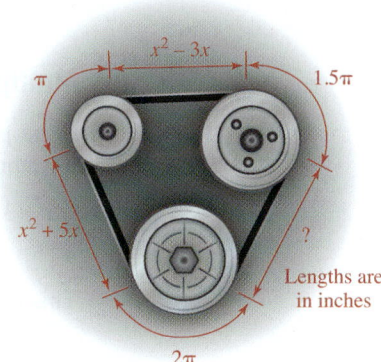

WRITING

111. How do you recognize like terms?

112. Explain why the vertical form used in algebra to add $2x^2 + 4x + 3$ and $5x^2 + 3x + 6$ is similar to the vertical form used in arithmetic to add 243 and 536.

113. Explain the error below.

$$7x^2y + 6x^2y = \cancel{13x^4y^2}$$

114. Explain the error below.

$$(12x^2 - 4) - (3x^2 - 1) = \cancel{12x^2 - 4 - 3x^2 - 1}$$
$$= \cancel{9x^2 - 5}$$

115. A student was asked to simplify $\frac{1}{6}x^2 - 3 + \frac{2}{3}x^2$. Explain the error below:

$$\cancel{6\left(\frac{1}{6}x^2 - 3 + \frac{2}{3}x^2\right) = x^2 - 18 + 4x^2}$$
$$= \cancel{5x^2 - 18}$$

116. Explain the error below.

Subtract $(2d^2 - d - 3)$ from $(d^2 - 9)$:

$$\cancel{(2d^2 - d - 3) - (d^2 - 9) = d^2 - d + 6}$$

REVIEW

117. What is the sum of the measures of the angles of a triangle?

118. What are the formulas for the area of a circle and the area of a triangle?

119. Graph: $y = -\frac{1}{2}x + 2$ **120.** Graph: $2x + 3y = 9$

CHALLENGE PROBLEMS

121. What polynomial must be added to $2x^2 - x + 3$ so that the sum is $6x^2 - 7x - 8$?

122. Is the sum of two trinomials always a trinomial? Explain why or why not.

SECTION 5.6

Multiplying Polynomials

OBJECTIVES

1. Multiply monomials.
2. Multiply a polynomial by a monomial.
3. Multiply binomials.
4. Multiply polynomials.

ARE YOU READY?

The following problems review some basic skills that are needed when multiplying polynomials.

1. Multiply: $5 \cdot 10a$

2. Multiply: $5(2x - 3)$

3. Multiply: $(4y + 8)7$

4. Multiply: $x^8 \cdot x^6$

5. Multiply: $-2(3)(5)$

6. Simplify by combining like terms: $9x^2 + 6x - 3x + 2$

We now discuss multiplying polynomials. We will begin with the simplest case—finding the product of two monomials.

1 Multiply Monomials.

To find the product of two monomials, such as $8x^2$ and $3x^4$, we use the commutative and associative properties of multiplication to reorder and regroup the factors.

$$(8x^2)(3x^4) = (8 \cdot 3)(x^2 \cdot x^4)$$ Group the coefficients together and the variables together.
$$= 24x^6$$ Multiply: $8 \cdot 3 = 24$. Simplify: $x^2 \cdot x^4 = x^{2+4} = x^6$.

This example illustrates the following rule.

Multiplying Two Monomials	To multiply two monomials, multiply the numerical factors (the coefficients) and then multiply the variable factors.

EXAMPLE 1   Multiply: **a.** $6r \cdot r$ **b.** $3t^4(-2t^5)$ **c.** $\left(\dfrac{1}{3}a^2b^3\right)(21ab^2)$ **d.** $-4y^5z^2(2y^3z^3)(3yz)$

Strategy   We will multiply the numerical factors and then multiply the variable factors.

Why   The commutative and associative properties of multiplication enable us to reorder and regroup the factors.

Solution   In each case, we will *multiply* the coefficients and *add* the exponents of the like bases.

a. $6r \cdot r = 6r^2$ Recall that $r = 1r$. Think: $6 \cdot 1 = 6$ and $r \cdot r = r^2$.

b. $(3t^4)(-2t^5) = -6t^9$ Think: $3(-2) = -6$ and $t^4 \cdot t^5 = t^{4+5} = t^9$.

c. $\left(\dfrac{1}{3}a^2b^3\right)(21ab^2) = 7a^3b^5$ Think: $\dfrac{1}{3} \cdot 21 = \dfrac{21}{3} = 7$, $a^2 \cdot a = a^3$, and $b^3 \cdot b^2 = b^5$.

d. $-4y^5z^2(2y^3z^3)(3yz) = -24y^9z^6$ Think: $-4(2)(3) = -24$, $y^5 \cdot y^3 \cdot y = y^9$, and $z^2 \cdot z^3 \cdot z = z^6$.

> **Caution**
>
> Only like terms can be added or subtracted.
> Like terms:
> $6r + r = 7r$
> Unlike terms:
> $3t^4 - 2t^5$ Does not simplify
>
> However, like and unlike terms may be multiplied.
> $6r \cdot r = 6r^2$
> $(3t^4)(-2t^5) = -6t^9$

Self Check 1   Multiply: **a.** $18t \cdot t$ **b.** $-10d^8(-6d^3)$
c. $(16y^{12})\left(\dfrac{1}{4}y^2\right)$ **d.** $(5a^3b^3)(-6a^3b^4)(ab)$

Now Try ▶   Problems 13, 19, and 21

2 Multiply a Polynomial by a Monomial.

We can use the distributive property to find the product of a monomial and a binomial such as $5x$ and $2x + 4$:

$$5x(2x + 4) = 5x(2x) + 5x(4)$$ Read as "5x times the quantity of 2x plus 4."
Distribute the multiplication by 5x.

$$= 10x^2 + 20x$$ Multiply the monomials.

This example suggests the following rule.

Multiplying Polynomials by Monomials   To multiply a monomial and a polynomial, multiply each term of the polynomial by the monomial.

EXAMPLE 2   Multiply: **a.** $3n^6(16n^{15} + n^{10})$ **b.** $3a^2(3a^2 - 5a + 2)$
c. $-2xz^3(6x^3z + x^2z^2 - xz^3 + 7z^4)$ **d.** $(-m^4 - 2.5)(4.1m^3)$

Strategy   To find each product, we will multiply each term of the polynomial by the monomial.

Why   We use the distributive property to multiply a monomial and a polynomial.

Solution   **a.** Multiply each term of the binomial $16n^{15} + n^{10}$ by the monomial $3n^6$.

$$3n^6(16n^{15} + n^{10}) = 3n^6(16n^{15}) + 3n^6(n^{10})$$ Distribute the multiplication by $3n^6$.
$$= 48n^{21} + 3n^{16}$$ Multiply the monomials.

b. Multiply each term of the trinomial $3a^2 - 5a + 2$ by the monomial $3a^2$.

$$3a^2(3a^2 - 5a + 2)$$
$$= 3a^2(3a^2) + 3a^2(-5a) + 3a^2(2)$$ Distribute the multiplication by $3a^2$.
$$= 9a^4 - 15a^3 + 6a^2$$ Multiply the monomials.

The rectangle below can be used to picture polynomial multiplication. The total area is $x(x + 2)$ and the sum of the two smaller areas is $x^2 + 2x$. Thus,

$$x(x + 2) = x^2 + 2x$$

	x^2	$2x$
x		

x 2

$x + 2$

c. Multiply each term of $6x^3z + x^2z^2 - xz^3 + 7z^4$ by the monomial $-2xz^3$.

$$-2xz^3(6x^3z + x^2z^2 - xz^3 + 7z^4)$$
$$= -2xz^3(6x^3z) - 2xz^3(x^2z^2) - 2xz^3(-xz^3) - 2xz^3(7z^4)$$
$$= -12x^4z^4 - 2x^3z^5 + 2x^2z^6 - 14xz^7 \qquad \text{Multiply the monomials.}$$

d. Multiply each term of the binomial $-m^4 - 2.5$ by the monomial $4.1m^3$.

$$(-m^4 - 2.5)(4.1m^3) = -m^4(4.1m^3) - 2.5(4.1m^3) \qquad \text{Distribute the multiplication by } 4.1m^3.$$

$$= -4.1m^7 - 10.25m^3 \qquad \text{Multiply the monomials.}$$

Self Check 2   Multiply: **a.** $22x(10x^5 + x^4)$
       b. $5c^2(4c^3 - 9c - 8)$
       c. $-s^2t^2(-s^4t^2 + s^3t^3 - s^2t^4 + 7s)$
       d. $(w^7 - 2w)6w^5$

Now Try   Problems 27, 33, and 35

EXAMPLE 3   Find a polynomial that represents the area of the parallelogram.

Strategy   We will multiply the length of the base of the parallelogram by its height.

Why   The area of a parallelogram is equal to the product of the length of its base and its height.

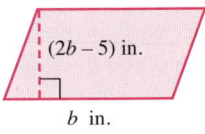
$(2b - 5)$ in.

b in.

Solution

A **parallelogram** is a four-sided figure whose opposite sides are parallel.

Area = **base · height**   This is the formula for the area of a parallelogram.

$$= b(2b - 5) \qquad b \text{ is the length of the base. Substitute } 2b - 5 \text{ for the height.}$$

$$= 2b^2 - 5b \qquad \text{Distribute the multiplication by } b.$$

The area of the parallelogram is $(2b^2 - 5b)$ square inches, which can be written as $(2b^2 - 5b)$ in.^{2}.

Self Check 3   Find a polynomial that represents the area of a rectangle with length n^3 meters and width $(3n^2 + 2n - 9)$ meters.

Now Try   Problem 37

---

3   Multiply Binomials.

The distributive property also can be used to multiply binomials. For example, to multiply $2a + 4$ and $3a + 5$, we think of $2a + 4$ as a single quantity and distribute it over each term of $3a + 5$.

$$(2a + 4)(3a + 5) = (2a + 4)3a + (2a + 4)5 \qquad \text{Read as "the quantity of } 2a + 4 \text{ times the quantity of } 3a + 5."}$$

$$= (2a + 4)3a + (2a + 4)5$$
$$= (2a)3a + (4)3a + (2a)5 + (4)5 \qquad \text{Distribute the multiplication by } 3a \text{ and by } 5.$$

$$= 6a^2 + 12a + 10a + 20 \qquad \text{Multiply the monomials.}$$

$$= 6a^2 + 22a + 20 \qquad \text{Combine like terms.}$$

In the third line of the solution, notice that each term of $3a + 5$ has been multiplied by each term of $2a + 4$. This example suggests the following rule.

Multiplying Two Binomials	To multiply two binomials, multiply each term of one binomial by each term of the other binomial, and then combine like terms.

EXAMPLE 4   Multiply: $(5x - 8)(x + 1)$

Strategy   To find the product, we will multiply $x + 1$ by $5x$ and by -8.

Why   To multiply two binomials, each term of one binomial must be multiplied by each term of the other binomial.

Solution

Notation

We will write the answers to polynomial multiplication in descending powers.

$$(5x - 8)(x + 1) = 5x(x + 1) - 8(x + 1)$$ Read as "the quantity of $5x - 8$ times the quantity of $x + 1$." Multiply $x + 1$ by $5x$ and multiply $x + 1$ by -8.

$$= 5x^2 + 5x - 8x - 8$$ Distribute the multiplication by $5x$. Distribute the multiplication by -8.

$$= 5x^2 - 3x - 8$$ Combine like terms.

Self Check 4   Multiply: $(9y + 3)(y - 4)$

Now Try ▶ Problem 45

The Language of Algebra

An **acronym** is an abbreviation of several words in such a way that the abbreviation itself forms a word. The *acronym* FOIL helps us remember the order to follow when multiplying two binomials: First, Outer, Inner, Last.

We can use a shortcut method, called the **FOIL method,** to multiply binomials. FOIL is an acronym for **F**irst terms, **O**uter terms, **I**nner terms, **L**ast terms. The FOIL method is a form of the distributive property. To use the FOIL method to multiply $2a + 4$ by $3a + 5$, we

1. Multiply the **F**irst terms $2a$ and $3a$ to obtain $6a^2$,
2. Multiply the **O**uter terms $2a$ and 5 to obtain $10a$,
3. Multiply the **I**nner terms 4 and $3a$ to obtain $12a$, and
4. Multiply the **L**ast terms 4 and 5 to obtain 20.

Then we simplify the resulting polynomial by combining like terms, if possible.

$$(2a + 4)(3a + 5) = 2a(3a) + 2a(5) + 4(3a) + 4(5)$$ Distribute $2a$ over $3a + 5$. Distribute 4 over $3a + 5$.

$$= 6a^2 + 10a + 12a + 20$$ Multiply the monomials.

$$= 6a^2 + 22a + 20$$ Combine like terms: $10a + 12a = 22a$.

EXAMPLE 5   Multiply: **a.** $(x + 5)(x + 7)$ **b.** $(3x + 4)(2x - 3)$

c. $\left(2r - \dfrac{1}{2}\right)\left(2r + \dfrac{5}{2}\right)$ **d.** $(3a^2 - 7b)(a^2 - b)$

Strategy   We will use the FOIL method.

Why   In each case we are to find the product of two binomials, and the FOIL method is a shortcut for multiplying two binomials.

Solution   **a.**

$$(x + 5)(x + 7) = x(x) + x(7) + 5(x) + 5(7)$$
$$= x^2 + 7x + 5x + 35 \qquad \text{Multiply the monomials.}$$
$$= x^2 + 12x + 35 \qquad \text{Combine like terms.}$$

b.

$$(3x + 4)(2x - 3) = 3x(2x) + 3x(-3) + 4(2x) + 4(-3)$$
$$= 6x^2 - 9x + 8x - 12 \qquad \text{Multiply the monomials.}$$
$$= 6x^2 - x - 12 \qquad \text{Combine like terms.}$$

c.

$$\left(2r - \frac{1}{2}\right)\left(2r + \frac{5}{2}\right) = 2r(2r) + 2r\left(\frac{5}{2}\right) - \frac{1}{2}(2r) - \frac{1}{2}\left(\frac{5}{2}\right)$$
$$= 4r^2 + 5r - r - \frac{5}{4} \qquad \text{Multiply the monomials.}$$
$$= 4r^2 + 4r - \frac{5}{4} \qquad \text{Combine like terms. We cannot clear the}$$
$$\text{fraction because we began with an}$$
$$\text{expression, not an equation.}$$

d.

$$(3a^2 - 7b)(a^2 - b) = 3a^2(a^2) + 3a^2(-b) - 7b(a^2) - 7b(-b)$$
$$= 3a^4 - 3a^2b - 7a^2b + 7b^2 \qquad \text{Multiply the monomials.}$$
$$= 3a^4 - 10a^2b + 7b^2 \qquad \text{Combine like terms.}$$

Self Check 5   Multiply: **a.** $(y + 3)(y + 1)$
b. $(2a - 1)(3a + 2)$
c. $\left(4x - \frac{1}{2}\right)\left(4x + \frac{3}{4}\right)$
d. $(5y^3 - 2b)(2y^3 - 7b)$

Now Try ▶   Problems 47, 51, and 55

Success Tip (sidebar)

The area of the large rectangle is given by $(x + 5)(x + 7)$. The sum of the areas of the smaller rectangles is $x^2 + 7x + 5x + 35$ or $x^2 + 12x + 35$. Thus,

$$(x + 5)(x + 7) = x^2 + 12x + 35$$

	5	$5x$	35
$x + 5$	x	x^2	$7x$
		x	7

$x + 7$

---

4 Multiply Polynomials.

To develop a general rule for multiplying any two polynomials, we will find the product of $2x + 3$ and $3x^2 + 3x + 5$. In the solution, the distributive property is used four times.

$$(2x + 3)(3x^2 + 3x + 5) = (2x + 3)3x^2 + (2x + 3)3x + (2x + 3)5 \qquad \text{Distribute } (2x + 3).$$
$$= (2x + 3)3x^2 + (2x + 3)3x + (2x + 3)5$$
$$= (2x)3x^2 + (3)3x^2 + (2x)3x + (3)3x + (2x)5 + (3)5 \qquad \text{Distribute.}$$
$$= 6x^3 + 9x^2 + 6x^2 + 9x + 10x + 15 \qquad \text{Multiply the monomials.}$$
$$= 6x^3 + 15x^2 + 19x + 15 \qquad \text{Combine like terms.}$$

In the third line of the solution, note that each term of $3x^2 + 3x + 5$ has been multiplied by each term of $2x + 3$. This example suggests the following rule.

Success Tip (sidebar)

In this section, you will see that every polynomial multiplication is basically a series of monomial multiplications. Then, if necessary, we combine like terms.

Multiplying Two Polynomials	To multiply two polynomials, multiply each term of one polynomial by each term of the other polynomial, and then combine like terms.

EXAMPLE 6   Multiply: $(7y + 3)(6y^2 - 8y + 1)$

Strategy   We will multiply each term of the trinomial, $6y^2 - 8y + 1$, by each term of the binomial, $7y + 3$.

Why   To multiply two polynomials, we must multiply each term of one polynomial by each term of the other polynomial.

Solution

$(\mathbf{7y} + \mathbf{3})(6y^2 - 8y + 1)$ Read as "the quantity of 7y + 3 times the quantity of $6y^2 - 8y + 1$."

$= \mathbf{7y}(6y^2) + \mathbf{7y}(-8y) + \mathbf{7y}(1) + \mathbf{3}(6y^2) + \mathbf{3}(-8y) + \mathbf{3}(1)$

$= 42y^3 - 56y^2 + 7y + 18y^2 - 24y + 3$ Multiply the monomials.

$= 42y^3 - 38y^2 - 17y + 3$ Combine like terms.

The FOIL method cannot be applied here—only to products of two binomials.

Self Check 6   Multiply: $(3a^2 - 1)(2a^4 - a^2 - a)$

Now Try ▶ Problem 59

It is often convenient to multiply polynomials using a vertical form similar to that used to multiply whole numbers.

EXAMPLE 7   Multiply using vertical form: **a.** $(3a^2 - 4a + 7)(2a + 5)$
b. $(6y^3 - 5y + 4)(-4y^2 - 3)$

Strategy   First, we will write one polynomial underneath the other and draw a horizontal line beneath them. Then, we will multiply each term of the upper polynomial by each term of the lower polynomial.

Why   *Vertical form* means to use an approach similar to that used in arithmetic to multiply two numbers.

Solution   **a.** Multiply:

$$\begin{array}{r} 3a^2 - 4a + 7 \\ 2a + 5 \\ \hline 15a^2 - 20a + 35 \\ 6a^3 - 8a^2 + 14a \quad\quad\quad \\ \hline 6a^3 + 7a^2 - 6a + 35 \end{array}$$

Multiply $3a^2 - 4a + 7$ by 5.
Multiply $3a^2 - 4a + 7$ by 2a. Line up like terms.
In each column, combine like terms.

Success Tip

Multiplying two polynomials in vertical form is much like multiplying two numbers in arithmetic.

$$\begin{array}{r} 347 \\ \times\ 25 \\ \hline 1,735 \\ +694\quad \\ \hline 8,675 \end{array}$$

b. With this method, it is often necessary to leave a space for a missing term to align like terms vertically.

Multiply:

$$\begin{array}{r} 6y^3 - 5y + 4 \\ -4y^2 - 3 \\ \hline -18y^3 \quad\quad + 15y - 12 \\ -24y^5 + 20y^3 - 16y^2 \quad\quad\quad\quad \\ \hline -24y^5 + 2y^3 - 16y^2 + 15y - 12 \end{array}$$

Multiply $6y^3 - 5y + 4$ by -3.
Multiply $6y^3 - 5y + 4$ by $-4y^2$.
Leave a space for any missing powers of y.
In each column, combine like terms.

The Language of Algebra

The pair of polynomials written below the first horizontal line, $-18y^3 + 15y - 12$ and $-24y^5 + 20y^3 - 16y^2$, are called **partial products**.

Self Check 7   Multiply using vertical form:
   a. $(3x + 2)(2x^2 - 4x + 5)$
   b. $(-2x^2 + 3)(2x^2 - 4x - 1)$

Now Try ▶ Problem 65

Multiplying Three Polynomials   ▼   To multiply three polynomials, multiply any two of them, and then multiply that result by the third polynomial.

EXAMPLE 8   Multiply: $-3a(4a + 1)(a - 7)$

Strategy   We will find the product of $4a + 1$ and $a - 7$ and then multiply that result by $-3a$.

Why   It is wise to perform the most difficult multiplication first. (In this case, that would be the product of the two binomials). Save the simpler multiplication by $-3a$ for last.

Solution

$$-3a(4a + 1)(a - 7) = -3a(4a^2 - 28a + a - 7)$$ Multiply the two binomials.

$$= -3a(4a^2 - 27a - 7)$$ Combine like terms within the parentheses: $-28a + a = -27a$.

$$= -12a^3 + 81a^2 + 21a$$ Distribute the multiplication by $-3a$.

Self Check 8   Multiply: $-2y(y + 3)(3y - 2)$

Now Try ▶ Problem 69

SECTION 5.6 ▶ STUDY SET

VOCABULARY

Fill in the blanks.

1. $(2x^3)(3x^4)$ is the product of two _____ and $(2a - 4)(3a + 5)$ is the product of two _____.

2. We read $(x + 7)(2x - 3)$ as "the _____ of $x + 7$ _____ the quantity of $2x - 3$."

3. In the acronym FOIL, F stands for _____ terms, O for _____ terms, I for _____ terms, and L for _____ terms.

4. $(2a - 4)(3a^2 + 5a - 1)$ is the product of a _____ and a _____.

CONCEPTS

Fill in the blanks.

5. **a.** To multiply two polynomials, multiply _____ term of one polynomial by _____ term of the other polynomial, and then combine like terms.

   b. When multiplying three polynomials, we begin by multiplying _____ two of them, and then we multiply that result by the _____ polynomial.

6. Label each arrow using one of the letters F, O, I, or L. Then fill in the blanks.

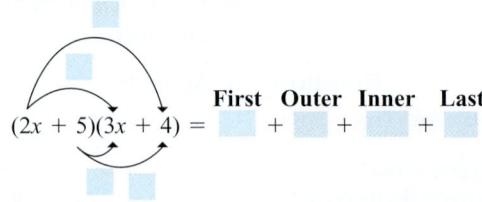

$$(2x + 5)(3x + 4) = \boxed{} + \boxed{} + \boxed{} + \boxed{}$$

First   Outer   Inner   Last

7. Simplify each polynomial by combining like terms.
   a. $6x^2 - 8x + 9x - 12$
   b. $5x^4 + 3ax^2 + 5ax^2 + 3a^2$

8. $(3a)(2a^2)$ can be classified as a monomial · monomial. Classify the following products by identifying the types of polynomial factors.
   a. $6x(x - 7)$
   b. $(9a + 1)(5a - 3)$
   c. $(c - d)(c^2 - c + d)$
   d. $6m(m^2 + 1)(m^2 - 1)$

NOTATION

Complete each solution.

9. $(9n^3)(8n^2) = (9 \cdot \boxed{})(\boxed{} \cdot n^2) = \boxed{}$

10. $7x(3x^2 - 2x + 5) = \boxed{}(3x^2) - \boxed{}(2x) + \boxed{}(5)$
$\qquad\qquad\qquad = \boxed{} - 14x^2 + 35x$

11. $(2x + 5)(3x - 2) = 2x(3x) - \boxed{}(2) + \boxed{}(3x) - \boxed{}(2)$
$\qquad\qquad\qquad = 6x^2 - \boxed{} + \boxed{} - 10$
$\qquad\qquad\qquad = 6x^2 + \boxed{} - 10$

12. $\qquad 3x^2 + 4x - 2$
$\qquad\qquad\quad \underline{2x + 3}$
$\qquad\quad \boxed{} + 12x - 6$
$\qquad \underline{6x^3 + 8x^2 - 4x \qquad\qquad}$
$\qquad\quad \boxed{} + 17x^2 + \boxed{} - 6$

GUIDED PRACTICE

Multiply. See Example 1.

13. $5m \cdot m$

14. $4s \cdot s$

15. $(3x^2)(4x^3)$

16. $(-2a^3)(11a^2)$

17. $(1.2c^3)(5c^3)$

18. $(2.5h^4)(2h^4)$

19. $(3b^2)(-2b)(4b^3)$

20. $(3y)(7y^2)(-y^4)$

21. $(2x^2y^3)(4x^3y^2)$

22. $(-5x^3y^6)(2x^2y^2)$

23. $(8a^5)\left(-\dfrac{1}{4}a^6\right)$

24. $\left(-\dfrac{2}{3}x^6\right)(9x^3)$

Multiply. See Example 2.

25. $3x(x + 4)$

26. $3a(a + 2)$

27. $-4t(t^2 - 7)$

28. $-6s(s^2 - 3)$

29. $-2x^3(3x^2 - x + 1)$

30. $-4b^3(2b^2 - 2b + 2)$

31. $\dfrac{5}{8}t^2(t^6 + 8t^2)$

32. $\dfrac{4}{9}a^2(9a^3 + a^2)$

33. $-4x^2z(3x^2 + z^2 + xz - 1)$

34. $-3x^2y(x^2 + y^2 + xy - 1)$

35. $(x^2 - 12x)(6x^{12})$

36. $(w^9 - 11w)(2w^7)$

Find a polynomial that represents the area of the parallelogram or rectangle. See Example 3.

37.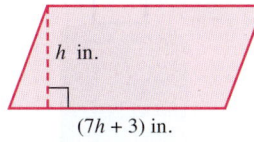
h in.
$(7h + 3)$ in.

38.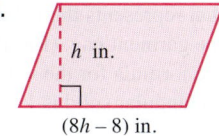
h in.
$(8h - 8)$ in.

39.
w ft
$(4w - 2)$ ft

40.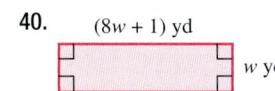
$(8w + 1)$ yd
w yd

Multiply. See Examples 4 and 5.

41. $(y + 3)(y + 5)$

42. $(a + 4)(a + 5)$

43. $(m + 6)(m - 9)$

44. $(n + 8)(n - 10)$

45. $(4y - 5)(y + 7)$

46. $(3x - 4)(x + 5)$

47. $(2x - 3)(6x - 5)$

48. $(5x - 3)(2x - 3)$

49. $(3.8y - 1)(2y - 1)$

50. $(2.6x - 3)(2x - 1)$

51. $\left(6m - \dfrac{2}{3}\right)\left(3m - \dfrac{4}{3}\right)$

52. $\left(8t - \dfrac{1}{2}\right)\left(4t - \dfrac{5}{2}\right)$

53. $(t^2 - 3)(t^2 - 4)$

54. $(s^3 - 6)(s^3 - 8)$

55. $(3a - 2b)(4a + b)$

56. $(2t + 3s)(3t - s)$

Multiply. See Example 6.

57. $(x + 2)(x^2 - 2x + 3)$

58. $(x - 5)(x^2 + 2x - 3)$

59. $(4t + 3)(t^2 + 2t + 3)$

60. $(3x + 1)(2x^2 - 3x + 1)$

61. $(x^2 + 6x + 7)(2x - 5)$

62. $(y^2 - 2y + 1)(4y + 8)$

63. $(r^2 - r + 3)(r^2 - 4r - 5)$

64. $(w^2 + w - 9)(w^2 - w + 3)$

Multiply using vertical form. See Example 7.

65. $\quad x^2 - 2x + 1$
$\quad\;\; \underline{x + 2}$

66. $\quad 5r^2 + r + 6$
$\quad\;\; \underline{2r - 1}$

67. $\quad 4x^2 + 3x - 4$
$\quad\;\; \underline{3x + 2}$

68. $\quad x^2 - x + 1$
$\quad\;\; \underline{x + 1}$

Multiply. See Example 8.

69. $4x(2x + 1)(x - 2)$

70. $5a(3a - 2)(2a + 3)$

71. $-3a(a + b)(a - b)$

72. $-2r(r + s)(r + s)$

73. $(-2a^2)(-3a^3)(3a - 2)$

74. $(3x)(-2x^2)(x + 4)$

75. $(x - 4)(x + 1)(x - 3)$

76. $(x + 6)(x - 2)(x - 4)$

TRY IT YOURSELF

Multiply.

77. $(5x - 2)(6x - 1)$

78. $(8x - 1)(3x - 7)$

79. $(3x^2 + 4x - 7)(2x^2)$

80. $(2y^2 - 7y - 8)(3y^3)$

81. $2(t + 4)(t - 3)$

82. $4(x + 7)(x - 6)$

83. $2a^2 + 3a + 1$
    $\underline{3a^2 - 2a + 4}$

84. $3y^2 + 2y - 4$
    $\underline{2y^2 - 4y + 3}$

85. $(t + 2s)(9t - 3s)$

86. $(4t - u)(3t + u)$

87. $\left(\dfrac{1}{2}a\right)(4a^4)(a^5)$

88. $(12b)\left(\dfrac{7}{6}b\right)(b^4)$

89. $\left(4a - \dfrac{5}{4}r\right)\left(4a + \dfrac{3}{4}r\right)$

90. $\left(5c - \dfrac{2}{5}t\right)\left(10c + \dfrac{1}{5}t\right)$

91. $(a + b)(a + b)$

92. $(m + n)(m + n)$

93. $(x + 6)(x^3 + 5x^2 - 4x - 4)$
94. $(x - 8)(x^3 - 4x^2 - 2x - 2)$
95. $9x^2(x^2 - 2x + 6)$

96. $4y^2(y^2 + 5y - 10)$

97. $4y(y + 3)(y + 7)$

98. $2t(t + 8)(t + 10)$

99. $0.3p^5(0.4p^4 - 6p^2)$

100. $0.5u^5(0.4u^6 - 0.5u^3)$

101. $8.2pq(2pq - 3p + 5q)$

102. $5.3ab(2ab + 6a - 3b)$

103. $(-3x + y)(x^2 - 8xy + 16y^2)$
104. $(3x - y)(x^2 + 3xy - y^2)$

Look Alikes . . .

Perform the indicated operations to simplify each expression, if possible.

105. **a.** $(x - 2) + (x^2 + 2x + 4)$ **b.** $(x - 2)(x^2 + 2x + 4)$

106. **a.** $(a + 3) + (a^2 - 3a + 9)$ **b.** $(a + 3)(a^2 - 3a + 9)$

107. **a.** $(6x^2z^5) - (-3xz^3)$ **b.** $(6x^2z^5)(-3xz^3)$

108. **a.** $(-5r^4t^2) - (2r^2t)$ **b.** $(-5r^4t^2)(2r^2t)$

109. **a.** $(2x^2 - x) - (3x^2 - 3x)$ **b.** $(2x^2 - x)(3x^2 - 3x)$

110. **a.** $(4.9a - b) - (2a + b)$ **b.** $(4.9a - b)(2a + b)$

111. **a.** $3a + (4a - 1) + (6a + 2)$
     b. $3a(4a - 1)(6a + 2)$

112. **a.** $\left(\dfrac{1}{2}c - 3d\right) + \left(\dfrac{3}{4}c + d\right)$ **b.** $\left(\dfrac{1}{2}c - 3d\right)\left(\dfrac{3}{4}c + d\right)$

APPLICATIONS

113. **Stamps.** Find a polynomial that represents the area of the stamp.

114. **Parking.** Find a polynomial that represents the total area of the van-accessible parking space and its access aisle.

$(x + 10)$ ft   2x ft

115. **Sunglasses.** An ellipse is an oval-shaped curve. The area of an ellipse is approximately $0.785lw$, where l is its length and w is its width. Find a polynomial that represents the approximate area of one of the elliptical-shaped lenses of the sunglasses.

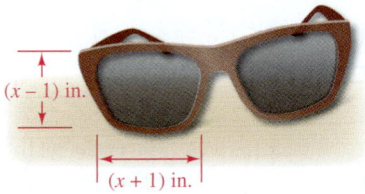

$(x - 1)$ in.

$(x + 1)$ in.

116. **Gardening.** Refer to the illustration below.

    a. Find the area of the region planted with corn, tomatoes, beans, and carrots. Add your answers to find the total area of the garden.

    b. Find the length and width of the garden. Multiply your answers to find its area.

    c. How do the answers from parts (a) and (b) for the area of the garden compare?

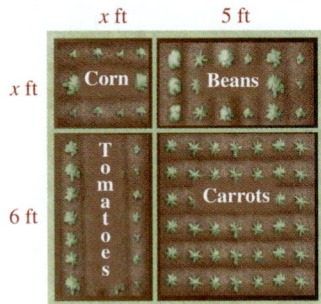

	x ft	5 ft
x ft	Corn	Beans
6 ft	Tomatoes	Carrots

117. **Luggage.** Find a polynomial that represents the volume of the garment bag. (Recall that the formula for the volume of a rectangular solid is $V = lwh$.)

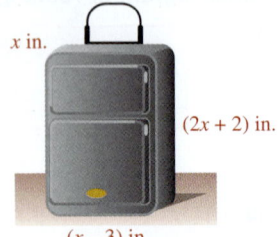

x in.

$(2x + 2)$ in.

$(x - 3)$ in.

118. Baseball. Find a polynomial that represents the volume within the batting cage.

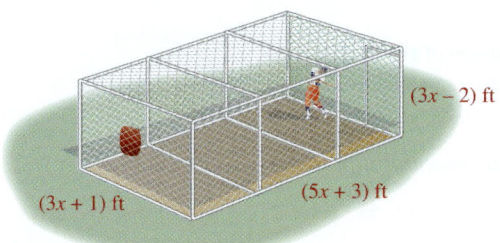

$(3x - 2)$ ft
$(3x + 1)$ ft
$(5x + 3)$ ft

WRITING

119. Is the product of a monomial and a monomial always a monomial? Explain.

120. Explain this diagram.

$$(5x + 6)(7x - 1)$$

121. Explain why the FOIL method cannot be used to find $(3x + 2)(4x^2 - x + 10)$.

122. Explain the error: $(x + 3)(x - 2) = x^2 - 6$

123. Explain why the vertical form used in algebra to multiply $2x^2 + 3x + 1$ and $3x + 2$ is similar to the vertical form used in arithmetic to multiply 231 and 32.

124. Would the OLIF method give the same result as the FOIL method when multiplying two binomials? Explain why or why not.

REVIEW

125. What is the slope of
   a. Line 1?   c. Line 3?
   b. Line 2?   d. the x-axis?

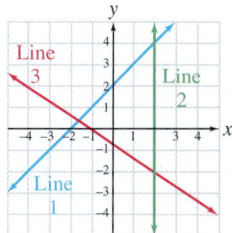

126. a. What is the y-intercept of Line 1?
   b. What is the x-intercept of Line 1?

CHALLENGE PROBLEMS

127. a. Find each of the following products.
   i. $(x - 1)(x + 1)$
   ii. $(x - 1)(x^2 + x + 1)$
   iii. $(x - 1)(x^3 + x^2 + x + 1)$

   b. Write a product of two polynomials such that the result is $x^5 - 1$.

128. Solve: $(y - 1)(y + 6) = (y - 3)(y - 2) + 8$

SECTION 5.7

Special Products

OBJECTIVES

1 Square a binomial.

2 Multiply the sum and difference of the same two terms.

3 Find higher powers of binomials.

4 Simplify expressions containing polynomial multiplication.

ARE YOU READY?

The following problems review some basic skills that are needed when finding special products.

1. Simplify by combining like terms: $x^2 + 2x + 2x + 4$

2. In each expression, identify the base and the exponent:
   a. $(x + 4)^2$
   b. $(x - 1)^3$

3. Multiply: $(a + 5)(a + 5)$ **4.** Multiply: $(a + 5)(a - 5)$

Certain products of binomials, called **special products,** occur so often that it is worthwhile to learn their forms.

1 Square a Binomial.

To develop a rule to find the *square of a binomial sum,* we consider $(x + y)^2$. We can use the definition of exponent and the procedure for multiplying two binomials to find the product.

$$(x + y)^2 = (x + y)(x + y) \quad \text{In } (x + y)^2 \text{, the base is } (x + y) \text{ and the exponent is 2.}$$
$$= x^2 + xy + xy + y^2 \quad \text{Multiply the binomials.}$$
$$= x^2 + 2xy + y^2 \quad \text{Combine like terms: } xy + xy = 1xy + 1xy = 2xy.$$

Note that the terms of the resulting trinomial are related to the terms of the binomial that was squared.

$$(x + y)^2 = x^2 + 2xy + y^2$$

└ The square of the second term, y.

└ Twice the product of the first and second terms, x and y.

└ The square of the first term, x.

To develop a rule to find the *square of a binomial difference,* we consider $(x - y)^2$.

$$(x - y)^2 = (x - y)(x - y) \qquad \text{In } (x - y)^2, \text{ the base is } (x - y) \text{ and the exponent is 2.}$$
$$= x^2 - xy - xy + y^2 \qquad \text{Multiply the binomials.}$$
$$= x^2 - 2xy + y^2 \qquad \text{Combine like terms: } -xy - xy = -1xy - 1xy = -2xy.$$

Again, the terms of the resulting trinomial are related to the terms of the binomial that was squared.

$$(x - y)^2 = x^2 - 2xy + y^2$$

└ The square of the second term, −y.

└ Twice the product of the first and second terms, x and −y.

└ The square of the first term, x.

The observations from these two examples illustrate the following **special-product rules.**

Squaring a Binomial

The **square of a binomial** is a trinomial, such that:

- Its first term is the square of the first term of the binomial.
- Its last term is the square of the second term of the binomial.
- Its middle term is twice the product of both terms of the binomial.

$$(A + B)^2 = A^2 + 2AB + B^2 \qquad (A - B)^2 = A^2 - 2AB + B^2$$

EXAMPLE 1 Find each square: **a.** $(t + 9)^2$ **b.** $(8a - 5)^2$ **c.** $(d + 0.5)^2$ **d.** $\left(c^3 - \dfrac{7}{2}d\right)^2$

Strategy To find each square of a binomial, we will use one of the special-product rules.

Why This approach is faster than using the FOIL method.

Solution **a.** $(t + 9)^2$ is the square of a binomial sum. The first term is t and the second term is 9.

$$(t + 9)^2 = \underbrace{t^2}_{\substack{\text{The square of} \\ \text{the first term, } t.}} + \underbrace{2(t)(9)}_{\substack{\text{Twice the product} \\ \text{of both terms.}}} + \underbrace{9^2}_{\substack{\text{The square of the} \\ \text{second term, 9.}}}$$

$$= t^2 + 18t + 81$$

b. $(8a - 5)^2$ is the square of a binomial difference. The first term is $8a$ and the second term is −5.

$$(8a - 5)^2 = \underbrace{(8a)^2}_{\substack{\text{The square of} \\ \text{the first term, } 8a.}} + \underbrace{2(8a)(-5)}_{\substack{\text{Twice the product} \\ \text{of both terms.}}} + \underbrace{(-5)^2}_{\substack{\text{The square of the} \\ \text{second term, −5.}}}$$

$$= 64a^2 - 80a + 25 \qquad \text{Use the power of a product rule: } (8a)^2 = 8^2 a^2 = 64a^2.$$

c. $(d + 0.5)^2$ is the square of a binomial sum. The first term is d and the second term is 0.5.

$$(d + 0.5)^2 = \underbrace{(d)^2}_{\substack{\text{The square of} \\ \text{the first term, } d.}} + \underbrace{2(d)(0.5)}_{\substack{\text{Twice the product} \\ \text{of both terms.}}} + \underbrace{(0.5)^2}_{\substack{\text{The square of the} \\ \text{second term, 0.5.}}}$$

$$= d^2 + d + 0.25$$

d. $\left(c^3 - \dfrac{7}{2}d\right)^2$ is the square of a binomial difference. The first term is c^3 and the second term is $-\dfrac{7}{2}d$.

$$\left(c^3 - \frac{7}{2}d\right)^2 = \underbrace{(c^3)^2}_{\substack{\text{The square of} \\ \text{the first term, } c^3.}} + \underbrace{2(c^3)\left(-\frac{7}{2}d\right)}_{\substack{\text{Twice the product} \\ \text{of both terms.}}} + \underbrace{\left(-\frac{7}{2}d\right)^2}_{\substack{\text{The square of the} \\ \text{second term, } -\frac{7}{2}d.}}$$

$$= c^6 - 7c^3d + \frac{49}{4}d^2 \quad \text{Use rules for exponents to find } (c^3)^2 \text{ and } \left(-\frac{7}{2}d\right)^2.$$

> **Self Check 1** Find each square: **a.** $(r + 6)^2$
> **b.** $(7g - 2)^2$ **c.** $(v + 0.8)^2$
> **d.** $\left(w^4 - \dfrac{3}{2}y\right)^2$
>
> **Now Try** ▶ Problems 9, 15, and 23

> ### Caution
>
> The square of a binomial is a *trinomial*. A common error when squaring a binomial is to forget the middle term of the product.
>
> $$(8a - 5)^2 \neq 64a^2 + 25$$
>
> Missing $-80a$

---

2 Multiply the Sum and Difference of the Same Two Terms.

A final special product that occurs often has the form $(A + B)(A - B)$. In these products, one binomial is the sum of two terms and the other binomial is the difference of the same two terms. To develop a rule to find such products, consider the following multiplication:

$$(x + y)(x - y) = x^2 - xy + xy - y^2 \quad \text{Multiply the binomials.}$$
$$= x^2 - y^2 \quad \text{Combine like terms: } -xy + xy = 0.$$

Note that when we combined like terms, we added opposites. This will always be the case for products of this type; the sum of the outer and inner products will be 0. The first and last products will be squares.

$$(x + y)(x - y) = x^2 - y^2$$

The square of the first term, x.
The square of the second term, y.

These observations suggest a third **special-product rule**.

> ### Success Tip
>
> We can use the FOIL method to find each of the special products discussed in this section. However, these forms occur so often, it is worthwhile to learn the special-product rules.

> **Multiplying the Sum and Difference of Two Terms**
>
> The product of the sum of two terms and difference of the same two terms is the square of the first term minus the square of the second term.
>
> $$(A + B)(A - B) = A^2 - B^2$$

EXAMPLE 2 Multiply: **a.** $(m + 2)(m - 2)$ **b.** $(3y + 4)(3y - 4)$ **c.** $\left(b - \dfrac{2}{3}\right)\left(b + \dfrac{2}{3}\right)$

d. $(t^4 - 6u)(t^4 + 6u)$

Strategy To find the product of each pair of binomials, we will use the special-product rule for the sum and difference of the same two terms.

Why This approach is faster than using the FOIL method.

Solution   **a.** $(m + 2)$ and $(m - 2)$ are the sum and difference of the same two terms, m and 2.

$$(m + 2)(m - 2) = \underbrace{m^2}_{\substack{\text{The square of the} \\ \text{first term, } m.}} - \underbrace{2^2}_{\substack{\text{The square of the} \\ \text{second term, 2.}}}$$

$$= m^2 - 4$$

b. $(3y + 4)$ and $(3y - 4)$ are the sum and difference of the same two terms, $3y$ and 4.

$$(3y + 4)(3y - 4) = \underbrace{(3y)^2}_{\substack{\text{The square of the} \\ \text{first term, } 3y.}} - \underbrace{4^2}_{\substack{\text{The square of the} \\ \text{second term, 4.}}}$$

$$= 9y^2 - 16$$

c. By the commutative property of multiplication, the special-product rule can be written with the factor containing the $-$ symbol first: $(A - B)(A + B) = A^2 - B^2$. Since $\left(b - \frac{2}{3}\right)$ and $\left(b + \frac{2}{3}\right)$ are the difference and sum of the same two terms, b and $\frac{2}{3}$, we have

$$\left(b - \frac{2}{3}\right)\left(b + \frac{2}{3}\right) = \underbrace{b^2}_{\substack{\text{The square of the} \\ \text{first term, } b.}} - \underbrace{\left(\frac{2}{3}\right)^2}_{\substack{\text{The square of the} \\ \text{second term, } \frac{2}{3}.}}$$

$$= b^2 - \frac{4}{9}$$

d. $(t^4 - 6u)$ and $(t^4 + 6u)$ are the difference and sum of the same two terms, t^4 and $6u$.

$$(t^4 - 6u)(t^4 + 6u) = \underbrace{(t^4)^2}_{\substack{\text{The square of the} \\ \text{first term, } t^4.}} - \underbrace{(6u)^2}_{\substack{\text{The square of the} \\ \text{second term, } 6u.}}$$

$$= t^8 - 36u^2$$

Self Check 2   Multiply: **a.** $(b + 4)(b - 4)$
b. $(5m + 9)(5m - 9)$ **c.** $\left(s - \frac{3}{4}\right)\left(s + \frac{3}{4}\right)$
d. $(c^3 + 2d)(c^3 - 2d)$

Now Try   Problems 25, 27, and 31

3 Find Higher Powers of Binomials.

When we find the third, fourth, or even higher powers of a binomial, we say that we are **expanding the binomial.** The special-product rules can be used in such cases. The result is an expression that has more terms than the original binomial.

EXAMPLE 3   Expand: $(x + 1)^3$

Strategy   We will use a special-product rule to find the third power of $x + 1$.

Why   Since $(x + 1)^3$ can be written as $(x + 1)(x + 1)^2$, we can use a special-product rule to find $(x + 1)^2$ quickly.

Solution

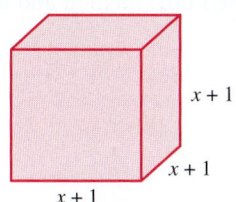

$x + 1$

$x + 1$

$x + 1$

$$(x + 1)^3 = (x + 1)(x + 1)^2 \qquad \text{Read as "the quantity of } x + 1, \text{ cubed."}$$
$$= (x + 1)(x^2 + 2x + 1) \qquad \text{Find } (x + 1)^2 \text{ using the rule for the square of a sum.}$$
$$= (x + 1)(x^2 + 2x + 1) \qquad \text{Multiply the binomial and the trinomial.}$$
$$= x(x^2) + x(2x) + x(1) + 1(x^2) + 1(2x) + 1(1) \qquad \begin{array}{l}\text{Multiply each term of} \\ x^2 + 2x + 1 \text{ by each} \\ \text{term of } x + 1.\end{array}$$
$$= x^3 + 2x^2 + x + x^2 + 2x + 1 \qquad \text{Multiply the monomials.}$$
$$= x^3 + 3x^2 + 3x + 1 \qquad \text{Combine like terms.}$$

Notation

$(x + 1)^3$ represents the volume of a cube with sides of length $x + 1$ units.

Self Check 3 Expand: $(n - 3)^3$

Now Try ▶ Problem 37

4 Simplify Expressions Containing Polynomial Multiplication.

We can use a modified version of the order of operations rule to simplify expressions that involve polynomial addition, subtraction, multiplication, and raising to a power.

Order of Operations with Polynomials

1. If possible, simplify any polynomials within parentheses by combining like terms.
2. Square (or expand) all polynomials raised to powers using the FOIL method or a special-product rule.
3. Perform all polynomial multiplications using the distributive property, the FOIL method, or a special-product rule.
4. Perform all polynomial additions and subtractions by combining like terms.

EXAMPLE 4 Simplify each expression:
a. $-8(y^2 - 2y + 3) - 4(2y^2 + y - 6)$
b. $(x + 1)(x - 2) + 3x(x + 3)$
c. $(3y - 2)^2 - (y - 5)(y + 5)$

Strategy We will follow the rules for the order of operations to simplify each expression.

Why If we don't follow the correct order of operations, we can obtain different results that are not equivalent.

Solution **a.** The two polynomials within the parentheses do not simplify further and no polynomials are raised to a power. To perform the multiplication, we will use the distributive property twice. Then we will combine like terms.

$$-8(y^2 - 2y + 3) - 4(2y^2 + y - 6) = -8y^2 + 16y - 24 - 8y^2 - 4y + 24 \qquad \text{Distribute.}$$
$$= -16y^2 + 12y \qquad \text{Add and subtract to combine like terms.}$$

b. The three polynomials within parentheses do not simplify further and no polynomials are raised to a power. To perform the multiplications, we use the FOIL method and the distributive property. Then we will combine like terms.

$$(x + 1)(x - 2) + 3x(x + 3)$$

$$= x^2 - x - 2 + 3x^2 + 9x \qquad \text{Use the FOIL method to find } (x + 1)(x - 2).$$
$$\text{Distribute the multiplication by } 3x.$$

$$= 4x^2 + 8x - 2 \qquad \text{Combine like terms.}$$

c. The three polynomials within parentheses do not simplify further. To square $3y - 2$, we use a special-product rule. To find the product of $(y - 5)(y + 5)$, we will use the special-product rule for the sum and difference of the same two terms. Then we will combine like terms.

$$(3y - 2)^2 - (y - 5)(y + 5)$$

$$= 9y^2 - 12y + 4 - (y^2 - 25) \qquad \text{Write } y^2 - 25 \text{ within parentheses}$$
$$\text{so that both terms are subtracted.}$$

$$= 9y^2 - 12y + 4 - y^2 + 25 \qquad \text{Change the sign of each term within}$$
$$(y^2 - 25) \text{ and drop the parentheses.}$$

$$= 8y^2 - 12y + 29 \qquad \text{Combine like terms.}$$

Self Check 4 Simplify each expression: **a.** $2(a^2 - 3a) + 5(a^2 + 2a)$
b. $(x - 4)(x + 6) + 5x(2x - 1)$
c. $(a + 9)(a - 9) - (2a - 4)^2$

Now Try ▶ Problems 45, 47, and 48

EXAMPLE 5 Find a polynomial that represents the area of the triangle.

Strategy We will multiply one-half, the length of the base, and the height of the triangle.

Why The area of a triangle is equal to one-half the product of the length of its base and its height.

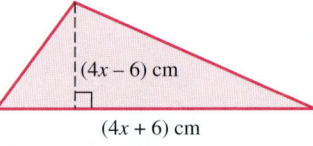
$(4x - 6)$ cm
$(4x + 6)$ cm

Solution We begin by substituting $4x + 6$ for the length of the base and $4x - 6$ for the height in the formula for the area of a triangle. It is wise to find $(4x + 6)(4x - 6)$ first, using a special-product rule, and then to multiply that result by $\frac{1}{2}$.

Success Tip

Recall that to multiply three polynomials, begin by multiplying any two of them. Then multiply that result by the third polynomial.

$$\text{Area} = \frac{1}{2} \cdot \textbf{base} \cdot \textbf{height} \qquad \text{This is the formula for the area of a triangle.}$$

$$= \frac{1}{2}(4x + 6)(4x - 6) \qquad \text{Substitute.}$$

$$= \frac{1}{2}(16x^2 - 36) \qquad \text{Use the special product rule for a sum}$$
$$\text{and difference of two terms.}$$

$$= 8x^2 - 18 \qquad \text{Distribute the multiplication by } \frac{1}{2}.$$

The area of the triangle is $(8x^2 - 18)$ square centimeters, which can be written as $(8x^2 - 18)$ cm^2.

Self Check 5 Find a polynomial that represents the area of a triangle with height $(12a - 2)$ ft and the length of the base $(12a + 2)$ ft.

Now Try ▶ Problem 53

SECTION 5.7 ► STUDY SET

VOCABULARY

Fill in the blanks.

1. Expressions of the form $(x + y)^2$, $(x - y)^2$, and $(x + y)(x - y)$ occur so frequently in algebra that they are called special _____.

2. $(2x + 3)^2$ is the _____ of a binomial and $(a + 6)(a - 6)$ is the product of the sum and difference of the _____ two terms.

CONCEPTS

3. Fill in the blanks to describe each special product.

   a. $(x + y)^2 = x^2 + 2xy + y^2$

      The _____ of the second term
      _____ the product of the first and second terms
      The square of the _____ term

   b. $(x + y)(x - y) = x^2 - y^2$

      The square of the _____ term
      The _____ of the first term

4. Consider the binomial $5x + 4$.

   a. What is the square of its first term?
   b. What is twice the product of its two terms?
   c. What is the square of its second term?

NOTATION

Complete each solution to find the product.

5. $(x + 4)^2 = \boxed{}^2 + 2(x)(\boxed{}) + \boxed{}^2$

   $= x^2 + \boxed{} + 16$

6. $(6r - 1)^2 = (\boxed{})^2 \boxed{} 2(6r)(1) + (-1)^2$

   $= \boxed{} - \boxed{} + 1$

7. $(s + 5)(s - 5) = \boxed{}^2 - \boxed{}^2$

   $= s^2 - \boxed{}$

8. True or false: $(t + 7)(t - 7) = (t - 7)(t + 7)$?

GUIDED PRACTICE

Find each product. See Example 1.

9. $(x + 1)^2$

10. $(y + 7)^2$

11. $(m - 6)^2$

12. $(b - 1)^2$

13. $(4x + 5)^2$

14. $(6y + 3)^2$

15. $(7m - 2)^2$

16. $(9b - 2)^2$

17. $(1 - 3y)^2$

18. $(1 - 4a)^2$

19. $(y + 0.9)^2$

20. $(d + 0.2)^2$

21. $(a^2 + b^2)^2$

22. $(c^2 + d^2)^2$

23. $\left(s + \dfrac{3}{4}\right)^2$

24. $\left(y - \dfrac{5}{3}\right)^2$

Find each product. See Example 2.

25. $(x + 3)(x - 3)$

26. $(y + 6)(y - 6)$

27. $(2p + 7)(2p - 7)$

28. $(5t + 4)(5t - 4)$

29. $(3n + 1)(3n - 1)$

30. $(5a + 4)(5a - 4)$

31. $\left(c + \dfrac{3}{4}\right)\left(c - \dfrac{3}{4}\right)$

32. $\left(m + \dfrac{4}{5}\right)\left(m - \dfrac{4}{5}\right)$

33. $(0.4 - 9m^2)(0.4 + 9m^2)$

34. $(0.3 - 2c^2)(0.3 + 2c^2)$

35. $(5 - 6g)(5 + 6g)$

36. $(6 - c^2)(6 + c^2)$

Expand each binomial. See Example 3.

37. $(x + 4)^3$

38. $(y + 2)^3$

39. $(n - 6)^3$

40. $(m - 5)^3$

41. $(2g - 3)^3$

42. $(3x - 2)^3$

43. $(a + b)^3$

44. $(c - d)^3$

Perform the operations. See Example 4.

45. $2(x^2 + 7x - 1) - 3(x^2 - 2x + 2)$

46. $2t(t + 2) + (t - 1)(t + 9)$

47. $(3x + 4)(2x - 2) - (2x + 1)(x + 3)$

48. $(5a - 1)^2 - (a - 8)(a + 8)$

49. $-5d(4d - 1)^2$

50. $-2h(7h - 2)^2$

51. $4d(d^2 + g^3)(d^2 - g^3)$

52. $8y(x^2 + y^2)(x^2 - y^2)$

Find a polynomial that represents the area of the figure. Leave π in your answer. See Example 5.

53.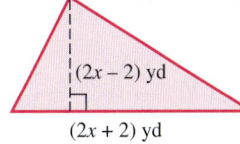
(2x − 2) yd
(2x + 2) yd

54.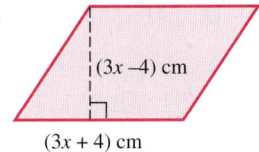
(3x −4) cm
(3x + 4) cm

55.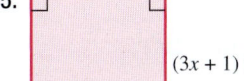
(3x + 1) ft
(3x + 1) ft

56.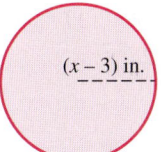
(x − 3) in.

TRY IT YOURSELF

Perform the operations.

57. $(2v^3 - 8)^2$

58. $(8x^4 - 3)^2$

59. $3x(2x + 3)(2x + 3)$

60. $4y(3y + 4)(3y + 4)$

61. $(4f + 0.4)(4f - 0.4)$

62. $(4t + 0.6)(4t - 0.6)$

63. $(r^2 + 10s)^2$

64. $(m^2 + 8n)^2$

65. $2(x + 3) + 4(x - 2)$

66. $3(y - 4) - 5(y + 3)$

67. $\left(d^4 + \dfrac{1}{4}\right)^2$

68. $\left(q^6 + \dfrac{1}{3}\right)^2$

69. $(d + 7)(d - 7)$

70. $(t + 2)(t - 2)$

71. $(2a - 3b)^2$

72. $(2x + 5y)^2$

73. $(n + 6)(n - 6)$

74. $(a + 12)(a - 12)$

75. $(m + 10)^2 - (m - 8)^2$
76. $(5y - 1)^2 - (y + 7)(y - 7)$
77. $(2m + n)^3$
78. $(p - 2q)^3$

79. $\left(5m - \dfrac{6}{5}\right)^2$

80. $\left(6m - \dfrac{7}{6}\right)^2$

81. $(r^2 - s^2)^2$

82. $(t^2 - u^2)^2$

83. $(x - 2)^2$

84. $(a + 2)^2$

85. $(r + 2)^2$

86. $(n + 10)^2$

87. $(n - 2)^4$
88. $(c + d)^4$
89. $5(y^2 - 2y - 6) + 6(2y^2 + 2y - 5)$
90. $(4b + 1)^2 - (b - 7)(b + 7)$
91. $(3x - 2)^2 + (2x + 1)^2$
92. $(4a - 3)^2 + (a + 6)^2$

93. $(f - 8)^2$

94. $(w - 9)^2$

95. $\left(6b + \dfrac{1}{2}\right)\left(6b - \dfrac{1}{2}\right)$

96. $\left(4h + \dfrac{2}{3}\right)\left(4h - \dfrac{2}{3}\right)$

97. $3y(y + 2) + (y + 1)(y - 1)$
98. $(x + y)(x - y) + x(x + y)$
99. $(6 - 2d^3)^2$

100. $(6 - 5p^2)^2$

101. $(2e + 1)^3$

102. $(3m - 2n)^3$

103. $(8x + 3)^2$

104. $(4b - 8)^2$

Look Alikes . . .

Perform the indicated operations.

105. a. $(xy)^2$ b. $(x + y)^2$

106. a. $(cd)^2$ b. $(c - d)^2$

107. a. $(2b^2d)^2$ b. $(2b^2 + d)^2$

108. a. $(mn)^3$ b. $(m + n)^3$

APPLICATIONS

109. **Playpens.** Find a polynomial that represents the area of the floor of the playpen.

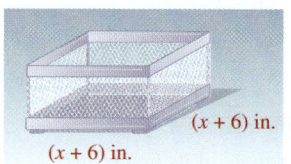

110. **Storage.** Find a polynomial that represents the volume of the cubicle.

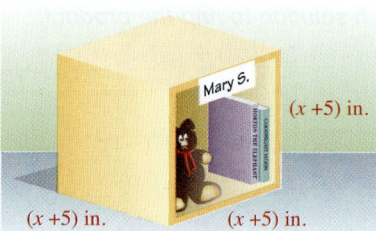

111. **Paper Towels.** The amount of space (volume) occupied by the paper on the roll of paper towels is given by the expression $\pi h(R + r)(R - r)$, where R is the outer radius and r is the inner radius. Perform the indicated multiplication.

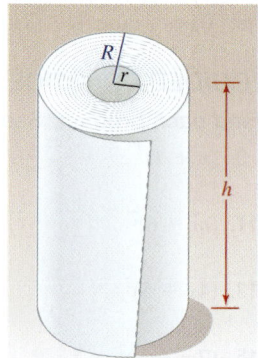

112. Signal Flags.   Refer to the illustration below. Find a polynomial that represents the area in blue of the maritime signal flag for the letter p. The dimensions are given in centimeters.

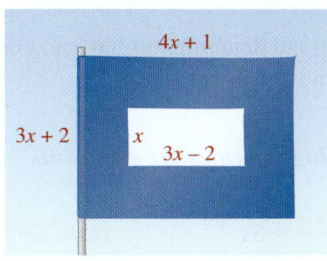

WRITING

113. What is a binomial? Explain how to square it.

114. Writing $(x + y)^2$ as $x^2 + y^2$ illustrates a common error. Explain.

115. We can find $(2x + 3)^2$ and $(5y - 6)^2$ using the FOIL method or using special product rules. Explain why the special product rules are faster.

116. a. Fill in the blanks: $(xy)^2$ is the _____ of a product and $(x + y)^2$ is the _____ of a sum.
   b. Explain why $(xy)^2 \neq (x + y)^2$.

REVIEW

117. Simplify: $\dfrac{30}{36}$

118. Add: $\dfrac{5}{12} + \dfrac{1}{4}$

119. Multiply: $\dfrac{7}{8} \cdot \dfrac{3}{5}$

120. Divide: $\dfrac{1}{3} \div \dfrac{4}{5}$

CHALLENGE PROBLEMS

121. a. Find two binomials whose product is a binomial.

   b. Find two binomials whose product is a trinomial.

   c. Find two binomials whose product is a four-term polynomial.

122. A special-product rule can be used to find $31 \cdot 29$.

$$31 \cdot 29 = (30 + 1)(30 - 1)$$
$$= 30^2 - 1^2$$
$$= 900 - 1$$
$$= 899$$

Use this method to find $52 \cdot 48$.

SECTION 5.8

OBJECTIVES

1 Divide a monomial by a monomial.

2 Divide a polynomial by a monomial.

3 Divide a polynomial by a polynomial.

Dividing Polynomials

ARE YOU READY?

 The following problems review some basic skills that are needed when dividing polynomials.

1. Simplify: **a.** $\dfrac{6}{15}$ **b.** $\dfrac{a^6}{a^4}$

2. Add: $\dfrac{a}{d} + \dfrac{b}{d}$

3. Divide: $24\overline{)864}$

4. Subtract: $\begin{array}{r} 2x^3 + 8x^2 \\ -(2x^3 + 6x^2) \end{array}$

In this section, we will conclude our study of operations with polynomials by discussing division of polynomials. To begin, we consider the simplest case, the quotient of two monomials.

1 **Divide a Monomial by a Monomial.**

To divide monomials, we can use the method for simplifying fractions or a method that involves a rule for exponents.

EXAMPLE 1 Divide the monomials: **a.** $\dfrac{21x^5}{7x^2}$ **b.** $\dfrac{10r^6s}{6rs^3}$

Strategy We can use the rules for simplifying fractions and/or the quotient rule for exponents.

Why We need to make sure that the numerator and denominator have no common factors other than 1. If that is the case, then the fraction is in *simplest form*.

Solution

> **Success Tip**
>
> In this section, you will see that regardless of the number of terms involved, every polynomial division is a series of monomial divisions.

By simplifying fractions

a. $\dfrac{21x^5}{7x^2} = \dfrac{3 \cdot \overset{1}{\cancel{7}} \cdot \overset{1}{\cancel{x}} \cdot \overset{1}{\cancel{x}} \cdot x \cdot x \cdot x}{\cancel{7} \cdot \cancel{x} \cdot \cancel{x}}$

$= 3x^3$

b. $\dfrac{10r^6s}{6rs^3} = \dfrac{\overset{1}{\cancel{2}} \cdot 5 \cdot \overset{1}{\cancel{r}} \cdot r \cdot r \cdot r \cdot r \cdot r \cdot \overset{1}{\cancel{s}}}{\underset{1}{\cancel{2}} \cdot 3 \cdot \underset{1}{\cancel{r}} \cdot \underset{1}{\cancel{s}} \cdot s \cdot s}$

$= \dfrac{5r^5}{3s^2}$

Using the rules for exponents

$\dfrac{21x^5}{7x^2} = 3x^{5-2}$ Divide the coefficients.

$= 3x^3$ Subtract the exponents.

$\dfrac{10r^6s}{6rs^3} = \dfrac{5}{3}r^{6-1}s^{1-3}$ Simplify $\frac{10}{6}$.

$= \dfrac{5}{3}r^5s^{-2}$ Subtract exponents.

$= \dfrac{5r^5}{3s^2}$ Move s^{-2} to the denominator and change the sign of the exponent.

Self Check 1 Divide the monomials: **a.** $\dfrac{30y^4}{5y^2}$ **b.** $\dfrac{8c^2d^6}{32c^5d^2}$

Now Try ▶ Problems 15 and 21

2 Divide a Polynomial by a Monomial.

Recall that to add two fractions with the same denominator, we add their numerators and keep their common denominator.

$$\dfrac{a}{d} + \dfrac{b}{d} = \dfrac{a+b}{d}$$

We can use this rule in reverse to divide polynomials by monomials.

Dividing a Polynomial by a Monomial

To divide a polynomial by a monomial, divide each term of the polynomial by the monomial.

If A, B, and D represent monomials, where $D \neq 0$, then

$$\dfrac{A+B}{D} = \dfrac{A}{D} + \dfrac{B}{D}$$

EXAMPLE 2 Divide: **a.** $\dfrac{9x^2 + 3x}{3x}$ **b.** $\dfrac{12a^4b^3 - 18a^3b^2 + 2a^2}{6a^2b^2}$

Strategy We will divide each term of the polynomial in the numerator by the monomial in the denominator.

Why A fraction bar indicates division of the numerator by the denominator.

Solution

a. Here, we have a binomial divided by a monomial.

$$\frac{9x^2 + 3x}{3x} = \frac{9x^2}{3x} + \frac{3x}{3x}$$

Divide each term of the numerator, $9x^2 + 3x$, by the denominator, $3x$.

$$= 3x^{2-1} + 1x^{1-1}$$

Do each monomial division. Divide the coefficients. Keep each base and subtract the exponents.

$$= 3x^1 + 1x^0$$

$$= 3x + 1$$

Recall that $x^0 = 1$.

Caution

A common error is to "remove" the $3x$'s in the expression incorrectly. Recall that only *factors* common to the numerator and denominator can be removed. Here, $3x$ is not a factor of the numerator—it is a *term*.

$$\frac{9x^2 + \overset{1}{\cancel{3x}}}{\underset{1}{\cancel{3x}}}$$

Check: We multiply the divisor, $3x$, and the quotient, $3x + 1$. The result should be the dividend, $9x^2 + 3x$.

$$3x(3x + 1) = 9x^2 + 3x$$ The answer checks.

b. Here, we have a trinomial divided by a monomial.

$$\frac{12a^4b^3 - 18a^3b^2 + 2a^2}{6a^2b^2} = \frac{12a^4b^3}{6a^2b^2} - \frac{18a^3b^2}{6a^2b^2} + \frac{2a^2}{6a^2b^2}$$

Divide each term of the numerator by the denominator, $6a^2b^2$.

$$= 2a^{4-2}b^{3-2} - 3a^{3-2}b^{2-2} + \frac{a^{2-2}}{3b^2}$$

Do each monomial division. Simplify: $\frac{2}{6} = \frac{1}{3}$.

$$= 2a^2b - 3a + \frac{1}{3b^2}$$

$b^{2-2} = b^0 = 1$ and $a^{2-2} = a^0 = 1$.

The Language of Algebra

The names of the parts of a division statement are

Dividend
$$\overset{\frown}{9x^2 + 3x} \over \underset{\text{Divisor}}{3x}} = \underset{\text{Quotient}}{3x + 1}$$

Recall that the variables in a polynomial must have whole-number exponents. Therefore, the result, $2a^2b - 3a + \frac{1}{3b^2}$, is not a polynomial because the last term has a variable in the denominator.

Check:

$$6a^2b^2\left(2a^2b - 3a + \frac{1}{3b^2}\right) = 12a^4b^3 - 18a^3b^2 + 2a^2$$ The answer checks.

Self Check 2   Divide: **a.** $\dfrac{50h^3 + 5h^2}{5h^2}$ **b.** $\dfrac{22s^5t^2 - s^4t^3 + 44s^2t^4}{11s^2t^2}$

Now Try ▶ Problems 25, 29, and 33

3   Divide a Polynomial by a Polynomial.

To divide a polynomial by a polynomial (other than a monomial), we use a method similar to long division in arithmetic.

EXAMPLE 3   Divide $x^2 + 5x + 6$ by $x + 2$.

Strategy   We will use the long division method. The dividend is $x^2 + 5x + 6$ and the divisor is $x + 2$.

Why   Since the divisor has more than one term, we must use the long division method to divide the polynomials.

Solution   We write the division using a long division symbol $\overline{)}$ and proceed as follows:

Step 1: $x + 2\overline{)x^2 + 5x + 6}$ with x above

Divide the first term of the dividend by the first term of the divisor: $\frac{x^2}{x} = x$. Write the result, x, above the long division symbol.

Notice how the instruction translates:

Divide $x^2 + 5x + 6$ by $x + 2$

$$x + 2\overline{)x^2 + 5x + 6}$$

Success Tip

Notice that this method is much like that used for division of whole numbers.

$$\begin{array}{r} 13 \\ 12\overline{)156} \\ -12\downarrow \\ \overline{036} \\ -36 \\ \overline{0} \end{array}$$

Hundreds ⌐
Tens ⌐
Ones ⌐

Success Tip

The long division method aligns like terms vertically.

$$\begin{array}{r} x + 3 \\ x + 2\overline{)x^2 + 5x + 6} \\ -(x^2 + 2x) \\ \overline{3x + 6} \\ -(3x + 6) \\ \overline{0} \end{array}$$

x^2-terms ⌐
x-terms ⌐
Constants ⌐

Step 2:
$$\begin{array}{r} x \\ x + 2\overline{)x^2 + 5x + 6} \\ x^2 + 2x \end{array}$$

Multiply each term of the divisor by x. Write the result, $x^2 + 2x$, under $x^2 + 5x$, and draw a line. Be sure to align the like terms.

Step 3:
$$\begin{array}{r} x \\ x + 2\overline{)x^2 + 5x + 6} \\ -(x^2 + 2x)\;\downarrow \\ \overline{3x + 6} \end{array}$$

Write parentheses around $x^2 + 2x$ so that both of its terms are subtracted. Subtract $x^2 + 2x$ from $x^2 + 5x$. Work column by column: $x^2 - x^2 = 0$ and $5x - 2x = 3x$.
Bring down the next term, 6.

Step 4:
$$\begin{array}{r} x + 3 \\ \textcircled{x} + 2\overline{)x^2 + 5x + 6} \\ -(x^2 + 2x) \\ \overline{\textcircled{3x} + 6} \end{array}$$

Divide the first term of $3x + 6$ by the first term of the divisor: $\frac{3x}{x} = 3$. Write $+ 3$ above the long division symbol to form the second term of the quotient.

Step 5:
$$\begin{array}{r} x + 3 \\ x + 2\overline{)x^2 + 5x + 6} \\ -(x^2 + 2x) \\ \overline{3x + 6} \\ 3x + 6 \end{array}$$

Multiply each term of the divisor by 3. Write the result, $3x + 6$, under $3x + 6$ and draw a line. Be sure to align the like terms.

Step 6:
$$\begin{array}{r} x + 3 \\ x + 2\overline{)x^2 + 5x + 6} \\ -(x^2 + 2x) \\ \overline{3x + 6} \\ -(3x + 6) \\ \overline{0} \end{array}$$

Write parentheses around $3x + 6$ so that both of its terms are subtracted. Subtract $3x + 6$ from $3x + 6$. Work vertically: $3x - 3x = 0$ and $6 - 6 = 0$. There are no more terms to bring down.

This is the remainder.

The quotient is $x + 3$ and the remainder is 0.

Step 7: Check the result by verifying that $(x + 2)(x + 3)$ is $x^2 + 5x + 6$.

$$(x + 2)(x + 3) = x^2 + 3x + 2x + 6 \qquad \text{Divisor} \cdot \text{quotient} = \text{dividend (if 0 remainder)}$$
$$= x^2 + 5x + 6 \qquad \text{The result checks.}$$

Self Check 3   Divide $x^2 + 7x + 12$ by $x + 3$.

Now Try ▶ Problem 37

Dividing a Polynomial by a Polynomial

To divide a polynomial by a polynomial (other than a monomial) use long division. If there is a remainder, write the result in the form: quotient $+ \frac{\text{remainder}}{\text{divisor}}$.

The long division method used in algebra can have a remainder just as long division in arithmetic often does.

EXAMPLE 4   Divide: $(6x^2 - 7x - 2) \div (2x - 1)$

Strategy   We will use the long division method. The dividend is $6x^2 - 7x - 2$ and the divisor is $2x - 1$.

Why   Since the divisor has more than one term, we must use the long division method to divide the polynomials.

Solution

Caution

The long-division method is a series of four steps:
- Divide
- Multiply
- Subtract
- Bring down the next term

Use parentheses to avoid sign errors when subtracting.

Step 1:

$$\begin{array}{r} 3x \\ 2x - 1 \overline{)6x^2 - 7x - 2} \end{array}$$

Divide the first term of the dividend by the first term of the divisor: $\frac{6x^2}{2x} = 3x$. Write the result, 3x, above the long division symbol.

Step 2:

$$\begin{array}{r} 3x \\ 2x - 1 \overline{)6x^2 - 7x - 2} \\ 6x^2 - 3x \end{array}$$

Multiply each term of the divisor by 3x. Write the result, $6x^2 - 3x$, under $6x^2 - 7x$, and draw a line.

Step 3:

$$\begin{array}{r} 3x \\ 2x - 1 \overline{)6x^2 - 7x - 2} \\ -(6x^2 - 3x) \downarrow \\ -4x - 2 \end{array}$$

Write parentheses around $6x^2 - 3x$ so that both of its terms are subtracted. Subtract $6x^2 - 3x$ from $6x^2 - 7x$. Work vertically: $6x^2 - 6x^2 = 0$ and $-7x - (-3x) = -7x + 3x = -4x$. Bring down the next term, -2.

Step 4:

$$\begin{array}{r} 3x - 2 \\ 2x - 1 \overline{)6x^2 - 7x - 2} \\ -(6x^2 - 3x) \\ -4x - 2 \end{array}$$

Divide the first term of $-4x - 2$ by the first term of the divisor: $\frac{-4x}{2x} = -2$. Write -2 above the long division symbol to form the second term of the quotient.

Step 5:

$$\begin{array}{r} 3x - 2 \\ 2x - 1 \overline{)6x^2 - 7x - 2} \\ -(6x^2 - 3x) \\ -4x - 2 \\ -4x + 2 \end{array}$$

Multiply each term of the divisor by -2. Write the result, $-4x + 2$, under $-4x - 2$, and draw a line.

Success Tip

The long division method for polynomials continues until the degree of the remainder is less than the degree of the divisor. Here, the remainder, -4, has degree 0. The divisor, $2x - 1$, has degree 1. Therefore, the division ends.

Step 6:

$$\begin{array}{r} 3x - 2 \\ 2x - 1 \overline{)6x^2 - 7x - 2} \\ -(6x^2 - 3x) \\ -4x - 2 \\ -(-4x + 2) \\ -4 \end{array}$$

Write $-4x + 2$ in parentheses so that both of its terms are subtracted. Subtract $-4x + 2$ from $-4x - 2$. Work vertically: $-4x - (-4x) = -4x + 4x = 0$ and $-2 - 2 = -4$. There are no more terms to bring down.

This is the remainder.

The quotient is $3x - 2$ and the remainder is -4. It is common to write the answer in *Quotient* $+ \frac{remainder}{divisor}$ form as either

$$3x - 2 + \frac{-4}{2x - 1} \qquad \text{or} \qquad 3x - 2 - \frac{4}{2x - 1}$$

Step 7: We can check the result using the fact that for any division:

$$\begin{array}{ccccc} \text{Divisor} \cdot & \text{quotient} & + \text{ remainder} & = & \text{dividend} \\ (2x - 1)(3x - 2) & + & (-4) & = & 6x^2 - 4x - 3x + 2 + (-4) \\ & & & = & 6x^2 - 7x - 2 \quad \text{The result checks.} \end{array}$$

Self Check 4 Divide: $(8x^2 + 6x - 3) \div (2x + 3)$. Check the result.

Now Try ▶ Problem 43

The division method works best when the terms of the divisor and the dividend are written in descending powers of the variable. If the powers in the dividend or divisor are not in descending order, we use the commutative property of addition to write them that way.

EXAMPLE 5 Divide $4x^2 + 2x^3 + 12 - 2x$ by $x + 3$.

Strategy We will write the dividend in descending powers of x and use the long division method.

Why It is easier to align like terms in columns when the powers of the variable are written in descending order.

Solution

$$
\begin{array}{r}
2x^2 - 2x + 4 \\
x + 3 \overline{)\, 2x^3 + 4x^2 - 2x + 12} \\
-(2x^3 + 6x^2) \\
\hline
-2x^2 - 2x \\
-(-2x^2 - 6x) \\
\hline
4x + 12 \\
-(4x + 12) \\
\hline
0
\end{array}
$$

The first division: $\frac{2x^3}{x} = 2x^2$.

The second division: $\frac{-2x^2}{x} = -2x$.

The third division: $\frac{4x}{x} = 4$.

Check: $(x + 3)(2x^2 - 2x + 4) = 2x^3 - 2x^2 + 4x + 6x^2 - 6x + 12$
$$= 2x^3 + 4x^2 - 2x + 12 \quad \text{The result checks.}$$

Self Check 5 Divide $x^2 - 10x + 6x^3 + 4$ by $2x - 1$.

Now Try Problem 45

When we write the terms of a dividend in descending powers, we must determine whether some powers of the variable are missing. If any are missing, insert **placeholder terms** with a coefficient of 0 or leave blank spaces for them. This keeps like terms in the same column, which is necessary when performing the subtraction in vertical form.

EXAMPLE 6 Divide: $\dfrac{27x^3 + 1}{3x + 1}$

Strategy The divisor is $3x + 1$. The dividend, $27x^3 + 1$, does not have an x^2-term or an x-term. We will insert a $0x^2$ term and a $0x$ term as placeholders, and use the long division method.

Why We insert placeholder terms so that like terms will be aligned in the same column when we subtract.

Solution

$$
\begin{array}{r}
9x^2 - 3x + 1 \\
3x + 1 \overline{)\, 27x^3 + 0x^2 + 0x + 1} \\
-(27x^3 + 9x^2) \\
\hline
-9x^2 + 0x \\
-(-9x^2 - 3x) \\
\hline
3x + 1 \\
-(3x + 1) \\
\hline
0
\end{array}
$$

The first division: $\frac{27x^3}{3x} = 9x^2$.

The second division: $\frac{-9x^2}{3x} = -3x$.

The third division: $\frac{3x}{3x} = 1$.

Check: $(3x + 1)(9x^2 - 3x + 1) = 27x^3 - 9x^2 + 3x + 9x^2 - 3x + 1$
$$= 27x^3 + 1 \quad \text{The result checks.}$$

Self Check 6 Divide: $\dfrac{x^2 - 9}{x - 3}$. Check the result.

Now Try Problem 49

EXAMPLE 7

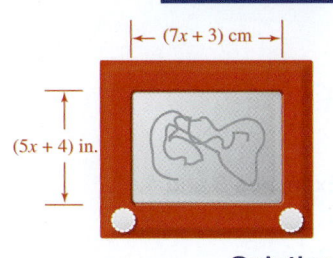

(7x + 3) cm

(5x + 4) in.

Toys. The area of an Etch A Sketch screen is represented by the polynomial $(35x^2 + 43x + 12)$ in.^{2}. If the width of the screen is $(5x + 4)$ inches, what polynomial represents its length?

Strategy We will find the length of the screen by dividing its area, $(35x^2 + 43x + 12)$ in.^{2}, by its width, $(5x + 4)$ inches.

Why Recall that the area of a rectangle is given by the formula $A = lw$. If we divide both sides of the formula by w, we see that $l = \frac{A}{w}$.

Solution

$$\text{Length} = \frac{\textbf{area}}{\textbf{width}}$$ This is the formula for the length of a rectangle.

$$= \frac{35x^2 + 43x + 12}{5x + 4}$$ Substitute $35x^2 + 43x + 12$ for the area and $5x + 4$ for the width.

To divide $35x^2 + 43x + 12$ by $5x + 4$, we use long division.

$$
\begin{array}{r}
7x + 3 \\
5x + 4{\overline{\smash{\big)}\,35x^2 + 43x + 12}} \\
-(35x^2 + 28x) \downarrow \\
\hline
15x + 12 \\
-(15x + 12) \\
\hline
0
\end{array}
$$

The first division: $\frac{35x^2}{5x} = 7x$.

The second division: $\frac{15x}{5x} = 3$.

The length of the Etch A Sketch screen is $(7x + 3)$ inches.

Self Check 7   **Televisions.** The area of the rectangular screen of a plasma TV is represented by the polynomial $(36x^2 - 51x - 21)$ in.^{2}. If the width of the screen is $(4x - 7)$ inches, what polynomial represents its length?

Now Try ▶ Problem 103

SECTION 5.8 ▶ STUDY SET

VOCABULARY

Fill in the blanks.

1. The expression $\dfrac{18x^7}{9x^4}$ is a monomial divided by a _____.

2. The expression $\dfrac{6x^3y - 4x^2y^2 + 8xy^3 - 2y^4}{2x^4}$ is a _____ divided by a monomial.

3. The expression $\dfrac{x^2 - 8x + 12}{x - 6}$ is a trinomial divided by a _____.

4.

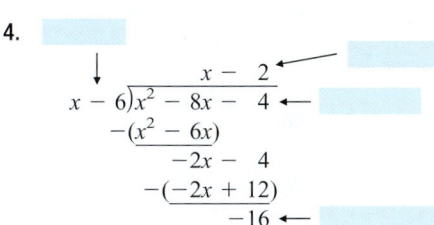

$$
\begin{array}{r}
x - 2 \\
x - 6{\overline{\smash{\big)}\,x^2 - 8x - 4}} \\
-(x^2 - 6x) \\
\hline
-2x - 4 \\
-(-2x + 12) \\
\hline
-16
\end{array}
$$

CONCEPTS

5. The long division method is a series of four steps that are repeated. Put them in the correct order:

   subtract   multiply   bring down   divide

6. In the following long divisions, find the answer to the subtraction that must be performed at this stage.

   a.
   $$
   \begin{array}{r}
   3x \\
   2x + 1{\overline{\smash{\big)}\,6x^2 + 9x - 10}} \\
   -(6x^2 + 3x)
   \end{array}
   $$

   b.
   $$
   \begin{array}{r}
   x \\
   x - 7{\overline{\smash{\big)}\,x^2 + 10x + 21}} \\
   -(x^2 - 7x)
   \end{array}
   $$

   c.
   $$
   \begin{array}{r}
   x \\
   x - 7{\overline{\smash{\big)}\,x^2 - 9x - 6}} \\
   -(x^2 - 7x)
   \end{array}
   $$

   d.
   $$
   \begin{array}{r}
   x \\
   x + 4{\overline{\smash{\big)}\,x^2 + 0x - 16}} \\
   -(x^2 + 4x)
   \end{array}
   $$

7. Fill in the blanks: To check an answer of a long division, we use the fact that

Divisor · [blank] + remainder = [blank]

8. Check to see whether the following result of a long division is correct.

$$\frac{x^2 + 4x - 20}{x - 3} = x + 7 + \frac{1}{x - 3}$$

NOTATION

Complete each solution.

9. $\dfrac{28x^5 - x^3 + 5x^2}{7x^2} = \dfrac{28x^5}{\boxed{}} - \dfrac{\boxed{}}{7x^2} + \dfrac{5x^2}{\boxed{}}$

$$= 4x^{\boxed{}} - \dfrac{x^{3-2}}{\boxed{}} + \dfrac{5x^{\boxed{}}}{7}$$

$$= \boxed{} - \dfrac{x}{7} + \boxed{}$$

10.
$$
\begin{array}{r}
\boxed{} + 2 \\
x + 2\overline{)x^2 + 4x + 5} \\
-(x^2 + \boxed{}) \\
\hline
\boxed{} + 5 \\
-(2x + 4) \\
\end{array}
$$

11. Write the polynomial $2x^2 - 1 + 5x^4$ in descending powers of x and insert placeholders for each missing term.

12. True or false: $6x + 4 + \dfrac{-3}{x + 2} = 6x + 4 - \dfrac{3}{x + 2}$

GUIDED PRACTICE

Divide the monomials. See Example 1.

13. $\dfrac{x^5}{x^2}$

14. $\dfrac{a^{12}}{a^8}$

15. $\dfrac{12h^8}{9h^6}$

16. $\dfrac{22b^9}{6b^6}$

17. $\dfrac{-3d^4}{15d^8}$

18. $\dfrac{-4x^3}{16x^5}$

19. $\dfrac{10s^2}{s^3}$

20. $\dfrac{16y^3}{y^4}$

21. $\dfrac{8x^3y^2}{40xy^6}$

22. $\dfrac{3y^3z}{18yz^6}$

23. $\dfrac{-16r^3y^2}{-4r^2y^7}$

24. $\dfrac{-35xz^6}{-7x^8z^2}$

Divide the polynomial by the monomial. See Example 2.

25. $\dfrac{6x + 3}{3}$

26. $\dfrac{8x + 4}{4}$

27. $\dfrac{a - a^3 + a^4}{a^4}$

28. $\dfrac{b^2 + b^3 - b^4}{b^4}$

29. $\dfrac{6h^{12} + 48h^9}{24h^{10}}$

30. $\dfrac{4x^{14} - 36x^8}{36x^{12}}$

31. $\dfrac{9s^8 - 18s^5 + 12s^4}{3s^3}$

32. $\dfrac{16b^{10} + 4b^6 - 20b^4}{4b^2}$

33. $\dfrac{7c^5 + 21c^4 - 14c^3 - 35c}{7c^2}$

34. $\dfrac{12r^{15} - 48r^{12} + r^{10} - 18r^8}{6r^{10}}$

35. $\dfrac{-25x^2y^3 + 30xy^2 - 5xy}{-5x^2y^2}$

36. $\dfrac{-30a^4b^4 - 15a^3b - 10a^2b^2}{-10a^2b^3}$

Perform each division. See Examples 3 and 4.

37. Divide $x^2 + 8x + 12$ by $x + 2$.

38. Divide $x^2 + 5x + 6$ by $x + 2$.

39. Divide $x^2 - 5x + 6$ by $x - 3$.

40. Divide $x^2 - 12x + 32$ by $x - 4$.

41. $\dfrac{2x^2 + 5x + 2}{2x + 3}$

42. $\dfrac{3x^2 - 8x + 3}{3x - 2}$

43. $\dfrac{6x^2 - 11x + 2}{3x - 1}$

44. $\dfrac{4x^2 + 6x - 1}{2x + 1}$

Perform each division. See Example 5.

45. $x + 2\overline{)3x + 2x^2 - 2}$

46. $x + 3\overline{)-x + 2x^2 - 21}$

47. $(3 + 11x + 10x^2) \div (5x + 3)$

48. $(6x + 1 + 9x^2) \div (3x + 1)$

Perform each division. See Example 6.

49. $(a^2 - 25) \div (a + 5)$

50. $(b^2 - 36) \div (b + 6)$

51. $(x^2 - 1) \div (x - 1)$

52. $(x^2 - 9) \div (x + 3)$

53. $\dfrac{4x^2 - 9}{2x + 3}$

54. $\dfrac{25x^2 - 16}{5x - 4}$

55. $\dfrac{81b^2 - 49}{9b - 7}$

56. $\dfrac{16t^2 - 121}{4t + 11}$

TRY IT YOURSELF

Perform each division.

57. Divide $y^2 + 13y + 13$ by $y + 1$.

58. Divide $z^2 + 7z + 14$ by $z + 3$.

59. $\dfrac{15a^8b^2 - 10a^2b^5}{5a^3b^2}$

60. $\dfrac{9a^4b^3 - 16a^3b^4}{12a^2b}$

61. $3x + 2\overline{)2 + 7x + 6x^3 + 10x^2}$

62. $3x - 2\overline{)4x - 4 + 6x^3 - x^2}$

63. $\dfrac{8x^9 - 32x^6}{4x^4}$

64. $\dfrac{30y^8 + 40y^7}{10y^6}$

65. $\dfrac{6a^2 + 5a - 6}{2a + 3}$

66. $\dfrac{3b^2 - 5b + 2}{3b - 2}$

67. $\dfrac{45m^{10}}{9m^5}$

68. $\dfrac{24n^{12}}{8n^4}$

69. $\dfrac{3b^2 + 11b + 6}{3b + 2}$

70. $\dfrac{8a^2 + 2a - 3}{2a - 1}$

71. $2x - 7 \overline{) -x - 21 + 2x^2}$

72. $2x - 1 \overline{) x - 2 + 6x^2}$

73. $\dfrac{x^3 + 1}{x + 1}$

74. $\dfrac{x^3 - 8}{x - 2}$

75. $\dfrac{-65rs^2}{15r^2s^5}$

76. $\dfrac{112uz^4}{-42u^3z^8}$

77. $\dfrac{-18w^6 - 9}{9w^4}$

78. $\dfrac{-40f^4 + 16}{8f^3}$

79. $\dfrac{9m - 6}{m}$

80. $\dfrac{10n - 6}{n}$

81. $\dfrac{y^3 + y}{y - 2}$

82. $\dfrac{a^3 + a}{a + 3}$

83. $\dfrac{5x^4 - 10x}{25x^3}$

84. $\dfrac{24x^7 - 32x^2}{16x^3}$

85. $3 + 4x \overline{) 3 - 5x^2 - 2x + 4x^3}$

86. $2x + 3 \overline{) 7x^2 - 3 + 4x + 2x^3}$

87. $(x^2 + 6x + 15) \div (x + 5)$

88. $(x^2 + 10x + 30) \div (x + 6)$

89. $\dfrac{12x^3y^2 - 8x^2y - 4x}{4xy}$

90. $\dfrac{12a^2b^2 - 8a^2b - 4ab}{4ab}$

91. $a - 5 \overline{) a^2 - 17a + 64}$

92. $b - 2 \overline{) b^2 - 4b + 6}$

93. $\dfrac{a^3 - 1}{a - 1}$

94. $\dfrac{y^3 + 8}{y + 2}$

95. $\dfrac{6x^3 + x^2 + 2x + 1}{3x - 1}$

96. $\dfrac{3y^3 - 4y^2 + 2y + 3}{y + 3}$

97. $\dfrac{8x^{17}y^{20}}{16x^{15}y^{30}}$

98. $\dfrac{21a^{30}b^{15}}{14a^{40}b^{12}}$

99. $(6m^2 - m - 40) \div (2m + 5)$

100. $(12d^2 - 20d + 3) \div (6d - 1)$

Look Alikes . . .

Perform the indicated operations.

101. a. $\dfrac{16x^2 - 16x - 5}{4x}$ **b.** $\dfrac{16x^2 - 16x - 5}{4x + 1}$

102. a. $\dfrac{9x^3 + 3x^2 + 4x + 4}{3x}$ **b.** $\dfrac{9x^3 + 3x^2 + 4x + 4}{3x + 2}$

APPLICATIONS

103. Furnace Filters. The area of the rectangular-shaped furnace filter is $(x^2 - 2x - 24)$ square inches. What expression represents its height?

$(x + 4)$ in.

104. Mini-Blinds. The area covered by the mini-blinds is $(3x^3 - 6x)$ square feet. What expression represents the height of the blinds?

$\longleftarrow 3x \text{ ft} \longrightarrow$

105. Pool. The rack shown in the illustration is used to set up the balls for a game of pool. If the perimeter of the rack, in inches, is given by the polynomial $6x^2 - 3x + 9$, what expression represents the approximate length of one side?

106. Communications. Telephone poles were installed every $(2x - 3)$ feet along a stretch of railroad track $(8x^3 - 6x^2 + 5x - 21)$ feet long. What expression represents the number of poles that were used?

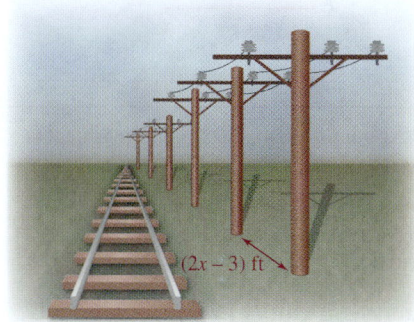

$(2x - 3)$ ft

WRITING

107. Explain how to check the following long division.

$$
\begin{array}{r}
x + 5 \\
3x + 5 \overline{) 3x^2 + 20x - 5} \\
-(3x^2 + 5x) \\
\hline
15x - 5 \\
-(15x + 25) \\
\hline
-30
\end{array}
$$

108. Explain the error: 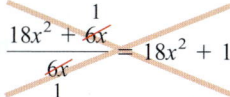 $\dfrac{18x^2 + 6x}{6x} = 18x^2 + 1$

109. How do you know when to stop the long division method when dividing polynomials?

110. When dividing $x^3 + 1$ by $x + 1$, why is it helpful to write $x^3 + 1$ as $x^3 + 0x^2 + 0x + 1$?

REVIEW

111. Write an equation of the line with slope $-\frac{11}{6}$ that passes through $(2, -6)$. Write the answer in slope–intercept form.

112. Solve $S = 2\pi rh + 2\pi r^2$ for h.

CHALLENGE PROBLEMS

Perform each division.

113. $\dfrac{6a^3 - 17a^2b + 14ab^2 - 3b^3}{2a - 3b}$

114. $(2x^4 + 3x^3 + 3x^2 - 5x - 3) \div (2x^2 - x - 1)$

115. $(x^6 + 2x^4 - 6x^2 - 9) \div (x^2 + 3)$

116. $\dfrac{6x^{6m}y^{6n} + 15x^{4m}y^{7n} - 24x^{2m}y^{8n}}{3x^{2m}y^n}$

117. $\dfrac{a^8 + a^6 - 4a^4 + 5a^2 - 3}{a^4 + 2a^2 - 3}$

118. $\dfrac{-17x^2 + 5x + x^4 + 2}{4x + x^2 - 1}$

5 ▶ Summary & Review

SECTION 5.1 ▶ Rules for Exponents

DEFINITIONS AND CONCEPTS	EXAMPLES
An **exponent** indicates repeated multiplication. It tells how many times the **base** is to be used as a factor. Exponent ⟶ *n* factors of *x* $x^n = x \cdot x \cdot x \cdot \cdots \cdot x$ Base	Identify the base and the exponent in each expression. $2^6 = 2 \cdot 2 \cdot 2 \cdot 2 \cdot 2 \cdot 2$ *2 is the base and 6 is the exponent.* $(-xy)^3 = (-xy)(-xy)(-xy)$ *−xy is the base and 3 is the exponent.* $5t^4 = 5 \cdot t \cdot t \cdot t \cdot t$ *t is the base and 4 is the exponent.* $8^1 = 8$ *8 is the base and 1 is the exponent.*
Rules for Exponents: If *m* and *n* represent integers and there are no divisions by 0, then **Product rule:** $x^m x^n = x^{m+n}$ **Quotient rule:** $\dfrac{x^m}{x^n} = x^{m-n}$ **Power rule:** $(x^m)^n = x^{m \cdot n} = x^{mn}$ **Power of a product rule:** $(xy)^m = x^m y^m$ **Power of a quotient rule:** $\left(\dfrac{x}{y}\right)^n = \dfrac{x^n}{y^n}$	Simplify each expression: $5^2 5^7 = 5^{2+7} = 5^9$ $\dfrac{t^7}{t^3} = t^{7-3} = t^4$ $(6^3)^7 = 6^{3 \cdot 7} = 6^{21}$ $(2p)^5 = 2^5 p^5 = 32p^5$ $\left(\dfrac{s}{4}\right)^4 = \dfrac{s^4}{4^4} = \dfrac{s^4}{256}$

REVIEW EXERCISES

1. Identify the base and the exponent in each expression.
   a. n^{12}
   b. $(2x)^6$
   c. $3r^4$
   d. $(y - 7)^3$

2. Write each expression in an equivalent form using an exponent.
   a. $m \cdot m \cdot m \cdot m \cdot m$
   b. $-3 \cdot x \cdot x \cdot x \cdot x$
   c. $(x + 8)(x + 8)$
   d. $\left(\dfrac{1}{2}pq\right)\left(\dfrac{1}{2}pq\right)\left(\dfrac{1}{2}pq\right)$

Simplify each expression. Assume there are no divisions by 0.

3. $7^4 \cdot 7^8$
4. $mmnn$
5. $(y^7)^3$
6. $(3x)^4$
7. $\dfrac{b^{12}}{b^3}$
8. $-b^3 b^4 b^5$
9. $(-16s^3)^2 s^4$
10. $(2.1x^2 y)^2$

11. $[(-9)^3]^5$
12. $(a^5)^3 (a^2)^4$
13. $\left(\dfrac{1}{2}x^2 x^3\right)^3$
14. $\left(\dfrac{x^7}{3xy}\right)^2$
15. $\dfrac{(m - 25)^{16}}{(m - 25)^4}$
16. $\dfrac{(5y^2 z^3)^3}{(yz)^5}$
17. $\dfrac{a^5 a^4 a^5}{a^2 a}$
18. $\dfrac{(cd)^9}{(cd)^4}$

Find an expression that represents the area or the volume of each figure, whichever is appropriate.

19.
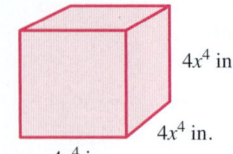
$4x^4$ in.
$4x^4$ in.
$4x^4$ in.

20.
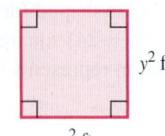
y^2 ft
y^2 ft

SECTION 5.2 ▶ Zero and Negative Exponents

DEFINITIONS AND CONCEPTS	EXAMPLES
Rules for exponents: For any nonzero real numbers x and y and any integers m and n,	Simplify each expression. Do not use negative exponents in the answer.
Zero exponent: $x^0 = 1$	$5^0 = 1$
Negative exponents: $x^{-n} = \dfrac{1}{x^n}$	$4^{-2} = \dfrac{1}{4^2} = \dfrac{1}{16}$ and $7c^{-6} = \dfrac{7}{c^6}$
Negative to positive rules: $\dfrac{1}{x^{-n}} = x^n \qquad \dfrac{x^{-m}}{y^{-n}} = \dfrac{y^n}{x^m}$	$\dfrac{1}{t^{-8}} = t^8$ and $\dfrac{2^{-4}}{x^{-6}} = \dfrac{x^6}{2^4} = \dfrac{x^6}{16}$
Negative exponents and reciprocals: $\left(\dfrac{x}{y}\right)^{-n} = \left(\dfrac{y}{x}\right)^n$	$\left(\dfrac{x}{10}\right)^{-3} = \left(\dfrac{10}{x}\right)^3 = \dfrac{10^3}{x^3} = \dfrac{1{,}000}{x^3}$

REVIEW EXERCISES

Simplify each expression. Do not use negative exponents in the answer.

21. x^0

22. $(3x^2y^2)^0$

23. $3x^0$

24. 10^{-3}

25. -5^{-2}

26. $\dfrac{t^4}{t^{10}}$

27. $\dfrac{8}{x^{-5}}$

28. $-6y^4y^{-5}$

29. $\dfrac{7^{-2}}{2^{-3}}$

30. $(x^{-3}x^{-4})^{-2}$

31. $\left(\dfrac{-3r^4r^{-3}}{r^{-3}r^7}\right)^3$

32. $\left(\dfrac{4z^4}{z^3}\right)^{-2}$

33. $\dfrac{3^{-2}c^3d^3}{2^{-3}c^2d^8}$

34. $\dfrac{t^{-30}}{t^{-60}}$

35. $\dfrac{w(w^{-3})^{-4}}{w^{-9}}$

36. $\left(\dfrac{4}{f^4}\right)^{-10}$

SECTION 5.3 ▶ Scientific Notation

DEFINITIONS AND CONCEPTS	EXAMPLES
A positive number is written in **scientific notation** when it is written in the form $N \times 10^n$, where $1 \le N < 10$ and n is an integer.	Write each number in scientific notation. $$32{,}500 = 3.25 \times 10^4 \qquad \text{and} \qquad 0.0025 = 2.5 \times 10^{-3}$$ 4 decimal places 3 decimal places Write each number in standard notation. $$1.91 \times 10^5 = 191{,}000 \qquad \text{and} \qquad 4.7 \times 10^{-6} = 0.0000047$$ 5 decimal places 6 decimal places
Scientific notation provides an easier way to perform computations involving very large or very small numbers.	Use scientific notation to perform the calculation: $$\dfrac{684{,}000{,}000}{456{,}000} = \dfrac{6.84 \times 10^8}{4.56 \times 10^5} = \dfrac{6.84}{4.56} \times \dfrac{10^8}{10^5} = 1.5 \times 10^3 \text{ or } 1{,}500$$

Write each number in scientific notation.

37. $720,000,000$

38. $9,370,000,000,000,000$

39. 0.00000000942

40. 0.00013

41. 0.018×10^{-2}

42. 853×10^3

Write each number in standard notation.

43. 1.26×10^5

44. 3.919×10^{-8}

45. 2.68×10^0

46. 5.76×10^1

Evaluate each expression by first writing each number in scientific notation. Express the result in scientific notation and standard notation.

47. $\dfrac{(0.000012)(0.000004)}{0.00000016}$

48. $\dfrac{(4,800,000)(20,000,000)}{600,000}$

49. World Population. As of January 2007, the world's population was estimated to be 6.57 billion. Write this number in standard notation and in scientific notation.

50. Atoms. The illustration shows a cross section of an atom. How many nuclei (plural for nucleus), placed end to end, would it take to stretch across the atom?

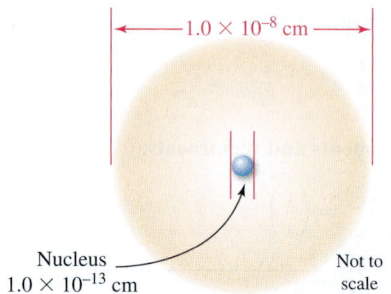

$\longleftarrow 1.0 \times 10^{-8}$ cm $\longrightarrow$

Nucleus
1.0×10^{-13} cm

Not to scale

SECTION 5.4 ▶ Polynomials

DEFINITIONS AND CONCEPTS	EXAMPLES
A **polynomial** is a single term or a sum of terms in which all variables have whole-number exponents and no variable appears in a denominator.	Polynomials: 32, $-5x^2y^3$, $7p^3 - 14q^3$, $4m^2 + 5m - 12$ Not Polynomials: $y^2 - y^{-5}$, $4x^3 - \dfrac{7}{x} + 3x$
A polynomial with exactly one term is called a **monomial.** A polynomial with exactly two terms is called a **binomial.** A polynomial with exactly three terms is called a **trinomial.**	*Monomials* $\qquad$ *Binomials* $\qquad$ *Trinomials* $3x^2 \qquad\qquad 2y^3 + 3y \qquad\qquad 3p^2 - 7p + 12$ $-12m^3n^2 \qquad 87t - 25 \qquad\quad 4p^2q^3 - 8p^2q^2 + 12p^2q$
The **coefficient of a term** is its numerical factor. The **degree of a term** of a polynomial in one variable is the value of the exponent on the variable. If a polynomial has more than one variable, the **degree of a term** is the sum of the exponents on the variables. The **degree of a nonzero constant** is 0.	*Term* $\qquad$ *Coefficient* $\qquad$ *Degree of the term* $6a^7 \qquad\qquad 6 \qquad\qquad\qquad 7$ $-7.3x^5y^4 \qquad -7.3 \qquad\qquad 5 + 4 = 9$ $32 \qquad\qquad 32 \qquad\qquad\qquad 0$
The **degree of a polynomial** is equal to the highest degree of any term of the polynomial.	*Polynomial* $\qquad\qquad$ *Degree of the polynomial* $7m^3 - 4m^2 + 5m - 12 \qquad\qquad 3$ $\dfrac{1}{2}a^4b + \dfrac{3}{4}a^3b^2 - \dfrac{2}{3}a^2b^4 \qquad 2 + 4 = 6$
To **evaluate a polynomial** for a given value, substitute the value for the variable and follow the rules for the order of operations.	Evaluate $3x^2 - 4x + 2$ for $x = 2$. $3x^2 - 4x + 2 = 3(2)^2 - 4(2) + 2$ $\quad$ Substitute 2 for each x. $\qquad\qquad\qquad = 3(4) - 8 + 2$ $\qquad$ Evaluate $(2)^2$ first. $\qquad\qquad\qquad = 6$

We can **graph equations defined by polynomials** such as $y = x^2 - 2$, $y = -x^2$, and $y = x^3 + 1$.

The graph of $y = x^2 - 2$ is a cup-shaped curve called a **parabola.**

Graph: $y = x^2 - 2$

Find several solutions of the equation, plot them on a rectangular coordinate system, and then draw a smooth curve passing through the points.

$y = x^2 - 2$

x	y	(x, y)
-2	2	$(-2, 2)$
-1	-1	$(-1, -1)$
0	-2	$(0, -2)$
1	-1	$(1, -1)$
2	2	$(2, 2)$

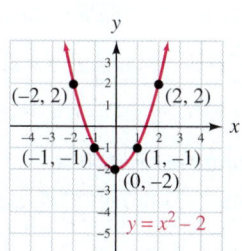

Select x-values. ⌐ ↑   ⌐ Plot points.
            Find y-values.

REVIEW EXERCISES

51. Consider the polynomial $3x^3 - x^2 + x + 10$.

   a. How many terms does the polynomial have?

   b. What is the lead term?

   c. What is the coefficient of each term?

   d. What is the constant term?

52. Find the degree of each polynomial and classify it as a monomial, binomial, trinomial, or none of these.

   a. $13x^7$ **b.** $-16a^2b$

   c. $5^3x + x^2$ **d.** $-3x^5 + x - 1$

   e. $9xy^2 + 21x^3y^3$ **f.** $4s^4 - 3s^2 + 5s + 4$

53. Evaluate $-x^5 - 3x^4 + 3$ for $x = 0$ and $x = -2$.

54. Diving.   The number of inches that the woman deflects the diving board is given by the polynomial $0.1875x^2 - 0.0078125x^3$ where x is the number of feet that she stands from the front anchor point of the board. Find the amount of deflection if she stands on the end of the diving board, 8 feet from the anchor point.

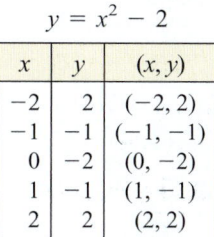

Construct a table of solutions like the one shown here and then graph the equation.

55. $y = x^2$

x	-2	-1	0	1	2
y					

56. $y = x^3 + 1$

x	-2	-1	0	1	2
y					

SECTION 5.5 ▶ **Adding and Subtracting Polynomials**

DEFINITIONS AND CONCEPTS	EXAMPLES
To **simplify a polynomial,** combine like terms.	Simplify: $3r^4 - 4r^3 + 7r^4 + 8r^2$ $= 10r^4 - 4r^3 + 8r^2$ Combine like terms. Think: $(3 + 7)r^4 = 10r^4$.
To **add polynomials,** combine their like terms.	Add: $(4x^2 + 9x + 4) + (3x^2 - 5x - 1)$ $= (4x^2 + 3x^2) + (9x - 5x) + (4 - 1)$ Group like terms. $= 7x^2 + 4x + 3$ Combine like terms.
To **subtract two polynomials,** change the signs of the terms of the polynomial being subtracted, drop the parentheses, and combine like terms.	Subtract: $(8a^3b - 4ab^2) - (-3a^3b + 9ab^2)$ $= 8a^3b - 4ab^2 + 3a^3b - 9ab^2$ Change the sign of each term of $-3a^3b + 9ab^2$ and drop the parentheses. $= 11a^3b - 13ab^2$ Combine like terms.

REVIEW EXERCISES

Simplify each polynomial and write the result in descending powers of one variable.

57. $6y^3 + 8y^4 + 7y^3 + (-8y^4)$

58. $4a^2b + 5 - 6a^3b - 3a^2b + 2a^3b + 1$

59. $\dfrac{5}{6}x^2 + \dfrac{1}{3}y^2 - \dfrac{1}{4}x^2 - \dfrac{3}{4}xy + \dfrac{2}{3}y^2$

60. $7.6c^5 - 2.1c^3 - 0.9c^5 + 8.1c^4$

Perform the operations.

61. $(2r^6 + 14r^3) + (23r^6 - 5r^3 + 5r)$

62. $(7.1a^2 + 2.2a - 5.8) - (3.4a^2 - 3.9a + 11.8)$

63. $(3r^3s + r^2s^2 - 3rs^3 - 3s^4) + (r^3s - 8r^2s^2 - 4rs^3 + s^4)$

64. $\left(\dfrac{7}{8}m^4 - \dfrac{1}{5}m^3\right) - \left(\dfrac{1}{4}m^4 + \dfrac{1}{5}m^3\right) - \dfrac{3}{5}m^3$

65. Find the difference when $(-3z^3 - 4z + 7)$ is subtracted from the sum of $(2z^2 + 3z - 7)$ and $(-4z^3 - 2z - 3)$.

66. Gardening.   Find a polynomial that represents the length of the wooden handle of the shovel.

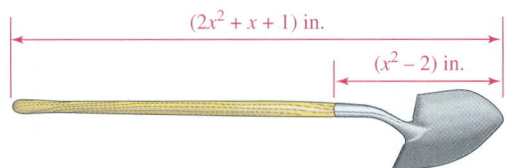

67. Add:

$$3x^2 + 5x + 2$$
$$x^2 - 3x + 6$$

68. Subtract:

$$20x^3 \qquad\quad + 12x$$
$$-(12x^3 + 7x^2 - \ \ 7x)$$

SECTION 5.6 ▶ **Multiplying Polynomials**

DEFINITIONS AND CONCEPTS	EXAMPLES
To **multiply two monomials,** multiply the numerical factors (the coefficients) and then multiply the variable factors.	Multiply: $(5p^6)(2p^5) = (5 \cdot 2)(p^6 \cdot p^5)$ Group the coefficients together and the variables together. $\qquad\qquad = 10p^{11}$ Think: $5 \cdot 2 = 10$ and $p^6 \cdot p^5 = p^{6+5} = p^{11}$.
To **multiply a monomial and a polynomial,** multiply each term of the polynomial by the monomial.	Multiply: $3r^2(2r^4 + 7r^2 - 4)$ $\quad = 3r^2(2r^4) + 3r^2(7r^2) + 3r^2(-4)$ Distribute the multiplication by $3r^2$. $\quad = 6r^6 + 21r^4 - 12r^2$ Multiply the monomials.

To **multiply two binomials,** use the *FOIL method:*
  F: First
  O: Outer
  I: Inner
  L: Last
	Multiply: $(3m + 4)(2m - 5) = 3m(2m) + 3m(-5) + 4(2m) + 4(-5)$
	$\qquad\qquad\qquad\qquad\qquad = 6m^2 - 15m + 8m - 20$ Multiply the monomials.
	$\qquad\qquad\qquad\qquad\qquad = 6m^2 - 7m - 20$ Combine like terms.

To **multiply two polynomials,** multiply each term of one polynomial by each term of the other polynomial and then combine like terms.	Multiply: $(a - b)(6a^2 - 4ab + b^2)$ $\quad = a(6a^2) + a(-4ab) + a(b^2) - b(6a^2) - b(-4ab) - b(b^2)$ $\quad = 6a^3 - 4a^2b + ab^2 - 6a^2b + 4ab^2 - b^3$ Multiply the monomials. $\quad = 6a^3 - 10a^2b + 5ab^2 - b^3$ Combine like terms.

When finding the **product of three polynomials,** begin by multiplying any two of them, and then multiply that result by the third polynomial.

Multiply: $-9x^4(x-1)(x-7) = -9x^4(x^2 - 7x - x + 7)$ Multiply the two binomials.

$$= -9x^4(x^2 - 8x + 7)$$ Combine like terms within the parentheses.

$$= -9x^6 + 72x^5 - 63x^4$$ Distribute the multiplication by $-9x^4$.

REVIEW EXERCISES

Multiply.

69. $(2x^2)(5x)$

70. $(-6x^4z^3)(x^6z^2)$

71. $5b^3 \cdot 6b^2 \cdot 4b^6$

72. $\frac{2}{3}h^5(3h^9 + 12h^6)$

73. $3n^2(3n^2 - 5n + 2)$

74. $x^2y(y^2 - xy)$

75. $2x(3x^4)(x + 2)$

76. $-a^2b^2(-a^4b^2 + a^3b^3 - ab^4 + 7a)$

77. $(x + 3)(x + 2)$

78. $(2x + 1)(x - 1)$

79. $(3t - 3)(2t + 2)$

80. $(3n^4 - 5n^2)(2n^4 - n^2)$

81. $-a^5(a^2 - b)(5a^2 + b)$

82. $6.6(a - 1)(a + 1)$

83. $\left(3t - \frac{1}{3}\right)\left(6t + \frac{5}{3}\right)$

84. $(5.5 - 6b)(2 - 4b)$

85. $(2a - 3)(4a^2 + 6a + 9)$

86. $(8x^2 + x - 2)(7x^2 + x - 1)$

87. Multiply using vertical form: $4x^2 - 2x + 1$
$$2x + 1$$

88. Refer to the illustration below. Find a polynomial that represents
   a. the perimeter of the base of the dishwasher.
   b. the area of the base of the dishwasher.

   c. the volume occupied by the dishwasher.

$3x$ in.

$(x + 6)$ in.

$(2x - 1)$ in.

SECTION 5.7 ▶ Special Products

DEFINITIONS AND CONCEPTS	EXAMPLES

The following **special products** occur so often that it is worthwhile to learn their forms.

Square of a binomial:
$$(A + B)^2 = A^2 + 2AB + B^2$$
This is the square of a binomial sum.

$$(A - B)^2 = A^2 - 2AB + B^2$$
This is the square of a binomial difference.

Multiplying the Sum and Difference of the Same Two Terms:
$$(A + B)(A - B) = A^2 - B^2$$

Multiply: $(n + 4)^2 = \underbrace{n^2}_{} + \underbrace{2(n)(4)}_{} + \underbrace{4^2}_{}$

   The square of   Twice the product   The square of the
   the first term, *n*.   of both terms.   second term, 4.

$$= n^2 + 8n + 16$$

Multiply: $(5a - 1)^2 = \underbrace{(5a)^2}_{} + \underbrace{2(5a)(-1)}_{} + \underbrace{(-1)^2}_{}$

   The square of   Twice the product   The square of the
   the first term, 5*a*.   of both terms.   second term, −1.

$$= 25a^2 - 10a + 1$$

Multiply: $(x + 8)(x - 8) = \underbrace{x^2}_{} - \underbrace{8^2}_{}$

   The square of   The square of the
   the first term, *x*.   second term, 8.

$$= x^2 - 64$$

Find each product.

89. $(a - 3)^2$

90. $(m + 2)^3$

91. $(x + 7)(x - 7)$

92. $(2x - 0.9)(2x + 0.9)$

93. $(2y + 1)^2$

94. $(y^2 + 1)(y^2 - 1)$

95. $(6r^2 + 10s)^2$

96. $-(8a - 3c)^2$

97. $80s(r^2 + s^2)(r^2 - s^2)$

98. $4b(3b - 4)^2$

99. $\left(t - \dfrac{3}{4}\right)^2$

100. $\left(x + \dfrac{4}{3}\right)^2$

Perform the operations.

101. $3(9x^2 + 3x + 7) - 2(11x^2 - 5x + 9)$

102. $(5c - 1)^2 - (c + 6)(c - 6)$

103. Graphic Arts. A Dr. Martin Luther King poster has his picture with a $\frac{1}{2}$-inch wide blue border around it. The length of the poster is $(x + 3)$ inches and the width is $(x - 1)$ inches. Find a polynomial that represents the area of the *picture* of Dr. King.

National Archives

104. Find a polynomial that represents the area of the triangle.

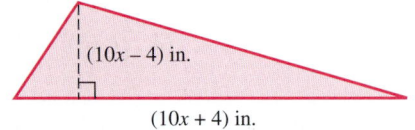
$(10x - 4)$ in.
$(10x + 4)$ in.

SECTION 5.8 ▶ Dividing Polynomials

DEFINITIONS AND CONCEPTS	EXAMPLES

To **divide monomials,** use the method for simplifying fractions and/or the quotient rule for exponents.

Divide the monomials:

$$\frac{8p^2q}{4pq^3} = \frac{2 \cdot \cancel{4} \cdot \cancel{p} \cdot p \cdot \cancel{q}}{\cancel{4} \cdot \cancel{p} \cdot \cancel{q} \cdot q \cdot q} \quad \text{or} \quad \frac{8p^2q}{4pq^3} = \frac{8}{4}p^{2-1}q^{1-3}$$

Keep each base and subtract the exponents.

$$= \frac{2p}{q^2} \qquad\qquad = 2p^1q^{-2}$$

$$= \frac{2p}{q^2}$$

Move q^{-2} to the denominator and change the sign of the exponent.

To **divide a polynomial by a monomial,** divide each term of the numerator by the denominator.

Divide: $\dfrac{9a^2b^4 - 12a^3b^6}{18ab^5} = \dfrac{9a^2b^4}{18ab^5} - \dfrac{12a^3b^6}{18ab^5}$

$$= \frac{a}{2b} - \frac{2a^2b}{3} \qquad \text{Perform each monomial division.}$$

Long division can be used to **divide a polynomial by a polynomial** (other than a monomial). The long division method is a series of four steps that are repeated: Divide, multiply, subtract, and bring down the next term.

When the division has a remainder, write the answer in the form: Quotient $+ \dfrac{\text{remainder}}{\text{divisor}}$.

Divide $4x^2 - 4x + 5$ by $2x + 1$.

$$2x + 1\overline{)\,4x^2 - 4x + 5}$$ with quotient $2x - 3$
$$-(4x^2 + 2x)$$
$$-6x + 5$$
$$-(-6x - 3)$$
$$8$$

The first division: $\frac{4x^2}{2x} = 2x$.

The second division: $\frac{-6x}{2x} = -3$.

The remainder is 8.

The result is: $2x - 3 + \dfrac{8}{2x + 1}$

The long division method works best when the terms of the divisor and the dividend are written in **descending powers of the variable.**

When the dividend has **missing terms,** insert such terms with a coefficient of 0, or leave a blank space.

Set up each long division.

$$\frac{5x + x^3 + 3 + 3x^2}{x + 1}$$

The terms of the dividend are not in descending powers of x.

$$x + 1 \overline{)x^3 + 3x^2 + 5x + 3}$$

$$\frac{x^2 - 9}{x - 3}$$

The dividend is missing a term.

$$x - 3 \overline{)x^2 + 0x - 9}$$

REVIEW EXERCISES

Divide. Do not use negative exponents in the answer.

105. $\dfrac{16n^8}{8n^5}$

106. $\dfrac{-14x^2y}{21xy^3}$

107. $\dfrac{a^{15} - 24a^8}{6a^{12}}$

108. $\dfrac{15a^5b + ab^2 - 25b}{5a^2b}$

109. $x - 1 \overline{)x^2 - 6x + 5}$

110. $\dfrac{2x^2 + 3 + 7x}{x + 3}$

111. $(15x^2 - 8x - 8) \div (3x + 2)$

112. Divide $25y^2 - 9$ by $5y + 3$.

113. $3x + 1 \overline{)-13x - 4 + 9x^3}$

114. $2x - 1 \overline{)6x^3 + x^2 + 1}$

115. Use multiplication to show that $(3y^2 + 11y + 6) \div (y + 3)$ is $3y + 2$.

116. **Bedding.** The area of a rectangular-shaped bed sheet is represented by the polynomial $(4x^3 + 12x^2 + x - 12)$ in.^{2}. If the width of the sheet is $(2x + 3)$ inches, find a polynomial that represents its length.

5 ▶ CHAPTER TEST

1. Fill in the blanks.

  a. In the expression y^{10}, the _____ is y and the _____ is 10.

  b. We call a polynomial with exactly one term a _____, with exactly two terms a _____, and with exactly three terms a _____.

  c. The _____ of a term of a polynomial in one variable is the value of the exponent on the variable.

  d. $(x + y)^2$, $(x - y)^2$, and $(x + y)(x - y)$ are called _____ products.

2. Use exponents to rewrite $2xxxyyyy$.

Simplify each expression. Do not use negative exponents in the answer.

3. $y^2(yy^3)$

4. $\left(\dfrac{1}{2}x^3\right)^5 (x^2)^3$

5. $3.5x^0$

6. $2y^{-5}y^2$

7. 5^{-3}

8. $\dfrac{(x + 1)^{15}}{(x + 1)^6}$

9. $\dfrac{(y^{-5})^{-4}}{yy^{-2}}$

10. $\left(\dfrac{a^2b^{-1}}{4a^3b^{-2}}\right)^3$

11. $\left(\dfrac{8}{m^6}\right)^{-2}$

12. $\dfrac{-6a}{b^{-9}}$

13. Find an expression that represents the volume of a cube that has sides of length $10y^4$ inches.

14. **Electricity.** One ampere (amp) corresponds to the flow of 6,250,000,000,000,000,000 electrons per second past any point in a direct current (DC) circuit. Write this number in scientific notation.

15. Write 9.3×10^{-5} in standard notation.

16. Evaluate $(2.3 \times 10^{18})(4.0 \times 10^{-15})$. Write the answer in scientific notation and standard notation.

17. Identify $x^4 + 8x^2 - 12$ as a monomial, binomial, or trinomial. Then complete the table.

Term	Coefficient	Degree

Degree of the polynomial:

18. Find the degree of the polynomial $3x^3y + 2x^2y^3 - 5xy^2 - 6y$.

19. Complete the table of solutions for $y = x^2 + 2$ and then graph the equation.

x	-2	-1	0	1	2
y					

20. Free Fall. A visitor standing on the rim of the Grand Canyon drops a rock over the side. The distance (in feet) that the rock is from the canyon floor t seconds after being dropped is given by the polynomial $-16t^2 + 5,184$. Find the position of the rock 18 seconds after being dropped. Explain your answer.

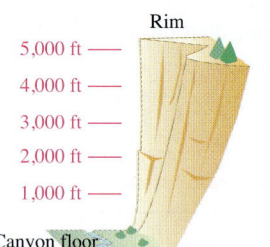

Rim

5,000 ft ——
4,000 ft ——
3,000 ft ——
2,000 ft ——
1,000 ft ——

Canyon floor

Simplify each polynomial.

21. $\frac{3}{5}x^2 + 6 + \frac{1}{4}x - 8 - \frac{1}{2}x^2 + \frac{1}{3}x$

22. $4a^2b + 5 - 6a^3b - 3a^2b + 2a^3b$

Perform the operations.

23. $(12.1h^3 - 9.9h^2 + 9.5) + (7.3h^3 - 1.2h^2 - 10.1)$

24. Subtract $b^3c - 3bc + 12$ from the sum of $6b^3c - 3bc$ and $b^3c - 2bc$.

25. Subtract:

$$-5y^3 + 4y^2 \qquad + 3$$
$$\underline{-(-2y^3 - 14y^2 + 17y - 32)}$$

26. Find a polynomial that represents the perimeter of the rectangle.

$(5a^2 + 3a - 1)$ in.

$(a - 9)$ in.

Multiply.

27. $(2x^3y^3)(5x^2y^8)$

28. $9b^3(8b^4)(-b)$

29. $3y^2(y^2 - 2y + 3)$

30. $0.6p^5(0.4p^6 - 0.9p^3)$

31. $\frac{3}{4}s^3t^9(s^4t^8 + 16st)$

32. $(x - 5)(3x + 4)$

33. $\left(6t + \frac{1}{2}\right)\left(2t - \frac{3}{2}\right)$

34. $(3.8m - 1)(2m - 1)$

35. $(a^3 - 6)(a^3 + 7)$

36. $(2x - 3)(x^2 - 2x + 4)$

37. $(1 + 10c)(1 - 10c)$

38. $(7b^3 - 3t)^2$

39. $(2.2a)(a + 5)(a - 3)$

40. Perform the operations: $(x + y)(x - y) + (x + y)^2$

Divide.

41. $\dfrac{6a^2 - 12b^2}{24ab}$

42. $\dfrac{x^2 + x - 6}{x + 3}$

43. $2x - 1\overline{)1 + x^2 + 6x^3}$

44. Find a polynomial that represents the width of a rectangle if its area is represented by the polynomial $(x^2 - 6x + 5)$ ft^2 and the length is $(x - 1)$ ft.

45. Use a check to determine whether $(5m^2 - 29m - 6) \div (5m + 1) = m - 6$.

46. Is $(a + b)^2 = a^2 + b^2$? Show why or why not.

Group Project

BINOMIAL MULTIPLICATION AND THE AREA OF RECTANGLES

Overview: In this activity, rectangles are used to illustrate binomial multiplication.

Instructions: Form groups of 2 or 3 students. Study the figure. The area of the largest rectangle (outlined in blue) is given by $(x + 2)(x + 3)$. Its area is also the sum of the areas of the four smaller rectangles: $x^2 + 3x + 2x + 6$. Thus,

$$(x + 2)(x + 3) = x^2 + 3x + 2x + 6$$
$$= x^2 + 5x + 6$$

Draw three similar models to represent the following products.

1. $(x + 4)(x + 5)$ 2. $(x + 8)^2$ 3. $x(x + 6)$ 4. $(2x + 1)^2$

Determine the missing number so that each rectangle below has the given area.

5. Area: $x^2 + 9x + 14$ 6. Area: $x^2 + 16x + 55$

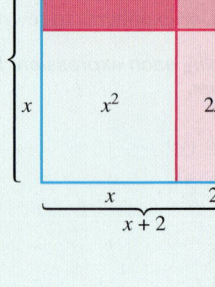

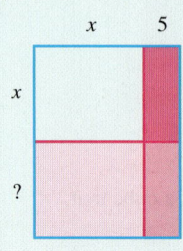

7. Draw a model that represents a rectangle with area $x^2 + 10x + 24$. Label it completely.

CUMULATIVE REVIEW Chapters 1–5

1. Use exponents to write the prime factorization of 270. [Section 1.2]

2. **a.** Use the variables a and b to state the commutative property of addition. [Section 1.4]

   b. Use the variables x, y, and z to state the associative property of multiplication. [Section 1.6]

Evaluate each expression.

3. $3 - 4[-10 - 4(-5)]$

   [Section 1.7]

4. $\dfrac{|-45| - 2(-5) + 1^5}{2 \cdot 9 - 2^4}$

   [Section 1.7]

Simplify each expression.

5. $27\left(\dfrac{2}{3}x\right)$ [Section 1.9]

6. $3x^2 + 2x^2 - 5x^2$ [Section 1.9]

Solve each equation.

7. $2 - (4x + 7) = 3 + 2(x + 2)$ [Section 2.2]

8. $\dfrac{2}{5}y + 3 = 9$ [Section 2.2]

9. **Candy Sales.** The circle graph shows how $6.5 billion in seasonal candy sales for 2009 was spent. Find the candy sales for Halloween. [Section 2.3]

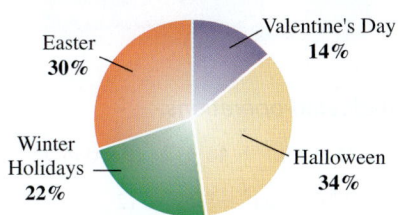

Source: National Confectioners Association

10. **Air Conditioning.** Find the volume of air contained in the duct. Round to the nearest tenth of a cubic foot. [Section 2.4]

11. **Angle of Elevation.** Find x. [Section 2.5]

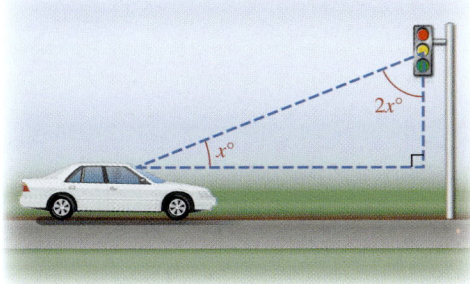

12. **Livestock Auction.** A farmer is going to sell one of her prize hogs at an auction and would like to make $6,000 after paying a 4% commission to the auctioneer. For what selling price will the farmer make this amount of money? [Section 2.5]

13. **Stock Market.** An investment club invested part of $45,000 in a high-yield mutual fund that earned 12% annual simple interest. The remainder of the money was invested in Treasury bonds that earned 6.5% simple annual interest. The two investments earned $4,300 in one year. How much was invested in each account? [Section 2.6]

14. Solve $-4x + 6 > 17$ and graph the solution set. Then describe the graph using interval notation. [Section 2.7]

Graph each equation.

15. $y = 3x$ [Section 3.2]   16. $x = -2$ [Section 3.3]

17. Find the slope of the line passing through $(6, -2)$ and $(-3, 2)$. [Section 3.4]

18. **Diabetes.** The graph below shows the number of people in the United States diagnosed with diabetes from 1996 to 2008. Find the rate of change in the number over this time span. [Section 3.4]

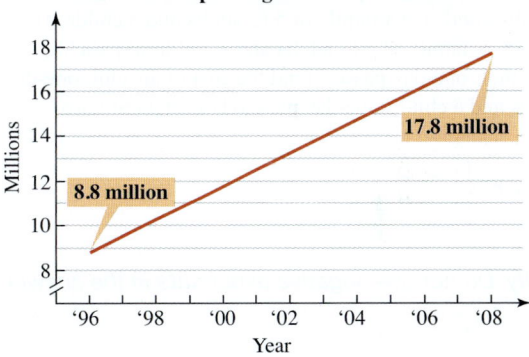

Source: U.S. Centers for Disease Control

19. Find the slope and y-intercept of the line. Then write the equation of the line. [Section 3.5]

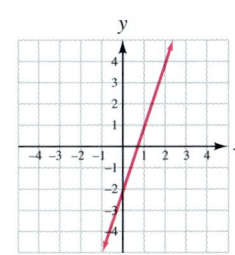

20. Without graphing, determine whether the graphs of $y = \frac{3}{2}x - 1$ and $2x + 3y = 10$ are parallel, perpendicular, or neither. [Section 3.5]

21. Write the equation of the line that passes through $(-2, 10)$ with slope -4. Write the result in slope–intercept form. [Section 3.6]

22. Is $(-2, 1)$ a solution of $2x - 3y \geq -6$? [Section 3.7]

23. If $f(x) = 2x^2 + 3x - 9$, find $f(-5)$. [Section 3.8]

24. Determine whether the graph below is the graph of a function. If it is not, find ordered pairs that show a value of x that is assigned more than one value of y. [Section 3.8]

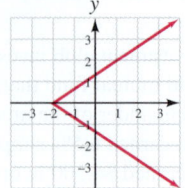

25. Is $\left(\dfrac{2}{3}, -1\right)$ a solution of the system $\begin{cases} y = -3x + 1 \\ 3x + 3y = -2 \end{cases}$?
    [Section 4.1]

26. Solve the system $\begin{cases} 3x + 2y = 14 \\ y = \dfrac{1}{4}x \end{cases}$ by graphing.

    [Section 4.1]

27. Solve the system $\begin{cases} 2b - 3a = 18 \\ a + 3b = 5 \end{cases}$ by substitution.

    [Section 4.2]

28. Solve the system $\begin{cases} 8s + 10t = 24 \\ 11s - 3t = -34 \end{cases}$ by elimination (addition).

    [Section 4.3]

29. **Vacations.**   One-day passes to Universal Studios Hollywood cost a family of 5 (2 adults and 3 children) $275. A family of 6 (3 adults and 3 children) paid $336 for their one-day passes. Find the cost of an adult one-day pass and a child's one-day pass to Universal Studios.
    [Section 4.4]

30. Graph: $\begin{cases} y \leq 2x - 1 \\ x + 3y > 6 \end{cases}$
    [Section 4.5]

Simplify. Do not use negative exponents in the answer.

31. $(-3x^2 y^4)^2$ [Section 5.1]

32. $(v^5)^2(v^3)^4$ [Section 5.1]

33. $ab^3 c^4 \cdot ab^4 c^2$ [Section 5.1]

34. $\left(\dfrac{4t^3 t^4 t^5}{3t^2 t^6}\right)^3$ [Section 5.1]

35. $(2y)^{-4}$ [Section 5.2]

36. $\dfrac{a^4 b^0}{a^{-3}}$ [Section 5.2]

37. -5^{-2} [Section 5.2]

38. $\left(\dfrac{a}{x}\right)^{-10}$ [Section 5.2]

Write each number in scientific notation.

39. 615,000 [Section 5.3]

40. 0.0000013 [Section 5.3]

41. Graph: $y = x^2$ [Section 5.4]

42. **Musical Instruments.**   The amount of deflection of the horizontal beam (in inches) is given by the polynomial $0.01875x^4 - 0.15x^3 + 1.2x$, where x is the distance (in feet) that the gong is hung from one end of the beam. Find the deflection if the gong is hung in the middle of the support.
    [Section 5.4]

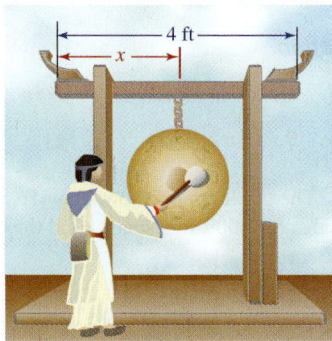

Perform the indicated operations.

43. $(4c^2 + 3c - 2) + (3c^2 + 4c + 2)$ [Section 5.5]

44. Subtract:
$$\begin{aligned} 17x^4 - 3x^2 - 65x - 12 \\ -(23x^4 + 14x^2 + 3x - 23) \end{aligned}$$
    [Section 5.5]

45. $(2t + 3s)(3t - s)$ [Section 5.6]

46. $3x(2x + 3)^2$ [Section 5.7]

47. $5x + 3 \overline{)11x + 10x^2 + 3}$ [Section 5.8]

48. $\dfrac{2x - 32}{16x}$ [Section 5.8]

Factoring and Quadratic Equations

6

©iStockphoto.com/Neustockimages

from Campus to Careers

Elementary School Teacher

It has been said that a teacher takes a hand, opens a mind, and touches a heart. That is certainly true for the thousands of dedicated elementary school teachers across the country. Elementary school teachers use their training in mathematics in many ways. Besides teaching math on a daily basis, they calculate student grades, analyze test results, and order instructional materials and supplies. They use measurement and geometry for designing bulletin board displays and they construct detailed schedules so that the classroom time is used wisely.

Problem 129 in **Study Set 6.1**, **problem 103** in **Study Set 6.3**, and **problem 17** in **Study Set 6.8** involve situations that an elementary school teacher might encounter on the job. The mathematical concepts discussed in this chapter can be used to solve those problems.

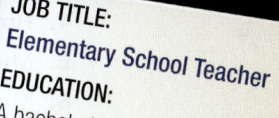

JOB TITLE:
Elementary School Teacher
EDUCATION:
A bachelor's degree and completion of an approved teacher training program
JOB OUTLOOK:
Varies from good to excellent, in some locations
ANNUAL EARNINGS:
U.S. median $51,180*
*Can vary greatly by region and experience.
FOR MORE INFORMATION:
www.bls.gov/oco/ocos069.htm

Reading an algebra textbook is different from reading a newspaper or a novel. Here are two ways that you should be reading this textbook.

SKIMMING FOR AN OVERVIEW: This is a quick way to look at material just before it is covered in class. It helps you become familiar with the new vocabulary and notation that will be used by your instructor in the lecture. It lays a foundation.

READING FOR UNDERSTANDING: This in-depth type of reading is done more slowly, with a pencil and paper at hand. Don't skip anything—every word counts! You should do just as much writing as you do reading. Highlight the important points and work each example. If you become confused, stop and reread the material until you understand it.

Now Try This ▶

Choose a section from this chapter and . . .
1. . . . quickly skim it. Write down any terms in bold face type and the titles of any properties, definitions, or strategies that are given in the colored boxes.
2. . . . work each *Self Check* problem. Your solutions should look like those in the *Examples*. Be sure to include your own "author notes" (the sentences in red to the right of each step of a solution).

SECTION 6.1

The Greatest Common Factor; Factoring by Grouping

OBJECTIVES

1 Find the greatest common factor of a list of terms.

2 Factor out the greatest common factor.

3 Factor by grouping.

ARE YOU READY?

The following problems review some basic skills that are needed when factoring expressions.

1. Find the prime factorization of 54.
2. Write $a \cdot a \cdot a \cdot a \cdot a \cdot a$ in an equivalent form using exponents.
3. Simplify: $-(x - 7)$
4. How many terms does the expression $3x^3 + x^2 + 4x + 8$ have?
5. Multiply: $4d(d^3 + 2d)$
6. Multiply and simplify: $x^2(x + 1) - 3(x + 1)$

In Chapter 5, we learned how to multiply polynomials. For example, to multiply $3x + 5$ by $4x$, we use the distributive property, as shown below.

$$4x(3x + 5) = 4x \cdot 3x + 4x \cdot 5$$
$$= 12x^2 + 20x$$

In this section, we reverse the previous steps and determine what factors were multiplied to obtain $12x^2 + 20x$. We call that process *factoring the polynomial.*

Multiplication: Given the factors, we find a polynomial.

$$4x(3x + 5) = 12x^2 + 20x$$

Factoring: Given a polynomial, we find the factors.

To **factor a polynomial** means to express it as a product of two (or more) polynomials. The first step when factoring a polynomial is to determine whether its terms have any common factors.

Success Tip

On the game show *Jeopardy!*, answers are revealed and contestants respond with the appropriate questions. Factoring is similar. Answers to multiplications are given. You are to respond by telling what factors were multiplied.

1   Find the Greatest Common Factor of a List of Terms.

To determine whether two or more integers have common factors, it is helpful to write them as products of prime numbers. For example, the prime factorizations of 42 and 90 are given below.

$$42 = \mathbf{2} \cdot \mathbf{3} \cdot 7 \qquad\qquad 90 = \mathbf{2} \cdot \mathbf{3} \cdot 3 \cdot 5$$

The highlighting shows that 42 and 90 have one factor of 2 and one factor of 3 in common. To find their *greatest common factor* (*GCF*), we multiply the common factors: $2 \cdot 3 = 6$. Thus, the GCF of 42 and 90 is 6.

The Greatest Common Factor (GCF)	The **greatest common factor (GCF)** of a list of integers is the largest common factor of those integers. To find the greatest common factor of two (or more) integers:
	1. Prime factor each number.
	2. Identify the common prime factors.
	3. The GCF is a product of all the common prime factors found in Step 2. If there are no common prime factors, the GCF is 1.

Recall from arithmetic that the factors of a number divide the number exactly, leaving no remainder. Therefore, the greatest common factor of two or more integers is the largest integer that divides each of them exactly.

EXAMPLE 1   Find the GCF of each list of numbers: **a.** 21 and 140   **b.** 24, 60, and 96 **c.** 9, 10, and 30

Strategy   We will prime factor each number in the list. Then we will identify the common prime factors and find their product.

Why   The product of the common prime factors is the GCF of the numbers in the list.

Solution

a. The prime factorization of each number is shown:

$$21 = 3 \cdot 7$$
$$140 = 2 \cdot 2 \cdot 5 \cdot 7 \quad \text{\color{red}This can be written as } 2^2 \cdot 5 \cdot 7.$$

Since the only prime factor common to 21 and 140 is 7, the GCF of 21 and 140 is 7.

b. To find the GCF of three numbers, we proceed in a similar way by first finding the prime factorization of each number in the list.

$$24 = \mathbf{2} \cdot \mathbf{2} \cdot 2 \cdot \mathbf{3} \quad \text{\color{red}This can be written as } 2^3 \cdot 3.$$
$$60 = \mathbf{2} \cdot \mathbf{2} \cdot \mathbf{3} \cdot 5 \quad \text{\color{red}This can be written as } 2^2 \cdot 3 \cdot 5.$$
$$96 = \mathbf{2} \cdot \mathbf{2} \cdot 2 \cdot 2 \cdot 2 \cdot \mathbf{3} \quad \text{\color{red}This can be written as } 2^5 \cdot 3.$$

The highlighting shows that 24, 60, and 96 have two factors of 2 and one factor of 3 in common. The GCF of 24, 60, and 96 is the product of their common prime factors.

$$\text{GCF} = \mathbf{2} \cdot \mathbf{2} \cdot \mathbf{3} = 2^2 \cdot 3^1 = 12$$

c. Since there are no prime factors common to 9, 10, and 30, their GCF is 1.

$$9 = 3 \cdot 3$$
$$10 = 2 \cdot 5$$
$$30 = 2 \cdot 3 \cdot 5$$

To find the greatest common factor of a list of terms, we can use the following approach.

Strategy for Finding the GCF	▼ 1. Write each coefficient as a product of prime factors.
	2. Identify the numerical and variable factors common to each term.
	3. Multiply the common numerical and variable factors identified in Step 2 to obtain the GCF. If there are no common factors, the GCF is 1.

EXAMPLE 2 Find the GCF of each list of terms: **a.** $12x^2$ and $20x$ **b.** $9a^5b^2$, $15a^4b^2$, and $90a^3b^3$

Strategy We will prime factor each coefficient of each term in the list. Then we will identify the numerical and variable factors common to each term and find their product.

Why The product of the common factors is the GCF of the terms in the list.

Solution **a.** *Step 1:* We write each coefficient, 12 and 20, as a product of prime factors. Recall that an exponent, as in x^2, indicates repeated multiplication.

$$12x^2 = \mathbf{2 \cdot 2} \cdot 3 \cdot \mathbf{x} \cdot x \qquad \text{This can be written as } 2^2 \cdot 3 \cdot x^2.$$
$$20x = \mathbf{2 \cdot 2} \cdot 5 \cdot \mathbf{x} \qquad \text{This can be written as } 2^2 \cdot 5 \cdot x.$$

Step 2: There are two common factors of 2 and one common factor of x.

Step 3: We multiply the common factors, 2, 2, and x, to obtain the GCF.

$$\text{GCF} = \mathbf{2 \cdot 2} \cdot x = 2^2 \cdot x = 4x$$

Success Tip

One way to identify common factors is to circle them:

$12x^2 = \boxed{2} \cdot \boxed{2} \cdot 3 \cdot \boxed{x} \cdot x$
$20x = \boxed{2} \cdot \boxed{2} \cdot 5 \cdot \boxed{x}$
$\text{GCF} = \boxed{2} \cdot \boxed{2} \cdot \boxed{x} = 4x$

Success Tip

The exponent on any variable in a GCF is the *smallest* exponent that appears on that variable in all of the terms under consideration.

b. *Step 1:* We write the coefficients, 9, 15, and 90, as products of primes. The exponents on the variables represent repeated multiplication.

$$9a^5b^2 = \mathbf{3} \cdot 3 \cdot \mathbf{a \cdot a \cdot a} \cdot a \cdot a \cdot \mathbf{b \cdot b} \qquad \text{This can be written as } 3^2 \cdot a^5 \cdot b^2.$$
$$15a^4b^2 = \mathbf{3} \cdot 5 \cdot \mathbf{a \cdot a \cdot a} \cdot a \cdot \mathbf{b \cdot b} \qquad \text{This can be written as } 3 \cdot 5 \cdot a^4 \cdot b^2.$$
$$90a^3b^3 = 2 \cdot \mathbf{3} \cdot 3 \cdot 5 \cdot \mathbf{a \cdot a \cdot a} \cdot \mathbf{b \cdot b} \cdot b \qquad \text{This can be written as } 2 \cdot 3^2 \cdot 5 \cdot a^3 \cdot b^3.$$

Step 2: The highlighting shows one common factor of 3, three common factors of a, and two common factors of b.

Step 3: $\text{GCF} = \mathbf{3} \cdot \mathbf{a \cdot a \cdot a} \cdot \mathbf{b \cdot b} = 3a^3b^2$

2 Factor Out the Greatest Common Factor.

The concept of greatest common factor is used to factor polynomials. For example, to factor $12x^2 + 20x$, we note that there are two terms, $12x^2$ and $20x$. We previously determined that the GCF of $12x^2$ and $20x$ is $4x$. With this in mind, we write each term of $12x^2 + 20x$ as a product of the GCF and one other factor. Then we apply the distributive property in reverse: $ab + ac = a(b + c)$.

$12x^2 + 20x = 4x \cdot 3x + 4x \cdot 5$ Write $12x^2$ and $20x$ as the product of the GCF, $4x$, and one other factor.

$= 4x(3x + 5)$ Write an expression so that the multiplication by $4x$ distributes over the sum of the terms $3x$ and 5.

We have found that the factored form of $12x^2 + 20x$ is $4x(3x + 5)$. This process is called **factoring out the greatest common factor.**

EXAMPLE 3   Factor: **a.** $8m + 24$ **b.** $35a^3b^2 - 14a^2b^3$ **c.** $3x^4 - 5x^3 + x^2$

Strategy   First, we will determine the GCF of the terms of the polynomial. Then we will write each term of the polynomial as the product of the GCF and one other factor.

Why   We can then use the distributive property to factor out the GCF.

Solution   **a.** Since the GCF of $8m$ and 24 is 8, we write $8m$ and 24 as the product of 8 and one other factor.

$8m + 24 = 8 \cdot m + 8 \cdot 3$ This step can be done mentally.

$= 8(m + 3)$ Factor out the GCF, 8.

To check, we multiply: $8(m + 3) = 8 \cdot m + 8 \cdot 3 = 8m + 24$. Since we obtain the original polynomial, $8m + 24$, the factorization is correct.

CAUTION   Remember to factor out the greatest common factor, not just a common factor. If we factored out 4 in the previous example, we would get

$8m + 24 = 4(2m + 6)$

However, the terms in red within parentheses have a common factor of 2, indicating that the factoring is not complete.

b. First, find the GCF of $35a^3b^2$ and $14a^2b^3$.

$35a^3b^2 = 5 \cdot 7 \cdot a \cdot a \cdot a \cdot b \cdot b$
$14a^2b^3 = 2 \cdot 7 \cdot a \cdot a \cdot b \cdot b \cdot b$ } The GCF is $7a^2b^2$.

$5\lfloor 35$ $2\lfloor 14$
$\quad 7$ $\quad 7$

Now, we write $35a^3b^2$ and $14a^2b^3$ as the product of the GCF, $7a^2b^2$, and one other factor.

$35a^3b^2 - 14a^2b^3 = 7a^2b^2 \cdot 5a - 7a^2b^2 \cdot 2b$ $\frac{35a^3b^2}{7a^2b^2} = 5a$ and $\frac{14a^2b^3}{7a^2b^2} = 2b$.

$= 7a^2b^2(5a - 2b)$ Factor out the GCF, $7a^2b^2$.

We check by multiplying: $7a^2b^2(5a - 2b) = 35a^3b^2 - 14a^2b^3$.

c. We factor out the GCF of the three terms, which is x^2.

$3x^4 - 5x^3 + x^2 = x^2(3x^2) - x^2(5x) + x^2(1)$ Write the last term, x^2, as $x^2(1)$.

$= x^2(3x^2 - 5x + 1)$ Factor out the GCF, x^2.

We check by multiplying: $x^2(3x^2 - 5x + 1) = 3x^4 - 5x^3 + x^2$.

Self Check 3   Factor: **a.** $6f + 36$ **b.** $24s^2t^2 - 42s^3t$
c. $y^6 - 10y^4 - y^3$

Now Try   Problems 41, 49, and 51

When asked to factor a polynomial whose leading coefficient is negative, factor out the *opposite of the GCF.*

EXAMPLE 4   Factor -1 from each polynomial: **a.** $-a^3 + 2a^2 - 4$ **b.** $6 - x$

Strategy   We will write each term of the polynomial as the product of -1 and one other factor.

Why   We can then use the distributive property to factor out the -1.

Solution   **a.** $-a^3 + 2a^2 - 4 = \mathbf{(-1)}a^3 + \mathbf{(-1)}(-2a^2) + \mathbf{(-1)}4$ *This step can be done mentally.*

$$= \mathbf{-1}(a^3 - 2a^2 + 4)$$ *Factor out -1.*
$$= -(a^3 - 2a^2 + 4)$$ *The 1 need not be written.*

We check by multiplying: $-(a^3 - 2a^2 + 4) = -a^3 + 2a^2 - 4$.

b. First, we will write the terms of $6 - x$ in descending order to determine whether the leading coefficient is negative.

$$6 - x = -x + 6$$ *Write in descending powers of x.*
$$= \mathbf{-1}(x) + \mathbf{(-1)}(-6)$$ *This step can be done mentally.*
$$= \mathbf{-1}(x - 6)$$ *Factor out -1.*
$$= -(x - 6)$$ *The 1 need not be written.*

> **Self Check 4**   Factor -1 from each polynomial: **a.** $-b^4 - 3b^2 + 2$
> **b.** $9 - t$
>
> **Now Try** ▶ Problems 57 and 61

> **Success Tip**
>
> This result suggests a quick way to factor out -1. Simply change the sign of each term of $-a^3 + 2a^2 - 4$ and write a $-$ symbol in front of the parentheses.

EXAMPLE 5   Factor out the opposite of the GCF in $-20m + 30$.

Strategy   First, we will determine the GCF of the terms of the polynomial. Then we will write each term of the polynomial as the product of the opposite of the GCF and one other factor.

Why   We can then use the distributive property to factor out the opposite of the GCF.

Solution   Since the GCF is 10, the opposite of the GCF is -10. We write each term of the polynomial as the product of -10 and another factor. Then we factor out -10.

$$-20m + 30 = \mathbf{(-10)}(2m) + \mathbf{(-10)}(-3)$$ *This step can be done mentally.*
$$= \mathbf{-10}(2m - 3)$$ *Note that the leading coefficient of the polynomial within the parentheses is positive.*

We check by multiplying: $-10(2m - 3) = -20m + 30$.

> **Self Check 5**   Factor out the opposite of the GCF in $-44c + 55$.
>
> **Now Try** ▶ Problem 67

> **Success Tip**
>
> It is standard practice to factor in such a way that the leading coefficient of the polynomial within the parentheses is positive.

EXAMPLE 6   Factor: $x(x + 4) + 3(x + 4)$

Strategy   We will identify the terms of the expression and find their GCF.

Why   We can then use the distributive property to factor out the GCF.

Solution   The expression has two terms: $\underline{x(x + 4)}$ + $\underline{3(x + 4)}$

The first term   The second term

The GCF of the two terms is the binomial $x + 4$, which can be factored out.

$x(x + 4) + 3(x + 4) = (x + 4)x + (x + 4)3$ Write each term as the product of $(x + 4)$ and one other factor.

$= (x + 4)(x + 3)$ Factor out the common factor, $(x + 4)$. Caution: Don't write $(x + 4)^2$.

Self Check 6   Factor: $2y(y - 1) + 7(y - 1)$

Now Try ▶   Problem 73

3 Factor by Grouping.

Although the terms of many polynomials don't have a common factor, other than 1, it is possible to factor some of them by arranging their terms in convenient groups. This method is called **factoring by grouping.**

EXAMPLE 7   Factor by grouping: **a.** $2x^3 + x^2 + 12x + 6$ **b.** $5c - 5d + cd - d^2$

Strategy   We will factor out a common factor from the first two terms and a common factor from the last two terms.

Why   This often produces a common binomial factor that can then be factored out.

Solution   **a.** Except for 1, there is no factor that is common to all four terms. However, the first two terms, $2x^3$ and x^2, have a common factor, x^2, and the last two terms, $12x$ and 6, have a common factor, 6.

$$\boxed{2x^3 + x^2}   +   \boxed{12x + 6}$$

$$x^2(2x + 1)   +   6(2x + 1)$$

We now see that $2x^3 + x^2$ and $12x + 6$ have a common binomial factor, $2x + 1$, which can be factored out.

$2x^3 + x^2 + 12x + 6 = x^2(2x + 1) + 6(2x + 1)$ Factor $2x^3 + x^2$ and $12x + 6$. Don't forget the blue + sign.

$= (2x + 1)(x^2 + 6)$ Factor out $2x + 1$. Caution: Don't write $(2x + 1)^2$.

We can check the factorization by multiplying. The result should be the original polynomial.

$(2x + 1)(x^2 + 6) = 2x^3 + 12x + x^2 + 6$

$= 2x^3 + x^2 + 12x + 6$ Rearrange the terms to get the original polynomial.

Caution

Don't think that
$5(c - d) + d(c - d)$ is in
factored form. It is a sum of two
terms. To be in factored form, the
result must be a product.

b. The first two terms have a common factor, 5, and the last two terms have a common factor, d.

$$5c - 5d + cd - d^2 = 5(c - d) + d(c - d)$$

Factor out 5 from $5c - 5d$
and d from $cd - d^2$.
Don't forget the blue + sign.

$$= (c - d)(5 + d)$$

Factor out the common binomial
factor, $c - d$. Caution:
Don't write $(c - d)^2$.

We can check by multiplying:

$$(c - d)(5 + d) = 5c + cd - 5d - d^2$$
$$= 5c - 5d + cd - d^2$$

Self Check 7   Factor by grouping: **a.** $3n^3 + 2n^2 + 9n + 6$
b. $7x - 7y + xy - y^2$

Now Try ▶ Problems 77 and 79

Factoring a Four-Termed Polynomial by Grouping

1. Group the terms of the polynomial so that the first two terms have a common factor and the last two terms have a common factor.
2. Factor out the common factor from each group.
3. Factor out the resulting common binomial factor. If there is no common binomial factor, regroup the terms of the polynomial and repeat steps 2 and 3.

By the multiplication property of 1, we know that 1 is a factor of every term. We can use this observation to factor certain polynomials by grouping.

EXAMPLE 8   Factor: **a.** $x^3 + 6x^2 + x + 6$ **b.** $x^2 - ax - x + a$

Strategy   We will follow the steps for factoring a four-termed polynomial.

Why   Since the terms of the polynomials do not have a common factor (other than 1), the only option is to attempt to factor these polynomials by grouping.

Solution   **a.** The first two terms, x^3 and $6x^2$, have a common factor of x^2. The only common factor of the last two terms, x and 6, is 1.

$$x^3 + 6x^2 + x + 6 = x^2(x + 6) + 1(x + 6)$$

Factor out x^2 from $x^3 + 6x^2$.
Factor out 1 from $x + 6$.
Don't forget the blue + sign.

$$= (x + 6)(x^2 + 1)$$

Factor out the common binomial
factor, $x + 6$. Caution:
Don't write $(x + 6)^2$.

Check the factorization by multiplying.

When we factor out −1 from the last two terms,

$$x^2 - ax \ \boxed{- x + a}$$

$$= x(x - a) - 1(x - a)$$

the signs of those terms change within the parentheses. The binomials within both sets of parentheses are then identical.

b. Since x is a common factor of the first two terms, we can factor it out.

$$\boxed{x^2 - ax} - x + a = x(x - a) - x + a \qquad \text{Factor out x from } x^2 - ax.$$

When factoring four terms by grouping, if the coefficient of the 3rd term is negative, we often factor out a negative coefficient from the last two terms. If we factor -1 from $-x + a$, a common binomial factor $x - a$ appears within the second set of parentheses, which we can factor out.

$$x^2 - ax \ \boxed{- x + a} = x(x - a) - 1(x - a) \qquad \text{To factor out −1, change the sign of}$$
$$\qquad\qquad\qquad\qquad\qquad\qquad\qquad -x \text{ and } a, \text{ and write −1 in front of the}$$
$$\qquad\qquad\qquad\qquad\qquad\qquad\qquad \text{parentheses.}$$

$$= (x - a)(x - 1) \qquad \text{Factor out the common factor, x − a.}$$
$$\qquad\qquad\qquad\qquad \text{Caution: Don't write } (x - a)^2.$$

Check by multiplying.

Self Check 8   Factor: **a.** $a^5 + 11a^4 + a + 11$
b. $b^2 - bc - b + c$

Now Try ▶ Problems 85 and 87

EXAMPLE 9   Factor: $5x^3 - 8 + 10x^2 - 4x$

Strategy   We will follow the steps for factoring by grouping.

Why   Since the four terms of the polynomial do not have a common factor (other than 1), we will attempt to factor it by grouping.

Solution   We note that the four terms of the polynomial do not have a common factor.

$$\underset{\substack{\text{No common factor}\\\text{(other than 1)}}}{\boxed{5x^3 - 8}} \ + \ \underset{\text{GCF = 2x}}{\boxed{10x^2 - 4x}} \qquad \text{We cannot factor the polynomial in its current form.}$$

An equivalent factorization, $(5x^2 - 4)(x + 2)$, results if the terms are arranged as $5x^3 - 4x + 10x^2 - 8$ or as $5x^3 - 4x - 8 + 10x^2$.

We will write the polynomial in descending powers of x and attempt to factor by grouping again.

$$5x^3 - 8 + 10x^2 - 4x = \boxed{5x^3 + 10x^2} \ \boxed{- 4x - 8}$$
$$= 5x^2(x + 2) - 4(x + 2) \qquad \text{Factor } 5x^2 \text{ from } 5x^3 + 10x^2 \text{ and}$$
$$\qquad\qquad\qquad\qquad\qquad\qquad\qquad -4 \text{ from } -4x - 8.$$

$$= (x + 2)(5x^2 - 4) \qquad \text{Factor out the GCF, x + 2.}$$

Self Check 9   Factor: $y^3 - 6 + 3y^2 - 2y$

Now Try ▶ Problem 93

The next example illustrates that when factoring a polynomial, we should *always look for a common factor first.*

EXAMPLE 10   Factor: $10k + 10m - 2km - 2m^2$

Strategy   Since all four terms have a common factor of 2, we factor it out first. Then we will factor the resulting polynomial by grouping.

Why   Factoring out the GCF first makes factoring by any method easier.

Solution   After factoring out 2 from all four terms, notice that within the parentheses, the first two terms have a common factor of 5, and the last two terms have a common factor of $-m$.

$$10k + 10m - 2km - 2m^2$$

$$= 2(5k + 5m - km - m^2)$$ Factor out the GCF, 2.

$$= 2[5(k + m) - m(k + m)]$$ Factor out 5 from $5k + 5m$. Factor out $-m$ from $-km - m^2$. This causes the signs of $-km$ and $-m^2$ to change within the second set of parentheses. Enclose the factoring by grouping process within brackets [ ].

$$= 2[(k + m)(5 - m)]$$ Factor out the common binomial factor, $k + m$.

$$= 2(k + m)(5 - m)$$ Drop the unnecessary brackets.

Check by multiplying.

> **Success Tip**
>
> Here is a factoring guideline: If a new pair of parentheses is produced in the factoring process, always see if the resulting expression within those parentheses can be factored further.

Self Check 10   Factor: $4t + 4s + 4tz + 4sz$

Now Try ▶ Problem 97

SECTION 6.1   **STUDY SET**

VOCABULARY

Fill in the blanks.

1. To _____ a polynomial means to express it as a product of two (or more) polynomials.

2. GCF stands for ____ ____ ____. When we write $2x + 4$ as $2(x + 2)$, we say that we have _____ out the GCF, 2.

3. To factor $m^3 + 3m^2 + 4m + 12$ by _____, we begin by writing $m^2(m + 3) + 4(m + 3)$.

4. The terms $x(x - 1)$ and $4(x - 1)$ have the common _____ factor $x - 1$.

CONCEPTS

5. Complete each factorization.
   a. $6x = 2 \cdot \boxed{} \cdot x$
   b. $35h^2 = 5 \cdot \boxed{} \cdot h \cdot \boxed{}$
   c. $18y^3z = 2 \cdot \boxed{} \cdot 3 \cdot \boxed{} \cdot y \cdot \boxed{} \cdot z$

6. a. Find the GCF of $30x^2$ and $105x^3$.
   $$30x^2 = 2 \cdot 3 \cdot 5 \cdot x \cdot x$$
   $$105x^3 = 3 \cdot 5 \cdot 7 \cdot x \cdot x \cdot x$$
   b. Find the GCF of $12a^2b^2$, $15a^3b$, and $75a^4b^2$.
   $$12a^2b^2 = 2 \cdot 2 \cdot 3 \cdot a \cdot a \cdot b \cdot b$$
   $$15a^3b = 3 \cdot 5 \cdot a \cdot a \cdot a \cdot b$$
   $$75a^4b^2 = 3 \cdot 5 \cdot 5 \cdot a \cdot a \cdot a \cdot a \cdot b \cdot b$$

7. a. Write a binomial such that the GCF of its terms is 2.

   b. Write a trinomial such that the GCF of its terms is x.

8. Check to determine whether each factorization is correct.
   a. $9y^3 + 5y^2 - 15y = 3y(3y^2 + 2y - 5)$
   b. $3s^3 + 2s^2 + 6s + 4 = (3s + 2)(s^2 + 2)$

Fill in the blanks to complete each factorization.

9. $2x^2 + 6x = \mathbf{2x} \cdot x + \mathbf{2x} \cdot 3$

$$= \boxed{} (\boxed{})$$

10. $3t^3 - t^2 + 15t - 5 = t^2(\mathbf{3t - 1}) + 5(\mathbf{3t - 1})$

$$= (\boxed{})(\boxed{})$$

11. Consider the polynomial $2k - 8 + hk - 4h$.
    a. How many terms does the polynomial have?
    b. Is there a common factor of all the terms, other than 1?
    c. What is the GCF of the first two terms and what is the GCF of the last two terms?

12. What is the first step in factoring $8y^2 - 16yz - 6y + 12z$?

NOTATION

Complete each factorization.

13. $8m^2 - 32m + 16 = \boxed{} (m^2 - 4m + 2)$

14. $10a^4 - 15a^3 = 5a \boxed{} (2a - 3)$

15. $b^3 - 6b^2 + 2b - 12 = \boxed{} (b - 6) + \boxed{} (b - 6)$
    $$= (\boxed{})(b^2 + 2)$$

16. $12 + 8n - 3m - 2mn = 4(3 + 2n) \boxed{} m(3 + 2n)$
    $$= (3 + 2n)(4 \boxed{} m)$$

GUIDED PRACTICE

Find the GCF of each list of numbers. **See Example 1.**

17. $6, 10$ **18.** $10, 15$

19. $18, 24$ **20.** $60, 72$

21. $14, 21, 42$ **22.** $16, 24, 48$

23. $40, 32, 24$ **24.** $28, 35, 21$

Find the GCF of each list of terms. **See Example 2.**

25. m^4, m^3 **26.** c^2, c^7

27. $15x, 25$ **28.** $9a, 21$

29. $20c^2, 12c$ **30.** $18r, 27r^3$

31. $18a^4, 9a^3, 27a^3$ **32.** $33m^5, 22m^6, 11m^5$

33. $24a^2, 16a^3b, 40ab$ **34.** $12r^2, 15rs, 9r^2s^2$

35. $6m^4n, 12m^3n^2, 9m^3n^3$ **36.** $15c^2d^4, 10c^2d, 40c^3d^3$

37. $4(x + 7), 9(x + 7)$ **38.** $2(y - 1), 5(y - 1)$

39. $4(p - t), p(p - t)$ **40.** $a(b + c), 3(b + c)$

Factor out the GCF. **See Example 3.**

41. $3x + 6$ **42.** $18x + 24$

43. $18m - 9$ **44.** $15s - 35$

45. $d^2 - 7d$ **46.** $a^2 + 9a$

47. $15c^3 + 25$ **48.** $33h^4 - 22$

49. $24a - 16a^2$ **50.** $18r - 30r^2$

51. $14x^2 - 7x - 7$ **52.** $27a^2 - 9a + 9$

53. $t^4 + t^3 + 2t^2$ **54.** $b^4 - b^3 - 3b^2$

55. $21x^2y^3 + 3xy^2$ **56.** $3x^2y^3 - 9x^4y^3$

Factor out −1 from each polynomial. **See Example 4.**

57. $-a - b$ **58.** $-x - 2y$

59. $-x^2 - x + 16$ **60.** $-t^2 - 9t + 1$

61. $5 - x$ **62.** $10 - m$

63. $9 - 4a$ **64.** $7 - 8b$

Factor each polynomial by factoring out the opposite of the GCF. **See Example 5.**

65. $-3x^2 - 6x$ **66.** $-4a^2 - 6a$

67. $-4a^2b + 12a^3$ **68.** $-25x^4 + 30x^2$

69. $-24x^4 - 48x^3 + 36x^2$ **70.** $-28a^5 - 42a^4 + 14a^3$

71. $-4a^3b^2 + 14a^2b^2 - 10ab^2$ **72.** $-30x^4y^3 + 24x^3y^2 - 60x^2y$

Factor. **See Example 6.**

73. $y(x + 2) + 3(x + 2)$

74. $r(t + v) + 3(t + v)$

75. $m(p - q) - 5(p - q)$

76. $ab(c - 7) - 12(c - 7)$

Factor by grouping. **See Example 7.**

77. $2x + 2y + ax + ay$ **78.** $bx + bz + 5x + 5z$

79. $rs - ru + 8sw - 8uw$ **80.** $12ab - 4ac + 3db - dc$

81. $7m^3 - 2m^2 + 14m - 4$ **82.** $9s^3 - 2s^2 + 36s - 8$

83. $5x^3 - x^2 + 10x - 2$ **84.** $6a^3 - a^2 + 18a - 3$

Factor by grouping. **See Example 8.**

85. $ab + ac + b + c$ **86.** $xy + 3y^2 + x + 3y$

87. $rs + 4s^2 - r - 4s$ **88.** $tx + tz - x - z$

89. $2ax + 2bx - 3a - 3b$ **90.** $rx + sx - ry - sy$

91. $mp - np - mq + nq$ **92.** $9p - 9q - mp + mq$

Factor by grouping. **See Example 9.**

93. $5m^3 + 6 + 5m^2 + 6m$

94. $4t^3 + 14 + 28t^2 + 2t$

95. $y^3 - 12 + 3y - 4y^2$

96. $h^3 - 8 + h - 8h^2$

Factor by grouping. Remember to factor out the GCF first. **See Example 10.**

97. $ax^3 - 2ax^2 + 5ax - 10a$

98. $x^3y^2 - 2x^2y^2 + 3xy^2 - 6y^2$

99. $6x^3 - 6x^2 + 12x - 12$

100. $3x^3 - 6x^2 + 15x - 30$

TRY IT YOURSELF

Factor.

101. $h^2(14 + r) + 5(14 + r)$

102. $x(y + 9) - 21(y + 9)$

103. $22a^3 - 33a^2$

104. $39r^3 + 26r^2$

105. $ax + bx - a - b$

106. $2xy + y^2 - 2x - y$

107. $15r^8 - 18r^6 - 30r^5$

108. $24cm - 12cn + 16c$

109. $27mp + 9mq - 9np - 3nq$

110. $4abc + 4ac^2 - 2bc - 2c^2$

111. $-60p^2t^2 - 80pt^3$

112. $-25x^5y^7 + 75x^3y^2$

113. $-2x + 5$

114. $-3x + 8$

115. $6x^2 - 2xy - 15x + 5y$

116. $m^3 + 5m^2 + m + 5$

117. $2x^3z - 4x^2z + 32xz - 64z$

118. $4a^2b + 12a^2 - 8ab - 24a$

119. $12uvw^3 - 54uv^2w^2$

120. $14xyz - 16x^2y^2z$

121. $x^3 + x^2 + x + 1$
122. $m^4 + m^3 + 2m + 2$
123. $-3r + 2s - 3$
124. $-6yz + 12xz + 5xy$

Look Alikes . . .

125. **a.** $5t^3 + 6t^2 + 15t + 18$ **b.** $3t^3 + 6t^2 + 15t + 18$

126. **a.** $x^2 + xy + x + y$ **b.** $2x^2 + 2xy + 2x + 2y$

APPLICATIONS

127. **Geometry.**   The dimensions of the rectangle shown below can be found by factoring the polynomial that represents its area. Find the polynomials that represent the length and the width of the rectangle.

Area = $(x^3 + 4x^2 + 5x + 20)$ ft^2

128. **Front Doors.**   Find a polynomial that represents the amount of surface on the front face of the door shown on the right that needs to be stained. Then factor the polynomial.

$2x$

x

$8x$

$4x + 4$

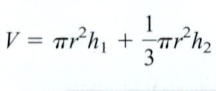

129. 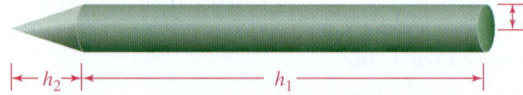 **from Campus to Careers**

Elementary School Teacher

A teacher is going to purchase a block of wax so that the children in her classroom can melt it down, add coloring, pour it into a mold, and make their own crayons. The amount of wax to make one crayon is given by the volume formula:

$$V = \pi r^2 h_1 + \frac{1}{3}\pi r^2 h_2$$

r

h_2 h_1

a. Rewrite the formula by factoring the expression on the right side.

b. Use the formula to estimate what size block of wax (in cubic inches) she needs to purchase if there are 20 students in her class, and each student will get 5 crayons. The dimensions for the crayon molds they will use are: $h_1 = 1\frac{5}{6}$ in., $h_2 = \frac{1}{2}$ in., and $r = \frac{1}{4}$ in. Round up to the nearest cubic inch.

130. **Interior Decorating.**   The expression $\pi rs + \pi Rs$ can be used to find the amount of material needed to make the lamp shade shown. Factor the expression.

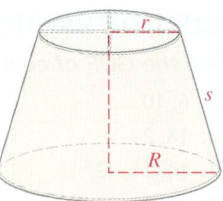

r

s

R

WRITING

131. Explain how to find the GCF of $32a^3$ and $16a^2$.
132. Explain this diagram.

Multiplication →

$$3x^2(5x^2 - 6x + 4) = 15x^4 - 18x^3 + 12x^2$$

← Factoring

133. Explain how factorizations of polynomials are checked. Give an example.
134. Explain the error.

Factor out the GCF: $\cancel{30a^3 - 12a^2 = 6a(5a^2 - 2a)}$

REVIEW

135. **Insurance Costs.**   A college student's good grades earned her a student discount on her car insurance premium. What was the percent of decrease, to the nearest percent, if her annual premium was lowered from $1,050 to $925?
136. **Calculating Grades.**   A student has test scores of 68%, 75%, and 79% in a government class. What must she score on the last exam to earn a B (80% or better) in the course?

CHALLENGE PROBLEMS

137. Factor: $6x^{4m}y^n + 21x^{3m}y^{2n} - 15x^{2m}y^{3n}$

138. Factor $ab - b^2 - bc + ac - bc - c^2$ by grouping.

©iStockphoto.com/Neustockimages

OBJECTIVES

1. Factor trinomials of the form $x^2 + bx + c$.
2. Factor trinomials of the form $x^2 + bx + c$ after factoring out the GCF.
3. Factor trinomials of the form $x^2 + bx + c$ using the grouping method.

Factoring Trinomials of the Form $x^2 + bx + c$

ARE YOU READY?

The following problems review some basic skills that are needed when factoring trinomials.

1. In $x^2 - 4x + 10$, what is the coefficient of the leading term?
2. Multiply: $(x + 8)(x - 1)$
3. What is the GCF of the terms of $2x^2 - 10x + 4$?
4. $x^2 - 4xy - 5y^2$ is a polynomial in how many variables?
5. Find two integers whose product is 8 and whose sum is 6.
6. Find two integers whose product is -15 and whose sum is 2.

In Chapter 5, we learned how to multiply binomials. For example, to multiply $x + 1$ and $x + 2$, we proceed as follows.

$$(x + 1)(x + 2) = x^2 + 2x + x + 2$$
$$= x^2 + 3x + 2$$

To *factor the trinomial* $x^2 + 3x + 2$, we will reverse the multiplication process and determine what factors were multiplied to obtain this result. Since the product of two binomials is often a trinomial, many trinomials factor into two binomials.

Multiplication: Given the binomial factors, we find a trinomial.

$$(x + 1)(x + 2) = x^2 + 3x + 2$$

Factoring: Given a trinomial, we find the binomial factors.

To begin the discussion of trinomial factoring, we consider trinomials of the form $x^2 + bx + c$, such as

$$x^2 + 8x + 15, \qquad y^2 - 13y + 12, \qquad a^2 + a - 20, \qquad \text{and} \qquad z^2 - 20z - 21$$

In each case, the **leading coefficient**—the coefficient of the squared variable—is 1.

The Language of Algebra

Recall that when a polynomial in one variable is written in descending powers of that variable, the coefficient of the first term is called the **leading coefficient**.

1. Factor Trinomials of the Form $x^2 + bx + c$.

To develop a method for factoring trinomials, we will find the product of $x + 6$ and $x + 4$ and make some observations about the result.

$$\begin{array}{c} \quad \mathbf{F} \qquad \mathbf{O} \qquad \mathbf{I} \qquad \mathbf{L} \\ (x + 6)(x + 4) = x \cdot x + x \cdot 4 + 6 \cdot x + 6 \cdot 4 \quad \text{Use the FOIL method.} \\ = x^2 + 4x + 6x + 24 \\ = x^2 + 10x + 24 \end{array}$$

First term — Middle term — Last term

The Language of Algebra

If a term of a trinomial is a number only, it is called a **constant term**.

$$x^2 + 10x + \underline{24}$$

Constant term

The result is a trinomial, where

- the first term, x^2, is the product of x and x
- the last term, 24, is the product of 6 and 4
- the coefficient of the middle term, 10, is the sum of 6 and 4

These observations suggest a strategy to use to factor trinomials that have 1 as the leading coefficient.

EXAMPLE 1 Factor: $x^2 + 8x + 15$

Strategy We will assume that $x^2 + 8x + 15$ is the product of two binomials and we will use a systematic method to find their terms.

Why Since the terms of $x^2 + 8x + 15$ do not have a common factor (other than 1), the only option available is to try to factor it as the product of two binomials.

Solution We represent the binomials using two sets of parentheses. Since the first term of the trinomial is x^2, we enter x and x as the first terms of its binomial factors.

$$x^2 + 8x + 15 = \left(x \;\boxed{}\right)\left(x \;\boxed{}\right) \quad \text{Because } x \cdot x \text{ will give } x^2.$$

The second terms of the binomials must be two integers whose product is 15 and whose sum is 8. Since the integers must have a positive product and a positive sum, we consider only pairs of positive integer factors of 15. The only such pairs, $1 \cdot 15$ and $3 \cdot 5$, are listed in the table. Then we find the sum of each pair and enter each result in the table.

Positive factors of 15	Sum of the positive factors of 15
$1 \cdot 15 = 15$	$1 + 15 = 16$
$3 \cdot 5 = 15$	$3 + 5 = 8$

List all of the pairs of positive integers that multiply to give 15.

Add each pair of factors.

Notation

By the commutative property of multiplication, the order of the binomial factors in a factorization does not matter. Thus, we can also write:

$x^2 + 8x + 15 = (x + 5)(x + 3)$

The second row of the table contains the correct pair of integers 3 and 5, whose product is 15 and whose sum is 8. To complete the factorization, we enter 3 and 5 as the second terms of the binomial factors.

$$x^2 + 8x + 15 = (x + 3)(x + 5)$$

We can check the factorization by multiplying:

$$(x + 3)(x + 5) = x^2 + 5x + 3x + 15$$
$$= x^2 + 8x + 15 \quad \text{This is the original trinomial.}$$

Self Check 1 Factor: $y^2 + 7y + 10$

Now Try Problem 15

EXAMPLE 2 Factor: $y^2 - 13y + 12$

Strategy We will assume that $y^2 - 13y + 12$ is the product of two binomials and we will use a systematic method to find their terms.

Why Since the terms of $y^2 - 13y + 12$ do not have a common factor (other than 1), the only option available is to try to factor it as the product of two binomials.

Solution We represent the binomials using two sets of parentheses. Since the first term of the trinomial is y^2, the first term of each binomial factor must be y.

$$y^2 - 13y + 12 = \left(y \;\boxed{}\right)\left(y \;\boxed{}\right) \quad \text{Because } y \cdot y \text{ will give } y^2.$$

The second terms of the binomials must be two integers whose product is 12 and whose sum is -13. Since the integers must have a positive product and a negative sum, we only consider pairs of negative integer factors of 12. The possible pairs are listed in the table.

Negative factors of 12	Sum of the negative factors of 12
$-1(-12) = 12$	$-1 + (-12) = -13$
$-2(-6) = 12$	$-2 + (-6) = -8$
$-3(-4) = 12$	$-3 + (-4) = -7$

You can stop listing the factors after finding the correct combination.

The Language of Algebra

Make sure you understand this vocabulary: **Many trinomials factor as the product of two binomials.**

Trinomial → Product of two binomials

$y^2 - 13y + 12 = (y - 1)(y - 12)$

The first row of the table contains the correct pair of integers -1 and -12, whose product is 12 and whose sum is -13. To complete the factorization, we enter -1 and -12 as the second terms of the binomial factors.

$$y^2 - 13y + 12 = (y - 1)(y - 12)$$

We check the factorization by multiplying:

$$(y - 1)(y - 12) = y^2 - 12y - y + 12$$
$$= y^2 - 13y + 12 \quad \text{This is the original trinomial.}$$

Self Check 2   Factor: $p^2 - 6p + 8$

Now Try ▶ Problem 19

EXAMPLE 3   Factor: $x^2 + x - 20$

Strategy   We will assume that $x^2 + x - 20$ is the product of two binomials and we will use a systematic method to find their terms.

Why   Since the terms of $x^2 + x - 20$ do not have a common factor (other than 1), the only option available is to try to factor it as the product of two binomials.

Solution   We represent the binomials using two sets of parentheses. Since the first term of the trinomial is x^2, the first term of each binomial factor must be x.

$$x^2 + x - 20 = \left(x \; \boxed{} \right)\left(x \; \boxed{} \right) \quad \text{Because } x \cdot x \text{ will give } x^2.$$

To determine the second terms of the binomials, we must find two integers whose product is -20 and whose sum is 1. Because the integers must have a negative product, their signs must be different. The possible pairs are listed in the table.

It is wise to follow an order when listing the factors in the table so that you don't skip the correct combination. Here, the first factors 1, 2, 4, 5, 10, and 20 are listed from least to greatest.

Factors of -20	Sum of the factors of -20
$1(-20) = -20$	$1 + (-20) = -19$
$2(-10) = -20$	$2 + (-10) = -8$
$4(-5) = -20$	$4 + (-5) = -1$
$5(-4) = -20$	$5 + (-4) = 1$
$10(-2) = -20$	$10 + (-2) = 8$
$20(-1) = -20$	$20 + (-1) = 19$

The fourth row of the table contains the correct pair of integers 5 and -4, whose product is -20 and whose sum is 1. To complete the factorization, we enter 5 and -4 as the second terms of the binomial factors.

$$x^2 + x - 20 = (x + 5)(x - 4)$$

Check the factorization by multiplying.

Self Check 3   Factor: $m^2 + m - 42$

Now Try ▶ Problem 23

EXAMPLE 4   Factor: $z^2 - 4z - 21$

Strategy   We will assume that $z^2 - 4z - 21$ is the product of two binomials and we will use a systematic method to find their terms.

Why   Since the terms of $z^2 - 4z - 21$ do not have a common factor (other than 1), the only option available is to try to factor it as the product of two binomials.

Solution   We represent the binomials using two sets of parentheses. Since the first term of the trinomial is z^2, the first term of each binomial factor must be z.

$$z^2 - 4z - 21 = \left(z \boxed{}\right)\left(z \boxed{}\right) \quad \text{Because } z \cdot z \text{ will give } z^2.$$

To determine the second terms of the binomials, we must find two integers whose product is -21 and whose sum is -4. Because the integers must have a negative product, their signs must be different. The possible pairs are listed in the table.

Factors of -21	Sum of the factors of -21
$1(-21) = -21$	$1 + (-21) = -20$
$3(-7) = -21$	$3 + (-7) = -4$
$7(-3) = -21$	$7 + (-3) = 4$
$21(-1) = -21$	$21 + (-1) = 20$

The second row of the table contains the correct pair of integers 3 and -7, whose product is -21 and whose sum is -4. To complete the factorization, we enter 3 and -7 as the second terms of the binomial factors.

$$z^2 - 4z - 21 = (z + 3)(z - 7)$$

Check by multiplying.

Self Check 4   Factor: $q^2 - 2q - 24$

Now Try ▶ Problem 29

The following guidelines are helpful when factoring trinomials.

Factoring Trinomials Whose Leading Coefficient Is 1

To factor a trinomial of the form $x^2 + bx + c$, find two numbers whose product is c and whose sum is b.

1. If c is positive, the numbers have the same sign.

2. If c is negative, the numbers have different signs.

Then write the trinomial as a product of two binomials. You can check by multiplying.

$$x^2 + bx + c = \left(x \boxed{}\right)\left(x \boxed{}\right)$$

The product of these numbers must be c and their sum must be b.

EXAMPLE 5   Factor: $-h^2 + 2h + 63$

Strategy   We will factor out -1 and then factor the resulting trinomial.

Why   It is easier to factor trinomials that have a positive leading coefficient.

Solution   After factoring out -1, we factor the trinomial within the parentheses.

$$
\begin{aligned}
-h^2 + 2h + 63 &= \mathbf{-1}(h^2 - 2h - 63) &&\text{Factor out } -1. \\
&= -(h^2 - 2h - 63) &&\text{The 1 need not be written.} \\
&= -(h + 7)(h - 9) &&\text{Factor } h^2 - 2h - 63.
\end{aligned}
$$

Check:
$$
\begin{aligned}
-(h + 7)(h - 9) &= -(h^2 - 9h + 7h - 63) &&\text{Multiply the binomials first.} \\
&= -(h^2 - 2h - 63) &&\text{Combine like terms.} \\
&= -h^2 + 2h + 63 &&\text{Drop the } - \text{ sign and change the} \\
&&&\text{sign of every term within the} \\
&&&\text{parentheses.}
\end{aligned}
$$

The result is the original trinomial.

Self Check 5   Factor: $-x^2 + 11x - 28$

Now Try ▶ Problem 33

We factor trinomials in two variables in a similar way.

EXAMPLE 6   Factor: $x^2 - 4xy - 5y^2$

Strategy   We will assume that $x^2 - 4xy - 5y^2$ is the product of two binomials and we will use a systematic method to find their terms.

Why   Since the terms of $x^2 - 4xy - 5y^2$ do not have a common factor (other than 1), the only option available is to try to factor it as the product of two binomials.

Solution   We represent the binomials using two sets of parentheses. Since the first term of the trinomial is x^2, the first term of each binomial factor must be x. Since the third term contains y^2, the last term of each binomial factor must contain y. To complete the factorization, we need to determine the coefficient of each y-term.

$$x^2 - 4xy - 5y^2 = \left(x \boxed{} y\right)\left(x \boxed{} y\right)$$ Because $x \cdot x$ will give x^2 and $y \cdot y$ will give y^2.

The trinomial $x^2 - 4xy - 5y^2$ is in two variables, x and y. It is written in **descending** powers of x and **ascending** powers of y.

The coefficients of y must be two integers whose product is -5 and whose sum is -4. Such a pair is 1 and -5. Instead of writing the first factor as $(x + 1y)$, we write it as $(x + y)$, because $1y = y$.

$$x^2 - 4xy - 5y^2 = (x + y)(x - 5y)$$

Check:
$$(x + y)(x - 5y) = x^2 - 5xy + xy - 5y^2$$
$$= x^2 - 4xy - 5y^2 \quad \text{This is the original trinomial.}$$

Self Check 6   Factor: $s^2 + 6st - 7t^2$

Now Try ▶ Problem 41

2 Factor Trinomials of the Form $x^2 + bx + c$ After Factoring Out the GCF.

If the terms of a trinomial have a common factor, it should be factored out first. The instruction "Factor" means for you to *factor the given expression completely*. A trinomial is **factored completely** when no factor can be factored further. Each factor of a completely factored expression will be prime.

EXAMPLE 7   Factor: $2x^4 + 26x^3 + 80x^2$

Strategy   We will factor out the GCF, $2x^2$, first. Then we will factor the resulting trinomial.

Why   The first step in factoring any polynomial is to factor out the GCF. Factoring out the GCF first makes factoring by any method easier.

Solution   We begin by factoring out the GCF, $2x^2$, from $2x^4 + 26x^3 + 80x^2$.

$$2x^4 + 26x^3 + 80x^2 = 2x^2(x^2 + 13x + 40)$$

Next, we factor $x^2 + 13x + 40$. The integers 8 and 5 have a product of 40 and a sum of 13, so the completely factored form of the given trinomial is

$$2x^4 + 26x^3 + 80x^2 = 2x^2(x + 8)(x + 5) \quad \text{The complete factorization must include } 2x^2.$$

Check:
$$2x^2(x + 8)(x + 5) = 2x^2(x^2 + 13x + 40)$$
$$= 2x^4 + 26x^3 + 80x^2 \quad \text{This is the original trinomial.}$$

Self Check 7   Factor: $4m^5 + 8m^4 - 32m^3$

Now Try ▶ Problem 47

Caution

When prime factoring 30, for example, you wouldn't stop here because 6 can be factored:

30 / 5 6

If a new pair of parentheses is produced in the factoring process, always see if the resulting expression within the parentheses can be factored further.

EXAMPLE 8   Factor: $-13g^2 + 36g + g^3$

Strategy   We will write the terms of the trinomial in descending powers of g.

Why   It is easier to factor a trinomial if its terms are written in descending powers of one variable.

Solution
$$-13g^2 + 36g + g^3 = g^3 - 13g^2 + 36g \quad \text{Rearrange the terms.}$$
$$= g(g^2 - 13g + 36) \quad \text{Factor out the GCF, } g.$$
$$= g(g - 9)(g - 4) \quad \text{Factor the trinomial.}$$

Caution

For multistep factorizations, don't forget to write the GCF in the final factored form.

Check the factorization by multiplying.

Self Check 8　Factor:　$-12t + t^3 + 4t^2$

Now Try ▶ Problem 59

If a trinomial with integer coefficients cannot be factored using only integers, it is called a **prime trinomial.**

EXAMPLE 9　Factor:　$x^2 + 2x + 3$

Strategy　We will assume that $x^2 + 2x + 3$ is the product of two binomials and we will use a systematic method to find their terms.

Why　Since the terms of $x^2 + 2x + 3$ do not have a common factor (other than 1), the only option available is to try to factor it as the product of two binomials.

Solution　To factor the trinomial, we must find two integers whose product is 3 and whose sum is 2. The possible factorizations are shown in the table.

The Language of Algebra

When a trinomial is not factorable using only integers, we say it is **prime** and that it does not factor **over** the integers.

Factors of 3	Sum of the factors of 3
$1(3) = 3$	$1 + 3 = 4$
$-1(-3) = 3$	$-1 + (-3) = -4$

Since there are no two integers whose product is 3 and whose sum is 2, the trinomial $x^2 + 2x + 3$ cannot be factored and is a *prime trinomial.*

Self Check 9　Factor:　$x^2 - 4x + 6$

Now Try ▶ Problem 63

3　Factor Trinomials of the Form $x^2 + bx + c$ Using the Grouping Method.

Another way to factor trinomials is to write them as equivalent four-termed polynomials and factor by grouping. To factor $x^2 + 8x + 15$ using this method, we proceed as follows.

1. First, identify a as the coefficient of the x^2 term, b as the coefficient of the x-term, and c as the last (constant) term. In this section, we are only factoring trinomials that have a leading coefficient of 1. Thus, a will always be equal to 1.

$$\left.\begin{array}{c} ax^2 + bx + c \\ \downarrow \quad \downarrow \quad \downarrow \\ 1x^2 + 8x + 15 \end{array}\right\} a = 1, b = 8, \text{ and } c = 15$$

Then, find the product ac, called the **key number:** $ac = \mathbf{1(15)} = 15$.

2. Next, find two integers whose product is the key number, 15, and whose sum is $b = \mathbf{8}$. Since the integers must have a positive product and a positive sum, we consider only positive factors of 15.

Key number = 15　　　　　$b = 8$

Positive factors of 15	Sum of the positive factors of 15
$1 \cdot 15 = 15$	$1 + 15 = 16$
$\mathbf{3 \cdot 5} = 15$	$\mathbf{3 + 5} = 8$

The second row of the table contains the correct pair of integers 3 and 5, whose product is the key number 15 and whose sum is $b = 8$.

3. Express the middle term, $8x$, of the trinomial as the *sum of two terms,* using the integers 3 and 5 found in step 2 as coefficients of the two terms.

$$x^2 + \textbf{8}x + 15 = x^2 + \textbf{3}x + \textbf{5}x + 15 \qquad \text{Express 8x as 3x + 5x.}$$

4. Factor the equivalent four-term polynomial by grouping:

$$x^2 + 3x + 5x + 15 = x(x + 3) + 5(x + 3) \qquad \text{Factor x out of } x^2 + 3x$$
$$\text{and 5 out of 5x + 15.}$$

$$= (x + 3)(x + 5) \qquad \text{Factor out x + 3.}$$

Check the factorization by multiplying.

The grouping method is an alternative to the method for factoring trinomials discussed earlier in this section. It is especially useful when the constant term, c, has many factors.

Factoring Trinomials of the Form $x^2 + bx + c$ Using Grouping

To factor a trinomial of the form $ax^2 + bx + c$, where $a = 1$:

1. Identify a, b, and c, and the key number, ac.

2. Find two integers whose product is the key number and whose sum is b.

3. Express the middle term, bx, as the sum (or difference) of two terms. Enter the two numbers found in step 2 as coefficients of x in the form shown below. Then factor the equivalent four-term polynomial by grouping.

$$x^2 + \boxed{}x + \boxed{}\,x + c$$

The product of these numbers must be ac, and their sum must be b.

4. Check the factorization using multiplication.

EXAMPLE 10 Factor by grouping: $x^2 + x - 20$

Strategy We will express the middle term, x, of the trinomial as the difference of two carefully chosen terms.

Why We want to produce an equivalent four-term polynomial that can be factored by grouping.

Solution Since $x^2 + x - 20 = \textbf{1}x^2 + \textbf{1}x - \textbf{20}$, we identify a as $\textbf{1}$, b as $\textbf{1}$, c as $-\textbf{20}$, and the key number ac as $\textbf{1}(-\textbf{20}) = -20$. We must find two integers whose product is -20 and whose sum is $b = \textbf{1}$. Since the integers must have a negative product, their signs must be different.

Key number $= -20$	$b = 1$
Factors of -20	**Sum of the factors of -20**
$1(-20) = -20$	$1 + (-20) = -19$
$2(-10) = -20$	$2 + (-10) = -8$
$4(-5) = -20$	$4 + (-5) = -1$
$\textbf{5}(-\textbf{4}) = -20$	$\textbf{5} + (-\textbf{4}) = 1$
$10(-2) = -20$	$10 + (-2) = 8$
$20(-1) = -20$	$20 + (-1) = 19$

Success Tip

We could also express the middle term as $-4x + 5x$. We obtain the same binomial factors, but in reverse order.

$$x^2 - 4x + 5x - 20$$
$$= x(x - 4) + 5(x - 4)$$
$$= (x - 4)(x + 5)$$

The fourth row of the table contains the correct pair of integers 5 and -4, whose product is -20 and whose sum is 1. They serve as the coefficients of $5x$ and $-4x$, the two terms that we use to represent the middle term, x, of the trinomial.

$$x^2 + x - 20 = x^2 + 5x - 4x - 20 \quad \text{Express the middle term, x, as 5x − 4x.}$$
$$= x(x + 5) - 4(x + 5) \quad \text{Factor x out of } x^2 + 5x \text{ and −4 out of −4x − 20.}$$
$$= (x + 5)(x - 4) \quad \text{Factor out x + 5.}$$

Check the factorization by multiplying.

Self Check 10   Factor by grouping: $m^2 + m - 42$

Now Try ▶ Problem 23

EXAMPLE 11   Factor by grouping: $x^2 - 4xy - 5y^2$

Strategy   We will express the middle term, $-4xy$, of the trinomial as the sum of two carefully chosen terms.

Why   We want to produce an equivalent four-term polynomial that can be factored by grouping.

Solution   In $1x^2 - 4xy - 5y^2$, we identify a as **1**, b as **−4**, c as **−5**, and the key number ac as $1(-5) = -5$. We must find two integers whose product is -5 and whose sum is $b = -4$. Such a pair is -5 and 1. They serve as the coefficients of $-5xy$ and $1xy$, the two terms that we use to represent the middle term, $-4xy$, of the trinomial.

Key number $= -5$ $b = -4$

Factors	Sum
$-5(1) = -5$	$-5 + 1 = -4$

$$x^2 - 4xy - 5y^2 = x^2 - 5xy + 1xy - 5y^2 \quad \text{Express the middle term, −4xy, as −5xy + 1xy. (1xy − 5xy could also be used.)}$$
$$= x(x - 5y) + y(x - 5y) \quad \text{Factor x out of } x^2 − 5xy \text{ and y out of 1xy − 5y}^2.$$
$$= (x - 5y)(x + y) \quad \text{Factor out x − 5y.}$$

Check the factorization by multiplying.

Self Check 11   Factor by grouping: $q^2 - 2qt - 24t^2$

Now Try ▶ Problem 41

EXAMPLE 12   Factor: $2x^3 - 20x^2 + 18x$

Strategy   We will factor out the GCF, $2x$, first. Then we will factor the resulting trinomial using the grouping method.

Why   The first step in factoring any polynomial is to factor out the GCF.

Solution   We begin by factoring out the GCF, $2x$, from $2x^3 - 20x^2 + 18x$.

$$2x^3 - 20x^2 + 18x = 2x(x^2 - 10x + 9)$$

To factor $x^2 - 10x + 9$ by grouping we identify a as 1, b as -10, and c as 9. We must find two integers whose product is the key number $ac = 1(9) = 9$ and whose sum is $b = -10$. Such a pair is -9 and -1.

Key number = 9	$b = -10$
Factors	**Sum**
$-9(-1) = 9$	$-9 + (-1) = -10$

$$x^2 - 10x + 9 = x^2 - 9x - 1x + 9 \quad \text{Express } -10x \text{ as } -9x - 1x.$$
$$(-1x - 9x \text{ could also be used.})$$
$$= x(x - 9) - 1(x - 9) \quad \text{Factor } x \text{ out of } x^2 - 9x \text{ and } -1 \text{ out of } -1x + 9.$$
$$= (x - 9)(x - 1) \quad \text{Factor out } x - 9.$$

The complete factorization of the original trinomial is

$$2x^3 - 20x^2 + 18x = 2x(x - 9)(x - 1) \quad \text{Don't forget to write the GCF, 2x.}$$

Check the factorization by multiplying.

Self Check 12 Factor: $3m^3 - 27m^2 + 24m$

Now Try ▶ Problem 47

SECTION 6.2 ▶ STUDY SET

VOCABULARY

Fill in the blanks.

1. The trinomial $x^2 - x - 12$ _____ as the product of two binomials: $(x - 4)(x + 3)$.
2. A _____ trinomial cannot be factored by using only integers.
3. The _____ coefficient of $x^2 - 3x + 2$ is 1.
4. A trinomial is factored _____ when no factor can be factored further.

CONCEPTS

Fill in the blanks.

5. **a.** Before attempting to factor a trinomial, be sure that it is written in _____ powers of a variable.
   b. Before attempting to factor a trinomial into two binomials, always factor out any _____ factors first.

6. $x^2 + x - 56 = \left(x \boxed{}\right)\left(x \boxed{}\right)$

   The product of these numbers must be ____, and their sum must be ____.

7. $x^2 + 5x + 3$ cannot be factored because we cannot find two integers whose product is ____ and whose sum is ____.

8. Complete the following table.

Factors of 8	Sum of the factors of 8
1(8)	
2( )	
−1(−8)	
(−4)	

9. Check to determine whether each factorization is correct.
   a. $x^2 - x - 20 = (x + 5)(x - 4)$
   b. $4a^2 + 12a - 16 = 4(a - 1)(a + 4)$

10. Find two integers whose
    a. product is 10 and whose sum is 7.
    b. product is 8 and whose sum is -6.
    c. product is -6 and whose sum is 1.
    d. product is -9 and whose sum is -8.

11. Consider a trinomial of the form $x^2 + bx + c$.
    a. If c is positive, what can be said about the two integers that should be chosen for the factorization?

    b. If c is negative, what can be said about the two integers that should be chosen for the factorization?

12. Fill in each blank to explain how to factor $x^2 + 7x + 10$ by grouping.

    We express the middle term, $7x$, as the sum of ____ terms:
    $$x^2 + 7x + 10 = x^2 + \boxed{}x + \boxed{}x + 10$$

    The product of these numbers must be ____, and their sum must be ____.

NOTATION

13. To factor a trinomial, a student made a table and circled the correct pair of integers, as shown. Complete the factorization of the trinomial.

    $(x \boxed{})(x \boxed{})$

Factors	Sum
1(−6)	−5
2(−3)	−1
③(−2)	①
6(−1)	5

14. To factor a trinomial by grouping, a student made a table and circled the correct pair of integers, as shown. Enter the correct coefficients for the first stage in the factorization process.

Key number = 16

Factors	Sum
$1 \cdot 16$	17
$2 \cdot 8$	10
$4 \cdot 4$	8

$$x^2 + \boxed{}x + \boxed{}x + 16$$

GUIDED PRACTICE

Factor. See Example 1 or Objective 1.

15. $x^2 + 3x + 2$

16. $y^2 + 4y + 3$

17. $z^2 + 7z + 12$

18. $x^2 + 7x + 10$

Factor each trinomial. See Example 2 or Example 10.

19. $m^2 - 5m + 6$

20. $n^2 - 7n + 10$

21. $t^2 - 11t + 28$

22. $c^2 - 9c + 8$

Factor. See Example 3 or Example 10.

23. $x^2 + 5x - 24$

24. $u^2 + u - 42$

25. $t^2 + 13t - 48$

26. $m^2 + 2m - 48$

Factor each trinomial. See Example 4 or Example 10.

27. $a^2 - 6a - 16$

28. $a^2 - 10a - 39$

29. $b^2 - 9b - 36$

30. $x^2 - 3x - 40$

Factor. See Example 5.

31. $-x^2 - 7x - 10$

32. $-x^2 + 9x - 20$

33. $-t^2 - t + 30$

34. $-t^2 - 15t + 34$

35. $-r^2 - 3r + 54$

36. $-d^2 - 2d + 63$

37. $-m^2 + 18m - 77$

38. $-n^2 + 14n - 33$

Factor. See Example 6 or Example 11.

39. $a^2 + 4ab + 3b^2$

40. $a^2 + 6ab + 5b^2$

41. $x^2 - 6xy - 7y^2$

42. $x^2 + 10xy - 11y^2$

43. $r^2 + rs - 2s^2$

44. $m^2 + mn - 6n^2$

45. $a^2 - 5ab + 6b^2$

46. $p^2 - 7pq + 10q^2$

Factor. See Example 7 or Example 12.

47. $2x^2 + 10x + 12$

48. $3y^2 - 21y + 18$

49. $6a^2 - 30a + 24$

50. $4b^2 + 12b - 16$

51. $5a^2 - 25a + 30$

52. $2b^2 - 20b + 18$

53. $-z^3 + 29z^2 - 100z$

54. $-m^3 + m^2 + 56m$

Write each trinomial in descending powers of one variable and factor. See Example 8.

55. $80 - 24x + x^2$

56. $y^2 + 100 + 25y$

57. $10y + 9 + y^2$

58. $x^2 - 13 - 12x$

59. $r^3 - 16r + 6r^2$

60. $u^3 - 12u - u^2$

61. $4r^2x + r^3 + 3rx^2$

62. $a^3 + 5ab^2 + 6a^2b$

Factor. See Example 9.

63. $u^2 + 10u + 15$

64. $v^2 + 9v + 15$

65. $r^2 + 2r - 4$

66. $r^2 - 9r - 12$

TRY IT YOURSELF

Choose the correct method from Section 6.1 or Section 6.2 to factor each of the following.

67. $5x + 15 + xy + 3y$

68. $ab + b + 2a + 2$

69. $26n^2 - 8n$

70. $40c^2 - 12c$

71. $a^2 - 4a - 5$

72. $t^2 - 5t - 50$

73. $-x^2 + 21x + 22$

74. $-r^2 + 14r - 45$

75. $4xy - 4x + 28y - 28$

76. $3xy - 3x + 15y - 15$

77. $12b^4 - 48b^3 - 36b^2$

78. $14n^5 - 42n^4 - 28n^3$

79. $r^2 - 9r + 18$

80. $y^2 - 17y + 72$

81. $-n^4 + 28n^3 + 60n^2$

82. $-c^5 + 16c^4 + 80c^3$

83. $x^2 + 4xy + 4y^2$

84. $m^2 - 8mn + 16n^2$

85. $a^2 - 4ab - 12b^2$

86. $p^2 + pq - 6q^2$

87. $4x^4 + 16x^3 + 16x^2$

88. $3a^4 + 30a^3 + 75a^2$

89. $a^2 - 46a + 45$

90. $r^2 - 37r + 36$

91. $r^2 - 2r + 4$

92. $m^2 + 3m - 20$

93. $t(x + 2) + 7(x + 2)$

94. $r(t - v) + 10(t - v)$

95. $s^4 + 11s^3 - 26s^2$

96. $x^4 + 14x^3 + 45x^2$

97. $15s^3 + 75$

98. $33g^4 - 99$

99. $-13y + y^2 - 14$

100. $-3a + a^2 + 2$

101. $2x^2 - 12x + 16$

102. $6t^2 - 18t - 24$

Look Alikes . . .

103. a. $x^2 - 10x + 24$ **b.** $x^2 - 10x - 24$

104. a. $x^2 - 5x + 6$ **b.** $x^2 - 5x - 6$

APPLICATIONS

105. Pets. The cage shown on the right is used for transporting dogs. Its volume is $(x^3 + 12x^2 + 27x)$ in.^{3}. The dimensions of the cage can be found by factoring. If the cage is longer than it is tall and taller than it is wide, find the polynomials that represent its length, width, and height.

Height
Width
Length

106. Photography. A picture cube is a clever way to display 6 photographs in a small amount of space. Suppose the surface area of the entire cube is given by the polynomial $(6s^2 + 12s + 6)$ in.^{2}. Find the polynomial that represents the length of an edge of the cube.

WRITING

107. Explain what it means when we say that a trinomial is the product of two binomials. Give an example.

108. Are $2x^2 - 12x + 16$ and $x^2 - 6x + 8$ factored in the same way? Explain.

109. When factoring $x^2 - 2x - 3$, one student got $(x - 3)(x + 1)$, and another got $(x + 1)(x - 3)$. Are both answers acceptable? Explain.

110. In the partial solution shown below, a student began to factor the trinomial. Write a note to the student explaining his mistake.

Factor: $x^2 - 2x - 63$

$(x - \quad)(x - \quad)$

111. Explain the error in the following factorization.

$x^3 + 8x^2 + 15x = x(x^2 + 8x + 15)$
$= (x + 3)(x + 5)$

112. Explain why the factorization is not complete.

$2y^2 - 12y + 16 = 2(y^2 - 6y + 8)$

REVIEW

Simplify each expression. Write each answer without negative exponents.

113. $\dfrac{x^{12}x^{-7}}{x^3x^4}$ **114.** $\dfrac{a^4a^{-2}}{a^2a^0}$

115. $(x^{-3}x^{-2})^2$ **116.** $\left(\dfrac{18a^2b^3c^{-4}}{3a^{-1}b^2c}\right)^{-3}$

CHALLENGE PROBLEMS

Factor completely.

117. $x^2 - \dfrac{6}{5}x + \dfrac{9}{25}$

118. $x^2 - 0.5x + 0.06$

119. $x^{2m} - 12x^m - 45$

120. $x^2(y + 1) - 3x(y + 1) - 70(y + 1)$

121. Find all positive integer values of c that make $n^2 + 6n + c$ factorable.

122. Find all integer values of b that make $x^2 + bx - 44$ factorable.

SECTION 6.3

Factoring Trinomials of the Form $ax^2 + bx + c$

OBJECTIVES

1 Factor trinomials using the trial-and-check method.

2 Factor trinomials after factoring out the GCF.

3 Factor trinomials using the grouping method.

ARE YOU READY?

The following problems review some basic skills that are needed when factoring trinomials.

1. In $3x^2 - x + 8$, what is the coefficient of the leading term?

2. Multiply: $5y \cdot y$

3. Multiply: $(2x - 5)(3x + 1)$

4. Find two integers whose product is -3. (There are two possible answers.)

5. Identify the coefficients of each term of $6x^2 - 3x + 7$.

6. Factor by grouping: $10b^2 + 15b - 2b - 3$

In this section, we will factor trinomials with leading coefficients other than 1, such as

$$2x^2 + 5x + 3, \qquad 6a^2 - 17a + 5, \qquad \text{and} \qquad 4b^2 + 8bc - 45c^2$$

We can use two methods to factor these trinomials. With the first method, we make educated guesses and then check them using multiplication. The correct factorization is determined through a process of elimination. The second method is an extension of factoring by grouping.

1 Factor Trinomials Using the Trial-and-Check Method.

EXAMPLE 1

Factor: $2x^2 + 5x + 3$

Strategy   We will assume that $2x^2 + 5x + 3$ is the product of two binomials and we will use a systematic method to find their terms.

Why   Since the terms of $2x^2 + 5x + 3$ do not have a common factor (other than 1), the only option available is to try to factor it as the product of two binomials.

Solution   We represent the binomials using two sets of parentheses. Since the first term of the trinomial is $2x^2$, we enter $2x$ and x as the first terms of the binomial factors.

$$\left(2x \;\boxed{}\;\right)\left(x \;\boxed{}\;\right)$$ Because $2x \cdot x$ will give $2x^2$.

> **The Language of Algebra**
>
> To **interchange** means to put each in the place of the other. We create all of the possible factorizations by *interchanging* the second terms of the binomials.
>
> $(2x + 1)(x + 3)$
>
> $(2x + 3)(x + 1)$

The second terms of the binomials must be two integers whose product is 3. Since the coefficients of the terms of $2x^2 + 5x + 3$ are positive, we consider only pairs of positive integer factors of 3. Since there is just one such pair, $1 \cdot 3$, we can enter 1 and 3 as the second terms of the binomials, or we can reverse the order and enter 3 and 1.

$(2x + 1)(x + 3)$ or $(2x + 3)(x + 1)$

The first possibility is incorrect, because when we find the outer and inner products and combine like terms, we obtain an incorrect middle term of $7x$.

Outer: 6x

$(2x + 1)(x + 3)$ Multiply and add to find the middle term: $6x + x = 7x$.
                     $7x$ is not the middle term we want.

Inner: x

The second possibility is correct, because it gives a middle term of $5x$.

Outer: 2x

$(2x + 3)(x + 1)$ Multiply and add to find the middle term: $2x + 3x = \boxed{5x}$.
                     $5x$ is the middle term we want.

Inner: 3x

Thus, the factorization is:

$$2x^2 + 5x + 3 = (2x + 3)(x + 1)$$

Check the factorization by multiplying:

$$(2x + 3)(x + 1) = 2x^2 + 2x + 3x + 3$$
$$= 2x^2 + 5x + 3 \qquad \text{This is the original trinomial.}$$

Self Check 1   Factor: $2x^2 + 5x + 2$

Now Try ▶ Problem 19

EXAMPLE 2

Factor: $6a^2 - 17a + 5$

Strategy   We will assume that $6a^2 - 17a + 5$ is the product of two binomials and we will use a systematic method to find their terms.

Why   Since the terms of $6a^2 - 17a + 5$ do not have a common factor (other than 1), the only option available is to try to factor it as the product of two binomials.

Solution   We represent the binomials using two sets of parentheses. Since the first term is $6a^2$, the first terms of the factors must be $6a$ and a or $3a$ and $2a$.

$$\left(6a \;\boxed{}\;\right)\left(a \;\boxed{}\;\right) \quad \text{or} \quad \left(3a \;\boxed{}\;\right)\left(2a \;\boxed{}\;\right)$$ Because $6a \cdot a$ or $3a \cdot 2a$ will give $6a^2$.

The second terms of the binomials must be two integers whose product is 5. Since the last term of $6a^2 - 17a + 5$ is positive and the coefficient of the middle term is negative, we consider only negative integer factors of the last term. Since there is just one such pair, $-1(-5)$, we can enter -1 and -5, or we can reverse the order and enter -5 and -1 as second terms of the binomials.

$(6a - 1)(a - 5)$ $-30a$ $-a$ $-30a - a = -31a$ Incorrect middle term

$(6a - 5)(a - 1)$ $-6a$ $-5a$ $-6a - 5a = -11a$ Incorrect middle term

$(3a - 1)(2a - 5)$ $-15a$ $-2a$ $-15a - 2a = \boxed{-17a}$ Correct middle term

$(3a - 5)(2a - 1)$ $-3a$ $-10a$ $-3a - 10a = -13a$ Incorrect middle term

Only the possibility shown in blue gives the correct middle term of $-17a$. Thus,

$$6a^2 - 17a + 5 = (3a - 1)(2a - 5)$$

We check by multiplying: $(3a - 1)(2a - 5) = 6a^2 - 17a + 5$.

Self Check 2   Factor: $6b^2 - 19b + 3$

Now Try   Problem 27

EXAMPLE 3   Factor: $3y^2 - 7y - 6$

Strategy   We will assume that $3y^2 - 7y - 6$ is the product of two binomials and we will use a systematic method to find their terms.

Why   Since the terms of $3y^2 - 7y - 6$ do not have a common factor (other than 1), the only option available is to try to factor it as the product of two binomials.

Solution   Since the first term is $3y^2$, the first terms of the binomial factors must be $3y$ and y.

$$\left(3y \;\boxed{}\;\right)\left(y \;\boxed{}\;\right)$$ Because $3y \cdot y$ will give $3y^2$.

The second terms of the binomials must be two integers whose product is -6. There are four such pairs: $1(-6)$, $-1(6)$, $2(-3)$, and $-2(3)$. When these pairs are entered, and then reversed as second terms of the binomials, there are eight possibilities to consider. Four of them can be discarded because they include a binomial whose terms have a common factor. If the terms of $3y^2 - 7y - 6$ do not have a common factor (other than 1), neither can any of its binomial factors.

Success Tip

If the terms of a trinomial do not have a common factor other than 1, the terms of each of its binomial factors will not have a common factor other than 1.

For 1 and -6: $(3y + 1)(y - 6)$ $-18y$ y $-18y + y = -17y$ Incorrect middle term   or $(3y - 6)(y + 1)$ See the Success Tip on the left. $3y - 6$ has a common factor of 3. This can't be the factorization.

For -1 *and* 6: $(3y - 1)(y + 6)$ or $(3y + 6)(y - 1)$

$18y$

$-y$

$18y - y = 17y$
Incorrect middle term

$3y + 6$ has a common factor of 3.
This can't be the factorization.

Success Tip

Reversing the signs within the binomial factors reverses the sign of the middle term. For example, the factors 2 and -3 give the middle term $-7y$, while -2 and 3 give the middle term $7y$.

For 2 *and* -3: $(3y + 2)(y - 3)$ or $(3y - 3)(y + 2)$

$-9y$

$2y$

$-9y + 2y = \boxed{-7y}$
Correct middle term

$3y - 3$ has a common factor of 3.
This can't be the factorization.

Success Tip

All eight possible factorizations are listed. In practice, you will often find the correct factorization without having to examine the entire list of possibilities.

For -2 *and* 3: $(3y - 2)(y + 3)$ or $(3y + 3)(y - 2)$

$9y$

$-2y$

$9y - 2y = 7y$
Incorrect middle term

$3y + 3$ has a common factor of 3.
This can't be the factorization.

Only the possibility shown in green gives the correct middle term of $-7y$. Thus,

$$3y^2 - 7y - 6 = (3y + 2)(y - 3)$$

Check the factorization by multiplying.

Self Check 3   Factor: $5t^2 - 23t - 10$

Now Try ▶ Problem 35

EXAMPLE 4   Factor: $4b^2 + 8bc - 45c^2$

Strategy   We will assume that $4b^2 + 8bc - 45c^2$ is the product of two binomials and we will use a systematic method to find their terms.

Why   Since the terms of $4b^2 + 8bc - 45c^2$ do not have a common factor (other than 1), the only option available is to try to factor it as the product of two binomials.

Solution   Since the first term is $4b^2$, the first terms of the binomial factors must be $4b$ and b or $2b$ and $2b$. Since the last term contains c^2, the second terms of the binomial factors must contain c.

Notation

The trinomial is in two variables, b and c. It is written in descending powers of b and ascending powers of c.

$$\left(4b \ \boxed{} \ c\right)\left(b \ \boxed{} \ c\right) \quad \text{or} \quad \left(2b \ \boxed{} \ c\right)\left(2b \ \boxed{} \ c\right)$$

Because $4b \cdot b$ or $2b \cdot 2b$ gives $4b^2$, and because $c \cdot c$ gives c^2.

The coefficients of c must be two integers whose product is -45. Since the coefficient of the last term is negative, the signs of the integers must be different. If we pick factors of $4b$ and b for the first terms, and -1 and 45 for the coefficients of c, the multiplication gives an incorrect middle term of $179bc$.

$180bc$

$(4b - c)(b + 45c)$ $180bc - bc = 179bc$
Incorrect middle term

$-bc$

Success Tip

When using the trial-and-check method to factor $ax^2 + bx + c$, here is one suggestion: If b is relatively small, test the factorizations that use the smaller factors of a and c first.

If we pick factors of $4b$ and b for the first terms, and 15 and -3 for the coefficients of c, the multiplication gives an incorrect middle term of $3bc$.

$$-12bc$$
$$(4b + 15c)(b - 3c) \quad -12bc + 15bc = \boxed{3bc}$$
$$15bc \qquad \text{Incorrect middle term}$$

If we pick factors of $2b$ and $2b$ for the first terms, and -5 and 9 for the coefficients of c, we have

$$18bc$$
$$(2b - 5c)(2b + 9c) \quad 18bc - 10bc = \boxed{8bc}$$
$$-10bc \qquad \text{Correct middle term}$$

which gives the correct middle term of $8bc$. Thus,

$$4b^2 + 8bc - 45c^2 = (2b - 5c)(2b + 9c)$$

Check the factorization by multiplying.

Self Check 4   Factor: $4x^2 + 4xy - 3y^2$

Now Try ▶ Problem 43

Because guesswork is often necessary, it is difficult to give specific rules for factoring trinomials with leading coefficients other than 1. However, the following hints are helpful when using the **trial-and-check method.**

Factoring Trinomials with Leading Coefficients Other Than 1

To factor trinomials with leading coefficients other than 1:

1. Factor out any GCF (including -1 if that is necessary to make a positive in a trinomial of the form $ax^2 + bx + c$).

2. Write the trinomial as a product of two binomials. The coefficients of the first terms of each binomial factor must be factors of a, and the last terms must be factors of c.

The product of these numbers must be a.

$$ax^2 + bx + c = (\square x \quad \square)(\square x \quad \square)$$

The product of these numbers must be c.

3. If c is positive, the signs within the binomial factors match the sign of b. If c is negative, the signs within the binomial factors are opposites.

4. Try combinations of the first terms and the second terms of the binomial factors until you find the one that gives the proper middle term in the trinomial. If no combination works, the trinomial is prime.

5. Check by multiplying.

---

2 Factor Trinomials After Factoring Out the GCF.

If the terms of a trinomial have a common factor, *the GCF (or the opposite of the GCF) should always be factored out first.*

EXAMPLE 5   Factor: $2x^2 - 8x^3 + 3x$

Strategy   We will write the trinomial in descending powers of x and factor out the common factor, $-x$.

Why   It is easier to factor trinomials that have a positive leading coefficient.

Solution   Write the trinomial in descending powers of x: $-8x^3 + 2x^2 + 3x$.

$$-8x^3 + 2x^2 + 3x = -x(8x^2 - 2x - 3) \quad \text{Factor out the opposite of the GCF, } -x.$$

We now factor $8x^2 - 2x - 3$. Its factorization has the form

$$\left(8x \ \boxed{}\right)\left(x \ \boxed{}\right) \quad \text{or} \quad \left(2x \ \boxed{}\right)\left(4x \ \boxed{}\right) \quad \begin{array}{l}\text{Because } 8x \cdot x \text{ or} \\ 4x \cdot 2x \text{ gives } 8x^2.\end{array}$$

> **The Language of Algebra**
>
> When asked to *factor* a polynomial, that means we should *factor completely*.

The second terms of the binomials must be two integers whose product is -3. There are two such pairs: $1(-3)$ and $-1(3)$. Since the coefficient of the middle term of the trinomial, $-2x$, is small, we pick the smaller factors of $8x^2$, which are $2x$ and $4x$, for the first terms and 1 and -3 for the second terms.

$$\overset{\displaystyle -6x}{(2x + 1)(4x \underset{\displaystyle 4x}{- 3})} \quad -6x + 4x = \boxed{-2x}$$

This factorization gives the correct middle term of $-2x$. Thus,

$$8x^2 - 2x - 3 = (2x + 1)(4x - 3)$$

> **Caution**
>
> For multistep factorization, don't forget to write the GCF (or its opposite) in the final factored form.

We can now give the complete factorization of the original trinomial.

$$-8x^3 + 2x^2 + 3x = -x(8x^2 - 2x - 3)$$
$$= -x(2x + 1)(4x - 3)$$
$$\hspace{4cm}\uparrow\underline{\hspace{4cm}} \text{ Don't forget this factor.}$$

Check the factorization by multiplying.

Self Check 5   Factor: $12y - 14y^3 + 22y^2$

Now Try ▶   Problem 53

---

3 Factor Trinomials Using the Grouping Method.

Another way to factor a trinomial of the form $ax^2 + bx + c$ is to write it as an equivalent four-termed polynomial and factor it by grouping. For example, to factor $2x^2 + 5x + 3$, we proceed as follows.

1. Identify the values of a, b, and c.

$$\left.\begin{array}{ccc} ax^2 & + bx & + c \\ \downarrow & \downarrow & \downarrow \\ 2x^2 & + 5x & + 3 \end{array}\right\} a = 2, b = 5, \text{ and } c = 3$$

Then, find the product ac, called the **key number**: $ac = 2(3) = 6$.

2. Next, find two integers whose product is $ac = 6$ and whose sum is $b = 5$. Since the integers must have a positive product and a positive sum, we consider only positive factors of 6.

Key number $= 6$ $b = 5$

Positive factors of 6	Sum of the positive factors of 6
$1 \cdot 6 = 6$	$1 + 6 = 7$
$2 \cdot 3 = 6$	$2 + 3 = 5$

The second row of the table contains the correct pair of integers 2 and 3, whose product is 6 and whose sum is 5.

3. Express the middle term, $5x$, of the trinomial as the *sum of two terms,* using the integers 2 and 3 found in step 2 as coefficients of the two terms.

$$2x^2 + \mathbf{5x} + 3 = 2x^2 + \mathbf{2x + 3x} + 3 \qquad \text{Express 5x as 2x + 3x.}$$

4. Factor the equivalent four-term polynomial by grouping:

$$2x^2 + 2x + 3x + 3 = 2x(x + 1) + 3(x + 1) \qquad \begin{array}{l}\text{Factor 2x out of } 2x^2 + 2x \\ \text{and 3 out of 3x + 3.}\end{array}$$

$$= (x + 1)(2x + 3) \qquad \text{Factor out x + 1.}$$

Check by multiplying.

Factoring by grouping is especially useful when the leading coefficient, a, and the constant term, c, have many factors.

Factoring Trinomials by Grouping

To factor a trinomial by grouping:

1. Factor out any GCF (including -1 if that is necessary to make a positive in a trinomial of the form $ax^2 + bx + c$).

2. Identify a, b, and c, and find the key number ac.

3. Find two integers whose product is the key number and whose sum is b.

4. Express the middle term, bx, as the sum (or difference) of two terms. Enter the two numbers found in step 3 as coefficients of x in the form shown below. Then factor the equivalent four-term polynomial by grouping.

$$ax^2 + \boxed{}x + \boxed{}x + c$$

The product of these numbers must be ac and their sum must be b.

5. Check the factorization by multiplying.

EXAMPLE 6   Factor by grouping: $\quad 10x^2 + 13x - 3$

Strategy   We will express the middle term, $13x$, of the trinomial as the sum of two carefully chosen terms.

Why   We want to produce an equivalent four-term polynomial that can be factored by grouping.

Solution   In $10x^2 + 13x - 3$, we have $a = 10, b = 13$, and $c = -3$. The key number is $ac = 10(-3) = -30$. We must find a factorization of -30 such that the sum of the factors is $b = 13$. The possible factor pairs are listed in the table. Since the factors must have a negative product, their signs must be different.

Key number = -30 $b = 13$

It is wise to follow an order when listing the factors in the table so that you don't skip the correct combination. Here, the first factors 1, 2, 3, 5, 6, 10, 15, and 30 are listed from least to greatest.

Factors of -30	Sum of the factors of -30
$1(-30) = -30$	$1 + (-30) = -29$
$2(-15) = -30$	$2 + (-15) = -13$
$3(-10) = -30$	$3 + (-10) = -7$
$5(-6) = -30$	$5 + (-6) = -1$
$6(-5) = -30$	$6 + (-5) = 1$
$10(-3) = -30$	$10 + (-3) = 7$
$15(-2) = -30$	$15 + (-2) = 13$
$30(-1) = -30$	$30 + (-1) = 29$

Notation

The middle term, $13x$, may be expressed as $15x - 2x$ or as $-2x + 15x$ when using factoring by grouping. The resulting factorizations will be equivalent.

The seventh row contains the correct pair of numbers 15 and -2, whose product is -30 and whose sum is 13. They serve as the coefficients of $15x$ and $-2x$, the two terms that we use to represent the middle term, $13x$, of the trinomial.

$$10x^2 + 13x - 3 = 10x^2 + 15x - 2x - 3 \quad \text{Express 13x as 15x - 2x.}$$

Finally, we factor by grouping.

$$10x^2 + 15x - 2x - 3 = 5x(2x + 3) - 1(2x + 3) \quad \text{Factor out 5x from } 10x^2 + 15x. $$
$$\text{Factor out } -1 \text{ from } -2x - 3.$$

$$= (2x + 3)(5x - 1) \quad \text{Factor out 2x + 3.}$$

So $10x^2 + 13x - 3 = (2x + 3)(5x - 1)$. Check the factorization by multiplying.

Self Check 6   Factor by grouping: $15a^2 + 17a - 4$

Now Try ▶ Problems 19, 27, and 35

EXAMPLE 7   Factor: $12x^5 - 17x^4 + 6x^3$

Strategy   We will factor out the GCF, x^3, first. Then we will factor the resulting trinomial using the grouping method.

Why   The first step in factoring any polynomial is to factor out the GCF.

Solution   The GCF of the three terms of the trinomial is x^3.

$$12x^5 - 17x^4 + 6x^3 = x^3(12x^2 - 17x + 6)$$

To factor $12x^2 - 17x + 6$, we must find two integers whose product is $12(6) = 72$ and whose sum is -17. Two such numbers are -8 and -9. They serve as the coefficients of $-8x$ and $-9x$, the two terms that we use to represent the middle term, $-17x$, of the trinomial.

Key number = 72 $b = -17$

Factors	Sum
$-8(-9) = 72$	$-8 + (-9) = -17$

$$12x^2 - \mathbf{17x} + 6 = 12x^2 - \mathbf{8x} - \mathbf{9x} + 6 \qquad \text{Express } -17x \text{ as } -8x - 9x.$$
$$\text{(}-9x - 8x \text{ could also be used.)}$$
$$= 4x(3x - 2) - 3(3x - 2) \qquad \text{Factor out } 4x \text{ and factor out } -3.$$
$$= (3x - 2)(4x - 3) \qquad \text{Factor out } 3x - 2.$$

The complete factorization of the original trinomial is

$$12x^5 - 17x^4 + 6x^3 = x^3(3x - 2)(4x - 3) \qquad \text{Don't forget to write the GCF, } x^3.$$

Check the factorization by multiplying.

Self Check 7 Factor: $21a^4 - 13a^3 + 2a^2$

Now Try ▶ Problem 53

VOCABULARY

Fill in the blanks.

1. The _____ coefficient of $3x^2 - x - 12$ is 3.
2. Given $5y^2 + 16y + 3 = (5y + 1)(y + 3)$. We say that $5y^2 + 16y + 3$ factors as the product of two _____.
3. The first terms of the binomial factors $(5y + 1)(y + 3)$ are ___ and ___. The second terms of the binomial factors are ___ and ___.
4. To factor $2m^2 + 11m + 12$ by _____, we write it as $2m^2 + 8m + 3m + 12$.

CONCEPTS

5. If $10x^2 - 27x + 5$ is to be factored as the product of two binomials, what are the possible *first terms* of the binomial factors?
6. Complete each sentence.

The product of these numbers must be ☐.
$$5x^2 + 6x - 8 = (\boxed{}x \ \boxed{})(\boxed{}x \ \boxed{})$$
The product of these numbers must be ☐.

7. **a.** Fill in the blanks. When factoring a trinomial, we write it in _____ powers of the variable. Then we factor out any _____ (including -1 if that is necessary to make the leading _____ positive).
   b. What is the GCF of the terms of $6s^4 + 33s^3 + 36s^2$?
   c. Factor out -1 from $-2d^2 + 19d - 8$.
8. Check to determine whether $(3t - 1)(5t - 6)$ is the correct factorization of $15t^2 - 19t + 6$.

A trinomial has been partially factored. Complete each statement that describes the type of integers we should consider for the blanks.

9. $5y^2 - 13y + 6 = \left(5y \ \boxed{}\right)\left(y \ \boxed{}\right)$
   Since the last term of the trinomial is positive and the middle term is negative, the integers must be _____ factors of 6.
10. $5y^2 + 13y + 6 = \left(5y \ \boxed{}\right)\left(y \ \boxed{}\right)$
    Since the last term of the trinomial is positive and the middle term is positive, the integers must be _____ factors of 6.
11. $5y^2 - 7y - 6 = \left(5y \ \boxed{}\right)\left(y \ \boxed{}\right)$
    Since the last term of the trinomial is negative, the signs of the integers will be _____.
12. $5y^2 + 7y - 6 = \left(5y \ \boxed{}\right)\left(y \ \boxed{}\right)$
    Since the last term of the trinomial is negative, the signs of the integers will be _____.
13. Complete the key number table.

Negative factors of 12	Sum of the negative factors of 12
$-1(-12)$	
$-2()$	
(-4)	

14. Complete the sentence to explain how to factor $3x^2 + 16x + 5$ by grouping.
    $$3x^2 + 16x + 5 = 3x^2 + \boxed{}x + \boxed{}x + 5$$
    The product of these numbers must be ☐ and their sum must be ☐.

NOTATION

15. **a.** Suppose we wish to factor $12b^2 + 20b - 9$ by grouping. Identify a, b, and c.
    b. What is the key number, ac?

16. To factor $6x^2 + 13x + 6$ by grouping, a student made a table and circled the correct pair of integers, as shown. Enter the correct coefficients for the first stage in the factorization process.

$ac = 36 \quad b = 13$

Factors	Sum
$1 \cdot 36$	37
$2 \cdot 18$	20
$3 \cdot 12$	15
$\boxed{4 \cdot 9}$	$\boxed{13}$
$6 \cdot 6$	12

$6x^2 + \boxed{}x + \boxed{}x + 6$

Complete each step of the factorization of the trinomial by grouping.

17. $12t^2 + 17t + 6 = 12t^2 + 9t + 8t + 6$

$ = \boxed{}(4t + 3) + \boxed{}(4t + 3)$

$ = (\boxed{})(3t + 2)$

18. $35t^2 - 11t - 6 = 35t^2 + 10t - 21t - 6$

$ = 5t(7t + 2) \boxed{} 3(7t + 2)$

$ = (\boxed{})(5t - 3)$

GUIDED PRACTICE

Factor. See Example 1 or Example 6.

19. $2x^2 + 3x + 1$ **20.** $3x^2 + 4x + 1$

21. $3a^2 + 10a + 3$ **22.** $2b^2 + 7b + 3$

23. $5x^2 + 7x + 2$ **24.** $7t^2 + 12t + 5$

25. $7x^2 + 18x + 11$ **26.** $5n^2 + 12n + 7$

Factor. See Example 2 or Example 6.

27. $4x^2 - 8x + 3$ **28.** $4z^2 - 13z + 3$

29. $8x^2 - 22x + 5$ **30.** $15a^2 - 28a + 5$

31. $15t^2 - 26t + 7$ **32.** $10x^2 - 9x + 2$

33. $6y^2 - 13y + 2$ **34.** $6y^2 - 43y + 7$

Factor. See Example 3 or Example 6.

35. $3x^2 - 2x - 21$ **36.** $3u^2 - 44u - 15$

37. $5m^2 - 7m - 6$ **38.** $5y^2 - 18y - 8$

39. $7y^2 + 55y - 8$ **40.** $7x^2 + 33x - 10$

41. $11y^2 + 7y - 4$ **42.** $13y^2 + 9y - 4$

Factor. See Example 4.

43. $6r^2 + rs - 2s^2$ **44.** $3m^2 + 5mn + 2n^2$

45. $4x^2 + 8xy + 3y^2$ **46.** $4b^2 + 15bc - 4c^2$

47. $8m^2 + 91mn + 33n^2$ **48.** $2m^2 + 17mn - 9n^2$

49. $15x^2 - xy - 6y^2$ **50.** $4a^2 - 15ab + 9b^2$

Factor. See Example 5 or Example 7.

51. $-26x + 6x^2 - 20$ **52.** $-28 + 6a^2 - 2a$

53. $15a + 8a^3 - 26a^2$ **54.** $16r - 40r^2 + 25r^3$

55. $2u^2 - 6v^2 - uv$ **56.** $6a^2 + 6b^2 - 13ab$

57. $36y^2 - 88y + 32$ **58.** $70a^2 - 95a + 30$

TRY IT YOURSELF

Factor. If an expression is prime, so indicate.

59. $6t^2 - 7t - 20$ **60.** $6w^2 + 13w + 5$

61. $15p^2 - 2pq - q^2$ **62.** $8c^2 - 10cd + 3d^2$

63. $4t^2 - 16t + 7$ **64.** $9x^2 - 32x + 15$

65. $130r^2 + 20r - 110$ **66.** $170h^2 - 210h - 260$

67. $8y^2 - 2y - 1$ **68.** $14y^2 + 11y + 2$

69. $18x^2 + 31x - 10$ **70.** $20y^2 - 93y - 35$

71. $-y^3 - 13y^2 - 12y$ **72.** $-2xy^2 - 8xy + 24x$

73. $10u^2 - 13u - 6$ **74.** $8m^2 + 5m - 10$

75. $-6x^4 + 15x^3 + 9x^2$ **76.** $-9y^4 - 3y^3 + 6y^2$

77. $6p^2 + pq - q^2$ **78.** $12m^2 - 11mn + 2n^2$

79. $30r^5 + 63r^4 - 30r^3$ **80.** $6s^5 - 26s^4 - 20s^3$

81. $16m^3n + 20m^2n^2 + 6mn^3$ **82.** $-28u^3v^3 + 26u^2v^4 - 6uv^5$

83. $3x^2 + x + 6$ **84.** $2u^2 + 3u + 25$

85. $-12y^2 - 12 + 25y$ **86.** $-10t^2 + 1 + 3t$

Choose the correct method from Sections 6.1, 6.2, or 6.3 to factor each of the following.

87. $m^2 + 3m - 28$ **88.** $-b^2 - 5b + 24$

89. $6a^3 + 15a^2$ **90.** $9x^4 + 27x^6$

91. $x^3 - 2x^2 + 5x - 10$ **92.** $x^3 - x^2 + 2x - 2$

93. $5y^2 + 3 - 8y$ **94.** $3t^2 + 7 - 10t$

95. $-2x^2 - 10x - 12$ **96.** $4y^2 + 36y + 72$

97. $12x^3y^3 - 18x^2y^3 + 15x^2y^2$ **98.** $15c^2d^3 - 25c^3d^2 - 10c^4d^4$

99. $a^2 - 7ab + 10b^2$ **100.** $x^2 - 13xy + 12y^2$

101. $9u^6 - 71u^5 - 8u^4$ **102.** $25n^8 - 49n^7 - 2n^6$

APPLICATIONS

103.  **from Campus to Careers**
Elementary School Teacher

The area of a teacher's desktop is represented by the trinomial $(4x^2 + 20x - 11)$ in.^{2}. Factor it to find the polynomials that represent its length and width.

104. Storage. The volume of an 8-foot-wide portable storage container is represented by the trinomial $(72x^2 + 120x - 400)$ ft^3. Its dimensions can be determined by factoring the trinomial. Find the polynomials that represent the height and the length of the container.

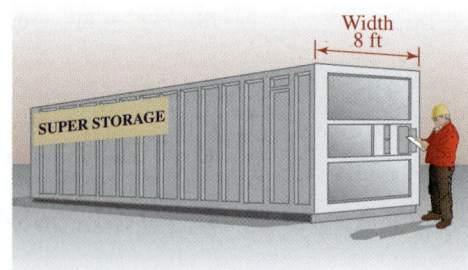

WRITING

105. Two students factor $2x^2 + 20x + 42$ and get two different answers: $(2x + 6)(x + 7)$ and $(x + 3)(2x + 14)$.

Do both answers check? Why don't they agree? Is either answer completely correct? Explain.

106. Why is the process of factoring $6x^2 - 5x - 6$ more complicated than the process of factoring $x^2 - 5x - 6$?

107. Suppose a factorization check of $(3x - 9)(5x + 7)$ gives a middle term $-24x$, but a middle term of $24x$ is actually needed. Explain how to quickly obtain the correct factorization.

108. Suppose we want to factor $2x^2 + 7x - 72$. Explain why $(2x - 1)(x + 72)$ is not a wise choice to try first.

REVIEW

Evaluate each expression.

109. -7^2 **110.** $(-7)^2$

111. 7^0 **112.** 7^{-2}

113. $\dfrac{1}{7^{-2}}$ **114.** $2 \cdot 7^2$

CHALLENGE PROBLEMS

Factor.

115. $6a^{10} + 5a^5 - 21$

116. $3x^4y^2 - 29x^2y + 56$

117. $8x^2(c^2 + c - 2) - 2x(c^2 + c - 2) - (c^2 + c - 2)$

118. Find all integer values of b that make $2x^2 + bx - 5$ factorable.

SECTION 6.4

Factoring Perfect-Square Trinomials and the Difference of Two Squares

OBJECTIVES

1 Recognize perfect-square trinomials.

2 Factor perfect-square trinomials.

3 Factor the difference of two squares.

ARE YOU READY?

The following problems review some basic skills that are needed when factoring binomials and trinomials.

1. Multiply: $(3y + 2)(3y + 2)$

2. Multiply: $(m + 9)(m - 9)$

3. Simplify: $(8d)^2$

4. Translate to symbols: the sum of x^2 and 25

In this section, we will discuss a method that can be used to factor two types of trinomials, called *perfect-square trinomials*. We also develop techniques for factoring a type of binomial called the *difference of two squares*.

1 Recognize Perfect-Square Trinomials.

We have seen that the square of a binomial is a trinomial. We also have seen that the special-product rules shown below can be used to find the square of a sum and the square of a difference quickly. The terms of the resulting trinomial are related to the terms of the binomial that was squared.

Success Tip

To prepare for this section, it would be helpful to review Section 5.7 Special Products.

$$(A + B)^2 \;=\; A^2 \;+\; 2AB \;+\; B^2$$

| This is the square of the first term of the binomial. | This is twice the product of A and B, or its opposite. | This is the square of the last term of the binomial. |

$$(A - B)^2 \;=\; A^2 \;-\; 2AB \;+\; B^2$$

Trinomials that are squares of a binomial are called **perfect-square trinomials.** Some examples are

$y^2 + 6y + 9$ Because it is the square of $(y + 3)$: $(y + 3)^2 = y^2 + 6y + 9$

$t^2 - 14t + 49$ Because it is the square of $(t - 7)$: $(t - 7)^2 = t^2 - 14t + 49$

$4m^2 - 20m + 25$ Because it is the square of $(2m - 5)$: $(2m - 5)^2 = 4m^2 - 20m + 25$

EXAMPLE 1   Determine whether the following are perfect-square trinomials: **a.** $x^2 + 10x + 25$ **b.** $c^2 - 12c - 36$ **c.** $25y^2 - 30y + 9$ **d.** $4t^2 + 18t + 81$

Strategy   We will compare each trinomial, term-by-term, to one of the special-product forms discussed in Section 5.7.

Why   If a trinomial matches one of these forms, it is a perfect-square trinomial.

Solution   **a.** To determine whether this is a perfect-square trinomial, we note that

$$x^2 + 10x + 25$$

| The first term is the square of x. | The middle term is twice the product of x and 5: $2 \cdot x \cdot 5 = 10x.$ | The last term is the square of 5. |

Thus, $x^2 + 10x + 25$ is a perfect-square trinomial.

b. To determine whether this is a perfect-square trinomial, we note that

$$c^2 - 12c - 36$$ The last term, −36, is not the square of a real number.

Since the last term is negative, $c^2 - 12c - 36$ is not a perfect-square trinomial.

The Language of Algebra

The expressions $25y^2$ and 9 are called *perfect squares* because $25y^2 = (5y)^2$ and $9 = 3^2$.

c. To determine whether this is a perfect-square trinomial, we note that

$$25y^2 - 30y + 9$$

| The first term is the square of 5y. | The middle term is the opposite of twice the product of 5y and 3: $-2(5y)(3) = -30y.$ | The last term is the square of 3. |

Thus, $25y^2 - 30y + 9$ is a perfect-square trinomial.

d. To determine whether this is a perfect-square trinomial, we note that

$$4t^2 + 18t + 81$$

The first term is the square of 2t.   The middle term is not twice the product of 2t and 9, because $2(2t)(9) = 36t$.   The last term is the square of 9.

Thus, $4t^2 + 18t + 81$ is not a perfect-square trinomial.

Self Check 1   Determine whether the following are perfect-square trinomials:

a. $y^2 + 4y + 4$ **b.** $b^2 - 6b - 9$

c. $4z^2 + 4z + 4$ **d.** $49x^2 - 28x + 16$

Now Try ▶ Problems 13 and 17

Recognizing Perfect-Square Trinomials

1. The first and last terms are squares of integers or monomials.
2. The middle term is twice the product of the expressions that are squared to produce the first and last terms, or its opposite.

2 Factor Perfect-Square Trinomials.

We can factor perfect-square trinomials using the methods previously discussed in Sections 6.2 and 6.3. However, in many cases, we can factor them more quickly by inspecting their terms and applying the special-product rules in reverse.

Factoring Perfect-Square Trinomials

Perfect-square trinomial		Square of a binomial	
$A^2 + 2AB + B^2$	$=$	$(A + B)^2$	Each of these trinomials factors
$A^2 - 2AB + B^2$	$=$	$(A - B)^2$	as the square of a binomial.

When factoring perfect-square trinomials, it is helpful to know the following perfect squares printed in red. The number 400, for example, is a perfect square, because $400 = 20^2$.

$1 = 1^2$	$25 = 5^2$	$81 = 9^2$	$169 = 13^2$	$289 = 17^2$
$4 = 2^2$	$36 = 6^2$	$100 = 10^2$	$196 = 14^2$	$324 = 18^2$
$9 = 3^2$	$49 = 7^2$	$121 = 11^2$	$225 = 15^2$	$361 = 19^2$
$16 = 4^2$	$64 = 8^2$	$144 = 12^2$	$256 = 16^2$	$400 = 20^2$

EXAMPLE 2   Factor: **a.** $x^2 + 20x + 100$ **b.** $9x^2 - 30xy + 25y^2$

Strategy   The terms of each trinomial do not have a common factor (other than 1). We will determine whether each is a perfect-square trinomial.

Why   If it is, we can factor it using a special-product rule in reverse.

Solution   **a.** $x^2 + 20x + 100$ is a perfect-square trinomial, because:

- The first term x^2 is the square of **x**.
- The last term 100 is the square of **10**.
- The middle term is twice the product of **x** and **10**: $2(x)(10) = 20x$.

To find the factorization, we match $x^2 + 20x + 100$ to the proper rule for factoring a perfect-square trinomial.

$$A^2 + 2 \ A \ B + B^2 = (A + B)^2$$
$$x^2 + 20x + 10^2 = x^2 + 2 \cdot x \cdot 10 + 10^2 = (x + 10)^2$$

Therefore, $x^2 + 20x + 10^2 = (x + 10)^2$. Check by finding $(x + 10)^2$.

b. $9x^2 - 30xy + 25y^2$ is a perfect-square trinomial, because:

- The first term $9x^2$ is the square of $3x$: $(3x)^2 = 9x^2$.
- The last term $25y^2$ is the square of $5y$: $(5y)^2 = 25y^2$.
- The middle term is the opposite of twice the product of $3x$ and $5y$: $-2(3x)(5y) = -30xy$.

We can use these observations to write the trinomial in one of the special-product forms that then leads to its factorization.

$$9x^2 - 30xy + 25y^2 = (3x)^2 - 2(3x)(5y) + (5y)^2$$
$$= (3x - 5y)^2$$

Therefore, $9x^2 - 30xy + 25y^2 = (3x - 5y)^2$. Check by finding $(3x - 5y)^2$.

Self Check 2   Factor: **a.** $x^2 + 18x + 81$ **b.** $16x^2 - 8xy + y^2$

Now Try ▶ Problems 21 and 29

EXAMPLE 3   Factor: $4a^3 - 4a^2 + a$

Strategy   We will factor out the GCF, a, first. Then we will factor the resulting perfect-square trinomial using a special-product rule in reverse.

Why   The first step in factoring any polynomial is to factor out the GCF.

Solution   The terms of $4a^3 - 4a^2 + a$ have the common factor a, which should be factored out first. Within the parentheses, we recognize $4a^2 - 4a + 1$ as a perfect-square trinomial of the form $A^2 - 2AB + B^2$, and factor it as such.

$$4a^3 - 4a^2 + a = a(4a^2 - 4a + 1) \quad \text{Factor out the GCF, } a.$$
$$= a(2a - 1)^2 \quad \text{Since } 4a^2 = (2a)^2, 1 = (1)^2, \text{ and } -4a = -2(2a)(1),$$
$$\text{} 4a^2 - 4a + 1 \text{ can be factored as a perfect-square}$$
$$\text{trinomial.}$$

Self Check 3   Factor: $49x^3 - 14x^2 + x$

Now Try ▶ Problem 33

3 **Factor the Difference of Two Squares.**

Recall the special-product rule for multiplying the sum and difference of the same two terms:

$$(A + B)(A - B) = A^2 - B^2$$

The binomial $A^2 - B^2$ is called a **difference of two squares,** because A^2 is the square of A and B^2 is the square of B. If we reverse this rule, we obtain a method for factoring a difference of two squares.

Factoring ⟶

$$A^2 - B^2 = (A + B)(A - B)$$

This pattern is easy to remember if we think of a difference of two squares as the square of a **F**irst quantity minus the square of a **L**ast quantity.

Factoring a Difference of Two Squares	To factor the square of a First quantity minus the square of a Last quantity, multiply the First plus the Last by the First minus the Last. $$F^2 - L^2 = (F + L)(F - L)$$

To factor differences of two squares, it will be helpful if you recognize the perfect squares listed in red on page 466.

EXAMPLE 4   Factor: **a.** $x^2 - 9$ **b.** $16 - b^2$ **c.** $n^2 - 45$ **d.** $a^2 + 81$

Strategy   The terms of each binomial do not have a common factor (other than 1). The only option available is to attempt to factor each as a difference of two squares.

Why   If a binomial is a difference of two squares, we can factor it using a special-product rule in reverse.

Solution   **a.** $x^2 - 9$ is the difference of two squares because it can be written as $x^2 - 3^2$. We can match it to the rule for factoring a difference of two squares to find the factorization.

$$F^2 - L^2 = (F + L)(F - L)$$
$$x^2 - 3^2 = (x + 3)(x - 3) \quad \text{9 is a perfect square: } 9 = 3^2.$$

Therefore, $x^2 - 9 = (x + 3)(x - 3)$.

Check by multiplying: $(x + 3)(x - 3) = x^2 - 9$.

b. $16 - b^2$ is the difference of two squares because $16 - b^2 = 4^2 - b^2$. Therefore,

$$16 - b^2 = (4 + b)(4 - b) \quad \text{16 is a perfect square: } 16 = 4^2.$$

Check by multiplying.

c. Since 45 is not a perfect square, $n^2 - 45$ cannot be factored using integers. It is *prime*.

d. $a^2 + 81$ can be written $a^2 + 9^2$, and is, therefore, the **sum of two squares.** We might attempt to factor $a^2 + 81$ as $(a + 9)(a + 9)$ or $(a - 9)(a - 9)$. However, the following checks show that neither product is $a^2 + 81$.

$$(a + 9)(a + 9) = a^2 \boxed{+ 18a} + 81 \qquad (a - 9)(a - 9) = a^2 \boxed{- 18a} + 81$$

In general, **the sum of two squares (with no common factor other than 1) cannot be factored using real numbers.** Thus, $a^2 + 81$ is *prime*.

Notation

By the commutative property of multiplication, the factors of a difference of two squares can be written in either order. For example, we can write:

$$x^2 - 9 = (x - 3)(x + 3)$$

The Language of Algebra

An expression of the form $A^2 + B^2$ is called the **sum of two squares,** whereas $(A + B)^2$ is the **square of a sum.** They are not equivalent because $(A + B)^2 \neq A^2 + B^2$.

Self Check 4   Factor: **a.** $c^2 - 4$
b. $121 - t^2$ **c.** $x^2 - 24$
d. $s^2 + 36$

Now Try ▶   Problems 37 and 45

Terms containing variables such as a^4, $25x^2$, and $4y^4$ are perfect squares, because they can be written as the square of a quantity. For example:

$$a^4 = (a^2)^2, \qquad 25x^2 = (5x)^2, \qquad \text{and} \qquad 4y^4 = (2y^2)^2$$

EXAMPLE 5   Factor: **a.** $25x^2 - 49$ **b.** $-121z^2 + 4y^4$

Strategy   In each case, the terms of the binomial do not have a common factor (other than 1). To factor them, we will write each binomial in a form that clearly shows it is a difference of two squares.

Why   We can then use a special-product rule in reverse to factor it.

Solution   **a.** We can write $25x^2 - 49$ in the form $(5x)^2 - 7^2$ and match it to the rule for factoring the difference of two squares:

$$\mathbf{F^2} \;-\; \mathbf{L^2} = (\,\mathbf{F} \;+\; \mathbf{L}\,)(\,\mathbf{F} \;-\; \mathbf{L}\,)$$
$$(\mathbf{5x})^2 \;-\; \mathbf{7^2} = (\mathbf{5x} \;+\; \mathbf{7})(\mathbf{5x} \;-\; \mathbf{7})$$

Therefore, $25x^2 - 49 = (5x + 7)(5x - 7)$. Check by multiplying.

Success Tip

Remember that a *difference of two squares* is a binomial. Each term is a square and the terms have different signs. The powers of the variables in the terms must be even.

b. If we reorder the terms, the resulting binomial, $4y^4 - 121z^2$ is obviously a difference of two squares. We can write $4y^4 - 121z^2$ in the form $(2y^2)^2 - (11z)^2$ and match it to the rule for factoring the difference of two squares:

$$\mathbf{F^2} \;-\; \mathbf{L^2} \;=\; (\,\mathbf{F} \;+\; \mathbf{L}\,)(\,\mathbf{F} \;-\; \mathbf{L}\,)$$
$$(\mathbf{2y^2})^2 \;-\; (\mathbf{11z})^2 = (\mathbf{2y^2} \;+\; \mathbf{11z})(\mathbf{2y^2} \;-\; \mathbf{11z})$$

Therefore, $-121z^2 + 4y^4 = 4y^4 - 121z^2 = (2y^2 + 11z)(2y^2 - 11z)$. Check by multiplying.

An alternate approach is first to factor out -1: $-(121z^2 - 4y^4)$. Then factor the difference of two squares within the parentheses to get an equivalent result: $-(11z + 2y^2)(11z - 2y^2)$.

Self Check 5   Factor: **a.** $16y^2 - 9$
 b. $9m^2 - 64n^4$ **c.** $-a^4 + 100$

Now Try ▸   Problems 49 and 53

EXAMPLE 6   Factor: $8x^2 - 8$

Strategy   We will factor out the GCF, 8, first. Then we will factor the resulting difference of two squares.

Why   The first step in factoring any polynomial is to factor out the GCF.

Solution
$$8x^2 - 8 = 8(x^2 - 1) \qquad \text{The GCF is 8.}$$
$$= 8(x + 1)(x - 1) \qquad \text{Think of } x^2 - 1 \text{ as } x^2 - 1^2 \text{ and}$$
$$\text{factor the difference of two squares.}$$

Check: $8(x + 1)(x - 1) = 8(x^2 - 1)$ Multiply the binomials first.
$$= 8x^2 - 8 \qquad \text{Distribute the multiplication by 8.}$$

Self Check 6   Factor: $2p^2 - 200$

Now Try ▸   Problem 57

Sometimes we must factor a difference of two squares more than once to factor a polynomial completely.

EXAMPLE 7 Factor: $x^4 - 16$

Strategy The terms of $x^4 - 16$ do not have a common factor (other than 1). To factor this binomial, we will write it in a form that clearly shows it is a difference of two squares.

Why We can then use a special-product rule in reverse to factor it.

Solution

$$x^4 - 16 = (x^2)^2 - 4^2 \qquad \text{Write } x^4 \text{ as } (x^2)^2 \text{ and } 16 \text{ as } 4^2.$$
$$= (x^2 + 4)(x^2 - 4) \qquad \text{Factor the difference of two squares.}$$
$$= (x^2 + 4)(x + 2)(x - 2) \qquad \text{Factor } x^2 - 4, \text{ which is itself a difference of two squares. The binomial } x^2 + 4 \text{ is a sum of two squares and does not factor further.}$$

Success Tip

Factoring a polynomial is complete when no factor can be factored further.

Self Check 7 Factor: $a^4 - 81$

Now Try ▶ Problem 63

SECTION 6.4 ▶ STUDY SET

VOCABULARY

Fill in the blanks.

1. $x^2 + 6x + 9$ is a _____-square trinomial because it is the square of the binomial $x + 3$.

2. The binomial $x^2 - 25$ is called a _____ of two squares and it factors as $(x + 5)(x - 5)$. The binomial $x^2 + 25$ is a _____ of two squares and since it does not factor using integers, it is _____.

CONCEPTS

Fill in the blanks.

3. Consider $25x^2 + 30x + 9$.
   a. The first term is the square of ▢.
   b. The last term is the square of ▢.
   c. The middle term is twice the product of ▢ and ▢.

4. Consider $49x^2 - 28xy + 4y^2$.
   a. The first term is the square of ▢.
   b. The last term is the square of ▢.
   c. The middle term is the opposite of twice the product of ▢ and ▢.

5. a. $x^2 + 2xy + y^2 = (▢ + ▢)^2$
   b. $x^2 - 2xy + y^2 = (x \ ▢ \ y)^2$
   c. $x^2 - y^2 = (x \ ▢ \ y)(▢ - ▢)$

6. a. $36x^2 = (▢)^2$ b. $100x^4 = (▢)^2$
   c. $4x^2 - 9 = (▢)^2 - (▢)^2$

7. List the squares of the integers from 1 through 20.

8. Use multiplication to determine if each factorization is correct.
   a. $9y^2 - 12y + 4 = (3y - 2)^2$
   b. $n^2 - 16 = (n + 8)(n - 8)$

NOTATION

Complete each factorization.

9. $x^2 + 10x + 25 = (x + 5)^▢$ 10. $9b^2 - 12b + 4 = (3b \ ▢ \ 2)^2$

11. $x^2 - 64 = (x \ ▢ \ 8)(x \ ▢ \ 8)$

12. $16t^2 - 49 = (4t + ▢)(4t - ▢)$

GUIDED PRACTICE

Determine whether each of the following is a perfect-square trinomial. See Example 1.

13. $x^2 + 18x + 81$ 14. $x^2 + 14x + 49$
15. $y^2 + 2y + 4$ 16. $y^2 + 4y + 16$
17. $9n^2 - 30n - 25$ 18. $9a^2 - 48a - 64$
19. $4y^2 - 12y + 9$ 20. $9y^2 - 30y + 25$

Factor. See Example 2.

21. $x^2 + 6x + 9$ 22. $x^2 + 10x + 25$

23. $b^2 + 2b + 1$ 24. $m^2 + 12m + 36$

25. $c^2 - 12c + 36$ 26. $d^2 - 10d + 25$

27. $9 + 4x^2 + 12x$ 28. $121 + 4x^2 + 44x$

29. $36m^2 + 60mn + 25n^2$ 30. $25a^2 + 30ab + 9b^2$

31. $81x^2 - 72xy + 16y^2$ **32.** $9x^2 - 48xy + 64y^2$

Factor. See Example 3.

33. $3u^2 - 18u + 27$ **34.** $3v^2 - 42v + 147$

35. $36x^3 + 12x^2 + x$ **36.** $4x^4 - 20x^3 + 25x^2$

Factor. If a polynomial can't be factored, write "prime."
See Example 4.

37. $x^2 - 4$ **38.** $x^2 - 9$

39. $x^2 - 16$ **40.** $x^2 - 25$

41. $36 - y^2$ **42.** $49 - w^2$

43. $t^2 - 25$ **44.** $h^2 - 144$

45. $a^2 + b^2$ **46.** $121a^2 + b^2$

47. $y^2 - 63$ **48.** $x^2 - 27$

Factor. See Example 5.

49. $25t^2 - 64$ **50.** $49d^2 - 16$

51. $81y^2 - 1$ **52.** $400z^2 - 1$

53. $9x^4 - y^2$ **54.** $4x^2 - z^4$

55. $-49d^4 + 16c^2$ **56.** $-121b^4 + 36a^2$

Factor. See Example 6.

57. $8x^2 - 32y^2$ **58.** $2a^2 - 200b^2$

59. $63a^2 - 7$ **60.** $20x^2 - 5$

Factor. See Example 7.

61. $81 - s^4$ **62.** $y^4 - 625$

63. $b^4 - 256$ **64.** $m^4n^4 - 16$

TRY IT YOURSELF

Factor.

65. $a^4 - 144b^2$ **66.** $81y^4 - 100z^2$

67. $9x^2y^2 + 30xy + 25$ **68.** $s^2t^2 - 20st + 100$

69. $16t^4 - 16s^4$ **70.** $2p^4 - 32q^4$

71. $t^2 - 20t + 100$ **72.** $r^2 + 24r + 144$

73. $9y^2 - 24y + 16$ **74.** $49z^2 - 14z + 1$

75. $z^2 - 64$ **76.** $25 + B^2$

77. $25m^4 - 25$ **78.** $9 - 9n^4$

79. $18a^5 + 84a^4b + 98a^3b^2$ **80.** $32b^6 + 80b^5c + 50b^4c^2$

81. $x^3 - 144x$ **82.** $g^3 - 121g$

83. $49t^2 - 28ts + 4s^2$ **84.** $81p^2 - 36pq + 4q^2$

85. $3m^4 - 3n^4$ **86.** $5a^4 - 80b^4$

87. $25m^2 + 70m + 49$ **88.** $25x^2 + 20x + 4$

89. $-100t^2 + 20t - 1$ **90.** $-81r^2 - 18r - 1$

91. $6x^4 - 6x^2y^2$ **92.** $4b^2y - 16c^2y$

93. $100a^2 + 81$ **94.** $25y^2 + 16$

95. $-169 + 25x^2$ **96.** $-196 + 49f^2$

Choose the correct method from Section 6.1, Section 6.2,
Section 6.3, or Section 6.4 to factor each of the following:

97. $x^2 + x - 42$ **98.** $rx - sx + r - s$

99. $x^2 - 9$ **100.** $3a^2 - 4a - 4$

101. $24a^3b - 16a^2b$ **102.** $20ns^2 - 60nu + 100n$

103. $-2r^2 + 28r - 80$ **104.** $10s - 39 + s^2$

105. $x^3 + 3x^2 + 4x + 12$ **106.** $2y^2 - 128z^2$

107. $4b^2 - 20b + 25$ **108.** $a^2 - 4ab - 12b^2$

APPLICATIONS

109. Genetics.   The Hardy–Weinberg equation, one of the fundamental concepts in population genetics, is $p^2 + 2pq + q^2 = 1$, where p represents the frequency of a certain dominant gene and q represents the frequency of a certain recessive gene. Factor the left side of the equation.

110. Signal Flags.   The maritime signal flag for the letter X is shown. Find the polynomial that represents the area of the shaded region and express it in factored form.

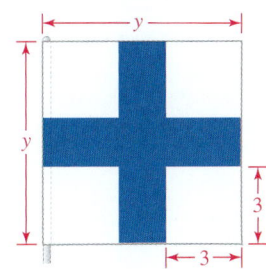

111. Physics. The illustration shows a time-sequence picture of a falling apple. Factor the expression, which gives the difference in the distance fallen by the apple during the time interval from t_1 to t_2 seconds.

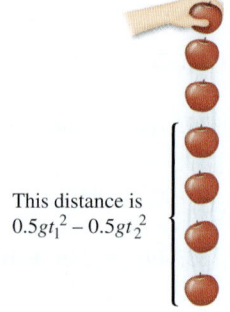

This distance is $0.5gt_1^2 - 0.5gt_2^2$

112. Darts. A circular dart board has a series of rings around a solid center, called the bullseye. To find the area of the outer black ring, we can use the formula $A = \pi R^2 - \pi r^2$. Factor the expression on the right side of the equation.

WRITING

113. When asked to factor $x^2 - 25$, one student wrote $(x + 5)(x - 5)$, and another student wrote $(x - 5)(x + 5)$. Are both answers correct? Explain.

114. Explain the error that was made in the following factorization:

$x^2 - 100 = (x + 50)(x - 50)$

115. Explain why the following factorization isn't complete.

$$x^4 - 625 = (x^2 + 25)(x^2 - 25)$$

116. Explain why $a^2 + 2a + 1$ is a perfect-square trinomial and why $a^2 + 4a + 1$ isn't a perfect-square trinomial.

REVIEW

Perform each division.

117. $\dfrac{-30c^2d^2 - 15c^2d - 10cd^2}{-10cd}$

118. $2a - 1\overline{)a - 2 + 6a^2}$

CHALLENGE PROBLEMS

119. For what value of c does $80x^2 - c$ factor as $5(4x + 3)(4x - 3)$?

120. Find all values of b so that $0.16x^2 + bxy + 0.25y^2$ is a perfect-square trinomial.

Factor completely.

121. $81x^6 + 36x^3y^2 + 4y^4$

122. $p^2 + p + \dfrac{1}{4}$

123. $c^2 + 1.6c + 0.64$

124. $x^{2n} - y^{4n}$

125. $(x + 5)^2 - y^2$

126. $\dfrac{1}{2} - 2a^2$

127. $c^2 - \dfrac{1}{16}$

128. $t^2 - \dfrac{9}{25}$

SECTION 6.5

Factoring the Sum and Difference of Two Cubes

OBJECTIVES

1 Factor the sum and difference of two cubes.

ARE YOU READY?

The following problems review some basic skills that are needed when factoring binomials and trinomials.

1. Multiply: $(x + 4)(x^2 - 4x + 16)$

2. Multiply: $(2h - 1)(4h^2 + 2h + 1)$

3. Evaluate: **a.** 3^3 **b.** 5^3

4. Explain why $x^2 - 2x + 4$ is prime.

In this section, we will discuss how to factor two types of binomials, called the *sum* and the *difference of two cubes*.

1 Factor the Sum and Difference of Two Cubes.

We have seen that the sum of two squares, such as $x^2 + 4$ or $25a^2 + 9b^2$, cannot be factored. However, the sum of two cubes and the difference of two cubes can be factored.

The sum of two cubes   *The difference of two cubes*

$$x^3 + 8$$ $$a^3 - 64b^3$$

This is   This is 2 cubed: This is   This is 4b cubed:
x cubed. $2^3 = 8$.   a cubed. $(4b)^3 = 64b^3$.

The Language of Algebra

The expression $x^3 + y^3$ is a **sum of two cubes**, whereas $(x + y)^3$ is the **cube of a sum.** If you expand $(x + y)^3$, you will see that $(x + y)^3 \neq x^3 + y^3$.

To find rules for factoring the sum of two cubes and the difference of two cubes, we need to find the products shown below. Note that each term of the trinomial is multiplied by each term of the binomial.

$$(x + y)(x^2 - xy + y^2) = x^3 - x^2y + xy^2 + x^2y - xy^2 + y^3$$
$$= x^3 + y^3 \quad \text{Combine like terms: } -x^2y + x^2y = 0 \text{ and } xy^2 - xy^2 = 0.$$

$$(x - y)(x^2 + xy + y^2) = x^3 + x^2y + xy^2 - x^2y - xy^2 - y^3$$
$$= x^3 - y^3 \quad \text{Combine like terms.}$$

These results justify the rules for factoring the **sum** or **difference of two cubes.** They are easier to remember if we think of a sum (or a difference) of two cubes as the cube of a **F**irst quantity plus (or minus) the cube of the **L**ast quantity.

Factoring the Sum and Difference of Two Cubes

To factor the cube of a First quantity plus the cube of a Last quantity, multiply the First plus the Last by the First squared, minus the First times the Last, plus the Last squared.

$$F^3 + L^3 = (F + L)(F^2 - FL + L^2)$$

To factor the cube of a First quantity minus the cube of a Last quantity, multiply the First minus the Last by the First squared, plus the First times the Last, plus the Last squared.

$$F^3 - L^3 = (F - L)(F^2 + FL + L^2)$$

To factor the sum or difference of two cubes, it's helpful to know the cubes of integers from 1 to 10 shown in red below. The number 216, for example, is a **perfect cube,** because $216 = 6^3$.

$1 = 1^3$	$27 = 3^3$	$125 = 5^3$	$343 = 7^3$	$729 = 9^3$
$8 = 2^3$	$64 = 4^3$	$216 = 6^3$	$512 = 8^3$	$1,000 = 10^3$

EXAMPLE 1   Factor: $x^3 + 8$

Strategy   We will write $x^3 + 8$ in a form that clearly shows it is the sum of two cubes.

Why   We can then use the rule for factoring the sum of two cubes.

Solution $x^3 + 8$ is the sum of two cubes because it can be written as $x^3 + 2^3$. We can match it to the rule for factoring the sum of two cubes to find its factorization.

Caution

A common error is to try to factor $x^2 - 2x + 4$. It is not a perfect square trinomial, because the middle term needs to be $-4x$. Furthermore, it cannot be factored by the methods of Section 6.2. It is prime.

$$\mathbf{F^3 + L^3 = (F + L)(F^2 - F \ L + L^2)} \quad \text{To write the trinomial factor:}$$
$$\quad \blacksquare \text{ Square the first term of the binomial factor.}$$
$$\quad \blacksquare \text{ Multiply the terms of the binomial factor.}$$
$$x^3 + 2^3 = (x + 2)(x^2 - x \cdot 2 + 2^2) \quad \blacksquare \text{ Square the last term of the binomial factor.}$$
$$= (x + 2)(x^2 - 2x + 4) \quad x^2 - 2x + 4 \text{ does not factor. It is prime.}$$

Therefore, $x^3 + 8 = (x + 2)(x^2 - 2x + 4)$. We can check by multiplying.

$$(x + 2)(x^2 - 2x + 4) = x^3 - 2x^2 + 4x + 2x^2 - 4x + 8$$
$$= x^3 + 8 \quad \text{This is the original binomial.}$$

Self Check 1 Factor: $h^3 + 27$

Now Try ▶ Problem 17

Terms containing variables such as a^6, $64b^3$, and $27m^6$ are also perfect cubes, because they can be written as the cube of a quantity:

$$a^6 = (a^2)^3, \qquad 64b^3 = (4b)^3, \qquad \text{and} \qquad 27m^6 = (3m^2)^3$$

EXAMPLE 2 Factor: $a^3 - 64b^3$

Strategy We will write $a^3 - 64b^3$ in a form that clearly shows it is the difference of two cubes.

Why We can then use the rule for factoring the difference of two cubes.

Solution $a^3 - 64b^3$ is the difference of two cubes because it can be written as $a^3 - (4b)^3$. We can match it to the rule for factoring the difference of two cubes to find its factorization.

> **The Language of Algebra**
>
> An expression of the form $a^3 - 64b^3$ is called the **difference of two cubes**, whereas $(a - 64b)^3$ is the **cube of a difference**. They are not equivalent because $(a - 64b)^3 \neq a^3 - 64b^3$.

$$\mathbf{F}^3 - \mathbf{L}^3 = (\mathbf{F} - \mathbf{L})(\mathbf{F}^2 + \mathbf{F}\ \mathbf{L} + \mathbf{L}^2)$$

To write the trinomial factor:
- Square the first term of the binomial factor.
- Multiply the terms of the binomial factor.
- Square the last term of the binomial factor.

$$a^3 - (4b)^3 = (a - 4b)[a^2 + a \cdot 4b + (4b)^2]$$
$$= (a - 4b)(a^2 + 4ab + 16b^2)$$

$a^2 + 4ab + 16b^2$ does not factor.

Therefore, $a^3 - 64b^3 = (a - 4b)(a^2 + 4ab + 16b^2)$. Check by multiplying.

Self Check 2 Factor: $8c^3 - 1$

Now Try ▶ Problem 37

You should memorize the rules for factoring the sum and the difference of two cubes. Note that the right side of each rule has the form

(a binomial)(a trinomial)

and that there is a relationship between the signs that appear in these forms.

The Sum of Two Cubes

The same sign

$$F^3 + L^3 = (F + L)(F^2 - FL + L^2)$$

Opposite signs   Always plus

The Difference of Two Cubes

The same sign

$$F^3 - L^3 = (F - L)(F^2 + FL + L^2)$$

Opposite signs   Always plus

> **Success Tip**
>
> An easy way to remember the sign patterns within the binomial and trinomial of these factoring forms is with the word "SOAP."
>
> **S**ame, **O**pposite, **A**lways **P**lus

If the terms of a binomial have a common factor, the GCF (or the opposite of the GCF) should always be factored out first.

EXAMPLE 3 Factor: $-2t^5 + 250t^2$

Strategy We will factor out the common factor, $-2t^2$. We can then factor the resulting binomial as a difference of two cubes.

Why The first step in factoring any polynomial is to factor out the GCF, or its opposite.

Solution $-2t^5 + 250t^2 = -2t^2(t^3 - 125)$ Factor out the opposite of the GCF, $-2t^2$.

$= -2t^2(t - 5)(t^2 + 5t + 25)$ Factor $t^3 - 125$.

Therefore, $-2t^5 + 250t^2 = -2t^2(t - 5)(t^2 + 5t + 25)$. Check by multiplying.

Self Check 3 Factor: $4c^3 + 4d^3$

Now Try ▶ Problem 43

SECTION 6.5 ▶ STUDY SET

VOCABULARY

Fill in the blanks.

1. $x^3 + 27$ is the _____ of two cubes and $a^3 - 125$ is the difference of two _____.

2. The factorization of $x^3 + 8$ is $(x + 2)(x^2 - 2x + 4)$. The first factor is a binomial and the second is a _____.

CONCEPTS

Fill in the blanks.

3. **a.** $F^3 + L^3 = \left(\boxed{} + \boxed{}\right)(F^2 - FL + L^2)$

   b. $F^3 - L^3 = \left(F \boxed{} L\right)\left(\boxed{} + FL + \boxed{}\right)$

4. $m^3 + 64$
        ↑   ↑
    This is   This is
    ☐ cubed. ☐ cubed.

5. $216n^3 - 125$
        ↑   ↑
    This is   This is
    ☐ cubed. ☐ cubed.

6. **a.** $x^3 + 64y^3 = (\boxed{})^3 + (\boxed{})^3$

   b. $8x^3 - 27 = (\boxed{})^3 - (\boxed{})^3$

7. List the first ten positive integer cubes.

8. $(x - 2)(x^2 + 2x + 4)$ is the factorization of what binomial?

9. Use multiplication to determine if the factorization is correct.

   $b^3 + 27 = (b + 3)(b^2 + 3b + 9)$

10. The factorization of $y^3 + 27$ is $(y + 3)(y^2 - 3y + 9)$. Is this factored completely, or does $y^2 - 3y + 9$ factor further?

NOTATION

Complete each factorization.

11. $a^3 + 8 = (a + 2)\left(a^2 - \boxed{} + 4\right)$

12. $x^3 - 1 = (x - 1)\left(x^2 + \boxed{} + 1\right)$

13. $b^3 + 27 = \left(\boxed{}\right)(b^2 - 3b + 9)$

14. $z^3 - 125 = (z - 5)\left(\boxed{} + 5z + \boxed{}\right)$

Give an example of each type of expression.

15. **a.** the sum of two cubes
    b. the cube of a sum

16. **a.** the difference of two cubes
    b. the cube of a difference

GUIDED PRACTICE

Factor. See Example 1.

17. $y^3 + 125$ 18. $b^3 + 216$

19. $a^3 + 64$ 20. $n^3 + 1$

21. $n^3 + 512$ 22. $t^3 + 729$

23. $8 + t^3$ 24. $27 + y^3$

25. $a^3 + 1{,}000b^3$
26. $8u^3 + w^3$
27. $125c^3 + 27d^3$
28. $64m^3 + 343n^3$

Factor. See Example 2.

29. $a^3 - 27$ 30. $r^3 - 8$

31. $m^3 - 343$ 32. $y^3 - 216$

33. $216 - v^3$ 34. $125 - t^3$

35. $8s^3 - t^3$ 36. $27a^3 - b^3$

37. $1{,}000a^3 - w^3$
38. $s^3 - 64t^3$
39. $64x^3 - 27y^3$
40. $27x^3 - 1{,}000y^3$

Factor. See Example 3.

41. $2x^3 + 2$

42. $8y^3 + 8$

43. $3d^3 + 81$

44. $2x^3 + 54$

45. $x^4 - 216x$

46. $x^5 - 125x^2$

47. $64m^3x - 8n^3x$

48. $16r^4 - 128rs^3$

TRY IT YOURSELF

Choose the correct method from Section 6.1 through Section 6.5 and factor completely.

49. $x^2 + 8x + 16$

50. $64p^3 - 27$

51. $9r^2 - 16s^2$

52. $-63 - 13x + 6x^2$

53. $xy - ty + sx - st$

54. $12p^2 + 14p - 6$

55. $4p^3 + 32q^3$

56. $56a^4 - 15a^3 + a^2$

57. $16c^3t^2 + 20c^2t^3 + 6ct^4$

58. $-t^2 - 9t + 1$

59. $36e^4 - 36$

60. $3(z + 4) - a(z + 4)$

61. $35a^3b^2 - 14a^2b^3 + 14a^3b^3$

62. $-y^2 - 15y + 34$

63. $36r^2 + 60rs + 25s^2$

64. $16u^2 - 16$

Look Alikes . . .

65. a. $x^2 - 1$ b. $x^3 - 1$

66. a. $x^2 - 64$ b. $x^3 - 64$

67. a. $x^2 + 2x$ b. $x^2 + 2x + 1$

68. a. $x^2 - 4x$ b. $x^2 - 4x + 4$

APPLICATIONS

69. Mailing Breakables. Write a polynomial that describes the amount of space in the larger box that must be filled with styrofoam chips if the smaller box containing a glass tea cup is to be placed within the larger box for mailing. Then factor the polynomial.

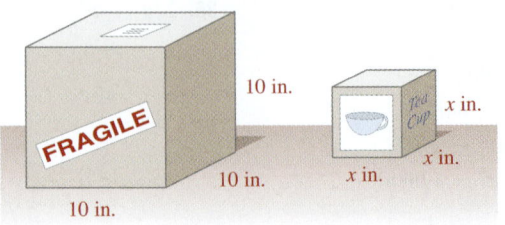

70. Melting Ice. In one hour, the block of ice shown below had melted to the size shown on the right. Write a polynomial that describes the volume of ice that melted away. Then factor the polynomial.

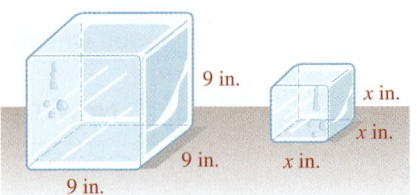

WRITING

71. Explain why $x^3 - 25$ is not a difference of two cubes.

72. Explain this diagram. Then draw a similar diagram for the difference of two cubes.

$$\underbrace{F^3 + L^3}_{\text{The same}} = (F + L)(F^2 - FL + L^2)$$

Opposite   Always plus

REVIEW

73. When expressed as a decimal, is $\frac{7}{9}$ a terminating or a repeating decimal?

74. Solve: $x + 20 = 4x - 1 + 2x$

75. Solve: $2x + 2 = \frac{2}{3}x - 2$

76. Check to determine whether 4 is a solution of $3(m - 8) + 2m = 4 - (m + 2)$.

CHALLENGE PROBLEMS

77. Consider: $x^6 - 1$. Write the binomial as a difference of two squares. Then factor.

78. What binomial multiplied by $(a^2b^2 + 7ab + 49)$ produces a difference of two cubes?

Factor.

79. $x^6 - y^9$

80. $\frac{125}{8}s^3 + \frac{1}{27}t^3$

81. $64x^{12} + y^{15}z^{18}$

82. $x^{3m} - y^{3n}$

SECTION 6.6

OBJECTIVE

1 Use a general strategy for factoring polynomials.

A Factoring Strategy

The following problems review some basic skills that are needed when factoring polynomials.

1. How many terms does each expression have?
    a. $x^4 - 5x^3 + x - 5$
    b. $16x - 5x^2 + 6x^3$

2. Do any of the factors in $(m^2 + 9)(m + 3)(m - 3)$ factor further?

3. Multiply: $3n(n + 5)(n - 14)$

4. What is the greatest common factor of the terms of $15c^3d^2 - 10cd^3$?

The factoring methods discussed so far will be used in the remaining chapters to simplify expressions and solve equations. In such cases, we must determine the factoring method—it will not be specified. This section will give you practice in selecting the appropriate factoring method to use given a randomly chosen polynomial.

1 Use a General Strategy for Factoring Polynomials.

The following strategy is helpful when factoring polynomials.

Steps for Factoring a Polynomial

1. Is there a common factor? If so, factor out the GCF, or the opposite of the GCF so that the leading coefficient is positive. Remember to include it in your final answer.

2. How many terms does the polynomial have?

    If it has **two terms,** look for the following problem types:
        a. The difference of two squares
        b. The sum of two cubes
        c. The difference of two cubes

    If it has **three terms,** look for the following problem types:
        a. A perfect-square trinomial
        b. If the trinomial is not a perfect square, use the trial-and-check-method or the grouping method.

    If it has **four or more terms,** try to factor by grouping.

3. Can any factors be factored further? If so, factor them completely.

4. Does the factorization check? Check by multiplying.

| **EXAMPLE 1** | Factor: $2x^4 - 162$ |

Strategy   We will answer the four questions listed in the *Steps for Factoring a Polynomial.*

Why   The answers to these questions help us determine which factoring techniques to use.

Solution   ***Is there a common factor?***   Yes. Factor out the GCF, which is 2.

$$2x^4 - 162 = 2(x^4 - 81)$$

How many terms does it have?   The polynomial within the parentheses, $x^4 - 81$, has two terms. It is a difference of two squares.

$$2x^4 - 162 = 2(x^4 - 81) \qquad \text{\color{red}{Think of } } x^4 - 81 \text{ \color{red}{as} } (x^2)^2 - 9^2.$$
$$= 2(x^2 + 9)(x^2 - 9) \qquad \text{\color{red}{Factor the difference of two squares.}}$$

Is it factored completely?   No. $x^2 - 9$ is also the difference of two squares and can be factored.

$$2x^4 - 162 = 2(x^4 - 81)$$
$$= 2(x^2 + 9)(x^2 - 9) \qquad \text{\color{red}{Think of } } x^2 - 9 \text{ \color{red}{as} } x^2 - 3^2.$$
$$= 2(x^2 + 9)(x + 3)(x - 3) \qquad \text{\color{red}{$x^2 + 9$ is a sum of two squares}}$$
$$\text{\color{red}{and does not factor.}}$$

Therefore, $2x^4 - 162 = 2(x^2 + 9)(x + 3)(x - 3)$.

Does it check?   Yes.

$$2(x^2 + 9)(x + 3)(x - 3) = 2(x^2 + 9)(x^2 - 9) \qquad \text{\color{red}{Multiply $(x + 3)(x - 3)$ first.}}$$
$$= 2(x^4 - 81) \qquad \text{\color{red}{Multiply $(x^2 + 9)(x^2 - 9)$.}}$$
$$= 2x^4 - 162 \qquad \text{\color{red}{This is the original polynomial.}}$$

Self Check 1   Factor: $11a^6 - 11a^2$

Now Try ▶ Problem 21

EXAMPLE 2   Factor: $-4c^5d^2 - 12c^4d^3 - 9c^3d^4$

Strategy   We will answer the four questions listed in the *Steps for Factoring a Polynomial*.

Why   The answers to these questions help us determine which factoring techniques to use.

Solution   *Is there a common factor?*   Yes. Factor out the opposite of the GCF, $-c^3d^2$, so that the leading coefficient is positive.

$$-4c^5d^2 - 12c^4d^3 - 9c^3d^4 = -c^3d^2(4c^2 + 12cd + 9d^2)$$

How many terms does it have?   The polynomial within the parentheses has three terms. It is a perfect-square trinomial because $4c^2 = (2c)^2$, $9d^2 = (3d)^2$, and $12cd = 2 \cdot 2c \cdot 3d$.

$$-4c^5d^2 - 12c^4d^3 - 9c^3d^4 = -c^3d^2(4c^2 + 12cd + 9d^2)$$
$$= -c^3d^2(2c + 3d)^2$$

Is it factored completely?   Yes. The binomial $2c + 3d$ does not factor further.

Therefore, $-4c^5d^2 - 12c^4d^3 - 9c^3d^4 = -c^3d^2(2c + 3d)^2$.

Does it check?   Yes.

$$-c^3d^2(2c + 3d)^2 = -c^3d^2(4c^2 + 12cd + 9d^2) \qquad \text{\color{red}{Use a special-product rule.}}$$
$$= -4c^5d^2 - 12c^4d^3 - 9c^3d^4 \qquad \text{\color{red}{This is the original polynomial.}}$$

Self Check 2   Factor: $-32h^4 - 80h^3 - 50h^2$

Now Try ▶ Problem 33

EXAMPLE 3   Factor: $y^4 - 3y^3 + y - 3$

Strategy   We will answer the four questions listed in the *Steps for Factoring a Polynomial.*

Why   The answers to these questions help us determine which factoring techniques to use.

Solution   *Is there a common factor?*   No. There is no common factor (other than 1).

How many terms does it have?   Since the polynomial has four terms, we will try factoring by grouping.

$$y^4 - 3y^3 + y - 3 = y^3(y - 3) + 1(y - 3) \qquad \text{Factor out } y^3 \text{ from } y^4 - 3y^3. \text{ Factor out 1 from } y - 3.$$

$$= (y - 3)(y^3 + 1)$$

> **Success Tip**
>
> Something as simple as counting the number of terms that a polynomial has is very important when determining how to factor it.

Is it factored completely?   No. We can factor $y^3 + 1$ as a sum of two cubes.

$$y^4 - 3y^3 + y - 3 = y^3(y - 3) + 1(y - 3)$$
$$= (y - 3)(y^3 + 1) \qquad \text{Think of } y^3 + 1 \text{ as } y^3 + 1^3.$$
$$= (y - 3)(y + 1)(y^2 - y + 1) \qquad y^2 - y + 1 \text{ does not factor further.}$$

Therefore, $y^4 - 3y^3 + y - 3 = (y - 3)(y + 1)(y^2 - y + 1)$.

Does it check?   Yes.

$$(y - 3)(y + 1)(y^2 - y + 1) = (y - 3)(y^3 + 1) \qquad \text{Multiply the last two factors.}$$
$$= y^4 + y - 3y^3 - 3 \qquad \text{Use the FOIL method.}$$
$$= y^4 - 3y^3 + y - 3 \qquad \text{This is the original polynomial.}$$

> **Self Check 3**   Factor: $b^4 + b^3 + 8b + 8$
>
> **Now Try** ▶ Problem 37

EXAMPLE 4   Factor: $32n - 4n^2 + 4n^3$

Strategy   We will answer the four questions listed in the *Steps for Factoring a Polynomial.*

Why   The answers to these questions help us determine which factoring techniques to use.

Solution   *Is there a common factor?*   Yes. When we write the terms in descending powers of n, we see that the GCF is $4n$.

$$4n^3 - 4n^2 + 32n = 4n(n^2 - n + 8)$$

How many terms does it have?
The polynomial within the parentheses has three terms. It is not a perfect-square trinomial because the last term, 8, is not a perfect square.

Negative factors of 8	Sum of the negative factors of 8
$-1(-8) = 8$	$-1 + (-8) = -9$
$-2(-4) = 8$	$-2 + (-8) = -10$

   To factor the trinomial $n^2 - n + 8$, we must find two integers whose product is 8 and whose sum is -1. As we see in the table, there are no such integers. Thus, $n^2 - n + 8$ is prime.

Is it factored completely?   Yes.

Therefore, $4n^3 - 4n^2 + 32n = 4n(n^2 - n + 8)$.   Remember to write the GCF, $4n$, from the first step.

Does it check? Yes.

$$4n(n^2 - n + 8) = 4n^3 - 4n^2 + 32n \quad \text{This is equivalent to the original polynomial.}$$

Self Check 4 Factor: $6m^2 - 54m + 6m^3$

Now Try Problem 45

EXAMPLE 5 Factor: $3y^3 - 4y^2 - 4y$

Strategy We will answer the four questions listed in the *Steps for Factoring a Polynomial*.

Why The answers to these questions help us determine which factoring techniques to use.

Solution *Is there a common factor?* Yes. The GCF is y.

$$3y^3 - 4y^2 - 4y = y(3y^2 - 4y - 4)$$

How many terms does it have? The polynomial within the parentheses has three terms. It is not a perfect-square trinomial because the first term, $3y^2$, is not a perfect square.

If we use grouping to factor $3y^2 - 4y - 4$, the key number is $ac = 3(-4) = -12$. We must find two integers whose product is -12 and whose sum is $b = -4$.

Key number $= -12$ $\qquad b = -4$

Factors of -12	Sum of the factors of -12
$2(-6) = -12$	$2 + (-6) = -4$

From the table, the correct pair is 2 and -6. They serve as the coefficients of $2y$ and $-6y$, the two terms that we use to represent the middle term, $-4y$, of the trinomial.

$$3y^2 - 4y - 4 = 3y^2 + 2y - 6y - 4 \quad \text{Express } -4y \text{ as } 2y - 6y.$$
$$= y(3y + 2) - 2(3y + 2) \quad \text{Factor } y \text{ from the first two terms and factor } -2 \text{ from the last two terms.}$$
$$= (3y + 2)(y - 2) \quad \text{Factor out } 3y + 2.$$

The trinomial $3y^2 - 4y - 4$ factors as $(3y + 2)(y - 2)$.

Is it factored completely? Yes. Because $3y + 2$ and $y - 2$ do not factor.

Therefore, $3y^3 - 4y^2 - 4y = y(3y + 2)(y - 2)$. Remember to write the GCF, y, from the first step.

Does it check? Yes.

$$y(3y + 2)(y - 2) = y(3y^2 - 4y - 4) \quad \text{Multiply the binomials.}$$
$$= 3y^3 - 4y^2 - 4y \quad \text{This is the original polynomial.}$$

Self Check 5 Factor: $6y^3 + 21y^2 - 12y$

Now Try Problem 67

SECTION 6.6   STUDY SET

VOCABULARY

Fill in the blanks.

1. To factor a polynomial means to express it as a _____ of two (or more) polynomials.

2. A polynomial is factored _____ when each factor is prime.

CONCEPTS

For each of the following polynomials, which factoring method would you use first?

3. $2x^5y - 4x^3y$

4. $9b^2 + 12y - 5$

5. $x^2 + 18x + 81$

6. $ax + ay - x - y$

7. $x^3 + 27$

8. $y^3 - 64$

9. $m^2 + 3mn + 2n^2$

10. $16 - 25z^2$

11. What is the first question that should be asked when using the strategy of this section to factor a polynomial?

12. Use multiplication to determine whether the factorization is correct.

$$5c^3d^2 - 40c^2d^3 + 35cd^4 = 5cd^2(c - 7d)(c - d)$$

NOTATION

Complete each factorization.

13. $6m^3 - 28m^2 + 16m = 2m(3m^2 - \boxed{} + 8)$
$$= 2m(3m - 2)(\boxed{} - 4)$$

14. $2a^3 + 3a^2 - 2a - 3$
$$= \boxed{}(2a + 3) - 1(\boxed{} + 3)$$
$$= (\boxed{})(a^2 - 1)$$
$$= (2a + 3)(a + 1)(\boxed{})$$

TRY IT YOURSELF

The following is a list of random factoring problems. Factor each expression. If an expression is not factorable, write "prime." See Examples 1–5.

15. $2b^2 + 8b - 24$

16. $32 - 2t^4$

17. $8p^3q^7 + 4p^2q^3$

18. $8m^2n^3 - 24mn^4$

19. $2 + 24y + 40y^2$

20. $6r^2 + 3rs - 18s^2$

21. $8x^4 - 8$

22. $t - 90 + t^2$

23. $14c - 147 + c^2$

24. $ab^2 - 4a + 3b^2 - 12$

25. $x^2 + 7x + 1$

26. $3a^3 + 24b^3$

27. $-2x^5 + 128x^2$

28. $16 - 40z + 25z^2$

29. $a^2c + a^2d^2 + bc + bd^2$

30. $6t^4 + 14t^3 - 40t^2$

31. $-9x^2 + 6x - 1$

32. $x^2y^2 - 2x^2 - y^2 + 2$

33. $-20m^3 - 100m^2 - 125m$

34. $5x^3y^3z^4 + 25x^2y^4z^2 - 35x^3y^2z^5$

35. $2c^2 - 5cd - 3d^2$

36. $125p^3 - 64y^3$

37. $p^4 - 2p^3 - 8p + 16$

38. $a^2 + 8a + 3$

39. $a^2(x - a) - b^2(x - a)$

40. $70p^4q^3 - 35p^4q^2 + 49p^5q^2$

41. $a^2b^2 - 144$

42. $-16x^4y^2z + 24x^5y^3z^4 - 15x^2y^3z^7$

43. $2x^3 + 10x^2 + x + 5$

44. $u^2 - 18u + 81$

45. $8v^2 - 14v^3 + v^4$

46. $28 - 3m - m^2$

47. $18a^2 - 6ab + 42ac - 14bc$

48. $81r^4 - 256$

49. $8a^2x^3 - 2b^2x$

50. $12x^2 + 14x - 6$

51. $6x^2 - 14x + 8$

52. $12x^2 - 12$

53. $4x^2y^2 + 4xy^2 + y^2$

54. $81r^4s^2 - 24rs^5$

55. $4m^5 + 500m^2$

56. $ae + bf + af + be$

57. $a^3 - 24 - 4a + 6a^2$

58. $6x^2 - x - 16$

59. $4x^2 + 9y^2$

60. $x^4y + 216xy^4$

61. $16a^5 - 54a^2$

62. $25x^2 - 16y^2$

63. $27x - 27y - 27z$

64. $12x^2 + 52x + 35$

65. $xy - ty + xs - ts$

66. $bc + b + cd + d$

67. $35x^8 - 2x^7 - x^6$

68. $x^3 - 25$

69. $5(x - 2) + 10y(x - 2)$

70. $16x^2 - 40x^3 + 25x^4$

71. $49p^2 + 28pq + 4q^2$

72. $16d^2 - 56dz + 49z^2$

73. $4t^2 + 36$

74. $r^5 + 3r^3 + 2r^2 + 6$

75. $m^2n^2 - 9m^2 + 3n^2 - 27$

76. $z^2 + 6yz^2 + 9y^2z^2$

77. Which factoring method do you find the most difficult? Why?

78. What four questions make up the factoring strategy for polynomials discussed in this section?

79. What does it mean to factor a polynomial?

80. How is a factorization checked?

REVIEW

81. Graph the real numbers -3, 0, 2, and $-\frac{3}{2}$ on a number line.

82. Graph the interval $(-2, 3]$ on a number line.

83. Graph: $y = \frac{1}{2}x + 1$ **84.** Graph: $y < 2 - 3x$

CHALLENGE PROBLEMS

Factor using rational numbers.

85. $x^6 - 4x^3 - 12$

86. $x(x - y) - y(y - x)$

87. $24 - x^3 + 8x^2 - 3x$

88. $25b^2 + 14b + \frac{49}{25}$

89. $x^9 + y^6$

90. $\frac{1}{4} - \frac{u^2}{81}$

91. $x^4 - 13x^2 + 36$ **92.** $x^4 - 2x^2 - 8$

93. $x^2y^2 - 6xy - 16$ **94.** $5x + 4y + 25x^2 - 16y^2$

SECTION 6.7

Solving Quadratic Equations by Factoring

OBJECTIVES

1 Define quadratic equations.

2 Solve quadratic equations using the zero-factor property.

3 Solve third-degree equations by factoring.

ARE YOU READY?

The following problems review some basic skills that are needed when solving quadratic equations.

1. Evaluate: $0 \cdot 5$

2. Fill in the blank: $8 \cdot \boxed{} = 0$

3. Solve: $x + 4 = 0$

4. Solve: $8x = 0$

5. Factor: $x^2 - x - 6$

6. Factor: $3n^2 - n - 2$

The factoring methods that we have discussed have many applications in algebra. In this section, we will use factoring to solve *quadratic equations*. These equations are different from those that we solved in Chapter 2. They contain a term in which the variable is raised to the second power, such as x^2 or t^2.

1 Define Quadratic Equations.

In a linear, or first degree equation, such as $2x + 3 = 8$, the exponent on the variable is an unwritten 1. A quadratic, or second degree equation, has a term in which the exponent on the variable is 2, and has no other terms of higher degree.

Quadratic Equations	A **quadratic equation** is an equation that can be written in the **standard form** $$ax^2 + bx + c = 0$$ where a, b, and c represent real numbers, and $a \neq 0$.

The Language of Algebra

Quadratic equations involve the square of a variable, not the 4th power as *quad* might suggest. Why is this? Because the origin of the word **quadratic** is the Latin word *quadratus*, meaning square.

Some examples of quadratic equations are

$$x^2 - 2x - 63 = 0, \quad x^2 - 25 = 0, \quad 6x^2 - 12x = 0, \quad \text{and} \quad 2x^2 + 3x = 2$$

The first three equations are in standard form. To write the fourth equation in standard form, we subtract 2 from both sides to get $2x^2 + 3x - 2 = 0$.

Even though it does not have an x-term, the equation $x^2 - 25 = 0$ is a quadratic equation because the definition allows b, the coefficient of x, to equal 0. And even though the equation $6x^2 - 12x = 0$ does not have a constant term, it is also a quadratic equation, because the definition allows c, the constant term, to equal 0. However, a quadratic equation must have a variable-squared term, because the definition requires that $a \neq 0$.

2 Solve Quadratic Equations Using the Zero-Factor Property.

To **solve a quadratic equation,** we find all values of the variable that make the equation true. The methods that we used to solve linear equations in Chapter 2 cannot be used to solve a quadratic equation, because we cannot isolate the variable on one side of the equation. However, we often can solve quadratic equations using factoring and the following property of real numbers.

The Zero-Factor Property	When the product of two real numbers is 0, at least one of them is 0. If a and b represent real numbers, and $$\text{if } ab = 0, \text{ then } a = 0 \text{ or } b = 0$$

EXAMPLE 1 Solve: $(4x - 1)(x + 6) = 0$

Strategy We will set $4x - 1$ equal to 0 and $x + 6$ equal to 0 and solve each equation.

Why If the product of $4x - 1$ and $x + 6$ is 0, then, by the zero-factor property, $4x - 1$ must equal 0, or $x + 6$ must equal 0.

Solution If $(4x - 1)(x + 6) = 0$ is to be a true statement, then either

$$4x - 1 = 0 \quad \text{or} \quad x + 6 = 0$$

Caution

It would not be helpful to multiply $(4x - 1)$ and $(x + 6)$. We want the left side of the equation to be in factored form so that we can use the zero-factor property.

Now we solve each of these linear equations using the methods from Chapter 2.

$$4x - 1 = 0 \qquad \text{or} \qquad x + 6 = 0$$
$$4x = 1 \quad \text{Add 1 to both sides.} \qquad x = -6 \quad \text{Subtract 6 from both sides.}$$
$$x = \frac{1}{4} \quad \text{Divide both sides by 4.}$$

The results must be checked separately to see whether each of them produces a true statement. We substitute $\frac{1}{4}$ and then -6 for x in the original equation and evaluate the left side.

The Language of Algebra

In the zero-factor property, the word **or** means one or the other or both. If the product of two numbers is 0, then one factor is 0, or the other factor is 0, or both factors can be 0.

Check $\frac{1}{4}$:

$$(4x - 1)(x + 6) = 0$$
$$\left[4\left(\frac{1}{4}\right) - 1\right]\left(\frac{1}{4} + 6\right) \stackrel{?}{=} 0$$
$$(1 - 1)\left(\frac{25}{4}\right) \stackrel{?}{=} 0$$
$$0\left(\frac{25}{4}\right) \stackrel{?}{=} 0 \quad \text{The factor } 4x - 1 \text{ is 0 when x is } \frac{1}{4}.$$
$$0 = 0 \quad \text{True}$$

Check -6:

$$(4x - 1)(x + 6) = 0$$
$$[4(-6) - 1](-6 + 6) \stackrel{?}{=} 0$$
$$(-24 - 1)(0) \stackrel{?}{=} 0 \quad \text{The factor x + 6 is 0 when x is −6.}$$
$$-25(0) \stackrel{?}{=} 0$$
$$0 = 0 \quad \text{True}$$

The resulting true statements indicate that $(4x - 1)(x + 6) = 0$ has two solutions: $\frac{1}{4}$ and -6. Recall from Chapter 2 that the *solution set* of an equation is the set of all numbers that make the equation true. Thus, the solution set is $\left\{-6, \frac{1}{4}\right\}$.

Self Check 1 Solve: $(x - 12)(5x + 6) = 0$

Now Try ▶ Problem 15

In Example 1, the left side of $(4x - 1)(x + 6) = 0$ is in factored form and the right side is 0, so we can immediately use the zero-factor property. However, to solve many quadratic equations, we must factor before using the zero-factor property.

EXAMPLE 2   Solve: $x^2 - 2x - 63 = 0$

Strategy   We will factor the trinomial on the left side of the equation and use the zero-factor property.

Why   To use the zero-factor property, we need one side of the equation to be factored completely and the other side to be 0.

Solution

$x^2 - 2x - 63 = 0$	This is the equation to solve.
$(x + 7)(x - 9) = 0$	Factor the trinomial, $x^2 - 2x - 63$.
$x + 7 = 0$ or $x - 9 = 0$	Set each factor equal to 0.
$x = -7$ $\mid$ $x = 9$	Solve each equation using the methods from Chapter 2.

Success Tip

When you see the word **solve** in this example, you probably think of steps from Chapter 2 such as combining like terms, distributing, or doing something to both sides. However, to solve this quadratic equation, we begin by factoring $x^2 - 2x - 63$.

To check the results, we substitute -7 and then 9 for x in the original equation and evaluate the left side.

Check -7:

$$x^2 - 2x - 63 = 0$$
$$(-7)^2 - 2(-7) - 63 \stackrel{?}{=} 0$$
$$49 - (-14) - 63 \stackrel{?}{=} 0$$
$$63 - 63 \stackrel{?}{=} 0$$
$$0 = 0 \quad \text{True}$$

Check 9:

$$x^2 - 2x - 63 = 0$$
$$(9)^2 - 2(9) - 63 \stackrel{?}{=} 0$$
$$81 - 18 - 63 \stackrel{?}{=} 0$$
$$63 - 63 \stackrel{?}{=} 0$$
$$0 = 0 \quad \text{True}$$

The solutions of $x^2 - 2x - 63 = 0$ are -7 and 9, and the solution set is $\{-7, 9\}$.

Self Check 2   Solve: $x^2 + 5x + 6 = 0$

Now Try ▶ Problem 27

The previous examples suggest the following strategy to solve quadratic equations by factoring.

The Factoring Method for Solving a Quadratic Equation

1. Write the equation in standard form: $ax^2 + bx + c = 0$ or $0 = ax^2 + bx + c$.
2. Factor completely.
3. Use the zero-factor property to set each factor equal to 0.
4. Solve each resulting equation.
5. Check the results in the original equation.

With this method, we factor *expressions* to solve *equations*.

EXAMPLE 3   Solve: $x^2 - 25 = 0$

Strategy   We will factor the binomial on the left side of the equation and use the zero-factor property.

Why   To use the zero-factor property, we need one side of the equation to be factored completely and the other side to be 0.

Solution

Notation

Although $x^2 - 25 = 0$ is missing a term involving x, it is a quadratic equation in standard $ax^2 + bx + c = 0$ form, where $a = 1$, $b = 0$, and $c = -25$.

We factor the difference of two squares on the left side of the equation and proceed as follows.

$$x^2 - 25 = 0 \qquad \text{This is the equation to solve.}$$

$$(x + 5)(x - 5) = 0 \qquad \text{Factor the difference of two squares, } x^2 - 25.$$

$$x + 5 = 0 \quad \text{or} \quad x - 5 = 0 \qquad \text{Set each factor equal to 0.}$$

$$x = -5 \qquad\qquad x = 5 \qquad \text{Solve each equation.}$$

Check each result by substituting it into the original equation.

Check -5:	*Check* **5**:
$x^2 - 25 = 0$	$x^2 - 25 = 0$
$(-5)^2 - 25 \stackrel{?}{=} 0$	$5^2 - 25 \stackrel{?}{=} 0$
$25 - 25 \stackrel{?}{=} 0$	$25 - 25 \stackrel{?}{=} 0$
$0 = 0$ True	$0 = 0$ True

The solutions of $x^2 - 25 = 0$ are -5 and 5, and the solution set is $\{-5, 5\}$.

Self Check 3   Solve: $x^2 - 49 = 0$

Now Try ▶ Problem 35

EXAMPLE 4   Solve: $6x^2 = 12x$

Strategy   We will subtract $12x$ from both sides of the equation to get 0 on the right side. Then we will factor the resulting binomial and use the zero-factor property.

Why   To use the zero-factor property, we need one side of the equation to be factored completely and the other side to be 0.

Solution   The equation is not in standard form, $ax^2 + bx + c = 0$. To get 0 on the right side, we proceed as follows.

Caution

A creative, but incorrect, approach to solve $6x^2 = 12x$ is to divide both sides by $6x$.

$$\frac{6x^2}{6x} = \frac{12x}{6x}$$

You will obtain $x = 2$; however, you will lose the second solution, 0.

$$6x^2 = 12x \qquad \text{This is the equation to solve.}$$

$$6x^2 - 12x = 12x - 12x \qquad \text{Use the subtraction property of equality to get 0 on the right side: Subtract } 12x \text{ from both sides.}$$

$$6x^2 - 12x = 0 \qquad \text{Combine like terms: } 12x - 12x = 0. \text{ Although it is missing a constant term, this is a quadratic equation in standard } ax^2 + bx + c = 0 \text{ form, where } a = 6, b = -12, \text{ and } c = 0.$$

To solve this equation, we factor the left side and proceed as follows.

$$6x(x - 2) = 0 \qquad \text{Factor out the GCF, } 6x.$$

$$6x = 0 \quad \text{or} \quad x - 2 = 0 \qquad \text{Set each factor equal to 0.}$$

$$x = \frac{0}{6} \qquad\qquad x = 2 \qquad \text{Solve each equation using the methods from Chapter 2.}$$

$$x = 0$$

The solutions are 0 and 2 and the solution set is $\{0, 2\}$. Check each solution in the original equation, $6x^2 = 12x$.

Self Check 4   Solve: $5x^2 = 25x$

Now Try ▶ Problem 47

EXAMPLE 5   Solve: $2x^2 - 2 = -3x$

Strategy   We will add $3x$ to both sides of the equation to get 0 on the right side. Then we will factor the resulting trinomial and use the zero-factor property.

Why   To use the zero-factor property, we need one side of the equation to be factored completely and the other side to be 0.

Solution   The equation is not in standard form, $ax^2 + bx + c = 0$. To get 0 on the right side, we proceed as follows.

$2x^2 - 2 = -3x$	This is the equation to solve.
$2x^2 + 3x - 2 = -3x + 3x$	Use the addition property of equality to get 0 on the right side: Add 3x to both sides.
$2x^2 + 3x - 2 = 0$	Combine like terms: $-3x + 3x = 0$. This equation is in standard form.
$(2x - 1)(x + 2) = 0$	Factor the trinomial.
$2x - 1 = 0 \quad$ or $\quad x + 2 = 0$	Set each factor equal to 0.
$2x = 1 \qquad\qquad x = -2$	Solve each equation using the methods
$x = \dfrac{1}{2}$	from Chapter 2.

The solutions are $\frac{1}{2}$ and -2 and the solution set is $\left\{-2, \frac{1}{2}\right\}$. Check each solution in the original equation, $2x^2 - 2 = -3x$.

Self Check 5   Solve: $3x^2 - 8 = -10x$

Now Try ▶ Problem 51

Unlike linear equations, quadratic equations have two solutions. In some cases, however, the two solutions are the same number.

EXAMPLE 6   Solve: $x(9x - 12) = -4$

Strategy   To write the equation in standard form, we will distribute the multiplication by x and add 4 to both sides. Then we will factor the resulting trinomial and use the zero-factor property.

Why   To use the zero-factor property, we need one side of the equation to be factored completely and the other side to be 0.

Solution

$x(9x - 12) = -4$	This is the equation to solve.
$9x^2 - 12x = -4$	Distribute the multiplication by x.
$9x^2 - 12x + 4 = -4 + 4$	To get 0 on the right side, add 4 to both sides.
$9x^2 - 12x + 4 = 0$	Combine like terms: $-4 + 4 = 0$. This equation is in standard form.
$(3x - 2)(3x - 2) = 0$	Factor the trinomial, $9x^2 - 12x + 4$.
$3x - 2 = 0 \quad$ or $\quad 3x - 2 = 0$	Set each factor equal to 0.
$3x = 2 \qquad\qquad 3x = 2$	Solve each equation using the methods
$x = \dfrac{2}{3} \qquad\qquad x = \dfrac{2}{3}$	from Chapter 2.

Caution

To use the zero-factor property, one side of the equation must be 0. In this example, it would be incorrect to set each factor equal to -4.

$x = -4$ ~~or~~ $9x - 12 = -4$

If the product of two numbers is -4, one of them does not have to be -4. For example, $2(-2) = -4$.

After solving both equations, we see that $\frac{2}{3}$ is a **repeated solution.** Thus, the solution set is $\left\{\frac{2}{3}\right\}$. Check this result by substituting it into the original equation.

Self Check 6 Solve: $x(4x + 12) = -9$

Now Try ▶ Problem 59

3 Solve Third-Degree Equations by Factoring.

Some equations involving polynomials with degrees higher than 2 also can be solved by using the factoring method. In such cases, we use an extension of the zero-factor property: When the product of two *or more* real numbers is 0, at least one of them is 0.

EXAMPLE 7   Solve: $6x^3 + 12x = 17x^2$

Strategy   This equation is not quadratic, because it contains a term involving x^3. However, we can solve it by using factoring. First we get 0 on the right side by subtracting $17x^2$ from both sides. Then we factor the polynomial on the left side and use an extension of the zero-factor property.

Why   To use the zero-factor property, we need one side of the equation to be factored completely and the other side to be 0.

Solution

The Language of Algebra

Since the highest degree of any term in $6x^3 + 12x = 17x^2$ is 3, it is called a **third-degree** equation. Note that it has three solutions.

$$6x^3 + 12x = 17x^2 \qquad \text{This is the equation to solve.}$$

$$6x^3 - 17x^2 + 12x = 17x^2 - 17x^2 \qquad \text{To get 0 on the right side, subtract } 17x^2$$
$$\text{from both sides.}$$

$$6x^3 - 17x^2 + 12x = 0 \qquad \text{Combine like terms: } 17x^2 - 17x^2 = 0.$$

$$x(6x^2 - 17x + 12) = 0 \qquad \text{Factor out the GCF, } x.$$

$$x(2x - 3)(3x - 4) = 0 \qquad \text{Factor the trinomial, } 6x^2 - 17x + 12.$$

If $x(2x - 3)(3x - 4) = 0$, then at least one of the factors is equal to 0.

$x = 0$ or	$2x - 3 = 0$ or	$3x - 4 = 0$ Set each factor equal to 0.
	$2x = 3$	$3x = 4$ Solve each equation.
	$x = \dfrac{3}{2}$	$x = \dfrac{4}{3}$

The solutions are 0, $\frac{3}{2}$, and $\frac{4}{3}$ and the solution set is $\left\{0, \frac{4}{3}, \frac{3}{2}\right\}$. Check each solution in the original equation, $6x^3 + 12x = 17x^2$.

Self Check 7 Solve: $10x^3 + x^2 = 2x$

Now Try ▶ Problem 63

SECTION 6.7 ▶ STUDY SET

VOCABULARY

Fill in the blanks.

1. $2x^2 + 3x - 1 = 0$ and $x^2 - 36 = 0$ are examples of _____ equations.

2. $ax^2 + bx + c = 0$ is called the _____ form of a quadratic equation.

3. The _____ property states that if the product of two numbers is 0, at least one of them is 0: If $ab = 0$, then $a = $ ▢ or $b = $ ▢.

4. Since the highest degree of any term in $x^3 - 5x^2 - 6x = 0$ is 3, it is called a _____-degree equation.

CONCEPTS

5. Which of the following are quadratic equations?
   a. $x^2 + 2x - 10 = 0$ b. $2x - 10 = 0$
   c. $x^2 = 15x$ d. $x^3 + x^2 + 2x = 0$

6. Write each equation in the standard form $ax^2 + bx + c = 0$.
   a. $x^2 + 2x = 6$ b. $x^2 = 5x$

   c. $3x(x - 8) = -9$ d. $4x^2 = 25$

7. Set $5x + 4$ equal to 0 and solve for x.

8. What step should be performed first to solve $x^2 - 6x - 16 = 0$?

9. What step (or steps) should be performed first before factoring is used to solve each equation?
   a. $x^2 + 7x = -6$
   b. $x(x + 7) = 3$

10. Check to determine whether the given number is a solution of the given quadratic equation.
    a. $x^2 - 4x = 0$; 4
    b. $x^2 - 2x - 7 = 0$; -2

NOTATION

Complete each solution to solve the equation.

11. $(x - 1)(x + 7) = 0$

   $x - 1 = \boxed{}$ or $\boxed{} = 0$

   $x = 1$ $\Big|$ $x = \boxed{}$

12. $7y^2 + 14y = 0$

   $\boxed{}(y + 2) = 0$

   $7y = 0$ $\boxed{}$ $y + 2 = 0$

   $y = \boxed{}$ $\Big|$ $y = -2$

13. $p^2 - p - 6 = 0$

   $(\boxed{} - 3)(p + 2) = 0$

   $\boxed{} = 0$ or $p + 2 = \boxed{}$

   $p = \boxed{}$ $\Big|$ $p = \boxed{}$

14. $4y^2 - 25 = 0$

   $(2y + \boxed{})(2y - \boxed{}) = 0$

   $2y + 5 = \boxed{}$ or $\boxed{} = 0$

   $2y = \boxed{}$ $\Big|$ $2y = 5$

   $y = -\frac{5}{2}$ $\Big|$ $y = \boxed{}$

GUIDED PRACTICE

Solve each equation. See Example 1.

15. $(x - 3)(x - 2) = 0$ 16. $(x + 2)(x + 3) = 0$

17. $(x + 7)(x - 7) = 0$ 18. $(x - 8)(x + 8) = 0$

19. $6x(2x - 5) = 0$ 20. $5x(5x + 7) = 0$

21. $-7a(3a + 10) = 0$ 22. $-6t(2t - 9) = 0$

23. $t(t - 6)(t + 8) = 0$ 24. $n(n + 1)(n - 6) = 0$

25. $(x - 1)(x + 2)(x - 3) = 0$ 26. $(x + 2)(x + 3)(x - 4) = 0$

Solve each equation. See Example 2.

27. $x^2 - 13x + 12 = 0$ 28. $x^2 + 7x + 6 = 0$

29. $x^2 - 4x - 21 = 0$ 30. $x^2 + 2x - 15 = 0$

31. $x^2 - 9x + 8 = 0$ 32. $x^2 - 14x + 45 = 0$

33. $a^2 + 8a + 15 = 0$ 34. $a^2 - 17a + 60 = 0$

Solve each equation. See Example 3.

35. $x^2 - 81 = 0$ 36. $x^2 - 36 = 0$

37. $t^2 - 25 = 0$ 38. $m^2 - 49 = 0$

39. $4x^2 - 1 = 0$ 40. $9y^2 - 1 = 0$

41. $9y^2 - 49 = 0$ 42. $16z^2 - 25 = 0$

Solve each equation. See Example 4.

43. $w^2 = 7w$ 44. $x^2 = 5x$

45. $s^2 = 16s$ 46. $p^2 = 20p$

47. $4y^2 = 12y$ 48. $5m^2 = 15m$

49. $3x^2 = -8x$ 50. $3s^2 = -4s$

Solve each equation. See Example 5.

51. $3x^2 + 5x = 2$ 52. $3x^2 + 14x = -8$

53. $2x^2 + x = 3$ 54. $2x^2 - 5x = -2$

55. $5x^2 + 1 = 6x$ 56. $6x^2 + 1 = 5x$

57. $2x^2 - 3x = 20$ 58. $2x^2 - 3x = 14$

Solve each equation. See Example 6.

59. $4r(r + 7) = -49$ 60. $5m(5m + 8) = -16$

61. $9a(a - 3) = 3a - 25$ 62. $3x(3x + 10) = 6x - 16$

Solve each equation. See Example 7.

63. $x^3 + 3x^2 + 2x = 0$ 64. $x^3 - 7x^2 + 10x = 0$

65. $k^3 - 27k - 6k^2 = 0$ 66. $j^3 - 22j - 9j^2 = 0$

TRY IT YOURSELF

Solve each equation.

67. $4x^2 = 81$ 68. $9y^2 = 64$

69. $x^2 - 16x + 64 = 0$

70. $h^2 + 2h + 1 = 0$

71. $(2s - 5)(s + 6) = 0$

72. $h(3h - 4)(h + 1) = 0$

73. $3b^2 - 30b = 6b - 60$

74. $2m^2 - 8m = 2m - 12$

75. $k^3 + k^2 - 20k = 0$

76. $n^3 - 6n^2 + 8n = 0$

77. $x^2 - 100 = 0$

78. $z^2 - 25 = 0$

79. $z(z - 7) = -12$

80. $p(p + 1) = 6$

81. $3y^2 - 14y - 5 = 0$

82. $4y^2 - 11y - 3 = 0$

83. $(x - 2)(x^2 - 8x + 7) = 0$

84. $(x - 1)(x^2 + 5x + 6) = 0$

85. $(n + 8)(n - 3) = -30$

86. $(2s + 5)(s + 1) = -1$

87. $x^3 - 6x^2 = -9x$

88. $m^3 - 8m^2 + 16m = 0$

89. $4a^2 + 1 = 8a + 1$

90. $3b^2 - 6 = 12b - 6$

91. $2b(6b + 13) = -12$

92. $5f(5f - 16) = -15$

93. $3a^3 + 4a^2 + a = 0$

94. $10b^3 - 15b^2 - 25b = 0$

95. $2x^3 = 2x(x + 2)$

96. $x^3 + 7x^2 = x^2 - 9x$

97. $-15x^2 + 2 + 7x = 0$

98. $-8x^2 + 3 - 10x = 0$

99. $4p^2 - 121 = 0$

100. $q^2 - \dfrac{1}{4} = 0$

101. $d(8d - 9) = -1$

102. $6n^3 - 6n = 0$

Look Alikes . . .

Factor the expression in part a and solve the equation in part b.

103. a. $x^2 + 4x - 21$ **b.** $x^2 + 4x - 21 = 0$

104. a. $4a^2 - 8a$ **b.** $4a^2 - 8a = 0$

105. a. $12n^2 - 5n - 2$ **b.** $12n^2 - 5n - 2 = 0$

106. a. $x^2 - 36$ **b.** $x^2 - 36 = 0$

WRITING

107. Explain the zero-factor property.

108. Find the error in the following solution.

$$x(x + 1) = 6$$

~~$x = 6$~~   or   ~~$x + 1 = 6$~~
~~$x = 5$~~

The solutions are 6 and 5.

109. A student solved $x^2 - 5x + 6 = 0$ and obtained two solutions: 2 and 3. Explain the error in his check.

Check: ~~$x^2 - 5x + 6 = 0$~~
~~$2^2 - 5(3) + 6 \overset{?}{=} 0$~~
$4 - 15 + 6 \overset{?}{=} 0$
$-5 = 0$ **False**

2 is not a solution.   3 is not a solution.

110. In this section, we solved quadratic equations by factoring. Did we always obtain two different solutions? Explain.

111. What is wrong with the step used to solve $x^2 = 2x$ shown below?

$$x^2 = 2x$$

$$\frac{x^2}{x} \,\, \frac{2x}{x}$$

$$x = 2$$

The solution is 2.

112. Explain the error in the following solution.

Factor: $x^2 - 5x + 6$

$(x - 2)(x - 3) = 0$

~~$x - 2 = 0$~~   or   ~~$x - 3 = 0$~~
$x = 2$ $x = 3$

The solutions are 2 and 3.

REVIEW

113. Exercise. A doctor advises a patient to exercise at least 15 minutes but less than 30 minutes per day. Use a compound inequality to express the range of these times t in minutes.

114. Snacks. A bag of peanuts is worth $0.30 less than the same size bag of cashews. Equal amounts of peanuts and cashews are used to make 40 bags of a mixture that is worth $1.05 per bag. How much is a bag of cashews worth?

CHALLENGE PROBLEMS

Solve each equation.

115. $x^4 - 625 = 0$

116. $2a^3 + a^2 - 32a - 16 = 0$

117. $(x - 3)^2 = 2x + 9$

118. $(x + 3)^2 = (2x - 1)^2$

<table>
<tr><td>

SECTION 6.8

OBJECTIVES

1 Solve problems involving geometric figures.

2 Solve problems involving consecutive integers.

3 Solve problems using the Pythagorean theorem.

4 Solve problems given the quadratic equation model.

</td></tr>
</table>

Applications of Quadratic Equations

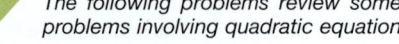

ARE YOU READY?

The following problems review some basic skills that are needed when solving application problems involving quadratic equations.

1. What is the formula for the area of a rectangle?

2. What is the formula for the area of a triangle?

3. Let x = a number. Write an algebraic expression that represents 1 more than the number.

4. How many sides does a triangle have?

In Chapter 2, we solved mixture, investment, and uniform motion problems. To model those situations, we used *linear equations* in one variable. We will now consider situations that are modeled by *quadratic equations*.

1 Solve Problems Involving Geometric Figures.

We can use the six-step problem-solving strategy and the factoring method for solving quadratic equations to find the dimensions of certain figures, given their area.

EXAMPLE 1

THE AMERICAN SWEETHEART

©Morgan Art Foundation Limited/Art Resource, NY

Painting.   In 2002, the pop art painting *The American Sweetheart,* by artist Robert Indiana, sold for $614,500. The area of the rectangular painting is 32 square feet. Find the dimensions of the painting if it is twice as long as it is wide.

Analyze

- The area of the painting is 32 ft^2.
- The length is twice as long as the width.
- Find the length and width (the dimensions).

Assign   Since the length is related to the width, let w = the width of the painting in feet. Then $2w$ = the length of the painting.

Form   To form an equation, we use the formula for the area of a rectangle, $A = lw$, where $A = 32$.

The area of the rectangle	equals	the length	times	the width.
32	=	2w	·	w

Solve

$$32 = 2w \cdot w$$ This is the equation to solve.

$$32 = 2w^2$$ Multiply 2w and w. This is a quadratic equation but it is not in standard form.

$$0 = 2w^2 - 32$$ To get 0 on the left side, subtract 32 from both sides.

$$0 = 2(w^2 - 16)$$ Factor out the GCF, 2.

$$0 = 2(w + 4)(w - 4)$$ Factor the difference of two squares, $w^2 - 16$.

$$w + 4 = 0 \quad \text{or} \quad w - 4 = 0$$ Since 2 cannot equal 0, discard that possibility. Set each factor that contains a variable equal to 0.

$$w = -4 \qquad\qquad w = 4$$ Solve each equation.

The Language of Algebra

When solving real-world application problems, we **discard** or **reject** any solutions of equations that do not make sense, such as a negative width of a painting.

State   The solutions of the equation are -4 and 4. Since w represents the width of the picture, and the width cannot be negative, we discard -4. Thus, the width of the picture is 4 feet and the length is $2 \cdot 4 = 8$ feet.

Check   A rectangle with dimensions 4 feet by 8 feet has an area of 32 ft^2, and the length is twice the width. The answers check.

> **Self Check 1**   **Geometry.**   A rectangle has an area of 55 square meters. Its length is 1 meter more than twice its width. Find the dimensions of the rectangle.
>
> **Now Try** ▶ Problem 13

EXAMPLE 2

Windmills.   The height of a triangular canvas sail of a windmill is 1 foot less than twice the length of its base. If the sail has an area of 22.5 ft^2, find the length of the base and the height.

Analyze

- The height is 1 ft less than twice the length of the base.
- The area is 22.5 ft^2.
- Find the length of the base and the height.

Assign   Since the height is related to the length of the base, we let $b =$ the length of the base of the sail in feet. Then $2b - 1 =$ the height of the sail in feet.

Form   To form an equation, we use the formula for the area of a triangle: $A = \frac{1}{2}bh$, where $A = 22.5$.

The area of the triangle	equals	one-half	times	the length of the base	times	the height.
22.5	$=$	$\frac{1}{2}$	$\cdot$	b	$\cdot$	$(2b - 1)$

Solve

$$22.5 = \frac{1}{2}b(2b - 1) \qquad \text{This is the equation to solve.}$$

$$\mathbf{2} \cdot 22.5 = \mathbf{2} \cdot \frac{1}{2}b(2b - 1) \qquad \begin{array}{l}\text{To clear the equation of the fraction, multiply} \\ \text{both sides by 2.}\end{array}$$

$$45 = b(2b - 1) \qquad \text{Multiply: } 2 \cdot 22.5 = 45 \text{ and } 2 \cdot \frac{1}{2} = 1.$$

$$45 = 2b^2 - b \qquad \begin{array}{l}\text{Distribute the multiplication by } b. \text{ This is a} \\ \text{quadratic equation but it is not in standard} \\ \text{form.}\end{array}$$

$$0 = 2b^2 - b - 45 \qquad \begin{array}{l}\text{To get 0 on the left side, subtract 45} \\ \text{from both sides.}\end{array}$$

$$0 = (2b + 9)(b - 5) \qquad \text{Factor the trinomial.}$$

$$2b + 9 = 0 \quad \text{or} \quad b - 5 = 0 \qquad \text{Set each factor equal to 0.}$$

$$2b = -9 \qquad\qquad b = 5 \qquad \text{Solve each equation.}$$

$$b \cancel{= -\frac{9}{2}}$$

Caution

A common error is to "distribute" the 2 incorrectly on the right side of the equation.

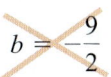

State   The solutions of the equation are $-\frac{9}{2}$ and 5. Since b represents the length of the base of the sail, and it cannot be negative, we discard $-\frac{9}{2}$. The length of the base is then 5 feet, and the height is $2(5) - 1 = 9$ feet.

Check   A triangle with height 9 feet and base 5 feet has area $\frac{1}{2}(9)(5) = 22.5 \text{ ft}^2$, and the height is 1 foot less than twice the base. The answers check.

> **Self Check 2**   **Sailing.**   The height of a triangular sail is 4 yards more than twice the length of the base. If the sail has an area of 15 yd², find the base and the height.
>
> **Now Try** ▶ Problem 19

2 Solve Problems Involving Consecutive Integers.

Consecutive integers are integers that follow one another, such as 15 and 16. When solving consecutive integer problems, if we let $x =$ the first integer, then:

- two consecutive integers are x and $x + 1$
- two consecutive even integers are x and $x + 2$
- two consecutive odd integers are x and $x + 2$

EXAMPLE 3

©Phil Anthony/iShutterstock.com

Women's Tennis.   In the 1998 Australian Open, sisters Venus and Serena Williams played against each other for the first time as professionals. Venus was victorious over her younger sister. At that time, their ages were consecutive integers whose product was 272. How old were Venus and Serena when they met in this match?

Analyze

- Venus is older than Serena.
- Their ages were consecutive integers.
- The product of their ages was 272.
- Find Venus' and Serena's ages when they played this match.

Assign   Let $x =$ Serena's age when she played in the 1998 Australian Open. Since their ages were consecutive integers, and since Venus is older, we let $x + 1 =$ Venus' age.

Form   The word *product* indicates multiplication.

Serena's age	times	Venus' age	was	272.
x	·	$(x + 1)$	=	272

Solve

$$x(x + 1) = 272$$
$$x^2 + x = 272 \qquad \text{Distribute the multiplication by } x. \text{ This is a quadratic equation but it is not in standard form.}$$
$$x^2 + x - 272 = 0 \qquad \text{Subtract 272 from both sides to make the right side 0.}$$
$$(x + 17)(x - 16) = 0 \qquad \text{Factor } x^2 + x - 272. \text{ Two numbers whose product is } -272 \text{ and whose sum is 1 are 17 and } -16.$$
$$x + 17 = 0 \quad \text{or} \quad x - 16 = 0 \qquad \text{Set each factor equal to 0.}$$
$$\cancel{x = -17} \qquad\qquad x = 16 \qquad \text{Solve each equation.}$$

State   The solutions of the equation are -17 and 16. Since x represents Serena's age, and it cannot be negative, we discard -17. Thus, Serena Williams was 16 years old and Venus Williams was $16 + 1 = 17$ years old when they played against each other for the first time as professionals.

Success Tip

The prime factorization of 272 is helpful in determining that $272 = 17 \cdot 16$.

$$272 = 17 \cdot 16$$

```
    2|272
    2|136
16{ 2| 68
    2| 34
       17
```

Check   Since 16 and 17 are consecutive integers, and since $16 \cdot 17 = 272$, the answers check.

Self Check 3   The product of two consecutive positive integers is 552. Find the integers.

Now Try ▶ Problem 23

3 Solve Problems Using the Pythagorean Theorem.

A **right triangle** is a triangle that has a 90° (right) angle. The longest side of a right triangle is the **hypotenuse,** which is the side opposite the right angle. The remaining two sides are the **legs** of the triangle. The **Pythagorean theorem** provides a formula relating the lengths of the three sides of a right triangle.

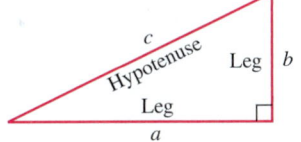

The Pythagorean Theorem   If a and b are the lengths of the legs of a right triangle and c is the length of the hypotenuse, then

$$a^2 + b^2 = c^2$$

In a right triangle, the sum of the squares of the lengths of the two legs is equal to the square of the length of the hypotenuse.

EXAMPLE 4   **Right Triangles.**   The longer leg of a right triangle is 3 units longer than the shorter leg. If the hypotenuse is 6 units longer than the shorter leg, find the lengths of the sides of the triangle.

Analyze   We begin by drawing a right triangle and labeling the legs and the hypotenuse.

Assign   Let a = the length of the shorter leg. Then the length of the hypotenuse is $a + 6$ and the length of the longer leg is $a + 3$.

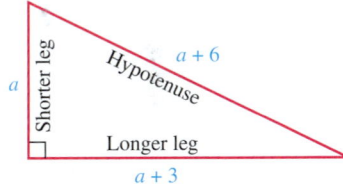

Form   By the Pythagorean theorem, we have

$\left(\begin{array}{c}\text{The length of}\\ \text{the shorter leg}\end{array}\right)^2$	plus	$\left(\begin{array}{c}\text{the length of}\\ \text{the longer leg}\end{array}\right)^2$	equals	$\left(\begin{array}{c}\text{the length of the}\\ \text{hypotenuse}\end{array}\right)^2$
a^2	$+$	$(a + 3)^2$	$=$	$(a + 6)^2$

Solve

$$a^2 + (a + 3)^2 = (a + 6)^2$$

$$a^2 + a^2 + 6a + 9 = a^2 + 12a + 36$$ Find $(a + 3)^2$ and $(a + 6)^2$. Don't forget the middle terms.

$$2a^2 + 6a + 9 = a^2 + 12a + 36$$ On the left side: $a^2 + a^2 = 2a^2$. This is a quadratic equation but it is not in standard form.

$$a^2 - 6a - 27 = 0$$ To get 0 on the right side, subtract a^2, $12a$, and 36 from both sides. This is a quadratic equation.

$$(a - 9)(a + 3) = 0$$ Factor the trinomial.

$$a - 9 = 0 \quad \text{or} \quad a + 3 = 0$$ Set each factor equal to 0.

$$a = 9 \quad | \quad \cancel{a = -3}$$ Solve each equation.

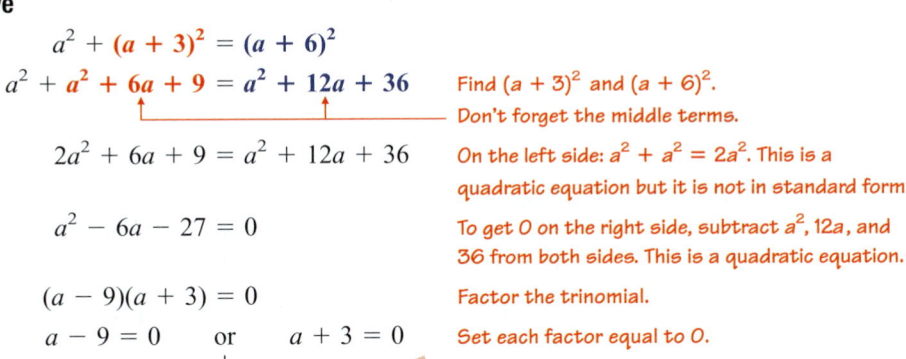

State   Since a side cannot have a negative length, we discard the solution -3. Thus, the shorter leg is 9 units long, the hypotenuse is $9 + 6 = 15$ units long, and the longer leg is $9 + 3 = 12$ units long.

Check   The longer leg, 12, is 3 units longer than the shorter leg, 9. The hypotenuse, 15, is 6 units longer than the shorter leg, 9, and the side lengths satisfy the Pythagorean theorem. So the results check.

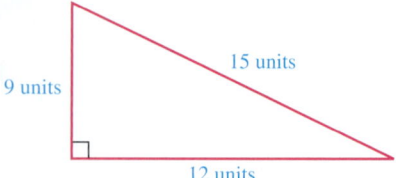

$$9^2 + 12^2 \stackrel{?}{=} 15^2$$
$$81 + 144 \stackrel{?}{=} 225$$
$$225 = 225$$

> **Self Check 4**   **Right Triangles.**   The longer leg of a right triangle is 7 inches longer than the shorter leg. If the hypotenuse is 9 inches longer than the shorter leg, find the lengths of the sides of the triangle.
>
> **Now Try** ▶   Problem 31

4   Solve Problems Given the Quadratic Equation Model.

A quadratic equation can be used to describe the height of an object that is projected upward, such as a ball thrown into the air or an arrow shot into the sky.

EXAMPLE 5   **College Pranks.**   A student uses rubber tubing to launch a water balloon from the roof of his dormitory. The height h (in feet) of the balloon, t seconds after being launched, is approximated by the formula $h = -16t^2 + 48t + 64$. After how many seconds will the balloon hit the ground?

Analyze   When the water balloon hits the ground, its height will be 0 feet.

Assign   To find the time that it takes for the balloon to hit the ground, we set h equal to 0, and solve the quadratic equation for t, the time.

Form $h = -16t^2 + 48t + 64$

$0 = -16t^2 + 48t + 64$ Substitute 0 for the height, h.
This is a quadratic equation.

Solve

$0 = -16t^2 + 48t + 64$	This is the equation to solve.
$0 = -16(t^2 - 3t - 4)$	Factor out the opposite of the GCF, -16.
$0 = -16(t + 1)(t - 4)$	Factor the trinomial.
$t + 1 = 0$ or $t - 4 = 0$	Since -16 cannot equal 0, discard that possibility. Set each factor that contains a variable equal to 0.
$t = -1$ | $t = 4$	Solve each equation.

> **Success Tip**
>
> Note that the common factor, -16, divides -16, 48, and 64 exactly:
>
> $\dfrac{-16}{-16} = 1$ $\dfrac{48}{-16} = -3$
>
> $\dfrac{64}{-16} = -4$

State   The equation has two solutions, -1 and 4. Since t represents time, and, in this case, time cannot be negative, we discard -1. The second solution, 4, indicates that the balloon hits the ground 4 seconds after being launched.

Check   Check this result by substituting 4 for t in $h = -16t^2 + 48t + 64$. You should get $h = 0$.

Self Check 5 **Archery.** An arrow is shot into the air from a balcony. The height h (in feet) of the tip of the arrow, t seconds after being shot, is approximated by $h = -16t^2 + 77t + 15$. After how many seconds will the arrow hit the ground?

Now Try ▶ Problem 37

SECTION 6.8 ▶ STUDY SET

VOCABULARY

Fill in the blanks.

1. Integers that follow one another, such as 6 and 7, are called _____ integers.
2. A _____ triangle is a triangle that has a 90° angle.
3. The longest side of a right triangle is the _____. The remaining two sides are the _____ of the triangle.
4. The _____ theorem is a formula that relates the lengths of the three sides of a right triangle.

CONCEPTS

5. A rectangle has an area of 40 in.². The length is 3 inches longer than the width. Which rectangle below meets these conditions?

   i.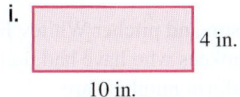
   4 in.
   10 in.

   ii.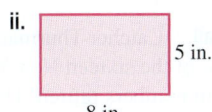
   5 in.
   8 in.

6. A triangle has an area of 15 ft². The height is 7 feet less than twice the length of the base. Which triangle below meets these conditions?

   i.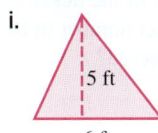
   5 ft
   6 ft

   ii.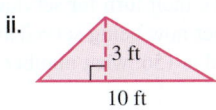
   3 ft
   10 ft

7. Multiply both sides of the equation by 2. ***Do not solve.***

   $10 = \frac{1}{2}b(b + 5)$

8. Fill in the blanks.
   a. If the length of the hypotenuse of a right triangle is c and the lengths of the other two legs are a and b, then _____ $= c^2$.
   b. In a right triangle, the sum of the _____ of the lengths of the two legs is equal to the square of the length of the _____.

9. a. What kind of triangle is shown?
   b. What are the lengths of the legs of the triangle?
   c. How much longer is the hypotenuse than the shorter leg?

   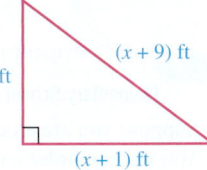
   $(x + 9)$ ft
   x ft
   $(x + 1)$ ft

10. A ball is thrown into the air. Its height h in feet, t seconds after being released, is given by the formula $h = -16t^2 + 24t + 6$. When the ball hits the ground, what is the value of h?

NOTATION

Complete the solution to solve the equation.

11. $0 = -16t^2 + 32t + 48$

    $0 = \boxed{}(t^2 - 2t - 3)$

    $0 = -16(t - 3)(t + \boxed{})$

    $t - 3 = \boxed{}$ or $t + 1 = \boxed{}$

    $t = \boxed{}$ | $t = \boxed{}$

12. Fill in the blanks.
    a. Consecutive integers can be represented by x and _____.
    b. Consecutive odd integers can be represented by x and _____.
    c. Consecutive even integers can be represented by x and _____.

APPLICATIONS

Geometry Problems

13. **Flags.** The length of the flag of Australia is twice as long as the width. Find the dimensions of an Australian flag if its area is 18 ft².

14. **Billiards.** Pool tables are rectangular, and their length is twice the width. Find the dimensions of a pool table if it occupies 50 ft² of floor space.

15. X-Rays. A rectangular-shaped x-ray film has an area of 80 square inches. The length is 2 inches longer than the width. Find its width and length.

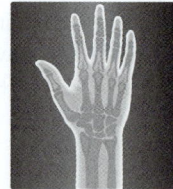

16. Insulation. The area of the rectangular slab of foam insulation in the illustration is 36 square meters. Find the dimensions of the slab.

w m

(2*w* + 1) m

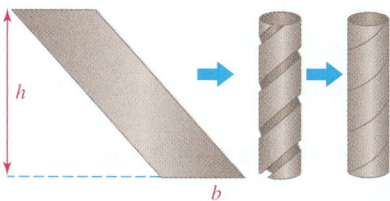

17. from **Campus to Careers**

Elementary School Teacher

Suppose you are an elementary school teacher. You want to order a rectangular bulletin board to mount on a classroom wall that has an area of 90 square feet. Fire code requirements allow for no more than 30% of a classroom wall to be covered by a bulletin board. If the length of the board is to be three times as long as the width, what are the dimensions of the largest bulletin board that meets fire code?

---

18. Tubing. Refer to the diagram below. A piece of cardboard in the shape of a parallelogram is twisted to form the tube. The parallelogram has an area of 60 square inches. If its height *h* is 7 inches more than the length of the base *b*, what is the length of the base? (*Hint:* The formula for the area of a parallelogram is $A = bh$.)

h

b

19. Jeans. The height of the triangular-shaped logo on a pair of jeans is 1 centimeter less than the length of its base. If the area of the logo is 15 square centimeters, find the length of the base and the height.

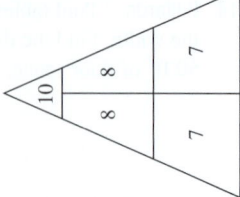

AMERICAN
jeans

20. Shuffleboard. The area of the numbered triangle on a shuffleboard court is 27 ft². Its height is 3 feet more than the length of the base. Find the length of the base and the height.

10   8   8   7   7

21. Sailboats. Refer to the diagram of a sail shown here. The length of the *luff* is 3 times longer than the length of the *foot* of the sail. Find the length of the foot and the length of the luff.

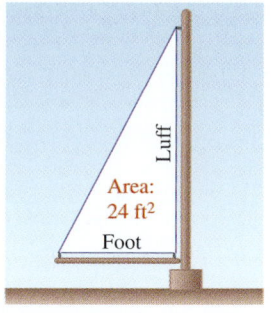

Luff

Area: 24 ft²

Foot

22. Designing Tents. The length of the base of the triangular sheet of canvas above the door of a tent is 2 feet more than twice its height. The area is 30 square feet. Find the height and the length of the base of the triangle.

h

Consecutive Integer Problems

23. NASCAR. The car numbers of drivers Kasey Kahne and Scott Riggs are consecutive positive integers whose product is 90. If Kahne's car number is the smaller, what is the number of each car?

24. Baseball. Catcher Thurman Munson and pitcher Whitey Ford are two of the sixteen New York Yankees who have had their uniform numbers retired. Their uniform numbers are consecutive integers whose product is 240. If Munson's was the smaller number, determine the uniform number of each player.

25. Customer Service. At a pharmacy, customers take a ticket to reserve their turn for service. If the product of the ticket number now being served and the next ticket number to be served is 156, what number is now being served?

NOW SERVING
?

PLEASE take a number

26. History. Delaware was the first state to enter the Union and Hawaii was the 50th. If we order the positions of entry for the rest of the states, we find that Kentucky entered the Union right after Vermont, and the product of their order-of-entry numbers is 210. Use the given information to complete these statements:

Kentucky was the ___ th state to enter the Union.

Vermont was the ___ th state to enter the Union.

27. Plotting Points. The x-coordinate and y-coordinate of a point in quadrant I are consecutive odd integers whose product is 143. The x-coordinate is less than the y-coordinate. Find the coordinates of the point.

28. Presidents. George Washington was born on 2-22-1732 (February 22, 1732). He died in 1799 at the age of 67. The month in which he died and the day of the month on which he died are consecutive even integers whose product is 168. When did Washington die?

Pythagorean Theorem Problems

29. High-Ropes Adventures Courses. A builder of a high-ropes adventure course wants to secure a pole by attaching a support cable from the anchor stake 8 yards from its base to a point 6 yards up the pole. How long should the cable be?

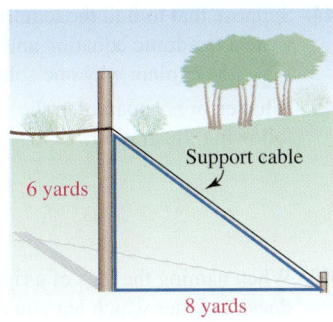

6 yards · Support cable · 8 yards

30. Wind Damage. A tree was blown over in a wind storm. Find x. Then find the height of the tree when it was standing upright.

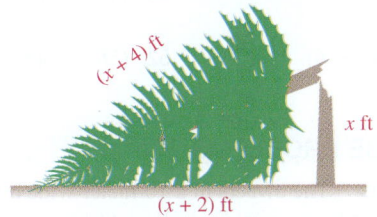

$(x + 4)$ ft · x ft · $(x + 2)$ ft

31. Moto X. Find x, the height of the landing ramp.

Take-off ramp · x · Landing ramp · 17 ft · 15 ft

32. Gardening Tools. The dimensions (in millimeters) of the teeth of a pruning saw blade are given in the illustration. Find each length.

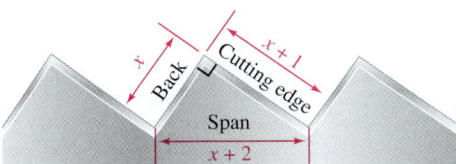

x · Back · $x + 1$ · Cutting edge · Span · $x + 2$

33. Boating. The inclined ramp of the boat launch shown in the next column is 8 meters longer than the rise of the ramp. The run is 7 meters longer than the rise. How long are the three sides of the ramp?

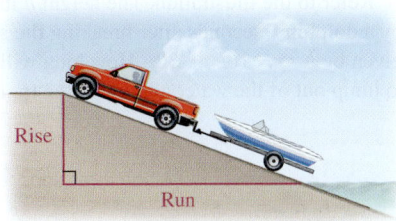

Rise · Run

34. Car Repairs. To create some space to work under the front end of a car, a mechanic drives it up steel ramps. A ramp is 1 foot longer than the back, and the base is 2 feet longer than the back of the ramp. Find the length of each side of the ramp.

Back · 90° · Base

Quadratic Equation Model Problems

35. Thrill Rides. At the peak of a roller coaster ride, a rider's sunglasses fly off his head. The height h (in feet) of the glasses, t seconds after he loses them, is given by $h = -16t^2 + 64t + 80$. After how many seconds will the glasses hit the ground? (*Hint:* Factor out -16.)

36. Parades. A celebrity on the top of a parade float is tossing pieces of candy to the people on the street below. The height h (in feet) of a piece of candy, t seconds after being thrown, is given by $h = -16t^2 + 16t + 32$. After how many seconds will the candy hit the ground? (*Hint:* Factor out -16.)

37. Softball. A pitcher can throw a fastball underhand at 63 feet per second (about 45 mph). If she throws a ball into the air with that velocity, its height h in feet, t seconds after being released, is given by $h = -16t^2 + 63t + 4$. After the ball is thrown, in how many seconds will it hit the ground? (*Hint:* Factor out -1.)

38. Officiating. Before a football game, a coin toss is used to determine which team will kick off. The height h (in feet) of a coin above the ground t seconds after being flipped up into the air is given by $h = -16t^2 + 22t + 3$. How long does a team captain have to call heads or tails if it must be done while the coin is in the air? (*Hint:* Factor out -1.)

39. Dolphins.   Refer to the illustration. The height h in feet reached by a dolphin t seconds after breaking the surface of the water is given by $h = -16t^2 + 32t$. How long will it take the dolphin to jump out of the water and touch the trainer's hand?

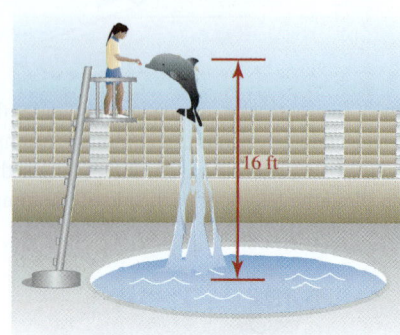

40. Exhibition Diving.   In Acapulco, Mexico, men diving from a cliff to the water 64 feet below are quite a tourist attraction. A diver's height h above the water (in feet), t seconds after diving, is given by $h = -16t^2 + 64$. How long does a dive last?

41. Choreography.   For the finale of a musical, 36 dancers are to assemble in a triangular-shaped series of rows, where each row has one more dancer than the previous row. The illustration shows the beginning of such a formation. The relationship between the number of rows r and the number of dancers d is given by $d = \frac{1}{2}r(r + 1)$. Determine the number of rows in the formation.

42. Crafts.   The illustration shows how a wall hanging can be created by stretching yarn from peg to peg across a wooden ring. The relationship between the number of pegs p placed evenly around the ring and the number of yarn segments s that criss-cross the ring is given by the formula $s = \frac{p(p-3)}{2}$. How many pegs are needed if the designer wants 27 segments to criss-cross the ring? (*Hint:* Multiply both sides of the equation by 2.)

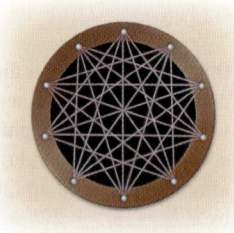

WRITING

43. A student was asked to solve the following problem: *The length of a rectangular room is 2 feet more than twice the width. If the area of the room is 60 square feet, find its dimensions.* Here is the student's solution:

Since $10 \cdot 6 = 60$, the length of the room is 10 feet and the width is 6 feet.

Explain why his solution is incorrect.

44. Suppose that to find the length of the base of a triangle, you write a quadratic equation and solve it to find $b = 6$ or $b = -8$. Explain why one solution should be discarded.

45. What error is apparent in the following illustration?

46. When naming the legs of a right triangle, explain why it doesn't matter which leg you label a and which leg you label b.

REVIEW

Find each special product.

47. $(5b - 2)^2$

48. $(2a + 3)^2$

49. $(s^2 + 4)^2$

50. $(m^2 - 1)^2$

51. $(9x + 6)(9x - 6)$

52. $(5b + 2)(5b - 2)$

CHALLENGE PROBLEMS

53. Pool Borders.   The owners of a 10-meter-wide by 25-meter-long rectangular swimming pool want to surround the pool with a crushed-stone border of uniform width. They have enough stone to cover 74 square meters. How wide should they make the border?

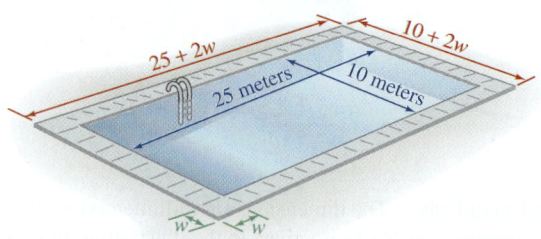

54. Find h.

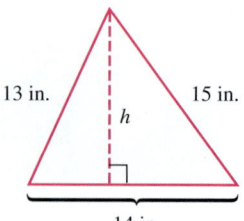

6 Summary & Review

SECTION 6.1 ▶ The Greatest Common Factor; Factoring by Grouping

DEFINITIONS AND CONCEPTS	EXAMPLES
Factoring is multiplication reversed. To **factor a polynomial** means to express it as a product of two (or more) polynomials.	Multiplication: Given the factors, we find a polynomial. ⟶ $$2x(5x + 3) = 10x^2 + 6x$$ ⟵ Factoring: Given a polynomial, we find the factors.
A natural number is in **prime-factored form** when it is written as the product of prime numbers.	The prime-factored form of 28 is $2 \cdot 2 \cdot 7 = 2^2 \cdot 7$.
To find the **greatest common factor, GCF,** of a list of terms 1. Write each coefficient as a product of prime factors. 2. Identify the numerical and variable factors common to each term. 3. Multiply the common numerical and variable factors identified in step 2 to obtain the GCF. If there are no common factors, the GCF is 1.	Find the GCF of $35x^4$, $63x^3$, and $42x^2$. $$35x^4 = 5 \cdot 7 \cdot x \cdot x \cdot x \cdot x$$ $$63x^3 = 3 \cdot 3 \cdot 7 \cdot x \cdot x \cdot x$$ $$42x^2 = 2 \cdot 3 \cdot 7 \cdot x \cdot x$$ $\text{GCF} = 7 \cdot x \cdot x = 7x^2$
The first step of factoring a polynomial is to see whether the terms of the polynomial have a common factor. If they do, **factor out the GCF.**	Factor: $35x^4 + 63x^3 - 42x^2$ $$= 7x^2(5x^2 + 9x - 6) \quad \text{Factor out the GCF, } 7x^2.$$ Use multiplication to check the factorization: $7x^2(5x^2 + 9x - 6) = 35x^4 + 63x^3 - 42x^2$ This is the original polynomial.
If a polynomial has four terms, try **factoring by grouping.** 1. Group the terms of the polynomial so that the first two terms have a common factor and the last two terms have a common factor. 2. Factor out the common factor from each group. 3. Factor out the resulting common binomial factor. If there is no common binomial factor, regroup the terms of the polynomial and repeat steps 2 and 3.	Factor: $ax - bx + ay - by$ $$= x(a - b) + y(a - b) \quad \text{Factor out } x \text{ from } ax - bx \text{ and } y \text{ from } ay - by.$$ $$= (a - b)(x + y) \quad \text{Factor out the common binomial factor, } (a - b).$$

REVIEW EXERCISES

Find the prime-factorization of each number.

1. 35
2. 96

Find the GCF of each list.

3. 28 and 35
4. $36a^4$, $54a^3$, and $126a^6$

Factor.

5. $3x + 9y$
6. $5ax^2 + 15a$
7. $7s^5 + 14s^3$
8. $\pi ab - \pi ac$
9. $24x^3 + 60x^2 - 48x$
10. $x^5y^3z^2 + xy^5z^3 - xy^3z^2$

11. $-5ab^2 + 10a^2b - 15ab$
12. $4(x - 2) - x(x - 2)$

Factor out -1.

13. $-a - 7$
14. $-4t^2 + 3t - 1$

Factor.

15. $2c + 2d + ac + ad$
16. $3xy + 18x - 5y - 30$
17. $2a^3 + 2a^2 - a - 1$
18. $4m^2n + 12m^2 - 8mn - 24m$

SECTION 6.2 ▶ Factoring Trinomials of the Form $x^2 + bx + c$

DEFINITIONS AND CONCEPTS	EXAMPLES
Many trinomials factor as the product of two binomials. To **factor a trinomial** of the form $x^2 + bx + c$, whose **leading coefficient is 1,** find two integers whose product is c and whose sum is b. $x^2 + bx + c = \left(x \,\boxed{}\right)\left(x \,\boxed{}\right)$ **The product of these numbers must be c and their sum must be b.** Use the FOIL method to check the factorization.	Factor: $p^2 + 7p + 12$ $= \left(p \,\boxed{}\right)\left(p \,\boxed{}\right)$ $= (p + 3)(p + 4)$ | **Positive factors of 12**	**Sum of positive factors of 12**
$1 \cdot 12 = 12$	$1 + 12 = 13$
$2 \cdot 6 = 12$	$2 + 6 = 8$
$3 \cdot 4 = 12$	$3 + 4 = 7$ | ***Check:*** $(p + 3)(p + 4) = p^2 + 4p + 3p + 12$ $= p^2 + 7p + 12$
Before factoring a trinomial, write it in **descending powers** of one variable. Also, factor out -1 if that is necessary to make the **leading coefficient positive.**	Factor: $7q - q^2 - 6$ $= -q^2 + 7q - 6$ Write the terms in descending powers of q. $= -(q^2 - 7q + 6)$ Factor out -1. $= -(q - 1)(q - 6)$ Factor the trinomial.
If a trinomial cannot be factored using only integers, it is called a **prime trinomial.**	$t^2 + 2t - 5$ is a prime trinomial because there are no two integers whose product is -5 and whose sum is 2.
The GCF should always be factored out first. A trinomial is **factored completely** when no factor can be factored further. Use multiplication to check the factorization.	Factor: $3m^3 - 6m^2 - 24m$ $= 3m(m^2 - 2m - 8)$ Factor out the GCF, 3m, first. $= 3m(m - 4)(m + 2)$ Factor the trinomial.
To factor a trinomial of the form $ax^2 + bx + c$ (where $a = 1$) by **grouping,** write it as a equivalent four-term polynomial: $x^2 + \boxed{}\,x + \boxed{}\,x + c$ **The product of these numbers must be ac, and their sum must be b.** Then factor the four-term polynomial by grouping. Use the FOIL method to check the factorization.	Factor by grouping: $p^2 + 7p + 12$ We must find two numbers whose product is $ac = 1(12) = 12$ and whose sum is $b = 7$. Two such numbers are 4 and 3. They serve as the coefficients of $4p$ and $3p$, the two terms that we use to represent the middle term, $7p$, of the trinomial. $p^2 + 7p + 12 = p^2 + 4p + 3p + 12$ Express 7p as 4p + 3p. $= p(p + 4) + 3(p + 4)$ Factor p out of $p^2 + 4p$ and 3 out of $3p + 12$. $= (p + 4)(p + 3)$ Factor out $(p + 4)$.

REVIEW EXERCISES

19. What is the leading coefficient of $x^2 + 8x - 9$?

20. Complete the table.

Factors of 6	Sum of the factors of 6
1(6)	
2(3)	
$-1(-6)$	
$-2(-3)$	

Factor.

21. $x^2 + 2x - 24$

22. $x^2 - 18x - 40$

23. $x^2 - 14x + 45$

24. $t^2 + 10t + 15$

25. $-y^2 + 15y - 56$

26. $10y + 9 + y^2$

27. $c^2 + 3cd - 10d^2$

28. $-3mn + m^2 + 2n^2$

29. Explain how we can check to determine whether $(x - 4)(x + 5)$ is the factorization of $x^2 + x - 20$.

30. Explain why $x^2 + 7x + 11$ is prime.

Factor.

31. $5a^5 + 45a^4 - 50a^3$

32. $-4x^2y - 4x^3 + 24xy^2$

SECTION 6.3 ▶ Factoring Trinomials of the Form $ax^2 + bx + c$

DEFINITIONS AND CONCEPTS	EXAMPLES
We can use the **trial-and-check method** to factor trinomials with **leading coefficients other than 1.** Write the trinomial as the product of two binomials and determine four integers. The product of these numbers must be a. $ax^2 + bx + c = (\boxed{\ }x\ \boxed{\ })(\boxed{\ }x\ \boxed{\ })$ The product of these numbers must be c. Use the FOIL method to check the factorization to obtain the correct middle term.	Factor: $2x^2 - 5x - 12$ Since the first term is $2x^2$, the first terms of the binomial factors must be $2x$ and x. $(2x\ \boxed{\ })(x\ \boxed{\ })$ Because 2x · x will give $2x^2$ The second terms of the binomials must be two integers whose product is -12. There are six such pairs: $12(-1)$, $6(-2)$, $4(-3)$, **3(−4)**, $2(-6)$, and $1(-12)$ The pair in blue gives the correct middle term when we use the FOIL method to check: Outer: −8x $(2x + 3)(x - 4)$ Combine like terms: −8x + 3x = −5x. Inner: 3x Correct middle term. Thus, $2x^2 - 5x - 12 = (2x + 3)(x - 4)$.
To factor $ax^2 + bx + c$ by **grouping,** write it as an equivalent four-term polynomial: $ax^2 + \boxed{\ }x + \boxed{\ }x + c$ The product of these numbers must be ac, and their sum must be b. Then factor the four-term polynomial by grouping. Use the FOIL method to check your work.	Factor by grouping: $2x^2 - 5x - 12$ We must find two numbers whose product is $ac = 2(-12) = -24$ and whose sum is $b = -5$. Two such numbers are -8 and 3. They serve as the coefficients of $-8x$ and $3x$, the two terms that we use to represent the middle term, $-5x$, of the trinomial. $2x^2 - 5x - 12 = 2x^2 - 8x + 3x - 12$ Express −5x as −8x + 3x. $= 2x(x - 4) + 3(x - 4)$ Factor out 2x and 3. $= (x - 4)(2x + 3)$ Factor out (x − 4).

REVIEW EXERCISES

Factor.

33. $2x^2 - 5x - 3$

34. $35y^2 + 11y - 10$

35. $-3x^2 + 13x + 30$

36. $-33p^2 - 6p + 18p^3$

37. $4b^2 - 17bc + 4c^2$

38. $7y^2 + 7y - 18$

39. Entertaining.   The rectangular area occupied by a table setting is $(12x^2 - x - 1)$ square inches. Factor the polynomial to find the binomials that represent the length and width of the table setting.

40. In the following work, a student began to factor $5x^2 - 8x + 3$. Explain his mistake.

$(5x -\ \ )(x +\ \ )$

SECTION 6.4 ▶ Factoring Perfect-Square Trinomials and the Difference of Two Squares

DEFINITIONS AND CONCEPTS	EXAMPLES
Trinomials that are squares of a binomial are called **perfect-square trinomials.** We can factor perfect-square trinomials by applying the special-product rules in reverse. $A^2 + 2AB + B^2 = (A + B)^2$ $A^2 - 2AB + B^2 = (A - B)^2$	Factor: $g^2 + 8g + 16$ and $m^2 - 18mn + 81n^2$ We match each trinomial to a special-product form shown in the left column. $g^2 + 8g + 16 = g^2 + 2 \cdot 4 \cdot g + 4^2 = (g + 4)^2$ $m^2 - 18mn + 81n^2 = m^2 - 2 \cdot m \cdot 9n + (9n)^2 = (m - 9n)^2$
To factor the **difference of two squares,** use the rule $F^2 - L^2 = (F + L)(F - L)$ It will be helpful to review the table of **squares of integers** shown on page 466.	Factor: $25b^2 - 36$ $= (5b)^2 - 6^2$ This is a difference of two squares. $= (5b + 6)(5b - 6)$
In general, the **sum of two squares** (with no common factor other than 1) cannot be factored using real numbers.	$x^2 + 100$ and $36y^2 + 49$ are prime polynomials.

REVIEW EXERCISES

Factor.

41. $x^2 + 10x + 25$

42. $9y^2 + 16 - 24y$

43. $-z^2 + 2z - 1$

44. $25a^2 + 20ab + 4b^2$

45. $x^2 - 9$

46. $49t^2 - 121y^2$

47. $x^2y^2 - 400$

48. $8at^2 - 32a$

49. $c^4 - 256$

50. $h^2 + 36$

SECTION 6.5 ▶ Factoring the Sum and Difference of Two Cubes

DEFINITIONS AND CONCEPTS	EXAMPLES
To factor the **sum** and **difference of two cubes,** use the following rules. $F^3 + L^3 = (F + L)(F^2 - FL + L^2)$ $F^3 - L^3 = (F - L)(F^2 + FL + L^2)$ It will be helpful to review the table of **cubes of integers** shown on page 473.	Factor: $p^3 + 64$ and $125a^3 - 27b^3$ We match each binomial to a factoring rule shown in the left column. $p^3 + 64 = p^3 + 4^3$ This is a sum of two cubes. $= (p + 4)(p^2 - p \cdot 4 + 4^2)$ $= (p + 4)(p^2 - 4p + 16)$ $125a^3 - 27b^3 = (5a)^3 - (3b)^3$ This is a difference of two cubes. $= (5a - 3b)[(5a)^2 + 5a \cdot 3b + (3b)^2]$ $= (5a - 3b)(25a^2 + 15ab + 9b^2)$

REVIEW EXERCISES

Factor.

51. $b^3 + 1$

52. $x^3 - 216$

53. $p^3 + 125q^3$

54. $16x^5 - 54x^2y^3$

SECTION 6.6 ▶ A Factoring Strategy

DEFINITIONS AND CONCEPTS	EXAMPLES
To factor a random polynomial, use the **factoring strategy** discussed in Section 6.6 on page 477. Remember that the instruction to factor means to **factor completely.** A polynomial is factored completely when no factor can be factored further.	Factor: $a^5 + 8a^2 + 4a^3 + 32$ ***Is there a common factor?*** No. There is no common factor (other than 1). ***How many terms does it have?*** Since the polynomial has four terms, try factoring by grouping. $a^5 + 8a^2 + 4a^3 + 32 = a^2(a^3 + 8) + 4(a^3 + 8)$ Factor a^2 from $a^5 + 8a^2$ and 4 from $4a^3 + 32$. $= (a^3 + 8)(a^2 + 4)$ Factor out $a^3 + 8$. ***Is it factored completely?*** No. We can factor $a^3 + 8$ as a sum of two cubes. $a^5 + 8a^2 + 4a^3 + 32$ $= (a^3 + 8)(a^2 + 4)$ $a^2 + 4$ is prime. $= (a + 2)(a^2 - 2a + 4)(a^2 + 4)$ $a^2 - 2a + 4$ is prime. ***Does it check?*** Use multiplication to check.

REVIEW EXERCISES

Factor.

55. $14y^3 + 6y^4 - 40y^2$ **56.** $5s^2t + 5s^2u^2 + 5tv + 5u^2v$ **61.** $2t^3 + 10$ **62.** $121p^2 + 36q^2$

57. $j^4 - 16$ **58.** $-3j^3 - 24$ **63.** $x^2z + 64y^2z + 16xyz$ **64.** $18c^3d^2 - 12c^3d - 24c^2d$

59. $400x + 400 - m^2x - m^2$ **60.** $12w^4 - 36w^3 + 27w^2$

SECTION 6.7 ▶ Solving Quadratic Equations by Factoring

DEFINITIONS AND CONCEPTS	EXAMPLES
A **quadratic equation** is an equation that can be written in the **standard form** $ax^2 + bx + c = 0$, where a, b, and c are real numbers and $a \neq 0$.	Examples of quadratic equations are: $5x^2 + 25x = 0$, $4a^2 - 9 = 0$, and $y^2 - 13y = 6$
The Zero-Factor Property If the product of two (or more) numbers is 0, then at least one of the numbers is 0.	If $(x + 2)(x - 3) = 0$ then, $x + 2 = 0$ or $x - 3 = 0$.
To use the **factoring method to solve a quadratic equation:** 1. Write the equation in standard form: $ax^2 + bx + c = 0$ or $0 = ax^2 + bx + c$ 2. Factor completely. 3. Use the *zero-factor property* to set each factor equal to 0. 4. Solve each resulting equation. 5. Check each result in the original equation.	Solve: $5x^2 + 25x = 0$ Solve: $4a^2 - 9 = 0$ $5x(x + 5) = 0$ $(2a + 3)(2a - 3) = 0$ $5x = 0$ or $x + 5 = 0$ $2a + 3 = 0$ or $2a - 3 = 0$ $x = 0$ | $x = -5$ $2a = -3$ | $2a = 3$ The solutions are 0 and -5. $a = -\frac{3}{2}$ | $a = \frac{3}{2}$ The solution set is $\{0, -5\}$. Check each result in the original equation. The solutions are $-\frac{3}{2}$ and $\frac{3}{2}$. The solution set is $\left\{-\frac{3}{2}, \frac{3}{2}\right\}$.

To use the zero-factor property to solve a quadratic equation, we need one side of the equation to be factored completely and the other side to be 0.

Solve: $5y^2 - 13y = 6$ This equation is not in standard form.

$5y^2 - 13y - 6 = 6 - 6$ To get 0 on the right side, subtract 6 from both sides.

$5y^2 - 13y - 6 = 0$ Do the subtraction.

$(5y + 2)(y - 3) = 0$ Factor the trinomial.

$5y + 2 = 0$ or $y - 3 = 0$ Set each factor equal to 0.

$5y = -2$ $y = 3$ Solve each equation.

$y = -\dfrac{2}{5}$

The solutions are $-\dfrac{2}{5}$ and 3. Check each result in the original equation.

REVIEW EXERCISES

Solve each equation by factoring.

65. $8x(x - 6) = 0$

66. $(4x - 7)(x + 1) = 0$

67. $x^2 + 2x = 0$

68. $x^2 - 9 = 0$

69. $144x^2 - 25 = 0$

70. $a^2 - 7a + 12 = 0$

71. $2t^2 + 28t + 98 = 0$

72. $2x - x^2 = -24$

73. $5a^2 - 6a + 1 = 0$

74. $2p^3 = 2p(p + 2)$

SECTION 6.8 ▶ **Applications of Quadratic Equations**

DEFINITIONS AND CONCEPTS	EXAMPLES

To solve application problems, use the **six-step problem-solving strategy:**

1. Analyze the problem

2. Assign a variable

3. Form an equation

4. Solve the equation

5. State the conclusion

6. Check the result

Find two consecutive positive integers whose product is 72.

Analyze   *Consecutive integers* are integers that follow each other. The word *product* indicates multiplication.

Assign   Let x = the smaller positive integer. Then $x + 1$ = the larger integer.

Form

The smaller integer	times	the larger integer	equals	72.
x	$\cdot$	$(x + 1)$	$=$	72

Solve

$x(x + 1) = 72$ This is the equation to solve.

$x^2 + x = 72$ Distribute the multiplication by x. This is a quadratic equation but it is not in standard form.

$x^2 + x - 72 = 0$ To get 0 on the right side, subtract 72 from both sides.

$(x + 9)(x - 8) = 0$ Factor the trinomial.

$x + 9 = 0$ or $x - 8 = 0$ Set each factor equal to 0.

$x = -9$ $x = 8$ Solve each linear equation.

State   Since we are looking for positive integers, the solution -9 must be discarded. Thus, the smaller integer is 8 and the larger integer is $x + 1 = 9$.

Check   The integers 8 and 9 are consecutive positive integers and their product is 72.

The Pythagorean Theorem:

If a and b are the lengths of the legs of a right triangle and c is the length of the hypotenuse, then

$$a^2 + b^2 = c^2$$

To show that a triangle with sides of 5, 12, and 13 units is a right triangle, we verify that $5^2 + 12^2 = 13^2$.

$$5^2 + 12^2 \overset{?}{=} 13^2$$
$$25 + 144 \overset{?}{=} 169$$
$$169 = 169 \quad \text{True}$$

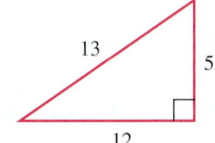

REVIEW EXERCISES

75. Sandpaper. A standard piece of sandpaper is 2 inches longer than it is wide. Find the dimensions of a piece of sandpaper if it has an area of 99 in².

76. Construction. The face of the triangular concrete panel has an area of 45 square meters, and its base is 3 meters longer than twice its height. Find the length of its base.

77. Fill in the blanks. If we let x = the first integer, then:
- two consecutive integers are x and _____
- two consecutive even integers are x and _____
- two consecutive odd integers are x and _____

78. Music Awards. The record for the most Grammy nominations in one year is held by Michael Jackson. Kanye West is currently in second place. The number of times Jackson and West were nominated are consecutive even integers whose product is 120. How many times was each artist nominated? (Source: *Wikipedia*)

79. Tightrope Walkers. A circus performer intends to walk up a taut cable shown in the illustration to a platform at the top of a pole. How high above the ground is the platform?

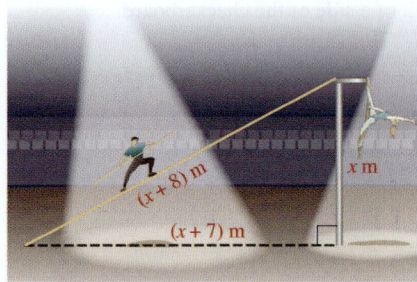

80. Ballooning. A hot-air balloonist accidentally dropped his camera overboard while traveling at a height of 1,600 ft. The height h in feet of the camera t seconds after being dropped is given by $h = -16t^2 + 1,600$. In how many seconds will the camera hit the ground?

---

6 ▶ CHAPTER TEST

1. Fill in the blanks.
- **a.** The letters GCF stand for _____ _____ _____.
- **b.** To factor a polynomial means to express it as a _____ of two (or more) polynomials.
- **c.** The _____ theorem provides a formula relating the lengths of the three sides of a right triangle.
- **d.** $y^2 - 25$ is a _____ of two squares.
- **e.** The trinomial $x^2 + x - 6$ factors as the product of two _____: $(x + 3)(x - 2)$.

2. a. Find the prime factorizations of 45 and 30.

   b. Find the greatest common factor of $45x^4$ and $30x^3$.

Factor. If an expression cannot be factored, write "prime."

3. $4x + 16$

4. $q^2 - 81$

5. $30a^2b^3 - 20a^3b^2 + 5ab$

6. $x^2 + 9$

7. $2x(x + 1) + 3(x + 1)$

8. $x^2 + 4x + 3$

9. $-x^2 + 9x + 22$

10. $60x^2 - 32x^3 + x^4$

11. $9a - 9b + ax - bx$

12. $2a^2 + 5a - 12$

13. $18x^2 + 60xy + 50y^2$

14. $x^3 + 8$

15. $60m^8 - 45m^6$

16. $3a^3 - 81$

17. $16x^4 - 81$

18. $a^3 + 5a^2 + a + 5$

19. $a^4 - 24 - 4a + 6a^3$

20. $3d - 4 + 10d^2$

21. $8m^2 - 800$

22. $36n^2 - 84n + 49$

23. $8r^2 - 14r + 3$

24. $t^2 - 6t + 10$

25. **Checkers.**   The area of a square checkerboard is represented by $(25x^2 - 40x + 16)$ in.^{2}. Find the polynomial that represents the length of a side of the checkerboard.

26. Factor $x^2 - 3x - 54$. Show a check of your answer.

Solve each equation.

27. $(x + 3)(x - 2) = 0$

28. $x^2 - 25 = 0$

29. $36x^2 = 6x$

30. $x^2 + 6x = -9$

31. $6x^2 + x - 1 = 0$

32. $a(a - 7) = 18$

33. $x^3 + 7x^2 = -6x$

34. **Driving Safety.**   All cars have a blind spot where it is difficult for the driver to see a car behind and to the right. The area of the rectangular blind spot shown is 54 ft^2. Its length is 3 feet longer than its width. Find its dimensions.

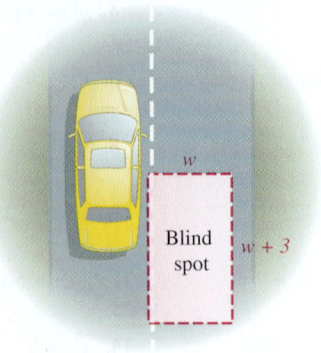

35. **Rocketry.**   The height h, in feet, of a toy rocket t seconds after being launched is given by $h = -16t^2 + 80t$. After how many seconds will the rocket hit the ground?

36. **ATV's.**   The area of a triangular-shaped safety flag on an all-terrain vehicle is 33 in.^{2}. Its height is 1 inch less than twice the length of the base. Find the length of the base and the height of the flag.

37. Find two consecutive positive integers whose product is 156.

38. Find the length of the hypotenuse of the right triangle shown.

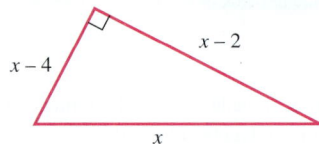

39. What is a quadratic equation? Give an example.

40. If the product of two numbers is 0, what conclusion can be drawn about the numbers?

Group Project

FACTORING MODELS

Overview In this activity, you will construct geometric models to find factorizations of several trinomials.

Instructions Form groups of 2 or 3 students.

1. Copy and cut out each of the following figures. On each figure, write its area.

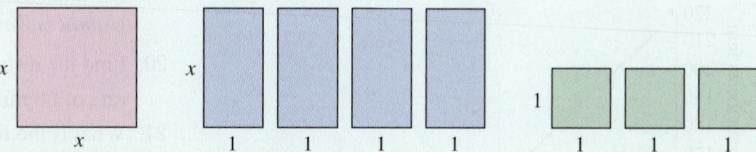

   Write a trinomial that represents the *sum* of the areas of the eight figures by combining any like terms: _____ + _____ + _____

2. Now assemble the eight figures to form the large rectangle shown below.

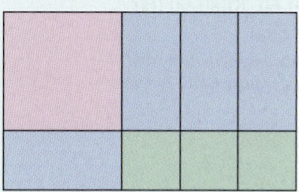

   Write an expression that represents the *length* of the rectangle: _____ + _____

   Write an expression that represents the *width* of the rectangle: _____ + _____

   Express the area of the rectangle as the product of its length and width:
   (_____)(_____)

3. The set of figures used in step 1 and the set of figures used in step 2 are the same. Therefore, the expressions for the areas must be equal. Set your answers from steps 1 and 2 equal to find the factorization of the trinomial $x^2 + 4x + 3$.

   $$\underset{\text{Answer from step 1}}{\underline{\hspace{3cm}}} = \underset{\text{Answer from step 2}}{\underline{\hspace{3cm}}}$$

4. Make a new model to find the factorization of $x^2 + 5x + 4$. (*Hint:* You will need to make one more 1-by-x figure and one more 1-by-1 figure.)

   $$\underline{\hspace{3cm}} = \underline{\hspace{3cm}}$$

5. Make a new model to find the factorization of $2x^2 + 5x + 2$. (*Hint:* You will need to make one more x-by-x figure.)

   $$\underline{\hspace{3cm}} = \underline{\hspace{3cm}}$$

CUMULATIVE REVIEW ▶▶ Chapters 1–6

1. **Heart Rates.** Refer to the graph. Determine the difference in the maximum heart beat rate for a 70-year-old as compared to someone half that age. [Section 1.1]

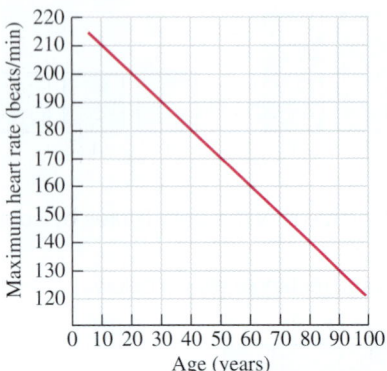

Based on data from *Cardiopulmonary Anatomy and Physiology: Essentials for Respiratory Care*, 2nd ed.

2. Find the prime factorization of 250. [Section 1.2]

3. Find the quotient: $\frac{16}{5} \div \frac{10}{3}$ [Section 1.2]

4. Write $\frac{124}{125}$ as a decimal. [Section 1.3]

5. Determine whether each statement is true or false. [Section 1.2]
   a. Every integer is a whole number.
   b. Every integer is a rational number.
   c. π is a real number.

6. Which division is undefined, $\frac{0}{5}$ or $\frac{5}{0}$? [Section 1.6]

Evaluate each expression.

7. $3 + 2[-1 - 4(5)]$ [Section 1.7]

8. $\frac{|-25| - 2(-5)}{9 - 2^4}$ [Section 1.7]

9. What is -3 cubed? [Section 1.7]

10. What is the value of x twenty-dollar bills? [Section 1.8]

11. Evaluate $\frac{-x - a}{y - b}$ for $x = -2$, $y = 1$, $a = 5$, and $b = 2$. [Section 1.8]

12. Identify the coefficient of each term in the expression $8x^2 - x + 9$. [Section 1.8]

Simplify each expression.

13. $-8y^2 - 5y^2 + 6$ [Section 1.9]

14. $3z + 2(y - z) + y$ [Section 1.9]

Solve each equation.

15. $-(3a + 1) + a = 2$ [Section 2.2]

16. $2 - (4x + 7) = 3 + 2(x + 2)$ [Section 2.2]

17. $\frac{3t - 21}{2} = t - 6$ [Section 2.2]

18. $-\frac{1}{3} - \frac{x}{5} = \frac{3}{2}$ [Section 2.2]

19. **Watermelons.** The heaviest watermelon on record weighed 270 pounds. If watermelon is 92% water by weight, what was its water weight? Round to the nearest pound. (Source: *Guinness World Records*) [Section 2.3]

20. Find the distance traveled by a truck traveling for $5\frac{1}{2}$ hours at a rate of 60 miles per hour. [Section 2.4]

21. What is the formula for simple interest? [Section 2.4]

22. **Geometry Tools.** A compass is adjusted so that the distance between the pointed ends is 2 inches. Then a circle is drawn. What will the area of the circle be? Round to the nearest tenth of a square inch. [Section 2.4]

23. Solve $A = P + Prt$ for t. [Section 2.4]

24. **History.** George Washington was the first president of the United States. John Adams was the second, and Thomas Jefferson was the third, and so on. Grover Cleveland was president two *different* times, as shown in the illustration. The sum of the numbers of Cleveland's presidencies is 46. Find these two numbers. [Section 2.5]

Grover Cleveland   Benjamin Harrison   Grover Cleveland

25. **Antique Shows.** A traveling antique show will be on the road for 17 weeks, visiting three cities. They will be in Los Angeles for 2 weeks longer than they will be in Las Vegas. Their stay in Dallas will be 1 week less than twice that in Las Vegas. How many weeks will they be in each city? [Section 2.5]

26. **Photographic Chemicals.** A photographer wishes to mix 6 liters of a 5% acetic acid solution with a 10% solution to get a 7% solution. How many liters of 10% solution must be added? [Section 2.6]

27. **Dried Fruits.** Dried apple slices cost $4.60 per pound, and dried banana chips sell for $3.40 per pound. How many pounds of each should be used to create a 10-pound mixture that sells for $4 per pound? [Section 2.6]

28. Solve: $-\frac{x}{2} + 4 > 5$. Write the solution set in interval notation and graph it. [Section 2.8]

29. Is $(-2, 5)$ a solution of $3x + 2y = 4$? [Section 3.2]

30. Graph: $y = 2x - 3$ [Section 3.2]

31. Is the graph of $x = 3$ a vertical or horizontal line? [Section 3.3]

32. If two lines are parallel, what can be said about their slopes? [Section 3.4]

33. **Encyclopedias.** The graph below approximates the total number of articles on the English-language edition of *Wikipedia* for the years 2005 through 2010. Find the rate of change in the number of articles over that time span. [Section 3.4] (Source: Wikipedia)

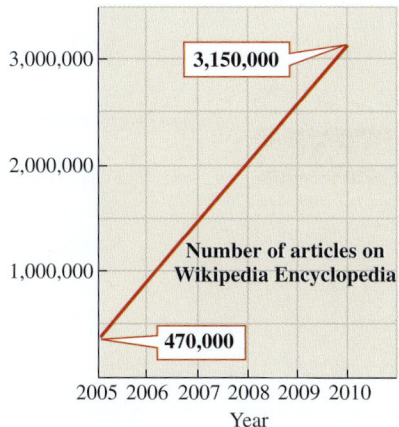

34. Find the slope and the y-intercept of the graph of $3x - 3y = 6$. [Section 3.5]

35. Find an equation of the line passing through $(-2, 5)$ and $(-3, -2)$. Write the equation in slope–intercept form. [Section 3.5]

36. Graph the line passing through $(-4, 1)$ that has slope -3. [Section 3.6]

37. Graph: $8x + 4y \geq -24$ [Section 3.7]

38. If $f(x) = 3x^2 - 2x + 1$, find $f(-2)$. [Section 3.8]

39. Is $\left(\frac{1}{2}, 1\right)$ a solution of the system $\begin{cases} 4x - y = 1 \\ 2x + y = 2 \end{cases}$? [Section 4.1]

40. Solve the system $\begin{cases} 3x - 2y = 6 \\ x - y = 1 \end{cases}$ by graphing. [Section 4.1]

41. Solve the system $\begin{cases} y = -4x + 1 \\ 4x - y = 5 \end{cases}$ by substitution. [Section 4.2]

42. Solve the system $\begin{cases} 5a + 3b = -8 \\ 2a + 9b = 2 \end{cases}$ by elimination (addition).
   [Section 4.3]

43. **Fundraising.** A Rotary Club held a city-wide recycling drive. They collected a total of 14 tons of newspaper and cardboard that earned them $356. They were paid $31 per ton for the newspaper and $18 per ton for the cardboard. How many tons of each did they collect? [Section 4.4]

44. Graph: $\begin{cases} 4x + 3y \geq 12 \\ y < 4 \end{cases}$ [Section 4.5]

Simplify each expression. Write each answer without negative exponents.

45. $-y^2(4y^3)$ [Section 5.1]

46. $\dfrac{(x^2y^5)^5}{(x^3y)^2}$ [Section 5.1]

47. $\dfrac{b^5}{b^{-2}}$ [Section 5.2]

48. $2x^0$ [Section 5.1]

49. Write 0.00009011 in scientific notation. [Section 5.3]

50. Write 1,700,000 in scientific notation. [Section 5.3]

51. Find the degree of $7y^3 + 4y^2 + y + 3$. [Section 5.4]

52. Graph: $y = x^3 + 2$ [Section 5.4]

Perform the operations.

53. $(x^2 - 3x + 8) - (3x^2 + x + 3)$ [Section 5.5]

54. $4b^3(2b^2 - 2b)$ [Section 5.6]

55. $(3x - 2)(x + 4)$ [Section 5.6]

56. $(y - 6)^2$ [Section 5.7]

57. $\dfrac{12a^2b^2 - 8a^2b - 4ab}{4ab}$ [Section 5.8]

58. $x - 3 \overline{)2x^2 - 5x - 3}$ [Section 5.8]

59. **Playpens.** Find an expression that represents the
   a. perimeter of the playpen. [Section 5.5]
   b. area of the floor of the playpen. [Section 5.6]
   c. volume of the playpen. [Section 5.6]

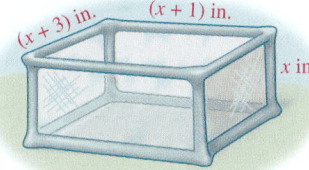

60. Find the GCF of $24x^5y^8$ and $54x^6y$. [Section 6.1]

Factor.

61. $9b^3 - 27b^2$ [Section 6.1]

62. $ax + bx + ay + by$ [Section 6.1]

63. $u^2 - 3 + 2u$ [Section 6.2]

64. $10x^2 + x - 2$ [Section 6.3]

65. $4a^2 - 12a + 9$ [Section 6.4]

66. $9z^2 - 1$ [Section 6.4]

67. $t^3 - 8$ [Section 6.5]

68. $3a^2b^2 - 6a^2 - 3b^2 + 6$ [Section 6.6]

Solve each equation.

69. $15s^2 - 20s = 0$ [Section 6.7]

70. $2x^2 - 5x = -2$ [Section 6.7]

71. $x^3 + 3x^2 + 2x = 0$ [Section 6.7]

72. **Camping.** The rectangular-shaped cooking surface of a small camping stove is 108 in.². If its length is 3 inches longer than its width, what are its dimensions? [Section 6.8]

Rational Expressions and Equations

7

©BananaStock/SuperStock

from Campus to Careers

Recreation Director

People of all ages enjoy participating in activities, such as arts and crafts, camping, sports, and the performing arts. Recreation directors plan, organize, and oversee these activities in local playgrounds, camps, community centers, religious organizations, theme parks, and tourist attractions. The job of recreation director requires mathematical skills such as budgeting, scheduling, and forecasting trends.

Problem 33 in **Study Set 7.7**, **problem 77** in **Study Set 7.8**, and **problem 57** in **Study Set 7.9** involve situations that a recreation director might encounter on the job. The mathematical concepts discussed in this chapter can be used to solve those problems.

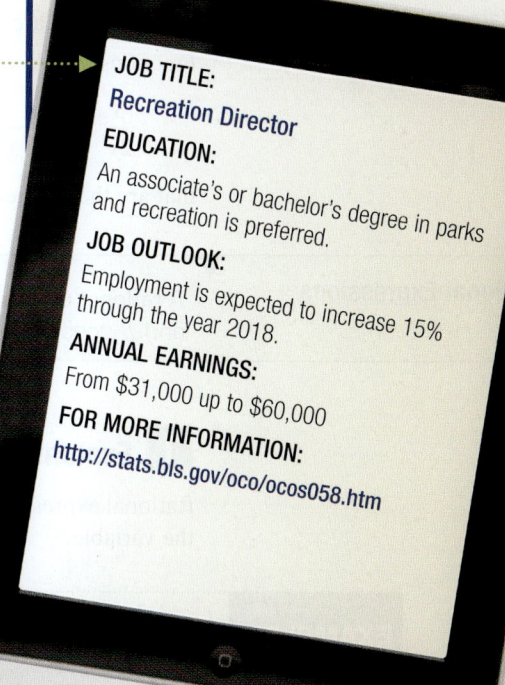

JOB TITLE:
Recreation Director

EDUCATION:
An associate's or bachelor's degree in parks and recreation is preferred.

JOB OUTLOOK:
Employment is expected to increase 15% through the year 2018.

ANNUAL EARNINGS:
From $31,000 up to $60,000

FOR MORE INFORMATION:
http://stats.bls.gov/oco/ocos058.htm

Study groups give students an opportunity to ask their classmates questions, share ideas, compare lecture notes, and review for tests. If something like this interests you, here are some suggestions.

GROUP SIZE: A study group should be small—from 3 to 6 people is best.

TIME AND PLACE: You should meet regularly in a place where you can spread out and talk without disturbing others.

GROUND RULES: The study group will be more effective if, early on, you agree on some rules to follow.

Now Try This ▶

Would you like to begin a study group? If so, you need to answer the following questions. Who will be in your group? Where will your group meet? How often will it meet? For how long will each session last? Will you have a group leader? What will be the leader's responsibilities? What will you try to accomplish each session? How will the members prepare for each meeting? Will you follow a set agenda each session? How will the members share contact information? When will you discuss ways to improve the study sessions?

SECTION 7.1

OBJECTIVES

1. Evaluate rational expressions.
2. Find numbers that cause a rational expression to be undefined.
3. Simplify rational expressions.
4. Simplify rational expressions that have factors that are opposites.

Simplifying Rational Expressions

ARE YOU READY?

The following problems review some basic skills that are needed when simplifying rational expressions.

1. Evaluate: **a.** $\dfrac{0}{7}$ **b.** $\dfrac{7}{0}$
2. Simplify: **a.** $\dfrac{18}{24}$ **b.** $\dfrac{5}{45}$
3. Factor: $12x - 8$
4. Factor: $a^2 - 16$
5. Factor: $5x^2 - 23x - 10$
6. Evaluate: $\dfrac{6}{-6}$

Fractions that are the quotient of two integers are *rational numbers*. Examples are $\frac{1}{2}$ and $\frac{9}{5}$. Fractions such as

$$\frac{3}{2y}, \qquad \frac{x}{x+2}, \qquad \text{and} \qquad \frac{2a^2 - 8a}{a^2 - 6a + 8}$$

that are the quotient of two polynomials are called **rational expressions.**

Rational Expressions	A **rational expression** is an expression of the form $\frac{A}{B}$, where A and B are polynomials and B does not equal 0.

1 Evaluate Rational Expressions.

Rational expressions can have different values depending on the number that is substituted for the variable.

EXAMPLE 1 Evaluate $\dfrac{2x - 1}{x^2 + 1}$ for $x = -3$ and for $x = 0$.

Strategy We will replace each x in the rational expression with the given value of the variable. Then we will evaluate the numerator and denominator separately, and simplify, if possible.

Why Recall from Chapter 1 that to *evaluate an expression* means to find its numerical value, once we know the value of its variable.

Solution

For x = −3:

$$\frac{2x - 1}{x^2 + 1} = \frac{2(-3) - 1}{(-3)^2 + 1}$$

$$= \frac{-6 - 1}{9 + 1}$$

$$= -\frac{7}{10}$$

For x = 0:

$$\frac{2x - 1}{x^2 + 1} = \frac{2(0) - 1}{(0)^2 + 1}$$

$$= \frac{0 - 1}{0 + 1}$$

$$= -1$$

Self Check 1 Evaluate $\frac{2x - 1}{x^2 + 1}$ for $x = 7$ and for $x = -2$.

Now Try ▶ Problems 13 and 21

2 Find Numbers That Cause a Rational Expression to Be Undefined.

The fraction bar in a rational expression indicates division. Since division by 0 is undefined, we must make sure that the denominator of a rational expression is not equal to 0.

EXAMPLE 2   Find all real numbers for which the expression is undefined: **a.** $\dfrac{7x}{x - 5}$

b. $\dfrac{3x - 2}{x^2 - x - 6}$ **c.** $\dfrac{8}{x^2 + 1}$ **d.** $\dfrac{x + 9.4}{12}$

Strategy   To find the real numbers for which each expression is undefined, we will find the values of the variable that make the *denominator* 0.

Why   We don't need to examine the numerator of the rational expression; it can be any value, including 0. It's a denominator of 0 that makes a rational expression undefined, because a denominator of 0 indicates division by 0.

Solution   **a.** The denominator of $\dfrac{7x}{x - 5}$ will be 0 if we replace x with 5.

$$\frac{7x}{x - 5} = \frac{7(5)}{5 - 5} = \frac{35}{0}$$

Since $\frac{35}{0}$ is undefined, the expression $\frac{7x}{x - 5}$ is undefined for $x = 5$.

b. $\dfrac{3x - 2}{x^2 - x - 6}$ will be undefined for values of x that make the denominator 0. To find these values, we set $x^2 - x - 6$ equal to 0, and solve for x.

$x^2 - x - 6 = 0$	Set the denominator of $\frac{3x - 2}{x^2 - x - 6}$ equal to 0.
$(x - 3)(x + 2) = 0$	To solve the quadratic equation, factor the trinomial.
$x - 3 = 0 \quad$ or $\quad x + 2 = 0$	Set each factor equal to 0.
$x = 3 \qquad\qquad\quad x = -2$	Solve each equation.

The Language of Algebra

Another way that Example 2 could be phrased is: State the **restrictions** on the variable. For $\frac{3x - 2}{x^2 - x - 6}$, we can state the restrictions by writing $x \neq 3$ and $x \neq -2$.

Since 3 and −2 make the denominator 0, the expression $\dfrac{3x - 2}{x^2 - x - 6}$ is undefined for $x = 3$ and $x = -2$. To check the results, we proceed as follows.

For x = 3:

$$\frac{3x - 2}{x^2 - x - 6} = \frac{3(3) - 2}{3^2 - 3 - 6}$$

$$= \frac{9 - 2}{9 - 3 - 6}$$

$$= \frac{7}{0} \quad \text{This expression is undefined.}$$

For x = −2:

$$\frac{3x - 2}{x^2 - x - 6} = \frac{3(-2) - 2}{(-2)^2 - (-2) - 6}$$

$$= \frac{-6 - 2}{4 + 2 - 6}$$

$$= \frac{-8}{0} \quad \text{This expression is undefined.}$$

c. No matter what real number is substituted for x, the denominator, $x^2 + 1$, will not be equal to 0. (A number squared plus 1 cannot equal 0.) Thus, no real numbers make $\frac{8}{x^2 + 1}$ undefined.

d. Since the denominator of $\frac{x + 9.4}{12}$ does not contain a variable, the denominator can never be equal to 0. Thus, no real numbers make the expression undefined.

Self Check 2 Find all real numbers for which the expression is undefined:

a. $\frac{x}{x + 9}$ **b.** $\frac{9x + 7}{x^2 - 25}$ **c.** $\frac{4 - x}{x^2 + 64}$ **d.** $\frac{x + 3.8}{100}$

Now Try ▶ Problems 23, 27, 31, and 33

---

3 Simplify Rational Expressions.

In Section 1.2, we simplified fractions by removing factors equal to 1. For example, to simplify $\frac{6}{15}$, we factor 6 and 15, and then remove the factor $\frac{3}{3}$.

$$\frac{6}{15} = \frac{2 \cdot 3}{5 \cdot 3} = \frac{2}{5} \cdot \frac{3}{3} = \frac{2}{5} \cdot 1 = \frac{2}{5}$$

To streamline this process, we can replace $\frac{3}{3}$ in $\frac{2 \cdot 3}{5 \cdot 3}$ with the equivalent fraction $\frac{1}{1}$.

$$\frac{6}{15} = \frac{2 \cdot 3}{5 \cdot 3} = \frac{2 \cdot \overset{1}{\cancel{3}}}{5 \cdot \underset{1}{\cancel{3}}} = \frac{2}{5} \qquad \text{We are removing } \frac{3}{3} = 1.$$

We can simplify rational expressions in a similar manner using a procedure that is based on the following property.

The Fundamental Property of Rational Expressions	If A, B, and C are polynomials, and B and C are not 0, $$\frac{AC}{BC} = \frac{A}{B}$$

A rational expression is **simplified** if its numerator and denominator have no common factors other than 1. To simplify a rational expression, follow these steps.

Simplifying Rational Expressions	1. Factor the numerator and denominator completely to determine their common factors. 2. Remove factors equal to 1 by replacing each pair of factors common to the numerator and denominator with the equivalent fraction $\frac{1}{1}$. 3. Multiply the remaining factors in the numerator and in the denominator.

EXAMPLE 3   Simplify: $\dfrac{21x^3}{14x^2}$

Strategy   We will write the numerator and denominator in factored form and then remove pairs of factors that are equal to 1.

Why   The rational expression is simplified when the numerator and denominator have no common factor other than 1.

Solution $\dfrac{21x^3}{14x^2} = \dfrac{3 \cdot 7 \cdot x \cdot x \cdot x}{2 \cdot 7 \cdot x \cdot x}$ To prepare to simplify the rational expression, factor the numerator and the denominator.

$$= \dfrac{3 \cdot \overset{1}{7} \cdot \overset{1}{\cancel{x}} \cdot \overset{1}{\cancel{x}} \cdot x}{2 \cdot \underset{1}{7} \cdot \underset{1}{\cancel{x}} \cdot \underset{1}{\cancel{x}}}$$

Simplify by replacing $\frac{7}{7}$ and $\frac{x}{x}$ with the equivalent fraction $\frac{1}{1}$. This removes the factor $\frac{7 \cdot x \cdot x}{7 \cdot x \cdot x}$, which is equal to 1.

$$= \dfrac{3x}{2}$$

Multiply the remaining factors in the numerator: $3 \cdot 1 \cdot 1 \cdot 1 \cdot x = 3x.$
Multiply the remaining factors in the denominator: $2 \cdot 1 \cdot 1 \cdot 1 = 2.$

We say that $\dfrac{21x^3}{14x^2}$ simplifies to $\dfrac{3x}{2}$. Since $\dfrac{21x^3}{14x^2}$ is undefined for $x = 0$, the expressions $\dfrac{21x^3}{14x^2}$ and $\dfrac{3x}{2}$ are equal only if $x \neq 0$. That is,

$$\dfrac{21x^3}{14x^2} = \dfrac{3x}{2} \quad \text{provided } x \neq 0$$

An alternate approach for Example 3 is to use rules for exponents to simplify rational expressions that are the quotient of two monomials.

$$\dfrac{21x^3}{14x^2} = \dfrac{3 \cdot \overset{1}{7} \cdot x^{3-2}}{2 \cdot \underset{1}{7}} = \dfrac{3x^1}{2} = \dfrac{3x}{2}$$

To divide exponential expressions with the same base, keep the base and subtract the exponents.

Self Check 3   Simplify: $\dfrac{32a^3}{24a}$

Now Try ▶ Problem 41

To simplify rational expressions, we often make use of the factoring methods discussed in Chapter 6.

EXAMPLE 4   Simplify: **a.** $\dfrac{30t - 6}{36}$ **b.** $\dfrac{x^2 + 13x + 12}{x^2 + 12x}$ **c.** $\dfrac{x^3 + x^2}{1 + x}$

Strategy   We will begin by factoring the numerator and denominator. Then we will remove any factors common to the numerator and denominator.

Why   We need to make sure that the numerator and denominator have no common factor other than 1. When this is the case, the rational expression is simplified.

Solution   **a.** $\dfrac{30t - 6}{36} = \dfrac{6(5t - 1)}{6 \cdot 6}$ To prepare to simplify the rational expression, factor the numerator: The GCF is 6. Factor the denominator.

$$= \dfrac{\overset{1}{\cancel{6}}(5t - 1)}{\underset{1}{\cancel{6}} \cdot 6}$$

Simplify by removing a factor equal to 1. Replace $\frac{6}{6}$ with $\frac{1}{1}$.

$$= \dfrac{5t - 1}{6}$$

Multiply the remaining factors in the numerator: $1 \cdot (5t - 1) = 5t - 1.$
Multiply the remaining factors in the denominator: $1 \cdot 6 = 6.$

b. $\dfrac{x^2 + 13x + 12}{x^2 + 12x} = \dfrac{(x + 1)(x + 12)}{x(x + 12)}$ To prepare to simplify the rational expression, factor the numerator. Factor the denominator: The GCF is x.

$$= \dfrac{(x + 1)(x \overset{1}{\cancel{+ 12}})}{x(x \underset{1}{\cancel{+ 12}})}$$

Simplify by replacing $\frac{x + 12}{x + 12}$ with the equivalent fraction $\frac{1}{1}$. This removes the factor $\frac{x + 12}{x + 12} = 1$.

$$= \dfrac{x + 1}{x}$$

This rational expression cannot be simplified further.

c. $\dfrac{x^3 + x^2}{1 + x} = \dfrac{x^2(x + 1)}{1 + x}$ To prepare to simplify the rational expression, factor the numerator. The GCF is x^2.

$= \dfrac{x^2\cancel{(x+1)}^{1}}{\cancel{1+x}_{1}}$ By the commutative property of addition, $x + 1 = 1 + x$. Simplify by removing a factor equal to 1. Replace $\frac{x+1}{1+x}$ with $\frac{1}{1}$.

$= \dfrac{x^2}{1}$ Multiply in the the numerator: $x^2 \cdot 1 = x^2$.

$= x^2$ Any number divided by 1 is itself.

> **Self Check 4** Simplify: **a.** $\dfrac{4t - 20}{12}$ **b.** $\dfrac{x^2 - x - 6}{x^2 - 3x}$ **c.** $\dfrac{2x^4 + 4x^3}{2 + x}$
>
> **Now Try** ▶ Problems 43 and 47

When a rational expression is simplified, the result is an **equivalent expression.** In Example 4b, for instance, this means that $\dfrac{x^2 + 13x + 12}{x^2 + 12x}$ and $\dfrac{x + 1}{x}$ have the same value for **all** values of x, except those that make either denominator 0. We can use that fact to perform an informal check of our work. If we let $x = 1$, for example, we see that the original rational expression and the simplified expression have the same value, 2.

The original expression	*The simplified expression*
$\dfrac{x^2 + 13x + 12}{x^2 + 12x} = \dfrac{(\mathbf{1})^2 + 13(\mathbf{1}) + 12}{(\mathbf{1})^2 + 12(\mathbf{1})}$	$\dfrac{x + 1}{x} = \dfrac{\mathbf{1} + 1}{\mathbf{1}}$
$= \dfrac{1 + 13 + 12}{1 + 12}$	$= \dfrac{2}{1}$
$= \dfrac{26}{13}$	$= 2$ Same result
$= 2$	

If the results are different, an error has been made, and the problem should be reworked.

EXAMPLE 5 Simplify: **a.** $\dfrac{3x^2 - 8x - 3}{2x^5 - 18x^3}$ **b.** $\dfrac{(x - y)^4}{x^2 - 2xy + y^2}$

Strategy We will begin by factoring the numerator and denominator using the methods discussed in Chapter 6. Then we will remove any factors common to the numerator and denominator.

Why We need to make sure that the numerator and denominator have no common factor other than 1. When this is the case, then the rational expression is simplified.

Solution **a.** $\dfrac{3x^2 - 8x - 3}{2x^5 - 18x^3} = \dfrac{(3x + 1)(x - 3)}{2x^3(x^2 - 9)}$ To prepare to simplify, factor the trinomial in the numerator. Factor the denominator: The GCF is $2x^3$.

$= \dfrac{(3x + 1)(x - 3)}{2x^3(x + 3)(x - 3)}$ In the denominator, factor the difference of two squares, $x^2 - 9$.

$= \dfrac{(3x + 1)\cancel{(x - 3)}^{1}}{2x^3(x + 3)\cancel{(x - 3)}_{1}}$ Simplify by replacing $\frac{x-3}{x-3}$ with the equivalent fraction $\frac{1}{1}$. This removes the factor $\frac{x-3}{x-3} = 1$.

$= \dfrac{3x + 1}{2x^3(x + 3)}$ It is not necessary to perform the multiplication $2x^3(x + 3)$ in the result. It is usually more convenient to leave the denominator in factored form.

b. $\dfrac{(x-y)^4}{x^2-2xy+y^2} = \dfrac{(x-y)^4}{(x-y)^2}$ To prepare to simplify, factor the perfect-square trinomial $x^2 - 2xy + y^2$ in the denominator.

$= \dfrac{(x-y)(x-y)(x-y)(x-y)}{(x-y)(x-y)}$ Write the repeated multiplication indicated by each exponent.

$= \dfrac{\overset{1}{(\cancel{x-y})}\overset{1}{(\cancel{x-y})}(x-y)(x-y)}{\underset{1}{(\cancel{x-y})}\underset{1}{(\cancel{x-y})}}$ Simplify by replacing each $\frac{x-y}{x-y}$ with $\frac{1}{1}$.

$= (x-y)^2$ Use an exponent to write the repeated multiplication in the numerator. It is not necessary to find $(x-y)^2$. The result can be presented in factored form.

Self Check 5 Simplify: **a.** $\dfrac{4x^2-4x-15}{8x^3-50x}$ **b.** $\dfrac{(a+3b)^5}{a^2+6ab+9b^2}$

Now Try ▶ Problems 51 and 53

CAUTION When simplifying rational expressions, we can remove only factors common to the entire numerator and denominator. *It is incorrect to remove any terms common to the numerator and denominator.*

$$\dfrac{\overset{1}{\cancel{x}}+1}{\underset{1}{\cancel{x}}} \qquad \dfrac{a^2-3a+\overset{1}{\cancel{2}}}{a+\underset{1}{\cancel{2}}} \qquad \dfrac{\overset{1}{\cancel{y^2}}-36}{\underset{1}{\cancel{y^2}}-y-7}$$

x is a term of $x+1$.   2 is a term of $a^2 - 3a + 2$ and a term of $a + 2$. y^2 is a term of $y^2 - 36$ and a term of $y^2 - y - 7$.

EXAMPLE 6 Simplify: $\dfrac{5(x+3)-5}{7(x+3)-7}$

Strategy We will begin by simplifying the numerator, $5(x+3) - 5$, and the denominator, $7(x+3) - 7$, separately. Then we will factor each result and remove any common factors.

Why We cannot remove $x + 3$ immediately because it is not a factor of the *entire* numerator and the *entire* denominator.

Solution

$\dfrac{5(x+3)-5}{7(x+3)-7} = \dfrac{5x+15-5}{7x+21-7}$ Use the distributive property in the numerator and in the denominator.

$= \dfrac{5x+10}{7x+14}$ Combine like terms: $15 - 5 = 10$ and $21 - 7 = 14$.

$= \dfrac{5(x+2)}{7(x+2)}$ To prepare to simplify, factor the numerator: The GCF is 5. Factor the denominator: The GCF is 7.

$= \dfrac{5\overset{1}{(\cancel{x+2})}}{7\underset{1}{(\cancel{x+2})}}$ Simplify by replacing $\frac{x+2}{x+2}$ with the equivalent fraction $\frac{1}{1}$. This removes the factor $\frac{x+2}{x+2} = 1$.

$= \dfrac{5}{7}$

The Language of Algebra

Some rational expressions cannot be simplified. For example, to attempt to simplify the following rational expression, we factor its numerator and denominator. Since there are no common factors, we say it **does not simplify** or that it is in **simplest form**.

$$\dfrac{x^2+x-2}{x^2+x} = \dfrac{(x+2)(x-1)}{x(x+1)}$$

Self Check 6 Simplify: $\dfrac{4(x-2)+4}{3(x-2)+3}$

Now Try ▶ Problem 55

4   Simplify Rational Expressions That Have Factors That Are Opposites.

If the terms of two polynomials are the same, except that they are opposite in sign, the polynomials are **opposites.** For example, the following pairs of polynomials are opposites.

$$2a - 1 \quad \text{and} \quad 1 - 2a \qquad\qquad -3x^2 - x + 5 \quad \text{and} \quad 3x^2 + x - 5$$

Compare terms: $2a$ and $-2a$; -1 and 1.   Compare terms: $-3x^2$ and $3x^2$; $-x$ and x; 5 and -5.

Notice that the sum of a polynomial and its opposite is 0.

$$(2a - 1) + (1 - 2a) = \mathbf{0} \qquad\qquad (-3x^2 - x + 5) + (3x^2 + x - 5) = \mathbf{0}$$

We have seen that the quotient of two real numbers that are opposites is always -1:

$$\frac{2}{-2} = -1 \qquad\qquad \frac{-78}{78} = -1 \qquad\qquad \frac{3.5}{-3.5} = -1$$

Likewise, the quotient of two polynomials that are opposites is always -1.

EXAMPLE 7   Simplify: $\dfrac{2a - 1}{1 - 2a}$

Strategy   We will rearrange the terms of the numerator, $2a - 1$, and factor out -1.

Why   This step is useful when the numerator and denominator contain factors that are opposites, such as $2a - 1$ and $1 - 2a$. It produces a common factor that can be removed.

Solution

$$\frac{2a - 1}{1 - 2a} = \frac{-1 + 2a}{1 - 2a} \qquad \text{Think of the numerator, } 2a - 1, \text{ as } 2a + (-1). \text{ Then change the order of the terms: } 2a + (-1) = -1 + 2a.$$

$$= \frac{-1(1 - 2a)}{1 - 2a} \qquad \text{To prepare to simplify: factor out } -1 \text{ from the two terms of the numerator.}$$

$$= \frac{-1(\overset{1}{\cancel{1 - 2a}})}{\underset{1}{\cancel{1 - 2a}}} \qquad \text{Simplify by replacing } \frac{1 - 2a}{1 - 2a} \text{ with the equivalent fraction } \frac{1}{1}. \text{ This removes the factor } \frac{1 - 2a}{1 - 2a} = 1.$$

$$= \frac{-1}{1} \qquad \text{Multiply the remaining factors in the numerator.}$$

$$= -1 \qquad \text{Any number divided by 1 is itself.}$$

Self Check 7   Simplify: $\dfrac{3p - 2}{2 - 3p}$

Now Try ▶   Problem 59

In general, we have this fact.

The Quotient of Opposites   The quotient of any nonzero polynomial and its opposite is -1.

For each of the following rational expressions, the numerator and denominator are opposites. Thus, each expression is equal to -1.

$$\frac{x - 6}{6 - x} = -1 \qquad\qquad \frac{2a - 9b}{9b - 2a} = -1 \qquad\qquad \frac{-3x^2 - x + 5}{3x^2 + x - 5} = -1$$

This fact can be used to simplify certain rational expressions by removing a factor equal to -1. If a factor of the numerator is the opposite of a factor of the denominator, we can replace them with the equivalent fraction $\frac{-1}{1}$, as shown in the following example.

EXAMPLE 8 Simplify: **a.** $\dfrac{y^2 - 1}{3 - 3y}$ **b.** $\dfrac{t + 8}{t - 8}$

Strategy We will begin by factoring the numerator and denominator. Then we look for common factors, or factors that are opposites, and remove them.

Why We need to make sure that the numerator and denominator have no common factor (or opposite factors) other than 1. When this is the case, then the rational expression is simplified.

Solution **a.** $\dfrac{y^2 - 1}{3 - 3y} = \dfrac{(y + 1)(y - 1)}{3(1 - y)}$ To prepare to simplify, factor the numerator, and factor the denominator.

$$= \dfrac{(y + 1)\overset{-1}{\cancel{(y - 1)}}}{3\underset{1}{\cancel{(1 - y)}}}$$ Since $y - 1$ and $1 - y$ are opposites, simplify by replacing $\frac{y-1}{1-y}$ with the equivalent fraction $\frac{-1}{1}$. This removes the factor $\frac{y-1}{1-y} = -1$.

$$= \dfrac{-(y + 1)}{3}$$ In the numerator, $-1 \cdot (y + 1)$ can be written as $-(y + 1)$. In the denominator, $3 \cdot 1 = 3$.

This result may be written in several other equivalent forms.

$$\dfrac{-(y + 1)}{3} = -\dfrac{y + 1}{3}$$ The $-$ symbol in $-(y + 1)$ can be written in the front of the fraction, and the parentheses can be dropped.

$$\dfrac{-(y + 1)}{3} = \dfrac{-y - 1}{3}$$ The $-$ symbol in $-(y + 1)$ represents a factor of -1. Distribute the multiplication by -1 in the numerator.

$$\dfrac{-(y + 1)}{3} = \dfrac{y + 1}{-3}$$ The $-$ symbol in $-(y + 1)$ can be applied to the denominator. However, we don't usually use this form.

b. The binomials $t + 8$ and $t - 8$ are not opposites because their first terms do not have opposite signs. Thus, $\dfrac{t + 8}{t - 8}$ does not simplify.

Caution

A $-$ symbol in front of a fraction may be applied to the numerator or to the denominator, but not to both:

$$-\dfrac{y + 1}{3} \neq \dfrac{-(y + 1)}{-3}$$

Self Check 8 Simplify: **a.** $\dfrac{m^2 - 100}{10m - m^2}$ **b.** $\dfrac{2x - 3}{2x + 3}$

Now Try ▶ Problem 63

SECTION 7.1 ▶ STUDY SET

VOCABULARY

Fill in the blanks.

1. A quotient of two polynomials, such as $\frac{x^2 + x}{x^2 - 3x}$, is called a _____ expression.

2. To simplify a rational expression, we remove common _____ of the numerator and denominator.

3. Because of the division by 0, the expression $\frac{8}{0}$ is _____.

4. The binomials $x - 15$ and $15 - x$ are called _____, because their terms are the same, except that they are opposite in sign.

CONCEPTS

5. When we simplify $\frac{x^2 + 5x}{4x + 20}$, the result is $\frac{x}{4}$. These equivalent expressions have the same value for all real numbers, except $x = -5$. Show that they have the same value for $x = 1$.

6. Determine whether each pair of polynomials are opposites. Write *yes* or *no*.
   a. $y + 7$ and $y - 7$
   b. $b - 20$ and $20 - b$
   c. $x^2 + 2x - 1$ and $-x^2 - 2x - 1$

7. Simplify each expression, if possible.

   a. $\dfrac{x - 8}{x - 8}$ b. $\dfrac{x - 8}{8 - x}$

   c. $\dfrac{x + 8}{8 + x}$ d. $\dfrac{x + 8}{x}$

8. Simplify each expression.

   a. $\dfrac{(x + 2)(x - 2)}{(x + 1)(x + 2)}$ b. $\dfrac{y(y - 2)}{9(2 - y)}$

   c. $\dfrac{(2m + 7)(m - 5)}{(2m + 7)}$ d. $\dfrac{x \cdot x}{x \cdot x(x - 30)}$

NOTATION

Complete the solution to simplify the rational expression.

9. $\dfrac{x^2 + 2x + 1}{x^2 + 4x + 3} = \dfrac{(x + 1)(\boxed{} + 1)}{(x + 3)(x + \boxed{})}$

$$= \dfrac{(x + 1)(\overset{1}{\cancel{x + 1}})}{(x + 3)\,\boxed{}}$$

$$= \dfrac{x + 1}{\boxed{}}$$

10. In the following table, a student's answers to four homework problems are compared with the answers in the back of the book. Are the answers equivalent?

Student's Answer	Book's answer	Equivalent?
$\dfrac{-3}{x + 3}$	$-\dfrac{3}{x + 3}$	
$\dfrac{-(x - 4)}{6x + 1}$	$\dfrac{-x + 4}{6x + 1}$	
$\dfrac{x + 7}{(x - 4)(x + 2)}$	$\dfrac{x + 7}{(x + 2)(x - 4)}$	
$\dfrac{x^2 + 6}{-x}$	$-\dfrac{x^2 + 6}{x}$	

GUIDED PRACTICE

Evaluate each expression for $x = 6$. See Example 1.

11. $\dfrac{x - 2}{x - 5}$ 12. $\dfrac{3x - 2}{x - 2}$

13. $\dfrac{x^2 - 4x - 12}{x^2 + x - 2}$ 14. $\dfrac{x^2 - 36}{x^3 - 1}$

15. $\dfrac{-x + 1}{x^2 - 5x - 6}$ 16. $\dfrac{-2x^2 - 3}{x - 6}$

Evaluate each expression for $y = -3$. See Example 1.

17. $\dfrac{y + 5}{3y - 2}$ 18. $\dfrac{2y + 9}{y^2 + 25}$

19. $-\dfrac{y}{y^2 - y + 6}$ 20. $-\dfrac{y^3}{3y^2 + 1}$

21. $\dfrac{y^2 + 9}{9 - y^2}$ 22. $\dfrac{-y - 11}{y^2 + 2y - 3}$

Find all real numbers for which the rational expression is undefined. See Example 2.

23. $\dfrac{15}{x - 2}$ 24. $\dfrac{5x}{x + 5}$

25. $\dfrac{x + 5}{8x}$ 26. $\dfrac{4x - 1}{6x}$

27. $\dfrac{15x + 2}{x^2 + 6}$ 28. $\dfrac{x^2 - 4x}{x^2 + 4}$

29. $\dfrac{x + 1}{2x - 1}$ 30. $\dfrac{-6x}{3x - 1}$

31. $\dfrac{x^2 - 6x}{9}$ 32. $\dfrac{x^3 - x^2}{15}$

33. $\dfrac{30x}{x^2 - 36}$ 34. $\dfrac{2x - 15}{x^2 - 49}$

35. $\dfrac{15}{x^2 + x - 2}$ 36. $\dfrac{x - 20}{x^2 + 2x - 8}$

37. $\dfrac{16}{20 - x}$ 38. $\dfrac{44}{57 - x}$

Simplify. See Example 3.

39. $\dfrac{45}{9a}$ 40. $\dfrac{48}{16y}$

41. $\dfrac{6x^4}{4x^2}$ 42. $\dfrac{9x^3}{6x}$

Simplify. See Example 4.

43. $\dfrac{6x + 3}{9}$ 44. $\dfrac{4x + 12}{16}$

45. $\dfrac{x + 3}{3x + 9}$ 46. $\dfrac{2x - 14}{x - 7}$

47. $\dfrac{x^2 - 4}{x^2 - 6x + 8}$ 48. $\dfrac{y^2 - 25}{y^2 - 3y - 10}$

49. $\dfrac{x^2 + 5x + 4}{x^2 + 4x}$ 50. $\dfrac{x^2 - 10x + 21}{x^2 - 3x}$

Simplify. See Example 5.

51. $\dfrac{m^2 - 2mn + n^2}{7m^2 - 7n^2}$ 52. $\dfrac{11c^2 - 11d^2}{c^2 - 2cd + d^2}$

53. $\dfrac{4b^2 + 4b + 1}{(2b + 1)^3}$ 54. $\dfrac{9y^2 - 12y + 4}{(3y - 2)^3}$

Simplify. See Example 6.

55. $\dfrac{10(c - 3) + 10}{3(c - 3) + 3}$ 56. $\dfrac{6(d + 3) - 6}{7(d + 3) - 7}$

57. $\dfrac{6(x + 3) - 18}{3x - 18}$ 58. $\dfrac{4(t - 1) + 4}{4t + 4}$

Simplify. See Example 7.

59. $\dfrac{2x - 7}{7 - 2x}$ 60. $\dfrac{18 - d}{d - 18}$

61. $\dfrac{3 - 4t}{8t - 6}$ 62. $\dfrac{5t - 1}{3 - 15t}$

Simplify. See Example 8.

63. $\dfrac{2 - a}{a^2 - a - 2}$

64. $\dfrac{4 - b}{b^2 - 5b + 4}$

65. $\dfrac{25 - 5m}{m^2 - 25}$

66. $\dfrac{36 - 6h}{h^2 - 36}$

TRY IT YOURSELF

Simplify. If an expression cannot be simplified, write "Does not simplify."

67. $\dfrac{a^3 - a^2}{a^4 - a^3}$

68. $\dfrac{2c^4 + 2c^3}{4c^5 + 4c^4}$

69. $\dfrac{4 - x^2}{x^2 - x - 2}$

70. $\dfrac{81 - y^2}{y^2 + 10y + 9}$

71. $\dfrac{6x - 30}{5 - x}$

72. $\dfrac{6t - 42}{7 - t}$

73. $\dfrac{x^2 + 3x + 2}{x^2 + x - 2}$

74. $\dfrac{x^2 + x - 6}{x^2 - x - 2}$

75. $\dfrac{15x^2 y}{5xy^2}$

76. $\dfrac{12xz}{4xz^2}$

77. $\dfrac{x^8 + 9x^7}{9 + x}$

78. $\dfrac{x^9 + 50x^8}{50 + x}$

79. $\dfrac{x(x - 8) + 16}{16 - x^2}$

80. $\dfrac{x^2 - 3(2x - 3)}{9 - x^2}$

81. $\dfrac{4c + 4d}{d + c}$

82. $\dfrac{a + b}{5b + 5a}$

83. $\dfrac{3x^2 - 27}{2x^2 - 5x - 3}$

84. $\dfrac{2x^2 - 8}{3x^2 - 5x - 2}$

85. $\dfrac{-3x^2 + 10x + 77}{x^2 - 4x - 21}$

86. $\dfrac{-2x^2 + 5x + 3}{x^2 + 2x - 15}$

87. $\dfrac{42c^3 d}{18cd^3}$

88. $\dfrac{49m^4 n^5}{35mn^6}$

89. $\dfrac{16a^2 - 1}{4a + 4}$

90. $\dfrac{25m^2 - 1}{5m + 5}$

91. $\dfrac{8u^2 - 2u - 15}{4u^4 + 5u^3}$

92. $\dfrac{6n^2 - 7n + 2}{3n^3 - 2n^2}$

93. $\dfrac{(2x + 3)^4}{4x^2 + 12x + 9}$

94. $\dfrac{(3y - 2)^5}{9y^2 - 12y + 4}$

95. $\dfrac{6a + 3(a + 2) + 12}{a + 2}$

96. $\dfrac{2y + 4(y - 1) - 2}{y - 1}$

97. $\dfrac{15x - 3x^2}{25y - 5xy}$

98. $\dfrac{18c - 2c^2}{81d - 9cd}$

99. $\dfrac{2x^2}{x + 2}$

100. $\dfrac{5y^2}{y + 5}$

101. $\dfrac{18 + 2x}{x^2 - 81}$

102. $\dfrac{12 + 6x}{x^2 - 4}$

APPLICATIONS

103. **Organ Pipes.** The number of vibrations n per second of an organ pipe is given by the formula $n = \dfrac{512}{L}$ where L is the length of the pipe in feet. How many times per second will a 6-foot pipe vibrate?

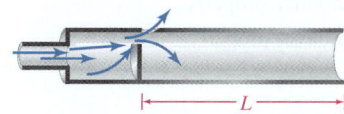

104. **Raising Turkeys.** The formula $T = \dfrac{2{,}000m}{m + 1}$ gives the number T of turkeys on a poultry farm m months after the beginning of the year. How many turkeys will there be on the farm by the end of July?

105. **Medical Dosages.** The formula $c = \dfrac{4t}{t^2 + 1}$ gives the concentration c (in milligrams per liter) of a certain dosage of medication in a patient's bloodstream t hours after the medication is administered. Suppose the patient received the medication at noon. Find the concentration of medication in his blood at the following times later that afternoon.

106. **Manufacturing.** If a company produces x child car seats, the average cost c (in dollars) to produce one car seat is given by the formula $c = \dfrac{50x + 50{,}000}{x}$. Find the company's average production cost if 1,000 are produced.

WRITING

107. Explain why $\dfrac{x - 7}{7 - x} = -1$.

108. Explain why $\dfrac{x - 3}{x + 4}$ is undefined for $x = -4$ but defined for $x = 3$.

109. Explain the error in the following work:

Simplify: $\dfrac{\cancel{x}}{x + 2} = \dfrac{\overset{1}{\cancel{x}}}{\cancel{x} + 2}$

$= \dfrac{1}{3}$

110. Explain the error in the following work:

Simplify: $\dfrac{30t - 6}{36} = \dfrac{\overset{1}{\cancel{2}} \cdot \overset{1}{\cancel{3}} \cdot 5 \cdot t - \overset{1}{\cancel{2}} \cdot \overset{1}{\cancel{3}}}{\underset{1}{\cancel{2}} \cdot 2 \cdot \underset{1}{\cancel{3}} \cdot \underset{1}{\cancel{3}}}$

$= 5t - 1$

111. Explain why there are no values for x for which $\dfrac{x - 7}{x^2 + 49}$ is undefined.

112. Write a rational expression that is not defined for $x = 5$. Then explain why that is so.

REVIEW

State each property using the variables a, b, and when necessary, c.

113. **a.** The associative property of addition

   b. The commutative property of multiplication

114. **a.** The distributive property

   b. The zero-factor property

CHALLENGE PROBLEMS

Simplify.

115. $\dfrac{(x^2 + 2x + 1)(x^2 - 2x + 1)}{(x^2 - 1)^2}$

116. $\dfrac{2x^2 + 2x - 12}{x^3 + 3x^2 - 4x - 12}$

117. $\dfrac{x^3 - 27}{x^3 - 9x}$

118. $\dfrac{b^3 + a^3}{a^2 - ab + b^2}$

119. $\dfrac{m^3 + 64}{m^3 + 4m^2 + 3m + 12}$

120. $\dfrac{s^3 + s^2 - 6s - 6}{s^3 + 1}$

SECTION 7.2

Multiplying and Dividing Rational Expressions

OBJECTIVES

1. Multiply rational expressions.

2. Divide rational expressions.

3. Convert units of measurement.

ARE YOU READY?

 The following problems review some basic skills that are needed when multiplying and dividing rational expressions.

1. Multiply: $\dfrac{3}{4} \cdot \dfrac{1}{5}$

2. What is the reciprocal of $\dfrac{8}{9}$?

3. Divide: $\dfrac{11}{16} \div \dfrac{7}{8}$

4. Factor: $x - x^2$

In this section, we will extend the rules for multiplying and dividing fractions to problems involving multiplication and division of rational expressions.

1 Multiply Rational Expressions.

Recall that to multiply fractions, we multiply their numerators and multiply their denominators. For example,

$$\frac{4}{7} \cdot \frac{3}{5} = \frac{4 \cdot 3}{7 \cdot 5}$$ Multiply the numerators and multiply the denominators.

$$= \frac{12}{35}$$

We use the same procedure to multiply rational expressions.

Multiplying Rational Expressions	To multiply two rational expressions, multiply their numerators and their denominators. Then, if possible, factor and simplify. For any two rational expressions, $\dfrac{A}{B}$ and $\dfrac{C}{D}$, $$\dfrac{A}{B} \cdot \dfrac{C}{D} = \dfrac{AC}{BD}$$

EXAMPLE 1   Multiply: **a.** $\dfrac{x+1}{x} \cdot \dfrac{9}{4x^2}$ **b.** $\dfrac{35x^3}{17y} \cdot \dfrac{y}{5x}$

Strategy   To find the product, we will use the rule for multiplying rational expressions. In the process, we must be prepared to factor the numerators and denominators so that any common factors can be removed.

Why   We want to give the result in simplified form, which requires that the numerator and denominator have no common factors other than 1.

Solution   **a.** $\dfrac{x+1}{x} \cdot \dfrac{9}{4x^2} = \dfrac{9(x+1)}{4x^3}$ Multiply the numerators.
Multiply the denominators.

Since the numerator and denominator do not share any common factors, $\dfrac{9(x+1)}{4x^3}$ cannot be simplified. We can leave the numerator in factored form, or we can distribute the multiplication by 9 and write the result as $\dfrac{9x+9}{4x^3}$.

b. $\dfrac{35x^3}{17y} \cdot \dfrac{y}{5x} = \dfrac{35x^3 \cdot y}{17y \cdot 5x}$ Multiply the numerators.
Multiply the denominators.

It is obvious that the numerator and denominator of $\dfrac{35x^3 \cdot y}{17y \cdot 5x}$ have several common factors, such as 5, x, and y. These common factors become more apparent when we factor the numerator and denominator completely.

$$\dfrac{35x^3 \cdot y}{17y \cdot 5x} = \dfrac{5 \cdot 7 \cdot x \cdot x \cdot x \cdot y}{17 \cdot y \cdot 5 \cdot x}$$ To prepare to simplify, factor $35x^3$.

$$= \dfrac{\overset{1}{\cancel{5}} \cdot 7 \cdot \overset{1}{\cancel{x}} \cdot x \cdot x \cdot \cancel{y}}{17 \cdot \underset{1}{\cancel{y}} \cdot \underset{1}{\cancel{5}} \cdot \underset{1}{\cancel{x}}}$$ Simplify by replacing $\frac{5}{5}$, $\frac{x}{x}$, and $\frac{y}{y}$ with the equivalent fraction $\frac{1}{1}$. This removes the factor $\frac{5 \cdot x \cdot y}{5 \cdot x \cdot y} = 1$.

$$= \dfrac{7x^2}{17}$$ Multiply the remaining factors in the numerator.
Multiply the remaining factors in the denominator.

> **Caution**
>
> When multiplying rational expressions, always write the result in simplest form by removing any factors common to the numerator and denominator.

Self Check 1   Multiply: **a.** $\dfrac{a+7}{a} \cdot \dfrac{6}{5a^3}$ **b.** $\dfrac{a^4}{8b} \cdot \dfrac{24b}{11a^3}$

Now Try ▶ Problems 13 and 17

EXAMPLE 2   Multiply: **a.** $\dfrac{x+3}{2x+4} \cdot \dfrac{6}{x^2-9}$ **b.** $\dfrac{8x^2-8x}{x^2+x-56} \cdot \dfrac{3x^2-22x+7}{x-x^2}$

Strategy   To find the product, we will use the rule for multiplying rational expressions. In the process, we need to factor the monomials, binomials, or trinomials that are not prime, so that any common factors can be removed.

Why   We want to give the result in simplified form, which requires that the numerator and denominator have no common factor other than 1.

Solution   **a.** $\dfrac{x+3}{2x+4} \cdot \dfrac{6}{x^2-9} = \dfrac{(x+3)6}{(2x+4)(x^2-9)}$ Multiply the numerators and multiply the denominators.

> **Notation**
>
> It is not necessary to multiply $(x+2)(x-3)$ in the denominator. When we add and subtract rational expressions in the next section, it is usually more convenient to leave the denominator in factored form as shown here.

$$= \dfrac{(x+3) \cdot 3 \cdot 2}{2(x+2)(x+3)(x-3)}$$ Factor 6. Factor out the GCF, 2, from $2x+4$. Factor the difference of two squares, x^2-9.

$$= \dfrac{\overset{1}{\cancel{(x+3)}} \cdot 3 \cdot \overset{1}{\cancel{2}}}{\underset{1}{\cancel{2}}(x+2)\underset{1}{\cancel{(x+3)}}(x-3)}$$ Simplify by replacing $\frac{x+3}{x+3}$ and $\frac{2}{2}$ with $\frac{1}{1}$. This removes the factor $\frac{2 \cdot (x+3)}{2 \cdot (x+3)} = 1$.

$$= \dfrac{3}{(x+2)(x-3)}$$ Multiply the remaining factors in the numerator.
Multiply the remaining factors in the denominator.

b. $\dfrac{8x^2 - 8x}{x^2 + x - 56} \cdot \dfrac{3x^2 - 22x + 7}{x - x^2}$

$= \dfrac{(8x^2 - 8x)(3x^2 - 22x + 7)}{(x^2 + x - 56)(x - x^2)}$ Multiply the numerators and multiply the denominators.

$= \dfrac{8x(x - 1)(3x - 1)(x - 7)}{(x + 8)(x - 7)x(1 - x)}$ To prepare to simplify, factor all four polynomials.

$$= \dfrac{\overset{1}{8x}(\overset{-1}{\cancel{x - 1}})(3x - 1)(\overset{1}{\cancel{x - 7}})}{(x + 8)(\underset{1}{\cancel{x - 7}})\underset{1}{x}(\underset{1}{\cancel{1 - x}})}$$ Simplify. Since $x - 1$ and $1 - x$ are opposites, replace $\frac{x-1}{1-x}$ with $\frac{-1}{1}$. This removes the factor $\frac{x-1}{1-x} = -1$.

$= \dfrac{-8(3x - 1)}{x + 8}$ Multiply the remaining factors in the numerator.
Multiply the remaining factors in the denominator.

The result can also be written as $-\dfrac{8(3x - 1)}{x + 8}$ or $-\dfrac{24x - 8}{x + 8}$.

> **Notation**
>
> We could distribute in the numerator and write the result as $\frac{-24x + 8}{x + 8}$. Check with your instructor to see which form of the result he or she prefers.

Self Check 2   Multiply: **a.** $\dfrac{3n - 9}{3n + 2} \cdot \dfrac{9n^2 - 4}{6}$

                b. $\dfrac{m^2 - 4m - 5}{2m - m^2} \cdot \dfrac{2m^2 - 4m}{3m^2 - 14m - 5}$

Now Try ▶ Problems 21 and 29

EXAMPLE 3   Multiply: **a.** $63x\left(\dfrac{1}{7x}\right)$ **b.** $5a\left(\dfrac{3a - 1}{a}\right)$

Strategy   We will write each of the monomials, $63x$ and $5a$, as rational expressions with denominator 1. (Remember, any number divided by 1 remains unchanged.) Then we will use the rule for multiplying rational expressions.

Why   Writing $63x$ and $5a$ over 1 is helpful during the multiplication process when we multiply numerators and multiply denominators.

Solution   **a.** $63x\left(\dfrac{1}{7x}\right) = \dfrac{63x}{1}\left(\dfrac{1}{7x}\right)$ Write $63x$ as a fraction: $63x = \frac{63x}{1}$.

$= \dfrac{63x \cdot 1}{1 \cdot 7 \cdot x}$ Multiply the numerators and multiply the denominators.

$= \dfrac{9 \cdot \overset{1}{\cancel{7}} \cdot \overset{1}{\cancel{x}} \cdot 1}{1 \cdot \underset{1}{\cancel{7}} \cdot \underset{1}{\cancel{x}}}$ Write $63x$ in factored form as $9 \cdot 7 \cdot x$.
Then simplify by removing a factor equal to 1: $\frac{7x}{7x}$.

$= 9$ Because $\frac{9}{1} = 9$

b. $5a\left(\dfrac{3a - 1}{a}\right) = \dfrac{5a}{1}\left(\dfrac{3a - 1}{a}\right)$ Write $5a$ as a fraction: $5a = \frac{5a}{1}$.

$= \dfrac{5\overset{1}{\cancel{a}}(3a - 1)}{1 \cdot \underset{1}{\cancel{a}}}$ Multiply the numerators and multiply the denominators.
Then simplify by removing a factor equal to 1: $\frac{a}{a}$.

$= 5(3a - 1)$

$= 15a - 5$ Distribute the multiplication by 5.

Self Check 3   Multiply: **a.** $36b\left(\dfrac{1}{6b}\right)$ **b.** $4x\left(\dfrac{x + 3}{x}\right)$

Now Try ▶ Problems 31 and 37

2 Divide Rational Expressions.

Recall that one number is the **reciprocal** of another if their product is 1. To find the reciprocal of a fraction, we invert its numerator and denominator. We have seen that to divide fractions, we multiply the first fraction by the reciprocal of the second fraction.

$$\frac{4}{7} \div \frac{3}{5} = \frac{4}{7} \cdot \frac{5}{3}$$ Invert $\frac{3}{5}$ and change the division to a multiplication.

$$= \frac{20}{21}$$ Multiply the numerators and multiply the denominators.

We use the same procedure to divide rational expressions.

Dividing Rational Expressions

To divide two rational expressions, multiply the first by the reciprocal of the second. Then, if possible, factor and simplify.

For any two rational expressions, $\frac{A}{B}$ and $\frac{C}{D}$, where $\frac{C}{D} \neq 0$,

$$\frac{A}{B} \div \frac{C}{D} = \frac{A}{B} \cdot \frac{D}{C} = \frac{AD}{BC}$$

EXAMPLE 4   Divide: **a.** $\dfrac{a}{13} \div \dfrac{17}{26}$ **b.** $\dfrac{9x}{35y} \div \dfrac{15x^2}{14}$

Strategy   We will use the rule for dividing rational expressions. After multiplying by the reciprocal, we will factor the monomials that are not prime, and remove any common factors of the numerator and denominator.

Why   We want to give the result in simplified form, which requires that the numerator and denominator have no common factor other than 1.

Solution   **a.** $\dfrac{a}{13} \div \dfrac{17}{26} = \dfrac{a}{13} \cdot \dfrac{26}{17}$ Multiply by the reciprocal of $\frac{17}{26}$.

$$= \frac{a \cdot 2 \cdot 13}{13 \cdot 17}$$ Multiply the numerators and multiply the denominators. Then, to prepare to simplify, factor 26 as $2 \cdot 13$.

$$= \frac{a \cdot 2 \cdot \overset{1}{\cancel{13}}}{\underset{1}{\cancel{13}} \cdot 17}$$ Simplify by removing common factors of the numerator and denominator.

$$= \frac{2a}{17}$$ Multiply the remaining factors in the numerator. Multiply the remaining factors in the denominator.

> **Caution**
>
> When dividing rational expressions, always write the result in simplest form, by removing any factors common to the numerator and denominator.

b. $\dfrac{9x}{35y} \div \dfrac{15x^2}{14} = \dfrac{9x}{35y} \cdot \dfrac{14}{15x^2}$ Multiply by the reciprocal of $\frac{15x^2}{14}$.

$$= \frac{3 \cdot 3 \cdot x \cdot 2 \cdot 7}{5 \cdot 7 \cdot y \cdot 3 \cdot 5 \cdot x \cdot x}$$ Multiply the numerators and multiply the denominators. Then, to prepare to simplify, factor 9, 35, 14, and $15x^2$.

$$= \frac{3 \cdot \overset{1}{\cancel{3}} \cdot \overset{1}{\cancel{x}} \cdot 2 \cdot \overset{1}{\cancel{7}}}{5 \cdot \underset{1}{\cancel{7}} \cdot y \cdot \underset{1}{\cancel{3}} \cdot 5 \cdot \underset{1}{\cancel{x}} \cdot x}$$ Simplify by removing factors equal to 1.

$$= \frac{6}{25xy}$$ Multiply the remaining factors in the numerator. Multiply the remaining factors in the denominator.

Self Check 4 Divide: $\dfrac{8a}{3b} \div \dfrac{16a^2}{9b^2}$

Now Try ▶ Problems 41 and 45

EXAMPLE 5 Divide: $\dfrac{x^2 + x}{3x - 15} \div \dfrac{(x + 1)^2}{6x - 30}$

Strategy To find the quotient, we will use the rule for dividing rational expressions. After multiplying by the reciprocal, we will factor the binomials that are not prime, and remove any common factors of the numerator and denominator.

Why We want to give the result in simplified form, which requires that the numerator and denominator have no common factor other than 1.

Solution

$$\frac{x^2 + x}{3x - 15} \div \frac{(x + 1)^2}{6x - 30}$$

$$= \frac{x^2 + x}{3x - 15} \cdot \frac{6x - 30}{(x + 1)^2} \qquad \text{Multiply by the reciprocal of } \frac{(x + 1)^2}{6x - 30}.$$

$$= \frac{x(x + 1) \cdot 2 \cdot 3(x - 5)}{3(x - 5)(x + 1)(x + 1)} \qquad \begin{array}{l}\text{Multiply the numerators and multiply the denominators.} \\ \text{Then, to prepare to simplify, factor the binomials. Write} \\ (x + 1)^2 \text{ as repeated multiplication.}\end{array}$$

$$= \frac{\overset{1}{x(\cancel{x + 1})} \cdot 2 \cdot \overset{1}{\cancel{3}}\overset{1}{(\cancel{x - 5})}}{\underset{1}{\cancel{3}}\underset{1}{(\cancel{x - 5})}\underset{1}{(\cancel{x + 1})}(x + 1)} \qquad \begin{array}{l}\text{Simplify by removing common factors of the numerator} \\ \text{and denominator.}\end{array}$$

$$= \frac{2x}{x + 1} \qquad \begin{array}{l}\text{Multiply the remaining factors in the numerator.} \\ \text{Multiply the remaining factors in the denominator.}\end{array}$$

The Language of Algebra

To find the reciprocal of $\frac{(x + 1)^2}{6x - 30}$, we invert it. To **invert** means to turn upside down: $\frac{6x - 30}{(x + 1)^2}$. Some amusement park thrill rides have giant loops where the riders become inverted.

Self Check 5 Divide: $\dfrac{z^2 - 9}{z^2 + 4z + 3} \div \dfrac{z^2 - 3z}{(z + 1)^2}$

Now Try ▶ Problem 51

EXAMPLE 6 Divide: $\dfrac{2x^2 - 3xy - 2y^2}{2x + y} \div (4y^2 - x^2)$

Strategy We begin by writing $4y^2 - x^2$ as a rational expression by inserting a denominator 1. Then we will use the rule for dividing rational expressions.

Why Writing $4y^2 - x^2$ over 1 is helpful when we invert its numerator and denominator to find its reciprocal.

Solution

$$\frac{2x^2 - 3xy - 2y^2}{2x + y} \div (4y^2 - x^2)$$

$$= \frac{2x^2 - 3xy - 2y^2}{2x + y} \div \frac{4y^2 - x^2}{1} \qquad \text{Write } 4y^2 - x^2 \text{ as a fraction with a denominator of 1.}$$

$$= \frac{2x^2 - 3xy - 2y^2}{2x + y} \cdot \frac{1}{4y^2 - x^2} \qquad \text{Multiply by the reciprocal of } \frac{4y^2 - x^2}{1}.$$

$$= \frac{(2x + y)(x - 2y) \cdot 1}{(2x + y)(2y + x)(2y - x)} \qquad \begin{array}{l}\text{Multiply the numerators and denominators. Then,} \\ \text{to simplify, factor } 2x^2 - 3xy - 2y^2 \text{ and } 4y^2 - x^2.\end{array}$$

$$= \frac{\overset{1}{\cancel{(2x+y)}}\overset{-1}{\cancel{(x-2y)}} \cdot 1}{\underset{1}{\cancel{(2x+y)}}(2y+x)\underset{1}{\cancel{(2y-x)}}}$$

Since $x - 2y$ and $2y - x$ are opposites, simplify by replacing $\frac{x-2y}{2y-x}$ with $\frac{-1}{1}$.

$$= \frac{-1}{2y+x}$$

Multiply the remaining factors in the numerator.
Multiply the remaining factors in the denominator.

$$= -\frac{1}{2y+x}$$

Write the − sign in front of the fraction.

Self Check 6 Divide : $(b-a) \div \frac{a^2 - b^2}{a^2 + ab}$

Now Try ▶ Problem 63

3 Convert Units of Measurement.

We can use the concepts discussed in this section to make conversions from one unit of measure to another. *Unit conversion factors* play an important role in this process. A **unit conversion factor** is a fraction that has a value of 1. For example, we can use the fact that 1 square yard = 9 square feet to form two unit conversion factors:

$$\frac{1 \text{ yd}^2}{9 \text{ ft}^2} = 1 \qquad \text{Read as "1 square yard per 9 square feet."} \qquad \frac{9 \text{ ft}^2}{1 \text{ yd}^2} = 1 \qquad \text{Read as "9 square feet per 1 square yard."}$$

Since a unit conversion factor is equal to 1, multiplying a measurement by a unit conversion factor does not change the measurement, it only changes the units of measure.

EXAMPLE 7

Carpeting. A roll of carpeting is 12 feet wide and 150 feet long. Find the number of square yards of carpeting on the roll.

Strategy We will begin by determining the number of square feet of carpeting on the roll. Then we will multiply that result by a unit conversion factor.

Why A properly chosen unit conversion factor can convert the number of square feet of carpeting on the roll to the number of square yards on the roll.

Solution When unrolled, the carpeting forms a rectangular shape with an area of $12 \cdot 150 = 1{,}800$ square feet. We will multiply 1,800 ft^2 by a unit conversion factor such that the units of ft^2 are removed and the units of yd^2 are introduced. Since $1 \text{ yd}^2 = 9 \text{ ft}^2$, we will use $\frac{1 \text{ yd}^2}{9 \text{ ft}^2}$.

$$\frac{1{,}800 \text{ ft}^2}{1 \text{ roll}} = \frac{1{,}800 \text{ ft}^2}{1 \text{ roll}} \cdot \frac{\mathbf{1 \text{ yd}^2}}{\mathbf{9 \text{ ft}^2}}$$

Multiply by a unit conversion factor that relates yd^2 to ft^2.

$$= \frac{1{,}800 \cancel{\text{ ft}^2}}{1 \text{ roll}} \cdot \frac{1 \text{ yd}^2}{9 \cancel{\text{ ft}^2}}$$

Remove the units of ft^2 that are common to the numerator and denominator.

$$= \frac{200 \text{ yd}^2}{1 \text{ roll}}$$

Divide 1,800 by 9 to get 200.

There are 200 yd^2 of carpeting on the roll.

Self Check 7 Convert 5,400 ft^2 to square yards.

Now Try ▶ Problem 67

EXAMPLE 8   **The Speed of Light.**   The speed with which light moves through space is about 186,000 miles per second. Express this speed in miles per minute.

Strategy   The speed of light can be expressed as $\frac{186{,}000 \text{ mi}}{1 \text{ sec}}$. We will multiply that fraction by a unit conversion factor.

Why   A properly chosen unit conversion factor can convert the number of miles traveled per second to the number of miles traveled per minute.

Solution   We will multiply $\frac{186{,}000 \text{ mi}}{1 \text{ sec}}$ by a unit conversion factor such that the units of seconds are removed and the units of minutes are introduced. Since 60 seconds = 1 minute, we will use $\frac{60 \text{ sec}}{1 \text{ min}}$.

$$\frac{186{,}000 \text{ mi}}{1 \text{ sec}} = \frac{186{,}000 \text{ mi}}{1 \text{ sec}} \cdot \frac{60 \text{ sec}}{1 \text{ min}} \quad \text{Multiply by a unit conversion factor that relates seconds to minutes.}$$

$$= \frac{186{,}000 \text{ mi}}{1 \text{ sec}} \cdot \frac{60 \text{ sec}}{1 \text{ min}} \quad \text{Remove the units of seconds that are common to the numerator and denominator.}$$

$$= \frac{11{,}160{,}000 \text{ mi}}{1 \text{ min}} \quad \text{Multiply 186,000 and 60 to get 11,160,000.}$$

The speed of light is about 11,160,000 miles per minute.

> **Success Tip**
>
> We can remove common units just as we remove factors that are common to the numerator and denominator when multiplying rational expressions.

Self Check 8   **Insects.**   A mosquito beats it wings about 600 times per second. How many times is that per minute?

Now Try ▶ **Problem 71**

SECTION 7.2 ▶ STUDY SET

VOCABULARY

Fill in the blanks.

1. The _____ of $\frac{x^2 + 6x + 1}{10x}$ is $\frac{10x}{x^2 + 6x + 1}$.

2. A _____ conversion factor is a fraction that is equal to 1, such as $\frac{3 \text{ ft}}{1 \text{ yd}}$.

CONCEPTS

Fill in the blanks.

3. **a.** To multiply rational expressions, multiply their _____ and multiply their _____. To divide two rational expressions, multiply the first by the _____ of the second. In symbols,

   b. $\frac{A}{B} \cdot \frac{C}{D} = \dfrac{}{}$ and $\frac{A}{B} \div \frac{C}{D} = \frac{A}{B} \cdot \dfrac{}{}$

Simplify each expression.

4. $\dfrac{(x + 7) \cdot 2 \cdot 5}{5(x + 1)(x + 7)(x - 9)}$

5. $\dfrac{y \cdot y \cdot y(15 - y)}{y(y - 15)(y + 1)}$

6. **a.** Write $3x + 5$ in fractional form.

   b. What is the reciprocal of $18x$?

7. Find the product of the rational expression and its reciprocal.

   $$\frac{3}{x + 2} \cdot \frac{x + 2}{3}$$

8. Use the fact that 1 tablespoon = 3 teaspoons to write two unit conversion factors.

NOTATION

9. What units are common to the numerator and denominator of the following product?

   $$\frac{45 \text{ ft}}{1} \cdot \frac{1 \text{ yd}}{3 \text{ ft}}$$

10. **a.** What fact is indicated by the unit conversion factor $\frac{1 \text{ day}}{24 \text{ hours}}$?

    b. Fill in the blank: $\frac{1 \text{ day}}{24 \text{ hours}} = \boxed{}$.

GUIDED PRACTICE

Multiply, and then simplify, if possible. **See Example 1.**

11. $\dfrac{3}{7} \cdot \dfrac{y}{2}$

12. $\dfrac{2}{7} \cdot \dfrac{z}{3}$

13. $\dfrac{y + 2}{y} \cdot \dfrac{3}{y^2}$

14. $\dfrac{4}{a + 1} \cdot \dfrac{a}{7}$

15. $\dfrac{35n}{12} \cdot \dfrac{16}{7n^2}$

16. $\dfrac{11m}{21} \cdot \dfrac{14}{55m^3}$

17. $\dfrac{2x^2y}{3xy} \cdot \dfrac{3xy^2}{2}$

18. $\dfrac{2x^2z}{z} \cdot \dfrac{5x}{z}$

Multiply, and then simplify, if possible. See Example 2.

19. $\dfrac{x+5}{5} \cdot \dfrac{x}{x+5}$

20. $\dfrac{a-9}{9} \cdot \dfrac{8a}{a-9}$

21. $\dfrac{2x+6}{x+3} \cdot \dfrac{3}{4x}$

22. $\dfrac{3y-9}{y-3} \cdot \dfrac{y}{3y^2}$

23. $\dfrac{(x+1)^2}{x+2} \cdot \dfrac{x+2}{x+1}$

24. $\dfrac{(y-3)^2}{y-5} \cdot \dfrac{y-5}{y-3}$

25. $\dfrac{x^2-x}{x} \cdot \dfrac{3x-6}{3-3x}$

26. $\dfrac{5z-10}{z+2} \cdot \dfrac{3}{6-3z}$

27. $\dfrac{x^2+x-6}{5x} \cdot \dfrac{5x-10}{x+3}$

28. $\dfrac{z^2+4z-5}{5z-5} \cdot \dfrac{5z}{z+5}$

29. $\dfrac{m^2-2m-3}{2m+4} \cdot \dfrac{m^2-4}{m^2+3m+2}$

30. $\dfrac{p^2-p-6}{3p-9} \cdot \dfrac{2p^2-5p-3}{p^2-3p}$

Multiply, and then simplify, if possible. See Example 3.

31. $7m\left(\dfrac{5}{m}\right)$

32. $9p\left(\dfrac{10}{p}\right)$

33. $15x\left(\dfrac{x+1}{5x}\right)$

34. $30t\left(\dfrac{t-7}{10t}\right)$

35. $12y\left(\dfrac{5y-8}{6y}\right)$

36. $16x\left(\dfrac{3x+8}{4x}\right)$

37. $24\left(\dfrac{3a-5}{2a}\right)$

38. $28\left(\dfrac{8-3t}{4t}\right)$

Divide, and then simplify, if possible. See Example 4.

39. $\dfrac{2}{y} \div \dfrac{4}{3}$

40. $\dfrac{3}{a} \div \dfrac{9}{5}$

41. $\dfrac{3a}{25} \div \dfrac{1}{5}$

42. $\dfrac{3y}{8} \div \dfrac{3}{2}$

43. $\dfrac{x^3}{18y} \div \dfrac{x}{6y}$

44. $\dfrac{21x}{z^2} \div \dfrac{7x^3}{z^5}$

45. $\dfrac{27p^4}{35q} \div \dfrac{9p}{21q}$

46. $\dfrac{12}{25s^5} \div \dfrac{10}{15s^2}$

Divide, and then simplify, if possible. See Example 5.

47. $\dfrac{9a-18}{28} \div \dfrac{9a^3}{35}$

48. $\dfrac{3x+6}{40} \div \dfrac{3x^2}{24}$

49. $\dfrac{x^2-4}{3x+6} \div \dfrac{2-x}{x+2}$

50. $\dfrac{x^2-9}{5x+15} \div \dfrac{3-x}{x+3}$

51. $\dfrac{x^2+7x}{5x-10} \div \dfrac{(x+7)^2}{15x-30}$

52. $\dfrac{x^2-10x}{7x+7} \div \dfrac{(x-10)^2}{35x+35}$

53. $\dfrac{m^2+m-20}{m} \div \dfrac{4-m}{m}$

54. $\dfrac{n^2+4n-21}{n} \div \dfrac{3-n}{n}$

55. $\dfrac{t^2+5t-14}{t} \div \dfrac{t-2}{t}$

56. $\dfrac{r^2+12r+11}{r} \div \dfrac{r+11}{r}$

57. $\dfrac{x^2-2x-35}{3x^2+27x} \div \dfrac{3x^2+17x+10}{18x^2+12x}$

58. $\dfrac{x^2-x-6}{2x^2+9x+10} \div \dfrac{x^2-25}{2x^2+15x+25}$

Divide, and then simplify, if possible. See Example 6.

59. $\dfrac{x^2-1}{3x-3} \div (x+1)$

60. $\dfrac{x^2-16}{x-4} \div (3x+12)$

61. $\dfrac{n^2-10n+9}{n-9} \div (n-1)$

62. $\dfrac{r^2-11r+18}{r-9} \div (r-2)$

63. $\dfrac{2r-3s}{12} \div (4r^2-12rs+9s^2)$

64. $\dfrac{3m+n}{18} \div (9m^2+6mn+n^2)$

65. $24n^2 \div \dfrac{18n^3}{n-1}$

66. $12m \div \dfrac{16m^2}{m+4}$

Complete each unit conversion. See Examples 7 and 8.

67. $\dfrac{150 \text{ yards}}{1} \cdot \dfrac{3 \text{ feet}}{1 \text{ yard}} = ?$

68. $\dfrac{60 \text{ inches}}{1} \cdot \dfrac{1 \text{ feet}}{12 \text{ inches}} = ?$

69. $\dfrac{6 \text{ pints}}{1} \cdot \dfrac{1 \text{ gallon}}{8 \text{ pints}} = ?$

70. $\dfrac{4 \text{ cups}}{1} \cdot \dfrac{1 \text{ gallon}}{16 \text{ cups}} = ?$

71. $\dfrac{30 \text{ miles}}{1 \text{ hour}} \cdot \dfrac{1 \text{ hour}}{60 \text{ minutes}} = ?$

72. $\dfrac{300 \text{ meters}}{3 \text{ months}} \cdot \dfrac{12 \text{ months}}{1 \text{ year}} = ?$

73. $\dfrac{30 \text{ meters}}{1 \text{ seconds}} \cdot \dfrac{60 \text{ seconds}}{1 \text{ minute}} = ?$

74. $\dfrac{288 \text{ inches}^2}{1 \text{ year}} \cdot \dfrac{1 \text{ feet}^2}{144 \text{ inches}^2} = ?$

TRY IT YOURSELF

Perform the operations and simplify, if possible.

75. $\dfrac{b^2-5b+6}{b^2-10b+16} \div \dfrac{b^2+2b}{b^2-6b-16}$

76. $\dfrac{m^2+m-6}{m^2-6m+9} \div \dfrac{m^2-4}{m^2-9}$

77. $\dfrac{5x+5}{25} \cdot \dfrac{5}{(x+1)^3}$

78. $\dfrac{7t-7}{28} \cdot \dfrac{4}{(t-1)^4}$

79. $\dfrac{6a^2}{a^2+6a+9} \cdot \dfrac{(a+3)^4}{4a^5}$

80. $\dfrac{9b^3}{b^2-8b+16} \cdot \dfrac{(b-4)^4}{15b^8}$

81. $\dfrac{36c^2-49d^2}{3d^3} \div \dfrac{12c+14d}{d^4}$

82. $\dfrac{25y^2-16z^2}{2yz} \div \dfrac{10y-8z}{y^2}$

83. $10h\left(\dfrac{5h-3}{2h}\right)$

84. $33r\left(\dfrac{5r+4}{11r}\right)$

85. $\dfrac{n^2 - 9}{n^2 - 3n} \div \dfrac{n + 3}{n^2 - n}$

86. $\dfrac{b^2 - b}{b + 2} \div \dfrac{b^2 - 2b}{b^2 - 4}$

87. $\dfrac{10r^2 s}{6rs^2} \cdot \dfrac{3r^3}{2rs}$

88. $\dfrac{3a^3 b}{25cd^3} \cdot \dfrac{5cd^2}{6ab}$

89. $\dfrac{7}{3p^3} \cdot \dfrac{p + 2}{p}$

90. $\dfrac{5t^2}{11} \cdot \dfrac{2t}{t - 5}$

91. $\dfrac{5x^2 + 13x - 6}{x + 3} \div \dfrac{5x^2 - 17x + 6}{x - 2}$

92. $\dfrac{3p^2 + 5p - 2}{p^3 + 2p^2} \div \dfrac{6p^2 + 13p - 5}{2p^3 + 5p^2}$

93. $\dfrac{4x^2 - 12xy + 9y^2}{x^3 y^2} \cdot \dfrac{x^3 y}{4x^2 - 9y^2}$

94. $\dfrac{ab^4}{25a^2 - 16b^2} \cdot \dfrac{25a^2 - 40ab + 16b^2}{a^2 b^4}$

95. $\dfrac{x - 2}{x} \cdot \dfrac{2x}{2 - x}$

96. $\dfrac{y - 3}{y} \cdot \dfrac{3y}{3 - y}$

Look Alikes . . .

97. a. $\dfrac{3x + 6}{4} \cdot \dfrac{4x + 8}{3}$ b. $\dfrac{3x + 6}{4} \div \dfrac{4x + 8}{3}$

98. a. $\dfrac{4a - 8}{5} \cdot \dfrac{5a - 10}{4}$ b. $\dfrac{4a - 8}{5} \div \dfrac{5a - 10}{4}$

99. a. $\dfrac{x^2 - 5x + 6}{2x - 4} \cdot \dfrac{2x - 6}{x - 2}$ b. $\dfrac{x^2 - 5x + 6}{2x - 4} \div \dfrac{2x - 6}{x - 2}$

100. a. $\dfrac{x^2 + 9x + 20}{9x + 36} \cdot \dfrac{9x + 45}{x + 4}$ b. $\dfrac{x^2 + 9x + 20}{9x + 36} \div \dfrac{9x + 45}{x + 4}$

APPLICATIONS

101. **Geometry.**   Find the area of the rectangle.

$\dfrac{x}{2x - 14}$ ft

$\dfrac{x^2 - 7x}{5}$ ft

102. **Motion.**   The table contains algebraic expressions for the rate an object travels and the time traveled at that rate. Complete the table.

Rate (mph)	Time (hr)	Distance (mi)
$\dfrac{k^2 + k - 6}{k - 3}$	$\dfrac{k^2 - 9}{k^2 - 4}$	

103. **Talking.**   According to the *Sacramento Bee* newspaper, the number of words an average man speaks a day is about 12,000. How many words does an average man speak in 1 year? (*Hint:* 365 days = 1 year.)

104. **Classroom Space.**   The recommended size of an elementary school classroom in the United States is approximately 900 square feet. Convert this to square yards.

105. **Natural Light.**   According to the University of Georgia School Design and Planning Laboratory, the basic classroom should have at least 72 square feet of windows for natural light. Convert this to square yards.

106. **Trucking.**   A cement truck holds 9 cubic yards of concrete. How many cubic feet of concrete does it hold? (*Hint:* 27 cubic feet = 1 cubic yard.)

107. **Bears.**   The maximum speed a grizzly bear can run is about 30 miles per hour. What is its maximum speed in miles per minute?

108. **Fuel Economy.**   Use the information that follows to determine the miles per fluid ounce of gasoline for city and for highway driving for the Ford Ranger. (*Hint:* 1 gallon = 128 fluid ounces.)

2010 Ford Ranger

Fuel Economy

Fuel Type	Regular
MPG (city)	**16**
MPG (highway)	**20**

109. **TV Trivia.**   On the comedy television series *Green Acres* (1965–1971), New York socialites Oliver Wendell Douglas (played by Eddie Albert) and his wife, Lisa Douglas (played by Eva Gabor), move from New York to purchase a 160-acre farm in Hooterville. Convert this to square miles. (*Hint:* 1 square mile = 640 acres.)

110. **Camping.**   The capacity of backpacks is usually given in cubic inches. Convert a backpack capacity of 5,400 cubic inches to cubic feet. (*Hint:* 1 cubic foot = 1,728 cubic inches.)

©Georgy Markov/Shutterstock.com

WRITING

111. Explain how to multiply rational expressions.

112. To divide rational expressions, you must first know how to multiply rational expressions. Explain why.

113. Explain why 60 miles per hour and 1 mile per minute are the same speed.

114. Explain why the unit conversion factor $\dfrac{1 \text{ ft}}{12 \text{ in.}}$ is equal to 1.

REVIEW

115. Hardware. A brace has a length that is 2 inches less than twice the width of the shelf that it supports. The brace is anchored to the wall 8 inches below the shelf. Find the width of the shelf and the length of the brace.

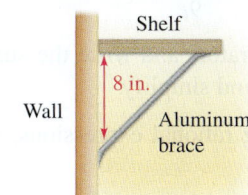

Shelf

8 in.

Wall

Aluminum brace

116. Solve $A = \frac{1}{2}h(b + d)$ for b.

CHALLENGE PROBLEMS

Perform the operations. Simplify, if possible.

117. $\dfrac{c^3 - 2c^2 + 5c - 10}{c^2 - c - 2} \cdot \dfrac{c^3 + c^2 - 5c - 5}{c^4 - 25}$

118. $\dfrac{x^3 - y^3}{x^3 + y^3} \div \dfrac{x^3 + x^2y + xy^2}{x^2y - xy^2 + y^3}$

119. $\dfrac{-x^3 + x^2 + 6x}{3x^3 + 21x^2} \div \left(\dfrac{2x + 4}{3x^2} \div \dfrac{2x + 14}{x^2 - 3x} \right)$

120. $\dfrac{x^2 - y^2}{2x^4 - 2x^3} \div \left(\dfrac{x - y}{2x^2} \div \dfrac{x + y}{x^2 + 2xy + y^2} \right)$

SECTION 7.3

OBJECTIVES

1 Add and subtract rational expressions that have the same denominator.

2 Find the least common denominator.

3 Build rational expressions into equivalent expressions.

Adding and Subtracting with Like Denominators; Least Common Denominators

ARE YOU READY?

The following problems review some basic skills that are needed when adding and subtracting rational expressions with like denominators.

1. Add: $\dfrac{5}{11} + \dfrac{3}{11}$

2. Subtract: $\dfrac{17}{21} - \dfrac{10}{21}$

3. Simplify: $x^2 + 8x - (3x - 1)$

4. Simplify: $\dfrac{x + 2}{x^2 - 5x - 14}$

5. Factor completely: $24x^2$

6. Factor: $5x + 5$

In this section, we extend the rules for adding and subtracting fractions to problems involving addition and subtraction of rational expressions.

1 Add and Subtract Rational Expressions That Have the Same Denominator.

Recall from Chapter 1 that to add (or subtract) fractions that have the same denominator, we add (or subtract) their numerators and write the sum (or difference) over the common denominator. For example,

The Language of Algebra

We can describe $\frac{3}{7}$ and $\frac{2}{7}$ as having the **same** denominator, **common** denominators, or **like** denominators.

$$\frac{3}{7} + \frac{2}{7} = \frac{3 + 2}{7} \qquad \text{and} \qquad \frac{18}{25} - \frac{9}{25} = \frac{18 - 9}{25}$$
$$= \frac{5}{7} \qquad\qquad\qquad\qquad = \frac{9}{25}$$

We use the same procedure to add and subtract rational expressions with like denominators.

Adding and Subtracting Rational Expressions That Have the Same Denominator

To add (or subtract) rational expressions that have same denominator, add (or subtract) their numerators and write the sum (or difference) over the common denominator. Then, if possible, factor and simplify.

If $\frac{A}{D}$ and $\frac{B}{D}$ are rational expressions,

$$\frac{A}{D} + \frac{B}{D} = \frac{A + B}{D} \qquad \text{and} \qquad \frac{A}{D} - \frac{B}{D} = \frac{A - B}{D}$$

EXAMPLE 1   Add: **a.** $\dfrac{x}{8} + \dfrac{3x}{8}$ **b.** $\dfrac{4s - 9}{9t} + \dfrac{7}{9t}$

Strategy   We will add the numerators and write the sum over the common denominator. Then, if possible, we will factor and simplify.

Why   This is the rule for adding rational expressions, such as these, that have the same denominator.

Solution   **a.** The given rational expressions have the same denominator, 8.

$$\frac{x}{8} + \frac{3x}{8} = \frac{x + 3x}{8}$$

$$= \frac{4x}{8} \qquad \text{Combine like terms in the numerator: } x + 3x = 4x.$$
$$\qquad\qquad \text{This result can be simplified.}$$

$$= \frac{\overset{1}{\cancel{4}} \cdot x}{2 \cdot \underset{1}{\cancel{4}}} \qquad \text{Factor 8 as } 4 \cdot 2. \text{ Then simplify by removing a factor equal to 1.}$$

$$= \frac{x}{2}$$

b. The given rational expressions have the same denominator, $9t$.

$$\frac{4s - 9}{9t} + \frac{7}{9t} = \frac{4s - 9 + 7}{9t} \qquad \text{Add the numerators. Write the sum over the common denominator, } 9t.$$

$$= \frac{4s - 2}{9t} \qquad \text{Combine like terms in the numerator: } -9 + 7 = -2.$$

To attempt to simplify the result, we factor the numerator to get $\dfrac{2(2s - 1)}{9t}$. Since the numerator and denominator do not have any common factors, $\dfrac{4s - 2}{9t}$ cannot be simplified. Thus,

$$\frac{4s - 9}{9t} + \frac{7}{9t} = \frac{4s - 2}{9t}$$

> **Self Check 1**   Add: **a.** $\dfrac{2x}{15} + \dfrac{4x}{15}$ **b.** $\dfrac{3m - 8}{23n} + \dfrac{2}{23n}$
>
> **Now Try** ▶   Problems 17 and 19

Caution

We *do not* add rational expressions by adding numerators and adding denominators!

$$\frac{x}{8} + \frac{3x}{8} \bcancel{=} \frac{4x}{16}$$

The same caution applies when subtracting rational expressions.

Notation

The numerator of the result may be written two ways:

Not factored   Factored

$$\frac{4s - 2}{9t} \qquad\qquad \frac{2(2s - 1)}{9t}$$

Check with your instructor to see which form he or she prefers.

EXAMPLE 2   Add: **a.** $\dfrac{3x + 21}{5x + 10} + \dfrac{8x + 1}{5x + 10}$ **b.** $\dfrac{x^2 + 9x - 7}{2x(x - 6)} + \dfrac{x^2 - 9x}{(x - 6)2x}$

Strategy   We will add the numerators and write the sum over the common denominator. Then, if possible, we will factor and simplify.

Why   This is the rule for adding rational expressions that have the same denominator.

Solution   **a.** $\dfrac{3x + 21}{5x + 10} + \dfrac{8x + 1}{5x + 10} = \dfrac{3x + 21 + 8x + 1}{5x + 10}$ Add the numerators. Write the sum over the common denominator, $5x + 10$.

$$= \frac{11x + 22}{5x + 10} \qquad \text{Combine like terms in the numerator:}$$
$$\qquad\qquad\qquad 3x + 8x = 11x \text{ and } 21 + 1 = 22.$$

$$= \frac{11(\cancel{x + 2})}{5(\cancel{x + 2})} \qquad \text{Factor the numerator: The GCF is 11. Factor}$$
$$\qquad\qquad\qquad \text{the denominator: The GCF is 5. Then simplify}$$
$$\qquad\qquad\qquad \text{by removing a factor equal to 1.}$$

$$= \frac{11}{5}$$

Caution

When adding or subtracting rational expressions, always write the result in simplest form by removing any factors common to the numerator and denominator.

b. By the commutative property of multiplication, $2x(x - 6) = (x - 6)2x$. Therefore, the denominators are the same. We add the numerators and write the sum over the common denominator.

$$\frac{x^2 + 9x - 7}{2x(x - 6)} + \frac{x^2 - 9x}{(x - 6)2x} = \frac{x^2 + 9x - 7 + x^2 - 9x}{2x(x - 6)}$$

$$= \frac{2x^2 - 7}{2x(x - 6)} \quad \begin{array}{l} \textcolor{red}{\text{Combine like terms in the numerator:}} \\ \textcolor{red}{x^2 + x^2 = 2x^2 \text{ and } 9x - 9x = 0.} \end{array}$$

Since the numerator, $2x^2 - 7$, does not factor, $\frac{2x^2 - 7}{2x(x - 6)}$ is in simplest form.

Self Check 2 Add: **a.** $\frac{m + 3}{3m - 9} + \frac{m - 9}{3m - 9}$ **b.** $\frac{c^2 - c}{(c - 1)(c + 2)} + \frac{c^2 - 10c}{(c + 2)(c - 1)}$

Now Try ▶ Problems 21 and 23

The method used to subtract rational expressions with like denominators is similar to the method used for adding such expressions.

EXAMPLE 3 Subtract: $\dfrac{x + 6}{x^2 + 4x - 5} - \dfrac{1}{x^2 + 4x - 5}$

Strategy We will subtract the numerators and write the sum over the common denominator. Then, if possible, we will factor and simplify.

Why This is the rule for subtracting rational expressions that have the same denominator.

Solution

$$\frac{x + 6}{x^2 + 4x - 5} - \frac{1}{x^2 + 4x - 5} = \frac{x + 6 - 1}{x^2 + 4x - 5} \quad \begin{array}{l} \textcolor{red}{\text{Subtract the numerators. Write the}} \\ \textcolor{red}{\text{difference over the common}} \\ \textcolor{red}{\text{denominator, } x^2 + 4x - 5.} \end{array}$$

$$= \frac{x + 5}{x^2 + 4x - 5} \quad \begin{array}{l} \textcolor{red}{\text{Combine like terms in the}} \\ \textcolor{red}{\text{numerator: } 6 - 1 = 5.} \end{array}$$

$$= \frac{\overset{1}{\cancel{x + 5}}}{\underset{1}{\cancel{(x + 5)}}(x - 1)} \quad \begin{array}{l} \textcolor{red}{\text{Factor the denominator. Then simplify}} \\ \textcolor{red}{\text{by removing a factor equal to 1.}} \end{array}$$

$$= \frac{1}{x - 1}$$

Self Check 3 Subtract: $\dfrac{n - 3}{n^2 - 16} - \dfrac{1}{n^2 - 16}$

Now Try ▶ Problem 29

Be very careful when subtracting rational expressions. There is great potential for making sign errors.

EXAMPLE 4 Subtract: **a.** $\dfrac{x^2 + 10x}{x + 3} - \dfrac{4x - 9}{x + 3}$ **b.** $\dfrac{x^2}{(x + 7)(x - 8)} - \dfrac{-x^2 + 14x}{(x + 7)(x - 8)}$

Strategy We will use the rule for subtracting rational expressions that have the same denominators. In both cases, it is important to note that the numerator of the second fraction has *two* terms.

Why We must make sure that the entire numerator (not just the first term) of the second fraction is subtracted.

Solution

a. To subtract the numerators, each term of $4x - 9$ must be subtracted from $x^2 + 10x$.

Caution

Don't make this common error by forgetting to write $4x - 9$ within parentheses.

$$\frac{x^2 + 10x - 4x - 9}{x + 3}$$

Notation

A fraction bar is a grouping symbol. Think of parentheses around the terms of the numerator and the denominator.

$$\frac{4x - 9}{x + 3} = \frac{(4x - 9)}{(x + 3)}$$

This $-$ symbol applies to the entire numerator $4x - 9$.

This numerator is written within parentheses to make sure that we subtract both of its terms.

$$\frac{x^2 + 10x}{x + 3} - \frac{4x - 9}{x + 3} = \frac{x^2 + 10x - (4x - 9)}{x + 3}$$

Subtract the numerators. Write the difference over the common denominator.

$$= \frac{x^2 + 10x - 4x + 9}{x + 3}$$

In the numerator, use the distributive property: $-(4x - 9) = -1(4x - 9) = -4x + 9.$

$$= \frac{x^2 + 6x + 9}{x + 3}$$

Combine like terms in the numerator: $10x - 4x = 6x$.

$$= \frac{(x + 3)(x + 3)}{x + 3}$$

To see if the result simplifies, factor the numerator.

$$= \frac{\overset{1}{\cancel{(x + 3)}}(x + 3)}{\underset{1}{\cancel{x + 3}}}$$

Simplify by removing a factor equal to 1.

$$= x + 3$$

b. We subtract the numerators and write the difference over the common denominator.

$$\frac{x^2}{(x + 7)(x - 8)} - \frac{-x^2 + 14x}{(x + 7)(x - 8)} = \frac{x^2 - (-x^2 + 14x)}{(x + 7)(x - 8)}$$

Write the second numerator within parentheses.

$$= \frac{x^2 + x^2 - 14x}{(x + 7)(x - 8)}$$

Use the distributive property: $-(-x^2 + 14x) = x^2 - 14x.$

$$= \frac{2x^2 - 14x}{(x + 7)(x - 8)}$$

In the numerator, combine like terms: $x^2 + x^2 = 2x^2$.

Success Tip

$-(-x^2 + 14x)$ means $-1(-x^2 + 14x)$. The multiplication by -1 changes the sign of each term within the parentheses:

$$-(-x^2 + 14x) = x^2 - 14x$$

In an attempt to simplify, we can factor $2x^2 - 14x$ as $2x(x - 7)$. However, the numerator and denominator have no common factors. The result is in simplest form.

Self Check 4 Subtract: **a.** $\dfrac{x^2 + 3x}{x - 1} - \dfrac{5x - 1}{x - 1}$

b. $\dfrac{3y^2}{(y + 3)(y - 3)} - \dfrac{-3y^2 + y}{(y + 3)(y - 3)}$

Now Try ▶ Problems 33 and 39

2 Find the Least Common Denominator.

We will now discuss two skills that are needed for adding and subtracting rational expressions that have unlike denominators. To begin, let's consider

$$\frac{11}{8x} + \frac{7}{18x^2}$$

To add these expressions, we must express them as equivalent expressions with a common denominator. The **least common denominator (LCD)** is usually the easiest one to use.

Finding the LCD

To find the LCD of a set of rational expressions:
1. Factor each denominator completely.
2. The LCD is a product that uses each different factor obtained in step 1 the greatest number of times it appears in any one factorization.

EXAMPLE 5 Find the LCD of each pair of rational expressions: **a.** $\dfrac{11}{8x}$ and $\dfrac{7}{18x^2}$

b. $\dfrac{20}{x}$ and $\dfrac{4x}{x-9}$

Strategy We will begin by factoring completely the denominator of each rational expression. Then we will form a product using each factor the greatest number of times it appears in any one factorization.

Why Since the LCD must contain the factors of each denominator, we need to write each denominator in factored form.

Solution **a.** $8x = \mathbf{2 \cdot 2 \cdot 2} \cdot x$ Prime factor 8.

$18x^2 = 2 \cdot \mathbf{3 \cdot 3} \cdot x \cdot x$ Prime factor 18. Factor x^2.

$$\begin{array}{l} 2\underline{|8} \quad\quad 2\underline{|18} \\ 2\underline{|4} \quad\quad 3\underline{|9} \\ 2 \quad\quad3 \end{array}$$

The factorizations of $8x$ and $18x^2$ contain the factors 2, 3, and x. The LCD of $\dfrac{11}{8x}$ and $\dfrac{7}{18x^2}$ should contain each factor of $8x$ and $18x^2$ the greatest number of times it appears in any one factorization.

The greatest number of times the factor 2 appears is three times.

The greatest number of times the factor 3 appears is twice.

The greatest number of times the factor x appears is twice.

$$\text{LCD} = \mathbf{2 \cdot 2 \cdot 2 \cdot 3 \cdot 3} \cdot x \cdot x$$
$$= 72x^2$$

The LCD for $\dfrac{11}{8x}$ and $\dfrac{7}{18x^2}$ is $72x^2$.

b. Since the denominators of $\dfrac{20}{x}$ and $\dfrac{4x}{x-9}$ are completely factored, the factor x appears once and the factor $x - 9$ appears once. Thus, the LCD is $x(x-9)$.

Self Check 5 Find the LCD of each pair of rational expressions:

a. $\dfrac{y+7}{6y^3}$ and $\dfrac{7}{75y}$ **b.** $\dfrac{a-3}{a+3}$ and $\dfrac{21}{a}$

Now Try ▶ Problems 41 and 47

EXAMPLE 6 Find the LCD of each pair of rational expressions:

a. $\dfrac{x}{7x+7}$ and $\dfrac{x-2}{5x+5}$ **b.** $\dfrac{6-x}{x^2+8x+16}$ and $\dfrac{15x}{x^2-16}$

Strategy We will begin by factoring completely each binomial and trinomial in the denominators of the rational expressions. Then we will form a product using each factor the greatest number of times it appears in any one factorization.

Why Since the LCD must contain the factors of each denominator, we need to write each denominator in factored form.

Solution **a.** Factor each denominator completely.

$$7x + 7 = \mathbf{7}(x+1)   \text{The GCF is 7.}$$
$$5x + 5 = \mathbf{5}(x+1)   \text{The GCF is 5.}$$

The factorizations of $7x + 7$ and $5x + 5$ contain the factors 7, 5, and $x + 1$. The LCD of $\dfrac{x}{7x+7}$ and $\dfrac{x-2}{5x+5}$ should contain each factor of $7x + 7$ and $5x + 5$ the greatest number of times it appears in any one factorization.

The greatest number of times the factor 7 appears is once.

The greatest number of times the factor 5 appears is once.

The greatest number of times the factor $x + 1$ appears is once.

$$\text{LCD} = \mathbf{7 \cdot 5} \cdot (x+1) = 35(x+1)$$

b. Factor each denominator completely.

$$x^2 + 8x + 16 = (x + 4)(x + 4) \quad \text{Factor the trinomial.}$$
$$x^2 - 16 = (x + 4)(x - 4) \quad \text{Factor the difference of two squares.}$$

The factorizations of $x^2 + 8x + 16$ and $x^2 - 16$ contain the factors $x + 4$ and $x - 4$.

The greatest number of times the factor x + 4 appears is twice.
The greatest number of times the factor x − 4 appears is once.

$$\text{LCD} = (x + 4)(x + 4)(x - 4) = (x + 4)^2(x - 4)$$

Self Check 6 Find the LCD: **a.** $\dfrac{x^3}{x^2 - 6x}$ and $\dfrac{25x}{2x - 12}$
b. $\dfrac{m + 1}{m^2 - 9}$ and $\dfrac{6m^2}{m^2 - 6m + 9}$

Now Try ▶ Problems 51 and 57

3 Build Rational Expressions into Equivalent Expressions.

Recall from Chapter 1 that writing a fraction as an equivalent fraction with a larger denominator is called **building the fraction.** For example, to write $\frac{3}{5}$ as an equivalent fraction with a denominator of 35, we multiply it by 1 in the form of $\frac{7}{7}$:

$$\frac{3}{5} = \frac{3}{5} \cdot \frac{7}{7} = \frac{21}{35} \quad \begin{array}{l}\text{Multiply the numerators.} \\ \text{Multiply the denominators.}\end{array}$$

It is important to note that multiplying $\frac{3}{5}$ by $\frac{7}{7}$ changes its appearance but not its value, because we are multiplying it by 1.

To add and subtract rational expressions with different denominators, we must write them as equivalent expressions having a common denominator. To do so, we build rational expressions.

Building Rational Expressions

To build a rational expression, multiply it by 1 in the form of $\frac{c}{c}$, where c is any nonzero number or expression.

EXAMPLE 7 Write each rational expression as an equivalent expression with the indicated denominator:
a. $\dfrac{7}{15n}$, denominator $30n^3$ **b.** $\dfrac{6x}{x + 4}$, denominator $(x + 4)(x - 4)$

Strategy We will begin by asking, "By what must we multiply the given denominator to get the required denominator?"

Why The answer to that question helps us determine the form of 1 to be used to build an equivalent rational expression.

Solution **a.** We need to multiply the denominator of $\frac{7}{15n}$ by $2n^2$ to obtain a denominator of $30n^3$. It follows that $\frac{2n^2}{2n^2}$ is the form of 1 that should be used to build an equivalent expression.

The Language of Algebra

We say that $\frac{7}{15n}$ and $\frac{14n^2}{30n^3}$ are **equivalent expressions** because they have the same value for all values of n, except those that make either denominator 0.

$$\frac{7}{15n} = \frac{7}{15n} \cdot \frac{2n^2}{2n^2} \quad \text{Multiply the given rational expression by 1, in the form of } \frac{2n^2}{2n^2}.$$
$$= \frac{14n^2}{30n^3} \quad \begin{array}{l}\text{Multiply the numerators.} \\ \text{Multiply the denominators.}\end{array}$$

b. We need to multiply the denominator of $\frac{6x}{x+4}$ by $x-4$ to obtain a denominator of $(x+4)(x-4)$. It follows that $\frac{x-4}{x-4}$ is the form of 1 that should be used to build an equivalent expression.

$$\frac{6x}{x+4} = \frac{6x}{x+4} \cdot \frac{x-4}{x-4}$$ Multiply the given rational expression by 1, in the form of $\frac{x-4}{x-4}$.

$$= \frac{6x(x-4)}{(x+4)(x-4)}$$ Multiply the numerators. Multiply the denominators.

$$= \frac{6x^2 - 24x}{(x+4)(x-4)}$$ In the numerator, distribute the multiplication by 6x. Leave the denominator in factored form.

To get this answer, we multiplied the factors in the numerator to obtain a polynomial in unfactored form: $6x^2 - 24x$. However, we left the denominator in factored form. This approach is beneficial in the next section when we add and subtract rational expressions with unlike denominators.

> **Self Check 7** Write each rational expression as an equivalent expression with the indicated denominator: **a.** $\frac{7}{20m^2}$, denominator $60m^3$
>
> **b.** $\frac{2c}{c+1}$, denominator $(c+1)(c+3)$
>
> **Now Try** ▶ Problems 65 and 67

EXAMPLE 8 Write $\frac{x+1}{x^2+6x}$ as an equivalent expression with a denominator of $x(x+6)(x+2)$.

Strategy We will begin by factoring the denominator of $\frac{x+1}{x^2+6x}$. Then we will compare the factors of $x^2 + 6x$ to those of $x(x+6)(x+2)$.

Why This comparison will enable us to answer the question, "By what must we multiply $x^2 + 6x$ to obtain $x(x+6)(x+2)$?"

Solution We factor the denominator to determine what factors are missing.

$$\frac{x+1}{x^2+6x} = \frac{x+1}{x(x+6)}$$ Factor out the GCF, x, from $x^2 + 6x$.

It is now apparent that we need to multiply the denominator by $x + 2$ to obtain a denominator of $x(x+6)(x+2)$. It follows that $\frac{x+2}{x+2}$ is the form of 1 that should be used to build an equivalent expression.

> **Success Tip**
>
> When building rational expressions, write the numerator of the result as a polynomial in unfactored form. Write the denominator in factored form.

$$\frac{x+1}{x^2+6x} = \frac{x+1}{x(x+6)} \cdot \frac{x+2}{x+2}$$ Multiply the given rational expression by 1, in the form of $\frac{x+2}{x+2}$.

$$= \frac{(x+1)(x+2)}{x(x+6)(x+2)}$$ Multiply the numerators. Multiply the denominators.

$$= \frac{x^2 + 3x + 2}{x(x+6)(x+2)}$$ In the numerator, use the FOIL method to multiply $(x+1)(x+2)$. Leave the denominator in factored form.

> **Self Check 8** Write $\frac{x-3}{x^2-4x}$ as an equivalent expression with a denominator of $x(x-4)(x+8)$.
>
> **Now Try** ▶ Problem 71

SECTION 7.3 ▶ STUDY SET

VOCABULARY

Fill in the blanks.

1. The rational expressions $\frac{7}{6n}$ and $\frac{n+1}{6n}$ have the common _____ $6n$.

2. The _____ common denominator of $\frac{x-8}{x+6}$ and $\frac{6-5x}{x}$ is $x(x+6)$.

3. To _____ a rational expression, we multiply it by a form of 1. For example: $\frac{2}{n^2} \cdot \frac{8n}{8n} = \frac{16n}{8n^3}$

4. $\frac{16n}{8n^3}$ and $\frac{2}{n^2}$ are _____ expressions. They have the same value for all values of n, except for $n = 0$.

CONCEPTS

Fill in the blanks.

5. To add or subtract rational expressions that have the same denominator, add or subtract the _____, and write the sum or difference over the common _____. In symbols, $\frac{A}{D} + \frac{B}{D} = $ _____ and $\frac{A}{D} - \frac{B}{D} = $ _____.

6. When adding or subtracting rational expressions, always write the result in _____ form by removing any factors common to the numerator and denominator.

7. The sum of two rational expressions is $\frac{4x+4}{5(x+1)}$. Factor the numerator and then simplify the result.

8. Factor each denominator completely.

   a. $\frac{17}{40x^2}$ b. $\frac{x+25}{2x^2-6x}$

9. Consider the following factorizations.

   $18x - 36 = 2 \cdot 3 \cdot 3 \cdot (x-2)$

   $3x - 6 = 3(x-2)$

   a. What is the greatest number of times the factor 3 appears in any one factorization?

   b. What is the greatest number of times the factor $x-2$ appears in any one factorization?

10. Fill in the blanks. To write $\frac{x}{x-9}$ as an equivalent rational expression with a denominator of $3x(x-9)$, we need to multiply the denominator by ___. It follows that $\frac{}{}$ is the form of 1 that should be used to build $\frac{x}{x-9}$.

NOTATION

Complete the solution.

11. $\dfrac{6a-1}{4a+1} + \dfrac{2a+3}{4a+1} = \dfrac{6a - 1 + (2a +)}{}$

   $= \dfrac{8a + }{4a+1}$

   $= \dfrac{2()}{4a+1}$

   $= $

12. The type of multiplication that is used to build rational expressions is shown below. Fill in the blanks.

   a. $\dfrac{4x}{5} \cdot \dfrac{2}{2} = \dfrac{}{10}$ b. $\dfrac{3}{t} \cdot \dfrac{t-2}{t-2} = \dfrac{}{t(t-2)}$

   c. $\dfrac{m+1}{m-3} \cdot \dfrac{m-5}{m-5} = \dfrac{}{(m-3)(m-5)}$

GUIDED PRACTICE

Add and simplify the result, if possible. See Example 1.

13. $\dfrac{9}{x} + \dfrac{2}{x}$ 14. $\dfrac{4}{s} + \dfrac{4}{s}$

15. $\dfrac{x}{18} + \dfrac{5}{18}$ 16. $\dfrac{7}{10} + \dfrac{3y}{10}$

17. $\dfrac{a-5}{3a^3} + \dfrac{5}{3a^3}$ 18. $\dfrac{b^3-8}{10b^4} + \dfrac{8}{10b^4}$

19. $\dfrac{x+3}{2y} + \dfrac{x+5}{2y}$ 20. $\dfrac{y+2}{10z} + \dfrac{y+4}{10z}$

Add and simplify the result, if possible. See Example 2.

21. $\dfrac{2}{r^2-3r-10} + \dfrac{r}{r^2-3r-10}$

22. $\dfrac{1}{h^2-4h-5} + \dfrac{h}{h^2-4h-5}$

23. $\dfrac{3x-5}{x-2} + \dfrac{6x-13}{x-2}$ 24. $\dfrac{8x-7}{x+3} + \dfrac{2x+37}{x+3}$

Subtract and simplify the result, if possible. See Example 3.

25. $\dfrac{2x}{25} - \dfrac{x}{25}$ 26. $\dfrac{16c}{11} - \dfrac{4c}{11}$

27. $\dfrac{m-1}{6m^2} - \dfrac{5}{6m^2}$ 28. $\dfrac{c+7}{4c^4} - \dfrac{3}{4c^4}$

29. $\dfrac{t}{t^2+t-2} - \dfrac{1}{t^2+t-2}$

30. $\dfrac{r}{r^2-2r-3} - \dfrac{3}{r^2-2r-3}$

31. $\dfrac{11w+6}{3w(w-9)} - \dfrac{11w}{3w(w-9)}$

32. $\dfrac{y+8}{2y(y-14)} - \dfrac{y}{2y(y-14)}$

Subtract and simplify the result, if possible. See Example 4.

33. $\dfrac{3y-2}{2y+6} - \dfrac{2y-5}{2y+6}$ 34. $\dfrac{5x+8}{3x+15} - \dfrac{3x-2}{3x+15}$

35. $\dfrac{6x^2}{3x+2} - \dfrac{11x+10}{3x+2}$ 36. $\dfrac{8a^2}{2a+5} - \dfrac{4a^2+25}{2a+5}$

37. $\dfrac{6x-5}{3xy} - \dfrac{3x-5}{3xy}$ 38. $\dfrac{7x+7}{5y} - \dfrac{2x+7}{5y}$

39. $\dfrac{2-p}{p^2-p} - \dfrac{-p+2}{p^2-p}$ 40. $\dfrac{2-7n}{n^2+5} - \dfrac{-7n+2}{n^2+5}$

Find the LCD of each pair of rational expressions. See Example 5.

41. $\dfrac{1}{2x}, \dfrac{9}{6x}$

42. $\dfrac{4}{9y}, \dfrac{11}{3y}$

43. $\dfrac{33}{15a^3}, \dfrac{9}{10a}$

44. $\dfrac{m-21}{12m^4}, \dfrac{m+1}{18m}$

45. $\dfrac{35}{3a^2b}, \dfrac{23}{a^2b^3}$

46. $\dfrac{27}{c^3d}, \dfrac{17}{2c^2d^3}$

47. $\dfrac{8}{c}, \dfrac{8-c}{c+2}$

48. $\dfrac{d^2-5}{d+9}, \dfrac{d-3}{d}$

Find the LCD of each pair of rational expressions. See Example 6.

49. $\dfrac{3x+1}{3x-3}, \dfrac{3x}{4x-4}$

50. $\dfrac{b+1}{5b-10}, \dfrac{b}{6b-12}$

51. $\dfrac{b-9}{4b+8}, \dfrac{b}{6}$

52. $\dfrac{b^2-b}{10b-15}, \dfrac{11b}{10}$

53. $\dfrac{6-k}{2k+4}, \dfrac{11}{8k}$

54. $\dfrac{5m+6}{4m+12}, \dfrac{7}{6m}$

55. $\dfrac{-2x}{x^2-1}, \dfrac{5x}{x+1}$

56. $\dfrac{7-y^2}{y^2-4}, \dfrac{y-49}{y+2}$

57. $\dfrac{4x-5}{x^2-4x-5}, \dfrac{3x+1}{x^2-25}$

58. $\dfrac{44}{s^2-9}, \dfrac{s+9}{s^2-s-6}$

59. $\dfrac{5n^2-16}{2n^2+13n+20}, \dfrac{3n^2}{n^2+8n+16}$

60. $\dfrac{4y+25}{y^2+10y+25}, \dfrac{y^2-7}{2y^2+17y+35}$

Build each rational expression into an equivalent expression with the given denominator. See Example 7.

61. $\dfrac{5}{r}$; $10r$

62. $\dfrac{4}{y}$; $7y$

63. $\dfrac{8}{x}$; x^2y

64. $\dfrac{7}{y}$; xy^2

65. $\dfrac{9}{4b}$; $12b^2$

66. $\dfrac{7}{6c}$; $30c^2$

67. $\dfrac{3x}{x+1}$; $(x+1)^2$

68. $\dfrac{5y}{y-2}$; $(y-2)^2$

Build each rational expression into an equivalent expression with the given denominator. See Example 8.

69. $\dfrac{x+9}{x^2+5x}$; $x^2(x+5)$

70. $\dfrac{a+11}{a^2+9a}$; $a^2(a+9)$

71. $\dfrac{t+5}{4t+8}$; $4(t+2)(t+9)$

72. $\dfrac{x+7}{3x-15}$; $3(x-5)(x+10)$

73. $\dfrac{y+3}{y^2-5y+6}$; $4y(y-2)(y-3)$

74. $\dfrac{3x-4}{x^2+3x+2}$; $8x(x+1)(x+2)$

75. $\dfrac{12-h}{h^2-81}$; $3(h+9)(h-9)$

76. $\dfrac{m^2}{m^2-100}$; $9(m+10)(m-10)$

TRY IT YOURSELF

Perform the operations. Then simplify, if possible.

77. $\dfrac{3t}{t^2-8t+7} - \dfrac{3}{t^2-8t+7}$

78. $\dfrac{10x}{x^2-2x+1} - \dfrac{10}{x^2-2x+1}$

79. $\dfrac{c}{c^2-d^2} - \dfrac{d}{c^2-d^2}$

80. $\dfrac{b}{b^2-4} - \dfrac{2}{b^2-4}$

81. $\dfrac{a^2+a}{4a^2-8a} + \dfrac{2a^2-7a}{4a^2-8a}$

82. $\dfrac{3b^2+16b}{6b^2+9b} + \dfrac{7b^2-b}{6b^2+9b}$

83. $\dfrac{17a}{2a+4} - \dfrac{7a}{2a+4}$

84. $\dfrac{10b}{3b-18} - \dfrac{4b}{3b-18}$

85. $\dfrac{8}{9-3x^2} - \dfrac{-6x+8}{9-3x^2}$

86. $\dfrac{5}{10-5t^2} - \dfrac{-15t+5}{10-5t^2}$

87. $\dfrac{11n}{(n+4)(n-2)} - \dfrac{4n-1}{(n-2)(n+4)}$

88. $\dfrac{1}{(t-1)(t+1)} - \dfrac{6-t}{(t+1)(t-1)}$

89. $\dfrac{5r-27}{3r^2-9r} + \dfrac{4r}{3r^2-9r}$

90. $\dfrac{9a}{5a^2+25a} + \dfrac{a+50}{5a^2+25a}$

91. $\dfrac{11}{36y} + \dfrac{9}{36y}$

92. $\dfrac{13}{24w} + \dfrac{17}{24w}$

93. $\dfrac{-4x}{3x^2-7x+2} - \dfrac{-3x-2}{3x^2-7x+2}$

94. $\dfrac{-3c}{5c^2-16c+3} - \dfrac{-2c-3}{5c^2-16c+3}$

95. $\dfrac{3x^2}{x+1} - \dfrac{-x+2}{x+1}$

96. $\dfrac{8b^2}{3b-2} - \dfrac{-b^2+4}{3b-2}$

Look Alikes . . .

97. a. $\dfrac{t}{12} + \dfrac{5t}{12}$ b. $\dfrac{t}{12} \cdot \dfrac{5t}{12}$ c. $\dfrac{t}{12} \div \dfrac{5t}{12}$

98. a. $\dfrac{x}{9} + \dfrac{2x}{9}$ b. $\dfrac{x}{9} \cdot \dfrac{2x}{9}$ c. $\dfrac{x}{9} \div \dfrac{2x}{9}$

99. a. $\dfrac{m+6}{5} - \dfrac{m+2}{5}$ b. $\dfrac{m+6}{5} \cdot \dfrac{m+2}{5}$

c. $\dfrac{m+6}{5} \div \dfrac{m+2}{5}$

100. **a.** $\dfrac{2r + 9}{2r} - \dfrac{1}{2r}$ **b.** $\dfrac{2r + 9}{2r} \cdot \dfrac{1}{2r}$

 c. $\dfrac{2r + 9}{2r} \div \dfrac{1}{2r}$

APPLICATIONS

101. **Geometry.** What is the difference of the length and width of the rectangle?

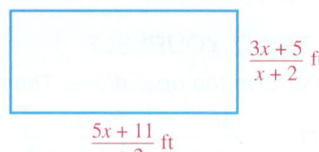

$\dfrac{3x + 5}{x + 2}$ ft

$\dfrac{5x + 11}{x + 2}$ ft

102. **Geometry.** What is the perimeter of the rectangle in Problem 101?

WRITING

103. Explain how to add fractions with the same denominator.

104. Explain how to find a least common denominator.

105. Explain the error in the following solution:

$$\frac{2x + 3}{x + 5} - \frac{x + 2}{x + 5} = \frac{2x + 3 - x + 2}{x + 5}$$

$$= \frac{x + 5}{x + 5}$$

$$= 1$$

106. Explain the error in the following solution:

$$\frac{y + 4}{y} - \frac{1}{y} = \frac{y + 4 - 1}{y + y}$$

$$= \frac{y + 3}{2y}$$

107. **a.** Explain why the LCD of $\frac{5}{h^2}$ and $\frac{3}{h}$ is h^2 and not h^3.

 b. Explain why the LCD of $\frac{1}{x}$ and $\frac{1}{x - 9}$ is $x(x - 9)$ and not $x - 9$.

108. Explain how multiplication by 1 is used to build a rational expression. Give an example.

REVIEW

Give the formula for . . .

109. **a.** simple interest

 b. the area of a triangle

 c. the perimeter of a rectangle

110. **a.** the slope of a line

 b. distance traveled

 c. the area of a circle

CHALLENGE PROBLEMS

Perform the operations. Simplify the result, if possible.

111. $\dfrac{3xy}{x - y} - \dfrac{x(3y - x)}{x - y} - \dfrac{x(x - y)}{x - y}$

112. $\dfrac{9t^3 - 12t^2}{27t^3 - 64} - \dfrac{-3t + 4}{27t^3 - 64}$

113. $\dfrac{2a^2 + 2}{a^3 + 8} + \dfrac{a^3 + a}{a^3 + 8}$

114. Find the LCD of

$$\frac{2}{a^3 + 8}, \; \frac{a}{a^2 - 4}, \; \text{and} \; \frac{2a + 5}{a^3 - 8}.$$

SECTION 7.4

Adding and Subtracting with Unlike Denominators

OBJECTIVES

1 Add and subtract rational expressions that have unlike denominators.

2 Add and subtract rational expressions that have denominators that are opposites.

ARE YOU READY?

The following problems review some basic skills that are needed when adding and subtracting rational expressions with unlike denominators.

1. Write $\dfrac{5}{4a}$ as an equivalent expression with a denominator of $16a^2$.

2. Factor completely: $36x^3$

3. Add: $\dfrac{6x}{9} + \dfrac{5x}{9}$

4. What is the opposite of $8 - t$?

We have discussed a method for finding the least common denominator (LCD) of two rational expressions. We have also built rational expressions into equivalent expressions having a given denominator. We will now use these skills to add and subtract rational expressions with unlike denominators.

1 Add and Subtract Rational Expressions That Have Unlike Denominators.

The following steps summarize how to add or subtract rational expressions that have different denominators.

Adding and Subtracting Rational Expressions That Have Unlike Denominators	1. Find the LCD. 2. Rewrite each rational expression as an equivalent expression with the LCD as the denominator. To do so, build each fraction using a form of 1 that involves any factor(s) needed to obtain the LCD. 3. Add or subtract the numerators and write the sum or difference over the LCD. 4. Simplify the result, if possible.

EXAMPLE 1   Add: $\dfrac{9x}{7} + \dfrac{3x}{5}$

Strategy   We will use the procedure for adding rational expressions that have unlike denominators. The first step is to determine the LCD.

Why   If we are to add (or subtract) rational expressions, their denominators must be the same. Since the denominators of these rational expressions are different, we cannot add them in their present form.

$$\underset{\text{sevenths}}{}\quad \dfrac{9x}{7} + \dfrac{3x}{5}\quad \underset{\text{fifths}}{}$$

Not the same number

Solution   **Step 1:**   The denominators are 7 and 5. The LCD is $7 \cdot 5 = 35$.

Step 2:   We need to multiply the denominator of $\frac{9x}{7}$ by 5 and we need to multiply the denominator of $\frac{3x}{5}$ by 7 to obtain the LCD, 35. It follows that $\frac{5}{5}$ and $\frac{7}{7}$ are the forms of 1 that should be used to write the equivalent rational expressions.

Caution

In Step 2, don't simplify $\frac{45x}{35}$ and $\frac{21x}{35}$, because that will take you back to the original rational expressions and you will lose the common denominator.

$$\dfrac{9x}{7} + \dfrac{3x}{5} = \dfrac{9x}{7} \cdot \dfrac{5}{5} + \dfrac{3x}{5} \cdot \dfrac{7}{7} \qquad \begin{array}{l}\text{Build the rational expressions so}\\ \text{that each has a denominator of 35.}\end{array}$$

$$= \dfrac{45x}{35} + \dfrac{21x}{35} \qquad \begin{array}{l}\text{Multiply the numerators: } 9x \cdot 5 = 45x \text{ and } 3x \cdot 7 = 21x.\\ \text{Multiply the denominators. Now the denominators are like.}\end{array}$$

Step 3:
$$= \dfrac{45x + 21x}{35} \qquad \begin{array}{l}\text{Add the numerators. Write the sum}\\ \text{over the common denominator, 35.}\end{array}$$

$$= \dfrac{66x}{35} \qquad \begin{array}{l}\text{Combine like terms in the numerator: } 45x + 21x = 66x.\end{array}$$

Step 4:   Since 66 and 35 have no common factor other than 1, the result cannot be simplified.

Self Check 1   Add: $\dfrac{y}{2} + \dfrac{6y}{7}$

Now Try ▶   Problem 13

EXAMPLE 2 Subtract: $\dfrac{13}{18b^2} - \dfrac{1}{24b}$

Strategy We will use the procedure for subtracting rational expressions that have unlike denominators. The first step is to determine the LCD.

Why If we are to subtract rational expressions, their denominators must be the same. Since the denominators of these rational expressions are different, we cannot subtract them in their present form.

Solution **Step 1:** To find the LCD, we form a product that uses each different factor of $18b^2$ and $24b$ the greatest number of times it appears in any one factorization.

$$\left. \begin{array}{l} 18b^2 = 2 \cdot 3 \cdot 3 \cdot b \cdot b \\ 24b = 2 \cdot 2 \cdot 2 \cdot 3 \cdot b \end{array} \right\} \quad \text{LCD} = 2 \cdot 2 \cdot 2 \cdot 3 \cdot 3 \cdot b \cdot b = 72b^2$$

Notation

When checking your answers with those in the back of the book, remember that the results can often be presented in several equivalent forms. For example, this result can also be expressed as

$$\dfrac{-3b + 52}{72b^2}$$

Step 2: We need to multiply $18b^2$ by 4 to obtain $72b^2$, and $24b$ by $3b$ to obtain $72b^2$. It follows that we should use $\dfrac{4}{4}$ and $\dfrac{3b}{3b}$ to build the equivalent rational expressions.

$$\dfrac{13}{18b^2} - \dfrac{1}{24b} = \dfrac{13}{18b^2} \cdot \dfrac{4}{4} - \dfrac{1}{24b} \cdot \dfrac{3b}{3b} \qquad \text{Build the rational expressions so that each has a denominator of } 72b^2.$$

$$= \dfrac{52}{72b^2} - \dfrac{3b}{72b^2} \qquad \text{Multiply the numerators. Multiply the denominators. Now the denominators are like.}$$

Step 3:
$$= \dfrac{52 - 3b}{72b^2} \qquad \text{Subtract the numerators. Write the difference over the common denominator, } 72b^2.$$

Step 4: Since $52 - 3b$ does not factor, the result cannot be simplified.

Self Check 2 Subtract: $\dfrac{5}{21z^2} - \dfrac{3}{28z}$

Now Try ▶ Problem 21

EXAMPLE 3 Add: $\dfrac{3}{2x + 18} + \dfrac{27}{x^2 - 81}$

Strategy We use the procedure for adding rational expressions when the denominators are binomials. The first step is to find the LCD.

Why Since the denominators are different, we cannot add these rational expressions in their present form.

Solution After factoring the denominators, we see that the greatest number of times each of the factors 2, $x + 9$, and $x - 9$ appear in any one of the factorizations is once.

$$\left. \begin{array}{l} 2x + 18 = 2(x + 9) \\ x^2 - 81 = (x + 9)(x - 9) \end{array} \right\} \quad \text{LCD} = 2(x + 9)(x - 9)$$

Since we need to multiply $2(x + 9)$ by $x - 9$ to obtain the LCD and $(x + 9)(x - 9)$ by 2 to obtain the LCD, $\dfrac{x - 9}{x - 9}$ and $\dfrac{2}{2}$ are the forms of 1 to use to build the equivalent rational expressions.

$$\frac{3}{2x + 18} + \frac{27}{x^2 - 81} = \frac{3}{2(x + 9)} + \frac{27}{(x + 9)(x - 9)}$$

Write each denominator in factored form.

$$= \frac{3}{2(x + 9)} \cdot \frac{x - 9}{x - 9} + \frac{27}{(x + 9)(x - 9)} \cdot \frac{2}{2}$$

Build the expressions so that each has a denominator of $2(x + 9)(x - 9)$.

Multiply the numerators to prepare to combine like terms later.

Don't multiply the denominators. Leave them in factored form to possibly simplify the result later.

$$= \frac{3x - 27}{2(x + 9)(x - 9)} + \frac{54}{2(x + 9)(x - 9)}$$

Multiply: $3(x - 9) = 3x - 27$ and $27 \cdot 2 = 54$. Now the denominators are like.

Although it is not required, the factors of each denominator are written in the same order.

$$= \frac{3x - 27 + 54}{2(x + 9)(x - 9)}$$

Add the numerators. Write the sum over the common denominator, $2(x + 9)(x - 9)$.

$$= \frac{3x + 27}{2(x + 9)(x - 9)}$$

Combine like terms in the numerator: $-27 + 54 = 27$.

Caution

Always write the result in simplest form by removing any factors common to the numerator and denominator.

$$= \frac{\overset{1}{3(x + 9)}}{2\underset{1}{(x + 9)}(x - 9)}$$

Factor the numerator. Then simplify the expression by removing a factor equal to 1.

$$= \frac{3}{2(x - 9)}$$

This is the result in simplest form.

Self Check 3   Add: $\dfrac{2}{5x + 25} + \dfrac{4}{x^2 - 25}$

Now Try ▶ Problem 29

EXAMPLE 4   Subtract: $\dfrac{x}{x - 1} - \dfrac{x - 6}{x - 4}$

Strategy   We use the same procedure for subtracting rational expressions when the denominators are binomials. The first step is to find the LCD.

Why   Since the denominators are different, we cannot subtract these rational expressions in their present form.

Solution   The denominators of $\frac{x}{x - 1}$ and $\frac{x - 6}{x - 4}$ are completely factored. The factor $x - 1$ appears once and the factor $x - 4$ appears once. Thus, the LCD $= (x - 1)(x - 4)$.

We need to multiply the first denominator by $x - 4$ to obtain the LCD and the second denominator by $x - 1$ to obtain the LCD. It follows that $\frac{x - 4}{x - 4}$ and $\frac{x - 1}{x - 1}$ are the forms of 1 to use to build the equivalent rational expressions.

$$\frac{x}{x - 1} - \frac{x - 6}{x - 4} = \frac{x}{x - 1} \cdot \frac{x - 4}{x - 4} - \frac{x - 6}{x - 4} \cdot \frac{x - 1}{x - 1}$$

Build the rational expressions so that each has a denominator of $(x - 1)(x - 4)$.

Multiply the numerators to prepare to combine like terms later.

Don't multiply the denominators. Leave them in factored form to possibly simplify later.

$$= \frac{x^2 - 4x}{(x - 1)(x - 4)} - \frac{x^2 - 7x + 6}{(x - 4)(x - 1)}$$

Multiply: $x(x - 4) = x^2 - 4x$.
Multiply: $(x - 6)(x - 1) = x^2 - 7x + 6$.
Now the denominators are like.

By the commutative property of multiplication, these are like denominators.

$$= \frac{x^2 - 4x - (x^2 - 7x + 6)}{(x - 1)(x - 4)}$$

Subtract the numerators. Remember the parentheses. Write the difference over the common denominator, $(x - 1)(x - 4)$.

$$= \frac{x^2 - 4x - x^2 + 7x - 6}{(x - 1)(x - 4)}$$

In the numerator, use the distributive property: $-(x^2 - 7x + 6) = -1(x^2 - 7x + 6) = -x^2 + 7x - 6$.

$$= \frac{3x - 6}{(x - 1)(x - 4)}$$

Combine like terms in the numerator: $x^2 - x^2 = 0$ and $-4x + 7x = 3x$.

Although, the numerator factors as $3(x - 2)$, the numerator and denominator do not have a common factor. Therefore, the result is in simplest form.

Self Check 4   Subtract: $\dfrac{x}{x + 9} - \dfrac{x - 7}{x + 8}$

Now Try ▶ Problem 37

EXAMPLE 5   Subtract: $\dfrac{m}{m^2 + 5m + 6} - \dfrac{2}{m^2 + 3m + 2}$

Strategy   We use the same procedure for subtracting rational expressions when the denominators are trinomials. The first step is to find the LCD.

Why   Since the denominators are different, we cannot subtract these rational expressions in their present form.

Solution   Factor each denominator and form the LCD.

$$\left. \begin{array}{l} m^2 + 5m + 6 = (m + 2)(m + 3) \\ m^2 + 3m + 2 = (m + 2)(m + 1) \end{array} \right\} \quad \text{LCD} = (m + 2)(m + 3)(m + 1)$$

Examining the factored forms, we see that the first denominator must be multiplied by $m + 1$, and the second must be multiplied by $m + 3$ to obtain the LCD. To build the expressions, we will use $\frac{m + 1}{m + 1}$ and $\frac{m + 3}{m + 3}$.

$$\frac{m}{m^2 + 5m + 6} - \frac{2}{m^2 + 3m + 2}$$

$$= \frac{m}{(m + 2)(m + 3)} - \frac{2}{(m + 2)(m + 1)}$$

Write each denominator in factored form.

$$= \frac{m}{(m + 2)(m + 3)} \cdot \frac{m + 1}{m + 1} - \frac{2}{(m + 2)(m + 1)} \cdot \frac{m + 3}{m + 3}$$

Build each expression so that it has a denominator of $(m + 2)(m + 3)(m + 1)$.

Multiply the numerators. ⟶

Leave the denominators in factored form.

$$= \frac{m^2 + m}{(m + 2)(m + 3)(m + 1)} - \frac{2m + 6}{(m + 2)(m + 1)(m + 3)}$$

Multiply: $m(m + 1) = m^2 + m$ and $2(m + 3) = 2m + 6$. Now the denominators are like.

By the commutative property of multiplication, these are like denominators.

$$= \frac{m^2 + m - (2m + 6)}{(m + 2)(m + 3)(m + 1)}$$

Subtract the numerators. Remember the parentheses. Write the difference over the common denominator, $(m + 2)(m + 3)(m + 1)$.

$$= \frac{m^2 + m - 2m - 6}{(m + 2)(m + 3)(m + 1)}$$

Use the distributive property: $-(2m + 6) = -1(2m + 6) = -2m - 6$.

$$= \frac{m^2 - m - 6}{(m + 2)(m + 3)(m + 1)}$$

Combine like terms in the numerator: $m - 2m = -m$.

$$= \frac{(m - 3)\overset{1}{\cancel{(m + 2)}}}{\underset{1}{\cancel{(m + 2)}}(m + 3)(m + 1)}$$ Factor the numerator and simplify the expression by removing a factor equal to 1.

$$= \frac{m - 3}{(m + 3)(m + 1)}$$

Self Check 5   Subtract: $\dfrac{b}{b^2 - 2b - 8} - \dfrac{6}{b^2 + b - 20}$

Now Try ▶   Problem 45

EXAMPLE 6   Add: $\dfrac{4b}{a - 5} + b$

Strategy   We will begin by writing the second addend, b, as $\frac{b}{1}$ and then find the LCD.

Why   To add b to the rational expression, $\frac{4b}{a - 5}$, we must rewrite b as a rational expression.

Solution   The LCD of $\frac{4b}{a - 5}$ and $\frac{b}{1}$ is $1(a - 5)$, or simply $a - 5$. Since we must multiply the denominator of $\frac{b}{1}$ by $a - 5$ to obtain the LCD, we will use $\frac{a - 5}{a - 5}$ to write an equivalent rational expression.

$$\frac{4b}{a - 5} + b = \frac{4b}{a - 5} + \frac{b}{1} \cdot \frac{a - 5}{a - 5}$$ Build $\frac{b}{1}$ so that it has a denominator of $a - 5$.

$$= \frac{4b}{a - 5} + \frac{ab - 5b}{a - 5}$$ Multiply numerators: $b(a - 5) = ab - 5b$. Multiply denominators: $1(a - 5) = a - 5$. Now the denominators are like.

$$= \frac{4b + ab - 5b}{a - 5}$$ Add the numerators. Write the sum over the common denominator.

$$= \frac{ab - b}{a - 5}$$ Combine like terms in the numerator: $4b - 5b = -b$.

Although the numerator factors as $b(a - 1)$, the numerator and denominator do not have a common factor. Therefore, the result is in simplest form.

Self Check 6   Add: $\dfrac{10y}{n + 4} + y$

Now Try ▶   Problem 47

> **Success Tip**
>
> Since the denominator of $\frac{4b}{a - 5}$ is the LCD, we *do not* need to build it by multiplying it by a form of 1. The step below is unnecessary.
>
> $$\frac{4b}{a - 5} \cdot \frac{1}{1}$$

---

2   Add and Subtract Rational Expressions That Have Denominators That Are Opposites.

Recall that two polynomials are **opposites** if their terms are the same but they are opposite in sign. For example, $x - 4$ and $4 - x$ are opposites. If we multiply one of these binomials by -1, the subtraction is reversed, and the result is the other binomial.

$$-1(x - 4) = -x + 4$$
$$= 4 - x$$ Write the expression with 4 first.

$$-1(4 - x) = -4 + x$$
$$= x - 4$$ Write the expression with x first.

These results suggest that when a polynomial is multiplied by -1, the result is its opposite. This fact can be used when adding or subtracting rational expressions whose denominators are opposites.

Rational Expressions with Opposite Denominators

When adding or subtracting two rational expressions whose denominators are opposites, multiply either expression by 1 in the form of $\frac{-1}{-1}$ to obtain a common denominator.

EXAMPLE 7

Subtract: $\dfrac{x}{x-7} - \dfrac{1}{7-x}$

Strategy We note that the denominators are opposites. Either can serve as the LCD; we will choose $x - 7$. To obtain a common denominator, we will multiply $\frac{1}{7-x}$ by $\frac{-1}{-1}$.

Why When $7 - x$ is multiplied by -1, the subtraction is reversed, and the result is $x - 7$.

Solution We must multiply the denominator of $\frac{1}{7-x}$ by -1 to obtain the LCD. It follows that $\frac{-1}{-1}$ should be the form of 1 that is used to write an equivalent rational expression.

Success Tip

Either denominator can serve as the LCD. However, it is common to have a result whose denominator is written in descending powers of the variable. Therefore, we chose $x - 7$, as opposed to $7 - x$, as the LCD.

$$\dfrac{x}{x-7} - \dfrac{1}{7-x} = \dfrac{x}{x-7} - \dfrac{1}{7-x} \cdot \dfrac{-1}{-1}$$

Build $\frac{1}{7-x}$ so that it has a denominator of $x - 7$.

$$= \dfrac{x}{x-7} - \dfrac{-1}{-7+x}$$

Multiply the numerators: $1(-1) = -1$. Multiply the denominators.

$$= \dfrac{x}{x-7} - \dfrac{-1}{x-7}$$

Rewrite the second denominator: $-7 + x = x - 7$. Now the denominators are like.

$$= \dfrac{x-(-1)}{x-7}$$

Subtract the numerators. Remember the parentheses. Write the difference over the common denominator, $x - 7$.

$$= \dfrac{x+1}{x-7}$$

Do the subtraction in the numerator: $x - (-1) = x + 1$.

The result does not simplify.

Self Check 7 Add: $\dfrac{n}{n-8} + \dfrac{12}{8-n}$

Now Try ▶ Problem 51

SECTION 7.4 ▶ STUDY SET

VOCABULARY

Fill in the blanks.

1. $\frac{x}{x-7}$ and $\frac{1}{x-7}$ have like denominators. $\frac{x+5}{x-7}$ and $\frac{4x}{x+7}$ have _____ denominators.

2. The polynomials $x - 3$ and $3 - x$ are _____.

CONCEPTS

3. Write each denominator in factored form.

   a. $\dfrac{x+1}{20x^2}$

   b. $\dfrac{3x^2 - 4}{x^2 + 4x - 12}$

4. The factorizations of the denominators of two rational expressions are given. Find the LCD.

   a. $12a = 2 \cdot 2 \cdot 3 \cdot a$

   $18a^2 = 2 \cdot 3 \cdot 3 \cdot a \cdot a$

   b. $x^2 - 36 = (x+6)(x-6)$

   $3x - 18 = 3(x-6)$

5. What is the LCD for $\frac{x-1}{x+6}$ and $\frac{1}{x+3}$?

6. The LCD for $\frac{1}{9n^2}$ and $\frac{37}{15n^3}$ is $3 \cdot 3 \cdot 5 \cdot n \cdot n \cdot n = 45n^3$.
   If we want to add these rational expressions, what form of 1 should be used
   a. to build $\frac{1}{9n^2}$?   b. to build $\frac{37}{15n^3}$?

Fill in the blanks.

7. To build $\frac{x}{x+2}$ so that it has a denominator of $5(x+2)$, we multiply it by 1 in the form of $\frac{\square}{\square}$.

8. To build $\frac{8x}{2-x}$ so that it has a denominator of $x-2$, we multiply it by 1 in the form of $\frac{\square}{\square}$.

NOTATION

Complete the solution.

9. $\dfrac{2}{5} + \dfrac{7}{3x} = \dfrac{2}{5} \cdot \dfrac{\square}{\square} + \dfrac{7}{3x} \cdot \dfrac{\square}{5}$

   $= \dfrac{6x}{\square} + \dfrac{35}{\square}$

   $= \dfrac{6x + \square}{15x}$

10. Are the student's answers and the book's answers equivalent?

Student's answer	Book's answer	Equivalent?
$\dfrac{m^2 + 2m}{(m-1)(m-4)}$	$\dfrac{m^2 + 2m}{(m-4)(m-1)}$	
$\dfrac{-5x^2 - 7}{4x(x+3)}$	$-\dfrac{5x^2 - 7}{4x(x+3)}$	
$\dfrac{-2x}{x-y}$	$-\dfrac{2x}{x-y}$	

GUIDED PRACTICE

Perform the operations. Simplify, if possible.
See Example 1.

11. $\dfrac{x}{3} + \dfrac{2x}{7}$ 12. $\dfrac{y}{4} + \dfrac{3y}{5}$

13. $\dfrac{7a}{8} + \dfrac{4a}{5}$ 14. $\dfrac{5t}{6} + \dfrac{4t}{7}$

Perform the operations. Simplify, if possible.
See Example 2.

15. $\dfrac{7}{m^2} - \dfrac{2}{m}$ 16. $\dfrac{6}{n^2} - \dfrac{2}{n}$

17. $\dfrac{3}{5p^2} - \dfrac{5}{10p}$ 18. $\dfrac{15}{16a} - \dfrac{3}{4a^2}$

19. $\dfrac{1}{6t} - \dfrac{11}{8t^3}$ 20. $\dfrac{3}{10a} - \dfrac{13}{15a^3}$

21. $\dfrac{1}{6c^4} - \dfrac{8}{9c^2}$ 22. $\dfrac{7}{8b^2} - \dfrac{5}{6b^3}$

Perform the operations. Simplify, if possible.
See Example 3.

23. $\dfrac{1}{2a+4} + \dfrac{5}{a^2-4}$ 24. $\dfrac{5}{p^2-9} + \dfrac{2}{3p+9}$

25. $\dfrac{2}{3a-2} + \dfrac{5}{9a^2-4}$ 26. $\dfrac{2}{5b-3} + \dfrac{5}{25b^2-9}$

27. $\dfrac{4}{a+2} - \dfrac{7}{a^2+4a+4}$ 28. $\dfrac{9}{b^2-2b+1} - \dfrac{2}{b-1}$

29. $\dfrac{6}{5m^2-5m} - \dfrac{3}{5m-5}$ 30. $\dfrac{9}{2c^2-2c} - \dfrac{5}{2c-2}$

Perform the operations. Simplify, if possible.
See Example 4.

31. $\dfrac{9}{t+3} + \dfrac{8}{t+2}$ 32. $\dfrac{2}{m-3} + \dfrac{7}{m-4}$

33. $\dfrac{3x}{2x-1} - \dfrac{2x}{2x+3}$ 34. $\dfrac{2y}{5y-1} - \dfrac{2y}{3y+2}$

35. $\dfrac{s+7}{s+3} - \dfrac{s-3}{s+7}$ 36. $\dfrac{t+5}{t-5} - \dfrac{t-5}{t+5}$

37. $\dfrac{3m}{m-2} - \dfrac{m-3}{m+5}$ 38. $\dfrac{2x}{x+2} - \dfrac{x+1}{x-3}$

Perform the operations. Simplify, if possible.
See Example 5.

39. $\dfrac{4}{s^2+5s+4} + \dfrac{s}{s^2+2s+1}$

40. $\dfrac{d}{d^2+6d+5} - \dfrac{3}{d^2+5d+4}$

41. $\dfrac{5}{x^2-9x+8} - \dfrac{3}{x^2-6x-16}$

42. $\dfrac{3}{t^2+t-6} + \dfrac{1}{t^2+3t-10}$

43. $\dfrac{2}{a^2+4a+3} + \dfrac{1}{a+3}$

44. $\dfrac{1}{c+6} + \dfrac{4}{c^2+8c+12}$

45. $\dfrac{8}{y^2-16} - \dfrac{7}{y^2-y-12}$

46. $\dfrac{6}{s^2-9} - \dfrac{5}{s^2-s-6}$

Perform the operations. Simplify, if possible.
See Example 6.

47. $\dfrac{9y}{x-4} + y$ 48. $\dfrac{9n}{m+4} + n$

49. $\dfrac{8}{x} + z$ 50. $\dfrac{2}{y} + z$

Perform the operations. Simplify, if possible.
See Example 7.

51. $\dfrac{7}{a-4} + \dfrac{5}{4-a}$ 52. $\dfrac{4}{b-6} + \dfrac{b}{6-b}$

53. $\dfrac{c}{7c-d} - \dfrac{d}{d-7c}$

54. $\dfrac{a}{5a-3b} - \dfrac{b}{3b-5a}$

TRY IT YOURSELF

Perform the operations and simplify, if possible.

55. $\dfrac{x-7}{x^2+4x-5} - \dfrac{x-9}{x^2+3x-10}$

56. $\dfrac{r}{r^2+5r+6} - \dfrac{2}{r^2+3r+2}$

57. $\dfrac{3d-3}{d-9} - \dfrac{3d}{9-d}$

58. $\dfrac{2x+2}{x-2} - \dfrac{2x}{2-x}$

59. $\dfrac{10}{x-1} + y$

60. $\dfrac{3}{s-8} + t$

61. $\dfrac{b}{b+1} - \dfrac{b-1}{b+2}$

62. $\dfrac{x}{x-2} - \dfrac{x+2}{x+3}$

63. $\dfrac{g}{g^2-4} + \dfrac{2}{4-g^2}$

64. $\dfrac{h}{h^2-49} + \dfrac{7}{49-h^2}$

65. $\dfrac{5y}{6} + \dfrac{5y}{3}$

66. $\dfrac{4x}{3} + \dfrac{x}{6}$

67. $\dfrac{1}{5x} + \dfrac{7x}{x+5}$

68. $\dfrac{10h}{h-3} + \dfrac{7}{9h}$

69. $\dfrac{11}{5x} - \dfrac{5}{6x}$

70. $\dfrac{5}{9y} - \dfrac{1}{4y}$

71. $\dfrac{x}{x+1} + \dfrac{x-1}{x}$

72. $\dfrac{t-2}{t} + \dfrac{t}{t+3}$

73. $\dfrac{y}{y-1} - \dfrac{4}{1-y}$

74. $\dfrac{1}{t-7} - \dfrac{t}{7-t}$

75. $\dfrac{n}{5} - \dfrac{n-2}{15}$

76. $\dfrac{m}{9} - \dfrac{m+1}{27}$

77. $\dfrac{y+2}{5y^2} + \dfrac{y+4}{15y}$

78. $\dfrac{x+3}{x^2} + \dfrac{x+5}{2x}$

79. $\dfrac{x}{x-2} + \dfrac{4+2x}{x^2-4}$

80. $\dfrac{y}{y+3} - \dfrac{2y-6}{y^2-9}$

81. $b - \dfrac{3}{a^2}$

82. $c - \dfrac{5}{3b}$

83. $\dfrac{7}{3a} + \dfrac{1}{a-2}$

84. $\dfrac{5}{9x} + \dfrac{4}{x+6}$

85. $\dfrac{3}{x^2} + \dfrac{17}{x}$

86. $\dfrac{7}{c} + \dfrac{14}{c^2}$

87. $\dfrac{x+2}{x+1} - 5$

88. $\dfrac{y+8}{y-8} - 4$

89. $\dfrac{4b}{3} - \dfrac{5b}{12}$

90. $\dfrac{21y}{12} - \dfrac{7y}{6}$

Look Alikes . . .

91. a. $\dfrac{5}{2x} + \dfrac{4x}{15}$

b. $\dfrac{5}{2x} \cdot \dfrac{4x}{15}$

92. a. $\dfrac{2a+4}{3} - \dfrac{9}{a+2}$

b. $\dfrac{2a+4}{3} \cdot \dfrac{9}{a+2}$

93. a. $\dfrac{t}{t-5} - \dfrac{t}{t^2-25}$

b. $\dfrac{t}{t-5} \div \dfrac{t}{t^2-25}$

94. a. $\dfrac{1}{m+2} - \dfrac{2}{m^2+4m+4}$

b. $\dfrac{1}{m+2} \div \dfrac{2}{m^2+4m+4}$

APPLICATIONS

95. Find the total height of the funnel.

96. Funnels. Refer to the illustration on the right. What is the difference between the diameter of the opening at the top of the funnel and the diameter of its spout?

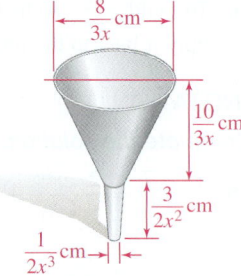

WRITING

97. Explain the error:

a. $\dfrac{3}{x} + \dfrac{8}{y} = \dfrac{\cancel{3+8}}{x\cancel{+}y}$

$= \dfrac{11}{x+y}$

b. $\dfrac{3}{x} + \dfrac{x}{3} = \dfrac{\overset{1}{\cancel{3}}}{\underset{1}{\cancel{x}}} + \dfrac{\overset{1}{\cancel{x}}}{\underset{1}{\cancel{3}}}$

$= 1 + 1$

$= 2$

98. Explain how to add two rational expressions with unlike denominators.

99. When will the LCD of two rational expressions be the product of the denominators of those rational expressions? Give an example.

100. Explain how multiplication by $\dfrac{-1}{-1}$ is used in this section.

REVIEW

101. Find the slope and y-intercept of the graph of $y = 8x + 2$.

102. Find the slope and y-intercept of the graph of $3x + 4y = -36$.

103. What is the slope of the graph of $y = 2$?

104. Is the graph of the equation $x = 0$ the x-axis or the y-axis?

CHALLENGE PROBLEMS

Perform the operations and simplify the result, if possible.

105. $\dfrac{a}{a-1} - \dfrac{2}{a+2} + \dfrac{3(a-2)}{a^2+a-2}$

106. $\dfrac{2x}{x^2-3x+2} + \dfrac{2x}{x-1} - \dfrac{x}{x-2}$

107. $\dfrac{1}{a+1} + \dfrac{a^2-7a+10}{2a^2-2a-4} \cdot \dfrac{2a^2-50}{a^2+10a+25}$

108. $1 - \dfrac{(x-2)^2}{(x+2)^2}$

SECTION 7.5

OBJECTIVES

1. Simplify complex fractions using division.
2. Simplify complex fractions using the LCD.

Simplifying Complex Fractions

ARE YOU READY?

The following problems review some basic skills that are needed when simplifying complex fractions.

1. What operation is indicated by the fraction bar in $\dfrac{56}{7}$?

2. Divide: $\dfrac{2}{9} \div \dfrac{4}{27}$

3. What is the LCD of $\dfrac{5}{2}$ and $\dfrac{3}{x}$?

4. Multiply: $10a\left(\dfrac{4}{5} - \dfrac{1}{a}\right)$

A **complex rational expression,** also called a **complex fraction,** is a rational expression whose numerator and/or denominator contains one or more rational expressions. The expression above the main fraction bar of a complex fraction is the numerator, and the expression below the main fraction bar is the denominator. Two examples of complex fractions are:

$$\dfrac{\dfrac{5x}{3}}{\dfrac{2x}{9}} \qquad \longleftarrow \text{ Numerator of complex fraction } \longrightarrow \qquad \dfrac{\dfrac{1}{2} - \dfrac{1}{x}}{\dfrac{x}{3} + \dfrac{1}{5}}$$

$\longleftarrow$ Main fraction bar $\longrightarrow$

$\longleftarrow$ Denominator of complex fraction $\longrightarrow$

In this section, we will discuss two methods for simplifying complex fractions. To **simplify a complex fraction** means to write it in the form $\dfrac{A}{B}$, where A and B are polynomials that have no common factors.

1 Simplify Complex Fractions Using Division.

One method for simplifying complex fractions uses the fact that the main fraction bar indicates division.

Simplifying Complex Fractions Method 1: Using Division

1. Add or subtract in the numerator and/or denominator so that the numerator is a single rational expression and the denominator is a single rational expression.
2. Perform the indicated division by multiplying the numerator of the complex fraction by the reciprocal of the denominator.
3. Simplify the result, if possible.

EXAMPLE 1

Simplify: $\dfrac{\dfrac{5x^2}{3}}{\dfrac{2x^3}{9}}$

Strategy We will perform the division indicated by the main fraction bar using the procedure for dividing rational expressions from Section 7.2.

Why We can skip the first step of method 1 and immediately divide because the numerator and the denominator of the complex fraction are already single rational expressions.

Solution

$$\frac{\dfrac{5x^2}{3}}{\dfrac{2x^3}{9}} = \frac{5x^2}{3} \div \frac{2x^3}{9}$$

Write the division indicated by the main fraction bar using a ÷ symbol.

$$= \frac{5x^2}{3} \cdot \frac{9}{2x^3}$$

To divide rational expressions, multiply the first by the reciprocal of the second.

$$= \frac{5x^2 \cdot 9}{3 \cdot 2x^3}$$

Multiply the numerators.
Multiply the denominators.

$$= \frac{5 \cdot \overset{1}{\cancel{x}} \cdot \overset{1}{\cancel{x}} \cdot \overset{1}{\cancel{3}} \cdot 3}{\underset{1}{\cancel{3}} \cdot 2 \cdot \underset{1}{\cancel{x}} \cdot \underset{1}{\cancel{x}} \cdot x}$$

Factor 9 as 3 · 3. Then simplify by removing factors equal to 1.

$$= \frac{15}{2x}$$

Multiply the remaining factors in the numerator.
Multiply the remaining factors in the denominator.

> **The Language of Algebra**
>
> The second step of this method could also be phrased: Perform the division by **inverting the denominator of the complex fraction and multiplying.**

Self Check 1   Simplify: $\dfrac{\dfrac{7y^3}{8}}{\dfrac{21y^2}{20}}$

Now Try ▶ Problem 17

In the next example, we must simplify the numerator and denominator of the complex fraction separately before the indicated division can be performed.

EXAMPLE 2   Simplify: $\dfrac{\dfrac{1}{2} - \dfrac{1}{x}}{\dfrac{x}{3} + \dfrac{1}{5}}$

Strategy We will simplify the expressions above and below the main fraction bar separately to write $\frac{1}{2} - \frac{1}{x}$ and $\frac{x}{3} + \frac{1}{5}$ as single rational expressions. Then we will perform the indicated division.

Why The numerator and the denominator of the complex fraction must be written as single rational expressions before dividing.

Solution To write the numerator as a single rational expression, we build $\frac{1}{2}$ and $\frac{1}{x}$ to have an LCD of $2x$, and then subtract. To write the denominator as a single rational expression, we build $\frac{x}{3}$ and $\frac{1}{5}$ to have an LCD of 15, and then add.

> **Success Tip**
>
> Notice that with this method, separate LCD's are found for the numerator and the denominator of the complex fraction.

$$\frac{\dfrac{1}{2} - \dfrac{1}{x}}{\dfrac{x}{3} + \dfrac{1}{5}} = \frac{\dfrac{1}{2} \cdot \dfrac{x}{x} - \dfrac{1}{x} \cdot \dfrac{2}{2}}{\dfrac{x}{3} \cdot \dfrac{5}{5} + \dfrac{1}{5} \cdot \dfrac{3}{3}}$$

← The LCD for the numerator is 2x. Build each fraction so that each has a denominator of 2x.

← The LCD for the denominator is 15. Build each fraction so that each has a denominator of 15.

$$= \frac{\dfrac{x}{2x} - \dfrac{2}{2x}}{\dfrac{5x}{15} + \dfrac{3}{15}}$$

Perform each of the four multiplications shown in color.

$$= \frac{\dfrac{x-2}{2x}}{\dfrac{5x+3}{15}}$$

Subtract in the numerator and add in the denominator of the complex fraction.

Now that the numerator and the denominator of the complex fraction are single rational expressions, we perform the indicated division.

$$\frac{\dfrac{x-2}{2x}}{\dfrac{5x+3}{15}} = \frac{x-2}{2x} \div \frac{5x+3}{15} \qquad \text{Write the division indicated by the main fraction bar using a} \div \text{symbol.}$$

$$= \frac{x-2}{2x} \cdot \frac{15}{5x+3} \qquad \text{Multiply by the reciprocal of } \tfrac{5x+3}{15}.$$

$$= \frac{15(x-2)}{2x(5x+3)} \qquad \text{Multiply the numerators. Multiply the denominators. Since the numerator and denominator have no common factor, the result does not simplify.}$$

$$= \frac{15x-30}{10x^2+6x} \qquad \text{Distribute 15 and distribute 2x.}$$

Self Check 2   Simplify: $\dfrac{\dfrac{1}{3}+\dfrac{1}{x}}{\dfrac{x}{5}-\dfrac{1}{2}}$

Now Try ▶ Problem 23

EXAMPLE 3   Simplify: $\dfrac{\dfrac{6}{x}+y}{\dfrac{6}{y}+x}$

Strategy   We will simplify the expressions above and below the main fraction bar separately to write $\frac{6}{x}+y$ and $\frac{6}{y}+x$ as single rational expressions. Then we will perform the indicated division.

Why   The numerator and the denominator of the complex fraction must be written as single rational expressions before dividing.

Solution   To write $\frac{6}{x}+y$ as a single rational expression, we build y into a fraction with a denominator of x and add. To write $\frac{6}{y}+x$ as a single rational expression, we build x into a fraction with a denominator of y and add.

$$\frac{\dfrac{6}{x}+y}{\dfrac{6}{y}+x} = \frac{\dfrac{6}{x}+\dfrac{y}{1}\cdot\dfrac{x}{x}}{\dfrac{6}{y}+\dfrac{x}{1}\cdot\dfrac{y}{y}} \qquad \begin{array}{l}\leftarrow \text{Write } y \text{ as } \tfrac{y}{1}.\text{ The LCD for the numerator is } x.\\ \text{Build } \tfrac{y}{1} \text{ so that it has a denominator of } x.\\ \leftarrow \text{Write } x \text{ as } \tfrac{x}{1}.\text{ The LCD for the denominator is } y.\\ \text{Build } \tfrac{x}{1} \text{ so that it has a denominator of } y.\end{array}$$

$$= \frac{\dfrac{6}{x}+\dfrac{xy}{x}}{\dfrac{6}{y}+\dfrac{xy}{y}} \qquad \text{Perform the two multiplications shown in color.}$$

$$= \frac{\dfrac{6+xy}{x}}{\dfrac{6+xy}{y}} \qquad \text{Add in the numerator and in the denominator of the complex fraction.}$$

Now that the numerator and the denominator of the complex fraction are single rational expressions, we can perform the division.

Success Tip

Simplifying using division (method 1) works well when a complex fraction is written, or can be easily written, as a quotient of two single rational expressions.

$$\frac{\dfrac{6 + xy}{x}}{\dfrac{6 + xy}{y}} = \frac{6 + xy}{x} \div \frac{6 + xy}{y}$$ Write the division indicated by the main fraction bar using a ÷ symbol.

$$= \frac{6 + xy}{x} \cdot \frac{y}{6 + xy}$$ Multiply by the reciprocal of $\frac{6 + xy}{y}$.

$$= \frac{y(6 + xy)}{x(6 + xy)}$$ Multiply the numerators. Multiply the denominators.

$$= \frac{y(6 + \cancel{xy})^{1}}{x(6 + \cancel{xy})_{1}}$$ Simplify the result by removing a factor equal to 1.

$$= \frac{y}{x}$$

Self Check 3   Simplify: $\dfrac{\dfrac{2}{a} - b}{\dfrac{2}{b} - a}$

Now Try ▶ **Problem 31**

2 Simplify Complex Fractions Using the LCD.

A second method for simplifying complex fractions uses the concepts of LCD and multiplication by a form of 1. The multiplication by 1 produces a simpler, equivalent expression, which will not contain rational expressions in its numerator or denominator.

Simplifying Complex Fractions
Method 2: Multiplying by the LCD

1. Find the LCD of all rational expressions within the complex fraction.
2. Multiply the complex fraction by 1 in the form $\frac{\text{LCD}}{\text{LCD}}$.
3. Perform the operations in the numerator and denominator. No rational expressions should remain within the complex fraction.
4. Simplify the result, if possible.

We will use method 2 to rework Example 2.

EXAMPLE 4   Simplify: $\dfrac{\dfrac{1}{2} - \dfrac{1}{x}}{\dfrac{x}{3} + \dfrac{1}{5}}$

Strategy   Using method 1 to simplify this complex fraction, we worked with $\frac{1}{2} - \frac{1}{x}$ and $\frac{x}{3} + \frac{1}{5}$ separately. With method 2, we will use the LCD of *all four* rational expressions within the complex fraction.

Why   Multiplying a complex fraction by 1 in the form of $\frac{\text{LCD}}{\text{LCD}}$ clears its numerator and denominator of fractions.

Solution   The denominators of all the rational expressions within the complex fraction are 2, x, 3, and 5. Thus, their LCD is $2 \cdot x \cdot 3 \cdot 5 = 30x$.

We now multiply the complex fraction by a factor equal to 1, using the LCD: $\frac{30x}{30x} = 1$.

$$\frac{\dfrac{1}{2} - \dfrac{1}{x}}{\dfrac{x}{3} + \dfrac{1}{5}} = \frac{\dfrac{1}{2} - \dfrac{1}{x}}{\dfrac{x}{3} + \dfrac{1}{5}} \cdot \frac{30x}{30x}$$

$$= \frac{\left(\dfrac{1}{2} - \dfrac{1}{x}\right)30x}{\left(\dfrac{x}{3} + \dfrac{1}{5}\right)30x}$$

← Multiply the numerators.

← Multiply the denominators.

$$= \frac{\dfrac{1}{2}(30x) - \dfrac{1}{x}(30x)}{\dfrac{x}{3}(30x) + \dfrac{1}{5}(30x)}$$

← In the numerator, distribute the multiplication by 30x.

← In the denominator, distribute the multiplication by 30x.

$$= \frac{15x - 30}{10x^2 + 6x}$$

Perform each of the four multiplications by 30x. Notice that no fractional expressions remain within the complex fraction.

To attempt to simplify the result, factor the numerator and denominator. Since they do not have a common factor, the result is in simplest form. This is the same result obtained in Example 2, which used method 1.

Self Check 4   Use method 2 to simplify: $\quad \dfrac{\dfrac{1}{4} - \dfrac{1}{x}}{\dfrac{x}{5} + \dfrac{1}{3}}$

Now Try ▶ Problem 37

Success Tip

With method 2, each term of the numerator and each term of the denominator of the complex fraction is multiplied by the LCD. Arrows can be helpful in showing this.

$$\frac{\left(\dfrac{1}{2} - \dfrac{1}{x}\right)}{\left(\dfrac{x}{3} + \dfrac{1}{5}\right)} \cdot \frac{30x}{30x}$$

Success Tip

When simplifying a complex fraction, the same result will be obtained regardless of the method used. See Example 2.

EXAMPLE 5   Simplify: $\quad \dfrac{\dfrac{1}{8} - \dfrac{1}{y}}{\dfrac{8 - y}{4y^2}}$

Strategy   Using method 1, we would work with $\frac{1}{8} - \frac{1}{y}$ and $\frac{8-y}{4y^2}$ separately. With method 2, we use the LCD of all three rational expressions within the complex fraction.

Why   Multiplying a complex fraction by 1 in the form of $\frac{\text{LCD}}{\text{LCD}}$ clears its numerator and denominator of fractions.

Solution   The denominators of all the rational expressions within the complex fraction are 8, y, and $4y^2$. Therefore, the LCD is $8y^2$ and we multiply the complex fraction by a factor equal to 1, using the LCD: $\frac{8y^2}{8y^2} = 1$.

$$\frac{\dfrac{1}{8} - \dfrac{1}{y}}{\dfrac{8 - y}{4y^2}} = \frac{\dfrac{1}{8} - \dfrac{1}{y}}{\dfrac{8 - y}{4y^2}} \cdot \frac{8y^2}{8y^2}$$

Success Tip

Notice that with this method, all the rational expressions in the complex fraction are considered to find one universal LCD.

$$= \frac{\left(\dfrac{1}{8} - \dfrac{1}{y}\right)8y^2}{\left(\dfrac{8 - y}{4y^2}\right)8y^2} \quad \leftarrow \text{Multiply the numerators.}$$

← Multiply the denominators.

$$= \frac{\dfrac{1}{8}(8y^2) - \dfrac{1}{y}(8y^2)}{\left(\dfrac{8 - y}{4y^2}\right)(8y^2)}$$

In the numerator, distribute the multiplication by $8y^2$.

$$= \frac{y^2 - 8y}{(8 - y)2}$$

Perform each of the three multiplications by $8y^2$.

$$= \frac{\overset{-1}{\cancel{y(y - 8)}}}{\underset{1}{\cancel{(8 - y)}}2}$$

In the numerator, factor out the GCF, y. Since $y - 8$ and $8 - y$ are opposites, simplify by replacing $\frac{y - 8}{8 - y}$ with $\frac{-1}{1}$.

$$= -\frac{y}{2}$$

The Language of Algebra

After multiplying a complex fraction by $\frac{\text{LCD}}{\text{LCD}}$ and performing the multiplications, the numerator and denominator of the complex fraction will be **cleared of fractions.**

Self Check 5   Simplify: $\dfrac{\dfrac{10 - n}{5n^2}}{\dfrac{1}{10} - \dfrac{1}{n}}$

Now Try ▶ **Problem 41**

EXAMPLE 6   Simplify: $\dfrac{1}{1 + \dfrac{1}{x + 1}}$

Strategy   Although either method can be used, we will use method 2 to simplify this complex fraction.

Why   Method 2 is often easier when the complex fraction contains a sum or difference.

Solution   The only rational expression within the complex fraction has the denominator $x + 1$. Therefore, the LCD is $x + 1$. We multiply the complex fraction by a factor equal to 1, using the LCD: $\frac{x + 1}{x + 1} = 1$.

Success Tip

Simplifying using the LCD (method 2) works well when the complex fraction has sums and/or differences in the numerator and/or denominator.

$$\frac{1}{1 + \dfrac{1}{x + 1}} = \frac{1}{1 + \dfrac{1}{x + 1}} \cdot \boxed{\frac{x + 1}{x + 1}}$$

$$= \frac{1(x + 1)}{\left(1 + \dfrac{1}{x + 1}\right)(x + 1)}$$

Multiply the numerators.
Multiply the denominators.

$$= \frac{1(x + 1)}{1(x + 1) + \dfrac{1}{x + 1}(x + 1)}$$

In the denominator, distribute the multiplication by $x + 1$.

$$= \frac{x + 1}{x + 1 + 1}$$

Perform each of the three multiplications by $x + 1$.

$$= \frac{x + 1}{x + 2}$$

Combine like terms in the denominator.

The result does not simplify.

Self Check 6   Simplify: $\dfrac{2}{\dfrac{1}{x+2}+2}$

Now Try ▶ Problem 47

SECTION **7.5** ▶ STUDY SET

VOCABULARY

Fill in the blanks.

1. The expression $\dfrac{\dfrac{2}{3}-\dfrac{1}{x}}{\dfrac{x-3}{4}}$ is called a _____ rational expression

   or a _____ fraction.

2. In a complex fraction, the numerator is above the _____ fraction bar and the _____ is below it.

CONCEPTS

Fill in the blanks.

3. Method 1: To simplify a complex fraction, write its numerator and denominator as _____ rational expressions. Then perform the indicated _____ by multiplying the numerator of the complex fraction by the _____ of the denominator.

4. Method 2: To simplify a complex fraction, find the LCD of ____ the rational expressions within the complex fraction. Multiply the complex fraction by 1 in the form ▢ .

5. Consider: $\dfrac{\dfrac{x-3}{4}}{\dfrac{1}{12}-\dfrac{x}{6}}$

   a. What is the numerator of the complex fraction? Is it a single rational expression?

   b. What is the denominator of the complex fraction? Is it a single rational expression?

6. Consider the complex fraction: $\dfrac{\dfrac{1}{y}-\dfrac{1}{3}}{\dfrac{5}{6}+\dfrac{1}{y}}$

   a. What is the LCD of all the rational expressions in the complex fraction?

   b. To simplify the complex fraction using method 2, it should be multiplied by what form of 1?

NOTATION

Fill in the blanks.

7. $\dfrac{\dfrac{12}{y^2}}{\dfrac{4}{y^3}}$ means $\dfrac{12}{y^2}\ \boxed{}\ \dfrac{4}{y^3}$

8. $\dfrac{\left(\dfrac{1}{5}-\dfrac{1}{a}\right)}{\left(\dfrac{a}{4}+\dfrac{2}{a}\right)}\cdot\dfrac{20a}{20a}=\dfrac{\dfrac{1}{5}(\quad)-\dfrac{1}{a}(\quad)}{\dfrac{a}{4}(\quad)+\dfrac{2}{a}(\quad)}$

   $=\dfrac{4a-\boxed{}}{\boxed{}+\boxed{}}$

GUIDED PRACTICE

Simplify each complex fraction. See Example 1.

9. $\dfrac{\dfrac{2}{3}}{\dfrac{3}{4}}$

10. $\dfrac{\dfrac{3}{5}}{\dfrac{2}{7}}$

11. $\dfrac{\dfrac{x}{2}}{\dfrac{6}{5}}$

12. $\dfrac{\dfrac{9}{4}}{\dfrac{7}{x}}$

13. $\dfrac{\dfrac{x}{y}}{\dfrac{1}{x}}$

14. $\dfrac{\dfrac{y}{x}}{\dfrac{x}{xy}}$

15. $\dfrac{\dfrac{n}{8}}{\dfrac{1}{n^2}}$

16. $\dfrac{\dfrac{1}{m}}{\dfrac{m^3}{15}}$

17. $\dfrac{\dfrac{4a}{11}}{\dfrac{6a^4}{55}}$

18. $\dfrac{\dfrac{14}{15m}}{\dfrac{21}{25m^6}}$

19. $\dfrac{\dfrac{x^4}{30}}{\dfrac{7x^2}{15}}$

20. $\dfrac{\dfrac{5x^2}{24}}{\dfrac{x^5}{56}}$

Simplify each complex fraction. See Examples 2 or 4.

21. $\dfrac{\dfrac{1}{2}+\dfrac{3}{4}}{\dfrac{3}{2}+\dfrac{1}{4}}$

22. $\dfrac{\dfrac{2}{3}-\dfrac{5}{2}}{\dfrac{2}{3}-\dfrac{3}{2}}$

23. $\dfrac{\dfrac{1}{4}+\dfrac{1}{y}}{\dfrac{y}{3}-\dfrac{1}{2}}$

24. $\dfrac{\dfrac{2}{x}-\dfrac{1}{3}}{\dfrac{2}{3}+\dfrac{x}{5}}$

25. $\dfrac{\dfrac{1}{y}-\dfrac{5}{2}}{\dfrac{3}{y}}$

26. $\dfrac{\dfrac{1}{6}-\dfrac{5}{s}}{\dfrac{2}{s}}$

27. $\dfrac{\dfrac{4}{c}-\dfrac{c}{6}}{\dfrac{2}{c}}$

28. $\dfrac{\dfrac{10}{n}-\dfrac{n}{4}}{\dfrac{8}{n}}$

Simplify each complex fraction. See Examples 3 or 5.

29. $\dfrac{\dfrac{2}{3}+1}{\dfrac{1}{3}+1}$

30. $\dfrac{\dfrac{3}{5}-2}{\dfrac{2}{5}-2}$

31. $\dfrac{\dfrac{1}{x}-3}{\dfrac{5}{x}+2}$

32. $\dfrac{\dfrac{1}{y}+3}{\dfrac{3}{y}-2}$

33. $\dfrac{\dfrac{2}{x}+2}{\dfrac{4}{x}+2}$

34. $\dfrac{\dfrac{3}{x}-3}{\dfrac{9}{x}-3}$

35. $\dfrac{\dfrac{3y}{x}-y}{y-\dfrac{y}{x}}$

36. $\dfrac{\dfrac{y}{x}+3y}{y+\dfrac{2y}{x}}$

Simplify each complex fraction. See Example 4.

37. $\dfrac{\dfrac{1}{6}-\dfrac{2}{x}}{\dfrac{1}{6}+\dfrac{1}{x}}$

38. $\dfrac{\dfrac{3}{4}+\dfrac{1}{y}}{\dfrac{5}{6}-\dfrac{1}{y}}$

39. $\dfrac{\dfrac{a}{7}-\dfrac{7}{a}}{\dfrac{1}{a}+\dfrac{1}{7}}$

40. $\dfrac{\dfrac{t}{9}-\dfrac{9}{t}}{\dfrac{1}{t}+\dfrac{1}{9}}$

Simplify each complex fraction. See Example 5.

41. $\dfrac{\dfrac{d^2}{4}+\dfrac{4d}{5}}{\dfrac{d+1}{2}}$

42. $\dfrac{\dfrac{d+2}{2}}{\dfrac{d}{3}-\dfrac{d}{4}}$

43. $\dfrac{\dfrac{2}{x}}{\dfrac{2}{y}-\dfrac{4}{x}}$

44. $\dfrac{\dfrac{2y}{3}}{\dfrac{2y}{3}-\dfrac{8}{y}}$

Simplify each complex fraction. See Example 6.

45. $\dfrac{\dfrac{1}{x+1}}{1+\dfrac{1}{x+1}}$

46. $\dfrac{\dfrac{1}{x-1}}{1-\dfrac{1}{x-1}}$

47. $\dfrac{\dfrac{x}{x+2}}{\dfrac{x}{x+2}+x}$

48. $\dfrac{\dfrac{2}{x-2}}{\dfrac{2}{x-2}-1}$

TRY IT YOURSELF

Simplify each complex fraction.

49. $\dfrac{\dfrac{1}{p}+\dfrac{1}{q}}{\dfrac{1}{p}}$

50. $\dfrac{\dfrac{m}{n}+1}{1-\dfrac{m}{n}}$

51. $\dfrac{\dfrac{40x^2}{20x}}{9}$

52. $\dfrac{\dfrac{18n^2}{6n}}{13}$

53. $\dfrac{\dfrac{1}{c}+\dfrac{1}{2}}{\dfrac{1}{c^2}-\dfrac{1}{4}}$

54. $\dfrac{\dfrac{1}{m}-\dfrac{1}{n}}{\dfrac{m}{n}-\dfrac{n}{m}}$

55. $\dfrac{\dfrac{1}{r+1}+1}{\dfrac{3}{r-1}+1}$

56. $\dfrac{5+\dfrac{1}{n+7}}{4-\dfrac{2}{n+7}}$

57. $\dfrac{\dfrac{b^2-81}{18a^2}}{\dfrac{4b-36}{9a}}$

58. $\dfrac{\dfrac{8x-64}{y}}{\dfrac{x^2-64}{y^2}}$

59. $\dfrac{\dfrac{10x}{x-3}}{\dfrac{6}{x-3}}$

60. $\dfrac{\dfrac{18a}{a-4}}{\dfrac{12}{a-4}}$

61. $\dfrac{4-\dfrac{1}{8h}}{12+\dfrac{3}{4h}}$

62. $\dfrac{12+\dfrac{1}{3b}}{12-\dfrac{1}{b^2}}$

63. $\dfrac{\dfrac{m}{n}+\dfrac{n}{m}}{\dfrac{m}{n}-\dfrac{n}{m}}$

64. $\dfrac{\dfrac{2a}{b}-\dfrac{b}{a}}{\dfrac{2a}{b}+\dfrac{b}{a}}$

65. $\dfrac{\dfrac{2}{c^2}}{\dfrac{1}{c}+\dfrac{5}{4}}$

66. $\dfrac{\dfrac{7}{s^2}}{\dfrac{1}{s}+\dfrac{10}{3}}$

67. $\dfrac{\dfrac{4t-8}{t^2}}{\dfrac{8t-16}{t^5}}$

68. $\dfrac{\dfrac{9m-27}{m^6}}{\dfrac{2m-6}{m^8}}$

69. $\dfrac{\dfrac{2}{s}-\dfrac{2}{s^2}}{\dfrac{4}{s^3}+\dfrac{4}{s^2}}$

70. $\dfrac{\dfrac{2}{x^3}-\dfrac{2}{x}}{\dfrac{4}{x}+\dfrac{8}{x^2}}$

71. $\dfrac{1+\dfrac{6}{t}+\dfrac{8}{t^2}}{1+\dfrac{1}{t}-\dfrac{12}{t^2}}$

72. $\dfrac{1-p+\dfrac{2}{p}}{\dfrac{6}{p^2}+\dfrac{1}{p}-1}$

73. $\dfrac{1}{\dfrac{1}{x}+\dfrac{1}{y}}$

74. $\dfrac{1}{\dfrac{b}{a}-\dfrac{a}{b}}$

75. $\dfrac{-\dfrac{25}{16x^2}}{\dfrac{15}{32x^5}}$

76. $\dfrac{\dfrac{21}{8g^3}}{\dfrac{35}{16g^8}}$

77. $\dfrac{3+\dfrac{3}{x-1}}{3-\dfrac{3}{x-1}}$

78. $\dfrac{2-\dfrac{2}{x+1}}{2+\dfrac{2}{x+1}}$

79. $\dfrac{1-\dfrac{9}{d^2}}{2+\dfrac{6}{d}}$

80. $\dfrac{1-\dfrac{16}{a^2}}{\dfrac{12}{a}+3}$

81. $\dfrac{\dfrac{1}{a^2b}-\dfrac{5}{ab}}{\dfrac{3}{ab}-\dfrac{7}{ab^2}}$

82. $\dfrac{\dfrac{3}{ab^2}+\dfrac{6}{a^2b}}{\dfrac{6}{a}-\dfrac{9}{b^2}}$

83. $\dfrac{m - \dfrac{1}{2m+1}}{1 - \dfrac{m}{2m+1}}$

84. $\dfrac{1 - \dfrac{r}{2r+1}}{r - \dfrac{1}{2r+1}}$

88. **Data Analysis.** Use the data in the table to find the average measurement for the three-trial experiment.

	Trial 1	Trial 2	Trial 3
Measurement	$\dfrac{k}{2}$	$\dfrac{k}{3}$	$\dfrac{k}{2}$

APPLICATIONS

85. **Slope.** We can use the slope formula shown below to find the slope of a line that passes through $\left(\frac{1}{2}, \frac{1}{3}\right)$ and $\left(\frac{3}{4}, \frac{5}{8}\right)$.
Simplify the complex fraction to find m.

$$m = \frac{\dfrac{5}{8} - \dfrac{1}{3}}{\dfrac{3}{4} - \dfrac{1}{2}}$$

86. **Pitching.** The earned run average (ERA) is a statistic that gives the average number of earned runs a pitcher allows. For a softball pitcher, this is based on a six-inning game. The formula for ERA is shown below. Simplify the complex fraction on the right side of the formula.

$$\text{ERA} = \frac{\text{earned runs}}{\dfrac{\text{innings pitched}}{6}}$$

87. **Electronics.** In electronic circuits, resistors are tiny components that limit the flow of an electric current. An important formula about two resistors in a circuit is shown below. Simplify the complex fraction on the right side of the formula.

$$\text{Total resistance} = \frac{1}{\dfrac{1}{R_1} + \dfrac{1}{R_2}}$$

(Recall that R_1 is read as R sub one.)

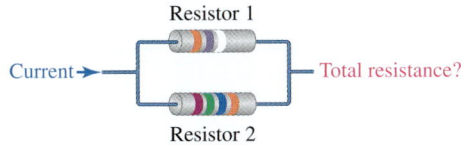

Current → Resistor 1 / Resistor 2 ← Total resistance?

WRITING

89. What is a complex fraction? Give several examples.

90. Explain how to use method 1 to simplify: $\dfrac{1 + \dfrac{1}{x}}{3 - \dfrac{1}{x}}$

91. Explain how to use method 2 to simplify the expression in Problem 90.

92. a. List an advantage and a disadvantage of using method 1 to simplify a complex fraction.
    b. List an advantage and a disadvantage of using method 2 to simplify a complex fraction.

REVIEW

Simplify each expression. Write each answer without negative exponents.

93. $(8x)^0$

94. $\left(-\dfrac{3r}{4r^3}\right)^4$

95. $\left(\dfrac{4x^3}{5x^{-3}}\right)^{-2}$

96. $\left(\dfrac{12xy^{-3}}{3x^{-2}y^2}\right)^{-2}$

CHALLENGE PROBLEMS

Simplify.

97. $\dfrac{\dfrac{h}{h^2 + 3h + 2}}{\dfrac{4}{h+2} - \dfrac{4}{h+1}}$

98. $\dfrac{\dfrac{2}{b^2-1} - \dfrac{3}{ab-a}}{\dfrac{3}{ab-a} - \dfrac{2}{b^2-1}}$

99. $a + \dfrac{a}{1 + \dfrac{a}{a+1}}$

100. $\dfrac{y^{-2}+1}{y^{-2}-1}$

SECTION 7.6

Solving Rational Equations

OBJECTIVES

1. Solve rational equations.
2. Solve for a specified variable in a formula.

ARE YOU READY?

The following problems review some basic skills that are needed when solving rational equations.

1. What is the LCD of $\dfrac{7}{2x}$, $\dfrac{1}{9x}$, and $\dfrac{17}{6x}$?

2. Multiply: $5(x-1)\left(\dfrac{x}{x-1}\right)$

3. Solve: $x^2 - x - 56 = 0$

4. Find all real numbers for which the rational expression $\dfrac{x+1}{x+8}$ is undefined.

In Chapter 2, we solved equations such as $\frac{1}{6}x + \frac{5}{2} = \frac{1}{3}$ by multiplying both sides by the LCD. With this approach, the equation that results is equivalent to the original equation, but easier to solve because it is cleared of fractions.

In this section, we will extend the fraction-clearing strategy to solve another type of equation, called a *rational equation*.

Rational Equations	A **rational equation** is an equation that contains one or more rational expressions.

Rational equations often have a variable in a denominator. Some examples of rational equations are:

$$\frac{2x}{3} = \frac{x}{6} + \frac{3}{2} \qquad\qquad \frac{2}{x} + \frac{1}{4} = \frac{5}{2x} \qquad\qquad \frac{11x}{x-5} = 6 + \frac{55}{x-5}$$

1 Solve Rational Equations.

To **solve a rational equation,** we find all the values of the variable that make the equation true. Any value of the variable that makes a denominator in a rational equation equal to 0 cannot be a solution of the equation. Such a number must be rejected, because division by 0 is undefined.

The goal when solving a rational equation is to use the multiplication property of equality to find an equivalent equation that we already know how to solve, such as a linear equation in one variable or a quadratic equation.

Strategy for Solving Rational Equations	1. Determine which numbers cannot be solutions of the equation.
	2. Multiply both sides of the equation by the LCD of all rational expressions in the equation. This clears the equation of fractions.
	3. Solve the resulting equation.
	4. Check all possible solutions in the original equation.

EXAMPLE 1 Solve: $\dfrac{2x}{3} = \dfrac{x}{6} + \dfrac{3}{2}$

Strategy We will use the multiplication property of equality to clear this rational equation of fractions by multiplying both sides by the LCD.

Why Equations that contain only integers are usually easier to solve than equations that contain fractions.

Solution There are no restrictions on x, because no value of x ever makes a denominator 0. Since the denominators are 3, 6, and 2, we multiply both sides of the equation by the LCD, 6.

$$\frac{2x}{3} = \frac{x}{6} + \frac{3}{2} \qquad\qquad \text{\textcolor{red}{This is the equation to solve.}}$$

Caution

Always enclose the left and right sides of an equation within parentheses when multiplying both sides by the LCD.

$$6\left(\frac{2x}{3}\right) = 6\left(\frac{x}{6} + \frac{3}{2}\right) \qquad \begin{array}{l}\text{\textcolor{red}{Multiply both sides of the equation}}\\\text{\textcolor{red}{by the LCD of } \frac{2x}{3}, \frac{x}{6}, \text{ and } \frac{3}{2}, \text{ which is 6.}}\end{array}$$

$$6\left(\frac{2x}{3}\right) = 6\left(\frac{x}{6}\right) + 6\left(\frac{3}{2}\right) \qquad \text{\textcolor{red}{Distribute the multiplication by 6.}}$$

$$2 \cdot \overset{1}{\cancel{3}}\left(\frac{2x}{\cancel{3}}\right)_{\!\!1} = \overset{1}{\cancel{6}}\left(\frac{x}{\cancel{6}}\right)_{\!\!1} + 2 \cdot \overset{1}{\cancel{3}}\left(\frac{3}{\cancel{2}}\right)_{\!\!1} \qquad \begin{array}{l}\text{\textcolor{red}{Perform the three multiplications by 6 by first removing}}\\\text{\textcolor{red}{common factors of the numerator and denominator.}}\\\text{\textcolor{red}{Try to do this step in your head.}}\end{array}$$

$$4x = x + 9$$ Simplify. Note that the fractions have been cleared. The result is a linear equation in one variable.

$$3x = 9$$ To eliminate x on the right side, subtract x from both sides.

$$x = 3$$ To undo the multiplication by 3, divide both sides by 3.

To check, we replace each x with 3 in the *original* equation.

$$\frac{2x}{3} = \frac{x}{6} + \frac{3}{2}$$

$$\frac{2(3)}{3} \stackrel{?}{=} \frac{3}{6} + \frac{3}{2}$$ Substitute 3 for x.

$$2 \stackrel{?}{=} \frac{1}{2} + \frac{3}{2}$$ Simplify: $\frac{2(\overset{1}{\cancel{3}})}{\cancel{3}} = 2$ and $\frac{3}{6} = \frac{1}{2}$.

$$2 = 2$$ Add: $\frac{1}{2} + \frac{3}{2} = \frac{4}{2} = 2$.

Since we obtain a true statement, 3 is the solution of $\frac{2x}{3} = \frac{x}{6} + \frac{3}{2}$. The solution set is $\{3\}$.

Self Check 1 Solve: $\frac{3x}{5} = \frac{x}{2} + \frac{1}{10}$

Now Try ▶ Problem 15

EXAMPLE 2 Solve: $\dfrac{2}{x} + \dfrac{1}{4} = \dfrac{5}{2x}$

Strategy This equation contains two rational expressions that have a variable in their denominators. We begin by asking, "What value(s) of x make either denominator 0?" Then we will clear the equation of fractions by multiplying both sides by the LCD.

Why If a number makes the denominator of a rational expression 0, that number cannot be a solution of the equation because division by 0 is undefined.

Solution If x is 0, the denominators of $\frac{2}{x}$ and $\frac{5}{2x}$ are 0 and the rational expressions would be undefined. Therefore, 0 cannot be a solution.

Since the denominators are x, 4, and $2x$, we multiply both sides of the equation by the LCD, $4x$, to clear the equation of fractions.

$$\frac{2}{x} + \frac{1}{4} = \frac{5}{2x}$$ This is the equation to solve.

$$4x\left(\frac{2}{x} + \frac{1}{4}\right) = 4x\left(\frac{5}{2x}\right)$$ Write each side of the equation within parentheses, and then multiply both sides by 4x.

$$4x\left(\frac{2}{x}\right) + 4x\left(\frac{1}{4}\right) = 4x\left(\frac{5}{2x}\right)$$ On the left side, distribute the multiplication by 4x.

$$\overset{1}{4x}\left(\frac{2}{\overset{}{\cancel{x}}}\right) + \overset{1}{\cancel{4}}x\left(\frac{1}{\overset{}{\cancel{4}}}\right) = 2 \cdot 2 \cdot \overset{1}{\cancel{x}}\left(\frac{5}{2 \cdot \overset{}{\cancel{x}}}\right)$$ On the right side, factor 4x as $2 \cdot 2 \cdot x$. Perform the three multiplications by 4x by first removing common factors of each numerator and denominator. Try to do this step in your head.

$$8 + x = 10$$ Simplify. Note that the fractions have been cleared. The result is a linear equation in one variable.

$$x = 2$$ To undo the addition of 8, subtract 8 from both sides.

The solution of $\frac{2}{x} + \frac{1}{4} = \frac{5}{2x}$ is 2. The solution set is $\{2\}$. Check by substituting 2 for each x in the *original* equation.

Self Check 2   Solve: $\dfrac{1}{6} + \dfrac{4}{3x} = \dfrac{5}{x}$

Now Try ▶   Problem 25

EXAMPLE 3   Solve: $y - \dfrac{12}{y} = 4$

Strategy   Since the only denominator is y, we will multiply both sides of the equation by y.

Why   Multiplying both sides by y will clear the equation of the fraction, $\dfrac{12}{y}$.

Solution   If y is 0, the denominator of $\dfrac{12}{y}$ is 0 and the fraction would be undefined. Therefore, 0 cannot be a solution.

$$y - \dfrac{12}{y} = 4 \qquad \text{This is the equation to solve.}$$

$$y\left(y - \dfrac{12}{y}\right) = y(4) \qquad \begin{array}{l}\text{Write each side of the equation within parentheses}\\\text{and then multiply both sides by the LCD, } y.\end{array}$$

$$y(y) - y\left(\dfrac{12}{y}\right) = y(4) \qquad \text{Distribute the multiplication by } y.$$

$$y^2 - 12 = 4y \qquad \begin{array}{l}\text{Simplify: } \overset{1}{y}\left(\dfrac{12}{\underset{1}{y}}\right) = 12. \text{ Note that the fraction has been cleared.}\end{array}$$

Note the y^2-term. The result is a quadratic equation.

> **Caution**
>
> By the multiplication property of equality, each term on both sides of the equation must be multiplied by the LCD. Here, it would be incorrect to multiply only the second term, $\frac{12}{y}$, by the LCD.
>
>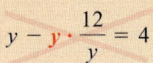

We can solve the resulting quadratic equation using the factoring method.

$$y^2 - 4y - 12 = 0 \qquad \text{Subtract } 4y \text{ from both sides to get 0 on the right side.}$$
$$(y - 6)(y + 2) = 0 \qquad \text{Factor the trinomial.}$$
$$y - 6 = 0 \quad \text{or} \quad y + 2 = 0 \qquad \text{Set each factor equal to 0.}$$
$$y = 6 \quad \mid \quad y = -2 \qquad \text{Solve each equation.}$$

There are two possible solutions, 6 and -2, to check in the original equation.

Check $y = 6$:

$$y - \dfrac{12}{y} = 4$$

$$6 - \dfrac{12}{6} \overset{?}{=} 4$$

$$6 - 2 \overset{?}{=} 4$$

$$4 = 4 \quad \text{True}$$

Check $y = -2$:

$$y - \dfrac{12}{y} = 4 \qquad \text{This is the original equation.}$$

$$-2 - \dfrac{12}{-2} \overset{?}{=} 4$$

$$-2 - (-6) \overset{?}{=} 4$$

$$4 = 4 \quad \text{True}$$

Thus, the solutions of $y - \dfrac{12}{y} = 4$ are 6 and -2. The solution set is $\{-2, 6\}$.

Self Check 3   Solve: $x - \dfrac{24}{x} = -5$

Now Try ▶   Problem 31

EXAMPLE 4

Solve: $\dfrac{11x}{x-5} = 6 + \dfrac{55}{x-5}$

Strategy Since both denominators are $x - 5$, we multiply both sides by the LCD, $x - 5$.

Why This will clear the equation of fractions.

Solution

If x is 5, the denominators of $\dfrac{11x}{x-5}$ and $\dfrac{55}{x-5}$ are 0, and the rational expressions are undefined. Therefore, 5 cannot be a solution of the equation.

$$\dfrac{11x}{x-5} = 6 + \dfrac{55}{x-5}$$

This is the equation to solve.

$$(x-5)\left(\dfrac{11x}{x-5}\right) = (x-5)\left(6 + \dfrac{55}{x-5}\right)$$

Write each side of the equation within parentheses and then multiply both sides by $x - 5$.

$$(x-5)^1\left(\dfrac{11x}{x-5}\right)_1 = (x-5)6 + (x-5)^1\left(\dfrac{55}{x-5}\right)_1$$

Distribute the multiplication by $x - 5$. Remove the common binomial factor $(x - 5)$ of the numerator and denominator.

$$11x = (x-5)6 + 55$$

Simplify. Note that the fractions have been cleared. The result is a linear equation in one variable.

$$11x = 6x - 30 + 55$$

To solve the resulting equation, distribute the 6.

$$11x = 6x + 25$$

Combine like terms: $-30 + 55 = 25$.

$$5x = 25$$

To eliminate $6x$ on the right side, subtract $6x$ from both sides.

$$x = 5$$

To undo the multiplication by 5, divide both sides by 5.

Earlier, we determined that 5 makes both denominators in the original equation 0. Therefore, 5 cannot be a solution. Since 5 is the only possible solution, and it must be rejected, it follows that $\dfrac{11x}{x-5} = 6 + \dfrac{55}{x-5}$ has *no solution*. The solution set is written as { } or $\varnothing$.

When solving an equation, a possible solution that does not satisfy the original equation is called an **extraneous solution.** In this example, 5 is an extraneous solution.

> **Caution**
>
> Even if you do not make an arithmetic or algebraic error when solving a rational equation, a possible solution may not check.

> **The Language of Algebra**
>
> **Extraneous** means not a vital part. Mathematicians speak of *extraneous* solutions. Rock groups don't want *extraneous* sounds (like feedback) coming from their amplifiers. Artists erase *extraneous* marks on their sketches.

Self Check 4 Solve $\dfrac{9x}{x-6} = 3 + \dfrac{54}{x-6}$, if possible.

Now Try ▶ Problem 41

EXAMPLE 5

Solve: $\dfrac{x+5}{x+3} + \dfrac{1}{x^2 + 2x - 3} = 1$

Strategy We will multiply both sides by the LCD of the two rational expressions in the equation. But first, we must factor the second denominator.

Why To determine the restrictions on the variable and to find the LCD, we need to write $x^2 + 2x - 3$ in factored form.

Solution

Since the trinomial $x^2 + 2x - 3$ factors as $(x + 3)(x - 1)$, we can write the given equation as:

$$\dfrac{x+5}{x+3} + \dfrac{1}{(x+3)(x-1)} = 1$$

If x is -3, the first denominator is 0. If x is -3 or 1, the second denominator is 0.

We see that -3 and 1 cannot be solutions of the equation, because they make rational expressions in the equation undefined.

Since the denominators are $x + 3$ and $(x + 3)(x - 1)$, we multiply both sides of the equation by the LCD, $(x + 3)(x - 1)$, to clear the fractions.

$$(x + 3)(x - 1)\left[\frac{x + 5}{x + 3} + \frac{1}{(x + 3)(x - 1)}\right] = (x + 3)(x - 1)[1]$$ Write each side within brackets [ ].

$$\overset{1}{(x + 3)}(x - 1)\frac{x + 5}{\underset{1}{x + 3}} + \overset{1}{(x + 3)}\overset{1}{(x - 1)}\frac{1}{\underset{1}{(x + 3)}\underset{1}{(x - 1)}} = (x + 3)(x - 1)[1]$$ Distribute the multiplication by $(x + 3)(x - 1)$ and remove common factors.

$$(x - 1)(x + 5) + 1 = (x + 3)(x - 1)$$ Simplify. The fractions are cleared.

The Language of Algebra

We say that the LCD is a **multiplier** that clears a rational equation of fractions.

To solve the resulting equation, we multiply the binomials on the left side and the right side, and proceed as follows.

$$x^2 + 4x - 5 + 1 = x^2 + 2x - 3$$ Find $(x - 1)(x + 5)$ and $(x + 3)(x - 1)$.

$$x^2 + 4x - 4 = x^2 + 2x - 3$$ Combine like terms: $-5 + 1 = -4$.

$$4x - 4 = 2x - 3$$ Subtract x^2 from both sides. The x^2-terms are eliminated. The result is a linear equation in one variable.

$$2x - 4 = -3$$ To eliminate $2x$ on the right side, subtract $2x$ from both sides.

$$2x = 1$$ To undo the subtraction of 4, add 4 to both sides.

$$x = \frac{1}{2}$$ To undo the multiplication by 2, divide both sides by 2.

A check will show that $\frac{1}{2}$ is the solution of the original equation.

Self Check 5 Solve: $\dfrac{1}{x + 3} + \dfrac{1}{x - 3} = \dfrac{5}{x^2 - 9}$

Now Try ▶ Problem 45

2 Solve for a Specified Variable in a Formula.

Many formulas are expressed as rational equations. To solve such formulas for a specified variable, we use the same steps, in the same order, as we do when solving rational equations having only one variable.

EXAMPLE 6 **Determining a Child's Dosage.** The formula $C = \dfrac{AD}{A + 12}$ is called **Young's rule.** It is a way to find the approximate child's dose C of a medication, where A is the age of the child in years and D is the recommended dosage for an adult. Solve the formula for D.

Strategy As we have done in the previous examples, we will begin by multiplying both sides of the equation by the LCD to clear it of the fraction.

Why To isolate D on the right side of the equation, we must first isolate the term AD on that side. That calls for clearing the right side of the denominator $A + 12$.

Solution

$$C = \frac{AD}{A + 12}$$ This is Young's rule.

$$(A + 12)(C) = (A + 12)\left(\frac{AD}{A + 12}\right)$$ Write each side of the formula within parentheses, and then multiply both sides by the LCD, $A + 12$.

Success Tip

It is often helpful to circle the variable in the formula that you are solving for:

$$C = \frac{A\boxed{D}}{A + 12}$$

$(A + 12)C = AD$ Simplify the right side: $(A + \overset{1}{12})\left(\dfrac{AD}{\cancel{A + 12}}\right)$.

$AC + 12C = AD$ Distribute the multiplication by C.

$\dfrac{AC + 12C}{A} = \dfrac{AD}{A}$ To undo the multiplication by A on the right side and isolate D, divide both sides by A.

$\dfrac{AC + 12C}{A} = D$ Simplify the right side: $\dfrac{A\overset{1}{D}}{\cancel{A}} = D.$

Solving Young's rule for D, we have $D = \dfrac{AC + 12C}{A}$.

Self Check 6   Solve $R = \dfrac{eS}{T - 10}$ for S.

Now Try ▶ Problem 49

EXAMPLE 7

Photography.   The design of a camera lens uses the formula $\frac{1}{f} = \frac{1}{p} + \frac{1}{q}$, where f is the focal length of the lens, p is the distance from the lens to the object, and q is the distance from the lens to the image. Solve the formula for q.

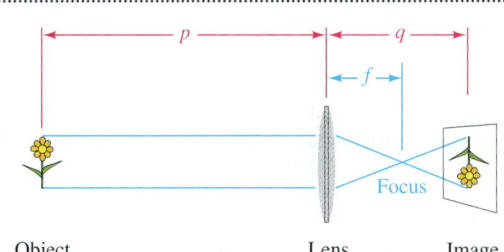

Strategy   We will begin by multiplying both sides of the equation by the LCD.

Why   It will be easier to isolate q if there are no fractions.

Solution

$$\frac{1}{f} = \frac{1}{p} + \frac{1}{q}$$ This is the given formula.

$$fpq\left(\frac{1}{f}\right) = fpq\left(\frac{1}{p} + \frac{1}{q}\right)$$ Write each side of the formula within parentheses and then multiply both sides by the LCD, fpq.

$$\overset{1}{\cancel{f}}pq\left(\frac{1}{\cancel{f}}\right) = f\overset{1}{\cancel{p}}q\left(\frac{1}{\cancel{p}}\right) + fp\overset{1}{\cancel{q}}\left(\frac{1}{\cancel{q}}\right)$$ Distribute the multiplication by fpq and then remove the common factors of each numerator and denominator.

$$pq = fq + fp$$ Simplify. Note that q is on both sides of the resulting equation.

Caution

A common error is to divide both sides of $pq = fq + fp$ by p to solve for q:

$$q = \frac{\cancel{fq} + fp}{\cancel{p}}$$

The formula is not solved for q because q appears on *both sides* of the equation.

If we subtract fq from both sides, all terms that contain q will be on the left side.

$pq - fq = fp$ Subtract fq from both sides.

$q(p - f) = fp$ Factor out the GCF, q, from the two terms on the left side.

$\dfrac{q(p - f)}{p - f} = \dfrac{fp}{p - f}$ To undo the multiplication by $(p - f)$ and isolate q, divide both sides by $p - f$.

$q = \dfrac{fp}{p - f}$ Simplify the left side: $\dfrac{q(p \overset{1}{\cancel{- f}})}{\cancel{p - f}} = q.$

Solving the formula for q, we have $q = \dfrac{fp}{p - f}$.

Self Check 7   Solve the formula in Example 7 for p.

Now Try ▶ Problem 51

VOCABULARY

Fill in the blanks.

1. Equations that contain one or more rational expressions, such as $\frac{x}{x + 2} = 4 + \frac{10}{x + 2}$, are called _____ equations.

2. To _____ a rational equation we find all the values of the variable that make the equation true.

3. To _____ a rational equation of fractions, multiply both sides by the LCD of all rational expressions in the equation.

4. When solving a rational equation, if we obtain a number that does not satisfy the original equation, the number is called an _____ solution.

CONCEPTS

5. Is 5 a solution of the given rational equation?

   a. $\frac{1}{x - 1} = 1 - \frac{3}{x - 1}$

   b. $\frac{x}{x - 5} = 3 + \frac{5}{x - 5}$

6. A student was asked to solve a rational equation. The first step of his solution is as follows:

   $$12x\left(\frac{5}{x} + \frac{2}{3}\right) = 12x\left(\frac{7}{4x}\right)$$

   a. What equation was he asked to solve?

   b. What LCD is used to clear the equation of fractions?

7. Consider the rational equation $\frac{x}{x - 3} = \frac{1}{x} + \frac{2}{x - 3}$.

   a. What values of x make a denominator 0?

   b. What values of x make a rational expression undefined?

   c. What numbers can't be solutions of the equation?

8. A student solved a rational equation and found 8 to be a possible solution. When she checked 8, she obtained $\frac{3}{0} = \frac{1}{0} + \frac{2}{3}$. What conclusion can be drawn?

By what should both sides of the equation be multiplied to clear it of fractions?

9. a. $\frac{1}{y} = 20 - \frac{5}{y}$ b. $\frac{x}{x^2 - 4} = \frac{4}{x - 2}$

10. a. $\frac{x}{5} = \frac{3x}{10} + \frac{7}{2x}$ b. $\frac{2x}{x - 6} = 4 + \frac{1}{x - 6}$

11. Perform each multiplication.

    a. $4x\left(\frac{3}{4x}\right)$ b. $(x + 6)(x - 2)\left(\frac{3}{x - 2}\right)$

12. Fill in the blanks.

    $$8x\left(\frac{3}{4x}\right) = 8x\left(\frac{1}{8x}\right) + 8x\left(\frac{5}{4}\right)$$
    $$\boxed{} = \boxed{} + \boxed{}$$

NOTATION

Complete the solution to solve the equation.

13. $\frac{2}{a} + \frac{1}{2} = \frac{7}{2a}$

    $$\boxed{}\left(\frac{2}{a} + \frac{1}{2}\right) = \boxed{}\left(\frac{7}{2a}\right)$$

    $$\boxed{}\left(\frac{2}{a}\right) + \boxed{}\left(\frac{1}{2}\right) = \boxed{}\left(\frac{7}{2a}\right)$$

    $$\boxed{} + a = \boxed{}$$

    $$4 + a - 4 = 7 - \boxed{}$$

    $$a = \boxed{}$$

14. Can $5x\left(\frac{2}{x} + \frac{4}{5}\right)$ be written as $5x \cdot \frac{2}{x} + \frac{4}{5}$? Explain.

GUIDED PRACTICE

Solve each equation and check the result. If an equation has no solution, so indicate. See Example 1.

15. $\frac{2}{3} = \frac{1}{2} + \frac{x}{6}$

16. $\frac{7}{4} = \frac{x}{8} + \frac{5}{2}$

17. $\frac{s}{12} - \frac{s}{2} = \frac{5s}{4}$

18. $\frac{n}{18} - \frac{n}{6} = \frac{4n}{3}$

19. $\frac{x}{18} = \frac{1}{3} - \frac{x}{2}$

20. $\frac{x}{4} = \frac{1}{2} - \frac{3x}{20}$

21. $\frac{b}{4} + \frac{1}{2} = \frac{b}{3} - \frac{1}{4}$

22. $\frac{n}{6} + \frac{2}{3} = \frac{n}{3} - \frac{1}{36}$

Solve each equation and check the result. If an equation has no solution, so indicate. See Example 2.

23. $\frac{5}{3k} + \frac{1}{k} = -2$

24. $\frac{3}{4h} + \frac{2}{h} = 1$

25. $\frac{1}{4} - \frac{5}{6} = \frac{1}{a}$

26. $\frac{5}{9} - \frac{1}{3} = \frac{1}{b}$

27. $\frac{1}{8} + \frac{2}{b} - \frac{1}{12} = 0$

28. $\frac{1}{14} + \frac{2}{n} - \frac{2}{21} = 0$

29. $\frac{4}{5} - \frac{1}{10x} = \frac{7}{15}$

30. $\frac{5}{14} - \frac{1}{2x} = \frac{3}{7}$

Solve each equation and check the result. If an equation has no solution, so indicate. See Example 3.

31. $x + \frac{8}{x} = 6$

32. $z - \frac{16}{z} = 6$

33. $\frac{10}{t} - t = 3$

34. $\frac{7}{p} - p = -6$

35. $\frac{20}{c} + c = -9$

36. $d = 4 + \frac{21}{d}$

37. $4 + \frac{15}{p} = 3p$

38. $2x = 6 + \frac{8}{x}$

Solve each equation and check the result. If an equation has no solution, so indicate. See Example 4.

39. $\frac{x}{x - 5} = 3 + \frac{5}{x - 5}$

40. $\dfrac{3}{y-2} = \dfrac{3}{y-2} + 1$

41. $\dfrac{a^2}{a+2} - a = \dfrac{4}{a+2}$

42. $\dfrac{z^2}{z+1} + 2 = \dfrac{1}{z+1}$

Solve each equation and check the result. If an equation has no solution, so indicate. See Example 5.

43. $\dfrac{x+6}{x+4} + \dfrac{1}{x^2+x-12} = 1$

44. $\dfrac{x+7}{x+2} + \dfrac{1}{x^2-3x-10} = 1$

45. $\dfrac{2x}{x^2+x-2} + \dfrac{2}{x+2} = 1$

46. $\dfrac{4x}{x^2+2x-3} + \dfrac{3}{x+3} = 1$

Solve each formula for the specified variable. See Example 6.

47. $h = \dfrac{2A}{b+d}$ for A

48. $T = \dfrac{3R}{M-n}$ for R

49. $I = \dfrac{E}{R+r}$ for r

50. $\dfrac{S}{k+h} = E$ for k

51. $\dfrac{5}{x} - \dfrac{4}{y} = \dfrac{5}{z}$ for x

52. $\dfrac{2}{c} + \dfrac{2}{d} = \dfrac{1}{h}$ for c

53. $\dfrac{1}{r} + \dfrac{1}{s} = \dfrac{1}{t}$ for r

54. $\dfrac{1}{x} - \dfrac{1}{y} = \dfrac{1}{z}$ for x

Solve each formula for the specified variable. See Example 7.

55. $\dfrac{P}{n} = rt$ for P

56. $\dfrac{F}{m} = a$ for F

57. $\dfrac{a}{b} = \dfrac{c}{d}$ for d

58. $\dfrac{pc}{s} = \dfrac{t}{r}$ for c

59. $\dfrac{1}{a} + \dfrac{1}{b} = 1$ for a

60. $\dfrac{1}{a} - \dfrac{1}{b} = 1$ for b

61. $F = \dfrac{L^2}{6d} + \dfrac{d}{2}$ for L^2

62. $H = \dfrac{J^3}{cd} - \dfrac{K^3}{d}$ for J^3

TRY IT YOURSELF

Solve each equation and check the result. If an equation has no solution, so indicate.

63. $\dfrac{1}{3} + \dfrac{2}{x-3} = 1$

64. $\dfrac{3}{5} + \dfrac{7}{x+2} = 2$

65. $\dfrac{7}{q^2-q-2} + \dfrac{1}{q+1} = \dfrac{3}{q-2}$

66. $\dfrac{3}{x-1} - \dfrac{1}{x+9} = \dfrac{18}{x^2+8x-9}$

67. $\dfrac{2}{3-t} = \dfrac{-t}{t+3}$

68. $\dfrac{n}{n+1} = \dfrac{6}{n+7}$

69. $\dfrac{1}{8} + \dfrac{2}{y} = \dfrac{1}{y} + \dfrac{1}{10}$

70. $\dfrac{7}{10} + \dfrac{4}{c} = \dfrac{1}{c} + \dfrac{11}{15}$

71. $4 - \dfrac{8}{x+1} = \dfrac{8x}{x+1}$

72. $\dfrac{x}{x-2} = \dfrac{2}{x-2} + 2$

73. $\dfrac{5a}{a+1} - 4 = \dfrac{3}{a+1}$

74. $\dfrac{4}{b-3} = \dfrac{b+5}{b-3} - 5$

75. $\dfrac{2}{y+1} + 5 = \dfrac{12}{y+1}$

76. $\dfrac{3}{p+6} - 2 = \dfrac{7}{p+6}$

77. $\dfrac{3}{x+1} = \dfrac{x-2}{x+1} + \dfrac{x-2}{2}$

78. $\dfrac{2}{x-1} + \dfrac{x-2}{3} = \dfrac{4}{x-1}$

79. $\dfrac{z-4}{z-3} = \dfrac{z+2}{z+1}$

80. $\dfrac{a+2}{a+8} = \dfrac{a-3}{a-2}$

81. $\dfrac{3}{x} + 2 = 3$

82. $\dfrac{2}{x} + 9 = 11$

83. $\dfrac{4}{y^2-4} = \dfrac{1}{y-2} + \dfrac{1}{y+2}$

84. $\dfrac{2w}{w^2-9} = \dfrac{1}{w+3} - \dfrac{4}{w-3}$

85. $\dfrac{3}{5d} + \dfrac{4}{3} = \dfrac{9}{10d}$

86. $\dfrac{2}{3d} + \dfrac{1}{4} = \dfrac{11}{6d}$

87. $\dfrac{n}{n^2-9} + \dfrac{n+8}{n+3} = \dfrac{n-8}{n-3}$

88. $\dfrac{7}{x-5} = \dfrac{40}{x^2-25} + \dfrac{3}{x+5}$

89. $\dfrac{3}{x-2} + \dfrac{1}{x} = \dfrac{6x+4}{x^2-2x}$

90. $\dfrac{x}{x-1} - \dfrac{12}{x^2-x} = \dfrac{-1}{x-1}$

91. $y + \dfrac{2}{3} = \dfrac{2y-12}{3y-9}$

92. $1 - \dfrac{3}{b} = \dfrac{-8b}{b^2+3b}$

93. $\dfrac{a-1}{7} - \dfrac{a-2}{14} = \dfrac{1}{2}$

94. $\dfrac{3x-1}{6} - \dfrac{x+3}{2} = \dfrac{3x+4}{3}$

Look Alikes . . .

For each expression, perform the indicated operations and then simplify, if possible. Solve each equation and check the result.

95. a. $\dfrac{a}{3} + \dfrac{3}{5} + \dfrac{a}{15}$

b. $\dfrac{a}{3} + \dfrac{3}{5} = \dfrac{a}{15}$

96. a. $\dfrac{1}{6x} - \dfrac{2}{x-6}$

b. $\dfrac{1}{6x} = \dfrac{2}{x-6}$

97. a. $\dfrac{x}{x-2} - \dfrac{1}{x-3}$

b. $\dfrac{x}{x-2} - \dfrac{1}{x-3} = 1$

98. a. $\dfrac{u^2 + 1}{u^2 - u} - \dfrac{u}{u - 1}$ **b.** $\dfrac{u^2 + 1}{u^2 - u} - \dfrac{u}{u - 1} = \dfrac{1}{u}$

APPLICATIONS

99. Medicine. Radioactive tracers are used for diagnostic work in nuclear medicine. The *effective half-life H* of a radioactive material in an organism is given by the formula $H = \dfrac{RB}{R + B}$ where R is the radioactive half-life and B is the biological half-life of the tracer. Solve the formula for R.

100. Chemistry. Charles's law describes the relationship between the volume and temperature of a gas that is kept at a constant pressure. It can be expressed as $\dfrac{V_1}{V_2} = \dfrac{T_1}{T_2}$ where V_1 and V_2 are variables representing two different volumes, and T_1 and T_2 are variables representing two different temperatures. (Recall that the notation V_1 is read as *V sub one*.) Solve for V_2.

101. Electronics. Most electronic circuits require resistors to make them work properly. Resistors are components that limit current. An important formula about resistors in a circuit is $\dfrac{1}{r} = \dfrac{1}{r_1} + \dfrac{1}{r_2}$. Solve for r.

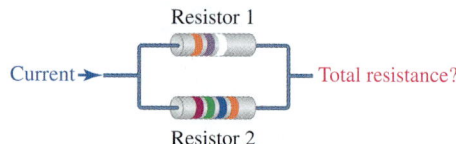

102. Mathematical Formulas. To quickly find the sum $\dfrac{1}{2} + \dfrac{1}{4} + \dfrac{1}{8} + \dfrac{1}{16} + \dfrac{1}{32} + \dfrac{1}{64} + \dfrac{1}{128}$, mathematicians use the formula $S = \dfrac{a(1 - r^n)}{1 - r}$. Solve the formula for a.

WRITING

103. Explain how the multiplication property of equality is used to solve rational equations. Give an example.

104. When solving rational equations, how do you know whether a solution is extraneous?

105. What is meant by clearing a rational equation of fractions? Give an example.

106. Explain the difference between the procedure used to simplify $\dfrac{1}{x} + \dfrac{1}{3}$ and the procedure used to solve $\dfrac{1}{x} + \dfrac{1}{3} = \dfrac{1}{2}$.

REVIEW

107. Uniforms. A cheerleading squad had their school mascot embroidered on the front of their uniform sweaters. They were charged $18.50 per sweater plus a one time setup fee of $75. If the project cost $445, how many sweaters were embroidered?

108. Geometry. The vertex angle of an isosceles triangle is 46°. Find the measure of each base angle.

CHALLENGE PROBLEMS

Solve each equation and check the result. If an equation has no solution, so indicate.

109. $\dfrac{x - 4}{x - 3} + \dfrac{x - 2}{x - 3} = x - 3$

110. $\dfrac{3}{x} = \dfrac{1 - \dfrac{1}{x}}{3 - \dfrac{7}{x}}$

111. $x^{-2} + 2x^{-1} + 1 = 0$

112. Engines. A formula that is used in the design and testing of diesel engines is $E = 1 - \dfrac{T_4 - T_1}{a(T_3 - T_2)}$. Solve the formula for T_1.

SECTION 7.7

Problem Solving Using Rational Equations

OBJECTIVES

1 Solve number problems.

2 Solve uniform motion problems.

3 Solve shared-work problems.

4 Solve investment problems.

ARE YOU READY?

The following problems review some basic skills that are needed when using rational equations to solve application problems.

1. Solve the uniform motion formula $d = rt$ for t.

2. Multiply: $\dfrac{1}{5} \cdot x$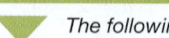

3. What is the simple interest formula?

4. What is the LCD for the fractions in the rational equation $\dfrac{x}{9} + \dfrac{x}{7} = 1$?

We will now use the six-step problem-solving strategy to solve application problems from a variety of areas, including banking, petroleum engineering, sports, and travel. In each case, we will use a rational equation to model the situation. We begin with an example in which we find an unknown number.

1 Solve Number Problems.

EXAMPLE 1   **Number Problem.**   If the same number is added to both the numerator and the denominator of the fraction $\frac{3}{5}$, the result is $\frac{4}{5}$. Find the number.

Analyze

- Begin with the fraction $\frac{3}{5}$.
- Add the same number to the numerator and to the denominator.
- The result is $\frac{4}{5}$.
- Find the number.

Assign   Let $n =$ the unknown number.

Form   To form an equation, add the unknown number to the numerator and to the denominator of $\frac{3}{5}$. Then set the result equal to $\frac{4}{5}$.

$$\frac{3+n}{5+n} = \frac{4}{5}$$

Solve   To solve this rational equation, we begin by clearing it of fractions.

$$\frac{3+n}{5+n} = \frac{4}{5}$$

$$5(5+n)\left(\frac{3+n}{5+n}\right) = 5(5+n)\left(\frac{4}{5}\right) \quad$$ Multiply both sides by the LCD, $5(5+n)$. Then remove common factors of the numerator and denominator.

$$5(3+n) = (5+n)4 \quad$$ Simplify. The fractions have been cleared.

$$15 + 5n = 20 + 4n \quad$$ Distribute the multiplication by 5 and by 4.

$$15 + n = 20 \quad$$ To isolate the variable term on the left side, subtract $4n$ from both sides.

$$n = 5 \quad$$ To undo the addition of 15, subtract 15 from both sides.

State   The number is 5.

Check   When we add 5 to both the numerator and denominator of $\frac{3}{5}$, we get

$$\frac{3+5}{5+5} = \frac{8}{10} = \frac{4}{5}$$

The result checks.

Self Check 1   **Number Problem.**   If the same number is added to both the numerator and denominator of the fraction $\frac{7}{9}$, the result is $\frac{8}{9}$. Find the number.

Now Try ▶ Problem 13

2 Solve Uniform Motion Problems.

The Language of Algebra

In uniform motion problems, the word **speed** is often used in place of the word **rate.** For example, we can say a car travels at a *rate* of 50 mph or its *speed* is 50 mph.

Recall that we use the distance formula $d = rt$ to solve motion problems. The relationship between distance, rate, and time can be expressed in another way by solving for t.

$$d = rt \quad$$ Distance = rate · time.

$$\frac{d}{r} = \frac{rt}{r} \quad$$ To undo the multiplication by r and isolate t, divide both sides by r.

$$\frac{d}{r} = t \quad$$ Simplify the right side: $\frac{rt}{r} = t$.

$$t = \frac{d}{r}$$

This result suggests an alternate form of the distance formula, time $= \frac{\text{distance}}{\text{rate}}$, that is used to solve the next example.

EXAMPLE 2

Runners. A coach can run 10 miles in the same amount of time as his best student-athlete can run 12 miles. If the student runs 1 mile per hour (mph) faster than the coach, find the running speeds of the coach and the student.

Analyze

- The coach runs 10 miles in the same time that the student runs 12 miles.
- The student runs 1 mph faster than the coach.
- Find the speed that each runs.

Assign Since the student's speed is 1 mph faster than the coach's, let $r =$ the speed that the coach can run. Then, $r + 1 =$ the speed that the student can run.

Form The expressions for the rates are entered in the Rate column of the table. The distances run by the coach and by the student are entered in the Distance column of the table.

Using $t = \frac{d}{r}$, we find that the time it takes the coach to run 10 miles, at a rate of r mph, is $\frac{10}{r}$ hours. Similarly, we find that the time it takes the student to run 12 miles, at a rate of $(r + 1)$ mph, is $\frac{12}{r + 1}$ hours. These expressions are entered in the Time column of the table.

©Exactostock/SuperStock

	Rate	·	Time	=	Distance
Coach	r		$\frac{10}{r}$		10
Student	$r + 1$		$\frac{12}{r + 1}$		12

Enter the information in these two columns first.

To get these entries, divide the distance by the rate to obtain an expression for the time: $t = \frac{d}{r}$.

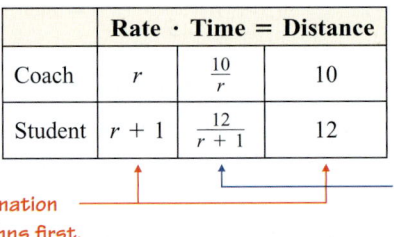

The time it takes the coach to run 10 miles	is the same as	the time it takes the student to run 12 miles.
$\frac{10}{r}$	$=$	$\frac{12}{r + 1}$

Solve To solve this rational equation, we begin by clearing it of fractions.

$$\frac{10}{r} = \frac{12}{r + 1}$$

$$\overset{1}{\cancel{r}}(r + 1)\left(\frac{10}{\cancel{r}}\right)_{1} = r(\overset{1}{\cancel{r + 1}})\left(\frac{12}{\cancel{r + 1}}\right)_{1}$$ Multiply both sides by the LCD, $r(r + 1)$. Then remove common factors of the numerator and denominator.

$$(r + 1)10 = 12r$$ Simplify. The fractions have been cleared.

$$10r + 10 = 12r$$ On the left side, distribute the multiplication by 10.

$$10 = 2r$$ To isolate the variable term on the right, subtract 10r from both sides.

$$5 = r$$ To undo the multiplication by 2, divide both sides by 2.

If $r = 5$, then $r + 1 = 6$.

State The coach's running speed is 5 mph and the student's running speed is 6 mph.

Check The coach will run 10 miles in $\frac{10 \text{ miles}}{5 \text{ mph}} = 2$ hours. The student will run 12 miles in $\frac{12 \text{ miles}}{6 \text{ mph}} = 2$ hours. The times are the same; the results check.

3 Solve Shared-Work Problems.

Problems in which two or more people (or machines) work together to complete a job are called *shared-work problems*. To solve such problems, we must determine the **rate of work** for each person (or machine) involved. For example, suppose it takes you 4 hours to clean your house. Your rate of work can be expressed as $\frac{1}{4}$ of the job is completed per hour. If someone else takes 5 hours to clean the same house, they complete $\frac{1}{5}$ of the job per hour. In general, a rate of work can be determined in the following way.

Rate of Work	If a job can be completed in t units of time, the rate of work can be expressed as: $\frac{1}{t}$ of the job is completed per unit of time.

To solve shared-work problems, we also must determine what fractional part of a job is completed. To do this, we use the formula

Work completed = rate of work · time worked   or $W = rt$

EXAMPLE 3 **Payroll.**   At the end of a pay period, it takes the president of a company 15 minutes to sign all of her employees' payroll checks. What fractional part of the job is completed if the president signs checks for 10 minutes?

Strategy   We will begin by finding the president's check-signing rate. Then we can use the formula $W = rt$ to find the part of the job that is completed.

Why   We know the time worked is 10 minutes. To use the work formula to find what part of the job is completed, we also need to know the president's work rate.

Solution   If all of the checks can be signed in 15 minutes, the president's work rate is $\frac{1}{15}$ job per minute. Substituting into the work formula, we have

$$W = rt$$

$$= \frac{1}{15} \cdot 10 \qquad \text{Substitute } \tfrac{1}{15} \text{ for } r, \text{ the work rate, and 10 for } t, \text{ the time worked.}$$

$$= \frac{10}{15} \qquad \text{Do the multiplication. Think: } \tfrac{1}{15} \cdot \tfrac{10}{1} = \tfrac{10}{15}.$$

$$= \frac{2}{3} \qquad \text{Simplify by removing the common factor of 5: } \frac{2 \cdot \overset{1}{\cancel{5}}}{3 \cdot \underset{1}{\cancel{5}}} = \tfrac{2}{3}.$$

In 10 minutes, the president will complete $\frac{2}{3}$ of the job of signing the payroll checks.

EXAMPLE 4

Filling a Tank. An inlet pipe can fill an oil storage tank in 7 days, and a second inlet pipe can fill the same tank in 9 days. If both pipes are used, how long will it take to fill the tank?

Analyze

- The first pipe can fill the tank in 7 days.
- The second pipe can fill the tank in 9 days.
- How long will it take the two pipes, working together, to fill the tank?

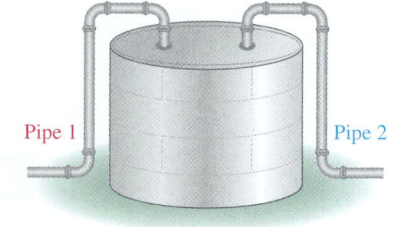

Pipe 1   Pipe 2

Assign Let x = the number of days it will take to fill the tank if both pipes are used.

Form It is helpful to organize the facts of the problem in a table. Since the pipes will be open for the same amount of time as they fill the tank, enter x as the time worked for each pipe.

The first pipe can fill the tank in 7 days; its rate working alone is $\frac{1}{7}$ of the job per day. The second pipe can fill the tank in 9 days; its rate working alone is $\frac{1}{9}$ of the job per day. To determine the work completed by each pipe, multiply the rate by the time.

	Rate	· Time	= Work completed
1st pipe	$\frac{1}{7}$	x	$\frac{x}{7}$
2nd pipe	$\frac{1}{9}$	x	$\frac{x}{9}$

Think: $\frac{1}{7} \cdot \frac{x}{1} = \frac{x}{7}$.

Think: $\frac{1}{9} \cdot \frac{x}{1} = \frac{x}{9}$.

Enter this information first.

Multiply to get each of these entries: $W = rt$.

In shared-work problems, the number 1 represents one whole job completed. So we have

The part of job done by 1st pipe	plus	part of job done by 2nd pipe	equals	1 job completed.
$\frac{x}{7}$	$+$	$\frac{x}{9}$	$=$	1

Solve

$$\frac{x}{7} + \frac{x}{9} = 1$$ This is a rational equation.

$$63\left(\frac{x}{7} + \frac{x}{9}\right) = 63(1)$$ Clear the equation of fractions by multiplying both sides by the LCD, 63.

$$63\left(\frac{x}{7}\right) + 63\left(\frac{x}{9}\right) = 63$$ On the left side, distribute the multiplication by 63.

$$9x + 7x = 63$$ Simplify the left side: $\overset{1}{7} \cdot 9\left(\frac{x}{\overset{}{7}}\right) = 9x$ and $7 \cdot \overset{1}{9}\left(\frac{x}{\overset{}{9}}\right) = 7x$.
The fractions have been cleared.

$$16x = 63$$ Combine like terms.

$$x = \frac{63}{16}$$ To undo the multiplication by 16 and isolate x, divide both sides by 16.

State If both pipes are used, it will take $\frac{63}{16}$ or $3\frac{15}{16}$ days to fill the tank.

Check To check, we use the work formula and multiply each rate by the time. In $\frac{63}{16}$ days, the first pipe fills $\frac{1}{7} \cdot \frac{63}{16} = \frac{9}{16}$ of the tank and the second pipe fills $\frac{1}{9} \cdot \frac{63}{16} = \frac{7}{16}$ of the tank. The sum of these efforts, $\frac{9}{16} + \frac{7}{16}$, is $\frac{16}{16}$ or 1 full tank. The result checks.

Self Check 4 **Mailing Flyers.** A school secretary can prepare a mass mailing of an informational flyer in 6 hours. A student worker would take 8 hours to prepare the mailing. How long will it take to prepare the mailing if they work together?

Now Try ▶ Problem 37

Strategy for Solving Shared-Work Problems

Equations that model shared-work problems involving two people (or machines) have the form

$$\frac{x}{a} + \frac{x}{b} = 1$$

where x represents the time they work together on the job, and a and b represent the respective times each worker needs to complete the job alone.

Example 4 can be solved in a different way by considering the amount of work done by each pipe in 1 day. As before, if we let $x =$ the number of days it will take to fill the tank if both inlet pipes are used, then together, in 1 day, they will complete $\frac{1}{x}$ of the job. If we add what the first pipe can do in 1 day to what the second pipe can do in 1 day, the sum is what they can do together in 1 day.

What the first inlet pipe can do in 1 day	plus	what the second inlet pipe can do in 1 day	equals	what they can do together in 1 day.
$\frac{1}{7}$	$+$	$\frac{1}{9}$	$=$	$\frac{1}{x}$

To solve the equation, begin by clearing it of fractions.

$$\frac{1}{7} + \frac{1}{9} = \frac{1}{x}$$

$$63x\left(\frac{1}{7} + \frac{1}{9}\right) = 63x\left(\frac{1}{x}\right)$$ Multiply both sides by the LCD, 63x.

$$9x + 7x = 63$$ Distribute the multiplication by 63x and simplify.

$$16x = 63$$ Combine like terms.

$$x = \frac{63}{16}$$ To isolate x, divide both sides by 16.

This is the same answer as the one obtained in Example 4.

4 Solve Investment Problems.

We have used the interest formula $I = Prt$ to solve investment problems. The relationships among interest, principal, rate, and time can be expressed in another way, by solving for P.

$$I = Prt$$ Interest = principal · rate · time.

$$\frac{I}{rt} = \frac{Prt}{rt}$$ To undo the multiplication by rt and isolate P, divide both sides by rt.

$$\frac{I}{rt} = P$$ Simplify the right side: $\dfrac{P\overset{1}{\cancel{r}}\overset{1}{\cancel{t}}}{\underset{1}{\cancel{r}}\underset{1}{\cancel{t}}} = P.$

$$P = \frac{I}{rt}$$ Reverse the sides of the equation so that P is on the left.

This alternate form of the interest formula, Principal $= \frac{\text{Interest}}{\text{rate} \cdot \text{time}}$, is used to solve the next example.

EXAMPLE 5   **Comparing Investments.**   An amount of money invested for one year in bonds will earn $120. At a bank, that same amount of money will only earn $75 interest, because the interest rate paid by the bank is 3% less than that paid by the bonds. Find the rate of interest paid by each investment.

Analyze

- The investment in bonds earns $120 in one year.
- The same amount of money, invested in a bank, earns $75 in one year.
- The interest rate paid by the bank is 3% less than that paid by the bonds.
- Find the bond's rate of interest and the bank's rate of interest.

Assign   Since the interest rate paid by the bank is 3% less than that paid by the bonds, let $r =$ the bond's rate of interest, and $r - 0.03 =$ the bank's interest rate. (Recall that $3\% = 0.03$.)

Form   If an investment earns $120 interest in 1 year at some rate r, we can use $P = \dfrac{I}{rt}$ to find that the principal invested was $\dfrac{120}{r}$ dollars. Similarly, if another investment earns $75 interest in 1 year at some rate $r - 0.03$, the principal invested was $\dfrac{75}{r - 0.03}$ dollars. We can organize the facts of the problem in a table.

	Principal $\cdot$	Rate	$\cdot$ Time =	Interest
Bonds	$\dfrac{120}{r}$	r	1	120
Bank	$\dfrac{75}{r - 0.03}$	$r - 0.03$	1	75

Divide to get each of these entries: $P = \dfrac{I}{rt}$.

Enter this information first.

The amount invested in the bonds	equals	the amount invested in the bank.
$\dfrac{120}{r}$	$=$	$\dfrac{75}{r - 0.03}$

Solve

$$\frac{120}{r} = \frac{75}{r - 0.03}$$ This is a rational equation.

$$\frac{1}{r}(r - 0.03)\left(\frac{120}{r}\right)\frac{}{1} = \frac{1}{r}(r - 0.03)\left(\frac{75}{r - 0.03}\right)\frac{}{1}$$ Multiply both sides by the LCD, $r(r - 0.03)$. Then remove common factors of the numerator and denominator.

$$(r - 0.03)120 = 75r$$ Simplify. The fractions have been cleared.

$$120r - 3.6 = 75r$$ On the left side, distribute the multiplication by 120.

$$45r - 3.6 = 0$$ To isolate the variable term on the left side, subtract 75r from both sides.

$$45r = 3.6$$ To undo the subtraction of 3.6, add 3.6 to both sides.

$$r = 0.08$$ To undo the multiplication by 45 and isolate r, divide both sides by 45.

If $r = 0.08$, then the bank's interest rate is given by $r - 0.03 = 0.05$.

State   The bonds pay 0.08, or 8%, interest. The bank's interest rate is 5%.

Check   The amount invested at 8% that will earn $120 interest in 1 year is $\dfrac{120}{(0.08)1} = \$1,500$.

The amount invested at 5% that will earn $75 interest in 1 year is $\dfrac{75}{(0.05)1} = \$1,500$. The amounts invested in the bonds and the bank are the same. The results check.

Self Check 5 **Comparing Investments.** An amount of money invested for one year in a certificate of deposit will earn $210. The same amount of money in a savings account will earn $70. If the certificate of deposit's interest rate is 2% more than the savings account's rate, find the interest rate of the savings account.

Now Try ▶ Problem 41

SECTION 7.7 ▶ STUDY SET

VOCABULARY

Fill in the blanks.

1. In this section, problems that involve:
   - moving vehicles are called uniform _____ problems.
   - depositing money are called _____ problems.
   - people completing jobs are called shared-_____ problems.
2. In the formula $W = rt$, the variable W stands for the _____ completed, r is the _____, and t is the _____.

CONCEPTS

3. Choose the equation that can be used to solve the following problem: *If the same number is added to the numerator and the denominator of the fraction $\frac{5}{8}$, the result is $\frac{2}{3}$. Find the number.*

   (i) $\dfrac{5}{8} + x = \dfrac{2}{3}$ (ii) $\dfrac{5 + x}{8} = \dfrac{2}{3}$

   (iii) $\dfrac{5 + x}{8 + x} = \dfrac{2}{3}$ (iv) $\dfrac{5}{8} = \dfrac{2 + x}{3 + x}$

4. Fill in the blank: If a job can be completed in t hours, then the rate of work can be expressed as $\frac{1}{}$ of the job is completed per hour.

5. a. It takes a night security officer 45 minutes to check each of the doors in an office building to make sure they are locked. What is the officer's rate of work?
   b. It takes an elementary school teacher 4 hours to make out the semester report cards. What part of the job does she complete in x hours?

6. **Hospitals.** An experienced employee can sterilize an operating room in 3 hours. It takes a new employee 5 hours to sterilize the same room. Select the best estimate below of the time it will take them to sterilize the room if they work together.

   - Less than 3 hours
   - Between 3 and 5 hours
   - More than 5 hours

7. a. Solve $d = rt$ for t.
   b. Solve $I = Prt$ for P.

8. Complete the table.

	r	$\cdot$	t	$= d$
Snowmobile	r			4
4×4 truck	$r - 5$			3

9. Complete the table.

	Rate	$\cdot$ Time	= Work completed
1st printer	$\frac{1}{15}$	x	
2nd printer	$\frac{1}{8}$	x	

10. Complete the table.

	P	$\cdot$	r	$\cdot t$	$= I$
City savings bank			r	1	50
Credit union			$r - 0.02$	1	75

NOTATION

11. Write $\frac{55}{9}$ days using a mixed number.
12. a. Write 9% as a decimal.
    b. Write 0.035 as a percent.

GUIDED PRACTICE

Solve each of these number problems. **See Example 1.**

13. If the same number is added to both the numerator and the denominator of $\frac{2}{5}$, the result is $\frac{2}{3}$. Find the number.
14. If the same number is subtracted from both the numerator and the denominator of $\frac{11}{13}$, the result is $\frac{3}{4}$. Find the number.
15. If the denominator of $\frac{3}{4}$ is increased by a number, and the numerator is doubled, the result is 1. Find the number.
16. If a number is added to the numerator of $\frac{7}{8}$, and the same number is subtracted from the denominator, the result is 2. Find the number.

17. If a number is added to the numerator of $\frac{3}{4}$, and twice as much is added to the denominator, the result is $\frac{4}{7}$. Find the number.

18. If a number is added to the numerator of $\frac{5}{7}$, and twice as much is subtracted from the denominator, the result is 8. Find the number.

19. The sum of a number and its reciprocal is $\frac{13}{6}$. Find the number.

20. The sum of the reciprocals of two consecutive even integers is $\frac{7}{24}$. Find each integer.

APPLICATIONS

21. **Cooking.**   If the same number is added to both the numerator and the denominator of the amount of butter used in the following recipe for toffee, the result is the amount of brown sugar to be used. Find the number.

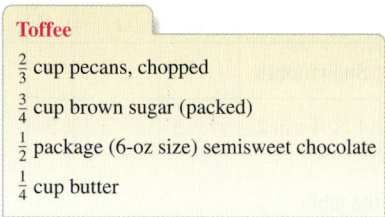

Toffee

$\frac{2}{3}$ cup pecans, chopped

$\frac{3}{4}$ cup brown sugar (packed)

$\frac{1}{2}$ package (6-oz size) semisweet chocolate

$\frac{1}{4}$ cup butter

22. **Tape Measures.**   If the same number is added to both the numerator and the denominator of the first measurement, the result is the second measurement. Find the number.

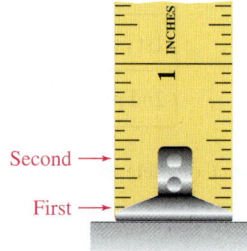

Second →
First →

23. **Tour De France.**   Maurice Garin of France won the first Tour de France bicycle road race in 1903. In 2005, American Lance Armstrong won his seventh consecutive Tour de France. Armstrong's average speed in 2005 was 10 mph faster than Garin's in 1903. In the time it took Garin to ride 80 miles, Armstrong could have ridden 130 miles. Find each cyclist's average speed.

24. **Physical Fitness.**   A woman can bicycle 28 miles in the same time as it takes her to walk 8 miles. She can ride 10 mph faster than she can walk. How fast can she walk?

25. **Packaging Fruit.**   The diagram below shows how apples are processed for market. Although the second conveyor belt is shorter, an apple spends the same amount of time on each belt because the second conveyor moves 1 foot per second slower than the first. Determine the speed of each conveyor belt.

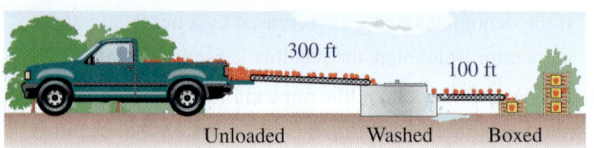

300 ft

100 ft

Unloaded   Washed   Boxed

26. **Comparing Travel.**   A plane can fly 300 miles in the same time as it takes a car to go 120 miles. If the car travels 90 mph slower than the plane, find the speed of the plane.

27. **Birds in Flight.**   Although flight speed is dependent upon the weather and the wind, in general, a Canada goose can fly about 10 mph faster than a great blue heron. In the same time that a Canada goose travels 120 miles, a great blue heron travels 80 miles. Find their flying speeds.

28. **Fast Cars.**   The top speed of a Dodge Charger SRT8 is 33 mph less than the top speed of a Chevrolet Corvette Z06. At their top speeds, a Corvette can travel 6 miles in the same time that a Charger can travel 5 miles. Find the top speed of each car.

©Jim West Alamy

29. **Wind Speed.**   When a plane flies downwind, the wind pushes the plane so that its speed is the *sum* of the speed of the plane in still air and the speed of the wind. Traveling upwind, the wind pushes against the plane so that its speed is the *difference* of the speed of the plane in still air and the speed of the wind. Suppose a plane that travels 255 mph in still air can travel 300 miles downwind in the same time as it takes to travel 210 miles upwind. Complete the following table and find the speed of the wind, represented by x.

	Rate · Time = Distance		
Downwind	$255 + x$		300
Upwind	$255 - x$		210

30. **Boating.**   A boat that travels 18 mph in still water can travel 22 miles downstream in the same time as it takes to travel 14 miles upstream. Find the speed of the current in the river. (See problem 29.)

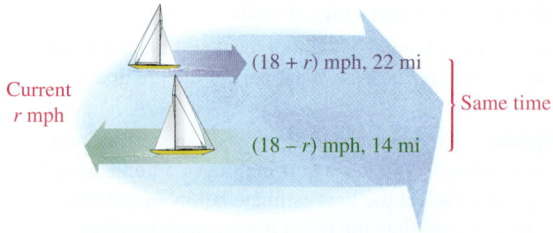

Current
r mph

$(18 + r)$ mph, 22 mi

$(18 - r)$ mph, 14 mi

Same time

31. **Roofing Houses.**   A homeowner estimates that it will take her 7 days to roof her house. A professional roofer estimates that he could roof the house in 4 days. How long will it take if the homeowner helps the roofer?

32. **Holiday Decorating.**   One crew can put up holiday decorations in the mall in 8 hours. A second crew can put up the decorations in 10 hours. How long will it take if both crews work together to decorate the mall?

33.

from Campus to Careers

Recreation Director

Suppose you are a recreation director at a summer camp. The water in the camp swimming pool was drained out for the winter and it is now time to refill the pool. One pipe can fill the empty pool in 12 hours and another can fill the empty pool in 18 hours. Suppose both pipes are opened at 8:00 A.M. and you have scheduled a swimming activity for 2:00 P.M. that day. Will the pool be filled by then?

34. Groundskeeping. It takes a groundskeeper 45 minutes to prepare a softball field for a game. It takes his assistant 55 minutes to prepare the same field. How long will it take if they work together to prepare the field?

35. Filling a Pool. One inlet pipe can fill an empty pool in 4 hours, and a drain can empty the pool in 8 hours. How long will it take the pipe to fill the pool if the drain is left open?

36. Sewage Treatment. A sludge pool is filled by two inlet pipes. One pipe can fill the pool in 15 days, and the other can fill it in 21 days. However, if no sewage is added, continuous waste removal will empty the pool in 36 days. How long will it take the two inlet pipes to fill an empty sludge pool?

37. Grading Papers. On average, it takes a teacher 30 minutes to grade a set of quizzes. It takes her teacher's aide twice as long to do the same grading. How long will it take if they work together to grade a set of quizzes?

38. Dog Kennels. It takes the owner/operator of a dog kennel 6 hours to clean all of the cages. It takes his assistant 2 hours more than that to clean the same cages. How long will it take if they work together?

39. Printers. It takes a printer 6 hours to print the class schedules for all of the students enrolled in a community college. A faster printer can print the schedules in 4 hours. How long will it take the two printers working together to print $\frac{3}{4}$ of the class schedules?

40. Office Work. In 5 hours, a secretary can address 100 envelopes. Another secretary can address 100 envelopes in 6 hours. How long would it take the secretaries, working together, to address 300 envelopes? (*Hint:* Think of addressing 300 envelopes as three 100-envelope jobs.)

41. Comparing Investments. An amount of money invested for 1 year in tax-free bonds will earn $300. In a certain credit union account, that same amount of money will only earn $200 interest in a year, because the interest paid is 2% less than that paid by the bonds. Find the rate of interest paid by each investment.

42. Comparing Investments. An amount of money invested for 1 year in a savings account will earn $1,500. That same amount of money invested in a mini-mall development will earn $6,500 interest in a year, because the interest paid is 10% more than that paid by the savings account. Find the rate of interest paid by each investment.

43. Comparing Investments. Two certificates of deposit (CDs) pay interest at rates that differ by 1%. Money invested for 1 year in the first CD earns $175 interest. The same principal invested in the second CD earns $200. Find the two rates of interest.

44. Comparing Interest Rates. Two bond funds pay interest at rates that differ by 2%. Money invested for 1 year in the first fund earns $315 interest. The same amount invested in the second fund earns $385. Find the lower rate of interest.

WRITING

45. In Example 4, one inlet pipe could fill an oil tank in 7 days, and another could fill the same tank in 9 days. We were asked to find how long it would take if both pipes were used. Explain why each of the following approaches is incorrect.

The time it would take to fill the tank

- is the *sum* of the lengths of time it takes each pipe to fill the tank: 7 days + 9 days = 16 days.
- is the *difference* in the lengths of time it takes each pipe to fill the tank: 9 days − 7 days = 2 days.
- is the *average* of the lengths of time it takes each pipe to fill the tank:

$$\frac{7 \text{ days} + 9 \text{ days}}{2} = \frac{16 \text{ days}}{2} = 8 \text{ days}.$$

46. Write a shared-work problem that can be modeled by the equation:

$$\frac{x}{3} + \frac{x}{4} = 1$$

REVIEW

47. Solve using substitution: $\begin{cases} x + y = 4 \\ y = 3x \end{cases}$

48. Solve using elimination (addition): $\begin{cases} 5x - 4y = 19 \\ 3x + 2y = 7 \end{cases}$

49. Use a check to determine whether $\frac{21}{5}$ is a solution of: $x + 20 = 4x - 1 + 2x$

50. Solve: $4x^2 + 8x = 0$

CHALLENGE PROBLEMS

51. River Tours. A river boat tour begins by going 60 miles upstream against a 5-mph current. There, the boat turns around and returns with the current. What still-water speed should the captain use to complete the tour in 5 hours?

52. Travel Time. A company president flew 680 miles one way in the corporate jet, but returned in a smaller plane that could fly only half as fast. If the total travel time was 6 hours, find the speeds of the planes.

53. Sales. A dealer bought some radios for a total of $1,200. She gave away 6 radios as gifts, sold the rest for $10 more than she paid for each radio, and broke even. How many radios did she buy?

54. Furnace Repairs. A repairman purchased several furnace-blower motors for a total cost of $210. If his cost per motor had been $5 less, he could have purchased one additional motor. How many motors did he buy at the regular rate?

<table>
<tr><td>

SECTION 7.8

OBJECTIVES

1 Write ratios and rates in simplest form.

2 Solve proportions.

3 Use proportions to solve problems.

4 Use proportions to solve problems involving similar triangles.

</td></tr>
</table>

Proportions and Similar Triangles

ARE YOU READY?

The following problems review some basic skills that are needed when working with ratios and proportions.

1. Simplify: $\dfrac{42}{54}$

2. Multiply: **a.** $bd \cdot \dfrac{a}{d}$ **b.** $bd \cdot \dfrac{c}{d}$

3. Solve: $16x = 136$

4. Solve: $x^2 - 7x = 18$

In this section, we will discuss a problem-solving tool called a *proportion*. A proportion is a type of rational equation that involves two *ratios* or two *rates*.

1 Write Ratios and Rates in Simplest Form.

Ratios are used to compare two numbers or two quantities measured in the same units. Here are some examples.

- To prepare fuel for a lawnmower, gasoline is mixed with oil in a 50-to-1 ratio.
- In the stock market, winning stocks might outnumber losers by a ratio of 7 to 4.
- Gold is combined with other metals in the ratio of 14 to 10 to make 14-karat jewelry.

Ratios	A **ratio** is the quotient of two numbers or the quotient of two quantities that have the same units.

There are three ways to write a ratio: as a fraction, using the word *to,* or with a colon. For example, the comparison of the number of winning stocks to the number of losing stocks mentioned earlier can be written as

$$\frac{7}{4}, \qquad 7 \text{ to } 4, \qquad \text{or} \qquad 7:4 \qquad \textit{Each of these forms can be read as "the ratio of 7 to 4."}$$

EXAMPLE 1   Translate each phrase into a ratio written in fractional form: **a.** The ratio of 5 to 9   **b.** 12 ounces to 2 pounds

Strategy   To translate, we need to identify the number (or quantity) before the word *to* and the number (or quantity) after it.

Why   The number before the word *to* is the numerator of the ratio and the number after it is the denominator.

Solution

a. The ratio of 5 *to* 9 is written $\dfrac{5}{9}$.

numerator

denominator

b. To write a ratio of two quantities with the same units, we must express 2 pounds in terms of ounces. Since 1 pound = 16 ounces, 2 pounds = 32 ounces. The ratio of 12 ounces to 32 ounces can be simplified so that no units appear in the final form.

$$\frac{12 \text{ ounces}}{32 \text{ ounces}} = \frac{3 \cdot \overset{1}{\cancel{4}} \cancel{\text{ ounces}}}{\underset{1}{\cancel{4}} \cdot 8 \cancel{\text{ ounces}}} = \frac{3}{8}$$

Caution
A ratio that is the quotient of two quantities having the same units should be simplified so that no units appear in the final answer.

Self Check 1 Translate each phrase into a ratio written in fractional form: **a.** The ratio of 15 to 2   **b.** 12 hours to 2 days

Now Try ▶ Problem 25

A quotient that compares quantities with different units is called a **rate**. For example, if the 495-mile drive from New Orleans to Dallas takes 9 hours, the average rate of speed is the quotient of the miles driven and the length of time the trip takes.

$$\text{Average rate of speed} = \frac{495 \text{ miles}}{9 \text{ hours}} = \frac{\overset{1}{\cancel{9}} \cdot 55 \text{ miles}}{\underset{1}{\cancel{9}} \cdot 1 \text{ hours}} = \frac{55 \text{ miles}}{1 \text{ hour}}$$

Rates	A **rate** is a quotient of two quantities that have different units.

2 Solve Proportions.

If two ratios or two rates are equal, we say that they are *in proportion*.

Proportion	A **proportion** is a mathematical statement that two ratios or two rates are equal.

Some examples of proportions are:

$$\frac{1}{2} = \frac{3}{6} \qquad \frac{3 \text{ waiters}}{7 \text{ tables}} = \frac{9 \text{ waiters}}{21 \text{ tables}} \qquad \frac{a}{b} = \frac{c}{d}$$

- The proportion $\frac{1}{2} = \frac{3}{6}$ can be read as "1 is to 2 as 3 is to 6."

- The proportion $\frac{3 \text{ waiters}}{7 \text{ tables}} = \frac{9 \text{ waiters}}{21 \text{ tables}}$ can be read as "3 waiters is to 7 tables as 9 waiters is to 21 tables."

- The proportion $\frac{a}{b} = \frac{c}{d}$ can be read as "a is to b as c is to d."

Each of the four numbers in a proportion is called a **term**. The first and fourth terms are called the **extremes,** and the second and third terms are called the **means.**

First term ⟶ $\dfrac{a}{b} = \dfrac{c}{d}$ ⟵ Third term

Second term ⟶   ⟵ Fourth term   *a* and *d* are the extremes. *b* and *c* are the means.

For the proportion $\frac{a}{b} = \frac{c}{d}$, we can show that the product of the extremes, ad, is equal to the product of the means, bc, by multiplying both sides of the proportion by bd, and observing that $ad = bc$.

$$\frac{a}{b} = \frac{c}{d}$$

$$\overset{1}{\cancel{bd}} \cdot \frac{a}{\underset{1}{\cancel{b}}} = \overset{1}{\cancel{bd}} \cdot \frac{c}{\underset{1}{\cancel{d}}} \qquad \text{To clear the fractions, multiply both sides by the LCD, } bd.$$
Remove common factors of the numerator and denominator.

$$ad = bc \qquad \text{Simplify: } \tfrac{b}{b} = 1 \text{ and } \tfrac{d}{d} = 1.$$

Since $ad = bc$, the product of the extremes equals the product of the means.

The same products ad and bc can be found by multiplying diagonally in the proportion $\frac{a}{b} = \frac{c}{d}$. We call ad and bc **cross products.**

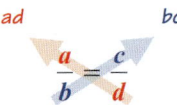

The Fundamental Property of Proportions	In a proportion, the product of the extremes is equal to the product of the means. If $\dfrac{a}{b} = \dfrac{c}{d}$, then $ad = bc$ and if $ad = bc$, then $\dfrac{a}{b} = \dfrac{c}{d}$.

EXAMPLE 2 Determine whether each equation is a proportion: **a.** $\dfrac{3}{7} = \dfrac{9}{21}$ **b.** $\dfrac{8}{3} = \dfrac{13}{5}$

Strategy We will check to see whether the product of the extremes is equal to the product of the means.

Why If the product of the extremes equals the product of the means, the equation is a proportion. If the cross products are not equal, the equation is not a proportion.

Solution **a.** The product of the extremes is $3 \cdot 21 = 63$. The product of the means is $7 \cdot 9 = 63$. Since the cross products are equal, $\dfrac{3}{7} = \dfrac{9}{21}$ is a proportion.

> **Caution**
>
> We cannot remove common factors "across" an $=$ symbol.
>
> $$\frac{\overset{1}{\cancel{3}}}{7} = \frac{9}{\underset{7}{\cancel{21}}}$$
>
> When this is done, the original proportion, $\frac{3}{7} = \frac{9}{21}$, which we found to be true, produces the false statement: $\frac{1}{7} = \frac{9}{7}$.

$3 \cdot 21 = 63 \qquad 7 \cdot 9 = 63$

$$\frac{3}{7} \diagup\!\!\!\!\diagdown \frac{9}{21} \qquad \text{Each cross product is 63.}$$

b. The product of the extremes is $8 \cdot 5 = 40$. The product of the means is $3 \cdot 13 = 39$. Since the cross products are not equal, the equation is not a proportion: $\dfrac{8}{3} \neq \dfrac{13}{5}$.

$8 \cdot 5 = 40 \qquad 3 \cdot 13 = 39$

$$\frac{8}{3} \diagup\!\!\!\!\diagdown \frac{13}{5} \qquad \text{One cross product is 40 and the other is 39.}$$

Self Check 2 Determine whether the equation $\dfrac{6}{13} = \dfrac{24}{53}$ is a proportion.

Now Try ▶ Problem 29

We have seen that a proportion contains four terms. If we know only three of the four terms of a proportion, we can use the fundamental property of proportions to find the value of the fourth term. This process is called **solving the proportion.**

EXAMPLE 3 Solve: $\dfrac{3}{2} = \dfrac{9}{x}$

Strategy To solve for x, we will set the cross products equal.

Why This equation is a proportion, and in a proportion the product of the extremes equals the product of the means.

Solution If $x = 0$, the denominator of $\frac{9}{x}$ is 0 and the fraction would be undefined. Therefore, 0 cannot be a solution.

<table>
<tr><td>Caution</td></tr>
</table>

Caution

Remember that a cross product is the product of the means or extremes of a proportion. It would be incorrect to try to calculate cross products to solve $\frac{12}{18} = \frac{4}{x} + \frac{1}{2}$ because there is more than one term on the right side of the equation. It is *not* a proportion.

$$\frac{3}{2} = \frac{9}{x}$$ This is the given proportion. Since it is a type of rational equation, we can solve it by multiplying both sides by the LCD, 2x. However, it is often easier to solve a proportion using the cross products.

$$3 \cdot x = 2 \cdot 9$$ Find each cross product and set them equal.

$$3x = 18$$ Do the multiplication.

$$\frac{3x}{3} = \frac{18}{3}$$ To isolate x, divide both sides by 3.

$$x = 6$$ Do the division.

Check: To check the result, we substitute 6 for x in $\frac{3}{2} = \frac{9}{x}$ and find the cross products.

$$3 \cdot 6 = 18 \qquad 2 \cdot 9 = 18$$

$$\frac{3}{2} \overset{?}{=} \frac{9}{6}$$ Each cross product is 18.

Since the cross products are equal, the solution of $\frac{3}{2} = \frac{9}{x}$ is 6. The solution set is $\{6\}$.

Self Check 3 Solve: $\frac{15}{x} = \frac{25}{40}$

Now Try ▶ Problem 35

EXAMPLE 4 Solve: $\frac{a}{2} = \frac{4}{a - 2}$

Strategy To solve for a, we will set the cross products equal.

Why Since this equation is a proportion, the product of the means equals the product of the extremes.

Solution If $a = 2$, the denominator of $\frac{4}{a - 2}$ is 0 and the fraction would be undefined. Therefore, 2 cannot be a solution.

Success Tip

Since proportions are rational equations, they can also be solved by multiplying both sides by the LCD. Here, an alternate approach is to multiply both sides by $2(a - 2)$:

$$\frac{a}{2} = \frac{4}{a - 2}$$ This is the given proportion. Since it is a type of rational equation, we can solve it by multiplying both sides by the LCD, $2(a - 2)$, as shown in the margin. However, it is often easier to solve a proportion using the cross products.

$$a(a - 2) = 2 \cdot 4$$ Find each cross product and set them equal. Don't forget to write the parentheses.

$$a^2 - 2a = 8$$ On the left side, distribute the multiplication by a. This is a quadratic equation. On the right side, multiply.

$$a^2 - 2a - 8 = 0$$ To get 0 on the right side of the equation, subtract 8 from both sides.

$$(a + 2)(a - 4) = 0$$ Factor $a^2 - 2a - 8$.

$$a + 2 = 0 \quad \text{or} \quad a - 4 = 0$$ Set each factor equal to 0.

$$a = -2 \quad | \quad a = 4$$ Solve each equation.

The solutions are -2 and 4. Verify this using a check.

Self Check 4 Solve: $\frac{6}{c} = \frac{c - 1}{5}$

Now Try ▶ Problem 47

3 Use Proportions to Solve Problems.

We can use proportions to solve many problems. If we are given a ratio (or rate) comparing two quantities, the words of the problem can be translated into a proportion, and we can solve it to find the unknown.

iStockPhoto.com/Graeme Gilmour

EXAMPLE 5

Grocery Shopping. If 6 apples cost $1.38, how much will 16 apples cost?

Analyze We know the cost of 6 apples; we are to find the cost of 16 apples.

Assign Let c = the cost of 16 apples.

Form If we compare the number of apples to their cost, the two ratios must be equal.

*6 apples is **to** $1.38 **as** 16 apples is **to** $c.*

Number of apples ⟶ $\dfrac{6}{1.38} = \dfrac{16}{c}$ ⟵ Number of apples
Cost ⟶ ⟵ Cost

The units can be written outside the proportion.

Solve We drop the units, find each cross product, set them equal, and then solve the resulting equation for c.

$$6 \cdot c = 1.38(16)$$ In a proportion, the product of the extremes equals the product of the means.

$$6c = 22.08$$ Multiply: $1.38(16) = 22.08$.

$$\dfrac{6c}{6} = \dfrac{22.08}{6}$$ To undo the multiplication by 6 and isolate c, divide both sides by 6.

$$c = 3.68$$ Recall that c represents the cost of 16 apples.

State Sixteen apples will cost $3.68.

Check We can use estimation to check the result. 16 apples are about 3 times as many as 6 apples, which cost $1.38. If we multiply $1.38 by 3, we get an estimate of the cost of 16 apples: $1.38 \cdot 3 = 4.14. The result, $3.68, seems reasonable.

The Language of Algebra

Remember that the word **to** separates the numerator and denominator of a ratio. If the units are written outside the ratio, we can write 6 apples is *to* $1.38 in fraction form as:

$$\dfrac{6}{1.38}$$

Self Check 5 **Concert Tickets.** If 9 tickets to a concert cost $112.50, how much will 15 tickets cost?

Now Try ▶ Problem 73

Be careful when solving problems using proportions. We must make sure that the units of both numerators are the same and the units of both denominators are the same. In Example 5, it would be incorrect to write

Cost of 6 apples ⟶ $\dfrac{1.38}{6} = \dfrac{16}{c}$ ⟵ 16 apples
6 apples ⟶ ⟵ Cost of 16 apples

EXAMPLE 6

Carousel ratio
1 inch:160 inches

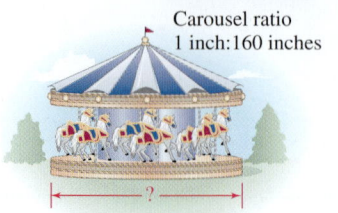

?

Miniatures. A **scale** is a ratio (or rate) that compares the size of a model, drawing, or map with the size of an actual object. The scale indicates that 1 inch on the model carousel is equivalent to 160 inches on the actual carousel. How wide should the model be if the actual carousel is 35 feet wide?

Analyze We are asked to determine the width of the miniature carousel if a ratio of 1 inch to 160 inches is used. We would like the width of the model to be given in inches, not feet, so we will express the 35-foot width of the actual carousel as $35 \cdot 12 = 420$ inches.

Assign Let w = the width of the model.

The Language of Algebra

Architects, interior decorators, landscapers, and automotive engineers are a few of the professionals who construct **scale** drawings or *scale* models of the projects they are designing.

Form   The ratios of the dimensions of the model to the corresponding dimensions of the actual carousel are equal.

*1 inch is **to** 160 inches **as** w inches is **to** 420 inches.*

Model $\longrightarrow$ $\dfrac{1}{160} = \dfrac{w}{420}$ $\longleftarrow$ Model
Actual size $\longrightarrow$ $\phantom{\dfrac{1}{160}}$ $\longleftarrow$ Actual size

Solve   We drop the units, find each cross product, set them equal, and then solve the resulting equation for w.

$420 = 160w$ In a proportion, the product of the extremes is equal to the product of the means.

$\dfrac{420}{160} = \dfrac{160w}{160}$ To undo the multiplication by 160 and isolate w, divide both sides by 160.

$2.625 = w$ Recall that w represents the width of the model.

State   The width of the miniature carousel should be 2.625 in., or $2\frac{5}{8}$ in.

Check   A width of $2\frac{5}{8}$ in. is approximately 3 in. When we write the ratio of the model's approximate width to the width of the actual carousel, we get $\frac{3}{420} = \frac{1}{140}$, which is about $\frac{1}{160}$. The answer seems reasonable.

Self Check 6   **Blueprints.**   The scale for a blueprint indicates that $\frac{1}{4}$ inch on the print is equivalent to 1 foot for the actual building. If the width of the building on the print is 7.5 inches, what is the width of the actual building?

Now Try ▶ Problem 89

When shopping, *unit prices* can be used to compare costs of different sizes of the same brand to determine the best buy. The **unit price** gives the cost per unit, such as cost per ounce, cost per pound, or cost per sheet. We can find the unit price of an item using a proportion.

EXAMPLE 7   **Comparison Shopping.**   Which size of toothpaste is the better buy?

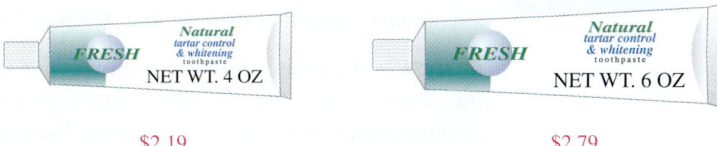

2.19 2.79

Strategy   We will find the unit price for each tube of toothpaste. Then we will identify which tube has the lower unit price.

Why   The better buy is the tube that has the lower unit price.

Solution

The Language of Algebra

A **unit price** indicates the cost of 1 unit of an item, such as 1 ounce of bottled water or 1 pound of hamburger. In advanced mathematics, we study unit circles—circles that have a radius of 1 unit.

To find the unit price for each tube, we let $x =$ the price of 1 ounce of toothpaste. Then we set up and solve the following proportions.

For the 4-ounce tube:

Price $\longrightarrow$ $\dfrac{2.19}{4} = \dfrac{x}{1}$ $\longleftarrow$ Price
Ounces $\longrightarrow$ $\phantom{\dfrac{2.19}{4}}$ $\longleftarrow$ Ounce

$2.19 = 4x$

$\dfrac{2.19}{4} = x$

$0.55 \approx x$ The unit price is approximately 0.55.

For the 6-ounce tube:

Price $\longrightarrow$ $\dfrac{2.79}{6} = \dfrac{x}{1}$ $\longleftarrow$ Price
Ounces $\longrightarrow$ $\phantom{\dfrac{2.79}{6}}$ $\longleftarrow$ Ounce

$2.79 = 6x$

$\dfrac{2.79}{6} = x$

$0.47 \approx x$ The unit price is approximately 0.47.

The price of 1 ounce of toothpaste from the 4-ounce tube is about 55¢. The price for 1 ounce of toothpaste from the 6-ounce tube is about 47¢. Since the 6-ounce tube has the lower unit price, it is the better buy.

> **Self Check 7**   **Comparison Shopping.**   Which is the better buy: 3 pounds of hamburger for $6.89 or 5 pounds for $12.49?
>
> **Now Try** ▶ Problem 93

4 Use Proportions to Solve Problems Involving Similar Triangles.

If two angles of one triangle have the same measures as two angles of a second triangle, the triangles have the same shape. Triangles with the same shape, but not necessarily the same size, are called **similar triangles.** In the following figure, $\triangle ABC \sim \triangle DEF$. (Read the symbol $\sim$ as "is similar to.")

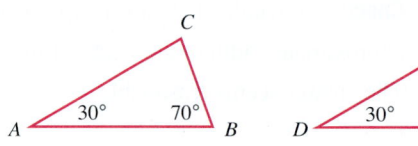

Property of Similar Triangles	If two triangles are **similar,** all pairs of corresponding sides are in proportion.

For the similar triangles previously shown, the following proportions are true.

$$\frac{AB}{DE} = \frac{BC}{EF}, \qquad \frac{BC}{EF} = \frac{CA}{FD}, \qquad \text{and} \qquad \frac{CA}{FD} = \frac{AB}{DE}$$ Read AB as "the length of segment AB."

EXAMPLE 8   **Finding the Height of a Tree.**   A tree casts a shadow 18 feet long at the same time as a woman 5 feet tall casts a shadow 1.5 feet long. Find the height of the tree.

Analyze   The figure shows the similar triangles determined by the tree and its shadow and the woman and her shadow. Since the triangles are similar, the lengths of their corresponding sides are in proportion. We can use this fact to find the height of the tree.

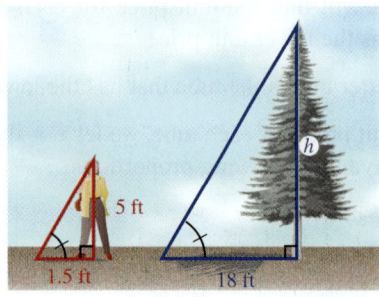

Each triangle has a right angle. Since the sun's rays strike the ground at the same angle, the angles highlighted with a tick mark have the same measure. Therefore, two angles of the smaller triangle have the same measures as two angles of the larger triangle; the triangles are similar.

> **Success Tip**
>
> Similar triangles do not have to be positioned the same. When they are placed differently, be careful to match their corresponding letters correctly. Here, $\triangle RST \sim \triangle MNO$.
>
>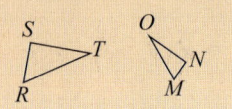

Assign   Let $h =$ the height of the tree.

Form   We can find h by solving the following proportion.

$$\frac{h}{5} = \frac{18}{1.5}$$ $\dfrac{\text{Height of the tree}}{\text{Height of the woman}} = \dfrac{\text{Length of shadow of the tree}}{\text{Length of shadow of the woman}}$

Solve

$1.5h = 5(18)$ In a proportion, the product of the extremes equals the product of the means.

$1.5h = 90$ Multiply: $5(18) = 90$.

$\dfrac{1.5h}{1.5} = \dfrac{90}{1.5}$ To undo the multiplication by 1.5 and isolate h, divide both sides by 1.5.

$h = 60$ Do the decimal division, $1.5\overline{)90}$, to get 60.

State   The tree is 60 feet tall.

Check $\dfrac{18}{1.5} = 12$ and $\dfrac{60}{5} = 12$. Since the ratios are the same, the result checks.

Self Check 8   **Shadows.**   Find the height of the tree in Example 8 if the woman is 5 feet 6 inches tall and her shadow is 1.5 feet long.

Now Try ▶   Problems 55 and 101

SECTION 7.8 ▶ **STUDY SET**

VOCABULARY

Fill in the blanks.

1. A _____ is the quotient of two numbers or the quotient of two quantities with the same units. A _____ is a quotient of two quantities that have different units.

2. A _____ is a mathematical statement that two ratios or two rates are equal.

3. In $\frac{50}{3} = \frac{x}{9}$, the terms 50 and 9 are called the _____ and the terms 3 and x are called the _____ of the proportion.

4. The _____ products for the proportion $\frac{5}{2} = \frac{6}{x}$ are $5x$ and 12.

5. Examples of _____ prices are $1.65 per gallon, 17¢ per day, and $50 per foot.

6. Two triangles with the same shape, but not necessarily the same size, are called _____ triangles.

CONCEPTS

7. Fill in the blanks: In a proportion, the product of the extremes is _____ to the product of the means. In symbols,

   If $\dfrac{a}{b} = \dfrac{c}{d}$, then ☐ = ☐.

8. Is 45 a solution of $\dfrac{5}{3} = \dfrac{75}{x}$?

9. **Snacks.**   In a sample of 25 bags of potato chips, 2 were found to be underweight. Complete the following proportion that could be used to find the number of underweight bags that would be expected in a shipment of 1,000 bags of potato chips.

   Number of bags → $\dfrac{25}{\boxed{}}$ = $\dfrac{\boxed{}}{\boxed{}}$ ← Number of bags
   Number underweight →   ← Number underweight

10. **Miniatures.**   A model of the Seattle Space Needle is to be made using a scale of 2 inches to 35 feet. Complete the following proportion to determine the height h of the model.

    $\dfrac{2}{35} = \dfrac{\boxed{}}{\boxed{}}$

605 ft

11. **Kleenex.**   Complete the following proportion that can be used to find the unit price of facial tissue if a box of 85 tissues sells for $2.19.

    Price → $\dfrac{\boxed{}}{85}$ = $\dfrac{x}{\boxed{}}$ ← Price
    Number of sheets →   ← Number of sheets

12. The two triangles shown in the following illustration are similar. Complete the proportion.

    $\dfrac{x}{\boxed{}} = \dfrac{\boxed{}}{10}$

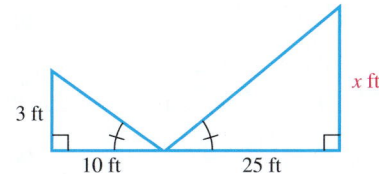

3 ft

10 ft   25 ft x ft

NOTATION

Complete the solution.

13. Solve for x: $\frac{12}{18} = \frac{x}{24}$

$$12 \cdot 24 = 18 \cdot \boxed{}$$

$$\boxed{} = 18x$$

$$\frac{288}{\boxed{}} = \frac{18x}{\boxed{}}$$

$$\boxed{} = x$$

14. Write the ratio of 25 to 4 in two other forms.

15. Fill in the blanks: The proportion $\frac{20}{1.6} = \frac{100}{8}$ can be read: 20 is to 1.6 ____ 100 is ___ 8.

16. Fill in the blank: We read $\triangle XYZ \sim \triangle MNO$ as: triangle XYZ is _____ to triangle MNO.

GUIDED PRACTICE

Translate each ratio into a fraction in simplest form. See Example 1.

17. 4 boxes to 15 boxes
18. 2 miles to 9 miles
19. 18 watts to 24 watts
20. 11 cans to 121 cans
21. 30 days to 24 days
22. 45 people to 30 people
23. 90 minutes to 3 hours
24. 20 inches to 2 feet
25. 8 quarts to 4 gallons
26. 6 feet to 12 yards
27. 6,000 feet to 1 mile
28. 5 tons to 4,000 pounds
    (*Hint:* 1 mi = 5,280 ft)
    (*Hint:* 1 ton = 2,000 lb)

Determine whether each equation is a true proportion. See Example 2.

29. $\frac{7}{3} = \frac{14}{6}$
30. $\frac{7}{16} = \frac{3}{7}$
31. $\frac{5}{8} = \frac{12}{19.4}$
32. $\frac{9}{32} = \frac{4.5}{16}$

Solve each proportion. See Example 3.

33. $\frac{2}{3} = \frac{x}{6}$
34. $\frac{3}{6} = \frac{x}{8}$
35. $\frac{63}{g} = \frac{9}{2}$
36. $\frac{27}{x} = \frac{9}{4}$
37. $\frac{x+1}{5} = \frac{3}{15}$
38. $\frac{x-1}{7} = \frac{2}{21}$
39. $\frac{5-x}{17} = \frac{13}{34}$
40. $\frac{4-x}{13} = \frac{11}{26}$
41. $\frac{15}{7b+5} = \frac{5}{2b+1}$
42. $\frac{8}{3n+6} = \frac{16}{3n-3}$
43. $\frac{8x}{3} = \frac{11x+9}{4}$
44. $\frac{3x}{16} = \frac{x+2}{5}$

Solve each proportion. See Example 4.

45. $\frac{2}{3x} = \frac{x}{6}$
46. $\frac{y}{4} = \frac{4}{y}$
47. $\frac{b-5}{3} = \frac{2}{b}$
48. $\frac{2}{q} = \frac{q-3}{2}$
49. $\frac{a-4}{a} = \frac{15}{a+4}$
50. $\frac{s}{s-5} = \frac{s+5}{24}$

51. $\frac{t+3}{t+5} = \frac{-1}{2t}$
52. $\frac{5h}{14h+3} = \frac{1}{h}$

Each pair of triangles is similar. Find the missing side length. See Example 8.

53.
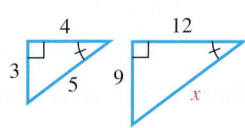

54.
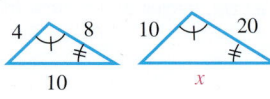

55.
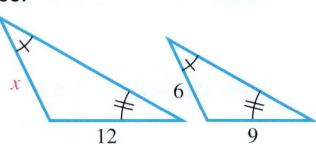

56.
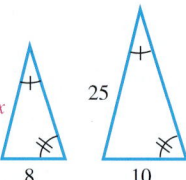

TRY IT YOURSELF

Solve each proportion.

57. $\frac{x-1}{x+1} = \frac{2}{3x}$
58. $\frac{2}{x+6} = \frac{-2x}{5}$
59. $\frac{x+1}{4} = \frac{3x}{8}$
60. $\frac{x-1}{9} = \frac{2x}{3}$
61. $\frac{y-4}{y+1} = \frac{y+3}{y+6}$
62. $\frac{r-6}{r-8} = \frac{r+1}{r-4}$
63. $\frac{c}{10} = \frac{10}{c}$
64. $\frac{-6}{r} = \frac{r}{-6}$
65. $\frac{m}{3} = \frac{4}{m+1}$
66. $\frac{n}{2} = \frac{5}{n+3}$
67. $\frac{3}{3b+4} = \frac{2}{5b-6}$
68. $\frac{2}{4d-1} = \frac{3}{2d+1}$

Look Alikes . . .

Solve each equation.

69. **a.** $-\frac{2}{5} = \frac{3}{4x}$
    b. $\frac{4}{x} - \frac{2}{5} = \frac{3}{4x}$

70. **a.** $\frac{1}{4} = \frac{2}{3a}$
    b. $\frac{5}{6a} + \frac{1}{4} = \frac{2}{3a}$

71. **a.** $\frac{3}{a-1} = \frac{8}{a}$
    b. $\frac{3}{a-1} + \frac{8}{a} = 3$

72. **a.** $\frac{4}{3x} = \frac{1}{3}$
    b. $\frac{4}{3x} - \frac{1}{3} = x$

APPLICATIONS

73. **Shopping for Clothes.** If shirts are on sale at two for $25, how much do five shirts cost?

74. **Mixing Perfume.** A perfume is to be mixed in the ratio of 3 drops of pure essence to 7 drops of alcohol. How many drops of pure essence should be mixed with 56 drops of alcohol?

75. **CPR.** A first aid handbook states that when performing cardiopulmonary resuscitation on an adult, the ratio of chest compressions to breaths should be 30:2. If 210 compressions were administered to an adult patient, how many breaths should have been given?

76. Cooking. A recipe for wild rice soup follows. Find the amounts of chicken broth, rice, and flour needed to make 15 servings.

> **Wild Rice Soup**
>
> *A sumptuous side dish with a nutty flavor*
>
> 3 cups chicken broth   1 cup light cream
>
> $\frac{2}{3}$ cup uncooked rice   2 tablespoons flour
>
> $\frac{1}{4}$ cup sliced onions $\frac{1}{8}$ teaspoon pepper
>
> $\frac{1}{2}$ cup shredded carrots
>
>   Serves: 6

77. from **Campus to Careers**

Recreation Director

A total of 966 boys and girls are members of a community recreation center.

a. If 504 are boys, how many members are girls?

b. Find the ratio of girls to boys who are members of the recreation center.

78. Gear Ratios. Write each ratio in two ways: as a fraction in simplest form and using a colon.

a. The number of teeth of the larger gear to the number of teeth of the smaller gear

b. The number of teeth of the smaller gear to the number of teeth of the larger gear

79. Computing a Paycheck. Billie earns $412 for a 40-hour week. If she missed 10 hours of work last week, how much did she get paid?

80. Waves. If the peak height to wavelength ratio is greater than 1:7, a wave becomes unstable and it breaks forward. (See the figure below.) What is the maximum height a wave with wavelength 637 feet can have before it breaks forward?

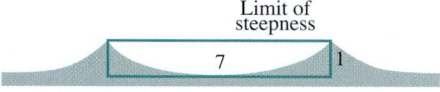

Limit of steepness

7   1

81. Twitter. According to a June, 2010 article in the Blog Herald, approximately 7,500 tweets are sent every 10 seconds. At this rate, about how many tweets are sent in one minute?

82. Engineering. A portion of a bridge is shown. Use the fact that $\frac{AB}{BC}$ is in proportion to $\frac{FE}{ED}$ to find FE.

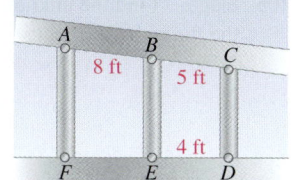

83. Nutrition. The table shows the nutritional facts about a 10-oz chocolate milkshake sold by a fast-food restaurant. Use the information to complete the table for the 16-oz shake. Round to the nearest unit when an answer is not exact.

	Calories	Fat (gm)	Protein (gm)
10-oz chocolate milkshake	355	8	9
16-oz chocolate milkshake			

84. Photo Enlargements. The 3-by-5 photo is to be blown up to the larger size. Find x.

5 in. $6\frac{1}{4}$ in.

3 in. x in.

85. Mixing Fuel. The instructions on a can of oil intended to be added to lawnmower gasoline are shown below. Are these instructions correct? (*Hint:* There are 128 ounces in 1 gallon.)

Recommended	Gasoline	Oil
50 to 1	6 gal	16 oz

86. Driver's Licenses. Of the 50 states, Alabama has one of the highest ratios of licensed drivers to residents. If the ratio is 399:500 and the population of Alabama is about 4,500,000, how many residents of that state have a driver's license?

87. Capture–Release Method. To estimate the ground squirrel population on his acreage, a farmer trapped, tagged, and then released a dozen squirrels. Two weeks later, the farmer trapped 35 squirrels and noted that 3 were tagged. Use this information to estimate the number of ground squirrels on his acreage.

88. Concrete. A 2:3 concrete mix means that for every two parts of sand, three parts of gravel are used. How much sand should be used in a mix composed of 25 cubic feet of gravel?

89. Model Railroads. An HO scale model railroad engine is 6 inches long. If the HO scale is 1 to 87, how long is a real engine, in inches? In feet?

90. Model Railroads. An N scale model railroad caboose is 4.5 inches long. If the N scale is 1 to 160, how long is a real caboose, in inches? In feet?

91. Blueprints. The scale for the drawing shown means that a $\frac{1}{4}$-inch length $\left(\frac{1}{4}''\right)$ on the drawing corresponds to an actual size of 1 foot (1'-0"). Suppose the length of the kitchen is $2\frac{1}{2}$ inches on the drawing. How long is the actual kitchen?

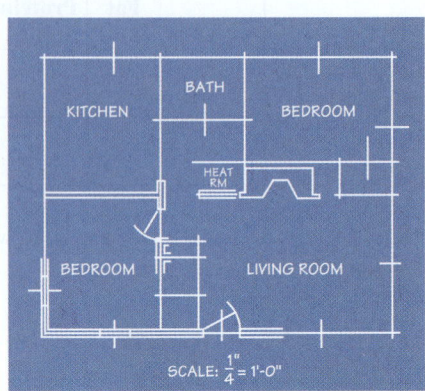

92. The Titanic. A 1:144 scale model of the *Titanic* is to be built. If the ship was 882 feet long, find the length of the model.

For each of the following purchases, determine the better buy. See Example 7.

93. Trumpet lessons: 45 minutes for $25 or 60 minutes for $35

94. Memory for a computer: 128 megabytes for $26 or 512 megabytes for $110

95. Business cards: 100 for $9.99 or 150 for $12.99

96. Dog food: 20 pounds for $7.49 or 44 pounds for $14.99

97. Soft drinks: 6-pack for $1.50 or a case (24 cans) for $6.25

98. Donuts: A dozen for $6.24 or a baker's dozen (13) for $6.65

99.

FAT-FREE PEACH YOGURT	FAT-FREE PEACH YOGURT
4.79	2.99
Six 4-OZ CARTONS	Four 4-OZ CARTONS

100.

AQUACLEAR WATER	AQUACLEAR WATER
1.79	4.49
12 8-OZ BOTTLES	24 12-OZ BOTTLES

101. Height of a Tree. A tree casts a shadow of 26 feet at the same time as a 6-foot man casts a shadow of 4 feet. Find the height of the tree.

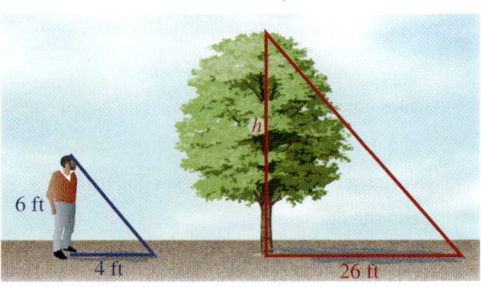

102. Height of a Building. A man places a mirror on the ground and sees the reflection of the top of a building, as shown. The two triangles in the illustration are similar. Find the height, h, of the building.

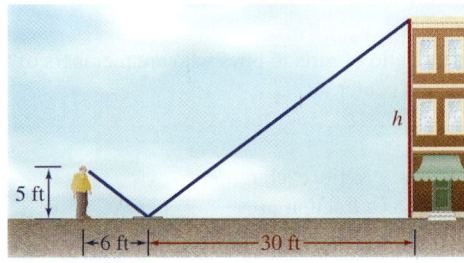

103. Surveying. To find the width of a river, a surveyor laid out the following similar triangles. Find w.

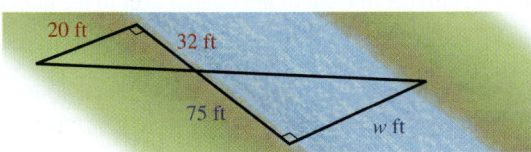

104. Flight Paths. An airplane ascends 100 feet as it flies a horizontal distance of 1,000 feet. How much altitude will it gain as it flies a horizontal distance of 1 mile? (*Hint:* 5,280 feet = 1 mile.)

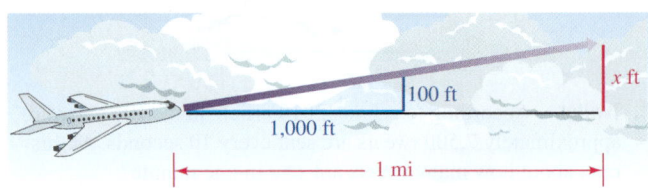

105. Slope. Find the unknown rise of the smaller slope triangle in the figure below.

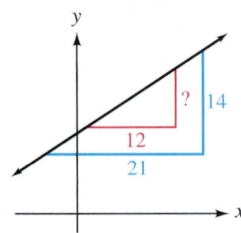

106. Washington, D.C. The Washington Monument casts a shadow of $166\frac{1}{2}$ feet at the same time as a 5-foot-tall tourist casts a shadow of $1\frac{1}{2}$ feet. Find the height of the monument.

WRITING

107. Explain how to solve the equation $\frac{7}{6} = \frac{2}{x}$ and how to simplify the expression $\frac{7}{6} \cdot \frac{2}{x}$.

108. Explain why the concept of cross products cannot immediately be used to solve the equation:

$$\frac{x}{3} - \frac{3x}{4} = \frac{1}{12}$$

109. What are similar triangles?

110. What is a unit price? Give an example.

REVIEW

111. Change $\frac{9}{10}$ to a percent.

112. Change $33\frac{1}{3}\%$ to a fraction.

113. Find 30% of 1,600.

114. Shopping. Maria bought a dress for 25% off the original price of $98. How much did the dress cost?

CHALLENGE PROBLEMS

115. Suppose $\frac{a}{b} = \frac{c}{d}$. Write three other proportions using a, b, c, and d.

116. a. Verify that $\frac{3}{5} = \frac{12}{20} = \frac{3 + 12}{5 + 20}$.

b. Is the following rule always true? Explain.

$$\frac{a}{b} = \frac{c}{d} = \frac{a + c}{b + d}$$

SECTION 7.9

Variation

OBJECTIVES

1 Solve direct variation problems.

2 Solve inverse variation problems.

ARE YOU READY?

The following problems review some basic skills that are needed when working with variation.

1. Find y if $y = 45x$ and $x = 10$.

2. Find d if $d = \frac{90}{n}$ and $n = 18$.

3. Solve: $26 = 4k$

4. Solve: $80 = \frac{k}{2.5}$

If the value of one quantity depends on the value of another quantity, we can often describe that relationship using the language of variation:

- The sales tax on an item *varies* as the price.
- The intensity of light *varies* as the distance from its source.
- The pressure exerted by water on an object *varies* as the depth of the object beneath the surface.

In this section, we will discuss two types of variation and see how to model them algebraically.

1 Solve Direct Variation Problems.

One type of variation, called **direct variation,** is represented by an equation of the form $y = kx$, where k is a nonzero constant. Two variables are said to *vary directly* if one is a constant multiple of the other.

Direct Variation	The words *y varies directly as x* or *y is directly proportional to x* mean that $$y = kx$$ for some nonzero constant k, called the **constant of variation.**

EXAMPLE 1   Suppose y varies directly as x. If $y = 12$ when $x = 4$, find y when $x = 6$.

Strategy   We will use the equation $y = kx$ to solve this problem.

Why   The words *varies directly* indicate that we should use the direct variation equation $y = kx$.

Solution   We can use the given pair of values of x and y to determine the constant of variation k.

$$y = kx \qquad \text{This is the equation that models direct variation.}$$
$$12 = k(4) \qquad \text{Substitute 4 for x and 12 for y.}$$
$$3 = k \qquad \text{To isolate k on the right side, divide both sides by 4.}$$
$$\qquad\qquad \text{This is the constant of variation.}$$

Since $y = kx$ and $k = 3$, we have $y = 3x$.

We can use $y = 3x$ to find other pairs of values of x and y. When $x = 6$, we see that

$$y = 3(6) = 18 \qquad \text{Substitute 6 for x.}$$

Thus, when $x = 6$, the value of y is 18.

Self Check 1   Suppose y varies directly as x. If $y = 24$ when $x = 3$, find y when $x = 5$.

Now Try ▶ Problem 25

Scientists have found that the distance a spring will stretch is directly proportional to the force applied to it. The more force that is applied to the spring, the more it will stretch.

We could use the equation $y = kx$ to model this direct variation. However, in application problems, the variables x and y are often replaced with letters that better describe the quantities involved. If we let d represent the distance the spring is stretched and f represent the force applied to it, this relationship can be represented by the equation

$$d = kf \qquad \text{where } k \text{ is the constant of variation}$$

Suppose that a 150-pound weight stretches a spring 18 inches. We can find the constant of variation for the spring by substituting 150 for f and 18 for d in the equation $d = kf$ and solving for k:

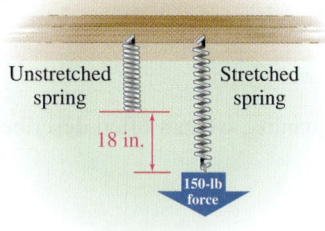

Unstretched spring   Stretched spring

18 in.

150-lb force

$$d = kf \qquad \text{This equation models direct variation.}$$
$$18 = k(150) \qquad \text{Substitute 18 for d and 150 for f.}$$
$$\frac{18}{150} = k \qquad \text{To isolate k on the right side, divide both sides by 150.}$$
$$\frac{3}{25} = k \qquad \text{Simplify the fraction: } \frac{18}{150} = \frac{\overset{1}{\cancel{6}} \cdot 3}{\underset{1}{\cancel{6}} \cdot 25} = \frac{3}{25}. \text{ This is the constant of variation.}$$

Thus, the equation that describes the relationship between the distance the spring will stretch and the amount of force applied to it is $d = \frac{3}{25}f$.

Once the value of k is known, other pairs of values can be found. To find the distance that the same spring will stretch when a lighter, 50-pound weight is used, we proceed as follows:

$$d = \frac{3}{25}f \qquad \text{This equation models direct variation for the specific spring used in this example.}$$

$$d = \frac{3}{25}(50) \qquad \text{Substitute 50 for f.}$$

$$d = 6 \qquad \text{Multiply: } \frac{3}{25}\left(\frac{50}{1}\right) = \frac{3 \cdot 2 \cdot \overset{1}{\cancel{25}}}{\underset{1}{\cancel{25}} \cdot 1} = 6.$$

The spring will stretch 6 inches when a 50-pound weight is used.

The table below shows some other possible values for f and d as determined by the equation $d = \frac{3}{25}f$. When these ordered pairs are graphed and a straight line is drawn through them, it is apparent that as the force f applied to a spring increases, the distance d that it stretches increases in a predictable way. Furthermore, the slope of the graph is $\frac{3}{25}$, which is the constant of variation.

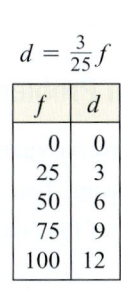

$$d = \frac{3}{25}f$$

f	d
0	0
25	3
50	6
75	9
100	12

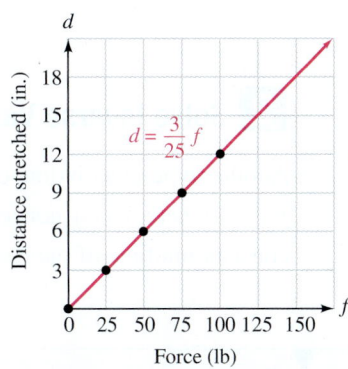

This straight-line graph shows that d varies directly as f.

We can use the following steps to solve variation problems.

Strategy for Solving Variation Problems

1. Translate the verbal model into an equation.

2. Substitute the values of a pair of variables into the equation from step 1 to determine the value of k.

3. Substitute the value of k into the equation from step 1.

4. Substitute the remaining given value into the equation from step 3 and solve for the unknown variable.

EXAMPLE 2

Geology. The weight of an object on Earth varies directly as its weight on the moon. If a rock weighed 5 pounds on the moon and 30 pounds on Earth, what would be the weight on Earth of a larger rock weighing 26 pounds on the moon?

Strategy We will follow the strategy for solving a direct variation problem.

Why In the words of the problem, the phrase *varies directly* indicates that a direct variation model should be used.

Solution

Step 1: We let e represent the weight (in pounds) of an object on Earth and m the weight (in pounds) of the object on the moon. Translating the words *weight on Earth varies directly as its weight on the moon,* we get the equation

$e = km$ This equation models direct variation.

Step 2: To find the constant of variation, k, we substitute 30 for e and 5 for m.

$e = km$

$30 = k(5)$ Substitute 30 for e and 5 for m.

$6 = k$ To isolate k, divide both sides by 5. This is the constant of variation.

Step 3: We now substitute the value of k into the equation from step 1. The equation describing the relationship between the weight of an object on Earth and on the moon is

$e = 6m$

Step 4: We can find the weight of the larger rock on Earth by substituting 26 for m in the equation from step 3.

$e = 6m$

$e = 6(26)$ Substitute 26 for m.

$e = 156$

The rock would weigh 156 pounds on Earth.

> **Self Check 2**   **Finding Distance.**   The distance that a car can go varies directly as the number of gallons of gasoline it consumes. If a car can go 288 miles on 12 gallons of gasoline, how far can it go on a full tank of 18 gallons?
>
> **Now Try** ▶ Problem 43

2 Solve Inverse Variation Problems.

Another type of variation, called **inverse variation,** is represented by an equation of the form $y = \frac{k}{x}$, where k is a nonzero constant. Two variables are said to *vary inversely* if one is a constant multiple of the reciprocal of the other.

Inverse Variation	The words *y varies inversely as x* or *y is inversely proportional to x* mean that $$y = \frac{k}{x}$$ for some nonzero constant k, called the **constant of variation**.

EXAMPLE 3   Suppose y varies inversely as x. If $y = 5$ when $x = 20$, find y when $x = 50$.

Strategy   We will use the equation $y = \frac{k}{x}$ to solve this problem.

Why   The words *varies inversely* indicate that we should use the inverse variation equation $y = \frac{k}{x}$.

Solution   We can use the given pair of values of x and y to determine the constant of variation, k, in $y = \frac{k}{x}$.

Success Tip

If we multiply both sides of $y = \frac{k}{x}$ by x, we get $xy = k$. Thus, for the inverse variation model, k is simply the product of one pair of values of x and y. (Assume $x \neq 0$.)

$$y = \frac{k}{x} \qquad \text{This is the equation that models inverse variation.}$$

$$5 = \frac{k}{20} \qquad \text{Substitute 20 for x and 5 for y.}$$

$$20 \cdot 5 = k \qquad \text{To isolate k on the right side, multiply both sides by 20.}$$

$$100 = k \qquad \text{This is the constant of variation.}$$

Since $y = \frac{k}{x}$ and $k = 100$, we have

$$y = \frac{100}{x} \qquad \text{Substitute 100 for k in } y = \frac{k}{x}.$$

We can use $y = \frac{100}{x}$ to find other pairs of values of x and y. When $x = 50$, we see that

$$y = \frac{100}{50} = 2 \qquad \text{Substitute 50 for x.}$$

Thus, when $x = 50$, the value of y is 2.

> **Self Check 3**   Suppose y varies inversely as x. If $y = 25$ when $x = 3$, find y when $x = 15$.
>
> **Now Try** ▶ Problem 33

The Language of Algebra

The phrase **"is inversely proportional to"** is also used to indicate inverse variation.

Suppose that the time (in hours) that it takes to paint a house is inversely proportional to the size of the painting crew. As the number of painters increases, the time that it takes to paint the house decreases. If n represents the number of painters and t represents the time it takes to paint the house, this relationship can be expressed by the equation

$$t = \frac{k}{n} \quad \text{where } k \text{ is the constant of variation}$$

If we know that a crew of 8 can paint the house in 12 hours, we can find the constant of variation by substituting 8 for n and 12 for t in the equation $t = \frac{k}{n}$ and solving for k:

$$\boldsymbol{t = \frac{k}{n}} \qquad \text{This equation models inverse variation.}$$

$$\boldsymbol{12 = \frac{k}{8}} \qquad \text{Substitute 8 for n and 12 for t.}$$

$$8 \cdot 12 = k \qquad \text{To isolate k on the right side, multiply both sides by 8.}$$

$$96 = k \qquad \text{This is the constant of variation.}$$

The equation describing the relationship between the size of the painting crew and the time it takes to paint the house is $t = \frac{96}{n}$. We can use this equation to find the time it will take a crew of any size to paint the house. For example, to find the time it would take a four-person crew, we substitute 4 for n in the equation $t = \frac{96}{n}$.

$$t = \frac{96}{\boldsymbol{n}} \qquad \text{This equation models inverse variation for the specific house used in this example.}$$

$$t = \frac{96}{\boldsymbol{4}} \qquad \text{Substitute 4 for n.}$$

$$t = 24 \qquad \text{Do the division.}$$

Success Tip

For any Inverse variation equation of the form $y = \frac{k}{x}$, where $k > 0$:

as x increases, y decreases

It would take a four-person crew 24 hours to paint the house.

The following table shows some possible values for n and t as determined by the equation $t = \frac{96}{n}$. When these ordered pairs are graphed and a smooth curve is drawn through them, it is clear that as the number of painters n increases, the time t decreases, in a predictable way.

$$t = \frac{96}{n}$$

n	t
2	48
3	32
4	24
6	16
8	12
12	8
16	6
24	4

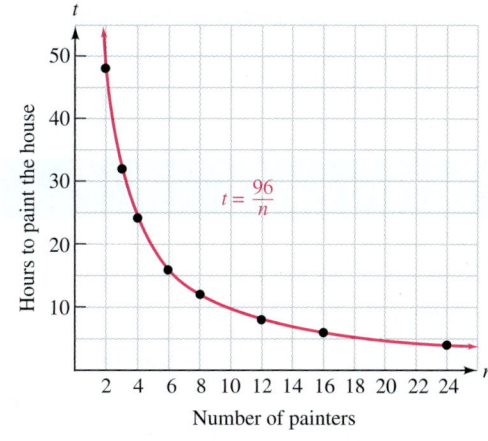

This curved graph shows that t varies inversely as n.

EXAMPLE 4   **Chemistry.** The volume occupied by a fixed weight of gas (held at a constant temperature) varies inversely as the pressure placed on it. If hydrogen gas occupies a volume of 22.5 cubic inches when placed under 3 pounds per square inch (psi) of pressure, find the volume occupied by the hydrogen gas when the pressure is 7.5 psi.

Strategy   We will follow the strategy for solving an inverse variation problem.

Why   In the words of the problem, the phrase *varies inversely* indicates that an inverse variation model should be used.

Solution   *Step 1:*   We let V represent the volume occupied by the gas and p represent the pressure. Translating the words *volume occupied by a gas varies inversely as the pressure,* we get the equation

$$V = \frac{k}{p} \qquad \text{This equation models inverse variation.}$$

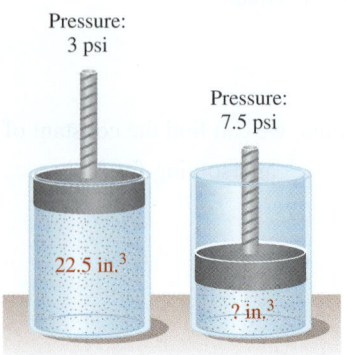

Pressure: 3 psi

Pressure: 7.5 psi

22.5 in.³

? in.³

The volume occupied by gas decreases as pressure increases.

Step 2:   To find the constant of variation, k, we substitute 22.5 for V and 3 for p.

$$V = \frac{k}{p}$$

$$22.5 = \frac{k}{3} \qquad \text{Substitute 3 for } p \text{ and 22.5 for } V.$$

$$67.5 = k \qquad \text{To isolate } k, \text{ multiply both sides by 3. This is the constant of variation.}$$

Step 3:   The equation describing the relationship between the volume occupied by the gas and the pressure placed on it is

$$V = \frac{67.5}{p} \qquad \text{Substitute 67.5 for } k \text{ in } V = \frac{k}{p}.$$

Step 4:   We can find the volume occupied by the gas when a pressure of 7.5 psi is placed on it by substituting 7.5 for p in the equation and evaluating the right side.

$$V = \frac{67.5}{p}$$

$$V = \frac{67.5}{7.5} \qquad \text{Substitute 7.5 for } p.$$

$$V = 9$$

The hydrogen gas will occupy a volume of 9 cubic inches when the pressure placed on it is 7.5 psi.

Self Check 4   **Chemistry.**   Find the volume occupied by the hydrogen gas in Example 4 when the pressure placed on it is 5 psi.

Now Try ▶ Problem 45

SECTION 7.9 ▷ STUDY SET

VOCABULARY

Fill in the blanks. Assume that k is a constant.

1. The equation $y = kx$ defines _____ variation and the equation $y = \frac{k}{x}$ defines _____ variation.

2. In $y = kx$ and in $y = \frac{k}{x}$, we call k the _____ of variation.

CONCEPTS

Tell whether each relationship suggests direct or inverse variation.

3. **Recycling.**   The amount of money you receive and the number of aluminum cans you return

4. **Karate.**   The force needed to break a board and the length of the board

5. **Tools.**   The force you must exert on the handle of a wrench to loosen a bolt and the length of the handle

6. **Anatomy.**   The volume of blood pumped from your heart each minute and your pulse rate

7. **Swimming Pools.**   The amount of chlorine needed in a pool and the amount of water in the pool

8. **Lightning.**   The time it takes you to hear the lightning after a strike and your distance from the strike

9. **Desserts.**   The number of servings you can get from a wedding cake and the size of the piece that is served

10. **Remodeling.**   The cost to remodel a house and the number of square feet to be added

Determine whether each graph represents direct variation or inverse variation.

11.

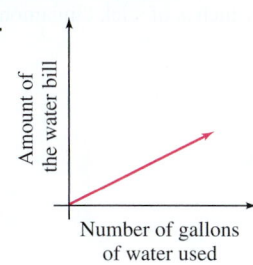

12.

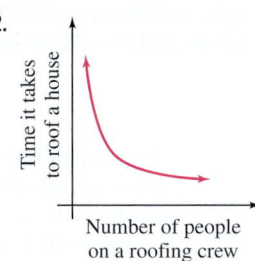

13.

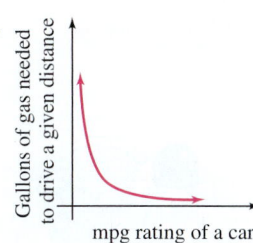

14.

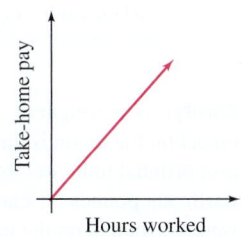

Complete each graph by sketching either a direct variation or an inverse variation.

15.

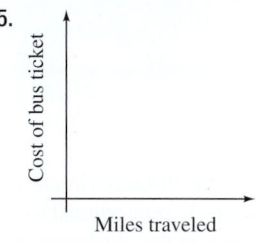

16.

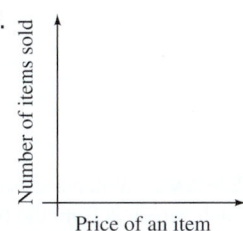

17.

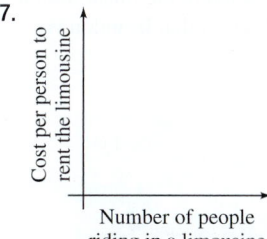

18.

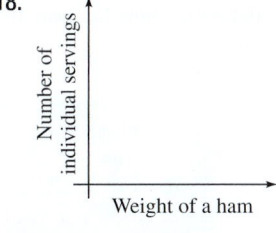

Write an equation to describe each variation. Use k for the constant of variation.

19. Swimming. When you swim underwater, the pressure p in your ears varies directly as the depth d at which you swim.

20. Farming. The number of bushels b of corn that a farmer harvests varies directly as a, the number of acres he plants.

21. Geology. The amount of dust d in a desert region varies inversely as the amount of rainfall r.

22. Word Processing. The number of words w that can be printed on an 8 in. by 11 in. piece of paper varies inversely as the size s of the font used.

NOTATION

23. Determine whether the equation describes direct variation.

   a. $y = kx$ **b.** $y = k + x$

   c. $y = \dfrac{k}{x}$ **d.** $m = kc$

24. Determine whether each equation describes inverse variation.

   a. $y = kx$ **b.** $y = \dfrac{k}{x}$

   c. $y = \dfrac{x}{k}$ **d.** $d = \dfrac{k}{g}$

GUIDED PRACTICE

Solve each direct variation problem. See Examples 1 and 2.

25. y varies directly as x. If $y = 10$ when $x = 2$, find y when $x = 7$.

26. A varies directly as z. If $A = 30$ when $z = 5$, find A when $z = 9$.

27. r varies directly as s. If $r = 21$ when $s = 7$, find r when $s = 12$.

28. h varies directly as m. If $h = 22$ when $m = 11$, find h when $m = 3$.

29. s varies directly as t. If $s = 1.2$ when $t = 4$, find s when $t = 30$.

30. y varies directly as x. If $y = 1.6$ when $x = 2$, find y when $x = 20$.

31. d is directly proportional to t. If $d = 21$ when $t = 6$, find d when $t = 4$.

32. b is directly proportional to c. If $b = 16$ when $c = 18$, find b when $c = 27$.

Solve each inverse variation problem. See Examples 3 and 4.

33. y varies inversely as x. If $y = 8$ when $x = 2$, find y when $x = 4$.

34. V varies inversely as p. If $V = 30$ when $p = 5$, find V when $p = 6$.

35. r varies inversely as t. If $r = 40$ when $t = 10$, find r when $t = 200$.

36. J varies inversely as v. If $J = 90$ when $v = 5$, find J when $v = 45$.

37. p varies inversely as x. If $p = 6$ when $x = 4$, find p when $x = 1.5$.

38. a varies inversely as d. If $a = 6$ when $d = 3$, find a when $d = 1.2$.

39. q is inversely proportional to s. If $q = 6$ when $s = 9$, find q when $s = 24$.

40. w is inversely proportional to s. If $w = 8$ when $s = 6$, find w when $s = 32$.

APPLICATIONS

Solve each direct variation problem. See Example 2.

41. Driving. The distance that a car can travel without refueling varies directly as the number of gallons of gasoline in the tank. If a car can go 360 miles on a full tank of gas (15 gallons), how far can it go on 7 gallons?

42. Gravity. The force of gravity acting on an object varies directly as the mass of the object. The force on a mass of 5 kilograms is 49 newtons. What is the force acting on a mass of 12 kilograms?

43. Medications. To fight ear infections in children, doctors often prescribe Ceclor. The recommended dose in milligrams is directly proportional to the child's body weight in pounds. The correct dosage for a 20-pound child is 124 milligrams. What would be the correct dosage for a 28-pound child?

44. Dosages. The recommended dose (in milligrams) of Demerol, a preoperative medication given to children, varies directly as the child's weight in pounds. The proper dosage for a child weighing 30 pounds is 18 milligrams. What would be the correct dosage for a child weighing 45 pounds?

Solve each inverse variation problem. **See Example 4.**

45. Traveling. The time it takes a car to travel a certain distance varies inversely as its rate of speed. If a certain trip takes 3 hours at 50 miles per hour, how long will the trip take at 60 miles per hour?

46. Geometry. For a fixed area, the length of a rectangle is inversely proportional to its width. A rectangle has a width of 12 feet and a length of 20 feet. If its width is increased to 12.5 feet, find the length that will maintain the same area.

47. Electricity. The current in an electric circuit varies inversely as the resistance. If the current in a circuit is 30 amps when the resistance is 4 ohms, what will the current be for a resistance of 15 ohms?

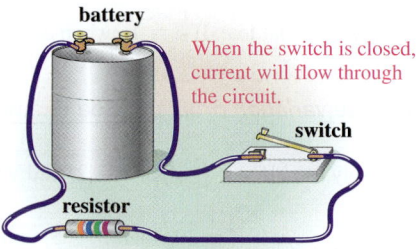

When the switch is closed, current will flow through the circuit.

48. Farming. The length of time a given number of bushels of corn will last when feeding cattle varies inversely as the number of animals. If a certain number of bushels will feed 25 cows for 10 days, how long will the feed last for 10 cows?

TRY IT YOURSELF

49. Pulleys. The speeds (in revolutions per minute) of two pulleys connected by a belt are inversely proportional to their diameters. If a pulley 24 inches in diameter, making 120 revolutions per minute, is belted to a second pulley 16 inches in diameter, how many revolutions per minute does the smaller pulley make?

50. Cider. For the following recipe, the number of inches of stick cinnamon to use varies directly as the number of servings of spiced cider to be made. How many inches of stick cinnamon are needed to make 36 servings?

Hot Spiced Cider

8 cups apple cider or apple juice
$\frac{1}{4}$ to $\frac{1}{2}$ cup packed brown sugar
6 inches stick cinnamon
1 teaspoon whole allspice
1 teaspoon whole cloves
8 thin orange wedges or slices (optional)
8 whole cloves (optional)   Makes 8 servings

51. Gravity. The weight of an object on the moon is directly proportional to its weight on Earth; six pounds on Earth weighs 1 pound on the moon. What would the scale register if the astronaut in the illustration were weighed on the moon?

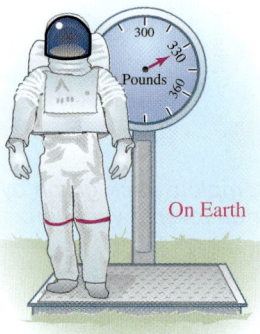

On Earth

52. Seesaws. When a seesaw is balanced, the distance (in feet) each person is from the fulcrum is inversely proportional to that person's weight. Use the information in the illustration to determine how far away from the fulcrum that Brandon is sitting.

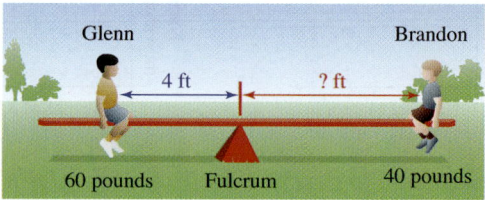

53. Hooke's Law. The distance a spring will stretch varies directly as the force applied to it. Suppose that a 15-kilogram weight stretches a spring 24 centimeters. Find the distance that the same spring will stretch when a heavier, 25-kilogram, weight is used.

54. Architecture. The total number of windows needed in the construction of an apartment building varies directly as the number of floors. If a 4-story building requires 176 windows, how many windows does an 11-story building require?

55. Architecture. If an office building is to have a fixed total floor space, the number of square feet of ground space it must occupy varies inversely as the number of floors. Suppose a 5-story building will occupy 163,000 square feet of ground space. How many square feet will a 25-story building occupy?

56. Hard Drives. The number of revolutions made by a computer hard drive varies directly as time. If it makes 16,200 revolutions in 3 minutes, how many revolutions will it make in 45 minutes?

57. **from Campus to Careers**

Recreation Director

Assume that the value of a stationary exercise bicycle varies inversely as its age. If a bike is worth $1,600 when it is 2 years old, find its value when it is 8 years old. How much does the bike depreciate over that 6-year period?

58. Chemistry. The volume occupied by a fixed weight of gas (held at a constant temperature) varies inversely as the pressure placed on it. If a nitrogen gas occupies a volume of 40 cubic meters under a pressure of 8 atmospheres, find the volume that the gas occupies when the pressure is changed to 6 atmospheres.

WRITING

59. Give examples of two quantities that vary directly and two quantities that do not.

60. What is the difference between direct variation and inverse variation?

61. What is a constant of variation?

62. Computer Printers. Is there a direct variation or an inverse variation between each pair of quantities? Explain. Draw a graph to support your answer.

    a. The time it takes to print a term paper and the speed of the printer.

    b. The time it takes to print a term paper and the length of the term paper.

REVIEW

Solve each equation.

63. $(a - 1)(a^2 + 5a + 6) = 0$

64. $(b - 2)(b^2 - 8b + 7) = 0$

65. $x^3 - 6x^2 - 27x = 0$

66. $6t^3 + 35t^2 - 6t = 0$

CHALLENGE PROBLEMS

67. Wind Energy. The power produced by a propeller wind turbine varies directly as the cube of the wind speed. If a wind speed of 15 mph produces 5.4 kilowatts of power, how much power would be produced by a 30-mph wind?

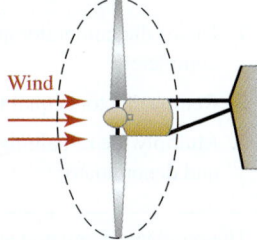

68. Gravity. The force with which the Earth attracts an object above the Earth's surface varies inversely as the square of the distance of the object from the center of the Earth. An object 4,000 miles from the center of the Earth is attracted with a force of 80 pounds. Find the force of attraction if the object were 5,000 miles from the center of the Earth.

69. Landing Aircraft. The runway distance required to land a single-engine airplane varies directly as the square of its touchdown speed. If a touchdown speed of 70 mph requires a landing distance of 1,470 feet, what landing distance is needed for a touchdown speed of 80 mph?

70. Construction. The time it takes to build a highway varies directly as the length of the road and inversely as the number of workers. It takes 100 workers 4 weeks to build 2 miles of highway. How long will it take 80 workers to build 10 miles of highway?

---

7 ▶ Summary & Review

SECTION 7.1 ▶ Simplifying Rational Expressions

DEFINITIONS AND CONCEPTS	EXAMPLES
A **rational expression** is an expression of the form $\frac{A}{B}$, where A and B are polynomials and B does not equal 0.	Rational expressions: $\dfrac{8}{7t}$, $\dfrac{a}{a-3}$, and $\dfrac{4x^2 - 16x}{x^2 - 6x + 8}$
To **evaluate a rational expression,** we substitute the values of its variables and simplify.	Evaluate $\frac{3x + 1}{x - 2}$ for $x = 5$. $\dfrac{3\mathbf{x} + 1}{\mathbf{x} - 2} = \dfrac{3(\mathbf{5}) + 1}{\mathbf{5} - 2} = \dfrac{16}{3}$ Substitute 5 for x.

To find the real numbers for which a **rational expression is undefined,** find the values of the variable that make the denominator 0.	For which real numbers is $\frac{11}{2x-3}$ undefined? $\quad 2x - 3 = 0$ Set the denominator equal to 0 and solve for x. $\quad\quad 2x = 3$ Add 3 to both sides. $\quad\quad\ x = \dfrac{3}{2}$ The expression is undefined for $x = \frac{3}{2}$.
To **simplify a rational expression:** 1. Factor the numerator and the denominator completely. 2. Remove factors equal to 1. 3. Multiply the remaining factors in the numerator and denominator.	Simplify: $\dfrac{x^2 - 4}{x^2 - 7x + 10} = \dfrac{(x+2)\overset{1}{\cancel{(x-2)}}}{(x-5)\underset{1}{\cancel{(x-2)}}}$ Factor and simplify. $\quad\quad\quad\quad\quad\quad = \dfrac{x+2}{x-5}$ Multiply the remaining factors in the numerator and the denominator.
The quotient of any nonzero expression and its **opposite** is -1.	$\dfrac{2t-3}{3-2t} = -1$ Because $2t-3$ and $3-2t$ are opposites.

REVIEW EXERCISES

1. Find the values of x for which the rational expression $\frac{x-1}{x^2-16}$ is undefined.

2. Evaluate $\frac{x^2-1}{x-5}$ for $x = -2$.

Simplify each rational expression, if possible. Assume that no denominators are zero.

3. $\dfrac{3x^2}{6x^3}$

4. $\dfrac{5xy^2}{2x^2y^2}$

5. $\dfrac{x^2}{x^2 + x}$

6. $\dfrac{a^2 - 4}{a + 2}$

7. $\dfrac{3p - 2}{2 - 3p}$

8. $\dfrac{8 - x}{x^2 - 5x - 24}$

9. $\dfrac{2x^2 - 16x}{2x^2 - 18x + 16}$

10. $\dfrac{x^2 + x - 2}{x^2 - x - 2}$

11. $\dfrac{x^2 - 2xy + y^2}{(x - y)^3}$

12. $\dfrac{4(t + 3) + 8}{3(t + 3) + 6}$

13. Explain the error in the following work: $\dfrac{x+1}{x} = \dfrac{\overset{1}{\cancel{x}}+1}{\underset{1}{\cancel{x}}} = \dfrac{2}{1} = 2.$

14. **Dosages.** Cowling's rule is a formula that can be used to determine the dosage of a prescription medication for children. If C is the proper child's dosage, D is an adult dosage, and A is the child's age in years, then $C = \frac{D(A + 1)}{24}$. Find the daily dosage of an antibiotic for an 11-year-old child if the adult daily dosage is 300 milligrams.

SECTION 7.2 ▶ Multiplying and Dividing Rational Expressions

DEFINITIONS AND CONCEPTS	EXAMPLES
To **multiply rational expressions,** multiply their numerators and multiply their denominators. $\dfrac{A}{B} \cdot \dfrac{C}{D} = \dfrac{AC}{BD}$ Then simplify, if possible.	Multiply: $\dfrac{4b}{b+2} \cdot \dfrac{7}{b} = \dfrac{4b \cdot 7}{(b+2)b}$ Multiply the numerators. Multiply the denominators. $\quad\quad\quad\quad\quad = \dfrac{4\overset{1}{\cancel{b}} \cdot 7}{(b+2)\underset{1}{\cancel{b}}}$ Simplify. $\quad\quad\quad\quad\quad = \dfrac{28}{b+2}$ Multiply the remaining factors in the numerator and the denominator.
To find the **reciprocal** of a rational expression, invert its numerator and denominator.	The reciprocal of $\dfrac{c}{c+7}$ is $\dfrac{c+7}{c}$.

| To **divide rational expressions,** multiply the first expression by the reciprocal of the second. $$\frac{A}{B} \div \frac{C}{D} = \frac{A}{B} \cdot \frac{D}{C} = \frac{AD}{BC}$$ Then simplify, if possible. | Divide: $\dfrac{t}{t+1} \div \dfrac{8}{t^2+t} = \dfrac{t}{t+1} \cdot \dfrac{t^2+t}{8}$ Multiply by the reciprocal. $$= \frac{t \cdot t \overset{1}{\cancel{(t+1)}}}{\underset{1}{\cancel{(t+1)}}8} \qquad \text{Factor and simplify.}$$ $$= \frac{t^2}{8} \qquad \text{Multiply the remaining factors in the numerator and the denominator.}$$ |
| A **unit conversion factor** is a fraction that has a value of 1. | $\dfrac{1\text{ yd}^2}{9\text{ ft}^2} = 1$ and $\dfrac{1\text{ mi}}{5{,}280\text{ ft}} = 1$ |

REVIEW EXERCISES

Multiply and simplify, if possible.

15. $\dfrac{3xy}{2x} \cdot \dfrac{4x}{2y^2}$ **16.** $56x\left(\dfrac{12}{7x}\right)$

17. $\dfrac{x^2-1}{x^2+2x} \cdot \dfrac{x}{x+1}$ **18.** $\dfrac{x^2+x}{3x-15} \cdot \dfrac{6x-30}{x^2+2x+1}$

Divide and simplify, if possible.

19. $\dfrac{3x^2}{5x^2y} \div \dfrac{6x}{15xy^2}$ **20.** $\dfrac{x^2-x-6}{1-2x} \div \dfrac{x^2-2x-3}{2x^2+x-1}$

21. Determine whether the given fraction is a unit conversion factor.

  a. $\dfrac{1\text{ ft}}{12\text{ in.}}$ **b.** $\dfrac{60\text{ min}}{1\text{ day}}$

  c. $\dfrac{2{,}000\text{ lb}}{1\text{ ton}}$ **d.** $\dfrac{1\text{ gal}}{4\text{ qt}}$

22. Traffic Signs. Convert the speed limit on the sign from miles per hour to miles per minute.

SPEED LIMIT **20** mph

SECTION 7.3 ▶ Adding and Subtracting with Like Denominators; Least Common Denominators

DEFINITIONS AND CONCEPTS	EXAMPLES
To **add (or subtract) rational expressions** that have the same denominator, add (or subtract) their numerators and write the sum (or difference) over their common denominator. $$\frac{A}{D} + \frac{B}{D} = \frac{A+B}{D} \qquad \frac{A}{D} - \frac{B}{D} = \frac{A-B}{D}$$ Then simplify, if possible.	Add: $\dfrac{2b}{3b-9} + \dfrac{b}{3b-9} = \dfrac{2b+b}{3b-9}$ Add the numerators and write the sum over the LCD, $3b-9$. $$= \frac{\overset{1}{\cancel{3}}b}{\underset{1}{\cancel{3}}(b-3)} \qquad \text{Factor and simplify.}$$ $$= \frac{b}{b-3}$$ Subtract: $\dfrac{x+1}{x} - \dfrac{x-1}{x} = \dfrac{x+1-(x-1)}{x}$ Don't forget the parentheses. $$= \frac{x+1-x+1}{x}$$ $$= \frac{2}{x} \qquad \text{Combine like terms.}$$
To find the **LCD** of several rational expressions, factor each denominator completely. Form a product using each different factor the greatest number of times it appears in any one factorization.	Find the LCD of $\dfrac{3}{x^3-x^2}$ and $\dfrac{x}{x^2-1}$. $$\left.\begin{array}{l} x^3-x^2 = x \cdot x \cdot (x-1) \\ x^2-1 = (x+1)(x-1) \end{array}\right\} \text{LCD} = x \cdot x \cdot (x-1)(x+1)$$

To **build an equivalent rational expression,** multiply the given expression by 1 in the form of $\frac{c}{c}$ where $c \neq 0$.

$$\frac{7}{4t} = \frac{7}{4t} \cdot \frac{3t}{3t} \qquad \text{and} \qquad \frac{x+1}{x-7} = \frac{x+1}{x-7} \cdot \frac{x-1}{x-1}$$

$$= \frac{21t}{12t^2} \qquad\qquad\qquad = \frac{(x+1)(x-1)}{(x-7)(x-1)}$$

$$= \frac{x^2 - 1}{x^2 - 8x + 7}$$

REVIEW EXERCISES

Add or subtract and simplify, if possible.

23. $\dfrac{13}{15d} - \dfrac{8}{15d}$

24. $\dfrac{x}{x+y} + \dfrac{y}{x+y}$

25. $\dfrac{3x}{x-7} - \dfrac{x-2}{x-7}$

26. $\dfrac{a}{a^2 - 2a - 8} + \dfrac{2}{a^2 - 2a - 8}$

Find the LCD of each pair of rational expressions.

27. $\dfrac{12}{x}, \dfrac{1}{9}$

28. $\dfrac{1}{2x^3}, \dfrac{5}{8x}$

29. $\dfrac{7}{m}, \dfrac{m+2}{m-8}$

30. $\dfrac{x}{5x+1}, \dfrac{5x}{5x-1}$

31. $\dfrac{6-a}{a^2 - 25}, \dfrac{a^2}{a-5}$

32. $\dfrac{4t+25}{t^2 + 10t + 25}, \dfrac{t^2 - 7}{2t^2 + 17t + 35}$

Build each rational expression into an equivalent fraction having the denominator shown in red.

33. $\dfrac{9}{a}, \mathbf{7a}$

34. $\dfrac{2y+1}{x-9}, \mathbf{x(x-9)}$

35. $\dfrac{b+7}{3b-15}, \mathbf{6(b-5)}$

36. $\dfrac{9r}{r^2 + 6r + 5}, \mathbf{(r+1)(r-4)(r+5)}$

SECTION 7.4 ▶ **Adding and Subtracting with Unlike Denominators**

DEFINITIONS AND CONCEPTS	EXAMPLES
To **add (or subtract) rational expressions** with unlike denominators: **1.** Find the LCD. **2.** Write each rational expression as an equivalent expression whose denominator is the LCD. **3.** Add (or subtract) the numerators and write the sum (or difference) over the LCD. **4.** Simplify the resulting rational expression if possible.	Add: $\dfrac{4x}{x} + \dfrac{2}{x-1} = \dfrac{4x}{x} \cdot \dfrac{x-1}{x-1} + \dfrac{2}{x-1} \cdot \dfrac{x}{x}$ The LCD is x(x − 1). $\begin{aligned}\text{Multiply} \rightarrow \\ \text{Don't multiply} \rightarrow \end{aligned} = \dfrac{4x(x-1)}{x(x-1)} + \dfrac{2x}{x(x-1)}$ Build so that each expression has a denominator of x(x − 1). $= \dfrac{4x^2 - 4x + 2x}{x(x-1)}$ Distribute the multiplication by 4x. $= \dfrac{4x^2 - 2x}{x(x-1)}$ Combine like terms. $= \dfrac{\overset{1}{2x}(2x-1)}{\underset{1}{x}(x-1)}$ Factor and simplify. $= \dfrac{2(2x-1)}{x-1}$
When a polynomial is multiplied by −1, the result is its opposite. This fact is used when adding or subtracting rational expressions whose **denominators are opposites.**	Add: $\dfrac{c}{c-4} + \dfrac{1}{4-c} = \dfrac{c}{c-4} + \dfrac{1}{4-c} \cdot \dfrac{-1}{-1}$ $= \dfrac{c}{c-4} + \dfrac{-1}{c-4}$ −1(4 − c) = c − 4 $= \dfrac{c-1}{c-4}$ Add the numerators. Write the sum over the LCD, c − 4.

REVIEW EXERCISES

Add or subtract and simplify, if possible.

37. $\dfrac{1}{7} - \dfrac{1}{a}$

38. $\dfrac{x}{x-1} + \dfrac{1}{x}$

39. $\dfrac{2t+2}{t^2+2t+1} - \dfrac{1}{t+1}$

40. $\dfrac{x+2}{2x} - \dfrac{2-x}{x^2}$

41. $\dfrac{6}{b-1} - \dfrac{b}{1-b}$

42. $\dfrac{8}{c} + 6$

43. $\dfrac{n+7}{n+3} - \dfrac{n-3}{n+7}$

44. $\dfrac{4}{t+2} - \dfrac{7}{(t+2)^2}$

45. $\dfrac{6}{a^2-9} - \dfrac{5}{a^2-a-6}$

46. $\dfrac{2}{3y-6} + \dfrac{3}{4y+8}$

47. Working on a homework assignment, a student added two rational expressions and obtained $\dfrac{-5n^3-7}{3n(n+6)}$. The answer given in the back of the book was $-\dfrac{5n^3+7}{3n(n+6)}$. Are the answers equivalent?

48. **Digital Video Cameras.**   Find the perimeter and the area of the LED screen of the camera.

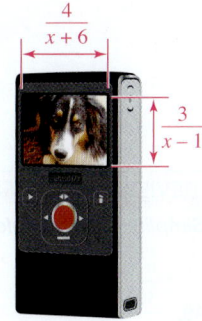

$\dfrac{4}{x+6}$

$\dfrac{3}{x-1}$

SECTION 7.5 ▶ ## Simplifying Complex Fractions

DEFINITIONS AND CONCEPTS	EXAMPLES
Complex fractions contain fractions in their numerators and/or their denominators.	Complex fractions: $\dfrac{\dfrac{2}{t}}{\dfrac{5}{4t}}$ and $\dfrac{\dfrac{3}{m}+\dfrac{m}{4}}{\dfrac{m}{2}}$
To simplify a complex fraction: **Method 1** Write the numerator and the denominator as single rational expressions and perform the indicated division.	Simplify: $\dfrac{\dfrac{3}{m}+\dfrac{m}{2}}{\dfrac{m}{4}} = \dfrac{\dfrac{3}{m}\cdot\dfrac{\mathbf{2}}{\mathbf{2}}+\dfrac{m}{2}\cdot\dfrac{\mathbf{m}}{\mathbf{m}}}{\dfrac{m}{4}}$ In the numerator, build to have an LCD of 2m. $= \dfrac{\dfrac{6}{2m}+\dfrac{m^2}{2m}}{\dfrac{m}{4}}$ In the numerator, multiply the fractions. $= \dfrac{\dfrac{6+m^2}{2m}}{\dfrac{m}{4}}$ Add the fractions in the numerator. $= \dfrac{6+m^2}{2m} \div \dfrac{m}{4}$ The main fraction bar indicates division. $= \dfrac{(6+m^2)\cdot \overset{1}{2}\cdot 2}{2m\cdot m}$ Multiply by the reciprocal of $\frac{m}{4}$. Factor 4 and simplify. $= \dfrac{12+2m^2}{m^2}$ Distribute the multiplication by 2 in the numerator.

Method 2

Determine the LCD of all the rational expressions in the complex fraction and multiply the complex fraction by 1, written in the form $\frac{LCD}{LCD}$.

Simplify: $\dfrac{\frac{3}{m}+\frac{m}{2}}{\frac{m}{4}} = \dfrac{\frac{3}{m}+\frac{m}{2}}{\frac{m}{4}}\cdot\dfrac{4m}{4m}$

The LCD for all the rational expressions is 4m.

$= \dfrac{\frac{3}{m}\cdot 4m + \frac{m}{2}\cdot 4m}{\frac{m}{4}\cdot 4m}$

In the numerator, distribute the multiplication by 4m.

$= \dfrac{12 + 2m^2}{m^2}$

Perform each multiplication by 4m.

REVIEW EXERCISES

Simplify each complex fraction.

49. $\dfrac{\frac{n^4}{30}}{\frac{7n}{15}}$

50. $\dfrac{\frac{r^2-81}{18s^2}}{\frac{4r-36}{9s}}$

53. $\dfrac{\frac{2}{x-1}+\frac{x-1}{x+1}}{\frac{1}{x^2-1}}$

54. $\dfrac{\frac{1}{x^2y}-\frac{5}{xy}}{\frac{3}{xy}-\frac{7}{xy^2}}$

51. $\dfrac{\frac{1}{y}+1}{\frac{1}{y}-1}$

52. $\dfrac{\frac{7}{a^2}}{\frac{1}{a}+\frac{10}{3}}$

SECTION 7.6 ▶ **Solving Rational Equations**

DEFINITIONS AND CONCEPTS

EXAMPLES

To **solve a rational equation** we use the multiplication property of equality to clear the equation of fractions. Use these steps:

1. Determine which numbers cannot be solutions.

2. Multiply both sides of the equation by the LCD of the rational expressions contained in the equation.

3. Solve the resulting equation.

4. Check all possible solutions in the *original* equation. A possible solution that does not satisfy the original equation is called an **extraneous solution**.

Solve: $\dfrac{y}{y-2}-1=\dfrac{1}{y}$ Since no denominators can be 0, $y \neq 2$ and $y \neq 0$.

$y(y-2)\left(\dfrac{y}{y-2}-1\right) = y(y-2)\left(\dfrac{1}{y}\right)$ The LCD is $y(y-2)$.

$y(y-2)\left(\dfrac{y}{y-2}\right) - y(y-2)1 = y(y-2)\left(\dfrac{1}{y}\right)$ Simplify.

$y\cdot y - y(y-2) = (y-2)\cdot 1$

$y^2 - y^2 + 2y = y - 2$ Multiply.

$2y = y - 2$ Combine like terms.

$y = -2$ Solve for y.

REVIEW EXERCISES

Solve each equation and check the result. If an equation has no solution, so indicate.

55. $\dfrac{3}{x}=\dfrac{2}{x-1}$

56. $\dfrac{a}{a-5}=3+\dfrac{5}{a-5}$

57. $\dfrac{2}{3t}+\dfrac{1}{t}=\dfrac{5}{9}$

58. $a=\dfrac{3a-50}{4a-24}-\dfrac{3}{4}$

59. $\dfrac{4}{x+2}-\dfrac{3}{x+3}=\dfrac{6}{x^2+5x+6}$

60. $\dfrac{3}{x+1}-\dfrac{x-2}{2}=\dfrac{x-2}{x+1}$

61. **Engineering.** The efficiency E of a Carnot engine is given by the following formula. Solve it for T_1.

$$E = 1 - \dfrac{T_2}{T_1}$$

62. Solve for y: $\dfrac{1}{x}=\dfrac{1}{y}+\dfrac{1}{z}$

SECTION 7.7 ▶ Problem Solving Using Rational Equations

DEFINITIONS AND CONCEPTS	EXAMPLES

DEFINITIONS AND CONCEPTS

To solve application problems, follow these steps:

1. Analyze the problem.
2. Assign a variable.
3. Form an equation.
4. Solve the equation.
5. State the conclusion.
6. Check the result.

Rate of Work: If a job can be completed in t units of time, the rate of work can be expressed as $\frac{1}{t}$ of the job is completed per unit of time.

Shared-work problems:

Work completed = rate of work · time worked

EXAMPLES

Washing Cars. Working alone, Carlos can wash the family SUV in 30 minutes. Victor, his brother, can wash the same SUV in 20 minutes working alone. How long will it take them if they wash the SUV together?

Analyze It takes Carlos 30 minutes and it takes Victor 20 minutes. How long will it take working together?

Assign Let x = the number of minutes it will take Carlos and Victor, working together, to wash the SUV.

Form Enter the data in a table.

	Rate ·	Time =	Work Completed
Carlos	$\frac{1}{30}$	x	$\frac{x}{30}$
Victor	$\frac{1}{20}$	x	$\frac{x}{20}$

The part of the job done by Carlos plus the part of the job done by Victor equals 1 job completed.

$$\frac{x}{30} + \frac{x}{20} = 1$$

Solve

$$60\left(\frac{x}{30} + \frac{x}{20}\right) = 60(1) \quad \text{Multiply both sides by the LCD, 60.}$$

$$60\left(\frac{x}{30}\right) + 60\left(\frac{x}{20}\right) = 60(1) \quad \text{On the left side, distribute the multiplication by 60.}$$

$$2x + 3x = 60 \quad \text{Perform each multiplication by 60.}$$

$$5x = 60 \quad \text{Combine like terms.}$$

$$x = \frac{60}{5} \quad \text{Divide both sides by 5.}$$

$$x = 12$$

State Working together, it will take Carlos and Victor 12 minutes to wash the family SUV.

Check In 12 minutes, Carlos will do $\frac{12}{30} = \frac{24}{60}$ of the job and Victor will do $\frac{12}{20} = \frac{36}{60}$ of the job. Together they will do $\frac{24}{60} + \frac{36}{60} = \frac{60}{60}$ or 1 whole job. The result checks.

Uniform motion problems: $\text{Time} = \dfrac{\text{distance}}{\text{rate}}$

See Example 2 in Section 7.7.

Investment problems: $\text{Principal} = \dfrac{\text{Interest}}{\text{rate} \cdot \text{time}}$

See Example 5 in Section 7.7.

REVIEW EXERCISES

63. Number Problems. If a number is subtracted from the denominator of $\frac{4}{5}$ and twice as much is added to the numerator, the result is 5. Find the number.

64. Exercise. A woman can bicycle 30 miles in the same time that it takes her to jog 10 miles. If she can ride 10 mph faster than she can jog, how fast can she jog?

65. House Cleaning. A maid can clean a house in 4 hours. What is her rate of work?

66. House Painting. If a homeowner can paint a house in 14 days and a professional painter can paint it in 10 days, how long will it take if they work together?

67. Investments. In 1 year, a student earned $100 interest on money she deposited at a savings and loan. She later learned that the money would have earned $120 if she had deposited it at a credit union, because the credit union paid 1% more interest at the time. Find the rate she received from the savings and loan.

68. Wind Speed. A plane flies 400 miles downwind in the same amount of time as it takes to travel 320 miles upwind. If the plane can fly at 360 mph in still air, find the velocity of the wind.

SECTION 7.8 ▶ Proportions and Similar Triangles

DEFINITIONS AND CONCEPTS	EXAMPLES
A **ratio** is the quotient of two numbers or two quantities with the same units. A **rate** is the quotient of two quantities with different units.	Ratios: $\dfrac{2}{3}$, $\dfrac{1}{50}$, and 2:3 Rates: $\dfrac{4 \text{ oz}}{6 \text{ lb}}$, $\dfrac{525 \text{ mi}}{15 \text{ hr}}$, and $\dfrac{\$1.95}{2 \text{ lb}}$

A **proportion** is a statement that two ratios or two rates are equal. In the proportion $\frac{a}{b} = \frac{c}{d}$, a and d are the **extremes** and b and c are the **means.**
 	
In any proportion, the product of the extremes is equal to the product of the means. (The **cross products** are equal.)
 	
To **solve a proportion,** set the product of the extremes equal to the product of the means and solve the resulting equation.
	A proportion: $\dfrac{4}{9} = \dfrac{28}{63}$ Extremes: 4 and 63
	Means: 9 and 28
	
	A proportion: 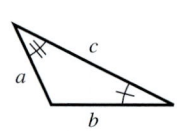 Cross product: **4 · 63** = 252
	Cross product: **9 · 28** = 252
	
	Solve the proportion: $\dfrac{3}{2} = \dfrac{x}{10}$
	
	   $3 \cdot 10 = 2 \cdot x$ Set the cross products equal.
	      $30 = 2x$
	      $15 = x$ Solve for x.

Triangles with the same shape but not necessarily the same size are called **similar triangles.** The lengths of the corresponding sides of two similar triangles are in proportion.	In these similar triangles: $\dfrac{a}{d} = \dfrac{b}{e} = \dfrac{c}{f}$ 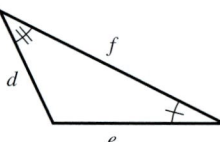
A **scale** is a ratio (or rate) that compares the size of a model to the size of an actual object. **Unit prices** can be used to compare costs of different sizes of the same brand to determine the best buy.	See Example 6 in Section 7.8. For the same item, a cost of $\dfrac{\$1.95}{1 \text{ lb}}$ is a better buy than a cost of $\dfrac{\$1.99}{1 \text{ lb}}$. See Example 7 in Section 7.8.

REVIEW EXERCISES

Determine whether each equation is a proportion.

69. $\dfrac{4}{7} = \dfrac{20}{34}$ **70.** $\dfrac{5}{7} = \dfrac{30}{42}$

Solve each proportion.

71. $\dfrac{3}{x} = \dfrac{6}{9}$ **72.** $\dfrac{x}{3} = \dfrac{x}{5}$

73. $\dfrac{x-2}{5} = \dfrac{x}{7}$ **74.** $\dfrac{2x}{x+4} = \dfrac{3}{x-1}$

75. Dentistry.   The diagram in the next column was displayed in a dentist's office. According to the diagram, if the dentist has 340 adult patients, how many will develop gum disease?

3 out of 4 adults will develop gum disease.

76. Utility Poles.   A telephone pole casts a shadow 12 feet long at the same time that a man 6 feet tall casts a shadow of 3.6 feet. How tall is the pole?

77. Porcelain Figurines.   A model of a flutist, standing and playing at a music stand, was made using a 1/12th scale. If the scale model is 5.5 inches tall, how tall is the flutist?

78. Comparison Shopping.   Which is the better buy for recordable compact discs: 150 for $60 or 250 for $98?

SECTION 7.9	Variation

DEFINITIONS AND CONCEPTS	EXAMPLES
The words *y* **varies directly** *as x* or *y* is **directly proportional** *to x* mean that $y = kx$ for some nonzero constant *k*, called the **constant of variation.**	The time *t* it takes you to order at a fast-food drive-through *varies directly* as the number *n* of cars ahead of you: $t = kn$.
The words *y* **varies inversely** *as x* or *y* is **inversely proportional** *to x* mean that $y = \frac{k}{x}$ for some nonzero constant *k*.	The time *t* it takes a person to read a book *varies inversely* as the person's reading rate *r*: $t = \frac{k}{r}$.

Strategy for Solving Variation Problems

1. Translate the verbal model into an equation.

2. Substitute a pair of values to find *k*.

3. Substitute the value of *k* into the variation equation.

4. Substitute the remaining given value into the equation from step 3 and answer the question.

Suppose *d* varies inversely as *h*. If $d = 5$ when $h = 4$, find *d* when $h = 10$.

1. The words *d varies inversely as h* translate to $d = \frac{k}{h}$.

2. If we substitute 5 for *d* and 4 for *h*, we have

$$5 = \frac{k}{4}$$

$$20 = k \qquad \text{Multiply both sides by 4.}$$

3. Since $k = 20$, the inverse variation equation is $d = \frac{20}{h}$.

4. To answer the final question, we substitute 10 for *h*.

$$d = \frac{20}{10} = 2$$

REVIEW EXERCISES

Write an equation to describe each variation. Use k for the constant of variation.

79. Fitness. The number *c* of calories burned while jogging varies directly as the time *t* spent jogging.

80. Guitars. The frequency *f* of a vibrating string varies inversely as the length *L* of the string.

81. Selling Fruit. The profit made by a strawberry farm varies directly as the number of baskets of strawberries sold. If a profit of $500 is made from the sale of 300 baskets, what is the profit when 1,200 baskets are sold?

82. *L* varies inversely as *w*. Find the constant of variation if $L = 30$ when $w = 20$.

83. Electricity. For a fixed voltage, the current in an electrical circuit varies inversely as the resistance in the ciruit. If a certain circuit has a current of 2.5 amps when the resistance is 150 ohms, find the current in the circuit when the resistance is 300 ohms.

84. Does the graph show direct or inverse variation?

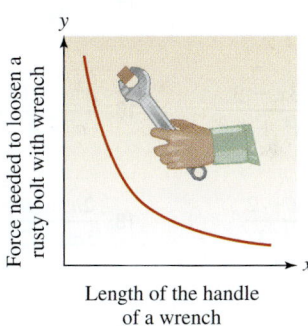

Length of the handle of a wrench

---

7 ▶ **CHAPTER TEST**

1. Fill in the blanks.

  a. A quotient of two polynomials, such as $\frac{x+7}{x^2 + 2x}$, is called a _____ expression.

  b. Two triangles with the same shape, but not necessarily the same size, are called _____ triangles.

  c. A _____ is a mathematical statement that two ratios or two rates are equal.

  d. To _____ a rational expression, we multiply it by a form of 1. For example, $\frac{2}{5x} \cdot \frac{8}{8} = \frac{16}{40x}$.

  e. To simplify $\frac{x-3}{(x+3)(x-3)}$, we remove common _____ of the numerator and denominator.

2. **Memory.** The formula $n = \frac{35 + 5d}{d}$ approximates the number of words n that a certain person can recall d days after memorizing a list of 50 words. How many words will the person remember in 1 week?

For what real numbers is each rational expression undefined?

3. $\frac{6x - 9}{5x}$

4. $\frac{x}{x^2 + x - 6}$

5. **The Internet.** A dial-up modem transmits up to 56K bits per second (K is an abbreviation for one thousand). Convert this to bits per minute.

6. Explain the error: $\frac{x + 5}{5} = \frac{\overset{1}{\cancel{x + 5}}}{\cancel{5}}$

$$= x + 1$$

Simplify each rational expression.

7. $\frac{48x^2y}{54xy^2}$

8. $\frac{7m - 49}{7 - m}$

9. $\frac{2x^2 - x - 3}{4x^2 - 9}$

10. $\frac{3(x + 2) - 3}{6x + 5 - (3x + 2)}$

Find the LCD of each pair of rational expressions.

11. $\frac{19}{3c^2d}$, $\frac{6}{c^2d^3}$

12. $\frac{4n + 25}{n^2 - 4n - 5}$, $\frac{6n}{n^2 - 25}$

Perform the operations. Simplify, if possible.

13. $\frac{12x^2y}{15xy} \cdot \frac{25y^2}{16x}$

14. $\frac{x^2 + 3x + 2}{3x + 9} \cdot \frac{x + 3}{x^2 - 4}$

15. $\frac{x - x^2}{3x^2 + 6x} \div \frac{3x - 3}{3x^3 + 6x^2}$

16. $\frac{a^2 - 16}{a - 4} \div (6a + 24)$

17. $\frac{3y + 7}{2y + 3} - \frac{-3y - 2}{2y + 3}$

18. $\frac{2n}{5m} - \frac{n}{2}$

19. $\frac{x + 1}{x} + \frac{x - 1}{x + 1}$

20. $\frac{a + 3}{a - 1} - \frac{a + 4}{1 - a}$

21. $\frac{9}{c - 4} + c$

22. $\frac{6}{t^2 - 9} - \frac{5}{t^2 - t - 6}$

Simplify each complex fraction.

23. $\dfrac{\dfrac{3m - 9}{8m}}{\dfrac{5m - 15}{32}}$

24. $\dfrac{\dfrac{3}{as^2} + \dfrac{6}{a^2s}}{\dfrac{6}{a} - \dfrac{9}{s^2}}$

Solve each equation. If an equation has no solution, so indicate.

25. $\frac{1}{3} + \frac{4}{3y} = \frac{5}{y}$

26. $\frac{9n}{n - 6} = 3 + \frac{54}{n - 6}$

27. $\frac{7}{q^2 - q - 2} + \frac{1}{q + 1} = \frac{3}{q - 2}$

28. $\frac{2}{3} = \frac{2c - 12}{3c - 9} - c$

29. $\frac{y}{y - 1} = \frac{y - 2}{y}$

30. $\frac{a}{a - 3} + \frac{4}{a + 3} = \frac{18}{a^2 - 9}$

31. Solve for B: $H = \frac{RB}{R + B}$

32. Solve for s: $\frac{1}{r} + \frac{1}{s} = \frac{1}{t}$

33. **Health Risks.** A medical newsletter states that a "healthy" waist-to-hip ratio for men is 19:20 or less. Does the patient shown in the illustration fall within the "healthy" range?

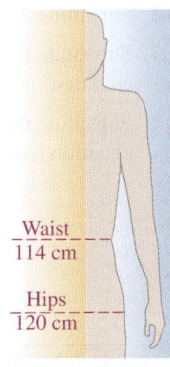

Waist
114 cm

Hips
120 cm

34. **Currency Exchange Rates.** Preparing for a visit to London, a New York resident exchanged 3,500 U.S. dollars for British pounds. (A pound is the basic monetary unit of Great Britain.) If the exchange rate was 100 U.S. dollars for 51 British pounds, how many British pounds did the traveler receive?

35. **TV Towers.** A television tower casts a shadow 114 feet long at the same time that a 6-foot-tall television reporter casts a shadow of 4 feet. Find the height of the tower.

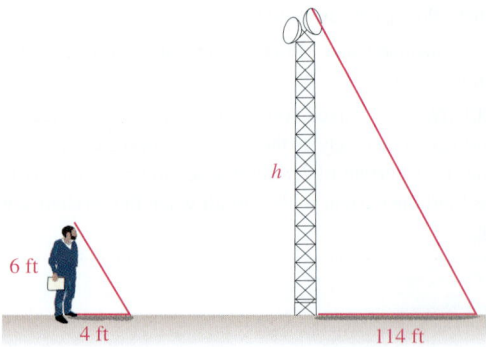

6 ft

h

4 ft   114 ft

36. **Comparison Shopping.** Which is the better buy for fabric softener: 80 sheets for $3.89 or 120 sheets for $6.19?

37. **Cleaning Highways.** One highway worker can pick up all the trash on a strip of highway in 7 hours, and his helper can pick up the trash in 9 hours. How long will it take them if they work together?

38. **Physical Fitness.** A man roller-blades at a rate 6 miles per hour faster than he jogs. In the same time it takes him to roller-blade 5 miles he can jog 2 miles. How fast does he jog?

39. **Number Problem.** If a number is subtracted from the numerator of $\frac{5}{8}$ and twice as much is added to the denominator, the result is $\frac{1}{4}$. Find the number.

40. Explain what it means to clear the following equation of fractions. Why is this a helpful first step in solving the equation?

$$\frac{u}{u-1} + \frac{1}{u} = \frac{u^2+1}{u^2-u}$$

41. Explain the difference between the procedure used to simplify $\frac{1}{x} + \frac{1}{4}$ and the procedure used to solve $\frac{1}{x} + \frac{1}{4} = \frac{1}{2}$.

42. **Pogo Sticks.** The force required to compress a spring varies directly with the change in the length of the spring. If a force of 130 pounds compresses the spring on the pogo stick 6.5 inches, how much force is required to compress the spring 5 inches?

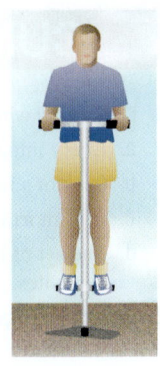

43. Assume that r varies inversely with s. If $r = 40$ when $s = 10$, find r when $s = 15$.

44. Tell whether each relationship suggests direct or inverse variation.

    a. The time it takes to complete a job and the number of workers on the job.

    b. The length of a beard and the time it has been growing.

Group Project

WHAT IS π?

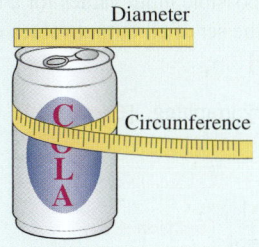

Diameter

Circumference

Overview In this activity, you will discover an important fact about the ratio of the circumference to the diameter of a circle.

Instructions Form groups of 2 or 3 students. With a piece of string or a cloth tape measure, find the circumference and the diameter of objects that are circular in shape. You can measure anything that is round: for example, a coin, the top of a can, a tire, or a wastepaper basket. Enter your results in a table, as shown below. Convert each measurement to a decimal, and then use a calculator to determine a decimal approximation of the ratio of the circumference C to diameter d.

Object	Circumference	Diameter	$\frac{C}{d}$ (approx.)
A quarter	$2\frac{15}{16}$ in. = 2.9375 in.	$\frac{15}{16}$ in. = 0.9375 in.	3.13333

Since early history, mathematicians have known that the ratio of the circumference to the diameter of a circle is the same for any size circle, approximately 3. Today, following centuries of study, we know that this ratio is exactly 3.141592653589. . . .

$$\frac{C}{d} = 3.141592653589\ldots$$

The Greek letter π (pi) is used to represent the ratio of circumference to diameter:

$$\pi = \frac{C}{d}, \quad \text{where } \pi = 3.141592653589\ldots$$

Are the ratios in your table numerically close to π? Give some reasons why they aren't exactly 3.141592653589 . . . in each case.

CUMULATIVE REVIEW ▶▶ Chapters 1–7

1. Determine whether each statement is true or false. [Section 1.3]

   a. Every integer is a whole number.

   b. 0 is not a rational number.

   c. π is an irrational number.

   d. The set of integers is the set of whole numbers and their opposites.

2. Insert the proper symbol, $<$ or $>$, in the blank to make a true statement.

   $$|2 - 4| \quad \underline{\quad} \quad -(-6) \quad \text{[Section 1.5]}$$

3. Evaluate: $9^2 - 3[45 - 3(6 + 4)]$ [Section 1.7]

4. Find the average (mean) test score of a student in a history class with scores of 80, 73, 61, 73, and 98. [Section 1.7]

5. Simplify: $8(c + 7) - 2(c - 3)$ [Section 1.9]

6. Solve: $\frac{4}{5}d = -4$ [Section 2.1]

7. Solve: $2 - 3(x - 5) = 4(x - 1)$ [Section 2.2]

8. **Grand King Size Beds.**   Because Americans are taller compared to 100 years ago, bed manufacturers are making larger models. Find the percent of increase in sleeping area of the new grand king size bed compared to the standard king size. [Section 2.3]

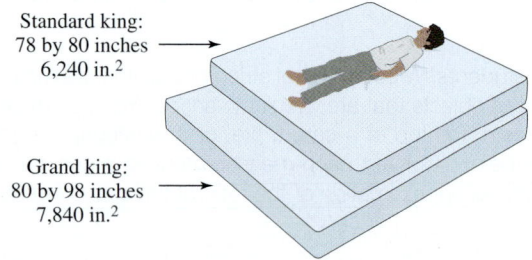

   Standard king: 78 by 80 inches 6,240 in.²

   Grand king: 80 by 98 inches 7,840 in.²

9. Solve $A - c = 2B + r$ for B. [Section 2.4]

10. Change 40°C to degrees Fahrenheit. [Section 2.4]

11. Find the volume of a pyramid that has a square base, measuring 6 feet on a side, and whose height is 20 feet. [Section 2.4]

12. **Blending Tea.**   One grade of tea (worth $6.40 per pound) is to be mixed with another grade (worth $4 per pound) to make 20 pounds of a mixture that will be worth $5.44 per pound. How much of each grade of tea must be used? [Section 2.6]

13. **Speed of a Plane.**   Two planes are 6,000 miles apart and their speeds differ by 200 mph. If they travel toward each other and meet in 5 hours, find the speed of the slower plane. [Section 2.6]

14. Solve $7x + 2 \geq 4x - 1$. Write the solution set in interval notation and graph it. [Section 2.7]

15. Graph: $y = 2x - 3$ [Section 3.2]

16. Find the slope of the line passing through $(-1, 3)$ and $(3, -1)$. [Section 3.4]

17. **Cutting Steel.**   The graph shows the amount of wear (in millimeters) on a cutting blade for a given length of a cut (in meters). Find the rate of change in the length of the cutting blade. [Section 3.4]

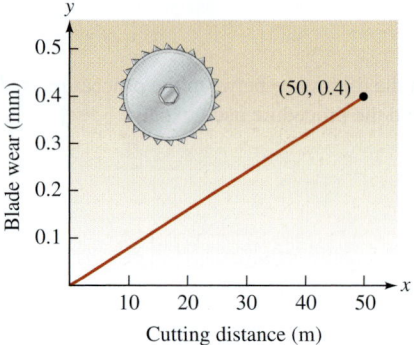

Cutting distance (m)

18. What is the slope of a line perpendicular to the line $y = -\frac{7}{8}x - 6$? [Section 3.5]

19. Write an equation of the line that has slope 3 and passes through the point $(1, 5)$. Write the answer in slope–intercept form. [Section 3.6]

20. Graph: $3x - 2y \leq 6$ [Section 3.7]

21. If $f(x) = -3x^2 - 6x$, find $f(-2)$. [Section 3.8]

22. Fill in the blanks. The set of all possible input values for a function is called the _____ and the set of all output values is called the _____. [Section 3.8]

23. Solve the system $\begin{cases} x + y = 1 \\ y = x + 5 \end{cases}$ by graphing. [Section 4.1]

24. Solve the system: $\begin{cases} x = 3y - 1 \\ 2x - 3y = 4 \end{cases}$ [Section 4.2]

25. Solve the system: $\begin{cases} 2x + 3y = -1 \\ 3x + 5y = -2 \end{cases}$ [Section 4.3]

26. **Poker.**   After a night of cards, a poker player finished with some red chips (worth $5 each) and some blue chips (worth $10 each). He received $190 when he cashed in the 23 chips. How many of each colored chip did he finish with? [Section 4.4]

Simplify each expression. Write each answer without using negative exponents.

27. x^4x^3 [Section 5.1]

28. $(x^2x^3)^5$ [Section 5.1]

29. $\left(\dfrac{y^3y}{2yy^2}\right)^3$ [Section 5.1]

30. $\left(\dfrac{-2a}{b}\right)^5$ [Section 5.1]

31. $(a^{-2}b^3)^{-4}$ [Section 5.2]

32. $\dfrac{9b^0b^3}{3b^{-3}b^4}$ [Section 5.2]

33. Write 290,000 in scientific notation. [Section 5.3]

34. What is the degree of the polynomial $5x^3 - 4x + 16$? [Section 5.4]

35. Graph: $y = -x^3$ [Section 5.4]

36. **Concentric Circles.**   The area of the ring between the two concentric circles of radius r and R is given by the formula

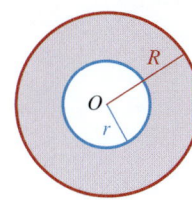

$$A = \pi(R + r)(R - r)$$

Do the multiplication on the right side of the equation.
[Section 5.7]

Perform the operations.

37. $(3x^2 - 3x - 2) + (3x^2 + 4x - 3)$ [Section 5.5]

38. $\left(\dfrac{1}{16}t^3 + \dfrac{1}{2}t^2 - \dfrac{1}{6}t\right) - \left(\dfrac{9}{16}t^3 + \dfrac{9}{4}t^2 - \dfrac{1}{12}t\right)$

[Section 5.5]

39. $(2x^2y^3)(3x^2y^2)$ [Section 5.6]

40. $(2y - 5)(3y + 7)$ [Section 5.6]

41. $-4x^2z(3x^2 - z)$ [Section 5.6]

42. $(3a - 4)^2$ [Section 5.7]

43. $\dfrac{6x + 9}{3}$ [Section 5.8]

44. $2x + 3\overline{)2x^3 + 7x^2 + 4x - 3}$ [Section 5.8]

Factor each polynomial completely.

45. $k^3t - 3k^2t$ 46. $2ab + 2ac + 3b + 3c$
[Section 6.1]   [Section 6.1]

47. $u^2 - 18u + 81$ 48. $-r^2 + 2 + r$
[Section 6.2]   [Section 6.2]

49. $u^2 + 10u + 15$ 50. $6x^2 - 63 - 13x$
[Section 6.2]   [Section 6.3]

51. $2a^2 - 200b^2$ 52. $b^3 + 125$
[Section 6.4]   [Section 6.5]

Solve each equation by factoring.

53. $5x^2 + x = 0$ 54. $6x^2 - 5x = -1$
[Section 6.7]   [Section 6.7]

55. **Cooking.**   The electric griddle shown has a cooking surface of 160 square inches. Find the length and the width of the griddle.
[Section 6.7]

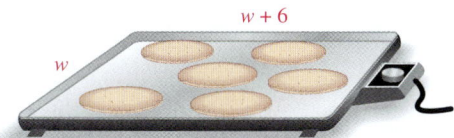

$w + 6$

w

56. For what values of x is the rational expression $\dfrac{3x^2}{x^2 - 25}$ undefined? [Section 7.1]

Perform the operations. Simplify, if possible.

57. $\dfrac{2x^2 - 8x}{x^2 - 6x + 8}$ 58. $\dfrac{x^2 - 16}{4 - x} \div \dfrac{3x + 12}{x^3}$
[Section 7.1]   [Section 7.2]

59. $\dfrac{8m^2}{2m + 5} - \dfrac{4m^2 + 25}{2m + 5}$ 60. $\dfrac{4}{x - 3} + \dfrac{5}{3 - x}$
[Section 7.3]   [Section 7.4]

61. $\dfrac{m}{m^2 + 5m + 6} - \dfrac{2}{m^2 + 3m + 2}$ [Section 7.4]

62. Simplify: $\dfrac{2 - \dfrac{2}{x + 1}}{2 + \dfrac{2}{x}}$ [Section 7.5]

Solve each equation.

63. $\dfrac{7}{5x} - \dfrac{1}{2} = \dfrac{5}{6x} + \dfrac{1}{3}$ [Section 7.6]

64. $\dfrac{u}{u - 1} + \dfrac{1}{u} = \dfrac{u^2 + 1}{u^2 - u}$ [Section 7.6]

65. **Draining a Tank.**   If one outlet pipe can drain a tank in 24 hours, and another pipe can drain the tank in 36 hours, how long will it take for both pipes to drain the tank?
[Section 7.7]

66. **Height of a Tree.**   A tree casts a shadow of 29 feet at the same time as a vertical yardstick casts a shadow of 2.5 feet. Find the height of the tree. [Section 7.8]

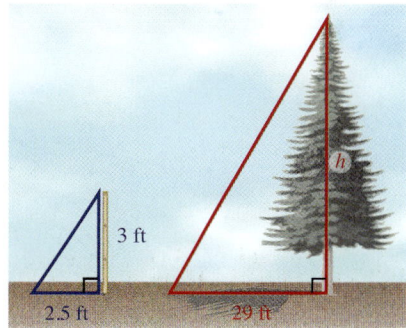

3 ft

h

2.5 ft

29 ft

67. **Forestry.**   For certain types of hardwood trees, the diameter of the tree trunk varies directly as the age of the tree. A 48-year old hardwood tree has a diameter of 16 inches. Find the diameter of a hardwood tree that is 84 years old.
[Section 7.9]

68. Suppose y varies inversely as x. If $y = 8$ when x is 6, find y when x is 64. [Section 7.9]

Radical Expressions and Equations

8

from Campus to Careers

Crime Scene Investigator

Crime scene investigators are often the first people on the scene after a crime has been committed. They are responsible for collecting evidence that can lead to criminal convictions. Most investigators spend some time each day in their offices and outside in the field. The job description calls for strong interest in the sciences and strong technical skills.

Problem 99 in **Study Set 8.1**, **problem 121** in **Study Set 8.4**, and **problem 95** in **Study Set 8.5** involve situations that a crime scene investigator might encounter on the job. The mathematical concepts discussed in this chapter can be used to solve those problems.

JOB TITLE:
Crime Scene Investigator
EDUCATION:
Some positions require a 4-year degree while others only high school graduation or a GED.
JOB OUTLOOK:
Employment is expected to increase between 9% and 17% through the year 2014.
ANNUAL EARNINGS:
Mean annual salary $65,860, mean hourly wage $31.66
FOR MORE INFORMATION:
bls.gov/oes/current/oes333021.htm

©Peter Coombs/Alamy

Final exams can be stressful for many students because the number of topics to study can seem overwhelming. Here are some suggestions to help reduce the stress and prepare you for the test.

GET ORGANIZED: Gather all of your notes, study sheets, homework assignments, and especially all of your returned tests to review.

TALK WITH YOUR INSTRUCTOR: Ask your instructor to list the topics that may appear on the final and those that won't be covered.

MANAGE YOUR TIME: Adjust your daily schedule one week before the final so that it includes extended periods of study time.

Now Try This ▶

1. Review your old tests. Make a list of the test problems that you are still unsure about and see a tutor or your instructor to get help.
2. Make a practice final exam that includes one or more of each type of problem that may appear on the test.
3. Make a detailed study plan. Determine when, where, and what you will study each day for one week before the final.

SECTION 8.1

An Introduction to Square Roots

OBJECTIVES

1. Find square roots of perfect squares.
2. Approximate irrational square roots.
3. Find square roots of variable expressions.
4. Use the Pythagorean theorem to solve problems.
5. Use the distance formula.

ARE YOU READY?

The following problems review some basic skills that are needed when working with square roots.

1. Evaluate: 6^2

2. Evaluate: $\left(\dfrac{4}{5}\right)^2$

3. Simplify: $(a^4)^2$

4. Evaluate: $(0.7)^2$

5. Evaluate: $9^2 + 3^2$

6. Evaluate: $\left[-3 - (-1)\right]^2$

Addition and subtraction are reverse operations, and so are multiplication and division. In this section, we will discuss another pair of reverse operations: raising a number to a power and finding the root of a number.

1 Find Square Roots of Perfect Squares.

When we raise a number to the second power, we are squaring it, or finding its **square.**

- The square of 5 is 25 because $5^2 = 25$.
- The square of -5 is 25, because $(-5)^2 = 25$.

We can reverse the squaring process to find **square roots** of numbers. For example, to find the square roots of 25, we ask ourselves "What number, when squared, is equal to 25?" There are two possible answers.

- 5 is a square root of 25, because $5^2 = 25$.
- -5 is also a square root of 25, because $(-5)^2 = 25$.

In general, we have the following definition.

The Definition of Square Root ▼   The number b is a **square root** of the number a if $b^2 = a$.

Every positive number has two square roots, one positive and one negative. For example, the two square roots of 9 are 3 and -3, and the two square roots of 144 are 12 and -12. The number 0 is the only real number with exactly one square root. In fact, it is its own square root, because $0^2 = 0$.

A **radical symbol** $\sqrt{}$ represents the **positive** or **principal square root** of a positive number. When reading this symbol, we usually drop the word *positive* (or *principal*) and simply say *square root.* Since 3 is the positive square root of 9, we can write

$$\sqrt{9} = 3 \qquad \text{\textcolor{red}{$\sqrt{9}$ represents the positive number whose square is 9.}}$$
$$\text{\textcolor{red}{Read as "the square root of 9 is 3."}}$$

The symbol $-\sqrt{}$ is used to represent the **negative square root** of a positive number. It is the opposite of the principal square root. Since -12 is the negative square root of 144, we can write

$$-\sqrt{144} = -12 \qquad \text{\textcolor{red}{Read as "the negative square root of 144 is -12" or "the opposite of the}}$$
$$\text{\textcolor{red}{square root of 144 is -12." The notation $-\sqrt{144}$ represents the negative}}$$
$$\text{\textcolor{red}{number whose square is 144.}}$$

If the number under the radical symbol is 0, we have $\sqrt{0} = 0$.

Square Root Notation

If a is a positive real number,

1. $\sqrt{a}$ represents the **positive** or **principal square root** of a. It is the positive number we square to get a.

2. $-\sqrt{a}$ represents the **negative square root** of a. It is the opposite of the principal square root of a: $-\sqrt{a} = -1 \cdot \sqrt{a}$.

3. The principal square root of 0 is 0: $\sqrt{0} = 0$.

The number or variable expression within (under) a radical symbol is called the **radicand,** and the radical symbol and radicand together are called a **radical.**

$$\underset{\textcolor{green}{\text{Radical}}}{\underbrace{\overset{\textcolor{red}{\text{Radical symbol}}}{\sqrt{16}}}} \;\textcolor{blue}{\leftarrow \text{Radicand}}$$

An algebraic expression containing a radical is called a **radical expression.** Some examples of radical expressions are

$$\sqrt{100}, \qquad \sqrt{2} + 3, \qquad \sqrt{x^2}, \qquad \text{and} \qquad \sqrt{\dfrac{a-1}{49}}$$

To evaluate (find the value of) square roots, you need to quickly recognize each of the following natural-number **perfect squares** shown in red:

$\mathbf{1} = 1^2$	$\mathbf{25} = 5^2$	$\mathbf{81} = 9^2$	$\mathbf{169} = 13^2$	$\mathbf{289} = 17^2$
$\mathbf{4} = 2^2$	$\mathbf{36} = 6^2$	$\mathbf{100} = 10^2$	$\mathbf{196} = 14^2$	$\mathbf{324} = 18^2$
$\mathbf{9} = 3^2$	$\mathbf{49} = 7^2$	$\mathbf{121} = 11^2$	$\mathbf{225} = 15^2$	$\mathbf{361} = 19^2$
$\mathbf{16} = 4^2$	$\mathbf{64} = 8^2$	$\mathbf{144} = 12^2$	$\mathbf{256} = 16^2$	$\mathbf{400} = 20^2$

EXAMPLE 1   Evaluate each square root: **a.** $\sqrt{16}$ **b.** $\sqrt{1}$ **c.** $\sqrt{0.36}$ **d.** $\sqrt{\dfrac{4}{9}}$ **e.** $-\sqrt{225}$

Strategy   In each case, we will determine what positive number, when squared, produces the radicand.

Why   The radical symbol $\sqrt{}$ indicates that the positive square root (principal square root) of the number written under it should be found.

Solution   a. $\sqrt{16} = 4$ Ask: What positive number, when squared, is 16? The answer is 4 because $4^2 = 16$.

b. $\sqrt{1} = 1$ Ask: What positive number, when squared, is 1? The answer is 1 because $1^2 = 1$.

c. $\sqrt{0.36} = 0.6$ Ask: What positive number, when squared, is 0.36? The answer is 0.6 because $(0.6)^2 = 0.36$.

d. $\sqrt{\dfrac{4}{9}} = \dfrac{2}{3}$ Ask: What positive number, when squared, is $\frac{4}{9}$? The answer is $\frac{2}{3}$ because $\left(\frac{2}{3}\right)^2 = \frac{4}{9}$.

e. $-\sqrt{225}$ is the opposite of the square root of 225. Since $\sqrt{225} = 15$, we have

$$-\sqrt{225} = -15 \quad \text{Because} -\sqrt{225} = -1 \cdot \sqrt{225} = -1 \cdot 15 = -15$$

Self Check 1   Evaluate each square root: a. $\sqrt{49}$ b. $\sqrt{121}$ c. $-\sqrt{0.09}$ d. $\sqrt{\dfrac{1}{25}}$

Now Try ▶ Problem 17

Not every number has a real-number square root. For example, to find $\sqrt{-4}$, we ask "What number, when squared, is equal to -4?" Since the square of a real number is always positive or 0, no real number squared is -4. Therefore, $\sqrt{-4}$ is not a real number. If we attempt to evaluate $\sqrt{-4}$ using a calculator, an error message is displayed.

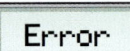

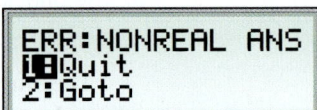

Scientific calculator   Graphing calculator

Since square roots of negative numbers such as $\sqrt{-4}$, $\sqrt{-1}$, and $\sqrt{-39}$ are not real numbers, they do not correspond to any point on the real number line. They are examples of **imaginary numbers.**

We summarize three important facts about square roots as follows.

Square roots

1. If a is a perfect square, then $\sqrt{a}$ is rational.

2. If a is a positive number that is not a perfect square, then $\sqrt{a}$ is irrational.

3. If a is a negative number, then $\sqrt{a}$ is not a real number.

A table of square root approximations

n	$\sqrt{n}$
1	1.000
2	1.414
3	1.732
4	2.000
5	2.236

2 Approximate Irrational Square Roots.

A number such as 16, 1, 0.36, $\frac{4}{9}$, or 225 that is the square of some rational number is called a **perfect square.** In Example 1, we saw that the square root of a perfect square is a rational number.

If a positive number is not a perfect square, its square root is irrational. For example, $\sqrt{3}$, $\sqrt{29}$, and $\sqrt{83}$ are irrational numbers because 3, 29, and 83 are not perfect squares. Since these square roots are irrational, they cannot be written as the quotient of two integers. Furthermore, their decimal representations are nonterminating and nonrepeating. We can find

Success Tip

Estimation is helpful when approximating square roots. For example, $\sqrt{83}$ must be a number between 9 and 10, because $\sqrt{81} < \sqrt{83} < \sqrt{100}$.

decimal approximations for them using the square root key $\sqrt{}$ on a calculator or from the table of square roots found in Appendix 2.

$$\sqrt{3} \approx 1.732 \qquad \text{Use a table. Read } \approx \text{ as "is approximately equal to."}$$

$$\sqrt{29} \approx 5.385164807 \qquad \text{Use a calculator. The number of decimal places displayed can vary slightly by model.}$$

$$\sqrt{83} \approx 9.11 \qquad \text{Use a calculator and round to the nearest hundredth.}$$

Remember that the values found for irrational square roots are only approximations. If we square 9.11, which is an approximation of $\sqrt{83}$, the result is not quite 83.

$$(9.11)^2 = 82.9921$$

Square roots appear in many applied problems.

EXAMPLE 2

Pendulums.   The **period of a pendulum** is the time required for the pendulum to swing back and forth to complete one cycle. To find the period P (in seconds) of a pendulum L feet long, we can use the formula

$$P = 1.11\sqrt{L} \qquad \text{Read } 1.11\sqrt{L} \text{ as "1.11 times the square root of L."}$$

Find the period of a clock pendulum that is 5 feet long. Round to the nearest tenth of a second.

Strategy   We will substitute 5 for L in the formula, use a calculator to approximate $\sqrt{5}$, and multiply that value by 1.11.

Why $1.11\sqrt{L}$ means $1.11 \cdot \sqrt{L}$.

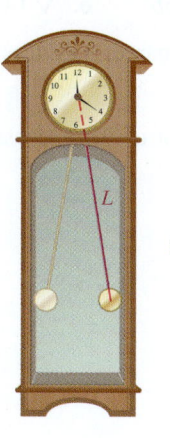

Solution
$$P = 1.11\sqrt{L}$$
$$P = 1.11\sqrt{5} \qquad \text{Substitute 5 for } L.$$
$$P \approx 2.482035455 \qquad \text{On a scientific calculator, enter } 1.11 \;\boxed{\times}\; 5 \;\boxed{\sqrt{}}\; \boxed{=}.$$
$$P \approx 2.5 \qquad \text{Round to the nearest tenth.}$$

The period is approximately 2.5 seconds.

Self Check 2   **Pendulums.**   Find the period of a pendulum that is 3 feet long. Round to the nearest tenth of a second.

Now Try ▶ Problem 101

EXAMPLE 3

Classify each square root as rational, irrational, or not a real number: **a.** $\sqrt{55}$ **b.** $\sqrt{-81}$ **c.** $-\sqrt{400}$

Strategy   We need to determine whether the radicand is positive or negative and whether it is a perfect square.

Why   If a positive number is a perfect square, its square root is rational. If a positive number is not a perfect square, its square root is irrational. The square root of a negative number is not a real number.

Solution   **a.** Since 55 is positive, but not a perfect square, $\sqrt{55}$ is an irrational number. If we use a calculator and round to two decimal places, we find that $\sqrt{55} \approx 7.42$.

b. $\sqrt{-81}$ is not a real number because it is the square root of a negative number.

c. Since $400 = (20)^2$, it is a perfect square and $-\sqrt{400}$ is rational: $-\sqrt{400} = -20$.

Self Check 3 Classify each square root as rational, irrational, or not a real number:
a. $\sqrt{-6}$ **b.** $-\sqrt{37}$ **c.** $\sqrt{\dfrac{16}{9}}$

Now Try ▶ Problem 51

3 Find Square Roots of Variable Expressions.

We have seen that the square root of a negative number is not a real number. To avoid negative radicands in this chapter, we will assume that **any variables within square root symbols do not represent negative real numbers.**

Variable expressions can be perfect squares. For example, $x^2, x^4, x^6, x^8,$ and x^{10} are perfect squares because

> **The Language of Algebra**
>
> We can state the restriction on the variables used in this chapter in another way: All variables represent **nonnegative** real numbers.

$$x^2 = (x^1)^2, \quad x^4 = (x^2)^2, \quad x^6 = (x^3)^2, \quad x^8 = (x^4)^2, \quad \text{and} \quad x^{10} = (x^5)^2$$

These examples illustrate an important fact: *even powers of a variable expression are perfect squares.* We can use this fact to simplify certain radical expressions.

EXAMPLE 4   Find each square root. **a.** $\sqrt{x^2}$ **b.** $\sqrt{a^4}$ **c.** $\sqrt{y^{16}}$ **d.** $\sqrt{25b^6}$ **e.** $\sqrt{144t^{14}}$

Strategy In each case, we will determine what expression, when squared, produces the radicand.

Why The radical symbol $\sqrt{}$ indicates that the positive square root of the expression written under it should be found.

Solution   **a.** $\sqrt{x^2} = x$ Ask: What expression, when squared, is x^2? The answer is x because $(x)^2 = x^2$.

b. $\sqrt{a^4} = a^2$ Ask: What expression, when squared, is a^4? The answer is a^2 because $(a^2)^2 = a^4$.

> **Caution**
>
> $\sqrt{y^{16}} \neq y^4$ because $(y^4)^2 \neq y^{16}$.
>
> Instead, $\sqrt{y^8} = y^4$ because $(y^4)^2 = y^8$.

c. $\sqrt{y^{16}} = y^8$ Ask: What expression, when squared, is y^{16}? The answer is y^8 because $(y^8)^2 = y^{16}$.

d. $\sqrt{25b^6} = 5b^3$ Ask: What expression, when squared, is $25b^6$? The answer is $5b^3$ because $(5b^3)^2 = 25b^6$.

e. $\sqrt{144t^{14}} = 12t^7$ Ask: What expression, when squared, is $144t^{14}$? The answer is $12t^7$ because $(12t^7)^2 = 144t^{14}$.

Self Check 4 Find each square root: **a.** $\sqrt{m^4}$ **b.** $\sqrt{b^2}$ **c.** $\sqrt{t^8}$ **d.** $\sqrt{81c^2}$

Now Try ▶ Problem 67

The results from Example 4 suggest that an easy way to find the square root of a variable raised to an even power is to *divide the exponent by 2.*

Divide 2 by 2   Divide 4 by 2   Divide 16 by 2

$$\sqrt{x^2} = x^1 = x \qquad \sqrt{a^4} = a^2 \qquad \sqrt{y^{16}} = y^8$$

4 Use the Pythagorean Theorem to Solve Problems.

The Pythagorean theorem relates the lengths of the sides of a right triangle.

<table>
<tr><td>The Pythagorean Theorem</td><td>If a and b are the lengths of the legs of a right triangle and c is the length of its hypotenuse,</td></tr>
</table>

The Pythagorean Theorem

If a and b are the lengths of the legs of a right triangle and c is the length of its hypotenuse,

$$a^2 + b^2 = c^2$$

In a right triangle, the square of the hypotenuse is equal to the sum of the squares of the other two sides.

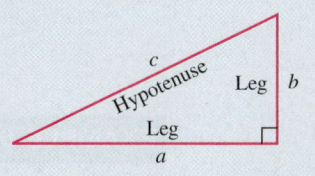

We call $a^2 + b^2 = c^2$ the **Pythagorean equation.**

When the lengths of two sides of a right triangle are given, we can use the Pythagorean theorem and the concept of square root to find the length of the third side of the triangle.

EXAMPLE 5   **Picture Frames.**   After nailing two pieces of molding together, a frame maker checks her work by making a diagonal measurement. What should she read on the yardstick if the sides of the frame form a right angle?

Analyze   The 15- and 20-inch sides of the frame in the figure are the legs and the diagonal is the hypotenuse of a right triangle. We need to find the length of the hypotenuse.

Assign   Let c represent the unknown length of the hypotenuse, as shown in the figure on the right.

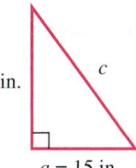

Form   We can use the Pythagorean theorem to form an equation.

$$a^2 + b^2 = c^2 \quad \text{This is the Pythagorean equation.}$$
$$15^2 + 20^2 = c^2 \quad \text{Substitute 15 for } a \text{ and 20 for } b.$$

Solve

$$225 + 400 = c^2 \quad \text{Evaluate the exponential expressions: } 15^2 = 225 \text{ and } 20^2 = 400.$$
$$625 = c^2 \quad \text{Do the addition.}$$

To find c, we ask "What number, when squared, is equal to 625?" There are two such numbers: the positive square root of 625 and the negative square root of 625. Since c represents the *length* of the hypotenuse, and it cannot be negative, it follows that c is the positive square root of 625.

$$\sqrt{625} = c$$
$$25 = c \quad \sqrt{625} = 25 \text{ because } (25)^2 = 625.$$

State   The diagonal measurement of the frame should be 25 inches.

Check   Sides of length 15, 20, and 25 inches satisfy the Pythagorean theorem.

$$15^2 + 20^2 \stackrel{?}{=} 25^2$$
$$225 + 400 \stackrel{?}{=} 625$$
$$625 = 625 \quad \text{True}$$

The Language of Algebra

If the square of one side of a triangle is equal to the sum of the squares of the other two sides, the triangle is a right triangle. This is called the **converse** of the Pythagorean theorem.

Self Check 5   **Tether ball.**   To make certain that a tether ball pole is vertical, a measurement is taken 24 inches up the pole and 10 inches along the ground. What measurement between the point on the pole and the point on the ground would guarantee the angle between the ground and pole to be a right angle?

Now Try ▶   Problem 73

When we use the Pythagorean theorem to find the length of a side of a right triangle, the solution is sometimes the square root of a number that is not a perfect square. In that case, we can use a calculator to *approximate* the square root.

EXAMPLE 6   Find the missing length of the side of the right triangle. Give the exact length and an approximation to the nearest hundredth.

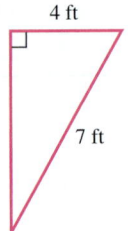

Strategy   We will use the Pythagorean theorem to find the missing length.

Why   The triangle is a right triangle, and we know the lengths of two of its sides.

Solution   If we substitute 4 for a, 7 for c, and let b represent the unknown length of the other leg in the Pythagorean equation, we have

$a^2 + b^2 = c^2$	
$4^2 + b^2 = 7^2$	The length of one leg is 4 ft. The length of the hypotenuse is 7 ft.
$16 + b^2 = 49$	Evaluate the exponential expressions: $4^2 = 16$ and $7^2 = 49$.
$b^2 = 33$	To isolate b^2, subtract 16 from both sides.
$b = \sqrt{33}$	If $b^2 = 33$, then b must be a square root of 33. Because b represents a length, it must be the positive square root of 33.
$b \approx 5.744562647$	Use a calculator to find an approximation of $\sqrt{33}$.
$b \approx 5.74$	Round to the nearest hundredth.

The length of the leg is exactly $\sqrt{33}$ feet, which is approximately 5.74 feet.

> **Caution**
>
> When using the Pythagorean equation $a^2 + b^2 = c^2$, we can let a represent the length of either leg of the right triangle. We then let b represent the length of the other leg. But c must always represent the length of the hypotenuse.

Self Check 6   Find the missing length of the side of the right triangle. Give the exact length and an approximation to the nearest hundredth.

Now Try ▶ Problems 81 and 85

Recall from geometry that an **isosceles triangle** has at least two sides of equal length. An **isosceles right triangle** is a right triangle with two legs of equal length and angles that measure 45°, 45°, and 90°.

EXAMPLE 7   **Roof Design.**   The gable end of a roof is an isosceles right triangle with a span of 48 feet. Find the distance from the eaves to the peak of the roof. Round to the nearest foot.

Analyze   The two equal sides of the isosceles triangle form the two legs of a right triangle, and the span of 48 feet is the length of the hypotenuse.

Assign   Let x represent the length (in feet) of each leg of the isosceles right triangle, which is the eaves-to-peak distance.

Form   We can use the Pythagorean theorem to form an equation.

$a^2 + b^2 = c^2$	This is the Pythagorean equation.
$x^2 + x^2 = 48^2$	Substitute x for a and b, and 48 for c.

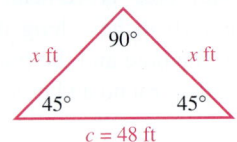

Solve	$2x^2 = 2{,}304$	Combine like terms: $x^2 + x^2 = 2x^2$. Evaluate: $48^2 = 2{,}304$.
	$x^2 = 1{,}152$	To isolate x^2, divide both sides by 2.
	$x = \sqrt{1{,}152}$	If $x^2 = 1{,}152$, then x must be a square root of 1,152. Because x represents a length, it must be the positive square root of 1,152.
	$x \approx 33.9411255$	Use a calculator to find an approximation of $\sqrt{1{,}152}$.
	$x \approx 34$	Round to the nearest foot.

The Language of Algebra

The word **isosceles** comes from the Greek word *isokelēs*, meaning equally legged.

State The eaves-to-peak distance is exactly $\sqrt{1,152}$ feet. Rounded to the nearest foot, the distance is approximately 34 feet.

Check For an eaves-to-peak distance of 34 feet, we have $(34)^2 + (34)^2 = 1,156 + 1,156 = 2,312$, which is approximately $(48)^2 = 2,304$. The result seems reasonable.

Self Check 7 **Geometry.** To find the area of an isosceles right triangle, Sarah must first find its base. If the equal sides of the triangle are each 13 cm long, what is the length of the base?

Now Try ▶ Problems 89 and 109

5 Use the Distance Formula.

We can use the Pythagorean theorem to develop a formula for finding the distance between two points with coordinates of (x_1, y_1) and (x_2, y_2) on a rectangular coordinate system. The distance d between the points is the length of the hypotenuse of the triangle in the figure. The two legs have lengths $x_2 - x_1$ and $y_2 - y_1$.

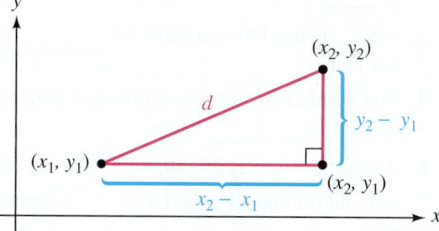

By the Pythagorean theorem, we have

$$d^2 = (x_2 - x_1)^2 + (y_2 - y_1)^2$$

Because d represents the distance between two points, it must be equal to the positive square root of $(x_2 - x_1)^2 + (y_2 - y_1)^2$. Thus,

$$d = \sqrt{(x_2 - x_1)^2 + (y_2 - y_1)^2}$$

We call this result the **distance formula**.

The Distance Formula	The distance between the points with coordinates (x_1, y_1) and (x_2, y_2) is given by $$d = \sqrt{(x_2 - x_1)^2 + (y_2 - y_1)^2}$$

EXAMPLE 8 Find the distance between $(-4, 5)$ and $(3, -1)$. Round to the nearest hundredth.

Strategy We will substitute the coordinates into the distance formula and solve for d.

Why In the formula, d represents the distance between the points.

Solution To use the distance formula, it doesn't matter which point we call (x_1, y_1) and which point we call (x_2, y_2). If we choose $(x_1, y_1) = (-4, 5)$ and $(x_2, y_2) = (3, -1)$, we have

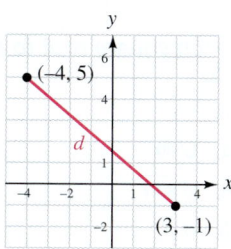

$d = \sqrt{(x_2 - x_1)^2 + (y_2 - y_1)^2}$ This is the distance formula.

$d = \sqrt{[3 - (-4)]^2 + (-1 - 5)^2}$ Substitute 3 for x_2, -4 for x_1, -1 for y_2, and 5 for y_1.

We will evaluate the entire expression under the radical symbol first, and then find the square root of that result.

$d = \sqrt{7^2 + (-6)^2}$ Perform each subtraction within the parentheses.

$d = \sqrt{49 + 36}$ Evaluate each exponential expression.

$d = \sqrt{85}$ The radical symbol acts like a grouping symbol. Do the addition under it first.

$d \approx 9.219544457$ Use a calculator to find an approximation of $\sqrt{85}$.

$d \approx 9.22$ Round to the nearest hundredth.

The distance between the two points is exactly $\sqrt{85}$ units, or approximately 9.22 units.

> **Self Check 8**   Find the distance between $(-6, 4)$ and $(-1, 2)$. Round to the nearest hundredth.
>
> **Now Try** ▶ Problem 93

SECTION 8.1 ▶ STUDY SET

VOCABULARY

Fill in the blanks.

1. The number b is a _____ root of the number a if $b^2 = a$.

2. The symbol $\sqrt{}$ is called a _____ symbol. It represents the _____ or principal square root of a number. The symbol $-\sqrt{}$ is used to represent the _____ square root of a number.

3. The number or variable expression under a radical symbol is called the _____ .

4. A number such as 25, 49, or $\frac{4}{81}$ that is the square of some rational number is called a _____ square.

5. The _____ theorem relates the lengths of the sides of a right triangle:

    $\boxed{} + \boxed{} = \boxed{}$

6. An _____ right triangle is shown here.

2 ft   2 ft
45°   45°

CONCEPTS

Fill in the blanks.

7. Every positive number has _____ square roots, one positive and one negative. The positive square root of 25 is _____ and the negative square root of 25 is _____. In symbols, we write

    $\sqrt{25} = \boxed{}$ and $-\sqrt{25} = \boxed{}$

8. To find the positive square root of 36, we ask "What positive number, when _____ , is equal to 36?"

9. The number 0 is the only real number with exactly one square root: $\sqrt{0} = \boxed{}$

10. $\sqrt{4} = 2$ because $(\boxed{})^2 = 4$ and $\sqrt{x^4} = x^2$ because $(\boxed{})^2 = x^4$.

11. Identify the radicand of each radical expression.

    a. $\sqrt{64}$ b. $-\sqrt{36y^4}$

    c. $2\sqrt{d}$ d. $\sqrt{\dfrac{16}{25}}$

12. Determine whether each statement is true or false.

    a. $-\sqrt{36} = \sqrt{-36}$ b. $\sqrt{m^{16}} = m^4$

13. Fill in the blanks: $\sqrt{29}$ must be a number between 5 and 6, because $\sqrt{\boxed{}} < \sqrt{29} < \sqrt{\boxed{}}$.

14. Evaluate: $\sqrt{(5-2)^2 + (8-4)^2}$

NOTATION

Complete the solution.

15. The legs of a right triangle measure 5 and 12 centimeters. Find the length of the hypotenuse.

$$a^2 + b^2 = c^2$$
$$\boxed{}^2 + 12^2 = c^2$$
$$25 + \boxed{} = c^2$$
$$\boxed{} = c^2$$
$$\sqrt{169} = \boxed{}$$
$$\boxed{} = c$$

The length of the hypotenuse is $\boxed{}$ cm.

16. Fill in the blanks. The distance formula is:

$$d = \sqrt{(x_2 - \boxed{})^2 + (\boxed{} - y_1)^2}$$

GUIDED PRACTICE

Evaluate each square root without using a calculator. See Example 1.

17. $\sqrt{64}$ 18. $\sqrt{25}$ 19. $\sqrt{36}$ 20. $\sqrt{1}$

21. $\sqrt{100}$ 22. $\sqrt{144}$ 23. $\sqrt{400}$ 24. $\sqrt{900}$

25. $-\sqrt{81}$ 26. $-\sqrt{36}$ 27. $\sqrt{1.21}$ 28. $\sqrt{1.69}$

29. $\sqrt{169}$ 30. $\sqrt{196}$ 31. $\sqrt{\dfrac{9}{25}}$ 32. $\sqrt{\dfrac{9}{49}}$

33. $-\sqrt{\dfrac{1}{64}}$ 34. $-\sqrt{\dfrac{1}{16}}$ 35. $-\sqrt{0.04}$ 36. $-\sqrt{0.64}$

37. $-\sqrt{289}$ 38. $-\sqrt{324}$ 39. $\sqrt{2,500}$ 40. $\sqrt{625}$

Use a calculator to approximate each square root to the nearest hundredth. See Example 2.

41. $\sqrt{3}$ 42. $\sqrt{2}$

43. $\sqrt{95}$ 44. $\sqrt{99}$

45. $\sqrt{428}$ 46. $\sqrt{844}$

47. $2\sqrt{3}$ 48. $3\sqrt{2}$

Classify each square root as rational, irrational, or not a real number. See Example 3.

49. $\sqrt{9}$ 50. $\sqrt{17}$

51. $\sqrt{-21}$ 52. $\sqrt{99}$

53. $\sqrt{33}$

54. $-\sqrt{\dfrac{81}{121}}$

55. $-\sqrt{0.25}$

56. $\sqrt{-47}$

Find each square root. All variables represent nonnegative real numbers. See Example 4.

57. $\sqrt{m^2}$ **58.** $\sqrt{b^2}$ **59.** $\sqrt{t^4}$ **60.** $\sqrt{a^6}$

61. $\sqrt{c^{10}}$ **62.** $\sqrt{r^8}$ **63.** $\sqrt{n^{12}}$ **64.** $\sqrt{t^{18}}$

65. $\sqrt{x^{36}}$ **66.** $\sqrt{y^{64}}$ **67.** $\sqrt{4y^2}$ **68.** $\sqrt{9b^2}$

69. $\sqrt{64b^4}$ **70.** $\sqrt{100s^{16}}$ **71.** $-\sqrt{49s^{14}}$ **72.** $-\sqrt{4a^4}$

Refer to the right triangle. See Example 5.

73. Find c if $a = 4$ and $b = 3$.
74. Find c if $a = 5$ and $b = 12$.
75. Find a if $b = 15$ and $c = 17$.
76. Find a if $b = 16$ and $c = 34$.
77. Find b if $a = 45$ and $c = 53$.
78. Find a if $b = 7$ and $c = 25$.
79. Find b if $c = 125$ and $a = 44$.
80. Find c if $a = 176$ and $b = 57$.

The lengths of two sides of a right triangle are given. Find the missing side length. Give the exact answer and an approximation to the nearest hundredth. See Example 6.

81. $a = 5$ cm and $c = 6$ cm
82. $a = 4$ in. and $c = 8$ in.
83. $a = 12$ m and $b = 8$ m
84. $a = 10$ ft and $b = 4$ ft
85. $a = 9$ in. and $b = 3$ in.
86. $a = 5$ mi and $b = 7$ mi
87. $b = 4$ in. and $c = 6$ in.
88. $a = 9$ mm and $c = 12$ mm

For each isosceles right triangle, find the length of the legs. Give an exact answer and an approximation to the nearest hundredth. See Example 7.

89.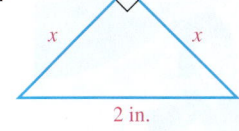

90.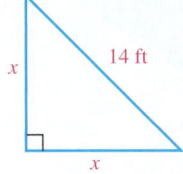

Find the distance between the two points. If an answer contains a radical, give an exact answer and an approximation to the nearest hundredth. See Example 8.

91. $(4, 6)$ and $(1, 2)$ **92.** $(9, 8)$ and $(3, 0)$
93. $(-2, -8)$ and $(3, 4)$ **94.** $(-5, -2)$ and $(7, 3)$
95. $(6, -5)$ and $(0, -4)$ **96.** $(-2, -5)$ and $(6, -6)$
97. $(9, -1)$ and $(7, -6)$ **98.** $(5, -6)$ and $(-1, 1)$

APPLICATIONS

If an answer contains a radical, give an exact answer and an approximation to the nearest hundredth.

99. from **Campus to Careers**

Crime Scene Investigator

Suppose you are a crime scene investigator and you need to determine whether the driver of a crashed car was speeding prior to an accident. You measure the car's skid marks on the pavement and find them to be 144 feet long. The formula $s = 4.5\sqrt{d}$ can be used to determine how fast the car was traveling. The variable s represents the speed (in mph) of the car and d the distance (in feet) of the skid when the driver hit the brakes. If the speed limit for the street was 35 mph, was the driver speeding? If so, by how many mph?

100. Free Fall. The time t (in seconds) that it takes an object dropped from a distance d (in feet) to fall to the ground is given by the formula $t = 0.25\sqrt{d}$. How long will it take a pair of sunglasses to fall to the ground if a sightseer drops them from the top of a 400-foot-tall observation tower?

101. Boating. To approximate the maximum speed (in knots) for certain types of boats, navigators use the formula $s = 1.34\sqrt{L}$, where s represents the maximum speed of the boat and L represents its length (in feet) at the waterline. Find the approximate maximum speed of the boat shown here.

102. Crossword Puzzles. If the area of a square is A square units, then the length of a side of the square is $\sqrt{A}$ units. The world's largest crossword puzzle can be purchased from the *Hammacher Schlemmer Gift Catalog* for $29.95. The crossword puzzle has a square shape and can be hung on a wall. If it covers 7,056 square inches of wall space, find the length of one of its sides. Answer in inches and also in feet.

103. Drafting. Among the tools used in drafting are the 30-60-90 and the 45-45-90 triangles shown.

a. Find the length of the hypotenuse of the 45-45-90 triangle if it is $\sqrt{2}$ times as long as a leg.

b. Find the length of the side opposite the 60° angle of the 30-60-90 triangle if it is $\sqrt{3}$ times as long as one-half of the hypotenuse.

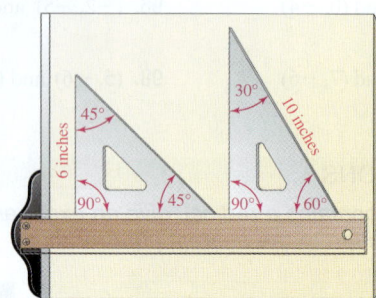

104. Gardening. A rectangular garden has sides of 28 and 45 feet. Find the length of a path that extends from one corner to the opposite corner.

105. Baseball. A baseball diamond is a square, with each side 90 feet long. How far is it from home plate to second base?

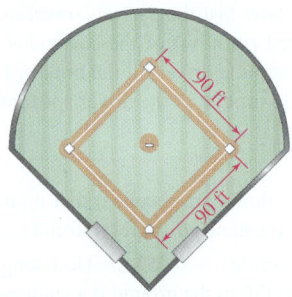

106. LCD Flat Panel HD-TV. The size of a television screen is the diagonal distance from the upper left to the lower right corner. What is the size of the screen shown in the illustration? Round to the nearest inch.

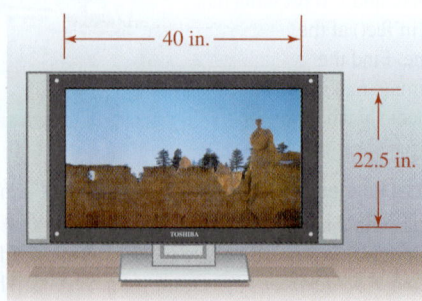

107. Carpentry. A carpenter's square is made of flat steel or aluminum and is shaped like an L. It is used for framing, roofing, and stairway work.

a. How long is each side of the carpenter's square shown in the next column? (The units are inches.)

b. What should be the diagonal measurement (shown in red) that will guarantee that the sides of the square form a 90° angle?

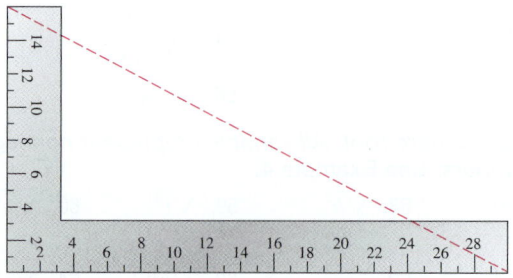

108. Boxing. The diagonal distance from corner to corner of a square boxing ring is 24 feet. Find the length of a side of the ring.

109. Football. On first down and ten, a quarterback tells his tight end to go out 6 yards, cut 45° to the right, and run 6 yards, as shown in the illustration. The tight end follows instructions, catches a pass, and is tackled immediately.

a. Find x.

b. Does he gain the necessary 10 yards for a first down?

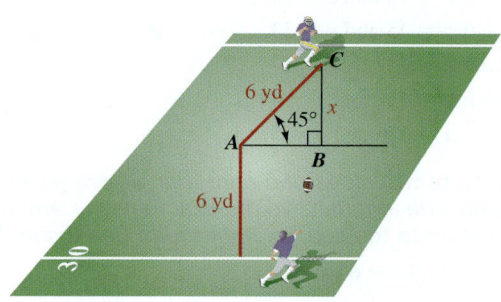

110. The Wizard of Oz. In the 1939 classic movie, the Scarecrow was in search of a brain. Once he received an honorary degree in "Thinkology" from the Wizard, he tried to impress his friends by stating: "*The sum of the square roots of any two sides of an isosceles triangle is equal to the square root of the remaining side.*" (You can watch this scene on YouTube.) What well-known mathematical fact was the Scarecrow attempting to recite? Explain the errors that he made. (Over 50 years later, in an episode of *The Simpsons,* Homer quotes the Scarecrow word-for-word after finding a pair of eyeglasses in a public restroom.)

111. Deck Designs. The plans for a patio deck shown in the illustration call for three redwood braces directly under the hot tub. Use the distance formula to find the length of each brace.

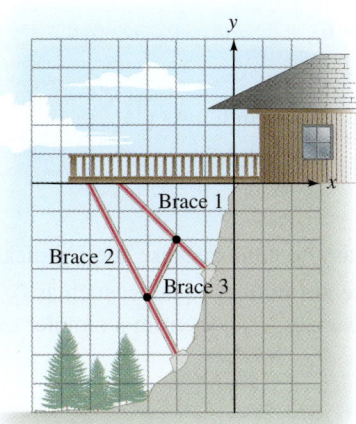

112. Navigation. A ship is sailing from Italy to Tunisia. The captain wants to travel a course that is always the same distance from a point on the coast of Sardinia as it is from a point on the coast of Sicily. (See the illustration below.) Determine if the ship is on the correct course if its current position is (9, 6).

WRITING

113. Consider the statement $\sqrt{26} \approx 5.1$. Explain why an $\approx$ symbol is used instead of an $=$ symbol.

114. Suppose you are told that $\sqrt{10} \approx 3.16$. Explain how another key on your calculator (besides the square root key $\sqrt{}$) could be used to see whether this is a reasonable approximation.

115. A calculator was used to find $\sqrt{-16}$. Explain the meaning of the message on the display.

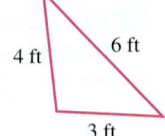

116. Explain why the Pythagorean theorem does not hold for this triangle.

$$3^2 + 4^2 = 6^2$$
$$9 + 16 \stackrel{?}{=} 36$$
$$25 \neq 36$$

REVIEW

117. Add: $(3s^2 - 3s - 2) + (3s^2 + 4s - 3)$

118. Subtract: $(3c^2 - 2c + 4) - (c^2 - 3c + 7)$

119. Multiply: $(3x - 2)(x + 4)$

120. Divide: $x^2 + 13x + 12$ by $x + 1$

CHALLENGE PROBLEMS

121. Shortcuts. Instead of walking on the sidewalk, students take a diagonal shortcut across a rectangular vacant lot, as shown in the illustration. How much distance do they save?

122. Evaluate: $\sqrt{\sqrt{\sqrt{256}}}$

123. Simplify $\sqrt{16x^{16n}}$, where n is a positive integer.

124. Graph $y = \sqrt{x}$ by first completing the table of solutions. Then plot the points and draw a smooth curve through them. Approximate when necessary to the nearest tenth.

x	0	1	3	4	5	9	12	16
y								

Simplifying Square Roots

ARE YOU READY?

The following problems review some basic skills that are needed when simplifying square roots.

1. Complete the factorization: $24 = \boxed{} \cdot 6$

2. Complete the factorization: $98 = \boxed{} \cdot 2$

3. Prime factor 300.

4. Find: $\sqrt{x^{12}}$

5. Find: $\sqrt{64x^6}$

6. Multiply: $x^6 \cdot x$

Square roots such as $\sqrt{9}$ and $\sqrt{16}$ are easy to find because their radicands are perfect squares. Square roots with radicands that are not perfect squares, such as $\sqrt{12}$ and $\sqrt{200}$, can be approximated using a calculator, or we can write them in a simpler exact form using the *product rule for square roots*.

1 Use the Product Rule to Simplify Square Roots.

To introduce the product rule for square roots, we will find $\sqrt{4 \cdot 25}$ and $\sqrt{4}\sqrt{25}$, and compare the results.

Notation
The product $\sqrt{4}\sqrt{25}$ also can be written as $\sqrt{4} \cdot \sqrt{25}$.

Square root of a product	*Product of square roots*
$\sqrt{4 \cdot 25} = \sqrt{100}$	$\sqrt{4}\sqrt{25} = 2 \cdot 5$
$= 10$	$= 10$

Read as "the square root of 4 times the square root of 25."

In each case, the answer is 10. Thus, $\sqrt{4 \cdot 25} = \sqrt{4}\sqrt{25}$. This result illustrates the *product rule for square roots*.

The Product Rule for Square Roots	The square root of the product of two nonnegative numbers is equal to the product of their square roots. For any nonnegative real numbers a and b, $$\sqrt{a \cdot b} = \sqrt{a}\sqrt{b}$$

A square-root radical expression is in **simplified form** when each of the following is true.

Simplified Form of a Square Root	1. Except for 1, the radicand has no perfect-square factors. 2. No fraction appears in the radicand. 3. No radical appears in the denominator of a fraction.

To simplify square roots, we follow these steps.

Simplifying Square Roots	1. Write the radicand as a product of the greatest perfect-square factor and one other factor. 2. Use the product rule for square roots to write the expression as a product of square roots. 3. Find the square root of the perfect-square factor.

To simplify square roots, we must often factor the radicand using two natural-number factors. Since one factor should be a perfect square, it is helpful to memorize the following natural-number **perfect squares:**

1, 4, 9, 16, 25, 36, 49, 64, 81, 100, 121, 144, 169, 196, 225

EXAMPLE 1

Simplify: $\sqrt{12}$

Strategy We begin by factoring 12 using one perfect-square factor and one non–perfect-square factor. Then we will use the product rule for square roots to simplify the expression.

Why Factoring the radicand in this way leads to a square root of a perfect square that we can easily evaluate.

Solution There are several possible factorizations of 12 to consider. They are

$$1 \cdot 12, \qquad 2 \cdot 6, \qquad \text{and} \qquad 4 \cdot 3$$

We will use $4 \cdot 3$ because this factorization contains the perfect square 4.

> **The Language of Algebra**
>
> When referring to a square root, the instruction **simplify** means to remove any perfect-square factors from the radicand.

$\sqrt{12} = \sqrt{4 \cdot 3}$ Ask: How can 12 be factored using a perfect square and one other factor? The answer is $4 \cdot 3$. Write the perfect-square factor first.

$= \sqrt{4}\sqrt{3}$ Use the product rule for square roots: The square root of a product is equal to the product of the square roots.

$= 2\sqrt{3}$ Evaluate the square root of the perfect square: $\sqrt{4} = 2$. Read as "2 times the square root of 3" or as "2 radical 3."

We say that $2\sqrt{3}$ is the simplified form of $\sqrt{12}$. Note that $2\sqrt{3}$ means $2 \cdot \sqrt{3}$.

> **The Language of Algebra**
>
> The instructions **simplify** and **approximate** do not mean the same thing.
>
> Simplify: $\sqrt{12} = 2\sqrt{3}$ (exact)
> Approximate:
> $\sqrt{12} \approx 3.464$ (not exact)

CAUTION To try to simplify $\sqrt{12}$, we could have factored 12 as $2 \cdot 6$. However, this factorization is not useful because neither of the resulting radical expressions is the square root of a perfect square.

$$\sqrt{12} = \sqrt{2 \cdot 6}$$
$$= \sqrt{2}\sqrt{6} \qquad \text{Use the product rule for square roots.}$$

In this case, neither radical expression simplifies. This example illustrates that we should always factor a radicand using a perfect square when simplifying a radical.

Self Check 1 Simplify: $\sqrt{45}$

Now Try ▶ Problem 15

EXAMPLE 2

Simplify: $\sqrt{200}$

Strategy We will factor 200 as $100 \cdot 2$ and then use the product rule for square roots to simplify the radical expression.

Why The radicand 200 has several factorizations that contain perfect squares:

$$4 \cdot 50, \qquad 25 \cdot 8, \qquad \text{and} \qquad 100 \cdot 2$$

The best one to choose is the one that contains the greatest perfect-square factor, which is 100.

Solution
$\sqrt{200} = \sqrt{100 \cdot 2}$ Factor 200 using its greatest perfect-square factor and one other factor.

$= \sqrt{100}\sqrt{2}$ Use the product rule for square roots: The square root of a product is equal to the product of the square roots.

$= 10\sqrt{2}$ Evaluate the square root of the perfect square: $\sqrt{100} = 10$.

To simplify $\sqrt{200}$, we could have factored 200 as $25 \cdot 8$. Even though 25 is not the greatest perfect-square factor of 200, we will still get the correct answer but it will take more steps:

$$\sqrt{200} = \sqrt{25 \cdot 8} \qquad \text{Factor 200 as 25 · 8.}$$
$$= \sqrt{25}\sqrt{8} \qquad \text{Use the product rule for square roots.}$$
$$= 5\sqrt{8} \qquad \text{Evaluate the square root of the perfect-square factor: } \sqrt{25} = 5.$$

Since the radicand 8 still contains the perfect-square factor 4, the expression $\sqrt{8}$ is not in simplest form. We continue simplifying:

$$\sqrt{200} = 5\sqrt{8}$$
$$= 5\sqrt{4 \cdot 2} \qquad \text{To simplify } \sqrt{8} \text{, factor 8 as 4 · 2.}$$
$$= 5\sqrt{4}\sqrt{2} \qquad \text{Use the product rule for square roots.}$$
$$= 5 \cdot 2 \cdot \sqrt{2} \qquad \text{Evaluate the square root of the perfect-square factor: } \sqrt{4} = 2.$$
$$= 10\sqrt{2} \qquad \text{Multiply 5 and 2 to get 10.}$$

Note that we obtain the same result.

Self Check 2 Simplify: $\sqrt{72}$

Now Try ▶ Problem 23

2 Use Prime Factorization to Simplify Square Roots.

When simplifying square roots, prime factorization can be useful in finding the greatest perfect-square factor of the radicand.

EXAMPLE 3 Simplify: **a.** $\sqrt{150}$ **b.** $\sqrt{95}$

Strategy In each case, the greatest perfect-square factor of the radicand (if there is one) is not obvious. Another approach is to find the prime factorization of the radicand and look for pairs of like factors.

Why Identifying a pair of like factors of the radicand leads to a square root of a perfect square that we can evaluate easily.

Solution **a.** $\sqrt{150} = \sqrt{2 \cdot 3 \cdot 5 \cdot 5}$ Write 150 in prime-factored form.

$$= \sqrt{2 \cdot 3}\sqrt{5 \cdot 5} \qquad \text{Group the pair of like factors together and use the product rule for square roots.}$$
$$= \sqrt{6} \cdot 5 \qquad \text{Evaluate the square root of the perfect square: } \sqrt{5 \cdot 5} = \sqrt{25} = 5.$$

$$
\begin{array}{r|r}
2 & 150 \\
\hline
3 & 75 \\
\hline
5 & 25 \\
\hline
 & 5
\end{array}
$$

$$= 5\sqrt{6} \qquad \text{Write the factor 5 first. This way, no misunderstanding can occur about exactly what is under the radical symbol.}$$

b. We prime factor 95 to get $95 = 5 \cdot 19$. Since the factorization does not contain a pair of like factors, 95 does not have a perfect-square factor. It follows that $\sqrt{95}$ cannot be simplified.

$$
\begin{array}{r|r}
5 & 95 \\
\hline
 & 19
\end{array}
$$

Self Check 3 Simplify: **a.** $\sqrt{140}$ **b.** $\sqrt{77}$

Now Try ▶ Problems 27 and 35

3 Simplify Square Roots of Variable Expressions.

Perfect squares such as x^2, x^4, and x^6, which are even powers of a variable, are used to simplify square roots whose radicands contain odd powers of a variable, such as x^3, x^5, and x^7. To perform such simplifications, it is helpful to memorize the following **perfect square forms:**

$$x^2, \ x^4, \ x^6, \ x^8, \ x^{10}, \ x^{12}, \ x^{14}, \ x^{16}, \dots$$

EXAMPLE 4 Simplify: **a.** $\sqrt{x^3}$ **b.** $\sqrt{a^9}$

Strategy We will factor each radicand as the product of the greatest possible even power of the variable and the variable to the first power.

Why Factoring an exponential expression raised to an odd power in this way leads to a square root of a perfect square that we can find easily.

Solution **a.** We note that x^2 is the greatest even power of x that is a factor of x^3. To simplify $\sqrt{x^3}$, we write x^3 as the product of x^2 and x and use the product rule for square roots.

> **Success Tip**
>
> To factor the radicand, remember the product rule for exponents: To multiply like bases, keep the base and add the exponents.
>
> $$x^2 \cdot x = x^{2+1} = x^3$$

$$\sqrt{x^3} = \sqrt{x^2 \cdot x} \qquad \text{Factor: } x^3 = x^2 \cdot x^1 = x^2 \cdot x. \text{ Think of the exponent 3 as the sum of 2 (which is even) and 1.}$$

$$= \sqrt{x^2}\sqrt{x} \qquad \text{Use the product rule: The square root of a product is equal to the product of the square roots.}$$

$$= x\sqrt{x} \qquad \text{Find the square root of the perfect square: } \sqrt{x^2} = x.$$

b. To simplify $\sqrt{a^9}$, we note that a^8 is the greatest even power of a that is a factor of a^9.

> **Caution**
>
> $\sqrt{a^9} \neq a^3$ because $(a^3)^2 \neq a^9$. (By the power rule for exponents, $(a^3)^2 = a^6$.)
> Similarly, $\sqrt{y^{25}} \neq y^5$ because $(y^5)^2 \neq y^{25}$.

$$\sqrt{a^9} = \sqrt{a^8 \cdot a} \qquad \text{Factor: } a^9 = a^8 \cdot a^1 = a^8 \cdot a. \text{ Think of the exponent 9 as the sum of 8 (which is even) and 1.}$$

$$= \sqrt{a^8}\sqrt{a} \qquad \text{Use the product rule: The square root of a product is equal to the product of the square roots.}$$

$$= a^4\sqrt{a} \qquad \text{Find the square root of the perfect square: } \sqrt{a^8} = a^4 \text{ because } (a^4)^2 = a^8.$$

Self Check 4 Simplify: **a.** $\sqrt{x^5}$ **b.** $\sqrt{b^{11}}$

Now Try ▶ Problems 39 and 41

EXAMPLE 5 Simplify: $4\sqrt{27m}$

Strategy To simplify this radical expression, we will factor $27m$ as $9 \cdot 3m$.

Why The greatest perfect-square factor of $27m$ is 9. Factoring the radicand in this way leads to a square root of a perfect square that we can evaluate easily.

Solution

$$4\sqrt{27m} = 4\sqrt{9 \cdot 3m} \qquad \text{Write } 27m \text{ as a product of its greatest perfect-square factor and one other factor: } 27m = 9 \cdot 3m.$$

> **Caution**
>
> When writing radical expressions such as $12\sqrt{3m}$, be sure to extend the radical symbol completely over $3m$, because $12\sqrt{3m}$ and $12\sqrt{3}m$ are not the same.

$$= 4 \cdot \sqrt{9}\sqrt{3m} \qquad \text{Use the product rule: The square root of a product is equal to the product of the square roots.}$$

$$= 4 \cdot 3\sqrt{3m} \qquad \text{Evaluate the square root of the perfect square: } \sqrt{9} = 3.$$

$$= 12\sqrt{3m} \qquad \text{Multiply 4 and 3 to get 12.}$$

Self Check 5 Simplify: $9\sqrt{32y}$

Now Try ▶ Problem 43

EXAMPLE 6   Simplify: **a.** $\sqrt{144a^7}$ **b.** $\dfrac{1}{6}\sqrt{72x^5}$

Strategy   We will determine the greatest perfect-square factor of the numerical part and the greatest perfect square factor of the variable part of the radicand separately.

Why   It is easier to determine the greatest perfect-square factor of the entire radicand if we consider the numerical and variable parts separately.

Solution   **a.** To simplify $\sqrt{144a^7}$, we look for the greatest perfect-square factor of $144a^7$. Because

- 144 is a perfect square and
- a^6 is a perfect square,

$144a^6$ is the greatest perfect-square factor of $144a^7$.

$$\sqrt{144a^7} = \sqrt{144a^6 \cdot a}$$ Write $144a^7$ as a product of its greatest perfect-square factor and one other factor: $144a^7 = 144a^6 \cdot a$.

$$= \sqrt{144a^6}\sqrt{a}$$ Use the product rule: The square root of a product is equal to the product of the square roots.

$$= 12a^3\sqrt{a}$$ Find the square root of the perfect square: $\sqrt{144a^6} = 12a^3$ because $(12a^3)^2 = 144a^6$.

b. To simplify $\dfrac{1}{6}\sqrt{72x^5}$, we look for the greatest perfect-square factor of $72x^5$. Because

- 36 is the greatest perfect-square factor of 72 and
- x^4 is the greatest perfect-square factor of x^5,

$36x^4$ is the greatest perfect-square factor of $72x^5$.

Caution

$$\sqrt{72x^5} = 6x^2\sqrt{2x}$$
↑
Simplified form: The radicand should not contain any variables with an exponent greater than 1.

$$\frac{1}{6}\sqrt{72x^5} = \frac{1}{6}\sqrt{36x^4 \cdot 2x}$$ Write $72x^5$ as a product of its greatest perfect-square factor and one other factor: $72x^5 = 36x^4 \cdot 2x$.

$$= \frac{1}{6}\sqrt{36x^4}\sqrt{2x}$$ Use the product rule: The square root of a product is equal to the product of the square roots.

$$= \frac{1}{6}6x^2\sqrt{2x}$$ Find the square root of the perfect square: $\sqrt{36x^4} = 6x^2$ because $(6x^2)^2 = 36x^4$.

$$= x^2\sqrt{2x}$$ Multiply: $\frac{1}{6} \cdot 6 = 1$.

Self Check 6   Simplify: **a.** $\sqrt{49p^9}$ **b.** $\dfrac{1}{5}\sqrt{75y^3}$

Now Try ▶ Problem 47

---

4   Use the Quotient Rule to Simplify Square Roots.

To introduce the quotient rule for square roots, we will find $\sqrt{\dfrac{100}{25}}$ and $\dfrac{\sqrt{100}}{\sqrt{25}}$ and compare the results.

Square root of a quotient	*Quotient of square roots*
$\sqrt{\dfrac{100}{25}} = \sqrt{4}$	$\dfrac{\sqrt{100}}{\sqrt{25}} = \dfrac{10}{5}$ Read as "the square root of 100 divided by the square root of 25."
$= 2$	$= 2$

Since the answer is 2 in each case, $\sqrt{\dfrac{100}{25}} = \dfrac{\sqrt{100}}{\sqrt{25}}$. This result illustrates the quotient rule for square roots.

The Quotient Rule for Square Roots	The square root of the quotient of two numbers is equal to the quotient of their square roots.
	For any positive real numbers a and b, $$\sqrt{\frac{a}{b}} = \frac{\sqrt{a}}{\sqrt{b}}$$

We use the quotient rule for square roots to simplify square roots in much the same way as we used the product rule.

EXAMPLE 7 Simplify: **a.** $\sqrt{\dfrac{4}{9}}$ **b.** $\sqrt{\dfrac{53}{121}}$ **c.** $\sqrt{\dfrac{108}{25}}$

Strategy In each case, the square root is not in simplified form because the radicand contains a fraction. To write each of these radical expressions in simplified form, we use the quotient rule for square roots.

Why Writing these expressions in $\dfrac{\sqrt{a}}{\sqrt{b}}$ form leads to square roots of perfect squares that we can evaluate easily.

Solution **a.** We have seen that $\sqrt{\dfrac{4}{9}} = \dfrac{2}{3}$, because $\left(\dfrac{2}{3}\right)^2 = \dfrac{4}{9}$. We also can use the quotient rule for square roots to simplify this expression.

$$\sqrt{\frac{4}{9}} = \frac{\sqrt{4}}{\sqrt{9}}$$ Use the quotient rule: The square root of a quotient is equal to the quotient of the square roots.

$$= \frac{2}{3}$$ Evaluate each square root: $\sqrt{4} = 2$ and $\sqrt{9} = 3$.

b. $\sqrt{\dfrac{53}{121}} = \dfrac{\sqrt{53}}{\sqrt{121}}$ Use the quotient rule: The square root of a quotient is equal to the quotient of the square roots.

$$= \frac{\sqrt{53}}{11}$$ In the denominator, evaluate: $\sqrt{121} = 11$.

c. $\sqrt{\dfrac{108}{25}} = \dfrac{\sqrt{108}}{\sqrt{25}}$ Use the quotient rule: The square root of a quotient is equal to the quotient of the square roots.

$$= \frac{\sqrt{36 \cdot 3}}{5}$$ To simplify $\sqrt{108}$, factor 108 using its greatest perfect-square factor, 36, and 3. In the denominator, evaluate: $\sqrt{25} = 5$.

$$= \frac{\sqrt{36}\sqrt{3}}{5}$$ In the numerator, use the product rule: The square root of a product is equal to the product of the square roots.

$$= \frac{6\sqrt{3}}{5}$$ Evaluate: $\sqrt{36} = 6$. The answer can also be written as $\dfrac{6}{5}\sqrt{3}$.

Self Check 7 Simplify: **a.** $\sqrt{\dfrac{36}{121}}$ **b.** $\sqrt{\dfrac{29}{64}}$ **c.** $\sqrt{\dfrac{72}{169}}$

Now Try ▶ Problems 55, 59, and 61

In the following example, a variable appears in the denominator of a radical expression. To avoid undefined situations such as this, we will assume *all variables represent positive real numbers* throughout this chapter.

EXAMPLE 8 Simplify: $\sqrt{\dfrac{44x^3}{81x}}$

Strategy Note that the radicand is a rational expression. To make the work easier, we will simplify the radicand before using the quotient rule for square roots.

Why After simplifying, our hope is that the numerator and/or the denominator of the resulting rational expression is a perfect square.

Solution

$$\sqrt{\dfrac{44x^3}{81x}} = \sqrt{\dfrac{44x^2}{81}}$$

Simplify the radicand by removing a factor of x that is common to the numerator and denominator: $\dfrac{44x^2 \cdot \overset{1}{\cancel{x}}}{81 \underset{1}{\cancel{x}}}$.

$$= \dfrac{\sqrt{44x^2}}{\sqrt{81}}$$

Use the quotient rule: The square root of a quotient is equal to the quotient of the square roots.

$$= \dfrac{\sqrt{4x^2}\sqrt{11}}{\sqrt{81}}$$

To simplify $\sqrt{44x^2}$, factor $44x^2$ using its greatest perfect-square factor: $\sqrt{4x^2 \cdot 11}$. Then use the product rule: The square root of a product is equal to the product of the square roots.

$$= \dfrac{2x\sqrt{11}}{9}$$

Find the square roots of the perfect squares: $\sqrt{4x^2} = 2x$ and $\sqrt{81} = 9$.

Success Tip

To simplify square roots, factor (or simplify) the radicand in such a way that perfect squares appear. Then use the product or quotient rule for square roots and find the square root of each of the perfect squares.

Self Check 8 Simplify: $\sqrt{\dfrac{99b^4}{16b^2}}$

Now Try ▶ Problem 69

SECTION 8.2 ▶ STUDY SET

VOCABULARY

Fill in the blanks.

1. "To _____ $\sqrt{20}$" means to write it as $2\sqrt{5}$.

2. The perfect-square factors of 200 are 4, 25, and 100. Of these, 100 is the _____ perfect-square factor of 200.

3. The expression $\sqrt{4 \cdot 3}$ is a square root of a _____ and $\sqrt{4}\sqrt{3}$ is a product of two _____ roots.

4. The expression $\sqrt{\dfrac{40}{9}}$ is a square root of a _____ and $\dfrac{\sqrt{40}}{\sqrt{9}}$ is a quotient of two _____ roots.

CONCEPTS

5. Fill in the blanks.

 a. The square root of the product of two positive numbers is equal to the _____ of their square roots. In symbols,

 $\sqrt{a \cdot b} = $ [ ]

 b. The square root of the quotient of two positive numbers is equal to the _____ of their square roots. In symbols,

 $\sqrt{\dfrac{a}{b}} = \dfrac{\quad}{\quad}$

6. Complete each list of perfect squares.

 a. 1, 4, 9, [ ], 25, 36, [ ], 64, 81, [ ], 121, [ ], . . .

 b. x^2, [ ], x^6, [ ], [ ], x^{12}, [ ], . . .

7. Which perfect square, 1, 4, 9, 16, 25, 36, 49, 64, 81, or 100, is the greatest perfect-square factor of the given number?

 a. 20   b. 98

 c. 54   d. 48

8. Complete each factorization using a perfect square.

 a. $24 = $ [ ] $\cdot 6$ b. $63 = $ [ ] $\cdot 7$

 c. $r^5 = $ [ ] $\cdot r$ d. $t^9 = $ [ ] $\cdot t$

9. To simplify $\sqrt{40}$, which one of the following factorizations should be used?

 $\sqrt{20 \cdot 2}$ $\sqrt{40 \cdot 1}$ $\sqrt{8 \cdot 5}$ $\sqrt{4 \cdot 10}$

10. To simplify $\sqrt{n^5}$, which one of the following factorizations should be used?

 $\sqrt{n^2 \cdot n^3}$ $\sqrt{n^3 \cdot n^2}$ $\sqrt{n^4 \cdot n}$

11. Use the square root of a product property to simplify each expression.

 a. $\sqrt{81 \cdot 2} = \sqrt{}\sqrt{} = \sqrt{}$

 b. $\sqrt{m^4 \cdot m} = \sqrt{}\sqrt{} = \sqrt{}$

12. Use the square root of a quotient property to simplify the expression.

 $\sqrt{\dfrac{17}{25}} = \dfrac{\sqrt{}}{\sqrt{}} = \dfrac{\sqrt{}}{}$

NOTATION

13. We can read $2\sqrt{6}$ as "2 _____ the square root of 6" or as "2 _____ 6."

14. Write each radical expression in better form.

a. $3 \cdot 2\sqrt{15}$ b. $\sqrt{7 \cdot 9}$

GUIDED PRACTICE

In the following problems, all variables represent positive real numbers.

Simplify. **See Example 1.**

15. $\sqrt{20}$ **16.** $\sqrt{18}$

17. $\sqrt{27}$ **18.** $\sqrt{28}$

19. $\sqrt{50}$ **20.** $\sqrt{54}$

21. $\sqrt{24}$ **22.** $\sqrt{88}$

Simplify. **See Example 3.**

23. $\sqrt{500}$ **24.** $\sqrt{700}$

25. $\sqrt{63}$ **26.** $\sqrt{75}$

Simplify. **See Example 3.**

27. $\sqrt{98}$ **28.** $\sqrt{128}$

29. $\sqrt{180}$ **30.** $\sqrt{147}$

31. $\sqrt{192}$ **32.** $\sqrt{216}$

33. $\sqrt{375}$ **34.** $\sqrt{405}$

35. $\sqrt{42}$ **36.** $\sqrt{55}$

37. $\sqrt{385}$ **38.** $\sqrt{182}$

Simplify. **See Example 4.**

39. $\sqrt{x^{11}}$ **40.** $\sqrt{n^3}$

41. $\sqrt{n^9}$ **42.** $\sqrt{t^7}$

Simplify. **See Example 5.**

43. $4\sqrt{12x}$ **44.** $9\sqrt{20y}$

45. $5\sqrt{54q}$ **46.** $7\sqrt{18k}$

Simplify. **See Example 6.**

47. $\sqrt{25t^3}$ **48.** $\sqrt{36p^3}$

49. $\sqrt{32x^5}$ **50.** $\sqrt{52y^5}$

51. $\frac{1}{5}\sqrt{75x^7}$ **52.** $\frac{1}{7}\sqrt{98y^7}$

53. $\frac{1}{3}\sqrt{18t^{11}}$ **54.** $\frac{1}{5}\sqrt{50m^{11}}$

Simplify. **See Example 7.**

55. $\sqrt{\dfrac{25}{9}}$ **56.** $\sqrt{\dfrac{36}{49}}$

57. $\sqrt{\dfrac{81}{64}}$ **58.** $\sqrt{\dfrac{121}{144}}$

59. $\sqrt{\dfrac{6}{121}}$ **60.** $\sqrt{\dfrac{15}{64}}$

61. $\sqrt{\dfrac{75}{16}}$ **62.** $\sqrt{\dfrac{32}{25}}$

Simplify. **See Example 8.**

63. $\sqrt{\dfrac{a^4}{4a}}$ **64.** $\sqrt{\dfrac{n^4}{81n}}$

65. $\sqrt{\dfrac{r^{10}}{225r}}$ **66.** $\sqrt{\dfrac{m^4}{225m}}$

67. $\sqrt{\dfrac{72x^3}{x}}$ **68.** $\sqrt{\dfrac{108b^3}{b}}$

69. $\sqrt{\dfrac{125n^5}{64n}}$ **70.** $\sqrt{\dfrac{72q^7}{25q^3}}$

TRY IT YOURSELF

Simplify.

71. $\sqrt{75t}$ **72.** $\sqrt{24s}$

73. $\sqrt{\dfrac{48}{81}}$ **74.** $\sqrt{\dfrac{27}{64}}$

75. $\sqrt{48}$ **76.** $\sqrt{32}$

77. $\sqrt{4k}$ **78.** $\sqrt{9p}$

79. $\sqrt{2d^{11}}$ **80.** $\sqrt{3x^9}$

81. $\sqrt{10b}$ **82.** $\sqrt{30s}$

83. $\sqrt{44}$ **84.** $\sqrt{60}$

85. $\sqrt{\dfrac{23}{64}}$ **86.** $\sqrt{\dfrac{35}{144}}$

87. $\frac{1}{6}\sqrt{72}$ **88.** $\frac{1}{2}\sqrt{28}$

89. $\sqrt{\dfrac{75q^2}{16q^4}}$ **90.** $\sqrt{\dfrac{128n^5}{81n^7}}$

91. $\sqrt{\dfrac{20}{49}}$ **92.** $\sqrt{\dfrac{50}{9}}$

93. $\sqrt{162}$ **94.** $\sqrt{486}$

95. $\frac{3}{2}\sqrt{16y^5}$ **96.** $\frac{4}{3}\sqrt{36y^5}$

97. $\sqrt{50x^4}$ **98.** $\sqrt{75y^4}$

99. $\sqrt{t^{15}}$ **100.** $\sqrt{m^{17}}$

Look Alikes . . .

101. a. $\sqrt{x^2}$ b. $\sqrt{x^3}$ c. $\sqrt{x^4}$ d. $\sqrt{x^5}$

e. $\sqrt{x^6}$ f. $\sqrt{x^7}$ g. $\sqrt{x^8}$ h. $\sqrt{x^9}$

102. a. $\sqrt{12}$ b. $\sqrt{27}$ c. $\sqrt{48}$ d. $\sqrt{75}$

103. a. $\sqrt{8}$ b. $\sqrt{18}$ c. $\sqrt{50}$ d. $\sqrt{72}$

104. a. $\sqrt{\dfrac{x^3}{4}}$ b. $\sqrt{\dfrac{x^3}{9}}$ c. $\sqrt{\dfrac{x^3}{16}}$ d. $\sqrt{\dfrac{x^3}{25}}$

APPLICATIONS

105. Studying Past Cultures. A coordinate system of 1 ft × 1 ft squares was constructed from string to record the location of objects found at a dig site. Use the distance formula (or the Pythagorean theorem) to determine the exact distance between a piece of pottery found at point *A* and an arrowhead found at point *B*. Express the answer in simplified radical form. Then approximate the distance to the nearest hundredth of a foot.

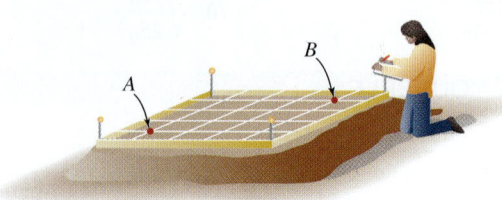

106. Square Dancing. A dance floor is a square with sides of length 12 yards. Use the Pythagorean theorem to find the exact length of the diagonal shown in the illustration. Express the answer in simplified radical form. Then approximate the length to the nearest yard.

107. Square Root Spiral. The pattern shown below is formed by a collection of right triangles, where the hypotenuse of one becomes a leg of the next. The length of each segment in red is 1 unit. Simplify any square roots that are not in simplified form.

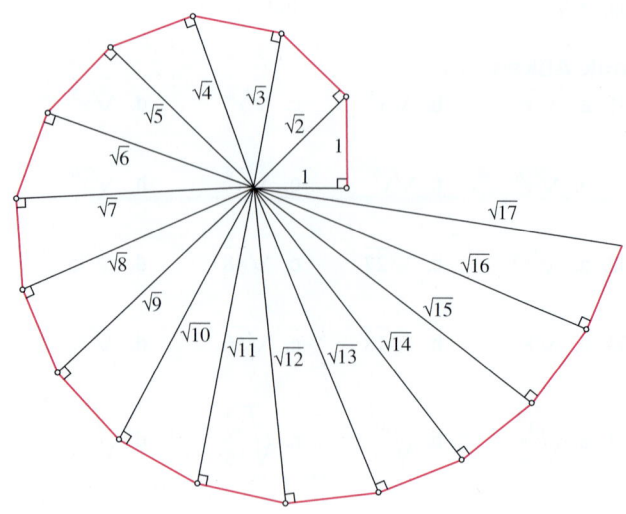

108. Amusement Park Rides. The time *t* (in seconds) that it takes a pirate ship ride to swing from one extreme to the other is given by the formula $t = \pi\sqrt{\frac{L}{32}}$. Use the value of $L = 54$ ft to find *t* and express it in simplified radical form. Leave π in the answer. Then approximate the time to the nearest tenth of a second.

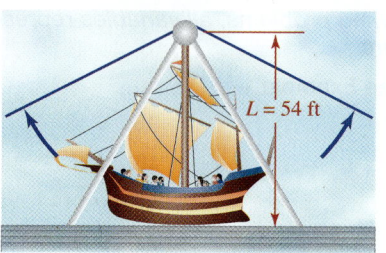

WRITING

109. Explain why each square root is not in simplified form.

a. $\sqrt{8}$ b. $\sqrt{\frac{7}{4}}$

110. Explain each property in words.

a. $\sqrt{a \cdot b} = \sqrt{a}\sqrt{b}$ b. $\sqrt{\frac{a}{b}} = \frac{\sqrt{a}}{\sqrt{b}}$

111. Explain why we cannot simplify $\sqrt{42}$.

112. Does $\sqrt{9x^9} = 3x^3$? Explain why or why not.

REVIEW

Solve each system.

113. $\begin{cases} y = 2x - 6 \\ 2x + y = 6 \end{cases}$ **114.** $\begin{cases} 2x + y = -2 \\ -2x - 3y = -6 \end{cases}$

115. $\begin{cases} 3x + 4y = -7 \\ 2x - y = -1 \end{cases}$ **116.** $\begin{cases} 2x + 3y = 8 \\ 3x - 2y = -1 \end{cases}$

CHALLENGE PROBLEMS

117. Simplify. All variables represent positive numbers.

$$\sqrt{\frac{196a^8b^5c^9}{50abc^8}}$$

118. Find the radicand: $\sqrt{} = 2a^4b\sqrt{7a}$

SECTION **8.3**

OBJECTIVES

1 Add and subtract square roots.

2 Simplify square roots in a sum or difference.

Adding and Subtracting Radical Expressions

ARE YOU READY?

The following problems review some basic skills that are needed when adding and subtracting radical expressions.

1. Which of the following are like terms: $7x, 3x^2, 9x, 2x^3$

2. Combine like terms: $15y + y$

3. Combine like terms: $8n - 2n$

4. Combine like terms: $4a + 8 + 5a - 9$

5. Simplify: $\sqrt{12}$

6. Simplify: $\sqrt{50h^2}$

We have learned how to combine like terms. In this section, we will use a similar procedure to combine *like radicals.*

1 Add and Subtract Square Roots.

Recall that we use the distributive property to simplify a sum or difference of like terms. For example,

$$3x + 5x = (3 + 5)x \qquad \text{and} \qquad 9y - 6y = (9 - 6)y$$
$$= 8x \qquad\qquad\qquad = 3y$$

In each case, we say that we have *combined like terms.*

The distributive property can also be used to simplify certain sums and differences of square roots. For example,

$$3\sqrt{2} + 5\sqrt{2} = (3 + 5)\sqrt{2} \qquad \text{and} \qquad 9\sqrt{15} - 6\sqrt{15} = (9 - 6)\sqrt{15}$$
$$= 8\sqrt{2} \qquad\qquad\qquad\qquad = 3\sqrt{15}$$

In the above examples, the terms $3\sqrt{2}$ and $5\sqrt{2}$ and the terms $9\sqrt{15}$ and $6\sqrt{15}$ are called **like radical terms.** In these examples, we say that we have *combined like radical terms,* or more simply, *combined like radicals.*

| **Like Radicals** | Square root radical terms are said to be **like radicals** when they have the same radicand. |

Like radicals

$9\sqrt{6}$ and $4\sqrt{6}$

↑ ↑

The same radicand

Unlike radicals

$8\sqrt{5}$ and $5\sqrt{3}$

↑ ↑

Different radicands

EXAMPLE 1

Simplify, if possible, by combining like radicals:

a. $4\sqrt{7} + 6\sqrt{7}$

b. $2\sqrt{m} - 3\sqrt{m}$

c. $5\sqrt{14} + \sqrt{14} - 9\sqrt{14}$

d. $\sqrt{10} + \sqrt{6}$

e. $8 + 4\sqrt{5}$

Strategy First, we will see if the radicands of the square roots are the same. If they are, we can use the distributive property in reverse to add (or subtract) like radicals.

Why We must check the radicands first because only square roots with the same radicand can be combined.

Solution

a. Since $4\sqrt{7}$ and $6\sqrt{7}$ are like radicals, we can add them.

$$4\sqrt{7} + 6\sqrt{7} = 10\sqrt{7} \qquad \text{To combine like radicals, think: } (4 + 6)\sqrt{7} = 10\sqrt{7}.$$

b. Since $2\sqrt{m}$ and $3\sqrt{m}$ are like radicals, we can subtract them.

$$2\sqrt{m} - 3\sqrt{m} = -1\sqrt{m} \qquad \text{Think: } (2 - 3)\sqrt{m} = -1\sqrt{m}.$$
$$= -\sqrt{m} \qquad \text{Because } -1\sqrt{m} = -\sqrt{m}.$$

c. Since each radicand is 14, the three radical terms are like radicals.

$$5\sqrt{14} + \sqrt{14} - 9\sqrt{14} = 5\sqrt{14} + 1\sqrt{14} - 9\sqrt{14} \qquad \text{Write } \sqrt{14} \text{ as } 1\sqrt{14}.$$
$$= -3\sqrt{14} \qquad \text{Think: } (5 + 1 - 9)\sqrt{14} = -3\sqrt{14}.$$

d. Since $\sqrt{10}$ and $\sqrt{6}$ have different radicands, they are unlike radicals and we cannot add them. The expression $\sqrt{10} + \sqrt{6}$ does not simplify.

CAUTION A common error is to add the radicands to simplify $\sqrt{10} + \sqrt{6}$ and get $\sqrt{16}$. However, the sum of two square roots does not equal the square root of the sum. To see why, we can approximate $\sqrt{10}$ with 3.16 and $\sqrt{6}$ with 2.45 and add. The sum, 5.61, is not equal to $\sqrt{16}$, which is 4.

$$\sqrt{10} + \sqrt{6} \neq \sqrt{16} \qquad \text{since} \qquad 3.16 + 2.45 \neq 4$$

e. Since 8 is not a radical expression, we cannot add 8 and $4\sqrt{5}$. The expression $8 + 4\sqrt{5}$ does not simplify.

Success Tip

Just as 4 apples plus 6 apples is 10 apples, we see that 4 square roots of 7 plus 6 square roots of 7 is equal to 10 square roots of 7.

$$4\sqrt{7} + 6\sqrt{7} = 10\sqrt{7}$$

Caution

We have discussed the product and quotient rules for square roots. There is no sum or difference rule for square roots. In general,

$$\sqrt{a} + \sqrt{b} \neq \sqrt{a + b}$$
$$\sqrt{a} - \sqrt{b} \neq \sqrt{a - b}$$

Self Check 1 Simplify, if possible: **a.** $20\sqrt{5} + 30\sqrt{5}$
b. $6\sqrt{b} - 4\sqrt{b}$ **c.** $\sqrt{21} + 4\sqrt{21} - 12\sqrt{21}$
d. $\sqrt{29} - \sqrt{2}$ **e.** $6 + 9\sqrt{6}$

Now Try ▶ Problems 9, 13, 15, 21, and 23

---

2 Simplify Square Roots in a Sum or Difference.

If a sum or difference involves unlike radicals, make sure that each one is written in simplified form. After doing so, like radicals may result that can be combined.

EXAMPLE 2 Perform the indicated operations: **a.** $\sqrt{18} + \sqrt{8}$ **b.** $\sqrt{108} - 8\sqrt{12} + 2\sqrt{27}$

Strategy Since the radicals in each part are unlike radicals, we cannot add or subtract them in their current form. However, we will simplify the radicals and hope that like radicals result.

Why Only like radicals can be combined.

Solution

a. Recall that to write a square root in simplified form we express the radicand as a product of the greatest perfect-square factor and one other factor. Then we use the product rule for square roots.

$$\sqrt{18} + \sqrt{8} = \sqrt{9 \cdot 2} + \sqrt{4 \cdot 2} \qquad \text{Factor 18 and 8 using their greatest perfect-square factors.}$$
$$= \sqrt{9}\sqrt{2} + \sqrt{4}\sqrt{2} \qquad \text{The square root of a product is equal to the product of the square roots.}$$
$$= 3\sqrt{2} + 2\sqrt{2} \qquad \text{Evaluate: } \sqrt{9} = 3 \text{ and } \sqrt{4} = 2.$$
$$= 5\sqrt{2} \qquad \text{To combine like radicals, think: } (3 + 2)\sqrt{2} = 5\sqrt{2}.$$

The Language of Algebra

The phrase **combining like radicals** refers to the operations of addition and subtraction.

b. When we write each radical in simplified form, the results are three like radicals.

$$\sqrt{108} - 8\sqrt{12} + 2\sqrt{27} = \sqrt{36 \cdot 3} - 8\sqrt{4 \cdot 3} + 2\sqrt{9 \cdot 3}$$ Factor using the greatest perfect squares.

$$= \sqrt{36}\sqrt{3} - 8\sqrt{4}\sqrt{3} + 2\sqrt{9}\sqrt{3}$$ The square root of a product is equal to the product of the square roots.

$$= 6\sqrt{3} - 8 \cdot 2\sqrt{3} + 2 \cdot 3\sqrt{3}$$ Evaluate: $\sqrt{36} = 6$, $\sqrt{4} = 2$, and $\sqrt{9} = 3$.

$$= 6\sqrt{3} - 16\sqrt{3} + 6\sqrt{3}$$ Do the multiplication.

$$= -4\sqrt{3}$$ Think: $(6 - 16 + 6)\sqrt{3} = -4\sqrt{3}$.

Self Check 2 Perform the indicated operations: **a.** $\sqrt{75} + \sqrt{12}$
b. $3\sqrt{50} - 6\sqrt{98} - \sqrt{128}$

Now Try ▶ Problems 25 and 33

EXAMPLE 3   Perform the indicated operations, if possible: **a.** $\sqrt{44x^2} + x\sqrt{99}$

b. $\sqrt{27x} - \dfrac{1}{2}\sqrt{20x}$

Strategy   Since the radicals in each part are unlike radicals, we cannot add or subtract them in their current form. However, we will simplify the radicals and hope that like radicals result.

Why   Only like radicals can be combined.

Solution   **a.** $\sqrt{44x^2} + x\sqrt{99}$

$$= \sqrt{4x^2 \cdot 11} + x\sqrt{9 \cdot 11}$$ Factor $44x^2$ and 99.

$$= \sqrt{4x^2}\sqrt{11} + x\sqrt{9}\sqrt{11}$$ The square root of a product is equal to the product of the square roots.

$$= 2x\sqrt{11} + 3x\sqrt{11}$$ Find the square roots: $\sqrt{4x^2} = 2x$ and $\sqrt{9} = 3$.

$$= 5x\sqrt{11}$$ To combine like radicals, think: $(2x + 3x)\sqrt{11} = 5x\sqrt{11}$.

> **The Language of Algebra**
>
> We say that the terms $2x\sqrt{11}$ and $3x\sqrt{11}$ have the same **radical factor,** namely $\sqrt{11}$.

b. $\sqrt{27x} - \dfrac{1}{2}\sqrt{20x} = \sqrt{9 \cdot 3x} - \dfrac{1}{2}\sqrt{4 \cdot 5x}$ Factor $27x$ and $20x$.

$$= \sqrt{9}\sqrt{3x} - \dfrac{1}{2} \cdot \sqrt{4}\sqrt{5x}$$ The square root of a product is equal to the product of the square roots.

$$= 3\sqrt{3x} - \dfrac{1}{2} \cdot 2\sqrt{5x}$$ Evaluate: $\sqrt{9} = 3$ and $\sqrt{4} = 2$.

$$= 3\sqrt{3x} - \sqrt{5x}$$ Multiply: $\dfrac{1}{2} \cdot 2 = 1$.

Since the remaining radicands are different, the radicals $3\sqrt{3x}$ and $\sqrt{5x}$ are unlike and the expression cannot be simplified further.

Self Check 3 Perform the indicated operations, if possible:

a. $\sqrt{24y^2} + y\sqrt{54}$ **b.** $\sqrt{75a} - \dfrac{1}{6}\sqrt{72a}$

Now Try ▶ Problems 37 and 41

When we discuss multiplying radicals in the next section, we will need to know how to simplify four-termed expressions like those in the following example.

EXAMPLE 4   Perform the indicated operations, if possible: **a.** $\sqrt{121} + 3\sqrt{11} + 5\sqrt{11} + \sqrt{225}$
b. $4\sqrt{y^2} - 6\sqrt{y} + 6\sqrt{y} - 9$

Strategy   We will examine each term of the expression looking for radicals that can be simplified. If there are any, we will simplify them.

Why   If there are any like radicals, we can combine them.

Solution   **a.** $\sqrt{121} + 3\sqrt{11} + 5\sqrt{11} + \sqrt{225}$

$= 11 + 3\sqrt{11} + 5\sqrt{11} + 15$ Evaluate: $\sqrt{121} = 11$ and $\sqrt{225} = 15$.

$= 8\sqrt{11} + 26$ Combine like radicals: $(3 + 5)\sqrt{11} = 8\sqrt{11}$.
    Combine like terms: $11 + 15 = 26$.

b. $4\sqrt{y^2} - 6\sqrt{y} + 6\sqrt{y} - 9$

$= 4y - 6\sqrt{y} + 6\sqrt{y} - 9$ Find the square root: $\sqrt{y^2} = y$.

$= 4y - 9$ Combine like radicals: $-6\sqrt{y} + 6\sqrt{y} = 0$.

Self Check 4   Perform the indicated operations.
a. $\sqrt{36} + 9\sqrt{6} + 3\sqrt{6} + 27$
b. $16\sqrt{c^2} + 40\sqrt{c} - 40\sqrt{c} - 10\sqrt{100}$

Now Try ▶ Problems 45 and 49

SECTION 8.3 ▶ STUDY SET

VOCABULARY

Fill in the blanks.

1. Square roots such as $3\sqrt{2}$ and $5\sqrt{2}$, that have the same radicand, are called _____ radicals.

2. To simplify $2\sqrt{2} + 3\sqrt{2}$, we _____ like radicals.

CONCEPTS

Determine whether each pair of radicals are like radicals.

3. **a.** $3\sqrt{3}, 4\sqrt{3}$ **b.** $9\sqrt{a}, 9\sqrt{7a}$

4. **a.** $6\sqrt{10}, 10\sqrt{6}$ **b.** $-\sqrt{5y}, \sqrt{5y}$

5. Fill in the blanks.

a. $5\sqrt{6} + 3\sqrt{6} = (\boxed{} + \boxed{})\sqrt{6} = \boxed{}\sqrt{6}$

b. $2\sqrt{n} - 9\sqrt{n} = (\boxed{} - \boxed{})\sqrt{n} = \boxed{}\sqrt{n}$

6. Write each radical in simplified form.

a. $\sqrt{18}$ **b.** $\sqrt{4x^3}$

NOTATION

7. Complete the solution to simplify the expression.

$9\sqrt{5} - 3\sqrt{20} = 9\sqrt{5} - 3\sqrt{\boxed{} \cdot 5}$

$= 9\sqrt{5} - 3\sqrt{\boxed{}}\sqrt{5}$

$= 9\sqrt{5} - 3 \cdot \boxed{}\sqrt{5}$

$= 9\sqrt{5} - \boxed{}\sqrt{5}$

$= \boxed{}\sqrt{5}$

8. Write each expression in simpler form.

a. $-1\sqrt{x}$ **b.** $1\sqrt{x}$

GUIDED PRACTICE

In the following problems, all variables represent nonnegative real numbers.

Perform the indicated operations, if possible, by combining like radicals. See Example 1.

9. $5\sqrt{7} + 4\sqrt{7}$ 10. $3\sqrt{10} + 4\sqrt{10}$

11. $14\sqrt{21} - 4\sqrt{21}$ 12. $7\sqrt{3} - 2\sqrt{3}$

13. $8\sqrt{n} + \sqrt{n}$ 14. $6\sqrt{m} + \sqrt{m}$

15. $\sqrt{3} + \sqrt{15}$ 16. $\sqrt{6} + 8\sqrt{13}$

17. $\sqrt{x} - 4\sqrt{x}$ 18. $\sqrt{t} - 9\sqrt{t}$

19. $2\sqrt{11} + 3\sqrt{11} + 5\sqrt{11}$ 20. $2\sqrt{5} + 6\sqrt{5} + 9\sqrt{5}$

21. $4\sqrt{2} + 4\sqrt{2} - 4\sqrt{2}$ 22. $9\sqrt{3} - 9\sqrt{3} + 9\sqrt{3}$

23. $7 + 2\sqrt{2}$ 24. $4 + 9\sqrt{3}$

Perform the indicated operations, if possible, by combining like radicals. See Example 2.

25. $\sqrt{12} + \sqrt{27}$ 26. $\sqrt{20} + \sqrt{45}$

27. $\sqrt{18} - \sqrt{8}$ 28. $\sqrt{32} - \sqrt{18}$

29. $\sqrt{12} - \sqrt{48}$ 30. $\sqrt{48} - \sqrt{75}$

31. $\sqrt{288} - 3\sqrt{200}$

32. $\sqrt{80} - \sqrt{245}$

33. $\sqrt{20} + \sqrt{45} + \sqrt{80}$

34. $\sqrt{48} + \sqrt{27} + \sqrt{75}$

35. $3\sqrt{200} - \sqrt{75} + \sqrt{48}$

36. $6\sqrt{20} + \sqrt{80} - \sqrt{125}$

Perform the indicated operations, if possible, by combining like radicals. See Example 3.

37. $\sqrt{72a} - \sqrt{98a}$

38. $\sqrt{25b} - \sqrt{49b}$

39. $\sqrt{18y} - \sqrt{27y}$

40. $\sqrt{49x} + \sqrt{8x}$

41. $\sqrt{2x^2} + \frac{1}{2}\sqrt{8x^2}$

42. $\sqrt{3y^2} + \frac{1}{2}\sqrt{12y^2}$

43. $\sqrt{2d^3} + \sqrt{8d^3}$

44. $\sqrt{3a^3} - \sqrt{12a^3}$

Perform the indicated operations, if possible, by combining like radicals. See Example 4.

45. $\sqrt{4} + \sqrt{2} + 3\sqrt{2} + \sqrt{9}$

46. $\sqrt{25} + 2\sqrt{5} - 3\sqrt{5} - \sqrt{36}$

47. $\sqrt{y^2} + 3\sqrt{y} - 5\sqrt{y} - 15$

48. $\sqrt{z^2} - 2\sqrt{z} + 7\sqrt{z} + 12$

49. $2\sqrt{t^2} + 5\sqrt{t} - 2\sqrt{t} - 5$

50. $4\sqrt{y^2} + 6\sqrt{y} - 6\sqrt{y} - 9$

51. $3\sqrt{a^2} + 3\sqrt{a} - 3\sqrt{a} - 1$

52. $6\sqrt{p^2} + 21\sqrt{p} - 4\sqrt{p} - 14$

TRY IT YOURSELF

Perform the indicated operations, if possible.

53. $8\sqrt{6} - 5\sqrt{2} - 3\sqrt{6}$

54. $3\sqrt{2} - 3\sqrt{15} - 4\sqrt{15}$

55. $\sqrt{24} + \sqrt{150} + \sqrt{240}$

56. $\sqrt{28} + \sqrt{63} + \sqrt{18}$

57. $12 + 2\sqrt{5}$

58. $60 + 20\sqrt{11}$

59. $-1 + 2\sqrt{r} - 3\sqrt{r}$

60. $-8 - 5\sqrt{c} + 4\sqrt{c}$

61. $-9\sqrt{21} + 6\sqrt{21}$

62. $-2\sqrt{5} + 8\sqrt{5}$

63. $-\sqrt{y} + \sqrt{y}$

64. $3\sqrt{t} - 3\sqrt{t}$

65. $2\sqrt{28} + \frac{1}{4}\sqrt{112}$

66. $4\sqrt{63} + \frac{1}{2}\sqrt{112}$

67. $15\sqrt{b^2} + 20\sqrt{b} - 3\sqrt{b} - \sqrt{16}$

68. $25\sqrt{m^2} + 10\sqrt{m} - 10\sqrt{m} - \sqrt{16}$

69. $5 + 3\sqrt{3} + 3\sqrt{3}$

70. $\sqrt{5} + 2 + 3\sqrt{5}$

71. $\sqrt{2} + \sqrt{3} + \sqrt{5}$

72. $\sqrt{3} + \sqrt{7} + \sqrt{11}$

73. $2\sqrt{45} + 2\sqrt{80}$

74. $3\sqrt{80} + 3\sqrt{125}$

75. $3\sqrt{54b^2} + 5\sqrt{24b^2}$

76. $3\sqrt{24x^4} + 2\sqrt{54x^4}$

77. $\sqrt{32x^5} - \frac{2}{3}\sqrt{18x^5}$

78. $\sqrt{27y^3} + \frac{3}{4}\sqrt{48y^3}$

79. $2\sqrt{80} - 3\sqrt{125}$

80. $3\sqrt{245} - 2\sqrt{180}$

81. $\sqrt{48} - \sqrt{8} + \sqrt{27} - \sqrt{32}$

82. $\sqrt{162} + \sqrt{50} - \sqrt{75} - \sqrt{108}$

83. $6\sqrt{40y} - 2\sqrt{360z}$

84. $3\sqrt{20x} + 2\sqrt{45y}$

Look Alikes . . .

85. a. $18\sqrt{n} + 18\sqrt{n}$ b. $18\sqrt{n} - 18\sqrt{n}$

86. a. $9\sqrt{6x} + \sqrt{6x}$ b. $9\sqrt{6x} - \sqrt{6x}$

87. a. $\sqrt{20} + \sqrt{20}$ b. $\sqrt{21} + \sqrt{21}$

88. a. $\sqrt{3} + \sqrt{12}$ b. $\sqrt{3} + \sqrt{13}$

APPLICATIONS

89. **Playground Equipment.** Find the exact total length of pipe necessary to construct the frame of the swing set. Then approximate the length to the nearest foot.

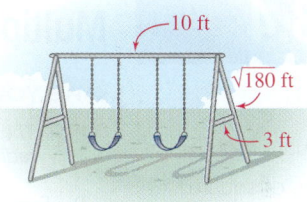

90. **Hardware.** Find the exact difference in the lengths of the "arms" of the door-closing device. Then approximate the difference to the nearest inch.

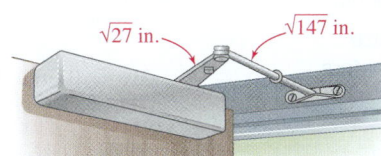

91. **Camping.** The length l of a center support pole for a tent is given by the formula $l = 0.5s\sqrt{3}$, where s is the length of the side of the tent. Find the exact total length of the four poles needed for the parents' and children's tents. Then approximate the total length to the nearest tenth of a foot.

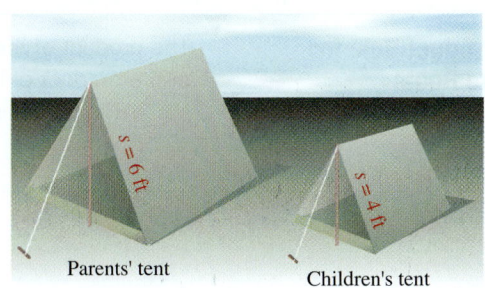

Parents' tent   Children's tent

92. Surveying. The length of each lot line of a swimming pool complex is shown. Find the exact perimeter of the irregular-shaped lot. Then approximate the perimeter to the nearest foot.

$10\sqrt{150}$ ft

DRESSING ROOM

SNACK BAR

LAWN

$7\sqrt{54}$ ft

TERRACE

POOL

$13\sqrt{24}$ ft

LANAI

$9\sqrt{96}$ ft

95. Explain why $\sqrt{6} + \sqrt{5}$ cannot be simplified further.

96. Is $2\sqrt{5} - \sqrt{5} = 2$? Explain.

97. Some of the instructions in this Study Set contain the sentence "All variables represent nonnegative real numbers." Why is this sentence necessary?

98. $\sqrt{2}$ and $\sqrt{50}$ are not like radicals. Can they be added? Explain.

REVIEW

Simplify. Write each answer without using negative exponents.

99. 3^{-2} **100.** $\dfrac{1}{3^{-2}}$ **101.** -3^2 **102.** 3^0

WRITING

93. Are $2\sqrt{3}$ and $3\sqrt{2}$ like radicals? Explain.

94. Does $\sqrt{9} + \sqrt{16} = \sqrt{25}$? Explain.

CHALLENGE PROBLEMS

103. Fill in the blank: $2y\sqrt{175y} - \sqrt{} = 0$

104. Add: $0.1\sqrt{50x^3} + \dfrac{1}{64}\sqrt{32x^3}$

SECTION 8.4

Multiplying and Dividing Radical Expressions

OBJECTIVES

1 Multiply square roots.

2 Find powers of square roots.

3 Multiply radical expressions.

4 Divide radical expressions.

5 Rationalize denominators.

ARE YOU READY?

The following problems review some basic skills that are needed when multiplying and dividing radical expressions.

Perform the indicated operations and simplify, if possible.

1. $2x \cdot 5x$ **2.** $(8a^6)^2$

3. $7t(6t + 2)$ **4.** $(2x + 3)(x - 1)$

5. $(n + 9)^2$ **6.** $(x + 2)(x - 2)$

We will now discuss methods used to multiply and divide square roots.

1 Multiply Square Roots.

We have used the product rule for square roots to write square roots in simplified form.

$$\sqrt{a \cdot b} = \sqrt{a} \cdot \sqrt{b}$$

We can use this rule in reverse to multiply square roots.

The Product Rule for Square Roots	The product of the square roots of two nonnegative numbers is equal to the square root of the product of those numbers. For any nonnegative real numbers a and b, $$\sqrt{a} \cdot \sqrt{b} = \sqrt{a \cdot b}$$

EXAMPLE 1   Multiply and then simplify, if possible: **a.** $\sqrt{3} \cdot \sqrt{2}$ **b.** $\sqrt{6x}\left(\sqrt{8x}\right)$

c. $\sqrt{a^3}\sqrt{5a^8}$

Strategy   To multiply the square roots, we will multiply their radicands and write the product within a square root symbol. Then, we will simplify the result, if possible.

Why   The product of the square roots of two nonnegative numbers is the square root of the product of those numbers.

Solution   **a.** $\sqrt{3} \cdot \sqrt{2} = \sqrt{3 \cdot 2} = \sqrt{6}$ Multiply the radicands, 3 and 2, to get 6, and write the square root of that product.

Since 6 does not have a perfect-square factor, $\sqrt{6}$ does not simplify.

b. $\sqrt{6x}\left(\sqrt{8x}\right) = \sqrt{6x \cdot 8x} = \sqrt{48x^2}$ Multiply the radicands, 6x and 8x, to get $48x^2$, and write the square root of that product.

Since $48x^2$ has a perfect-square factor of $16x^2$, we can simplify $\sqrt{48x^2}$.

$$\sqrt{48x^2} = \sqrt{16x^2 \cdot 3}$$ Factor $48x^2$ as $16x^2 \cdot 3$.

$$= \sqrt{16x^2}\sqrt{3}$$ The square root of a product is the product of the square roots.

$$= 4x\sqrt{3}$$ Find the square root of the perfect square: $\sqrt{16x^2} = 4x$.

c. $\sqrt{a^3}\sqrt{5a^8} = \sqrt{a^3 \cdot 5a^8} = \sqrt{5a^{11}}$ Multiply the radicands: $a^3 \cdot 5a^8 = 5a^{8+3} = 5a^{11}$. Then write the square root of that product.

Since a^{11} has a perfect-square factor of a^{10}, we can simplify $\sqrt{5a^{11}}$.

$$\sqrt{5a^{11}} = \sqrt{a^{10} \cdot 5a}$$ Factor $5a^{11}$ as $a^{10} \cdot 5a$.

$$= \sqrt{a^{10}}\sqrt{5a}$$ The square root of a product is the product of the square roots.

$$= a^5\sqrt{5a}$$ Find the square root of the perfect square: $\sqrt{a^{10}} = a^5$.

Self Check 1   Multiply and simplify, if possible: **a.** $\sqrt{5} \cdot \sqrt{3}$
b. $\sqrt{10y}\left(\sqrt{2y}\right)$ **c.** $\sqrt{5b^3}\sqrt{2b^6}$

Now Try ▶ Problems 13, 17, and 19

Notation

Multiplication of radicals can be shown in several ways:

$$\sqrt{3} \cdot \sqrt{2} = \sqrt{3}\left(\sqrt{2}\right) = \sqrt{3}\sqrt{2}$$

Success Tip

Here, the product rule for square roots is used in two ways.

To multiply:

$$\sqrt{a^3}\left(\sqrt{5a^8}\right) = \sqrt{a^3 \cdot 5a^8}$$

and to simplify:

$$\sqrt{a^{10} \cdot 5a} = \sqrt{a^{10}}\sqrt{5a}$$

EXAMPLE 2   Multiply and simplify, if possible: **a.** $6\sqrt{5} \cdot 7$ **b.** $2\sqrt{3} \cdot \sqrt{12}$ **c.** $3\sqrt{6} \cdot 4\sqrt{5}$

Strategy   We will use the commutative and associative properties of multiplication to reorder and regroup the factors in each expression. Then we will multiply and simplify the result, if possible.

Why   We want to write the factors that are integers together and the factors that are square roots together so that we can find their products separately.

Solution   **a.** $6\sqrt{5} \cdot 7 = 6 \cdot 7 \cdot \sqrt{5} = 42\sqrt{5}$ Multiply the integer factors, 6 and 7, to get 42. The result does not simplify.

b. $2\sqrt{3} \cdot \sqrt{12} = 2\sqrt{36}$ Multiply the radicands, 3 and 12, to get 36, and write the square root of that product.

$$= 2 \cdot 6$$ Simplify the result: $\sqrt{36} = 6$.

$$= 12$$ Multiply 2 and 6 to get 12.

Caution

Note the difference.

If 7 is *not* a radicand:

$$6\sqrt{5} \cdot 7 = 42\sqrt{5}$$

If 7 is a radicand:

$$6\sqrt{5} \cdot \sqrt{7} = 6\sqrt{35}$$

c. $3\sqrt{6} \cdot 4\sqrt{5} = 3 \cdot 4 \cdot \sqrt{6} \cdot \sqrt{5}$ Write the integer factors together and the square root factors together.

$$= 12\sqrt{6 \cdot 5}$$ Multiply 3 and 4 to get 12, and multiply $\sqrt{6}$ and $\sqrt{5}$.

$$= 12\sqrt{30}$$ Multiply 6 and 5 to get 30.

Since 30 does not have a perfect-square factor, the result does not simplify.

Self Check 2   Multiply and simplify, if possible: **a.** $2\sqrt{2} \cdot 6$
b. $9\sqrt{5} \cdot \sqrt{20}$ **c.** $7\sqrt{10} \cdot 8\sqrt{3}$

Now Try ▶ Problems 21 and 25

---

2 Find Powers of Square Roots.

By definition, when the square root of a positive number is squared, the result is that positive number. In symbols, we have the following fact.

The Square of a Square Root	For any nonnegative real number a, $$\left(\sqrt{a}\right)^2 = a$$

EXAMPLE 3   Find: **a.** $\left(\sqrt{5}\right)^2$ **b.** $\left(\sqrt{x+1}\right)^2$ **c.** $\left(2\sqrt{7}\right)^2$

Strategy   We will use the rule $\left(\sqrt{a}\right)^2 = a$ to find each power.

Why   Each expression involves the square of the square root of a positive number.

Solution   **a.** $\left(\sqrt{5}\right)^2 = 5$ The square of the square root of a positive number is that number.

Success Tip

Since $\left(\sqrt{a}\right)^2 = \sqrt{a} \cdot \sqrt{a}$, it also follows that $\sqrt{a} \cdot \sqrt{a} = a$, if a is a positive real number.

b. $\left(\sqrt{x+1}\right)^2 = x + 1$ If $x > 0$, then x represents a positive number, and $x + 1$ is positive. The square of the square root of a positive number is that number.

c. We can use the power of a product rule for exponents to evaluate $\left(2\sqrt{7}\right)^2$.

$$\left(2\sqrt{7}\right)^2 = 2^2\left(\sqrt{7}\right)^2$$ Raise each factor of the product $2\sqrt{7}$ to the second power.

$$= 4 \cdot 7$$ Simplify: $2^2 = 4$ and $\left(\sqrt{7}\right)^2 = 7$.

$$= 28$$ Multiply 4 and 7 to get 28.

Self Check 3   Find: **a.** $\left(\sqrt{11}\right)^2$ **b.** $\left(\sqrt{3y+5}\right)^2$ **c.** $\left(5\sqrt{2}\right)^2$

Now Try ▶ Problems 29, 33, and 35

---

3 Multiply Radical Expressions.

To multiply radical expressions that contain more than one term, we use the same methods that were used to multiply polynomials with more than one term.

EXAMPLE 4   Multiply and simplify, if possible: **a.** $\sqrt{7}\left(\sqrt{7} - 3\right)$ **b.** $2\sqrt{2x}\left(\sqrt{x} + 3\sqrt{2}\right)$

Strategy   As with polynomials, we will multiply each term within the parentheses by the term outside the parentheses.

Why   This is an application of the distributive property.

Solution   **a.** $\sqrt{7}\left(\sqrt{7} - 3\right) = \sqrt{7}\sqrt{7} - \sqrt{7} \cdot 3$ Distribute the multiplication by $\sqrt{7}$.

$$= 7 - 3\sqrt{7}$$ Multiply: $\sqrt{7}\sqrt{7} = \left(\sqrt{7}\right)^2 = 7$.
Write $\sqrt{7} \cdot 3$ as $3\sqrt{7}$.

> **Caution**
>
> A common error is to "simplify" incorrectly by subtracting 7 and 3:
>
> $$7 - 3\sqrt{7} = 4\sqrt{7}$$

b. $2\sqrt{2x}\left(\sqrt{x} + 3\sqrt{2}\right) = 2\sqrt{2x}\sqrt{x} + 2\sqrt{2x} \cdot 3\sqrt{2}$ Distribute the multiplication by $2\sqrt{2x}$.

$$= 2\sqrt{2x^2} + 6\sqrt{4x}$$ Multiply radicands and write the square root of each product.

$$= 2\sqrt{x^2 \cdot 2} + 6\sqrt{4 \cdot x}$$ To simplify $\sqrt{2x^2}$, factor $2x^2$.
To simplify $\sqrt{4x}$, factor $4x$.

$$= 2\sqrt{x^2}\sqrt{2} + 6\sqrt{4}\sqrt{x}$$ The square root of a product is the product of the square roots.

$$= 2x\sqrt{2} + 6 \cdot 2\sqrt{x}$$ Find the square roots of the perfect squares: $\sqrt{x^2} = x$ and $\sqrt{4} = 2$.

$$= 2x\sqrt{2} + 12\sqrt{x}$$ To simplify the last term, multiply: $6 \cdot 2$ to get 12.

Since $2x\sqrt{2}$ and $12\sqrt{x}$ are unlike radicals, we cannot add them. This result does not simplify.

> **Self Check 4**   Multiply and simplify, if possible: **a.** $\sqrt{3}\left(\sqrt{3} + 16\right)$
> **b.** $4\sqrt{5n}\left(\sqrt{n} - 2\sqrt{5}\right)$
>
> **Now Try** ▶ Problems 37 and 41

EXAMPLE 5   Multiply and simplify, if possible: **a.** $\left(\sqrt{7} + \sqrt{3}\right)\left(\sqrt{2} - \sqrt{5}\right)$
b. $\left(\sqrt{3x} - 5\right)\left(\sqrt{3x} + 2\right)$

Strategy   As with binomials, we will multiply each term within the first set of parentheses by each term within the second set of parentheses.

Why   This is an application of the FOIL method for multiplying binomials.

Solution

$$\text{a. } \left(\sqrt{7} + \sqrt{3}\right)\left(\sqrt{2} - \sqrt{5}\right) = \sqrt{7}\sqrt{2} - \sqrt{7}\sqrt{5} + \sqrt{3}\sqrt{2} - \sqrt{3}\sqrt{5}$$

$$= \sqrt{14} - \sqrt{35} + \sqrt{6} - \sqrt{15}$$ Do the multiplication.

Since there are no like radicals, we cannot simplify the result.

b. $\left(\sqrt{3x} - 5\right)\left(\sqrt{3x} + 2\right) = \sqrt{3x}\sqrt{3x} + 2\sqrt{3x} - 5\sqrt{3x} - 5 \cdot 2$ Multiply each term of $\sqrt{3x} - 5$ by each term of $\sqrt{3x} + 2$.

$$= 3x + 2\sqrt{3x} - 5\sqrt{3x} - 10 \quad \sqrt{3x}\sqrt{3x} = \left(\sqrt{3x}\right)^2 = 3x.$$

$$= 3x - 3\sqrt{3x} - 10 \quad \text{Combine like radicals:}$$
Think: $(2 - 5)\sqrt{3x}.$

Self Check 5   Find the product and simplify, if possible:
a. $\left(\sqrt{6} + \sqrt{5}\right)\left(\sqrt{11} - \sqrt{7}\right)$
b. $\left(\sqrt{5a} - 2\right)\left(\sqrt{5a} + 3\right)$

Now Try ▸ Problems 47 and 49

The special product formulas from Chapter 4 can be used to multiply two-term expressions containing radicals.

EXAMPLE 6   Find each product and simplify, if possible: **a.** $\left(\sqrt{6} + \sqrt{y}\right)^2$
b. $\left(\sqrt{7} + \sqrt{2}\right)\left(\sqrt{7} - \sqrt{2}\right)$

Strategy   We will use a special-product rule to find each product.

Why   It is often easier to find such products using special-product rules than it is using the FOIL method.

Solution   **a.** Recall that the square of the sum of two terms is the square of the first term, plus twice the product of both terms, plus the square of the last term:
$(A + B)^2 = A^2 + 2AB + B^2$.

$$\left(\sqrt{6} + \sqrt{y}\right)^2 = \left(\sqrt{6}\right)^2 + 2 \cdot \sqrt{6}\sqrt{y} + \left(\sqrt{y}\right)^2$$

$$= 6 + 2\sqrt{6y} + y \qquad \left(\sqrt{6}\right)^2 = 6 \text{ and } \left(\sqrt{y}\right)^2 = y.$$

> **Caution**
>
> A common error when squaring a two-term expression is simply to square the first term and square the last term and forget the middle term of the product.
>
> $\left(\sqrt{6} + \sqrt{y}\right)^2 = \left(\sqrt{6}\right)^2 + \left(\sqrt{y}\right)^2$
> $= 6 \quad + \quad y$

b. Recall that the product of the sum and difference of two terms is the square of the first term minus the square of the second term: $(A + B)(A - B) = A^2 - B^2$.

$$\left(\sqrt{7} + \sqrt{2}\right)\left(\sqrt{7} - \sqrt{2}\right) = \left(\sqrt{7}\right)^2 - \left(\sqrt{2}\right)^2$$

$$= 7 - 2 \qquad \left(\sqrt{7}\right)^2 = 7 \text{ and } \left(\sqrt{2}\right)^2 = 2.$$

$$= 5 \qquad \text{Do the subtraction.}$$

Self Check 6   Find each product and simplify, if possible:
a. $\left(\sqrt{15} - \sqrt{n}\right)^2$
b. $\left(\sqrt{3} + \sqrt{5}\right)\left(\sqrt{3} - \sqrt{5}\right)$

Now Try ▸ Problems 53 and 61

4 Divide Radical Expressions.

We have used the quotient rule for square roots to simplify square roots of quotients:

$$\sqrt{\frac{a}{b}} = \frac{\sqrt{a}}{\sqrt{b}}$$

We can use this rule in reverse to divide square roots.

The Quotient Rule for Square Roots

The quotient of the square roots of two numbers is equal to the square root of the quotient of the two numbers.

For any positive real numbers a and b,

$$\frac{\sqrt{a}}{\sqrt{b}} = \sqrt{\frac{a}{b}}$$

EXAMPLE 7   Divide and simplify, if possible: **a.** $\dfrac{\sqrt{39}}{\sqrt{3}}$ **b.** $\dfrac{\sqrt{40}}{\sqrt{5}}$ **c.** $\dfrac{\sqrt{18x^5}}{\sqrt{2x^3}}$

Strategy   To divide the square roots, we will divide their radicands and write the quotient within a square root symbol. Then, we will simplify the result, if possible.

Why   Since each denominator contains a square root, these expressions are not in simplified form.

Solution   **a.** $\dfrac{\sqrt{39}}{\sqrt{3}} = \sqrt{\dfrac{39}{3}}$ The quotient of the square roots is equal to the square root of the quotient.

$\qquad\qquad = \sqrt{13}$ Divide the radicands, 39 by 3, to get 13, and write the square root of that quotient. The result does not simplify.

Caution

Recall that a radical expression is *not* in simplified form if the denominator contains a radical.

b. $\dfrac{\sqrt{40}}{\sqrt{5}} = \sqrt{\dfrac{40}{5}}$ The quotient of the square roots is equal to the square root of the quotient.

$\qquad = \sqrt{8}$ Divide the radicands and write the square root of that quotient. Since 8 has a perfect-square factor of 4, we can simplify.

$\qquad = \sqrt{4 \cdot 2}$ To simplify the radical, factor 8 as 4 · 2.

$\qquad = \sqrt{4}\sqrt{2}$ Use the product rule.

$\qquad = 2\sqrt{2}$ Evaluate: $\sqrt{4} = 2$.

c. $\dfrac{\sqrt{18x^5}}{\sqrt{2x^3}} = \sqrt{\dfrac{18x^5}{2x^3}}$ The quotient of the square roots is equal to the square root of the quotient.

$\qquad = \sqrt{9x^2}$ Simplify the radicand: $\dfrac{18x^5}{2x^3} = \dfrac{2 \cdot 9 \cdot \cancel{x} \cdot \cancel{x} \cdot \cancel{x} \cdot x \cdot x}{2 \cdot \cancel{x} \cdot \cancel{x} \cdot \cancel{x}} = 9x^2$.

$\qquad = 3x$ Find the square root of $9x^2$.

Self Check 7   Divide and simplify, if possible: **a.** $\dfrac{\sqrt{42}}{\sqrt{7}}$ **b.** $\dfrac{\sqrt{90}}{\sqrt{2}}$ **c.** $\dfrac{\sqrt{125d^5}}{\sqrt{5d}}$

Now Try ▶ Problems 65 and 71

5   Rationalize Denominators.

The radical expression $\dfrac{1}{\sqrt{2}}$ is not in simplified form because a radical appears in the denominator of the fraction. There is a way to change the denominator of the fraction from a radical representing an irrational number to an expression representing a rational number. In this process, called **rationalizing the denominator,** we multiply the fraction by a form of 1 and use the fact that $\sqrt{a} \cdot \sqrt{a} = a$.

EXAMPLE 8   Rationalize the denominator: $\dfrac{1}{\sqrt{2}}$

Strategy   We look at the denominator of the fraction and ask, "By what must we multiply $\sqrt{2}$ to obtain a rational number?" The answer is $\sqrt{2}$, because $\sqrt{2} \cdot \sqrt{2} = 2$. Then we multiply the fraction by 1 in the form $\dfrac{\sqrt{2}}{\sqrt{2}}$.

Why   This will result in a new fraction that has the rational number 2 for its denominator.

Solution

The Language of Algebra

Since $\sqrt{2}$ is an irrational number, $\dfrac{1}{\sqrt{2}}$ has an *irrational* denominator. Since 2 is a rational number, $\dfrac{\sqrt{2}}{2}$ has a *rational* denominator.

$$\frac{1}{\sqrt{2}} = \frac{1}{\sqrt{2}} \cdot \frac{\sqrt{2}}{\sqrt{2}}$$

To build an equivalent fraction, multiply by 1 in the form $\dfrac{\sqrt{2}}{\sqrt{2}} = 1$.

$$= \frac{\sqrt{2}}{2}$$

Multiply the numerators: $1 \cdot \sqrt{2} = \sqrt{2}$.
Multiply the denominators: $\sqrt{2} \cdot \sqrt{2} = \sqrt{4} = 2$.

Thus, $\dfrac{1}{\sqrt{2}} = \dfrac{\sqrt{2}}{2}$. These fractions are equivalent (they represent the same number) but the second one does not have an irrational number in the denominator.

Self Check 8   Rationalize the denominator: $\dfrac{11}{\sqrt{6}}$

Now Try   Problem 73

One reason for rationalizing denominators is historical. The concept of rationalizing the denominator remains with us from the days when all arithmetic was done by hand. It makes evaluation by long division easier. For example, to estimate the decimal value of $\dfrac{1}{\sqrt{2}}$, we can divide 1 by 1.414213, or we can estimate the decimal value of $\dfrac{\sqrt{2}}{2}$, by dividing 1.414213 by 2. The results are approximately the same, but the second division is much easier.

$$\frac{1}{\sqrt{2}} \approx \frac{1}{1.414213} \approx 0.707107 \qquad \frac{\sqrt{2}}{2} \approx \frac{1.414213}{2} \approx 0.707107$$

With the calculators and computers that are now available, either of these divisions are rather simple.

Rationalizing a denominator is important today because it gives us a standard form in which we can write radical expressions. This helps when we compare such radicals and enables us to combine many radical expressions.

The procedure for rationalizing square root denominators is as follows.

Rationalizing Denominators

To **rationalize a square root denominator,** multiply the numerator and denominator of the given fraction by the square root that appears in its denominator, or by a square root that makes a perfect-square radicand in the denominator.

EXAMPLE 9   Simplify each expression: **a.** $\sqrt{\dfrac{5}{3}}$ **b.** $\dfrac{4}{\sqrt{24x}}$

Strategy   In each case, we will rationalize a denominator.

Why   The expression will then be written in simplest form.

Solution   **a.** The expression $\sqrt{\dfrac{5}{3}}$ is not in simplified form, because the radicand is a fraction. To write it in simplified form, we use the quotient rule for square roots and then rationalize the denominator.

$$\sqrt{\dfrac{5}{3}} = \dfrac{\sqrt{5}}{\sqrt{3}}$$ The square root of a quotient is equal to the quotient of the square roots.

$$= \dfrac{\sqrt{5}}{\sqrt{3}} \cdot \dfrac{\sqrt{3}}{\sqrt{3}}$$ To build an equivalent fraction, multiply by $\dfrac{\sqrt{3}}{\sqrt{3}} = 1$.

$$= \dfrac{\sqrt{15}}{3}$$ Multiply the numerators and multiply the denominators: $\sqrt{3} \cdot \sqrt{3} = \sqrt{9} = 3$.

b. We could begin by multiplying $\dfrac{4}{\sqrt{24x}}$ by $\dfrac{\sqrt{24x}}{\sqrt{24x}}$. However, to work with smaller numbers, we simplify $\sqrt{24x}$ first and then rationalize the denominator.

$$\dfrac{4}{\sqrt{24x}} = \dfrac{4}{2\sqrt{6x}}$$ $\sqrt{24x} = \sqrt{4 \cdot 6x} = \sqrt{4}\sqrt{6x} = 2\sqrt{6x}$.

$$= \dfrac{4}{2\sqrt{6x}} \cdot \dfrac{\sqrt{6x}}{\sqrt{6x}}$$ To build an equivalent fraction, multiply by $\dfrac{\sqrt{6x}}{\sqrt{6x}} = 1$.

$$= \dfrac{4\sqrt{6x}}{2 \cdot 6x}$$ Multiply the numerators and multiply the denominators.

$$= \dfrac{\overset{1}{\cancel{2}} \cdot \overset{1}{\cancel{2}}\sqrt{6x}}{\underset{1}{\cancel{2}} \cdot \underset{1}{\cancel{2}} \cdot 3x}$$ Factor 4 as $2 \cdot 2$ and $6x$ as $2 \cdot 3x$. Simplify by removing the common factors.

$$= \dfrac{\sqrt{6x}}{3x}$$ The expression does not simplify any further.

Self Check 9   Rationalize each denominator: **a.** $\sqrt{\dfrac{6}{7}}$ **b.** $\dfrac{15}{\sqrt{50y}}$

Now Try   Problems 77 and 79

We will now discuss a method to rationalize denominators of fractions that have two terms.

One-term denominators   *Two-term denominators*

$$\dfrac{1}{\sqrt{2}}, \quad \dfrac{\sqrt{5}}{\sqrt{3}}, \quad \dfrac{4}{\sqrt{24x}} \qquad\qquad \dfrac{5}{\sqrt{6} - 1}, \quad \dfrac{3}{\sqrt{7} + 2}$$

To rationalize the denominator of $\dfrac{5}{\sqrt{6} - 1}$, we multiply the numerator and denominator by $\sqrt{6} + 1$, because the product $\left(\sqrt{6} - 1\right)\left(\sqrt{6} + 1\right)$ contains no radicals. Radical expressions such as $\sqrt{6} - 1$ and $\sqrt{6} + 1$ are said to be **conjugates** of each other. Notice that they differ only in the sign separating the terms.

EXAMPLE 10   Rationalize each denominator: **a.** $\dfrac{2}{\sqrt{6} - 1}$ **b.** $\dfrac{\sqrt{2} - \sqrt{y}}{\sqrt{3} + \sqrt{y}}$

Strategy   To rationalize each two-term denominator, we will multiply the expression by a form of 1 that uses the conjugate of the denominator.

Why   Since the product of a radical expression and its conjugate contains no radicals, this step clears the denominator of radicals.

Solution   **a.** The conjugate of $\sqrt{6} - 1$ is $\sqrt{6} + 1$.

$$\frac{2}{\sqrt{6} - 1} = \frac{2}{\sqrt{6} - 1} \cdot \frac{\sqrt{6} + 1}{\sqrt{6} + 1}$$ To build an equivalent fraction, multiply by $\frac{\sqrt{6} + 1}{\sqrt{6} + 1} = 1$.

$$= \frac{2(\sqrt{6} + 1)}{(\sqrt{6} - 1)(\sqrt{6} + 1)}$$ Multiply the numerators and multiply the denominators.

$$= \frac{2(\sqrt{6} + 1)}{(\sqrt{6})^2 - 1^2}$$ In the denominator, use a special-product formula: $(A - B)(A + B) = A^2 - B^2$.

$$= \frac{2(\sqrt{6} + 1)}{6 - 1}$$ After evaluating $(\sqrt{6})^2$, the denominator no longer contains a radical.

$$= \frac{2(\sqrt{6} + 1)}{5}$$ In the denominator, subtract: $6 - 1 = 5$.

$$= \frac{2\sqrt{6} + 2}{5}$$ In the numerator, distribute the multiplication by 2.

> **Success Tip**
>
> It is best to leave the numerator, $2(\sqrt{6} + 1)$, in factored form until we are sure that the denominator does not contain a factor of 2 or $\sqrt{6} + 1$.

b. The conjugate of the denominator, $\sqrt{3} + \sqrt{y}$, is $\sqrt{3} - \sqrt{y}$.

$$\frac{\sqrt{2} - \sqrt{y}}{\sqrt{3} + \sqrt{y}} = \frac{\sqrt{2} - \sqrt{y}}{\sqrt{3} + \sqrt{y}} \cdot \frac{\sqrt{3} - \sqrt{y}}{\sqrt{3} - \sqrt{y}}$$ To build an equivalent fraction, multiply by $\frac{\sqrt{3} - \sqrt{y}}{\sqrt{3} - \sqrt{y}} = 1$.

$$= \frac{(\sqrt{2} - \sqrt{y})(\sqrt{3} - \sqrt{y})}{(\sqrt{3} + \sqrt{y})(\sqrt{3} - \sqrt{y})}$$ Multiply the numerators and multiply the denominators.

$$= \frac{\sqrt{6} - \sqrt{2y} - \sqrt{3y} + y}{(\sqrt{3})^2 - (\sqrt{y})^2}$$ In the numerator, use the FOIL method. In the denominator, use a special-product formula: $(A + B)(A - B) = A^2 - B^2$.

$$= \frac{\sqrt{6} - \sqrt{2y} - \sqrt{3y} + y}{3 - y}$$ After simplifying the denominator, it no longer contains radicals.

Self Check 10   Rationalize the denominator: **a.** $\dfrac{5}{\sqrt{7} + 2}$

b. $\dfrac{\sqrt{x} - \sqrt{7}}{\sqrt{x} - \sqrt{2}}$

Now Try ▶ Problems 81 and 87

SECTION 8.4 ▶ STUDY SET

VOCABULARY

Fill in the blanks.

1. The _____ of $\frac{1}{\sqrt{3}}$ is an irrational number.

2. To _____ the denominator of $\frac{4}{\sqrt{5}}$, we multiply the fraction by $\frac{\sqrt{5}}{\sqrt{5}}$.

3. The _____ of $1 + \sqrt{2}$ is $1 - \sqrt{2}$.

4. $\frac{3}{\sqrt{7} + 2}$ has a _____ -term denominator.

CONCEPTS

Fill in the blanks. All variables represent nonnegative numbers.

5. **a.** $\sqrt{a} \cdot \sqrt{b} = \sqrt{}$ **b.** $\dfrac{\sqrt{a}}{\sqrt{b}} = \sqrt{}$

   c. $(\sqrt{a})^2 = $ **d.** $\sqrt{b} \cdot \sqrt{b} = $

6. **a.** To rationalize the denominator of $\dfrac{4}{\sqrt{3}}$, we multiply the fraction by $\dfrac{\sqrt{3}}{\sqrt{3}}$, which is a form of ☐.

   b. To rationalize the denominator of $\dfrac{5}{\sqrt{6}+3}$, we multiply the fraction by ☐.

7. Explain why each expression is not in simplified radical form.

   a. $\sqrt{\dfrac{3}{4}}$ **b.** $\dfrac{1}{\sqrt{10}}$

8. Perform each operation, if possible.

   a. $\sqrt{2}+\sqrt{3}$ **b.** $\sqrt{2}\cdot\sqrt{3}$

   c. $\sqrt{2}-\sqrt{3}$ **d.** $\dfrac{\sqrt{2}}{\sqrt{3}}$

NOTATION

Complete each solution.

9. $7\sqrt{2}\cdot 4\sqrt{3} = 7\cdot\boxed{}\cdot\sqrt{2}\cdot\boxed{}$

   $\qquad\qquad = 28\sqrt{\boxed{}}$

10. $\left(\sqrt{x}+\sqrt{2}\right)\left(\sqrt{x}-\sqrt{2}\right)$

    $= \sqrt{x}\,\boxed{} - \sqrt{x}\cdot\sqrt{2} + \sqrt{2}\cdot\boxed{} - \sqrt{2}\cdot\sqrt{2}$

    $= \boxed{} - \sqrt{\boxed{}} + \sqrt{2x} - \boxed{}$

    $= x - \boxed{}$

11. $\dfrac{5}{\sqrt{7}} = \dfrac{5}{\sqrt{7}}\cdot\dfrac{\sqrt{\boxed{}}}{\sqrt{\boxed{}}}$

    $= \dfrac{5\sqrt{7}}{\boxed{}}$

12. $\dfrac{4}{\sqrt{x}+1} = \dfrac{4}{\sqrt{x}+1}\cdot\dfrac{\sqrt{x}\,\boxed{}\,1}{\sqrt{x}\,\boxed{}\,1}$

    $= \dfrac{4\left(\sqrt{x}-1\right)}{\left(\sqrt{x}\right)^{\boxed{}}-1}$

    $= \dfrac{4\sqrt{x}-\boxed{}}{\boxed{}-\boxed{}}$

GUIDED PRACTICE

In the following problems, all variables represent nonnegative real numbers.

Multiply and simplify, if possible. See Example 1.

13. $\sqrt{3}\cdot\sqrt{5}$ 14. $\sqrt{2}\cdot\sqrt{7}$

15. $\sqrt{5}\cdot\sqrt{10}$ 16. $\sqrt{2}\cdot\sqrt{6}$

17. $\sqrt{5d}\cdot\sqrt{8d}$ 18. $\sqrt{3b}\cdot\sqrt{15b}$

19. $\sqrt{5x^3}\sqrt{x^5}$ 20. $\sqrt{a^7}\sqrt{2a^3}$

Multiply and simplify, if possible. See Example 2.

21. $3\sqrt{5}\cdot 5$ 22. $4\sqrt{3}\cdot 7$

23. $10\sqrt{5}\cdot\sqrt{15}$ 24. $16\sqrt{2}\cdot\sqrt{6}$

25. $5\sqrt{3}\cdot 2\sqrt{5}$ 26. $2\sqrt{5}\cdot 5\sqrt{2}$

27. $2\sqrt{6}\left(3\sqrt{3}\right)$ 28. $7\sqrt{5}\left(3\sqrt{10}\right)$

Find each power. See Example 3.

29. $\left(\sqrt{6}\right)^2$ 30. $\left(\sqrt{11}\right)^2$

31. $\left(\sqrt{y}\right)^2$ 32. $\left(\sqrt{m}\right)^2$

33. $\left(\sqrt{2b+7}\right)^2$ 34. $\left(\sqrt{3t+8}\right)^2$

35. $\left(2\sqrt{3}\right)^2$ 36. $\left(3\sqrt{5}\right)^2$

Multiply and simplify, if possible. See Example 4.

37. $\sqrt{2}\left(\sqrt{2}+1\right)$ 38. $\sqrt{5}\left(\sqrt{5}+2\right)$

39. $3\sqrt{3}\left(\sqrt{27}-\sqrt{2}\right)$ 40. $2\sqrt{2}\left(\sqrt{8}-\sqrt{3}\right)$

41. $\sqrt{x}\left(\sqrt{3x}-2\right)$ 42. $\sqrt{y}\left(\sqrt{y}+5\right)$

43. $\sqrt{m}\left(\sqrt{8m}+9\right)$ 44. $\sqrt{n}\left(\sqrt{50n}-4\right)$

Multiply and simplify, if possible. See Example 5.

45. $\left(\sqrt{2}+1\right)\left(\sqrt{2}-1\right)$ 46. $\left(\sqrt{3}-1\right)\left(\sqrt{3}+1\right)$

47. $\left(\sqrt{2}-\sqrt{3}\right)\left(\sqrt{3}+\sqrt{5}\right)$ 48. $\left(\sqrt{3}+\sqrt{5}\right)\left(\sqrt{5}-\sqrt{2}\right)$

49. $\left(\sqrt{2x}+3\right)\left(\sqrt{8x}-6\right)$ 50. $\left(\sqrt{5y}-3\right)\left(\sqrt{20y}+6\right)$

51. $\left(2\sqrt{7}-x\right)\left(3\sqrt{2}+x\right)$

52. $\left(4\sqrt{2}-\sqrt{x}\right)\left(\sqrt{x}+2\sqrt{3}\right)$

Find each product and simplify, if possible. See Example 6.

53. $\left(5+\sqrt{3}\right)^2$ 54. $\left(4+\sqrt{2}\right)^2$

55. $\left(a+\sqrt{7}\right)^2$ 56. $\left(t+\sqrt{6}\right)^2$

57. $\left(\sqrt{5}-\sqrt{m}\right)^2$ 58. $\left(\sqrt{3}-\sqrt{v}\right)^2$

59. $\left(\sqrt{6}+\sqrt{3}\right)\left(\sqrt{6}-\sqrt{3}\right)$ 60. $\left(\sqrt{8}+\sqrt{7}\right)\left(\sqrt{8}-\sqrt{7}\right)$

61. $\left(\sqrt{11}-y\right)\left(\sqrt{11}+y\right)$ 62. $\left(\sqrt{2}-s\right)\left(\sqrt{2}+s\right)$

63. $\left(\sqrt{7c}-3\right)\left(\sqrt{7c}+3\right)$ 64. $\left(\sqrt{5r}+4\right)\left(\sqrt{5r}-4\right)$

Simplify. See Example 7.

65. $\dfrac{\sqrt{60}}{\sqrt{6}}$ 66. $\dfrac{\sqrt{115}}{\sqrt{5}}$

67. $\dfrac{\sqrt{a^7}}{\sqrt{a^3}}$ 68. $\dfrac{\sqrt{b^{11}}}{\sqrt{b^5}}$

69. $\dfrac{\sqrt{18x}}{\sqrt{25x}}$ 70. $\dfrac{\sqrt{27y}}{\sqrt{16y}}$

71. $\dfrac{\sqrt{27x}}{\sqrt{75x^3}}$ 72. $\dfrac{\sqrt{50z}}{\sqrt{98z^3}}$

Rationalize each denominator and simplify. See Example 8.

73. $\dfrac{1}{\sqrt{3}}$ 74. $\dfrac{1}{\sqrt{5}}$

75. $\dfrac{4}{\sqrt{19}}$ 76. $\dfrac{9}{\sqrt{2}}$

Rationalize each denominator and simplify. See Example 9.

77. $\sqrt{\dfrac{12}{5}}$

78. $\sqrt{\dfrac{24}{7}}$

79. $\dfrac{6}{\sqrt{27}}$

80. $\dfrac{4}{\sqrt{20}}$

Rationalize each denominator and simplify. See Example 10.

81. $\dfrac{3}{\sqrt{3}-1}$

82. $\dfrac{5}{\sqrt{7}-1}$

83. $\dfrac{5}{\sqrt{3}+\sqrt{2}}$

84. $\dfrac{3}{\sqrt{3}-\sqrt{2}}$

85. $\dfrac{\sqrt{5}+\sqrt{7}}{\sqrt{2}-\sqrt{5}}$

86. $\dfrac{\sqrt{7}-\sqrt{2}}{\sqrt{5}+\sqrt{2}}$

87. $\dfrac{\sqrt{3}-\sqrt{a}}{\sqrt{5}+\sqrt{a}}$

88. $\dfrac{\sqrt{y}-\sqrt{2}}{\sqrt{y}+\sqrt{2}}$

TRY IT YOURSELF

Perform the operations and simplify, if possible, or rationalize the denominator.

89. $\dfrac{3}{\sqrt{32}}$

90. $\dfrac{5}{\sqrt{18}}$

91. $\sqrt{x}\left(\sqrt{14x}+\sqrt{2}\right)$

92. $3\sqrt{z}\left(\sqrt{z}-\sqrt{5}\right)$

93. $\dfrac{\sqrt{3}}{5+\sqrt{x}}$

94. $\dfrac{\sqrt{2}}{3-\sqrt{b}}$

95. $\left(10\sqrt{x}\right)^2$

96. $\left(\sqrt{5y}+2\right)^2$

97. $\sqrt{\dfrac{13}{7}}$

98. $\sqrt{\dfrac{3}{11}}$

99. $\left(3p+\sqrt{5}\right)^2$

100. $\left(6t-\sqrt{3}\right)^2$

101. $\dfrac{8}{\sqrt{7}+2}$

102. $\dfrac{9}{\sqrt{11}+3}$

103. $\sqrt{6}\sqrt{11}$

104. $\sqrt{5}\sqrt{13}$

105. $\dfrac{\sqrt{12x^3}}{\sqrt{27x}}$

106. $\dfrac{\sqrt{32}}{\sqrt{98x^2}}$

107. $\left(\sqrt{15}+\sqrt{2}\right)\left(\sqrt{15}-\sqrt{2}\right)$

108. $\left(\sqrt{14x}+\sqrt{3}\right)\left(\sqrt{14x}-\sqrt{3}\right)$

109. $\dfrac{\sqrt{9}}{\sqrt{2x}}$

110. $\dfrac{\sqrt{4}}{\sqrt{3z}}$

111. $7\sqrt{3}\left(2\sqrt{5}\right)$

112. $3\sqrt{5}\left(4\sqrt{7}\right)$

113. $\dfrac{\sqrt{3}}{7+\sqrt{3}}$

114. $\dfrac{\sqrt{r}-3}{\sqrt{r}+3}$

115. $\sqrt{6n^4}\cdot\sqrt{10n^5}$

116. $\sqrt{5m^2}\cdot\sqrt{10m^5}$

Look Alikes . . .

117. a. $3\sqrt{5}\cdot4\sqrt{5}$ b. $3\sqrt{5}+4\sqrt{5}$

118. a. $\sqrt{32a}\cdot\sqrt{2a}$ b. $\sqrt{32a}-\sqrt{2a}$

119. a. $\left(2\sqrt{x}+2\right)\left(\sqrt{x}-6\right)$ b. $2\sqrt{x}+2+\sqrt{x}-6$

120. a. $\left(2\sqrt{x}\right)^2$ b. $\left(2+\sqrt{x}\right)^2$

APPLICATIONS

121. **from Campus to Careers**
Crime Scene Investigator
The overhead view of a living room below shows where evidence was found during a criminal investigation. Find the area of the rectangle. (*Hint:* Use the grid marks and the Pythagorean theorem to find the length and the width of the rectangle first.)

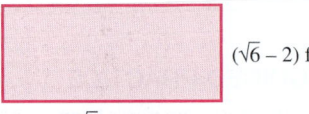
1 ft x 1 ft grid

122. **Geometry.** Find the perimeter and the area of the rectangle.
$(\sqrt{6}-2)$ ft
$(\sqrt{6}+2)$ ft

123. **Tuning Guitars.** The formula $f=0.772\sqrt{\dfrac{T}{u}}$ can be used to calculate the pitch of a string on a Fender Stratocaster electric guitar. In the formula, f is related to the vibration of the string, T represents the tension in the string, and u is related to the thickness of the string. Rewrite the right side of the formula by rationalizing the denominator.

124. **The Perfect Smile.** Throughout history, **golden rectangles** have been considered the most pleasing to the human eye. They can be seen in art, architecture, and nature. The ratio of the length to the width of any golden rectangle is $\dfrac{2}{\sqrt{5}-1}$. Some cosmetic dentists feel the front two teeth should form a golden rectangle. Rationalize the denominator of this ratio.

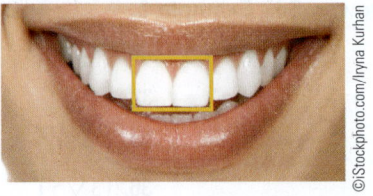

WRITING

125. Explain why each of the following expressions is not in simplified radical form. Then simplify it. Finally, use a calculator to approximate its value. (*Hint:* See page 622.)

	Why isn't it in simplified form?	Simplified form	Approximation
$\dfrac{3}{\sqrt{2}}$			
$\dfrac{\sqrt{18}}{2}$			
$\sqrt{\dfrac{9}{2}}$			

126. Explain why multiplying $\dfrac{9}{\sqrt{5}}$ by $\dfrac{\sqrt{5}}{\sqrt{5}}$ does not change its value.

127. Is the conjugate of $\sqrt{11} - 8$ the same as the opposite of $\sqrt{11} - 8$? Explain.

128. Explain what happens if we try to rationalize $\dfrac{5}{1 + \sqrt{7}}$ using $\dfrac{1 + \sqrt{7}}{1 + \sqrt{7}}$.

REVIEW

129. Hot Dog Eating Champion. On July 4, 2010, Joey Chestnut won Nathan's Annual Contest by eating 54 hot dogs and buns in 10 minutes. At this rate, how many hot dogs and buns would he eat in 30 seconds?

130. Peanut Butter. It takes about 720 peanuts to make 1 pound of peanut butter. How many peanuts would it take to make 12 ounces of peanut butter? (*Hint:* 1 pound = 16 ounces.)

CHALLENGE PROBLEMS

131. Rationalize the denominator of the reciprocal of $\dfrac{\sqrt{y} - 2\sqrt{x}}{2\sqrt{x} + \sqrt{y}}$. All variables represent positive numbers.

132. Rationalize the *numerator:* $\dfrac{\sqrt{7} + 1}{\sqrt{6} - 4}$

SECTION 8.5

Solving Radical Equations

OBJECTIVES

1 Use the squaring property of equality.

2 Solve equations containing one square root.

3 Identify radical equations that have no solutions.

4 Solve radical equations by squaring a binomial.

5 Solve equations containing two square roots.

6 Solve problems modeled by a radical equation.

ARE YOU READY?

The following problems review some basic skills that are needed when solving radical equations.

1. Find: **a.** $\left(\sqrt{x}\right)^2$ **b.** $\left(\sqrt{3x + 1}\right)^2$ **2.** Solve: $2x - 4 = 16$

3. Solve: $x^2 - 6x - 27 = 0$ **4.** Multiply: $(x - 4)^2$

When we solve equations containing fractions, we first clear them of the fractions by multiplying both sides by the LCD. To solve equations containing radicals, the objective is much the same. The first step is to clear them of the radicals. To do this, we will use a new property of equality called the *squaring property*.

1 Use the Squaring Property of Equality.

A **radical equation** is an equation that contains a variable in a radicand. Some examples are

$$\sqrt{x} = 6, \qquad \sqrt{2x - 1} = 9, \qquad \text{and} \qquad \sqrt{x + 12} = 3\sqrt{x + 4}$$

To **solve a radical equation,** we find all the values of the variable that make the equation true. In these examples and throughout this section, we will assume that all radicands represent nonnegative numbers.

The goal when solving a radical equation is to use the *squaring property of equality* to find an equivalent equation that we already know how to solve, such as a linear equation in one variable or a quadratic equation.

Squaring Property of Equality	If two numbers are equal, their squares are equal. For any real numbers a and b, if $a = b$, then $a^2 = b^2$.

When we use the squaring property of equality to solve radical equations, it produces expressions of the form $\left(\sqrt{a}\right)^2$. We have seen that when this expression is simplified, the radical symbol is removed.

The Square of a Square Root	For any nonnegative real number a, $$(\sqrt{a})^2 = a$$

Here are some examples of the square of a square root. Notice how squaring such an expression *removes the square root symbol*.

$$\left(\sqrt{n}\right)^2 = n, \qquad \left(\sqrt{x-3}\right)^2 = x - 3, \qquad \text{and} \qquad \left(\sqrt{5a+8}\right)^2 = 5a + 8$$

EXAMPLE 1   Solve: $\sqrt{x} = 6$

Strategy   We will square both sides of the equation.

Why   Squaring both sides will produce, on the left side, the expression $(\sqrt{x})^2$ that simplifies to x. This step clears the equation of the square root.

Solution

$$\sqrt{x} = 6 \qquad \text{\color{red}This is the equation to solve.}$$

$$\left(\sqrt{x}\right)^2 = (6)^2 \qquad \text{\color{red}Use the squaring property of equality and square both sides.}$$

$$x = 36 \qquad \text{\color{red}Simplify: } (\sqrt{x})^2 = x \text{ and } (6)^2 = 36. \text{ The square root has been removed.}$$

> **The Language of Algebra**
>
> When we square both sides of an equation, we are **raising both sides** to the second power.

We now check the result, 36, in the original equation.

$$\sqrt{x} = 6$$

$$\sqrt{36} \stackrel{?}{=} 6 \qquad \text{\color{red}Substitute 36 for x.}$$

$$6 = 6 \qquad \text{\color{red}True}$$

Since we obtain a true statement, 36 is the solution of $\sqrt{x} = 6$. The solution set is $\{36\}$.

Self Check 1   Solve: $\sqrt{y} = 10$

Now Try ▶   Problem 11

CAUTION   If we square both sides of an equation, *the resulting equation may not have the same solutions as the original one.* For example, consider

$$x = 2$$

The only solution of this equation is 2. However, if we square both sides, we obtain $(x)^2 = (2)^2$, which simplifies to

$$x^2 = 4$$

This new equation has solutions 2 and -2, because $2^2 = 4$ and $(-2)^2 = 4$.

The equations $x = 2$ and $x^2 = 4$ are not equivalent equations because they do not have the same solutions. The solution -2 satisfies $x^2 = 4$ but it does not satisfy $x = 2$. We see that squaring both sides of an equation can produce an equation with solutions that don't satisfy the original one. Therefore, **we must check each possible solution in the original equation.**

2 Solve Equations Containing One Square Root.

To solve a radical equation containing one square root, we follow these steps.

Strategy for Solving Radical Equations Containing Square Roots	1. Isolate a radical term on one side of the equation.
	2. Square both sides of the equation.
	3. Solve the resulting equation.
	4. Check the possible solutions in the original equation. This step is required.

EXAMPLE 2   Solve: $\sqrt{2x - 1} = 9$

Strategy   Since the radical is already isolated on one side, we can move to Step 2 of the strategy for solving radical equations and square both sides of the equation.

Why   Squaring both sides will produce, on the left side, the expression $\left(\sqrt{2x - 1}\right)^2$ that simplifies to $2x - 1$. This step clears the equation of the radical.

Solution

$$\sqrt{2x - 1} = 9 \qquad \text{This is the equation to solve.}$$

$$\left(\sqrt{2x - 1}\right)^2 = (9)^2 \qquad \text{Use the squaring property of equality and square both sides.}$$

$$2x - 1 = 81 \qquad \text{Simplify: } \left(\sqrt{2x - 1}\right)^2 = 2x - 1 \text{ and } (9)^2 = 81. \text{ The radical has been removed. The result is a linear equation in one variable.}$$

$$2x = 82 \qquad \text{To solve the resulting equation, add 1 to both sides.}$$

$$x = 41 \qquad \text{To isolate x, divide both sides by 2.}$$

The Language of Algebra

Possible solutions also are called **potential** or **proposed** solutions.

Check the possible solution, 41, in the original equation.

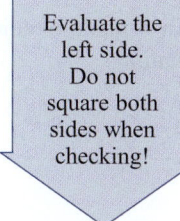

Evaluate the left side. Do not square both sides when checking!

$$\sqrt{2x - 1} = 9 \qquad \text{This is the original equation.}$$

$$\sqrt{2(41) - 1} \stackrel{?}{=} 9 \qquad \text{Substitute 41 for x.}$$

$$\sqrt{82 - 1} \stackrel{?}{=} 9 \qquad \text{Do the multiplication within the radical.}$$

$$\sqrt{81} \stackrel{?}{=} 9 \qquad \text{Do the subtraction within the radical.}$$

$$9 = 9 \qquad \text{True}$$

The solution of $\sqrt{2x - 1} = 9$ is 41. The solution set is written $\{41\}$.

Self Check 2   Solve: $\sqrt{3x - 5} = 2$

Now Try ▶ Problem 19

EXAMPLE 3   Solve: $a = \sqrt{a^2 + 3a - 3}$

Strategy   Since the radical is isolated on the right side, we will square both sides.

Why   Squaring both sides will produce, on the right side, the expression $\left(\sqrt{a^2 + 3a - 3}\right)^2$ that simplifies to $a^2 + 3a - 3$. This step clears the equation of the radical.

Solution

$$a = \sqrt{a^2 + 3a - 3} \qquad \text{This is the equation to solve.}$$

$$(a)^2 = \left(\sqrt{a^2 + 3a - 3}\right)^2 \qquad \text{Square both sides.}$$

$$a^2 = a^2 + 3a - 3 \qquad \text{Simplify: } \left(\sqrt{a^2 + 3a - 3}\right)^2 = a^2 + 3a - 3. \text{ The radical has been removed.}$$

$$a^2 - a^2 = a^2 + 3a - 3 - a^2 \qquad \text{To solve the resulting equation, eliminate } a^2 \text{ from the left side by subtracting } a^2 \text{ from both sides.}$$

$$0 = 3a - 3 \qquad \text{Combine like terms: } a^2 - a^2 = 0. \text{ Notice that the } a^2 \text{ terms drop out on both sides of the equation. The result is a linear equation, not a quadratic equation.}$$

$$3 = 3a \qquad \text{To isolate the variable term, 3a, add 3 to both sides.}$$

$$1 = a \qquad \text{To isolate a, divide both sides by 3.}$$

Check:

$$a = \sqrt{a^2 + 3a - 3}$$ This is the original equation.

$$1 \stackrel{?}{=} \sqrt{(1)^2 + 3(1) - 3}$$ Substitute 1 for a.

$$1 \stackrel{?}{=} \sqrt{1 + 3 - 3}$$ Evaluate the expression within the radical.

$$1 \stackrel{?}{=} \sqrt{1}$$ Do the addition and subtraction within the radical to get 1.

$$1 = 1$$ True

The solution of $a = \sqrt{a^2 + 3a - 3}$ is 1. The solution set is $\{1\}$.

Self Check 3   Solve: $b = \sqrt{b^2 - 2b + 24}$

Now Try ▶ Problem 27

---

3 **Identify Radical Equations That Have No Solutions.**

We will now consider radical equations in which the radical term containing a variable is not by itself (isolated) on one side of the equation.

EXAMPLE 4   Solve: $\sqrt{x + 2} + 5 = 3$

Strategy   Since 5 is outside the square root symbol, there are two terms on the left side of the equation. To isolate the radical, we will subtract 5 from both sides.

Why   This will put the equation in a form in which we can square both sides to clear the radical.

Solution

$$\sqrt{x + 2} + 5 = 3$$ This is the equation to solve.

$$\sqrt{x + 2} + 5 - 5 = 3 - 5$$ To isolate the radical, subtract 5 from both sides.

$$\sqrt{x + 2} = -2$$ The square root term is by itself on one side of the equation.

We now square both sides to clear the equation of the radical.

$$\left(\sqrt{x + 2}\right)^2 = (-2)^2$$ Square both sides.

$$x + 2 = 4$$ Simplify: $\left(\sqrt{x + 2}\right)^2 = x + 2$ and $(-2)^2 = 4$.

$$x = 2$$ To solve the resulting equation, subtract 2 from both sides.

> **Success Tip**
>
> It is apparent that $\sqrt{x + 2} = -2$ has no solution. Since $\sqrt{x + 2}$ is the principal square root, and cannot be negative, there is no real number x that could make $\sqrt{x + 2}$ equal to a negative number.

Check:

$$\sqrt{x + 2} + 5 = 3$$ This is the original equation.

$$\sqrt{2 + 2} + 5 \stackrel{?}{=} 3$$ Substitute 2 for x.

$$\sqrt{4} + 5 \stackrel{?}{=} 3$$ Do the addition within the radical.

$$2 + 5 \stackrel{?}{=} 3$$ Find the square root.

$$7 = 3$$ False

Since the result is a false statement, 2 is not a solution of the original equation and must be discarded. The original equation, $\sqrt{x + 2} + 5 = 3$, has *no solution*. The solution set is the empty set, written as $\varnothing$ or $\{ \ \}$.

Self Check 4   Solve: $\sqrt{x - 2} + 8 = 1$

Now Try ▶ Problem 39

Example 4 shows that squaring both sides of an equation can lead to possible solutions that do not satisfy the original equation. We call such numbers **extraneous solutions.** In that example, we found that 2 is an extraneous solution of $\sqrt{x+2} + 5 = 3$.

4 Solve Radical Equations by Squaring a Binomial.

Sometimes, after clearing an equation of a radical, the result is a quadratic equation.

EXAMPLE 5   Solve: $\sqrt{3-x} - x = -3$

Strategy   Since $-x$ is outside the radical, there are two terms on the left side of the equation. To isolate the radical, we will add x to both sides.

Why   This will put the equation in a form in which we can square both sides to clear the radical.

Solution

$$\sqrt{3-x} - x = -3 \qquad \text{This is the equation to solve.}$$

$$\sqrt{3-x} - x + x = x - 3 \qquad \begin{array}{l}\text{Add } x \text{ to both sides. On the right side, it is}\\ \text{best to add the } x\text{-term in front of } -3.\end{array}$$

$$\sqrt{3-x} = x - 3 \qquad \text{The radical is now isolated.}$$

We then square both sides to clear the equation of the radical.

$$\left(\sqrt{3-x}\right)^2 = (x-3)^2 \qquad \text{Square both sides.}$$

$$3 - x = x^2 - 6x + 9 \qquad \begin{array}{l}\text{Simplify: } \left(\sqrt{3-x}\right)^2 = 3 - x. \text{ On the right, use a special-}\\ \text{product formula to find } (x-3)^2. \text{ Don't forget the middle}\\ \text{term, } -6x. \text{ This is a quadratic equation.}\end{array}$$

To solve the resulting quadratic equation, we write it in standard form so that the left side is 0.

$$3 = x^2 - 5x + 9 \qquad \text{To eliminate } x \text{ on the left side, add } x \text{ to both sides.}$$

$$0 = x^2 - 5x + 6 \qquad \text{To get 0 on the left side, subtract 3 from both sides.}$$

$$0 = (x-3)(x-2) \qquad \text{Factor the trinomial } x^2 - 5x + 6.$$

$$x - 3 = 0 \quad \text{or} \quad x - 2 = 0 \qquad \text{Set each factor equal to 0.}$$

$$x = 3 \qquad\qquad x = 2 \qquad \text{Solve each equation.}$$

Success Tip

Even if you are absolutely certain that no algebraic mistakes were made when solving a radical equation, you must still check your solutions. Squaring both sides can introduce extraneous solutions that must be discarded.

Since there are two possible solutions, we check each one in the original equation.

The check for 3:

$$\sqrt{3-x} - x = -3$$
$$\sqrt{3-3} - 3 \overset{?}{=} -3$$
$$\sqrt{0} - 3 \overset{?}{=} -3$$
$$0 - 3 \overset{?}{=} -3$$
$$-3 = -3 \qquad \text{True}$$

The check for 2:

$$\sqrt{3-x} - x = -3$$
$$\sqrt{3-2} - 2 \overset{?}{=} -3$$
$$\sqrt{1} - 2 \overset{?}{=} -3$$
$$1 - 2 \overset{?}{=} -3$$
$$-1 = -3 \qquad \text{False}$$

Since a true statement results when 3 is substituted for x, 3 is a solution. Since a false statement results when 2 is substituted for x, 2 is an extraneous solution. The solution set is $\{3\}$.

Self Check 5   Solve: $\sqrt{x+4} - x = -2$

Now Try ▶ Problem 43

5 Solve Equations Containing Two Square Roots.

To solve an equation containing two square roots, we want to have one radical on the left side and one radical on the right side.

EXAMPLE 6   Solve: $\sqrt{x + 12} = 3\sqrt{x + 4}$

Strategy   We will square both sides to clear the equation of both radicals.

Why   We can square both sides immediately since each square root term is by itself on one side of the equation. This step will clear the equation of both radicals.

Solution

> **Caution**
>
> Recall that to raise a product to a power, we raise each factor of the product to that power:
>
> $$(xy)^n = x^n y^n$$
>
> Use this rule for exponents to simplify $\left(3\sqrt{x + 4}\right)^2$. A common error is to forget to square 3.

$$\sqrt{x + 12} = 3\sqrt{x + 4} \qquad \text{This is the equation to solve.}$$

$$\left(\sqrt{x + 12}\right)^2 = \left(3\sqrt{x + 4}\right)^2 \qquad \text{Square both sides.}$$

$$x + 12 = 3^2\left(\sqrt{x + 4}\right)^2 \qquad \text{On the left, simplify: } \left(\sqrt{x + 12}\right)^2 = x + 12. \text{ On the right, raise each factor of the product } 3\sqrt{x + 4} \text{ to the second power. Don't forget to square 3.}$$

$$x + 12 = 9(x + 4) \qquad \text{Simplify: } 3^2 = 9 \text{ and } \left(\sqrt{x + 4}\right)^2 = x + 4. \text{ The square roots have been removed. This is a linear equation.}$$

$$x + 12 = 9x + 36 \qquad \text{To solve the resulting equation, first distribute the multiplication by 9.}$$

$$12 = 8x + 36 \qquad \text{To eliminate } x \text{ on the left side, subtract } x \text{ from both sides.}$$

$$-24 = 8x \qquad \text{To isolate the variable term, } 8x, \text{ subtract 36 from both sides.}$$

$$-3 = x \qquad \text{To isolate } x, \text{ divide both sides by 8.}$$

Check:

$$\sqrt{x + 12} = 3\sqrt{x + 4} \qquad \text{This is the original equation.}$$

$$\sqrt{-3 + 12} \stackrel{?}{=} 3\sqrt{-3 + 4} \qquad \text{Substitute } -3 \text{ for } x.$$

$$\sqrt{9} \stackrel{?}{=} 3\sqrt{1} \qquad \text{Evaluate the expressions within the radicals.}$$

$$3 \stackrel{?}{=} 3 \cdot 1 \qquad \text{Find each square root.}$$

$$3 = 3 \qquad \text{True}$$

The solution of $\sqrt{x + 12} = 3\sqrt{x + 4}$ is -3. The solution set is $\{-3\}$.

Self Check 6   Solve: $\sqrt{x - 4} = 2\sqrt{x - 16}$

Now Try ▶   Problem 55

6 Solve Problems Modeled by a Radical Equation.

Radical equations can be used to model many real-world situations.

EXAMPLE 7   **Bridges.**   The time t in seconds that it takes an object to fall d feet is given by the formula $t = \frac{\sqrt{d}}{4}$. To find the height of the Benjamin Franklin Bridge in Philadelphia, a man stands in the center section and drops a coin into the water. If it takes the coin 3 seconds to hit the water, how high above the water is the bridge?

Strategy   We will substitute 3 for t in the formula and solve for d.

Why   Since the coin fell for 3 seconds, $t = 3$. The height of the bridge above the water is the same as the distance d that the coin fell.

Solution	$t = \dfrac{\sqrt{d}}{4}$	This is the equation that models the situation.
	$3 = \dfrac{\sqrt{d}}{4}$	Substitute 3 for t, the time in seconds.
	$12 = \sqrt{d}$	To clear the equation of the fraction, multiply both sides by 4.
	$(12)^2 = \left(\sqrt{d}\right)^2$	To clear the equation of the radical, square both sides.
	$144 = d$	Simplify each side.

The bridge is 144 feet above the water. Check this result in the original equation.

Self Check 7 **Wells.** To find the depth of a well, a farmer dropped a rock down it. If the rock fell straight downward for 2 seconds before he heard a splash, how deep was the water level?

Now Try ▶ Problem 91

SECTION 8.5 ▷ STUDY SET

VOCABULARY

Fill in the blanks.

1. $\sqrt{x + 1} = 3$ and $\sqrt{x} - 2 = 10$ are examples of _____ equations.

2. To _____ the radical in $\sqrt{x} + 1 = 6$ means to get $\sqrt{x}$ by itself on one side of the equation.

3. To _____ a radical equation, we find all the values of the variable that make the equation true.

4. Squaring both sides of an equation can lead to possible solutions that do not satisfy the original equation. We call such numbers _____ solutions.

CONCEPTS

Fill in the blanks.

5. The squaring property of equality states that if two numbers are equal, their _____ are equal. If $a = b$, then $a^2 = $ ▢ .

6. To clear $\sqrt{x + 1} = 5$ of the radical, we _____ both sides of the equation.

7. Use properties of algebra to isolate each radical term on one side of the equation. DO NOT SOLVE THE EQUATION.
   a. $\sqrt{x - 4} - 1 = 2$ b. $8 = \sqrt{x} - x$

8. Simplify each expression.
   a. $\left(\sqrt{x}\right)^2$ b. $\left(\sqrt{2x + 3}\right)^2$
   c. $\left(4\sqrt{x}\right)^2$ d. $\left(\sqrt{n^2 - 2n + 6}\right)^2$

NOTATION

Complete each solution.

9. Solve: $\sqrt{x - 3} = 5$

   $\left(\sqrt{x - 3}\right)^{▢} = 5^{▢}$

   ▢ $= 25$

   $x = $ ▢

10. Check to determine whether -1 is a solution of the equation.

$$2\sqrt{3x + 4} = \sqrt{5x + 9}$$

$$2\sqrt{3(\;\;)} + 4 \overset{?}{=} \sqrt{5(\;\;)} + 9$$

$$2\sqrt{\;\;\;\; + 4} \overset{?}{=} \sqrt{\;\;\;\; + 9}$$

$$2\sqrt{1} \overset{?}{=} \sqrt{\;\;\;\;}$$

$$2 \cdot \;\;▢\;\; \overset{?}{=} 2$$

$$▢ = 2$$

Thus, ▢ is the solution of $2\sqrt{3x + 4} = \sqrt{5x + 9}$.

GUIDED PRACTICE

Solve each equation. See Example 1.

11. $\sqrt{x} = 3$ 12. $\sqrt{x} = 5$

13. $\sqrt{y} = 12$ 14. $\sqrt{y} = 20$

15. $\sqrt{2a} = 4$ 16. $\sqrt{3a} = 9$

17. $\sqrt{4n} = 6$ 18. $\sqrt{4n} = 8$

Solve each equation. See Example 2.

19. $\sqrt{x + 3} = 2$ 20. $\sqrt{x + 2} = 3$

21. $\sqrt{5 - T} = 10$ 22. $\sqrt{3 - T} = 2$

23. $\sqrt{6x + 19} = 7$ 24. $\sqrt{2x + 6} = 4$

25. $\sqrt{5x - 5} = 5$ 26. $\sqrt{4x - 7} = 7$

Solve each equation. **See Example 3.**

27. $x = \sqrt{x^2 - 2x + 16}$ **28.** $x = \sqrt{x^2 + 3x - 21}$

29. $c = \sqrt{c^2 - 3c + 39}$ **30.** $2s = \sqrt{4s^2 - 2s + 10}$

31. $\sqrt{9q^2 - 5q + 10} = 3q$ **32.** $\sqrt{m^2 - 7m + 21} = m$

33. $\sqrt{4m^2 + 6m + 6} = -2m$ **34.** $\sqrt{9t^2 + 4t + 20} = -3t$

Solve each equation. **See Example 4.**

35. $\sqrt{x} = -6$ **36.** $\sqrt{x} = -7$

37. $\sqrt{r} + 4 = 0$ **38.** $\sqrt{r} + 1 = 0$

39. $\sqrt{2x + 7} + 4 = 1$ **40.** $\sqrt{15a + 3} + 5 = 4$

41. $\sqrt{6 - 2b} - 7 = -9$ **42.** $\sqrt{3 - 12t} - 2 = -3$

Solve each equation. **See Example 5.**

43. $\sqrt{3a + 7} - a = 3$ **44.** $-1 = \sqrt{2a - 2} - a$

45. $b = \sqrt{5b + 1} - 1$ **46.** $1 = \sqrt{t + 1} - t$

47. $y = 9 + \sqrt{y - 3}$ **48.** $m = \sqrt{m - 7} + 9$

49. $\sqrt{15 - 3t} + 5 = t$ **50.** $\sqrt{1 - 8s} - s = 4$

Solve each equation. **See Example 6.**

51. $\sqrt{10 - 3x} = \sqrt{2x + 20}$ **52.** $\sqrt{1 - 2x} = \sqrt{x + 10}$

53. $\sqrt{3c - 8} = \sqrt{c}$ **54.** $\sqrt{2x} = \sqrt{x + 8}$

55. $5\sqrt{a} = \sqrt{10a + 15}$ **56.** $\sqrt{17m - 4} = 4\sqrt{m}$

57. $2\sqrt{3x + 4} = \sqrt{5x + 9}$ **58.** $\sqrt{3x + 6} = 2\sqrt{2x - 11}$

Solve each equation.

59. $5\sqrt{x} - 11 = 9$ **60.** $5\sqrt{x} - 6 = 4$

61. $x = \sqrt{x^2 - 15} + 3$ **62.** $v = \sqrt{v^2 - 16} + 2$

63. $\sqrt{24 + 10n} - n = 4$ **64.** $\sqrt{7 + 6y} - y = 2$

65. $\sqrt{3t - 9} = \sqrt{t + 1}$ **66.** $\sqrt{a - 3} = \sqrt{2a - 8}$

67. $\sqrt{a} + 2 = 9$ **68.** $\sqrt{a} - 1 = 0$

69. $-2 = 2\sqrt{x} - 12$ **70.** $-1 = 2\sqrt{x} - 7$

71. $1 = \sqrt{x^2 + x + 4} - x$ **72.** $-1 = \sqrt{x^2 - 4x + 9} - x$

73. $\sqrt{9y} = 2y + 1$ **74.** $\sqrt{d + 1} = d + 1$

75. $\frac{1}{2}\sqrt{4t^2 + 2t + 20} = t$ **76.** $\frac{1}{3}\sqrt{9x^2 + 4x + 4} = x$

77. $-\sqrt{x} = -2$ **78.** $-\sqrt{x} = -12$

79. $10 - \sqrt{s} = 7$ **80.** $-4 = 6 - \sqrt{x}$

81. $\sqrt{4x - 2} = \sqrt{3x + 5}$ **82.** $2\sqrt{5y - 1} = \sqrt{2y + 14}$

83. $\sqrt{x + 1} = 7$ **84.** $\sqrt{x - 2} = 8$

85. $b = \sqrt{2b - 2} + 1$ **86.** $c = \sqrt{5c + 1} - 1$

87. $6 + \sqrt{m} - m = 0$ **88.** $6 + \sqrt{3m} - m = 0$

89. $\sqrt{2a^2 - 3a - 4} = a$ **90.** $\sqrt{2b^2 - 5b - 36} = b$

APPLICATIONS

91. Niagara Falls.   The time t in seconds that it takes an object to fall d feet is given by the formula $t = \frac{\sqrt{d}}{4}$. The time it took a stuntman to go over the Niagara Falls in a barrel was 3.25 seconds. Find the height of the waterfall.

92. Monuments.   Gabby Street, a professional baseball player of the 1920s, was known for once catching a ball dropped from the top of the Washington Monument in Washington, D.C. If the ball fell for slightly less than 6 seconds before it was caught, find the approximate height of the monument. (See Problem 91.)

93. Pendulums.   The time t (in seconds) required for a pendulum of length L feet to swing through one back-and-forth cycle, called its period, is given by the formula $t = 1.11\sqrt{L}$. The Foucault pendulum in Chicago's Museum of Science and Industry is used to demonstrate the rotation of the Earth. It completes one cycle in 8.91 seconds. To the nearest tenth of a foot, how long is the pendulum?

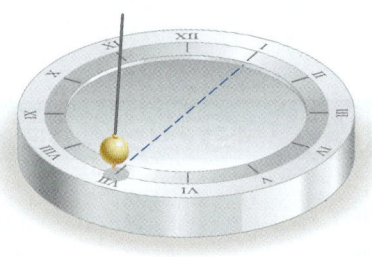

94. Power Usage.   The current I (in amperes), the resistance R (in ohms), and the power P (in watts) are related by the formula $I = \sqrt{\frac{P}{R}}$. Find the power (to the nearest watt) used by a space heater that draws 7 amps when the resistance is 10.2 ohms.

95. ▶   *from* **Campus to Careers**

Crime Scene Investigator

Investigators often do tests on guns used in the commission of a crime. The formula $v = 891\sqrt{h}$ gives the velocity v (in ft/sec) of a certain size bullet fired into a block of wood. The variable h represents the height in feet that the block is raised after the collision. If a gun is known to fire such a bullet at 1,100 ft/sec, how high will the block be raised in a test?

©Peter Coombs/Alamy

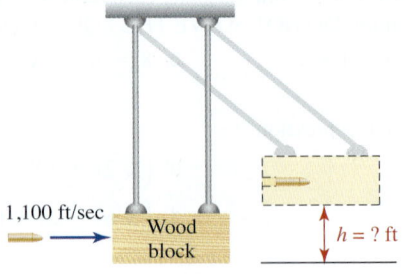

96. Road Safety. The formula $s = k\sqrt{d}$ relates the speed s (in mph) of a car and the distance d (in feet) of the skid when a driver hits the brakes. On wet pavement, $k = 3.24$. How far will a car skid if it is going 55 mph? Round to the nearest foot.

97. Highway Design. A highway curve banked at 8° will accommodate traffic traveling at speed s (in mph) if the radius of the curve is r (feet), according to the equation $s = 1.45\sqrt{r}$. If highway engineers expect traffic to travel at 65 mph, what radius should they specify (to the nearest foot)?

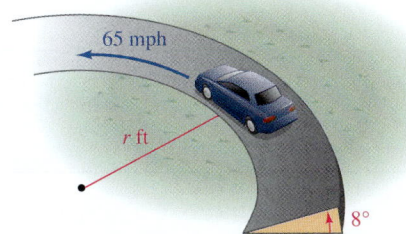

98. Geometry. The radius of a cone with volume V and height h is given by the formula $r = \sqrt{\dfrac{3V}{\pi h}}$. Solve the equation for V.

WRITING

99. Explain why a check is necessary when solving radical equations.

100. How would you know, without solving it, that the equation $\sqrt{x - 9} = -4$ has no solutions?

101. Explain the error in the following partial solution.

Solve: $\sqrt{1 - 4y} = y + 2$
$$\left(\sqrt{1 - 4y}\right)^2 = (y + 2)^2$$
$$1 - 4y = y^2 + 4$$

102. Explain the error in the following partial solution.

Solve: $\sqrt{a + 2} - 5 = 4$
$$\left(\sqrt{a + 2} - 5\right)^2 = (4)^2$$

REVIEW

Solve each equation.

103. $\dfrac{1}{2} + \dfrac{x}{5} = \dfrac{3}{4}$

104. $\dfrac{1}{3} + \dfrac{c}{5} = -\dfrac{3}{2}$

105. $\dfrac{2}{5}x + 1 = \dfrac{1}{3} + x$

106. $\dfrac{2}{3}y + 2 = \dfrac{1}{5} + y$

CHALLENGE PROBLEMS

Solve each equation. (Hint: You have to square both sides of the equation twice.)

107. $\sqrt{n} + 3 = \sqrt{n + 21}$

108. $\sqrt{m} + 6 = \sqrt{m + 72}$

109. $\sqrt{1 - t} + \sqrt{t + 9} = 4$

110. $2\sqrt{x + 9} - 3 = \sqrt{4x + 3}$

111. $\sqrt{\sqrt{x + 2}} - 3 = 0$

112. $\sqrt{\sqrt{x - 4}} - 2 = 0$

SECTION 8.6

Higher-Order Roots and Rational Exponents

OBJECTIVES

1 Find cube roots of perfect cubes.

2 Find higher-order roots.

3 Simplify higher-order roots.

4 Evaluate expressions of the form $a^{1/n}$.

5 Evaluate expressions of the form $a^{m/n}$.

ARE YOU READY?

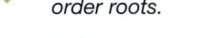

The following problems review some basic skills that are needed when working with higher-order roots.

1. Evaluate: **a.** 5^3
  b. 3^5

2. Evaluate: **a.** $(-4)^3$
  b. $(-5)^3$

3. Express using a positive exponent: x^{-3} **4.** Evaluate: $\sqrt{9}$

In this section, we will consider higher-order roots such as cube roots and fourth roots. We will also extend the definition of exponents to include rational (fractional) exponents.

1 Find Cube Roots of Perfect Cubes.

When we raise a number to the third power, we are cubing it, or finding its **cube.** We can reverse the cubing process to find **cube roots** of numbers. To find the cube root of 8, we ask, "What number, when cubed, is equal to 8?" It follows that 2 is a cube root of 8, because $2^3 = 8$.

In general, we have this definition.

The Definition of Cube Root	The number b is a **cube root** of the number a if $b^3 = a$.

All real numbers have one real-number cube root. A positive number has a positive cube root, a negative number has a negative cube root, and the cube root of 0 is 0.

Cube Root Notation	The **cube root of a** is written as $\sqrt[3]{a}$. By definition, $$\sqrt[3]{a} = b \quad \text{if} \quad b^3 = a$$

Earlier, we determined that the cube root of 8 is 2. In symbols, we can write: $\sqrt[3]{8} = 2$. The number 3 is called the **index**, 8 is called the **radicand**, and the entire expression is called a **radical.**

Index → $\sqrt[3]{8}$ ← Radicand   Read as "the cube root of 8."

Radical

To evaluate cube roots, you need to quickly recognize each of the following **perfect cubes** shown in blue.

$$1 = 1^3 \qquad 27 = 3^3 \qquad 125 = 5^3 \qquad 343 = 7^3 \qquad 729 = 9^3$$
$$8 = 2^3 \qquad 64 = 4^3 \qquad 216 = 6^3 \qquad 512 = 8^3 \qquad 1{,}000 = 10^3$$

EXAMPLE 1   Evaluate each cube root: **a.** $\sqrt[3]{27}$ **b.** $\sqrt[3]{-64}$ **c.** $-\sqrt[3]{125}$

Strategy   In each case, we will determine what number, when cubed, produces the radicand.

Why   The symbol $\sqrt[3]{\ }$ indicates that the cube root of the number written under it should be found.

Solution   **a.** $\sqrt[3]{27} = 3$ Ask: What number, when cubed, is 27? The answer is 3 because $3^3 = 27$.

b. $\sqrt[3]{-64} = -4$ Read as "the cube root of -64." Ask: What number, when cubed, is -64? The answer is -4 because $(-4)^3 = -64$.

We have seen that the *square root* of a negative number is not a real number. In this example, we see that the *cube root* of a negative number is a real number.

c. $-\sqrt[3]{125}$ is read as "the opposite of the cube root of 125." Since $\sqrt[3]{125} = 5$, we have

$$-\sqrt[3]{125} = -5$$ Because $-\sqrt[3]{125} = -1 \cdot \sqrt[3]{125} = -1 \cdot 5 = -5$.

Self Check 1   Evaluate each cube root: **a.** $\sqrt[3]{1}$ **b.** $\sqrt[3]{-8}$ **c.** $-\sqrt[3]{216}$

Now Try ▶ Problems 13, 17, and 23

2 Find Higher-Order Roots.

Just as there are square roots and cube roots, there are also fourth roots, fifth roots, and so on. In general, we have the following definition.

*n*th Roots of *a*	The *n*th root of *a* is written as $\sqrt[n]{a}$, and
	$$\sqrt[n]{a} = b \quad \text{if} \quad b^n = a$$

In the previous definition, *n* is called the **index** of the radical. If *n* is an odd natural number, $\sqrt[n]{a}$ represents an **odd root**. If *n* is even, and if *a* is nonnegative, $\sqrt[n]{a}$ represents an **even root**.

The Language of Algebra

The plural of the word *index* is indices.

When the index *n* of $\sqrt[n]{a}$ is an even natural number (square root, fourth root, sixth root, and so on), the radicand *a* must be positive or zero for the root to be a real number. Like a square root symbol, the symbol $\sqrt[4]{}$ represents the **positive** or **principal fourth root**. The symbol $-\sqrt[4]{}$ represents a negative root.

To evaluate fourth and fifth roots, you need to quickly recognize each of the following:

Perfect fourth powers		*Perfect fifth powers*	
$1 = 1^4$	$256 = 4^4$	$1 = 1^5$	$1,024 = 4^5$
$16 = 2^4$	$625 = 5^4$	$32 = 2^5$	
$81 = 3^4$		$243 = 3^5$	

EXAMPLE 2   Evaluate each root: **a.** $\sqrt[4]{81}$ **b.** $\sqrt[5]{32}$ **c.** $\sqrt[5]{-1}$ **d.** $\sqrt[4]{-16}$

Strategy   In each case, we will determine what number, when raised to the fourth (or fifth) power, produces the radicand.

Why   The symbols $\sqrt[4]{}$ and $\sqrt[5]{}$ indicate that the fourth or fifth root of the number written under it should be found.

Solution

a. $\sqrt[4]{81} = 3$ Read as "the fourth root of 81." Since the index is even, ask: What positive number, raised to the 4th power, is 81? The answer is 3 because $3^4 = 81$.

b. $\sqrt[5]{32} = 2$ Read as "the fifth root of 32." Since the index is odd, ask: What number, raised to the 5th power, is 32? The answer is 2 because $2^5 = 32$.

c. $\sqrt[5]{-1} = -1$ Read as "the fifth root of −1." Ask: What number, raised to the 5th power, is −1? The answer is −1 because $(-1)^5 = -1$.

d. $\sqrt[4]{-16}$ is not a real number because no real number raised to the fourth power is -16.

Self Check 2   Evaluate each root: **a.** $\sqrt[4]{16}$ **b.** $\sqrt[5]{-1,024}$
c. $\sqrt[4]{-1}$

Now Try   Problems 31, 33, and 35

We can find even roots of expressions that contain variables, provided these variable expressions do not represent negative numbers. When finding fourth roots of variable expressions, it is helpful to memorize the following forms:

Perfect fourth powers:

$$x^4 = (x^1)^4, \quad x^8 = (x^2)^4, \quad x^{12} = (x^3)^4, \quad x^{16} = (x^4)^4, \quad \text{and} \quad x^{20} = (x^5)^4$$

EXAMPLE 3   Find each root. Assume that each variable represents a nonnegative number: **a.** $\sqrt[4]{m^4}$
b. $\sqrt[4]{16x^{12}}$

Strategy   In each case, we will determine what positive expression, when raised to the fourth power produces the radicand.

Why   The symbol $\sqrt[4]{}$ indicates that the positive fourth root of the expression written under it should be found.

Solution   **a.** $\sqrt[4]{m^4} = m$ ⟵ Read as "the fourth root of m^4." Ask: What expression, raised to the 4th power, is m^4? The answer is m because $(m)^4 = m^4$.

b. $\sqrt[4]{16x^{12}} = 2x^3$ ⟵ Ask: What expression, raised to the 4th power, is $16x^{12}$? The answer is $2x^3$ because $(2x^3)^4 = 16x^{12}$.

Self Check 3   Find each root. Each variable represents a nonnegative number:
a. $\sqrt[4]{n^8}$ **b.** $\sqrt[4]{81x^{16}}$

Now Try ▶ Problems 41 and 45

We can also find odd roots of expressions that contain variables. When finding cube and fifth roots of variable expressions, it is helpful to memorize the following forms:

Perfect cubes:
$$x^3 = (x^1)^3, \quad x^6 = (x^2)^3, \quad x^9 = (x^3)^3, \quad x^{12} = (x^4)^3, \quad \text{and} \quad x^{15} = (x^5)^3$$

Perfect fifth powers:
$$x^5 = (x^1)^5, \quad x^{10} = (x^2)^5, \quad x^{15} = (x^3)^5, \quad x^{20} = (x^4)^5, \quad \text{and} \quad x^{25} = (x^5)^5$$

EXAMPLE 4   Find each root: **a.** $\sqrt[3]{y^3}$ **b.** $\sqrt[3]{-64x^9}$ **c.** $\sqrt[5]{x^{10}}$

Strategy   In each case, we will determine what expression, when raised to the third (or fifth) power produces the radicand.

Why   The symbols $\sqrt[3]{}$ and $\sqrt[5]{}$ indicate that the third (or fifth) root of the expression written under it should be found.

Solution   **a.** $\sqrt[3]{y^3} = y$ ⟵ Read as "the cube root of y cubed." Ask: What expression, when cubed, is y^3? The answer is y because $(y)^3 = y^3$.

b. $\sqrt[3]{-64x^9} = -4x^3$ ⟵ Ask: What expression, when cubed, is $-64x^9$? The answer is $-4x^3$ because $(-4x^3)^3 = -64x^9$.

c. $\sqrt[5]{x^{10}} = x^2$ ⟵ Read as "the fifth root of x to the tenth power." Ask: What expression, raised to the 5th power, is x^{10}? The answer is x^2 because $(x^2)^5 = x^{10}$.

Self Check 4   Find each root: **a.** $\sqrt[3]{s^{15}}$ **b.** $\sqrt[3]{-1,000b^3}$ **c.** $\sqrt[5]{x^{15}}$

Now Try ▶ Problems 49 and 51

3   Simplify Higher-Order Roots.

The product and quotient rules for square roots can be generalized to apply to higher-order roots.

The Product and Quotient Rules for Radicals	If $\sqrt[n]{a}$ and $\sqrt[n]{b}$ are real numbers,

$$\sqrt[n]{a \cdot b} = \sqrt[n]{a}\sqrt[n]{b} \qquad \sqrt[n]{\frac{a}{b}} = \frac{\sqrt[n]{a}}{\sqrt[n]{b}} \qquad (b \neq 0)$$

Read as "the nth root of a times b" and "the nth root of a divided by b."

EXAMPLE 5 Simplify: **a.** $\sqrt[3]{54}$ **b.** $\sqrt[4]{405}$ **c.** $\sqrt[3]{-\dfrac{32}{125}}$

Strategy To simplify means to remove all perfect cube (or perfect fourth power) factors from the radicands. To do so, we will factor each radicand, use the product rule for radicals, and then simplify.

Why Factoring the radicand in this way leads to a cube root of a perfect cube (or a fourth root of a perfect fourth power) that we can evaluate easily.

Solution **a.** From the list of perfect cubes on page 656, we see that 27 is the greatest perfect cube factor of the radicand 54.

$$\sqrt[3]{54} = \sqrt[3]{27 \cdot 2}$$ Ask: How can 54 be factored using a perfect cube and one other factor? The answer is 27 · 2. Write the perfect-cube factor first.

$$= \sqrt[3]{27}\sqrt[3]{2}$$ The cube root of a product is equal to the product of the cube roots.

$$= 3\sqrt[3]{2}$$ Find the cube root of the perfect cube: $\sqrt[3]{27} = 3$.

We say that $3\sqrt[3]{2}$ is the simplified form of $\sqrt[3]{54}$. Note that $3\sqrt[3]{2}$ means $3 \cdot \sqrt[3]{2}$.

b. The greatest perfect fourth power factor of 405 (if there indeed is one) is not obvious. To determine it, we can write 405 in prime-factored form.

$$\sqrt[4]{405} = \sqrt[4]{3 \cdot 3 \cdot 3 \cdot 3 \cdot 5}$$ Prime-factor 405.

$$= \sqrt[4]{3 \cdot 3 \cdot 3 \cdot 3}\sqrt[4]{5}$$ Use the product rule for radicals. Write 3 · 3 · 3 · 3 within one fourth root because it is a perfect-fourth power: 3 · 3 · 3 · 3 = 81.

$$= 3\sqrt[4]{5}$$ Evaluate: $\sqrt[4]{3 \cdot 3 \cdot 3 \cdot 3} = \sqrt[4]{81} = 3$.

```
5 | 405
3 | 81
3 | 27
3 | 9
    3
```

c. We can use the quotient rule for radicals to simplify this expression.

$$\sqrt[3]{-\frac{32}{125}} = \frac{\sqrt[3]{-32}}{\sqrt[3]{125}}$$ The cube root of the quotient is equal to the quotient of the cube roots.

$$= \frac{\sqrt[3]{-8 \cdot 4}}{5}$$ Factor −32 as −8 · 4. In the denominator, $\sqrt[3]{125} = 5$.

$$= \frac{\sqrt[3]{-8}\sqrt[3]{4}}{5}$$ The cube root of a product is equal to the product of the cube roots.

$$= \frac{-2\sqrt[3]{4}}{5}$$ Find the cube root: $\sqrt[3]{-8} = -2$.

$$= -\frac{2\sqrt[3]{4}}{5}$$ Write the − sign in front of the fraction.

Self Check 5 Simplify: **a.** $\sqrt[3]{128}$ **b.** $\sqrt[4]{176}$ **c.** $\sqrt[3]{\dfrac{40}{343}}$

Now Try ▶ Problems 57, 61, and 67

4   Evaluate Expressions of the Form $a^{1/n}$.

It is possible to raise numbers to fractional powers. To give meaning to rational (fractional) exponents, we first consider $\sqrt{7}$. Because $\sqrt{7}$ is the positive number whose square is 7, we have

$$\left(\sqrt{7}\right)^2 = 7$$

The Language of Algebra

Rational exponents are also called **fractional exponents**.

We now consider the notation $7^{1/2}$, which is read as "7 to the one-half power." If rational exponents are to follow the same rules as integer exponents, the square of $7^{1/2}$ must be 7, because

$$(7^{1/2})^2 = 7^{(1/2) \cdot 2} \quad \text{Keep the base and multiply the exponents.}$$
$$= 7^1 \quad \text{Do the multiplication: } \tfrac{1}{2} \cdot 2 = 1.$$
$$= 7$$

Since the square of $7^{1/2}$ and the square of $\sqrt{7}$ are both equal to 7, we define $7^{1/2}$ to be $\sqrt{7}$. Similarly,

$$7^{1/3} = \sqrt[3]{7}, \qquad 7^{1/4} = \sqrt[4]{7}, \qquad \text{and} \qquad 7^{1/5} = \sqrt[5]{7}$$

In general, we have the following definition.

Definition of $a^{1/n}$

If n represents a positive integer greater than 1 and $\sqrt[n]{a}$ represents a real number,

$$a^{1/n} = \sqrt[n]{a} \quad \text{Read as "a to the } \tfrac{1}{n} \text{ power is equal to the nth root of a."}$$

We can use this definition to evaluate exponential expressions that have rational exponents with a numerator of 1.

EXAMPLE 6   Evaluate: **a.** $64^{1/3}$ **b.** $9^{1/2}$ **c.** $(-8)^{1/3}$ **d.** $-16^{1/4}$

Strategy   We will identify the base and the exponent of the exponential expression so that we can write the exponential expression in an equivalent radical form.

Why   We know how to evaluate square roots, cube roots, and fourth roots.

Solution   **a.** In the exponential expression $64^{1/3}$, the base is 64 and the exponent is $\tfrac{1}{3}$. The base is the same as the radicand of the corresponding radical. The denominator of the rational exponent is the same as the radical's index.

Root

$$64^{1/3} = \sqrt[3]{64} = 4 \qquad \text{Read as "64 to the one-third power." Because the denominator of}$$
the exponent is 3, find the cube root of the base, 64.

Base

Caution

Consider $-16^{1/4}$. Because of the lack of parentheses, the exponent $\tfrac{1}{4}$ only applies to the base 16, not -16.

b. $9^{1/2} = \sqrt{9} = 3$ Because the denominator of the exponent is 2, find the square root of the base, 9.

c. $(-8)^{1/3} = \sqrt[3]{-8} = -2$ Read as "negative eight to the one-third power." Because the denominator of the exponent is 3, find the cube root of the base, -8.

d. $-16^{1/4} = -\sqrt[4]{16} = -2$ Read as "the opposite of 16 to the one-fourth power." Because the denominator of the exponent is 4, find the fourth root of the base, 16.

Self Check 6   Evaluate: **a.** $144^{1/2}$ **b.** $(-27)^{1/3}$ **c.** $-81^{1/4}$

Now Try ▶   Problems 69, 71, and 73

5 Evaluate Expressions of the Form $a^{m/n}$.

We can extend the definition of $a^{1/n}$ to cover fractional exponents for which the numerator is not 1. For example, because $4^{3/2}$ can be written as $(4^{1/2})^3$, we have

$$4^{3/2} = (4^{1/2})^3 = \left(\sqrt{4}\right)^3 = 2^3 = 8 \qquad \text{Read } 4^{3/2} \text{ as "4 to the three-halves power."}$$

Because $4^{3/2}$ can also be written as $(4^3)^{1/2}$, we have

$$4^{3/2} = (4^3)^{1/2} = 64^{1/2} = \sqrt{64} = 8$$

In general, $a^{m/n}$ can be written as $(a^{1/n})^m$ or as $(a^m)^{1/n}$. Since $(a^{1/n})^m = \left(\sqrt[n]{a}\right)^m$ and $(a^m)^{1/n} = \sqrt[n]{a^m}$, we make the following definition.

Definition of $a^{m/n}$	If m and n represent positive integers ($n \neq 1$) and $\sqrt[n]{a}$ represents a real number, $$a^{m/n} = \left(\sqrt[n]{a}\right)^m \quad \text{and} \quad a^{m/n} = \sqrt[n]{a^m}$$

We read the first definition given above as "a to the $\frac{m}{n}$ power is equal to the nth root of a, raised to the mth power."

We can use this definition to evaluate exponential expressions that have rational exponents with a numerator that is not 1. To avoid large numbers, we usually find the root of the base first and then calculate the power using the relationship $a^{m/n} = \left(\sqrt[n]{a}\right)^m$.

EXAMPLE 7   Evaluate: **a.** $125^{4/3}$ **b.** $81^{3/4}$ **c.** $-25^{3/2}$ **d.** $(-27)^{2/3}$

Strategy   We will identify the base and the exponent of the exponential expression so that we can write the exponential expression in an equivalent radical form.

Why   We know how to evaluate square roots, cube roots, and fourth roots.

Solution   **a.** In the exponential expression $125^{4/3}$, the base is 125 and the exponent is $\frac{4}{3}$. The base is the same as the radicand of the corresponding radical. The denominator of the rational exponent is the same as the radical's index. The numerator of the rational exponent indicates the power to which the radical base is raised.

$$\underset{\text{Base}}{\underbrace{125}}\,^{\overset{\text{Power}}{\overset{\text{Root}}{4/3}}} = \left(\sqrt[3]{125}\right)^4 = (5)^4 = 625$$

Read as "125 to the four-thirds power." Because the exponent is 4/3, find the cube root of the base, 125, to get 5. Then find the fourth power of 5.

Caution

We can also evaluate $x^{m/n}$ using $\sqrt[n]{x^m}$; however, the resulting radicand is often extremely large. For example,

$$81^{3/4} = \sqrt[4]{81^3}$$
$$= \sqrt[4]{531,441}$$
$$= 27$$

b. $81^{3/4} = \left(\sqrt[4]{81}\right)^3 = (3)^3 = 27$ Because the exponent is 3/4, find the fourth root of the base, 81, to get 3. Then find the third power of 3.

c. For the exponential expression $-25^{3/2}$, the base is 25, not -25.

$$-25^{3/2} = -\left(\sqrt{25}\right)^3 = -(5)^3 = -125$$

Because the exponent is 3/2, find the square root of the base, 25, to get 5. Then find the third power of 5.

d. Because of the parentheses, the base of the exponential expression $(-27)^{2/3}$ is -27.

$$(-27)^{2/3} = \left(\sqrt[3]{-27}\right)^2 = (-3)^2 = 9$$

Because the exponent is 2/3, find the cube root of the base, -27, to get -3. Then find the second power of -3.

Self Check 7 Evaluate: **a.** $100^{3/2}$ **b.** $(-8)^{2/3}$ **c.** $-32^{4/5}$

Now Try ▶ Problems 81, 85, and 91

We define negative rational exponents in the following way.

Definition of $a^{-m/n}$	If $a^{m/n}$ is a nonzero real number,
	$$a^{-m/n} = \frac{1}{a^{m/n}}$$ **Read as "a to the negative $\frac{m}{n}$ power is equal to 1 over a to the $\frac{m}{n}$ power."**

From the definition, we see that another way to write $a^{-m/n}$ is to write its reciprocal and change the sign of its exponent.

EXAMPLE 8 Evaluate: **a.** $27^{-2/3}$ **b.** $36^{-1/2}$ **c.** $-625^{-3/4}$

Strategy We will identify the base and the exponent of the exponential expression so that we can write the given exponential expression in an equivalent radical form.

Why We know how to evaluate square roots, cube roots, and fourth roots.

Solution **a.** In the exponential expression $27^{-2/3}$, the base is 27 and the exponent is $-\frac{2}{3}$. Because the exponent is $-2/3$, we write the reciprocal of $27^{-2/3}$ and change the sign of the exponent. Then we find the cube root of the base, 27, to get 3. Finally, we find the second power of 3.

Caution

A negative exponent does not indicate a negative number. For example,

$$27^{-2/3} = \frac{1}{9}$$

$$36^{-1/2} = \frac{1}{6}$$

Reciprocal; change sign

$$27^{-2/3} = \frac{1}{27^{2/3}} = \frac{1}{\left(\sqrt[3]{27}\right)^2} = \frac{1}{(3)^2} = \frac{1}{9}$$ Read as "27 to the negative two-thirds power."

Root

Power

b. $36^{-1/2} = \dfrac{1}{36^{1/2}} = \dfrac{1}{\sqrt{36}} = \dfrac{1}{6}$

c. In $-625^{-3/4}$, the base is 625, not -625.

$$-625^{-3/4} = -\frac{1}{625^{3/4}} = -\frac{1}{\left(\sqrt[4]{625}\right)^3} = -\frac{1}{(5)^3} = -\frac{1}{125}$$

Self Check 8 Evaluate: **a.** $49^{-1/2}$ **b.** $8^{-4/3}$ **c.** $-16^{-1/4}$

Now Try ▶ Problems 93, 97, and 101

SECTION 8.6 ▶ STUDY SET

VOCABULARY

Fill in the blanks.

1. We read $\sqrt[3]{8}$ as "the _____ root of 8," $\sqrt[4]{16}$ as "the _____ root of 16," and $\sqrt[5]{-1}$ as the _____ root of -1.

2. For the radical $\sqrt[3]{27}$, the _____ is 3 and the _____ is 27.

3. "To _____ $\sqrt[3]{54}$" means to write it as $3\sqrt[3]{2}$.

4. The expression $25^{3/2}$ has a _____ (fractional) exponent. We read it as "25 to the three-_____ power."

CONCEPTS

Fill in the blanks.

5. a. $\sqrt[3]{64} = 4$ because $()^3 = 64$.

   b. $\sqrt[4]{16x^4} = 2x$ because $()^4 = 16x^4$.

   c. $\sqrt[5]{-32} = -2$ because $(-2)^5 = $.

6. a. Write the first seven perfect cubes.

   $1, , , , , , 343$

   b. Write the first five perfect fourth powers.

   $1, , , 256, $

7. a. $\sqrt[n]{a \cdot b} = $

   b. $\sqrt[n]{\dfrac{a}{b}} = $

8. a. $x^{1/n} = $

   b. $x^{m/n} = ()^m$

   c. $x^{-m/n} = \dfrac{1}{}$

9. a. To simplify $\sqrt[3]{24}$, which of the following expressions should be used?

   $\sqrt[3]{12 \cdot 2} \qquad \sqrt[3]{4 \cdot 6} \qquad \sqrt[3]{8 \cdot 3} \qquad \sqrt[3]{24 \cdot 1}$

   b. To simplify $\sqrt[4]{48}$, which of the following expressions should be used?

   $\sqrt[4]{24 \cdot 2} \qquad \sqrt[4]{16 \cdot 3} \qquad \sqrt[4]{8 \cdot 6} \qquad \sqrt[4]{48 \cdot 1}$

10. a. Find the cube root of the perfect cube to simplify $\sqrt[3]{\mathbf{8}}\sqrt[3]{3}$.

    b. Find the fourth root of the perfect fourth power to simplify $\sqrt[4]{\mathbf{16}}\sqrt[4]{3}$.

NOTATION

Write each exponential expression in an equivalent form involving a radical. Do not evaluate it.

11. a. $25^{1/2}$ b. $(-27)^{2/3}$

12. a. $36^{-1/2}$ b. $16^{-5/4}$

GUIDED PRACTICE

Evaluate. See Example 1.

13. $\sqrt[3]{8}$ 14. $\sqrt[3]{27}$

15. $\sqrt[3]{0}$ 16. $\sqrt[3]{1}$

17. $\sqrt[3]{-125}$ 18. $\sqrt[3]{-1}$

19. $\sqrt[3]{-64}$ 20. $\sqrt[3]{-27}$

21. $-\sqrt[3]{-1}$ 22. $-\sqrt[3]{343}$

23. $-\sqrt[3]{-27}$ 24. $-\sqrt[3]{64}$

25. $\sqrt[3]{729}$ 26. $\sqrt[3]{512}$

27. $\sqrt[3]{\dfrac{1}{125}}$ 28. $\sqrt[3]{\dfrac{1}{1,000}}$

Evaluate. See Example 2.

29. $\sqrt[4]{16}$ 30. $\sqrt[4]{81}$

31. $\sqrt[4]{256}$ 32. $-\sqrt[4]{625}$

33. $\sqrt[4]{-625}$ 34. $\sqrt[4]{-16}$

35. $\sqrt[5]{-243}$ 36. $\sqrt[5]{-1}$

37. $\sqrt[5]{32}$ 38. $\sqrt[5]{-32}$

39. $-\sqrt[4]{\dfrac{1}{81}}$ 40. $\sqrt[4]{\dfrac{1}{256}}$

Find each root. See Example 3.

41. $\sqrt[4]{m^8}$ 42. $\sqrt[4]{n^{20}}$

43. $\sqrt[4]{x^{24}}$ 44. $\sqrt[4]{x^{12}}$

45. $\sqrt[4]{81a^4}$ 46. $\sqrt[4]{625b^{16}}$

47. $\sqrt[4]{16b^{12}}$ 48. $\sqrt[4]{81h^{16}}$

Find each root. See Example 4.

49. $\sqrt[5]{y^5}$ 50. $\sqrt[3]{m^3}$

51. $\sqrt[3]{27a^6}$ 52. $\sqrt[3]{-64b^9}$

53. $\sqrt[5]{x^{15}}$ 54. $\sqrt[5]{t^{10}}$

55. $\sqrt[5]{32v^{30}}$ 56. $\sqrt[5]{32x^5}$

Simplify. See Example 5.

57. $\sqrt[3]{24}$ 58. $\sqrt[3]{32}$

59. $\sqrt[3]{-128}$ 60. $\sqrt[3]{-250}$

61. $\sqrt[4]{162}$ 62. $\sqrt[4]{80}$

63. $\sqrt[3]{72}$ 64. $\sqrt[3]{189}$

65. $\sqrt[3]{\dfrac{250}{27}}$ 66. $\sqrt[3]{\dfrac{54}{125}}$

67. $\sqrt[5]{64}$ 68. $\sqrt[5]{160}$

Evaluate. See Example 6.

69. $81^{1/2}$ 70. $100^{1/2}$

71. $8^{1/3}$ 72. $27^{1/3}$

73. $\left(\dfrac{1}{16}\right)^{1/4}$ 74. $\left(\dfrac{1}{81}\right)^{1/4}$

75. $-144^{1/2}$ 76. $-400^{1/2}$

77. $(-125)^{1/3}$ 78. $(-8)^{1/3}$

79. $\left(\dfrac{9}{64}\right)^{1/2}$ 80. $\left(\dfrac{4}{49}\right)^{1/2}$

Evaluate. See Example 7.

81. $16^{3/2}$ 82. $25^{3/2}$

83. $27^{2/3}$ 84. $8^{4/3}$

85. $(-125)^{2/3}$ 86. $(-64)^{2/3}$

87. $4^{5/2}$ 88. $9^{5/2}$

89. $\left(\dfrac{1}{16}\right)^{3/4}$ 90. $\left(-\dfrac{243}{32}\right)^{3/5}$

91. $-32^{3/5}$ 92. $-32^{2/5}$

Evaluate. See Example 8.

93. $8^{-1/3}$ 94. $81^{-1/2}$

95. $25^{-3/2}$ 96. $9^{-3/2}$

97. $16^{-5/4}$ 98. $16^{-3/4}$

99. $(-27)^{-2/3}$ 100. $8^{-5/3}$

101. $-81^{-3/2}$ 102. $-125^{-2/3}$

103. $32^{-3/5}$ 104. $(-32)^{-2/5}$

TRY IT YOURSELF

Look Alikes . . .

105. Evaluate.

a. $\sqrt{64}$ b. $\sqrt[3]{64}$ c. $\sqrt{-64}$ d. $\sqrt[3]{-64}$

106. Evaluate.

a. $\sqrt{81}$ b. $\sqrt[4]{81}$ c. $\sqrt{-81}$ d. $\sqrt[4]{-81}$

107. Simplify each root.

a. $\sqrt{32}$ b. $\sqrt[3]{32}$

c. $\sqrt[4]{32}$ d. $\sqrt[5]{32}$

108. Evaluate.

a. $16^{1/2}$ b. $16^{3/4}$

c. $16^{-1/4}$ d. $16^{-5/2}$

APPLICATIONS

109. Windmills.   The power generated by a windmill is related to the speed of the wind by the formula $S = \sqrt[3]{\dfrac{P}{0.02}}$, where S is the speed of the wind (in mph) and P is the power (in watts). Find the speed of the wind when the windmill is producing 20 watts of power.

110. Depreciation.   The formula $r = 1 - \sqrt[n]{\dfrac{S}{C}}$ gives the annual depreciation rate r of an item that had an original cost of C dollars and has a useful life of n years and a salvage value of S dollars. Use the information in the illustration to find (in percent) the annual depreciation rate for the new piece of sound equipment. (In the memo, $81K = $81,000.)

> **OFFICE MEMO**
>
> **To:** Purchasing Dept.
> **From:** Bob Kinsell, Engineering Dept. *BK*
> **Re:** New sound board
>
> We recommend you purchase the new Sony sound board @ $81K. This equipment does become obsolete quickly but we figure we can use it for 4 yrs. A college would probably buy it from us then. I bet we could get around $16K for it.

111. Holiday Decorations.   Find the length s of each string of colored lights used to decorate an evergreen tree in the manner shown if $s = (r^2 + h^2)^{1/2}$.

$h = 24$ ft

s

$r = 10$ ft

112. Visibility.   The distance d in miles that a person in an airplane can see to the horizon on a clear day is given by the formula $d = 1.22a^{1/2}$, where a is the altitude of the plane in feet. Find d. Round to the nearest tenth of a mile.

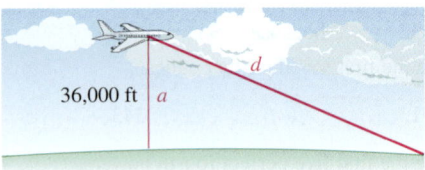

36,000 ft a d

113. Speakers.   The formula $A = V^{2/3}$ can be used to find the area A of one face of a cube if its volume V is known. Find the amount of floor space occupied by a speaker if it is a cube with a volume of 216 cubic inches.

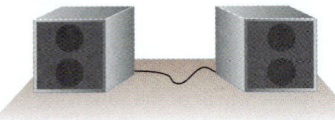

114. Medical Tests.   Before a series of X-rays are taken, a patient is injected with a special contrast mixture that highlights obstructions in his blood vessels. The amount of the original dose of contrast material remaining in the patient's bloodstream h hours after it is injected is given by $h^{-3/2}$. How much of the contrast material remains in the patient's bloodstream 4 hours after the injection?

WRITING

115. Explain why a negative number can have a real number for its cube root yet cannot have a real number for its fourth root.

116. What is a rational exponent? Give several examples.

117. Explain this statement: In the expression $16^{3/2}$, the number $3/2$ requires that two operations be performed on 16.

118. To evaluate $27^{4/3}$, which expression would be easier to use, $\left(\sqrt[3]{27}\right)^4$ or $\sqrt[3]{27^4}$? Explain.

REVIEW

Graph each equation.

119. $x = 3$

120. $y = -3$

121. $2x - y = -4$

122. $y = 4x - 4$

CHALLENGE PROBLEMS

123. Graph $y = \sqrt[3]{x}$ by first completing the table of solutions. Then plot the points and draw a smooth curve through them.

x	-27	-8	-1	0	1	8	27
y							

124. Evaluate: $\left(\dfrac{25}{49}\right)^{-1.5}$

8 ▶ Summary & Review

DEFINITIONS AND CONCEPTS	EXAMPLES
The number b is a **square root** of a if $b^2 = a$. The **positive** or **principal square root** of a positive number a, written as $\sqrt{a}$, is the positive square root of a. The **negative square root** of a is written as $-\sqrt{a}$.	3 is a square root of 9 because $3^2 = 9$. $\sqrt{49} = 7$ because $7^2 = 49$. $\sqrt{25} = 5$ because $5^2 = 25$. $\sqrt{16a^6} = 4a^3$ because $(4a^3)^2 = 16a^6$.
The expression within a **radical symbol** $\sqrt{}$ is called a **radicand**.	In $\sqrt{49}$, the radicand is 49. In $\sqrt{16a^6}$, the radicand is $16a^6$.
A number such as 16, 1, 0.36, $\frac{4}{9}$, or 225, that is the square of some rational number, is called a **perfect square**. The square root of a perfect square is a **rational number**. If a positive number is not a perfect square, its square root is an **irrational number**. Square roots of negative numbers are **not real numbers**.	▪ Since 81 is a perfect square, $\sqrt{81}$ is a rational number: $\sqrt{81} = 9$. ▪ Since 5 is not a perfect square, $\sqrt{5}$ is irrational. Its decimal value is approximately 2.24. ▪ Since no real number squared is -4, $\sqrt{-4}$ is not a real number.
Pythagorean theorem: If a and b are the lengths of the legs of a right triangle and c is the length of the hypotenuse, then $a^2 + b^2 = c^2$. We call $a^2 + b^2 = c^2$ the **Pythagorean equation.**	Find the unknown length of the third side of the right triangle. $a^2 + b^2 = c^2$ This is the Pythagorean equation. $6^2 + b^2 = 10^2$ Substitute 6 for a and 10 for c. $36 + b^2 = 100$ Evaluate: $6^2 = 36$ and $10^2 = 100$. $b^2 = 64$ To isolate b^2, subtract 36 from both sides. $b = \sqrt{64}$ Since b must be positive, find the positive square root of 64. $b = 8$ Find the square root of 64. The length of the third side of the triangle is 8 ft.
Distance formula: The distance between points with coordinates (x_1, y_1) and (x_2, y_2) is given by $d = \sqrt{(x_2 - x_1)^2 + (y_2 - y_1)^2}$	Find the distance between $(2, 0)$ and $(-2, 3)$. $d = \sqrt{(x_2 - x_1)^2 + (y_2 - y_1)^2}$ This is the distance formula. $d = \sqrt{(-2 - 2)^2 + (3 - 0)^2}$ Substitute. $d = \sqrt{(-4)^2 + 3^2}$ Do the subtraction. $d = \sqrt{25}$ Evaluate within the radical symbol. $d = 5$ Find the square root of 25.

REVIEW EXERCISES

1. Fill in the blanks: $\sqrt{36} = 6$ because $(\ \ )^2 = \ \ $

2. Determine whether the statement is true or false:

$-\sqrt{4} = \sqrt{-4}$

Find each square root. DO NOT USE A CALCULATOR.

3. $\sqrt{25}$ **4.** $\sqrt{49}$ **5.** $-\sqrt{144}$

6. $\sqrt{\dfrac{16}{81}}$ **7.** $\sqrt{0.64}$ **8.** $\sqrt{1}$

Use a calculator to approximate each expression to the nearest hundredth.

9. $\sqrt{21}$ **10.** $-2\sqrt{7}$

11. Determine whether each number is rational, irrational, or not a real number: $\sqrt{-2}$, $\sqrt{68}$, $\sqrt{81}$, $\sqrt{3}$.

12. Road Signs. To find the maximum velocity a car can safely travel around a curve without skidding, we can use the formula $v = \sqrt{2.5r}$, where v is the velocity in mph and r is the radius of the curve in feet. How should the road sign be labeled if it is to be posted in front of a curve with a radius of 360 feet?

Simplify. All variables represent positive real numbers.

13. $\sqrt{x^2}$ **14.** $\sqrt{4b^4}$ **15.** $-\sqrt{y^{12}}$ **16.** $\sqrt{9h^{16}}$

Refer to the right triangle.

17. Find c where $a = 6$ and $b = 8$.

18. Find b where $a = 8$ and $c = 17$.

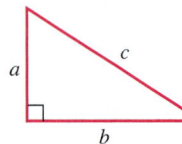

19. Theater Seating. How much higher is the seat at the top of the incline than the one at the bottom?

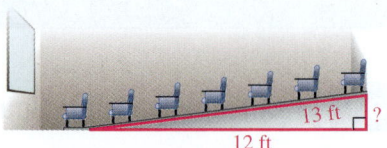

20. Triangles. The length of the hypotenuse of an isosceles right triangle is 3 inches. Find the length of a leg of the triangle. Round to the nearest hundredth of an inch.

Find the distance between the points. If an answer is not exact, round to the nearest hundredth.

21. $(-7, 12), (-4, 8)$ **22.** $(-15, -3), (-10, -16)$

SECTION 8.2 ▶ Simplifying Square Roots

DEFINITIONS AND CONCEPTS	EXAMPLES
Simplified form of a radical: **1.** Except for 1, the radicand has no perfect-square factors. **2.** No fraction appears in the radicand. **3.** No radical appears in the denominator.	These radicals are *not* in simplified form: $\sqrt{18m^3}$ $\sqrt{\dfrac{5}{64}}$ $\dfrac{1}{\sqrt{3}}$ The radicand has a perfect-square factor of $9m^2$. The radicand is a fraction. A radical appears in the denominator.
The **product and quotient rules for square roots** can be used to simplify radical expressions. $$\sqrt{a \cdot b} = \sqrt{a}\sqrt{b}$$ $$\sqrt{\dfrac{a}{b}} = \dfrac{\sqrt{a}}{\sqrt{b}} \quad (b \neq 0)$$	Simplify: $\sqrt{18m^3} = \sqrt{9m^2 \cdot 2m}$ Write $18m^3$ as a product of its greatest perfect-square factor, $9m^2$, and one other factor. $= \sqrt{9m^2}\sqrt{2m}$ The square root of a product is equal to the product of the square roots. $= 3m\sqrt{2m}$ Find the square root of the perfect square: $\sqrt{9m^2} = 3m$. Simplify: $\sqrt{\dfrac{5}{64}} = \dfrac{\sqrt{5}}{\sqrt{64}}$ The square root of a quotient is equal to the quotient of the square roots. $= \dfrac{\sqrt{5}}{8}$ In the denominator, evaluate: $\sqrt{64} = 8$.

REVIEW EXERCISES

Simplify. All variables represent positive real numbers.

23. $\sqrt{32}$

24. $\sqrt{x^5}$

25. $\sqrt{80x^2}$

26. $-2\sqrt{63}$

27. $\sqrt{250t^3}$

28. $\sqrt{\dfrac{16}{25}}$

29. $\sqrt{\dfrac{60}{49}}$

30. $\sqrt{\dfrac{242x^4}{169x^2}}$

31. Fitness Equipment. The length of the sit-up board can be found using the Pythagorean theorem. Find its length. Express the answer in simplified radical form. Then express your result as a decimal approximation rounded to the nearest tenth.

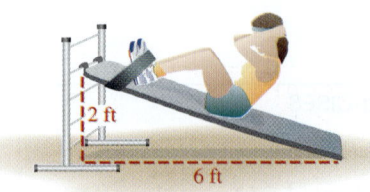

32. Determine whether the statement is true or false: $\sqrt{x^9} = x^3$.

SECTION 8.3 ▶ Adding and Subtracting Radical Expressions

DEFINITIONS AND CONCEPTS	EXAMPLES
Square root radicals are called **like radicals** when they have the same radicand.	Like radicals: $4\sqrt{2}$ and $5\sqrt{2}$ Unlike radicals: $3\sqrt{6}$ and $7\sqrt{3}$ *The same radicand* *Different radicands*
Radical expressions can be added or subtracted if they contain like radicals. To **combine like radicals** we use the distributive property in reverse.	Add: $3\sqrt{7} + 5\sqrt{7} = 8\sqrt{7}$ *Think: $(3 + 5)\sqrt{7}$.* Subtract: $8\sqrt{2y} - 2\sqrt{2y} = 6\sqrt{2y}$ *Think: $(8 - 2)\sqrt{2y}$.*
If a sum or difference involves unlike radicals, make sure that each one is written in simplified form. After doing so, like radicals may result that can be combined.	Add: $\sqrt{12} + \sqrt{75} = \sqrt{4}\sqrt{3} + \sqrt{25}\sqrt{3}$ *Simplify $\sqrt{12}$ and $\sqrt{75}$.* $= 2\sqrt{3} + 5\sqrt{3}$ *$\sqrt{4} = 2$ and $\sqrt{25} = 5$.* $= 7\sqrt{3}$ *Combine like radicals: $2 + 5 = 7$.*

REVIEW EXERCISES

Perform the operations, if possible. All variables represent nonnegative real numbers.

33. $\sqrt{10} + \sqrt{10}$ **34.** $6\sqrt{x} - \sqrt{x}$

35. $\sqrt{2} + \sqrt{8} - \sqrt{18}$ **36.** $\sqrt{3} + 4 + \sqrt{27} - 7$

37. $5\sqrt{28} - 3\sqrt{63}$ **38.** $3\sqrt{5y^3} - 5y\sqrt{20y}$

39. Explain why we cannot add $3\sqrt{5}$ and $5\sqrt{3}$.

40. Gardening. Find the difference in the lengths of the two wires used to secure the tree.

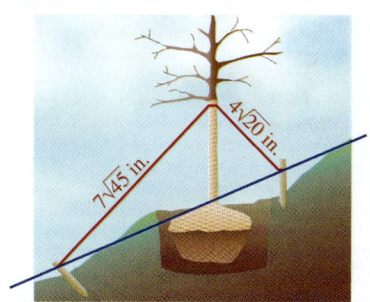

SECTION 8.4 ▶ Multiplying and Dividing Radical Expressions

DEFINITIONS AND CONCEPTS	EXAMPLES
To **multiply square roots:** The product of two square roots is equal to the square root of the product. $$\sqrt{a} \cdot \sqrt{b} = \sqrt{a \cdot b}$$	Multiply: $3\sqrt{5} \cdot 2\sqrt{7} = 6\sqrt{5 \cdot 7} = 6\sqrt{35}$ *Multiply the radicands, 5 and 7, to get 35, and write the square root of that product.* Multiply and simplify: $\sqrt{2a^5} \cdot \sqrt{6a} = \sqrt{2a^5 \cdot 6a}$ $= \sqrt{12a^6}$ Since $12a^6$ has a perfect-square factor of $4a^6$, we can simplify the result: $= \sqrt{4a^6 \cdot 3}$ $= 2a^3\sqrt{3}$
The Square of a Square Root: For any nonnegative real number a, $(\sqrt{a})^2 = a$.	$(\sqrt{5})^2 = 5$ $(\sqrt{z + 1})^2 = z + 1$ $(3\sqrt{a})^2 = 9a$
Radical expressions with more than one term are multiplied as if they were polynomials.	Multiply: $2\sqrt{x}(3 - 4\sqrt{x}) = 2\sqrt{x} \cdot 3 - 2\sqrt{x} \cdot 4\sqrt{x}$ *Distribute $2\sqrt{x}$.* $= 6\sqrt{x} - 8\sqrt{x^2}$ *$2 \cdot 3 = 6$ and $2 \cdot 4 = 8$.* $= 6\sqrt{x} - 8x$ *Simplify: $\sqrt{x^2} = x$.*

Use the FOIL method to multiply two radical expressions, each having two terms.	Multiply: $(\sqrt{a} - 2)(\sqrt{a} + 6) = \overset{F}{\sqrt{a} \cdot \sqrt{a}} + \overset{O}{6\sqrt{a}} - \overset{I}{2\sqrt{a}} - \overset{L}{2 \cdot 6}$ $= \sqrt{a^2} + 4\sqrt{a} - 12$ *Combine like radicals.* $= a + 4\sqrt{a} - 12$ *Simplify:* $\sqrt{a^2} = a$.
To **divide square roots:** The quotient of two square roots is equal to the square root of the quotient. $\dfrac{\sqrt{a}}{\sqrt{b}} = \sqrt{\dfrac{a}{b}}$ $(b \neq 0)$	Divide: $\dfrac{\sqrt{70}}{\sqrt{10}} = \sqrt{\dfrac{70}{10}} = \sqrt{7}$ *Divide the radicands, 70 by 10, to get 7, and write the square root of that quotient. The result does not simplify.*
To **rationalize a square root denominator,** multiply the numerator and denominator of the given fraction by the square root that appears in its denominator, or by a square root that makes a perfect-square radicand in the denominator.	Rationalize the denominator: $\dfrac{3}{\sqrt{5}} = \dfrac{3}{\sqrt{5}} \cdot \dfrac{\sqrt{5}}{\sqrt{5}}$ *Build an equivalent fraction by multiplying by a form of 1.* $= \dfrac{3\sqrt{5}}{5}$ *Multiply the numerators. Multiply the denominators:* $\sqrt{5} \cdot \sqrt{5} = \sqrt{25} = 5$.
To **rationalize a two-term denominator,** multiply the expression by a form of 1 that uses the **conjugate** of the denominator. $\dfrac{3 - \sqrt{a}}{2 + \sqrt{a}}$ ←— Two terms	Rationalize the denominator: $\dfrac{3 - \sqrt{a}}{2 + \sqrt{a}} = \dfrac{3 - \sqrt{a}}{2 + \sqrt{a}} \cdot \dfrac{2 - \sqrt{a}}{2 - \sqrt{a}}$ $2 - \sqrt{a}$ *is the conjugate of* $2 + \sqrt{a}$. $= \dfrac{6 - 3\sqrt{a} - 2\sqrt{a} + \sqrt{a^2}}{4 - \sqrt{a^2}}$ $= \dfrac{6 - 5\sqrt{a} + a}{4 - a}$ *Simplify.*

REVIEW EXERCISES

Perform the operations. All variables represent nonnegative real numbers.

41. $\sqrt{2}\sqrt{3}$

42. $\left(-5\sqrt{5}\right)^2$

43. $\left(3\sqrt{3x^3}\right)\left(4\sqrt{6x^2}\right)$

44. $\left(\sqrt{15} + 3x\right)^2$

45. $\sqrt{2}\left(\sqrt{8} - \sqrt{18}\right)$

46. $\left(\sqrt{3} + \sqrt{5}\right)\left(\sqrt{3} - \sqrt{5}\right)$

47. $\left(\sqrt{x}\right)^2$

48. $\left(\sqrt{t-1}\right)^2$

49. Find the area of the rectangle at the right. Express the answer in simplified radical form. Then express your result as a decimal approximation to the nearest tenth.

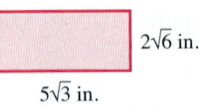

$2\sqrt{6}$ in.

$5\sqrt{3}$ in.

50. Patio Tiles. The length of the diagonal of a square tile is 1 foot. It can be shown using the Pythagorean theorem that the length of one side of the tile is $\dfrac{1}{\sqrt{2}}$ foot. Rationalize the denominator of this expression.

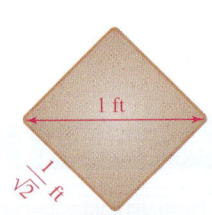

1 ft

$\frac{1}{\sqrt{2}}$ ft

Rationalize the denominator. All variables represent positive real numbers.

51. $\dfrac{9}{\sqrt{7}}$

52. $\sqrt{\dfrac{3}{a}}$

53. $\dfrac{16\sqrt{3}}{\sqrt{5}}$

54. $\dfrac{11}{\sqrt{75}}$

55. $\dfrac{1}{\sqrt{8x}}$

56. $\dfrac{\sqrt{7}}{\sqrt{7} - \sqrt{2}}$

57. $\dfrac{\sqrt{a}}{\sqrt{a} + 1}$

58. $\dfrac{2 + \sqrt{b}}{\sqrt{b} - 3}$

SECTION 8.5 ▶ Solving Radical Equations

DEFINITIONS AND CONCEPTS	EXAMPLES
The **squaring property of equality:** If $a = b$, then $a^2 = b^2$. The **square of a square root:** For any nonnegative real number a, $$\left(\sqrt{a}\right)^2 = a$$ To **solve radical equations** containing square roots: 1. Isolate a radical term on one side of the equation. 2. Square both sides of the equation. 3. Solve the resulting equation. 4. Check the possible solutions in the original equation. Discard any **extraneous solutions.**	Solve: $\quad \sqrt{x+2} - 3 = 5$ $\sqrt{x+2} - 3 + 3 = 5 + 3 \quad$ To isolate the radical, add 3 to both sides. $\sqrt{x+2} = 8 \quad$ Do the addition. $\left(\sqrt{x+2}\right)^2 = (8)^2 \quad$ Use the squaring property of equality and square both sides. $x + 2 = 64 \quad$ Simplify the square of a square root: $\left(\sqrt{x+2}\right)^2 = x+2$. $x = 62 \quad$ To isolate x, subtract 2 from both sides. Check: $\quad \sqrt{x+2} - 3 = 5 \quad$ The original equation. $\sqrt{62+2} - 3 \stackrel{?}{=} 5 \quad$ Substitute 62 for x. $\sqrt{64} - 3 \stackrel{?}{=} 5 \quad$ Evaluate the left side. $5 = 5 \quad$ True Since we obtain a true statement, 62 is the solution of $\sqrt{x+2} - 3 = 5$.

REVIEW EXERCISES

Solve each equation and check each result.

59. $\sqrt{x} = 9$

60. $\sqrt{2x + 10} = 10$

61. $\sqrt{3x + 4} + 5 = 3$

62. $\sqrt{b + 12} = 3\sqrt{b + 4}$

63. $\sqrt{p^2 + 8p + 13} = p + 3$

64. $\sqrt{24 + 10y} - y = 4$

65. $5 + \sqrt{3n + 3} = n$

66. $\sqrt{a - 2} = a - 8$

67. Ferris Wheels. The distance d in feet that an object will fall in t seconds is given by the formula $t = \sqrt{\dfrac{d}{16}}$. If a person drops their sunglasses from the top of a Ferris wheel and it takes the glasses 2 seconds to hit the ground, how tall is the Ferris wheel?

68. The Horizon. On a clear day, the approximate distance d (in miles) that a person can see to the horizon is given by the formula $d = \sqrt{1.5h}$, where h is the height (in feet) of the person's eyes above the ground. Suppose the view of a worker at the top of a cell phone tower extends 12 miles to the horizon. Estimate the height of the tower.

SECTION 8.6 ▶ Higher-Order Roots and Rational Exponents

DEFINITIONS AND CONCEPTS	EXAMPLES
The **cube root** of a is denoted by $\sqrt[3]{a}$. By definition, $\sqrt[3]{a} = b$ if $b^3 = a$.	$\sqrt[3]{64} = 4 \qquad$ because $\qquad 4^3 = 64$.
Just as there are square roots and cube roots, there are also fourth roots, fifth roots, and so on. The ***n*th root of a** is written $\sqrt[n]{a}$, and $\sqrt[n]{a} = b$ if $b^n = a$. The number n is called the **index** of the radical.	$\sqrt[4]{81} = 3 \qquad$ because $\qquad 3^4 = 81 \qquad$ The index is 4. $\sqrt[5]{-32m^{10}} = -2m^2 \qquad$ because $\qquad (-2m^2)^5 = -32m^{10} \qquad$ The index is 5.
When n is even, we say that the radical $\sqrt[n]{a}$ is an **even root.** When n is odd, we say that the radical $\sqrt[n]{a}$ is an **odd root.**	$\sqrt[4]{16}$ is an even root. $\sqrt[4]{-12}$ is not a real number. No real number raised to the fourth power is -12. $\sqrt[3]{125}$ is an odd root.

The product and quotient rules for square roots can be generalized to apply to higher-order roots. They can be used to simplify radical expressions.

$$\sqrt[n]{a \cdot b} = \sqrt[n]{a}\,\sqrt[n]{b}$$

$$\sqrt[n]{\dfrac{a}{b}} = \dfrac{\sqrt[n]{a}}{\sqrt[n]{b}} \qquad \left(\sqrt[n]{b} \neq 0\right)$$

Simplify: $\sqrt[4]{32a} = \sqrt[4]{16 \cdot 2a}$ Write 32a as a product of its greatest perfect fourth power factor, 16, and one other factor.

$$= \sqrt[4]{16}\,\sqrt[4]{2a}$$ The fourth root of a product is equal to the product of the fourth roots.

$$= 2\sqrt[4]{2a}$$ Evaluate: $\sqrt[4]{16} = 2$.

Simplify: $\sqrt[3]{\dfrac{26}{27}} = \dfrac{\sqrt[3]{26}}{\sqrt[3]{27}}$ The cube root of a quotient is equal to the quotient of the cube roots.

$$= \dfrac{\sqrt[3]{26}}{3}$$ In the denominator, evaluate: $\sqrt[3]{27} = 3$.

To evaluate exponential expressions involving fractional exponents, use the **rules for rational exponents** to write the expressions in an equivalent radical form.

$$x^{1/n} = \sqrt[n]{x} \qquad x^{m/n} = \left(\sqrt[n]{x}\right)^m = \sqrt[n]{x^m}$$

$$x^{-m/n} = \dfrac{1}{x^{m/n}}$$

Evaluate: $(-125)^{1/3} = \sqrt[3]{-125} = -5$

Evaluate: $16^{3/4} = \left(\sqrt[4]{16}\right)^3 = (2)^3 = 8$

Evaluate: $8^{-2/3} = \dfrac{1}{8^{2/3}} = \dfrac{1}{\left(\sqrt[3]{8}\right)^2} = \dfrac{1}{(2)^2} = \dfrac{1}{4}$

REVIEW EXERCISES

Evaluate each root.

69. $\sqrt[3]{-27}$

70. $-\sqrt[3]{125}$

71. $\sqrt[4]{81}$

72. $\sqrt[5]{32}$

73. $\sqrt[3]{0}$

74. $\sqrt[5]{-1}$

75. $\sqrt[3]{\dfrac{1}{64}}$

76. $\sqrt[4]{256}$

Simplify. All variables represent nonnegative real numbers.

77. $\sqrt[3]{x^3}$

78. $\sqrt[3]{27y^6}$

79. $\sqrt[4]{16a^{12}}$

80. $\sqrt[5]{b^{20}}$

81. $\sqrt[3]{54}$

82. $\sqrt[4]{80}$

83. $\sqrt[3]{-\dfrac{16}{343}}$

84. $\sqrt[3]{\dfrac{56}{125}}$

Evaluate.

85. $49^{1/2}$

86. $(-1,000)^{1/3}$

87. $36^{3/2}$

88. $\left(\dfrac{8}{27}\right)^{2/3}$

89. $4^{-3/2}$

90. $-81^{5/4}$

91. Dentistry. The fractional amount of painkiller remaining in the system of a patient h hours after the original dose was injected into her gums is given by $h^{-3/2}$. How much of the original dose is in the patient's system 16 hours after the injection?

92. Explain why $(-64)^{1/2}$ is not a real number.

8 ▸ CHAPTER TEST

1. Fill in the blanks.

  a. The symbol $\sqrt{}$ is called a _____ symbol.

  b. The _____ of $\sqrt{25a^2}$ is $25a^2$.

  c. The _____ theorem relates the lengths of the sides of a right triangle.

  d. We read $\sqrt[3]{8}$ as "the _____ _____ of 8."

  e. For the expression $\sqrt[5]{32}$, the _____ is 5.

2. Fill in the blanks: $\sqrt{16} = 4$ because $()^2 = $.

Find each square root.

3. $\sqrt{100}$

4. $-\sqrt{\dfrac{64}{9}}$

5. $\sqrt{0.25}$

6. $\sqrt{1}$

7. Electronics. The current I (in amperes), the resistance R (in ohms), and the power P (in watts) are related by the formula $I = \sqrt{\dfrac{P}{R}}$. Find the current I drawn by an electrical appliance that uses 980 watts of power when the resistance is 20 ohms.

8. **Ladders.** A 26-foot ladder reaches a point on a wall 24 feet above the ground. How far from the wall is the ladder's base?

9. Determine whether each number is rational, irrational, or not a real number: $\sqrt{19}$, $\sqrt{-16}$, $\sqrt{144}$.

10. Find the distance between $(-4, 5)$ and $(2, 11)$. Express the result in simplified radical form. Then approximate the distance to the nearest hundredth.

Simplify each expression. All variables represent positive real numbers.

11. $\sqrt{4x^2}$

12. $\sqrt{54x^3}$

13. $\sqrt{\dfrac{50}{49}}$

14. $\sqrt{\dfrac{18a^6}{2a}}$

Perform each operation, if possible, and simplify. All variables represent nonnegative real numbers.

15. $8\sqrt{5} + \sqrt{5}$

16. $16\sqrt{11} - 4\sqrt{11} - 9\sqrt{11}$

17. $\sqrt{12b^4} + \sqrt{27b^4}$

18. $10\sqrt{3} \cdot \sqrt{2}$

19. $\sqrt{3}\left(\sqrt{8} + \sqrt{6}\right)$

20. $x\sqrt{50x} - \sqrt{200x^3}$

21. $\left(\sqrt{2} + \sqrt{3}\right)\left(\sqrt{2} - \sqrt{3}\right)$

22. $\left(-2\sqrt{8x}\right)\left(3\sqrt{12x^4}\right)$

23. $\left(5\sqrt{x} - 1\right)\left(\sqrt{x} - 4\right)$

24. $\left(2\sqrt{3t}\right)^2$

25. $\left(\sqrt{x+1}\right)^2$

26. $\left(\sqrt{x} + 1\right)^2$

27. **Sewing.** A corner of fabric is folded over to form a collar and stitched down. From the dimensions given in the figure, determine the exact number of inches of stitching that must be made.

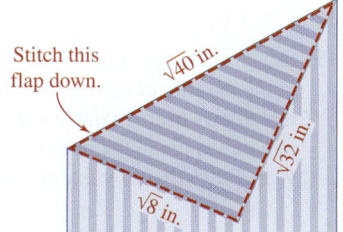

Stitch this flap down. $\sqrt{40}$ in. $\sqrt{32}$ in. $\sqrt{8}$ in.

28. Find the unknown length of a leg of the isosceles right triangle. Give an exact answer in simplified form and an approximation to the nearest hundredth.

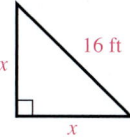

16 ft x x

Rationalize each denominator. All variables represent positive real numbers.

29. $\dfrac{2}{\sqrt{7}}$

30. $\dfrac{\sqrt{3}}{\sqrt{x} - 8}$

Solve each equation and check the result.

31. $\sqrt{x} = 15$

32. $\sqrt{2 - x} - 2 = 6$

33. $\sqrt{3x + 9} = 2\sqrt{x + 1}$

34. $x = \sqrt{x - 1} + 1$

35. $\sqrt{7t + 4} + 1 = 0$

36. $\sqrt{m - 5} = m - 11$

37. **Carpentry.** A carpenter uses a tape measure to see whether the wall he just put up is perfectly square with the floor. Explain what mathematical concept he is applying. If the wall is positioned correctly, what should the measurement on the tape read?

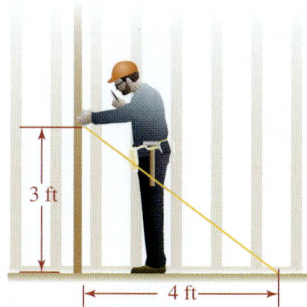

3 ft   4 ft

38. **Curving Test Scores.** Some professors curve their students' scores when the test results in a class are lower than he or she expected. One method to curve the grades on a 100-point test is to use the formula $N = 10\sqrt{s}$, where s is the original score and N is the new score. Find the original test score of a student if her new "curved" score is 80.

39. Explain why we cannot add $3\sqrt{2}$ and $2\sqrt{3}$.

40. Explain why $\sqrt{-49}$ is not a real number.

Find each root, if possible. All variables represent nonnegative real numbers.

41. $\sqrt[3]{-125}$

42. $\sqrt[4]{-256}$

43. $\sqrt[3]{\dfrac{1}{64}}$

44. $\sqrt[5]{-32x^{10}}$

Simplify. All variables represent nonnegative real numbers.

45. $\sqrt[4]{81x^4}$

46. $\sqrt[3]{88}$

Evaluate.

47. $144^{1/2}$

48. $8^{2/3}$

49. $(-27)^{-4/3}$

50. $-9^{-1/2}$

Group Project

A Spiral of Roots

Overview: In this activity, you will gain a better understanding of square roots and get additional practice using the Pythagorean theorem.

Instructions: Form groups of 2 students. You will need an $8\frac{1}{2} \times 11$ inch piece of white, unlined paper, a 3×5 inch index card (to trace a right angle), a ruler, and a pencil.

Draw an isosceles right triangle with legs 1 inch long near the right edge of the paper. Use the Pythagorean theorem to determine the length of the hypotenuse.

Next, construct a second right triangle using the hypotenuse of the first triangle as one leg. Draw a second leg that is 1 inch long. Find the length of the hypotenuse of Triangle 2 using the Pythagorean theorem. Continue creating right triangles, using the previous hypotenuse as one leg and drawing a second leg that is 1 inch long. For each right triangle created, use the Pythagorean theorem to find the length of its hypotenuse. Make a list of the lengths of each hypotenuse, and look for a pattern.

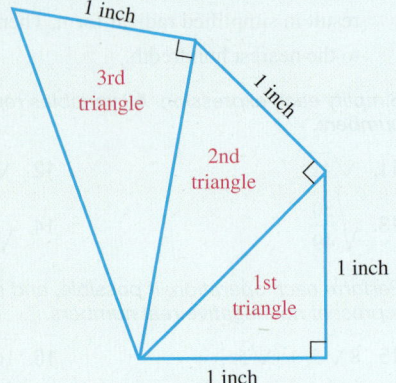

CUMULATIVE REVIEW ▶▶ Chapters 1–8

1. Determine whether each statement is true or false. [Section 1.3]
   a. All whole numbers are integers.
   b. π is a rational number.
   c. A real number is either rational or irrational.

2. Evaluate: $\dfrac{-3(3+2)^2 - (-5)}{17 - |-22|}$ [Section 1.7]

3. Simplify: $3p - 6(p + z) + p$ [Section 1.9]

4. Solve: $2 - (4x + 7) = 3 + 2(x + 2)$ [Section 2.2]

5. **Backpacks.** Pediatricians advise that children should not carry more than 20% of their own body weight in a backpack. According to this warning, how much weight can a fifth-grade girl who weighs 85 pounds safely carry in her backpack? [Section 2.3]

6. **Surface Area.** The total surface area A of a box with dimensions l, w, and h is given by the formula $A = 2lw + 2wh + 2lh$. If $A = 202$ square inches, $l = 9$ inches, and $w = 5$ inches, find h. [Section 2.4]

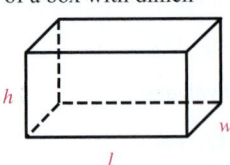

7. **Search and Rescue.** Two search and rescue teams leave base at the same time, looking for a lost boy. The first team, on foot, heads north at 2 mph and the other, on horseback, south at 4 mph. How long will it take them to search a distance of 21 miles between them? [Section 2.6]

8. **Blending Coffee.** A store sells regular coffee for $8 a pound and gourmet coffee for $14 a pound. Using 40 pounds of the gourmet coffee, the owner makes a blend to put on sale for $10 a pound. How many pounds of regular coffee should he use? [Section 2.6]

9. Solve: $3 - 3x \geq 6 + x$. Graph the solution set. Then describe the graph using interval notation. [Section 2.7]

10. Is $(-6, -7)$ a solution of $4x - 3y = -4$? [Section 3.1]

Graph each equation.

11. $y = \dfrac{1}{2}x$ [Section 3.2]

12. $3x - 4y = 12$ [Section 3.3]

13. $x = 5$ [Section 3.3]

14. $y = 2x^2 - 3$ [Section 5.4]

15. **Skype.** The line graph below shows the approximate growth in the number of subscribers to Skype, a software application that allows users to make video calls over the Internet. Find the rate of change in the number of subscribers to Skype during the years 2007–2010 by finding the slope of the line. [Section 3.4]

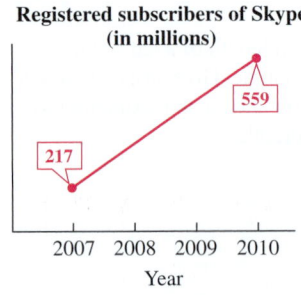

Registered subscribers of Skype (in millions)

Source: businessinsider.com

16. What is the slope of the line defined by each equation? [Section 3.5]
   a. $y = 3x - 7$
   b. $2x + 3y = -10$

17. Find the equation of the line passing through $(-2, 5)$ and $(4, 8)$. Write the answer in slope–intercept form. [Section 3.6]

18. If $f(x) = 2x^2 - 3x + 1$, find $f(-3)$. [Section 3.8]

19. **Boating.** The graph shows the vertical distance from a point on the tip of a propeller to the centerline as the propeller spins. Is this the graph of a function? [Section 3.8]

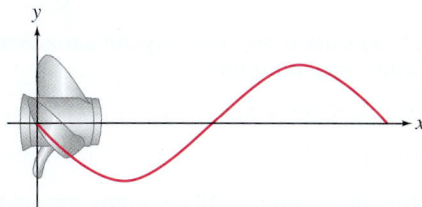

20. Solve the system $\begin{cases} x + y = 4 \\ y = x + 6 \end{cases}$ by graphing. [Section 4.1]

Solve each system of equations.

21. $\begin{cases} x = y + 4 \\ 2x + y = 5 \end{cases}$ [Section 4.2]

22. $\begin{cases} 3s + 4t = 5 \\ 2s - 3t = -8 \end{cases}$ [Section 4.3]

23. **Financial Planning.** In investing \$6,000 of a couple's money, a financial planner put some of it into a savings account paying 6% annual interest. The rest was invested in a riskier mini-mall development plan paying 12% annually. The combined interest earned for the first year was \$540. How much money was invested at each rate? Use two variables to solve this problem. [Section 4.4]

24. Graph: $\begin{cases} 3x + 2y \geq 6 \\ x + 3y \leq 6 \end{cases}$ [Section 4.5]

Simplify. Use only positive exponents in your answers.

25. $(x^5)^2(x^7)^3$ [Section 5.1]

26. $\left(\dfrac{a^3b}{c^4}\right)^5$ [Section 5.1]

27. $4^{-3} \cdot 4^{-2} \cdot 4^5$ [Section 5.2]

28. $(2a^{-2}b^3)^{-4}$ [Section 5.2]

29. **Astronomy.** The **parsec,** a unit of distance used in astronomy, is 3×10^{16} meters. The distance to Betelgeuse, a star in the constellation Orion, is 1.6×10^2 parsecs. Use scientific notation to express this distance in meters. [Section 5.3]

30. Write 0.0000000043 in scientific notation. [Section 5.3]

Perform the operations.

31. $(3a^2 - 2a + 4) - (a^2 - 3a + 7)$ [Section 5.5]

32. $0.3p^5(0.4p^4 - 6p^2)$ [Section 5.6]

33. $(-3t + 2s)(2t - 3s)$ [Section 5.6]

34. $(4b - 8)^2$ [Section 5.7]

35. $\left(6b + \dfrac{1}{2}\right)\left(6b - \dfrac{1}{2}\right)$ [Section 5.7]

36. $x + 2\overline{)2x^2 + 3x - 2}$ [Section 5.8]

Factor completely, if possible.

37. $12x^2y - 6xy^2 + 9xy^3$ [Section 6.1]

38. $2x^2 + 2xy - 3x - 3y$ [Section 6.1]

39. $x^2 + 7x + 10$ [Section 6.2]

40. $6a^2 - 7a - 20$ [Section 6.3]

41. $6 + 3x^2 + x$ [Section 6.3]

42. $25a^2 - 70ab + 49b^2$ [Section 6.4]

43. $a^3 + 8b^3$ [Section 6.5]

44. $2x^5 - 32x$ [Section 6.6]

Solve each equation.

45. $x^2 + 3x + 2 = 0$ [Section 6.7]

46. $5x^2 = 10x$ [Section 6.7]

47. $6x^2 - x = 2$ [Section 6.7]

48. $a^2 - 25 = 0$ [Section 6.7]

49. **Children's Stickers.** A rectangular-shaped sticker has an area of 20 cm². The width is 1 cm shorter than the length. Find the length of the sticker. [Section 6.8]

50. Simplify: $\dfrac{x^2 + 2x + 1}{x^2 - 1}$ [Section 7.1]

Perform the operations. Simplify, if possible.

51. $\dfrac{p^2 - p - 6}{3p - 9} \div \dfrac{p^2 + 6p + 9}{p^2 - 9}$ [Section 7.2]

52. $\dfrac{12x^2}{7 - x} \cdot \dfrac{x - 7}{20x^3}$ [Section 7.2]

53. $\dfrac{13}{15a} - \dfrac{8}{15a}$ [Section 7.3]

54. $\dfrac{x + 2}{x + 5} - \dfrac{x - 3}{x + 7}$ [Section 7.4]

55. $\dfrac{1}{6b^4} - \dfrac{8}{9b^2}$ [Section 7.4]

56. $\dfrac{\dfrac{1}{x} + \dfrac{1}{y}}{\dfrac{1}{x} - \dfrac{1}{y}}$ [Section 7.5]

57. Solve: $\dfrac{7}{a^2 - a - 2} + \dfrac{1}{a + 1} = \dfrac{3}{a - 2}$ [Section 7.6]

58. **Filling a Pool.** An inlet pipe can fill an empty swimming pool in 5 hours, and another inlet pipe can fill the pool in 4 hours. How long will it take both pipes to fill the pool? [Section 7.7]

59. **Online Sales.** A company found that, on average, it made 9 online sales transactions for every 500 hits on its Internet Web site. If the company's Web site had 360,000 hits in one year, how many sales transactions did it have that year? [Section 7.8]

60. **Cafeterias.** The length of time the punch in a dispenser at a children's summer camp will last varies inversely as the number of children at the camp. If the punch in the dispenser is consumed by 45 children in 3 days, how long will the punch last for 30 children? [Section 7.9]

61. Cargo Space. How wide a piece of plywood can be stored diagonally in the back of the van? [Section 8.1]

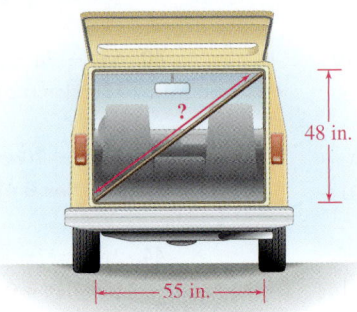

48 in.

?

55 in.

62. Explain why $\sqrt{-4}$ is not a real number. [Section 8.1]

Simplify. All variables represent nonnegative real numbers.

63. $\sqrt{64}$ [Section 8.1]   **64.** $\sqrt{32x^3}$ [Section 8.2]

65. $-\sqrt[3]{-27m^6}$ [Section 8.6]   **66.** $\sqrt[4]{81t^4}$ [Section 8.6]

Perform each operation and simplify. All variables represent nonnegative real numbers.

67. $\sqrt{48} - \sqrt{8} + \sqrt{27} - \sqrt{32}$ [Section 8.3]

68. $\left(\sqrt{y} - 4\right)\left(\sqrt{y} - 5\right)$ [Section 8.4]

Rationalize the denominator. All variables represent positive real numbers.

69. $\dfrac{4}{\sqrt{5}}$ [Section 8.4]   **70.** $\sqrt{\dfrac{17}{2x}}$ [Section 8.4]

71. Solve: $\sqrt{6x + 19} - 5 = 2$ [Section 8.5]

72. Evaluate: $16^{3/2}$ [Section 8.6]

Quadratic Equations

9

©Mark Richards/PhotoEdit

from Campus to Careers

Graphic Designer

Graphic designers combine their artistic talents with their mathematical skills to create attractive advertisements, photography, packaging, and even Web pages. They perform arithmetic computations with fractions and decimals to determine proper font sizes and page margins. They apply algebraic concepts such as graphing and equation solving to plan page layouts. And they use formulas to create production schedules and calculate costs.

Problem 89 in **Study Set 9.1** and **problem 77** in **Study Set 9.3** involve situations that a graphic designer might encounter on the job. The mathematical concepts discussed in this chapter can be used to solve those problems.

JOB TITLE:
Graphic Designer

EDUCATION:
A bachelor's degree is required for most entry-level positions.

JOB OUTLOOK:
Employment is expected to increase between 7% and 13% through the year 2018.

ANNUAL EARNINGS:
Median annual salary for staff-level designers is $45,000.

FOR MORE INFORMATION:
www.bls.gov/oco/0C05090.htm

Before moving on to a new mathematics course, it's worthwhile to take some time to reflect on your effort and performance in this course.

Now Try This ▶

As this course draws to a close, here are some questions to ask yourself.

1. How was my attendance?
2. Was I organized? Did I have the right materials?
3. Did I follow a regular schedule?
4. Did I pay attention in class and take good notes?
5. Did I spend the appropriate amount of time on homework?
6. How did I prepare for tests? Did I have a test-taking strategy?
7. Was I part of a study group? If not, why not? If so, was it worthwhile?
8. Did I ever seek extra help from a tutor or from my instructor?
9. In what topics was I the strongest? In what topics was I the weakest?
10. If I had it to do over, would I do anything differently?

SECTION 9.1

Solving Quadratic Equations: The Square Root Property

OBJECTIVES

1. Use the square root property to solve equations of the form $x^2 = c$.

2. Use the square root property to solve equations of the form $(ax + b)^2 = c$.

3. Solve problems modeled by quadratic equations.

ARE YOU READY?

The following problems review some basic skills that are needed when solving quadratic equations.

1. Factor: $x^2 - 64$

2. How many square roots does 9 have? What are they?

3. Simplify: $\sqrt{75}$

4. Rationalize the denominator: $\sqrt{\dfrac{5}{7}}$

5. Simplify: $\sqrt{\dfrac{23}{16}}$

6. Approximate to the nearest hundredth: $\sqrt{51}$

Recall that a **quadratic equation** can be written in the form $ax^2 + bx + c = 0$, where a, b, and c represent real numbers and $a \neq 0$. Some examples of quadratic equations are

$$x^2 - x - 6 = 0, \qquad 4x^2 + 4x + 1 = 0, \qquad \text{and} \qquad x^2 - 16 = 0$$

We have solved quadratic equations like these using factoring in combination with the zero-factor property. To review this method, let's solve $x^2 - 16 = 0$.

$$x^2 - 16 = 0$$
$$(x + 4)(x - 4) = 0 \qquad \text{\color{red}{Factor the difference of two squares.}}$$
$$x + 4 = 0 \quad \text{or} \quad x - 4 = 0 \qquad \text{\color{red}{Set each factor equal to 0.}}$$
$$x = -4 \qquad\qquad x = 4 \qquad \text{\color{red}{Solve each equation.}}$$

The solutions are -4 and 4.

1 **Use the Square Root Property to Solve Equations of the Form $x^2 = c$.**

We will now solve $x^2 - 16 = 0$ in another way. This time, we will ignore the zero-factor property condition that requires 0 on one side of the equation. Instead, we will add 16 to both sides to isolate x^2.

$$x^2 - 16 = 0$$
$$x^2 = 16 \qquad \text{\color{red}{Add 16 to both sides.}}$$

We see that x must be a number whose square is 16. Therefore, x must be a square root of 16. Since *every positive number has two square roots,* one positive and one negative, we have

$$x = \sqrt{16} \quad \text{or} \quad x = -\sqrt{16}$$
$$x = 4 \qquad\quad\ \ \ |\qquad x = -4$$

As with the factor method, we find that the solutions of $x^2 - 16 = 0$ are 4 and -4. This approach illustrates how the following *square root property* can be used to solve quadratic equations that consist of a squared quantity and a constant.

The Square Root Property of Equations	For any nonnegative real number c, if $x^2 = c$, then $$x = \sqrt{c} \quad \text{or} \quad x = -\sqrt{c}$$

We can write the conclusion of the square root property ($x = \sqrt{c}$ or $x = -\sqrt{c}$) in more compact form, called **double-sign notation:**

$$x = \pm\sqrt{c}$$ This is formally read as "x equals the positive or negative square root of c." However, it is often read more informally as "x equals plus or minus the square root of c."

EXAMPLE 1   Solve: **a.** $x^2 - 5 = 0$ **b.** $x^2 = 12$ **c.** $3a^2 + 1 = 11$ **d.** $n^2 = -4$

Strategy   We will use properties of equality to isolate the *squared term* on one side of the equation. Then we will use the square root property to isolate the variable itself.

Why   To solve the original equation, we want to find a simpler equivalent equation of the form **a variable = ± a number**, whose solutions are obvious.

Solution   **a.** We can use the addition property of equality to isolate x^2.

$$x^2 - 5 = 0 \quad \text{This is the equation to solve.}$$
$$x^2 = 5 \quad \text{Add 5 to both sides.}$$

Now we use the square root property to isolate x.

$$x = \sqrt{5} \quad \text{or} \quad x = -\sqrt{5}$$
$$x = \pm\sqrt{5} \qquad\qquad \text{Write the result using double-sign notation.}$$

To verify that $\sqrt{5}$ and $-\sqrt{5}$ are solutions, we substitute each one into the original equation.

The check for $\sqrt{5}$:	*The check for* $-\sqrt{5}$:
$x^2 - 5 = 0$	$x^2 - 5 = 0$
$(\sqrt{5})^2 - 5 \stackrel{?}{=} 0$	$(-\sqrt{5})^2 - 5 \stackrel{?}{=} 0$
$5 - 5 \stackrel{?}{=} 0$	$5 - 5 \stackrel{?}{=} 0$
$0 = 0$ True	$0 = 0$ True

Since each statement is true, the solutions of $x^2 - 5 = 0$ are $\sqrt{5}$ and $-\sqrt{5}$, or $\pm\sqrt{5}$. The solution set is written as $\{-\sqrt{5}, \sqrt{5}\}$ or $\{\pm\sqrt{5}\}$.

b. Since x^2 is already isolated, we simply apply the square root property.

$$x^2 = 12 \qquad\qquad\qquad \text{This is the equation to solve.}$$

$$x = \sqrt{12} \quad \text{or} \quad x = -\sqrt{12} \qquad \text{To isolate x, use the square root property.}$$
$$x = \pm\sqrt{12} \qquad\qquad\qquad \text{Use double-sign notation.}$$
$$x = \pm 2\sqrt{3} \qquad\qquad\qquad \text{Simplify the radical: } \sqrt{12} = \sqrt{4 \cdot 3} = \sqrt{4}\sqrt{3} = 2\sqrt{3}.$$

Verify that $2\sqrt{3}$ and $-2\sqrt{3}$ are solutions by checking each in the original equation.

c. First, we will use properties of equality to isolate a^2. Then, we will use the square root property for a.

$$3a^2 + 1 = 11 \qquad \text{This is the equation to solve.}$$

$$3a^2 = 10 \qquad \text{To isolate the term } 3a^2, \text{ subtract 1 from both sides.}$$

$$a^2 = \frac{10}{3} \qquad \text{To isolate } a^2, \text{ divide both sides by 3.}$$

$$a = \pm\sqrt{\frac{10}{3}} \qquad \text{To isolate } a, \text{ use the square root property: } a = \sqrt{\frac{10}{3}} \text{ or } a = -\sqrt{\frac{10}{3}}.$$
$$\text{Use double-sign notation.}$$

The Language of Algebra

The *exact* solutions are $\pm\frac{\sqrt{30}}{3}$. To the nearest hundredth, the *approximate* solutions are ± 1.83.

To rationalize the denominator of the result, we proceed as follows:

$$a = \pm\sqrt{\frac{10}{3}} = \pm\frac{\sqrt{10}}{\sqrt{3}} \cdot \frac{\sqrt{3}}{\sqrt{3}} = \pm\frac{\sqrt{30}}{3}$$

Verify that $\frac{\sqrt{30}}{3}$ and $-\frac{\sqrt{30}}{3}$ are solutions by checking each in the original equation.

d. $n^2 = -4 \qquad \text{This is the equation to solve.}$

$n = \pm\sqrt{-4} \qquad \text{Use the square root property. Write the result using double-sign notation.}$

Since the square root of -4 is not a real number, the equation has no real-number solutions.

Self Check 1 Solve: **a.** $x^2 - 21 = 0$ **b.** $b^2 = 54$
c. $5m^2 - 1 = 6$ **d.** $x^2 = -25$

Now Try ▶ Problems 17, 19, 23, and 25

2 Use the Square Root Property to Solve Equations of the Form $(ax + b)^2 = c$.

We can extend the square root property to solve equations that involve the square of a binomial and a constant.

EXAMPLE 2 Solve: **a.** $(x - 3)^2 = 36$ **b.** $(x + 1)^2 = 50$ **c.** $(2s - 4)^2 = 7$

Strategy Instead of a variable squared on the left side of the equation, we have a quantity squared. We still use the square root property to solve each equation.

Why We want to eliminate the square on the binomial, so that we can eventually isolate the variable on one side of the equation.

Solution **a.**

$$(x - 3)^2 = 36 \qquad \text{This is the equation to solve.}$$

$$x - 3 = \sqrt{36} \quad \text{or} \quad x - 3 = -\sqrt{36} \qquad \text{Use the square root property.}$$

$$x - 3 = \pm\sqrt{36} \qquad \text{Write the result using double-sign notation.}$$

$$x - 3 = \pm 6 \qquad \text{Evaluate: } \sqrt{36} = 6.$$

$$x = 3 \pm 6 \qquad \text{To isolate } x, \text{ add 3 to both sides. It is standard practice to write the 3 in front of the } \pm \text{ symbol.}$$

Caution

It might be tempting to square the binomial on the left side of $(x - 3)^2 = 36$. However, that causes unnecessary, additional steps to be used to solve the equation in another way.

$(x - 3)^2 = 36$
$x^2 - 6x + 9 = 36$

We read 3 ± 6 as "3 plus or minus 6." To find the solutions, we perform the calculation using a plus symbol $+$ and then using a minus symbol $-$.

$$x = 3 + 6 \qquad \text{or} \qquad x = 3 - 6$$
$$x = 9 \qquad \qquad \qquad x = -3$$

To verify that 9 and -3 are solutions, we substitute each one into the original equation.

The check for 9:	**The check for −3:**
$(x - 3)^2 = 36$	$(x - 3)^2 = 36$
$(9 - 3)^2 \stackrel{?}{=} 36$	$(-3 - 3)^2 \stackrel{?}{=} 36$
$(6)^2 \stackrel{?}{=} 36$	$(-6)^2 \stackrel{?}{=} 36$
$36 = 36$ True	$36 = 36$ True

The solutions are 9 and −3. The solution set is written as $\{-3, 9\}$.

The Language of Algebra

The ± symbol is often seen in political polls. A candidate with 48% support (±4%) could be between 48 + 4 = 52% and 48 − 4 = 44%.

b. $(x + 1)^2 = 50$ This is the equation to solve.

$x + 1 = \pm\sqrt{50}$ By the square root property, $x + 1 = \sqrt{50}$ or $x + 1 = -\sqrt{50}$. Use double-sign notation.

$x + 1 = \pm 5\sqrt{2}$ $\sqrt{50} = \sqrt{25 \cdot 2} = \sqrt{25}\sqrt{2} = 5\sqrt{2}$.

$x = -1 \pm 5\sqrt{2}$ To isolate x, subtract 1 from both sides (or add −1 to both sides). It is standard practice to write the −1 in front of the ± symbol. Read as "−1 plus or minus 5 times $\sqrt{2}$."

Use a check to verify that the solutions are $-1 + 5\sqrt{2}$ and $-1 - 5\sqrt{2}$. The solution set is written as $\left\{-1 + 5\sqrt{2},\ -1 - 5\sqrt{2}\right\}$ or $\left\{-1 \pm 5\sqrt{2}\right\}$.

Caution

Since 2 is not a common factor of the entire numerator, it would be incorrect to simplify the solutions as shown:

$$s = \frac{\overset{2}{\cancel{4}} \pm \sqrt{7}}{\underset{1}{\cancel{2}}}$$

c. $(2s - 4)^2 = 7$ This is the equation to solve.

$2s - 4 = \pm\sqrt{7}$ By the square root property, $2s - 4 = \sqrt{7}$ or $2s - 4 = -\sqrt{7}$. Use double-sign notation.

$2s = 4 \pm \sqrt{7}$ To isolate the variable term 2s, add 4 to both sides.

$s = \dfrac{4 \pm \sqrt{7}}{2}$ To isolate s, divide both sides by 2. Read as "4 plus or minus $\sqrt{7}$, divided by 2."

Use a check to verify that the solutions are $\dfrac{4 + \sqrt{7}}{2}$ and $\dfrac{4 - \sqrt{7}}{2}$.

Self Check 2   Solve each equation: **a.** $(x - 2)^2 = 64$
b. $(x + 3)^2 = 98$
c. $(3r - 1)^2 = 3$

Now Try ▶ Problems 27, 31, and 33

CAUTION Equations like $(x - 3)^2 = -16$ and $(5y + 6)^2 = -4$ have no real-number solutions because no real number squared is negative.

If one side of an equation factors as the square of a binomial and the other side is a constant, we can use the methods of Example 2 to solve the equation.

EXAMPLE 3   Solve: $x^2 + 16x + 64 = 2$

Strategy   We will attempt to factor the trinomial on the left side of the equation. Our hope is that it factors as the square of a binomial.

Why   If the equation can be written in the form $(ax + b)^2 = c$, we can use the square root property to solve it.

Solution   Since $x^2 + 16x + 64 = x^2 + 2 \cdot x \cdot 8 + 8^2$, it is a perfect-square trinomial and factors as $(x + 8)^2$.

The Language of Algebra

Recall that trinomials that are squares of binomials are called **perfect-square trinomials**.

$x^2 + 16x + 64 = 2$ This is the equation to solve.

$(x + 8)^2 = 2$ Factor the perfect-square trinomial $x^2 + 16x + 64$.

$x + 8 = \pm\sqrt{2}$ By the square root property, $x + 8 = \sqrt{2}$ or $x + 8 = -\sqrt{2}$. Use double-sign notation.

$x = -8 \pm \sqrt{2}$ To isolate x, subtract 8 from both sides. It is standard practice to write the −8 in front of the ± symbol.

As an informal check, we can use a calculator to approximate $-8 + \sqrt{2}$ and $-8 - \sqrt{2}$ to the nearest tenth. Then we can substitute each approximation into the original equation.

The check for $-8 + \sqrt{2} \approx -6.6$:

$$x^2 + 16x + 64 = 2$$
$$(-6.6)^2 + 16(-6.6) + 64 \stackrel{?}{=} 2$$
$$1.96 \approx 2$$

The check for $-8 - \sqrt{2} \approx -9.4$:

$$x^2 + 16x + 64 = 2$$
$$(-9.4)^2 + 16(-9.4) + 64 \stackrel{?}{=} 2$$
$$1.96 \approx 2$$

In each case, the sides are approximately equal. These results suggest that $-8 + \sqrt{2}$ and $-8 - \sqrt{2}$ are the exact solutions, and -6.6 and -9.4 are approximate solutions.

Self Check 3   Solve: $x^2 - 14x + 49 = 11$. Approximate the solutions to the nearest tenth.

Now Try ▶ Problem 35

3 Solve Problems Modeled by Quadratic Equations.

The equation-solving methods discussed in this section can be used to solve a variety of real-world applications that are modeled by quadratic equations.

EXAMPLE 4   **Movie Stunts.**   In a scene for an action movie, a stuntwoman falls from the top of a 95-foot-tall building into a 10-foot-tall airbag directly below her on the ground. The formula $d = 16t^2$ gives the distance d in feet that she falls in t seconds. For how many seconds will she fall before making contact with the airbag?

Strategy   We will calculate the distance the woman falls, substitute that value into the formula $d = 16t^2$, and solve for t.

Why   The variable t represents the unknown time the woman free-falls.

Solution   The woman will fall a distance of $95 - 10 = 85$ feet before making contact with the airbag. To find the number of seconds that the fall will last, we substitute 85 for d in the formula and solve for t, the time.

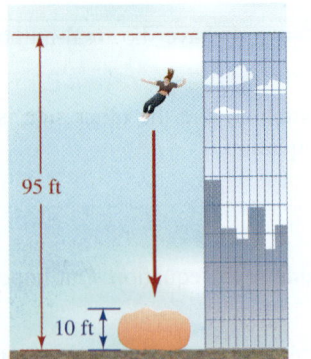

95 ft

10 ft

$$d = 16t^2 \quad \text{This is the formula that models free-falling objects.}$$
$$85 = 16t^2 \quad \text{Substitute 85 for } d\text{, the distance in feet, that the stuntwoman falls.}$$

The resulting quadratic equation is easily solved by the square root property.

$$\frac{85}{16} = t^2 \quad \text{To isolate } t^2\text{, divide both sides by 16.}$$

$$\pm\sqrt{\frac{85}{16}} = t \quad \text{To isolate } t\text{, use the square root property. Write the result using double-sign notation.}$$

$$\pm\frac{\sqrt{85}}{\sqrt{16}} = t \quad \text{Use the quotient rule for square roots: The square root of a quotient is the quotient of square roots.}$$

$$\pm\frac{\sqrt{85}}{4} = t \quad \text{Evaluate: } \sqrt{16} = 4. \text{ Since } 85 = 5 \cdot 17, \text{ we cannot simplify } \sqrt{85}.$$

The stuntwoman will fall for exactly $\frac{\sqrt{85}}{4}$ seconds before making contact with the airbag. We can use a calculator to find that this is approximately 2.3 seconds. We will discard the other solution, $-\frac{\sqrt{85}}{4}$, because a negative time does not make sense in this example.

Self Check 4	**Falling Objects.** If a penny is dropped from a bridge that is 48 feet above a river, how long would it take for it to hit the ground?
Now Try ▶	Problem 79

SECTION 9.1 ▶ **STUDY SET**

VOCABULARY

Fill in the blanks.

1. $x^2 - 15 = 0$ is an example of a _____ equation.
2. $x^2 + 6x + 9$ is a perfect-_____ trinomial because $x^2 + 6x + 9 = (x + 3)^2$.

CONCEPTS

Fill in the blanks.

3. The square root property of equations: If $x^2 = c$, then $x = $ ____ or $x = $ _____.

4. a. If $x^2 = 5$, then $x = \pm$ ____.

   b. If $(x - 2)^2 = 7$, then $x - 2 = \pm$ ____.

5. Use a property of equality to isolate the variable on the left side of the equation.

   a. $x + 9 = \pm\sqrt{2}$

   b. $6x = 3 \pm \sqrt{2}$

6. Rationalize the denominator: $x = \pm\sqrt{\dfrac{7}{2}}$

7. Is $2\sqrt{5}$ a solution of $x^2 = 20$?

8. Is -4 a solution of $(x - 1)^2 = 25$?

NOTATION

9. Write the statement $x = \sqrt{6}$ or $x = -\sqrt{6}$ using a $\pm$ symbol (double-sign notation).

10. Fill in the blanks: $2 \pm \sqrt{3}$ is read as "Two _____ or _____ the square root of three."

GUIDED PRACTICE

Use the square root property to solve each equation, if possible. See Example 1.

11. $x^2 - 36 = 0$

12. $x^2 - 4 = 0$

13. $x^2 = \dfrac{49}{16}$

14. $x^2 = \dfrac{81}{121}$

15. $5x^2 = 125$

16. $4x^2 = 16$

17. $x^2 - 6 = 0$

18. $x^2 - 7 = 0$

19. $m^2 = 20$

20. $n^2 = 32$

21. $t^2 = 72$

22. $n^2 = 75$

23. $2x^2 + 8 = 23$

24. $3m^2 + 5 = 18$

25. $x^2 = -81$

26. $y^2 = -100$

Use the square root property to solve each equation. See Example 2.

27. $(x + 1)^2 = 25$

28. $(x - 1)^2 = 49$

29. $(x - 2)^2 = 8$

30. $(x + 2)^2 = 50$

31. $(s + 9)^2 = 63$

32. $(t - 11)^2 = 45$

33. $(5c - 10)^2 - 6 = 0$

34. $(4n + 8)^2 - 17 = 0$

Use the square root property to solve each equation. Approximate each solution to the nearest tenth. See Example 3.

35. $x^2 + 2x + 1 = 10$

36. $x^2 + 8x + 16 = 6$

37. $x^2 - 18x + 81 = 7$

38. $x^2 - 14x + 49 = 19$

39. $a^2 - 6a + 9 = 40$

40. $b^2 - 10b + 25 = 90$

41. $m^2 + 4m + 4 = 75$

42. $m^2 + 16m + 64 = 80$

TRY IT YOURSELF

Use the square root property to solve each equation, if possible.

43. $(x + 12)^2 = 27$

44. $(m + 1)^2 = 32$

45. $m^2 = 98$

46. $n^2 = 99$

47. $4x^2 = 400$

48. $3m^2 = 27$

49. $(3x + 1)^2 - 18 = 0$

50. $(6y + 5)^2 - 72 = 0$

51. $b^2 - 12b + 36 = 2$

52. $a^2 - 18a + 81 = 5$

53. $b^2 - 17 = 0$

54. $a^2 - 26 = 0$

55. $6r^2 - 3 = 4$

56. $2w^2 - 9 = 12$

57. $(y - 15)^2 - 8 = 0$

58. $(y - 5)^2 - 12 = 0$

59. $t^2 = \dfrac{1}{144}$

60. $d^2 = \dfrac{1}{9}$

61. $4(t - 7)^2 - 12 = 0$ 62. $2(t - 6)^2 - 22 = 0$

63. $h^2 + 25 = 0$ 64. $4r^2 + 16 = 0$

65. $5x^2 + 1 = 18$ 66. $7x^2 + 3 = 6$

67. $(x + 2)^2 = 81$ 68. $(x + 3)^2 = 16$

69. $x^2 - 14 = 0$ 70. $x^2 - 46 = 0$

71. $(8y + 9)^2 = 44$ 72. $(6y + 13)^2 = 99$

73. $\frac{1}{2}a^2 + 6 = 4$ 74. $\frac{1}{3}m^2 + 9 = 3$

Look Alikes . . .

Solve the equation in part a using factoring. Solve the equation in part b using the square root property.

75. a. $b^2 - 9 = 0$ b. $b^2 - 8 = 0$
76. a. $5x^2 - 125 = 0$ b. $5x^2 - 60 = 0$
77. a. $x^2 - 14x + 49 = 0$ b. $x^2 - 14x + 49 = 6$

78. a. $r^2 = 16$ b. $r^2 = 17$

APPLICATIONS

For Exercises 79–84, use the formula $d = 16t^2$, where d is the distance an object falls (in feet) and t is the time (in seconds) that it has been falling. **See Example 4.**

79. **Lighthouses.** The 144-foot-tall Tybee Island Lighthouse is located near Savannah, Georgia. If an object is dropped from the top of the lighthouse, how long would it take for it to hit the ground?

80. **Skyscrapers.** A downtown office building on the north side of the Chicago River is 784 feet tall. If an object is dropped from the top of the building, how long would it take for it to hit the ground?

81. **Science History.** Legend has it that Galileo Galilei (1564–1642) dropped two objects having different weights from the leaning tower of Pisa in order to prove that they fall at the same rate. If a steel ball is dropped from the lowest side of the tower, and falls 183 feet, how long will it take to hit the ground? Round to the nearest tenth of a second.

82. **Studying Microgravity.** NASA's Glenn Research Center in Cleveland, Ohio, has a 435-foot drop tower that begins on the surface and descends into Earth like a mineshaft. How long will it take a sealed container to fall 435 feet? Round to the nearest tenth of a second.

83. **Daredevils.** In 1873, Henry Bellini combined a tightrope walk with a leap into the Niagara River below, where he was picked up by a boat. If the rope was 200 feet above the water, for how many seconds did he fall before hitting the water? Round to the nearest tenth of a second.

84. **Roller Coasters.** Soaring to a height of 420 feet, the *Top Thrill Dragster* at Cedar Point Amusement Park in Sandusky, Ohio, is one of the tallest and fastest roller coasters in the world. If a rider accidentally dropped a camera as the coaster reached its highest point, how long would it take the camera to hit the ground? Round to the nearest tenth.

85. **Pro Wrestling.** A WWE (World Wrestling Entertainment) ring is square in shape and has an area of 400 ft². Find the length of a side of the ring. (*Hint:* Use the formula for the area of a square, $A = s^2$.)

86. **Chess.** A tournament chessboard is square in shape and has an area of 441 in.². Find the length of a side of the board. (*Hint:* Use the formula for the area of a square, $A = s^2$.)

87. **Escape Velocity.** The speed at which a rocket must be fired for it to leave the earth's gravitational attraction is called the *escape velocity*. If the formula for the escape velocity v, in miles per hour, is given by $\frac{v^2}{2g} = R$, where $g = 78{,}545$ and $R = 3{,}960$, find v. Round to the nearest mi/hr.

A launch speed of v_e results in this path.

A launch speed slightly less than v_e results in this path

88. **Carousels.** In 1999, the city of Lancaster, Pennsylvania, considered installing a classic Dentzel carousel in an abandoned downtown building. After learning that the circular carousel (like that shown below) would occupy 2,376 square feet of floor space the proposal was determined to be impractical because of the large remodeling costs. Find the radius of the carousel to the nearest tenth of one foot. (*Hint:* Use the formula for the area of a circle, $A = \pi r^2$.)

89.

from Campus To Careers

Graphic Designer

©Mark Richards/PhotoEdit

A **golden rectangle** is one of the most visually appealing of all geometric forms. It is often used in art, architecture, and graphics because it is a shape that seems "right" to the eye. The length of a golden rectangle is 1.618 times as long as its width. A graphic artist wants to design a poster with an area of approximately 275 square inches using the dimensions of a golden rectangle. Find the length and the width of the poster. Round to the nearest inch.

SAVE A LIFE

ADOPT A PET
SPONSORED BY
YOUR LOCAL HUMANE SOCIETY

$1.618x$

x

90. Marlins. The formula

$$w = \frac{Lg^2}{800}$$

is used by sports fishermen to estimate the weight w (in pounds) of a blue marlin that is L inches long with a girth of g inches. (See the figure.) Find the girth of a 360-pound marlin that is 144 inches long. Round to the nearest inch.

Girth

Length

WRITING

91. Explain why the equation $x^2 + 16 = 0$ has no real-number solutions.

92. Explain why the notation $6 \pm \sqrt{2}$ represents two real numbers.

93. Explain the error in the following work.

a. Solve: $x^2 = 7$

$x = \sqrt{7}$

b. Solve: $x^2 = 28$

$x = \pm\sqrt{28}$

$x = 2 \pm \sqrt{7}$

94. Explain two methods for solving the equation $x^2 - 100 = 0$.

REVIEW

Solve each equation.

95. $\sqrt{5x - 6} = 2$

96. $\sqrt{6x + 1} + 2 = 7$

97. $2\sqrt{x} = \sqrt{5x - 16}$

98. $\sqrt{22y + 86} = y + 9$

CHALLENGE PROBLEMS

Solve each equation.

99. $25\left(x + \dfrac{1}{3}\right)^2 = 144$

100. $100(0.2x - 1.2)^2 = 64$

SECTION 9.2

Solving Quadratic Equations: Completing the Square

OBJECTIVES

1 Complete the square to write perfect-square trinomials.

2 Solve quadratic equations with leading coefficients of 1 by completing the square.

3 Solve quadratic equations with leading coefficients other than 1 by completing the square.

ARE YOU READY?

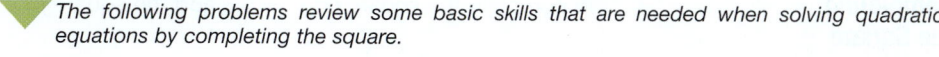

The following problems review some basic skills that are needed when solving quadratic equations by completing the square.

1. Factor: $x^2 - 10x + 25$

2. What is the coefficient of the middle term of $x^2 + 6x - 14$?

3. Find one-half of 12. Then square that result.

4. Find $\dfrac{1}{2} \cdot 3$. Then square that result.

5. Solve $x - 6 = \pm\sqrt{7}$ for x.

6. Add: $2 + \dfrac{37}{4}$

In Section 9.1, we used the square root property to solve equations such as $(x - 3)^2 = 36$ and $(x + 1)^2 = 50$, whose left side is a binomial squared and right side is a constant. We also solved equations such as $x^2 + 16x + 64 = 2$ in a similar way by first factoring the perfect-square trinomial on the left side.

In this section, we will discuss a procedure that enables us to solve quadratic equations such as $x^2 + 4x = -3$, whose left side is not a perfect-square trinomial. To make the left side a perfect-square trinomial, we will use a procedure called *completing the square*.

1 Complete the Square to Write Perfect-Square Trinomials.

In Section 5.7, we learned how to square binomials quickly using special-product rules. Two examples are shown below.

The square of a binomial		*The perfect-square trinomial result*
$(x + 4)^2$	$=$	$x^2 + 8x + 16$
$(x - 5)^2$	$=$	$x^2 - 10x + 25$

In both results, there is a relationship between the coefficient of x and the constant (third) term. In $x^2 + 8x + 16$, for example, the coefficient of x is 8 and the constant term is 16. Note that the constant, 16, is the square of one-half the coefficient of x. Similarly, in $x^2 - 10x + 25$, the constant, 25, is the square of one-half the coefficient of x, which is -10.

Now, let's generalize. Consider the following perfect-square trinomials (with leading coefficients of 1) and their factored forms.

$$x^2 + 2bx + b^2 = (x + b)^2 \qquad \text{and} \qquad x^2 - 2bx + b^2 = (x - b)^2$$

In each of these perfect-square trinomials, the third term is the square of one-half of the coefficient of x.

- In $x^2 + \mathbf{2b}x + b^2$, the coefficient of x is $\mathbf{2b}$. If we find $\frac{1}{2} \cdot \mathbf{2b}$, which is b, and square it, we get the third term, b^2.
- In $x^2 - \mathbf{2b}x + b^2$, the coefficient of x is $\mathbf{-2b}$. If we find $\frac{1}{2}(\mathbf{-2b})$, which is $-b$, and square it, we get the third term: $(-b)^2 = b^2$.

We can use these observations to change certain binomials into perfect-square trinomials. For example, to change $x^2 + 12x$ into a perfect-square trinomial, we find one-half of the coefficient of x, square the result, and add the square to $x^2 + 12x$.

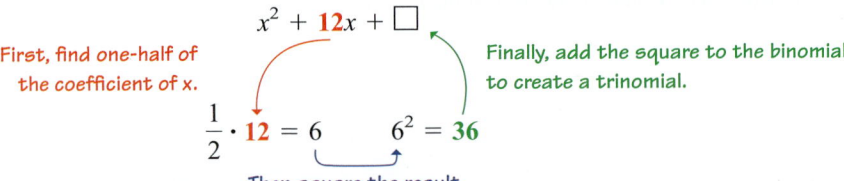

First, find one-half of the coefficient of x.

Finally, add the square to the binomial to create a trinomial.

$$\frac{1}{2} \cdot \mathbf{12} = 6 \qquad 6^2 = \mathbf{36}$$

Then square the result.

We obtain the perfect-square trinomial $x^2 + 12x + 36$ that factors as $(x + 6)^2$. By adding 36 to $x^2 + 12x$, we **completed the square** on $x^2 + 12x$.

Completing the Square

To complete the square on $x^2 + bx$, add the square of one-half of the coefficient of x:

$$x^2 + bx + \left(\frac{1}{2}b\right)^2$$

EXAMPLE 1 Complete the square and factor the resulting perfect-square trinomial: **a.** $x^2 + 6x$ **b.** $x^2 - 5x$ **c.** $x^2 + \frac{7}{3}x$

Strategy We will add the square of one-half of the coefficient of x to the given binomial.

Why Adding such a term will change the binomial into a perfect-square trinomial that will factor.

Solution **a.** In $x^2 + \mathbf{6}x$, the coefficient of x is $\mathbf{6}$. One-half of $\mathbf{6}$ is 3, and $3^2 = 9$. If we add 9 to $x^2 + 6x$, it becomes a perfect-square trinomial.

$$x^2 + 6x + \mathbf{9} \qquad \text{To complete the square: } \tfrac{1}{2} \cdot 6 = 3 \text{ and } 3^2 = 9. \text{ Add 9 to the binomial.}$$

The Language of Algebra

When we add 9 to $x^2 + 6x$, we say that we have **completed the square** on $x^2 + 6x$. For that reason, binomials such as $x^2 + 6x$ are often called **incomplete squares**.

The resulting trinomial factors as $(x + 3)^2$. We can check using multiplication.

Check: $(x + 3)^2 = (x + 3)(x + 3) = x^2 + 3x + 3x + 9 = x^2 + 6x + 9$

b. In $x^2 - 5x$, the coefficient of x is -5. One-half of -5 is $-\frac{5}{2}$, and $\left(-\frac{5}{2}\right)^2 = \frac{25}{4}$. If we add $\frac{25}{4}$ to $x^2 - 5x$, it becomes a perfect-square trinomial.

$$x^2 - 5x + \frac{25}{4} \qquad \text{To complete the square: } \tfrac{1}{2}(-5) = -\tfrac{5}{2} \text{ and } \left(-\tfrac{5}{2}\right)^2 = \tfrac{25}{4}.$$
$$\text{Add } \tfrac{25}{4} \text{ to the binomial.}$$

Caution

When we complete the square on a binomial, we are not writing an equivalent trinomial expression. Since the result is a completely different polynomial, it would be incorrect to use an = symbol between the two.

~~$x^2 - 5x = x^2 - 5x + \dfrac{25}{4}$~~

The resulting trinomial factors as $\left(x - \frac{5}{2}\right)^2$. We can check using multiplication.

Check: $\left(x - \dfrac{5}{2}\right)^2 = \left(x - \dfrac{5}{2}\right)\left(x - \dfrac{5}{2}\right) = x^2 - \dfrac{5}{2}x - \dfrac{5}{2}x + \dfrac{25}{4} = x^2 - 5x + \dfrac{25}{4}$

c. In $x^2 + \frac{7}{3}x$, the coefficient of x is $\frac{7}{3}$. One-half of $\frac{7}{3}$ is $\frac{7}{6}$ and $\left(\frac{7}{6}\right)^2 = \frac{49}{36}$. If we add $\frac{49}{36}$ to $x^2 + \frac{7}{3}x$, it becomes a perfect-square trinomial.

$$x^2 + \frac{7}{3}x + \frac{49}{36} \qquad \text{To complete the square: } \tfrac{1}{2}\left(\tfrac{7}{3}\right) = \tfrac{7}{6} \text{ and } \left(\tfrac{7}{6}\right)^2 = \tfrac{49}{36}. \text{ Add } \tfrac{49}{36} \text{ to the binomial.}$$

The resulting trinomial factors as $\left(x + \frac{7}{6}\right)^2$. Check using multiplication.

Self Check 1 Complete the square for each expression and factor the resulting perfect-square trinomial: **a.** $y^2 - 8y$
b. $y^2 + 3y$ **c.** $x^2 + \frac{7}{5}x$

Now Try ▶ Problems 13, 15, and 23

2 Solve Quadratic Equations with Leading Coefficients of 1 by Completing the Square.

We can use completing the square to solve quadratic equations.

EXAMPLE 2 Solve by completing the square: $x^2 + 4x = 12$

Strategy We will use the addition property of equality and add the square of one-half of the coefficient of x to both sides of the equation.

Why This will create a perfect-square trinomial on the left side that will factor as the square of a binomial. Then we can use the square root property to solve for x.

Solution In $x^2 + 4x$, the coefficient of x is 4. One-half of 4 is 2, and $2^2 = 4$. If we add 4 to $x^2 + 4x$, it becomes a perfect-square trinomial.

$$x^2 + 4x \qquad\ = 12 \qquad\qquad \text{This is the equation to solve.}$$
$$x^2 + 4x + 4 = 12 + 4 \qquad \text{To complete the square on the left side, add 4 to both sides.}$$
$$(x + 2)^2 = 16 \qquad\qquad \text{Factor the perfect-square trinomial. Add on the right side.}$$
$$x + 2 = \pm\sqrt{16} \qquad \text{By the square root property, } x + 2 = \sqrt{16} \text{ or } x + 2 = -\sqrt{16}.$$
$$\text{Write the result using double-sign notation.}$$
$$x + 2 = \pm 4 \qquad\qquad \text{Evaluate: } \sqrt{16} = 4.$$
$$x = -2 \pm 4 \qquad \text{To isolate } x, \text{ subtract 2 from both sides.}$$

Success Tip

We also could solve $x^2 + 4x = 12$ by writing it in the form $x^2 + 4x - 12 = 0$ and using factoring:

$$(x + 6)(x - 2) = 0$$
$$x = -6 \quad \text{or} \quad x = 2$$

$$x = -2 + 4 \quad \text{or} \quad x = -2 - 4$$

To find the solutions, perform the calculation using a + symbol and then using a − symbol.

$$x = 2 \qquad\qquad x = -6$$

The solutions are 2 and -6 and the solution set is $\{-6, 2\}$. Verify this by substituting each one in the original equation.

Self Check 2 Solve by completing the sqaure: $x^2 + 4x = 32$

Now Try ▶ Problem 25

EXAMPLE 3 Solve by completing the square: $x^2 - 8x - 5 = 0$. Approximate the solutions to the nearest hundredth.

Strategy We will use the addition property of equality and add 5 to both sides. Then we will proceed as in Example 2.

Why To prepare to complete the square, we need to isolate the variable terms, x^2 and $-8x$, on the left side of the equation and the constant term on the right side.

Solution
$$x^2 - 8x - 5 = 0 \quad \text{This is the equation to solve.}$$
$$x^2 - 8x = 5 \quad \text{Add 5 to both sides so that the constant term is on the right side.}$$

In $x^2 - \mathbf{8}x$, the coefficient of x is $-\mathbf{8}$. One-half of $-\mathbf{8}$ is -4, and $(-4)^2 = 16$. If we add 16 to $x^2 - 8x$, it becomes a perfect-square trinomial.

$$x^2 - 8x + \mathbf{16} = 5 + \mathbf{16} \quad \text{To complete the square on the left side, add 16 to both sides.}$$
$$(x - 4)^2 = 21 \quad \text{Factor the perfect-square trinomial. Add on the right side.}$$
$$x - 4 = \pm\sqrt{21} \quad \text{By the square root property, } x - 4 = \sqrt{21} \text{ or } x - 4 = -\sqrt{21}. \text{ Use double-sign notation to show this.}$$
$$x = 4 \pm \sqrt{21} \quad \text{To isolate } x, \text{ add 4 to both sides.}$$

Success Tip

Example 2 but not Example 3 can be solved by factoring. These observations illustrate that the method of completing the square can be used to solve any quadratic equation.

The exact solutions of $x^2 - 8x - 5 = 0$ are $4 + \sqrt{21}$ and $4 - \sqrt{21}$.

We also can approximate each solution by using the decimal approximation of $\sqrt{21}$, which is 4.582575695:

$$4 + \sqrt{21} \approx 4 + 4.582575695 \qquad 4 - \sqrt{21} \approx 4 - 4.582575695$$
$$\approx 8.58 \qquad\qquad\qquad \approx -0.58$$

To the nearest hundredth, the solutions are 8.58 and -0.58. These approximations can be used to check the exact solutions informally by substituting each of them into the original equation.

Self Check 3 Solve by completing the square: $x^2 - 10x - 4 = 0$. Approximate the solutions to the nearest hundredth.

Now Try ▶ Problem 37

EXAMPLE 4 Solve by completing the square: $a^2 - 7a = 2$. Approximate the solutions to the nearest hundredth.

Strategy We will use the addition property of equality and add the square of one-half of the coefficient of a to both sides of the equation.

Why This will create a perfect-square trinomial on the left side that will factor as the square of a binomial. Then we can use the square root property to solve for a.

Solution

In $a^2 - 7a$, the coefficient of a is -7. One-half of -7 is $-\frac{7}{2}$, and $\left(-\frac{7}{2}\right)^2 = \frac{49}{4}$. If we add $\frac{49}{4}$ to $a^2 - 7a$, it becomes a perfect-square trinomial.

$$a^2 - 7a = 2 \qquad \text{This is the equation to solve.}$$

$$a^2 - 7a + \frac{49}{4} = 2 + \frac{49}{4} \qquad \begin{array}{l}\text{To complete the square on the left side,}\\ \text{add } \frac{49}{4} \text{ to both sides.}\end{array}$$

$$\left(a - \frac{7}{2}\right)^2 = \frac{2}{1} \cdot \frac{4}{4} + \frac{49}{4} \qquad \begin{array}{l}\text{On the left side, factor the perfect-square trinomial. On}\\ \text{the right side, prepare to add by writing 2 as } \frac{2}{1} \text{ and}\\ \text{building it so that its denominator is 4.}\end{array}$$

$$\left(a - \frac{7}{2}\right)^2 = \frac{8}{4} + \frac{49}{4} \qquad \text{On the right side, multiply: } \frac{2}{1} \cdot \frac{4}{4} = \frac{8}{4}.$$

$$\left(a - \frac{7}{2}\right)^2 = \frac{57}{4} \qquad \begin{array}{l}\text{On the right side, add the numerators.}\\ \text{Write the sum, 57, over the common denominator, 4.}\end{array}$$

$$a - \frac{7}{2} = \pm\sqrt{\frac{57}{4}} \qquad \begin{array}{l}\text{Use the square root property. Write the result using}\\ \text{double-sign notation.}\end{array}$$

$$a - \frac{7}{2} = \pm\frac{\sqrt{57}}{2} \qquad \text{Use the quotient rule to simplify: } \sqrt{\frac{57}{4}} = \frac{\sqrt{57}}{\sqrt{4}} = \frac{\sqrt{57}}{2}.$$

$$a = \frac{7}{2} \pm \frac{\sqrt{57}}{2} \qquad \text{To isolate } a, \text{ add } \frac{7}{2} \text{ to both sides.}$$

$$a = \frac{7 \pm \sqrt{57}}{2} \qquad \begin{array}{l}\text{Write the sum (and difference) over}\\ \text{the common denominator 2.}\end{array}$$

The exact solutions of $a^2 - 7a = 2$ are $\frac{7 + \sqrt{57}}{2}$ and $\frac{7 - \sqrt{57}}{2}$. If we approximate them to the nearest hundredth, we have

$$\frac{7 + \sqrt{57}}{2} \approx 7.27 \qquad \text{and} \qquad \frac{7 - \sqrt{57}}{2} \approx -0.27$$

Self Check 4 Solve by completing the square: $b^2 - 3b = 3$. Approximate the solutions to the nearest hundredth.

Now Try ▶ Problem 41

CAUTION Completing the square sometimes leads to equations such as $(x - 1)^2 = -9$ and $(y + 6)^2 = -25$. Recall from Section 9.1 that equations like these have no real-number solutions because no real number squared is negative.

3 Solve Quadratic Equations with Leading Coefficients Other Than 1 by Completing the Square.

The method of completing the square can be used to solve any quadratic equation. However, when the coefficient of the squared variable (called the **leading coefficient**) is not 1, we must make it 1 before we can complete the square.

EXAMPLE 5 Solve by completing the square: $2x^2 + 5x + 1 = 0$

Strategy We will use the division property of equality and divide both sides by 2 so that the coefficient of x^2 is 1.

Why We create a leading coefficient that is 1 so that we can complete the square to solve the equation.

Solution

$$2x^2 + 5x + 1 = 0$$

This is the equation to solve.

$$x^2 + \frac{5}{2}x + \frac{1}{2} = 0$$

To make the leading coefficient 1, divide both sides by 2, term-by-term: $\frac{2x^2}{2} + \frac{5x}{2} + \frac{1}{2} = \frac{0}{2}$.

$$x^2 + \frac{5}{2}x = -\frac{1}{2}$$

Subtract $\frac{1}{2}$ from both sides so that the constant term is on the right side.

$$x^2 + \frac{5}{2}x + \frac{25}{16} = -\frac{1}{2} + \frac{25}{16}$$

Complete the square: $\frac{1}{2}\left(\frac{5}{2}\right) = \frac{5}{4}$ and $\left(\frac{5}{4}\right)^2 = \frac{25}{16}$. Add $\frac{25}{16}$ to both sides.

$$\left(x + \frac{5}{4}\right)^2 = -\frac{1}{2} \cdot \frac{8}{8} + \frac{25}{16}$$

On the left side, factor the perfect-square trinomial. On the right, prepare to add by building $-\frac{1}{2}$ so that its denominator is 16.

$$\left(x + \frac{5}{4}\right)^2 = -\frac{8}{16} + \frac{25}{16}$$

On the right side, multiply: $-\frac{1}{2} \cdot \frac{8}{8} = -\frac{8}{16}$.

$$\left(x + \frac{5}{4}\right)^2 = \frac{17}{16}$$

On the right side, add the numerators. Write the result, 17, over the common denominator 16.

$$x + \frac{5}{4} = \pm\sqrt{\frac{17}{16}}$$

Use the square root property. Write the result using double-sign notation.

$$x + \frac{5}{4} = \pm\frac{\sqrt{17}}{4}$$

Use the quotient rule to simplify: $\sqrt{\frac{17}{16}} = \frac{\sqrt{17}}{\sqrt{16}} = \frac{\sqrt{17}}{4}$.

$$x = -\frac{5}{4} \pm \frac{\sqrt{17}}{4}$$

To isolate x, subtract $\frac{5}{4}$ from both sides.

$$x = \frac{-5 \pm \sqrt{17}}{4}$$

Write the sum (and difference) over the common denominator 4.

The exact solutions are $\frac{-5 \pm \sqrt{17}}{4}$. Check each one or its approximation in the original equation.

> **Success Tip**
>
> We also could make the leading coefficient 1 by multiplying both sides of the equation by $\frac{1}{2}$:
>
> $$\frac{1}{2}(2x^2 + 5x + 1) = \frac{1}{2} \cdot 0$$

Self Check 5   Solve by completing the square: $3x^2 + x - 1 = 0$

Now Try ▶ Problem 49

To solve a quadratic equation in x by completing the square, we follow these steps.

Completing the Square to Solve a Quadratic Equation in x

1. If the coefficient of x^2 is 1, go to step 2. If it is not 1, make it 1 by dividing both sides of the equation by the coefficient of x^2.

2. Get all variable terms on one side of the equation and constants on the other side.

3. Complete the square by finding one-half of the coefficient of x, squaring the result, and adding the square to both sides of the equation.

4. Factor the perfect-square trinomial as the square of a binomial.

5. Solve the resulting equation using the square root property.

6. Check your answers in the original equation.

EXAMPLE 6   Solve by completing the square: $4x^2 - 3 = 24x$. Approximate the solutions to the nearest hundredth.

Strategy   We will use the addition and subtraction properties of equality to get the variable terms on one side of the equation and the constant term on the other. Then we will use the division property of equality and divide both sides by 4 so that the coefficient of x^2 is 1.

Why This will create a leading coefficient of 1 so that we can complete the square to solve the equation.

Solution

$$4x^2 - 3 = 24x$$

This is the equation to solve.

$$4x^2 - 24x = 3$$

To have both variable terms on the left side, subtract 24x from both sides. To have the constant term on the right, add 3 to both sides.

$$x^2 - 6x = \frac{3}{4}$$

To make the coefficient of the x^2 term 1, divide both sides by 4: $\frac{4x^2}{4} - \frac{24x}{4} = \frac{3}{4}$.

Caution

A common error is to add a constant to one side of an equation to complete the square and forget to add it to the other side.

$$x^2 - 6x + 9 = \frac{3}{4} + 9$$

Complete the square: $\frac{1}{2}(-6) = -3$ and $(-3)^2 = 9$. Add 9 to both sides.

$$(x - 3)^2 = \frac{39}{4}$$

On the left side, factor. On the right side, express 9 as $\frac{36}{4}$ and add to $\frac{3}{4}$ to get $\frac{39}{4}$.

$$x - 3 = \pm\sqrt{\frac{39}{4}}$$

Use the square root property. Write the result using double-sign notation.

$$x - 3 = \pm\frac{\sqrt{39}}{2}$$

Use the quotient rule to simplify: $\sqrt{\frac{39}{4}} = \frac{\sqrt{39}}{\sqrt{4}} = \frac{\sqrt{39}}{2}$.

$$x = 3 \pm \frac{\sqrt{39}}{2}$$

To isolate x, add 3 to both sides.

$$x = \frac{6}{2} \pm \frac{\sqrt{39}}{2}$$

To write the solutions in compact form, express 3 as a fraction with denominator 2: $3 = \frac{3}{1} \cdot \frac{2}{2} = \frac{6}{2}$.

$$x = \frac{6 \pm \sqrt{39}}{2}$$

Write the sum (and difference) over the common denominator 2.

The exact solutions are $\frac{6 \pm \sqrt{39}}{2}$. Check each one in the original equation. We can approximate the solutions using a calculator. To the nearest hundredth, we have

$$\frac{6 + \sqrt{39}}{2} \approx 6.12 \qquad \frac{6 - \sqrt{39}}{2} \approx -0.12$$

Self Check 6 Solve by completing the square: $4d^2 - 1 = 32d$. Approximate the solutions to the nearest hundredth.

Now Try ▶ Problem 59

SECTION 9.2 > STUDY SET

VOCABULARY

Fill in the blanks.

1. When we add 9 to $x^2 + 6x$, we say that we have completed the _____ on $x^2 + 6x$.

2. The _____ coefficient of $5x^2 - 2x + 7$ is 5 and the _____ term is 7.

CONCEPTS

3. Find one-half of the given number and square the result.
   a. 6   b. −5

4. Factor each perfect-square trinomial.
   a. $x^2 + 14x + 49$ b. $x^2 - 2x + 1$

5. Fill in the blank: To complete the square on $x^2 + 8x$, add the square of _____ of the coefficient of x.

6. Fill in the blanks to complete the square. Then factor the resulting perfect-square trinomial.
   a. $x^2 + 8x + \boxed{} = (x + \boxed{})^2$

   b. $x^2 - 9x + \boxed{} = \left(x - \boxed{}\right)^2$

7. What is the first step to solve the equation by completing the square? **Do not solve.**

   a. $x^2 + 9x + 7 = 0$

   b. $4x^2 + 5x - 16 = 0$

8. Check to determine whether $3 + \sqrt{7}$ is a solution of $x^2 - 6x + 2 = 0$.

9. Determine whether each statement is true or false.

   a. Any quadratic equation can be solved by the factoring method.

   b. Any quadratic equation can be solved by completing the square.

10. Simplify: $\sqrt{\dfrac{19}{4}}$

NOTATION

11. Translate to mathematical symbols: *the square of one-half of nine.* Then evaluate the expression.

12. Fill in the blanks: $\dfrac{5}{3} \pm \dfrac{\sqrt{17}}{3} = \dfrac{\boxed{}}{\boxed{}}$

GUIDED PRACTICE

Complete the square and factor the resulting perfect-square trinomial. See Example 1.

13. $x^2 + 2x$

14. $x^2 + 12x$

15. $x^2 - 4x$

16. $x^2 - 14x$

17. $a^2 - 7a$

18. $b^2 + 11b$

19. $x^2 + x$

20. $x^2 - x$

21. $b^2 - \dfrac{2}{3}b$

22. $t^2 - \dfrac{6}{5}t$

23. $x^2 - \dfrac{5}{2}x$

24. $x^2 - \dfrac{7}{6}x$

Solve each equation by completing the square. See Example 2.

25. $x^2 + 4x = 5$

26. $x^2 + 6x = 7$

27. $g^2 - 2g = 15$

28. $s^2 - 5s = 14$

29. $x^2 + 6x = -8$

30. $x^2 + 8x = -12$

31. $k^2 - 8k = -12$

32. $p^2 - 4p = -3$

Solve each equation by completing the square. Approximate the solutions to the nearest hundredth. See Example 3.

33. $s^2 - 4s - 3 = 0$

34. $t^2 - 6t + 3 = 0$

35. $x^2 - 2x - 17 = 0$

36. $x^2 + 10x - 7 = 0$

37. $x^2 + 8x - 6 = 0$

38. $x^2 + 6x - 2 = 0$

39. $x^2 + 6x + 4 = 0$

40. $x^2 + 8x + 6 = 0$

Solve each equation by completing the square. Approximate the solutions to the nearest hundredth. See Example 4.

41. $x^2 - 7x = 5$

42. $x^2 + 5x = 7$

43. $b^2 - 5b = 10$

44. $b^2 - 3b = 5$

45. $t^2 + 3t = 20$

46. $a^2 + a = 3$

47. $x^2 + x = 9$

48. $x^2 + 7x = 6$

Solve each equation by completing the square. See Example 5.

49. $2x^2 - 7x - 3 = 0$

50. $3x^2 - 5x - 1 = 0$

51. $4a^2 - 9a + 1 = 0$

52. $4a^2 - 11a + 1 = 0$

53. $5b^2 + 3b - 4 = 0$

54. $2x^2 + 9x + 3 = 0$

55. $3x^2 + 5x - 5 = 0$

56. $5b^2 + 7b - 2 = 0$

Solve each equation by completing the square. See Example 6.

57. $4x^2 - 13 = 24x$

58. $4d^2 - 9 = 16d$

59. $9a^2 - 5 = -18a$

60. $9r^2 - 7 = 36r$

61. $4t^2 + 11 = 48t$

62. $4x^2 + 13 = 40x$

63. $16x^2 + 3 = -64x$

64. $16b^2 + 1 = -32b$

Solve each equation by completing the square. Approximate the solutions to the nearest hundredth. See Example 6.

65. $2x^2 = -6x - 1$

66. $2x^2 + 3 = 10x$

67. $3x^2 - 4 = -2x$

68. $3x^2 = -4x + 3$

69. $4x^2 + 12x = 6$

70. $3x^2 - 6x = 5$

71. $6m^2 - 8m - 3 = 0$

72. $4t^2 - 8t - 1 = 0$

TRY IT YOURSELF

Solve each equation by completing the square.

73. $x^2 = -7x - 2$

74. $y^2 = -3y + 6$

75. $x^2 - 2x - 4 = 0$

76. $x^2 - 4x - 2 = 0$

77. $9x^2 - 36x - 1 = 0$

78. $16y^2 - 32y + 1 = 0$

79. $x^2 - 12x = -35$

80. $x^2 + 10x = -24$

81. $n^2 - 9n = 5$

82. $a^2 + 7a = 4$

83. $3n^2 - 8n = -4$

84. $3a^2 - 11a = 4$

85. $x^2 - 2x - 5 = 0$

86. $x^2 + 4x + 1 = 0$

87. $a^2 - 4a + 7 = 0$

88. $b^2 - 10b + 26 = 0$

89. $\frac{1}{2}t^2 - 1 = -\frac{5}{4}t$

90. $\frac{1}{2}s^2 - 4 = \frac{3}{4}s$

91. $3x^2 - 18x - 21 = 0$

92. $5x^2 - 20x - 25 = 0$

93. $2x^2 + 24x + 44 = 0$

94. $2x^2 + 32x + 6 = 0$

Look Alikes . . .

Solve each equation in part a using factoring. Solve the equation in part b by completing the square.

95. a. $y^2 + 5y + 4 = 0$
 b. $y^2 + 5y + 3 = 0$

96. a. $x^2 + 7x - 8 = 0$
 b. $x^2 + 7x - 9 = 0$

97. a. $2a^2 - 8a = 0$
 b. $2a^2 - 8a = 1$

98. a. $8x^2 + 2x - 3 = 0$
 b. $8x^2 + 2x - 2 = 0$

APPLICATIONS

99. **Geometry.** Completing the square can be applied to geometric figures. For example, the area of the figure in part (a) is $x^2 + x + x + x + x$, or simply $x^2 + 4x$. Note that the figure is not a square because it is missing a corner. Determine the number of small squares that were added in part (b) to "complete the square." Then fill in the blanks.

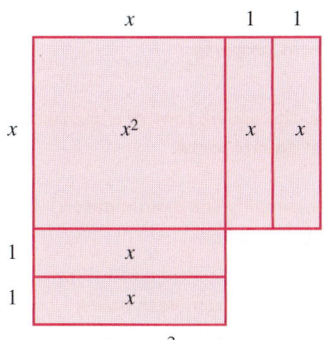

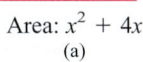

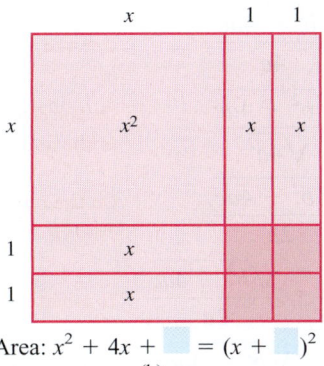

Area: $x^2 + 4x$ Area: $x^2 + 4x + \boxed{\ } = (x + \boxed{\ })^2$
      (a)   (b)

100. **Geometry.** Draw a figure similar to that in part (a) of Problem 99 that has area $x^2 + 8x$. Then draw part (b) to determine the number of small squares that must be added to "complete the square." Fill in the blanks:

Area: $x^2 + 8x + \boxed{\ } = (x + \boxed{\ })^2$

WRITING

101. Give an example of a perfect-square trinomial. Why do you think the word "perfect" is used to describe it?

102. Explain why completing the square on $x^2 + 5x$ is more difficult than completing the square on $x^2 + 4x$.

103. Find the error in the following solution.

Solve: $x^2 + 8x = 5$
$x^2 + 8x + 16 = 5$
$(x + 4)^2 = 5$
$x + 4 = \pm\sqrt{5}$
$x = -4 \pm \sqrt{5}$

104. Try to solve $x^2 - 4x = -40$ by completing the square and then explain why it has no real-number solutions.

REVIEW

Look Alikes . . .

Perform the indicated operation and simplify when possible.

105. $\dfrac{x+3}{x-3} \cdot \dfrac{x+1}{x^2-9}$

106. $\dfrac{x+3}{x-3} \div \dfrac{x+1}{x^2-9}$

107. $\dfrac{x+3}{x-3} + \dfrac{x+1}{x^2-9}$

108. $\dfrac{x+3}{x-3} - \dfrac{x+1}{x^2-9}$

CHALLENGE PROBLEMS

Solve each equation by completing the square.

109. $0.2x^2 + 0.4x + 0.1 = 0$

110. $0.03x^2 + 0.06x = 0.02$

111. $x(x + 3) - \dfrac{1}{2} = -2$

112. $x[(x - 2) + 3] = 3\left(x - \dfrac{2}{9}\right)$

Solving Quadratic Equations: The Quadratic Formula

OBJECTIVES

1 Use the quadratic formula to solve quadratic equations.

2 Identify quadratic equations with no real-number solutions.

3 Determine the most efficient method to use to solve a quadratic equation.

4 Solve problems modeled by quadratic equations.

ARE YOU READY?

The following problems review some basic skills that are needed when solving quadratic equations using the quadratic formula.

1. Evaluate: $\sqrt{5^2 - 4(4)(-6)}$

2. Simplify: $\sqrt{45}$

3. How many terms does $2x^2 - x + 7$ have? What is the coefficient of each term?

4. Evaluate: $\dfrac{-5 \pm 11}{8}$

5. Classify $\sqrt{-28}$ as rational, irrational, or not a real number.

6. Approximate $\dfrac{6 + \sqrt{3}}{2}$ to the nearest hundredth.

We can solve any quadratic equation by completing the square, but the work is often lengthy and involved. In this section, we will develop a formula that will enable us to solve quadratic equations with much less effort.

1 Use the Quadratic Formula to Solve Quadratic Equations.

To develop a formula that will produce the solutions of any given quadratic equation, we start with a quadratic equation in **standard form**, $ax^2 + bx + c = 0$, where $a > 0$. We can solve for x by completing the square.

$$ax^2 + bx + c = 0$$

$$\frac{ax^2}{a} + \frac{bx}{a} + \frac{c}{a} = \frac{0}{a}$$ Divide both sides by a so that the coefficient of x^2 is 1.

$$x^2 + \frac{b}{a}x + \frac{c}{a} = 0$$ Simplify: $\frac{ax^2}{a} = x^2$. Write $\frac{bx}{a}$ as $\frac{b}{a}x$.

$$x^2 + \frac{b}{a}x = -\frac{c}{a}$$ Subtract $\frac{c}{a}$ from both sides so that only the variable terms are on the left side of the equation and the constant is on the right side.

The Language of Algebra

Your instructor may ask you to *derive* the quadratic formula. That means to solve $ax^2 + bx + c = 0$ for x, using the series of steps shown here.

We can complete the square on $x^2 + \frac{b}{a}x$ by adding the square of one-half of the coefficient of x. Since the coefficient of x is $\frac{b}{a}$, we have $\frac{1}{2} \cdot \frac{b}{a} = \frac{b}{2a}$ and $\left(\frac{b}{2a}\right)^2 = \frac{b^2}{4a^2}$.

$$x^2 + \frac{b}{a}x + \frac{b^2}{4a^2} = -\frac{c}{a} + \frac{b^2}{4a^2}$$ To complete the square, add $\frac{b^2}{4a^2}$ to both sides.

$$x^2 + \frac{b}{a}x + \frac{b^2}{4a^2} = -\frac{4ac}{4aa} + \frac{b^2}{4a^2}$$ On the right side, build $-\frac{c}{a}$ by multiplying it by $\frac{4a}{4a}$. Now the fractions on that side have the common denominator $4a^2$.

$$\left(x + \frac{b}{2a}\right)^2 = \frac{b^2 - 4ac}{4a^2}$$ On the left side, factor the perfect-square trinomial. On the right side, add the fractions. In the numerator, write $-4ac + b^2$ as $b^2 - 4ac$.

$$x + \frac{b}{2a} = \pm\sqrt{\frac{b^2 - 4ac}{4a^2}}$$ Use the square root property.

$$x + \frac{b}{2a} = \pm\frac{\sqrt{b^2 - 4ac}}{\sqrt{4a^2}}$$ On the right side, the square root of a quotient is the quotient of square roots.

$$x + \frac{b}{2a} = \pm\frac{\sqrt{b^2 - 4ac}}{2a}$$ On the right side, simplify the denominator. Since $a > 0$, $\sqrt{4a^2} = 2a$.

$$x = -\frac{b}{2a} \pm \frac{\sqrt{b^2 - 4ac}}{2a}$$ To isolate x, subtract $\frac{b}{2a}$ from both sides.

$$x = \frac{-b \pm \sqrt{b^2 - 4ac}}{2a}$$ Combine the fractions. Write the sum (and difference) over the common denominator $2a$.

This result is called the **quadratic formula.** To develop this formula, we assumed that a was positive. If a is negative, similar steps are used, and we obtain the same result. This formula is very useful and should be memorized.

The Quadratic Formula

The solutions of $ax^2 + bx + c = 0$, with $a \neq 0$, are given by

$$x = \frac{-b \pm \sqrt{b^2 - 4ac}}{2a}$$

Read as "x equals the opposite of b plus or minus the square root of b squared minus 4ac, all over 2a."

The quadratic formula is a compact way of representing two solutions:

$$x = \frac{-b + \sqrt{b^2 - 4ac}}{2a} \qquad \text{or} \qquad x = \frac{-b - \sqrt{b^2 - 4ac}}{2a}$$

EXAMPLE 1   Use the quadratic formula to solve: $4x^2 + x - 3 = 0$

Strategy   We will compare the given equation to the standard form of a quadratic equation $ax^2 + bx + c = 0$ to identify a, b, and c.

Why   To use the quadratic formula, we need to know what numbers to substitute for a, b, and c in $x = \frac{-b \pm \sqrt{b^2 - 4ac}}{2a}$.

Solution

$4x^2 + x - 3 = 0$ This is the equation to solve.

$4x^2 + 1x - 3 = 0$ Write x as 1x.

$ax^2 + bx + c = 0$ This is standard form.

Caution

Make sure to include the correct sign when determining a, b, and c. In this example, $c = -3$.

We see that $a = 4$, $b = 1$, and $c = -3$. To find the solutions of $4x^2 + x - 3 = 0$, we substitute these numbers into the formula and evaluate the right side.

$$x = \frac{-b \pm \sqrt{b^2 - 4ac}}{2a}$$ This is the quadratic formula.

Caution

When writing the quadratic formula, be careful to draw the fraction bar so that it includes the entire numerator. Do not write

$$x = -b \pm \frac{\sqrt{b^2 - 4ac}}{2a}$$

$$x = \frac{-1 \pm \sqrt{1^2 - 4(4)(-3)}}{2(4)}$$ Substitute 4 for a, 1 for b, and −3 for c.

$$x = \frac{-1 \pm \sqrt{1 - (-48)}}{8}$$ Evaluate the power and multiply within the radical. Multiply in the denominator: 2(4) = 8.

$$x = \frac{-1 \pm \sqrt{49}}{8}$$ Evaluate the expression within the radical: 1 − (−48) = 1 + 48 = 49.

$$x = \frac{-1 \pm 7}{8}$$ Evaluate: $\sqrt{49} = 7$.

To find the first solution, evaluate the expression using the $+$ symbol. To find the second solution, evaluate the expression using the $-$ symbol.

$$x = \frac{-1 + 7}{8} \qquad \text{or} \qquad x = \frac{-1 - 7}{8}$$

$$x = \frac{6}{8} \qquad\qquad\qquad x = \frac{-8}{8}$$

$$x = \frac{3}{4} \qquad\qquad\qquad x = -1$$

Verify that $\frac{3}{4}$ and -1 are the solutions by substituting them into the original equation. The solution set is $\left\{-1, \frac{3}{4}\right\}$.

Self Check 1   Use the quadratic formula to solve: $3x^2 + x - 2 = 0$

Now Try ▶   Problem 17

EXAMPLE 2   Use the quadratic formula to solve $5x^2 + 1 = 5x$. Approximate the solutions to the nearest hundredth.

Strategy   We will use the subtraction property of equality to get 0 on the right side of the equation. Then we will compare the resulting equation to $ax^2 + bx + c = 0$ to identify a, b, and c.

Why   We need to know a, b, and c to use the quadratic formula.

Solution

$$5x^2 + 1 = 5x \qquad \text{This is the equation to solve.}$$

$$5x^2 - 5x + 1 = 0 \qquad \text{To get 0 on the right side, subtract 5x from both sides.}$$

$$ax^2 + bx + c = 0 \qquad \text{This is standard form.}$$

We see that $a = 5$, $b = -5$, and $c = 1$. To find the solutions, we substitute these values into the quadratic formula and evaluate the right side.

$$x = \frac{-b \pm \sqrt{b^2 - 4ac}}{2a} \qquad \text{This is the quadratic formula.}$$

$$x = \frac{-(-5) \pm \sqrt{(-5)^2 - 4(5)(1)}}{2(5)} \qquad \text{Substitute 5 for } a, -5 \text{ for } b, \text{ and 1 for } c.$$

$$x = \frac{5 \pm \sqrt{25 - 20}}{10} \qquad \text{Simplify: } -(-5) = 5. \text{ Evaluate the power and multiply within the radical. Multiply in the denominator.}$$

$$x = \frac{5 \pm \sqrt{5}}{10} \qquad \text{Subtract within the radical: } 25 - 20 = 5.$$

This fraction does not simplify further. Thus, the exact solutions are $\frac{5 \pm \sqrt{5}}{10}$. We can use a calculator to approximate them. To the nearest hundredth, we have

$$\frac{5 + \sqrt{5}}{10} \approx 0.72 \qquad\qquad \frac{5 - \sqrt{5}}{10} \approx 0.28$$

Success Tip

Example 1 but not Example 2 can be solved by factoring. These observations illustrate that the quadratic formula can be used to solve any quadratic equation.

When each approximation is substituted into the original equation, and the calculations are made, notice that the sides are approximately equal. This suggests that the results, $\frac{5 + \sqrt{5}}{10}$ and $\frac{5 - \sqrt{5}}{10}$, are reasonable.

$$5x^2 + 1 = 5x \qquad\qquad\qquad 5x^2 + 1 = 5x$$

$$5(0.72)^2 + 1 \overset{?}{=} 5(0.72) \qquad\qquad 5(0.28)^2 + 1 \overset{?}{=} 5(0.28)$$

$$3.592 \approx 3.6 \qquad\qquad\qquad 1.392 \approx 1.4$$

Self Check 2   Use the quadratic formula to solve: $4x^2 + 2 = 7x$. Approximate the solutions to the nearest hundredth.

Now Try ▶   Problem 31

EXAMPLE 3   Use the quadratic formula to solve: $3t^2 - 4 = -2t$

Strategy   We will use the addition property of equality to get 0 on the right side of the equation. Then we will compare the resulting equation to $at^2 + bt + c = 0$ to identify a, b, and c.

Why   We need to know a, b, and c to use the quadratic formula.

Solution

$$3t^2 - 4 = -2t \qquad \text{This is the equation to solve.}$$

$$\underset{\uparrow}{3t^2} + \underset{\uparrow}{2t} - \underset{\uparrow}{4} = 0 \qquad \text{To get 0 on the right side, add 2t to both sides.}$$

$$at^2 + bt + c = 0 \qquad \text{This is standard form using the variable t.}$$

Caution

Since the variable in the equation is t, not x, we must change the variable in the quadratic formula to reflect this:

$$t = \frac{-b \pm \sqrt{b^2 - 4ac}}{2a}$$

We see that $a = 3$, $b = 2$, and $c = -4$. To find the solutions, we substitute these values into the quadratic formula and evaluate the right side.

$$t = \frac{-b \pm \sqrt{b^2 - 4ac}}{2a} \qquad \text{This is the quadratic formula, where x is replaced by t.}$$

$$t = \frac{-2 \pm \sqrt{2^2 - 4(3)(-4)}}{2(3)} \qquad \text{Substitute 3 for a, 2 for b, and } -4 \text{ for c.}$$

$$t = \frac{-2 \pm \sqrt{4 - (-48)}}{6} \qquad \begin{array}{l}\text{Evaluate the power and multiply within the radical.}\\ \text{Multiply in the denominator: } 2(3) = 6.\end{array}$$

$$t = \frac{-2 \pm \sqrt{52}}{6} \qquad \text{Subtract within the radical: } 4 - (-48) = 4 + 48 = 52.$$

$$t = \frac{-2 \pm 2\sqrt{13}}{6} \qquad \text{Simplify: } \sqrt{52} = \sqrt{4}\sqrt{13} = 2\sqrt{13}.$$

This fraction can be simplified. We will factor the numerator and denominator and remove a common factor.

Caution

To simplify expressions such as $\frac{-2 \pm 2\sqrt{13}}{6}$, factor the numerator and denominator first. Then remove a common factor of the *entire numerator*. Some common errors are shown:

$$\frac{\overset{1}{\cancel{-2}} \pm 2\sqrt{13}}{\underset{1}{2 \cdot 3}} \qquad \frac{-2 \pm \overset{1}{\cancel{2}}\sqrt{13}}{\underset{1}{2 \cdot 3}}$$

$$t = \frac{2\left(-1 \pm \sqrt{13}\right)}{2 \cdot 3} \qquad \begin{array}{l}\text{Factor out the GCF, 2, from the two terms in the numerator.}\\ \text{In the denominator, factor 6 as } 2 \cdot 3.\end{array}$$

$$t = \frac{\overset{1}{\cancel{2}}\left(-1 \pm \sqrt{13}\right)}{\underset{1}{\cancel{2}} \cdot 3} \qquad \begin{array}{l}\text{Simplify the fraction by removing the common factor 2}\\ \text{in the numerator and denominator: } \frac{2}{2} = 1.\end{array}$$

$$t = \frac{-1 \pm \sqrt{13}}{3}$$

Thus, the exact solutions of $3t^2 - 4 = -2t$ are $\frac{-1 \pm \sqrt{13}}{3}$. Check each of them (or their approximations) in the original equation.

Self Check 3   Use the quadratic formula to solve: $2s^2 - 1 = -2s$

Now Try ▶ Problem 37

To solve a quadratic equation in x using the quadratic formula, we follow these steps.

Solving a Quadratic Equation in x Using the Quadratic Formula

1. Write the equation in standard form: $ax^2 + bx + c = 0$.

2. Identify a, b, and c.

3. Substitute the values for a, b, and c in the quadratic formula

$$x = \frac{-b \pm \sqrt{b^2 - 4ac}}{2a}$$

and evaluate the right side to obtain the solutions.

2 Identify Quadratic Equations with No Real-Number Solutions.

The next example shows that some quadratic equations have no real-number solutions.

EXAMPLE 4 Use the quadratic formula to solve $x^2 - x + 6 = 0$.

Strategy We will compare the given equation to the standard form of a quadratic equation $ax^2 + bx + c = 0$ to identify a, b, and c.

Why We need to know a, b, and c to use the quadratic formula.

Solution

Success Tip

When identifying a and b, recall that the term x^2 has an understood coefficient 1 and the term $-x$ has an understood coefficient -1.

$$x^2 = 1x^2 \quad \text{and} \quad -x = -1x$$

$$x^2 - x + 6 = 0 \quad \text{This is the equation to solve.}$$

$$1x^2 - 1x + 6 = 0 \quad \text{Write } x^2 \text{ as } 1x^2 \text{ and write } -x \text{ as } -1x.$$

$$ax^2 + bx + c = 0 \quad \text{This is standard form.}$$

We see that $a = 1$, $b = -1$, and $c = 6$ and substitute these values into the quadratic formula.

$$x = \frac{-b \pm \sqrt{b^2 - 4ac}}{2a}$$

$$x = \frac{-(-1) \pm \sqrt{(-1)^2 - 4(1)(6)}}{2(1)} \quad \text{Substitute 1 for } a, -1 \text{ for } b, \text{ and 6 for } c.$$

$$x = \frac{1 \pm \sqrt{1 - 24}}{2} \quad \begin{array}{l}\text{Evaluate the power and multiply within the radical.} \\ \text{Multiply in the denominator: } 2(1) = 2.\end{array}$$

$$x = \frac{1 \pm \sqrt{-23}}{2} \quad \text{Subtract within the radical: } 1 - 24 = -23.$$

Caution

When you obtain a result such as $x = \frac{1 \pm \sqrt{-19}}{2}$, don't forget to write, *"No real-number solutions."*

Since $\sqrt{-23}$ is not a real number, $x^2 - x + 6 = 0$ has no real-number solutions.

Self Check 4 Use the quadratic formula to solve: $x^2 - x + 1 = 0$

Now Try ▶ Problem 49

3 Determine the Most Efficient Method to Solve a Quadratic Equation.

We have discussed four methods that are used to solve quadratic equations. The following table shows some advantages and disadvantages of each method.

Method	Advantages	Disadvantages	Examples
Factoring and the zero-factor property	When each factor is set equal to 0, the resulting equations are usually easy to solve.	Some polynomials may be difficult to factor and others impossible.	$x^2 - 2x - 24 = 0$ $4a^2 + a = 0$
Square root property	It is the fastest way to solve equations of the form $ax^2 = n$ or $(ax + b)^2 = n$.	It only applies to equations that are in these forms.	$x^2 = 27$ $(2y + 3)^2 = 25$
Completing the square*	It can be used to solve any quadratic equation. It works well with equations of the form $x^2 + bx = n$, where b is even.	It involves more steps than the other methods. The algebra can be cumbersome if the leading coefficient is not 1 or the coefficient of the middle term is odd or a fraction.	$t^2 - 14t = 9$ $x^2 + 4x + 1 = 0$
Quadratic formula	It can be used to solve any quadratic equation.	It involves several calculations where sign errors can be made. Often the result must be simplified.	$x^2 + 3x - 33 = 0$ $4s^2 - 10s + 5 = 0$

*The quadratic formula is just a condensed version of completing the square and is usually easier to use. However, you need to know how to complete the square because it is used in more advanced mathematics courses.

4   Solve Problems Modeled by Quadratic Equations.

The equation-solving methods discussed in this section can be used to solve a variety of real-world applications that are modeled by quadratic equations.

EXAMPLE 5

Sailing.   The height of a triangular sail is 4 feet more than the length of the base. If the sail has an area of 30 square feet, find the length of its base and the height.

Analyze
- The height of the sail is 4 feet more than the length of the base.
- The area of the sail is 30 ft².
- Find the length of the base and height of the sail.

Assign   If we let b = the length of the base (in feet) of the triangular sail, then $b + 4 =$ the height of the sail (in feet).

Form   We can use the formula for the area of a triangle, $A = \frac{1}{2}bh$, to form an equation.

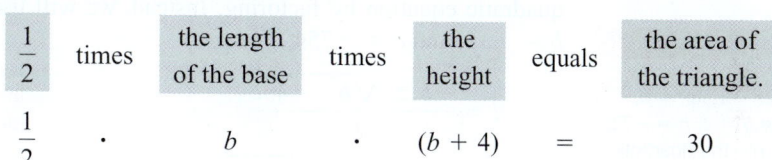

$$\frac{1}{2} \cdot b \cdot (b + 4) = 30$$

Solve

$$\frac{1}{2}b(b + 4) = 30$$

$$b(b + 4) = 60 \qquad \text{To clear the equation of the fraction, multiply both sides by 2.}$$

$$b^2 + 4b = 60 \qquad \text{Distribute the multiplication by } b. \text{ This is a quadratic equation.}$$

Since the coefficient of the b-term is the even number 4, this quadratic equation can be solved quickly by completing the square.

$$b^2 + 4b = 60$$

$$b^2 + 4b + 4 = 60 + 4 \qquad \text{Complete the square: } \tfrac{1}{2}(4) = 2 \text{ and } (2)^2 = 4. \\ \text{Add 4 to both sides.}$$

$$(b + 2)^2 = 64 \qquad \text{On the left side, factor the perfect-square trinomial.} \\ \text{On the right side, add: } 60 + 4 = 64.$$

$$b + 2 = \pm\sqrt{64} \qquad \text{Use the square root property. Use double-sign notation.}$$

$$b = -2 \pm 8 \qquad \text{To isolate } b, \text{ subtract 2 from both sides. Evaluate:} \\ \sqrt{64} = 8.$$

$$b = -2 + 8 \qquad \text{or} \qquad b = -2 - 8 \qquad \text{To find the solutions, perform the calculation} \\ \text{using a + symbol and then using a − symbol.}$$

$$b = 6 \qquad \qquad \cancel{b = -10} \qquad \text{Discard the solution } -10. \text{ The length} \\ \text{of the base cannot be negative.}$$

State   The length of the base of the sail is 6 feet. Since the height is given by $b + 4$, the height of the sail is $6 + 4 = 10$ feet.

Check   A height of 10 feet is 4 feet more than the length of the base, which is 6 feet. Also, the area of the triangle is $\frac{1}{2}(6)(10) = 30$ ft². The results check.

Success Tip

It is usually easier to clear quadratic equations of fractions before attempting to solve them.

Success Tip

This equation also could be solved by writing it in the form $b^2 + 4b - 60 = 0$ and factoring: $(b + 10)(b - 6) = 0$. Or, the quadratic formula could be used, where $a = 1$, $b = 4$, and $c = -60$.

Self Check 5   **Gardening.**   A rectangular garden is 4 feet longer than it is wide. If the garden has an area of 96 square feet, find the garden's length and width.

Now Try ▶   Problem 77

EXAMPLE 6

Televisions. A television's screen size is measured diagonally. For the 42-inch plasma television shown in the illustration, the screen's height is 16 inches less than its length. What are the length and height of the screen? Round to the nearest tenth.

Analyze A sketch of the screen shows that two adjacent sides and the diagonal form a right triangle. The length of the hypotenuse is 42 inches.

Assign If we let l = the length of the screen in inches, then $l - 16$ represents the height of the screen in inches.

Form We can use the Pythagorean theorem to form an equation.

$$a^2 + b^2 = c^2 \qquad \text{This is the Pythagorean equation.}$$

$$l^2 + (l - 16)^2 = 42^2 \qquad \text{Substitute } l \text{ for } a, l - 16 \text{ for } b, \text{ and } 42 \text{ for } c.$$

$$l^2 + l^2 - 32l + 256 = 1{,}764 \qquad \text{Find } (l-16)^2 \text{ and } 42^2.$$

$$2l^2 - 32l - 1{,}508 = 0 \qquad \begin{array}{l}\text{To get 0 on the right side of the equation,}\\ \text{subtract 1,764 from both sides.}\end{array}$$

$$l^2 - 16l - 754 = 0 \qquad \text{Divide both sides of the equation by 2: } \tfrac{2l^2}{2} - \tfrac{32l}{2} - \tfrac{1{,}508}{2} = \tfrac{0}{2}.$$

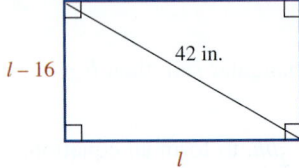

Solve Because of the large constant term, -754, we will not attempt to solve this quadratic equation by factoring. Instead, we will use the quadratic formula, with $a = 1$, $b = -16$, and $c = -754$.

$$l = \frac{-b \pm \sqrt{b^2 - 4ac}}{2a} \qquad \begin{array}{l}\text{In the quadratic formula,}\\ \text{replace the variable } x \text{ with } l.\end{array}$$

$$l = \frac{-(-16) \pm \sqrt{(-16)^2 - 4(1)(-754)}}{2(1)} \qquad \text{Substitute 1 for } a, -16 \text{ for } b, \text{ and } -754 \text{ for } c.$$

$$l = \frac{16 \pm \sqrt{256 - (-3{,}016)}}{2} \qquad \begin{array}{l}\text{Evaluate the power and multiply within the radical.}\\ \text{Multiply in the denominator: } 2(1) = 2.\end{array}$$

$$l = \frac{16 \pm \sqrt{3{,}272}}{2} \qquad \begin{array}{l}\text{Subtract within the radical:}\\ 256 - (-3{,}016) = 256 + 3{,}016 = 3{,}272.\end{array}$$

We can use a calculator to approximate each one to the nearest tenth. The negative solution is discarded because the length of the screen cannot be negative.

$$\frac{16 + \sqrt{3{,}272}}{2} \approx 36.6 \qquad \text{or} \qquad \cancel{\frac{16 - \sqrt{3{,}272}}{2} \approx -20.6}$$

State The length of the television screen is approximately 36.6 inches. Since the height is $l - 16$, the height is approximately $36.6 - 16$ or 20.6 inches.

Check The sum of the squares of the lengths of the sides is $(36.6)^2 + (20.6)^2 = 1{,}763.92$. The square of the length of the hypotenuse is $42^2 = 1{,}764$. Since these are approximately equal, the results seem reasonable.

Success Tip

To solve $2l^2 - 32l - 1{,}508 = 0$, we can substitute $a = 2$, $b = -32$, and $c = -1{,}508$ into the quadratic formula. Or, we can divide both sides by 2 and solve the equivalent equation $l^2 - 16l - 754 = 0$, and use smaller values: $a = 1$, $b = -16$, and $c = -754$.

Self Check 6 **Laptops.** Find the height and the length of the screen of a laptop if the diagonal measurement is 15.4 inches and its length is 4.5 inches more than its height. Round to the nearest hundredth.

Now Try ▶ Problem 79

SECTION 9.3 ▶ **STUDY SET**

VOCABULARY

Fill in the blanks.

1. The general _____ equation is $ax^2 + bx + c = 0$, where $a \neq 0$.

2. Complete the quadratic formula.

$$x = \frac{\rule{2cm}{0.4pt}}{\rule{1cm}{0.4pt}}$$

CONCEPTS

3. Write each equation in $ax^2 + bx + c = 0$ form.
   a. $x^2 + 2x = -5$ b. $3x^2 = -2x + 1$

4. For each quadratic equation, find a, b, and c.
   a. $x^2 + 5x + 6 = 0$ b. $8x^2 - x = 10$

5. Divide both sides of $2x^2 - 4x + 8 = 0$ by 2, and then find a, b, and c.

6. Evaluate each expression.

   a. $\dfrac{-2 \pm \sqrt{2^2 - 4(1)(-8)}}{2(1)}$

   b. $\dfrac{-(-1) \pm \sqrt{(-1)^2 - 4(2)(-4)}}{2(2)}$

7. a. How many terms does the expression $10 \pm 15\sqrt{2}$ have?
   b. What common factor do the terms have?

8. Simplify each expression.

   a. $\dfrac{-1 \pm \sqrt{45}}{2(7)}$

   b. $\dfrac{-(-4) \pm \sqrt{(-4)^2 - 4(2)(-9)}}{2(2)}$

9. A student used the quadratic formula to solve an equation and obtained

   $$x = \frac{-3 \pm \sqrt{15}}{2}$$

   a. How many solutions does the equation have?
   b. What are they?
   c. Approximate each to the nearest hundredth.

10. Match each quadratic equation with the best method for solving it. Each answer can be used only once.
    a. $x^2 + 7x - 8 = 0$ i. Square root method
    b. $x^2 + 19x - 2 = 0$ ii. Factoring method
    c. $x^2 = 21$ iii. Complete the square
    d. $x^2 + 10x = 5$ iv. Quadratic formula

NOTATION

Complete the solution.

11. Solve: $x^2 - 5x - 6 = 0$

$$x = \frac{-b \pm \sqrt{b^2 - 4ac}}{2a}$$

$$x = \frac{-(\ \ ) \pm \sqrt{(-5)^2 - 4(1)(\ \ )}}{2(\ \ )}$$

$$x = \frac{5 \pm \sqrt{25 + \boxed{}}}{2}$$

$$x = \frac{5 \pm \sqrt{\boxed{}}}{2}$$

$$x = \frac{\boxed{} \pm 7}{2}$$

$$x = \frac{5 + \boxed{}}{2} = \boxed{} \quad \text{or} \quad x = \frac{5 - \boxed{}}{2} = \boxed{}$$

12. Fill in the blanks: To read $\dfrac{-b \pm \sqrt{b^2 - 4ac}}{2a}$, we say, "the _____ of b, plus or _____ the square root of b _____ minus $4ac$, all _____ $2a$."

GUIDED PRACTICE

Use the quadratic formula to solve each equation. See Example 1.

13. $x^2 + 5x + 6 = 0$ 14. $x^2 + 5x + 4 = 0$

15. $x^2 + 7x + 12 = 0$ 16. $x^2 + 8x + 15 = 0$

17. $4x^2 + 3x - 1 = 0$ 18. $2x^2 + 3x - 2 = 0$

19. $6x^2 + 5x - 6 = 0$ 20. $4x^2 + 4x - 3 = 0$

21. $3x^2 - 5x - 2 = 0$ 22. $3x^2 - 8x - 3 = 0$

23. $5x^2 - 13x - 6 = 0$ 24. $2x^2 - x - 1 = 0$

Use the quadratic formula to solve each equation. Approximate the solutions to nearest hundredth. See Example 2.

25. $x^2 + 1 = -3x$ 26. $x^2 + 2 = -5x$

27. $x^2 - 4 = -7x$ 28. $x^2 - 2 = -3x$

29. $3x^2 - x = 3$ 30. $5x^2 - 3x = 1$

31. $7x^2 - 3x = 1$ 32. $4x^2 - x = 2$

33. $4x^2 = 7x - 2$ 34. $3x^2 = 5x + 4$

35. $2x^2 = 7 - 9x$ **36.** $6x^2 = 3 - 11x$

Use the quadratic formula to solve each equation. See Example 3.

37. $2n^2 + 10n + 11 = 0$ **38.** $2m^2 + 8m + 5 = 0$

39. $2d^2 - 6d + 1 = 0$ **40.** $2s^2 + 2s - 5 = 0$

41. $3x^2 - 8x + 2 = 0$ **42.** $3x^2 - 6x - 2 = 0$

43. $4x^2 - 12x = -1$ **44.** $4x^2 + 12x = 3$

45. $x^2 = 1 - 2x$ **46.** $x^2 = 4 + 2x$

47. $a^2 + 4a = 3$ **48.** $d^2 + 6d = 5$

Use the quadratic formula to solve each equation. See Example 4.

49. $3m^2 + 5 = 2m$ **50.** $2n^2 + 3n = -3$

51. $7x^2 - x = -8$ **52.** $9x^2 - 2x = -4$

TRY IT YOURSELF

Use the most efficient method to solve each equation. Give the exact solutions, and approximations to the nearest hundredth when appropriate.

53. $(2y - 1)^2 = 25$ **54.** $(y + 1)^2 = 64$

55. $2x^2 + x = 5$ **56.** $2x^2 - x + 2 = 0$

57. $x^2 - 12x + 35 = 0$ **58.** $m^2 + 14m + 49 = 0$

59. $r^2 + 9r + 1 = 0$ **60.** $b^2 = 18$

61. $5x^2 - 2x = 8$ **62.** $6x^2 - 6x - 1 = 0$

63. $x^2 + 3x = 9$ **64.** $x^2 + 7x = -1$

65. $7x^2 + 6x + 4 = 0$ **66.** $3x^2 - x = 1$

67. $4c^2 + 16c = 0$ **68.** $2t^2 - 162 = 0$

69. $3y^2 = 18$ **70.** $25x - 50x^2 = 0$

71. $x^2 - 4x = 29$ **72.** $a^2 + 20a = 3$

73. $t^2 - 24t + 144 = 0$ **74.** $3a^2 = 2a + 3$

75. $x^2 - 63 = 0$ **76.** $y^2 - 80 = 0$

APPLICATIONS

77.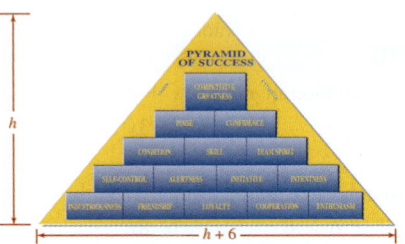

from Campus to Careers

Graphic Designer

A poster that shows former UCLA basketball coach John Wooden's *Pyramid of Success* has an area of 80 square inches. The base of the triangular-shaped poster is 6 inches longer than the height. Find the length of the base and the height of the poster.

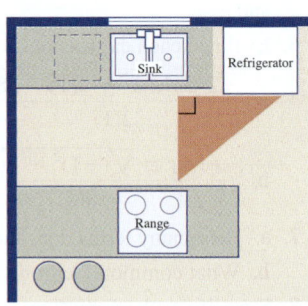

78. Kitchen Floor Plans. To minimize the number of steps that a cook must take when preparing meals, designers carefully plan the *kitchen work triangle* (the area between the sink, refrigerator, and range). The leg of the work triangle next to the refrigerator shown is 2 feet longer than the other leg, and the area covered is 30 ft². Find the length of each leg of the triangle. Round to the nearest tenth.

79. Earthquakes. After a powerful earthquake, a store owner nailed 52-inch-long boards across a broken display window. Find the length and height of the window, if the height is 10 inches less than the length. Round to the nearest tenth.

80. The Abacus. The Chinese abacus shown consists of a frame, parallel wires, and beads that are moved to perform arithmetic computations. The frame is 21 centimeters longer than it is high. Find its height and length.

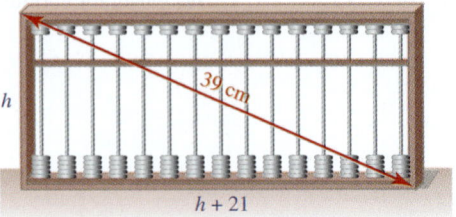

81. **Comics.** A comic strip occupies 100 square centimeters of space in a newspaper. The length of the rectangular space is 4 centimeters more than twice its width. Find its dimensions. Round to the nearest tenth.

82. **Badminton.** A badminton court occupies 880 square feet of floor space of a high school gym. If the length of the court is 16 feet less than three times its width, find its dimensions.

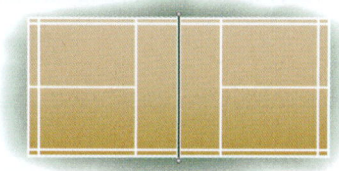

83. **Investing.** We can use the formula $A = P(1 + r)^2$ to find the amount $\$A$ that $\$P$ will become when invested at an annual rate of r for 2 years. What interest rate is needed to make $\$5,000$ grow to $\$5,724.50$ in 2 years?

84. **Retailing.** When a wholesaler sells n compact disc players, his revenue R is given by the formula $R = 150n - \frac{1}{2}n^2$. How many players would he have to sell to receive $\$11,250$? (*Hint:* Multiply both sides of the equation by -2.)

WRITING

85. Do you agree with the following statement? Explain. *The quadratic formula is the easiest method to use to solve quadratic equations.*

86. In this section, we used diagrams like the following when solving quadratic equations. What is its purpose?

$$5x^2 + 6x + 7 = 0$$
$$\uparrow \qquad \uparrow \qquad \uparrow$$
$$ax^2 + bx + c = 0$$

87. Write the quadratic formula in words.

88. At times, certain types of solutions to applied problems are discarded. Explain. Give an example of such a situation.

89. Solve $x^2 - 8x + 16 = 0$ using the quadratic formula. Explain why the equation has only one solution.

90. Find the error in the following work.
Solve: $x^2 + 4x - 5 = 0$

$$x = -4 \pm \frac{\sqrt{16 - 4(1)(-5)}}{2}$$

REVIEW

Solve each equation for the specified variable.

91. $A = p + prt$; for r

92. $F = \dfrac{GMm}{d^2}$; for M

93. $\dfrac{1}{r} = \dfrac{1}{r_1} + \dfrac{1}{r_2}$; for r

94. $2E = \dfrac{T - t}{9}$; for t

CHALLENGE PROBLEMS

Use the quadratic formula to solve each equation.

95. $\dfrac{2}{3}x^2 - \dfrac{1}{3} = \dfrac{5}{9}x$

96. $\dfrac{1}{x - 1} = \dfrac{1}{4} - \dfrac{2}{x}$

97. $\dfrac{(n + 5)(2n + 1)}{2} = 3.5$

98. $2x^2 + \sqrt{5}x - 3 = 0$

Solve each equation and approximate each solution to the nearest tenth.

99. $2.4x^2 - 9.5x + 6.2 = 0$

100. $-1.7x^2 + 0.5x + 0.9 = 0$

101. **Decking.** The owner of a pool wants to surround it with a concrete deck of uniform width (shown in gray). If he can afford 368 square feet of decking, how wide can he make the deck?

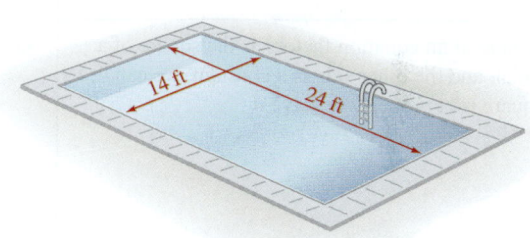

102. **Metal Fabrication.** A square piece of tin, 12 inches on a side, is to have four equal squares cut from its corners, as shown. If the edges are then to be folded up to make a box with a floor area of 64 square inches, find the depth of the box.

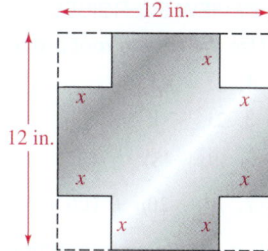

9 ▶ Summary & Review

DEFINITIONS AND CONCEPTS	EXAMPLES
We can use the **square root property** to solve equations of the form $x^2 = c$, where $c > 0$. The two solutions are $x = \sqrt{c}$ or $x = -\sqrt{c}$ We can write $x = \sqrt{c}$ or $x = -\sqrt{c}$ in more compact form using **double-sign notation**: $x = \pm\sqrt{c}$ Read as "x equals the positive or negative square root of c."	Solve: $x^2 = 27$ $x = \sqrt{27}$ or $x = -\sqrt{27}$ Use the square root property. $x = \pm\sqrt{27}$ Use double-sign notation. $x = \pm 3\sqrt{3}$ Simplify: $\sqrt{27} = \sqrt{9 \cdot 3} = 3\sqrt{3}$. The exact solutions are $\pm 3\sqrt{3}$. Verify this by substituting each into the original equation.
We can expand the square root property to solve equations that involve the square of a binomial and a constant.	Solve: $(x - 3)^2 = 5$ $x - 3 = \pm\sqrt{5}$ Use the square root property. Write the result using double-sign notation. $x = 3 \pm \sqrt{5}$ To isolate x, add 3 to both sides. The exact solutions are $3 + \sqrt{5}$ and $3 - \sqrt{5}$. If we approximate the solutions to the nearest hundredth, we have $3 + \sqrt{5} \approx 3 + 2.236067978 \approx 5.24$ Use a calculator. $3 - \sqrt{5} \approx 3 - 2.236067978 \approx 0.76$
If one side of an equation factors as the square of a binomial, and the other side is a constant, we can use the square root property to solve it.	Solve: $x^2 - 22x + 121 = 25$ $(x - 11)^2 = 25$ Factor the perfect-square trinomial. $x - 11 = \pm\sqrt{25}$ Use the square root property. $x = 11 \pm 5$ Add 11 to both sides and simplify $\sqrt{25}$. $x = 11 + 5$ or $x = 11 - 5$ $x = 16$ | $x = 6$ The solutions are 6 and 16.

REVIEW EXERCISES

Use the square root property to solve each equation.

1. $x^2 = 64$ **2.** $t^2 - 8 = 0$

3. $2x^2 - 1 = 149$ **4.** $(x - 1)^2 = 25$

5. $(9x - 8)^2 = 40$ **6.** $4(x - 2)^2 - 9 = 0$

7. $p^2 - 20p + 100 = 9$ **8.** $9m^2 + 6m + 1 = 6$

Use the square root property to find all real-number solutions of each equation. Round each solution to the nearest hundredth.

9. $x^2 = 12$ **10.** $(x - 1)^2 = 55$

11. $m^2 + 36 = 0$ **12.** $(2x - 3)^2 = -8$

13. Cliff Divers.   The La Quebrada Cliff Divers of Acapulco, Mexico, perform daily for the public by diving 148 feet from ocean-side cliffs into the sea below. Find the length of time of a dive. Round to the nearest one tenth of one second. (*Hint:* Use the formula $d = 16t^2$.)

14. Rubik's Cube.   The area of one of the square faces of a Rubik's cube is $\frac{81}{16}$ in.^{2}. Find the length of one side of a Rubik's cube. Express your answer as a mixed number. (*Hint:* Use the formula for the area of a square, $A = s^2$.)

SECTION 9.2 ▶ Solving Quadratic Equations: Completing the Square

DEFINITIONS AND CONCEPTS	EXAMPLES
To **complete the square** on $x^2 + bx$, add the square of one-half of the coefficient of x. $$x^2 + bx + \left(\frac{1}{2}b\right)^2$$	Complete the square on $x^2 + 12x$ and factor the resulting perfect-square trinomial. $$x^2 + 12x \mathbf{+ 36} \quad \text{\color{red}The coefficient of x is 12. To complete the square:}$$ $$\text{\color{red}$\frac{1}{2} \cdot 12 = 6$ and $6^2 = 36$. Add 36 to the binomial.}$$ This trinomial factors as $(x + 6)^2$. We can check using multiplication.

To **solve a quadratic equation in x by completing the square:** **1.** If necessary, divide both sides of the equation by the coefficient of x^2 to make its coefficient 1. **2.** Get all variable terms on one side of the equation and all constants on the other side. **3.** Complete the square. **4.** Factor the perfect-square trinomial. **5.** Solve the resulting equation by using the square root property. **6.** Check your answers in the original equation.	Solve: $3x^2 - 12x + 6 = 0$ $$\frac{3x^2}{\mathbf{3}} - \frac{12}{\mathbf{3}}x + \frac{6}{\mathbf{3}} = \frac{0}{\mathbf{3}} \quad \text{\color{red}To make the leading coefficient 1, divide both sides by 3, term-by-term.}$$ $$x^2 - 4x + 2 = 0 \quad \text{\color{red}Do the divisions.}$$ $$x^2 - 4x = -2 \quad \text{\color{red}Subtract 2 from both sides so that the constant term, -2, is on the right side.}$$ $$x^2 - 4x \mathbf{+ 4} = -2 \mathbf{+ 4} \quad \text{\color{red}The coefficient of x is -4. To complete the square: $\frac{1}{2}(-4) = -2$ and $(-2)^2 = 4$. Add 4 to both sides.}$$ $$(x - 2)^2 = 2 \quad \text{\color{red}Factor the perfect-square trinomial on the left side. Add on the right side.}$$ $$x - 2 = \pm\sqrt{2} \quad \text{\color{red}Use the square root property.}$$ $$x = 2 \pm \sqrt{2} \quad \text{\color{red}To isolate x, add 2 to both sides.}$$ The exact solutions are $2 + \sqrt{2}$ and $2 - \sqrt{2}$. We can approximate each solution. To the nearest hundredth: $$2 + \sqrt{2} \approx 3.41 \qquad 2 - \sqrt{2} \approx 0.59$$

REVIEW EXERCISES

Complete the square to make each expression a perfect-square trinomial. Then factor.

15. $x^2 + 4x$

16. $t^2 - 5t$

Solve each quadratic equation by completing the square.

17. $x^2 - 8x + 15 = 0$

18. $x^2 = -5x + 14$

19. $x^2 + 2x = 5$

20. $4x^2 - 16x = 7$

21. $2x^2 - 2x - 1 = 0$

22. $3x^2 + 5x + 2 = 0$

Solve each quadratic equation by completing the square. Approximate the solutions to the nearest hundredth.

23. $x^2 + 4x + 1 = 0$

24. $x^2 - 7x = 5$

SECTION 9.3 ▶ Solving Quadratic Equations: The Quadratic Formula

DEFINITIONS AND CONCEPTS	EXAMPLES

To **solve a quadratic equation in x using the quadratic formula:**

1. Write the equation in standard form:

$$ax^2 + bx + c = 0$$

2. Identify a, b, and c.

3. Substitute the values for a, b, and c in the quadratic formula

$$x = \frac{-b \pm \sqrt{b^2 - 4ac}}{2a}$$

and evaluate the right side to obtain the solutions.

Use the quadratic formula to solve: $3x^2 - 2x = 2$

$$3x^2 - 2x - 2 = 0 \qquad \text{To get 0 on the right side, subtract 2 from both sides.}$$

Here, $a = 3$, $b = -2$, and $c = -2$.

$$x = \frac{-b \pm \sqrt{b^2 - 4ac}}{2a} \qquad \text{This is the quadratic formula.}$$

$$x = \frac{-(-2) \pm \sqrt{(-2)^2 - 4(3)(-2)}}{2(3)} \qquad \begin{array}{l}\text{Substitute 3 for } a, \\ -2 \text{ for } b, \text{ and } -2 \text{ for } c.\end{array}$$

$$x = \frac{2 \pm \sqrt{4 - (-24)}}{6} \qquad \begin{array}{l}\text{Evaluate within the radical.} \\ \text{Multiply in the denominator.}\end{array}$$

$$x = \frac{2 \pm \sqrt{28}}{6} \qquad \text{Add the opposite: } 4 - (-24) = 4 + 24 = 28.$$

$$x = \frac{2 \pm 2\sqrt{7}}{6} \qquad \text{Simplify: } \sqrt{28} = \sqrt{4}\sqrt{7} = 2\sqrt{7}.$$

$$x = \frac{\overset{1}{\cancel{2}}(1 \pm \sqrt{7})}{\underset{1}{\cancel{2} \cdot 3}} \qquad \begin{array}{l}\text{Factor out the GCF, 2, in the numerator. In the} \\ \text{denominator, factor 6. Remove the common} \\ \text{factor, 2.}\end{array}$$

$$x = \frac{1 \pm \sqrt{7}}{3} \qquad \text{The exact solutions are } \frac{1 \pm \sqrt{7}}{3}.$$

A **strategy for solving quadratic equations** is given on page 696.

A suggested method for solving each quadratic equation is given.

Factor method	*Square root property*
$x^2 - 3x - 18 = 0$	$(x - 1)^2 = 18$
Complete the square	*Quadratic formula*
$x^2 + 6x - 11 = 0$	$3x^2 - 9x + 1 = 0$

even coefficient

REVIEW EXERCISES

Write each equation in $ax^2 + bx + c = 0$ form and find a, b, and c.

25. $x^2 + 2x = -5$

26. $6x^2 = 2x + 1$

Use the quadratic formula to find all real-number solutions of each equation.

27. $x^2 - 2x - 15 = 0$ **28.** $6x^2 = 7x + 3$

29. $p^2 - 4 = 2p$ **30.** $x^2 + 7 = 6x$

31. $3x^2 + 3x = 1$ **32.** $5x^2 + x = 1$

33. $7x^2 - x + 2 = 0$ **34.** $2x^2 + 6x = 5$

Use the most efficient method to find all real-number solutions of each equation.

35. $4x^2 + 16x = 0$ **36.** $(y + 3)^2 = 16$

37. $3g^2 - 81 = 0$ **38.** $3x^2 - 6x = -1$

39. $2x^2 + 2x - 5 = 0$ **40.** $a^2 = 4a - 4$

41. $(2x - 5)^2 = 64$ **42.** $a^2 - 2a + 5 = 0$

43. Use the quadratic formula to solve $3x^2 + 2x - 2 = 0$. Give the exact solutions and then approximate the solutions to the nearest hundredth.

44. Security Gates. The length of the frame for an iron gate is 14 feet longer than the width. A diagonal cross brace is 26 feet long. Find the width and length of the gate frame.

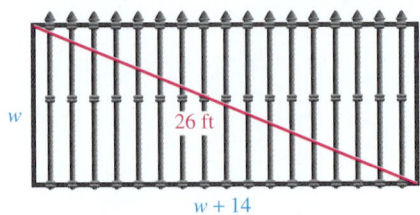

45. The Grand Canyon. The depth of the Grand Canyon at the South Rim is almost one mile. Suppose a visitor standing on the rim tosses a rock upward over the canyon. The time t (in seconds) that it takes for the rock to hit the bottom of the canyon can be found by solving the quadratic equation $0 = -16t^2 + 8t + 5{,}040$. Find t.

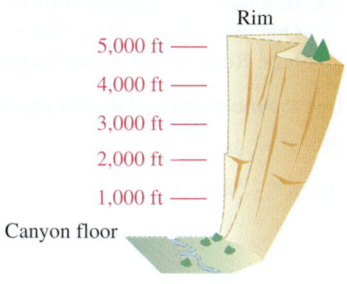

Rim
5,000 ft —
4,000 ft —
3,000 ft —
2,000 ft —
1,000 ft —
Canyon floor

46. Geometry. The triangle shown has an area of 30 square inches. Find its height. Round to the nearest tenth.

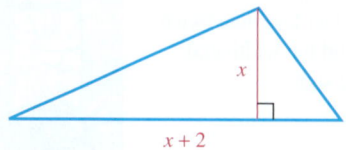

x

$x + 2$

9 ▸ CHAPTER TEST

1. Fill in the blanks.

  a. A _____ equation can be written in the form $ax^2 + bx + c = 0$, where a, b, and c represent real numbers and $a \neq 0$.

  b. $x^2 + 8x + 16$ is a perfect-_____ trinomial because $x^2 + 8x + 16 = (x + 4)^2$.

  c. When we add 25 to $x^2 + 10x$, we say we have _____ the square on $x^2 + 10x$.

  d. We read $3 \pm \sqrt{2}$ as "three _____ or _____ the square root of two."

  e. The _____ coefficient of $3x^2 + 8x - 9$ is 3 and the _____ term is -9.

2. Write the statement $x = \sqrt{5}$ or $x = -\sqrt{5}$ using double-sign notation.

Solve each equation by the square root method.

3. $x^2 = 17$

4. $r^2 - 48 = 0$

5. $(x - 2)^2 = 3$

6. $4y^2 - 20 = 5$

7. $t^2 = \dfrac{1}{49}$

8. $x^2 + 16x + 64 = 24$

9. Explain why the equation $m^2 + 49 = 0$ has no real-number solutions.

10. Check to determine whether $4\sqrt{2}$ is a solution of $n^2 - 32 = 0$.

Complete the square and factor the resulting perfect-square trinomial.

11. $x^2 - 14x$

12. $c^2 - 7c$

13. $x^2 + x$

14. $a^2 - \dfrac{5}{3}a$

15. Complete the square to solve $a^2 + 2a - 4 = 0$. Give the exact solutions and then approximate them to the nearest hundredth.

16. Complete the square to solve $a^2 + a = 3$.

17. Complete the square to solve $m^2 - 4m + 10 = 0$.

18. Complete the square to solve: $2x^2 = 3x + 2$

Use the quadratic formula to solve each equation.

19. $2x^2 - 5x - 12 = 0$

20. $5x^2 + 11x = -3$

21. $4n^2 - 12n + 1 = 0$

22. $7t^2 = -6t - 4$

23. Solve $3x^2 - 2x - 2 = 0$ using the quadratic formula. Give the exact solutions, and then approximate them to the nearest hundredth.

24. Check to determine whether $1 + \sqrt{5}$ is a solution of $x^2 - 2x - 4 = 0$.

25. Archery. The area of a circular archery target is 5,026 cm^2. What is the radius of the target? Round to the nearest centimeter.

26. St. Louis. On October 28, 1965, workers "topped out" the final section of the Gateway Arch in St. Louis, Missouri. It is the tallest national monument in the United States at 630 feet. If a worker dropped a tool from that height, how long would it take to reach the ground? Round to the nearest tenth.

27. New York City. The rectangular Samsung sign in Times Square is a full color LED screen that has an area of 2,665 ft². Its height is 17 feet less than twice its width. Find the width and height of the sign.

28. Geometry. The hypotenuse of a right triangle is 8 feet long. One leg is 4 feet longer than the other. Find the lengths of the legs. Round to the nearest tenth.

Use the most efficient method to solve each equation.

29. $x^2 - 4x = -2$

30. $(3b + 1)^2 = 16$

31. $u^2 - 24 = 0$

32. $6n^2 - 36n = 0$

Group Project

Overview: In this activity, you will learn how to predict the type of solutions that a quadratic equation has, before solving it.

Instructions: Form groups of 2 or 3 students. The expression $b^2 - 2ac$ is called the **discriminant** of the quadratic equation $ax^2 + bx + c = 0$. It provides us with information about the solutions of a quadratic equation.

- If $b^2 - 4ac > 0$, the equation has two different real-number solutions.
- If $b^2 - 4ac = 0$, the equation has one repeated rational-number solution.
- If $b^2 - 4ac < 0$, the equation has two different nonreal solutions.

Find the value of the discriminant for each of the following equations. Then, use the answer to predict the type of solutions that the equation has. Next, solve the equation. Do the solutions match your predictions about them?

1. $x^2 - 2x - 1 = 0$ 2. $x^2 - 4x + 4 = 0$ 3. $x^2 + 2x + 2 = 0$

CUMULATIVE REVIEW ▶▶ Chapters 1–9

1. Determine whether each statement is true or false. [Section 1.3]
   a. Every rational number can be written as a ratio of two integers.
   b. The set of real numbers corresponds to all points on the number line.
   c. The whole numbers and their opposites form the set of integers.

2. **Driving Safety.** In cold-weather climates, salt is spread on roads to keep snow and ice from bonding to the pavement. This allows snowplows to remove built-up snow quickly. According to the graph, when is the accident rate the highest? [Section 1.4]

3. Evaluate: $-4 + 2[-7 - 3(-9)]$ [Section 1.7]

4. Evaluate: $\left| \frac{4}{5} \cdot 10 - 12 \right|$ [Section 1.7]

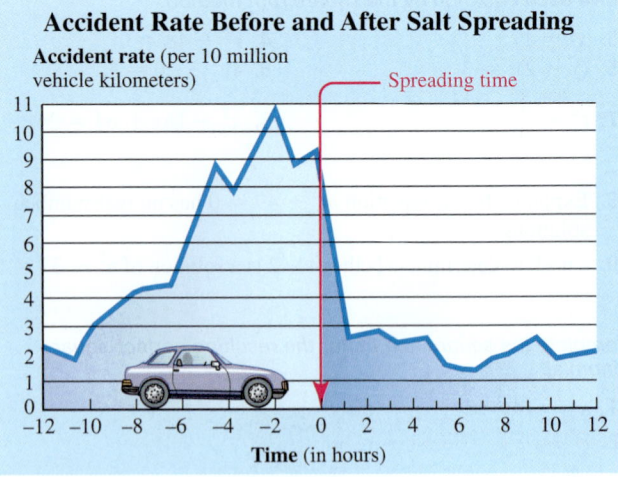

Accident Rate Before and After Salt Spreading

Based on data from a study done in Europe by the Salt Institute

5. Evaluate $(x - a)^2 + (y - b)^2$ for $x = -2$, $y = 1$, $a = 5$, and $b = -3$. [Section 1.8]

6. Simplify: $3p - 6(p - 9) + p$ [Section 1.9]

7. Solve $\frac{5}{6}k = 10$ and check the result. [Section 2.2]

8. Solve $-(3a + 1) + a = 2$ and check the result. [Section 2.2]

9. **Loose Change.** The Coinstar machines that are in many grocery stores count unsorted coins and print out a voucher that can be exchanged for cash at the checkout stand. However, to use this service, a processing fee is charged. If a boy turned in a jar of coins worth \$60 and received a voucher for \$54.12, what was the processing fee (expressed as a percent) charged by Coinstar? [Section 2.3]

10. Solve $T = 2r + 2t$ for r. [Section 2.4]

11. **Selling a Home.** At what price should a home be listed if the owner wants to make \$330,000 on its sale after paying a 4% real estate commission? [Section 2.5]

12. **Business Loans.** Last year, a women's professional organization made two small-business loans totaling \$28,000 to young women beginning their own businesses. The money was lent at 7% and 10% simple interest rates. If the annual income the organization received from these loans was \$2,560, what was each loan amount? [Section 2.6]

13. Solve $5x + 7 < 2x + 1$ and graph the solution set. Then use interval notation to describe the solution. [Section 2.7]

14. Check to determine whether $(-5, -3)$ is a solution of $2x - 3y = -1$. [Section 3.1]

Graph each equation or inequality.

15. $y = -x + 2$ [Section 3.2]   16. $2y - 2x = 6$ [Section 3.3]

17. $y = -3$ [Section 3.3]   18. $y < 3x$ [Section 3.7]

19. Find the slope of the line passing through $(-2, -2)$ and $(-12, -8)$. [Section 3.4]

20. **TV News.** The line graph in red below approximates the evening news viewership on all networks for the years 1995–2009. Find the rate of decrease over this period of time. [Section 3.4]

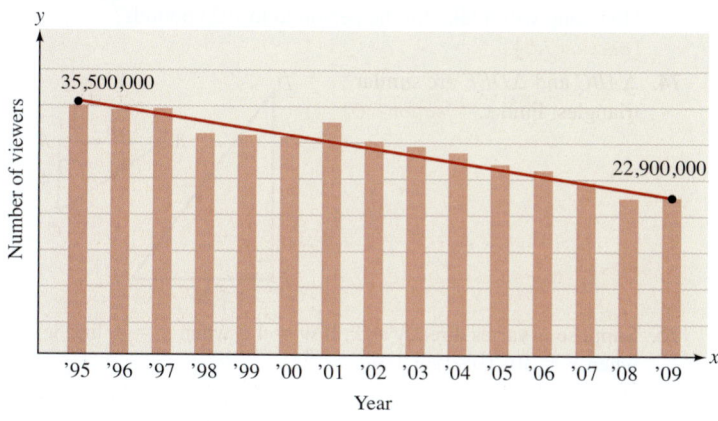

Source: The State of the News Media, 2010

21. What is the slope of the line defined by $4x + 5y = 6$? [Section 3.5]

22. Write the equation of the line whose graph has slope -2 and y-intercept $(0, 1)$. [Section 3.5]

23. Are the graphs of $y = 4x + 9$ and $x + 4y = -10$ parallel, perpendicular, or neither? [Section 3.5]

24. Write the equation of the line whose graph has slope $\frac{1}{4}$ and passes through the point $(8, 1)$. Write the equation in slope–intercept form. [Section 3.6]

25. Graph the line passing through $(-2, -1)$ and having slope $\frac{4}{3}$. [Section 3.6]

26. If $f(x) = 3x^2 + 3x - 8$, find $f(-1)$. [Section 3.8]

27. Find the domain and range of the relation: $\{(1, 8), (4, -3), (-4, 2), (5, 8)\}$ [Section 3.8]

28. Is this the graph of a function? [Section 3.8]

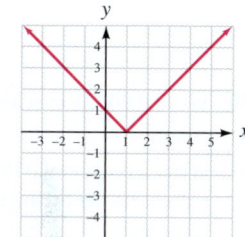

29. Solve using the graphing method. [Section 4.1]

$$\begin{cases} x + y = 1 \\ y = x + 5 \end{cases}$$

30. Solve using the substitution method.

$$\begin{cases} y = 2x + 5 \\ x + 2y = -5 \end{cases}$$ [Section 4.2]

31. Solve using the elimination (addition) method.

$$\begin{cases} \dfrac{3}{5}s + \dfrac{4}{5}t = 1 \\ -\dfrac{1}{4}s + \dfrac{3}{8}t = 1 \end{cases}$$ [Section 4.3]

32. **Aviation.** With the wind, a plane can fly 3,000 miles in 5 hours. Against the wind, the trip takes 6 hours. Find the airspeed of the plane (the speed in still air). Use two variables to solve this problem. [Section 4.4]

33. **Mixing Candy.** How many pounds of each candy must be mixed to obtain 48 pounds of candy that would be worth \$4.50 per pound? Use two variables to solve this problem. [Section 4.4]

34. Solve the system of linear inequalities.

$$\begin{cases} 3x + 4y > -7 \\ 2x - 3y \geq 1 \end{cases}$$ [Section 4.5]

Simplify each expression. Write each answer without using parentheses or negative exponents.

35. $y^3(y^2y^4)$ [Section 5.1]

36. $\left(\dfrac{b^2}{3a}\right)^3$ [Section 5.1]

37. $\dfrac{10a^4a^{-2}}{5a^2a^0}$

[Section 5.2]

38. $\left(\dfrac{21x^{-2}y^2z^{-2}}{7x^3y^{-1}}\right)^{-2}$

[Section 5.2]

39. **Five-Card Poker.** The odds against being dealt the hand shown are about 2.6×10^6 to 1. Express 2.6×10^6 using standard notation. [Section 5.3]

40. Write 0.00073 in scientific notation. [Section 5.3]

41. Graph: $y = x^3 - 2$ [Section 5.4]

42. Write a polynomial that represents the perimeter of the rectangle. [Section 5.5]

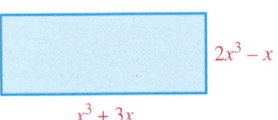

$2x^3 - x$

$x^3 + 3x$

Perform the operations.

43. $4(4x^3 + 2x^2 - 3x - 8) - 5(2x^3 - 3x + 8)$
[Section 5.5]

44. $(-2a^3)(3a^2)$ [Section 5.6]

45. $(2b - 1)(3b + 4)$ [Section 5.6]

46. $(3x + y)(2x^2 - 3xy + y^2)$ [Section 5.6]

47. $(2x + 5y)^2$ [Section 5.7]

48. $(9m^2 - 1)(9m^2 + 1)$ [Section 5.7]

49. $\dfrac{12a^3b - 9a^2b^2 + 3ab}{6a^2b}$ [Section 5.8]

50. $x - 3\overline{)2x^2 - 3 - 5x}$ [Section 5.8]

Factor each expression completely.

51. $6a^2 - 12a^3b + 36ab$
[Section 6.1]

52. $2x + 2y + ax + ay$
[Section 6.1]

53. $x^2 - 6x - 16$
[Section 6.2]

54. $30y^5 + 63y^4 - 30y^3$
[Section 6.3]

55. $t^4 - 16$
[Section 6.4]

56. $b^3 + 125$
[Section 6.5]

Solve each equation by factoring.

57. $3x^2 + 8x = 0$
[Section 6.7]

58. $15x^2 - 2 = 7x$
[Section 6.7]

59. **Geometry.** The triangle shown has an area of 22.5 square inches. Find its height. [Section 6.8]

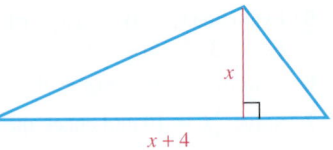

x

$x + 4$

60. For what value is $\dfrac{x}{x + 8}$ undefined? [Section 7.1]

Simplify each expression.

61. $\dfrac{3x^2 - 27}{x^2 + 3x - 18}$

[Section 7.1]

62. $\dfrac{a - 15}{15 - a}$

[Section 7.1]

Perform the operations and simplify when possible.

63. $\dfrac{x^2 - x - 6}{2x^2 + 9x + 10} \div \dfrac{x^2 - 25}{2x^2 + 15x + 25}$ [Section 7.2]

64. $\dfrac{1}{s^2 - 4s - 5} + \dfrac{s}{s^2 - 4s - 5}$ [Section 7.3]

65. $\dfrac{x + 5}{xy} - \dfrac{x - 1}{x^2y}$

[Section 7.4]

66. $\dfrac{x}{x - 2} + \dfrac{3x}{x^2 - 4}$

[Section 7.4]

Simplify each complex fraction.

67. $\dfrac{\dfrac{9m - 27}{m^6}}{\dfrac{2m - 6}{m^8}}$

[Section 7.5]

68. $\dfrac{\dfrac{5}{y} + \dfrac{4}{y + 1}}{\dfrac{4}{y} - \dfrac{5}{y + 1}}$

[Section 7.5]

Solve each equation.

69. $\dfrac{2p}{3} - \dfrac{1}{p} = \dfrac{2p - 1}{3}$ [Section 7.6]

70. $\dfrac{7}{q^2 - q - 2} + \dfrac{1}{q + 1} = \dfrac{3}{q - 2}$ [Section 7.6]

71. Solve the formula $\dfrac{1}{a} + \dfrac{1}{b} = 1$ for a. [Section 7.6]

72. **Roofing.** A homeowner estimates that it will take him 7 days to roof his house. A professional roofer estimates that he can roof the house in 4 days. How long will it take if the homeowner helps the roofer? [Section 7.7]

73. **Losing Weight.** If a person cuts his or her daily calorie intake by 100, it will take 350 days for that person to lose 10 pounds. How long will it take for the person to lose 25 pounds? [Section 7.8]

74. $\triangle ABC$ and $\triangle DEF$ are similar triangles. Find x. [Section 7.8]

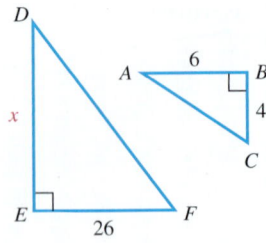

75. Suppose w varies directly as x. If $w = 1.2$ when $x = 4$, find w when $x = 30$. [Section 7.9]

76. Gears.   The speed of a gear varies inversely with the number of teeth. If a gear with 10 teeth makes 3 revolutions per second, how many revolutions per second will a gear with 25 teeth make? [Section 7.9]

Simplify each radical expression. All variables represent positive numbers.

77. $\sqrt{100x^2}$
[Section 8.1]

78. $-\sqrt{18b^3}$
[Section 8.2]

Perform the indicated operation.

79. $3\sqrt{24} + \sqrt{54}$
[Section 8.3]

80. $\left(\sqrt{2} + 1\right)\left(\sqrt{2} - 3\right)$
[Section 8.4]

Rationalize the denominator.

81. $\dfrac{8}{\sqrt{10}}$ [Section 8.4]

82. $\dfrac{\sqrt{2}}{3 - \sqrt{a}}$ [Section 8.4]

Solve each equation.

83. $\sqrt{6x + 1} + 2 = 7$
[Section 8.5]

84. $\sqrt{3t + 7} = t + 3$
[Section 8.5]

Simplify each radical expression. All variables represent positive numbers.

85. $\sqrt[3]{\dfrac{27m^3}{8n^6}}$ [Section 8.6]

86. $\sqrt[4]{16}$ [Section 8.6]

Evaluate each expression.

87. $25^{3/2}$ [Section 8.6]

88. $(-8)^{-4/3}$ [Section 8.6]

Solve each equation.

89. $t^2 = 75$
[Section 9.1]

90. $(6y + 5)^2 - 72 = 0$
[Section 9.1]

91. Storage Cubes.   The diagonal distance across the face of each of the stacking cubes is 15 inches. What is the height of the entire storage arrangement? Round to the nearest tenth of an inch.
[Section 9.1]

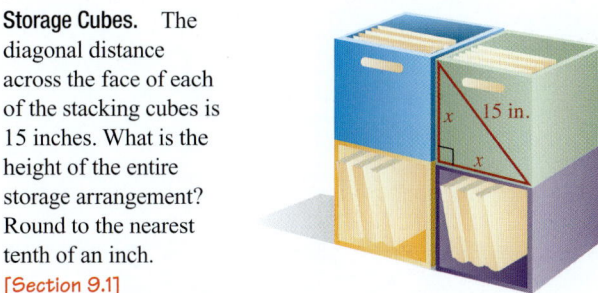

92. Solve $x^2 + 8x + 12 = 0$ by completing the square.
[Section 9.2]

93. Solve $4x^2 - x - 2 = 0$ using the quadratic formula. Give the exact solutions, and then approximate each to the nearest hundredth.
[Section 9.3]

94. Quilts.   According to the *Guinness Book of World Records 1998,* the world's largest quilt was made by the Seniors' Association of Saskatchewan, Canada, in 1994. If the length of the rectangular quilt is 11 feet less than twice its width and it has an area of 12,865 ft^2, find its width and length.
[Section 9.3]

APPENDIX 1 Statistics

In this appendix, you will learn about

1. The Mean
2. The Median
3. The Mode

Statistics is a branch of mathematics that deals with the analysis of numerical data. In statistics, three types of averages are commonly used as **measures of central tendency** of a set of values: the *mean,* the *median,* and the *mode.*

1 The Mean

We have previously discussed the mean of a set of numbers.

The Mean

The mean of several values is the sum of those values divided by the number of values.

$$\text{Mean} = \frac{\text{sum of the values}}{\text{number of values}}$$

EXAMPLE 1 **Physiology.** As part of a class project, a student measured ten people's reaction time to a visual stimulus. Their reaction times (in seconds) were

0.36, 0.24, 0.23, 0.41, 0.28, 0.25, 0.20, 0.28, 0.39, 0.26

Find the mean reaction time.

Solution To find the mean, we add the values and divide by the number of values.

$$\begin{aligned}\text{Mean} &= \frac{0.36 + 0.24 + 0.23 + 0.41 + 0.28 + 0.25 + 0.20 + 0.28 + 0.39 + 0.26}{10}\\ &= \frac{2.9}{10}\\ &= 0.29\end{aligned}$$

The mean reaction time is 0.29 second.

Self Check 1 Find the mean of 2.3, 4.1, 5.2, 6.3, 3.7, 5.1, 4.6, 5.3

Now Try Problem 6

EXAMPLE 2 **Banking.** When the mean (average) daily balance of a checking account falls below $500 in any week, the customer must pay a $20 service charge. What minimum balance must a customer have on Friday to avoid a service charge?

Security Savings Bank		
Day	**Date**	**Daily balance**
Mon	5/09	$670.70
Tues	5/10	$540.19
Wed	5/11	−$60.39
Thurs	5/12	$475.65
Fri	5/13	

Analyze We can find the mean (average) daily balance for the week by adding the daily balances and dividing by 5. If the mean is $500 or more, there will be no service charge.

Assign We can let x = the minimum balance needed on Friday.

Form To form an equation, we translate the words into mathematical symbols.

The sum of the five daily balances divided by 5 is $500.

$$\frac{670.70 + 540.19 + (-60.39) + 475.65 + x}{5} = 500$$

Solve

$$\frac{670.70 + 540.19 + (-60.39) + 475.65 + x}{5} = 500$$

$$\frac{1{,}626.15 + x}{5} = 500 \qquad \text{Simplify the numerator.}$$

$$5\left(\frac{1{,}626.15 + x}{5}\right) = 5(500) \qquad \text{Multiply both sides by 5.}$$

$$1{,}626.15 + x = 2{,}500$$

$$x = 873.85 \qquad \text{Subtract 1,626.15 from both sides.}$$

State On Friday, the account balance must be at least $873.85 to avoid a service charge.

Check Check the result by adding the five daily balances and dividing by 5.

Self Check 2 **Grading.** To receive a grade of A in math, Lindsey must have a mean (average) of at least 90 on five exams. On the first four exams, she received scores of 94, 97, 80, and 87. What is the lowest score she can get on her final exam and still earn an A?

Now Try Problem 13

2 The Median

The Median

The **median** of several values is the middle value. To find the median of several values:

1. Arrange the values in increasing order.
2. If there are an odd number of values, choose the middle value.
3. If there are an even number of values, add the middle two values and divide by 2.

EXAMPLE 3   **Finding the Median.**   In Example 1, the following values were the reaction times of ten people to a visual stimulus.

$$0.36, 0.24, 0.23, 0.41, 0.28, 0.25, 0.20, 0.28, 0.39, 0.26$$

Find the median of these values.

Solution   To find the median, we first arrange the values in increasing order:

$$0.20, 0.23, 0.24, 0.25, 0.26, 0.28, 0.28, 0.36, 0.39, 0.41$$

Because there are an even number of values, the median is the sum of the middle two values, 0.26 and 0.28, divided by 2. Thus, the median is

$$\text{Median} = \frac{0.26 + 0.28}{2} = 0.27$$

The median reaction time is 0.27 second.

Self Check 3   For the values in Self Check 1, find the median.

Now Try ▶ Problem 4

3   The Mode

The Mode	The **mode** of several values is the value that occurs most often.

EXAMPLE 4   **Finding the Mode.**   Find the mode of the following values.

$$0.36, 0.24, 0.23, 0.41, 0.28, 0.25, 0.20, 0.28, 0.39, 0.26$$

Solution   Since the value 0.28 occurs most often, it is the mode.

Self Check 4   Find the mode of 12, 16, 13, 13, 12, 15, 14, 13

Now Try ▶ Problem 5

If two different numbers in a distribution tie for occurring most often, there are two modes, and the distribution is called **bimodal.**

Although the mean is probably the most common measure of average, the median and the mode are frequently used. For example, workers' salaries are usually compared to the median (average) salary. To say that the modal (average) shoe size is 10 means that a shoe size of 10 occurs more often than any other shoe size.

APPENDIX 1   STUDY SET

PRACTICE

In Problems 1–3, use the following set of values: 7, 5, 9, 10, 8, 6, 6, 7, 9, 12, 9.

1. Find the mean.
2. Find the median.
3. Find the mode.

In Problems 4–6, use the following set of values: 8, 12, 23, 12, 10, 16, 26, 12, 14, 8, 16, 23.

4. Find the median.
5. Find the mode.
6. Find the mean.
7. Find the mean, median, and mode of the following values: 24, 27, 30, 27, 31, 30, and 27.

8. Find the mean, median, and mode of the following golf scores: 85, 87, 88, 82, 85, 91, 88, and 88.

APPLICATIONS

9. **Football.** The gains and losses made by a running back on seven plays were −8 yd, 2 yd, −6 yd, 6 yd, 4 yd, −7 yd, and −5 yd. Find his average (mean) yards per carry.

10. **Sales.** If a clerk had the sales shown for one week, find the mean of her daily sales.

Monday	$1,525
Tuesday	$ 785
Wednesday	$1,628
Thursday	$1,214
Friday	$ 917
Saturday	$1,197

11. **Viruses.** The table gives the approximate lengths (in centimicrons) of the viruses that cause five common diseases. Find the mean length of the viruses.

Polio	2.5
Influenza	105.1
Pharyngitis	74.9
Chicken pox	137.4
Yellow fever	52.6

12. **Salaries.** Ten workers in a small business have monthly salaries of $2,500, $1,750, $2,415, $3,240, $2,790, $3,240, $2,650, $2,415, $2,415, and $2,650. Find the average (mean) salary.

13. **Job Testing.** To be accepted into a police training program, a recruit must have an average (mean) score of 85 on a battery of four tests. If a candidate scored 78 on the oral test, 91 on the physical test, and 87 on the psychological test, what is the lowest score she can obtain on the written test and be accepted into the program?

14. **Gas Mileage.** Mileage estimates for four cars owned by a small business are shown. If the business buys a fifth car, what must its mileage average be so that the five-car fleet averages 20.8 mpg?

Model	City mileage (mpg)
Chevrolet Lumina	20.3
Jeep Cherokee	14.1
Ford Contour	28.2
Dodge Caravan	16.9

15. **Sport Fishing.** The weights (in pounds) of the trophy fish caught one week in Catfish Lake were 4, 7, 4, 3, 3, 5, 6, 9, 4, 5, 8, 13, 4, 5, 4, 6, and 9. Find the median and modal averages of the fish caught.

16. **Salaries.** Find the median and mode of the ten salaries given in Problem 12.

17. **Fuel Efficiency.** The ten most fuel-efficient cars in 2009, based on manufacturers' estimated city and highway average miles per gallon (mpg), are shown in the table below.
   a. Find the mean, median, and mode of the city mileage.

   b. Find the mean, median, and mode of the highway mileage.

Model	mpg city/hwy
Toyota Prius	50/49
Honda Civic Hybrid	40/45
Honda Insight	40/43
Ford Fusion Hybrid	41/36
Mercury Milan Hybrid	41/36
VW Jetta TDI	30/41
Nissan Altima Hybrid	35/33
Toyota Camry Hybrid	33/34
Toyota Yaris	29/36
Toyota Corolla	26/35

Source: edmonds.com

18. **Nutrition.** Refer to the table below.
   a. Find the mean number of calories in one serving of the meats shown.
   b. Find the median.
   c. Find the mode.

NUTRITIONAL COMPARISONS Per 3.5 oz. serving of cooked meat	
Species	**Calories**
Bison	143
Beef (Choice)	283
Beef (Select)	201
Pork	212
Chicken (Skinless)	190
Sockeye Salmon	216

Source: The National Bison Association

WRITING

19. Explain why the mean of two numbers is halfway between the numbers.
20. Can the mean, median, and mode of a distribution be the same number? Explain.
21. Must the mean, median, and mode of a distribution be the same number? Explain.
22. Can the mode of a distribution be greater than the mean? Explain.

n	n^2	$\sqrt{n}$	n^3	$\sqrt[3]{n}$	n	n^2	$\sqrt{n}$	n^3	$\sqrt[3]{n}$
1	1	1.000	1	1.000	51	2,601	7.141	132,651	3.708
2	4	1.414	8	1.260	52	2,704	7.211	140,608	3.733
3	9	1.732	27	1.442	53	2,809	7.280	148,877	3.756
4	16	2.000	64	1.587	54	2,916	7.348	157,464	3.780
5	25	2.236	125	1.710	55	3,025	7.416	166,375	3.803
6	36	2.449	216	1.817	56	3,136	7.483	175,616	3.826
7	49	2.646	343	1.913	57	3,249	7.550	185,193	3.849
8	64	2.828	512	2.000	58	3,364	7.616	195,112	3.871
9	81	3.000	729	2.080	59	3,481	7.681	205,379	3.893
10	100	3.162	1,000	2.154	60	3,600	7.746	216,000	3.915
11	121	3.317	1,331	2.224	61	3,721	7.810	226,981	3.936
12	144	3.464	1,728	2.289	62	3,844	7.874	238,328	3.958
13	169	3.606	2,197	2.351	63	3,969	7.937	250,047	3.979
14	196	3.742	2,744	2.410	64	4,096	8.000	262,144	4.000
15	225	3.873	3,375	2.466	65	4,225	8.062	274,625	4.021
16	256	4.000	4,096	2.520	66	4,356	8.124	287,496	4.041
17	289	4.123	4,913	2.571	67	4,489	8.185	300,763	4.062
18	324	4.243	5,832	2.621	68	4,624	8.246	314,432	4.082
19	361	4.359	6,859	2.668	69	4,761	8.307	328,509	4.102
20	400	4.472	8,000	2.714	70	4,900	8.367	343,000	4.121
21	441	4.583	9,261	2.759	71	5,041	8.426	357,911	4.141
22	484	4.690	10,648	2.802	72	5,184	8.485	373,248	4.160
23	529	4.796	12,167	2.844	73	5,329	8.544	389,017	4.179
24	576	4.899	13,824	2.884	74	5,476	8.602	405,224	4.198
25	625	5.000	15,625	2.924	75	5,625	8.660	421,875	4.217
26	676	5.099	17,576	2.962	76	5,776	8.718	438,976	4.236
27	729	5.196	19,683	3.000	77	5,929	8.775	456,533	4.254
28	784	5.292	21,952	3.037	78	6,084	8.832	474,552	4.273
29	841	5.385	24,389	3.072	79	6,241	8.888	493,039	4.291
30	900	5.477	27,000	3.107	80	6,400	8.944	512,000	4.309
31	961	5.568	29,791	3.141	81	6,561	9.000	531,441	4.327
32	1,024	5.657	32,768	3.175	82	6,724	9.055	551,368	4.344
33	1,089	5.745	35,937	3.208	83	6,889	9.110	571,787	4.362
34	1,156	5.831	39,304	3.240	84	7,056	9.165	592,704	4.380
35	1,225	5.916	42,875	3.271	85	7,225	9.220	614,125	4.397
36	1,296	6.000	46,656	3.302	86	7,396	9.274	636,056	4.414
37	1,369	6.083	50,653	3.332	87	7,569	9.327	658,503	4.431
38	1,444	6.164	54,872	3.362	88	7,744	9.381	681,472	4.448
39	1,521	6.245	59,319	3.391	89	7,921	9.434	704,969	4.465
40	1,600	6.325	64,000	3.420	90	8,100	9.487	729,000	4.481
41	1,681	6.403	68,921	3.448	91	8,281	9.539	753,571	4.498
42	1,764	6.481	74,088	3.476	92	8,464	9.592	778,688	4.514
43	1,849	6.557	79,507	3.503	93	8,649	9.644	804,357	4.531
44	1,936	6.633	85,184	3.530	94	8,836	9.695	830,584	4.547
45	2,025	6.708	91,125	3.557	95	9,025	9.747	857,375	4.563
46	2,116	6.782	97,336	3.583	96	9,216	9.798	884,736	4.579
47	2,209	6.856	103,823	3.609	97	9,409	9.849	912,673	4.595
48	2,304	6.928	110,592	3.634	98	9,604	9.899	941,192	4.610
49	2,401	7.000	117,649	3.659	99	9,801	9.950	970,299	4.626
50	2,500	7.071	125,000	3.684	100	10,000	10.000	1,000,000	4.642

ARE YOU READY? 1.1 (page 2)

1. 210 **2.** 1,750 **3.** 1,092 **4.** 9

SELF CHECKS 1.1

1. 350 calories **2.** The product of 22 and 11 equals 242.
3. $u = 500 - p$ (Answers may vary depending on the variables used.) **4.** 480 calories **5.**

m	c
8	80
75	750

STUDY SET SECTION 1.1 (page 7)

1. sum, difference, product, quotient **3.** constant
5. equation; expression **7.** horizontal **9. a.** Equation
b. Algebraic Expression **11. a.** Algebraic Expression
b. Equation **13.** Addition, multiplication, division; t
15.

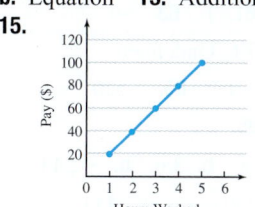

17. is not equal to
19. $5 \cdot 6, 5(6)$ **21.** $4x$
23. $2w$ **25.** $\frac{32}{x}$ **27.** $\frac{55}{5}$
29. 15-year-old machinery is worth
$35,000. **31.** $250
33. The product of 8 and 2 equals 16.
35. The difference of 11 and
9 equals 2.
37. The sum of x and 2 equals 10.
39. The quotient of 66 and 11 equals 6. **41.** $p = 100 - d$
43. $7d = h$ **45.** $s = 3c$ **47.** $w = e + 1,200$
49. $p = r - 600$ **51.** $\frac{l}{4} = m$ **53.** 390, 400, 405
55. 1,300; 1,200; 1,100 **57.** 12 **59.** 2
61. a. The number of calories burned is the product of 3 and the
number of minutes cleaning. **b.** $c = 3m$
c. 30, 60, 90, 120, 150, 180

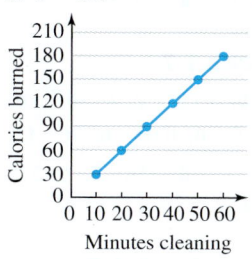

ARE YOU READY? 1.2 (page 9)

1. 1 **2.** 150 **3.** Yes **4.** four-fifths

SELF CHECKS 1.2

1. $189 = 3 \cdot 3 \cdot 3 \cdot 7$ **2.** $\frac{10}{27}$ **3.** $\frac{12}{25}$ **4.** $\frac{15}{24}$ **5. a.** $\frac{3}{7}$
b. In simplest form **6.** $\frac{9}{5}$ **7.** $\frac{13}{240}$ **8. a.** 28 **b.** $\frac{4}{5}$ **9.** $\frac{81}{8} = 10\frac{1}{8}$
10. $6\frac{11}{24}$

STUDY SET SECTION 1.2 (page 20)

1. multiplied **3.** prime-factored **5.** equivalent
7. least or lowest **9. a.** 1 **b.** a **c.** $\frac{a \cdot c}{b \cdot d}$ **d.** $\frac{a \cdot d}{b \cdot c}$ **e.** $\frac{a + b}{d}$
f. $\frac{a - b}{d}$ **11. a.** 1 **b.** 1 **13. a.** $\frac{5}{5}, \frac{25}{30}$ **b.** 2, 7, $\frac{2}{7}$
15. $3 \cdot 5 \cdot 5$ **17.** $2 \cdot 2 \cdot 7$ **19.** $3 \cdot 3 \cdot 3 \cdot 3$ **21.** $3 \cdot 3 \cdot 13$
23. $2 \cdot 2 \cdot 5 \cdot 11$ **25.** $2 \cdot 3 \cdot 11 \cdot 19$ **27.** $\frac{5}{48}$ **29.** $\frac{21}{55}$ **31.** $\frac{15}{8}$
33. $\frac{42}{25}$ **35.** $\frac{3}{9}$ **37.** $\frac{24}{54}$ **39.** $\frac{35}{5}$ **41.** $\frac{35}{7}$ **43.** $\frac{1}{3}$ **45.** $\frac{6}{7}$
47. $\frac{3}{8}$ **49.** Simplest form **51.** $\frac{2}{3}$ **53.** $\frac{4}{25}$ **55.** $\frac{6}{5}$ **57.** $\frac{4}{7}$
59. $\frac{5}{24}$ **61.** $\frac{22}{35}$ **63.** $\frac{41}{45}$ **65.** $\frac{5}{12}$ **67.** 24 **69.** 4 **71.** $\frac{7}{9}$
73. $\frac{7}{20}$ **75.** $32\frac{2}{3}$ **77.** $2\frac{1}{2}$ **79.** $\frac{5}{9}$ **81.** $5\frac{19}{48}$ **83.** $\frac{19}{15}$
85. 70 **87.** $13\frac{3}{4}$ **89.** $\frac{1}{2}$ **91.** $\frac{14}{5}$ **93.** $\frac{8}{5}$ **95.** $\frac{3}{35}$ **97.** $\frac{9}{4}$
99. $\frac{3}{25}$ **101.** $1\frac{9}{11}$ **103.** $\frac{1}{7}$ **105.** 0 **107.** 14 **109. a.** $\frac{55}{63}$
b. $\frac{1}{63}$ **c.** $\frac{4}{21}$ **d.** $\frac{28}{27}$ **111. a.** $\frac{7}{32}$ in. **b.** $\frac{3}{32}$ in.
113. $63\frac{1}{8}$ in. **115.** $40\frac{1}{2}$ in. **121.** Variables

ARE YOU READY? 1.3 (page 22)

1. 28 **2.** -10 **3.** a decimal (3.5 GPA)
4. a fraction ($\frac{3}{4}$ cup of sugar)

SELF CHECKS 1.3

1. Natural numbers: 45; whole numbers: 45; integers: 45, -2;
rational numbers: 0.1, $-\frac{2}{7}$, 45, -2, $\frac{13}{4}$, $-6\frac{7}{8}$; irrational numbers:
none; real numbers: all **2. a.** $>$ **b.** $<$ **c.** $>$ **d.** $<$
3. **4. a.** 100 **b.** 4.7
c. $\sqrt{2}$

STUDY SET SECTION 1.3 (page 29)

1. whole **3.** number line **5.** signed **7.** inequality
9. terminating; repeating **11.** decimal **13. a.** $-$$15 million
b. $\frac{5}{16}$ in. or $+\frac{5}{16}$ in. **15. a.** -20 **b.** $\frac{2}{3}$ **17.** -14 and -4
19. square, root **21.** is approximately equal to **23.** Greek
25. $-4, -5$
27.

	5	0	-3	$\frac{7}{8}$	0.17	$-9\frac{1}{4}$	$\sqrt{2}$	π
Real	✓	✓	✓	✓	✓	✓	✓	✓
Irrational							✓	✓
Rational	✓	✓	✓	✓	✓	✓		
Integer	✓	✓	✓					
Whole	✓	✓						
Natural	✓							

29. True   **31.** False   **33.** True   **35.** True   **37.** >   **39.** <
41. >   **43.** <   **45.** >   **47.** <   **49.** 0.625   **51.** $0.0\overline{3}$
53. $0.01\overline{6}$ **55.** 0.42
57.

59.

61. 83   **63.** $\frac{4}{3}$ **65.** 11   **67.** 6.1   **69.** >
71. <   **73.** =   **75.** =   **77.** <   **79.** >
81. Natural, whole, integers: 9; rational: $9, \frac{15}{16}, 3\frac{1}{8}, 1.765$; irrational: $2\pi, 3\pi, \sqrt{89}$; real: all   **83.** iii
85. a. 2006; −$90 billion   **b.** 2009; −$45 billion   **93.** $\frac{4}{9}$
95. $2\frac{5}{23}$

ARE YOU READY? **1.4** (page 32)

1. 3, 5; −5 has the larger absolute value.   **2.** 7.17   **3.** 621   **4.** $\frac{2}{15}$

SELF CHECKS **1.4**

1. a. −60   **b.** −13.18   **c.** $-\frac{11}{12}$ **2. a.** −24   **b.** 1.73   **c.** $\frac{2}{5}$
3. 569   **4.** 139   **5.** −5   **6.** 11

STUDY SET SECTION **1.4** (page 37)

1. sum   **3.** commutative, associative   **5. a.** 6   **b.** −9.2
7. a. Negative   **b.** Positive   **9. a.** $1 + (−5)$ **b.** $−80.5 + 15$
c. $20 + 4$ **d.** $3 + 2.1$ **11.** Step 1: Commutative Property of
Addition; Step 2: Associative Property of Addition
13. a. $x + y = y + x$ **b.** $(x + y) + z = x + (y + z)$ **15.** −9
17. −17   **19.** −74   **21.** −10.3   **23.** $-\frac{17}{12}$ **25.** $-\frac{7}{20}$ **27.** −3
29. 39   **31.** 0   **33.** 2.25   **35.** $-\frac{4}{15}$ **37.** $\frac{3}{8}$ **39.** 16   **41.** −15
43. −21   **45.** 195   **47.** 215   **49.** −112   **51.** $1\frac{2}{3}$ **53.** 15.4
55. 9   **57.** 1   **59.** 70   **61.** −6.6   **63.** $-\frac{1}{8}$ **65.** −14   **67.** 0
69. −26   **71.** 0.67   **73.** −5   **75.** −22.1   **77.** 2,167   **79.** −1.7
81. 68   **83.** $-\frac{15}{28}$ **85.** −0.9   **87. a.** 27   **b.** 3   **c.** −27   **d.** −3
89. a. $\frac{13}{18}$ **b.** $-\frac{5}{18}$ **c.** $-\frac{13}{18}$ **d.** $\frac{5}{18}$ **91.** 2,150 m
93. Yang: −8, Woods: −5, Westwood: −3, McIlroy: −3
95. 1,242.86   **97.** −$99,650,000   **99.** −1, 3
101. 79 feet above sea level   **103.** $6,276 million   **107.** True
109. −9 and 3

ARE YOU READY? **1.5** (page 40)

1. −6, 15   **2.** $22 − 6$ **3.** 12   **4.** −9

SELF CHECKS **1.5**

1. a. 1   **b.** y **c.** −500   **2. a.** −57   **b.** 2.9   **c.** $\frac{5}{12}$ **3. a.** 2.7
b. −2.7   **4.** −8   **5.** 41°F   **6.** −11 ft

STUDY SET SECTION **1.5** (page 44)

1. Subtraction   **3.** range   **5. a.** −12   **b.** $\frac{1}{5}$ **c.** −2.71   **d.** 0
7. +, 9, 10   **9.** $−10 + (−8) + (−23) + 5 + 34$ **11. a.** $−6 \ominus (−4)$
b. $7 + (−3) \ominus 5 \ominus (−2)$ **13. a.** $1 − (−7) = 8$ **b.** $−(−2) = 2$
c. $−|−3| = −3$ **d.** $2 − 6 = −4$ **15.** 55   **17.** x **19.** −25
21. $-\frac{3}{16}$ **23.** −3   **25.** −10   **27.** 11   **29.** −6   **31.** 2
33. 40   **35.** 5   **37.** 12   **39.** −6.9   **41.** −2.31   **43.** $-\frac{1}{2}$ **45.** $-\frac{5}{12}$
47. 22   **49.** −25   **51.** −7   **53.** −1   **55.** 256   **57.** 0   **59.** −2.1
61. $\frac{47}{56}$ **63.** 3   **65.** −47.5   **67.** 149   **69.** −171   **71.** 4.63

73. $-\frac{19}{12}$ **75.** −11   **77.** −88   **79.** −1.1   **81.** −50   **83.** −3.5
85. $-\frac{5}{16}$ **87.** 1   **89.** −13   **91. a.** −53   **b.** −47   **93. a.** $-\frac{13}{18}$
b. $-\frac{7}{18}$ **95.** 160°F   **97.** 21 points   **99.** Orlando: −115,000
passengers; Ft Lauderdale: −50,000 passengers   **101.** 1,030 ft
103. an increase of 315°F   **105.** −179, −206   **111.** $2 \cdot 3 \cdot 5$
113. True

ARE YOU READY? **1.6** (page 47)

1. 14, 6.75   **2.** Different signs   **3.** $\frac{5}{16}$ **4.** 3.7

SELF CHECKS **1.6**

1. a. −60   **b.** −15   **c.** −11.18   **d.** $-\frac{2}{5}$ **2. a.** 120   **b.** $\frac{1}{12}$
3. a. −300   **b.** 18   **4. a.** $-\frac{16}{15}$ **b.** $\frac{16}{15}$ **c.** $-\frac{1}{27}$ **5. a.** 7   **b.** −3
c. −0.2   **d.** $-\frac{6}{5}$ **6.** The house depreciated $2,250 per year.
7. a. Undefined   **b.** 0

STUDY SET SECTION **1.6** (page 55)

1. product, quotient   **3.** associative   **5. a.** positive   **b.** negative
7. a. −3, 3, −9   **b.** 0, 8, 0   **9. a.** a **b.** 1   **c.** 0
d. Undefined   **11. a.** Positive   **b.** Negative   **c.** Negative
13. a. $8 \cdot 5$ **b.** $(−2 \cdot 6)9$ **c.** $\frac{1}{5}$ **d.** 1   **15. a.** NEG
b. Not possible to tell   **c.** POS   **d.** NEG   **17.** $−4(−5) = 20$
19. −4   **21.** −16   **23.** −60   **25.** −66   **27.** −0.48   **29.** $-\frac{1}{4}$
31. 7   **33.** 54   **35.** 9   **37.** −441   **39.** 2.4   **41.** $\frac{1}{12}$
43. 66   **45.** −720   **47.** −861   **49.** −216   **51.** $\frac{9}{7}$; 1
53. $-\frac{1}{13}$; 1   **55.** 10   **57.** 3   **59.** −17   **61.** 1   **63.** −9
65. −0.005   **67.** $-\frac{5}{12}$ **69.** $\frac{15}{4}$ **71.** 0   **73.** Undefined
75. −4   **77.** −1   **79.** 16   **81.** 0   **83.** −4.7   **85.** −520
87. $\frac{1}{24}$ **89.** −11   **91.** $1\frac{1}{2}$ **93.** 30.24   **95.** $-\frac{3}{8}$
97. $-\frac{3}{20}$ **99.** 30.3   **101.** $\frac{15}{16}$ **103. a.** 1.8   **b.** 3.6   **c.** −2.43
d. −3   **105.** −67   **107.** 6   **109.** −$8,000 per year   **111.** −72°
113. a. ii   **b.** 36 lb   **115.** −51°F   **117.** (160), (250), (800)
119. a. 5, −10   **b.** 2.5, −5   **c.** 7.5, −15   **d.** 10, −20
125. −5   **127.** $1.08\overline{3}$

ARE YOU READY? **1.7** (page 58)

1. Subtraction, multiplication   **2.** Addition, multiplication,
division   **3.** 81   **4.** −125

SELF CHECKS **1.7**

1. a. 12^3 **b.** $2 \cdot 9^4$ **c.** $(−30)^2$ **d.** y^6 **e.** $8b^3c$ **2. a.** 32
b. $-\frac{27}{64}$ **c.** 9   **d.** 0.09   **e.** 36   **f.** −125   **3.** −625   **4. a.** 35
b. −66   **c.** 27   **d.** 256   **5. a.** 216   **b.** −35   **6.** −130   **7.** −1
8. 1,003   **9.** 2

STUDY SET SECTION **1.7** (page 65)

1. base, exponent, power   **3.** exponent   **5.** order
7. a. Subtraction   **b.** Division   **c.** Addition   **d.** Power
9. a. Parentheses, brackets, braces, absolute value symbols, fraction
bar   **b.** Innermost: parentheses; outermost: brackets   **11. a.** −5
b. 5   **13.** 3, 9, 27, 54, −73   **15.** 8^3 **17.** $7^3 12^2$ **19.** x^3 **21.** $r^4 s^2$
23. 49   **25.** 216   **27.** 625   **29.** 0.01   **31.** $-\frac{1}{64}$ **33.** $\frac{8}{27}$
35. 36, −36   **37.** 64, −64   **39.** −17   **41.** 30   **43.** 43
45. 8   **47.** −34   **49.** −118   **51.** −44   **53.** 0   **55.** −148

57. 100   **59.** 53   **61.** -86 **63.** -392 **65.** 3   **67.** $\frac{1}{2}$ **69.** 0
71. $-\frac{8}{9}$ **73.** 13   **75.** 2   **77.** -31 **79.** 11   **81.** 1   **83.** -32
85. 86   **87.** -8 **89.** -19 **91.** -500 **93.** -376 **95.** 12
97. 39   **99.** Undefined   **101.** -275 **103.** -54 **105.** $\frac{1}{8}$ **107.** 10
109. -1 **111. a.** 22   **b.** -13 **113. a.** -40 **b.** -10
115. 2^2 square units, 3^2 square units, 4^2 square units   **117.** 12 min
119. a. \$11,875   **b.** \$95   **121.** 81 in.   **127.** $-17, -5$

ARE YOU READY? 1.8 (page 68)

1. $9x$ **2.** m **3.** $-t$ **4.** $\frac{7}{8}y$ **5.** Quotient   **6.** Difference
7. Product   **8.** Sum

SELF CHECKS 1.8

1. $1, -12, 3, -4$ **2. a.** Factor   **b.** Term   **3. a.** $t - 80$ **b.** $\frac{2}{3}T$
c. $(2a - 15)^2$ **4.** $(m + 15)$ min   **5.** $s =$ amount donated to
scholarship fund in dollars; $900 - s =$ amount donated to building
fund in dollars   **6.** $x =$ number of votes received by the challenger;
$3x - 55 =$ number of votes received by the incumbent
7. Daughter's: x; Kayla's: $x + 5$; Son's: $x + 2$ **8.** $\frac{h}{24}$ days
9. a. \$300   **b.** \$100t **c.** \$1,000$(x - 4)$ **10. a.** 17   **b.** -18
c. 0   **11.** 96 ft

STUDY SET SECTION 1.8 (page 75)

1. expressions   **3.** terms   **5.** coefficient   **7.** $7, 14, 21, 7w$
9. $12 - h$ **11.** $x + 20$ **13. a.** $b - 15$
b. $p + 15$ **15.** $5, 25, 45$ **17. a.** $8y$ **b.** $2cd$ **c.** Commutative
19. a. 4   **b.** $3, 11, -1, 9$ **21.** Term   **23.** Factor   **25.** $l + 15$
27. $50x$ **29.** $\frac{w}{7}$ **31.** $P + \frac{2}{3}p$ **33.** $k^2 - 2{,}005$ **35.** $2a - 1$
37. $\frac{1{,}000}{n}$ **39.** $2p + 90$ **41.** $3(35 + h + 300)$ **43.** $p - 680$
45. $4d - 15$ **47.** $2(200 + t)$ **49.** $|a - 2|$ **51.** $0.1d$ or $\frac{1}{10}d$
53. Three-fourths of r **55.** 50 less than t
57. The product of x, y, and z **59.** Twice m, increased by 5
61. $x + 2$ **63.** $36 - x$ **65.** $60h$ **67.** $\frac{i}{12}$ **69.** \$8x
71. $49x$¢   **73.** \$2t **75.** \$25$(x + 2)$ **77.** 2   **79.** 13   **81.** 20
83. -12 **85.** -5 **87.** $-\frac{1}{5}$ **89.** 17   **91.** 36   **93.** 255
95. 8   **97.** $-1, -2, -28$ **99.** $41, 11, 2$ **101.** $150, -450$
103. $0, 0, 5$ **105. a.** $x + 7^2$ **b.** $(x + 7)^2$ **107. a.** $4(x + 2)$
b. $4x + 2$ **109. a.** Let $x =$ weight of the Element, $2x - 340 =$
weight of the Hummer   **b.** 6,400 lb   **111. a.** Let $x =$ age of Apple;
$x + 80 =$ age of IBM; $x - 9 =$ age of Dell
b. IBM: 112 yr; Dell: 23 yr   **117.** 60   **119.** $\frac{8}{27}$

ARE YOU READY? 1.9 (page 78)

1. The terms are in a different order.   **2.** The position of the
parentheses is different.   **3.** 30, 30; same result   **4.** Different
variable factors (x and y); the same coefficient (4)

SELF CHECKS 1.9

1. a. $(5x + 3x) + 1$ **b.** $-15a$ **2. a.** $54s$ **b.** $48u$ **c.** m **d.** $8y$
3. a. $7m + 14$ **b.** $-640x - 240$ **c.** $4y + 9$ **4. a.** $10x - 5$
b. $9y + 36$ **c.** $-c + 22$ **5. a.** $-2x - 8$ **b.** $54c - 108d$
c. $-1.4r - 3.5s + 5.6$ **6.** $5x - 18$ **7. a.** $-2y$ and $7y$
b. $5p^2$ and $17p^2$; -12 and 2   **8. a.** $8x$ **b.** $-3y$ **c.** $0.5s^4$
d. Doesn't simplify   **e.** $\frac{6}{7}c$ **9. a.** $8h$ **b.** $10h$ **c.** h **d.** $-h$
10. $7y^2 + 19y - 6$ **11.** $21y - 8$

STUDY SET SECTION 1.9 (page 86)

1. simplify   **3.** distributive   **5.** like   **7. a.** $4, 9, 36$
b. Associative Property of Multiplication   **9. a.** $+$ **b.** $-$ **c.** $-$
d. $+$ **11. a.** $10x$ **b.** Can't be simplified   **c.** $-42x$
d. Can't be simplified   **e.** $18x$ **f.** $3x + 5$ **13. a.** $6(h - 4)$
b. $-(z + 16)$ **15.** $(8 + 7) + a$ **17.** $11y$ **19.** $8d(2 \cdot 6)$
21. $t + 4$ **23.** $12t$ **25.** $-35q$ **27.** $11.2x$ **29.** $60c$ **31.** g
33. $5x$ **35.** $5x + 15$ **37.** $-12x - 27$ **39.** $9x + 10$
41. $0.4x + 1.6$ **43.** $36c - 42$ **45.** $-78c + 18$ **47.** $30t + 90$
49. $4a - 1$ **51.** $24t + 16$ **53.** $2w - 4$ **55.** $56y + 32$
57. $5a - 7.5b + 2.5$ **59.** $-x + 7$ **61.** $5.6y - 7$
63. $3x$ and $-2x$ **65.** $-3m^3$ and $-m^3$ **67.** $10x$ **69.** $20b^2$
71. $28y$ **73.** $\frac{4}{5}t$ **75.** r **77.** $-s^3$ **79.** $-6y - 10$ **81.** $-2x + 5$
83. $9m^2 + 6m - 4$ **85.** $4x^2 - 3x + 9$ **87.** $7z - 25$
89. $s^2 - 12$ **91.** $-\frac{5}{8}x$ **93.** $-3.6c$ **95.** $-96m$ **97.** 0   **99.** $0.4r$
101. $6y$ **103.** $-41r + 130$ **105.** $63m$ **107.** $12c + 34$
109. $300t$ **111.** $8x - 9$ **113.** $-10r$ **115.** $-20r$ **117.** $3a$
119. $9r - 16$ **121.** Doesn't simplify   **123.** $c - 13$ **125.** $a^3 - 8$
127. a. $70x$ **b.** $14x + 10$ **129.** $12x$ in.   **133.** 2

CHAPTER 1 REVIEW (page 89)

1. 1 hr; 100 cars   **2.** 100   **3.** 7 P.M.   **4.** 12 A.M. (midnight)
5. The difference of 15 and 3 equals 12.
6. The sum of 15 and 3 equals 18.
7. The quotient of 15 and 3 equals 5.
8. The product of 15 and 3 equals 45.   **9. a.** $4 \cdot 9; 4(9)$
b. $\frac{9}{3}$ **10. a.** $8b$ **b.** Prt **11. a.** Equation
b. Expression   **12.** $10, 15, 25$ **13. a.** $2 \cdot 12, 3 \cdot 8$ (Answers may
vary)   **b.** $2 \cdot 2 \cdot 6$ (Answers may vary)   **c.** $1, 2, 3, 4, 6, 8, 12, 24$
14. Equivalent   **15.** $2 \cdot 3^3$ **16.** $3 \cdot 7^2$ **17.** $5 \cdot 7 \cdot 11$
18. Prime   **19.** $\frac{4}{7}$ **20.** $\frac{4}{3}$ **21.** $\frac{40}{64}$ **22.** $\frac{36}{64}$ **23.** 90
24. 210   **25.** $\frac{7}{64}$ **26.** $\frac{5}{21}$ **27.** $\frac{16}{45}$ **28.** $3\frac{1}{4}$ **29.** $\frac{2}{5}$ **30.** $\frac{5}{22}$
31. $\frac{59}{60}$ **32.** $\frac{5}{18}$ **33.** $52\frac{1}{2}$ million   **34.** $\frac{17}{96}$ in.   **35. a.** 0
b. $\{\ldots, -2, -1, 0, 1, 2, \ldots\}$ **36.** -206 ft   **37. a.** $<$
b. $>$ **38. a.** $\frac{7}{10}$ **b.** $\frac{14}{3}$ **39.** 0.004   **40.** $0.\overline{772}$
41.

$$\begin{array}{c}
-\frac{17}{4} \qquad\qquad\qquad 0.333\ldots\ \frac{7}{8}\ \sqrt{2} \qquad \pi \quad 3.75 \\
\underset{-5\ \ -4\ \ -3\ \ -2\ \ -1\ \ \ 0\ \ \ 1\ \ \ 2\ \ \ 3\ \ \ 4\ \ \ 5}{\longleftarrow\!\!\!\bullet\!-\!\!-\!-\!-\!\bullet\!-\!-\!\bullet\!-\bullet\!-\!-\!\bullet\!\!\!\longrightarrow}
\end{array}$$

42. Natural: 8; whole: 0, 8; integers: $0, -12, 8$; rational: $-\frac{4}{5}$, 99.99,
$0, -12, 4\frac{1}{2}, 0.666\ldots, 8$; irrational: $\sqrt{2}$; real: all
43. False   **44.** False   **45.** True   **46.** True   **47.** $>$ **48.** $<$
49. -82 **50.** 12   **51.** -7 **52.** 0   **53.** -11 **54.** -12.3
55. $-\frac{3}{16}$ **56.** 11   **57. a.** Commutative Property of Addition
b. Associative Property of Addition   **c.** Addition Property of
Opposites (Inverse Property of Addition)   **d.** Addition Property of
0 (Identity Property of Addition)   **58.** 118°F   **59. a.** -10 **b.** 3
60. a. $\frac{9}{16}$ **b.** -4 **61.** -19 **62.** $-\frac{14}{15}$ **63.** 5   **64.** 5.7
65. -10 **66.** -29 **67.** 65,233 ft; $65{,}233 + (-36{,}205) = 29{,}028$
68. 287 B.C.; (-287); $-287 + 75 = -212$ **69.** -56 **70.** 1
71. 12   **72.** -12 **73.** 6.36   **74.** -2 **75.** $-\frac{2}{15}$ **76.** 0
77. High: 3, low: -4.5 **78. a.** Associative Property of
Multiplication   **b.** Commutative Property of Multiplication
c. Multiplication Property of 1 (Identity Property of Multiplication)
d. Inverse Property of Multiplication   **79.** -1 **80.** -17 **81.** 3
82. $-\frac{6}{5}$ **83.** Undefined   **84.** -4.5 **85.** $0, 18, 0$ **86.** $-\$360$
87. a. 8^5 **b.** $9\pi r^2$ **88. a.** 81   **b.** $-\frac{8}{27}$ **c.** 32   **d.** 50
89. 17   **90.** -36 **91.** -169 **92.** 23   **93.** -420 **94.** $-\frac{7}{19}$
95. 113   **96.** Undefined   **97. a.** $(-9)^2 = 81$ **b.** $-9^2 = -81$

98. $20   **99. a.** 3   **b.** 1   **100. a.** 16, −5, 25   **b.** $\frac{1}{2}$, 1
101. $h + 25$ **102.** $3s − 15$ **103.** $\frac{1}{2}t − 6$
104. $|2 − a^2|$ **105.** $n + 4$ **106.** $b − 4$ **107.** $10d$
108. $x − 5$ **109.** 30, $10d$ **110.** 0, 19, −16   **111.** 40
112. −36   **113.** $150a$ **114.** $(9 + 1) + 7y$ **115.** $(2.7 \cdot 10)b$
116. $2x^2 + x$ **117.** $−28w$ **118.** $24x$ **119.** $2.08f$ **120.** r
121. $5x + 15$ **122.** $−2x − 3 + y$ **123.** $3c − 6$
124. $12.6c + 29.4$ **125.** $9p$ **126.** $−7m$ **127.** $4n$
128. $−p − 18$ **129.** $0.1k^2$ **130.** $8a^3 − 1$ **131.** w
132. $4h − 15$ **133.** $4.6t − 7.7$ **134. a.** x **b.** $−x$
c. $4x + 1$ **d.** $4x − 1$

CHAPTER 1 TEST (page 98)

1. a. equivalent   **b.** product   **c.** reciprocal   **d.** like terms
e. undefined   **2. a.** $24   **b.** 5 hr   **3.** 3, 20, 70
4. $2 \cdot 2 \cdot 3 \cdot 3 \cdot 5 = 2^2 \cdot 3^2 \cdot 5$ **5.** $\frac{2}{5}$ **6.** $\frac{3}{2} = 1\frac{1}{2}$ **7.** $\frac{27}{35}$ **8.** $6\frac{11}{15}$
9. a. $4\frac{1}{4}$ lb   **b.** $3.57   **10.** $0.8\overline{3}$
11. a.

b. Natural numbers: 2; whole numbers: 0, 2; integers: 0, 2, −3;
rational numbers: $−1\frac{1}{4}$, 0, −3.75, 2, $\frac{7}{2}$, 0.5, −3; irrational numbers:
$\sqrt{2}$; real numbers: all   **12. a.** True   **b.** False   **c.** True   **d.** True
e. True   **13. a.** >   **b.** <   **c.** <   **d.** >
14. A gain of 0.6 of a rating point   **15.** −7.6   **16.** −2   **17.** $\frac{3}{8}$
18. a. −6   **b.** $−6 + (−4) = −10$ **19. a.** 14
b. $14(−0.9) = −12.6$ **20.** −30   **21.** −2.44   **22.** 0   **23.** $−\frac{27}{125}$
24. 0   **25.** −3   **26.** 50   **27.** 14   **28.** a net loss of 78 in.
29. −$1,275   **30. a.** $−12 + (97 + 3)$ **b.** $2x + 14$ **c.** $−2(5)m$
d. 1   **e.** $15x$ **31. a.** 9^5 **b.** $3x^2z^3$ **32.** 4, 17, −59   **33.** 170
34. −12   **35.** −99   **36.** −351   **37.** 20   **38.** −60   **39.** $2w − 7$
40. a. $x − 2$ **b.** $25q¢$ **41.** 3   **42.** 1, −6, −1, 10   **43.** $−20x$
44. $224t$ **45.** $−4a + 4$ **46.** $−5.9d^3$ **47.** $14x + 3$
48. $3m^2 + 2m − 4$ **49. a.** True   **b.** False   **c.** False   **d.** False
e. True   **f.** False

ARE YOU READY? 2.1 (page 102)

1. −   **2.** +   **3.** 7   **4.** ·   **5.** $\frac{2}{5}$ **6.** $−\frac{8}{9}$

SELF CHECKS 2.1

1. Yes   **2.** 49   **3. a.** 33   **b.** 49   **4. a.** $\frac{29}{15}$ **b.** −0.5   **5.** 72
6. a. 6   **b.** $−\frac{11}{21}$ **7. a.** 11   **b.** −25.1   **8.** 12

STUDY SET SECTION 2.1 (page 110)

1. equation   **3.** solve   **5.** equivalent   **7. a.** $x + 6$ **b.** Neither
c. No   **d.** Yes   **9. a.** c, c **b.** c, c **11. a.** x **b.** y **c.** t
d. h **13.** 5, 5, 50, 50, $\overset{?}{=}$, 45, 50   **15. a.** Is possibly equal to
b. Yes   **17.** No   **19.** No   **21.** No   **23.** No   **25.** Yes   **27.** No
29. No   **31.** Yes   **33.** Yes   **35.** Yes   **37.** 71   **39.** 18   **41.** −12
43. 3   **45.** $\frac{11}{10}$ **47.** −2.3   **49.** 45   **51.** −48   **53.** 20   **55.** $−\frac{5}{42}$
57. 4   **59.** 0.5   **61.** −18   **63.** $−\frac{4}{21}$ **65.** 7   **67.** $\frac{8}{9}$ **69.** 0
71. −0.9   **73.** 15   **75.** 20   **77.** $−\frac{1}{25}$ **79.** 21   **81.** 0   **83.** −2.64
85. 1   **87.** 4   **89.** 13   **91.** 2.5   **93.** $−\frac{8}{3}$ **95.** $−\frac{21}{16}$ **97.** −5
99. −200   **101.** $\frac{4}{3}$ **103.** 4   **105. a.** $\frac{13}{20}$ **b.** $\frac{17}{20}$ **c.** $\frac{15}{2}$ **d.** $\frac{3}{40}$
107. 65°   **109.** $6,000,000   **115.** 0   **117.** $45 − x$

ARE YOU READY? 2.2 (page 112)

1. −12   **2.** $2a$ **3.** $−7m + 18$ **4.** $3x$ **5.** $24n$ **6.** 8

SELF CHECKS 2.2

1. 7   **2.** −36   **3.** −3.9   **4. a.** 1   **b.** −11   **5.** −16   **6.** $−\frac{5}{2}$
7. 6,000   **8.** 0   **9.** All real numbers; the equation is an identity.
10. No solution; the equation is a contradiction.

STUDY SET SECTION 2.2 (page 120)

1. equation   **3.** identity   **5. a.** subtraction, multiplication
b. addition, division   **7.** No   **9. a.** 6   **b.** 10
11. 7, 7, 2, 2, 14, $\overset{?}{=}$, 28, 21, 14   **13.** −9   **15.** −5   **17.** 18   **19.** 16
21. −4   **23.** 2.9   **25.** $\frac{11}{5}$ **27.** −6   **29.** −21   **31.** 1   **33.** $\frac{2}{15}$
35. 6   **37.** 5   **39.** 200   **41.** −4   **43.** −1   **45.** All real numbers
47. No solution   **49.** $\frac{1}{4}$ **51.** 12   **53.** −1   **55.** 6   **57.** 1   **59.** 30
61. No solution   **63.** −11   **65.** 7   **67.** −11   **69.** $\frac{9}{2}$ **71.** $−\frac{12}{5}$
73. −0.25   **75.** −7   **77.** $\frac{10}{3}$ **79.** 5   **81.** 1,000   **83.** −3   **85.** $\frac{27}{5}$
87. 200   **89.** −11   **91.** 3   **93.** $\frac{52}{9}$ **95.** 0.04   **97.** 80   **99.** $−\frac{5}{2}$
101. −20   **103.** No solution   **105.** All real numbers   **107.** −6
109. −6   **111. a.** $x − 20$ **b.** −5   **113. a.** $0.4 − 0.2x$ **b.** 0
119. Commutative property of multiplication   **121.** Associative
property of addition

ARE YOU READY? 2.3 (page 122)

1. 61   **2.** 2   **3.** 0.5, 0.75   **4.** 0.27   **5.** 3.72   **6.** 49.5

SELF CHECKS 2.3

1. 2.24   **2.** 570 million   **3.** 1.45%   **4.** 21%   **5.** $120   **6.** $600

STUDY SET SECTION 2.3 (page 127)

1. Percent   **3.** multiplication, is   **5.** $\frac{51}{100}$, 0.51, 51%
7. amount, percent, base   **9. a.** 639   **b.** 639, what, 3,618
11. a. 0.35   **b.** 0.085   **c.** 1.5   **d.** 0.0275   **e.** 0.0925   **f.** 0.015
13. 312   **15.** 46.2   **17.** 300   **19.** 1,464   **21.** 26%   **23.** 2.5%
25. 0.48 oz   **27. a.** $1,102.6 billion   **b.** $715.2 billion   **29.** $10.45
31. $24.20   **33.** 60%, 40%   **35.** 19%   **37.** No (66%)
39. 120 children   **41. a.** 5 g; 25%   **b.** 20 g   **43. a.** 11%   **b.** 46%
45. 12%   **47. a.** 6%   **b.** 23.5 mpg   **49.** $75   **51.** $300
53. $95,000   **55.** $25,600   **61.** $\frac{12}{5} = 2\frac{2}{5}$ **63.** No

ARE YOU READY? 2.4 (page 130)

1. a. one   **b.** three   **2.** d **3.** x **4.** $8c − cx$ **5.** Yes   **6.** 1

SELF CHECKS 2.4

1. $1,510.50   **2.** 2.5 yr   **3.** 1.25 min   **4.** −283°F   **5.** 230 ft
6. 62.83 in.   **7.** 1,357.2 m^3 **8.** $c = r − p$ **9.** $a = \frac{2A}{r^2}$
10. $c = \frac{B − 4d}{3}$ **11.** $y = 4 − \frac{1}{3}x$ or $y = −\frac{1}{3}x + 4$
12. $s = \frac{A + 3xy}{xy}$

STUDY SET SECTION 2.4 (page 137)

1. formula   **3.** volume   **5. a.** $d = rt$ **b.** $r = c + m$
c. $p = r − c$ **d.** $I = Prt$ **7.** 11,176,920 mi, 65,280 ft
9. Ax, Ax, B, B, B **11. a.** $\pi r^2 h$ **b.** The radius of a cylinder;
the height of a cylinder   **13.** $240 million   **15.** $931   **17.** 3.5%
19. $6,000   **21.** 2.5 mph   **23.** 4.5 hours   **25.** 185°C

27. $-454°F$ **29.** 20 in. **31.** $1{,}885 \text{ mm}^3$ **33.** $c = r - m$
35. $b = P - a - c$ **37.** $h = \frac{3V}{B}$ **39.** $R = \frac{E}{I}$ **41.** $r = \frac{T - 2t}{2}$
43. $x = \frac{C - By}{A}$ **45.** $y = -\frac{2}{7}x + 3$ **47.** $y = \frac{9}{2}x + 4$
49. $m = \frac{T - 4ab}{4ab}$ **51.** $r = \frac{G + g}{4g}$ **53.** $c = 3A - a - b$
55. $y = -3x + 9$ **57.** $m = \frac{2K}{v^2}$ **59.** $r = \frac{C}{2\pi}$ **61.** $M = 4.2B + 19.8$
63. $f = \frac{s}{w}$ **65.** $y = \frac{1}{3}x + 3$ **67.** $b = \frac{2A}{h} - d$ or $b = \frac{2A - hd}{h}$
69. $a^2 = c^2 - b^2$ **71.** $c = \frac{72 - 8w}{7}$ **73.** $b = \frac{m - 70 - at}{t}$
75. $l = \frac{V}{wh}$ **77.** $t = T - 18E$ **79.** $r^2 = \frac{s}{4\pi}$ **81. a.** $R = A - ab$
b. $a = \frac{A - R}{b}$ **83. a.** $h = \frac{S - 4lw}{2w}$ **b.** $l = \frac{S - 2wh}{4w}$
85. Horsepower $= \frac{\text{RPM} \cdot \text{Torque}}{5{,}252}$ **87.** 168.4, 192.8 **89.** 14 in.
91. 50 in. **93.** 25 in., 2.5 in. **95.** 18.1 in.^2 **97.** $2{,}463 \text{ ft}^2$
99. $3{,}150 \text{ cm}^2$ **101.** 6 in. **103.** 8 ft **105.** 348 ft^3 **107.** 254 in.^2
109. $R = \frac{L - 2D - 3.25r}{3.25}$ or $R = \frac{L - 2D}{3.25} - r$ **115.** 137.76
117. 15%

ARE YOU READY? 2.5 (page 142)

1. $4x + 4$ **2.** 49 **3.** $39.15 **4.** $0.28x$ **5.** $P = 2l + 2w$
6. $180°$ **7.** $2x - 8$ **8.** 0.06

SELF CHECKS 2.5

1. 16 mi, 18 mi, 20 mi, and 22 mi **2.** 47 shirts **3.** $2,762.77
4. Page 579 **5.** 5 ft by 11 ft **6.** 12 cm, 12 cm

STUDY SET SECTION 2.5 (page 147)

1. consecutive **3.** vertex, base **5.** $17, x + 2, 3x$ **7.** $0.03x$
9. $180°$ **11. a.** $x + 1$ **b.** $x + 2$ **c.** $x + 2, x + 4$ **13.** 4 ft, 8 ft
15. Day 1: 102 mi; day 2: 108 mi; day 3: 114 mi; day 4: 120 mi
17. 7.3 ft, 10.7 ft **19.** *Guitar Hero:* $2.99; *Call of Duty:*
World at War: Zombies II: $9.99; *Tom Tom USA:* $39.99
21. 250 calories in ice cream, 600 calories in pie **23.** 7 hr
25. 580 mi **27.** 20 hr **29.** $50,000 **31.** $240 **33.** $5,250
35. Ronaldo: 15 goals; Mueller: 14 goals
37. *Friends:* 236 episodes; *Leave It to Beaver:* 234 episodes
39. July 22, 24, 26 **41.** Width: 27 ft; length: 78 ft
43. 21 in. by 30.25 in. **45.** 7 ft, 7 ft; 11 ft **47.** $20°$
49. At steering column: $42.5°$; at seat support: $70°$;
at pedal gear: $67.5°$ **51.** $x = 11°$; $22°, 68°$
53. Maximum stride angle: $106°$ **59.** -24 **61.** $-\frac{40}{37}$

ARE YOU READY? 2.6 (page 151)

1. $400 **2.** 135 mi **3.** 3.6 gal of antifreeze **4.** $19.60
5. $14,000 **6.** $3x$

SELF CHECKS 2.6

1. $2,400 ar 2%, $1,800 at 3% **2.** 3.5 hr **3.** 1.5 hr **4.** 10 gal
5. 40 lb **6.** 12 iPods, 4 skins, 24 cards

STUDY SET SECTION 2.6 (page 158)

1. investment, motion **3.** $30{,}000 - x$ **5.** $r - 150$
7. $35t; t; 45t; 80; 35t + 45t = 80$
9. a. $0.25x, 0.50(6), 6 + x, 0.30(6 + x)$,
$0.25x + 0.50(6) = 0.30(6 + x)$
b. $10 - x, 0.06x, 0.03(10 - x), 0.05(10)$,
$0.03(10 - x) + 0.06x = 0.05(10)$ **11.** 0.06, 0.152 **13.** 4
15. 6,000 **17.** $15,000 at 4%; $10,000 at 7% **19.** Silver: $1,500;
gold: $2,000 **21.** $26,000 **23.** 822: $9,000; 721: $6,000 **25.** $4,900
27. Credit union: $13,500; stocks: $4,500 **29.** 2 hr

31. $\frac{1}{4}$ hr $= 15$ min **33.** 1 hr **35.** 4 hr **37.** 55 mph
39. 50 gal **41.** 4%: 5 gal; 1%: 10 gal
43. 32 ounces of 8%; 32 ounces of 22% **45.** 6 gal **47.** 50 lb
49. 20 scoops **51.** 15 **53.** $4.25 **55.** 17 **57.** 90
59. 40 pennies, 20 dimes, 60 nickels
61. 2-pointers: 50; 3-pointers: 4 **67.** $-36a - 48b + 384$
69. $30t + 6$

ARE YOU READY? 2.7 (page 162)

1. is less than **2.** True **3.**
4. $0 < 10$

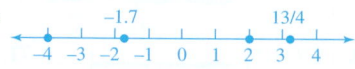

SELF CHECKS 2.7

1. Yes **2.** $[0, \infty)$
3. $(-\infty, 1)$
4. a. $[-200, \infty)$
b. $(-\infty, 12)$
5. $(-3, \infty)$
6. $(-\infty, 2.1)$
7. $\left[\frac{13}{4}, \infty\right)$
8. $\left(-\infty, \frac{21}{2}\right)$
9. $[-2, 1)$
10. $[-4, 0]$ **11.** 84% or better

STUDY SET SECTION 2.7 (page 171)

1. inequality **3.** interval **5. a.** both **b.** positive
c. negative **7.** $x > 32$ **9. a.** $-1, -\infty$ **b.** $2, \infty$ **11. a.** $\le$
b. ∞ **c.** [or] **d.** $>$ **13.** 5, 5, 12, 4, 4, 3 **15. a.** Yes **b.** No
17. a. No **b.** Yes **19.** $(-\infty, 5)$

21. $(-3, 1]$ **23.** $(3, \infty)$

25. $[10, \infty)$ **27.** $(-\infty, 48]$

29. $[3, \infty)$ **31.** $[2, \infty)$

33. $(-\infty, 6)$ **35.** $(7, \infty)$

37. $(-\infty, -5]$ **39.** $(-\infty, 0.4]$

41. $(2.4, \infty)$ **43.** $(-5, \infty)$

45. $\left[-\frac{5}{3}, \infty\right)$ **47.** $\left(\frac{5}{4}, \infty\right)$

49. $(-\infty, 15]$

51. $[-2, 3)$

53. $\left(-\frac{7}{4}, 2\right)$

55. $(7, 10)$

57. $[-10, 0]$

59. $[2, 3]$

61. $(-3, 6]$

63. $\left(-\infty, \frac{3}{2}\right]$

65. $(-\infty, 12]$

67. $(-2, 1]$

69. $(-\infty, 1.5]$

71. $\left(-\infty, \frac{1}{8}\right]$

73. $(-\infty, 0)$

75. $\left[\frac{9}{4}, \infty\right)$

77. $(-5, -2)$

79. $(-\infty, 2]$

81. $(-\infty, -27)$

83. $\left(-\infty, \frac{17}{21}\right]$

85. $[-13, \infty)$

87. $(6, \infty)$

89. $[-32, 48]$

91. $(-\infty, 1.5]$

93. $\left[-\frac{3}{8}, \infty\right)$

95. $\left(-\infty, \frac{1}{2}\right)$

97. $(-\infty, -1]$

99. a. $\left(\frac{1}{8}, \infty\right)$ **b.** $\frac{1}{8}$

101. a. $[5, \infty)$ **b.** $[5, 12)$

103. 98% or better   **105.** More than 27 mpg   **107.** 19 ft or less   **109.**
More than 5 ft   **111.** 40 or less   **113.** 12.5 in. or less
115. 26, 27, 28, 29, 30   **119.** 1, −3, 6

CHAPTER 2 REVIEW (page 175)

1. Yes   **2.** No   **3.** No   **4.** No   **5.** Yes   **6.** Yes
7. equation   **8.** True   **9.** 21   **10.** 32   **11.** −20.6
12. 107   **13.** 24   **14.** $\frac{16}{21}$ **15.** −9   **16.** −7.8   **17.** 0
18. $-\frac{16}{5}$ **19.** 2   **20.** −30.6   **21.** 30   **22.** −19   **23.** 4
24. 1   **25.** $\frac{5}{4}$ **26.** $\frac{47}{13}$ **27.** 6   **28.** $-\frac{22}{75}$ **29.** 5   **30.** 1
31. Identity; all real numbers   **32.** Contradiction; no solution
33. a. Percent   **b.** discount   **c.** commission   **34.** 192.4
35. 142.5   **36.** 12%   **37. a.** 28.8%   **b.** 221 million
38. $26.74   **39.** No   **40.** $450   **41.** $150   **42.** 1,567%
43. $176   **44.** $11,800   **45.** 8 min   **46.** 4.5%
47. 1,949°F   **48. a.** 168 in.   **b.** 1,440 in.²   **c.** 4,320 in.³
49. 76.5 m²   **50.** 144 in.²   **51. a.** 50.27 cm
b. 201 cm²   **52.** 9.4 ft³   **53.** 381.70 in.³   **54.** 120 ft³
55. $h = \frac{A}{2\pi r}$ **56.** $G = 3A - 3BC + K$ **57.** $t = \frac{4C}{s} + d$
58. $y = \frac{3}{4}x + 4$ **59.** 8 ft   **60.** 200 signatures   **61.** $2,500,000
62. Labonte: 43; Petty: 45
63. 24.875 in. × 29.875 in. $\left(24\frac{7}{8}\text{ in.} \times 29\frac{7}{8}\text{ in.}\right)$

64. 76.5°, 76.5°   **65.** $16,000 at 7%, $11,000 at 9%   **66.** 20 min
67. $1\frac{2}{3}$ hr = 1 hr 40 min   **68.** TV celebrities: 12 autographs;
movie stars: 4 autographs   **69.** 10 lb of each   **70.** 2 gal

71. $(-\infty, 1)$

72. $(-\infty, 12]$ **73.** $\left(\frac{5}{4}, \infty\right)$

74. $[3, \infty)$ **75.** $(-\infty, 40]$

76. $(7, \infty)$ **77.** $(6, 11)$

78. $\left(-\frac{7}{2}, \frac{3}{2}\right]$

79. $2.40\text{ g} \le w \le 2.53\text{ g}$ **80.** $0\text{ in.} < l \le 48\text{ in.}$; 48 inches or less

CHAPTER 2 TEST (page 182)

1. a. solve   **b.** Percent   **c.** circumference   **d.** inequality
e. multiplication, equality   **2.** No   **3.** 2   **4.** 22   **5.** −5
6. $\frac{12}{7}$ **7.** 1,336   **8.** All real numbers (an identity)   **9.** $\frac{7}{4}$ **10.** 55
11. 0   **12.** 0   **13.** −4   **14.** No solution (a contradiction)
15. 12.16   **16.** $76,000   **17.** 6%   **18.** $30   **19.** $295
20. −10°C   **21.** 66,480 ft   **22.** 393 in.³   **23.** $h = \frac{V}{\pi r^2}$
24. $r = \frac{A - P}{Pt}$ **25.** $c = 4A - a - b - d$ **26.** $y = \frac{2}{3}x - 3$
27. 20 in.²   **28.** Programming: 22 min; commercials: 8 min
29. $40.55   **30.** 80 balcony seats, 800 floor seats   **31.** $120,000
32. 380 mi, 280 mi   **33.** Green: 16 lb; herbal: 4 lb   **34.** 412, 413
35. $\frac{3}{5}$ hr   **36.** 10 liters   **37.** $\frac{1}{3}$ hr = 20 min   **38.** 68°   **39.** $5,250
40. No   **41.** $[-3, \infty)$

42. $(-\infty, 6.4)$ **43.** $[-7, 4)$

44. $(-\infty, -13)$

45. $(-\infty, 5]$ **46.** 180 words

ARE YOU READY? 3.1 (page 186)

1.

2. a. 8   **b.** −3.5

3. I, II, III, IV   **4.** $4\frac{1}{2}, -3\frac{2}{3}$

SELF CHECK 3.1

1.

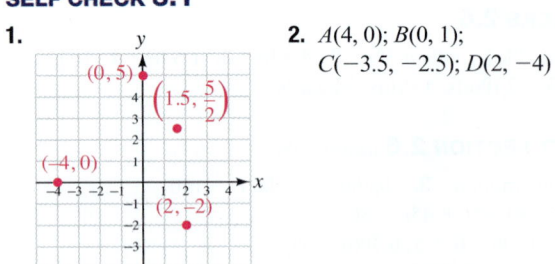

2. $A(4, 0)$; $B(0, 1)$;
$C(-3.5, -2.5)$; $D(2, -4)$

3. a. 30 min before and 85 min after taping began   **b.** 200
c. 40 min before taping began

STUDY SET SECTION **3.1** (page 191)

1. ordered   **3.** axis, axis, origin   **5.** rectangular
7. a. origin, left, up   **b.** origin, right, down   **9. a.** I and II
b. II and III   **c.** IV   **d.** The y-axis
11. (3, 5) is an ordered pair, $3(5) = 3 \cdot 5$
13. Yes   **15.** Horizontal
17.

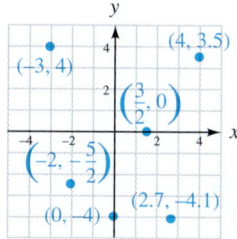

19. (4, 3), (0, 4), (−5, 0), (−4, −5), (3, −3)
21. a. 60 beats/min   **b.** 10 min
23. a. 5 min and 50 min after starting   **b.** 20 min
25. a. 2 hr   **b.** −1,000 ft   **27. a.** It ascends (rises) 500 ft
b. −500 ft   **29.** Rivets: (−60, 0), (−20, 0), (20, 0), (60, 0); welds:
(−40, 30), (0, 30), (40, 30); anchors: (−60, −30), (60, −30)
31. (G, 2), (G, 3), (G, 4)   **33. a.** 8 teeth   **b.** It represents the
patient's left side.   **35. a.** 60°; 4 ft   **b.** 30°; 4 ft   **37.** 10 square units
39.

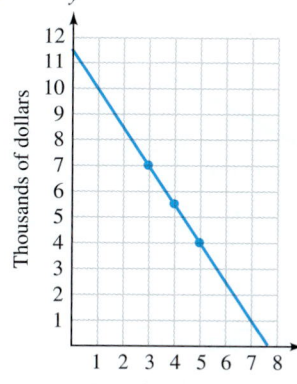

a. 20 mi.   **b.** 6 gal
c. 35 mi

41.

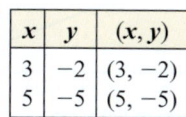

a. A 3-yr-old copier
is worth $7,000.   **b.** $1,000
c. 6 yr
47. $h = \frac{3(AC + T)}{2}$ or $h = \frac{3AC + 3T}{2}$
49. −1

ARE YOU READY? **3.2** (page 195)

1. False   **2.** 5   **3.** 2   **4.** $y = -\frac{4}{5}x - 3$ **5.** 3   **6.** −5

SELF CHECK **3.2**

1. Yes   **2.** (−2, −10)   **3.**

x	y	(x, y)
3	−2	(3, −2)
5	−5	(5, −5)

4.

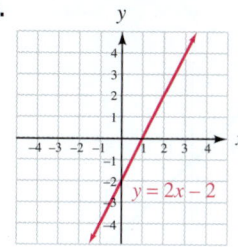

5.

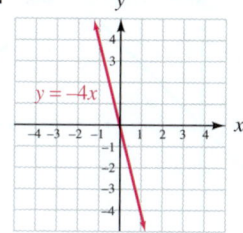

6.

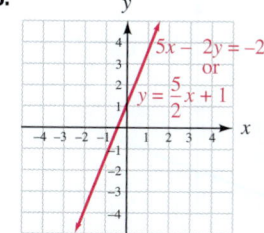

7. $c = 15 + 10n$

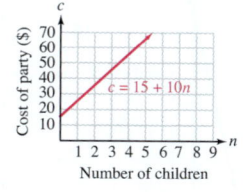

STUDY SET SECTION **3.2** (page 203)

1. two   **3.** table   **5.** linear   **7. a.** 2   **b.** Yes   **c.** No
d. Infinitely many   **9.** solution, point
11. a. −5, 0, 5 (Answers may vary)
b. −10, 0, 10 (Answers may vary)   **13.** 6, −2, 2, 6
15. a. 1's   **b.** The exponent on x is not 1.   **17.** Yes
19. No   **21.** Yes   **23.** Yes   **25.** No   **27.** No   **29.** 11
31. 4   **33.** 13   **35.** $-\frac{8}{7}$
37.

x	y	(x, y)
8	12	(8, 12)
6	8	(6, 8)

39.

x	y	(x, y)
−5	−13	(−5, −13)
−1	−1	(−1, −1)

41.

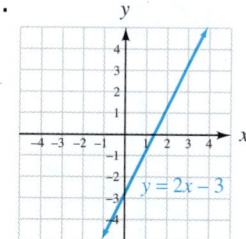

43.

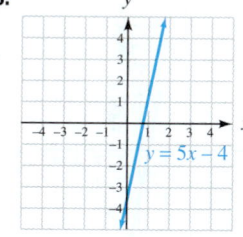

45.

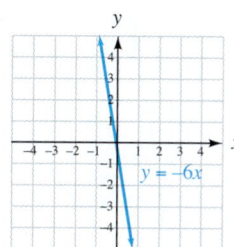

47.

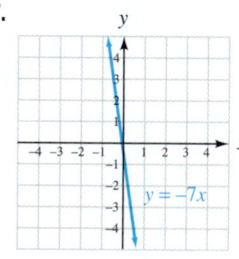

49.

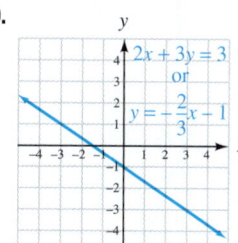

51.

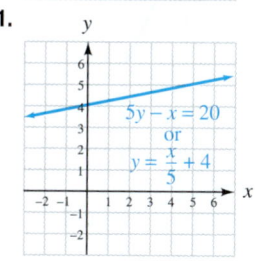

53.

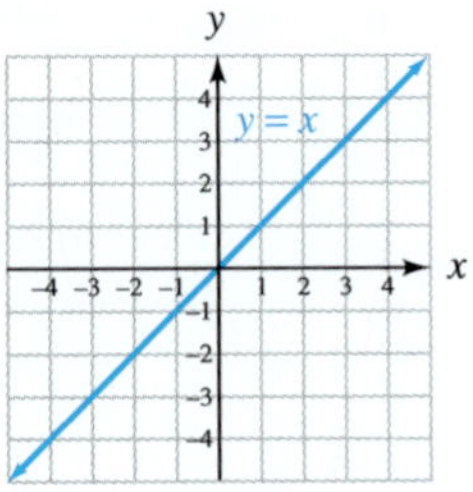

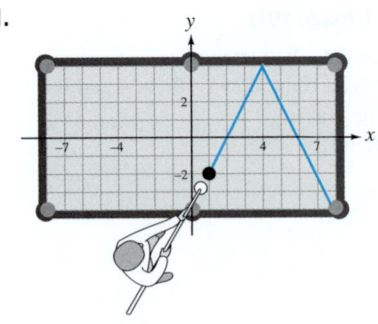

y = x

55.

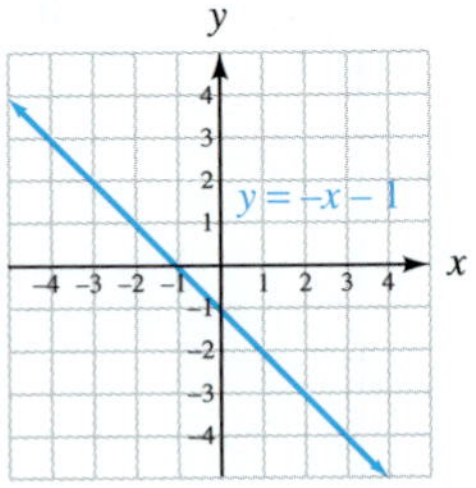

y = −x − 1

57.

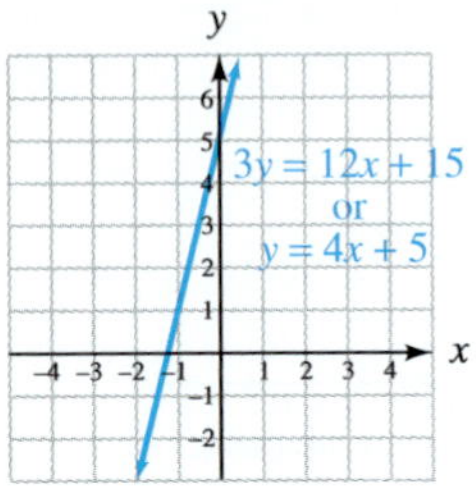

3y = 12x + 15
or
y = 4x + 5

59.

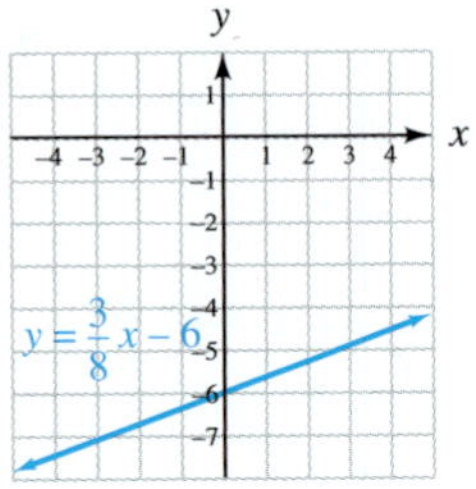

y = 3/8 x − 6

61.

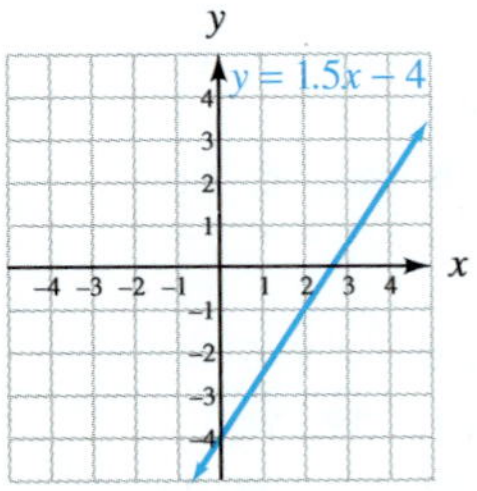

y = 1.5x − 4

63.

8x + 4y = 16
or
y = −2x + 4

65.

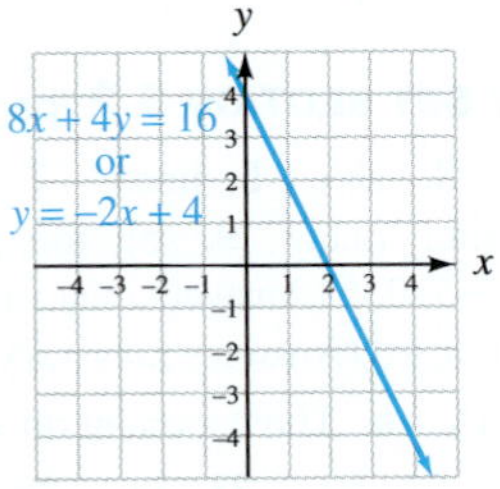

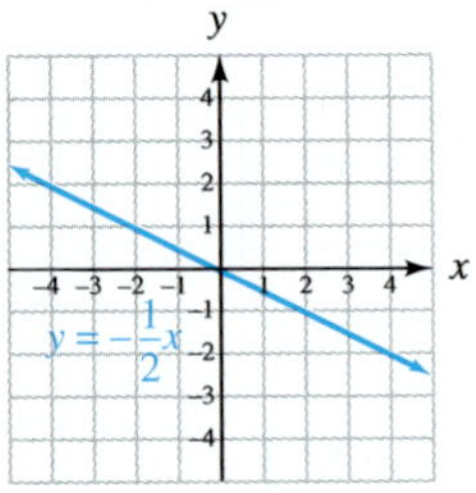

y = −1/2 x

67.

y = 5/6 x − 5

69.

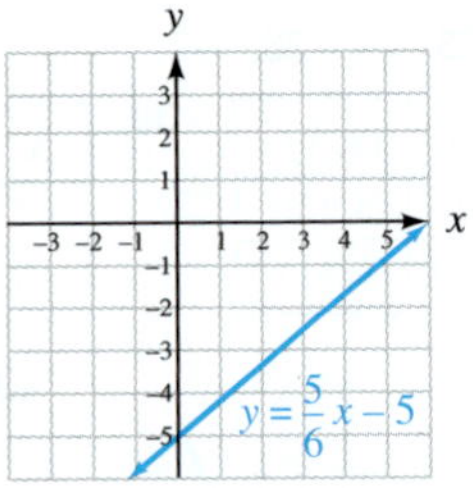

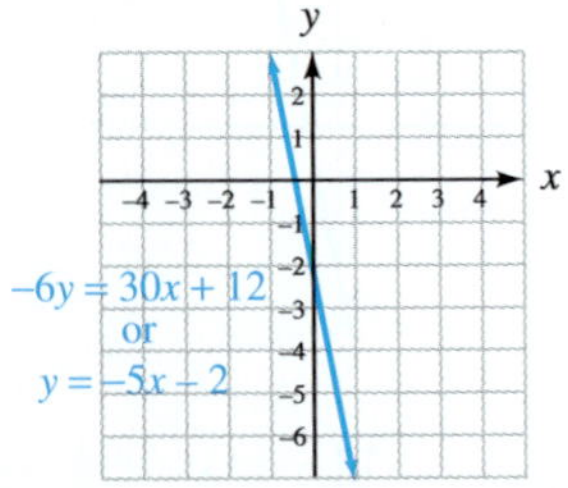

−6y = 30x + 12
or
y = −5x − 2

71.

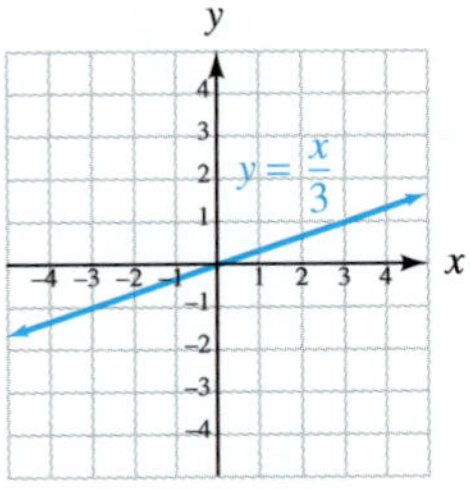

y = x/3

73.

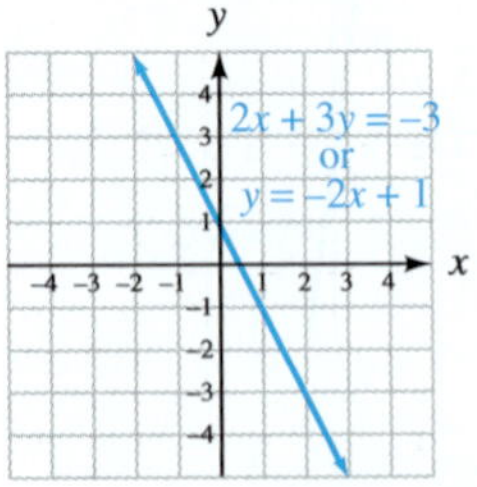

2x + 3y = −3
or
y = −2x + 1

75.

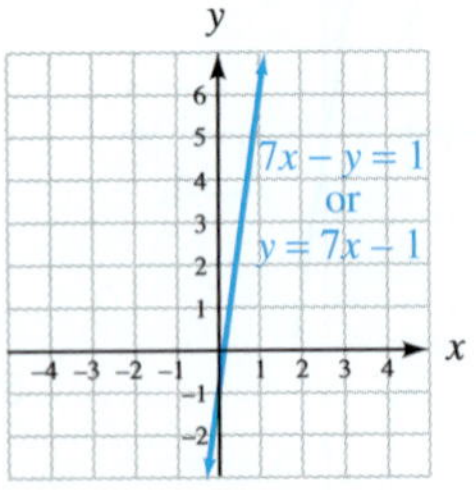

7x − y = 1
or
y = 7x − 1

77.

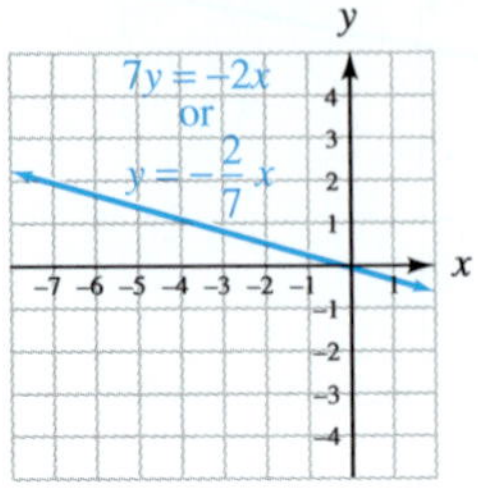

7y = −2x
or
y = −2/7 x

79.

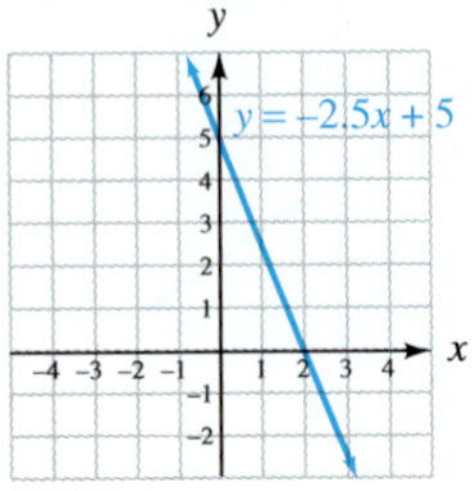

y = −2.5x + 5

81.

83. About 125 hr

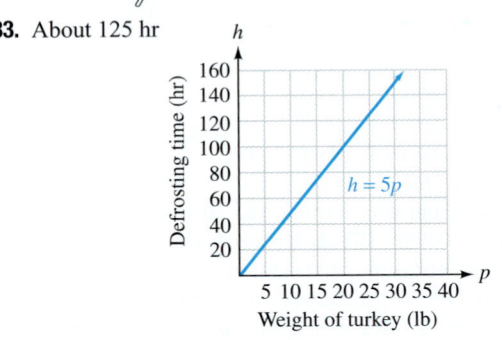

Defrosting time (hr)
h = 5p
Weight of turkey (lb)

85. About 3 oz

Amount of polish left in bottle (oz)
A = −0.02n + 16
Number of sprays

87. About $100

Average cost of NFL ticket ($)
p = 2.7t + 20
Years after 1990

89. About 180 tickets

Tickets sold
n = −20p + 300
Price of ticket ($)

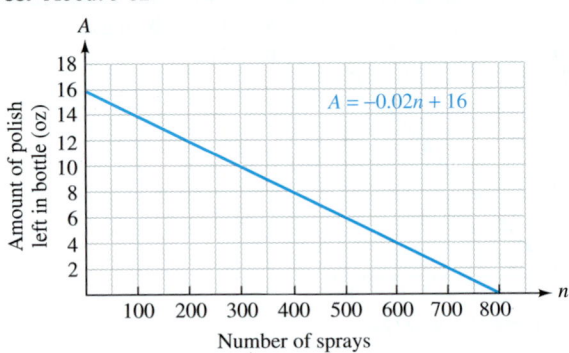

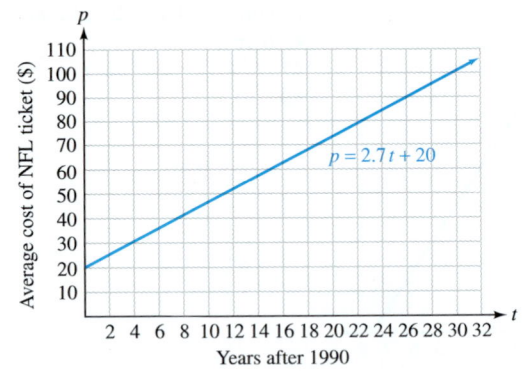

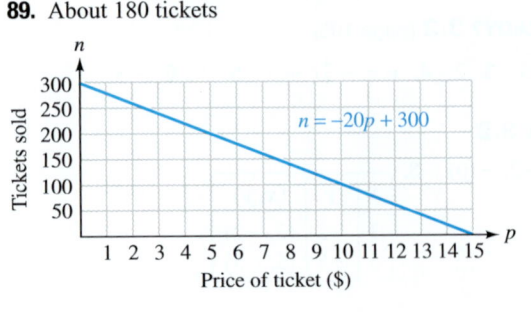

91. 2030

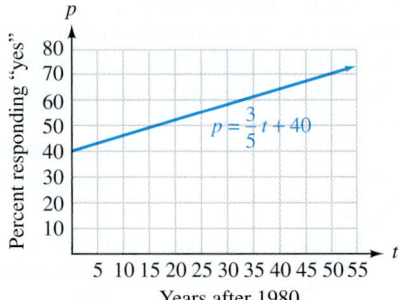

101. $5 + 4c$ or $4c + 5$ **103.** 904.8 ft³

ARE YOU READY? 3.3 (page 206)

1.

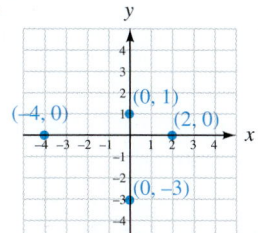

2. 0 **3.** The origin (0, 0) **4.** 5

SELF CHECK 3.3

1. $(-1, 0); (0, -3)$

2.

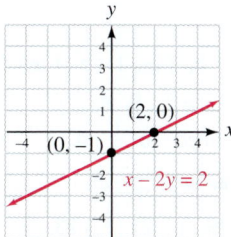

3.

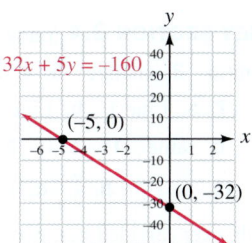

4.

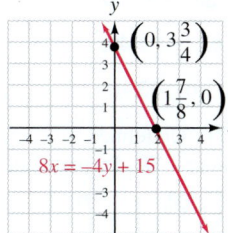

5.

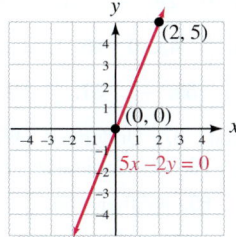

6.

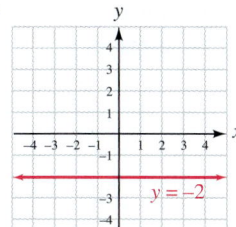

7.

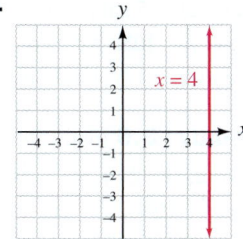

STUDY SET SECTION 3.3 (page 214)

1. x-intercept **3.** horizontal, vertical **5. a.** 0, y **b.** 0, x
7. a. y-intercept: (0, 80,000); \$80,000
b. x-intercept: (30, 0); 30 years after purchase **9.** $y = 0$; $x = 0$
11. x-intercept: (4, 0), y-intercept: (0, 3)
13. x-intercept: $(-5, 0)$, y-intercept: $(0, -4)$
15. No x-intercept, y-intercept: (0, 2)

17. x-intercept: $\left(-2\frac{1}{2}, 0\right)$; y-intercept: $\left(0, \frac{2}{3}\right)$ (Answers may vary)
19. (3, 0); (0, 8) **21.** (4, 0); (0, −14) **23.** $(-2, 0); \left(0, -\frac{10}{3}\right)$
25. $\left(\frac{3}{2}, 0\right); (0, 9)$

27.

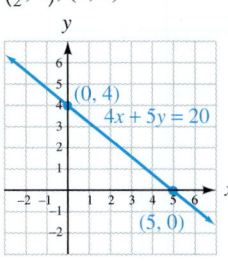

29.

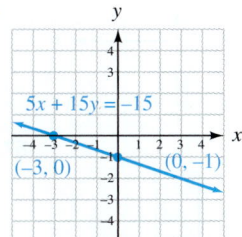

31.

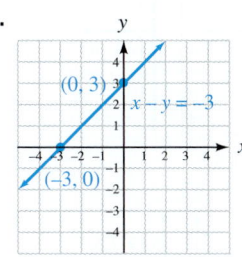

33.

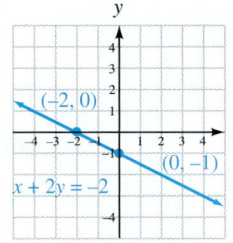

35.

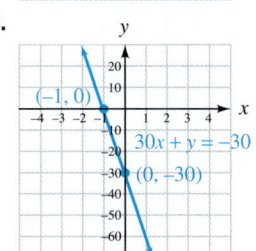

37.

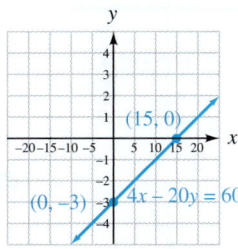

39.

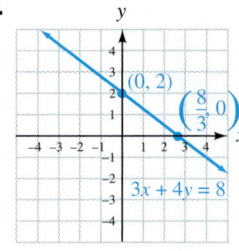

41.

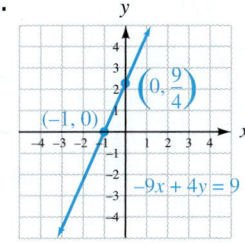

43.

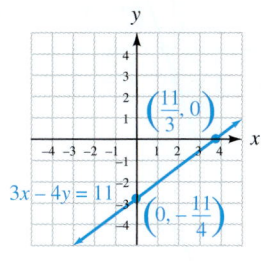

45.

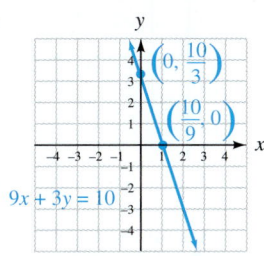

47.

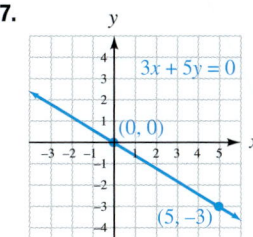

49.

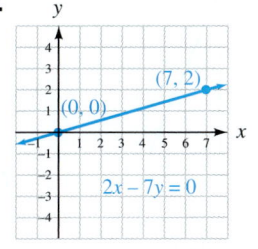

51.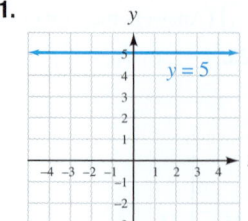

53.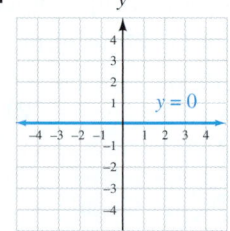

55.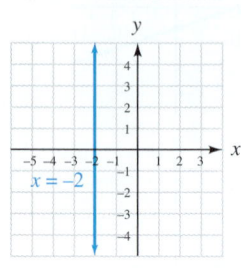

57.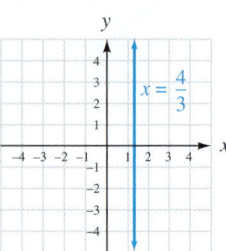

59. $y = 2$

61. $x = 1.5$

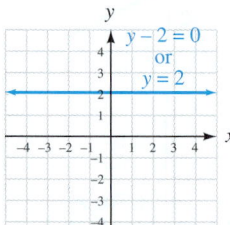

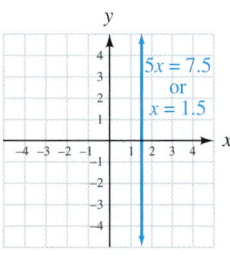

63.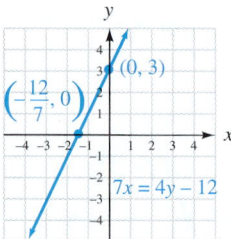

65.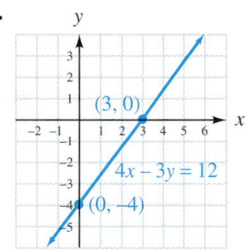

67.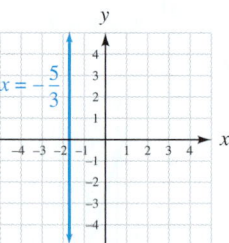

69.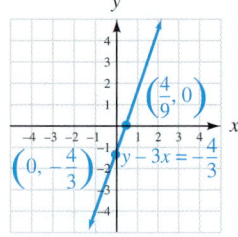

71.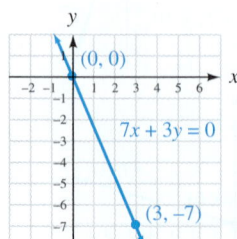

73.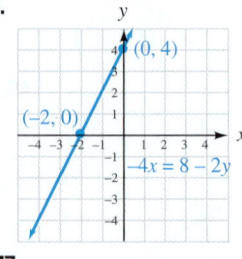

75.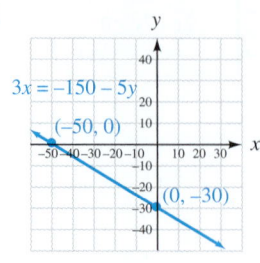

77.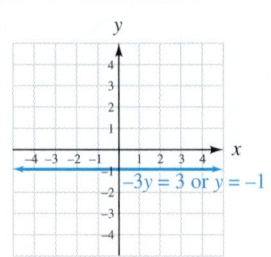

79. a. About $-270°C$ **b.** 0 milliliters

81. The g-intercept is $(0, 5)$: Before any cups of water have been served from the bottle, it contains 5 gallons of water. The c-intercept is $\left(106\frac{2}{3}, 0\right)$: The bottle will be empty after $106\frac{2}{3}$ six-ounce cups of water have been served from it.

83.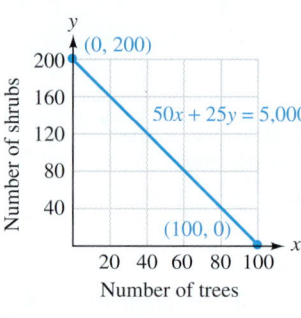

a. If only shrubs are purchased, he can buy 200.

b. If only trees are purchased, he can buy 100.

89. $\frac{1}{5}$ **91.** $2x - 6$

ARE YOU READY? 3.4 (page 217)

1. $\frac{3}{5}$ **2.** undefined   **3.** -1 **4.** $\frac{5}{6}$

SELF CHECK 3.4

1. $-\frac{1}{2}$ **2.** 5   **3.** $-\frac{5}{2}$ **4.** 0   **5.** Undefined slope   **6.** $\frac{5}{12}$

7. An increase of 650 million bushels per year   **8.** Neither

9. $-\frac{13}{4}$

STUDY SET SECTION 3.4 (page 225)

1. slope, ratio   **3.** change   **5. a.** Line 2   **b.** Line 1   **c.** Line 4

d. Line 3   **7. a.** Line 1   **b.** Line 1   **c.** Line 2   **9. a.** $\frac{1}{3}$

b. $\frac{4}{12} = \frac{1}{3}$ **c.** same   **11. a.** 0   **b.** Undefined   **c.** $\frac{1}{4}$ **d.** 2

13. 40%   **15. a.** $-\frac{1}{6}$ **b.** $\frac{8}{7}$ **c.** 1   **17. a.** $m = \frac{y_2 - y_1}{x_2 - x_1}$

b. sub, sub, over (divided by), two, one   **19.** 1   **21.** $\frac{2}{3}$ **23.** $\frac{4}{3}$

25. -2 **27.** 0   **29.** $-\frac{1}{5}$ **31.** $\frac{1}{2}$ **33.** 1   **35.** -3 **37.** $\frac{5}{4}$

39. $-\frac{1}{2}$ **41.** $\frac{3}{5}$ **43.** 0   **45.** Undefined   **47.** $-\frac{2}{3}$ **49.** -4.75

51. 0   **53.** $\frac{7}{5}$ **55.** $-\frac{2}{5}$ **57.** $\frac{3}{4}$ **59.** 0   **61.** 0   **63.** 0

65. Undefined   **67.** Undefined   **69.** 0   **71.** Undefined

73. Parallel   **75.** Perpendicular   **77.** Neither   **79.** Perpendicular

81. Parallel   **83.** Neither   **85.** $\frac{5}{9}$ **87.** $-\frac{2}{3}$ **89.** -1 **91.** $\frac{1}{2}$

93. $-\frac{2}{5}$ **95.** $\frac{1}{20}$; 5%   **97.** $\frac{3}{25}$; 12%   **99.** Front: $\frac{3}{2}$; side: $\frac{3}{5}$

101. a decrease of 875 gal per hour (-875 gal per hr)

103. 319 lb per yr   **105.** An increase of 325 students per year

111. 40 lb licorice; 20 lb gumdrops

ARE YOU READY? 3.5 (page 230)

1. a. $3x, -6$ **b.** 3   **2.** $y = -\frac{2}{5}x + 3$ **3.** True   **4.** $\frac{3}{1}$

5. the y-axis   **6.** True

SELF CHECK 3.5

1. a. $m = -9; (0, 4)$ **b.** $m = \frac{1}{11}; (0, -2)$ **c.** $m = -5; \left(0, -\frac{7}{2}\right)$

2. $y = x - 12$ **3.** $y = -\frac{3}{2}x + 2$

4.

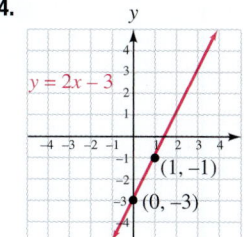

5.

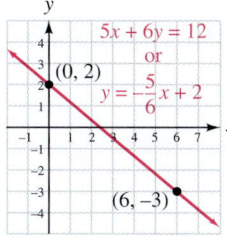

6. Neither   **7.** $c = -10.50p + 4,500$

STUDY SET SECTION **3.5** (page 237)

1. slope–intercept   **3. a.** No   **b.** No   **c.** Yes   **d.** No
5. a. $y = 2x + 8$ **b.** $y = -5x - 3$ **c.** $y = \frac{1}{3}x - 1$
d. $y = \frac{9}{5}x + 4$ **7.** $-2x, 5y, 5, 5, 5, -\frac{2}{5}, 3, -\frac{2}{5}, 0, 3$ **9.** $-2, -3$
11. $4, (0, 2)$ **13.** $-5, (0, -8)$ **15.** $25, (0, -9)$ **17.** $-1, (0, 11)$
19. $\frac{1}{2}, (0, 6)$ **21.** $\frac{1}{4}, \left(0, -\frac{1}{2}\right)$ **23.** $-5, (0, 0)$ **25.** $1, (0, 0)$
27. $0, (0, -2)$ **29.** $0, \left(0, -\frac{2}{5}\right)$ **31.** $-1, (0, 8)$ **33.** $\frac{1}{6}, (0, -1)$
35. $-\frac{3}{2}, (0, 1)$ **37.** $-\frac{2}{3}, (0, 2)$ **39.** $\frac{3}{5}, (0, -3)$ **41.** $1, \left(0, -\frac{11}{6}\right)$
43. $y = 5x - 3$ **45.** $y = -3x + 6$

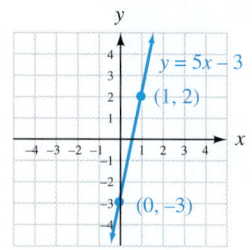

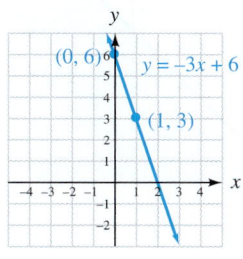

47. $y = \frac{1}{4}x - 2$ **49.** $y = -\frac{8}{3}x + 5$

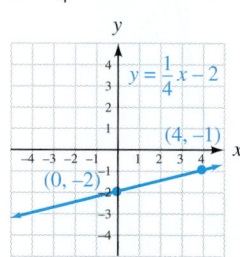

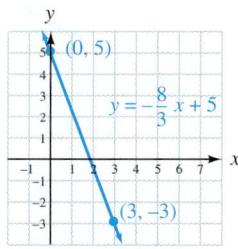

51. $y = \frac{6}{5}x$ **53.** $y = -2x + \frac{1}{2}$

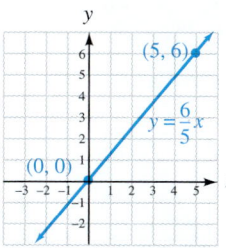

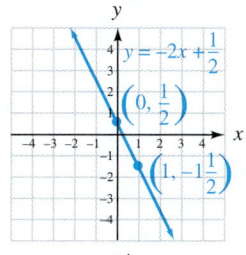

55. $y = 5x - 1$ **57.** $y = -2x + 3$ **59.** $y = \frac{4}{5}x - 2$
61. $y = -\frac{5}{3}x + 2$
63. $3, (0, 3)$ **65.** $\frac{1}{2}, (0, 2)$

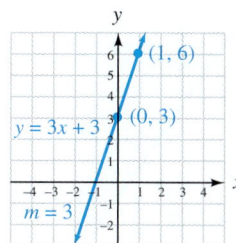

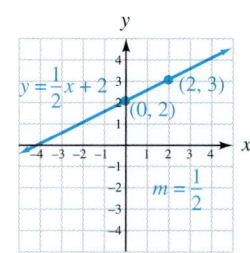

67. $-3, (0, 0)$ **69.** $-4, (0, -4)$

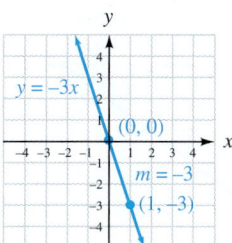

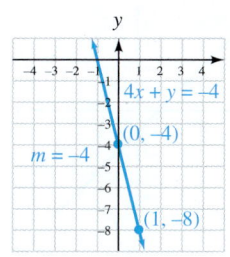

71. $-\frac{3}{4}, (0, 4)$ **73.** $2, (0, -1)$

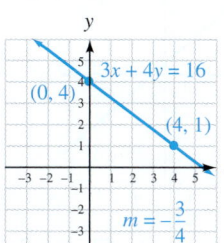

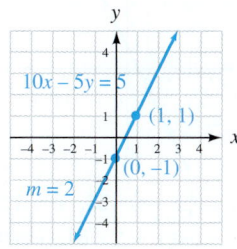

75. Parallel   **77.** Perpendicular   **79.** Parallel   **81.** Neither
83. Perpendicular   **85.** Parallel   **87.** Perpendicular   **89.** Neither
91. a. $c = 2,000h + 5,000$ **b.** \$21,000   **93.** $F = 5t - 10$
95. $c = -20m + 500$ **97.** $c = 5x + 20$
99. a. $c = 14.95m + 629.99$ **b.** \$988.79
101. a. When there are no head waves, the ship can travel at
18 knots.   **b.** $-\frac{1}{2}$ knot/ft   **c.** $y = -\frac{1}{2}x + 18$
105. 42 ft, 45 ft, 48 ft, 51 ft

ARE YOU READY? **3.6** (page 240)

1. $\frac{4}{5}$ **2.** $x + 5$ **3.** $y = 6x - 44$ **4.** $\frac{29}{4}$

SELF CHECK **3.6**

1. $y = -2x + 5$ **2.** $y = -\frac{10}{13}x + \frac{2}{13}$ **3. a.** $y = 2$ **b.** $x = -1$
4.   **5.** $F = \frac{9}{5}C + 32$ **6.** $c = 5n + 12$

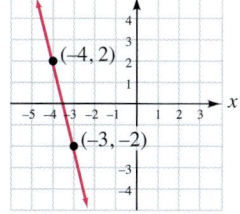

STUDY SET SECTION **3.6** (page 246)

1. point–slope, sub, times, minus, one   **3. a.** point–slope
b. slope–intercept   **5. a.** $(-2, -3)$ **b.** $\frac{5}{6}$ **c.** $y + 3 = \frac{5}{6}(x + 2)$
7. $(67, 170), (79, 220)$ **9.** $5, -1, +, 2, 3$
11. point–slope, slope–intercept   **13.** $y - 1 = 3(x - 2)$
15. $y + 1 = \frac{4}{5}(x + 5)$ **17.** $y = 2x - 1$ **19.** $y = -5x - 37$
21. $y = -3x$ **23.** $y = \frac{1}{5}x - 1$ **25.** $y = -\frac{4}{3}x + 4$
27. $y = -\frac{11}{6}x - \frac{7}{3}$ **29.** $y = 2x + 5$ **31.** $y = -\frac{1}{2}x + 1$
33. $y = 5$ **35.** $y = \frac{1}{10}x + \frac{1}{2}$ **37.** $x = -8$ **39.** $y = \frac{1}{2}x$
41. $x = 4$ **43.** $y = 5$

45.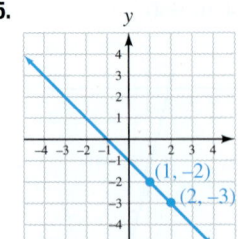

47.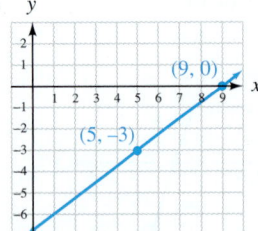

49.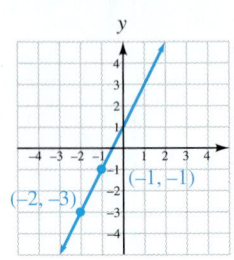

51.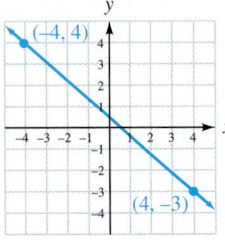

53. $y = \frac{1}{4}x - \frac{5}{4}$ **55.** $y = 12$ **57.** $y = -\frac{1}{4}x + \frac{7}{8}$

59. $y = -\frac{2}{3}x + 2$ **61.** $y = 8x + 4$ **63.** $x = -3$ **65.** $y = 7x - 11$

67. $y = -4x - 9$ **69.** $y = \frac{2}{7}x - 2$ **71.** $y = \frac{1}{10}x$ **73.** $x = -\frac{1}{8}$

75. $y = 1.7x - 2.8$ **77.** $h = 3.9r + 28.9$

79. $y = -\frac{2}{5}x + 4$, $y = -7x + 70$, $x = 10$

81. a. $y = -40m + 920$ **b.** 440 yd³ **83.** $l = \frac{25}{4}r + \frac{1}{4}$

85. a. $y = -\frac{3}{10}x + \frac{283}{10}$ or $y = -0.3x + 28.3$ **b.** 16.3 gal

91. 17 in. by 39 in.

ARE YOU READY? 3.7 (page 249)

1. False **2.** True **3.**

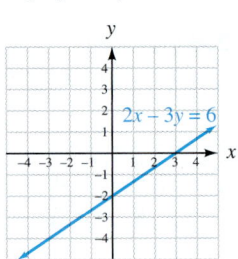

4. a. Below **b.** On **c.** Above

SELF CHECK 3.7

1. a. Not a solution **b.** Solution **c.** Solution **d.** Solution

2.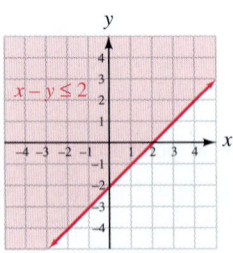

3.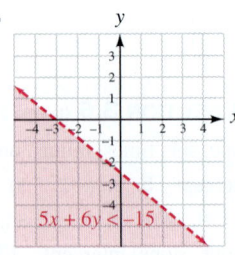

4.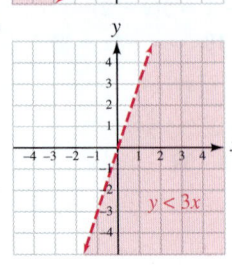

5.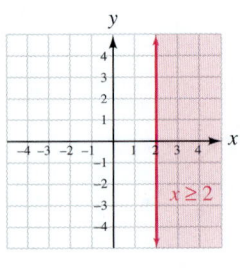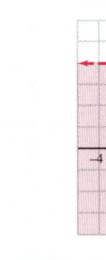

(a)   (b)

6. $x + 15y \le 150$

STUDY SET SECTION 3.7 (page 255)

1. inequality **3.** satisfies **5.** half-planes **7.** Yes
9. dashed, solid **11.** The half-plane opposite that in which the test point lies **13. a.** Yes **b.** No **c.** No **d.** Yes
15. a. Is less than **b.** Is greater than or equal to
c. Is less than or equal to **d.** Is possibly greater than **17.** $=, <$
19. Yes **21.** No **23.** No **25.** Yes

27.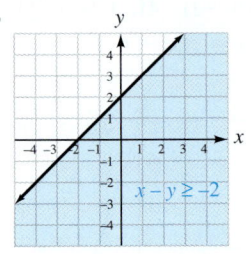

29.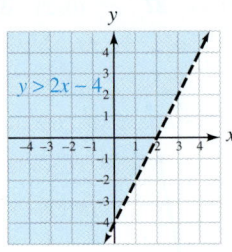

31.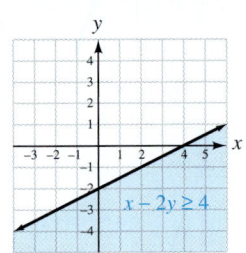

33.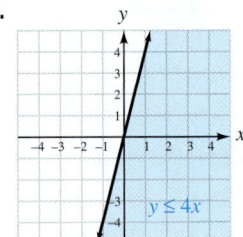

35.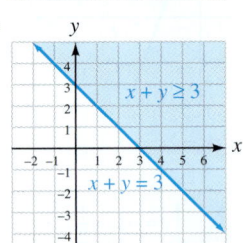

37.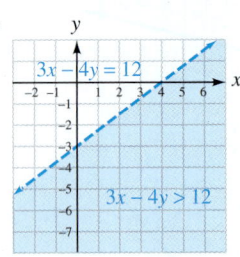

39.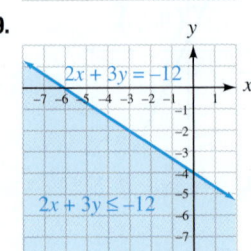

41.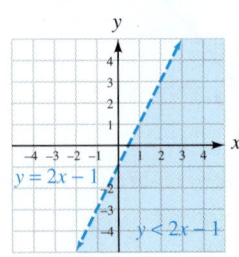

43.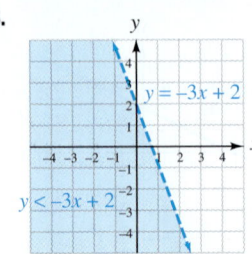

45.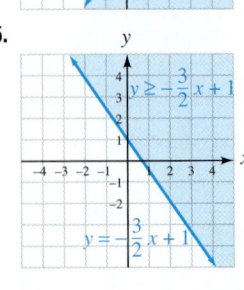

47.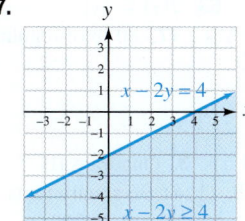

49.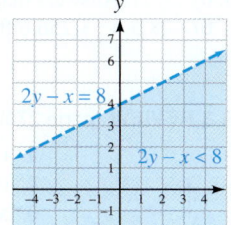

51.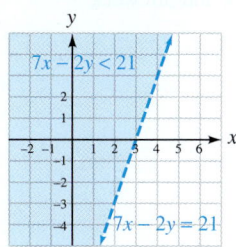

53.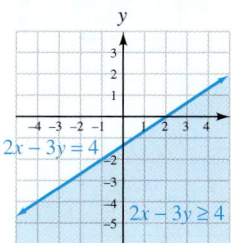

55.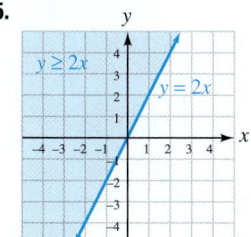

57.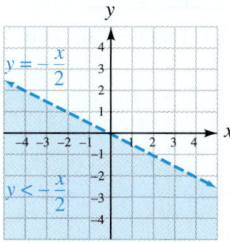

59.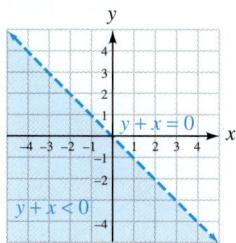

61.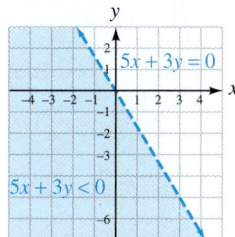

63.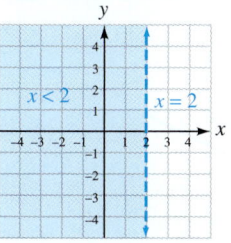

65.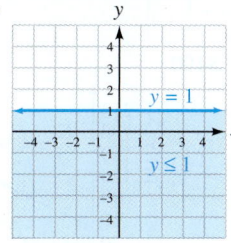

67.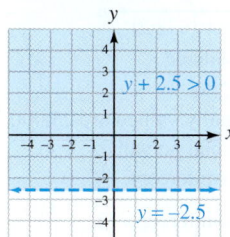

69.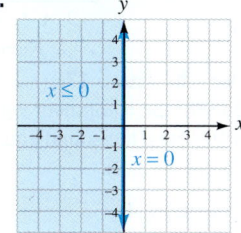

71. a. 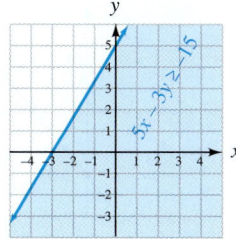   **b.**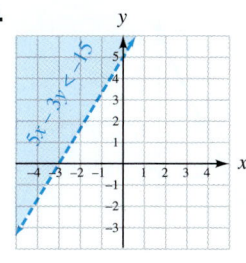

73. a. 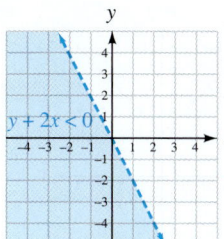   **b.**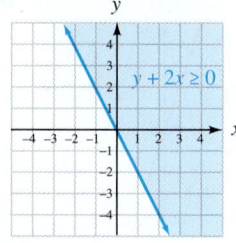

75. No   **77.** (2, 7), (4, 6), (9, 2); answers may vary.

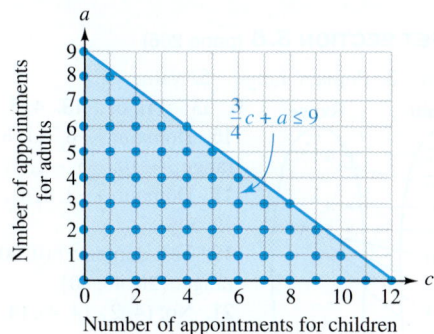

79. (10, 10), (20, 10), (10, 20); Answers may vary

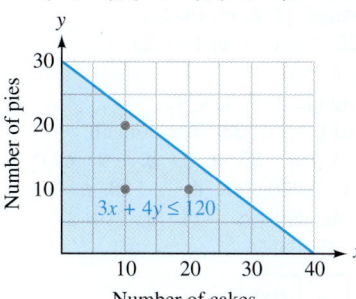

81. (40, 30), (30, 40), (40, 20); Answers may vary

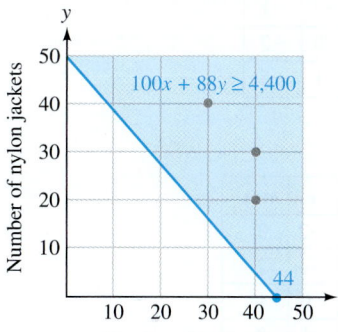

87. $t = \dfrac{A - P}{Pr}$

89. $15x + 22$

ARE YOU READY? 3.8 (page 258)

1. 17   **2.** (3, 5), (3, 0)   **3.** (8, 4), (−3, 4)   **4. a.** 5   **b.** 2

SELF CHECK 3.8

1. Domain: $\{-5, -1, 6, 8\}$; range: $\{-5, 2, 10\}$
2. a. Not a function, $-1 \rightarrow 4$ and $-1 \rightarrow 5$ **b.** Function; domain: $\{-6, 4, 5\}$; range: $\{-6, 5, 8\}$ **c.** Function; domain: $\{1, 3, 4, 9\}$; range: $\{4, 9\}$ **3. a.** -5 **b.** 5   **c.** 3   **4.** 3, 79

5.

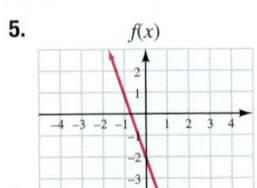

$f(x) = -3x - 2$

6.

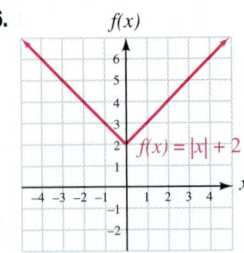

$f(x) = |x| + 2$

7. a. Function   **b.** Not a function   **8.** $140

STUDY SET SECTION 3.8 (page 266)

1. relation   **3.** domain, range   **5.** value

7.

Domain	Range
1992	4.25
1994	
1996	4.75
1998	5.15
2000	
2002	
2004	
2006	
2008	6.55
2010	7.25

9. 33 **11.** of **13.** 4, 5, 4, 5
15. Domain: $\{-6, -1, 6, 8\}$; range: $\{-10, -5, -1, 2\}$
17. Domain: $\{-8, 0, 6\}$; range: $\{9, 50\}$
19. Yes; domain: $\{10, 20, 30\}$; range: $\{20, 40, 60\}$
21. No; (4, 2), (4, 4), (4, 6) (Answers may vary)
23. Yes; domain: $\{1, 2, 3, 4, 5\}$; range: $\{7, 8, 15, 16, 23\}$
25. No; $(-1, 0), (-1, 2)$
27. No; (3, 4), (3, −4) or (4, 3), (4, −3)
29. Yes; domain: $\{-3, 1, 5, 6\}$; range: $\{-8, 0, 4, 9\}$
31. No, (3, 4), (3, −4) or (4, 3), (4, −3)   **33.** Yes; domain: $\{-2, -1, 0, 1\}$; range: $\{7, 10, 13, 16\}$ **35. a.** 3 **b.** −9 **c.** 0 **d.** 199 **37. a.** 0.32 **b.** 18 **c.** 2,000,000 **d.** $\frac{1}{32}$ **39. a.** 7 **b.** 14 **c.** 0 **d.** 1 **41. a.** 0 **b.** 990 **c.** −24 **d.** 210 **43. a.** 36 **b.** 0 **c.** 9 **d.** 4 **45.** 1.166

47.

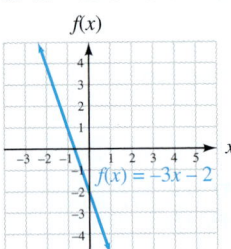

$f(x) = -3x - 2$

x	$f(x)$
−2	4
−1	1
0	−2
1	−5

49.

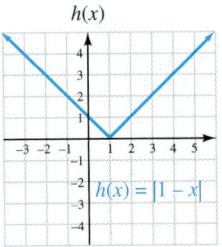

$h(x) = |1 - x|$

x	$h(x)$
−2	3
−1	2
0	1
1	0
2	1
3	2
4	3

51.

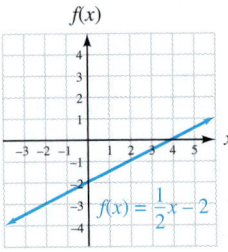

$f(x) = \frac{1}{2}x - 2$

53.

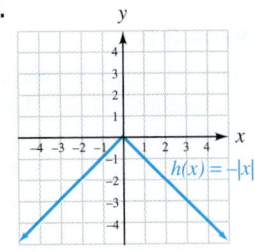

$h(x) = -|x|$

55. Yes   **57.** No; (3, 4), (3, −1) (Answers may vary)
59. No; (0, 2), (0, −4) (Answers may vary)

61. No; (3, 0), (3, 1) (Answers may vary)   **63.** $f(x) = |x|$
65. $900   **67.** 78.5 ft², 1,256.6 ft²   **69.** Yes   **77.** 80 lb of regular coffee

CHAPTER 3 REVIEW (page 270)

1.

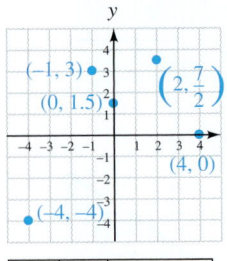

$(-1, 3)$, $(0, 1.5)$, $\left(2, \frac{7}{2}\right)$, $(4, 0)$, $(-4, -4)$

2. (158, 21.5)   **3.** Quadrant III
4. (0, 0)   **5.** (1, 4); 36 square units
6. a. 2,500; week 2   **b.** 1,000
c. 1st week and 5th week
7. Yes

8.

x	y	(x, y)
−2	−6	(−2, −6)
−8	3	(−8, 3)

9. $y = x^2 + 1$ and $y - x^3 = 0$ **10. a.** True   **b.** False

11.

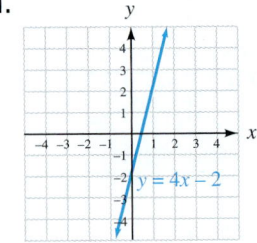

$y = 4x - 2$

12.

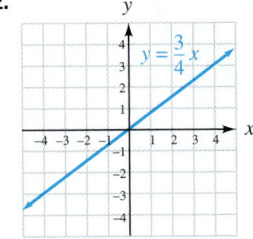

$y = \frac{3}{4}x$

13.

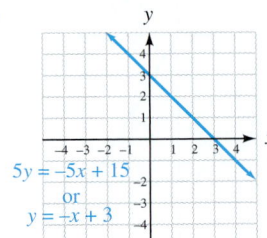

$5y = -5x + 15$ or $y = -x + 3$

14.

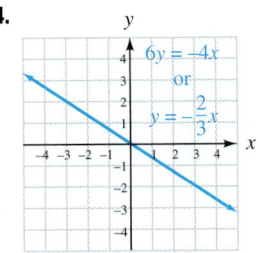

$6y = -4x$ or $y = -\frac{2}{3}x$

15. About $190

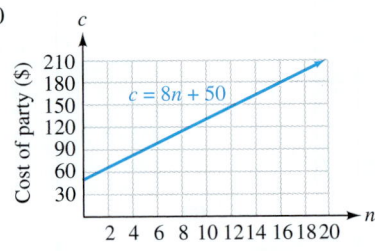

$c = 8n + 50$

Cost of party ($) vs. Number of children

16. a. False   **b.** True   **17.** (−3, 0), (0, 2.5)
18. (0, 25,000); the equipment was originally valued at $25,000. (10, 0); in 10 years, the sound equipment had no value.
19. x-intercept: (−2, 0); y-intercept: (0, 4)
20. x-intercept: $\left(\frac{13}{5}, 0\right)$; y-intercept: $\left(0, -\frac{13}{4}\right)$

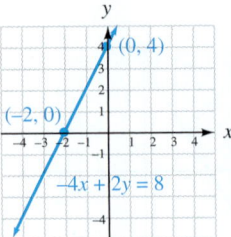

$(0, 4)$, $(-2, 0)$, $-4x + 2y = 8$

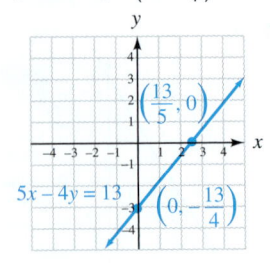

$\left(\frac{13}{5}, 0\right)$, $5x - 4y = 13$, $\left(0, -\frac{13}{4}\right)$

21. 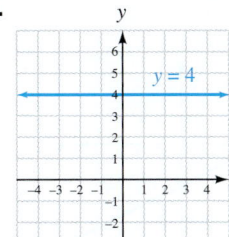 **22.**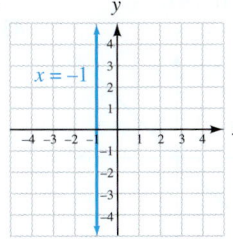

23. $\frac{1}{4}$ **24.** $-\frac{7}{8}$ **25.** -7 **26.** $-\frac{3}{2}$

27.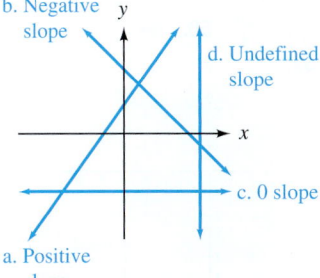

28. $\frac{3}{4}$ **29.** 8.3%   **30.** 1.5 gal/yr   **31.** They are neither.

32. $-\frac{7}{5}$ **33.** $m = \frac{3}{4}$; y-intercept: $(0, -2)$

34. $m = -4$; y-intercept: $(0, 0)$ **35.** $m = \frac{1}{8}$; y-intercept: $(0, 10)$

36. $m = -\frac{7}{5}$; y-intercept: $\left(0, -\frac{21}{5}\right)$

37. $y = -4x - 1$ **38.** $y = \frac{3}{2}x - 3$

39. $m = 3$; y-intercept: $(0, -5)$

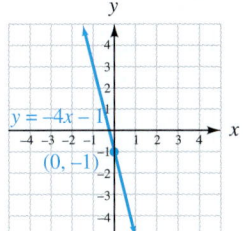 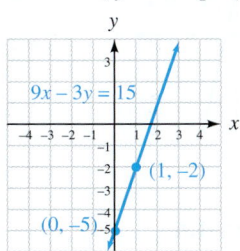

40. a. $c = 300w + 75{,}000$ **b.** 90,600 copies   **41.** Parallel
42. Perpendicular
43. $y = 3x + 2$ **44.** $y = -\frac{1}{2}x - 3$

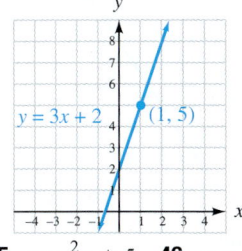 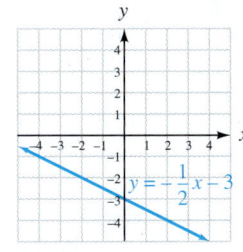

45. $y = \frac{2}{3}x + 5$ **46.** $y = -8$ **47.** $f = -35x + 450$
48. a. $P = \frac{3}{2}t + 310$ **b.** 400 parts per million
49. a. Yes   **b.** Yes   **c.** Yes   **d.** No   **50.** $=, >$
51. 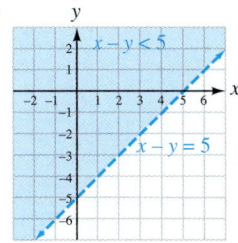   **52.**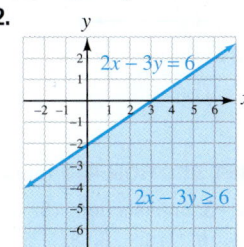

53. 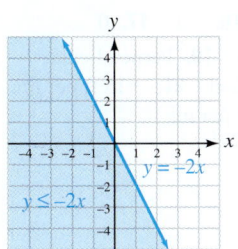   **54.**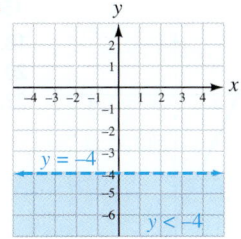

55. a. True   **b.** False   **c.** False
56. $(2, 4)$, $(5, 3)$, $(6, 2)$; answers may vary

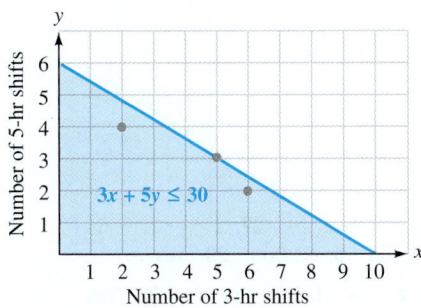

57. Domain: $\{-5, 0, 4, 7\}$; range: $\{-11, -3, 4, 9\}$
58. Domain: $\{-6, 1, 2, 15\}$; range: $\{-8, -2, 9\}$
59. Yes; domain: $\{1, 4, 8\}$; range: $\{0, 6, 9\}$
60. Yes; domain: $\{2, 3, 5, 6\}$; range: $\{1, 4\}$
61. Yes; domain: $\{3, 5, 7, 9\}$; range: $\{9, 25, 49, 81\}$
62. No; $(-1, 2)$, $(-1, 4)$ **63.** Yes; domain: $\{-1, 0, 1, 2\}$; range: $\{6\}$
64. No; $(4, 4)$, $(4, 6)$ **65.** domain, range   **66.** $f(x)$ **67.** -3
68. 0   **69.** 21   **70.** $-\frac{7}{4}$ **71.** -5 **72.** 37   **73.** -2
74. -8 **75.** No; $(1, 0.5)$, $(1, 4)$, (answers may vary)   **76.** Yes
77.

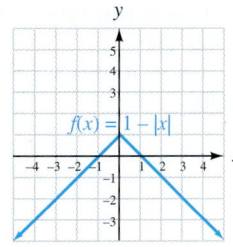

x	$f(x)$
0	1
1	0
3	-2
-1	0
-3	-2

78. $1{,}004.8$ in.³

CHAPTER 3 TEST (page 279)

1. a. axis, axis   **b.** solution   **c.** linear   **d.** slope   **e.** function
2. 10 dogs   **3.** 60 dogs   **4.** 1 day before and the 3rd day of the holiday   **5.** 50 dogs were in the kennel when the holiday began.
6. 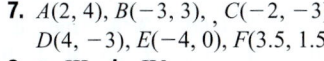   **7.** $A(2, 4)$, $B(-3, 3)$, $C(-2, -3)$, $D(4, -3)$, $E(-4, 0)$, $F(3.5, 1.5)$
8. a. III   **b.** IV
9. Yes
10.

x	y	(x, y)
2	1	$(2, 1)$
-6	3	$(-6, 3)$

11. a. False   **b.** True   **12.**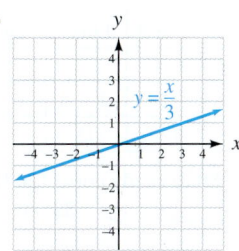
13. x-intercept: $(3, 0)$;
y-intercept: $(0, -2)$

14.

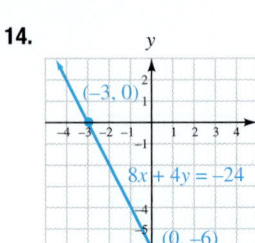

15. $\frac{8}{7}$ **16.** -1 **17.** 0
18. 10% **19.** Perpendicular
20. Parallel **21.** -15 ft per mi
22. 25 ft per mi

23.

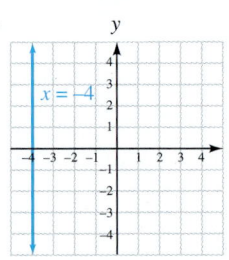

24.

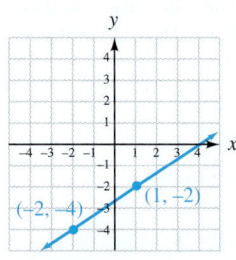

25. $m = -\frac{1}{2}$; $(0, 4)$ **26.** $y = 7x + 19$ **27.** $y = -2x - 5$
28. a. $v = -1{,}500x + 15{,}000$ **b.** $3{,}000 **29.** Yes
30. $y = -\frac{1}{5}T + 41$ **31. a.** Yes **b.** No **c.** No **32.** Yes

33.

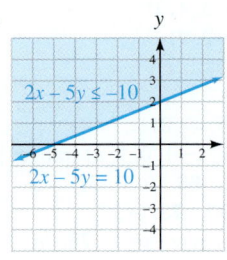

34. Domain: $\{-4, 0, 1, 5\}$;
range: $\{-8, 3, 12\}$
35. Yes; domain: $\{1, 2, 3, 4\}$;
range: $\{1, 2, 3, 4\}$
36. No; $(-3, 9)$, $(-3, -7)$;
37. Yes; domain: $\{6, 7, 8, 9, 10\}$;
range: $\{5\}$ **38.** No; $(2, 6)$, $(2, 2)$
39. No; $(2, 3.5)$, $(2, -3.5)$; (answers
may vary) **40.** No; $(-2, 2)$,
$(-2, -1)$; (answers may vary)
41. -13 **42.** 756
43. $C(45) = 28.50$; it costs $28.50 to
make 45 calls.

44.

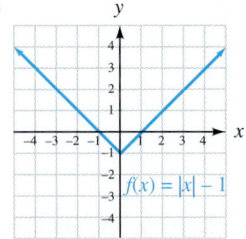

CUMULATIVE REVIEW CHAPTERS 1–3 (page 281)

1. $2^2 \cdot 3^3$ **2.** 0.004 **3. a.** True **b.** True **c.** True **4.** -15
5. -0.77 **6.** -945 **7.** 30 **8.** 2 **9.** 32 **10.** $500 - x$
11. $3, -2$ **12. a.** $2x + 8$ **b.** $-2x + 8$ **13.** $4a + 10$
14. $-63t$ **15.** $4b^2$ **16.** 0 **17.** 4 **18.** $-160a$ **19.** $-3y$
20. $7x - 12$ **21.** 6 **22.** 2.9 **23.** 9 **24.** -19 **25.** $\frac{1}{7}$
26. 1 **27.** $-\frac{55}{6}$ **28.** No solution, contradiction **29.** -99
30. $-\frac{1}{4}$ **31.** 1,100 **32.** $h = \frac{S - 2\pi r^2}{2\pi r}$ **33.** $3\frac{1}{8}$ in., $\frac{39}{64}$ in.^{2}
34. 45°

35.

	% acid	Liters	Amount of acid
50% solution	0.50	x	$0.50x$
25% solution	0.25	$13 - x$	$0.25(13 - x)$
30% mixture	0.30	13	$0.30(13)$

36. 7.5 hr **37.** 80 lb candy corn, 120 lb gumdrops
38. $(-\infty, 48]$ **39.** $(0, \infty)$

40. I and II **41.** No
42.

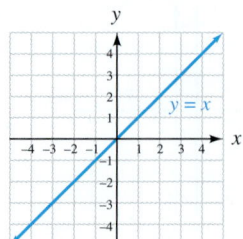

43.

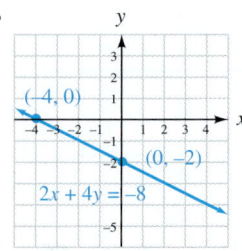

44. 0 **45.** $-\frac{10}{7}$ **46.** $\frac{7}{12}$ **47.** $\frac{2}{3}$, $(0, 2)$ **48.** $y = -2x + 1$
49. $y + 9 = -\frac{7}{8}(x - 2)$; $y = -\frac{7}{8}x - \frac{29}{4}$ **50.** Yes

51.

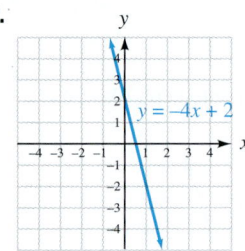

52.

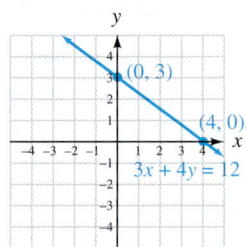

53. 78 **54.** No

ARE YOU READY? 4.1 (page 284)

1. Not a solution
2.

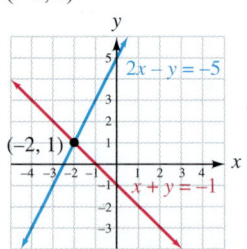

3.

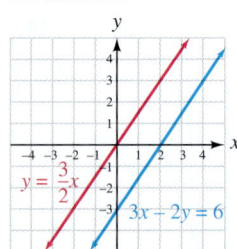

4. Parallel

SELF CHECKS 4.1

1. Not a solution
2. $(-2, 1)$ **3.** No solution

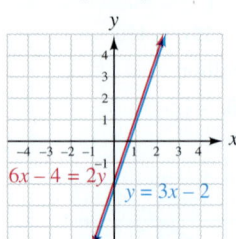

4. Infinitely many solutions

5. No solution

STUDY SET SECTION 4.1 (page 290)

1. system **3.** intersection **5.** consistent, inconsistent
7. a. True **b.** True **9. a.** $-5, 2$ **b.** $3, 3, (0, -2)$ **11.** No
solution; independent **13.** A solution **15.** A solution

17. Not a solution   **19.** Not a solution   **21.** Not a solution
23. A solution   **25.** (3, 2)

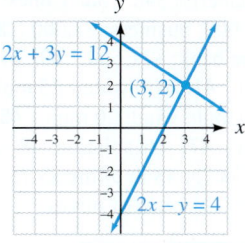

27. (−1, 5)

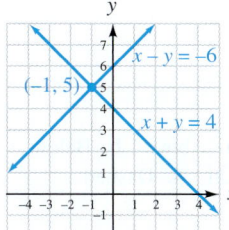

29. No solution

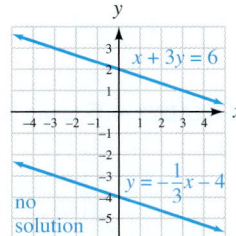

31. No solution

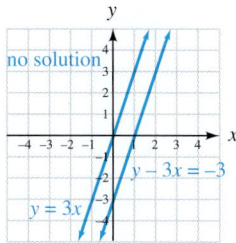

33. Infinitely many solutions

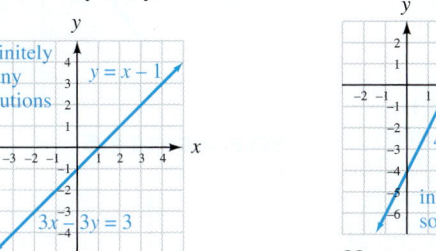

35. Infinitely many solutions

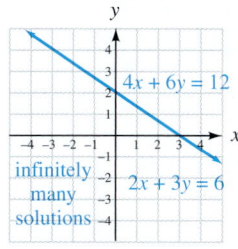

37. 1 solution   **39.** Infinitely many solutions   **41.** No solution
43. 1 solution   **45.** (1, 3)
47. No solution

49. (−2, 0)

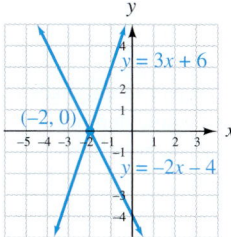

51. No solution

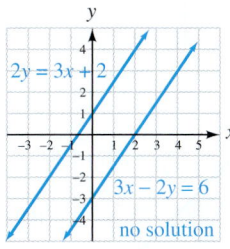

53. (3, −1)

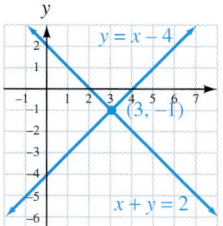

55. (3, 0)

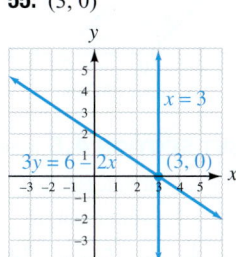

57. (−4, 0)

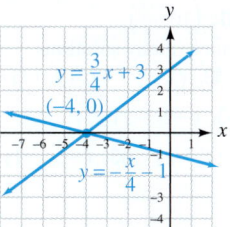

59. (4, −6)

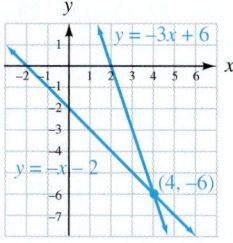

61. (5, −2)

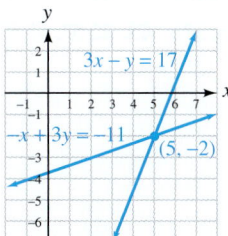

63. (1, 1)

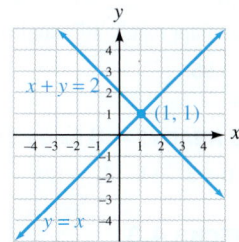

65. Infinitely many solutions

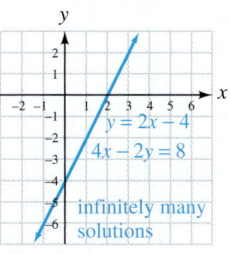

67. (−6, 1)

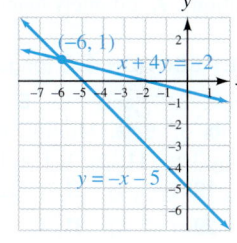

69. (−2, −3)

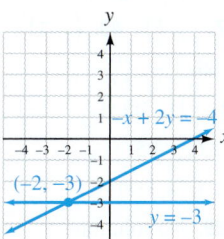

71. (4, −4)

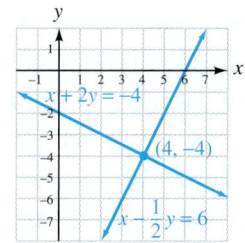

73. May '09; about 70 million   **75. a.** Houston, New Orleans, St. Augustine   **b.** St. Louis, Memphis, New Orleans
c. New Orleans   **77. a.** The incumbent; 7%   **b.** November 2
c. The challenger; 3   **79.** (6, 4), (6, 8), (12, 4), (12, 8); yes
87. $[-3, \infty)$

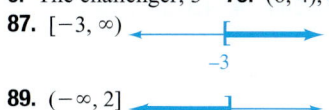

89. $(-\infty, 2]$

ARE YOU READY? 4.2 (page 294)

1. 1   **2.** $y = 2x + 3$ **3.** 18   **4.** 17   **5.** $3x$

SELF CHECKS 4.2

1. (3, 1)   **2.** (−8, 2)   **3.** (5, 0)   **4.** $\left(\frac{18}{7}, \frac{8}{7}\right)$ **5.** (−3, −3)
6. No solution   **7.** Infinitely many solutions

STUDY SET SECTION 4.2 (page 301)

1. substituting   **3.** $y = -3x$ **5.** $x + 3(x - 4) = 8$
7. Substitute 3 for a in the second equation.   **9. a.** No   **b.** ii
11. $3x, 4, -2, -2, -6, -2, -6$ **13.** (2, 4)   **15.** (3, 0)

17. $(-10, 2)$ **19.** $(2, -10)$ **21.** $(3, 2)$ **23.** $(-5, 5)$
25. $(-4, -6)$ **27.** $(3, -2)$ **29.** $(-6, 4)$ **31.** $\left(10, \frac{15}{2}\right)$
33. No solution **35.** No solution **37.** Infinitely many solutions
39. Infinitely many solutions **41.** $(3, -2)$ **43.** $(1, 1)$ **45.** $(9, 11)$
47. No solution **49.** $\left(\frac{1}{3}, \frac{2}{3}\right)$ **51.** $\left(\frac{1}{5}, 4\right)$ **53.** $(-2, -2)$
55. $\left(\frac{1}{2}, \frac{1}{3}\right)$ **57.** $\left(\frac{2}{3}, -\frac{1}{3}\right)$ **59.** $\left(-4, \frac{5}{4}\right)$ **61.** $(4, -2)$ **63.** $(-5, -1)$
65. $(4, 2)$ **67.** $(-4, -9)$ **69.** $(-2, 3)$
71. Infinitely many solutions **73.** $(-10, -24)$ **75.** $\left(\frac{1}{2}, 2\right)$
77. $\left(-1, \frac{2}{3}\right)$ **79.** Infinitely many solutions **81.** $(-4, -1)$
83. Angle of approach: 40°; angle of departure: 37° **85.** 22.5°, 67.5°
93. $3^3 \cdot 7$ **95.** $\frac{2}{3}$

ARE YOU READY? 4.3 (page 304)

1. 6 **2.** 0 **3.** -1 **4.** $-12x - 36y = -4$

SELF CHECKS 4.3

1. $(2, 4)$ **2.** $(-3, -3)$ **3.** $(0, 5)$ **4.** $(-5, 6)$ **5.** $\left(-\frac{22}{9}, \frac{10}{9}\right)$
6. No solution **7.** Infinitely many solutions

STUDY SET SECTION 4.3 (page 311)

1. opposites **3.** $7y$ and $-7y$ **5. a.** $5a = -4$ **b.** $-4y = 1$
7. a. -2 **b.** 3 **9. a.** Multiply both sides by 15. **b.** Multiply
both sides by 10. **11.** $2x, 1, 1, 4, 1, 4$ **13.** $(3, 2)$ **15.** $(-2, -3)$
17. $(0, 8)$ **19.** $(-3, 4)$ **21.** $(-12, 1)$ **23.** $(3, 11)$ **25.** $(-2, 7)$
27. $(1, 1)$ **29.** $(-2, 5)$ **31.** $(2, -1)$ **33.** $\left(\frac{5}{2}, -\frac{1}{2}\right)$ **35.** $\left(-\frac{1}{2}, -\frac{2}{3}\right)$
37. No solution **39.** No solution **41.** Infinitely many solutions
43. Infinitely many solutions **45.** $(2, 3)$ **47.** $\left(\frac{3}{4}, \frac{1}{3}\right)$ **49.** $\left(\frac{1}{3}, 3\right)$
51. $(3, -2)$ **53.** Infinitely many solutions **55.** $\left(\frac{10}{3}, \frac{10}{3}\right)$
57. $(6, -2)$ **59.** $(3, 0)$ **61.** $(10, 9)$ **63.** $(1, -1)$ **65.** $\left(\frac{7}{25}, -\frac{1}{25}\right)$
67. $(6, 8)$ **69.** $(4, -2)$ **71.** $\left(\frac{13}{75}, \frac{14}{75}\right)$ **73.** $(2.6, -4.5)$
75. $(0, -2)$ **77.** $(12, -9)$ **79.** $\left(1, -\frac{5}{2}\right)$ **81.** $(-4, -1)$
83. $(-4, 5)$ **85.** $(-4, -5)$ **87.** $(-5, 10)$ **89.** No solution
91. 1991 **93.** After 24 days **99.** $y = -\frac{11}{6}x - \frac{7}{3}$ **101.** -80

ARE YOU READY? 4.4 (page 314)

1. $P = 2l + 2w$ **2.** $38.85 **3.** $2,200 **4.** 420 mi
5. 0.96 oz alcohol

SELF CHECKS 4.4

1. 4 ft, 8 ft **2.** 101°, 79° **3.** Length: 55 ft; width: 20 ft
4. Practice cone: $0.50, portable goal: $20 **5.** $2,500 at 9%,
$7,500 at 10% **6.** Boat: 10 mph; current: 2 mph **7.** 1%: 40 L;
4%: 20 L **8.** Planting mix: 120 yd³; topsoil: 80 yd³

STUDY SET SECTION 4.4 (page 323)

1. complementary, supplementary **3.** $x + y = 20, y = 2x - 1$
5. $x + y = 180, y = x - 25$ **7.** $5x + 2y = 15$
9. Downstream: $(x + c)$, upstream: $(x - c)$ mph **11. a.** $(x + y)$ mL
b. 33% **13.** 20°, 70° **15.** 50°, 130° **17.** Upper: 22 ft,
lower: 29 ft **19.** President: $400,000; vice president: $227,300
21. $\angle 1$: 115°; $\angle 2$: 65° **23.** Length: 96 ft, width: 70 ft
25. Length: 15 m, width: 10 m **27.** Printer: $2; copier: $15

29. $29.50 for a 10 × 14; $21.00 for an 8 × 10
31. Elvis: 29¢; Liberty: 34¢ **33.** Cones: 85, sundaes: 63
35. Nursing: $2,000; business: $3,000 **37.** International fund:
$18,500; offshore bank: $21,500 **39.** 4% account: $11,000;
biotech: $11,000 **41.** Still water: 25 mph, current: 5 mph
43. Still air: 180 mph, wind: 20 mph **45.** 8 gal 6% salt water, 24
gal 2% salt water **47.** 4% solution: 48 oz; 12% solution: 80 oz
49. 52 lb of the $8.75, 48 lb of the $3.75 **51.** $\frac{20}{3} = 6\frac{2}{3}$ pints of
mushrooms; $\frac{40}{3} = 13\frac{1}{3}$ pints of olives
55. $(-\infty, 4)$
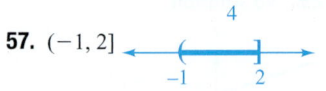
57. $(-1, 2]$

ARE YOU READY? 4.5 (page 328)

1. True **2.**
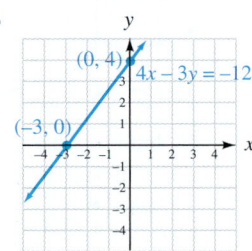
3. a. Above **b.** Below
c. On **4.** Dashed

SELF CHECKS 4.5

1.
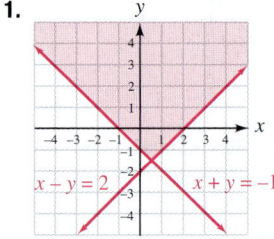
2.
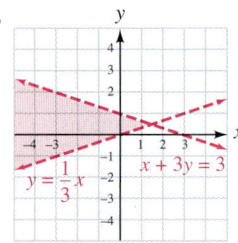

3.
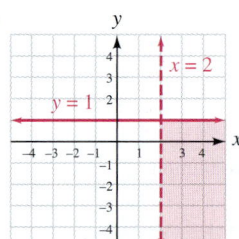
4.
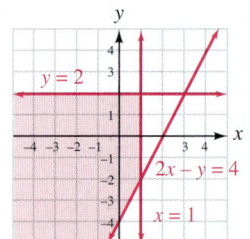

5.
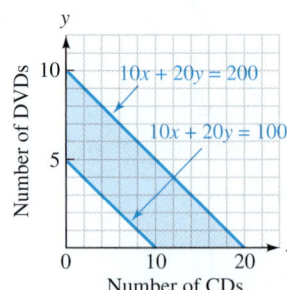

Any ordered pair in the shaded
region with whole number
coordinates is a possible
combination.

STUDY SET SECTION 4.5 (page 334)

1. inequalities **3.** intersection **5. a.** $3x - y = 5$ **b.** Dashed
7. Slope: $4 = \frac{4}{1}$, y-intercept: $(0, -3)$ **9. a.** No **b.** Above
11. a. Yes **b.** No **c.** No **13. a.** ii **b.** iii **c.** iv **d.** i

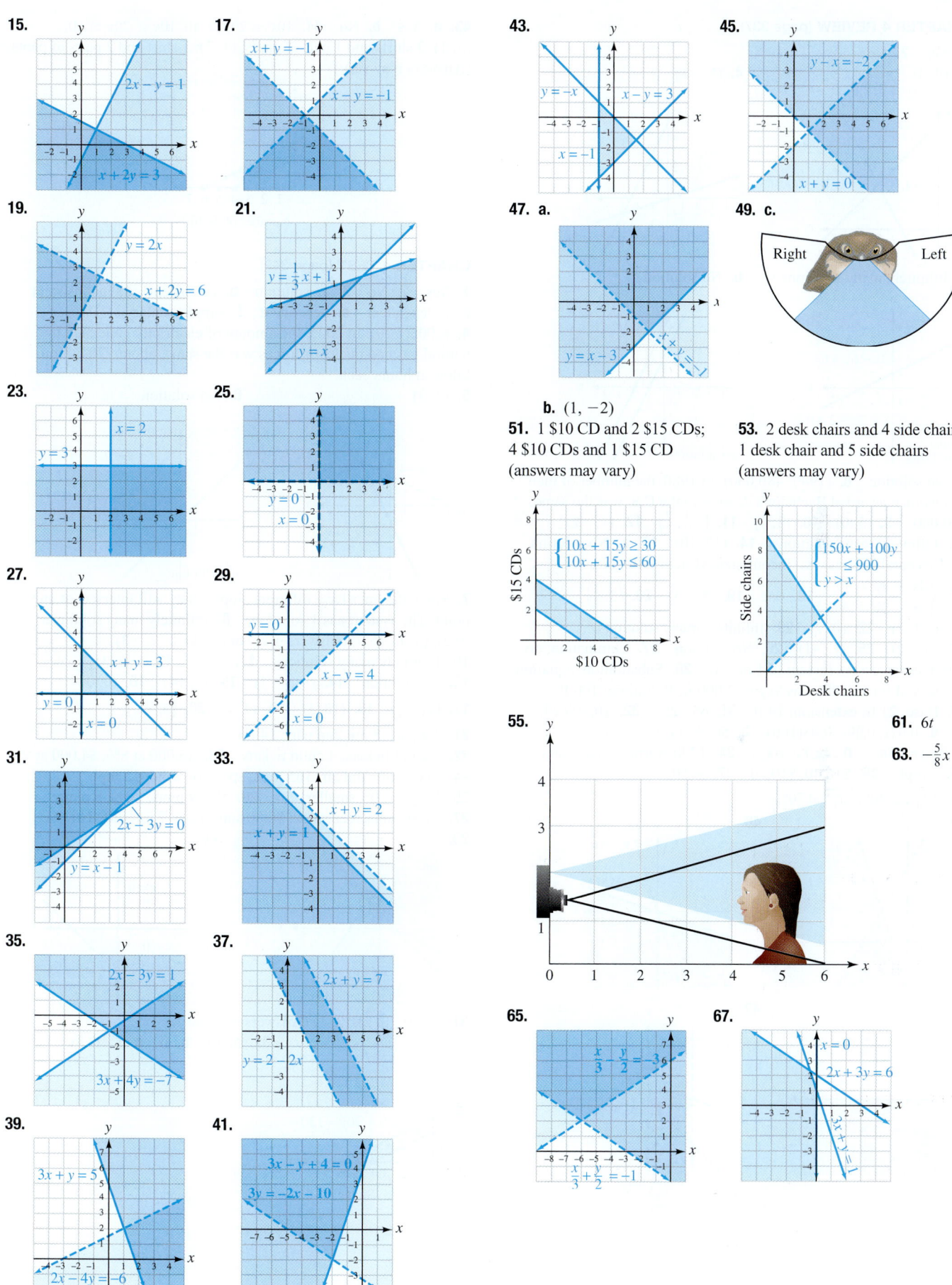

15. $2x - y = 1$; $x + 2y = 3$

17. $x + y = -1$; $x - y = -1$

43. $y = -x$; $x - y = 3$; $x = -1$

45. $y - x = -2$; $x + y = 0$

19. $y = 2x$; $x + 2y = 6$

21. $y = \frac{1}{3}x + 1$; $y = x$

47. a. $y = x - 3$; $x + y = -1$

b. $(1, -2)$

49. c. Right Left

23. $x = 2$; $y = 3$

25. $y = 0$; $x = 0$

51. 1 $10 CD and 2 $15 CDs; 4 $10 CDs and 1 $15 CD (answers may vary)

$\begin{cases} 10x + 15y \geq 30 \\ 10x + 15y \leq 60 \end{cases}$

$15 CDs / $10 CDs

53. 2 desk chairs and 4 side chairs; 1 desk chair and 5 side chairs (answers may vary)

$\begin{cases} 150x + 100y \leq 900 \\ y > x \end{cases}$

Side chairs / Desk chairs

27. $x + y = 3$; $y = 0$; $x = 0$

29. $y = 0$; $x - y = 4$; $x = 0$

31. $2x - 3y = 0$; $y = x - 1$

33. $x + y = 2$; $x + y = 1$

55.

61. $6t$

63. $-\frac{5}{8}x$

35. $2x - 3y = 1$; $3x + 4y = -7$

37. $2x + y = 7$; $y = 2 - 2x$

39. $3x + y = 5$; $2x - 4y = -6$

41. $3x - y + 4 = 0$; $3y = -2x - 10$

65. $\frac{x}{3} - \frac{y}{2} = -3$; $\frac{x}{3} + \frac{y}{2} = -1$

67. $x = 0$; $2x + 3y = 6$; $3x + y = 1$

CHAPTER 4 REVIEW (page 337)

1. Yes **2.** No
3. (4, 3) **4.** (3, −1)

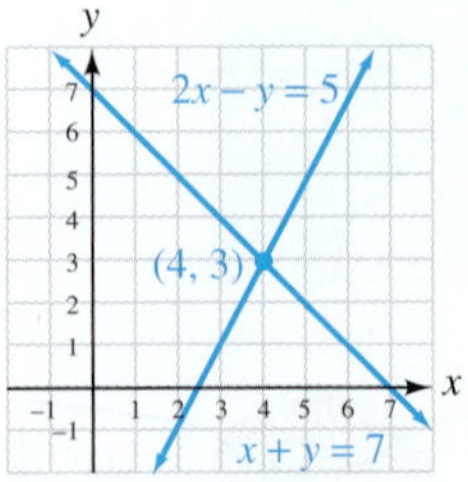

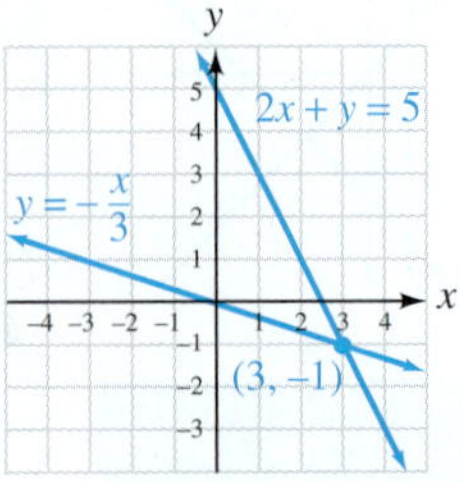

5. Infinitely many solutions **6.** No solution

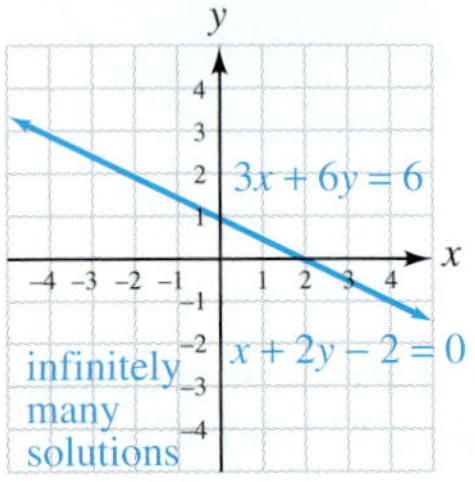

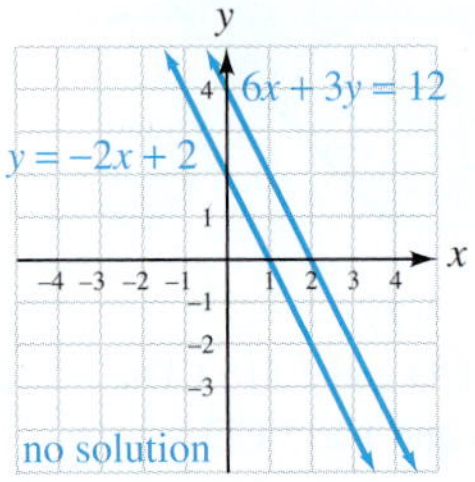

7. No solution **8.** (1980: 480,000); In 1980, the number of men and women awarded Bachelor's degrees in the U.S. was the same, 480,000. **9.** (5, 0) **10.** (3, 3) **11.** $\left(-\frac{1}{2}, \frac{7}{2}\right)$ **12.** (1, −2) **13.** Infinitely many solutions **14.** (12, 10) **15. a.** No solution **b.** Two parallel lines **c.** Inconsistent system **16.** one **17.** $\begin{cases} 4x + 2y = 7 \\ 5x - 3y = -6 \end{cases}$ **18.** one **19.** (3, −5) **20.** $\left(3, \frac{1}{2}\right)$ **21.** (−1, 7) **22.** (0, 9) **23.** Infinitely many solutions **24.** (−5, 2) **25.** (1, −1) **26.** No solution **27.** Elimination; no variables have a coefficient of 1 or −1. **28.** Substitution; equation 1 is solved for x. **29.** Las Vegas: 2,000 ft; Baltimore: 100 ft **30.** Base: 21 ft; extension: 14 ft **31.** 65°, 25° **32.** 10,800 yd² **33. a.** $0.02x$, $0.09y$, $0.08(100)$ **b.** $5(s + w)$, $7(s - w)$ **c.** $0.11x$, $0.06y$ **d.** $4x$, $8y$, $10(5)$ **34.** 12 lb worms, 18 lb bears **35.** 3 mph **36.** \$16.40, \$10.20 **37.** \$750 **38.** $13\frac{1}{3}$ gal 40%, $6\frac{2}{3}$ gal 70%

39.

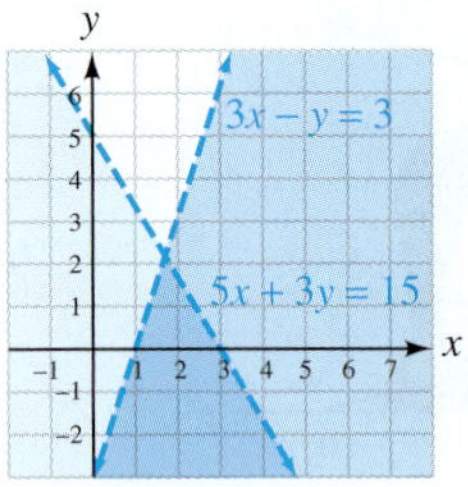

40.

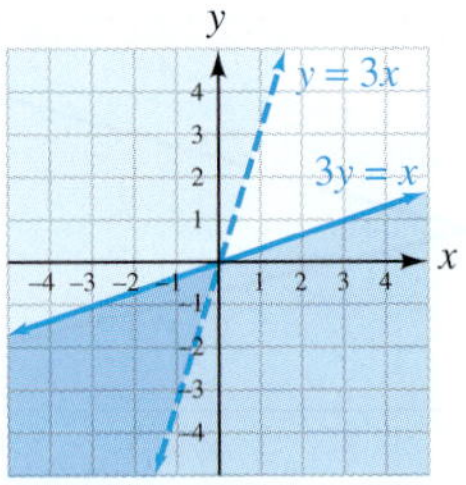

41.

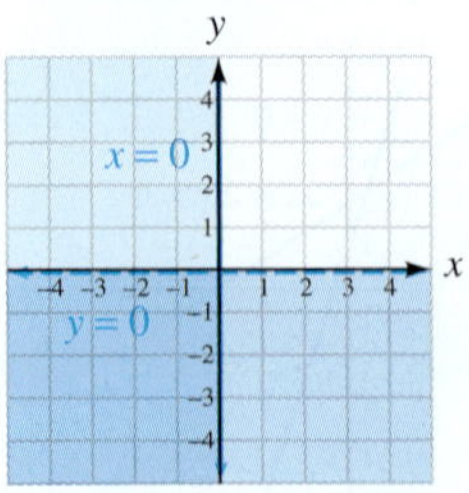

42.

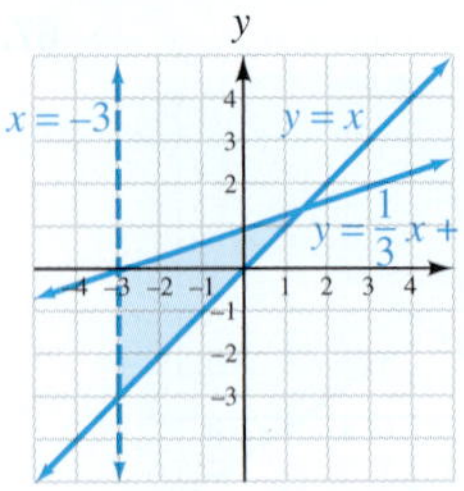

43. a. Yes **b.** No **44.** $10x + 20y \geq 40$, $10x + 20y \leq 60$; (3, 1): 3 shirts and 1 pair of pants; (1, 2): 1 shirt and 2 pairs of pants (Answers may vary)

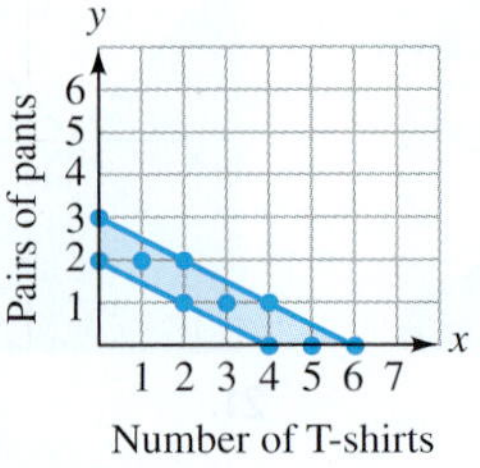

CHAPTER 4 TEST (page 342)

1. Yes **2.** No **3. a.** solution **b.** consistent **c.** inconsistent **d.** independent **e.** dependent **f.** supplementary
4. (2005, 770); In 2005, the amount of electricity generated by natural gas and nuclear sources was the same, about 770 billion kilowatt hours each.
5. (2, 3) **6.** No solution

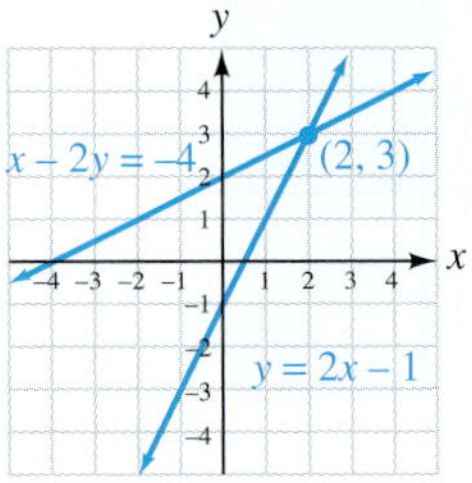

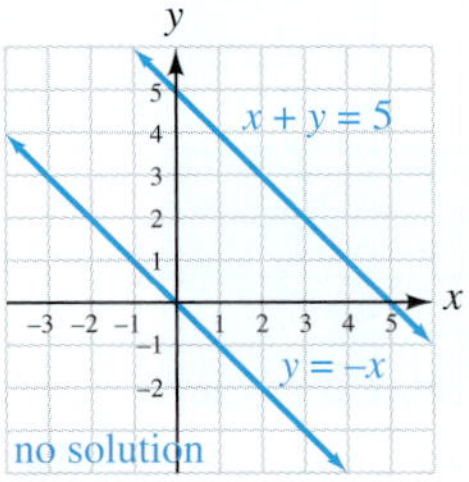

7. Since the lines have different slopes, they will intersect at one point. The system has 1 solution. **8.** Infinitely many solutions; (0,2), (3, 1), (6,0) (answers may vary) **9.** (−2, −3) **10.** Infinitely many solutions **11.** (2, 4) **12.** (−3, 3) **13.** No solution **14.** (−1, −1) **15.** (5, 14) **16.** (0, 0) **17.** $\left(\frac{1}{2}, -\frac{1}{4}\right)$ **18.** (10, 5) **19.** (3, 3) **20.** $\left(3, -\frac{1}{2}\right)$ **21.** 1st part: 8 mi, 2nd part: 14 mi **22.** 3 adult tickets; 4 child tickets **23.** \$6,000 at 8%, \$4,000 at 9% **24.** Speed of calm air: 165 mph, speed of wind: 15 mph **25.** Larger: 70°, smaller: 20° **26.** 5%: 4 pints; 20%: 8 pints **27.** \$1.50 sunscreen: 3 oz; \$0.80 sunscreen: 7 oz **28.** No

29.

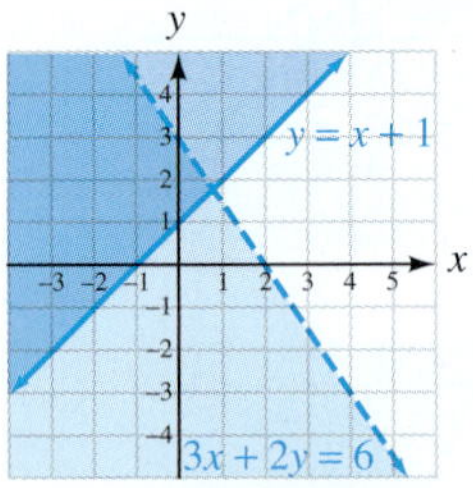

30.

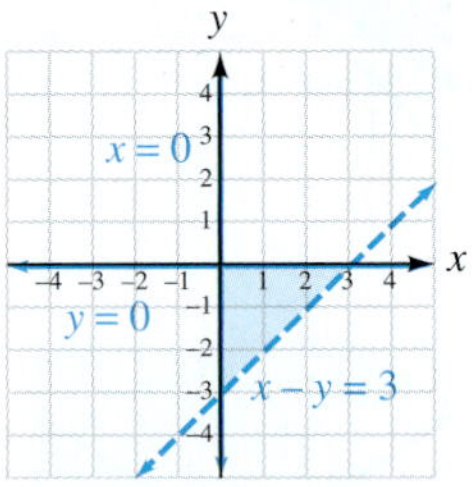

31. (1, 2), (2, 2), (3, 1) (Answers may vary)

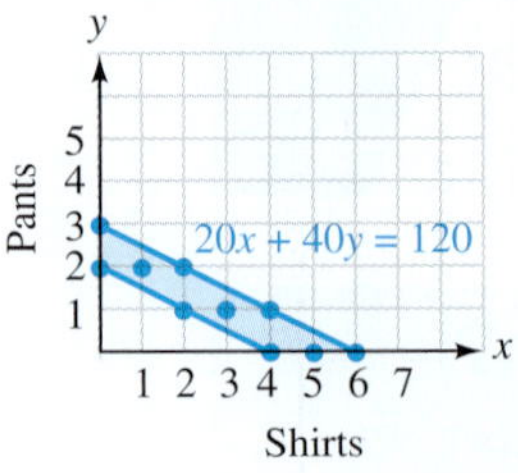

32. a. iii **b.** ii **c.** i **d.** iv

CUMULATIVE REVIEW CHAPTERS 1–4 (page 344)

1. sum, difference, product, quotient **2.** $2^2 \cdot 5^2$ **3.** $\frac{5}{8}$ **4.** $\frac{22}{35}$
5. Irrational **6.**

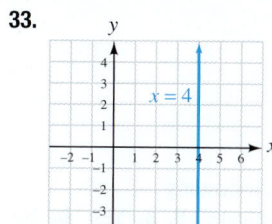

7. $0.\overline{6}$ **8.** Associative property of multiplication **9.** 0 **10.** 1
11. 4 **12.** -2 **13.** $250 - x$ **14.** $10d$ cents **15.** r **16.** $18x$
17. $3d - 11$ **18.** $-78c + 18$ **19.** 13 **20.** 41 **21.** $\frac{10}{9}$
22. -24 **23.** 140 children **24.** 1,941 teragrams **25.** $h = \frac{2A}{b + B}$
26. Men: 116,000; women: 103,000 **27.** 20 lb of $3.80 candy;
10 lb of $4.40 candy **28.** $(-\infty, -14)$

29. (Ninth Avenue, 44th Street) **30.** 38 units **31.** II **32.** No
33.

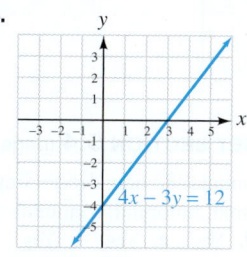

34.

35. $\frac{1}{2}$ **36.** 0 **37.** $\frac{2}{3}$ **38.** Perpendicular **39.** Slope: 2; $(-1, 0)$,
$(0, 2)$ **40.** A decrease of 1.6% per year **41.** $y = \frac{2}{3}x + 5$
42. $y = \frac{3}{4}x + \frac{11}{2}$ **43.** $y = 4$ **44.**

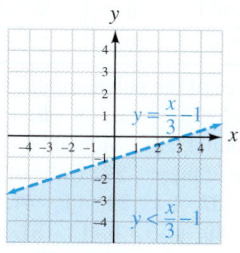

45. 1 **46.** No: $(1, 2), (1, -2)$ (answers may vary)
47. $(-6, 1)$

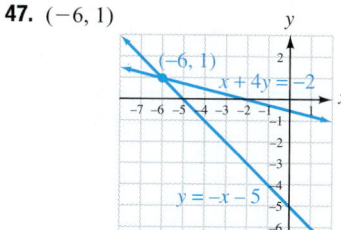

48.

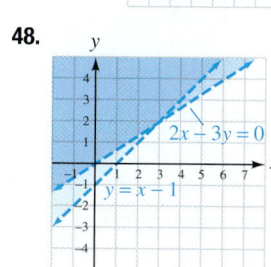

49. $(4, 1)$ **50.** $\left(\frac{2}{3}, \frac{3}{2}\right)$ **51.** Noodles: 2 servings, rice: 3 servings
52. $3,000 at 10%, $5,000 at 12%

ARE YOU READY? 5.1 (page 348)

1. a. 15 **b.** 125 **2. a.** 64 **b.** 12 **3.** x **4.** 4

SELF CHECKS 5.1

1. a. Base: 16; exponent: 2 **b.** Base: y; exponent: 4 **c.** Base: $3y$; exponent: 4 **d.** Base: m; exponent: 15 **2. a.** $9a^2b^4$ **b.** $(9a)^2$

c. $(x + y)^5$ **3. a.** 7^{15} **b.** x^6 **c.** $(y - 1)^{10}$ **d.** s^8t^7 **4.** a^{12} in.^{2}
5. a. 55^{25} **b.** a^2 **c.** $8.9t$ **d.** $b^{11}c^3$ **6.** b **7. a.** 4^{30} **b.** y^{10}
8. a. a^{21} **b.** a^{17} **9. a.** $16t^4$ **b.** $c^{18}d^{24}$ **c.** $-\frac{1}{27}a^3b^{15}$
10. $c^{10}d^{12}$ **11.** $64h^6$ **12. a.** $\frac{x^3}{343}$ **b.** $\frac{16x^{12}}{81y^8}$

STUDY SET SECTION 5.1 (page 356)

1. exponential **3. a.** $3x, 3x, 3x, 3x$ **b.** $(-5y)^3$
5. a. Subtract **b.** Add **c.** Multiply **d.** Multiply
7. a. $2x^2$ **b.** x^4 **c.** Doesn't simplify **d.** x^3 **9.** x^6, 18
11. a. power **b.** quantity, quantity **c.** sixth, ninth
13. a. Base: 4, exponent: 3 **b.** Base: 4, exponent: 3
c. Base: -4, exponent: 3 **15. a.** Base: $-3x$, exponent: 2
b. Base: x, exponent: 2 **c.** Base: $-3x$, exponent: 2
17. a. Base: m, exponent: 12 **b.** Base: $9m$, exponent: 12
c. Base: m, exponent: 12 **19.** $(4t)^4$ **21.** $-4t^5$ **23.** $\left(\frac{t}{2}\right)^3$
25. $(x - y)^2$ **27.** 5^7 **29.** b^6 **31.** $(y - 2)^7$ **33.** a^5b^6
35. a^{10} mi^2 **37.** x^9 ft^3 **39.** 8^8 **41.** x^{12} **43.** $(3.7p)^5$
45. c^2d^6 **47.** y^4 **49.** a **51.** 3^8 **53.** $(-4.3)^{24}$ **55.** m^{500}
57. y^{15} **59.** x^{25} **61.** p^{25} **63.** t^{18} **65.** u^{14} **67.** $36a^2$
69. $625y^4$ **71.** $-8r^6s^9$ **73.** $-\frac{1}{243}y^{10}z^{20}$ **75.** ab^4 **77.** $r^{13}s^3$
79. $216k^3$ **81.** $9q^2$ **83.** $\frac{a^3}{b^3}$ **85.** $\frac{64a^4}{121b^{10}}$ **87.** $\frac{x^{10}}{y^{15}}$ **89.** y^9
91. 15^3 **93.** t^7 **95.** $(k - 2)^{14}$ **97.** c^2d^5 **99.** y^{15} **101.** s^2
103. $-216a^9b^6$ **105.** $\frac{243m^{20}}{32n^{25}}$ **107.** $a^{21}b^{21}$ **109.** n^{33}
111. $36h^2$ **113.** x^3y^4 **115.** $\frac{m^4}{81}$ **117. a.** a^6 **b.** a^9 **c.** $2a^3$
119. a. b^9 **b.** b^{20} **c.** b **121. a.** $25x^2$ ft^2 **b.** $9a^2\pi$ ft^2
123. $\frac{1}{8,192}$ **127.** c **129.** d

ARE YOU READY? 5.2 (page 359)

1. $\frac{1}{2}$ **2.** $\frac{1}{x^2}$ **3. a.** 5 **b.** $\frac{1}{w}$ **4. a.** m, 0 **b.** 7, -2

SELF CHECKS 5.2

1. a. 1 **b.** $-5d$ **c.** 1 **2. a.** $\frac{1}{64}$ **b.** $\frac{1}{x^5}$ **c.** $-\frac{1}{27}$ **d.** $\frac{1}{16}$
3. a. $\frac{12}{h^9}$ **b.** $-\frac{1}{16}$ **4. a.** w^5 **b.** $\frac{64}{25}$ **c.** $-\frac{8a^7}{h^6}$ **5.** $\frac{81}{c^2}$ **6. a.** t^4
b. $\frac{1}{a^5}$ **c.** $\frac{1}{n^{20}}$ **d.** $\frac{64c^6}{d^3}$ **e.** $\frac{8}{c^{12}}$ **7. a.** $\frac{1}{a^6}$ **b.** $\frac{81x^2}{y^3}$ **c.** $\frac{d^{15}}{c^{18}}$

STUDY SET SECTION 5.2 (page 367)

1. negative **3.** zero
5.

Expression	Base	Exponent
4^{-2}	4	-2
$6x^{-5}$	x	-5
$\left(\frac{3}{y}\right)^{-8}$	$\frac{3}{y}$	-8
-7^{-1}	7	-1
$(-2)^{-3}$	-2	-3
$10a^0$	a	0

7.

x	3^x
2	9
1	3
0	1
-1	$\frac{1}{3}$
-2	$\frac{1}{9}$

9. a. 3 **b.** 6 **c.** sign, 3, 2 **d.** reciprocal, 2 **11.** y^8, -40, 40
13. 1 **15.** 1 **17.** 2 **19.** $\frac{5}{2}$ **21.** $\frac{1}{4}$ **23.** $\frac{1}{6}$ **25.** $\frac{1}{b^5}$ **27.** $-\frac{1}{5}$ **29.** $\frac{1}{2}$
31. $\frac{8}{9}$ **33.** $\frac{15}{g^6}$ **35.** $\frac{5}{x^3}$ **37.** $-\frac{1}{27}$ **39.** $-\frac{1}{64}$ **41.** 125 **43.** $8s$
45. $\frac{3}{16}$ **47.** $-\frac{4p^{10}}{d}$ **49.** 36 **51.** 8 **53.** $\frac{d^8}{c^8}$ **55.** $\frac{m^4}{81}$ **57.** y^6
59. b^7 **61.** $\frac{1}{y}$ **63.** $\frac{1}{h^7}$ **65.** $\frac{1}{x^{12}}$ **67.** $\frac{1}{b^8}$ **69.** $\frac{36s^8}{t^{14}}$ **71.** $\frac{x^9}{64}$ **73.** y^3
75. $\frac{1}{a^6}$ **77.** $\frac{9a^2}{2b^2}$ **79.** $\frac{y^{14}}{z^{10}}$ **81.** $\frac{8b^3}{a^{12}}$ **83.** r^{20} **85.** $\frac{b^2}{a^5}$ **87.** $-\frac{1}{1,000}$

89. 1   **91.** $\frac{64s^2}{81t^4}$ **93.** $\frac{32v^{25}}{u^{10}}$ **95.** $\frac{9}{y^8}$ **97.** $-15y$ **99.** $\frac{h^{20}}{16}$ **101.** $\frac{1}{x^6}$

103. $\frac{c^{12}}{d^{27}}$ **105.** 15   **107.** $\frac{9}{4g^2}$ **109.** $\frac{125}{d^6}$ **111.** $\frac{x^{28}}{y^{20}}$ **113.** $\frac{32x^{15}}{y^{10}}$

115. t^{10} **117.** $-\frac{4t^2}{s^5}$ **119.** $\frac{1}{x^3}$ **121. a.** $\frac{1}{8}$ **b.** $-\frac{1}{8}$ **c.** $-\frac{1}{8}$ **d.** $\frac{1}{8}$

123. a. $\frac{4x}{y^2}$ **b.** $\frac{1}{16x^2y^2}$ **c.** $\frac{4y}{x^2}$ **d.** $\frac{xy}{16}$

125.

Type of Sound	Intensity
Front row rock concert	10^{-1}
Normal conversation	10^{-6}
Vacuum cleaner	10^{-4}
Military jet takeoff	10^{2}
Whisper	10^{-10}

129. $-\frac{3}{2}$ **131.** $y = \frac{3}{4}x - 5$

ARE YOU READY? 5.3 (page 369)

1. 100   **2.** 4,528   **3.** $\frac{1}{10}$ **4.** 0.0622

SELF CHECKS 5.3

1. a. 4,880,000   **b.** 0.0098   **2. a.** 9.3×10^7 **b.** 9.055×10^{-5}
c. 8.5×10^{-2} **3.** 1.0414×10^9 **4.** 3.3×10^{-22} g

STUDY SET SECTION 5.3 (page 374)

1. scientific, standard   **3.** right, left   **5. a.** positive   **b.** negative
7. a. 7.7   **b.** 5.0   **c.** 8   **9. a.** $(5.1 \times 1.5)(10^9 \times 10^{22})$
b. $\frac{8.8}{2.2} \times \frac{10^{30}}{10^{19}}$ **11.** 1, 10, integer   **13.** 230   **15.** 812,000
17. 0.00115   **19.** 0.000976   **21.** 6,001,000   **23.** 2.718
25. 0.06789   **27.** 0.00002   **29.** 2.3×10^4 **31.** 1.7×10^6
33. 6.2×10^{-2} **35.** 5.1×10^{-6} **37.** 5.0×10^9
39. 3.0×10^{-7} **41.** 9.09×10^8 **43.** 3.45×10^{-2}
45. 9.0×10^0 **47.** 1.1×10^1 **49.** 1.718×10^{18}
51. 1.23×10^{-14} **53.** 7.3×10^5 **55.** 2.018×10^{17}
57. 7.3×10^{-5} **59.** 3.602×10^{-19} **61.** 7.14×10^5; 714,000
63. 4.032×10^{-3}; 0.004032   **65.** 4.0×10^{-4}; 0.0004
67. 3.0×10^4; 30,000   **69.** 4.3×10^{-3}; 0.0043
71. 3.08×10^{-2}; 0.0308   **73.** 2.0×10^5; 200,000
75. 7.5×10^{-11}; 0.000000000075   **77.** $9.038030748 \times 10^{15}$
79. $1.734152992 \times 10^{-12}$ **81.** 2.57×10^{13} mi
83. 197,000,000 mi²; 109,000,000,000,000,000 mi²;
14,600,000 mi²   **85.** 4.5×10^{-10} oz   **87.** g, x, u, v, i, m, r
89. 3.3×10^{-1} km/sec   **91.** 3.09936×10^{16} ft
93. 3.04×10^{11} dollars
95. 1.0×10^6, 1.0×10^9, 1.0×10^{12}, 1.0×10^{15}, 1.0×10^{18}
101. 5   **103.** $c = 30t + 45$

ARE YOU READY? 5.4 (page 376)

1. 3 terms   **2.** 6   **3.** $5b^4$ **4.** -26

SELF CHECKS 5.4

1. a. A trinomial in one variable of degree 2 written in descending
powers of x; terms: x^2, $4x$, -16; coefficients: 1, 4, -16; degree of
terms: 2, 1, 0   **b.** A binomial in two variables of degree 7 written in
descending powers of s and ascending powers of t; terms: $-14s^5t$, s^4t^3;
coefficients: -14, 1; degree of terms: 6, 7   **2.** 33   **3.** 55 cans   **4.** 16

5.
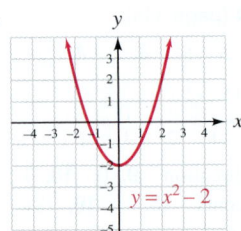

6.
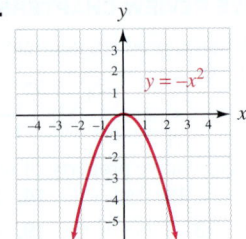

7.
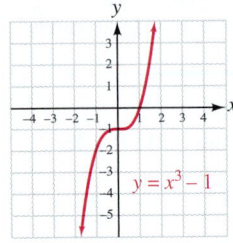

STUDY SET SECTION 5.4 (page 382)

1. polynomial   **3.** one, descending, two, ascending
5. monomial, binomial, trinomial   **7.** evaluate   **9. a.** Yes   **b.** No
c. No   **d.** Yes   **e.** Yes   **f.** Yes   **11.**

Term	Coefficient	Degree
$8x^2$	8	2
x	1	1
-7	-7	0

Degree of the polynomial: 2

13.

Term	Coefficient	Degree
$8a^6b^3$	8	9
$-27ab$	-27	2

Degree of the polynomial: 9

15. a. $5x^3 + 3x^2 + x - 9$ **b.** $x^2 - 2xy + y^2$ **17.** Binomial
19. Trinomial   **21.** Monomial   **23.** Binomial   **25.** Trinomial
27. None of these   **29.** None of these   **31.** Trinomial   **33.** 4th
35. 2nd   **37.** 1st   **39.** 4th   **41.** 12th   **43.** 0th   **45.** 18th
47. 3rd   **49. a.** 3   **b.** 13   **51. a.** -6 **b.** -8 **53. a.** 7   **b.** 34
55. a. -11.6 **b.** -40.2 **57. a.** 28   **b.** 4   **59. a.** 2   **b.** 0
61. 72   **63.** 19   **65.** -35 **67.** -257

69.
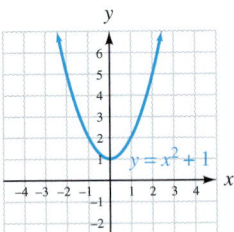

71.
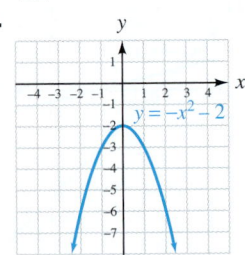

73.
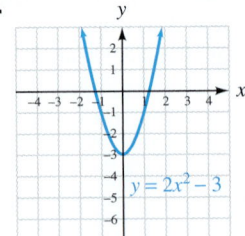

75.
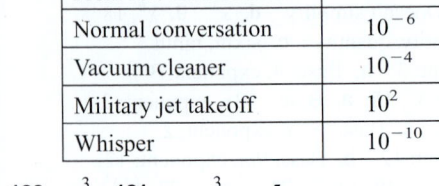

77.
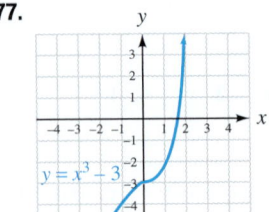

79.
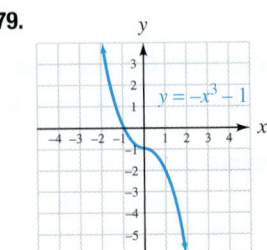

81. 91 cantaloupes **83.** 63 ft **85.** 28.6 billion downloads
87.

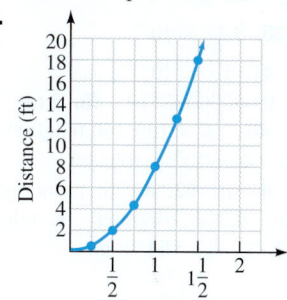

93. $[-3, \infty)$ **95.** x^{18} **97.** y^9

ARE YOU READY? 5.5 (page 385)

1. $17x$ **2.** $1.2a^3$ **3.** $5t^2u$ **4.** Does not simplify
5. $-4b^2 + 9b - 1$ **6.** -8

SELF CHECKS 5.5

1. a. $9m^4$ **b.** x **c.** $1.1s^3t + 0.3s^2t$ **d.** $\frac{11}{8}c^5 - \frac{4}{15}d^5 - 8$
2. a. $7a^2 + 5a - 1$ **b.** $\frac{9}{4}b^3 - \frac{3}{10}b - 3$ **c.** $11x^2 - 2xy - 2y^2$
3. $(29h^2 - 15h)$ in. **4.** $6q^2 - 8q + 2$ **5. a.** $7a^3 - 4a^2 + 12$
b. $x^2y - 8x - 8y$ **6. a.** $3p^2 - 8p + 15$
b. $-12m^3 + 6m^2 + 9m$ **7.** $0.6q^2 - 0.3q$
8. $\$(7{,}000x + 220{,}000)$

STUDY SET SECTION 5.5 (page 391)

1. polynomials **3.** Like **5.** combine **7. a.** $5x^2$ **b.** $14m^3$
c. $7a^3b$ **d.** $6cd + 4c^2d$ **9. a.** $-5x^2 + 8x - 23$
b. $5y^4 - 3y^2 + 7$ **11.** $4x^2, 2x, 1, 10x^2, 4$ **13.** $12t^2$
15. $20x^2 - 19x$ **17.** $x^2 + x$ **19.** $\frac{13}{15}x^2 - \frac{1}{8}x$ **21.** $1.3x^3$
23. $2st$ **25.** $-ab$ **27.** $-4x^3y + x^2y + 5$ **29.** $5q^2 - 4q - 5$
31. $y^3 + \frac{19}{20}y^2 + \frac{1}{3}$ **33.** $0.7p - 0.9q$ **35.** $7x^2 + xy + 2y^2$
37. $(3x^2 + 6x - 2)$ yd **39.** $(7x^2 + 5x + 6)$ mi **41.** $5x^2 + x + 11$
43. $-3a^2 + 7a + 7$ **45.** $10z^3 + z - 2$
47. $-x^3y^2 + 4x^2y + 5x + 6$ **49.** $2a^2 + a - 3$
51. $-5h^3 + 5h^2 + 30$ **53.** $\frac{1}{24}s^8 - \frac{19}{20}s^7$ **55.** $b^2 + 4ab - 2$
57. $x^2 + 6x + 2$ **59.** $4s^2 - 4s + 7$ **61.** $9a^3 - 8a^2 + 28a - 11$
63. $0.6x^3 + 1.2x^2 + 1.3x - 0.3$ **65.** $-9x^2 - 3x + 1$
67. $t^3 + 3t^2 + 6t - 5$ **69.** $13a^2 + a$ **71.** $3y^5 - 6y^4 + 1.2$
73. $10r^4 - 4r$ **75.** $-0.14f^2 + 0.25f + 2.09$ **77.** $\frac{5}{4}r^4 + \frac{11}{9}r^2 - 2$
79. $9c^2 - 6c - 14$ **81.** $19.4h^3 + 11h^2$ **83.** $-5r^2t - 9rt + 30$
85. $6x^2 + x - 5$ **87.** $\frac{7}{12}c^2 - \frac{1}{2}cd + d^2$ **89.** $7x + 4$
91. $-1.3t^2 + 0.7t + 0.6$ **93.** $-48u^3$ **95.** $5d^2 + 14d$
97. $7x^3y^2 - 2x^2y - 2x + 15$ **99.** $3x + 1$
101. $9x^3 + 2x^2 - 7x + 6$ **103. a.** $-19x^2 + 3x + 10$
b. $3x^2 - 9x - 10$ **105. a.** $(x^2 - 8x + 12)$ ft **b.** $(x^2 + 2x - 8)$ ft
107. $(2a^2 + 6a + 5)$ in. **109. a.** $(22t + 20)$ ft **b.** 108 ft
117. $180°$ **119.**

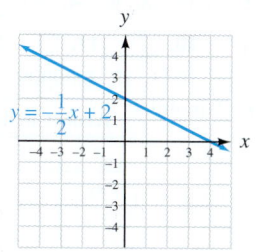

ARE YOU READY? 5.6 (page 394)

1. $50a$ **2.** $10x - 15$ **3.** $28y + 56$ **4.** x^{14} **5.** -30
6. $9x^2 + 3x + 2$

SELF CHECKS 5.6

1. a. $18t^2$ **b.** $60d^{11}$ **c.** $4y^{14}$ **d.** $-30a^7b^8$ **2. a.** $220x^6 + 22x^5$
b. $20c^5 - 45c^3 - 40c^2$ **c.** $s^6t^4 - s^5t^5 + s^4t^6 - 7s^3t^2$
d. $6w^{12} - 12w^6$ **3.** $(3n^5 + 2n^4 - 9n^3)$ m^2 **4.** $9y^2 - 33y - 12$
5. a. $y^2 + 4y + 3$ **b.** $6a^2 + a - 2$ **c.** $16x^2 + x - \frac{3}{8}$
d. $10y^6 - 39by^3 + 14b^2$ **6.** $6a^6 - 5a^4 - 3a^3 + a^2 + a$
7. a. $6x^3 - 8x^2 + 7x + 10$ **b.** $-4x^4 + 8x^3 + 8x^2 - 12x - 3$
8. $-6y^3 - 14y^2 + 12y$

STUDY SET SECTION 5.6 (page 400)

1. monomials, binomials **3.** first, outer, inner, last **5. a.** each,
each **b.** any, third **7. a.** $6x^2 + x - 12$ **b.** $5x^4 + 8ax^2 + 3a^2$
9. $8, n^3, 72n^5$ **11.** $2x, 5, 5, 4x, 15x, 11x$ **13.** $5m^2$ **15.** $12x^5$
17. $6c^6$ **19.** $-24b^6$ **21.** $8x^5y^5$ **23.** $-2a^{11}$ **25.** $3x^2 + 12x$
27. $-4t^3 + 28t$ **29.** $-6x^5 + 2x^4 - 2x^3$ **31.** $\frac{5}{8}t^8 + 5t^4$
33. $-12x^4z - 4x^2z^3 - 4x^3z^2 + 4x^2z$ **35.** $6x^{14} - 72x^{13}$
37. $(7h^2 + 3h)$ in.^{2} **39.** $(4w^2 - 2w)$ ft^2 **41.** $y^2 + 8y + 15$
43. $m^2 - 3m - 54$ **45.** $4y^2 + 23y - 35$ **47.** $12x^2 - 28x + 15$
49. $7.6y^2 - 5.8y + 1$ **51.** $18m^2 - 10m + \frac{8}{9}$ **53.** $t^4 - 7t^2 + 12$
55. $12a^2 - 5ab - 2b^2$ **57.** $x^3 - x + 6$
59. $4t^3 + 11t^2 + 18t + 9$ **61.** $2x^3 + 7x^2 - 16x - 35$
63. $r^4 - 5r^3 + 2r^2 - 7r - 15$ **65.** $x^3 - 3x + 2$
67. $12x^3 + 17x^2 - 6x - 8$ **69.** $8x^3 - 12x^2 - 8x$
71. $-3a^3 + 3ab^2$ **73.** $18a^6 - 12a^5$ **75.** $x^3 - 6x^2 + 5x + 12$ **77.**
$30x^2 - 17x + 2$ **79.** $6x^4 + 8x^3 - 14x^2$ **81.** $2t^2 + 2t - 24$
83. $6a^4 + 5a^3 + 5a^2 + 10a + 4$ **85.** $9t^2 + 15st - 6s^2$ **87.** $2a^{10}$
89. $16a^2 - 2ar - \frac{15}{16}r^2$ **91.** $a^2 + 2ab + b^2$
93. $x^4 + 11x^3 + 26x^2 - 28x - 24$ **95.** $9x^4 - 18x^3 + 54x^2$
97. $4y^3 + 40y^2 + 84y$ **99.** $0.12p^9 - 1.8p^7$
101. $16.4p^2q^2 - 24.6p^2q + 41pq^2$
103. $-3x^3 + 25x^2y - 56xy^2 + 16y^3$ **105. a.** $x^2 + 3x + 2$
b. $x^3 - 8$ **107. a.** Does not simplify **b.** $-18x^3z^8$
109. a. $-x^2 + 2x$ **b.** $6x^4 - 9x^3 + 3x^2$ **111. a.** $13a + 1$
b. $72a^3 + 6a^2 - 6a$ **113.** $(6x^2 + x - 1)$ cm^2
115. $(0.785x^2 - 0.785)$ in.^{2} **117.** $(2x^3 - 4x^2 - 6x)$ in.^{3}
125. a. 1 **b.** Undefined **c.** $-\frac{2}{3}$ **d.** 0

ARE YOU READY? 5.7 (page 403)

1. $x^2 + 4x + 4$ **2. a.** Base: $x + 4$; exponent: 2 **b.** Base: $x - 1$;
exponent: 3 **3.** $a^2 + 10a + 25$ **4.** $a^2 - 25$

SELF CHECKS 5.7

1. a. $r^2 + 12r + 36$ **b.** $49g^2 - 28g + 4$ **c.** $v^2 + 1.6v + 0.64$
d. $w^8 - 3w^4y + \frac{9}{4}y^2$ **2. a.** $b^2 - 16$ **b.** $25m^2 - 81$ **c.** $s^2 - \frac{9}{16}$
d. $c^6 - 4d^2$ **3.** $n^3 - 9n^2 + 27n - 27$ **4. a.** $7a^2 + 4a$
b. $11x^2 - 3x - 24$ **c.** $-3a^2 + 16a - 97$ **5.** $(72a^2 - 2)$ ft^2

STUDY SET SECTION 5.7 (page 409)

1. products **3. a.** square, Twice, first **b.** second, square
5. $x, 4, 4, 8x$ **7.** $s, 5, 25$ **9.** $x^2 + 2x + 1$ **11.** $m^2 - 12m + 36$
13. $16x^2 + 40x + 25$ **15.** $49m^2 - 28m + 4$ **17.** $1 - 6y + 9y^2$
19. $y^2 + 1.8y + 0.81$ **21.** $a^4 + 2a^2b^2 + b^4$ **23.** $s^2 + \frac{3}{2}s + \frac{9}{16}$
25. $x^2 - 9$ **27.** $4p^2 - 49$ **29.** $9n^2 - 1$ **31.** $c^2 - \frac{9}{16}$

33. $0.16 - 81m^4$ **35.** $25 - 36g^2$ **37.** $x^3 + 12x^2 + 48x + 64$
39. $n^3 - 18n^2 + 108n - 216$ **41.** $8g^3 - 36g^2 + 54g - 27$
43. $a^3 + 3a^2b + 3ab^2 + b^3$ **45.** $-x^2 + 20x - 8$
47. $4x^2 - 5x - 11$ **49.** $-80d^3 + 40d^2 - 5d$ **51.** $4d^5 - 4dg^6$
53. $(2x^2 - 2)$ yd^2 **55.** $(9x^2 + 6x + 1)$ ft^2 **57.** $4v^6 - 32v^3 + 64$
59. $12x^3 + 36x^2 + 27x$ **61.** $16f^2 - 0.16$
63. $r^4 + 20r^2s + 100s^2$ **65.** $6x - 2$ **67.** $d^8 + \frac{1}{2}d^4 + \frac{1}{16}$
69. $d^2 - 49$ **71.** $4a^2 - 12ab + 9b^2$ **73.** $n^2 - 36$ **75.** $36m + 36$
77. $8m^3 + 12m^2n + 6mn^2 + n^3$ **79.** $25m^2 - 12m + \frac{36}{25}$
81. $r^4 - 2r^2s^2 + s^4$ **83.** $x^2 - 4x + 4$ **85.** $r^2 + 4r + 4$
87. $n^4 - 8n^3 + 24n^2 - 32n + 16$ **89.** $17y^2 + 2y - 60$
91. $13x^2 - 8x + 5$ **93.** $f^2 - 16f + 64$ **95.** $36b^2 - \frac{1}{4}$
97. $4y^2 + 6y - 1$ **99.** $36 - 24d^3 + 4d^6$
101. $8e^3 + 12e^2 + 6e + 1$ **103.** $64x^2 + 48x + 9$
105. a. x^2y^2 **b.** $x^2 + 2xy + y^2$ **107. a.** $4b^4d^2$
b. $4b^4 + 4b^2d + d^2$ **109.** $(x^2 + 12x + 36)$ in.^{2}
111. $\pi hR^2 - \pi hr^2$ **117.** $\frac{5}{6}$ **119.** $\frac{21}{40}$

ARE YOU READY? 5.8 (page 411)

1. a. $\frac{2}{5}$ **b.** a^2 **2.** $\frac{a+b}{d}$ **3.** 36 **4.** $2x^2$

SELF CHECKS 5.8

1. a. $6y^2$ **b.** $\frac{d^4}{4c^3}$ **2. a.** $10h + 1$ **b.** $2s^3 - \frac{s^2t}{11} + 4t^2$ **3.** $x + 4$
4. $4x - 3 + \frac{6}{2x+3}$ **5.** $3x^2 + 2x - 4$ **6.** $x + 3$ **7.** $(9x + 3)$ in.

STUDY SET SECTION 5.8 (page 417)

1. monomial **3.** binomial **5.** Divide, multiply, subtract, bring
down **7.** quotient, dividend **9.** $7x^2, x^3, 7x^2, 5, 2, 7, 2, 2, 4x^3, 5, 7$
11. $5x^4 + 0x^3 + 2x^2 + 0x - 1$ **13.** x^3 **15.** $\frac{4h^2}{3}$ **17.** $-\frac{1}{5d^4}$
19. $\frac{10}{s}$ **21.** $\frac{x^2}{5y^4}$ **23.** $\frac{4r}{y^5}$ **25.** $2x + 1$ **27.** $\frac{1}{a^3} - \frac{1}{a} + 1$
29. $\frac{h^2}{4} + \frac{2}{h}$ **31.** $3s^5 - 6s^2 + 4s$ **33.** $c^3 + 3c^2 - 2c - \frac{5}{c}$
35. $5y - \frac{6}{x} + \frac{1}{xy}$ **37.** $x + 6$ **39.** $x - 2$ **41.** $x + 1 + \frac{-1}{2x+3}$
43. $2x - 3 + \frac{-1}{3x-1}$ **45.** $2x - 1$ **47.** $2x + 1$ **49.** $a - 5$
51. $x + 1$ **53.** $2x - 3$ **55.** $9b + 7$ **57.** $y + 12 + \frac{1}{y+1}$
59. $3a^5 - \frac{2b^3}{a}$ **61.** $2x^2 + 2x + 1$ **63.** $2x^5 - 8x^2$
65. $3a - 2$ **67.** $5m^5$ **69.** $b + 3$ **71.** $x + 3$ **73.** $x^2 - x + 1$
75. $-\frac{13}{3rs^3}$ **77.** $-2w^2 - \frac{1}{w^4}$ **79.** $9 - \frac{6}{m}$
81. $y^2 + 2y + 5 + \frac{10}{y-2}$ **83.** $\frac{x}{5} - \frac{2}{5x^2}$ **85.** $x^2 - 2x + 1$
87. $x + 1 + \frac{10}{x+5}$ **89.** $3x^2y - 2x - \frac{1}{y}$ **91.** $a - 12 + \frac{4}{a-5}$
93. $a^2 + a + 1$ **95.** $2x^2 + x + 1 + \frac{2}{3x-1}$ **97.** $\frac{x^2}{2y^{10}}$
99. $3m - 8$ **101. a.** $4x - 4 - \frac{5}{4x}$ **b.** $4x - 5$ **103.** $(x - 6)$ in.
105. $(2x^2 - x + 3)$ in. **111.** $y = -\frac{11}{6}x - \frac{7}{3}$

CHAPTER 5 REVIEW (page 420)

1. a. Base n, exponent 12 **b.** Base $2x$, exponent 6 **c.** Base r,
exponent 4 **d.** Base $y - 7$, exponent 3 **2. a.** m^5 **b.** $-3x^4$
c. $(x + 8)^2$ **d.** $\left(\frac{1}{2}pq\right)^3$ **3.** 7^{12} **4.** m^2n^2 **5.** y^{21} **6.** $81x^4$
7. b^9 **8.** $-b^{12}$ **9.** $256s^{10}$ **10.** $4.41x^4y^2$ **11.** $(-9)^{15}$
12. a^{23} **13.** $\frac{1}{8}x^{15}$ **14.** $\frac{x^{12}}{9y^2}$ **15.** $(m - 25)^{12}$ **16.** $125yz^4$
17. a^{11} **18.** c^5d^5 **19.** $64x^{12}$ in.^{3} **20.** y^4 ft^2 **21.** 1 **22.** 1
23. 3 **24.** $\frac{1}{1,000}$ **25.** $-\frac{1}{25}$ **26.** $\frac{1}{t^6}$ **27.** $8x^5$ **28.** $-\frac{6}{y}$

29. $\frac{8}{49}$ **30.** x^{14} **31.** $-\frac{27}{r^9}$ **32.** $\frac{1}{16z^2}$ **33.** $\frac{8c}{9d^5}$ **34.** t^{30}
35. w^{22} **36.** $\frac{f^{40}}{4^{10}}$ **37.** 7.2×10^8 **38.** 9.37×10^{15}
39. 9.42×10^{-9} **40.** 1.3×10^{-4} **41.** 1.8×10^{-4}
42. 8.53×10^5 **43.** 126,000 **44.** 0.00000003919 **45.** 2.68
46. 57.6 **47.** 3.0×10^{-4}; 0.0003 **48.** 1.6×10^8; 160,000,000
49. 6,570,000,000; 6.57×10^9 **50.** $1.0 \times 10^5 = 100,000$
51. a. 4 **b.** $3x^3$ **c.** $3, -1, 1, 10$ **d.** 10 **52. a.** 7th, monomial
b. 3rd, monomial **c.** 2nd, binomial **d.** 5th, trinomial **e.** 6th,
binomial **f.** 4th, none of these **53.** $3, -13$ **54.** 8 in.
55.

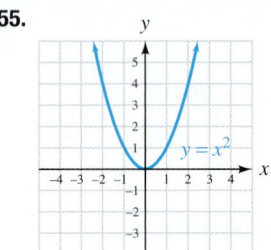

56.

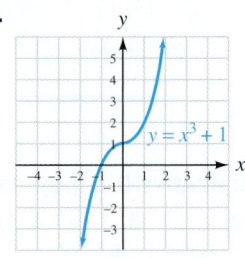

x	-2	-1	0	1	2
y	4	1	0	1	4

x	-2	-1	0	1	2
y	-7	0	1	2	9

57. $13y^3$ **58.** $-4a^3b + a^2b + 6$ **59.** $\frac{7}{12}x^2 - \frac{3}{4}xy + y^2$
60. $6.7c^5 + 8.1c^4 - 2.1c^3$ **61.** $25r^6 + 9r^3 + 5r$
62. $3.7a^2 + 6.1a - 17.6$ **63.** $4r^3s - 7r^2s^2 - 7rs^3 - 2s^4$
64. $\frac{5}{8}m^4 - m^3$ **65.** $-z^3 + 2z^2 + 5z - 17$ **66.** $(x^2 + x + 3)$ in.
67. $4x^2 + 2x + 8$ **68.** $8x^3 - 7x^2 + 19x$ **69.** $10x^3$
70. $-6x^{10}z^5$ **71.** $120b^{11}$ **72.** $2h^{14} + 8h^{11}$
73. $9n^4 - 15n^3 + 6n^2$ **74.** $x^2y^3 - x^3y^2$ **75.** $6x^6 + 12x^5$
76. $a^6b^4 - a^5b^5 + a^3b^6 - 7a^3b^2$ **77.** $x^2 + 5x + 6$
78. $2x^2 - x - 1$ **79.** $6t^2 - 6$ **80.** $6n^8 - 13n^6 + 5n^4$
81. $-5a^9 + 4a^7b + a^5b^2$ **82.** $6.6a^2 - 6.6$ **83.** $18t^2 + 3t - \frac{5}{9}$
84. $24b^2 - 34b + 11$ **85.** $8a^3 - 27$
86. $56x^4 + 15x^3 - 21x^2 - 3x + 2$ **87.** $8x^3 + 1$
88. a. $(6x + 10)$ in. **b.** $(2x^2 + 11x - 6)$ in.^{2}
c. $(6x^3 + 33x^2 - 18x)$ in.^{3} **89.** $a^2 - 6a + 9$
90. $m^3 + 6m^2 + 12m + 8$ **91.** $x^2 - 49$ **92.** $4x^2 - 0.81$
93. $4y^2 + 4y + 1$ **94.** $y^4 - 1$ **95.** $36r^4 + 120r^2s + 100s^2$
96. $-64a^2 + 48ac - 9c^2$ **97.** $80r^4s - 80s^5$
98. $36b^3 - 96b^2 + 64b$ **99.** $t^2 - \frac{3}{2}t + \frac{9}{16}$ **100.** $x^2 + \frac{8}{3}x + \frac{16}{9}$
101. $5x^2 + 19x + 3$ **102.** $24c^2 - 10c + 37$ **103.** $(x^2 - 4)$ in.^{2}
104. $(50x^2 - 8)$ in.^{2} **105.** $2n^3$ **106.** $-\frac{2x}{3y^2}$ **107.** $\frac{a^3}{6} - \frac{4}{a^4}$
108. $3a^3 + \frac{b}{5a} - \frac{5}{a^2}$ **109.** $x - 5$ **110.** $2x + 1$
111. $5x - 6 + \frac{4}{3x+2}$ **112.** $5y - 3$ **113.** $3x^2 - x - 4$
114. $3x^2 + 2x + 1 + \frac{2}{2x-1}$
115. $(y + 3)(3y + 2) = 3y^2 + 11y + 6$ **116.** $(2x^2 + 3x - 4)$ in.

CHAPTER 5 TEST (page 427)

1. a. base, exponent **b.** monomial, binomial, trinomial
c. degree **d.** special **2.** $2x^3y^4$ **3.** y^6 **4.** $\frac{1}{32}x^{21}$ **5.** 3.5 **6.** $\frac{2}{y^3}$
7. $\frac{1}{125}$ **8.** $(x + 1)^9$ **9.** y^{21} **10.** $\frac{b^3}{64a^3}$ **11.** $\frac{m^{12}}{64}$ **12.** $-6ab^9$
13. $1,000y^{12}$ in.^{3} **14.** 6.25×10^{18} **15.** 0.000093
16. 9.2×10^3; 9,200 **17.** Trinomial

Term	Coefficient	Degree
x^4	1	4
$8x^2$	8	2
-12	-12	0

Degree of the polynomial: 4

18. 5th degree **19.**

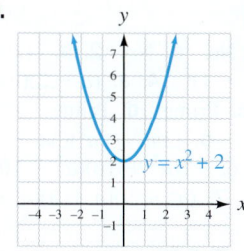

x	-2	-1	0	1	2
y	6	3	2	3	6

20. 0 ft; the rock hits the canyon floor 18 seconds after being dropped.
21. $\frac{1}{10}x^2 + \frac{7}{12}x - 2$ **22.** $-4a^3b + a^2b + 5$
23. $19.4h^3 - 11.1h^2 - 0.6$ **24.** $6b^3c - 2bc - 12$
25. $-3y^3 + 18y^2 - 17y + 35$ **26.** $(10a^2 + 8a - 20)$ in.
27. $10x^5y^{11}$ **28.** $-72b^8$ **29.** $3y^4 - 6y^3 + 9y^2$
30. $0.24p^{11} - 0.54p^8$ **31.** $\frac{3}{4}s^7t^{17} + 12s^4t^{10}$
32. $3x^2 - 11x - 20$ **33.** $12t^2 - 8t - \frac{3}{4}$
34. $7.6m^2 - 5.8m + 1$ **35.** $a^6 + a^3 - 42$
36. $2x^3 - 7x^2 + 14x - 12$ **37.** $1 - 100c^2$
38. $49b^6 - 42b^3t + 9t^2$ **39.** $2.2a^3 + 4.4a^2 - 33a$
40. $2x^2 + 2xy$ **41.** $\frac{a}{4b} - \frac{b}{2a}$ **42.** $x - 2$
43. $3x^2 + 2x + 1 + \frac{2}{2x - 1}$ **44.** $(x - 5)$ ft
45. Yes; $(5m + 1)(m - 6) = 5m^2 - 29m - 6$
46. No; $(a + b)^2 = a^2 + 2ab + b^2$

CUMULATIVE REVIEW CHAPTERS 1–5 (page 429)

1. $2 \cdot 3^3 \cdot 5$ **2. a.** $a + b = b + a$ **b.** $(xy)z = x(yz)$ **3.** -37
4. 28 **5.** $18x$ **6.** 0 **7.** -2 **8.** 15 **9.** \$2.21 billion
10. 1.2 ft^3 **11.** $30°$ **12.** \$6,250 **13.** Mutual fund: \$25,000;
bonds: \$20,000 **14.** $\left(-\infty, -\frac{11}{4}\right)$

15. 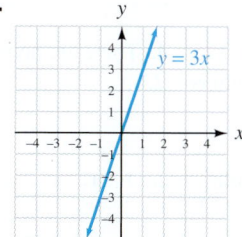 **16.**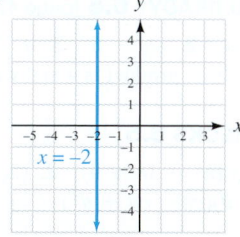

17. $-\frac{4}{9}$ **18.** 0.75 million people per year or $\frac{3}{4}$ million people per year **19.** $m = 3, (0, -2); y = 3x - 2$ **20.** Perpendicular
21. $y = -4x + 2$ **22.** No **23.** 26 **24.** Not a function; $(1, 2)$, $(1, -2)$; answers may vary. **25.** No
26. $(4, 1)$

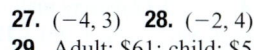

27. $(-4, 3)$ **28.** $(-2, 4)$
29. Adult: \$61; child: \$51

30.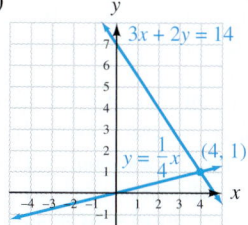
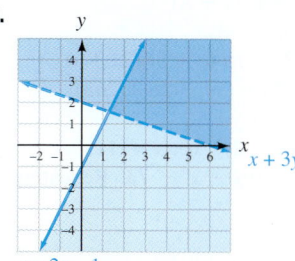

31. $9x^4y^8$ **32.** v^{22}
33. $a^2b^7c^6$ **34.** $\frac{64t^{12}}{27}$
35. $\frac{1}{16y^4}$ **36.** a^7 **37.** $-\frac{1}{25}$
38. $\frac{x^{10}}{a^{10}}$ **39.** 6.15×10^5
40. 1.3×10^{-6}

41.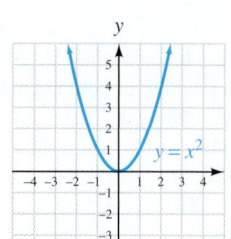

42. 1.5 in. **43.** $7c^2 + 7c$
44. $-6x^4 - 17x^2 - 68x + 11$
45. $6t^2 + 7st - 3s^2$
46. $12x^3 + 36x^2 + 27x$
47. $2x + 1$ **48.** $\frac{1}{8} - \frac{2}{x}$

ARE YOU READY? 6.1 (page 432)

1. $2 \cdot 3 \cdot 3 \cdot 3 = 2 \cdot 3^3$ **2.** a^6 **3.** $-x + 7$ **4.** 4
5. $4d^4 + 8d^2$ **6.** $x^3 + x^2 - 3x - 3$

SELF CHECKS 6.1

1. a. 2 **b.** 1 **c.** 15 **2. a.** $11c$ **b.** $21s^2t^2$ **3. a.** $6(f + 6)$
b. $6s^2t(4t - 7s)$ **c.** $y^3(y^3 - 10y - 1)$ **4. a.** $-(b^4 + 3b^2 - 2)$
b. $-(t - 9)$ **5.** $-11(4c - 5)$ **6.** $(y - 1)(2y + 7)$
7. a. $(3n + 2)(n^2 + 3)$ **b.** $(x - y)(7 + y)$
8. a. $(a + 11)(a^4 + 1)$ **b.** $(b - c)(b - 1)$ **9.** $(y + 3)(y^2 - 2)$
10. $4(t + s)(1 + z)$

STUDY SET SECTION 6.1 (page 440)

1. factor **3.** grouping **5. a.** 3 **b.** $7, h$ **c.** $3, y, y$
7. a. $2x + 4$ (Answers may vary) **b.** $x^3 + x^2 + x$ (Answers may vary) **9.** $2x, x + 3$ **11. a.** 4 **b.** No **c.** $2; h$ **13.** 8
15. $b^2, 2, b - 6$ **17.** 2 **19.** 6 **21.** 7 **23.** 8 **25.** m^3
27. 5 **29.** $4c$ **31.** $9a^3$ **33.** $8a$ **35.** $3m^3n$ **37.** $x + 7$
39. $p - t$ **41.** $3(x + 2)$ **43.** $9(2m - 1)$ **45.** $d(d - 7)$
47. $5(3c^3 + 5)$ **49.** $8a(3 - 2a)$ **51.** $7(2x^2 - x - 1)$
53. $t^2(t^2 + t + 2)$ **55.** $3xy^2(7xy + 1)$ **57.** $-(a + b)$
59. $-(x^2 + x - 16)$ **61.** $-(-5 + x)$ or $-(x - 5)$
63. $-(-9 + 4a)$ or $-(4a - 9)$ **65.** $-3x(x + 2)$
67. $-4a^2(b - 3a)$ **69.** $-12x^2(2x^2 + 4x - 3)$
71. $-2ab^2(2a^2 - 7a + 5)$ **73.** $(x + 2)(y + 3)$
75. $(p - q)(m - 5)$ **77.** $(x + y)(2 + a)$ **79.** $(s - u)(r + 8w)$
81. $(7m - 2)(m^2 + 2)$ **83.** $(5x - 1)(x^2 + 2)$ **85.** $(b + c)(a + 1)$
87. $(r + 4s)(s - 1)$ **89.** $(2x - 3)(a + b)$ **91.** $(m - n)(p - q)$
93. $(m + 1)(5m^2 + 6)$ **95.** $(y^2 + 3)(y - 4)$
97. $a(x - 2)(x^2 + 5)$ **99.** $6(x^2 + 2)(x - 1)$
101. $(14 + r)(h^2 + 5)$ **103.** $11a^2(2a - 3)$ **105.** $(a + b)(x - 1)$
107. $3r^5(5r^3 - 6r - 10)$ **109.** $3(3p + q)(3m - n)$
111. $-20pt^2(3p + 4t)$ **113.** $-(2x - 5)$ **115.** $(3x - y)(2x - 5)$
117. $2z(x - 2)(x^2 + 16)$ **119.** $6uvw^2(2w - 9v)$
121. $(x + 1)(x^2 + 1)$ **123.** $-(3r - 2s + 3)$
125. a. $(5t + 6)(t^2 + 3)$ **b.** $3(t^3 + 2t^2 + 5t + 6)$
127. $(x^2 + 5)$ ft; $(x + 4)$ ft **129. a.** $V = \pi r^2\left(h_1 + \frac{1}{3}h_2\right)$
b. A 40 in.^{3} block of wax is needed. **135.** 12%

ARE YOU READY? 6.2 (page 443)

1. 1 **2.** $x^2 + 7x - 8$ **3.** 2 **4.** 2 **5.** 2 and 4 **6.** -3 and 5

SELF CHECKS 6.2

1. $(y + 2)(y + 5)$ **2.** $(p - 2)(p - 4)$ **3.** $(m + 7)(m - 6)$
4. $(q + 4)(q - 6)$ **5.** $-(x - 4)(x - 7)$ **6.** $(s + 7t)(s - t)$
7. $4m^3(m + 4)(m - 2)$ **8.** $t(t - 2)(t + 6)$ **9.** Prime trinomial
10. $(m + 7)(m - 6)$ **11.** $(q + 4t)(q - 6t)$
12. $3m(m - 8)(m - 1)$

STUDY SET SECTION 6.2 (page 452)

1. factors   **3.** leading   **5. a.** descending   **b.** common   **7.** 3, 5
9. a. No   **b.** Yes   **11. a.** They are both positive or both negative.
b. One will be positive, the other negative.   **13.** $+3, -2$
15. $(x + 2)(x + 1)$ **17.** $(z + 4)(z + 3)$ **19.** $(m - 3)(m - 2)$
21. $(t - 7)(t - 4)$ **23.** $(x + 8)(x - 3)$ **25.** $(t - 3)(t + 16)$
27. $(a - 8)(a + 2)$ **29.** $(b - 12)(b + 3)$ **31.** $-(x + 5)(x + 2)$
33. $-(t + 6)(t - 5)$ **35.** $-(r + 9)(r - 6)$
37. $-(m - 7)(m - 11)$ **39.** $(a + 3b)(a + b)$
41. $(x - 7y)(x + y)$ **43.** $(r + 2s)(r - s)$ **45.** $(a - 3b)(a - 2b)$
47. $2(x + 3)(x + 2)$ **49.** $6(a - 4)(a - 1)$ **51.** $5(a - 3)(a - 2)$
53. $-z(z - 4)(z - 25)$ **55.** $(x - 4)(x - 20)$ **57.** $(y + 9)(y + 1)$
59. $r(r - 2)(r + 8)$ **61.** $r(r + 3x)(r + x)$ **63.** Prime
65. Prime   **67.** $(x + 3)(5 + y)$ **69.** $2n(13n - 4)$
71. $(a - 5)(a + 1)$ **73.** $-(x - 22)(x + 1)$ **75.** $4(y - 1)(x + 7)$
77. $12b^2(b^2 - 4b - 3)$ **79.** $(r - 3)(r - 6)$
81. $-n^2(n - 30)(n + 2)$ **83.** $(x + 2y)(x + 2y) = (x + 2y)^2$
85. $(a - 6b)(a + 2b)$ **87.** $4x^2(x + 2)(x + 2) = 4x^2(x + 2)^2$
89. $(a - 45)(a - 1)$ **91.** Prime   **93.** $(x + 2)(t + 7)$
95. $s^2(s + 13)(s - 2)$ **97.** $15(s^3 + 5)$ **99.** $(y - 14)(y + 1)$
101. $2(x - 2)(x - 4)$ **103. a.** $(x - 4)(x - 6)$
b. $(x - 12)(x + 2)$ **105.** $(x + 9)$ in., x in., $(x + 3)$ in.   **113.** $\frac{1}{x^2}$
115. $\frac{1}{x^{10}}$

ARE YOU READY? 6.3 (page 454)

1. 3   **2.** $5y^2$ **3.** $6x^2 - 13x - 5$ **4.** $1(-3)$ and $-1(3)$
5. $6, -3, 7$ **6.** $(2b + 3)(5b - 1)$

SELF CHECKS 6.3

1. $(2x + 1)(x + 2)$ **2.** $(6b - 1)(b - 3)$ **3.** $(5t + 2)(t - 5)$
4. $(2x + 3y)(2x - y)$ **5.** $-2y(7y + 3)(y - 2)$
6. $(3a + 4)(5a - 1)$ **7.** $a^2(7a - 2)(3a - 1)$

STUDY SET SECTION 6.3 (page 462)

1. leading   **3.** $5y, y, 1, 3$ **5.** $10x$ and x, $5x$ and $2x$
7. a. descending, GCF, coefficient   **b.** $3s^2$ **c.** $-(2d^2 - 19d + 8)$
9. negative   **11.** different   **13.** $-13; -6, -8; -3, -7$
15. a. $12, 20, -9$ **b.** -108 **17.** $3t, 2, 4t + 3$
19. $(2x + 1)(x + 1)$ **21.** $(3a + 1)(a + 3)$ **23.** $(5x + 2)(x + 1)$
25. $(7x + 11)(x + 1)$ **27.** $(2x - 3)(2x - 1)$
29. $(4x - 1)(2x - 5)$ **31.** $(5t - 7)(3t - 1)$ **33.** $(6y - 1)(y - 2)$
35. $(3x + 7)(x - 3)$ **37.** $(5m + 3)(m - 2)$ **39.** $(7y - 1)(y + 8)$
41. $(11y - 4)(y + 1)$ **43.** $(3r + 2s)(2r - s)$
45. $(2x + 3y)(2x + y)$ **47.** $(8m + 3n)(m + 11n)$
49. $(5x + 3y)(3x - 2y)$ **51.** $2(3x + 2)(x - 5)$
53. $a(2a - 5)(4a - 3)$ **55.** $(2u + 3v)(u - 2v)$
57. $4(9y - 4)(y - 2)$ **59.** $(2t - 5)(3t + 4)$
61. $(3p - q)(5p + q)$ **63.** $(2t - 1)(2t - 7)$
65. $10(13r - 11)(r + 1)$ **67.** $(4y + 1)(2y - 1)$
69. $(18x - 5)(x + 2)$ **71.** $-y(y + 12)(y + 1)$ **73.** Prime
75. $-3x^2(2x + 1)(x - 3)$ **77.** $(3p - q)(2p + q)$
79. $3r^3(5r - 2)(2r + 5)$ **81.** $2mn(4m + 3n)(2m + n)$
83. Prime   **85.** $-(4y - 3)(3y - 4)$ **87.** $(m + 7)(m - 4)$
89. $3a^2(2a + 5)$ **91.** $(x - 2)(x^2 + 5)$ **93.** $(5y - 3)(y - 1)$
95. $-2(x + 2)(x + 3)$ **97.** $3x^2y^2(4xy - 6y + 5)$
99. $(a - 5b)(a - 2b)$ **101.** $u^4(9u + 1)(u - 8)$
103. $(2x + 11)$ in., $(2x - 1)$ in.   **109.** -49 **111.** 1   **113.** 49

ARE YOU READY? 6.4 (page 464)

1. $9y^2 + 12y + 4$ **2.** $m^2 - 81$ **3.** $64d^2$ **4.** $x^2 + 25$

SELF CHECKS 6.4

1. a. Yes   **b.** No   **c.** No   **d.** No   **2. a.** $(x + 9)^2$
b. $(4x - y)^2$ **3.** $x(7x - 1)^2$ **4. a.** $(c + 2)(c - 2)$
b. $(11 + t)(11 - t)$ **c.** Prime   **d.** Prime
5. a. $(4y + 3)(4y - 3)$ **b.** $(3m + 8n^2)(3m - 8n^2)$
c. $(10 + a^2)(10 - a^2)$ or $-1(a^2 + 10)(a^2 - 10)$
6. $2(p + 10)(p - 10)$ **7.** $(a^2 + 9)(a + 3)(a - 3)$

STUDY SET SECTION 6.4 (page 470)

1. perfect   **3. a.** $5x$ **b.** 3   **c.** $5x, 3$ **5. a.** x, y **b.** $-$
c. $+, x, y$ **7.** 1, 4, 9, 16, 25, 36, 49, 64, 81, 100, 121, 144, 169,
196, 225, 256, 289, 324, 361, 400   **9.** 2   **11.** $+, -$ **13.** Yes
15. No   **17.** No   **19.** Yes   **21.** $(x + 3)^2$ **23.** $(b + 1)^2$
25. $(c - 6)^2$ **27.** $(2x + 3)^2$ **29.** $(6m + 5n)^2$ **31.** $(9x - 4y)^2$
33. $3(u - 3)^2$ **35.** $x(6x + 1)^2$ **37.** $(x + 2)(x - 2)$
39. $(x + 4)(x - 4)$ **41.** $(6 + y)(6 - y)$ **43.** $(t + 5)(t - 5)$
45. Prime   **47.** Prime   **49.** $(5t + 8)(5t - 8)$
51. $(9y + 1)(9y - 1)$ **53.** $(3x^2 + y)(3x^2 - y)$
55. $(4c + 7d^2)(4c - 7d^2)$ or $-(7d^2 + 4c)(7d^2 - 4c)$
57. $8(x + 2y)(x - 2y)$ **59.** $7(3a + 1)(3a - 1)$
61. $(9 + s^2)(3 + s)(3 - s)$ **63.** $(b^2 + 16)(b + 4)(b - 4)$
65. $(a^2 + 12b)(a^2 - 12b)$ **67.** $(3xy + 5)^2$
69. $16(t^2 + s^2)(t + s)(t - s)$ **71.** $(t - 10)^2$
73. $(3y - 4)^2$ **75.** $(z + 8)(z - 8)$ **77.** $25(m^2 + 1)(m + 1)(m - 1)$
79. $2a^3(3a + 7b)^2$ **81.** $x(x + 12)(x - 12)$ **83.** $(7t - 2s)^2$
85. $3(m^2 + n^2)(m + n)(m - n)$ **87.** $(5m + 7)^2$ **89.** $-(10t - 1)^2$
91. $6x^2(x + y)(x - y)$ **93.** Prime   **95.** $(5x + 13)(5x - 13)$ or
$-(13 + 5x)(13 - 5x)$ **97.** $(x + 7)(x - 6)$ **99.** $(x + 3)(x - 3)$
101. $8a^2b(3a - 2)$ **103.** $-2(r - 10)(r - 4)$ **105.** $(x + 3)(x^2 + 4)$
107. $(2b - 5)^2$ **109.** $(p + q)^2$ **111.** $0.5g(t_1 + t_2)(t_1 - t_2)$
117. $3cd + \frac{3c}{2} + d$

ARE YOU READY? 6.5 (page 472)

1. $x^3 + 64$ **2.** $8h^3 - 1$ **3. a.** 27   **b.** 125   **4.** There are no two
integers whose product is 4 and whose sum is -2.

SELF CHECKS 6.5

1. $(h + 3)(h^2 - 3h + 9)$ **2.** $(2c - 1)(4c^2 + 2c + 1)$
3. $4(c + d)(c^2 - cd + d^2)$

STUDY SET SECTION 6.5 (page 475)

1. sum, cubes   **3. a.** F, L **b.** $-, F^2, L^2$ **5.** $6n, 5$ **7.** 1, 8, 27,
64, 125, 216, 343, 512, 729, 1,000   **9.** No   **11.** $2a$ **13.** $b + 3$
15. a. $x^3 + 8$ (Answers may vary.)   **b.** $(x + 8)^3$
17. $(y + 5)(y^2 - 5y + 25)$ **19.** $(a + 4)(a^2 - 4a + 16)$
21. $(n + 8)(n^2 - 8n + 64)$ **23.** $(2 + t)(4 - 2t + t^2)$
25. $(a + 10b)(a^2 - 10ab + 100b^2)$
27. $(5c + 3d)(25c^2 - 15cd + 9d^2)$ **29.** $(a - 3)(a^2 + 3a + 9)$
31. $(m - 7)(m^2 + 7m + 49)$ **33.** $(6 - v)(36 + 6v + v^2)$
35. $(2s - t)(4s^2 + 2st + t^2)$ **37.** $(10a - w)(100a^2 + 10aw + w^2)$
39. $(4x - 3y)(16x^2 + 12xy + 9y^2)$ **41.** $2(x + 1)(x^2 - x + 1)$
43. $3(d + 3)(d^2 - 3d + 9)$ **45.** $x(x - 6)(x^2 + 6x + 36)$
47. $8x(2m - n)(4m^2 + 2mn + n^2)$ **49.** $(x + 4)^2$
51. $(3r + 4s)(3r - 4s)$ **53.** $(x - t)(y + s)$
55. $4(p + 2q)(p^2 - 2pq + 4q^2)$ **57.** $2ct^2(4c + 3t)(2c + t)$
59. $36(e^2 + 1)(e + 1)(e - 1)$ **61.** $7a^2b^2(5a - 2b + 2ab)$
63. $(6r + 5s)^2$ **65. a.** $(x + 1)(x - 1)$
b. $(x - 1)(x^2 + x + 1)$ **67. a.** $x(x + 2)$ **b.** $(x + 1)^2$
69. $(1,000 - x^3)$ in.^{3}; $(10 - x)(100 + 10x + x^2)$ **73.** Repeating
75. -3

ARE YOU READY? 6.6 (page 477)

1. a. 4   **b.** 3   **2.** No   **3.** $3n^3 - 27n^2 - 210n$ **4.** $5cd^2$

SELF CHECKS 6.6

1. $11a^2(a^2 + 1)(a + 1)(a - 1)$ **2.** $-2h^2(4h + 5)^2$
3. $(b + 1)(b + 2)(b^2 - 2b + 4)$ **4.** $6m(m^2 + m - 9)$
5. $3y(2y - 1)(y + 4)$

STUDY SET SECTION 6.6 (page 481)

1. product   **3.** Factor out the GCF   **5.** Perfect-square trinomial
7. Sum of two cubes   **9.** Trinomial factoring   **11.** Is there a
common factor?   **13.** $14m, m$ **15.** $2(b + 6)(b - 2)$
17. $4p^2q^3(2pq^4 + 1)$ **19.** $2(2y + 1)(10y + 1)$
21. $8(x^2 + 1)(x + 1)(x - 1)$ **23.** $(c + 21)(c - 7)$ **25.** Prime
27. $-2x^2(x - 4)(x^2 + 4x + 16)$ **29.** $(c + d^2)(a^2 + b)$
31. $-(3x - 1)^2$ **33.** $-5m(2m + 5)^2$ **35.** $(2c + d)(c - 3d)$
37. $(p - 2)^2(p^2 + 2p + 4)$ **39.** $(x - a)(a + b)(a - b)$
41. $(ab + 12)(ab - 12)$ **43.** $(x + 5)(2x^2 + 1)$
45. $v^2(v^2 - 14v + 8)$ **47.** $2(3a - b)(3a + 7c)$
49. $2x(2ax + b)(2ax - b)$ **51.** $2(3x - 4)(x - 1)$
53. $y^2(2x + 1)^2$ **55.** $4m^2(m + 5)(m^2 - 5m + 25)$
57. $(a + 6)(a + 2)(a - 2)$ **59.** Prime
61. $2a^2(2a - 3)(4a^2 + 6a + 9)$ **63.** $27(x - y - z)$
65. $(x - t)(y + s)$ **67.** $x^6(7x + 1)(5x - 1)$
69. $5(x - 2)(1 + 2y)$ **71.** $(7p + 2q)^2$ **73.** $4(t^2 + 9)$
75. $(n + 3)(n - 3)(m^2 + 3)$
81.

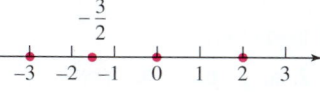

83.

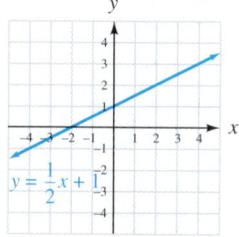

ARE YOU READY? 6.7 (page 482)

1. 0   **2.** 0   **3.** -4 **4.** 0   **5.** $(x - 3)(x + 2)$
6. $(3n + 2)(n - 1)$

SELF CHECKS 6.7

1. $12, -\frac{6}{5}$ **2.** $-2, -3$ **3.** $-7, 7$ **4.** $0, 5$ **5.** $\frac{2}{3}, -4$ **6.** $-\frac{3}{2}$
repeated   **7.** $0, \frac{2}{5}, -\frac{1}{2}$

STUDY SET SECTION 6.7 (page 487)

1. quadratic   **3.** zero-factor, 0, 0   **5. a.** Yes   **b.** No   **c.** Yes
d. No   **7.** $-\frac{4}{5}$ **9. a.** Add 6 to both sides.   **b.** Distribute the
multiplication by x and subtract 3 from both sides.
11. $0, x + 7, -7$ **13.** $p, p - 3, 0, 3, -2$ **15.** $3, 2$ **17.** $-7, 7$
19. $0, \frac{5}{2}$ **21.** $0, -\frac{10}{3}$ **23.** $0, 6, -8$ **25.** $1, -2, 3$
27. $12, 1$ **29.** $-3, 7$ **31.** $8, 1$ **33.** $-3, -5$ **35.** $-9, 9$
37. $-5, 5$ **39.** $-\frac{1}{2}, \frac{1}{2}$ **41.** $-\frac{7}{3}, \frac{7}{3}$ **43.** $0, 7$ **45.** $0, 16$
47. $0, 3$ **49.** $0, -\frac{8}{3}$ **51.** $-2, \frac{1}{3}$ **53.** $-\frac{3}{2}, 1$ **55.** $\frac{1}{5}, 1$
57. $-\frac{5}{2}, 4$ **59.** $-\frac{7}{2}$ repeated   **61.** $\frac{5}{3}$ repeated   **63.** $0, -1, -2$
65. $0, 9, -3$ **67.** $-\frac{9}{2}, \frac{9}{2}$ **69.** 8 repeated   **71.** $\frac{5}{2}, -6$ **73.** $2, 10$

75. $0, -5, 4$ **77.** $-10, 10$ **79.** $3, 4$ **81.** $-\frac{1}{3}, 5$ **83.** $2, 7, 1$
85. $-3, -2$ **87.** $0, 3$ repeated   **89.** $0, 2$ **91.** $-\frac{2}{3}, -\frac{3}{2}$
93. $0, -1, -\frac{1}{3}$ **95.** $0, -1, 2$ **97.** $\frac{2}{3}, -\frac{1}{5}$ **99.** $-\frac{11}{2}, \frac{11}{2}$ **101.** $\frac{1}{8}, 1$
103. a. $(x + 7)(x - 3)$ **b.** $-7, 3$ **105. a.** $(4n + 1)(3n - 2)$
b. $-\frac{1}{4}, \frac{2}{3}$ **113.** 15 min $\le t <$ 30 min

ARE YOU READY? 6.8 (page 490)

1. $A = lw$ **2.** $A = \frac{1}{2}bh$ **3.** $x + 1$ **4.** 3

SELF CHECKS 6.8

1. Width: 5 m; length: 11 m   **2.** Base: 3 yd; height: 10 yd
3. 23 and 24   **4.** 8 in., 15 in., and 17 in.   **5.** 5 sec

STUDY SET SECTION 6.8 (page 495)

1. consecutive   **3.** hypotenuse, legs   **5.** ii.   **7.** $20 = b(b + 5)$
9. a. A right triangle   **b.** x ft; $(x + 1)$ ft   **c.** 9 ft
11. $-16, 1, 0, 0, 3, -1$ **13.** Width: 3 ft; length: 6 ft   **15.** 8 in.,
10 in.   **17.** 3 ft by 9 ft   **19.** Base: 6 cm; height: 5 cm
21. Foot: 4 ft; luff: 12 feet   **23.** Kahne: 9; Riggs: 10   **25.** 12
27. $(11, 13)$ **29.** 10 yd   **31.** 8 ft   **33.** 5 m, 12 m, 13 m
35. 5 sec   **37.** 4 sec   **39.** 1 sec   **41.** 8   **47.** $25b^2 - 20b + 4$
49. $s^4 + 8s^2 + 16$ **51.** $81x^2 - 36$

CHAPTER 6 REVIEW (page 499)

1. $5 \cdot 7$ **2.** $2^5 \cdot 3$ **3.** 7   **4.** $18a^3$ **5.** $3(x + 3y)$
6. $5a(x^2 + 3)$ **7.** $7s^3(s^2 + 2)$ **8.** $\pi a(b - c)$
9. $12x(2x^2 + 5x - 4)$ **10.** $xy^3z^2(x^4 + y^2z - 1)$
11. $-5ab(b - 2a + 3)$ **12.** $(x - 2)(4 - x)$ **13.** $-(a + 7)$
14. $-(4t^2 - 3t + 1)$ **15.** $(c + d)(2 + a)$ **16.** $(y + 6)(3x - 5)$
17. $(a + 1)(2a^2 - 1)$ **18.** $4m(n + 3)(m - 2)$ **19.** 1
20.

Factors of 6	Sum of the factors of 6
1(6)	7
2(3)	5
−1(−6)	−7
−2(−3)	−5

21. $(x + 6)(x - 4)$
22. $(x - 20)(x + 2)$
23. $(x - 5)(x - 9)$
24. Prime
25. $-(y - 8)(y - 7)$
26. $(y + 9)(y + 1)$
27. $(c + 5d)(c - 2d)$
28. $(m - 2n)(m - n)$ **29.** Multiply   **30.** There are no two
integers whose product is 11 and whose sum is 7.
31. $5a^3(a + 10)(a - 1)$ **32.** $-4x(x + 3y)(x - 2y)$
33. $(2x + 1)(x - 3)$ **34.** $(7y + 5)(5y - 2)$
35. $-(3x + 5)(x - 6)$ **36.** $3p(6p + 1)(p - 2)$
37. $(4b - c)(b - 4c)$ **38.** Prime   **39.** $(4x + 1)$ in., $(3x - 1)$ in.
40. The signs of the second terms must be negative.   **41.** $(x + 5)^2$
42. $(3y - 4)^2$ **43.** $-(z - 1)^2$ **44.** $(5a + 2b)^2$
45. $(x + 3)(x - 3)$ **46.** $(7t + 11y)(7t - 11y)$
47. $(xy + 20)(xy - 20)$ **48.** $8a(t + 2)(t - 2)$
49. $(c^2 + 16)(c + 4)(c - 4)$ **50.** Prime
51. $(b + 1)(b^2 - b + 1)$ **52.** $(x - 6)(x^2 + 6x + 36)$
53. $(p + 5q)(p^2 - 5pq + 25q^2)$
54. $2x^2(2x - 3y)(4x^2 + 6xy + 9y^2)$ **55.** $2y^2(3y - 5)(y + 4)$
56. $5(t + u^2)(s^2 + v)$ **57.** $(j^2 + 4)(j + 2)(j - 2)$
58. $-3(j + 2)(j^2 - 2j + 4)$ **59.** $(x + 1)(20 + m)(20 - m)$
60. $3w^2(2w - 3)^2$ **61.** $2(t^3 + 5)$ **62.** Prime   **63.** $z(x + 8y)^2$
64. $6c^2d(3cd - 2c - 4)$ **65.** $0, 6$ **66.** $\frac{7}{4}, -1$ **67.** $0, -2$
68. $-3, 3$ **69.** $-\frac{5}{12}, \frac{5}{12}$ **70.** $3, 4$ **71.** -7 repeated
72. $6, -4$ **73.** $1, \frac{1}{5}$ **74.** $0, -1, 2$ **75.** Width: 9 in.;
length: 11 in.   **76.** 15 m   **77.** $x + 1; x + 2; x + 2$

78. Jackson: 12 nominations; West: 10 nominations   **79.** 5 m
80. 10 sec

CHAPTER 6 TEST (page 505)

1. a. greatest, common, factor   **b.** product   **c.** Pythagorean
d. difference   **e.** binomials   **2. a.** $45 = 3^2 \cdot 5; 30 = 2 \cdot 3 \cdot 5$
b. $15x^3$ **3.** $4(x + 4)$ **4.** $(q + 9)(q - 9)$
5. $5ab(6ab^2 - 4a^2b + 1)$ **6.** Prime   **7.** $(x + 1)(2x + 3)$
8. $(x + 3)(x + 1)$ **9.** $-(x - 11)(x + 2)$ **10.** $x^2(x - 30)(x - 2)$
11. $(a - b)(9 + x)$ **12.** $(2a - 3)(a + 4)$ **13.** $2(3x + 5y)^2$
14. $(x + 2)(x^2 - 2x + 4)$ **15.** $15m^6(4m^2 - 3)$
16. $3(a - 3)(a^2 + 3a + 9)$ **17.** $(4x^2 + 9)(2x + 3)(2x - 3)$
18. $(a + 5)(a^2 + 1)$ **19.** $(a + 6)(a^3 - 4)$
20. $(5d + 4)(2d - 1)$ **21.** $8(m + 10)(m - 10)$ **22.** $(6n - 7)^2$
23. $(4r - 1)(2r - 3)$ **24.** Prime   **25.** $(5x - 4)$ in.
26. $(x - 9)(x + 6); x^2 + 6x - 9x - 54 = x^2 - 3x - 54$
27. $-3, 2$ **28.** $-5, 5$ **29.** $0, \frac{1}{6}$ **30.** -3 repeated   **31.** $\frac{1}{3}, -\frac{1}{2}$
32. $9, -2$ **33.** $0, -1, -6$ **34.** 6 ft by 9 ft   **35.** 5 sec
36. Base: 6 in.; height: 11 in.   **37.** 12, 13   **38.** 10
39. A quadratic equation is an equation that can be written in the form $ax^2 + bx + c = 0; x^2 - 2x + 1 = 0$. (Answers may vary.)
40. At least one of them is 0.

CUMULATIVE REVIEW CHAPTERS 1–6 (page 508)

1. About 35 beats/min difference   **2.** $2 \cdot 5^3$ **3.** $\frac{24}{25}$ **4.** 0.992
5. a. False   **b.** True   **c.** True   **6.** $\frac{5}{0}$ **7.** -39 **8.** -5
9. -27 **10.** $\$20x$ **11.** 3   **12.** $8, -1, 9$ **13.** $-13y^2 + 6$
14. $3y + z$ **15.** $-\frac{3}{2}$ **16.** -2 **17.** 9   **18.** $-\frac{55}{6}$ **19.** 248 lb
20. 330 mi   **21.** $I = Prt$ **22.** 12.6 in.^{2} **23.** $t = \frac{A - P}{Pr}$
24. 22nd president, 24th president   **25.** Los Angeles: 6 wk; Las Vegas: 4 wk; Dallas: 7 wk   **26.** 4 L   **27.** 5 lb apple slices, 5 lb banana chips   **28.** $(-\infty, -2)$ **29.** Yes
30. 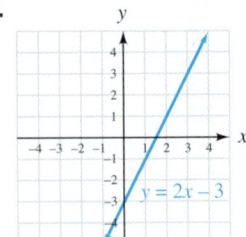   **31.** Vertical line   **32.** They are the same.   **33.** An increase of 536,000 articles per year   **34.** 1; (0, −2)   **35.** $y = 7x + 19$
36. 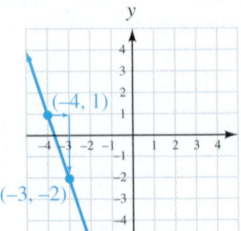   **37.**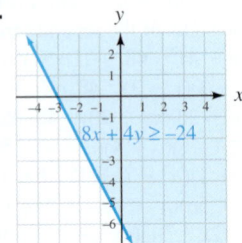
38. 17   **39.** Yes   **40.**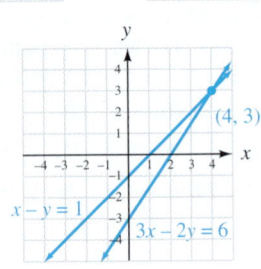

41. $\left(\frac{3}{4}, -2\right)$ **42.** $\left(-2, \frac{2}{3}\right)$ **43.** Newspaper: 8 tons; cardboard: 6 tons   **44.** 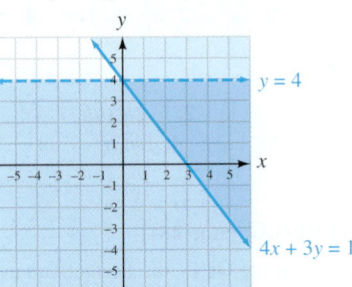   **45.** $-4y^5$
46. x^4y^{23} **47.** b^7 **48.** 2   **49.** 9.011×10^{-5} **50.** 1.7×10^6
51. 3   **52.** 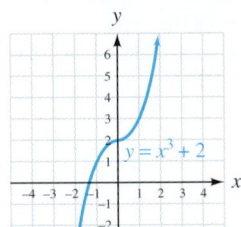   **53.** $-2x^2 - 4x + 5$
54. $8b^5 - 8b^4$
55. $3x^2 + 10x - 8$
56. $y^2 - 12y + 36$
57. $3ab - 2a - 1$
58. $2x + 1$
59. a. $(4x + 8)$ in.
b. $(x^2 + 4x + 3)$ in.^{2}
c. $(x^3 + 4x^2 + 3x)$ in.^{3}
60. $6x^5y$ **61.** $9b^2(b - 3)$ **62.** $(x + y)(a + b)$
63. $(u + 3)(u - 1)$ **64.** $(2x + 1)(5x - 2)$ **65.** $(2a - 3)^2$
66. $(3z + 1)(3z - 1)$ **67.** $(t - 2)(t^2 + 2t + 4)$
68. $3(b^2 - 2)(a + 1)(a - 1)$ **69.** $0, \frac{4}{3}$ **70.** $\frac{1}{2}, 2$ **71.** $0, -1, -2$
72. 9 in. by 12 in.

ARE YOU READY? 7.1 (page 512)

1. a. 0   **b.** Undefined   **2. a.** $\frac{3}{4}$ **b.** $\frac{1}{9}$ **3.** $4(3x - 2)$
4. $(a + 4)(a - 4)$ **5.** $(5x + 2)(x - 5)$ **6.** -1

SELF CHECKS 7.1

1. $\frac{13}{50}, -1$ **2. a.** -9 **b.** $-5, 5$ **c.** None   **d.** None   **3.** $\frac{4a^2}{3}$
4. a. $\frac{t - 5}{3}$ **b.** $\frac{x + 2}{x}$ **c.** $2x^3$ **5. a.** $\frac{2x + 3}{2x(2x + 5)}$ **b.** $(a + 3b)^3$
6. $\frac{4}{3}$ **7.** -1 **8. a.** $-\frac{m + 10}{m}$ **b.** Does not simplify

STUDY SET SECTION 7.1 (page 519)

1. rational   **3.** undefined   **5.** $\frac{6}{24} = \frac{1}{4}$ **7. a.** 1   **b.** -1 **c.** 1
d. Does not simplify   **9.** $x, 1, x + 1, x + 3$ **11.** 4   **13.** 0
15. Undefined   **17.** $-\frac{2}{11}$ **19.** $\frac{1}{6}$ **21.** Undefined   **23.** 2
25. 0   **27.** None   **29.** $\frac{1}{2}$ **31.** None   **33.** $-6, 6$ **35.** $-2, 1$
37. 20   **39.** $\frac{5}{a}$ **41.** $\frac{3x^2}{2}$ **43.** $\frac{2x + 1}{3}$ **45.** $\frac{1}{3}$ **47.** $\frac{x + 2}{x - 4}$
49. $\frac{x + 1}{x}$ **51.** $\frac{m - n}{7(m + n)}$ or $\frac{m - n}{7m + 7n}$ **53.** $\frac{1}{2b + 1}$ **55.** $\frac{10}{3}$
57. $\frac{2x}{x - 6}$ **59.** -1 **61.** $-\frac{1}{2}$ **63.** $-\frac{1}{a + 1}$ **65.** $-\frac{5}{m + 5}$ **67.** $\frac{1}{a}$
69. $-\frac{x + 2}{x + 1}$ **71.** -6 **73.** $\frac{x + 1}{x - 1}$ **75.** $\frac{3x}{y}$ **77.** x^7 **79.** $\frac{4 - x}{4 + x}$ or $-\frac{x - 4}{x + 4}$ **81.** 4   **83.** $\frac{3(x + 3)}{2x + 1}$ or $\frac{3x + 9}{2x + 1}$ **85.** $-\frac{3x + 11}{x + 3}$ **87.** $\frac{7c^2}{3d^2}$
89. Does not simplify   **91.** $\frac{2u - 3}{u^3}$ **93.** $(2x + 3)^2$ **95.** 9
97. $\frac{3x}{5y}$ **99.** Does not simplify   **101.** $\frac{2}{x - 9}$ **103.** $85\frac{1}{3}$
105. 2, 1.6, and 1.2 milligrams per liter
113. a. $(a + b) + c = a + (b + c)$ **b.** $ab = ba$

ARE YOU READY? 7.2 (page 522)

1. $\frac{3}{20}$ **2.** $\frac{9}{8}$ **3.** $\frac{11}{14}$ **4.** $x(1 - x)$

SELF CHECKS 7.2

1. a. $\frac{6(a+7)}{5a^4}$ **b.** $\frac{3a}{11}$ **2. a.** $\frac{(n-3)(3n-2)}{2}$ **b.** $-\frac{2(m+1)}{3m+1}$
3. a. 6 **b.** $4x+12$ **4.** $\frac{3b}{2a}$ **5.** $\frac{z+1}{z}$ **6.** $-a$ **7.** 600 yd^2
8. 36,000 beats per minute

STUDY SET SECTION 7.2 (page 528)

1. reciprocal **3. a.** numerators, denominators, reciprocal
b. AC, BD, D, C **5.** $-\frac{y^2}{y+1}$ **7.** 1 **9.** ft **11.** $\frac{3y}{14}$
13. $\frac{3(y+2)}{y^3}$ or $\frac{3y+6}{y^3}$ **15.** $\frac{20}{3n}$ **17.** x^2y^2 **19.** $\frac{x}{5}$
21. $\frac{3}{2x}$ **23.** $x+1$ **25.** $-(x-2)$ or $-x+2$ **27.** $\frac{(x-2)^2}{x}$
29. $\frac{(m-2)(m-3)}{2(m+2)}$ **31.** 35 **33.** $3x+3$ **35.** $10y-16$
37. $\frac{36a-60}{a}$ **39.** $\frac{3}{2y}$ **41.** $\frac{3a}{5}$ **43.** $\frac{x^2}{3}$ **45.** $\frac{9p^3}{5}$ **47.** $\frac{5(a-2)}{4a^3}$
49. $-\frac{x+2}{3}$ **51.** $\frac{3x}{x+7}$ **53.** $-(m+5)$ or $-m-5$ **55.** $t+7$
57. $\frac{2(x-7)}{x+9}$ **59.** $\frac{1}{3}$ **61.** 1 **63.** $\frac{1}{12(2r-3s)}$ **65.** $\frac{4(n-1)}{3n}$
67. 450 ft **69.** $\frac{3}{4}$ gal **71.** $\frac{1}{2}$ mi per min **73.** 1,800 m per min
75. $\frac{b-3}{b}$ **77.** $\frac{1}{(x+1)^2}$ **79.** $\frac{3(a+3)^2}{2a^3}$ **81.** $\frac{d(6c-7d)}{6}$
83. $25h-15$ **85.** $n-1$ **87.** $\frac{5r^3}{2s^2}$
89. $\frac{7(p+2)}{3p^4}$ or $\frac{7p+14}{3p^4}$ **91.** $\frac{x-2}{x-3}$ **93.** $\frac{2x-3y}{y(2x+3y)}$ **95.** -2
97. a. $(x+2)^2$ **b.** $\frac{9}{16}$ **99. a.** $\frac{(x-3)^2}{x-2}$ **b.** $\frac{x-2}{4}$ **101.** $\frac{x^2}{10}$ ft^2
103. 4,380,000 **105.** 8 yd^2
107. $\frac{1}{2}$ mi per min **109.** $\frac{1}{4}$ mi^2 **115.** $w=6$ in., $l=10$ in.

ARE YOU READY? 7.3 (page 531)

1. $\frac{8}{11}$ **2.** $\frac{1}{3}$ **3.** x^2+5x+1 **4.** $\frac{1}{x-7}$ **5.** $2\cdot2\cdot2\cdot3\cdot x\cdot x$
6. $5(x+1)$

SELF CHECKS 7.3

1. a. $\frac{2x}{5}$ **b.** $\frac{3m-6}{23n}$ or $\frac{3(m-2)}{23n}$ **2. a.** $\frac{2}{3}$ **b.** $\frac{2c^2-11c}{(c-1)(c+2)}$ or
$\frac{c(2c-11)}{(c-1)(c+2)}$ **3.** $\frac{1}{n+4}$ **4. a.** $x-1$ **b.** $\frac{6y^2-y}{(y+3)(y-3)}$
5. a. $150y^3$ **b.** $a(a+3)$ **6. a.** $2x(x-6)$
b. $(m+3)(m-3)^2$ **7. a.** $\frac{21m}{60m^3}$ **b.** $\frac{2c^2+6c}{(c+1)(c+3)}$
8. $\frac{x^2+5x-24}{x(x-4)(x+8)}$

STUDY SET SECTION 7.3 (page 538)

1. denominator **3.** build **5.** numerators, denominator,
$A+B, D, A-B, D$ **7.** $\frac{4}{5}$ **9. a.** Twice **b.** Once
11. 3, $4a+1$, 2, $4a+1$, 2 **13.** $\frac{11}{x}$ **15.** $\frac{x+5}{18}$ **17.** $\frac{1}{3a^2}$
19. $\frac{x+4}{y}$ **21.** $\frac{1}{r-5}$ **23.** 9 **25.** $\frac{x}{25}$ **27.** $\frac{m-6}{6m^2}$ **29.** $\frac{1}{t+2}$
31. $\frac{2}{w(w-9)}$ **33.** $\frac{1}{2}$ **35.** $2x-5$ **37.** $\frac{1}{y}$ **39.** 0 **41.** $6x$
43. $30a^3$ **45.** $3a^2b^3$ **47.** $c(c+2)$ **49.** $12(x-1)$
51. $12(b+2)$ **53.** $8k(k+2)$ **55.** $(x+1)(x-1)$
57. $(x+1)(x+5)(x-5)$ **59.** $(2n+5)(n+4)^2$ **61.** $\frac{50}{10r}$
63. $\frac{8xy}{x^2y}$ **65.** $\frac{27b}{12b^2}$ **67.** $\frac{3x^2+3x}{(x+1)^2}$ **69.** $\frac{x^2+9x}{x^2(x+5)}$
71. $\frac{t^2+14t+45}{4(t+2)(t+9)}$ **73.** $\frac{4y^2+12y}{4y(y-2)(y-3)}$ **75.** $\frac{36-3h}{3(h+9)(h-9)}$
77. $\frac{3}{t-7}$ **79.** $\frac{1}{c+d}$ **81.** $\frac{3}{4}$ **83.** $\frac{5a}{a+2}$ **85.** $\frac{2x}{3-x^2}$

87. $\frac{7n+1}{(n+4)(n-2)}$ **89.** $\frac{3}{r}$ **91.** $\frac{5}{9y}$ **93.** $-\frac{1}{3x-1}$ **95.** $3x-2$
97. a. $\frac{t}{2}$ **b.** $\frac{5t^2}{144}$ **c.** $\frac{1}{5}$ **99. a.** $\frac{4}{5}$ **b.** $\frac{m^2+8m+12}{25}$ **c.** $\frac{m+6}{m+2}$
101. $\frac{2x+6}{x+2}$ ft **109. a.** $I=Prt$ **b.** $A=\frac{1}{2}bh$ **c.** $P=2l+2w$

ARE YOU READY? 7.4 (page 540)

1. $\frac{20a}{16a^2}$ **2.** $2\cdot2\cdot3\cdot3\cdot x\cdot x\cdot x$ **3.** $\frac{11x}{9}$ **4.** $t-8$

SELF CHECKS 7.4

1. $\frac{19y}{14}$ **2.** $\frac{20-9z}{84z^2}$ **3.** $\frac{2}{5(x-5)}$ **4.** $\frac{6x+63}{(x+9)(x+8)}$
5. $\frac{b+3}{(b+2)(b+5)}$ **6.** $\frac{14y+ny}{n+4}$ **7.** $\frac{n-12}{n-8}$

STUDY SET SECTION 7.4 (page 546)

1. unlike **3. a.** $2\cdot2\cdot5\cdot x\cdot x$ **b.** $(x-2)(x+6)$
5. $(x+6)(x+3)$ **7.** $\frac{5}{5}$ **9.** $3x, 5, 15x, 15x, 35$ **11.** $\frac{13x}{21}$
13. $\frac{67a}{40}$ **15.** $\frac{7-2m}{m^2}$ **17.** $\frac{6-5p}{10p^2}$ **19.** $\frac{4t^2-33}{24t^3}$
21. $\frac{3-16c^2}{18c^4}$ **23.** $\frac{a+8}{2(a+2)(a-2)}$ **25.** $\frac{6a+9}{(3a+2)(3a-2)}$
27. $\frac{4a+1}{(a+2)^2}$ **29.** $\frac{6-3m}{5m(m-1)}$ **31.** $\frac{17t+42}{(t+3)(t+2)}$
33. $\frac{2x^2+11x}{(2x-1)(2x+3)}$ **35.** $\frac{14s+58}{(s+3)(s+7)}$ **37.** $\frac{2m^2+20m-6}{(m-2)(m+5)}$
39. $\frac{s^2+8s+4}{(s+4)(s+1)(s+1)}$ **41.** $\frac{2x+13}{(x-8)(x-1)(x+2)}$ **43.** $\frac{1}{a+1}$
45. $\frac{1}{(y+3)(y+4)}$ **47.** $\frac{5y+xy}{x-4}$ **49.** $\frac{8+xz}{x}$ **51.** $\frac{2}{a-4}$ **53.** $\frac{c+d}{7c-d}$
55. $\frac{1}{(x-1)(x-2)}$ **57.** $\frac{6d-3}{d-9}$ **59.** $\frac{xy-y+10}{x-1}$
61. $\frac{2b+1}{(b+1)(b+2)}$ **63.** $\frac{1}{g+2}$ **65.** $\frac{5y}{2}$ **67.** $\frac{35x^2+x+5}{5x(x+5)}$
69. $\frac{41}{30x}$ **71.** $\frac{2x^2-1}{x(x+1)}$ **73.** $\frac{y+4}{y-1}$ **75.** $\frac{2n+2}{15}$ **77.** $\frac{y^2+7y+6}{15y^2}$
79. $\frac{x+2}{x-2}$ **81.** $\frac{a^2b-3}{a^2}$ **83.** $\frac{10a-14}{3a(a-2)}$ **85.** $\frac{17x+3}{x^2}$
87. $\frac{-4x-3}{x+1}$ or $-\frac{4x+3}{x+1}$ **89.** $\frac{11b}{12}$ **91. a.** $\frac{75+8x^2}{30x}$ **b.** $\frac{2}{3}$
93. a. $\frac{t^2+4t}{(t-5)(t+5)}$ **b.** $t+5$ **95.** $\frac{20x+9}{6x^2}$ cm
101. 8; $(0, 2)$ **103.** 0

ARE YOU READY? 7.5 (page 549)

1. Division **2.** $\frac{3}{2}$ **3.** $2x$ **4.** $8a-10$

SELF CHECKS 7.5

1. $\frac{5y}{6}$ **2.** $\frac{10x+30}{6x^2-15x}$ **3.** $\frac{b}{a}$ **4.** $\frac{15x-60}{12x^2+20x}$ **5.** $-\frac{2}{n}$ **6.** $\frac{2x+4}{2x+5}$

STUDY SET SECTION 7.5 (page 555)

1. complex, complex **3.** single, division, reciprocal
5. a. $\frac{x-3}{4}$, yes **b.** $\frac{1}{12}-\frac{x}{6}$; no **7.** $\div$ **9.** $\frac{8}{9}$ **11.** $\frac{5x}{12}$ **13.** $\frac{x^2}{y}$
15. $\frac{n^3}{8}$ **17.** $\frac{10}{3a^3}$ **19.** $-\frac{x^2}{14}$ **21.** $\frac{5}{7}$ **23.** $\frac{3y+12}{4y^2-6y}$ **25.** $\frac{2-5y}{6}$
27. $\frac{24-c^2}{12}$ **29.** $\frac{5}{4}$ **31.** $\frac{1-3x}{5+2x}$ **33.** $\frac{1+x}{2+x}$ **35.** $\frac{3-x}{x-1}$
37. $\frac{x-12}{x+6}$ **39.** $a-7$ **41.** $\frac{5d^2+16d}{10d+10}$ **43.** $\frac{y}{x-2y}$ **45.** $\frac{1}{x+2}$
47. $\frac{1}{x+3}$ **49.** $\frac{q+p}{q}$ **51.** $18x$ **53.** $\frac{2c}{2-c}$ **55.** $\frac{r-1}{r+1}$ **57.** $\frac{b+9}{8a}$
59. $\frac{5x}{3}$ **61.** $\frac{32h-1}{96h+6}$ **63.** $\frac{m^2+n^2}{m^2-n^2}$ **65.** $\frac{8}{4c+5c^2}$ **67.** $\frac{t^3}{2}$
69. $\frac{s^2-s}{2+2s}$ **71.** $\frac{t+2}{t-3}$ **73.** $\frac{xy}{y+x}$ **75.** $-\frac{10x^3}{3}$ **77.** $\frac{x}{x-2}$

79. $\dfrac{d-3}{2d}$ **81.** $\dfrac{b-5ab}{3ab-7a}$ **83.** $2m-1$ **85.** $\dfrac{7}{6}$ **87.** $\dfrac{R_1R_2}{R_2+R_1}$

93. 1 **95.** $\dfrac{25}{16x^{12}}$

ARE YOU READY? 7.6 (page 557)

1. $18x$ **2.** $5x$ **3.** $-7, 8$ **4.** -8

SELF CHECKS 7.6

1. 1 **2.** 22 **3.** $3, -8$ **4.** No solution, 6 is extraneous **5.** $\dfrac{5}{2}$
6. $S = \dfrac{RT-10R}{e}$ **7.** $p = \dfrac{fq}{q-f}$

STUDY SET SECTION 7.6 (page 564)

1. rational **3.** clear **5. a.** Yes **b.** No **7. a.** $3, 0$
b. $3, 0$ **c.** $3, 0$ **9. a.** y **b.** $(x+2)(x-2)$ **11. a.** 3
b. $3x+18$ **13.** $2a, 2a, 2a, 2a, 2a, 4, 7, 4, 3$ **15.** 1 **17.** 0
19. $\dfrac{3}{5}$ **21.** 9 **23.** $-\dfrac{4}{3}$ **25.** $-\dfrac{12}{7}$ **27.** -48 **29.** $\dfrac{3}{10}$ **31.** $2, 4$
33. $-5, 2$ **35.** $-4, -5$ **37.** $3, -\dfrac{5}{3}$ **39.** No solution; 5 is
extraneous **41.** No solution; -2 is extraneous **43.** $\dfrac{5}{2}$ **45.** $0, 3$
47. $A = \dfrac{h(b+d)}{2}$ **49.** $r = \dfrac{E-IR}{I}$ **51.** $x = \dfrac{5yz}{5y+4z}$
53. $r = \dfrac{st}{s-t}$ **55.** $P = nrt$ **57.** $d = \dfrac{bc}{a}$ **59.** $a = \dfrac{b}{b-1}$
61. $L^2 = 6dF - 3d^2$ **63.** 6 **65.** 1 **67.** $-1, 6$ **69.** -40
71. No solution; -1 is extraneous **73.** 7 **75.** 1 **77.** $-4, 3$
79. 1 **81.** 3 **83.** No solution; 2 is extraneous **85.** $\dfrac{9}{40}$ **87.** 0
89. -3 **91.** $1, 2$ **93.** 7 **95. a.** $\dfrac{2a+3}{5}$ **b.** $-\dfrac{9}{4}$
97. a. $\dfrac{x^2-4x+2}{(x-2)(x-3)}$ **b.** 4 **99.** $R = \dfrac{HB}{B-H}$ **101.** $r = \dfrac{r_1r_2}{r_2+r_1}$
107. 20

ARE YOU READY? 7.7 (page 566)

1. $t = \dfrac{d}{r}$ **2.** $\dfrac{x}{5}$ **3.** $I = Prt$ **4.** 63

SELF CHECKS 7.7

1. 9 **2.** 4 mph **3.** $\dfrac{3}{4}$ of the job **4.** $3\dfrac{3}{7}$ hr **5.** 1%

STUDY SET SECTION 7.7 (page 573)

1. motion, investment, work **3.** iii **5. a.** $\dfrac{1}{45}$ of the job per minute
b. $\dfrac{x}{4}$ **7. a.** $t = \dfrac{d}{r}$ **b.** $P = \dfrac{I}{rt}$ **9.** $\dfrac{x}{15}, \dfrac{x}{8}$ **11.** $6\dfrac{1}{9}$ days **13.** 4
15. 2 **17.** 5 **19.** $\dfrac{2}{3}$ or $\dfrac{3}{2}$ **21.** 8 **23.** Garin: 16 mph; Armstrong:
26 mph **25.** 1st: $1\dfrac{1}{2}$ ft per sec; 2nd: $\dfrac{1}{2}$ ft per sec **27.** Canada
goose: 30 mph; great blue heron: 20 mph **29.** $\dfrac{300}{255+x}, \dfrac{210}{255-x}$;
45 mph **31.** $2\dfrac{6}{11}$ days **33.** No, after the pipes are opened, the
swimming is scheduled to take place in 6 hours. It takes 7.2 hr
(7 hr 12 min) to fill the pool. **35.** 8 hr **37.** 20 min
39. $1\dfrac{4}{5}$ hr $= 1.8$ hr **41.** Credit union: 4%; bonds: 6%
43. 7% and 8% **47.** $(1, 3)$ **49.** Yes

ARE YOU READY? 7.8 (page 576)

1. $\dfrac{7}{9}$ **2. a.** ab **b.** bc **3.** 8.5 **4.** $-2, 9$

SELF CHECKS 7.8

1. a. $\dfrac{15}{2}$ **b.** $\dfrac{1}{4}$ **2.** No **3.** 24 **4.** $-5, 6$ **5.** $187.50 **6.** 30 ft
7. 3 lb for $6.89 **8.** 66 ft

STUDY SET SECTION 7.8 (page 583)

1. ratio, rate **3.** extremes, means **5.** unit **7.** equal, ad, bc
9. $2, 1,000, x$ **11.** $2.19, 1$ **13.** $x, 288, 18, 18, 16$ **15.** as, to
17. $\dfrac{4}{15}$ **19.** $\dfrac{3}{4}$ **21.** $\dfrac{5}{4}$ **23.** $\dfrac{1}{2}$ **25.** $\dfrac{1}{2}$ **27.** $\dfrac{25}{22}$ **29.** Yes
31. No **33.** 4 **35.** 14 **37.** 0 **39.** $-\dfrac{3}{2}$ **41.** -2 **43.** -27
45. $2, -2$ **47.** $6, -1$ **49.** $-1, 16$ **51.** $-\dfrac{5}{2}, -1$ **53.** 15
55. 8 **57.** $-\dfrac{1}{3}, 2$ **59.** 2 **61.** $-\dfrac{27}{2}$ **63.** $-10, 10$ **65.** $-4, 3$
67. $\dfrac{26}{9}$ **69. a.** $-\dfrac{15}{8}$ **b.** $\dfrac{65}{8}$ **71. a.** $\dfrac{8}{5}$ **b.** $\dfrac{2}{3}, 4$ **73.** $62.50
75. 14 breaths **77. a.** 462 **b.** $\dfrac{11}{12}$; 11:12 **79.** $309
81. 45,000 tweets **83.** 568, 13, 14 **85.** Not exactly, but close
87. 140 **89.** 522 in.; 43.5 ft **91.** 10 ft **93.** 45 min for $25
95. 150 for $12.99 **97.** 6-pack for $1.50 **99.** Four 4-oz cartons
101. 39 ft **103.** $46\dfrac{7}{8}$ ft **105.** 8 **111.** 90% **113.** 480

ARE YOU READY? 7.9 (page 587)

1. 450 **2.** 5 **3.** $\dfrac{13}{2}$ **4.** 200

SELF CHECKS 7.9

1. 40 **2.** 432 mi **3.** 5 **4.** 13.5 in.³

STUDY SET SECTION 7.9 (page 592)

1. direct, inverse **3.** Direct **5.** Inverse **7.** Direct **9.** Inverse
11. Direct **13.** Inverse
15. 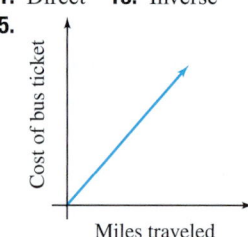 **17.**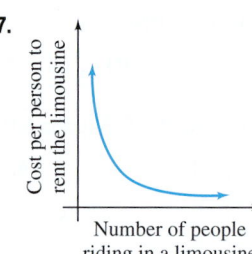
19. $p = kd$ **21.** $d = \dfrac{k}{r}$ **23. a.** Yes **b.** No **c.** No **d.** Yes
25. 35 **27.** 36 **29.** 9 **31.** 14 **33.** 4 **35.** 2 **37.** 16
39. 2.25 or $\dfrac{9}{4}$ **41.** 168 mi **43.** 173.6 mg **45.** $2\dfrac{1}{2}$ hr
47. 8 amps **49.** 180 **51.** 55 lb **53.** 40 cm **55.** 32,600 ft²
57. $400; $1,200 **63.** $1, -2, -3$ **65.** $0, -3, 9$

CHAPTER 7 REVIEW (page 595)

1. $4, -4$ **2.** $-\dfrac{3}{7}$ **3.** $\dfrac{1}{2x}$ **4.** $\dfrac{5}{2x}$ **5.** $\dfrac{x}{x+1}$ **6.** $a-2$ **7.** -1
8. $-\dfrac{1}{x+3}$ **9.** $\dfrac{x}{x-1}$ **10.** Does not simplify **11.** $\dfrac{1}{x-y}$ **12.** $\dfrac{4}{3}$
13. x is not a common factor of the numerator and the denominator; x
is a term of the numerator. **14.** 150 mg **15.** $\dfrac{3x}{y}$ **16.** 96
17. $\dfrac{x-1}{x+2}$ **18.** $\dfrac{2x}{x+1}$ **19.** $\dfrac{3y}{2}$ **20.** $-x-2$ **21. a.** Yes
b. No **c.** Yes **d.** Yes **22.** $\dfrac{1}{3}$ mi per min **23.** $\dfrac{1}{3d}$ **24.** 1
25. $\dfrac{2x+2}{x-7}$ **26.** $\dfrac{1}{a-4}$ **27.** $9x$ **28.** $8x^3$ **29.** $m(m-9)$
30. $(5x+1)(5x-1)$ **31.** $(a+5)(a-5)$ **32.** $(2t+7)(t+5)^2$
33. $\dfrac{63}{7a}$ **34.** $\dfrac{2xy+x}{x(x-9)}$ **35.** $\dfrac{2b+14}{6(b-5)}$ **36.** $\dfrac{9r^2-36r}{(r+1)(r-4)(r+5)}$
37. $\dfrac{a-7}{7a}$ **38.** $\dfrac{x^2+x-1}{x(x-1)}$ **39.** $\dfrac{1}{t+1}$ **40.** $\dfrac{x^2+4x-4}{2x^2}$
41. $\dfrac{b+6}{b-1}$ **42.** $\dfrac{6c+8}{c}$ **43.** $\dfrac{14n+58}{(n+3)(n+7)}$ **44.** $\dfrac{4t+1}{(t+2)^2}$
45. $\dfrac{1}{(a+3)(a+2)}$ **46.** $\dfrac{17y-2}{12(y-2)(y+2)}$ **47.** Yes
48. $\dfrac{14x+28}{(x+6)(x-1)}$ units, $\dfrac{12}{(x+6)(x-1)}$ square units **49.** $\dfrac{n^3}{14}$

50. $\frac{r+9}{8s}$ **51.** $\frac{1+y}{1-y}$ **52.** $\frac{21}{3a+10a^2}$ **53.** x^2+3 **54.** $\frac{y-5xy}{3xy-7x}$

55. 3 **56.** No solution; 5 is extraneous **57.** 3 **58.** 2, 4

59. 0 **60.** $-4, 3$ **61.** $T_1 = \frac{T_2}{1-E}$ **62.** $y = \frac{xz}{z-x}$ **63.** 3

64. 5 mph **65.** $\frac{1}{4}$ of the job per hr **66.** $5\frac{5}{6}$ days **67.** 5%

68. 40 mph **69.** No **70.** Yes **71.** $\frac{9}{2}$ **72.** 0 **73.** 7

74. $4, -\frac{3}{2}$ **75.** 255 **76.** 20 ft **77.** 5 ft 6 in. **78.** 250 for $98

79. $c = kt$ **80.** $f = \frac{k}{L}$ **81.** $2,000 **82.** 600 **83.** 1.25 amps

84. Inverse variation

CHAPTER 7 TEST (page 603)

1. a. rational **b.** similar **c.** proportion **d.** build **e.** factors
2. 10 words **3.** 0 **4.** $-3, 2$ **5.** 3,360,000 or 3,360K bits per
minute **6.** 5 is not a common factor of the numerator, and therefore
cannot be removed. 5 is a term of the numerator. **7.** $\frac{8x}{9y}$ **8.** -7

9. $\frac{x+1}{2x+3}$ **10.** 1 **11.** $3c^2d^3$ **12.** $(n+1)(n+5)(n-5)$

13. $\frac{5y^2}{4}$ **14.** $\frac{x+1}{3(x-2)}$ **15.** $-\frac{x^2}{3}$ **16.** $\frac{1}{6}$ **17.** 3 **18.** $\frac{4n-5mn}{10m}$

19. $\frac{2x^2+x+1}{x(x+1)}$ **20.** $\frac{2a+7}{a-1}$ **21.** $\frac{c^2-4c+9}{c-4}$ **22.** $\frac{1}{(t+3)(t+2)}$

23. $\frac{12}{5m}$ **24.** $\frac{a+2s}{2as^2-3a^2}$ **25.** 11 **26.** No solution; 6 is extraneous

27. 1 **28.** 1, 2 **29.** $\frac{2}{3}$ **30.** -10; 3 is extraneous

31. $B = \frac{HR}{R-H}$ **32.** $s = \frac{rt}{r-t}$ **33.** Yes **34.** 1,785

35. 171 ft **36.** 80 sheets for $3.89 **37.** $3\frac{15}{16}$ hr **38.** 4 mph

39. 2 **40.** We multiply both sides of the equation by the LCD of
the rational expressions appearing in the equation. The resulting

equation is easier to solve. **41.** To simplify $\frac{1}{x} + \frac{1}{4}$, we build each

fraction to have the LCD of $4x$. To solve $\frac{1}{x} + \frac{1}{4} = \frac{1}{2}$, we multiply

both sides by the LCD $4x$ to eliminate the denominators and clear
the equation of fractions. **42.** 100 lb **43.** $\frac{80}{3}$ or $26\frac{2}{3}$
44. a. Inverse **b.** Direct

CUMULATIVE REVIEW CHAPTERS 1–7 (page 606)

1. a. False **b.** False **c.** True **d.** True **2.** $<$ **3.** 36
4. 77 **5.** $6c + 62$ **6.** -5 **7.** 3 **8.** About 26%
9. $B = \frac{A-c-r}{2}$ **10.** 104°F **11.** 240 ft³ **12.** 12 lb of the
$6.40 tea and 8 lb of the $4 tea **13.** 500 mph
14. $[-1, \infty)$ **15.**

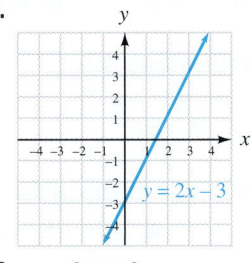

$y = 2x - 3$

16. -1 **17.** 0.008 mm/m **18.** $\frac{8}{7}$ **19.** $y = 3x + 2$
20.

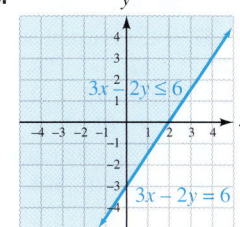

$3x + 2y \le 6$

$3x - 2y = 6$

21. 0 **22.** domain, range

23.

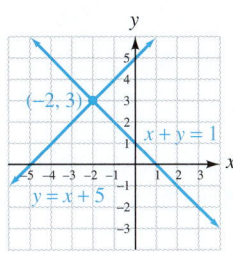

$(-2, 3)$

$x + y = 1$

$y = x + 5$

24. $(5, 2)$ **25.** $(1, -1)$ **26.** Red:
8, blue: 15 **27.** x^7 **28.** x^{25}
29. $\frac{y^3}{8}$ **30.** $-\frac{32a^5}{b^5}$ **31.** $\frac{a^8}{b^{12}}$
32. $3b^2$ **33.** 2.9×10^5 **34.** 3

35.

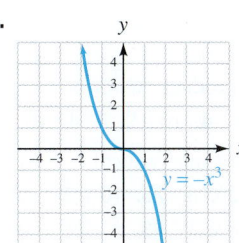

$y = -x^3$

36. $A = \pi R^2 - \pi r^2$
37. $6x^2 + x - 5$
38. $-\frac{1}{2}t^3 - \frac{7}{4}t^2 - \frac{1}{12}$ **39.** $6x^4y^5$
40. $6y^2 - y - 35$
41. $-12x^4z + 4x^2z^2$
42. $9a^2 - 24a + 16$
43. $2x + 3$ **44.** $x^2 + 2x - 1$
45. $k^2t(k - 3)$
46. $(b + c)(2a + 3)$ **47.** $(u - 9)^2$

48. $-(r - 2)(r + 1)$ **49.** Prime **50.** $(2x - 9)(3x + 7)$
51. $2(a + 10b)(a - 10b)$ **52.** $(b + 5)(b^2 - 5b + 25)$
53. $0, -\frac{1}{5}$ **54.** $\frac{1}{3}, \frac{1}{2}$ **55.** 16 in., 10 in. **56.** 5, -5 **57.** $\frac{2x}{x-2}$
58. $-\frac{x^3}{3}$ **59.** $2m - 5$ **60.** $-\frac{1}{x-3}$ **61.** $\frac{m-3}{(m+3)(m+1)}$
62. $\frac{x^2}{(x+1)^2}$ **63.** $\frac{17}{25}$ **64.** 2 **65.** $14\frac{2}{5}$ hr **66.** 34.8 ft **67.** 28 in.
68. $\frac{3}{4} = 0.75$

ARE YOU READY? 8.1 (page 610)

1. 36 **2.** $\frac{16}{25}$ **3.** a^8 **4.** 0.49 **5.** 90 **6.** 4

SELF CHECKS 8.1

1. a. 7 **b.** 11 **c.** -0.3 **d.** $\frac{1}{5}$ **2.** 1.9 sec **3. a.** Not a real
number **b.** Irrational **c.** Rational **4. a.** m^2 **b.** b **c.** t^4
d. $9c$ **5.** 26 in. **6.** $\sqrt{55}$ cm ≈ 7.42 cm **7.** About 18.4 cm
8. $\sqrt{29} \approx 5.39$

STUDY SET SECTION 8.1 (page 618)

1. square **3.** radicand **5.** Pythagorean, a^2, b^2, c^2 **7.** two,
$5, -5, 5, -5$ **9.** 0 **11. a.** 64 **b.** $36y^4$ **c.** d **d.** $\frac{16}{25}$
13. 25, 36 **15.** 5, 144, 169, c, 13, 13 **17.** 8 **19.** 6 **21.** 10
23. 20 **25.** -9 **27.** 1.1 **29.** 13 **31.** $\frac{3}{5}$ **33.** $-\frac{1}{8}$
35. -0.2 **37.** -17 **39.** 50 **41.** 1.73 **43.** 9.75 **45.** 20.69
47. 3.46 **49.** Rational **51.** Not a real number **53.** Irrational
55. Rational **57.** m **59.** t^2 **61.** c^5 **63.** n^6 **65.** x^{18}
67. $2y$ **69.** $8b^2$ **71.** $-7s^7$ **73.** 5 **75.** 8 **77.** 28
79. 117 **81.** $\sqrt{11}$ cm ≈ 3.32 cm **83.** $\sqrt{208}$ m ≈ 14.42 m
85. $\sqrt{90}$ in. ≈ 9.49 in. **87.** $\sqrt{20}$ in. ≈ 4.47 in.
89. $\sqrt{2}$ in. ≈ 1.41 in. **91.** 5 **93.** 13 **95.** $\sqrt{37} \approx 6.08$
97. $\sqrt{29} \approx 5.39$ **99.** 54 mph; 19 mph **101.** 10.72 knots
103. a. $6\sqrt{2}$ in. ≈ 8.49 in. **b.** $5\sqrt{3}$ in. ≈ 8.66 in.
105. $\sqrt{16,200}$ ft ≈ 127.28 ft **107. a.** 16 in., 30 in.
b. 34 in. **109. a.** $\sqrt{18}$ yd ≈ 4.24 yd
b. Yes: 6 yd + 4.24 yd > 10 yd
111. Brace 1: $\sqrt{18}$ m ≈ 4.24 m; Brace 2: $\sqrt{45}$ m ≈ 6.71 m;
Brace 3: $\sqrt{5}$ m ≈ 2.24 m **117.** $6s^2 + s - 5$
119. $3x^2 + 10x - 8$

ARE YOU READY? 8.2 (page 622)

1. 4 **2.** 49 **3.** $2^2 \cdot 3 \cdot 5^2$ **4.** x^6 **5.** $8x^3$ **6.** x^7

SELF CHECKS 8.2

1. $3\sqrt{5}$ **2.** $6\sqrt{2}$ **3. a.** $2\sqrt{35}$ **b.** Cannot be simplified
4. a. $x^2\sqrt{x}$ **b.** $b^5\sqrt{b}$ **5.** $36\sqrt{2y}$ **6. a.** $7p^4\sqrt{p}$ **b.** $y\sqrt{3y}$
7. a. $\frac{6}{11}$ **b.** $\frac{\sqrt{29}}{8}$ **c.** $\frac{6\sqrt{2}}{13}$ **8.** $\frac{3b\sqrt{11}}{4}$

STUDY SET SECTION 8.2 (page 628)

1. simplify **3.** product, square **5. a.** product, $\sqrt{a}\sqrt{b}$
b. quotient, $\frac{\sqrt{a}}{\sqrt{b}}$ **7. a.** 4 **b.** 49 **c.** 9 **d.** 16 **9.** $\sqrt{4 \cdot 10}$
11. a. $81, 2, 9, 2$ **b.** m^4, m, m^2, m **13.** times, radical
15. $2\sqrt{5}$ **17.** $3\sqrt{3}$ **19.** $5\sqrt{2}$ **21.** $2\sqrt{6}$ **23.** $10\sqrt{5}$
25. $3\sqrt{7}$ **27.** $7\sqrt{2}$ **29.** $6\sqrt{5}$ **31.** $8\sqrt{3}$ **33.** $5\sqrt{15}$
35. Cannot be simplified **37.** Cannot be simplified **39.** $x^5\sqrt{x}$
41. $n^4\sqrt{n}$ **43.** $8\sqrt{3x}$ **45.** $15\sqrt{6q}$ **47.** $5t\sqrt{t}$ **49.** $4x^2\sqrt{2x}$
51. $x^3\sqrt{3x}$ **53.** $t^5\sqrt{2t}$ **55.** $\frac{5}{3}$ **57.** $\frac{9}{8}$ **59.** $\frac{\sqrt{6}}{11}$
61. $\frac{5\sqrt{3}}{4}$ **63.** $\frac{a\sqrt{a}}{2}$ **65.** $\frac{r^4\sqrt{r}}{15}$ **67.** $6x\sqrt{2}$ **69.** $\frac{5n^2\sqrt{5}}{8}$
71. $5\sqrt{3t}$ **73.** $\frac{4\sqrt{3}}{9}$ **75.** $4\sqrt{3}$ **77.** $2\sqrt{k}$ **79.** $d^5\sqrt{2d}$
81. Cannot be simplified **83.** $2\sqrt{11}$ **85.** $\frac{\sqrt{23}}{8}$ **87.** $\sqrt{2}$
89. $\frac{5\sqrt{3}}{4q}$ **91.** $\frac{2\sqrt{5}}{7}$ **93.** $9\sqrt{2}$ **95.** $6y^2\sqrt{y}$ **97.** $5x^2\sqrt{2}$
99. $t^7\sqrt{t}$ **101. a.** x **b.** $x\sqrt{x}$ **c.** x^2 **d.** $x^2\sqrt{x}$ **e.** x^3
f. $x^3\sqrt{x}$ **g.** x^4 **h.** $x^4\sqrt{x}$ **103. a.** $2\sqrt{2}$ **b.** $3\sqrt{2}$ **c.** $5\sqrt{2}$
d. $6\sqrt{2}$ **105.** $\sqrt{18}$ ft $= 3\sqrt{2}$ ft; 4.24 ft
107. $\sqrt{4} = 2, \sqrt{8} = 2\sqrt{2}, \sqrt{9} = 3, \sqrt{12} = 2\sqrt{3}, \sqrt{16} = 4$
113. $(3, 0)$ **115.** $(-1, -1)$

ARE YOU READY? 8.3 (page 631)

1. $7x, 9x$ **2.** $16y$ **3.** $6n$ **4.** $9a - 1$ **5.** $2\sqrt{3}$ **6.** $5h\sqrt{2}$

SELF CHECKS 8.3

1. a. $50\sqrt{5}$ **b.** $2\sqrt{b}$ **c.** $-7\sqrt{21}$ **d.** Cannot be simplified
e. Cannot be simplified **2. a.** $7\sqrt{3}$ **b.** $-35\sqrt{2}$ **3. a.** $5y\sqrt{6}$
b. $5\sqrt{3a} - \sqrt{2a}$ **4. a.** $12\sqrt{6} + 33$ **b.** $16c - 100$

STUDY SET SECTION 8.3 (page 634)

1. like **3. a.** Yes **b.** No **5. a.** $5, 3, 8$ **b.** $2, 9, -7$
7. $4, 4, 2, 6, 3$ **9.** $9\sqrt{7}$ **11.** $10\sqrt{21}$ **13.** $9\sqrt{n}$
15. Cannot be simplified **17.** $-3\sqrt{x}$ **19.** $10\sqrt{11}$ **21.** $4\sqrt{2}$
23. Cannot be simplified **25.** $5\sqrt{3}$ **27.** $\sqrt{2}$ **29.** $-2\sqrt{3}$
31. $-18\sqrt{2}$ **33.** $9\sqrt{5}$ **35.** $30\sqrt{2} - \sqrt{3}$ **37.** $-\sqrt{2a}$
39. $3\sqrt{2y} - 3\sqrt{3y}$ **41.** $2x\sqrt{2}$ **43.** $3d\sqrt{2d}$ **45.** $5 + 4\sqrt{2}$
47. $y - 2\sqrt{y} - 15$ **49.** $2t + 3\sqrt{t} - 5$ **51.** $3a - 1$
53. $5\sqrt{6} - 5\sqrt{2}$ **55.** $7\sqrt{6} + 4\sqrt{15}$ **57.** Cannot be
simplified **59.** $-1 - \sqrt{r}$ **61.** $-3\sqrt{21}$ **63.** 0 **65.** $5\sqrt{7}$
67. $15b + 17\sqrt{b} - 4$ **69.** $5 + 6\sqrt{3}$ **71.** Cannot be simplified
73. $14\sqrt{5}$ **75.** $19b\sqrt{6}$ **77.** $2x^2\sqrt{2x}$ **79.** $-7\sqrt{5}$
81. $7\sqrt{3} - 6\sqrt{2}$ **83.** $12\sqrt{10y} - 12\sqrt{10z}$ **85. a.** $36\sqrt{n}$
b. 0 **87. a.** $4\sqrt{5}$ **b.** $2\sqrt{21}$ **89.** $\left(16 + 24\sqrt{5}\right)$ ft ≈ 70 ft
91. $10\sqrt{3}$ ft ≈ 17.3 ft **99.** $\frac{1}{9}$ **101.** -9

ARE YOU READY? 8.4 (page 636)

1. $10x^2$ **2.** $64a^{12}$ **3.** $42t^2 + 14t$ **4.** $2x^2 + x - 3$
5. $n^2 + 18n + 81$ **6.** $x^2 - 4$

SELF CHECKS 8.4

1. a. $\sqrt{15}$ **b.** $2y\sqrt{5}$ **c.** $b^4\sqrt{10b}$ **2. a.** $12\sqrt{2}$ **b.** 90
c. $56\sqrt{30}$ **3. a.** 11 **b.** $3y + 5$ **c.** 50 **4. a.** $3 + 16\sqrt{3}$
b. $4n\sqrt{5} - 40\sqrt{n}$ **5. a.** $\sqrt{66} - \sqrt{42} + \sqrt{55} - \sqrt{35}$
b. $5a + \sqrt{5a} - 6$ **6. a.** $15 - 2\sqrt{15n} + n$ **b.** -2
7. a. $\sqrt{6}$ **b.** $3\sqrt{5}$ **c.** $5d^2$ **8.** $\frac{11\sqrt{6}}{6}$ **9. a.** $\frac{\sqrt{42}}{7}$ **b.** $\frac{3\sqrt{2y}}{2y}$
10. a. $\frac{5\sqrt{7} - 10}{3}$ **b.** $\frac{x + \sqrt{2x} - \sqrt{7x} - \sqrt{14}}{x - 2}$

STUDY SET SECTION 8.4 (page 644)

1. denominator **3.** conjugate **5. a.** $a \cdot b$ **b.** $\frac{a}{b}$ **c.** a **d.** b
7. a. The radicand is a fraction. **b.** There is a radical in the
denominator. **9.** $4, \sqrt{3}, 6$ **11.** $7, 7, 7$ **13.** $\sqrt{15}$ **15.** $5\sqrt{2}$
17. $2d\sqrt{10}$ **19.** $x^4\sqrt{5}$ **21.** $15\sqrt{5}$ **23.** $50\sqrt{3}$ **25.** $10\sqrt{15}$
27. $18\sqrt{2}$ **29.** 6 **31.** y **33.** $2b + 7$ **35.** 12 **37.** $2 + \sqrt{2}$
39. $27 - 3\sqrt{6}$ **41.** $x\sqrt{3} - 2\sqrt{x}$ **43.** $2m\sqrt{2} + 9\sqrt{m}$ **45.** 1
47. $\sqrt{6} + \sqrt{10} - 3 - \sqrt{15}$ **49.** $4x - 18$
51. $6\sqrt{14} + 2x\sqrt{7} - 3x\sqrt{2} - x^2$ **53.** $28 + 10\sqrt{3}$
55. $a^2 + 2a\sqrt{7} + 7$ **57.** $5 - 2\sqrt{5m} + m$ **59.** 3
61. $11 - y^2$ **63.** $7c - 9$ **65.** $\sqrt{10}$ **67.** a^2 **69.** $\frac{3\sqrt{2}}{5}$ **71.** $\frac{3}{5x}$
73. $\frac{\sqrt{3}}{3}$ **75.** $\frac{4\sqrt{19}}{19}$ **77.** $\frac{2\sqrt{15}}{5}$ **79.** $\frac{2\sqrt{3}}{3}$ **81.** $\frac{3\sqrt{3} + 3}{2}$
83. $5\sqrt{3} - 5\sqrt{2}$ **85.** $-\frac{\sqrt{10} + 5 + \sqrt{14} + \sqrt{35}}{3}$
87. $\frac{\sqrt{15} - \sqrt{3a} - \sqrt{5a} + a}{5 - a}$ **89.** $\frac{3\sqrt{2}}{8}$ **91.** $x\sqrt{14} + \sqrt{2x}$
93. $\frac{5\sqrt{3} - \sqrt{3x}}{25 - x}$ **95.** $100x$ **97.** $\frac{\sqrt{91}}{7}$ **99.** $9p^2 + 6p\sqrt{5} + 5$
101. $\frac{8\sqrt{7} - 16}{3}$ **103.** $\sqrt{66}$ **105.** $\frac{2x}{3}$ **107.** 13 **109.** $\frac{3\sqrt{2x}}{2x}$
111. $14\sqrt{15}$ **113.** $\frac{7\sqrt{3} - 3}{46}$ **115.** $2n^4\sqrt{15n}$ **117. a.** 60
b. $7\sqrt{5}$ **119. a.** $2x - 10\sqrt{x} - 12$ **b.** $3\sqrt{x} - 4$
121. Length: $\sqrt{8^2 + 2^2} = \sqrt{68} = 2\sqrt{17}$ ft;
width: $\sqrt{4^2 + 1^2} = \sqrt{17}$ ft; area: 34 ft² **123.** $f = 0.772\frac{\sqrt{Tu}}{u}$
129. 2.7 hot dogs and buns

ARE YOU READY? 8.5 (page 647)

1. a. x **b.** $3x + 1$ **2.** 10 **3.** $-3, 9$ **4.** $x^2 - 8x + 16$

SELF CHECKS 8.5

1. 100 **2.** 3 **3.** 12 **4.** No solution **5.** 5, 0 is extraneous
6. 20 **7.** 64 ft

STUDY SET SECTION 8.5 (page 653)

1. radical **3.** solve **5.** squares, b^2 **7. a.** $\sqrt{x - 4} = 3$
b. $x + 8 = \sqrt{x}$ **9.** $2, 2, x - 3, 28$ **11.** 9 **13.** 144 **15.** 8
17. 9 **19.** 1 **21.** -95 **23.** 5 **25.** 6 **27.** 8 **29.** 13 **31.** 2
33. -1 **35.** No solution **37.** No solution **39.** No solution
41. No solution **43.** $-2, -1$ **45.** $0, 3$ **47.** 12, 7 is extraneous
49. 5, 2 is extraneous **51.** -2 **53.** 4 **55.** 1 **57.** -1 **59.** 16
61. 4 **63.** $-2, 4$ **65.** 5 **67.** 49 **69.** 25 **71.** 3 **73.** $\frac{1}{4}, 1$
75. No solution, -10 is extraneous **77.** 4 **79.** 9 **81.** 7
83. 36 **85.** 1, 3 **87.** 9, 4 is extraneous **89.** 4, -1 is extraneous
91. 169 ft **93.** 64.4 ft **95.** About 1.5 ft **97.** 2,010 ft **103.** $\frac{5}{4}$
105. $\frac{10}{9}$

ARE YOU READY? 8.6 (page 655)

1. a. 125 **b.** 243 **2. a.** -64 **b.** -125 **3.** $\frac{1}{x^3}$ **4.** 3

SELF CHECKS 8.6

1. a. 1 **b.** -2 **c.** -6 **2. a.** 2 **b.** -4 **c.** Not a real number
3. a. n^2 **b.** $3x^4$ **4. a.** s^5 **b.** $-10b$ **c.** x^3 **5. a.** $4\sqrt[3]{2}$
b. $2\sqrt[4]{11}$ **c.** $\frac{2\sqrt[3]{5}}{7}$ **6. a.** 12 **b.** -3 **c.** -3 **7. a.** 1,000
b. 4 **c.** -16 **8. a.** $\frac{1}{7}$ **b.** $\frac{1}{16}$ **c.** $-\frac{1}{2}$

STUDY SET SECTION 8.6 (page 662)

1. cube, fourth, fifth **3.** simplify **5. a.** 4 **b.** $2x$ **c.** -32
7. a. $\sqrt[n]{a}\sqrt[n]{b}$ **b.** $\frac{\sqrt[n]{a}}{\sqrt[n]{b}}$ **9. a.** $\sqrt[3]{8\cdot3}$ **b.** $\sqrt[4]{16\cdot3}$
11. a. $\sqrt{25}$ **b.** $\left(\sqrt[3]{-27}\right)^2$ **13.** 2 **15.** 0 **17.** -5
19. -4 **21.** 1 **23.** 3 **25.** 9 **27.** $\frac{1}{5}$ **29.** 2 **31.** 4
33. Not a real number **35.** -3 **37.** 2 **39.** $-\frac{1}{3}$ **41.** m^2
43. x^6 **45.** $3a$ **47.** $2b^3$ **49.** y **51.** $3a^2$ **53.** x^3 **55.** $2v^6$
57. $2\sqrt[3]{3}$ **59.** $-4\sqrt[3]{2}$ **61.** $3\sqrt[4]{2}$ **63.** $2\sqrt[3]{9}$ **65.** $\frac{5\sqrt[3]{2}}{3}$
67. $2\sqrt[5]{2}$ **69.** 9 **71.** 2 **73.** $\frac{1}{2}$ **75.** -12 **77.** -5 **79.** $\frac{3}{8}$
81. 64 **83.** 9 **85.** 25 **87.** 32 **89.** $\frac{1}{8}$ **91.** -8 **93.** $\frac{1}{2}$
95. $\frac{1}{125}$ **97.** $\frac{1}{32}$ **99.** $\frac{1}{9}$ **101.** $-\frac{1}{729}$ **103.** $\frac{1}{8}$ **105. a.** 8 **b.** 4
c. Not a real number **d.** -4 **107. a.** $4\sqrt{2}$ **b.** $2\sqrt[3]{4}$ **c.** $2\sqrt[4]{2}$
d. 2 **109.** 10 mph **111.** 26 ft **113.** 36 in.^{2}
119.

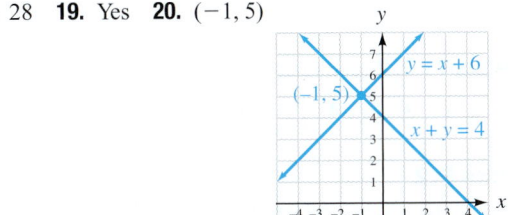

121.

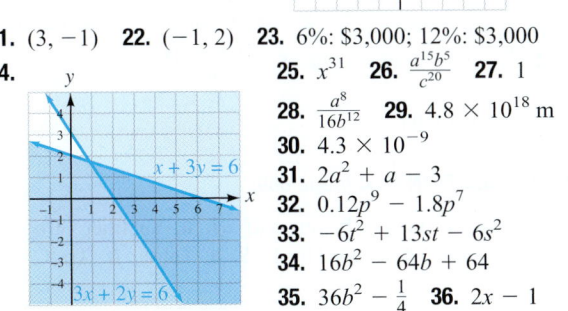

CHAPTER 8 REVIEW (page 665)

1. 6, 36 **2.** False **3.** 5 **4.** 7 **5.** -12 **6.** $\frac{4}{9}$ **7.** 0.8
8. 1 **9.** 4.58 **10.** -5.29 **11.** Not a real number, irrational,
rational, irrational **12.** 30 mph **13.** x **14.** $2b^2$ **15.** $-y^6$
16. $3h^8$ **17.** 10 **18.** 15 **19.** 5 ft **20.** 2.12 in. **21.** 5
22. $\sqrt{194}\approx13.93$ **23.** $4\sqrt{2}$ **24.** $x^2\sqrt{x}$ **25.** $4x\sqrt{5}$
26. $-6\sqrt{7}$ **27.** $5t\sqrt{10t}$ **28.** $\frac{4}{5}$ **29.** $\frac{2\sqrt{15}}{7}$ **30.** $\frac{11x\sqrt{2}}{13}$
31. $2\sqrt{10}$ ft; 6.3 ft **32.** False **33.** $2\sqrt{10}$ **34.** $5\sqrt{x}$ **35.** 0
36. $-3+4\sqrt{3}$ **37.** $\sqrt{7}$ **38.** $-7y\sqrt{5y}$ **39.** The radicands
are different. **40.** $13\sqrt{5}$ in. **41.** $\sqrt{6}$ **42.** 125
43. $36x^2\sqrt{2x}$ **44.** $15+6x\sqrt{15}+9x^2$ **45.** -2 **46.** -2
47. x **48.** $t-1$ **49.** $30\sqrt{2}$ in.^{2}; 42.4 in.^{2} **50.** $\frac{\sqrt{2}}{2}$
51. $\frac{9\sqrt{7}}{7}$ **52.** $\frac{\sqrt{3a}}{a}$ **53.** $\frac{16\sqrt{15}}{5}$ **54.** $\frac{11\sqrt{3}}{15}$ **55.** $\frac{\sqrt{2x}}{4x}$
56. $\frac{7+\sqrt{14}}{5}$ **57.** $\frac{a-\sqrt{a}}{a-1}$ **58.** $\frac{b+5\sqrt{b}+6}{b-9}$ **59.** 81 **60.** 45
61. No solution **62.** -3 **63.** -2 **64.** $-2,4$ **65.** 11, 2 is
extraneous **66.** 9, 4 is extraneous **67.** 64 ft **68.** 96 ft
69. -3 **70.** -5 **71.** 3 **72.** 2 **73.** 0 **74.** -1
75. $\frac{1}{4}$ **76.** 4 **77.** x **78.** $3y^2$ **79.** $2a^3$ **80.** b^4 **81.** $3\sqrt[3]{2}$
82. $2\sqrt[4]{5}$ **83.** $-\frac{2\sqrt[3]{2}}{7}$ **84.** $\frac{2\sqrt[3]{7}}{5}$ **85.** 7 **86.** -10 **87.** 216
88. $\frac{4}{9}$ **89.** $\frac{1}{8}$ **90.** -243 **91.** $\frac{1}{64}$ of the original dose **92.** No
real number squared is -64.

CHAPTER 8 TEST (page 670)

1. a. radical **b.** radicand **c.** Pythagorean **d.** cube, root
e. index **2.** 4, 16 **3.** 10 **4.** $-\frac{8}{3}$ **5.** 0.5 **6.** 1 **7.** 7 amps
8. 10 ft **9.** Irrational, not a real number, rational
10. $6\sqrt{2}\approx8.49$ **11.** $2x$ **12.** $3x\sqrt{6x}$ **13.** $\frac{5\sqrt{2}}{7}$ **14.** $3a^2\sqrt{a}$
15. $9\sqrt{5}$ **16.** $3\sqrt{11}$ **17.** $5b^2\sqrt{3}$ **18.** $10\sqrt{6}$
19. $2\sqrt{6}+3\sqrt{2}$ **20.** $-5x\sqrt{2x}$ **21.** -1 **22.** $-24x^2\sqrt{6x}$
23. $5x-21\sqrt{x}+4$ **24.** $12t$ **25.** $x+1$ **26.** $x+2\sqrt{x}+1$
27. $\left(6\sqrt{2}+2\sqrt{10}\right)$ in. **28.** $8\sqrt{2}$ ft ≈11.31 ft
29. $\frac{2\sqrt{7}}{7}$ **30.** $\frac{\sqrt{3x}+8\sqrt{3}}{x-64}$ **31.** 225 **32.** -62 **33.** 5
34. 2, 1 **35.** No solution **36.** 9, 4 is extraneous **37.** The
Pythagorean theorem; 5 ft **38.** 64 **39.** They are not like radicals,
the radicands are different. **40.** No real number squared is equal
to -49. **41.** -5 **42.** Not a real number **43.** $\frac{1}{4}$ **44.** $-2x^2$
45. $3x$ **46.** $2\sqrt[3]{11}$ **47.** 12 **48.** 4 **49.** $\frac{1}{81}$ **50.** $-\frac{1}{3}$

CUMULATIVE REVIEW CHAPTERS 1–8 (page 672)

1. a. True **b.** False **c.** True **2.** 14 **3.** $-2p-6z$
4. -2 **5.** 17 lb **6.** 4 in. **7.** 3.5 hr **8.** 80
9. $\left(-\infty,-\frac{3}{4}\right]$ 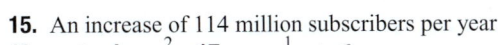 **10.** No

11.

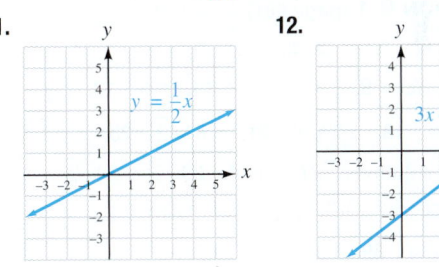

12.

13.

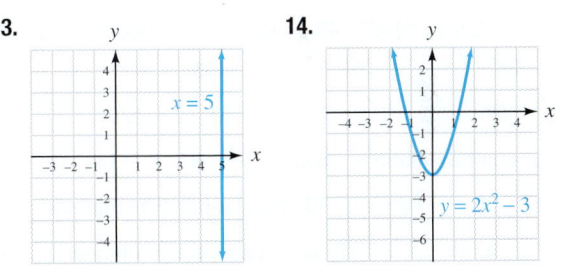

14.

15. An increase of 114 million subscribers per year
16. a. 3 **b.** $-\frac{2}{3}$ **17.** $y=\frac{1}{2}x+6$
18. 28 **19.** Yes **20.** $(-1,5)$

21. $(3,-1)$ **22.** $(-1,2)$ **23.** 6%: \$3,000; 12%: \$3,000
24.
25. x^{31} **26.** $\frac{a^{15}b^5}{c^{20}}$ **27.** 1
28. $\frac{a^8}{16b^{12}}$ **29.** 4.8×10^{18} m
30. 4.3×10^{-9}
31. $2a^2+a-3$
32. $0.12p^9-1.8p^7$
33. $-6t^2+13st-6s^2$
34. $16b^2-64b+64$
35. $36b^2-\frac{1}{4}$ **36.** $2x-1$

37. $3xy(4x - 2y + 3y^2)$ **38.** $(x + y)(2x - 3)$
39. $(x + 5)(x + 2)$ **40.** $(3a + 4)(2a - 5)$ **41.** Prime
42. $(5a - 7b)^2$ **43.** $(a + 2b)(a^2 - 2ab + 4b^2)$
44. $2x(x^2 + 4)(x + 2)(x - 2)$ **45.** $-1, -2$ **46.** $0, 2$
47. $\frac{2}{3}, -\frac{1}{2}$ **48.** $-5, 5$ **49.** 5 cm **50.** $\frac{x+1}{x-1}$
51. $\frac{(p-3)(p+2)}{3(p+3)}$ **52.** $-\frac{3}{5x}$ **53.** $\frac{1}{3a}$ **54.** $\frac{7x+29}{(x+5)(x+7)}$
55. $\frac{3-16b^2}{18b^4}$ **56.** $\frac{y+x}{y-x}$ **57.** 1 **58.** $2\frac{2}{9}$ hr **59.** 6,480
60. $4\frac{1}{2}$ days **61.** 73 in. **62.** No real number squared is -4.
63. 8 **64.** $4x\sqrt{2x}$ **65.** $3m^2$ **66.** $3t$ **67.** $7\sqrt{3} - 6\sqrt{2}$
68. $y - 9\sqrt{y} + 20$ **69.** $\frac{4\sqrt{5}}{5}$ **70.** $\frac{\sqrt{34x}}{2x}$ **71.** 5 **72.** 64

ARE YOU READY? 9.1 (page 676)

1. $(x + 8)(x - 8)$ **2.** Two; 3 and -3 **3.** $5\sqrt{3}$ **4.** $\frac{\sqrt{35}}{7}$
5. $\frac{\sqrt{23}}{4}$ **6.** 7.14

SELF CHECKS 9.1

1. a. $\pm\sqrt{21}$ **b.** $\pm 3\sqrt{6}$ **c.** $\pm\frac{\sqrt{35}}{5}$ **d.** No real-number
solutions **2. a.** $10, -6$ **b.** $-3 \pm 7\sqrt{2}$ **c.** $\frac{1 \pm \sqrt{3}}{3}$
3. $7 \pm \sqrt{11}$; 10.3, 3.7 **4.** $\sqrt{3}$ sec ≈ 1.7 sec

STUDY SET SECTION 9.1 (page 681)

1. quadratic **3.** $\sqrt{c}, -\sqrt{c}$ **5. a.** $x = -9 \pm \sqrt{2}$
b. $x = \frac{3 \pm \sqrt{2}}{6}$ **7.** Yes **9.** $x = \pm\sqrt{6}$ **11.** ± 6 **13.** $\pm\frac{7}{4}$
15. ± 5 **17.** $\pm\sqrt{6}$ **19.** $\pm 2\sqrt{5}$ **21.** $\pm 6\sqrt{2}$ **23.** $\pm\frac{\sqrt{30}}{2}$
25. No real-number solutions **27.** $-6, 4$ **29.** $2 \pm 2\sqrt{2}$
31. $-9 \pm 3\sqrt{7}$ **33.** $\frac{10 \pm \sqrt{6}}{5}$ **35.** $-1 \pm \sqrt{10}$; 2.2, -4.2
37. $9 \pm \sqrt{7}$; 11.6, 6.4 **39.** $3 \pm 2\sqrt{10}$; 9.3, -3.3
41. $-2 \pm 5\sqrt{3}$; 6.7, -10.7 **43.** $-12 \pm 3\sqrt{3}$ **45.** $\pm 7\sqrt{2}$
47. ± 10 **49.** $\frac{-1 \pm 3\sqrt{2}}{3}$ **51.** $6 \pm \sqrt{2}$ **53.** $\pm\sqrt{17}$
55. $\pm\frac{\sqrt{42}}{6}$ **57.** $15 \pm 2\sqrt{2}$ **59.** $\pm\frac{1}{12}$ **61.** $7 \pm \sqrt{3}$
63. No real-number solutions **65.** $\pm\frac{\sqrt{85}}{5}$ **67.** $7, -11$
69. $\pm\sqrt{14}$ **71.** $\frac{-9 \pm 2\sqrt{11}}{8}$ **73.** No real-number solutions
75. a. ± 3 **b.** $\pm 2\sqrt{2}$ **77. a.** 7 is a repeated solution.
b. $7 \pm \sqrt{6}$ **79.** 3 sec **81.** 3.4 sec **83.** 3.5 sec **85.** 20 ft
87. 24,941 mi/hr **89.** Width: 13 in.; length: 21 in. **95.** 2
97. 16

ARE YOU READY? 9.2 (page 683)

1. $(x - 5)^2$ **2.** 6 **3.** 36 **4.** $\frac{9}{4}$ **5.** $x = 6 \pm \sqrt{7}$ **6.** $\frac{45}{4}$

SELF CHECKS 9.2

1. a. $y^2 - 8y + 16 = (y - 4)^2$ **b.** $y^2 + 3y + \frac{9}{4} = \left(y + \frac{3}{2}\right)^2$
c. $x^2 + \frac{7}{5}x + \frac{49}{100} = \left(x + \frac{7}{10}\right)^2$ **2.** $-8, 4$ **3.** $5 \pm \sqrt{29}$; 10.39,
-0.39 **4.** $\frac{3 \pm \sqrt{21}}{2}$; 3.79, -0.79 **5.** $\frac{-1 \pm \sqrt{13}}{6}$
6. $\frac{8 \pm \sqrt{65}}{2}$; 8.03, -0.03

STUDY SET SECTION 9.2 (page 689)

1. square **3. a.** 9 **b.** $\frac{25}{4}$ **5.** one-half **7. a.** Subtract 7 from
both sides **b.** Divide both sides by 4 **9. a.** False **b.** True

11. $\left(\frac{1}{2} \cdot 9\right)^2 = \frac{81}{4}$ **13.** $x^2 + 2x + 1 = (x + 1)^2$
15. $x^2 - 4x + 4 = (x - 2)^2$ **17.** $a^2 - 7a + \frac{49}{4} = \left(a - \frac{7}{2}\right)^2$
19. $x^2 + x + \frac{1}{4} = \left(x + \frac{1}{2}\right)^2$ **21.** $b^2 - \frac{2}{3}b + \frac{1}{9} = \left(b - \frac{1}{3}\right)^2$
23. $x^2 - \frac{5}{2}x + \frac{25}{16} = \left(x - \frac{5}{4}\right)^2$ **25.** $-5, 1$ **27.** $-3, 5$
29. $-2, -4$ **31.** 2, 6 **33.** $2 \pm \sqrt{7}$; 4.65, -0.65
35. $1 \pm 3\sqrt{2}$; 5.24, -3.24 **37.** $-4 \pm \sqrt{22}$; 0.69, -8.69
39. $-3 \pm \sqrt{5}$; $-0.76, -5.24$ **41.** $\frac{7 \pm \sqrt{69}}{2}$; 7.65, -0.65
43. $\frac{5 \pm \sqrt{65}}{2}$; 6.53, -1.53 **45.** $\frac{-3 \pm \sqrt{89}}{2}$; 3.22, -6.22
47. $\frac{-1 \pm \sqrt{37}}{2}$; 2.54, -3.54 **49.** $\frac{7 \pm \sqrt{73}}{4}$ **51.** $\frac{9 \pm \sqrt{65}}{8}$
53. $\frac{-3 \pm \sqrt{89}}{10}$ **55.** $\frac{-5 \pm \sqrt{85}}{6}$ **57.** $-\frac{1}{2}, \frac{13}{2}$ **59.** $\frac{-3 \pm \sqrt{14}}{3}$
61. $\frac{12 \pm \sqrt{133}}{2}$ **63.** $\frac{-8 \pm \sqrt{61}}{4}$ **65.** $\frac{-3 \pm \sqrt{7}}{2}$; $-0.18, -2.82$
67. $\frac{-1 \pm \sqrt{13}}{3}$; 0.87, -1.54 **69.** $\frac{-3 \pm \sqrt{15}}{2}$; 0.44, -3.44
71. $\frac{4 \pm \sqrt{34}}{6}$; 1.64, -0.31 **73.** $\frac{-7 \pm \sqrt{41}}{2}$ **75.** $1 \pm \sqrt{5}$
77. $\frac{6 \pm \sqrt{37}}{3}$ **79.** 7, 5 **81.** $\frac{9 + \sqrt{101}}{2}$ **83.** $2, \frac{2}{3}$ **85.** $1 \pm \sqrt{6}$
87. No real-number solutions **89.** $\frac{-5 \pm \sqrt{57}}{4}$ **91.** $-1, 7$
93. $-6 \pm \sqrt{14}$ **95. a.** $-1, -4$ **b.** $\frac{-5 \pm \sqrt{13}}{2}$ **97. a.** 0, 4
b. $\frac{4 \pm 3\sqrt{2}}{2}$ **99.** 4; 4, 2 **105.** $\frac{x+1}{(x-3)^2}$ **107.** $\frac{x^2 + 7x + 10}{x^2 - 9}$

ARE YOU READY? 9.3 (page 692)

1. 11 **2.** $3\sqrt{5}$ **3.** 3; 2, $-1, 7$ **4.** $\frac{3}{4}, -2$ **5.** Not a real number
6. 3.87

SELF CHECKS 9.3

1. $\frac{2}{3}, -1$ **2.** $\frac{7 \pm \sqrt{17}}{8}$; 1.39, 0.36 **3.** $\frac{-1 \pm \sqrt{3}}{2}$
4. No real-number solutions **5.** Width: 8 ft; length: 12 ft
6. Height: 8.40 in.; length: 12.90 in.

STUDY SET SECTION 9.3 (page 699)

1. quadratic **3. a.** $x^2 + 2x + 5 = 0$ **b.** $3x^2 + 2x - 1 = 0$
5. $1, -2, 4$ **7. a.** 2 **b.** 5 **9. a.** 2 **b.** $\frac{-3 + \sqrt{15}}{2}, \frac{-3 - \sqrt{15}}{2}$
c. $0.44, -3.44$ **11.** $-5, -6, 1, 24, 49, 5, 7, 6, 7, -1$
13. $-2, -3$ **15.** $-3, -4$ **17.** $\frac{1}{4}, -1$ **19.** $\frac{2}{3}, -\frac{3}{2}$ **21.** $2, -\frac{1}{3}$
23. $3, -\frac{2}{5}$ **25.** $\frac{-3 \pm \sqrt{5}}{2}$; $-0.38, -2.62$
27. $\frac{-7 \pm \sqrt{65}}{2}$; 0.53, -7.53 **29.** $\frac{1 \pm \sqrt{37}}{6}$; 1.18, -0.85
31. $\frac{3 \pm \sqrt{37}}{14}$; 0.65, -0.22 **33.** $\frac{7 \pm \sqrt{17}}{8}$; 1.39, 0.36
35. $\frac{-9 \pm \sqrt{137}}{4}$; 0.68, -5.18 **37.** $\frac{-5 \pm \sqrt{3}}{2}$ **39.** $\frac{3 \pm \sqrt{7}}{2}$
41. $\frac{4 \pm \sqrt{10}}{3}$ **43.** $\frac{3 \pm 2\sqrt{2}}{2}$ **45.** $-1 \pm \sqrt{2}$ **47.** $-2 \pm \sqrt{7}$
49. No real-number solutions **51.** No real-number solutions
53. $-2, 3$ **55.** $\frac{-1 \pm \sqrt{41}}{4}$; $-1.85, 1.35$ **57.** 5, 7
59. $\frac{-9 \pm \sqrt{77}}{2}$; $-0.11, -8.89$ **61.** $\frac{1 \pm \sqrt{41}}{5}$; 1.48, -1.08
63. $\frac{-3 \pm 3\sqrt{5}}{2}$; 1.85, -4.85 **65.** No real-number solutions
67. $-4, 0$ **69.** $\pm\sqrt{6}$; ± 2.45 **71.** $2 \pm \sqrt{33}$; 7.74, -3.74
73. 12 **75.** $\pm 3\sqrt{7}$; ± 7.94 **77.** 16 in., 10 in. **79.** 41.4 in.,
31.4 in. **81.** 6.1 cm by 16.3 cm **83.** 7% **91.** $r = \frac{A - p}{pt}$
93. $r = \frac{r_1 r_2}{r_2 + r_1}$

ARE YOU READY? 9.4 (available online)

1. No real number squared is equal to -16 **2.** $5x + 12$
3. $-3x - 9$ **4.** $45x^2 + 54x$ **5.** $14x^2 + 19x - 3$
6. $x = \pm 2\sqrt{6}$

SELF CHECKS 9.4 (available online)

1. a. $9i$ **b.** $-i\sqrt{11}$ **c.** $2i\sqrt{7}$ **d.** $\frac{3i\sqrt{3}}{10}$ **2. a.** $-18 + 0i$
b. $0 + 6i$ **c.** $1 + 2i\sqrt{6}$ **3. a.** $1 + i$ **b.** $-3 + 5i$ **c.** $25 - 4i$
4. a. $36 + 12i$ **b.** $7 - 22i$ **c.** $40 + 0i$ **5. a.** $\frac{20}{17} + \frac{5}{17}i$
b. $\frac{7}{10} - \frac{11}{10}i$ **6. a.** $0 \pm 10i$ **b.** $3 \pm 4i\sqrt{3}$ **7.** $-1 \pm i\sqrt{2}$

STUDY SET SECTION 9.4 (available online)

1. complex. real, imaginary   **3. a.** $\sqrt{-1}$ **b.** $2, -1$ **5. a.** $8i$
b. $2i$ **7.** $\frac{6+i}{6+i}$ **9. a.** True   **b.** True   **c.** False   **d.** True
11. a. $i\sqrt{7}$ **b.** $2i\sqrt{3}$ **13.** $3i$ **15.** $i\sqrt{7}$ or $\sqrt{7}i$ **17.** $2i\sqrt{6}$
or $2\sqrt{6}i$ **19.** $-4i\sqrt{2}$ or $-4\sqrt{2}i$ **21.** $45i$ **23.** $\frac{5}{3}i$
25. $12 + 0i$ **27.** $0 + 10i$ **29.** $6 + 4i$ **31.** $-9 - 7i$
33. $8 - 2i$ **35.** $3 - 5i$ **37.** $14 - 13i$ **39.** $12 - 9i$
41. $6 - 3i$ **43.** $-25 - 25i$ **45.** $12 + 5i$ **47.** $13 - i$
49. $2 + i$ **51.** $\frac{8}{53} - \frac{28}{53}i$ **53.** $-\frac{5}{13} + \frac{12}{13}i$ **55.** $\frac{31}{50} - \frac{17}{50}i$
57. $0 \pm 3i$ **59.** $0 \pm 2i\sqrt{2}$ **61.** $-3 \pm i$ **63.** $11 \pm 5i\sqrt{3}$
65. $\frac{3}{2} \pm \frac{\sqrt{7}}{2}i$ **67.** $-\frac{1}{4} \pm \frac{\sqrt{7}}{4}i$ **69.** $\frac{15}{26} - \frac{3}{26}i$ **71.** $15 + 2i$
73. $0 + i$ **75.** $1 + 8i$ **77.** $\frac{5}{13} - \frac{12}{13}i$ **79.** $9 - 8i$
81. $-12 - 16i$ **83.** $3 + 4i$ **85.** $-\frac{1}{4} \pm \frac{\sqrt{39}}{4}i$ **87.** $4 \pm 3i\sqrt{5}$
89. $-1 \pm i$ **91.** $0 \pm 6i$ **93.** $0 \pm \frac{4}{3}i$ **95.** $-\frac{1}{3} \pm \frac{\sqrt{2}}{3}i$
97. $18.45 - 2.18i$ **103.** $\frac{\sqrt{7}}{7}$ **105.** $\frac{8\sqrt{x}+16}{x-4}$

ARE YOU READY? 9.5 (available online)

1. $-1, 3$ **2.** -10 **3.** $1, -4, 6$ **4.** 2.7

SELF CHECKS 9.5 (available online)

1. y-intercept: $(0, 8)$; x-intercepts: $(-2, 0), (-4, 0)$ **2.** $(-3, -1)$
3.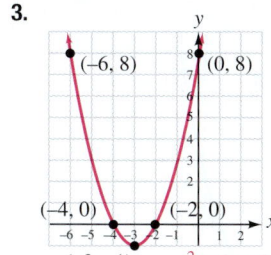
4.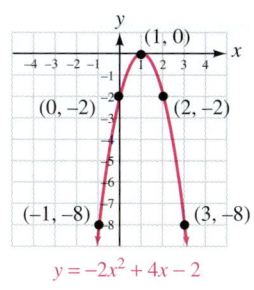
5.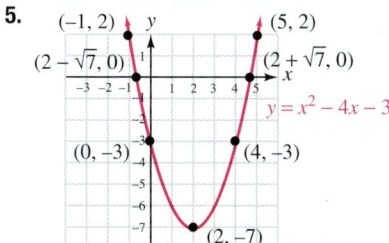

STUDY SET SECTION 9.5 (available online)

1. quadratic, parabola   **3.** intercepts, intercept   **5.** $<, 0$ **7. a.** 0
b. x **9. a.** Parabola   **b.** $(1, 0), (3, 0)$ **c.** $(0, -3)$ **d.** $(2, 1)$
e. It is a vertical line through $(2, 1)$.
11. $(2, -2), (-1, 1)$ **13.** Two; $-3, 1$
15. a. $2, 4, -8$
b. -1
17. $(4, 0), (2, 0)$; $(0, 8)$

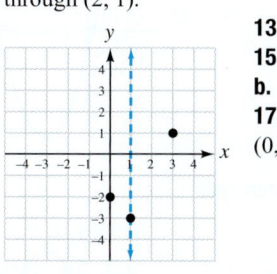

19. $(-3, 0), (-7, 0); (0, -21)$ **21.** $(1, -1)$ **23.** $(3, 1)$
25. 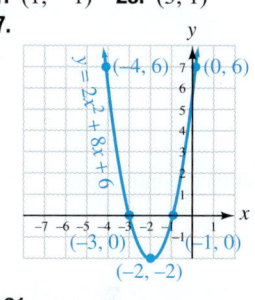   **27.**

29.   **31.**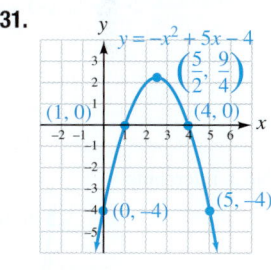

33.   **35.**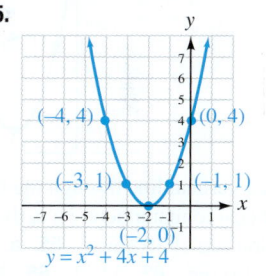

37.   **39.**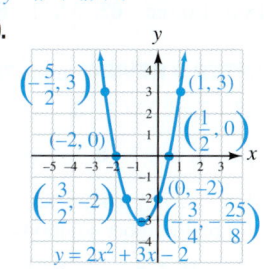

41.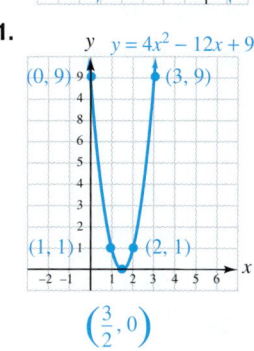

43. Irrational x-intercepts

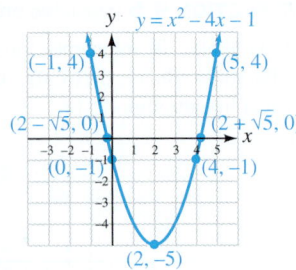

45. Irrational x-intercepts

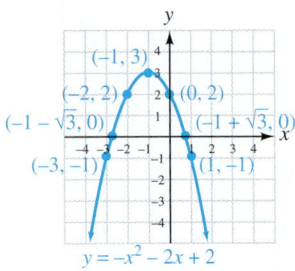

47. Irrational x-intercepts

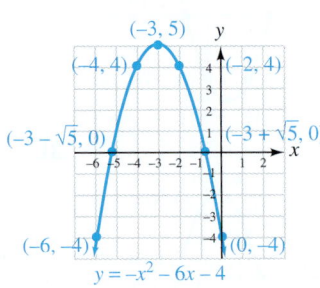

49. No x-intercepts

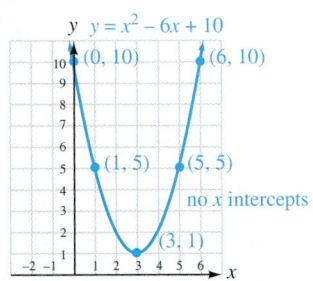

51. It is a vertical line through the body of the butterfly.
53. The cost to manufacture a carburetor is lowest ($100) for a production run of 30 units. **55. a.** 14 ft **b.** 0.25 sec and 1.75 sec **c.** 18 ft; 1.0 sec **63.** $3\sqrt{2}$ **65.** $3z$

CHAPTER 9 REVIEW (page 702)

1. ± 8 **2.** $\pm 2\sqrt{2}$ **3.** $\pm 5\sqrt{3}$ **4.** $-4, 6$ **5.** $\frac{8 \pm 2\sqrt{10}}{9}$

6. $\frac{7}{2}, \frac{1}{2}$ **7.** $13, 7$ **8.** $\frac{-1 \pm \sqrt{6}}{3}$ **9.** $\pm 2\sqrt{3}; \pm 3.46$

10. $1 \pm \sqrt{55}; -6.42, 8.42$ **11.** No real-number solutions
12. No real-number solutions **13.** 3.0 sec **14.** $2\frac{1}{4}$ in.
15. $x^2 + 4x + 4 = (x + 2)^2$ **16.** $t^2 - 5t + \frac{25}{4} = \left(t - \frac{5}{2}\right)^2$

17. $3, 5$ **18.** $2, -7$ **19.** $-1 \pm \sqrt{6}$ **20.** $\frac{4 \pm \sqrt{23}}{2}$ **21.** $\frac{1 \pm \sqrt{3}}{2}$

22. $-1, -\frac{2}{3}$ **23.** $-2 \pm \sqrt{3}; -0.27, -3.73$

24. $\frac{7 \pm \sqrt{69}}{2}; -0.65, 7.65$ **25.** $x^2 + 2x + 5 = 0; 1, 2, 5$

26. $6x^2 - 2x - 1 = 0; 6, -2, -1$ **27.** $5, -3$ **28.** $\frac{3}{2}, -\frac{1}{3}$

29. $1 \pm \sqrt{5}$ **30.** $3 \pm \sqrt{2}$ **31.** $\frac{-3 \pm \sqrt{21}}{6}$ **32.** $\frac{-1 \pm \sqrt{21}}{10}$

33. No real-number solutions **34.** $\frac{-3 \pm \sqrt{19}}{2}$ **35.** $0, -4$

36. $1, -7$ **37.** $\pm 3\sqrt{3}$ **38.** $\frac{3 \pm \sqrt{6}}{3}$ **39.** $\frac{-1 \pm \sqrt{11}}{2}$

40. 2 is a repeated solution **41.** $\frac{13}{2}, -\frac{3}{2}$
42. No real-number solutions **43.** $\frac{-1 \pm \sqrt{7}}{3}; -1.22, 0.55$
44. 10 ft, 24 ft **45.** 18 sec **46.** 6.8 in. **47.** $5i$ **48.** $3i\sqrt{2}$
49. $-7i$ **50.** $\frac{3}{8}i$ **51.** Real numbers, imaginary numbers
52. a. True **b.** True **c.** False **d.** False **53.** $3 - 6i$
54. $-1 + 7i$ **55.** $0 - 19i$ **56.** $0 + i$ **57.** $8 - 2i$
58. $3 - 5i$ **59.** $3 + 6i$ **60.** $9 + 7i$ **61.** $-\frac{5}{13} + \frac{12}{13}i$

62. $\frac{15}{26} - \frac{3}{26}i$ **63.** $0 \pm 3i$ **64.** $0 \pm \frac{4\sqrt{3}}{3}i$ **65.** $2 \pm 2i\sqrt{6}$

66. $-3 \pm 3i\sqrt{6}$ **67.** $-1 \pm i$ **68.** $\frac{3}{4} \pm \frac{\sqrt{7}}{4}i$

69. a. $(-3, 0), (1, 0)$ **b.** $(0, -3)$ **c.** $(-1, -4)$ **d.** A vertical line through $(-1, -4)$ **70.** $(-2, -3)$ **71.** $(1, 5)$; upward
72. $(3, 16)$; downward **73.** $(-5, 0), (-1, 0); (0, 5)$
74. No x-intercepts; $(0, 3)$
75.

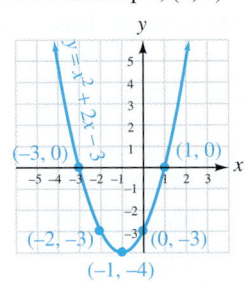

76.

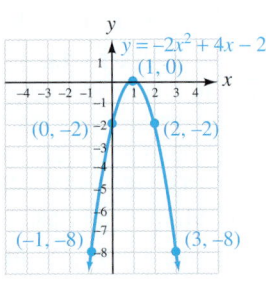

77. Irrational x-intercepts

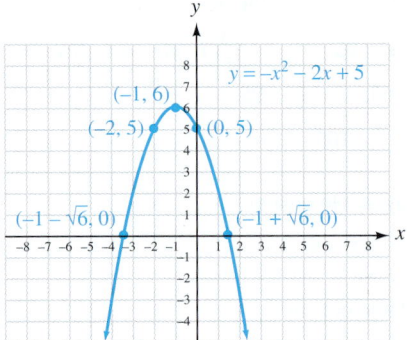

78. Irrational x-intercepts

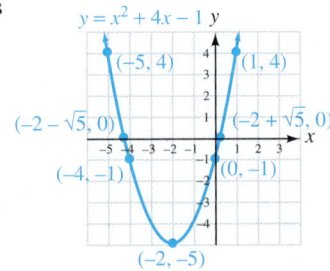

79. $2; -2, 1; 1; -3$; no **80.** The maximum profit of $16,000 is obtained from the sale of 400 units.

CHAPTER 9 TEST (page 705)

1. a. quadratic **b.** square **c.** completed **d.** plus, minus
e. leading, constant **2.** $x = \pm\sqrt{5}$ **3.** $\pm\sqrt{17}$ **4.** $\pm 4\sqrt{3}$
5. $2 \pm \sqrt{3}$ **6.** $\pm\frac{5}{2}$ **7.** $\pm\frac{1}{7}$ **8.** $-8 \pm 2\sqrt{6}$

9. $m = \pm\sqrt{-49}$, and $\sqrt{-49}$ is not a real number.
10. It is a solution. **11.** $x^2 - 14x + 49 = (x - 7)^2$
12. $c^2 - 7c + \frac{49}{4} = \left(c - \frac{7}{2}\right)^2$ **13.** $x^2 + x + \frac{1}{4} = \left(x + \frac{1}{2}\right)^2$
14. $a^2 - \frac{5}{3}a + \frac{25}{36} = \left(a - \frac{5}{6}\right)^2$ **15.** $-1 \pm \sqrt{5}; -3.24, 1.24$

16. $\dfrac{-1 \pm \sqrt{13}}{2}$ **17.** No real-number solutions   **18.** $2, -\dfrac{1}{2}$

19. $-\dfrac{3}{2}, 4$ **20.** $\dfrac{-11 \pm \sqrt{61}}{10}$ **21.** $\dfrac{3 \pm 2\sqrt{2}}{2}$

22. No real-number solutions   **23.** $\dfrac{1 \pm \sqrt{7}}{3}$; $-0.55, 1.22$

24. It is a solution   **25.** 40 cm   **26.** 6.3 seconds   **27.** 41 ft, 65 ft

28. 3.3 ft, 7.3 ft   **29.** $2 \pm \sqrt{2}$ **30.** $1, -\dfrac{5}{3}$ **31.** $\pm 2\sqrt{6}$ **32.** $0, 6$

33. $10i$ **34.** $-3i\sqrt{2}$ **35.** $1 + i$ **36.** $-1 + 12i$ **37.** $14 - 8i$
38. $\dfrac{5}{13} - \dfrac{12}{13}i$ **39.** $0 \pm 10i$ **40.** $-3 \pm i$ **41.** $0 \pm \dfrac{5}{4}i$

42. $-\dfrac{1}{3} \pm \dfrac{\sqrt{2}}{3}i$ **43.** The most air conditioners sold in a week

(18) occurred when 3 ads were run.   **44.** $<, >$

45.

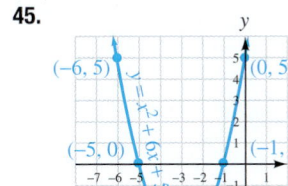

46.

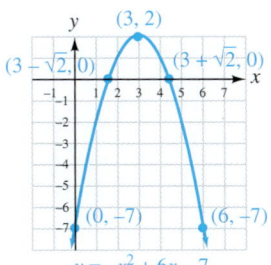

CUMULATIVE REVIEW CHAPTERS 1–9 (page 706)

1. a. True   **b.** True   **c.** True   **2.** 2 hours before salt is spread
3. 36   **4.** 4   **5.** 65   **6.** $-2p + 54$ **7.** 12   **8.** $-\dfrac{3}{2}$

9. 9.8%   **10.** $r = \dfrac{T - 2t}{2}$ **11.** $343,750
12. $8,000 at 7%, $20,000 at 10%
13. $(-\infty, -2)$ **14.** Yes

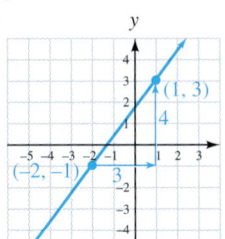

15.

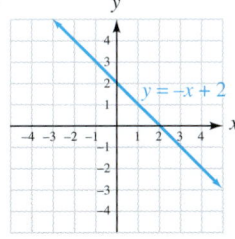

16.

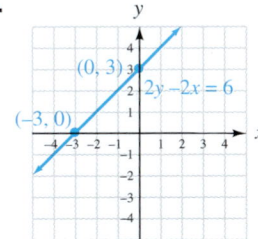

17.

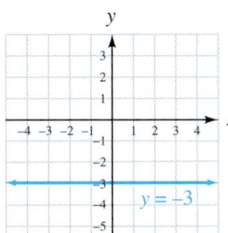

18.

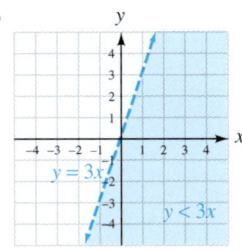

19. $\dfrac{3}{5}$ **20.** A decrease of 900,000 viewers per year   **21.** $-\dfrac{4}{5}$
22. $y = -2x + 1$ **23.** Perpendicular   **24.** $y = \dfrac{1}{4}x - 1$
25.

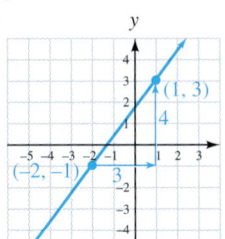

26. -8
27. Domain: $\{-4, 1, 4, 5\}$;
range: $\{-3, 2, 8\}$ **28.** Yes

29. $(-2, 3)$

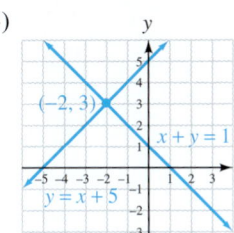

30. $(-3, -1)$
31. $(-1, 2)$ **32.** 550 mph
33. 36 lb of hard candy,
12 lb soft candy

34.

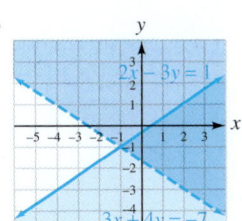

35. y^9 **36.** $\dfrac{b^6}{27a^3}$ **37.** 2
38. $\dfrac{x^{10}z^4}{9y^6}$ **39.** 2,600,000
40. 7.3×10^{-4}

41.

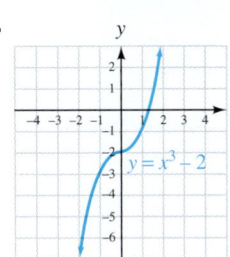

42. $6x^3 + 4x$
43. $6x^3 + 8x^2 + 3x - 72$
44. $-6a^5$ **45.** $6b^2 + 5b - 4$
46. $6x^3 - 7x^2y + y^3$
47. $4x^2 + 20xy + 25y^2$
48. $81m^4 - 1$ **49.** $2a - \dfrac{3}{2}b + \dfrac{1}{2a}$
50. $2x + 1$
51. $6a(a - 2a^2b + 6b)$
52. $(x + y)(2 + a)$

53. $(x + 2)(x - 8)$ **54.** $3y^3(5y - 2)(2y + 5)$
55. $(t^2 + 4)(t + 2)(t - 2)$ **56.** $(b + 5)(b^2 - 5b + 25)$
57. $0, -\dfrac{8}{3}$ **58.** $\dfrac{2}{3}, -\dfrac{1}{5}$ **59.** 5 in.   **60.** -8 **61.** $\dfrac{3(x + 3)}{x + 6}$
62. -1 **63.** $\dfrac{x - 3}{x - 5}$ **64.** $\dfrac{1}{s - 5}$ **65.** $\dfrac{x^2 + 4x + 1}{x^2y}$ **66.** $\dfrac{x^2 + 5x}{x^2 - 4}$
67. $\dfrac{9m^2}{2}$ **68.** $\dfrac{9y + 5}{4 - y}$ **69.** 3   **70.** 1   **71.** $a = \dfrac{b}{b - 1}$
72. $2\dfrac{6}{11}$ days   **73.** 875 days   **74.** 39   **75.** 9
76. 1.2 revolutions per second   **77.** $10x$ **78.** $-3b\sqrt{2b}$
79. $9\sqrt{6}$ **80.** $-1 - 2\sqrt{2}$ **81.** $\dfrac{4\sqrt{10}}{5}$ **82.** $\dfrac{3\sqrt{2} + \sqrt{2a}}{9 - a}$
83. 4   **84.** $-2, -1$ **85.** $\dfrac{3m}{2n^2}$ **86.** 2   **87.** 125   **88.** $\dfrac{1}{16}$
89. $\pm 5\sqrt{3}$ **90.** $\dfrac{-5 \pm 6\sqrt{2}}{6}$ **91.** 21.2 in.   **92.** $-2, -6$
93. $\dfrac{1 \pm \sqrt{33}}{8}$; $-0.59, 0.84$ **94.** 83 ft $\times$ 155 ft   **95.** $7i$
96. $3i\sqrt{6}$ **97.** $1 + 5i$ **98.** $16 - 2i$ **99.** $6 - 17i$
100. $1 - i$ **101.** $0 \pm 4i$ **102.** $2 \pm i$
103.

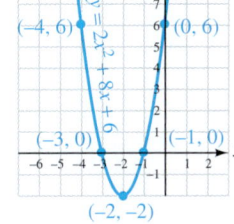

104. About 4,000 rpm

STUDY SET APPENDIX (page A-3)

1. 8   **3.** 9   **5.** 12   **7.** 28, 27, 27   **9.** -2 yd   **11.** 74.5
centimicrons   **13.** 84   **15.** 5 lb, 4 lb   **17. a.** Mean: 36.5 mpg;
median: 37.5 mpg; bimodal: 40 mpg and 41 mpg   **b.** Mean:
38.8 mpg; median: 36 mpg; mode: 36 mpg

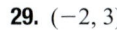
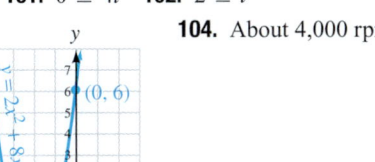

INDEX

Page numbers beginning with 9- refer to online content available at www.cengage.com/tussy.